Braconidae of the Middle East (Hymenoptera)

Braconidae of the Middle East (Hymenoptera)

Taxonomy, Distribution, Biology, and Biocontrol Benefits of Parasitoid Wasps

Edited by

Neveen Samy Gadallah
Entomology Department, Faculty of Science, Cairo University, Giza, Egypt

Hassan Ghahari
Department of Plant Protection, Yadegar-e Imam Khomeini (RAH) Shahre Rey Branch, Islamic Azad University, Tehran, Iran

Scott Richard Shaw
Department of Ecosystem Science and Management, University of Wyoming, Laramie, WY, United States

ELSEVIER

ACADEMIC PRESS
An imprint of Elsevier

Academic Press is an imprint of Elsevier
125 London Wall, London EC2Y 5AS, United Kingdom
525 B Street, Suite 1650, San Diego, CA 92101, United States
50 Hampshire Street, 5th Floor, Cambridge, MA 02139, United States
The Boulevard, Langford Lane, Kidlington, Oxford OX5 1GB, United Kingdom

Notices
Knowledge and best practice in this field are constantly changing. As new research and experience broaden our understanding, changes in research methods, professional practices, or medical treatment may become necessary.

Practitioners and researchers must always rely on their own experience and knowledge in evaluating and using any information, methods, compounds, or experiments described herein. In using such information or methods they should be mindful of their own safety and the safety of others, including parties for whom they have a professional responsibility.

To the fullest extent of the law, neither the Publisher nor the authors, contributors, or editors, assume any liability for any injury and/or damage to persons or property as a matter of products liability, negligence or otherwise, or from any use or operation of any methods, products, instructions, or ideas contained in the material herein.

ISBN: 978-0-323-96099-1

For information on all Academic Press publications visit our website at
https://www.elsevier.com/books-and-journals

Publisher: Nikki P. Levy
Acquisitions Editor: Anna Valutkevich
Editorial Project Manager: Ivy Dawn Torre
Production Project Manager: Swapna Srinivasan
Cover Designer: Vicky Pearson

Typeset by TNQ Technologies

This book is dedicated to all the people who have had roles in discovering the Braconidae of the Middle East and advancing their use in applied biological pest suppression.

Zele albiditarsus Curtis, 1832 (♀, Euphorinae)
[photo courtesy of H. Jonkman and C. van Achterberg]

Contents

31. Subfamily Sigalphinae Haliday, 1833

*Neveen Samy Gadallah, Hassan Ghahari,
Scott Richard Shaw and Donald L.J. Quicke*

32. Diversity of Braconidae in the Middle East with an emphasis on Iran

*Hassan Ghahari, Neveen Samy Gadallah,
Scott Richard Shaw and Donald L.J. Quicke*

List of contributors

Sergey A. Belokobylskij, Zoological Institute Russian Academy of Sciences, Universitetskaya nab. 1, St. Petersburg, Russia

Maximilian Fischer, Naturhistorisches Museum, Zoologische Abteilung, Wien, Austria

Neveen Samy Gadallah, Entomology Department, Faculty of Science, Cairo University, Giza, Egypt

Hassan Ghahari, Department of Plant Protection, Yadegar-e Imam Khomeini (RAH) Shahre Rey Branch, Islamic Azad University, Tehran, Iran

Ilgoo Kang, Department of Entomology, Louisiana State University Agricultural Center, Baton Rouge, LA, United States

Nickolas G. Kavallieratos, Department of Crop Science, Agricultural University of Athens, Athens, Greece

Rebecca N. Kittel, Laboratory of Insect Biodiversity and Ecosystems Science, Graduate School of Agricultural Science, Kobe University, Kobe, Japan

Francisco Javier Peris-Felipo, Basel, Switzerland

Donald L.J. Quicke, Integrative Ecology Laboratory, Department of Biology, Faculty of Science, Chulalongkorn University, Pathumwan, Bangkok, Thailand

Michael J. Sharkey, Department of Entomology, University of Kentucky, Lexington, KY, United States

Scott Richard Shaw, UW Insect Museum, Department of Ecosystem Science and Management, University of Wyoming, Laramie, WY, United States

James B. Whitfield, Department of Entomology, University of Illinois at Urbana-Champaign, Urbana, IL, United States

Acknowledgments

The second editor of this book (HG) wishes to express his sincere gratitude and appreciation to several colleagues who kindly identified Braconidae specimens of the Iranian fauna. Some of them have passed away; however, I do not forget their invaluable roles in documenting the fauna of Iranian Braconidae: Karl-Johan Hedqvist (Sweden; 1917–2009), Vladimir Ivanovich Tobias (Russia, 1929–2011), Jenö Papp (Hungary; 1933–2017), and Maximilian Fischer (Austria; 1929–2019). Also thanked for their support are Kees van Achterberg (Netherlands), Ahmet Beyarslan (Turkey), Elena Davidian (Russia), Özlem Çetin Erdoğan (Turkey), Neveen S. Gadallah (Egypt), Nickolas G. Kavallieratos (Greece), Francisco Javier Peris Felipo (Spain), and Donald L.J. Quicke (UK).

The editors would also like to express their sincere thanks to colleagues who reviewed and commented upon different chapters and gave us insightful comments: Ahmet Beyarslan (Braconinae, Doryctinae, Microgastrinae, Opiinae, Rogadinae), Mark R. Shaw (Pampolinae, Sigalphinae), Michael J. Sharkey (Brachistinae, Sigalphinae), Manfred Mackauer (Aphidiinae), Julia Stigenberg (Euphorinae), Francisco J. Peris Felipo (Microgastrinae), Donald L. J. Quicke (Charmontinae, Helconinae, Microgastrinae), and James B. Whitfield (Introduction, Microtypinae, Proteropinae).

We are indebted to Kees van Achterberg (Netherlands) who replied to numerous scientific questions and taxonomical ambiguities, supplied some images, and provided several necessary papers and data. Additional thanks are also due to Mark R. Shaw (UK), Ahmet Beyarslan (Turkey), Andy Austin (Australia), James B. Whitfield (USA), Robert A. Wharton (USA), Jose Fernandez Triana (Canada), and Anatoly Kotenko (Ukraine) for supplying necessary papers. Alex Wild (University of Texas, USA) and Irinel E. Popescu (Alexandru Ioan Cuza University, Romania) are thanked for providing the images used on the front and back covers and some others, and also M. Hajilari (Iran) for assisting of the layout of the front cover. We would like to extend our sincerest thanks to Vladyslav Mirutenko (Uzhhorod National University, Ukraine), Pierre Moulet (Museum Requien, France), and Jian Huang (Fujian Agriculture and Forestry University, China) for translating of Russian, French, and Chinese data.

This work was supported by Islamic Azad University (Yadegar-e Imam Khomeini (RAH) Shahre Rey Branch), USDA National Institute of Food and Agriculture (Wyoming Agricultural Experiment Station; WYO-612–20), and Cairo University (Giza, Egypt).

Finally, we would like to thank everyone else who provided us with advice, support, and assistance throughout this study, and the staff at Elsevier publishing, especially Anna Valutkevich, Ivy Dawn C. Torre, and Swapna Srinivasan.

Dendrosoter protuberans (Nees, 1834) (Doryctinae), ♀, lateral habitus. *Photo prepared by S.R. Shaw.*

Heterogamus dispar (Haliday, 1833) (Rogadinae), ♂, dorsal habitus. *Photo prepared by S.R. Shaw.*

Chapter 1

Introduction to the Braconidae of the Middle East

Neveen Samy Gadallah[1], Hassan Ghahari[2], Scott Richard Shaw[3] and Donald L.J. Quicke[4]

[1]Entomology Department, Faculty of Science, Cairo University, Giza, Egypt; [2]Department of Plant Protection, Yadegar-e Imam Khomeini (RAH) Shahre Rey Branch, Islamic Azad University, Tehran, Iran; [3]UW Insect Museum, Department of Ecosystem Science and Management, University of Wyoming, Laramie, WY, United States; [4]Integrative Ecology Laboratory, Department of Biology, Faculty of Science, Chulalongkorn University, Pathumwan, Bangkok, Thailand

The Braconidae is one of the largest and most beneficial families of the insect order Hymenoptera (sawflies, wasps, ants, bees), which itself is one of the most species-rich orders of insects. The Hymenoptera comprises over 153,000 extant and described species, as well as more than 2400 extinct species classified into 132 families (Aguiar et al., 2013). The largest hymenopteran superfamily, in terms of species described to date, is the Ichneumonoidea, and this book deals with the second largest of the three extant ichneumonoid families, the Braconidae.

Members of the Hymenoptera play many ecological roles. The bees and many wasps are familiar as important pollinators for most flowering plants, and some bees are well known for producing honey. Ants are almost ubiquitous and are a major group of predators of other insects. Most of the rest of the order are known as "wasps" which can be confusing to lay people as they often associate the word only with the social stinging vespid wasp species. In the broad sense of the word, wasps are encountered in most terrestrial and even some freshwater habitats, and there are many economically significant wasp species. It is among the "wasps" that topic of this book, Braconidae, falls (Borror et al., 1989; Hanson & Gauld, 2006; Naumann & Carne, 1991). Hymenopterans collectively display diverse lifestyles (Austin & Dowton, 2000), including socially organized systems, specialized predators and parasitoids (valuable for keeping population of insects and other arthropods in check), and phytophagous species that feed directly on plants or on fungal tissues (Austin & Dowton, 2000; Goulet & Huber, 1993).

The braconid wasps are examples of what are termed parasitoid wasps. As the word "parasitoid" implies, it has something to do with parasitism, but not in the same way as mosquitoes or tapeworms are parasites. For both parasites and parasitoids, there is a phase of their life when both they and their host are both alive at the same time; in this case, the larva of the parasitoid wasp feeds on the living host's tissues. The distinction is that parasites have not evolved to kill their hosts, quite the opposite in fact, though they can make their host very sick. One host always provides all the food needed by its parasitoid (or parasitoids) until it is killed. However, parasitoids inevitably kill their hosts, and this means that they have a quite different effect on host population dynamics, which is precisely why many parasitoid wasps, especially Braconidae species, are of great importance to both natural habitats and agro-ecosystems (Hochberg & Ives, 2000; Hoy et al., 1983; Zuparko, 2008).

The order Hymenoptera was traditionally divided into two suborders, the Symphyta (sawflies and woodwasps which have broad waists) and the Apocrita characterized by having a narrow "wasp waist." This system no longer used formally because the former are rendered paraphyletic by the latter (wasps evolved from sawflies and woodwasp-like ancestors). The apocritans are further divided into the Parasitica, most of which are parasitoids of other insects (Sharkey, 2007), and the Aculeata (stinging wasps and allies). In older literature, the name Terebrantia is sometimes used instead of Parasitica and refers to the long, protruding ovipositor of many parasitoid species.

The parasitoid wasps comprise a number of superfamilies but (a bit confusingly to the novice) actually includes some nonparasitoid lineages as well. The most well-known nonparasitic Parasitica are the phytophagous gall wasps (Cynipoidea). This is an example of the evolution of "secondary" plant feeding. The transition back to phytophagy has actually occurred on several independent occasions among the parasitoid wasps and occurs only rarely in the Braconidae.

In the Hymenoptera, the parasitoid lifestyle appears to have first evolved in Jurassic period (Rasnitsyn & Quicke,

2002). Several of the modern extant families appeared in the Cretaceous period. Fossil wasps of various kinds have been discovered from both amber and sedimentary deposits of Australia, Asia, Europe, North America, and South America (Zuparko, 2008). The family Braconidae appears to date from early Cretaceous (assuming *Eobracon* is properly assigned to family). The family Braconidae diversified extensively in the Middle to Late Cretaceous and Early Tertiary Periods, when flowering plants and their associated holometabolous herbivores radiated (herbivores such as caterpillars are among the main hosts for braconid parasitoids) (Basibuyuk et al., 1999; Ghahari, Yu, & van Achterberg, 2006; Rasnitsyn, 1983; Whitfield, 2002).

Estimating the true total number of Hymenoptera species is difficult as many species still remain undiscovered. The large number of unnamed species in insect collections demonstrates that the actual diversity is vastly larger than what has been published so far. Conservative estimates suggest the existence of over 600,000 species of Hymenoptera (Heraty & Darling, 2009), and the total could easily exceed one million (Rasplus et al., 2010; Sharkey, 2007). That could mean that only about 10% of hymenopteran species have been formally classified and given scientific names. Much higher numbers reaching 2.5 million species or higher, have even been proposed (Grissell, 1999; Stork, 1997); Sharkey et al. (2021) have even suggested that the single superfamily Ichneumonoidea could eventually turn out to comprise one million species. Clearly, much taxonomic (and biological) work still needs to be accomplished.

Most of the different parasitoid biology strategies occurring in the Hymenoptera are exhibited by various species of Braconidae (with the notable exception of hyperparasitism which is quite rare in braconid species). The most obviously differing feeding behaviors are ectoparasitism and endoparasitism. Ectoparasitoids are those larvae which feed and develop attached to the host on its outside, while endoparasitoids are those which feed inside the host's body. It should be noted that many endoparasitoid species have a last instar larva that exits the host and may feed externally for a time, but in these examples other aspects of the biology are largely determined by the earlier endoparasitoid phase. Both ectoparasitoids and endoparasitoid species are common in the Braconidae, with endoparasitoidism having arisen independently on several occasions. The ancestral biology of the family is still unknown, although widely presumed to have been ectoparasitoid in behavior (Quicke, 1997; Zuparko, 2008).

Another way of categorizing parasitic behaviors involves whether the species exhibits idiobiosis or koinobiosis. Idiobionts are those species which use a venom to paralyze the host during oviposition and paralysis is permanent (the host does not develop further); while koinobionts comprise those species that do not permanently paralyze the host, and the host continues to feed, develop, and molt (Godfray, 1994; Zuparko, 2008). Idiobiont species must develop solely by consuming the host biomass that is available to them at the time the parasitoid egg hatches, while koinobionts often delay their own feeding and development for some time, allowing the host to feed and grow to a much larger size before rapidly consuming it. While most ectoparasitoid species are idiobionts, and most koinobionts are endoparasitoids, this is not always the case (some species are endoparasitic idiobionts or ectoparasitic koinobionts) (Beckage et al., 1993; Hawkins & Sheehan, 1994; Hochberg & Ives, 2000; Quicke, 2015). Among the Braconidae (the focus of this book), most species are either ectoparasitic idiobionts or endoparasitic koinobionts, with the exception of just one subfamily (Shaw, 1983).

Several means of defense against parasitism have evolved through both morphological and physiological adaptations by the host (Gauld & Bolton, 1988; Godfray, 1994; Quicke, 1997). Most insects have an encapsulation response in which "foreign objects" in their bodies, such as a parasitoid egg, get coated in a thick layer of hemocytes which hardens to form a melanized capsule, thus killing the intruder. Endoparasitoids have to overcome encapsulation, usually either by general or specific suppression of the host immune response. This protective suppression involves complicated interactions of chemicals either injected into the host by the adult female wasp (venoms) or components released by the developing parasitoid larva within the host (Quicke & Butcher, 2021). In other cases, the wasp may avoid the host immune system entirely by inserting the egg into a nerve ganglion or muscle tissue, thereby evading the host's blood. One lineage of braconids and two of ichneumonids, a long time ago evolved mutualistic relationships with particular insect viruses that also helped to protect them. These viruses are no longer viruses in the strict sense because the beneficial parts of their DNA are now encoded within the wasp's genome, and the "virus" particles are generated within the female wasp's genital tract. These are termed polydnaviruses and are transferred to the host insect during oviposition, allowing delivery of genes that affect host function (Herniou et al., 2013; Lapointe et al., 2007).

Insect parasitoids have to search for their hosts in a dynamic heterogeneous environment. They must assess if potential host species are nearby, how many there may be, and how they are distributed in space. Since for parasitoids there is a direct link between host encounter rate and the production of offspring, natural selection acts strongly on parasitoid searching efficiency (Godfray, 1994; Hassell, 1980; Vet, 2001). Interactions between insect parasitoids and their arthropod hosts characteristically result in the premature death of the hosts and are obligatory for the development of the parasitic insects. This places strong selective pressure on the hosts to avoid detection by parasitoids, and on the parasitoids themselves to increase encounter rates with suitable hosts. To confront the challenge of locating the often-inconspicuous, well-hidden hosts, parasitoids have developed various sophisticated searching strategies that depend on a vast array of environmental cues (Turlings et al., 1993; Waage, 1979; Waage & Greathead, 1986).

Parasitoids are a substantial component of higher trophic levels in terrestrial ecosystems, and a major source of mortality in herbivorous insects (DeBach, 1974; Hawkins, 1994). Operating at higher trophic levels suggests an intrinsic vulnerability to local extirpation and possible extinction, and this may be exacerbated by the high degree of trophic specialization displayed by many species (Purvis et al., 2000). Indeed, there is evidence that parasitoids are highly sensitive to several extinction threats, including human-caused processes and activities such as habitat alteration or destruction, pesticide usage, introduction of exotic species, pollution, and climate change (Kruess & Tscharntke, 1994; Mayhew et al., 2009; Stireman et al., 2005). Once female parasitoids have located their hosts, they have the capacity to distinguish between parasitized and nonparasitized hosts, a behavior known as "discrimination ability" (van Alphen & Visser, 1990). This behavior can occur at three levels: (1) self-discrimination, (2) conspecific discrimination, and (3) heterospecific discrimination (Mackauer, 1990). This ability has been observed in many species of hymenopteran parasitoids (Vinson, 1976). It is particularly important in the case of potential biocontrol agents, since these are expected to be efficient in host searching and to have the ability to discriminate between parasitized and nonparasitized hosts (van Lenteren et al., 1978). The latter helps females to avoid superparasitism, and reduces the time and energy spent in searching behavior (Ayala et al., 2018; Mackauer, 1990).

Braconid wasps play important roles in regulating the populations of many herbivorous insects, and thus they are essential for the maintenance of ecological processes and shaping the diversity of the biosphere (e.g., by selectively feeding on caterpillars they can shape the distribution and abundance of the host insect's food plants) (Ghahari, Yu, & van Achterberg, 2006; Hanson & Gauld, 2006; LaSalle & Gauld, 1993; Ruiz-Guerra et al., 2015). These capable parasitoids recognize their hosts by various means during the searching period, such as chemodetection of kairomones released by plants as a result of the host's feeding behavior, host silk, and host frass (Powell, Pennacchio, Poppy, & Tremblay, 1998; Sheehan et al., 1993). Braconids are often host specific (especially koinobionts), making them an attractive group for use in biological control programs and as indicators of ecosystem health (Matthews, 1974; Ruiz-Guerra et al., 2015). The most common hosts selected by braconid wasps are species of Lepidoptera, Coleoptera, and Diptera (Quicke, 2015; Wharton, 1993); however, hosts from several other insect orders are utilized by particular groups, most notably by members of Euphorinae.

Diptera are used as hosts predominantly by members of two braconid subfamilies, the Alysiinae (Figs. 1.1 and 1.2) and the closely related Opiinae. Both of these are exclusively endoparasitoids as well as being exclusively parasitoids of flies. Thus being an endoparasitoid of flies is a rather specialist way of life and may constrain the wasps to that group of hosts.

Lepidopteran caterpillars (e.g., Fig. 1.3) of economic importance may be classified into pest injury guilds based on their method of injury (Klem & Zaspel, 2019). This is typical of pest Lepidoptera in agricultural settings (although exceptions exist), and larval feeding is the most common cause of economic damage. The larvae of pest Lepidoptera species may affect yield by causing direct damage to crops, such as by feeding on corn ears or fruit-boring, or by causing indirect damage and weakening the plant. Most lepidopterans are phytophagous and individual species depend on a limited range of plant species (within one or a

FIGURE 1.1 *Heterolexis* sp. (♀, Alysiinae) at rest on a leaf. *Photo courtesy of Stephen A. Marshall, University of Guelph.*

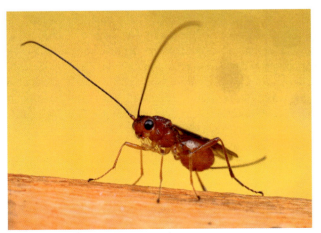

FIGURE 1.2 *Phaenocarpa* sp. (♀, Alysiinae). *Photo courtesy of Alex Wild, Texas State Museum.*

FIGURE 1.3 Caterpillar of *Euphydryas maturna* (Linnaeus, 1758) (Nymphalidae) parasitized by *Cotesia* sp. (Microgastrinae) whose larvae emerged and spun their silk cocoons below the remains of the host. *Photo courtesy of Irinel E. Popescu, Alexandru Ioan Cuza University, Romania.*

few plant families) for larval growth (Carter, 1984; Honda, 1995; Sathe & Pandharbale, 2008). There are numerous efficient larval parasitoids within the family Braconidae, which suppress the population density of many lepidopteran pests (Quicke, 2015; Shaw, 1990; Sree & Varma, 2015).

Caterpillars which have been attacked by endoparasitoid wasps not only ultimately die, but also develop more slowly and usually consume much less plant food. Therefore parasitism has a twofold benefit: reducing the numbers of pests that can go on to breed and produce a new pest generation, and reducing the damage done *in situ*.

The Ichneumonidae are regarded as the sister group to the braconids (Belshaw et al., 1998; Dowton et al., 2002; Li et al., 2016; Quicke & van Achterberg, 1990; Quicke et al., 1999; Sharanowski et al., 2011, 2021; Sharkey & Wahl, 1992; Shi et al., 2005; Wei et al., 2010); although the most recent study finds the sister-group of Ichneumonidae to be the Braconidae + Trachypetidae combined (Quicke et al., 2020). One of the best-studied groups of Braconidae is the subfamily Aphidiinae, which was for a long time treated as a separate family, Aphidiidae, and much of the older literature appears under that family name (Chen & van Achterberg, 2019).

Braconids are often encountered, as they are widely distributed, being well represented in nearly all parts of the world (Yu et al., 2016). They are among the most beneficial parasitic wasps and are of great value as biological control agents against pests that cause economic damage to agriculture, horticulture, and forestry (Austin & Dowton, 2000; Chen and van Achterberg, 2019; Wharton, 1993). Braconidae are considered as one of the principal groups of natural enemies of phytophagous insects (Chen & van Achterberg, 2019; Shaw, 1995), and hosts from more than 120 families are attacked (Wharton, 1993; Yu et al., 2016). Additionally, early life stages of many hosts are attacked by

koinobiont braconid species (Gauld, 1988) and this has important implications for being the most effective parasitoids for biocontrol, as these can kill their hosts prior to the stage causing the most economic damage to crops (Austin & Dowton, 2000). Aside from their importance as biological control agents, braconids are excellent models for many other studies, including studies in biodiversity and conservation (Lewis & Whitfield, 1999); studies on host-parasitoid interactions, and studies of physiology, behavior, and evolutionary biology (Clausen, 1942, 1954; Strand, 2000; Strand & Obrycki, 1996; Whitfield, 1998). Braconids are also used as model organisms in many evolutionary studies of parasitic development (Belshaw & Quicke, 2002; Gauld, 1988; Jervis et al., 2001; Quicke and Belshaw, 1999; Shaw, 1985; Whitfield, 1992; Zaldívar-Riverón et al., 2008), studies of morphological adaptation and convergence (Belshaw et al., 2003; Quicke & Belshaw, 1999), and exploration of polydnavirus evolution (Bracovirus, BV) (Bezier et al., 2009; Whitfield, 1997). Based on phylogenetic studies conducted by Jones et al. (2009), about 18,000 braconid species (along with at least 26,000 estimated undescribed species), are BV-carrying Braconidae. All of these belong to five subfamilies forming a monophyletic group, the microgastroid complex (Murphy, Banks, Whitfield, & Austin, 2008; Whitfield, 1997, 2002).

Based on head and mouth part structure, members of Braconidae have been separated into two taxonomically informal groups: "cyclostomes" with more or less concave and glabrous labrum, and a partly depressed clypeus (forming a circular or oval "hypoclypeal depression") (Fig. 1.9a), and the "noncyclostomes," characterized by having a flat or convex clypeus, as well as a flat and setose labrum (Fig. 1.9b) (Maetô, 1987; Quicke and van Achterberg, 1990; Tobias, 1967; van Achterberg, 1984a; Wharton, 1993). Cyclostome braconids have a wide range of behaviors but most of them are idiobiont ectoparasitoids of concealed-feeding hosts, while the noncyclostomes are mostly koinobiont endoparasitoids of various hosts (often exposed-feeders). The hypothetical relationships between these two groups differ based on the characters used for phylogenetic analysis, whether based upon morphological data (Quicke and van Achterberg, 1990; van Achterberg, 1984a); molecular data (Belshaw and Quicke, 2002; Belshaw et al., 1998, 2000; Dowton et al., 1998; Sharanowski et al., 2021); or both morphological and molecular data combined (Dowton et al., 2002; Pitz et al., 2007; Shi et al., 2005; Zaldívar-Riverón et al., 2006).

The inability to fully resolve relationships between all braconid subfamilies is considered one of the greatest problems for ongoing systematic studies of this family (Chen & van Achterberg, 2019). This issue is attributed to the enormous biological diversification of this group of wasps (Whitfield, 1992; Zaldívar-Riverón et al., 2006). The number of recognized subfamilies ranges from 17 to 50, depending on the author, with no universally accepted single classification (Wharton, 2000; Wharton & van Achterberg, 2000). Braconid-host relationships are summarized by several authors, including Belokobylskij and Tobias (1986), Quicke and van Achterberg (1990), Shaw (1997), Shaw and Huddleston (1991), van Achterberg (1984a), and Whitfield et al. (2018).

A wide range of parasitic lifestyles are exhibited by the Braconidae (Austin & Dangerfield, 1998). They are either ectoparasitic or endoparasitic and may either induce permanent paralysis of the host during oviposition, thus stopping host development (idiobiosis) (Askew & Shaw, 1986; Gupta, 1988; Wharton, 1993), or they may allow the host to continue its development (koinobiosis) (Askew & Shaw, 1986). Most braconids are either koinobiont endoparasitoids or idiobiont ectoparasitoids (Gauld, 1988; Shaw & Huddleston, 1991). The exceptions are few. Shaw (1983) documented ectoparasitic koinobiosis in the subfamily Rhysipolinae (as Rogadinae *s.l.*), and a few tropical Braconinae are known to be idiobiont endoparasitoids within lepidopteran pupae (van Achterberg, 1984b). Thus virtually all ectoparasitic braconids attack concealed hosts, while endoparasitoid braconids use both concealed hosts and exposed ones (Wharton, 1993).

The majority of braconid subfamilies are comprised of solitary parasitoids (Wharton, 1993). However, gregarious development, with more than one individual wasp developing in or on a single host is widespread (Fig. 1.3). It is true of many smaller ectoparasitoids such as various species of *Bracon* (Braconinae) and *Hormius* (Hormiinae) but also occurs in some subfamilies of koinobiont endoparasitoids. In fact, recent detailed observations of rearing tropical braconids are revealing gregariousness to be more widespread than previously thought (Sharkey et al., 2021). Gregarious koinobiosis is common among many genera of Microgastrinae and Macrocentrinae (Clausen, 1940). However, it comes about in different ways in these two subfamilies. In the former, the female wasp oviposits many eggs into a single host, and these each develop to produce an adult. In the second case, the gregariousness is the result of polyembryony. That is, the female macrocentrine oviposits only one or two eggs into a host caterpillar, but the resulting embryos keep dividing and dividing to produce many genetically identical siblings.

A few rare braconid species are seed-feeders or gall-forming herbivores (Austin & Dangerfield, 1998; Centrella & Shaw, 2010; Chavarría et al., 2009; Infante et al., 1995; Macêdo & Monteiro, 1989; Wharton & Hanson, 2005; Zaldívar-Riverón et al., 2007). This biology is thus far known best among New World species, although it is likely to be more widespread. An intermediate biology, entomophytophagy, in which the wasp larva first consumes a gall-forming insect host but then continues its development by feeding on the plant gall tissues was recently described (Ranjith et al., 2016). This all goes to illustrate a good degree of developmental plasticity in some groups.

Earlier studies on the systematics of the Braconidae were based only on morphological data, sometimes combined with a little biological data (Quicke and van Achterberg, 1990; Quicke et al., 1999; Sharkey & Wahl, 1992; Wharton, 2000). Deeper understanding of phylogeny and evolution of the family has resulted from studies using either molecular data alone (Belshaw et al., 1998; Jasso-Martínez et al., 2020; Sharanowski et al., 2011, 2021; Wei et al., 2010), or molecular and morphological data combined (Dowton et al., 2002; Quicke et al., 1999; Shi et al., 2005; Zaldívar-Riverón et al., 2006). With modern molecular techniques such as use of ultraconserved elements and whole genomes, it is to be expected that within the near future we will have a fully resolved and well-supported phylogeny of the family. This in turn will allow us to understand how many biological traits have evolved.

Although the major biological feature of braconid wasps is parasitoidism, they may also play some other either incidental or important ecological roles. Flower pollination is one of the crucial events in the life cycle of many flowering plants (Faegri and van der Pijl, 1979; Proctor et al., 1996). Pollinator diversity varies between habitats, both in species richness and in the total number of individuals recorded (McGregor, 1976; Mudri-Stojnić et al., 2012). It involves 67% of species of flowering plants and a relatively high diversity of insect taxa (Forup et al., 2008). On the other hand, 35% of crop production worldwide and 70% of major global crop species rely on animal pollination (Steffan-Dewenter & Westphal, 2008). Although the most numerous group of pollinating insects in all investigated localities are bees (Hymenoptera: Apoidea) (Mudri-Stojnić et al., 2012), some members of the superfamily Ichneumonoidea (Braconidae + Trachypetidae + Ichneumonidae) have minor roles as pollinators in some ecosystems (Figs. 1.4 and 1.5), as well as some other insect taxa such as Coleoptera, Diptera, and Lepidoptera (Free, 1993; Ghahari, 2017; Pena et al., 2002; Richards, 1978). Surely these parasitoids are not specialized pollinators; however, while feeding on nectar or while searching for hidden hosts in developing flower seeds, pollen grains stick to their bodies and so are transferred to other flowers.

Especially in Iran and other, largely arid, Mediterranean ecosystem countries, the prevalence and success of braconids and other insect parasitoids is likely to be affected greatly by the availability of fluids to drink. Therefore floral and extrafloral nectar are likely to be extremely important resources for adult wasp survival.

Iran, also historically called "Persia," is the world's 17th largest country and is the second largest country in the

FIGURE 1.4 *Agathis malvacearum* Latreille, 1805 (♀, Agathidinae) feeding from an Asteraceae flower. *Photo courtesy of Stephen A. Marshall, University of Guelph.*

FIGURE 1.5 *Bracon* sp. (♀, Braconinae) on a flower. *Photo courtesy of I.E. Popescu, Alexandru Ioan Cuza University, Romania.*

Middle East, spanning 1,623,779 km^2, occupying the largest part of the Iranian plateau. Iran is situated in Western Asia and is bordered to the north by the Caspian Sea, to the east by Afghanistan, to the south by the Persian Gulf and the Gulf of Oman, and to the west by Turkey and Iraq. Its significant geostrategic importance results from its central location in Eurasia and its proximity to the Strait of Hormuz (Barthold, 1984; Fisher, 1968; Zehzad et al., 2002). Iran's climate is diverse, ranging from arid and semiarid, to subtropical along the Caspian Coast, and the forested coniferous in the north (Mansouri et al., 2014). Although Iran is part of the Palaearctic region, it is located near two other biogeographic regions (Afrotropical and Indo-Australian), as well as near the Caucasus Mountains. Thus, Iran encompasses considerable topographic diversity and features three main climatic zones including arid and semiarid regions, Mediterranean climate (mainly in the western Zagros Mountains) and humid and semihumid regions (mainly in the Caspian) (Hakimzadeh Khoei et al., 2011; Tavassoli, 2016; Zehzad et al., 2002).

From the floristic point of view, the diverse and various climatic and geological conditions throughout Iran, from the Alborz and Zagros Mountains, Caspian Sea in the north, and Persian Gulf and Oman Gulf in the south of Iran, have resulted in the evolution of distinctive floristic regions (Sanjerehei, 2019). The country is divided into five floristic regions: the Hyrcanian, Irano-Turanian, Zagros, Khalijio-Omanian, and Arasbaran floristic zones (Trugobov & Mobayen, 1970). The varieties in landscapes and weather conditions across Iran (Zehzad et al., 2002) have resulted in rich environmental conditions and high diversity of flora within the Iranian Plateau (Ghahreman & Attar, 1999), resulting also in a high species richness of insects.

The Middle East is a region comprising the vast majority of the western Asia and all Egypt (most in North Africa), or in other words, it represents the lands around the southern and eastern shores of the Mediterranean Sea. The central part of this general area was formerly called the Near East, so the term Middle East has come as a replacement of the formerly used term "Near East" in the early 20th century. Most of the Middle East countries (13 out of 16) are part of the Arab World, from which Egypt, Iran, and Turkey are the most populous countries in the region, while Saudi Arabia is the largest in terms of area. The Middle East region is generally characterized by its hot, arid climate, with several major rivers providing irrigation to support agriculture in limited areas. The arid climate and the great dependence on the fossil fuel industry have a negative impact on the region that may cause climate change.

The following countries also are traditionally included in the Middle East region: Bahrain, Cyprus, Egypt, Iraq, Israel and Palestine, Jordan, Kuwait, Lebanon, Oman, Qatar, Saudi Arabia, Syria, Turkey, United Arab Emirates, and Yemen.

The first attempt to list all the recorded species of Iranian Braconidae was done by Farahbakhsh (1961) including only six species (*Apanteles glomerata* (L.), *Habrobracon brevicornis* (Wesmael), *Habrobracon hebetor* (Say), *Chelonus contractor* (Nees von Esenbeck), *Phanerotoma* sp. and *Rogas* sp.). Modarres Awal (1997, 2012) has listed 57 (in addition to nine unknown species) and 114 known species, respectively (in two separate families, Aphidiidae and Braconidae). Fallahzadeh and Saghaei (2010), and Farahani et al. (2016) recorded 202 and 780 species, respectively. Yu et al. (2016) cataloged 804 species and subspecies based on 246 references, and the last catalogs have been published by Samin et al. (2018a,b) reporting 834 and 861 species and subspecies, respectively. In addition to the previously mentioned checklists on Iranian Braconidae, several other checklists were published on all the subfamilies of Iranian Braconidae (Barahoei et al., 2014; Beyarslan et al., 2017; Cortés et al., 2016; Fischer et al., 2011; Gadallah and Ghahari, 2013a,b, 2015, 2016, 2017; Gadallah et al., 2015a,b, 2016a,b, 2019; Ghahari, 2016; Khajeh et al., 2014; Sedighi & Madjidzadeh, 2015).

The present book attempts to include all braconid wasp species (Hymenoptera: Braconidae) recorded from any of the Middle East countries, with special reference to Iran, as being in particular rich and diverse with braconid species. We follow the classification of Chen and van Achterberg (2019) in the taxonomic arrangement of the family Braconidae. In this volume, we document the presence of 30 braconid subfamilies in the Middle East (Agathidinae, Alysiinae, Aphidiinae, Brachistinae, Braconinae, Cardiochilinae, Cenocoelinae, Charmontinae, Cheloninae, Dirrhopinae, Doryctinae, Euphorinae, Exothecinae, Gnamptodontinae, Helconinae, Homolobinae, Hormiinae, Ichneutinae, Macrocentrinae, Microgastrinae, Microtypinae, Miracinae, Opiinae, Orgilinae, Pambolinae, Proteropinae, Rhysipolinae, Rhyssalinae, Rogadinae, and Sigalphinae). In each subfamily chapter, an introductory section is provided, articulating the number of species, genera, and tribes worldwide as well as in the Iranian fauna. Also provided is a subfamily diagnosis differentiating it from other braconid subfamilies, notes on the biology of its members, and discussion of the hypothesized phylogenetic

relationships with other subfamilies. Following this general subfamily information, we summarize the information included in the following checklist of species for that subfamily. This summary information includes the number of species recorded from the Middle East countries, local distribution of species in the Iranian fauna in particular, their worldwide distribution, and their host records. For general distribution of species worldwide, we follow Yu et al. (2016), supplemented by other available, more recent references. Following the checklist of recorded species for each subfamily is a separate list of any species newly excluded from the Iranian fauna, if any, with the reasons for excluding them. Additionally, identification key for 30 subfamilies of the Middle East Braconidae is constructed; for wing venation, we follow van Achterberg's terminology (1993) (Fig. 1.7).

In each chapter, we represent the list of species recorded from Middle East countries (Bahrain, Cyprus, Egypt, Iran, Iraq, Israel and Palestine, Jordan, Kuwait, Lebanon, Libya, Oman, Qatar, Saudi Arabia, Syria, Turkey, United Arab Emirates, and Yemen), as well as compare the fauna of Iran with those of the other 16 Middle East countries together with seven adjacent ones to Iran (Afghanistan, Armenia, Azerbaijan, Kazakhstan, Pakistan, Russia, and Turkmenistan) having land and sea borders with Iran (Fig. 1.6). Additionally, the shared species between Iran and the mentioned countries is discussed. These comparisons should be interpreted with caution because of

FIGURE 1.6 Map of Iran in relation to Middle East and adjacent countries.

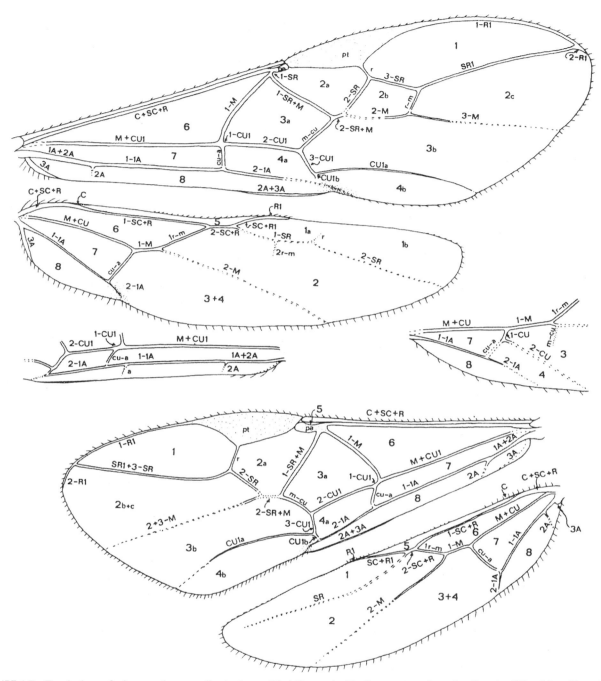

FIGURE 1.7 Terminology of wing venation according to the modified Comstock—Needham system. *A*, analis; *C*, costa; *CU*, cubitus; *M*, media; *R*, radius; *SC*, subcosta; *SR*, sectio radii (or RS of "radial sector"); *a*, transverse anal vein; *cu-a*, transverse cubito-anal vein; *m-cu*, transverse medio-cubital vein; *r*, transverse radial vein; *r—m*, transverse radio-medial vein; *pa*, parastigma; *pt.*, pterostigma. Cells: 1: marginal cell; 2: submarginal cell; 3: discal cell; 4: subdis-cal cell; 5: costal cell; 6: basal cell; 7: subbasal cell; 8: plical cell or (if protruding) lobe; a, b, and c indicate first, second and third cell, respectively. *Adapted from van Achterberg, C. (1993). Illustrated key to the subfamilies of the Braconidae (Hymenoptera: Ichneumonoidea).* Zoologische Verhandelingen, 283, 1—189.

differences in the states of knowledge of the braconid faunas of these different countries, but they are offered to provide baseline data for future studies and analyses. In the last chapter, we tabulate the species diversity of Braconidae of the Middle East by subfamily (comprising the number of genera and species for each) (Table 32.1), highlighting the species so far only known from the Middle Eastern countries (Tables 32.3—32.8). Because of the importance of Braconidae for biological control of agricultural and forest pests in Iran, host information for parasitoid species (parasitoid–host relationships) that is provided throughout the chapters is synthesized in Table 32.9. Additionally, species diversity of the Iranian Braconidae is discussed, including the number of species of each subfamily (Table 32.2), the species that are currently only known from Iran (Table 32.4), the total number of reported species of Iranian Braconidae by province (Fig. 32.2). In order to simply summarize the Iranian Braconidae fauna, all the species treated in this book are listed in appendix 1 at the end of the book. Many more species are expected to exist, for which more field collecting is needed in some of the more poorly studied areas of Iran. However, this catalog of currently known species should prove useful for researchers and students interested in this important wasp group in Iran, as well as in other Middle Eastern countries and neighboring countries to Iran.

Key to the subfamilies of Braconidae in the Middle East

1. Mandibles exodont (outward-facing), teeth directed away from longitudinal body axis and tips not meeting when mandibles are closed (Fig. 1.8a) .... **Alysiinae** Leach, 1815
- Mandibles normal, teeth directed toward each other and tips meeting or overlapping when mandibles are closed (Fig. 1.8b) 2
2. Metasomal tergum 2 + 3 basally with a narrow polished and shining transverse or semicircular area, and posteriorly often with a crenulate sulcus (Fig. 1.8c); minute species less than 2.0 mm in body length **Gnamptodontinae** Fischer, 1970
- Metasomal tergum 2 + 3 not as above, without a polished transverse or semicircular area basally (Fig. 1.8d), or polished area larger and triangular, and body size larger than 2.0 m 3

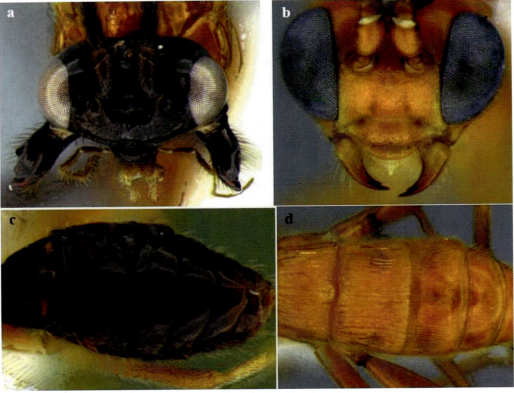

FIGURE 1.8 (a) Frontal view of *Gnathopleura* sp. head (Alysiinae) (showing exodont mandibles); (b) Frontal view of *Zele* sp. head (Euphorinae) (showing mandibles facing inwards); (c) Dorso-lateral view of *Gnamptodon* sp. metasoma (Gnamptodontinae); (d) Dorsal view of *Aleiodes* sp. metasoma (Rogadinae). *Photos prepared by S.R. Shaw.*

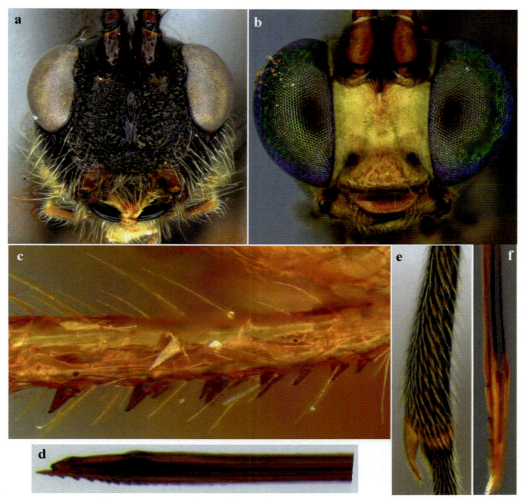

FIGURE 1.9   (a) Frontal view of *Odontobracon* sp. head (Doryctinae) (showing cyclostome oral area); (b) Frontal view of *Zele* sp. head (Euphorinae) (showing non-cyclostome head); (c) Front tibia of *Gymnobracon* sp. (Doryctinae); (d) Ovipositor of *Odontobracon* sp. (Doryctinae); (e) Front tibia of *Cyanopterus* sp. (Braconinae); (f) Ovipositor of a braconine species. *Photos prepared by S.R. Shaw.*

3. Mandibles, when closed, remote from lower margin of clypeus and thus a conspicuous oral cavity is present between mandibles and clypeus; labrum normally visible, shining, and often concave (cyclostomes) (Fig. 1.9a) [note: some Opiinae can be keyed by either couplet] 4

- Mandibles, when closed, close to lower margin of clypeus and concealing labrum, without a conspicuous oral cavity or depression (noncyclostomes) (Fig. 1.9b) 12

4. Fore tibia with a line of stout peg-like setae along anterior surface (Fig. 1.9c), or (if difficult to see setae in minute specimens) then fore wing with vein 2-SR partly or completely absent (Fig. 1.10a), or males with a pterostigma on anterior hind wing margin (Fig. 1.10c); ovipositor well-sclerotized and distinctly blackened apically, dorsal valve of ovipositor tip with a subapical double nodus (Fig. 1.9d) .... **Doryctinae** Foerster, 1863

- Fore tibia with only normal hair-like setae, lacking stout peg-like setae (Fig. 1.9e); fore wing with vein 2-SR present (Fig. 1.10b); males lacking a pterostigma on the hind wing; ovipositor variable, but often not as well-sclerotized and not blackened apically, dorsal valve with only a single nodus, or none (Fig. 1.9f) 5

5. Epicnemial carina absent (Fig. 1.11b); occipital carina absent (Fig. 1.11c), either entirely or at least dorsally 6

- Epicnemial carina present (Fig. 1.11a); occipital carina present dorsally (Fig. 1.11d) 8

6. Occipital carina completely absent (Fig. 1.11c) .... **Braconinae** Nees, 1811

- Occipital carina present laterally and ventrally, absent only dorsally (Fig. 1.12c) 7

FIGURE 1.10 (a) Fore wing of *Heterospilus* sp. (♀) (Doryctinae) (showing absence of vein 2-SR); (b) Fore and hind wings of *Aleiodes* sp. (♂) (Rogadinae) (showing the presence of 2-SR of fore wing and the absence of pterostigma on hind wing); (c) Hind wing of male *Heterospilus* sp. (Doryctinae) (showing pterostigma). *Photos prepared by S.R. Shaw.*

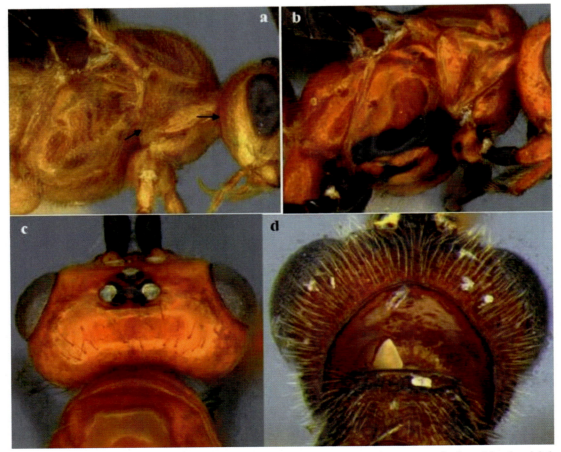

FIGURE 1.11 (a) Lateral view of head (part) and mesosoma of *Aleiodes* sp. (Rogadinae) (showing presence of epicnemial and occipital carinae); (b) Lateral view of head (part) and mesosoma of a braconine species (Braconinae) (showing absence of epicnemial and occipital carinae); (c) Dorsal view of head of *Vipio* sp. (Braconinae) (showing absence of occipital carina); (d) Dorsal view of head *Heterogamus tatianae* Telenga (Rogadinae) (showing the complete occipital carina). *Photos prepared by S.R. Shaw.*

FIGURE 1.12   (a) Fore and hind wings of an opiine species (Opiinae) (Showing the elongate second submarginal cell, and short vein r); (b) Fore and part of hind wing of an opiine species (Opiinae) with elongate pterostigma, vein r arising near base of pterostigma); (c) Posterior view of head of *Opius* sp. (Opiinae) (showing the absence of occipital carina dorsally and its presence laterally); (d) Fore and hind wings of *Colastes braconius* Haliday (Exothecinae) (showing the normal second submarginal cell, and insertion of vein r near to middle of pterostigma); (e) Fore and hind wings of *Shawiana catenator* (Haliday) (Exothecinae) (showing normal pterostigma, and insertion of vein r near to middle of pterostigma). *Photos prepared by S.R. Shaw.*

7. Second submarginal cell of fore wing often elongated and narrow apically, vein 3-SR quite long, often as much as 6× longer than vein r (Fig. 1.12a and b); or pterostigma greatly elongated and narrow, vein r arising from near base of pterostigma (Fig. 1.12b) .... **Opiinae** Blanchard, 1845 (in part)
- Second submarginal cell of fore wing not unusually elongated or narrow apically, vein 3-SR not so long, much less than 6× longer than vein r (Fig. 1.12d and e); pterostigma of normal dimensions and never greatly elongated or narrow; vein r arising nearer to middle of pterostigma, never at extreme base (Fig. 1.12e) .... **Exothecinae** Foerster, 1863
8. Spiracles of metasomal terga 2 + 3 located ventrally on the laterotergites, below lateral edges of the dorsal surface of the metasoma (Fig. 1.13c, and d) .... **Rhyssalinae** Foerster, 1863
- Spiracles of metasomal terga 2 + 3 located on the dorsal surface, near but above the lateral edges of the dorsal surface of the metasoma (Fig. 1.13a and b) 9
9. Propodeum with spines or tubercles on lateral corners (Fig. 1.13e and f), if absent or small, then first metasomal tergum greatly widened posteriorly; metasomal terga 2 + 3 well-sclerotized, entirely smooth and shining .... **Pambolinae** Marshall, 1885
- Propodeum lacking spines or tubercles on lateral corners; first metasomal tergum not greatly widened posteriorly; metasomal terga 2 + 3 either extensively desclerotized, or well-sclerotized and not entirely smooth, often with extensive areas of fine to coarse microsculpture 10

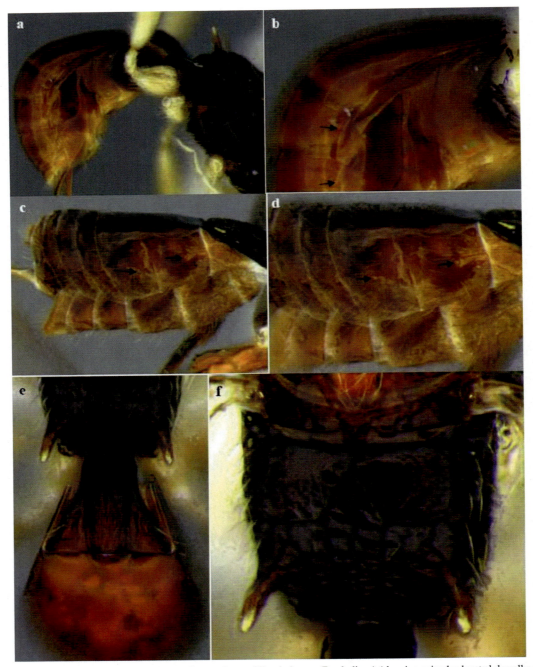

FIGURE 1.13 (a) Lateral view of metasoma and part of propodeum of *Pambolus* sp. (Pambolinae) (showing spiracles located dorsally on tergum 2 & tergum 3); (b) Lateral view of metasoma (part) of *Pambolus* sp. (Pambolinae); (c) Lateral view of metasoma of *Dolopsidea* sp. (Rhyssalinae) (showing spiracles located on laterotergites of tergum 2 & tergum 3); (d) Lateral view of part of metasoma of *Dolopsidea* sp. (Rhyssalinae); (e) Dorsal view of parts of propodeum and metasoma of *Pambolus* sp. (Pambolinae); (f) Propodeum of *Pampolus* sp. *Photos prepared by S.R. Shaw.*

10. Dorsal carinae of first metasomal tergum widely separated posteriorly (Fig. 1.14a); metasomal terga 2 + 3 mostly membranous and desclerotized (Fig. 1.14b) .... **Hormiinae** Foerster, 1863
- Dorsal lateral carinae of first metasomal tergum not widely separated, usually joining posteriorly to form a median carina (Fig. 1.14d); metasomal terga 2 + 3 not mostly membranous, mostly sclerotized at least weakly, and most commonly strongly sclerotized and rigid (Fig. 1.14d) 11

FIGURE 1.14   (a) Dorsolateral view of metasoma of *Hormius* sp. (Hormiinae) (showing widely separated dorsal carinae of tergum 1); (b) Dorsal view of metasoma of *Hormius* sp. (Hormiinae) (showing the membranous tergum 1 and tergum 2); (c) Dorsolateral view of metasoma of female *Aleiodes* sp. (Rogadinae) (showing short ovipositor); (d) Dorsolateral view of metasoma of *Aleiodes* sp.; (e) Dorsolateral view of metasoma of *Rhysipolis* sp. (Rhysipolinae) (showing long ovipositor); (f) Host caterpillar mummified by *Aleiodes* sp. *Photos prepared by S.R. Shaw.*

11. Metasomal terga 2—6 smooth and weakly sclerotized, commonly collapsing irregularly in dried specimens; first and second metasomal terga lacking a well-developed or long median carina (Fig. 1.14e); ovipositor longer than ½ hind tibia length and sometimes much longer (twice hind tibia length or longer) (Fig. 1.14e); parasitoids of gracillariid leaf-mining larvae, uncommon .... **Rhysipolinae** Belokobylskij, 1984
- Metasomal terga 2—6 well-sclerotized, variously sculptured, sometimes carapace-like (Fig. 1.14d); metasomal dorsum solid and well-preserved in dried specimens; first and second metasomal terga usually with a well-developed or long median carina (Fig. 1.14d); ovipositor usually shorter than ½ hind tibia length (Fig. 1.14c); endoparasitoids of various Lepidoptera larvae, pupating inside the mummified caterpillar remains and resulting in a well-preserved host mummy (Fig. 1.14f); diverse and commonly encountered species .... **Rogadinae** Foerster, 1863
12. First metasomal spiracle located within a membranous or poorly sclerotized area of the first laterotergite (Fig. 1.15c and d), or spiracle not visible in air-dried specimens due to shrinking of the membrane 13
- First metasomal spiracle located within a well-sclerotized area of first tergite laterally (Fig. 1.15a and b) 15
13. Antenna with 18—49 flagellomeres (Fig. 1.16a); fore wing vein SR1 strongly curved towards anterior margin of wing (Fig. 1.16c and d); maxillary palpus 6-segmented (Fig. 1.16b) .... **Cardiochilinae** Ashmead, 1900

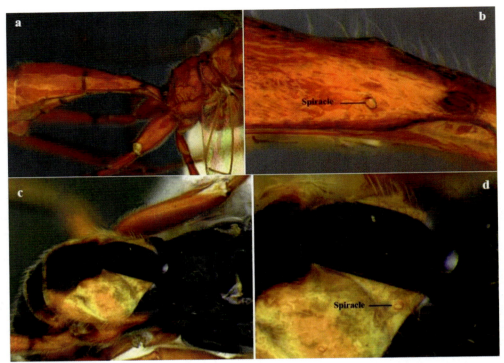

FIGURE 1.15   (a) Dorsolateral view of parts of mesosoma and metasoma of *Zele* sp. (Euphorinae, Meteorini) (showing the presence of spiracle within a well-sclerotized tergum 1); (b) Lateral view of tergum 1 of *Zele* sp. (Euphorinae, Meteorini) (focusing on spiracle); (c) Dorsolateral view of part of mesosoma and metasoma of *Sathon* sp. (Microgastrinae) (showing the presence of spiracle within a membranous area of tergum 1); (d) Lateral view of first metasomal segment of *Sathon* sp. (Microgastrinae) (focusing on spiracle). *Photos prepared by S.R. Shaw.*

FIGURE 1.16   (a) Lateral view of head and antenna of a cardiochiline species (Cardiochilinae); (b) Lateral view of mouthparts of a cardiochiline species (Cardiochilinae) (showing the six-segmented maxillary palpus, the head should be positioned correctly to count the number of segments); (c) Fore and hind wings of a cardiochiline species (Cardiochilinae); (d) Fore wing (part) of a cardiochiline species (focused to show the conspicuously curved SR1 toward wing margin). *Photos prepared by S.R. Shaw.*

FIGURE 1.17   (a) Fore and hind wings of *Microplitis* sp. (Microgastrinae); (b) Fore wing (part) of *Microplitis* sp. (Microgastrinae) (showing the SR1 straight or slightly curved, reduced apically); (c) Posterior view of *Microplitis* sp. head (Microgastrinae); (d) Posterior view of head (part) of *Microplitis* sp. (Microgastrinae) (showing five-segmented maxillary palpus). *Photos prepared by S.R. Shaw.*

- Antenna with 12—16 flagellomeres; fore wing vein SR1 straight or only slightly curved, often reduced apically (Fig. 1.17a and b); maxillary palpus 5-segmented (Fig. 1.17c and d) [note: microgastrines have the longitudinal antennal sensilla arranged in two even ranks per flagellomere, thus the 16-segmented flagellum may superficially appear to have 32 segments) 14

14. Antenna with 16 flagellomeres; first metasomal tergite variable in shape but never extremely narrow or linear medially (Fig. 1.18a and b) [parasitoids of Lepidoptera larvae with various feeding habits but quite commonly associated with large exposed-feeding caterpillars] .... **Microgastrinae** Foerster, 1863

- Antenna with 12 flagellomeres; first metasomal tergite extremely narrow and linear medially (Fig. 1.18c) [minute species which are parasitoids of small leaf-mining Lepidoptera larvae of the families Nepticulidae, Heliozelidae, and Lyonetiidae] .... **Miracinae** Viereck, 1918

15. First metasomal tergite flat and narrow, with spiracle situated behind middle of segment, and tergum noticeably narrowed posteriorly (beyond spiracle) (Fig. 1.18d); rarely encountered minute species, 2 mm or less (Fig. 1.18g), which are parasitoids of small leaf-mining Lepidoptera larvae of the family Nepticulidae... **Dirrhopinae** van Achterberg, 1984

- First metasomal tergite not as above, variously shaped but commonly convex or wide, not narrowing posteriorly, and with the spiracle situated near middle of segment, or anterior of middle (Fig. 1.15a and b); size and behavior variable but often much larger than 2 mm long 16

16. Hind trochantellus laterally with an aggregation of short, thick spines (Fig. 1.18f); occipital carina absent (Fig. 1.18e) .... **Macrocentrinae** Foerster, 1863

- Hind trochantellus lacking short, thick spines; occipital carina variable but most commonly present (Fig. 1.11d), at least laterally 17

17. Metasoma attached very high on propodeum (Fig. 1.19a), widely separated from bases of hind coxae .... **Cenocoeliinae** Szépligeti, 1901

- Metasoma attached low (Fig. 1.19b), at posterior base of propodeum, and very close to bases of hind coxa 18

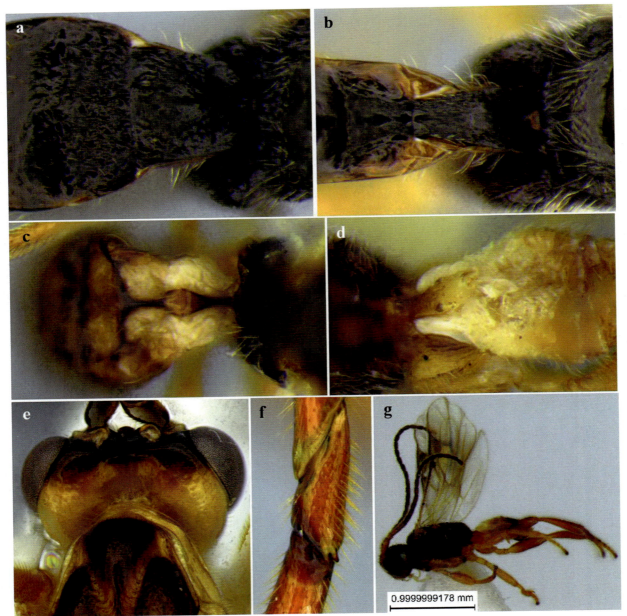

FIGURE 1.18 (a) Metasomal tergum 1 and part of tergum 2 of *Cotesia* sp. (Microgastrinae); (b) Metasomal tergum 1 and part of tergum 2 of *Glyptapanteles* sp. (Microgastrinae); (c) Metasomal tergum 1 of *Mirax* sp. (Miracinae) (showing tergum 1 extremely narrow and linear medially); (d) Metasomal tergum 1 of *Dirrhope* sp. (Dirrhopinae) (showing situation of small spiracle behind the middle of the segment); (e) Posterior view of head of *Dolichozele* sp. (Macrocentrinae) (showing absence of occipital carina); (f) Hind trochantellus of *Dolichozele* sp. (Macrocentrinae) (showing short thick spines); (g) Lateral habitus of *Dirrhope* sp. *Photos prepared by S.R. Shaw.*

18. Fore wing with marginal cell long and quite narrow, vein SR1 straight and meeting anterior wing margin well before wing tip (Fig. 1.19c and f); second submarginal cell usually present and quite small, triangular (Fig. 1.19f) or quadrate shaped (Fig. 1.19c) .... **Agathidinae** Haliday, 1833
- Fore wing venation different; marginal cell usually much longer (Fig. 1.19d), or if short, then SR1 vein is curved (Fig. 1.19e) or RS is incomplete and does not reach the wing margin (Figs. 1.19g and 1.20a); second submarginal cell not as above, either absent (Fig. 1.20a) or, if present, then much larger and variously shaped (Fig. 1.19d) 19

FIGURE 1.19   (a) Lateral habitus of female *Cenocoelius* sp. (Cenocoeliinae) (showing high attachment of metasoma on propodeum); (b) Lateral habitus of female *Aridelus* sp. (Euphorinae) (showing low attachment of metasoma on propodeum); (c) Fore and hind wings of *Cremnops* sp. (Agathidinae) (showing SR1 meeting wing margin before tip); (d) Fore and hind wings of *Chelonus* sp. (Cheloninae) (showing SR1 meeting wing tip); (e) Fore and hind wings of *Peristenus* sp. (Euphorinae) (showing the short, strongly curved SR1); (f) Fore and hind wings of *Zelomorpha* sp. (Agathidinae) (showing SR1 meeting wing margin before tip); (g) Fore wing (part) of *Adelius* sp. (Cheloninae) (showing the incomplete SR1). *Photos prepared by S.R. Shaw.*

19. Metasomal terga 1-3 forming a rigid, shell-like carapace covering most of the abdomen (Fig. 1.20c), and having at most one flexible joint (most often lacking any flexible joints, although sculptured lines indicating the points of fusion may be present) (Fig. 1.20b); fore wing with a closed second submarginal cell (Fig. 1.19d) 20
- Metasomal dorsum not carapace-like (Fig. 1.20d), or metasoma having more than three dorsally visible terga, or fore wing lacking a closed second submarginal cell (cross-vein r-m absent) (Figs. 1.19g and 1.20a) 21

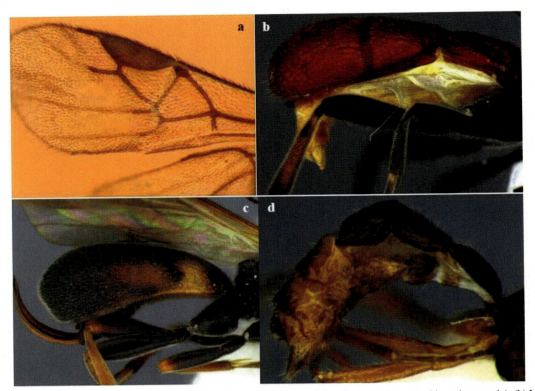

FIGURE 1.20   (a) Fore wing of *Adelius* sp. (Cheloninae: Adeliini) (showing the incomplete SR1+3-SR, not reaching wing margin); (b) Lateral view of metasoma of *Sigalphus* sp. (Sigalphinae) (showing membranous, flexible joint separating tergum 1 and tergum 2); (c) Lateral view of metasoma of *Chelonus* sp. (Cheloninae) (showing the fusion of tergum 1-3 forming a carapace); (d) Lateral view of metasoma of *Aphidius* sp. (Aphidiinae) (showing metasoma not forming a carapace). *Photos prepared by S.R. Shaw.*

20. Metasomal carapace completely rigid and lacking any flexible joints (Fig. 1.20c), at most with two transverse lines or grooves but often lacking any visible segmentation dorsally (Fig. 1.21a); often quite common and abundant species seen on leafy vegetation where they search for Lepidoptera eggs to parasitize .... **Cheloninae** Foerster, 1863 (in part, most species)
- Metasomal carapace with membranous, flexible joint separating tergum 1 and tergum 2 (Fig. 1.20b); relatively rare species seldom encountered) .... **Sigalphinae** Haliday, 1833
21. Metasomal terga 1-3 fused into a single flat plate making up more than half the dorsum of the metasoma (Fig. 1.21b); fore wing with SR1+3-SR incomplete and ending well before wing margin (Figs. 1.19d and 1.20a), minute rare species just a few millimeters long (Fig. 1.21c), associated with leaf-mining Lepidoptera larvae .... **Cheloninae** Foerster, 1863 (in part, Tribe Adeliini)
- Metasomal terga 1-3 not fused into a single flat plate making up more than half the dorsum of the metasoma; fore wing with SR1 usually complete and reaching wing margin, or (more rarely), if incomplete, then arising from other venation, not directly off pterostigma; size variable but often much larger (many common species key this way) 22
22. Fore wing with vein 1-M strongly and often abruptly curved near intersection with vein SR1 (Fig. 1.21d–f); parasitoids of immature plant-feeding Hymenoptera (sawfly larvae) 23
- Fore wing with vein 1-M either straight or just gradually curved over more of its length, never abruptly curved near intersection with vein SR1 (Fig. 1.21d); parasitoids of various insects but not sawflies 24
23. Fore wing with vein 1-M strongly bent apically and abruptly curved near intersection with vein 1SR + M (Fig. 1.21d and e); hind wing vein SR mostly desclerotized, clear, and not reaching wing margin (Fig. 1.21f); anterior tentorial pits normal and small (Fig. 1.22a) .... **Ichneutinae** Foerster, 1863

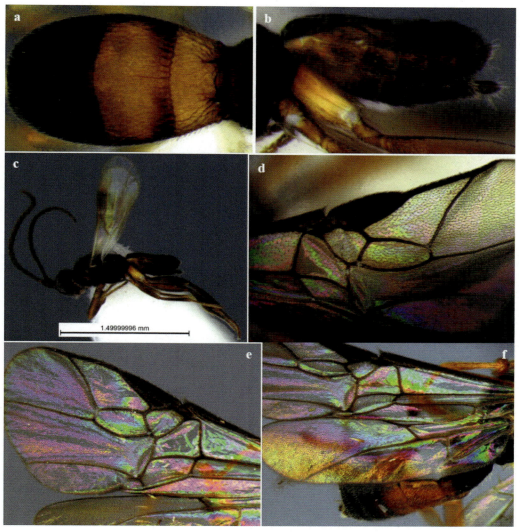

FIGURE 1.21   (a) Dorsal view of metasoma of *Phanerotoma* sp. (Cheloninae) (showing two metasomal lines); (b) Lateral view of metasoma of *Adelius* sp. (Cheloninae); (c) Lateral habitus of *Adelius* sp. (Cheloninae); (d) Fore wing (part) of *Ichneutes* sp. (Ichneutinae); (e) Fore wing of *Ichneutes* sp. (Ichneutinae); (f) Fore and hind wings of *Ichneutes* sp. (Ichneutinae) (showing desclerotized vein SR1 of hind wing). *Photos prepared by S.R. Shaw.*

- Fore wing with vein 1-M bent apically (Fig. 1.22c), but not so abruptly curved as above; hind wing vein SR sclerotized, pigmented, and reaching wing margin (Fig. 1.22d); anterior tentorial pits very large (Fig. 1.22b) .... **Proteropinae** van Achterberg, 1976
24. Metasomal terga weakly sclerotized (Fig. 1.20d); fused tergum 2 + 3 with a flexible joint, able to bend at line of fusion of tergum 2 with tergum 3 (Fig. 1.20d); scutellar sulcus smooth (Fig. 1.23a); hind wing with veins CU and cu-a absent, therefore lacking a small closed cell in the basal posterior area of hind wing (Fig. 1.23c); parasitoids of aphids, pupating inside the mummified aphid remains (Fig. 1.23b) .... **Aphidiinae** Haliday, 1833
- Metasomal terga well sclerotized; tergum 2 and tergum 3 solidly fused, metasoma not able to bend at this joint; scutellar sulcus usually crenulate or otherwise coarsely sculptured, or, more rarely, if smooth, then hind wing with veins CU and cu-a present; hind wing subbasal cell present; parasitoids of various insect hosts but not associated with aphids; pupating externally in a silken cocoon outside the host remains 25

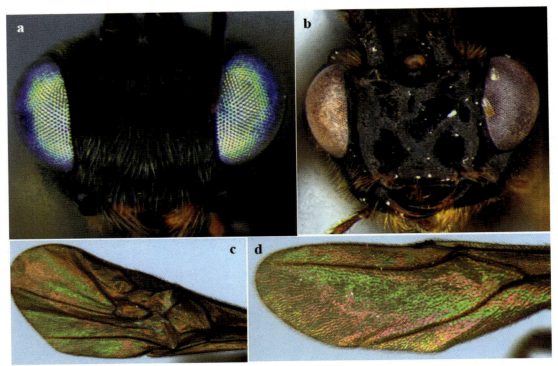

FIGURE 1.22   (a) Frontal view of head of *Ichneutes* sp. (Ichneutinae) (showing normal tentorial pits); (b) Frontal view of head of *Proterops* sp. (Proteropinae) (showing large tentorial pits); (c) Fore wing of *Proterops* sp. (Proteropinae) (showing the slightly curved vein 1M); (d) Hind wing of *Proterops* sp. (Proteropinae) (showing vein SR1 sclerotized and reaching wing margin). *Photos prepared by S.R. Shaw.*

25. Second submarginal cell of fore wing often elongated and narrow apically, vein 3-SR quite long, often as much as 6× longer than vein r (Fig. 1.12a and b); or pterostigma greatly elongated and narrow, vein r arising from near base of pterostigma .... **Opiinae** Blanchard, 1845 (in part)
- Second submarginal cell of fore wing absent, or, if present then not unusually elongated or narrow apically, vein 3-SR not so long, much less than 6× longer than vein r; pterostigma of normal dimensions and never greatly elongated or narrow; vein r arising nearer to middle of pterostigma, never at extreme base 26
26. Second submarginal cell of forewing absent (Fig. 1.19e); occipital carina partly or entirely absent, at least absent dorso-medially 27
- Second submarginal cell of forewing present (as in Fig. 1.10b), or, if absent (more rarely), then occipital carina entirely present 28
27. Apical segment of SR1 vein of fore wing long and straight, extending most of the distance toward wing tip and ending on anterior wing margin at a position that is closer to the wing tip than to the pterostigma (Fig. 1.23e); first subdiscal cell of fore wing closed apically by a short cross-vein, therefore separated from second subdiscal cell (Fig. 1.23d); ovipositor long and straight (Fig. 1.23f); parasitoids of concealed-feeding Lepidoptera larvae, common in woodland habitats .... **Orgilinae** Ashmead, 1900
- Apical segment of SR1 vein of fore wing short and curved, ending on wing margin at a position that is closer to the pterostigma than to the wing tip (Fig. 1.24c); first subdiscal cell of fore wing open apically and (narrowly) confluent with the second subdiscal cell (cross-vein CU1b absent); ovipositor short and curved (Fig. 1.24a and b); parasitoids of various nymphal or adult insects but not associated with Lepidoptera larvae (more rarely encountered species) .... **Euphorinae** Foerster, 1863 (in part)
28. Fore wing with a closed second submarginal cell (cross-vein r-m present) 29
- Fore wing lacking a closed second submarginal cell (cross-vein r-m absent) 33

FIGURE 1.23 (a) Dorsal view of mesosoma of *Aphidius* sp. (Aphidiinae) (showing smooth scutellar sulcus); (b) Dried remains of an aphid mummified by *Aphidius* sp. (Aphidiinae); (c) Fore and hind wings of *Praon* sp. (Aphidiinae) (showing the reduction of hind wing venation); (d) Fore wing (part) of *Orgilus* sp. (Orgilinae) (showing the closed first subdiscal cell); (e) Fore wing of *Orgilus* sp. (Orgilinae) (showing the long, straight SR1); (f) Lateral habitus of female *Orgilus* sp. (Orgilinae) (showing the long, straight ovipositor). *Photos prepared by S.R. Shaw.*

29. Fore wing with four-sided second submarginal cell that is distinctively trapezoidal in shape (Fig. 1.24d); metasoma with tergum 1 narrowly petiolate basally and broadening abruptly beyond the middle of the segment (Fig. 1.24f); parasitoid cocoon commonly suspended from substrate by a short to long silk thread (Fig. 1.24e) .... **Euphorinae** Foerster, 1863 (in part, tribe Meteorini)
- Fore wing with second submarginal cell (cross-vein r-m absent) differently shaped; metasoma with tergum 1 of various shapes but often broad basally, if narrow basally then narrow over its entire length; not broadening abruptly beyond the middle of the segment; parasitoid cocoon not suspended from substrate by a silk thread 30
30. Second submarginal cell of fore wing large and more or less quadrate (Fig. 1.25a and c) (more common species) 31
- Second submarginal cell of fore wing small and either rounded or triangular 32
31. Hind tibia inner apical spur about half as long as basitarsus (Fig. 1.25e); often night-active species attracted to lights and with large ocelli (Fig. 1.25b) .... **Homolobinae** van Achterberg, 1979
- Hind tibia inner apical spur much shorter than half basitarsus length (Fig. 1.25f); day-active species associated with dead trees, not attracted to lights, and with small (normal-sized) ocelli .... **Helconinae** Foerster, 1863

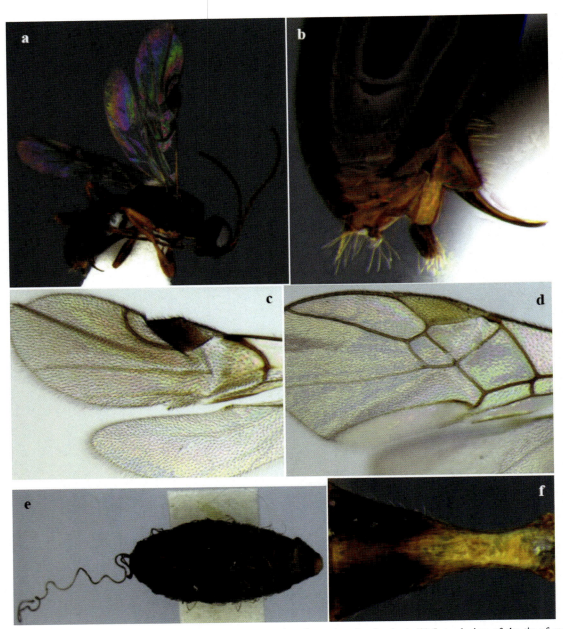

FIGURE 1.24 (a) Lateral habitus of *Leiophron* sp. (Euphorinae) (showing short, curved ovipositor); (b) Lateral view of the tip of metasoma of *Leiophron* sp. (Euphorinae) (showing the short, curved ovipositor); (c) Fore wing of *Leiophron* sp. (Euphorinae) (showing short, curved vein SR1); (d) Fore wing of *Meteorus* sp. (Euphorinae) (showing four-sided second submarginal cell); (e) parasitoid cocoon of *Meteorus* sp. (Euphorinae) (showing a short to long silk thread suspended from substrate); (f) Dorsal view of first metasomal segment of *Meteorus* sp. (Euphorinae) (showing petiolate tergum 1).

32. Metasoma with tergum 1 broad and more or less parallel-sided (Fig. 1.26a); second submarginal cell of fore wing small and triangular (Fig. 1.26c); ovipositor long and straight (Fig. 1.26b) .... **Microtypinae** Szépligeti, 1908
- Metasoma with tergum 1 narrowly tubular, elongate, and petiolate (Fig. 1.26d); second submarginal cell of fore wing small and either rounded or triangular (Fig. 1.26e) (rarely encountered species) .... **Euphorinae** Foerster, 1863 (in part, tribe Helorimorphini)
33. First three metasomal terga forming a rigid dorsal carapace, which covers most of the metasoma (Fig. 1.27a–c) (normally at most three metasomal segments are visible in dorsal view) .... **Brachistinae** Foerster, 1863 (in part)

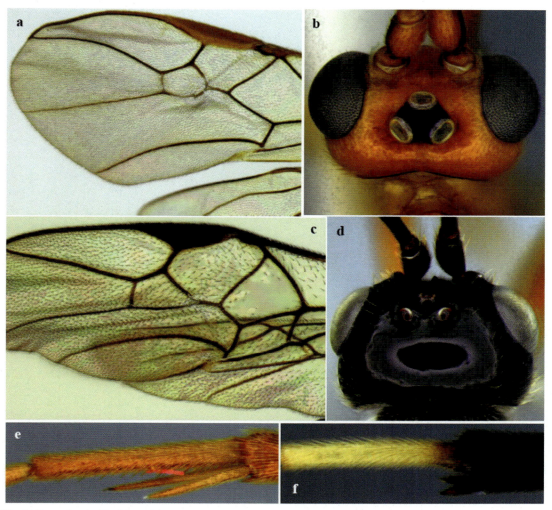

FIGURE 1.25 (a) Fore wing (part) of *Homolobus* sp. (Homolobinae) (showing large, quadrate second submarginal cell); (b) Dorsal view of head of *Homolobus* sp. (Homolobinae) (showing large ocelli); (c) Fore wing (part) of *Wroughtonia* sp. (Helconinae) (showing large second submarginal cell); (d) Dorsal view of head of *Wroughtonia* sp. (Helconinae) (showing smaller ocelli); (e) Apex of hind tibia and metatarsus of *Homolobus* sp. (Homolobinae) (showing long inner hind tibial spur); (f) Tip of hind tibia and metatarsus of *Wroughtonia* sp. (Helconinae) (showing short hind tibial spurs). *Photos prepared by S.R. Shaw.*

- First three metasomal terga not forming a rigid dorsal carapace; more than three metasomal segments are visible in dorsal view 34

34. Fore wing with apical segment of vein SR1 strongly curved over its length, RS reaching wing margin well before the wing tip (ending closer to pterostigma than to wing apex) (Fig. 1.24c), or, more rarely, SR1 absent (Fig. 1.27d); first metasomal tergum often narrow and petiolate (Fig. 1.24f); ovipositor short and curved (Fig. 1.24a and b) .... **Euphorinae** Foerster, 1863 (in part)

- Fore wing with apical segment of vein SR1 straight over most its length, or at least straight apically, and SR1 reaching wing margin near the wing tip (Fig. 1.28a); first metasomal tergum not petiolate (Fig. 1.28c); ovipositor long and usually quite straight (Fig. 1.28b) 35

35. Females with ovipositor quite long (Fig. 1.28b), distinctly much longer than the entire metasoma length (excluding ovipositor); fore wing with vein 1-SR + M emerging from the base of a small parastigma just basal from the larger pterostigma (Fig. 1.28a); parastigma not enlarged in males .... **Charmontinae** van Achterberg, 1979

- Females with ovipositor shorter, less than or equal to the metasoma length (excluding ovipositor) (Fig. 1.28d); fore wing with vein 1-SR + M emerging from the middle of a larger parastigma just basal from the larger pterostigma; parastigma distinctly enlarged in males (Fig. 1.28e) .... **Brachistinae** Foerster, 1863 (in part, Tribe Blacini)

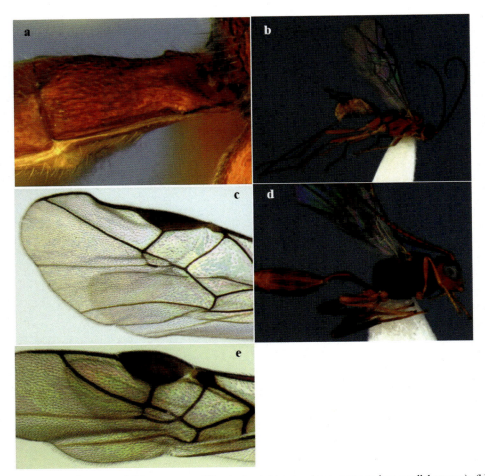

FIGURE 1.26 (a) First metasomal tergum of *Microtypus* sp. (Microtypinae) (showing large, more or less parallel tergum); (b) Lateral habitus of *Microtypus* sp. (Microtypinae) (showing long and straight ovipositor); (c) Fore wing (part) of *Microtypus* sp. (Microtypinae) (showing small and triangular second submarginal cell); (d) Lateral habitus of *Aridelus* sp. (Euphorinae: Helorimorphini) (showing the long, tubular and petiolate tergum 1); (e) Fore wing (part) of *Aridelus* sp. (Euphorinae: Helorimorphini) (showing triangular second submarginal cell). *Photos prepared by S.R. Shaw.*

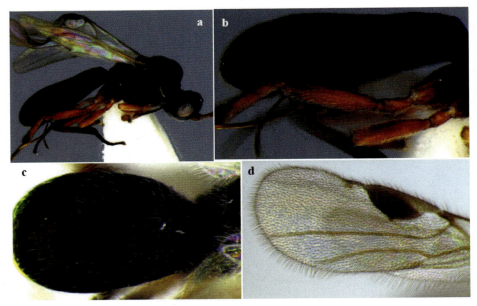

FIGURE 1.27 (a) Lateral habitus of *Schizoprymnus* sp. (Brachistinae) (showing metasomal carapace); (b) Lateral view of metasoma of *Schizoprymnus* sp. (Brachistinae) (showing metasomal carapace); (c) Dorsal view of metasoma of *Triaspis* sp. (Brachistinae) (showing metasomal carapace); (d) Fore wing of *Leiophron* (*Euphoiriella*) sp. (Euphorinae) (showing the absence of vein SR1). *Photos prepared by S.R. Shaw.*

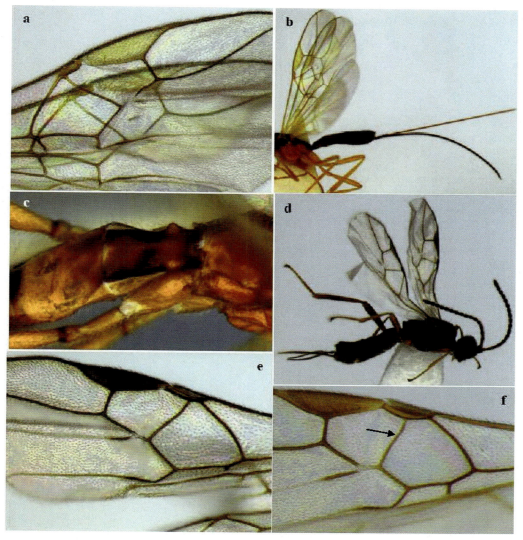

FIGURE 1.28 (a) Fore wing of *Charmon cruentatus* (Charmontinae); (b) Lateral habitus of *Charmon cruentatus* (Charmontinae) (showing ovipositor longer than metasoma); (c) Dorso-lateral view of mesosoma (part) and metasoma (part) of *Charmon* sp. (showing tergum 1 broad, not petiolate); (d) Lateral habitus of *Blacus caduceus* (Brachistinae) (showing ovipositor shorter than or as long as metasoma); (e) Fore wing of male *Blacus caduceus* (Brachistinae) (showing large parastigma); (f) Fore wing (part) of *Blacus ruficornis* (Brachistinae) (showing vein 1-SR + M emerging from middle of parastigma). *Photos prepared by S.R. Shaw.*

References

Aguiar, A. P., Deans, A. R., Engel, M. S., Forgage, M., Huber, J. T., Jennings, J. T., Johnson, N. F., Lelej, A. S., Longino, J. F., Lohrmann, V., Miko, I., Ohl, M., Rasmussen, C., Taeger, A., & Yu, D. S. K. (2013). Order Hymenoptera Linnaeus, 1758. In Z. Q. Zhang (Ed.), Animal biodiversity: An outline of higher-level classification and survey of taxonomic richness. *Zootaxa, 3705*(1), 51–62.

Askew, R. R., & Shaw, M. R. (1986). Parasitoid communities: Their size, structure and development. In J. Waage, & D. Greathead (Eds.), *Vol. xviii. Insect parasitoids* (pp. 225–264). London Academic Press, 389 pp.

Austin, A., & Dangerfield, P. C. (1998). Biology of the *Mesostoa kerri* Austin and Wharton (Insecta: Hymenoptera: Braconidae: Mesostoinae), an endemic Australian wasp that causes stem galls on *Banksia marginata* Cav. *Australian Journal of Botany, 46*, 559–569.

Austin, A., & Dowton, M. (2000). *Hymenoptera: evolution, biodiversity, and biological control.* Canberra: CSIRO, 512 pp.

Ayala, A., Pérez-Lachaud, G., Toledo, J., Liedo, P., & Montoya, P. (2018). Host acceptance by three native braconid parasitoid species attacking larvae of the Mexican fruit fly, *Anastrepha ludens* (Diptera, Tephritidae). *Journal of Hymenoptera Research, 63*, 33–49.

Barahoei, H., Rakhshani, E., Nader, E., Starý, P., Kavallieratos, N. G., Tomanović, Z., & Mehrparvar, M. (2014). Checklist of Aphidiinae parasitoids (Hymenoptera: Braconidae) and their host aphid associations in Iran. *Journal of Crop Protection, 3*(2), 199–232.

Barthold, V. V. (1984). *An historical geography of Iran.* Princeton University Press, 308 pp.

Basibuyuk, H. H., Rasnitsyn, A. P., van Achterberg, C., Fitton, M. G., & Quicke, D. L. J. (1999). A new, putatively primitive Cretaceous fossil braconid subfamily from New Jersey amber (Hymenoptera, Braconidae). *Zoologica Scripta, 28*(1&2), 211−214.

Beckag, N. E., Thompson, S. N., & Federic, B. A. (1993). *Parasitoids and pathogens of insects.* Academic Press. vol. 1: 364 pp., vol. 2: 294 pp.

Belokobylskij, S. A., & Tobias, V. I. (1986). Doryctinae. *Opred. Fauna SSSR, 145*, 21−72.

Belshaw, R., & Quicke, D. L. J. (2002). Robustness of ancestral state estimates: Evolution of life history strategy in ichneumonoid parasitoids. *Systematic Biology, 50*(3), 450−477.

Belshaw, R., Fitton, M., Hernion, E., Gimeno, C., & Quicke, D. L. J. (1998). A phylogenetic reconstruction of the Ichneumonoidea (Hymenoptera) based on the D2 variable region of the 28S ribosomal RNA. *Sytematic Entomology, 23*(2), 109−123.

Belshaw, R., Dowton, M., Quicke, D. L. J., & Austin, A. (2000). Estimating ancestral geographical distributions: A gondwanan origin for aphid parasitoids? *Proceedings of the Royal Society of London B, 267*, 491−496.

Belshaw, R., Grafen, A., & Quicke, D. L. J. (2003). Inferring life history from ovipositor morphology in parasitoid wasps using phylogenetic discriminant analysis. *Zoological Journal of the Linnean Society, 139*, 213−228.

Beyarslan, A., Gadallah, N. S., & Ghahari, H. (2017). An annotated catalogue of the Iranian Microtypinae and Rogadinae (Hymenoptera: Braconidae). *Zootaxa, 4291*(1), 99−116.

Bezier, A., Annaheim, M., Herbiniere, J., Wetterwald, C., Gyapay, G., Bernard-Samain, S., Wincker, P., Roditi, I., Heller, M., Belghazi, M., Pfister-Wilhem, R., Periquet, G., Dupuy, C., Huguet, E., Volkoff, A. N., Lanzrein, B., & Drezen, J.-M. (2009). Polydnaviruses of braconid wasps derive from an ancestral nudivirus. *Science, 323*, 926−930.

Borror, D. J., Triplehorn, C. A., & Johnson, N. F. (1989). *An introduction to the study of insects* (6th ed.). Philadelphia, Pennsylvania: Saunders College Publishing, 875 pp.

Carter, D. J. (1984). *Pest Lepidoptera of Europe: with special reference to the British Isles.* Junk, 431 pp.

Centrella, M., & Shaw, S. R. (2010). A new species of phytophagous braconid, *Allorhogas minimus* (Hymenoptera: Braconidae: Doryctinae) reared from fruit galls on *Miconia longifolia* (Melastomataceae) in Costa Rica. *International Journal of Tropical Insect Science, 30*(2), 1−6.

Chavarría, L., Hanson, P. E., Marsh, P. M., & Shaw, S. R. (2009). A phytophagous braconid, *Allorhogas conostegia* n. sp. (Hymenoptera: Braconidae), in the fruits of *Conostegia xalapensis* (Bonpl.) D. Don (Melastomataceae). *Journal of Natural History, 43*, 2677−2689.

Chen, X. X., & van Achterberg, C. (2019). Systematics, phylogeny, and evolution of braconid wasps: 30 years of progress. *Annual Review of Entomology, 64*, 1−24.

Clausen, C. P. (1940). *Entomophagous insects.* New York: McGraw-Hill, 688 pp.

Clausen, C. P. (1942). The relationship of taxonomy to biological control. *Journal of Economic Entomology, 35*, 744−748.

Clausen, C. P. (1954). The egg-larval host relationship among the parasitic Hymenoptera. *Bolletino del Laboratorio di Zoologia Generale e Agraria della Facoltá Agraria in Portici, 33*, 119−133.

Cortés, E., Ameri, A., Rakhshani, E., & Peris-Felipo, F. J. (2016). Subfamily Alysiinae Leach, 1815 (Hymenoptera: Braconidae) from Iran: New records and updated list. *Journal of Insect Biodiversity and Systematics, 2*(4), 411−418.

DeBach, P. (1974). *Biological control by natural enemies.* UK: Cambridge University Press, 440 pp.

Dowton, A., Austin, A. D., & Antolin, M. F. (1998). Evolutionary relationships among the microgastroid wasps (Hymenoptera: Braconidae): combined analysis of 16S and 28S rDNA genes. *Molecular Phylogenetics and Evolution, 10*(3), 129−150.

Dowton, M., Belshaw, R., Austin, A. D., & Quicke, D. L. J. (2002). Simultaneous molecular and morphological analysis of braconid relationships (Insecta: Hymenoptera: Braconidae) indicates independent mt-tRNA gene inversions within a single wasp family. *Journal of Molecular Evolution, 54*, 210−226.

Faegri, K., & van der Pijl, L. (1979). *The principles of pollination ecology* (3rd ed.). Oxford: Pergamon Press, 244 pp.

Fallahzadeh, M., & Saghaei, N. (2010). Checklist of Braconidae (Insecta: Hymenoptera) from Iran. *Munis Entomology & Zoology, 5*(1), 170−186.

Farahani, S., Talebi, A. A., & Rakhshani, E. (2016). Iranian Braconidae (Insecta: Hymenoptera: Ichneumonoidea): Diversity, distribution and host association. *Journal of Insect Biodiversity and Systematics, 2*(1), 1−92.

Farahbakhsh, G. (1961). Family Braconidae (Hymenoptera), p. 124. In G. Farahbakhsh (Ed.), *A checklist of economically important insects and other enemies of plants and agricultural products in Iran.* The Ministry of Agriculture, Department Plant Protection, 153 pp.

Fisher, W. B. (1968). *The Cambridge history of Iran* (Vol. 1). Cambridge University Press, 804 pp.

Fischer, M., Lashkari Bod, A., Rakhshani, E., & Talebi, A. A. (2011). Alysiinae from Iran (Insecta: Hymenoptera: Braconidae: Alysiinae). *Annalen des Naturhistorischen Museums in Wien B, 112*, 115−132.

Forup, M. L., Henson, K. S. E., Craze, P. G., & Memmott, J. (2008). The restoration of ecological interactions: Plant−pollinator networks on ancient and restored heathlands. *Journal of Applied Ecology, 45*, 742−752.

Free, J. B. (1993). *Insect pollination of crops* (2nd ed.). London: Academic Press, 684 pp.

Gadallah, N. S., & Ghahari, H. (2013a). An annotated catalogue of the Iranian Agathidinae and Brachistinae (Hymenoptera: Braconidae). *Linzer Biologische Beiträge, 45*(2), 1873−1901.

Gadallah, N. S., & Ghahari, H. (2013b). An annotated catalogue of the Iranian Cheloninae (Hymenoptera: Braconidae). *Linzer Biologische Beiträge, 45*(2), 1921−1943.

Gadallah, N. S., & Ghahari, H. (2015). An annotated catalogue of the Iranian Braconinae (Hymenoptera: Braconidae). *Entomofauna, 36*, 121−176.

Gadallah, N. S., & Ghahari, H. (2016). An updated checklist of the Iranian Miracinae, Pambolinae and Sigalphinae (Hymenoptera: Braconidae). *Orsis, 30*, 51−61.

Gadallah, N. S., & Ghahari, H. (2017). An annotated catalogue of the Iranian Doryctinae and Exothecinae (Hymenoptera: Braconidae). *Transactions of the American Entomological Society, 143*, 669−691.

Gadallah, N. S., Ghahari, H., & Kavallieratos, N. G. (2019). An annotated catalogue of the Iranian Charmontinae, Ichneutinae, Macrocentrinae and Orgilinae (Hymenoptera: Braconidae). *Journal of the Entomological Research Society, 21*(3), 333−354.

Gadallah, N. S., Ghahari, H., Fischer, M., & Peris-Felipo, F. J. (2015a). An annotated catalogue of the Iranian Alysiinae (Hymenoptera: Braconidae). *Zootaxa, 3974*(1), 1−28.

Gadallah, N. S., Ghahari, H., & Peris-Felipo, F. J. (2015b). Catalogue of the Iranian Microgastrinae (Hymenoptera: Braconidae). *Zootaxa, 4043*(1), 1−69.

Gadallah, N. S., Ghahari, H., Peris-Felipo, F. J., & Fischer, M. (2016a). Updated checklist of Iranian Opiinae (Hymenoptera: Braconidae). *Zootaxa, 4066*(1), 1−40.

Gadallah, N. S., Ghahari, H., & van Achterberg, C. (2016b). An annotated catalogue of the Iranian Euphorinae, Gnamptodontinae, Helconinae, Hormiinae and Rhysipolinae (Hymenoptera: Braconidae). *Zootaxa, 4072*(1), 1−38.

Gauld, I. D. (1988). Evolutionary patterns of host utilization by ichneumonoid parasitoids (Hymenoptera: Ichneumonidae and Braconidae). *Biological Journal of the Linnean Society, 35,* 331−377.

Gauld, I. D., & Bolton, B. B. (1988). *The Hymenoptera.* London: British Museum (Natural History) and Oxford University Press, 352 pp.

Ghahari, H. (2016). Five new records of Iranian Braconidae (Hymenoptera: Ichnemonoidea) for Iran and annotated catalogue of the subfamily Homolobinae. *Wuyi Science Journal, 32,* 35−43.

Ghahari, H. (2017). *Pollinator insects and honey bee.* Karaj: Publication of Jame Elmi Karbordi University, 312 pp. (in Persian).

Ghahari, H., Yu, D. S., & van Achterberg, C. (2006). *World bibliography of the family Braconidae (Hymenoptera: Ichneumonoidea) (1964−2003)* (Vol. 8). NNM Technical Bulletin, 293 pp.

Ghahreman, A., & Attar, F. (1999). Biodiversity of plant species in Iran. In *The vegetation of Iran, plant species, red data of Iran, endemic species, rare species, species threatened by extinction* (Vol. 1)Tehran, Iran: Central Herbarium of Tehran University, 1176 pp.

Godfray, H. C. J. (1994). *Parasitoid and behavioral ecology.* New Jersey: Princeton University Press, 473 pp.

Goulet, H., & Huber, J. (1993). *Hymenoptera of the world: an Identification guide to families* (Vol. vii). Ottawa, Canada: Research Branch, Agriculture Canada, 668 pp.

Grissell, E. E. (1999). Hymenopteran Biodiversity: Some alien notions. *American Entomology, 45*(4), 235−244.

Gupta, V. K. (1988). Advances in parasitic Hymenoptera research. In *Proceedings of the 2nd conference on the taxonomy and biology of parasitic Hymenoptera* (pp. 19−21). Gainesville, Fl: University of Florida. November 1987, EJ Brill, The Netherlands, 546 pp.

Hakimzadeh Khoei, M., Kaya, M., & Altindag, A. (2011). New records of Rotifers from Iran with biogeographic considerations. *Turkish Journal of Zoology, 35,* 395−402.

Hanson, P., & Gauld, I. (2006). Hymenoptera de la region tropical. *Memoirs of the American Entomological Institute, 77,* 1−994.

Hassell, M. P. (1980). Foraging strategies, population models and biological control: A case study. *Journal of Animal Ecology, 49,* 603−628.

Hawkins, B. A. (1994). *Pattern and process in host-parasitoid interactions* (Vol. X). Cambridge: Cambridge University Press, 190 pp.

Hawkins, B. A., & Sheehan, W. (1994). *Parasitoid community ecology.* Oxford, UK: Oxford University Press, 516 pp.

Heraty, J. M., & Darling, D. C. (2009). Fossil Eucharitidae and Perilampidae (Hymenoptera: Chalcidoidea) from Baltic amber. *Zootaxa, 2306,* 1−16.

Herniou, E. A., Huguet, E., Theze, J., Bezier, A., Periquet, G., & Drezen, J. M. (2013). When parasitic wasp hijacked viruses: Genomic and functional evolution on polydnaviruses. *Philosophical Transactions of the Royal Society B, 368*(1626), 1−12.

Hochberg, M. E., & Ives, A. R. (2000). *Parasitoid population biology.* Princeton, NJ, USA: Princeton University Press, 366 pp.

Honda, K. (1995). Chemical basis of differential oviposition by lepidopterous insects. *Archives of Insect Biochemistry and Physiology, 30,* 1−23.

Hoy, M. A., Cuningham, G. L., & Knutson, L. (1983). *Biological control of pets by mites.* Berkeley: University of California Press, 185 pp.

Infante, F., Hanson, P., & Wharton, R. A. (1995). Phytophagy in the genus *Monitoriella* (Hymenoptera: Braconidae) with description of new species. *Annals of the Entomological Society of America, 88*(4), 406−415.

Jasso-Martínez, J. M., Quicke, D. L. J., Belokobylskij, S. A., Meza-Lázaro, R. M., & Zaldívar-Riverón, A. (2020). Phylogenomics of the lepidopteran endoparasitoid wasp subfamily Rogadinae (Hymenoptera: Braconidae) and related subfamilies. *Systematic Entomology, 46,* 83−95.

Jervis, M. A., Heimpel, G. E., Ferns, P. N., Harvey, J. A., & Kidd, N. A. C. (2001). Life-history strategies in parasitoid wasps: A comparative analysis of 'ovigeny'. *Journal of Animal Ecology, 70,* 442−458.

Jones, O. R., Purvis, A., Baumgart, E., & Quicke, D. L. J. (2009). Using taxonomic revision data to estimate the geographic and taxonomic distribution of undescribed species richness in the Braconidae (Hymenoptera: Ichneumonoidea). *Insect Conservation and Diversity, 2,* 204−212.

Khajeh, N., Rakhshani, E., Peris-Felipo, F. J., & Žikić, V. (2014). Contributions to the Opiinae (Hymenoptera: Braconidae) of eastern Iran with updated checklist of Iranian species. *Zootaxa, 3784*(2), 131−147.

Klem, C. C., & Zaspel, J. (2019). Pest injury guilds, Lepidoptera, and placing fruit-piercing moths in context: A review. *Annals of the Entomological Society of America, 112*(5), 421−432.

Kruess, A., & Tscharntke, T. (1994). Habitat fragmentation, species loss, and biological control. *Science, 264,* 1581−1584.

Lapointe, R., Tanaka, K., Barney, W. E., Whitfield, J. B., Banks, J. C., Béliveau, C., Stoltz, D. B., Webb, B. A., & Cusson, M. (2007). Genomic and morphological features of a banchine polydnavirus: Comparison with bracoviruses and ichnoviruses. *Journal of Virology, 81,* 6491−6501.

LaSalle, J., & Gauld, I. D. (1993). *Hymenoptera and biodiversity.* New York: Oxford University Press, 368 pp.

Lewis, C. N., & Whitfield, J. B. (1999). Braconid wasps (Hymenoptera: Braconidae) diversity in forest plots under different silvicultural methods. *Environmental Entomology, 28,* 986−997.

Li, Q., Wei, S. J., Tang, P., Wu, Q., Shi, M., Sharkey, M. J., & Chen, X.-X. (2016). Multiple lines of evidence from mitochondrial genomes resolve phylogeny and evolution of parasitic wasp in Braconidae. *Genome Biology and Evolution, 8*(9), 2651−2662.

Macêdo, M. V., & Monteiro, R. T. (1989). Seed predation by a braconid wasp, *Allorhogas* sp. (Hymenoptera). *Journal of the New York Entomological Society, 97,* 359−362.

Mackauer, M. (1990). Host discrimination and larval competition in solitary endoparasitoids. In M. Mackauer, L. E. Ehler, & J. Roland (Eds.), *Vol. XVIII. Critical issues in biological control* (pp. 41−62). Intercept, Andover, 330 pp.

Maetô, K. (1987). A comparative morphology of the male internal reproductive organs of the family Braconidae (Hymenoptera: Ichneumonoidea). *Kontyu, 55,* 32−42.

Mansouri, R., Ganavati, E., & Servati, M. R. (2014). The assessment of geomorphological landscapes and geotourism potential role of the Ilam province with respect to sustainable development. *Geographical Data, 23,* 5−12.

Matthews, R. W. (1974). Biology of Braconidae. *Annual Review of Entomology, 19*, 15—32.

Mayhew, P. J., Dytham, C., Shaw, M. R., & Fraser, S. E. M. (2009). Collections of ichneumonid wasps (Subfamilies Diacritinae, Diplazontinae, Pimplinae and Poemeniinae) from woodlands near York and their implications for conservation planning. *Naturalist, 134*, 3—24.

McGregor, S. E. (1976). *Insect pollination of cultivated crop plants* (Vol. 496). Washington, DC: USDA Agriculture Handbook, 411 pp.

Modarres Awal, M. (1997). Family Braconidae (Hymenoptera). In M. Modarres Awal (Ed.), *List of agricultural pests and their natural enemies in Iran* (2nd ed., pp. 265—267). Ferdowsi University of Mashhad Press, 429 pp.

Modarres Awal, M. (2012). Family Braconidae (Hymenoptera). In M. Modarres Awal (Ed.), *List of agricultural pests and their natural enemies in Iran* (3rd ed., pp. 483—486). Ferdowsi University of Mashhad Press, 759 pp.

Mudri-Stojnić, S., Andrić, A., Józan, Z., & Vujic, A. (2012). Pollinator diversity (Hymenoptera and Diptera) in semi-natural habitats in Serbia during summer. *Archives of Biological Sciences, Belgrade, 64*(2), 777—786.

Murphy, N., Banks, J. C., Whitfield, J. B., & Austin, A. D. (2008). Phylogeny of the microgastroid complex of subfamilies of braconid wasps (Hymenoptera) based on sequence data from seven genes, with an improved estimate of the time of origin of the lineage. *Molecular Phylogenetics and Evolution, 47*, 378—395.

Naumann, I. D., & Carne, P. B. (1991). *The insects of Australia*. London: Melbourne University Press, Melbourne, Australia and UCL Press Limited, 1137 pp.

Pena, J. E., Sharp, J. L., & Wysoki, M. (2002). *Tropical fruit pests and pollinators. Biology, economic importance, natural enemies and control*. CABI Publishing, 430 pp.

Pitz, K., Dowling, A. P. G., Sharanowski, B. J., Boring, C. A. B., Seltmann, K. C., & Sharkey, M. J. (2007). Phylogenetic relationships among the Braconidae (Hymenoptera: Ichneumonoidea) as proposed by Shi et al.: A reassessment. *Molecular Phylogenetics and Evolution, 43*, 338—343.

Powell, W., Pennacchio, F., Poppy, G. M., & Tremblay, E. (1998). Strategies involved in the location of hosts by the parasitoid *Aphidius ervi* Haliday (Hymenoptera: Braconidae, Aphidiinae). *Biological Control, 11*, 104—112.

Proctor, M., Yeo, P., & Lack, A. (1996). *The natural history of pollination*. Timber Press, 487 pp.

Purvis, A., Jones, K. E., & Mace, G. M. (2000). Extinction. *Bioessays, 22*, 1123—1133.

Quicke, D. L. J. (1997). *Parasitic wasps*. London: Chapman & Hall, 470 pp.

Quicke, D. L. J. (2015). *The braconid and ichneumonid parasitoid wasps: biology, systematics, evolution and ecology*. Chichester: Wiley Blackwell, 688 pp.

Quicke, D. L. J., Austin, A. D., Fagan-Jeffries, E. P., Hebert, P. D. N., & Butcher, B. A. (2020). Recognition of the Trachypetidae stat.n. as a new extant family of Ichneumonoidea (Hymenoptera), based on molecular and morphological evidence. *Systematic Entomology*. https://doi.org/10.1111/syen.12426

Quicke, D. L. J., & Belshaw, R. (1999). Incongruence between morphological data sets: An example from the evolution of endoparasitism among parasitic wasps (Hymenoptera: Braconidae). *Systematic Biology, 48*, 436—454.

Quicke, D. L. J., & Butcher, B. A. (2021). Review of venoms of non-polydnavirus carrying ichneumonoid wasps. *Biology, 10*, 50. https://doi.org/10.3390/biology10010050

Quicke, D. L. J., & van Achterberg, C. (1990). Phylogeny of the subfamilies of the family Braconidae. *Zoologica Scripta, 21*, 403—416.

Quicke, D. L. J., Basibuyuk, H. H., Fitton, M. G., & Rasnitsyn, A. P. (1999). Morphological palaeontological and molecular aspects of ichneumonoid phylogeny (Hymenoptera, Insecta). *Zoologica Scripta, 28*(1—2), 175—202.

Ranjith, A. P., Quicke, D. L. J., Saleem, U. K. A., Butcher, B. A., Zaldívar-Riverón, A., & Nasser, M. (2016). Entomophytophagy in an Indian braconid "parasitoid" wasp (Hymenoptera): Specialized larval morphology, biology and description of a new species. *PLoS One, 11*(6), e0156997.

Rasnitsyn, A. P. (1983). Ichneumonoidea (Hymenoptera) from the lower cretaceous of Mongolia. *Contributions of the American Entomological Institute, 20*, 259—265.

Rasnitsyn, A. P., & Quicke, D. L. J. (2002). *History of insects*. Dordrecht: Kluwer Academic, 560 pp.

Rasplus, J.-Y., Villemant, C., Paiva, M. R., Delvare, G., & Roques, A. (2010). Hymenoptera. *BioRisk, 4*, 669—776.

Richards, A. J. (1978). *The pollination of flowers by insects*. Academic Press, 213 pp.

Ruiz-Guerra, B., López-Acosta, J. C., Zaldívar-Riverón, A., & Velázquez-Rosas, N. (2015). Braconidae (Hymenoptera: Ichneumonoidea) abundance and richness in four types of land use and preserved rain forest in southern Mexico. *Revista Mexicana de Biodiversidad, 86*, 164—171.

Samin, N., Coronado-Blanco, J. M., Kavallieratos, N. G., Fischer, M., & Sakenin, H. (2018a). Recent findings on Braconidae (Hymenoptera: Ichneumonoidea) of Iran with an updated checklist. *Acta Biologica Turcica, 31*(4), 160—173.

Samin, N., Coronado-Blanco, J. M., Fischer, M., van Achterberg, C., Sakenin, H., & Davidian, E. (2018b). Updated checklist of Iranian Braconidae (Hymenoptera: Ichneumonoidea) with twenty-three new records. *Natura Somogyiensis, 32*, 21—36.

Sanjerehei, M. M. (2019). *Life forms of plant species and floristic regions in Iran*. Tafakkor Talaei, 203 pp.

Sathe, T. V., & Pandharbale, A. R. (2008). *Forest pest Lepidoptera*. Manglam Publications, 186 pp.

Sedighi, S., & Madjdzadeh, M. (2015). Updated checklist of Iranian Euphorinae (Hymenoptera: Ichneumonoidea: Braconidae). *Biharean Biologist, 9*(2), 98—104.

Sharanowski, B. J., Dowling, A. P. G., & Sharkey, M. J. (2011). Molecular phylogenetics of Braconidae (Hymenoptera: Ichneumonoidea), based on multiple nuclear genes, and implications for classification. *Systematic Entomology, 36*(3), 549—572.

Sharanowski, B. J., Ridenbaugh, R. D., Piekarski, P. K., Broad, G. R., Burke, G. R., Deans, A. R., Lemmon, A. R., Moriarty Lemmon, E. C., Diehl, G. J., Whitfield, J. B., & Hines, H. M. (2021). Phylogenomics of Ichneumonoidea (Hymenoptera) and implications for evolution of mode of parasitism and viral endogenization. *Molecular Phylogenetics and Evolution, 56*. https://doi.org/10.1016/j.ympev.2020.107023

Sharkey, M. (2007). Phylogeny and classification of Hymenoptera. *Zootaxa, 1668*, 521—548.

Sharkey, M. J., & Wahl, D. B. (1992). Cladistics of the Ichneumonoidea (Hymenoptera). *Journal of Hymenoptera Research, 1*, 15—24.

Sharkey, M. J., Janzen, D. H., Hallwachs, W., Chapman, E. G., Smith, M. A., Dapkey, T., Brown, A., Ratnasigham, S., Naik, S., Manjunat, R., Perez, K., Milton, M., Hebert, P. D. N., Shaw, S. R., Kittel, R. N., Solis, A., Metz, M., Goldstein, P. Z., Brown, J. W., Quicke, D. L. J., van Achterberg, C., Brown, B. V., & Burns, J. M. (2021). Minimalist revision and description of 411 new species in 11 subfamilies of Costa Rican braconid parasitic wasps, including host records. *Zootaxa, 1013*, 1−665.

Shaw, M. R. (1983). On [e] evolution of endoparasitism: the biology of some genera of Rogadinae (Braconidae). *Contributions of the American Entomological Institute, 20*, 307−328.

Shaw, M. R. (1990). Parasitoids of European butterflies and their study. In O. Kudrna (Ed.), *Butterflies of Europe Vol. 2. Introduction to Lepidopterology* (pp. 449−479). Germany: Aula-Verlag Wiesbaden, 557 pp.

Shaw, M. R. (1997). *Rearing parasitic Hymenoptera*. Orpington: Amateur Entomologist's Society, 45 pp.

Shaw, M. R., & Huddleston, T. (1991). Classification and biology of braconid wasps (Hymenoptera: Braconidae). *Handbooks for the Identification of British Insects, 7*, 1−126.

Shaw, S. R. (1985). A phylogenetic study of the subfamilies Meteorinae and Euphorinae (Hymenoptera: Braconidae). *Entomography, 3*, 277−370.

Shaw, S. R. (1995). Chapter 12.2, Braconidae. In P. Hanson, & I. D. Gauld (Eds.), *The Hymenoptera of Costa Rica* (pp. 431−463). Oxford University Press, 893 pp.

Sheehan, W., Wackers, F. L., & Lewis, W. J. (1993). Discrimination of previously searched, host-free sites by *Microplitis croceipes* (Hymenoptera: Braconidae). *Journal of Insect Behavior, 6*, 323−331.

Shi, M., Chen, X. X., & van Achterberg, C. (2005). Molecular phylogeny of the Aphidiinae (Hymenoptera: Braconidae) based on the DNA sequences of the 16S rRNA and ATPase 6 genes. *European Journal of Entomology, 102*(2), 133−138.

Sree, K. S., & Varma, A. (2015). *Biocontrol of lepidopteran pests* (Vol. X). Springer International Publishing, 344 pp.

Steffan-Dewenter, I., & Westphal, C. (2008). The interplay of pollinator diversity, pollination services and landscape change. *Journal of Applied Ecology, 45*, 737−741.

Stireman, J. O., Dyer, L. A., Janzen, D. H., Singer, M. S., Lill, J. T., Marquis, R. J., Ricklefs, R. E., Gentry, G. L., Hallwachs, W., Coley, P. D., Barone, J. A., Greeney, H. F., Connahs, H., Barbosa, P., Morais, H. C., & Diniz, I. R. (2005). Climatic unpredictability and parasitism of caterpillars: Implications of global warming. *Proceedings of the National Academy of Sciences of the United State of America, 102*(48), 17384−17387.

Stork, N. E. (1997). Measuring global biodiversity and its decline. In M. L. Reaka-Kudla, D. E. Wilson, & E. O. Wilson (Eds.), *Biodiversity II. Understanding and protecting our biological resources* (pp. 41−68). Washington, DC: Joseph Henry Press, 560 pp.

Strand, M. R. (2000). Developmental traits and life-history evolution in parasitoids. In M. E. Hochberg, & A. R. Ives (Eds.), *Parasitoid population biology* (pp. 139−160). Princeton, New Jersey: Princeton University Press, 366 pp.

Strand, M. R., & Obrycki, J. J. (1996). Host specificity of insect parasitoids and predators. *Bioscience, 46*, 422−429.

Tavassoli, M. (2016). *Urban structure in hot arid environments. XXXVIII.* Springer International Publishing Switzerland, 241 pp.

Tobias, V. I. (1967). A review of the classification, phylogeny and evolution of the family Braconidae (Hymenoptera). *Entomologicheskoye Obozreniye, 46*, 645−669.

Tregobov, V., & Mobayen, S. (1970). *Guide pour la carte de la vegetation naturelle de l'Iran* (Vol. 14). Bulletin: University de Tehran, 21 pp.

Turlings, T. C. L., Wäckers, F. L., Vet, L. E. M., Lewis, W. J., & Tumlinson, J. H. (1993). Learning of host-finding cues by hymenopterous parasitoids. In D. R. Papaj, & A. C. Lewis (Eds.), *Insect learning. Ecology and evolutinary perspectives* (pp. 51−78). Springer XIII, 398 pp.

van Achterberg, C. (1984a). Essay on the phylogeny of Braconidae (Hymenoptera: Ichneumonoidea). *Entomologisk Tidskrift, 105*, 41−58.

van Achterberg, C. (1984b). Revision of the genera of Braconini with first and second metasomal tergites immovably joined (Hymenoptera, Braconidae, Braconinae). *Tijdschrift voor Entomologie, 127*, 137−164.

van Achterberg, C. (1993). Illustrated key to the subfamilies of the Braconidae (Hymenoptera: Ichneumonoidea). *Zoologische Verhandelingen, 283*, 1−189.

van Alphen, J. J. M., & Visser, M. E. (1990). Superparasitism as an adaptive strategy for insect parasitoids. *Annual Review of Entomology, 35*, 59−79.

van Lenteren, J. C., Bakker, K., & van Alphen, J. J. M. (1978). How to analyze host discrimination. *Ecological Entomology, 3*, 71−75.

Vet, L. E. M. (2001). Parasitoid searching efficiency links behaviour to population processes. *Applied Entomology and Zoology, 36*(4), 399−408.

Vinson, S. B. (1976). Host selection by insect parasitoids. *Annual Review of Entomology, 21*, 109−134.

Waage, J. K. (1979). Foraging for patchily-distributed hosts by the parasitoid, *Nemiritis canescens*. *Journal of Animal Ecology, 48*, 353−371.

Waage, J., & Greathead, D. J. (1986). *Insect parasitoids*. London: Academic Press, 389 pp.

Wei, S. J., Shi, M., Sharkey, M. J., van Achterberg, C., & Chen, X. X. (2010). Comparative microgenomics of Braconidae (Insecta, Hymenoptera) and the phylogenetic utility of mitochondrial genomes with special reference to Holometabolous insects. *BMC Genomics, 11*(371), 1−16.

Wharton, R. A. (1993). Bionomics of the Braconidae. *Annual Review of Entomology, 38*, 121−143.

Wharton, R. A. (2000). Can braconid classification be reconstructed to facilitate portrayal of relationships? In A. D. Austin, & M. Dowton (Eds.), *Hymenoptera: evolution, biodiversity, and biological control* (pp. 143−153). Canberra: CSIRO, 512 pp.

Wharton, R. A., & Hanson, P. E. (2005). Gall wasps in the family Braconidae (Hymenoptera). In A. Raman, W. C. Schaefer, & T. M. Withers (Eds.), *Vol. xxi. Biology, ecology, and evolution of gall-inducing arthropods* (pp. 495−505). Enfield, New Hampshire: Science Publishers, 817 pp.

Wharton, R. A., & van Achterberg, C. (2000). Family group names in Braconidae (Hymenoptera: Ichneumonoidea). *Journal of Hymenoptera Research, 9*, 254−270.

Whitfield, J. B. (1992). The phylogenetic origin of endoparasitism in the cyclostome lineages of Braconidae (Hymenoptera). *Systematic Entomology, 17*, 273−286.

Whitfield, J. B. (1997). Molecular and morphological data suggest a common origin for the polydnaviruses among braconid wasps. *Naturwissenschaften, 84*, 502−507.

Whitfield, J. B. (1998). Phylogeny and evolution of host-parasitoid interactions in Hymenoptera. *Annual Review of Entomology, 43*, 129−151.

Whitfield, J. B. (2002). Estimating the age of the polydnavirus/braconid wasp symbiosis. *Proceedings of the National Academy of Sciences of the United States of America, 99*(11), 7508−7513.

Whitfield, J. B., Austin, A. D., & Fernandez-Triana, J. L. (2018). Systematics, biology, and evolution of microgastrine parasitoid wasps. *Annual Review of Entomology, 63*, 389−406.

Yu, D. S., van Achterberg, C., & Horstmann, K. (2016). *Taxapad 2016, Ichneumonoidea 2015, Database on flash-drive. Ottawa, Ontario, Canada.*

Zaldívar-Riverón, A., Belokobylskij, S. A., León-Regagnon, V., Martínez, J. J., Briceño, R., & Quicke, D. L. J. (2007). A single origin of gall association in a group of parasitic wasps with disparate morphologies. *Molecular Phylogenetics and Evolution, 44*, 981−992.

Zaldívar-Riverón, A., Mori, M., & Quicke, D. L. J. (2006). Systematics of the cyclostomes subfamilies of braconid parasitic wasps (Hymenoptera, Ichneumonoidea): A simultaneous molecular and morphological Bayesian approach. *Molecular Phylogenetics and Evolution, 38*(1), 130−145.

Zaldívar-Riverón, A., Belokobylskij, S. A., León-Regagnon, V., Briceno, G. R., & Quicke, D. L. J. (2008). Molecular phylogeny and historical biogeography of the cosmopolitan parasitic wasp subfamily Doryctinae (Hymenoptera: Braconidae). *Invertebrate Systematics, 22*(3), 345−363.

Zehzad, B., Kiabi, B. H., & Madjnoonian, H. (2002). The natural areas and landscape of Iran: An overview. *Zoology in the Middle East, 26*, 7−10.

Zuparko, R. L. (2008). Parasitic Hymenoptera (Parasitica). In J. L. Capinera (Ed.), *Encyclopedia of entomology* (pp. 174−262). Dordrecht, CCLII: Springer, 4346 pp.

Coccygidium transcaspicum (Kokujev, 1902) (Agathidinae), ♀, lateral habitus. *Photo prepared by S.R. Shaw.*

Earinus gloriatorius (Panzer, 1809) (Agathidinae), ♀, lateral habitus. *Photo prepared by S.R. Shaw.*

Chapter 2

Subfamily Agathidinae Haliday, 1833

Scott Richard Shaw[1], Neveen Samy Gadallah[2], Michael J. Sharkey[3], Hassan Ghahari[4] and Donald L.J. Quicke[5]

[1]UW Insect Museum, Department of Ecosystem Science and Management, University of Wyoming, Laramie, WY, United States; [2]Entomology Department, Faculty of Science, Cairo University, Giza, Egypt; [3]UW Insect Museum, Department of Entomology, University of Kentucky, Lexington, KY, United States; [4]Department of Plant Protection, Yadegar-e Imam Khomeini (RAH) Shahre Rey Branch, Islamic Azad University, Tehran, Iran; [5]Integrative Ecology Laboratory, Department of Biology, Faculty of Science, Chulalongkorn University, Pathumwan, Bangkok, Thailand

Introduction

The cosmopolitan Agathidinae is a moderately large subfamily of Braconidae with at least 1213 species classified into 61 genera and seven tribes (Agathidini Haliday, 1833, Agathirsini Sharkey, 2017, Cremnoptini Sharkey, 1992, Disophrini Sharkey, 1992, Earinini Sharkey, 1992, Lytopylini Sharkey, 2017, and Mesocoelini van Achterberg, 1990) (Chen & van Achterberg, 2019; Sharkey & Chapman, 2017b; Yu et al., 2016). It is estimated that an additional 2000–3000 new species are yet to be named and awaiting description worldwide (Sharkey & Chapman, 2017a; Sharkey et al., 2006, 2009). The genus *Agathis* Latreille, 1804 is currently the most diverse and cosmopolitan genus of Agathidinae (tribe Agathidini), with about 153 species worldwide (Yu et al., 2016). Previous studies have indicated at least 86 agathidine species in the West Palaearctic region and with majority being from the genera *Agathis* and *Bassus* Fabricius (Çetin Erdoğan & Beyarslan, 2016; Yu et al., 2016). Following the recent reclassification of many species, the genus *Bassus* is interpreted as a rather small group of about 10 species as redefined by Sharkey et al. (2009). Most of the Palaearctic species formerly as *Bassus* are now treated as members of either *Aerophilus*, *Therophilus*, or *Zosteragathis*. In other words, *Bassus* has historically been used as a dumping-ground for many distantly related species and new combinations have not been made for all the Palaearctic fauna (outside those treated here).

In a recent study of Southeast Asian species, the tribe Agathidini was revised by Sharkey and Chapman (2017a), and 10 new genera were proposed based on molecular analysis, thus raising the total number of genera in the subfamily to 61. As a result of these new genera, many new combinations were proposed (Sharkey & Chapman, 2017a). A phylogenetic analysis conducted by Sharkey and Chapman (2017b) suggested that the genus *Bassus* may still be polyphyletic. However, that hypothesis of polyphyly was not supported by morphological characters, so the current concept of the genus was retained (Sharkey & Chapman, 2017b).

There were various nomenclatural difficulties concerning the subfamily Agathidinae in the period between 1917 and 1948 (e.g., Muesebeck, 1927; Simmonds, 1947), during which years the subfamily was often called Braconinae, until the clarification of the name *Bracon* by the International Commission of Zoological Nomenclature's opinion 162 (1945) fixed the name Agathidinae for this group (Shaw & Huddleston, 1991). Sharkey (1985) applied the name *Bassus* to the genus formerly named *Microdus*.

Although the detailed biology of many species is still unknown (Nixon, 1986; van Achterberg, 1993, 2011; van Achterberg & Long, 2010), it is thought that most agathidines are koinobiont endoparasitoids of lepidopteran larvae (Nixon, 1986; Sharkey, 1992; Sharkey & Chapman, 2017b; Sharkey et al., 2006, 2009; van Achterberg, 1993, 2011). Various families, including Blastobasidae, Coleophoridae, Gelechiidae, Noctuidae, Oecophoridae, Pyralidae, Sesiidae, Tineidae, and Tortricidae, are known to serve as hosts for Agathidinae (Yu et al., 2016).

Agathidine species are primarily solitary parasitoids with one known exception, the gregarious New World species *Zelomorpha gregaria* (Sharkey et al., 2006, 2009). Both exposed and concealed larvae (Sharkey et al., 2006, 2009; van Achterberg, 2011) serve as hosts largely depending on the genus of Agathidinae. Those attacking larvae concealed in plant tissues tend to have long ovipositors while species utilizing exposed-feeding hosts have shorter ovipositors. Most agathidines attack an early instar larva but do not kill the host until it reaches the final instar and is ready to construct a cocoon (Sharkey & Chapman, 2017b).

Braconidae of the Middle East (Hymenoptera). https://doi.org/10.1016/B978-0-323-96099-1.00020-0

Agathidines lay their eggs into the first and second instars of their hosts, often inserting the egg into a ventral ganglion or a lateral lobe of the protocerebrum. The egg increases in size before hatching and the caudate first-instar larva, which has a discernible tracheal system but no spiracles, floats freely in the hemocoel. The first instar larva has sharp mandibles used to fight with any competing larvae, and it remains as a first instar during the feeding and development period of the host, diapausing in this stage if the host diapauses while partly grown (Balduf, 1966; Dondale, 1954; Quednau, 1970).

Members of the subfamily Agathidinae are easily recognized from other braconid subfamilies by their distinctive wing venation: the fore wing has extremely narrow and somewhat long marginal cell; vein CU1b is absent; and the vein m-cu of the fore wing is more or less diverging posteriorly. Other distinguishing features of agathidines include the pronotum having a deep pit on each side. The propodeum is variable, with sculpture ranging from strongly and often symmetrically areolate or carinate to completely smooth. The metasoma of agathidines is usually shiny and smooth but the first two terga are sculptured in some genera. The spiracles of the first metasomal tergum situated on the dorsal plate (not on the membrane) (Nixon, 1986).

Checklists of Regional Agathidinae. Fallahzadeh and Saghaei (2010): four species in four genera (without precise localities); Gadallah and Ghahari (2013): 28 species in nine genera; Farahani et al. (2014): 29 species in eight genera; Farahani et al. (2016): 33 species in nine genera; Yu et al. (2016): 33 species in eight genera; Samin, Coronado-Blanco, Kavallieratos et al. (2018): 33 species in eight genera; Samin, Coronado-Blanco, Fischer et al. (2018): 35 species in nine genera including *Lepton nigrum* (Nees, 1812) under Agathidinae). In the following inventory, we include 80 species in 12 genera and six tribes for the Middle Eastern fauna (Agathidini, Cremnoptini, Disophrini, Earinini, Lyptopylini, and Mesocoelini). *Agathis taurica* Telenga is here recorded for the first time for the Iranian fauna. Here we follow Sharkey and Chapman (2017b) and Chen and van Achterberg (2019) for the higher classification of the subfamily, as well as Yu et al. (2016) for the global distribution of species.

Key to genera of the subfamily Agathidinae in the Middle East (modified from Sharkey et al., 2009)

1. Fore and mid-claws bifid 2
— Fore and mid-claws simple or with a basal lobe 5
2. Hind trochantellus with one or two distinct carinae (fore tibial spur about as long as basitarsus, ending in a long thin style .... *Coccygidium* Saussure
— Hind trochantellus without carinae 3
3. Posterolateral margins of frons not bordered with carinae .... *Euagathis* Szépligeti
— Posterolateral margins of frons bordered with cainae 4
4. Median and lateral carinae of frons lamellate; ovipositor hardly protruded, much shorter than half length of metasoma5
— Median and lateral carinae of frons not lamellate, ridgelike; ovipositor at least as long as metasoma ....
.... *Cremnops* Foerster
5. Lateral carina of frons with posterior ends directed toward middle ocellus .... *Troticus* Brullé
— Lateral carina of frons with posterior ends directed toward lateral ocelli .... *Disophrys* Foerster
6. Fore and mid-claws simple .... *Bassus* Fabricius
— Fore and mid-claws with a basal lobe 7
7. Vein 1-SR+M of fore wing mostly or entirely absent; notauli present either partly or completely 8
— Vein 1-SR+M of fore wing present and complete; notauli absent .... *Earinus* Wesmael
8. First middle tergite most strigate or rarely smooth; second submarginal cell of fore wing usually present 9
— First middle tergite mostly granulate or coriaceous; second submarginal cell of fore wing absent ....
.... *Camptothlipsis* Enderlein
9. Third tergum completely smooth; pair of carinae on first tergum not prominent; hind coxal cavities open to metasomal foramen or narrowly closed and positioned partly above ventral margin of metasomal foramen 10
— Third tergum partly or completely sculptured, sculpture often confined to narrow line along transverse depression; pair of carinae on first tergum prominent; hind coxal cavities closed and positioned completely below the metasomal foramen; ventral margin of metasomal foramen with a strong, relatively straight transverse carina .... *Aerophilus* Szépligeti
10. Median area of T3 usually densely striated in anterior half or more, sometimes with other sculpture, rarely smooth
.... *Lytopylus* Foerster
— Median area of T3 smooth or (rarely) coriarious 11
11. Mouth parts long, galea distinctly longer than wide; gena usually elongate .... *Agathis* Latreille
— Mouth parts short, galea not longer than wide; gena not elongate .... *Therophilus* Wesmael

List of species of the subfamily Agathidinae recorded in the Middle East

Subfamily Agathidinae Haliday, 1833

Tribe Agathidini Haliday, 1833

Genus *Aerophilus* Szépligeti, 1902

Aerophilus persicus (Farahani & Talebi, 2014)

Catalogs with Iranian records: Farahani et al. (2014, 2016 as *Lyptopylus persicus* Farahani and Talebi); Samin, Coronado-Blanco, Kavallieratos et al. (2018), Samin, Coronado-Blanco, Fischer et al., (2018 as *L. persicus*).
Distribution in Iran: Guilan, Mazandaran (Farahani et al., 2014 as *L. persicus*).
Distribution in the Middle East: Iran.
Extralimital distribution: None.
Host records: Unknown.

Aerophilus rufipes (Nees von Esenbeck, 1812)

Catalogs with Iranian records: Gadallah and Ghahari (2013 as *Lyptophylus rufipes* (Nees)), Farahani et al. (2014, 2016 as *L. rufipes*), Samin, Coronado-Blanco, Kavallieratos et al. (2018), Samin, Coronado-Blanco, Fischer et al. (2018 as *L. rufipes*).
Distribution in Iran: East Azarbaijan (Ranjbar Aghdam & Fathipour, 2010, as *Bassus rufipes* (Nees, 1812)); Farahani et al., 2014), Razavi Khorasan (Shojai et al., 2000, 2002 as *Braunsia rufipes*), Tehran (Shojai, 1989, as *Braunsia rufipes*), West Azarbaijan (Akbarzadeh Shoukat et al., 2015; Shojai et al., 2000, 2002 as *Braunsia rufipes*).
Distribution in the Middle East: Iran.
Extralimital distribution: Argentina, Armenia, Australia, Austria, Azerbaijan, Belgium, Bulgaria, former Czechoslovakia, Finland, France, Georgia, Germany, Hungary, Italy, Japan, Kazakhstan, Korea, Kyrgyzstan, Lithuania, Moldova, Netherlands, Poland, Romania, Russia, Slovakia, Sweden, Switzerland, Turkmenistan, Uruguay, United States of America, Ukraine, United Kingdom.
Host records: Summarized by Yu et al. (2016) as being a parasitoid of a wide range of insect species belonging to the order Lepidoptera. In Iran, it was reared from the following tortricids: *Cydia pomonella* (L.) (Ranjbar Aghdam & Fathipour, 2010; Shojai et al., 2000, 2002), *Spilonota ocellana* Denis and Schiffermüller (Shojai 1989), and *Lobesia botrana* Denis and Schiffermüller (Akbarzadeh Shoukat et al., 2015).

Genus *Agathis* Latreille, 1804

Agathis anglica Marshall, 1885

Catalogs with Iranian records: Gadallah and Ghahari (2013), Farahani et al. (2014, 2016), Samin, Coronado-Blanco, Kavallieratos et al. (2018), Samin, Coronado-Blanco, Fischer et al. (2018).
Distribution in Iran: Golestan (Ghahari, Fischer, Sakenin et al., 2011; Samin et al., 2011, Samin, Ghahari et al., 2015), Guilan (Ghahari, Fischer, & Tobias, 2012), Mazandaran (Ghahari, Fischer, Sakenin et al., 2011), West Azarbaijan (Samin et al., 2014).
Distribution in the Middle East: Cyprus (Simbolotti & van Achterberg, 1999), Iran (see references above), Israel—Palestine (Papp, 2012), Syria (Yu et al., 2016), Turkey (Beyarslan et al., 2002; Çetin Erdoğan, 2013, 2014; Çetin Erdoğan & Beyarslan, 2001, 2009; Çetin Erdoğan et al., 2009; Guclü & Ozbek, 2002).
Extralimital distribution: Albania, Armenia, Austria, Azerbaijan, Bulgaria, China, Croatia, Finland, France, Germany, Greece, Hungary, Italy, Kazakhstan, Mongolia, Montenegro, Morocco, Netherlands, Poland, Romania, Russia, Spain, Sweden, Switzerland, Taiwan, Tajikistan, Ukraine, United Kingdom, former Yugoslavia.
Host records: Summarized by Shenefelt (1970) and Yu et al. (2016) as being a parasitoid of the following coleophorids *Coleophora adjunctella* Hodgkinson, *Coleophora albitarsella* Zeller, *Coleophora argentula* (Stephens), *Coleophora discordella* Zeller, *Coleophora laricella* (Hübner), *Coleophora lusciniaepennella* (Treitschke), and *Coleophora salicorniae* Heinemann and Wocke; the following crambids *Loxostege sticticalis* (L.), and *Pyrausta aurata* (Scopoli); the depressariids *Agonopterix nervosa* (Zeller), and *Agonopterix pallorella* (Zeller); the gelechiids *Aproaerema anthyllidella* (Hübner), *Nothris verbascella* (Denis and Schiffermüller), *Pexicopia malvella* (Hübner), and *Teleiodes saltuum* (Zeller). It was also reported by Nixon (1986) as being a parasitoid of the tortricid *Epinotia mercuriana* (Frölich).

Agathis assimilis Kokujev, 1895

Catalogs with Iranian records: Gadallah and Ghahari (2013), Farahani et al. (2014, 2016), Samin, Coronado-Blanco, Kavallieratos et al. (2018), Samin, Coronado-Blanco, Fischer et al. (2018).
Distribution in Iran: Guilan (Ghahari, Fischer, & Tobias, 2012), Semnan (Ghahari, Gadallah et al., 2009).
Distribution in the Middle East: Iran (Ghahari, Gadallah et al., 2009; Ghahari, Fischer, & Tobias, 2012), Turkey (Çetin Erdoğan & Beyarslan, 2001; Çetin Erdoğan et al., 2009; Zettel & Beyarslan, 1992).
Extralimital distribution: Austria, Azerbaijan, Bulgaria, Croatia, France, Germany, Hungary, Italy, Kazakhstan, Korea, Lithuania, Macedonia, Moldova, Mongolia, Montenegro, Netherlands, Norway, Poland, Russia, Switzerland, Tajikistan, Ukraine, United Kingdom, Uzbekistan.

Host records: Recorded by Telenga (1955) as being a parasitoid of the coleophorid *Coleophora astragalella* Zeller.

Agathis berkei Çetin Erdoğan, 2010
Distribution in the Middle East: Turkey (Çetin Erdoğan, 2010).
Extralimital distribution: None.
Host records: Unknown.

Agathis breviseta Nees von Esenbeck, 1812
Catalogs with Iranian records: Gadallah and Ghahari (2013), Farahani et al. (2016), Samin, Coronado-Blanco, Kavallieratos et al. (2018), Samin, Coronado-Blanco, Fischer et al. (2018).
Distribution in Iran: Lorestan (Ghahari, Fischer, Papp et al., 2012), Semnan (Naderian et al., 2012).
Distribution in the Middle East: Iran (Ghahari, Fischer, Papp et al., 2012; Naderian et al., 2012), Turkey (Nixon, 1986).
Extralimital distribution: Azerbaijan, Belgium, Bulgaria, Czech Republic, Finland, France, Georgia, Germany, Greece, Hungary, Ireland, Italy, Kazakhstan, Latvia, Lithuania, Moldova, Mongolia, Netherlands, Poland, Russia, Slovakia, Slovenia, Sweden, Switzerland, Tajikistan, Ukraine, United Kingdom.
Host records: Summarized by Shenefelt (1970) and Yu et al. (2016) as being a parasitoid of the following coleophorids: *Coleophora conyzae* Zeller, *Coleophora follicularis* (Vallot), *Coleophora trochilella* (Duponchel), *Coleophora albitarsella* Zeller, *Coleophora inulae* Wocke, and *Coleophora lutipennella* Zeller; the crambid *Pyrausta purpuralis* (L.); the following: depressariids *Agonopterix broennoeensis* (Strand), and *Agonopterix kaekeritziana* (L.); the following gelechiids: *Chrysoesthia drurella* (Fabricius), *Dichomeris marginella* (Fabricius), and *Monochroa cytisella* (Curtis); and the following tortricids: *Aethes rutilana* (Hübner), and *Spilonota ocellana* (Denis and Schiffermüller).

Agathis fischeri Zettel and Beyarslan, 1992
Distribution in the Middle East: Turkey (Çetin Erdoğan, 2013; Çetin Erdoğan et al., 2009; Simbolotti & van Achterberg, 1999; Zettel & Beyarslan, 1992).
Extralimital distribution: None.
Host records: Unknown.

Agathis fulmeki Fischer, 1957
Catalogs with Iranian records: Gadallah and Ghahari (2013), Farahani et al. (2014, 2016), Samin, Coronado-Blanco, Kavallieratos et al. (2018), Samin, Coronado-Blanco, Fischer et al. (2018).
Distribution in Iran: East Azarbaijan, West Azarbaijan (Ghahari & Fischer, 2011a), Golestan (Ghahari, Fischer,

Sakenin et al., 2011), Guilan (Ghahari, Fischer, & Tobias, 2012), Kuhgiloyeh and Boyerahmad (Samin, van Achterberg et al., 2015), Semnan (Samin, Fischer et al., 2015).
Distribution in the Middle East: Iran (see references above), Turkey (Çetin Erdoğan, 2013; Çetin Erdoğan & Beyarslan, 2001; Çetin Erdoğan et al., 2009; Guclü & Öbek, 2002).
Extralimital distribution: Austria, Bulgaria, Croatia, France, Greece, Hungary, Mongolia, Montenegro, Morocco, Serbia, Spain.
Host records: Unknown.

Agathis fuscipennis (Zetterstedt, 1838)
Catalogs with Iranian records: Gadallah and Ghahari (2013), Farahani et al. (2014, 2016), Samin, Coronado-Blanco, Kavallieratos et al. (2018), Samin, Coronado-Blanco, Fischer et al. (2018).
Distribution in Iran: Alborz, Guilan, Qazvin, Tehran (Farahani et al., 2014), Lorestan (Ghahari, Fischer, Papp et al., 2012), West Azarbaijan (Ghahari, Fischer, Çetin Erdoğan, Beyarslan et al., 2009).
Distribution in the Middle East: Iran (see references above), Turkey (Çetin Erdoğan, 2013; Çetin Erdoğan & Beyarslan, 2001, 2009; Simbolotti & van Achterberg, 1999; Zettel & Beyarslan, 1992).
Extralimital distribution: Armenia, Austria, Bosnia-Herzegovina, Bulgaria, Croatia, Finland, France, Germany, Greece, Hungary, Ireland, Italy, Kazakhstan, Korea, Latvia, Lithuania, North Macedonia, Mongolia, Montenegro, Netherlands, Norway, Poland, Russia, Serbia, Spain, Sweden, Switzerland, Tajikistan, Tunisia, United Kingdom.
Plant associations in Iran: Alfalfa fields (*Medicago sativa*, Fabaceae) (Ghahari, Fischer, Çetin Erdoğan, Beyarslan et al., 2009).
Host records: Summarized by Yu et al. (2016) as being a parasitoid of the following coleophorids: *Coleophora* spp.; the epermeniid *Ochromolopis ictella* (Hübner); the following gelechiids: *Apoaerema anthyllidella* (Hübner), *Caryocolum saginella* (Zeller), *Chrysoesthia drurella* (Fabricius), *Chrysoesthia sexguttella* (Thunberg), *Scrobipalpa gallicella* (Constant), and *Thiotricha subocellea* (Stephens); the heliodinid *Heliodines roesella* (L.); and the following tortricids *Argyroploce arbutella* (L.) and *Spilonota ocellana* (Stephens). In Iran, *Agathis fuscipennis* has been reared from larvae of the gelechiid *Tuta absoluta* (Meyrick). Its possible role as a biological control agent for management of tomato pests remains to be studied (Loni et al., 2011).

Agathis glaucoptera Nees von Esenbeck, 1834
Catalogs with Iranian records: Gadallah and Ghahari (2013), Farahani et al. (2014, 2016), Samin, Coronado-Blanco, Kavallieratos et al. (2018), Samin, Coronado-Blanco, Fischer et al. (2018).

Distribution in Iran: Ardabil, West Azarbaijan (Ghahari & Fischer, 2011a), Golestan (Samin, Ghahari et al., 2015), Guilan (Ghahari & Fischer, 2011b; Sakenin et al., 2012), Mazandaran (Ghahari, Fischer, Sakenin et al., 2011).
Distribution in the Middle East: Iran (see references above), Turkey (Çetin Erdoğan et al., 2009; Guclü & Özbek, 2002; Nixon, 1986; Simbolotti & van Achterberg, 1999; Zettel & Beyarslan, 1992).
Extralimital distribution: Azerbaijan, France, Germany, Hungary, Italy, Kazakhstan, North Macedonia, Russia, Spain, Ukraine, former Yugoslavia.
Host records: Unknown.

Agathis gracilipes Hellén, 1956
Distribution in the Middle East: Turkey (Çetin Erdoğan et al., 2009).
Extralimital distribution: Finland, Tajikistan, Turkmenistan.
Host records: Unknown.

Agathis griseifrons Thomson, 1895
Catalogs with Iranian records: No catalog.
Distribution in Iran: Guilan (Gadallah et al., 2018).
Distribution in the Middle East: Iran (Gadallah et al., 2018), Turkey (Çetin Erdoğan, 2013; Çetin Erdoğan et al., 2009; Simbolotti & van Achterberg, 1999).
Extralimital distribution: Albania, Andorra, Armenia, Austria, Azerbaijan, Bulgaria, Finland, France, Greece, Hungary, Ireland, Italy, Kazakhstan, Korea, Lithuania, Mongolia, Netherlands, Norway, Poland, Romania, Russia, Serbia, Slovenia, Sweden, Switzerland, Ukraine, United Kingdom.
Host records: Recorded by Nixon (1986) as being a parasitoid of the crambid Pyrausta aurata (Scopoli).

Agathis jordanicola Koçak and Kemal, 2013
Distribution in the Middle East: Jordan (Nixon, 1986).
Extralimital distribution: None.
Host records: Unknown.

Agathis levis Abdinbekova, 1970
Catalogs with Iranian records: No catalog.
Distribution in Iran: Mazandaran (Kian et al., 2018, 2020).
Distribution in the Middle East: Iran.
Extralimital distribution: Armenia, Azerbaijan, Hungary.
Host records: Unknown.

Agathis lugubris (Foerster, 1863)
Catalogs with Iranian records: Gadallah and Ghahari (2013), Farahani et al. (2014, 2016), Samin, Coronado-Blanco, Kavallieratos et al. (2018), Samin, Coronado-Blanco, Fischer et al. (2018).
Distribution in Iran: Golestan (Ghahari, Fischer, Çetin Erdoğan et al., 2010), Ilam (Ghahari et al., 2011b), Kermanshah (Ghahari & Fischer, 2012; Samin, 2015), Zanjan (Ghahari & Beyarslan, 2017).
Distribution in the Middle East: Iran (see references above), Turkey (Çetin Erdoğan, 2013; Çetin Erdoğan & Beyarslan, 2001, 2009; Çetin Erdoğan et al., 2009; Guclü & Özbek, 2002; Zettel & Beyarslan, 1992;).
Extralimital distribution: Former Czechoslovakia, Finland, Germany, Greece, Hungary, Ireland, Mongolia, Netherlands, Norway, Poland, Switzerland, Ukraine, United Kingdom.
Host records: Agathis lugubris has been recorded by Simbolotti and van Achterberg (1999) as being a parasitoid of the following coleophorids: Coleophora alticolella Zeller and Coleophora glaucicolella Wood. In Iran, it was reared from Coleophora sp. (Ghahari & Beyarslan, 2017).

Agathis luteotegula van Achterberg, 2011
Distribution in the Middle East: United Arab Emirates (van Achterberg, 2011).
Extralimital distribution: None.
Host records: Unknown.

Agathis malvacearum Latreille, 1805
Catalogs with Iranian records: Gadallah and Ghahari (2013), Farahani et al. (2014, 2016), Samin, Coronado-Blanco, Kavallieratos et al. (2018), Samin, Coronado-Blanco, Fischer et al. (2018).
Distribution in Iran: Chaharmahal and Bakhtiari, Khuzestan (Samin, van Achterberg et al., 2015), Guilan (Ghahari, Fischer, Çetin Erdoğan et al., 2010), Lorestan (Ghahari, Fischer, Papp et al., 2012), Mazandaran (Ghahari, Fischer, Sakenin et al., 2011), Qazvin (Ghahari et al., 2011a).
Distribution in the Middle East: Iran (see references above), Israel—Palestine (Papp, 2012), Turkey (Beyarslan et al., 2006; Çetin Erdoğan, 2013, 2014; Çetin Erdoğan & Beyarslan, 2001, 2009).
Extralimital distribution: Nearctic (Canada, United States of America), Palaearctic [Adjacent countries to Iran: Armenia, Azerbaijan, Kazakhstan, Russia].
Host records: Summarized by Yu et al. (2016) as being a parasitoid of the following coleophorids: Coleophora galbulipennella Zeller, and Coleophora graminicolella Heinemann; the following gelechiids: Metzneria aestivella (Zeller), Metzneria lappella (L.), and Pexicopia malvella (Hübner); the pterophorid Hellinsa didactylites (Ström); and the tortricid Retinia resinella (L.).

Agathis mediator (Nees von Esenbeck, 1812)
Catalogs with Iranian records: No catalog.
Distribution in Iran: Fars (Samin, Fischer et al., 2019 as Bassus mediator (Nees)).
Distribution in the Middle East: Iran (Samin, Fischer et al., 2019), Israel—Palestine (Bodenheimer, 1930).

Extralimital distribution: Armenia, Azerbaijan, Czech Republic, Finland, France, Germany, Hungary, Italy, Kazakhstan, Moldova, Netherlands, Poland, Russia, Sweden, Switzerland, Ukraine, United Kingdom.

Host records: Recorded by Yu et al. (2016) as being a parasitoid of the following coleophorids: *Coleophora anatipennella* (Hübner), *Coleophora flavipennella* (Duponchel), *Coleophora hemerobiella* (Scopoli), *Coleophora kuehnella* (Goeze), *Coleophora lutipennella* (Zeller), and *Coleophora pennella* (Denis and Schiffermüller); and the gelechiid *Scrobipalpa ocellatella* (Boyd).

Agathis melanotegula van Achterberg, 2011
Distribution in the Middle East: United Arab Emirates (van Achterberg, 2011).
Extralimital distribution: None.
Host records: Unknown.

Agathis melpomene Nixon, 1986
Catalogs with Iranian records: Gadallah and Ghahari (2013), Farahani et al. (2016), Samin, Coronado-Blanco, Kavallieratos et al. (2018), Samin, Coronado-Blanco, Fischer et al. (2018).
Distribution in Iran: Kermanshah (Ghahari & Fischer, 2012), Mazandaran (Ghahari, Fischer, Çetin Erdoğan et al., 2010; Ghahari & Fischer, 2011b; Sakenin et al., 2012), Northern Iran (no specific locality cited) (Sakenin et al., 2011).
Distribution in the Middle East: Iran (see references above), Turkey (Çetin Erdoğan & Beyarslan, 2001; Çetin Erdoğan et al., 2009; Simbolotti & van Achterberg, 1999; Zettel & Beyarslan, 1992).
Extralimital distribution: Bulgaria, Hungary, Mongolia.
Host records: Unknown.

Agathis montana Shestakov, 1932
Catalogs with Iranian records: Gadallah and Ghahari (2013), Farahani et al. (2014, 2016), Samin, Coronado-Blanco, Kavallieratos et al. (2018), Samin, Coronado-Blanco, Fischer et al. (2018).
Distribution in Iran: Ardabil (Ghahari & Fischer, 2011a), Golestan (Ghahari, Fischer, Sakenin et al., 2011; Samin, Ghahari et al., 2015).
Distribution in the Middle East: Iran (see references above), Israel−Palestine (Papp, 2012), Turkey (Belshaw & Quicke, 2002; Çetin Erdoğan et al., 2009; Güçlü & Özbek, 2002, 2007; Simbolotti & van Achterberg, 1999; Zettel & Beyarslan, 1992).
Extralimital distribution: Andorra, Armenia, Azerbaijan, Bulgaria, China, Czech Republic, France, Greece, Hungary, Kazakhstan, Korea, Kyrgyzstan, North Macedonia, Moldova, Mongolia, Montenegro, Poland, Russia, Serbia, Switzerland, Ukraine, United Kingdom.
Host records: In Turkey, this species has been recorded by Güçlu and Özbek (2007) as being a parasitoid of the tortrix moth *Pandemis cerasana* (Hübner). It has been also reared from the crambid *Pyrausta aurata* (Scopoli) (Tobias, 1976, 1986).

Agathis nigra Nees von Esenbeck, 1812
Catalogs with Iranian records: Fallahzadeh and Saghaei (2010), Gadallah and Ghahari (2013), Farahani et al. (2014, 2016), Samin, Coronado-Blanco, Kavallieratos et al. (2018), Samin, Coronado-Blanco, Fischer et al. (2018).
Distribution in Iran: Golestan (Ghahari, Fischer, Sakenin et al., 2011), Guilan (Ghahari, Fischer, & Tobias, 2012), Iran (no specific locality cited) (Hellén, 1956; Shenefelt, 1970; Simbolotti and van Achterberg, 1999; Tobias, 1986).
Distribution in the Middle East: Iran (see references above), Israel−Palestine (Papp, 1970), Turkey (Çetin Erdoğan & Beyarslan, 2001, 2009; Çetin Erdoğan et al., 2009; Güçlü & Özbek, 2002; Zettel & Beyarslan, 1992).
Extralimital distribution: Austria, Belgium, Bulgaria, Croatia, former Czechoslovakia, Finland, France, Germany, Greece, Hungary, Italy, Kazakhstan, Korea, Latvia, Lithuania, North Macedonia, Moldova, Mongolia, Morocco, Netherlands, Poland, Russia, Serbia, Slovenia, Spain, Sweden, Switzerland, Ukraine, United Kingdom.
Host records: Summarized by Yu et al. (2016) as being a parasitoid of the coleophorid *Coleophora* spp.; the crambids *Anania coronata* (Hufnagel) and *Pyrausta aurata* (Scopoli); the following gelechiids: *Apodea bifractella* (Duponchel), *Isophrictis striatella* (Denis and Schiffermüller), *Metzeneria lappella* (L.), *Metzeneria metzneriella* (Stainton), and *Scrobipalpa atriplicella* (Fischer); the pyralid *Ortholepis betulae* (Goeze); and the following tortricids: *Acleris guercinana* (Zeller), *Cochylis roseana* Rosy Conch, and *Ptycholonia lecheana* (L.).

Agathis pedias Nixon, 1986
Catalogs with Iranian records: Samin, Coronado-Blanco, Fischer et al. (2018).
Distribution in Iran: Lorestan (Samin, Coronado-Blanco, Fischer et al., 2018).
Distribution in the Middle East: Iran.
Extralimital distribution: Algeria, Greece, Italy, Morocco, Portugal, Spain.
Host records: Unknown.

Agathis persephone Nixon, 1986
Distribution in the Middle East: Turkey (Çetin Erdoğan, 2014).

Extralimital distribution: Croatia, France.
Host records: Unknown.

Agathis pumila (Ratzeburg, 1844)
Catalogs with Iranian records: Farahani et al. (2014, 2016), Samin, Coronado-Blanco, Kavallieratos et al. (2018), Samin, Coronado-Blanco, Fischer et al. (2018).
Distribution in Iran: Guilan (Ghahari, Fischer, & Tobias, 2012, as *Bassus pumilus* (Ratzeburg, 1844)).
Distribution in the Middle East: Iran.
Extralimital distribution: Austria, Bulgaria, Canada (introduced in some parts), China (introduced), Czech Republic, Germany, Italy, Japan, Mongolia, Netherlands, Poland, Russia, Slovakia, Switzerland, United States of America, Ukraine, United Kingdom.
Host records: Summarized by Yu et al. (2016) as being a parasitoid of the argyrestiid *Argyresthia laevigatella* (Heydenreich); the coleophorid *Coleophora laricella* (Hübner); and the tortricid *Zeiraphera griseana* (Hübner).

Agathis rostrata Tobias, 1963
Catalogs with Iranian records: Gadallah and Ghahari (2013), Farahani et al. (2016), Samin, Coronado-Blanco, Kavallieratos et al. (2018), Samin, Coronado-Blanco, Fischer et al. (2018).
Distribution in Iran: Lorestan (Ghahari, Fischer, Papp et al., 2012).
Distribution in the Middle East: Iran.
Extralimital distribution: Germany, Italy, Kazakhstan, Lithuania, Mongolia, Russia, Sweden, United Kingdom.
Host records: Unknown.

Agathis rubens Tobias, 1963
Distribution in the Middle East: Turkey (Çetin Erdoğan & Beyarslan, 2004; Çetin Erdoğan et al., 2009).
Extralimital distribution: Kazakhstan (Tobias, 1986).
Host records: Unknown.

Agathis rufipalpis Nees von Esenbeck, 1812
Catalogs with Iranian records: Gadallah and Ghahari (2013), Farahani et al. (2014, 2016), Samin, Coronado-Blanco, Kavallieratos et al. (2018), Samin, Coronado-Blanco, Fischer et al. (2018).
Distribution in Iran: East Azarbaijan (Ghahari, Gadallah et al., 2009), Fars, Razavi Khorasan (Ghahari, 2020), Kermanshah (Ghahari & Fischer, 2012), Tehran (Ghahari & Gadallah, 2021).
Distribution in the Middle East: Iran (see references above), Israel—Palestine (Papp, 2012), Turkey (Çetin Erdoğan, 2013, 2014; Çetin Erdoğan & Beyarslan, 2001, 2009; Çetin Erdoğan et al., 2009; Simbolotti & van Acheterberg, 1999; Zettel & Beyarslan, 1992).
Extralimital distribution: Belgium, Bulgaria, Croatia, former Czechoslovakia, Finland, France, Germany, Hungary, Ireland, Italy, Macedonia, Mongolia, Montenegro, Netherlands, Norway, Poland, Portugal, Romania, Slovenia, Spain, Sweden, Switzerland, United Kingdom.
Host records: Summarized by Yu et al. (2016) as being a parasitoid of the coleophorid *Coleophora alcyonipennella* (Kollar); the crambid, *Pyrausta aurata* (Scopoli); the depressariid *Agonopterix kaekeritziana* (L.); and the gelechiid *Chrysosthia drurella* (Fabricius). In Iran, this species has been reared from the coleophorid *Coleophora serratella* (L.) (Ghahari & Gadallah, 2021).

Agathis semiaciculata Ivanov, 1899
Catalogs with Iranian records: Gadallah and Ghahari (2013), Farahani et al. (2014, 2016), Samin, Coronado-Blanco, Kavallieratos et al. (2018), Samin, Coronado-Blanco, Fischer et al. (2018).
Distribution in Iran: Golestan (Sakenin et al., 2012), Guilan (Ghahari & Fischer, 2011b), Qazvin (Samin, 2015).
Distribution in the Middle East: Iran.
Extralimital distribution: Azerbaijan, Bulgaria, China, Croatia, France, Georgia, Germany, Greece, Italy, Japan, Kazakhstan, Moldova, Mongolia, Poland, Russia, Switzerland, Turkmenistan, Ukraine.
Host records: Recorded by Marczak and Buszko (1994) as being a parasitoid of the coleophorid *Coleophora onobrychiella* Zeller.

Agathis syngenesiae Nees von Esenbeck, 1812
Catalogs with Iranian records: Gadallah and Ghahari (2013), Farahani et al. (2014, 2016), Samin, Coronado-Blanco, Kavallieratos et al. (2018), Samin, Coronado-Blanco, Fischer et al. (2018).
Distribution in Iran: Fars (Samin, van Achterberg et al., 2015), Hamadan, Ilam (Ghahari, Fischer, Hedqvist et al., 2010), Lorestan (Ghahari, Fischer, Papp et al., 2012).
Distribution in the Middle East: Iran (see references above), Turkey (Beyarslan et al., 2006; Čapek & Hofmann, 1997; Çetin Erdoğan & Beyarslan, 2001; Çetin Erdoğan et al., 2009; Fahringer, 1922; Güçlü & Özbek, 2002; Nixon, 1986; Simbolotti & van Achterberg, 1999; Zettel & Beyarslan, 1992).
Extralimital distribution: Azerbaijan, Croatia, Denmark, Finland, France, Germany, Greece, Hungary, Italy, Kazakhstan, Malta, Mongolia, Netherlands, Poland, Portugal, Russia, Spain, Sweden, Tajikistan, Uzbekistan, former Yugoslavia.
Host records: Unknown.

Agathis tatarica Telenga, 1933
Catalogs with Iranian records: No catalog.
Distribution in Iran: West Azarbaijan (Naderian et al., 2020).

Distribution in the Middle East: Iran (Naderian et al., 2020), Turkey (Çetin Erdoğan, 2013).
Extralimital distribution: Kazakhstan, North Macedonia, Mongolia, Russia, former Yugoslavia.
Host records: Unknown.

Agathis taurica Telenga, 1955
Catalogs with Iranian records: This species is a new record for the fauna of Iran.
Distribution in Iran: Guilan province, Astara, Asgar-Abad, 1♀, September 2015.
Distribution in the Middle East: Iran (new record), Turkey (Çetin Erdoğan et al., 2009; Nixon, 1986; Zettel & Beyarslan, 1992).
Extralimital distribution: Armenia, Bulgaria, Croatia, Ukraine.
Host records: Unknown.

Agathis tibialis Nees von Esenbeck, 1812
Catalogs with Iranian records: Gadallah and Ghahari (2013), Farahani et al. (2014, 2016), Samin, Coronado-Blanco, Kavallieratos et al. (2018), Samin, Coronado-Blanco, Fischer et al. (2018).
Distribution in Iran: Kermanshah (Ghahari & Fischer, 2012), Qazvin (Ghahari et al., 2011a), Semnan (Ghahari & Gadallah, 2013).
Distribution in the Middle East: Iran (see references above), Turkey (Kohl, 1905).
Extralimital distribution: Austria, Azerbaijan, Croatia, Czech Republic, Denmark, Finland, France, Germany, Greece, Hungary, Italy, Kazakhstan, Kyrgyzstan, Lithuania, North Macedonia, Moldova, Mongolia, Montenegro, Netherlands, Poland, Russia, Serbia, Slovenia, Sweden, Switzerland, Ukraine, United Kingdom, Uzbekistan.
Host records: Summarized by Yu et al. (2016) as being a parasitoid of the coleophorids *Coleophora astragalella* Zeller, and *Coleophora cracella* (Vallot); and the following gelechiids: *Apodia bifractella* (Duponchel), *Metzneria lappella* (L.), and *Ptocheuusa paupella* (Zeller).

Agathis turanica Kukojev, 1903
Distribution in the Middle East: Israel—Palestine (Papp, 2012).
Extralimital distribution: Kazakhstan, Turkmenistan, Uzbekistan.
Host records: Unknown.

Agathis umbellatarum Nees von Esenbeck, 1812
Catalogs with Iranian records: Gadallah and Ghahari (2013), Farahani et al. (2014, 2016), Samin, Coronado-Blanco, Kavallieratos et al. (2018), Samin, Coronado-Blanco, Fischer et al. (2018).

Distribution in Iran: Golestan (Ghahari, Fischer, Sakenin et al., 2011), Guilan (Ghahari, Fischer, Çetin Erdoğan et al., 2010), Ilam (Ghahari et al., 2011b), Kermanshah (Ghahari & Fischer, 2012), Tehran (Farahani et al., 2014).
Distribution in the Middle East: Cyprus (Nixon, 1986), Iran (see references above), Israel—Palestine (Papp, 2012), Turkey (Beyarslan et al., 2002, 2006; Çetin Erdoğan, 2013; Çetin Erdoğan & Beyarslan, 2001; Çetin Erdoğan et al., 2009; Nixon, 1986; Zettel & Beyarslan, 1992).
Extralimital distribution: Algeria, Azerbaijan, Bulgaria, Croatia, France, Germany, Greece, Hungary, Italy, Kazakhstan, Kyrgyzstan, Macedonia, Malta, Moldova, Mongolia, Portugal, Russia, Spain, Tajikistan, Tunisia, Turkmenistan, Ukraine, Uzbekistan, former Yugoslavia.
Host records: Summarized by Yu et al. (2016) as being a parasitoid of the depressariid *Depressaria* sp.; and the gelechiids *Metzneria aestivella* (Zeller), and *Metzneria lappella* (L.).

Agathis varipes Thomson, 1895
Catalogs with Iranian records: Farahani et al. (2016), Samin, Coronado-Blanco, Kavallieratos et al. (2018), Samin, Coronado-Blanco, Fischer et al. (2018).
Distribution in Iran: West Azarbaijan (Samin et al., 2014).
Distribution in the Middle East: Iran (Samin et al., 2014), Turkey (Çetin Erdoğan, 2013; Çetin Erdoğan et al., 2009; Simbolotti & van Achetrberg, 1999; Zettel & Beyarslan, 1992).
Extralimital distribution: Austria, Finland, Germany, Greece, Hungary, Italy, Kazakhstan, North Macedonia, Mongolia, Netherlands, Norway, Russia, Serbia, Slovakia, Sweden, Switzerland, Tajikistan, Ukraine, United Kingdom, Uzbekistan.
Host records: Summarized by Yu et al. (2016) as being a parasitoid of the following gelechiids: *Apodia bifractella* (Duponchel), and *Metzneria lappella* (L.); and the pyralid *Eurhodope cirrigella* (Zincken).

Agathis zaisanica Tobias, 1963
Catalogs with Iranian records: No catalog.
Distribution in Iran: Mazandaran (Naderian et al., 2020).
Distribution in the Middle East: Iran (Naderian et al., 2020), Turkey (Çetin Erdoğan et al., 2009).
Extralimital distribution: Hungary, Kazakhstan, Mongolia, Russia.
Host records: Unknown.

Genus Bassus Fabricius, 1804
Bassus beyarslani Çetin Erdoğan, 2005
Distribution in the Middle East: Turkey (Çetin Erdoğan, 2005).
Extralimital distribution: None.
Host records: Unknown.

Bassus calculator (Fabricius, 1798)
Catalogs with Iranian records: No catalog.
Distribution in Iran: Mazandaran (Samin, Fischer et al., 2019).
Distribution in the Middle East: Iran (Samin, Fischer et al., 2019), Turkey (Çetin Erdoğan & Beyarslan, 2006).
Extralimital distribution: Austria, Denmark, Finland, France, Georgia, Germany, Hungary, Italy, Latvia, Netherlands, Norway, Poland, Romania, Russia, Slovakia, Sweden, Switzerland, Ukraine, United Kingdom.
Host records: Summarized by Yu et al. (2016) as being a parasitoid of the following tineids: *Archinemapogon yildizae* Koçak, *Morophaga choragella* Denis and Schiffermüller, *Scardia boletella* (Fabricius), and *Triaxomera parasitella* (Hübner). Several records of Coleoptera are almost certainly errors.

Bassus tergalis Alexeev, 1971
Distribution in the Middle East: Israel—Palestine (Papp, 2012).
Extralimital distribution: Mongolia, Turkmenistan.
Host records: Unknown.

Genus *Camptothlipsis* Enderlein, 1920
Camptothlipsis arabica Ghramh, 2012
Distribution in the Middle East: Saudi Arabia (Ghramh, 2012).
Extralimital distribution: None.
Host records: Unknown.

Camptothlipsis armeniaca (Telenga, 1955)
Catalogs with Iranian records: Gadallah and Ghahari (2013), Farahani et al. (2014, 2016), Samin, Coronado-Blanco, Kavallieratos et al. (2018), Samin, Coronado-Blanco, Fischer et al. (2018).
Distribution in Iran: Alborz (Farahani et al., 2014), East Azarbaijan (Ghahari, Fischer, Çetin Erdoğan, Beyarslan et al., 2009, as *Bassus armeniacus* (Telenga, 1955)), Guilan (Ghahari, Fischer, Çetin Erdoğan et al., 2010, as *Baeognatha armeniaca* Telenga, 1955), Kermanshah (Ghahari & Fischer, 2012 under *B. armeniaca*).
Distribution in the Middle East: Iran (see references above), Turkey (Nixon, 1986; Çetin Erdoğan & Beyarslan, 2001; Zettel & Beyarslan, 1992).
Extralimital distribution: Armenia, Austria, Azerbaijan, Hungary, Italy, Kazakhstan, Moldova, Switzerland, Ukraine.
Host records: Recorded by Yu et al. (2016) as being a parasitoid of the following gelechiids: *Anarsia eleagnella* Kuznetzov, *Anarsia lineatella* Zeller, and *Recurvia nanella* (Denis and Schiffermüller); and the tortricid *Grapholita funebrana* (Treitschke).

Camptothlipsis breviantennalis van Achterberg, 2011
Distribution in the Middle East: Yemen (van Achterberg, 2011).
Extralimital distribution: None.
Host records: Unknown.

Camptothlipsis fuscistigmalis van Achterberg, 2011
Distribution in the Middle East: Yemen (van Achterberg, 2011).
Extralimital distribution: None.
Host records: Unknown.

Camptothlepsis luteostigmalis van Achterberg, 2011
Distribution in the Middle East: Yemen (van Achterberg, 2011).
Extralimital records: None.
Host records: Recorded by van Achterberg (2011) as being a parasitoid of the crambid *Antigastra catalaunalis* (Duponchel).

Tribe Cremnoptini Sharkey, 1992
Genus *Cremnops* Foerster, 1863
Cremnops desertor (Linnaeus, 1758)
Catalogs with Iranian records: Farahani et al. (2014, 2016), Samin, Coronado-Blanco, Kavallieratos et al. (2018), Samin, Coronado-Blanco, Fischer et al. (2018).
Distribution in Iran: East Azarbaijan (Samin, Beyarslan et al., 2020), Guilan, Mazandaran (Farahani et al., 2014).
Distribution in the Middle East: Iran (Farahani et al., 2014; Samin, Beyarslan et al., 2020), Turkey (Çetin Erdoğan & Beyarslan, 2009; Zettel & Beyarslan, 1992).
Extralimital distribution: Europe, Nearctic, Oriental, Palaearctic [Adjacent countries to Iran: Armenia, Azerbaijan, Russia].
Host records: Recorded by Yu et al. (2016) as being a parasitoid of the following crambids: *Anania hortulata* (L.) and *Ostrinia nubilalis* (Hübner); the noctuid *Euxoa triaena* Kozhanchikov; the pyralid *Palpita machaeralis* (Walker); the sesiid *Synathedon speciformis* (Denis and Schiffermüller); and the tortricid *Cydia pomonella* (L.). In Iran, this species was reared from the tortricid *Cydia pomonella* (L.) (Samin, Beyarslan et al., 2020).

Cremnops monochroa Szépligeti, 1913
Distribution in the Middle East: Yemen (Madl, 2007).
Extralimital distribution: Madagascar, South Africa, Tanzania (Yu et al., 2016).
Host records: Summarized by Yu et al. (2016) as a parasitoid of the crambid *Epascestria pustulalis* (Hübner).

Cremnops richteri Hedwig, 1957
Catalogs with Iranian records: Fallahzadeh and Saghaei (2010), Gadallah and Ghahari (2013), Farahani et al. (2014, 2016), Samin, Coronado-Blanco, Kavallieratos et al. (2018), Samin, Coronado-Blanco, Fischer et al. (2018).
Distribution in Iran: Sistan and Baluchestan (Hedwig, 1957).
Distribution in the Middle East: Iran.
Extralimital distribution: None.
Host records: Unknown.

Tribe Disophrini Sharkey, 1992
Genus Coccygidium Saussure, 1892
Coccygidium arabicum Ghramh, 2011
Distribution in the Middle East: Saudi Arabia (Ghramh, 2011 as C. arabica).
Extralimital distribution: None.
Host records: Unknown.

Coccygidium hebabi Ghramh, 2013
Distribution in the Middle East: Saudi Arabia (Ghramh, 2013).
Extralimital distribution: None.
Host records: Unknown.

Coccygidium luteum (Brullé, 1846)
Distribution in the Middle East: Yemen (including Socotra) (van Achterberg, 2011).
Extralimital distribution: Cameroon, Democratic Republic of Congo, Ethiopia, Guinea, Kenya, Madagascar, Mauritius, Mozambique, Namibia, Niger, Nigeria, Rodriques Island, Réunion, Senegal, Seychelles, Somalia, South Africa, Tanzania, Uganda.
Host records: Summarized by Yu et al. (2016) as being a parasitoid of the crambid Prophantis sp.; the erebid Crypsotidia mesosema Hampson; the nctuids Condica capensis (Guenée), Spodoptera exempta (Walker), and Spodoptera exigua (Hübner); and the tortricid Leguminivora ptychora (Meyrick).

Coccygidium maculatum van Achterberg, 2011
Distribution in the Middle East: United Arab Emirates (van Achterberg, 2011).
Extralimital distribution: None.
Host records: Unknown.

Coccygidium melleum (Roman, 1910)
Distribution in the Middle East: Egypt (Mahmoud et al., 2009), Oman, United Arab Emirates, Yemen (van Achterberg, 2011).
Extralimital distribution: Democratic Republic of Congo, Kenya, Somalia, Sudan.

Host records: Recorded by Bashir and Venkatraman (1968) as being a parasitoid of the noctuid Spodoptera exigua (Hübner), as well as Sesamia sp. (van Achterberg, 2011).

Coccygidium rugiferum van Achterberg, 2011
Distribution in the Middle East: Yemen (van Achterberg, 2011).
Extralimital distribution: None.
Host records: Unknown.

Coccygidium transcaspicum (Kokujev, 1902)
Catalogs with Iranian records: Fallahzadeh and Saghaei (2010), Gadallah and Ghahari (2013), Farahani et al. (2014, 2016), Samin, Coronado-Blanco, Kavallieratos et al. (2018), Samin, Coronado-Blanco, Fischer et al. (2018).
Distribution in Iran: Iran (no specific locality cited) (Hedwig, 1957; Sharkey, 1998; Telenga, 1955; Tobias, 1986).
Distribution in the Middle East: Iran (see references above), Israel−Palestine (Papp, 2012).
Extralimital distribution: Japan, Turkmenistan, Uzbekistan.
Host records: Unknown.

Genus Disophrys Foerster, 1863
Disophrys angitemporalis van Achterberg, 2011
Distribution in the Middle East: United Arab Emirates (van Achterberg, 2011).
Extralimital distribution: None.
Host records: Unknown.

Disophrys caesa (Klug, 1835)
Catalogs with Iranian records: Fallahzadeh and Saghaei (2010), Gadallah and Ghahari (2013), Farahani et al. (2014, 2016), Samin, Coronado-Blanco, Kavallieratos et al. (2018), Samin, Coronado-Blanco, Fischer et al. (2018).
Distribution in Iran: Guilan, Tehran (Farahani et al., 2014), Lorestan (Ghahari, Fischer, Papp et al., 2012), Mazandaran (Farahani et al., 2014; Kian et al., 2020 as Disophrys initiator (Fonscolombe, 1846)), Sistan and Baluchestan (Hedwig, 1957), Iran (no specific locality cited) (Kreichbaumer, 1898, as Disophrys anthracina (Kriechbaumer, 1898)).
Distribution in the Middle East: Iran (see references above), Israel−Palestine (Papp, 2012), Turkey (Çetin Erdoğan, 2013; Çetin Erdoğan & Beyarslan, 2001, 2009; Fahringer, 1922; Güclü & Özbek, 2002; Zettel & Beyarslan, 1992).
Extralimital distribution: Algeria, Armenia, Azerbaijan, Bulgaria, Croatia, France, Germany, Hungary, Italy, Morocco, Portugal, Romania, Russia, Serbia, Spain, Switzerland.
Host records: Unknown.
Comments: van Achterberg (2011) believes that Disophrys caesa and Disophrys inculcatrix should be synonymized with Disophrys initiator.

Disophrys dissors Kokujev, 1903

Catalogs with Iranian records: Gadallah and Ghahari (2013), Farahani et al. (2014, 2016), Samin, Coronado-Blanco, Kavallieratos et al. (2018), Samin, Coronado-Blanco, Fischer et al. (2018).

Distribution in Iran: Guilan (Ghahari, 2018—around rice fields), Mazandaran (Ghahari et al., 2009a).

Distribution in the Middle East: Iran (Ghahari, 2018; Ghahari, Fischer, Çetin Erdoğan, Tabari et al., 2009), Israel—Palestine (Papp, 2012).

Extralimital distribution: Azerbaijan, Hungary, North Macedonia, Moldova, Serbia, Turkmenistan, Ukraine.

Host records: Unknown.

Disophrys inculcatrix (Kriechbaumer, 1898)

Catalogs with Iranian records: Gadallah and Ghahari (2013), Farahani et al. (2016), Samin, Coronado-Blanco, Kavallieratos et al. (2018), Samin, Coronado-Blanco, Fischer et al. (2018).

Distribution in Iran: Fars (Lashkari Bod et al., 2011).

Distribution in the Middle East: Iran.

Extralimital distribution: Azerbaijan, Hungary, Russia, Ukraine.

Host records: Unknown.

Disophrys punctifera van Achterberg, 2011

Distribution in the Middle East: Yemen (van Achterberg, 2011).

Extralimital distribution: None.

Host records: Unknown.

Genus *Euagathis* Szépligeti, 1900

Euagathis indica Enderlein, 1920

Catalogs with Iranian records: No catalog.

Distribution in Iran: Hormozgan (Ameri et al., 2016).

Distribution in the Middle East: Iran.

Extralimital distribution: China, India, Sri Lanka.

Host records: Unknown.

Genus *Troticus* Brullé, 1846

Troticus ovalis (Fahringer, 1937)

Distribution in the Middle East: Egypt (van Achterberg et al., 2008).

Extralimital distribution: Italy (Telenga, 1955).

Host records: Recorded by van Achterberg et al. (2008) as being a parasitoid of the lasiocampid *Streblote repanda aegyptiaca* Bang-Haas.

Tribe Earinini Sharkey, 1992

Genus *Earinus* Wesmael, 1837

Earinus elator (Fabricius, 1804)

Catalogs with Iranian records: Gadallah and Ghahari (2013), Farahani et al. (2014, 2016), Samin, Coronado-Blanco, Kavallieratos et al. (2018), Samin, Coronado-Blanco, Fischer et al. (2018).

Distribution in Iran: East Azarbaijan (Ghahari & Fischer, 2011a; Samin, Beyarslan et al., 2020), Golestan (Ghahari, Fischer, Sakenin et al., 2011; Samin, Ghahari et al., 2015).

Distribution in the Middle East: Iran (see references above), Turkey (Güclü & Ozbek, 2002).

Extralimital distribution: Austria, Azerbaijan, Belgium, Croatia, former Czechoslovakia, Finland, France, Germany, Hungary, Ireland, Italy, Japan, Korea, Latvia, Lithuania, Moldova, Netherlands, Norway, Poland, Russia, Serbia, Slovakia, Sweden, Switzerland, United Kingdom.

Host records: Summarized by Yu et al. (2016) as being a parasitoid of the following geometrids: *Alsophila aescularia* (Denis and Schiffermüller), and *Lycia hirtaria* (Clerck); the gracillariid *Gracillaria syringella* (Fabricius); the following noctuids: *Acontia lucida* (Hufnagel), *Agrochola circellaris* (Hufnagel), *Agrochola lota* (Clerck), *Atethmia ambusta* Denis and Schiffermüller, *Atethmia centrago* (Haworth), *Conistra vaccini* (Denis and Schiffermüller), *Dichonia convergens* (Denis and Schiffermüller), *Lithophane ornitopus* (Hufnagel), and *Orthosia stabilis* (Fabricius). In Iran, this species has been reared from the gracillariid *Caloptilia rufipennella* (Hübner) (Samin, Beyarslan et al., 2020).

Earinus gloriatorius (Pazner, 1809)

Catalogs with Iranian records: No catalog.

Distribution in Iran: Tehran (Ghahari & Beyarslan, 2019).

Distribution in the Middle East: Iran.

Extralimital distribution: Austria, Azerbaijan, Belgium, Bulgaria, Czech Republic, Estonia, Finland, France, Germany, Hungary, Ireland, Latvia, Lithuania, Moldova, Mongolia, Netherlands, Norway, Poland, Russia, Slovakia, Sweden, Switzerland, Ukraine, United Kingdom.

Host records: Recorded by Yu et al. (2016) as being a parasitoid of the following coleophorids: *Coleophora follicularis* (Vallot) and *Coleophora laricella* (Hübner); the following depressariids: *Agonopterix cileilla* (Stainton) and *Agonopterix heracliana* (L.); the following gelechiids: *Anacampsis populella* (Clerck) and *Anacampsis timidella* (Wocke); the geometrid *Eupithecia pimpinellata* (Hübner); the noctuid *Orthosia stabilis* (Fabricius); and the following tortricids: *Acleris variegata* (Denis and Schiffermüller), and *Pammene germmana* (Hübner). In Iran, this species has been reared from the tortricid *Acleris variegana* (Denis and Schiffermüller, 1775) (Ghahari & Beyarslan, 2019).

Tribe Lytopylini Sharkey, 2017

Genus *Lytopylus* Foerster, 1863

Lytopylus brevitarsis van Achterberg, 2011

Distribution in the Middle East: Yemen (van Achterberg, 2011).

Extralimital distribution: None.
Host records: Unknown.

Tribe Mesocoelini van Achterberg, 1990
Genus *Therophilus* Wesmael, 1837
Therophilus breviscutum van Achterberg, 2011
Distribution in the Middle East: Yemen (van Achterberg, 2011).
Extralimital distribution: None.
Host records: Unknown.

Therophilus cingulipes (Nees von Esenbeck, 1812)
Catalogs with Iranian records: No catalog.
Distribution in Iran: Isfahan (Samin, Fischer et al., 2019, as *Bassus cingulipes* Nees), Zanjan (Ghahari & Gadallah, 2021).
Distribution in the Middle East: Iran (Ghahari & Gadallah, 2021; Samin, Fischer et al., 2019), Turkey (Çetin Erdoğan & Beyarslan, 2006).
Extralimital distribution: Austria, Azerbaijan, Belgium, Bulgaria, China, Croatia, Czech Republic, Finland, France, Germany, Greece, Hungary, Ireland, Italy, Japan, Kazakhstan, Korea, Lithuania, Moldova, Mongolia, Netherlands, Poland, Russia, Slovenia, Switzerland, United Kingdom, former Yugoslovia.
Host records: Summarized by Yu et al. (2016) as being a parasitoid of the following coleophorids: *Coleophora follicularis* (Vallot) and *Coleophora frischella* (L.); the gelechiids *Aproaerema anthyllidella* (Hübner), *Caryocolum fraternella* (Douglas), and *Metzneria aestivella* (Zeller); the geometrid *Eupithecia intricata* (Zetterstedt); and the following tortricids: *Aethes francillana* (Fabricius), *Cydia laricana* (Busck), *Exapate duratella* (von Heyden), *Phalonidia curvistrigana* (Stainton), *Spilonota lariciana* (Heinemann), and *Spilonota ocellana* (Denis and Schiffermüller). In Iran, this species has been reared from the tortricid *Spilonota ocellana* (Denis and Schiffermüller) (Ghahari & Gadallah, 2021).

Therophilus clausthalianus (Ratzeburg, 1844)
Catalogs with Iranian records: Gadallah and Ghahari (2013) as *Bassus clausthalianus*, Farahani et al. (2014, 2016) as *Bassus clausthalianus*, Samin, Coronado-Blanco, Kavallieratos et al. (2018), Samin, Coronado-Blanco, Fischer et al. (2018).
Distribution in Iran: Ilam (Ghahari et al., 2011b as *Bassus clausthalianus*).
Distribution in the Middle East: Iran.
Extralimital distribution: Austria, Belgium, Croatia, former Czechoslovakia, Finland, France, Germany, Greece, Hungary, Ireland, Italy, Kazakhstan, Mongolia, Netherlands, Russia, Slovakia, Slovenia, Sweden, Switzerland, United Kingdom, former Yugoslavia.

Host records: Summarized by Yu et al. (2016) as being a parasitoid of the following depressariids: *Agonopterix atomella* (Denis and Schiffermüller), and *Agonopterix scopariella* (Heinemann); and the following tortricids *Archips oporana* (L.), *Dichelia histrionana* (Frörich), *Dichrorampha acuminatana* (Lienig and Zeller), *Epiblema scutulana* (Denis and Schiffermüller), *Epinotia sordidana* (Hübner), *Epinotia tedella* (Clerck), *Pammene gallicana* (Guenée), and *Spilonota ocellana* (Denis and Schiffermüller).

Therophilus conspicuus (Wesmael, 1837)
Catalogs with Iranian records: No catalog.
Distribution in Iran: Guilan (Samin, Fischer et al., 2019 as *Bassus conspicuus* (Wesmael)).
Distribution in the Middle East: Iran (Samin, Fischer et al., 2019), Turkey (Çetin Erdoğan, 2013; Çetin Erdoğan & Beyarslan, 2006).
Extralimital distribution: Belgium, Canary Islands, China, Croatia, Czech Republic, Finland, France, Germany, Greece, Hungary, Ireland, Italy, Japan, Korea, Netherlands, Poland, Russia, Serbia, Slovenia, Spain, Sweden, Switzerland, United States of America, United Kingdom.
Host records: Summarized by Yu et al. (2016) as being a parasitoid of the crambid *Eudonia lacustrata* (Panzer); and the following tortricids: *Cydia pomonella* (L.), *Grapholita molesta* (Busck), *Gypsonoma nitidulana* (Lienig and Zeller), *Pammene regiana* (Zeller), *Phalonidia manniana* (Fischer), and *Rhopobota ustamaculana* (Curtis).

Therophilus dimidiator (Nees von Esenbeck, 1834)
Catalogs with Iranian records: Gadallah and Ghahari (2013) as *Bassus dimidiator*, Farahani et al. (2014, 2016) as *Bassus dimidiator*, Samin, Coronado-Blanco, Kavallieratos et al. (2018), Samin, Coronado-Blanco, Fischer et al. (2018).
Distribution in Iran: East Azarbaijan (Ghahari & Fischer, 2011a), Kordestan (Samin, 2015), Tehran (Ghahari, Fischer, Sakenin et al., 2011).
Distribution in the Middle East: Iran (see references above), Israel—Palestine (Simbolotti & van Achterberg, 1992), Turkey (Güclü & Ozbek, 2002).
Extralimital distribution: Canada, Nearctic, Oriental, Palaearctic [Adjacent countries to Iran: Armenia, Azerbaijan, Kazakhstan, Russia].
Host records: Summarized by Yu et al. (2016) as being a parasitoid of the agonoxenid *Blastodacna atra* (Haworth); the coleophorid *Coleophora spinella* (Schrank); the following gelechiids: *Recurvaria leucatella* (Clerck), and *Recurvaria nanella* (Denis and Schiffermueller); the following tortricids: *Acleris forsskaleana* (L.), *Acleris variana* (Fernald), *Aleimma loeflingiana* (L.), *Archips crataegana* (Hübner), *Archips rosana* (L.), *Archips xylosteana* (L.), *Argyrotaenia velutinana* (Walker),

Choristoneura rosaceana (Harris), *Croesia bergmanniana* (L.), *Cydia latiferreana* (Walsingham), *Dichelia histrionana* (Frölich), *Epiblema scutulana* (Denis and Schiffermüller), *Epinotia tetraquetrana* (Haworth), *Grapholita interstinctana* (Clemens), *Grapholita molesta* (Busck), *Hedya nubiferana* (Haworth), *Pandemis heparana* (Denis and Schiffermüller), *Spilonota ocellana* (Denis and Schiffermüller), and *Tortrix viridana* L., and the yponomeutid *Yponomeuta malinellus* (Zeller).

Therophilus graecus (Simbolotti & van Achterberg, 1992)

Distribution in the Middle East: Turkey (Çetin Erdoğan & Beyarslan, 2006).
Extralimital distribution: Greece (Simbolotti & van Achterberg, 1992).
Host records: Unknown.

Therophilus linguarius (Nees von Esenbeck, 1812)

Catalogs with Iranian records: Gadallah and Ghahari (2013) as *Bassus linguarius*, Farahani et al. (2014, 2016) as *Bassus linguarius*, Samin, Coronado-Blanco, Kavallieratos et al. (2018), Samin, Coronado-Blanco, Fischer et al. (2018).
Distribution in Iran: Ardabil (Ghahari, Gadallah et al., 2009), Guilan (Ghahari & Beyarslan, 2017), Lorestan (Ghahari, Fischer, Papp et al., 2012), Semnan (Ghahari & Gadallah, 2015).
Distribution in the Middle East: Iran (see references above), Turkey (Çetin Erdoğan & Beyarslan, 2001).
Extralimital distribution: Armenia, Austria, Belgium, Bulgaria, Croatia, Finland, France, Germany, Hungary, Italy, Kazakhstan, Mongolia, Netherlands, Poland, Slovenia, Spain, Switzerland, United Kingdom, former Yugoslavia.
Host records: Recorded by Čapek and Hofmann (1997) as being a parasitoid of *Coleophora* sp. In Iran, this species has been reared from the coleophorid *Coleophora serratella* (L.) (Ghahari & Beyarslan, 2017).

Therophilus longiscutum van Achterberg, 2011

Distribution in the Middle East: Yemen (van Achterberg, 2011).
Extralimital distribution: None.
Host records: Unknown.

Therophilus nigrator van Achterberg, 2011

Distribution in the Middle East: Yemen (van Achterberg, 2011).
Extralimital distribution: None.
Host records: Unknown.

Therophilus nugax (Reinhard, 1867)

Catalogs with Iranian records: No catalog.
Distribution in Iran: East Azarbaijan (Samin, Sakenin Chelav et al., 2020a).
Distribution in the Middle East: Iran (Samin, Sakenin Chelav et al., 2020a), Israel–Palestine (Papp, 2012), Turkey (Beyarslan et al., 2002).
Extralimital distribution: Azerbaijan, Bulgaria, former Czechoslovakia, France, Germany, Hungary, Italy, Poland, Russia, Switzerland, Ukraine, United Kingdom.
Host records: Recorded by Čapek and Hofmann (1997) as being a parasitoid of the tortricid *Cochylis roseana* (Howarth).

Therophilus rugulosus (Nees von Esenbeck, 1834)

Distribution in the Middle East: Israel–Palestine (Papp, 2012), Turkey (Çetin Erdoğan & Beyarslan, 2006).
Extralimital distribution: Andorra, Azerbaijan, Azores, Belgium, Bulgaria, Finland, France, Germany, Greece, Hungary, Ireland, Italy, Korea, Netherlands, Poland, Sweden, Switzerland, United Kingdom.
Host records: Summarized by Yu et al. (2016) as being a parasitoid of the following curculionids *Dryocoetes autographus* (Ratzeberg), and *Dryocoetes villosus* (Fabricius); the blastobasid *Blastobasis vittata* (Wollaston); the coleophorid *Coleophora* sp.; the gelechiid *Recurvaria nanella* (Denis and Schiffermüller), and the tineids *Nemapogon cloacella* (Howarth), *Nemaxera betulinella* (Paykull), and *Triaxomera parasitella* (Hübner).

Therophilus sulciferus van Achterberg, 2011

Distribution in the Middle East: Yemen (van Achterberg, 2011).
Extralimital distribution: None.
Host records: Unknown.

Therophilus tegularis (Thomson, 1895)

Catalogs with Iranian records: No catalog.
Distribution in Iran: Hamadan (Samin et al., 2019a).

Distribution in the Middle East: Iran (Samin, Coronado-Blanco et al., 2019), Turkey (Çetin Erdoğan, 2013).
Extralimital distribution: Bulgaria, France, Germany, Hungary, Italy, Netherlands, Norway, Poland, Sweden, United Kingdom, former Yugoslavia.
Host records: Unknown.

Therophilus tumidulus (Nees von Esenbeck, 1812)

Catalogs with Iranian records: Gadallah and Ghahari (2013), Farahani et al. (2014, 2016), Samin, Coronado-Blanco, Kavallieratos et al. (2018), Samin, Coronado-Blanco, Fischer et al. (2018).
Distribution in Iran: Golestan (Ghahari, Fischer, Sakenin et al., 2011), Kermanshah (Ghahari & Fischer, 2012), Kuhgiloyeh and Boyerahmad (Samin, van Achterberg et al., 2015, under *Bassus tumidulus* (Nees, 1812)), Lorestan (Ghahari, Fischer, Papp et al., 2012), Mazandaran (Ghahari, Fischer, Çetin Erdoğan et al., 2010).
Distribution in the Middle East: Iran (see references above), Turkey (Beyarslan et al., 2002, 2006; Çetin Erdoğan & Beyarslan, 2001;).
Extralimital distribution: Europe, Oriental, Palaearctic [Adjacent countries to Iran: Azerbaijan, Kazakhstan, Russia].
Host records: Summarized by Yu et al. (2016) as being a parasitoid of the dipressariid *Agonopterix atomella* (Denis and Schiffermüller); the gelechiid *Ptocheuusa inopella* (Zeller); the momphid *Mompha epilobiella* (Denis and Schiffermüller); and the following tortricids: *Cydia* spp., *Dichrorampha acuminatana* (Lienig and Zeller), *Epiblema cirsiana* (Zeller), *Epiblema scutulana* (Denis and Schiffermüller), *Gypsonoma aceriana* (Duponchel), *Gypsonoma minutana* (Hübner), *Lathronympha strigana* (Fabricius), *Lobesia botrana* (Denis and Schiffermüller), *Lobesia euphorbianus* (Freyer), *Rhopobata ustomaculana* (Curtis), and *Sparganothis pilleriana* (Denis and Schiffermüller).

Therophilus zaykovi (Nixon, 1986)

Distribution in the Middle East: Turkey (Çetin Erdoğan & Beyarslan, 2006).
Extralimital distribution: Bulgaria, Germany, Hungary.
Host records: Unknown.

Species excluded from the fauna of the Middle East

Cremnops testaceus Pérez, 1907, nomen dubium

Distribution in the Middle East: Saudi Arabia (Shenefelt, 1970).
Extralimital distribution: None.
Host records: Unknown.
Comments: The type of this species has been lost and it is doubtful that it is a species of the genus *Cremnops*. It is a nomen dubium and should be ignored unless someone finds the type (van Achterberg, 2011). At this time, the species should not be included in the counts.

Conclusion

Eighty species of the subfamily Agathidinae (6.6% of the world species) in 12 genera and six tribes (Agathidini Haliday, 1833, Cremnoptini Sharkey, 1992, Disophrini Sharkey, 1992, Earinini Sharkey, 1992, Lytopylini Sharkey, 2017, and Mesocoelini van Achterberg, 1990) have been reported from the Middle East countries. The genus *Agathis* is the most diverse, with 35 recorded species. In total, 47 species of Agathidinae (3.8% of the world species) in 10 genera and five tribes have been reported from Iran so far (being the richest in the Middle East countries): Agathidini (30 species, four genera), Cremnoptini (two species, one genus), Disophrini (five species, three genera), Earinini (two species, one genus), and Mesocoelini (eight species, one genus). The genus *Agathis* with 26 species is more diverse than the other genera, representing 55.3% of the Iranian agathidine species; followed by *Therophilus* (eight species), *Disophrys* (three species), *Aerophilus*, *Cremnops*, *Earinus* (each with two species), *Bassus*, *Camptothlipsis*, *Coccygidium*, and *Euagathis* (each with one species). Two species, *Aerophilus persicus* and *Cremnops richter* are so far only known from Iran (endemic or subendemic to Iran). Iranian Agathidinae have been recorded from 24 provinces, which Guilan with 19 species has the highest number of species; followed by Mazandaran (12 species), Golestan (10 species), and Lorestan (nine species) (Fig. 2.1). Host species have been discovered for only eight agathidine species. Those hosts belong to three families of Lepidoptera: Coleophoridae (three species), Gracillariidae (one species), and Tortricidae (six species). Comparison of the Agathidinae fauna of Iran with Middle East and adjacent countries indicates the fauna of Russia with 63 recorded species in 12 genera (Belokobylskij & Lelej, 2019) is more diverse than Iran; followed by Turkey (41 species), Kazakhstan (37 species), Azerbaijan (30 species), Armenia and Israel-Palesine (both with 15 species), Yemen (13 species), Turkmenistan (11 species), United Arab Emirates (five species), Saudi Arabia (four species), Afghanistan, Cyprus, Egypt, Pakistan (each with two species), and Jordan, Oman, and Syria (each with one species). No species have been recorded from Bahrain, Iraq, Kuwait, Lebanon, and Qatar (Yu et al., 2016). Additionally, among the 23 countries of the Middle East and adjacent to Iran, Turkey and Russia share the largest number of species with Iran (with 33 and 32 species, respectively), followed by Azerbaijan, Kazakhstan (both with 25 species), Armenia (18 species), Israel-Palesine (nine species), Turkmenistan (five species), Cyprus (two species), and Syria (one species). There is no agathidine species shared between Iran and 14 other countries.

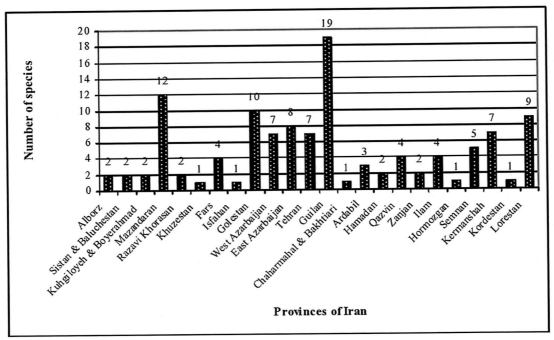

FIGURE 2.1 Number of reported species of Iranian Agathidinae by province.

References

Akbarzadeh Shoukat, G., Safaralizadeh, M., Ranjbar Aghdam, H., & Aramideh, S. (2015). Study on the parasitoid wasps belonging to the superfamily Ichneumonoidea on grape berry moth, *Lobesia botrana* (Lep.: Tortricidae) in Urmia vyneyards. In *Proceedings of the 1st Iranian International Congress of Entomology, 29—31 August 2015* (p. 115). Iranian Research Institute of Plant Protection.

Ameri, A., Talebi, A. A., Rakhshani, E., Mousavi, S. E., & Asakri Siahoyei, M. (2016). First record of the genus *Euagathis* Szépligeti, 1900 (Hym., Braconidae, Agathidinae) representing by *Euagathis indica* Enderlein, 1920 from the Palaearctic region. In *Proceedings of the 22nd Iranian plant protection congress, 27—30 August 2018* (p. 479). Gorgan University of Agricultural Sciences and Natural Resources.

Balduf, W. V. (1966). Life of Acroasis *rubrifasciella* (Lepidoptera: Phycitidae), its main parasite, *Agathis calcarata. Annals of Entomological Society of America, 59*, 1038—1049.

Bashir, M. O., & Venkatraman, T. V. (1968). Insect parasite complex of berseem armyworm *Spodoptera exigua* (Hübner) (Lepidoptera: Noctuidae). *Entomophaga, 13*, 151—158.

Belokobylskij, S. A., & Lelej, A. S. (2019). Annotated catalogue of the Hymenoptera of Russia. Volume II. Apocrita: Parasitica. *Proceedings of the Zoological Institute of the Russian Academy of Sciences*, (8), 594 pp.

Belshaw, R., & Quicke, D. L. J. (2002). Robustness of ancestral state estimates, evolution of life history strategy in ichneumonoid parasitoids. *Systematic Biology, 51*(3), 450—477.

Beyarslan, A., Inanç, F., Çetin Erdoğan, Ö., & Aydoğdu, M. (2002). Braconidae von den tuerkischen Inseln Imbros und Tenedos (Hymenoptera, Braconidae: Agathidinae, Braconinae, Cheloninae, Microgastrinae). *Entomofauna, 23*(15), 173—188.

Beyarslan, A., Yurtcan, M., Çetin Erdoğan, Ö., & Aydoğdu, M. (2006). A study on Braconidae and Ichneumonidae from ganos mountain (Thrace region, Turkey) (Hymenoptera Braconidae, Ichneumonidae). *Linzer Biologische Beitraege, 38/1*, 409—422.

Bodenheimer, F. S. (1930). Die Schädlingsfauna Palästinas. *Monographien zur Angewandten Entomologie, 10*, 438.

Čapek, M., & Hofmann, C. (1997). The Braconidae (Hymenoptera) in the collections of the Musée cantonal de Zoologie, Lausanne. *Litterae Zoologicae (Lausanne), 2*, 25—162.

Çetin Erdoğan, Ö. (2005). *Bassus beyarslani*, sp. n. (Hymenoptera, Braconidae, Agathidinae) from Turkey. *Biologia (Bratislava), 60/2*, 129—132.

Çetin Erdoğan, Ö. (2010). A new species, *Agathis berkei* sp. n., from Eastern Anatolia, Turkey (Hymenoptera, Braconidae, Agathidinae). *Turkish Journal of Zoology, 34*(2), 177—180.

Çetin Erdoğan, Ö. (2013). Contribution to the knowledge of Agathidinae fauna of the Eastern Anatolia region of Turkey. *Turkish Journal of Zoology, 37*(2), 195—199.

Çetin Erdoğan, Ö. (2014). A contribution to the subfamily Agathidinae (Hymenoptera: Braconidae) in Dyarbakir, Mardin, and Sanliurfa provinces of Turkey. *Turkish Journal of Zoology, 38*(1), 101—103.

Çetin Erdoğan, Ö., & Beyarslan, A. (2001). The Agathidinae (Hymenoptera: Braconidae) fauna of the Marmara region. *Turkish Journal of Zoology, 25*(3), 257—268.

Çetin Erdoğan, Ö., & Beyarslan, A. (2004). First record of Agathis rubens Tobias from Turkey (Hymenoptera: Braconidae: Agathidinae). *Acta Entomologica Slovenica, 12*(2), 253—254.

Çetin Erdoğan, Ö., & Beyarslan, A. (2006). New records of endoparasitoid *Bassus* Fabricius, 1804 (Hymenoptera: Braconidae: Agathidinae) species from Turkey. *Phytoparasitica, 34*(4), 350−353.

Çetin Erdoğan, Ö., & Beyarslan, A. (2009). A study on the Agathidinae (Braconidae: Hymanoptera) species from East Black Sea region. *Turkiye Entomoloji Dergisi, 33*(1), 73−80.

Çetin Erdoğan, Ö., & Beyarslan, A. (2016). Faunistic survey on the Agathidinae (Hymenoptera: Braconidae) in the central Anatolia region with a new record for the Turkish fauna: *Bassus tegularis* (Thomson, 1895). *Turkish Journal of Zoology, 40*, 448−453.

Çetin Erdoğan, Ö., van Achterberg, C., & Beyarslan, A. (2009). On the zoogeographical distribution of the genus *Agathis* Latreille, 1804 (Hymenoptera: Braconidae: Agathidinae) in Turkey. *Journal of the Entomological Research Society, 11*(1), 17−25.

Chen, X. X., & van Achterberg, C. (2019). Systematics, phylogeny, and evolution of braconid wasps: 30 years of progress. *Annual Review of Entomology, 64*, 1−24.

Dondale, C. D. (1954). Biology of *Agathis laticinctus* (Cress) (Hymenoptera, Braconidae), a parasite of eye-spotted bud moth, in Nova. Scotia. *The Canadian Entomologist, 86*, 40−44.

Fallahzadeh, M., & Saghaei, N. (2010). Checklist of Braconidae (Insecta: Hymenoptera) from Iran. *Munis Entomology and Zoology, 5*(1), 170−186.

Farahani, S., Talebi, A. S., Rakhshani, E., van Achterberg, C., & Sharkey, M. (2014). A contribution to the knowledge of Agathidinae (Hymenoptera: Braconidae) from Iran with description of a new species. *Biologia, 69*(2), 228−235.

Farahani, S., Talebi, A. A., & Rakhshani, E. (2016). Iranian Braconidae (Insecta: Hymenoptera: Ichneumonoidea): diversity, distribution and host association. *Journal of Insect Biodiversity and Systematics, 2*(1), 1−92.

Gadallah, N. S., & Ghahari, H. (2013). An annotated catalogue of the Iranian Agathidinae and Brachistinae (Hymenoptera: Braconidae). *Linzer Biologische Beitraege, 45*(2), 1873−1901.

Gadallah, N. S., Ghahari, H., Papp, J., & Beyarslan, A. (2018). New records of Braconidae (Hymenoptera) from Iran. *Wuyi Science Journal, 34*, 43−48.

Ghahari, H. (2018). Species diversity of the parasitoids in rice fields of northern Iran, especially parasitoids of rice stem borer. *Journal of Animal Environment, 9*(4), 289−298 (in Persian, English summary).

Ghahari, H. (2020). A study on the fauna of predator and parasitoid arthropods in saffon fields (*Crocus sativus* L.). *Journal of Saffron Research, 7*(2), 203−215 (in Persian, English summary).

Ghahari, H., & Beyarslan, A. (2017). A faunistic study on Braconidae (Hymenoptera: Ichneumonoidea) from Iran. *Natura Somogyiensis, 30*, 39−46.

Ghahari, H., & Beyarslan, A. (2019). A faunistic study on Braconidae (Hymenoptera: Ichneumonoidea) from Iran, and in Memoriam Dr. Jenő Papp (20 May 1933-11 December 2017). *Acta Biologica Turcica, 32*(4), 248−254.

Ghahari, H., & Fischer, M. (2011a). A contribution to the Braconidae (Hymenoptera: Ichneumonoidea) from north-western Iran. *Calodema, 134*, 1−6.

Ghahari, H., & Fischer, M. (2011b). A study on the Braconidae (Hymenoptera: Ichneumonoidea) from some regions of northern Iran. *Entomofauna, 32*, 181−196.

Ghahari, H., & Fischer, M. (2012). A faunistic survey on the braconid wasps (Hymenoptera: Braconidae) from Kermanshah province, Iran. *Entomofauna, 33*, 305−312.

Ghahari, H., & Gadallah, N. S. (2015). A faunistic study on the Braconidae (Hymenoptera) from some regions of Semnan, Iran. *Entomofauna, 36*, 177−184.

Ghahari, H., & Gadallah, N. S. (2021). *Additional records to the braconid fauna (Hymenoptera: Ichneumonoidea) of Iran, with new host reports.* Entomological News (in press).

Ghahari, H., Fischer, M., Çetin Erdoğan, Ö., Tabari, M., Ostovan, H., & Beyarslan, A. (2009). A contribution to Braconidae (Hymenoptera) from rice fields and surrounding grasslands of northern Iran. *Munis Entomology and Zoology, 4*(2), 432−435.

Ghahari, H., Fischer, M., Çetin Erdoğan, Ö., Beyarslan, A., Hedqvist, K. J., & Ostovan, H. (2009). Faunistic note on the Braconidae (Hymenoptera: Ichneumonoidea) in Iranian alfalfa fields and surrounding grasslands. *Entomofauna, 30*, 437−444.

Ghahari, H., Gadallah, N. S., Çetin Erdoğan, Ö., Hedqvist, K. J., Fischer, F., Beyarslan, A., & Ostovan, H. (2009). Faunistic note on the Braconidae (Hymenoptera: Ichneumonoidea) in Iranian cotton fields and surrounding grasslands. *Egyptian Journal of Biological Pest Control, 19*(2), 115−118.

Ghahari, H., Fischer, M., Çetin Erdoğan, Ö., Beyarslan, A., & Ostovan, H. (2010). A contribution to the braconid wasps (Hymenoptera: Braconidae) from the forests of northern Iran. *Linzer Biologische Beitraege, 42*(1), 621−634.

Ghahari, H., Fischer, M., Hedqvist, K. J., Çetin Erdoğan, Ö., van Achterberg, C., & Beyarslan, A. (2010). Some new records of Braconidae (Hymenoptera) for Iran. *Linzer Biologische Beitraege, 42*(2), 1395−1404.

Ghahari, H., Fischer, M., & Papp, J. (2011a). A study on the Braconidae (Hymenoptera: Ichneumonoidea) from Qazvin province, Iran. *Entomofauna, 32*, 197−208.

Ghahari, H., Fischer, M., & Papp, J. (2011b). A study on the Braconidae (Hymenoptera: Ichneumonoidea) from Ilam province, Iran. *Calodema, 160*, 1−5.

Ghahari, H., Fischer, M., Sakenin, H., & Imani, S. (2011). A contribution to the Agathidinae, Alysinae, Aphidiinae, Braconinae, Microgastrinae and Opiinae (Hymenoptera: Braconidae) from cotton fields and surrounding grasslands of Iran. *Linzer Biologische Beitraege, 43*(2), 1269−1276.

Ghahari, H., Fischer, M., & Tobias, V. (2012). A study on the Braconidae (Hymenoptera: Ichneumonoidea) from Guilan province, Iran. *Entomofauna, 33*, 317−324.

Ghahari, H., Fischer, M., Papp, J., & Tobias, V. (2012). A contribution to the knowledge of braconids (Hymenoptera: Braconidae) from Lorestan province Iran. *Entomofauna, 33*, 65−72.

Ghramh, H. A. (2011). Records of the genus *Coccygidium* Saussure (Hymenoptera: Braconidae: Agathidinae), with description of a new species from Saudi Arabia. *African Journal of Biotechnology, 10*(42), 8481−8483.

Ghramh, H. A. (2012). Records of the genus *Camptothlipsis* Enderlein (Hymenoptera: Braconidae: Agathidinae) from Saudi Arabia, with description of a new species. *Journal of Entomology and Nematology, 4*(4), 38−41.

Ghramh, H. A. (2013). Description of a new species of the genus *Coccygidium* de Saussure, 1892 (Hymenoptera: Braconidae: Agathidinae) from Saudi Arabia. *Egyptian Academic Journal of Biological Sciences, 4*(1), 145−148.

Güçlü, C., & Özbek, H. (2002). The subfamily Agathidinae (Hymenoptera: Braconidae) of Erzurum province. *Journal of the Entomological Research Society, 4*(2), 7−19.

Güçlü, C., & Özbek, H. (2007). *Agathis montana* Shestakov (Hymenoptera: Braconidae), a new parasitoid of *Pandemis cerasana* Huebner (Lepidoptera: Tortricidae) in Turkey. *Entomological News, 118*(5), 534.

Hedwig, K. (1957). Ichneumoniden und Braconiden aus den Iran 1954 (Hymenoptera). *Jahresheft des Vereins für Vaterländische Naturkunde, 112*(1), 103−117.

Hellén, W. (1956). Zur Kenntniss der Agathidinen Finnlands (Hymenoptera, Braconidae). *Notulae Entomologicae, 36,* 116−125.

Kian, N., Goldasteh, S., Farahani, S., & van Achterberg, C. (2018). First record of *Agathis levis* Abdinbekova, 1970 (Braconidae: Agathidinae) from Iran. *Biharean Biologist, 12*(2), 111−113.

Kian, N., Goldasteh, S., & Farahani, S. (2020). A survey on abundance and species diversity of Braconid wasps in forest of Mazandaran province. *Journal of Entomological Research, 12*(1), 61−69 (in Persian, English summary).

Kohl, F. F. (1905). Hymenoptera. In A. Pencher, & E. Zederbauer (Eds.), *Ergebnisse einer naturwissenschaftlichen Reise zur Naturhistorischen Museum in Wien* (vol. 20, pp. 220−246).

Kriechbaumer, J. (1898). Über *Diophrys caesa* Klug und inculcatrix auct. Nebst einer neuen Art dieser Gattung. *Entomologische Nachrichten, 24*(1), 181−185.

Lashkari Bod, A., Rakhshani, E., Talebi, A. A., Lozan, A., & Žikič, V. (2011). A contribution to the knowledge of Braconidae (Hym., Ichneumonoidea) of Iran. *Biharean Biologist, 5*(2), 147−150.

Loni, A., Rossi, E., & van Achterberg, C. (2011). First report of *Agathis fuscipennis* in Europe as parasitoid of the tomato leafminer *Tuta absoluta. Bulletin of Insectology, 64*(1), 115−117.

Madl, M. (2007). Review of the Agathidinae (Hymenoptera: Braconidae) of the Malagasy subregion. *Linzer Biologische Beitraege, 39*(2), 993−1007.

Mahmoud, S. M., El-Heneidy, A. H., Gadallah, N. S., & Saleh, R. (2009). Survey and abundance of common ichneumonoid parasitoid species in Suez Canal Region, Egypt. *Egyptian Journal of Biological Pest Control, 19*(2), 185−190.

Marczak, P., & Buszko, J. (1994). Braconid wasps (Hymenoptera, Braconidae) reared from mining Lepidoptera. *Wiadomosci Entomologiszne, 12*(4), 259−272.

Muesebeck, C. F. W. (1927). A revision of the parasitic wasps of the subfamily Braconinae occurring in America north of Mexico. *Proceedings of the United States National Museum, 69*(16), 1−73.

Naderian, H., Ghahari, H., & Asgari, S. (2012). Species diversity of natural enemies in corn fields and surrounding grasslands of Semnan province, Iran. *Calodema, 217,* 1−8.

Naderian, H., Penteado-Dias, A. M., Sakenin Chelav, H., & Samin, N. (2020). A faunistic study on Braconidae and Ichneumonidae (Hymenoptera, Ichneumonoidea) of Iran. *Calodema, 844,* 1−9.

Nixon, G. E. J. (1986). A revision of the European Agathidinae (Hymenoptera: Braconidae). *Bulletin of the British Museum (Natural History), Entomology series, 52*(3), 183−242.

Papp, J. (1970). A contribution to the braconid fauna of Israel (Hymenoptera). *Israel Journal of Entomology, 5,* 63−76.

Papp, J. (2012). A contribution to the braconid fauna of Israel (Hymenoptera: Braconidae), 3. *Israel Journal of Entomology, 41−42,* 165−219.

Quednau, F. W. (1970). Notes on life history fecundity, longevity and attack pattern of the *Agathis punila* (Hymenoptera, Braconidae), a parasite of the larch casebearer. *The Canadian Entomologist, 102,* 736−745.

Ranjbar Aghdam, H., & Fathipour, Y. (2010). Fist report of parasitoid wasps, *Ascogaster quadridentata* and *Bassus rufipes* (Hym.: Braconidae) on codling moth (Lep.: Tortricidae) larvae from Iran. *Journal of Entomological Society of Iran, 30*(1), 55−58.

Sakenin, H., Fischer, M., Samin, N., Imani, S., Papp, J., Ghahari, H., & Rastegar, J. (2011). A faunistic survey on the Braconidae wasps (Hymenoptera: Braconidae) from northern Iran. In *Proceedings of global conference on entomology, 5−9 March 2011* (p. 123). Thailand: Chiang Mai.

Sakenin, H., Naderian, H., Samin, N., Rastegar, J., Tabari, M., & Papp, J. (2012). On a collection of Braconidae (Hymenoptera) from northern Iran. *Linzer Biologische Beitraege, 44*(2), 1319−1330.

Samin, N. (2015). A faunistic study on the Braconidae of Iran (Hymenoptera: Ichneumonoidea). *Arquivos Entomoloxicos, 13,* 339−345.

Samin, N., Sakemin, H., Imani, S., & Shojai, M. (2011). A study on the Braconidae (Hymenoptera) of Khorasan province and vicinity, Northeastern Iran. *Phagea, 39*(4), 137−143.

Samin, N., Ghahari, H., Gadallah, N. S., & Davidian, E. (2014). A study on the Braconidae (Hymenoptera: Ichneumonoidea) from West Azarbaijan province, northern Iran. *Linzer Biologische Beiträge, 46*(2), 1447−1478.

Samin, N., Ghahari, H., Gadallah, N. S., & Monaem, R. (2015). A study on the braconid wasps (Hymenoptera: Ichneumonoidea: Braconidae) from Golestan province, northern Iran. *Linzer Biologische Beitraege, 47*(1), 731−739.

Samin, N., van Achterberg, C., & Ghahari, H. (2015). A faunistic study of Braconidae (Hymenoptera: Ichneumonoidea) from southern Iran. *Linzer Biologische Beitraege, 47*(2), 1801−1809.

Samin, N., Fischer, M., & Ghahari, H. (2015). A contribution to the study on the fauna of Braconidae (Hymenoptera, Ichneumonoidea) from the province of Semnan, Iran. *Arquivos Entomoloxicos, 13,* 429−433.

Samin, N., Coronado-Blanco, J. M., Kavallieratos, N. G., Fischer, M., & Sakenin, H. (2018). Recent findings on Braconidae (Hymenoptera: Ichneumonoidea) of Iran with an updated checklist. *Acta Biologica Turcica, 31*(4), 160−173.

Samin, N., Coronado-Blanco, J. M., Fischer, M., van Achterberg, C., Sakenin, H., & Davidian, E. (2018). Updated checklist of Iranian Braconidae (Hymenoptera: Ichneumonoidea) with twenty-three new records. *Natura Somogyiensis, 32,* 21−36.

Samin, N., Coronado-Blanco, J. M., Hosseini, A., Fischer, M., & Sakenin Chelav, H. (2019). A faunistic study on the braconid wasps (Hymenoptera: Braconidae) of Iran. *Natura Somogyiensis, 33,* 75−80.

Samin, N., Fischer, M., Sakenin, H., Coronado-Blanco, J. M., & Tabari, M. (2019). A faunistic study on Agathidinae, Alysiinae, Doryctinae, Helconinae, Microgastrinae, and Rogadinae (Hymenoptera: Braconidae), with eight new country records. *Calodema, 734,* 1−7.

Samin, N., Sakenin Chelav, H., Ahmad, Z., Penteado-Dias, A. M., & Samiuddin, A. (2020). A faunistic study on the family Braconidae (Hymenoptera: Ichneumonoidea) from Iran. *Scientific Bulletin of Uzhhorod National University (Series: Biology), 48,* 14−19.

Samin, N., Beyarslan, A., Ranjith, A. P., Ahmad, Z., Sakenin Chelav, H., & Hosseini Boldaji, S. A. (2020). A faunistic study on Braconidae (Hymenoptera: Ichneumonoidea) from Ardebil and East Azarbayjan provinces, Northwestern Iran. *Egyptian Journal of Plant Protection Research Institute, 3*(4), 955−963.

Sharkey, M. J. (1985). Notes on the genera *Bassus* Fabricius and *Agathis* Latreille, with a description of *Bassus arthurellus* n. sp. (Hymenoptera: Braconidae). *The Canadian Entomologist, 117,* 1497−1502.

Sharkey, M. J. (1992). Cladistics and tribal classification of the Agathidinae (Hymenoptera: Braconidae). *Journal of the Natural History, 26*, 425–447.

Sharkey, M. J. (1998). Agathidinae, pp. 520–531. In P. A. Lehr (Ed.), *Neuropteroidea, Mecoptera, Hymenoptera: vol. 4. Key to the insects of Russian far East* (pp. 1–708). Part 3.

Sharkey, M. J., & Chapman, E. (2017a). Ten new genera of Agathidini (Hymenoptera, Braconidae, Agathidinae) from Southeast Asia. *ZooKeys, 660*, 107–150.

Sharkey, M. J., & Chapman, E. (2017b). Phylogeny of the Agathidinae (Hymenoptera: Braconidae) with a revised tribal classification and the description of a new genus. *Proceedings of the Entomological Society of Washington, 119* (Special Issue), 823–842.

Sharkey, M. J., Laurenne, N. M., Sharanowski, B., Quicke, D. L. J., & Murray, D. (2006). Revision of the Agathidinae (Hymenoptera: Braconidae) withcomparisons and dynamic alignments. *Cladistics, 22*, 546–567.

Sharkey, M. J., Yu, D. S., van Noort, S., Sltmann, K., & Penev, L. (2009). Revision of the Oriental genera of Agathidinae (Hymenoptera, Braconidae) with an emphasis on Thailand including interactive keys to genera published in three different formats. *ZooKeys, 21*, 19–54.

Shaw, M. R., & Huddleston, T. (1991). Classification and biology of braconid wasps (Hymenoptera: Braconidae). In *Handbooks for the Identification of British Insects* (Vol. 7, pp. 1–126).

Shenefelt, R. D. (1970). *Hymenopterorum Catalogus (nova editio). Pars 6* (pp. 307–428).

Shojai, M. (1989). *Entomology (ethology, social life and natural enemies) (biological control)* (2nd ed., Vol. III, p. 406). University of Tehran Publication.

Shojai, M., Esmaili, M., Ostovan, H., Khodaman, A., Daniali, M., Hosseini, M., Assadi, Y., Sadighfar, M., Korosh-Najad, A., Nasrollahi, A., Labbafi, Y., Azma, M., Ghavam, F., & Honarbakhsh, S. (2000). Integrated pest management of codling moth and other important pests of Pomoidea fruit trees. *Journal of Agricultural Sciences, 6*(2), 15–45 (in Persian, English summary).

Shojai, M., Ostovan, H., Hosseini, M., Sadighfar, M., Khodaman, A., Labbafi, Y., Nasrollahi, A., Ghavam, F., & Honarbakhsh, S. (2002). Biocenotic potentials of apple orchards IPM in organic crop production programme. *Journal of Agricultural Sciences, 8*(1), 1–27 (in Persian, English summary).

Simbolotti, G., & van Achterberg, C. (1992). Revision of the West Palaearctic species of the genus *Bassus* (Hymenoptera: Braconidae). *Zoologische Verhandelingen, 281*, 80 pp.

Simbolotti, G., & van Achterberg, C. (1999). *Revision of the West Palaearctic species of the genus Agathis Latreille (Hymenoptera:* *Braconidae: Agathidinae)* (Vol. 325). Zoologische Verhandlingen Leiden, 167 pp.

Simmonds, F. J. (1947). The biology of the parasites of *Loxostege sticticalis* L., in north America- *Bracon vulgaris* (cress.) (Braconidae, Agathidinae). *Bulletin of Entomological Research, 38*, 145–155.

Telenga, N. A. (1955). Braconidae, subfamily Microgasterinae, subfamily Agathinae. *Fauna USSR, Hymenoptera, 5*(4), 311 (Translated from Russian by Israel Program for Scientific Translation, Jerusalem, 1964).

Tobias, V. I. (1976). Braconids of the Caucasus (Hymenoptera: Braconidae). In *Opred. Faune USSR* (Vol. 110, p. 286). Leningrad: Nauka Press.

Tobias, V. I. (1986). Introduction, pp. 1–21. Rogadinae, pp. 72–85. Gnaptodontinae, Braconinae, Telengainae, pp. 85–149. Helconinae, Brachistinae, pp. 150–180. Euphorinae, pp. 181–250. Macrocentrinae. Xiphozelinae, pp. 250–263. Homolobinae, Orgilinae (Mimagathidinae, Microtypinae), pp. 263–274. Sigalphinae, Agathidinae, pp. 274–291. Ichneutinae, pp. 291–293. Cheloninae, pp. 293–335. Acaeliinae, Cardiochilinae, Microgastrinae, Miracinae, Supplement, pp. 336–501. In G. S. Medvedev (Ed.), *Opredelitel Nasekomych Evrospeiskoi Tsasti SSSR 3, Peredpontdatokrylye 4. Opr. Faune SSSR, 145, 1–501* [Keys to the insects of the European part of USSR. Hymenoptera] [English translation. Lebanon, USA].

van Achterberg, C. (1993). Illustrated key to the subfamilies of the Braconidae (Hymenoptera: Ichneumonoidea). *Zoologische Verhandelingen, 283*, 198.

van Achterberg, C. (2011). Order Hymenoptera, family Braconidae. The subfamily Agathidinae from the United Arab Emirates, with a review of the fauna of the Arabian Peninsula. *Arthropod fauna of the UAE, 4*, 286–352.

van Avhterberg, C., Karam, H. H., & Ramadan, H. M. (2008). Redescription of *Troticus ovalis* (Fahringer) comb. nov., its first host record and a note on *T. melanopterus* Cameron (Hymenoptera: Braconidae: Agathidinae). *Zoologische Mededelingen, 82*(42), 479–484.

van Achterberg, C., & Long, K. D. (2010). Revision of the Agathidinae (Hymenoptera, Braconidae) of Vietnam, with the description of forty-two new species and three new genera. *ZooKeys, 54*, 1–184.

Yu, D. S., van Achterberg, C., & Horstmann, K. (2016). *Taxapad 2016, Ichneumonoidea 2015, Database on flash-drive*. Ottawa, Ontario, Canada.

Zettel, H., & Beyarslan, A. (1992). Uber Agathidinae aus der Türkei (Hymenoptera: Braconidae). *Entomofauna, Zeitschruft Fur Entomologie, 13*(5), 121–132.

Phaenocarpa sp. (Alysiinae), ♀, lateral exodont mandible, and lateral habitus. *Photo prepared by S.R. Shaw.*

Alysia manducator (Panzer, 1799) (Alysiinae), ♀, lateral habitus. *Photo prepared by S.R. Shaw.*

Chapter 3

Subfamily Alysiinae Leach, 1815

Neveen Samy Gadallah[1], Francisco Javier Peris-Felipo[2], Hassan Ghahari[3] and Scott Richard Shaw[4]

[1]Entomology Department, Faculty of Science, Cairo University, Giza, Egypt; [2]Basel, Switzerland; [3]Department of Plant Protection, Yadegar-e Imam Khomeini (RAH) Shahre Rey Branch, Islamic Azad University, Tehran, Iran; [4]UW Insect Museum, Department of Ecosystem Science and Management, University of Wyoming, Laramie, WY, United States

Introduction

The subfamily Alysiinae Leach, 1815 is a monophyletic group on the basis of such distinctive apomorphic characters as the shape and position of the exodont mandibles (the mandibles do not touch each other even when closed) (Belokobylskij & Kostromina, 2011; van Achterberg, 1993), and by the absence of the occipital and prepectal carinae.

The monophyly of this subfamily is supported by molecular phylogenetic studies done by many authors (Gimeno et al., 1997; Zaldivar-Riverón et al., 2006). Based on molecular analyses, Alysiinae is sister-group to Opiinae (Belshaw et al., 1998; Dowton et al., 1998; Gimeno et al., 1997; Quicke & van Achterberg, 1990; Shi et al., 2005; Zaldívar-Riverón et al., 2004, 2006).

Alysiines are very large, cosmopolitan subfamily of medium- to small-sized braconid wasps that occur throughout the world (Fischer et al., 2011). They are often very common and frequently seen wherever decaying organic matter occurs (Peris-Felipo & Jiménez-Peydró, 2011). About 2442 species in 107 genera have been described worldwide within Alysiinae (Peris-Felipo, Belokobylskij, Jiménez-Peydró, 2014b; Yu et al., 2016), which are divided into two large and polymorphic tribes, Alysiini Leach, 1815 and Dacnusini Foerster, 1863 (Chen & van Achterberg, 2019; Peris-Felipo, Belokobylskij, Falcó-Garí et al., 2014; Peris-Felipo, Belokobylskij, Jiménez-Peydró, 2014; Shenefelt, 1974; Yu et al., 2016). Morphologically, the two tribes differ from each other by the presence (Alysiini) or absence (Dacnusini) of the forewing vein cuqu 2 (r-m or second radiomedial); accordingly, Alysiini has three submarginal (radiomedial) cells while Dacnusini have only two (Peris-Felipo, Belokobylskij, Jiménez-Peydró, 2014).

Tribe Dacnusini species are parasitoids of leaf and stem mining dipterans belonging to 13 families, especially those of the families Agromyzidae, Chloropidae, and Ephydridae (Kostromina et al., 2016; Yu et al., 2016). On the other hand, those of the tribe Alysiini are recorded to attack a wide range of dipteran hosts of cyclorrhaphous Diptera (predominantly from the families Agromyzidae, Anthomyiidae, Calliphoridae, Drosophilidae, Muscidae, Sarcophagidae, and Tephritidae) (Belokobylskij, 2005; Berry, 2007; Fischer & Beyarslan, 2012; Yu et al., 2016). They are larval-pupal endoparasitoids that finish their development and emerge from fly puparia (Belokobylskij, 2005; Belokobylskij & Tobias, 1997). From the economic point of view, they can play a potential role in the regulation of such dipteran pests (Berry, 2007).

Checklists of Regional Alysiinae. Khajeh et al. (2014) listed 55 species in 13 genera; Gadallah et al. (2015) cataloged 78 species in 15 genera; Farahani et al. (2016), 85 species in 18 genera; Yu et al. (2016), 90 species in 19 genera; Cortés et al. (2016), 108 species in 18 genera, and finally Samin, Coronado-Blanco, Kavallieratos et al. (2018) and Samin, Coronado-Blanco, Fischer et al. (2018) listed 93 and 94 species, respectively, in 20 genera. Samin, Fischer et al. (2019) added 12 alysiine species as new records for the Iranian fauna. In the present chapter, the subfamily Alysiinae is represented by 213 species in 32 genera and two tribe (Alysiini and Dacnusini) in the Middle East, of which eight species are only known from Iran (endemic). The number of genera has been reduced because of the synonymizations carried out during different studies (e.g., *Adelphenaldis* Fischer, 2003 = *Aspilota* (*Eusynaldis*) Foerster, 1863, and *Synaldis* Foerster, 1863 = *Dinotrema* (*Synaldis*) Foerster, 1863) (Peris-Felipo & Belokobylskj, 2020; Zhu et al., 2017). Here we follow Yu et al. (2016) and Chen and van Achterberg (2019) for classification, and Yu et al. (2016) for the global distribution of species, and in other situations the related references are given.

This chapter is dedicated to the late Maximilian Fischer (June 7, 1929–June 15, 2019; see Ghahari et al., 2020) who had invaluable cooperation with H. Ghahari for identification of several specimens of Iranian Braconidae.

Key to genera of the subfamily Alysiinae of Iran (modified from Tobias & Jakimavicius, 1986)

1. Fore wing with three submarginal cells (tribe Alysiini) 2
— Fore wing with two submarginal cells (tribe Dacnusini) 19
2. Wings well developed 3
— Wings distinctly reduced, or at most slightly exceeding propodeum 18
3. Hind wing narrow, without closed cells .... *Dinotrema* Foerster
— Hind wing wider, with one to two closed cells 4
4. Metasoma smooth, only first tergite sculptured; mandible tridentate, sometimes middle tooth with an additional dentiform projection above 5
— Metasoma with first to third tergites sculptured; 3-SR of fore wing as long as 2-SR vein .... *Trachyusa* Ruthe
5. Second flagellomere longer than first flagellomere 6
— First flagellomere longer than second; notauli developed; marginal cell reaching wing apex 11
6. Pterostigma short, triangular, vein r originating from its middle or posterior to middle; sometimes pterostigma fairly narrow, then precoxal sulcus smooth; notauli smooth above; Vein 3-SR of fore wing not longer than vein 2-SR 7
— Pterostigma cuneate or parallel-sided, vein r originating anterior to its middle; sometimes pterostigma entirely fused with metacarpus 8
7. Precoxal sulcus deep, crenulate; propodeum without distinct longitudinal ridge .... *Alysia* Latreille
— Precoxal sulcus weak or not developed, smooth or weakly sculptured; propodeum with longitudinal ridge; legs somewhat darkened .... *Pentapleura* Foerster
8. Vein 3-SR of fore wing not longer than or very slightly longer than first submarginal cell; metasoma slightly compressed, first tergite broadened apically; pterostigma cuneate .... *Tanycarpa* Foerster
— Vein 3-SR of fore wing much longer than first submarginal cell; metasoma usually somewhat distinctly compressed, first tergite parallel-sided or slightly broadened apically 9
9. Mandible with distinct and curved transverse carina; lower mandibular tooth very large, obtuse; pterostigma usually distinctly separated from metacarpus, linear, rarely narrow cuneate .... *Orthostigma* Ratzeburg
— Mandible without distinct and curved transverse carina; lower mandibular denticle small, of different shapes; pterostigma often fused with metacarpus 10
10. Notauli indistinct, only may be somewhat distinct anteriorly .... *Aspilota* Foerster
— Notauli distinct .... *Carinthilota* Fischer
11. Postscutellum without denticle, if sometimes with small denticle, then temple short and maxillary palp longer than head height 12
— Postscutellum with denticle; vein 3-M of fore wing usually originating from middle of outside of subdiscal cell, or slightly developed; vein 3-SR usually not longer than first submarginal cell 16
12. First flagellomere not longer than or barely longer than second flagellomere 13
— First flagellomere longer than second flagellomere; second section of radial vein of fore wing not longer, usually shorter than vein 2-SR; propodeum usually with somewhat developed longitudinal ridge .... *Cratospila* Foerster
13. Vein 3-SR of fore wing longer than vein 2-SR, sometimes vein 2-SR not developed 14
— Vein 3-SR of fore wing not longer than vein 2-SR; first and second submarginal and discal cells separate .... *Idiasta* Foerster
14. Pterostigma relatively short, not cuneate and linear; vein r originating usually posterior to its middle; second flagellomere longer than first flagellomere; ovipositor usually not less than half as long as metasoma 15
— Pterostigma narrow, cuneate or linear; vein r originate anterior to its middle, rarely from middle; second flagellomere not longer than or slightly longer than first flagellomere; ovipositor short .... *Adelurola* Strand
15. First subdiscal cell closed .... *Phaenocarpa* Foerster
— First subdiscal cell wide opened .... *Asobara* Foerster
16. Second flagellomere longer than first flagellomere 17
— Second flagellomere much shorter than first flagellomere .... *Alloea* Haliday

17. Fore wing with very small second submarginal cell; upper margin of middle mandibular tooth serrate .... *Angelovia* Zykova
— Fore wing with normally develop second submarginal cell; upper margin of middle mandibular tooth not serrate *Idiasta* Foerster

18. Notauli absent; mesoscutum slightly developed compared to pronotum and propodeum; scutellum not separated; first flagellomere shorter than second flagellomere .... *Pseudopezomachus* Mantero
— Notauli distinct; mesoscutum well-developed; scutellum separated .... *Alloea* Haliday

19. Apical metasomal tergites with rather numerous, widely scattered setae; second tergite longitudinally rugose, if rarely smooth, then pterostigma wide; vein SR1 of fore wing uniformly curved; mandible either 4- or if 3-dentate, then middle tooth strongly developed 20
— Apical metasomal tergites with only one row of setae on posterior margin; second tergite smooth or weakly sculptured at base; if longitudinally rugose, then mandible 3-dentate or sides of mesoscutum with dense setae, radially divergent from a small tubercle 24

20. Pterostigma short and wide; marginal cell somewhat reduced 21
— Pterostigma narrower and longer; marginal cell usually longer, if relatively short, then body more than or slightly more than 3.0 mm, and head smooth .... *Aristelix* Nixon

21. Female metasoma compressed at apex, at least 2.5× as long as its maximum width; mandible with large and acute middle tooth, 3-dentate, of if 4-dentate, then a small additional denticle above large middle one; metasomal tergites uniformly pubescent 22
— Metasoma not more than 2.0× as long as its maximum width in both sexes; metasoma adpressed; mandible wide and obtuse, 4-dentate, fourth tooth faint .... *Epimicta* Foerster

22. Vein SR1 of fore wing uniformly curved; first metasomal tergite more than 2.0× as long as wide; mandible elongate, with acuminate middle tooth, and a small tooth above it; second metasomal tergite sculptured, rarely smooth; antenna pubescent, with distinct rhinaria .... *Coelinidea* Viereck
— Vein SR1 of fore wing unevenly curved 23

23. First metasomal tergite thin, more than 2.0× as long as wide; mandible 4-dentate, third tooth relatively small; metasoma distinctly compressed; head quadrate dorsally, smooth; second metasomal tergite only longitudinally rugose at base *Coelinius* Nees
— First metasomal tergite massive, less than 2.0× as long as wide; mandible with very large and long third tooth; metasoma only compressed apically; head distinctly transverse, punctate .... *Polemochartus* Schulz

24. Mandible 4-dentate, sometimes third tooth as faint dentiform process on middle tooth; in most species sides of propodeum densely pubescent, hairs appressed, with lustrous tubercle, surrounded by radially divergent hairs; sides of mesosoma below always with a furrow .... *Chorebus* Haliday
— Mandible usually 3-dentate; sides of propodeum with sparse hairs, if somewhat densely hairy, then hairs obliquely downwardly directed to hind coxa and not appressed 25

25. Vein m-cu of fore wing merging with second submarginal cell .... *Exotela* Foerster
— Vein m-cu of fore wing usually merging with first submarginal cell (in some species of *Dacnusa* interstitial, but then pterostigma is wide, and the furrow in lower side of mesosoma not developed) 26

26. Head massive, roughly 1.5× as wide as mesosoma, with broad temple, at least as long as eye; mandible large, with first tooth widely deflected sidewards and wide but slightly projecting, almost bidentate .... *Protodacnusa* Griffiths
— Head and mandible less massive 29

27. First subdiscal cell open, at least in its lower outer angle .... *Coloneura* Foerster
— First subdiscal cell closed 28

28. Pterostigma either very long and thin, reaching beyond middle of marginal cell, or characterized by strong sexual dimorphism; sides of metanotum and propodeum sometimes fairly evenly pubescent, hairs thin, uniformly distributed; precoxal sulcus often smoothened .... *Dacnusa* Haliday
— Pterostigma usually less long and similar in both sexes; sides of metanotum and propodeum with sparse hairs; precoxal sulcus distinct, rugose .... *Exotela* Foerster

List of species of the subfamily Alysiinae recorded in the Middle East

Subfamily Alysiinae Leach, 1815

Tribe Alysiini Leach, 1815

Genus *Adelurola* Strand, 1928

Adelurola amplidens (Fischer, 1966)

Catalogues with Iranian records: Farahani et al. (2016), Cortés et al. (2016), Yu et al. (2016), Samin, Coronado-Blanco, Kavallieratos et al. (2018), Samin, Coronado-Blanco, Fischer et al. (2018).
Distribution in Iran: Hormozgan, Kermanshah (Peris-Felipo, Yari, van Achterberg et al., 2016), Kerman (Safahani et al., 2017).
Distribution in the Middle East: Iran (Peris-Felipo, Yari, van Achterberg et al., 2016; Safahani et al., 2017), Iraq (Fischer, 1966; Peris-Felipo, Yari, van Achterberg et al., 2016).
Extralimital distribution: None.
Host records: Recorded by Peris-Felipo, Yari, van Achterberg et al. (2016) as being a parasitoid of the anthomyiid *Pegomya hyoscyami* (Panzer).
Plant associations in Iran: *Medicago sativa* (Fabaceae) (Peris-Felipo, Yari, van Achterberg et al., 2016), *Anethum graveolens* (Apiaceae) (Safahani et al., 2017).

Adelurola florimela (Haliday, 1838)

Catalogues with Iranian records: No catalog.
Distribution in Iran: Fars (Samin, Fischer et al., 2019).
Distribution in the Middle East: Iran.
Extralimital distribution: Austria, former Czechoslovakia, Finland, Georgia, Germany, Hungary, Ireland, Italy, Japan, Latvia, Lithuania, Netherlands, Norway, Poland, Russia, Slovenia, Spain, Sweden, Switzerland, United Kingdom, former Yugoslavia.
Host records: Recorded by Königsmann (1959) as being a parasitoid of the anthomyiid *Pegomya solennis* (Meigen).

Genus *Alloea* Haliday, 1833

Alloea contracta (Haliday, 1833)

Catalogues with Iranian records: Khajeh et al. (2014), Gadallah et al. (2015), Farahani et al. (2016), Cortés et al. (2016), Yu et al. (2016), Samin, Coronado-Blanco, Kavallieratos et al., 2018, Samin, Coronado-Blanco, Fischer et al., 2018.
Distribution in Iran: Golestan (Samin, Ghahari et al., 2015), Mazandaran (Ghahari, Fischer, Hedqvist et al., 2010).
Distribution in the Middle East: Iran.
Extralimital distribution: Belgium, Faroe Islands, Germany, Hungary, Iceland, Ireland, Mongolia, Netherlands, Poland, Spain, Sweden, United Kingdom.
Host records: Unknown.

Genus *Alysia* Latreille, 1804

Alysia (*Alysia*) *alticola* (Ashmead, 1890)

Catalogues with Iranian records: No catalog.
Distribution in Iran: Fars (Samin, Fischer et al., 2019).
Distribution in the Middle East: Iran.
Extralimital distribution: Argentina, Armenia, Bulgaria, Canada, Georgia, Greenland, Ireland, Kazakhstan, Mongolia, Poland, Russia, Serbia, United States of America, United Kingdom.
Host records: Recorded by Wharton (1984) as being a parasitoid of the calliphorids *Calliphora lilaea* (Walker) and *Calliphora terraenovae* Robineau-Desvoidy.

Alysia (*Alysia*) *frigida* Haliday, 1838

Catalogues with Iranian records: No catalog.
Distribution in Iran: Khuzestan (Samin, Fischer et al., 2019).
Distribution in the Middle East: Iran.
Extralimital distribution: Canada, China, former Czechoslovakia, Finland, Georgia, Germany, Hungary, Lithuania, Mongolia, Norway, Poland, Romania, Russia, Sweden, Switzerland, United States of America, United Kingdom.
Host records: Recorded by Wharton (1986) as being a parasitoid of the anthomyiid *Pegomya* sp.

Alysia (*Alysia*) *incongrua* Nees von Esenbeck, 1834

Catalogues with Iranian records: No catalogue.
Distribution in Iran: East Azarbaijan (Samin, Fischer et al., 2019).
Distribution in the Middle East: Iran.
Extralimital distribution: Armenia, Austria, Bulgaria, former Czechoslovakia, Denmark, Finland, Georgia, Germany, Hungary, Ireland, Italy, Poland, Russia, Spain, Switzerland, United Kingdom.
Host records: Unknown.

Alysia (*Alysia*) *lucicola* Haliday, 1838

Catalogues with Iranian records: No catalogue.
Distribution in Iran: Hamadan (Samin, Fischer et al., 2019).
Distribution in the Middle East: Iran (Samin, Fischer et al., 2019), Turkey (Fischer & Beyarslan, 2012).
Extralimital distribution: Armenia, Austria, Bulgaria, Canada, Czech Republic, Georgia, Germany, Hungary, Ireland, Italy, Kyrgyzstan, Netherlands, Norway, Poland, Russia, Spain, Sweden, Switzerland, United States of America, Ukraine, United Kingdom.
Host records: Recorded by Wharton (1986) as being a parasitoid of the Heleomyzid *Suillia apicalis* (Loew).

Alysia (*Alysia*) *luciella* Stelfox, 1941

Catalogues with Iranian records: No catalogue.
Distribution in Iran: East Azarbaijan (Samin, Fischer et al., 2019).

Distribution in the Middle East: Iran (Samin, Fischer et al., 2019), Turkey (Fischer & Beyarslan, 2012).

Extralimital distribution: Armenia, Bulgaria, Canada, Czech Republic, Georgia, Ireland, Mongolia, Norway, Russia, Sweden, United States of America, Ukraine, United Kingdom.

Host records: Unknown.

Alysia (*Alysia*) *manducator* (Panzer, 1799)

Catalogues with Iranian records: No catalogue.

Distribution in Iran: West Azarbaijan (Samin, Fischer et al., 2019).

Distribution in the Middle East: Iran (Samin, Fischer et al., 2019), Israel–Palestine (Papp, 2012), Turkey (Fischer & Beyarslan, 2012).

Extralimital distribution: Armenia, Australia (introduced), Austria, Belgium, Bulgaria, China, Croatia, former Czechoslovakia, Finland, France, Georgia, Germany, Greenland, Hungary, Iceland, Ireland, Italy, Mongolia, Montenegro, Netherlands, New Zealand (introduced), Norway, Poland, Portugal, Romania, Russia, Serbia, Slovenia, Spain, Sweden, Switzerland, South Africa (introduced), United Kingdom, United States of America, Uruguay, Uzbekistan.

Host records: Summarized by Yu et al. (2016) as being a parasitoid of the anthomyiids *Delia antiqua* (Meigen), *D. floralis* (Fallén), and *D. radicum* (L.); the calliphorids *Calliphora vicina* Robineau-Desvoidy, *Chrysomya albiceps* (Wiedemann), *Lucilia caeser* (L.), and *L. sericata* (Meigen); and the muscids *Hydrotaea dentipes* (Fabricius), *Musca domestica* L., *Muscina stabulanus* Fallén, and *Stomoxys calcitrans* (L.).

Alysia (*Alysia*) *truncator* (Nees von Esenbeck, 1812)

Catalogues with Iranian records: No catalogue.

Distribution in Iran: Fars (Samin, Fischer et al., 2019).

Distribution in the Middle East: Iran.

Extralimital distribution: Armenia, Canada, Czech Republic, France, Georgia, Germany, Hungary, Ireland, Italy, Korea, Poland, Russia, Switzerland, United States of America, United Kingdom.

Host records: Summarized by Yu et al. (2016) as being a parasitoid of the agromyzid *Agromyza macquarti* Loew; and the anthomyiids *Delia florilega* (Zetterstedt), *D. platura* (Meigen), and *D. radicum* (L.).

Alysia (*Anarcha*) *fuscipennis* Haliday, 1838

Catalogues with Iranian records: This species is a new record for the fauna of Iran.

Distribution in Iran: Mazandaran province, Ramsar, Dalkhani forest, 2♀, 1♂, July 2016.

Distribution in the Middle East: Iran (new record), Turkey (Fischer & Beyarslan, 2012).

Extralimital distribution: Austria, Belgium, Bulgaria, Czech Republic, France, Georgia, Germany, Hungary, Ireland, Italy, Netherlands, Norway, Poland, Russia, Serbia, Sweden, Switzerland, United Kingdom.

Host records: Unknown.

Alysia (*Anarcha*) *rufidens* Nees von Esenbeck, 1834

Catalogues with Iranian records: Farahani et al. (2016), Cortés et al. (2016), Yu et al. (2016), Samin, Coronado-Blanco, Kavallieratos et al. (2018), Samin, Coronado-Blanco, Fischer et al. (2018).

Distribution in Iran: West Azarbaijan (Samin et al., 2014).

Distribution in the Middle East: Iran (Samin et al., 2014), Israel–Palestine (Papp, 2012), Turkey (Fischer & Beyarslan, 2012).

Extralimital distribution: Austria, Belgium, Bulgaria, Germany, Greece, Hungary, Ireland, Netherlands, Norway, Russia, Spain, Sweden, Switzerland, United States of America.

Host records: Summarized by Yu et al. (2016) as being a parasitoid of the anthomyiids *Delia floralis* (Fallén), and *D. quadripila* (Stein); and the tephritids *Ensina sonchi* L., and *Tephritis arnica* (L.).

Alysia (*Anarcha*) *tipulae* (Scopoli, 1763)

Catalogues with Iranian records: No catalogue.

Distribution in Iran: Lorestan (Ghahari & Beyarslan, 2019).

Distribution in the Middle East: Iran.

Extralimital distribution: Armenia, Austria, Belgium, Bulgaria, Croatia, Czech Republic, France, Georgia, Germany, Hungary, Ireland, Italy, Korea, Lithuania, Mongolia, Netherlands, Norway, Poland, Romania, Russia, Slovenia, Spain, Sweden, Switzerland, Ukraine, United Kingdom, former Yugoslavia.

Host records: Recorded by Kolarov and Bechev (1995) as being a parasitoid of the mycetophilid *Mycetophila* sp.

Genus *Angelovia* Zaykov, 1980

Angelovia elipsocubitalis Zaykov, 1980

Catalogues with Iranian records: Gadallah et al. (2015), Farahani et al. (2016), Cortés et al. (2016), Yu et al. (2016), Samin, Coronado-Blanco, Kavallieratos et al. (2018), Samin, Coronado-Blanco, Fischer et al. (2018).

Distribution in Iran: Golestan (Ghahari & Fischer, 2011b).

Distribution in the Middle East: Iran.

Extralimital distribution: Bulgaria, Greece.

Host records: Unknown.

Genus *Aphaereta* Foerster, 1863

Aphaereta brevis Tobias, 1962

Catalogues with Iranian records: No catalogue.

Distribution in Iran: Northern Khorasan (Sakenin et al., 2020).

Distribution in the Middle East: Iran.

Extralimital distribution: Afghanistan, Bulgaria, former Czechoslovakia, Hungary, Korea, Russia, Serbia, Spain.

Host records: Unknown.

Aphaereta difficilis Nixon, 1939

Catalogues with Iranian records: Khajeh et al. (2014), Gadallah et al. (2015), Farahani et al. (2016), Cortés et al. (2016), Yu et al. (2016), Samin, Coronado-Blanco, Kavallieratos et al. (2018), Samin, Coronado-Blanco, Fischer et al. (2018).

Distribution in Iran: Ilam (Ghahari et al., 2011c).

Distribution in the Middle East: Iran (Ghahari et al., 2011c), Israel−Palestine (Papp, 2012).

Extralimital distribution: Austria, Bulgaria, France, Germany, Greece, Hungary, Korea, Moldova, Morocco, Poland, Romania, Russia, Serbia, Spain, Switzerland, Tunisia, Uzbekistan.

Host records: Summarized by Yu et al. (2016) as being a parasitoid of the anthomyiids *Delia radicum* (L.) and *Fucellia tergina* (Zetterstedt).

Aphaereta falcigera Graham, 1960

Distribution in the Middle East: Israel−Palestine (Papp, 2012).

Extralimital distribution: Austria, Czech Republic, Germany, Hungary, Ireland, Korea, Spain, United Kingdom.

Host records: Unknown.

Aphaereta minuta (Nees von Esenbeck, 1811)

Catalogues with Iranian records: Khajeh et al. (2014), Gadallah et al. (2015), Farahani et al. (2016), Cortés et al. (2016), Yu et al. (2016), Samin, Coronado-Blanco, Kavallieratos et al. (2018), Samin, Coronado-Blanco, Fischer et al. (2018).

Distribution in Iran: Isfahan (Ghahari et al., 2011b as *Asobara minuta*; Khajeh et al., 2014), Tehran (Khajeh et al., 2014).

Distribution in the Middle East: Iran (Ghahari et al., 2011b; Khajeh et al., 2014), Turkey (Fischer & Beyarslan, 2012).

Extralimital distribution: Austria, Azores, Bulgaria, Croatia, Czech Republic, Faroe Islands, Finland, France, Germany, Greece, Greenland, Hungary, Iceland, Ireland, Italy, Korea, Moldova, Morocco, Netherlands, Norway, Poland, Portugal, Romania, Russia, Serbia, Slovenia, Spain, Sweden, Switzerland, Ukraine, United Kingdom, Uzbekistan.

Host records: Summarized by Yu et al. (2016) as being a parasitoid the anthomyiids *Anthomyia pluvialis* (L.), *Botanophila phrenione* (Séguy), *Delia antiqua* (Meigen), *Delia radicum* (L.), and *Fucellia* sp.; the calliphorids *Calliphora vicina* Robineau-Desvoidy, *Lucilia illustri* Meigen, and *Lucilia sericata* (Meigen); the coelopid *Coelopa* sp.; the drosophilid *Drosophila hydei* Sturtevant; the fanniid *Fannia canicularis* (L.); the muscids *Musca autumnalis* De Geer, *M. domestica* L., *M. stabulans* Fallén, *Muscina levida* (Harris), and *Potamia littoralis* Robineau-Desvoidy; the sarcophagids *Ravinia premix* (Harris), and *Sarcophaga haemorrhoidalis* Fallén; the scatophagid *Scatophaga stercoraria* (L.); the sepsids *Orygma luctuosum* Meigen, and *Themira putris* (L.); and the tephritid *Ceratitis capitata* (Wiedemann).

Aphaereta pallipes (Say, 1828)

Catalogues with Iranian records: No catalogue.

Distribution in Iran: Golestan (Ghahari & Beyarslan, 2019).

Distribution in the Middle East: Iran.

Extralimital distribution: Australia (introduced), United States of America; Nearctic, Neotropical, Oceanic, and Palaearctic regions.

Host records: Summarized by Yu et al. (2016) as being a parasitoid of the agromyzid *Calycomya jucunda* (Wulp); the anthomyiids *Adia cinerella* Fallén, *Delia antiqua* (Meigen), *Delia platura* (Meigen), and *Delia radicum* (L.); the chloropid *Chlorops* sp.; the fanniid *Fannia canicularis* (L.); the muscids *Haematobia irritans* (L.), *Musca autumnalis* De Geer, *Musca domestica* L., *Muscina levida* (Harris), and *Neomyia cornicina* (Fabricius); the sarcophagids *Blaesoxipha kellyi* Aldrich, *Helicobia rapax* (Walker), *Oxysarcodexia ventricosa* (Wulp), *Ravinia* spp., and *Sarcophaga* spp.; the sciomyzid *Atrichomelina pubera* (Loew); and the ulidiid *Tritoxa flexa* (Wiedemann).

Aphaereta tenuicornis Nixon, 1939

Distribution in the Middle East: Turkey (Fischer & Beyarslan, 2012).

Extralimital distribution: Austria, Czech Republic, Finland, Germany, Hungary, Iceland, Ireland, Netherlands, Poland, Russia, Spain, Switzerland, United Kingdom.

Host records: Summarized by Yu et al. (2016) as being a parasitoid of the anthomyiids *Botanophila phrenione* (Séguy), and *Delia radicum* (L.).

Genus *Asobara* Foerster, 1863

Asobara tabida (Nees von Esenbeck, 1834)

Catalogues with Iranian records: This species is a new record for the fauna of Iran.

Distribution in Iran: Guilan province, Talesh, 3♀, July 2015, ex. *Drosophila melanogaster* Meigen (Diptera: Drosophilidae).

Distribution in the Middle East: Iran (new record), Turkey (Beyarslan & Inanç, 1992; Fischer & Beyarslan, 2012; Kraaijeveld & van der Wel, 1994).

Extralimital distribution: Austria, Azores, Bulgaria, Canada, Cape Verde Islands, China, Czech Republic, France, Germany, Greece, Hungary, Ireland, Italy, Japan, Korea, Malaysia, Moldova, Morocco, Netherlands, New Zealand, Poland, Romania, Russia, Serbia, Spain, Sweden, Switzerland, Tunisia, United Kingdom, Uzbekistan.
Host records: Summarized by Yu et al. (2016) as being a parasitoid of the cecidomyiid *Fabomyia medicaginis* (Rübsaamen); and the drosophilids *Drosophila ambigua* Pomini, *D. auraria* Peng, *D. busckii* Coquillett, *D. funebris* (Fabricius), *D. kuntzei* Duda, *D. melanogaster* Meigen, *D. obscura* Fallén, *D. simulans* Sturtevant, *D. subobscura* Collin, and *D. suzukii* (Matsumura). In Iran, this species has been reared from *D. melanogaster* (present work).

Asobara vanharteni van Achterberg, 2019
Distribution in the Middle East: Yemen (Peris-Felipo et al., 2019).
Extralimital distribution: None.
Host records: Unknown.

Genus *Aspilota* Foerster, 1863

Aspilota (*Aspilota*) *alfalfae* Fischer, Lashkari Bod, Rakhshani and Talebi, 2011
Catalogues with Iranian records: Khajeh et al. (2014), Gadallah et al. (2015), Farahani et al. (2016), Cortés et al. (2016), Yu et al. (2016), Samin, Coronado-Blanco, Kavallieratos et al. (2018), Samin, Coronado-Blanco, Fischer et al. (2018).
Distribution in Iran: Fars (Fischer et al., 2011; Khajeh et al., 2014; Peris-Felipo, Ameri, et al., 2016), Kerman (Safahani et al., 2017), Khuzestan (Peris-Felipo, Ameri, et al., 2016).
Distribution in the Middle East: Iran.
Extralimital distribution: None.
Host records: Unknown.
Plant associations in Iran: *Medicago sativa* (Fabaceae) (Fischer et al., 2011; Khajeh et al., 2014; Peris-Felipo, Ameri, et al., 2016; Safahani et al., 2017).

Aspilota (*Aspilota*) *breviantennata* Tobias, 1962
Distribution in the Middle East: Israel–Palestine (Papp, 2012).
Extralimital distribution: Czech Republic, Hungary, Poland, Russia, Slovakia.
Host records: Unknown.

Aspilota (*Aspilota*) *delicata* Fischer, 1973
Catalogues with Iranian records: Khajeh et al. (2014), Gadallah et al. (2015), Farahani et al. (2016), Cortés et al.

(2016), Yu et al. (2016), Samin, Coronado-Blanco, Kavallieratos et al. (2018), Samin, Coronado-Blanco, Fischer et al. (2018).
Distribution in Iran: Ardabil (Ghahari, Fischer et al., 2011), Hormozgan, Kerman (Peris-Felipo, Ameri, et al., 2016), Iran (no specific locality cited) (Peris-Felipo, 2013; Peris-Felipo, Belokobylskij, Falcó-Garí et al., 2014).
Distribution in the Middle East: Iran.
Extralimital distribution: Austria, Greece, Hungary, Spain.
Host records: Unknown.

Aspilota (*Aspilota*) *flagellaris* Fischer, 1973
Catalogues with Iranian records: No catalogue.
Distribution in Iran: West Azerbaijan (Sakenin et al., 2020).
Distribution in the Middle East: Iran.
Extralimital distribution: Austria, Hungary, Korea, Spain.
Host records: Unknown.

Aspilota (*Aspilota*) *flagimilis* Fischer, 1996
Catalogues with Iranian records: Cortés et al. (2016).
Distribution in Iran: Isfahan (Peris-Felipo, Ameri et al., 2016).
Distribution in the Middle East: Iran.
Extralimital distribution: Spain.
Host records: Unknown.
Plant associations in Iran: *Chenopodium* sp. (Amaranthaceae) (Peris-Felipo, Ameri et al., 2016).

Aspilota (*Aspilota*) *fuscicornis* (Haliday, 1838)
Catalogues with Iranian records: No catalogue.
Distribution in Iran: Khuzestan (Samin, Coronado-Blanco et al., 2019).
Distribution in the Middle East: Iran.
Extralimital distribution: Austria, Belgium, former Czechoslovakia, Faroe Islands, Finland, Germany, Greece, Hungary, Iceland, Ireland, Italy, Korea, Lithuania, Mongolia, Netherlands, Poland, Portugal, Romania, Russia, Spain, Sweden, United Kingdom.
Host records: Summarized by Yu et al. (2016) as being a parasitoid of the phorid *Spiniphora bergenstammi* (Mik); and the tephritid *Euleia heraclei* (L.).

Aspilota (*Eusynaldis*) *globipes* (Fischer, 1962)
Catalogues with Iranian records: No catalogue.
Distribution in Iran: Guilan (Ghahari & Beyarslan, 2019 as *Adelphenaldis globipes* (Fischer, 1962)).
Distribution in the Middle East: Iran.
Extralimital distribution: Bulgaria, China, Czech Republic, Georgia, Hungary, Italy, Netherlands, Poland, Russia, Spain, Sweden.
Host records: Unknown.

Aspilota (Aspilota) insolita (Tobias, 1962)
Catalogues with Iranian records: Cortés et al. (2016).
Distribution in Iran: Kerman (Peris-Felipo, Ameri et al., 2016).
Distribution in the Middle East: Iran.
Extralimital distribution: Former Czechoslovakia, Hungary, Russia, Spain.
Host records: Unknown.

Aspilota (Aspilota) isfahanensis Peris-Felipo, 2016
Catalogues with Iranian records: Cortés et al. (2016), Yu et al. (2016), Samin, Coronado-Blanco, Kavallieratos et al. (2018), Samin, Coronado-Blanco, Fischer et al. (2018).
Distribution in Iran: Isfahan (Peris-Felipo, Yari, Rakhshani et al. 2016, Peris-Felipo, Ameri, et al., 2016), Kerman (Safahani et al., 2017).
Distribution in the Middle East: Iran.
Extralimital distribution: None.
Plant associations in Iran: *Chenopodium* sp. (Amaranthaceae) (Peris-Felipo, Yari, Rakhshani et al. 2016), and *Triticum aestivum* (Poaceae) (Safahani et al., 2017).

Aspilota (Aspilota) latitemporata Fischer, 1976
Catalogues with Iranian records: Cortés et al. (2016).
Distribution in Iran: Fars, Khuzestan (Peris-Felipo, Ameri et al., 2016).
Distribution in the Middle East: Iran (Peris-Felipo, Ameri et al., 2016), Israel−Palestine (Papp, 2012).
Extralimital distribution: Austria, Hungary.
Host records: Unknown.

Aspilota (Aspilota) nidicola Hedqvist, 1972
Catalogues with Iranian records: Cortés et al. (2016).
Distribution in Iran: Fars (Peris-Felipo, Ameri et al., 2016).
Distribution in the Middle East: Iran.
Extralimital distribution: Sweden.
Host records: Erroneously recorded by Hedqvist (1972) as being a parasitoid of the vespid wasp *Vespula vulgaris* (L.). It remains possible, however, that this species may parasitize fly species that sometimes invade this vespid's nest.

Aspilota (Aspilota) ruficornis (Nees von Esenbeck, 1834)
Catalogues with Iranian records: No catalogue.
Distribution in Iran: Ardabil (Samin, Beyarslan et al., 2020).
Distribution in the Middle East: Iran (Samin, Beyarslan et al., 2020), Turkey (Fischer & Beyarslan, 2012).
Extralimital distribution: Austria, Belgium, Croatia, former Czechoslovakia, Denmark, Germany, Hungary, Montenegro, Netherlands, Poland, Romania, Serbia, Sweden, Switzerland, United Kingdom.
Host records: Recorded by Györfi (1943) as being a parasitoid of the syrphid *Eristalis tenax* (L.).

Aspilota (Aspilota) stenogaster Stelfox and Graham, 1951
Distribution in the Middle East: Turkey (Fischer & Beyarslan, 2012).
Extralimital distribution: Czech Republic, Hungary, Japan, Korea, Norway, Russia, United Kingdom.
Host records: Unknown.

Genus *Carinthilota* Fischer, 1975
Carinthilota vechti van Achterberg, 1988
Catalogues with Iranian records: Samin, Coronado-Blanco, Fischer et al. (2018).
Distribution in Iran: Kordestan (Sakenin et al., 2018).
Distribution in the Middle East: Iran.
Extralimital distribution: Hungary, Netherlands, Poland, Russia.
Host records: Unknown.

Genus *Chasmodon* Haliday, 1838
Chasmodon apterus (Nees von Esenbeck, 1812)
Distribution in the Middle East: Israel−Palestine (Aubert, 1966).
Extralimital distribution: Austria, Belgium, Bulgaria, Czech Republic, Denmark, Finland, France, Germany, Hungary, Iceland, Ireland, Italy, Latvia, Moldova, Montenegro, Netherlands, Norway, Poland, Russia, Sweden, Switzerland, Ukraine, United Kingdom.
Host records: Summarized by Yu et al. (2016) as being a parasitoid of the agromyzid *Chromatomyia syngenesia* Hardy; the cecidomyiid *Contarinia pyrivora* (Riley); the chloropids *Iscinella frit* (L.) and *Oscinella vastator* (Curtis); the drosophilid *Scaptomyza pallida* (Zetterstedt); and the opomyzids *Geomyza tripunctata* (Fallén), and *Opomyza germinationis* (L.).

Genus *Cratospila* Foerster, 1863
Cratospila circe (Haliday, 1838)
Catalogues with Iranian records: No catalogue.
Distribution in Iran: Qazvin (Samin, Sakenin Chelav et al., 2020).
Distribution in the Middle East: Iran (Samin, Sakenin Chelav et al., 2020), Turkey (Fischer & Beyarslan, 2012).
Extralimital distribution: Austria, China, Czech Republic, France, Germany, Hungary, Ireland, Italy, Korea, Malaysia, Mongolia, Poland, Russia, Serbia, Slovenia, Spain, Switzerland, United Kingdom.
Host records: Unknown.

Genus *Dinotrema* Foerster, 1863
Dinotrema (Dinotrema) amoenidens (Fischer, 1973)
Catalogues with Iranian records: Gadallah et al. (2015), Farahani et al. (2016), Cortés et al. (2016), Yu et al. (2016),

Samin, Coronado-Blanco, Kavallieratos et al. (2018), Samin, Coronado-Blanco, Fischer et al. (2018).
Distribution in Iran: Golestan (Samin et al., 2011; Samin, Ghahari et al., 2015), Iran (no specific locality cited) (Peris-Felipo, 2013; Peris-Felipo, Belokobylskij, Jiménez-Peydró, 2014).
Distribution in the Middle East: Iran.
Extralimital distribution: Austria, Bosnia-Herzegovina, China, Denmark, Greece, Hungary, Italy, Mongolia, Poland, Romania, Russia.
Host records: Unknown.

Dinotrema (Dinotrema) amparoae Peris-Felipo, 2013
Catalogues with Iranian records: Cortés et al. (2016).
Distribution in Iran: Hormozgan (Cortés et al., 2016).
Distribution in the Middle East: Iran.
Extralimital distribution: Spain.
Host records: Unknown.

Dinotrema (Synaldis) argamani (Fischer, 1993)
Distribution in the Middle East: Israel—Palestine (Fischer, 1993; Papp, 2012).
Extralimital distribution: None.
Host records: Unknown.

Dinotrema (Dinotrema) borzhomii Tobias, 2004
Catalogues with Iranian records: Cortés et al. (2016).
Distribution in Iran: Kerman (Cortés et al., 2016).
Distribution in the Middle East: Iran.
Extralimital distribution: Georgia.
Host records: Unknown.

Dinotrema (Dinotrema) concinnum Haliday, 1838
Catalogues with Iranian records: Gadallah et al. (2015), Farahani et al. (2016), Yu et al. (2016), Samin, Coronado-Blanco, Kavallieratos et al. (2018), Samin, Coronado-Blanco, Fischer et al. (2018).
Distribution in Iran: Chaharmahal and Bakhtiari (Ghahari & Beyarslan, 2017), Qazvin (Ghahari, Fischer et al., 2011), Iran (no specific locality cited) (Peris-Felipo, 2013; Peris-Felipo, Belokobylskij, Jiménez-Peydró, 2014).
Distribution in the Middle East: Iran (Ghahari & Beyarslan, 2017; Ghahari, Fischer et al., 2011; Peris-Felipo, 2013; Peris-Felipo, Belokobylskij, Jiménez-Peydró, 2014), Israel—Palestine (Papp, 2012), Turkey (Peris-Felipo, Belokobylskij, Jiménez-Peydró, 2014).
Extralimital distribution: Afghanistan, Austria, Bulgaria, Croatia, Faroe Islands, Germany, Greece, Hungary, Iceland, Ireland, Italy, Mongolia, Netherlands, Poland, Russia, Spain, Sweden, Switzerland, Tunisia, United Kingdom.
Host records: Recorded by Bignell (1901) as being a parasitoid of the fanniid *Fannia canicularis* (L.).

Dinotrema (Synaldis) concolor (Nees von Esenbeck, 1812)
Catalogues with Iranian records: Khajeh et al. (2014) as *Synaldis concolor*, Gadallah et al. (2015) as *S. concolor*, Farahani et al. (2016), Cortés et al. (2016), Yu et al. (2016), Samin, Coronado-Blanco, Kavallieratos et al. (2018), Samin, Coronado-Blanco, Fischer et al. (2018).
Distribution in Iran: East Azarbaijan (Ghahari & Gadallah, 2021; Rastegar et al., 2012), Fars (Fischer et al., 2011; Khajeh et al., 2014; Lashkari Bod et al., 2010, 2011), Kerman (Ghotbi Ravandi et al., 2015), Qazvin (Ghahari, Fischer et al., 2011), East of Iran (no specific locality cited) (Yari et al., 2014).
Distribution in the Middle East: Iran (see references above), Turkey (Fischer & Beyarslan, 2012).
Extralimital distribution: Afghanistan, Austria, Bulgaria, Czech Republic, France, Germany, Greece, Hungary, Iceland, Ireland, Italy, Korea, Lithuania, Mongolia, Netherlands, Norway, Poland, Russia, Serbia, Spain, Switzerland, United Kingdom.
Host records: Recorded by Austin (1934) and Hussey (1960) as being a parasitoid of the phorid *Megaselia nigra* (Meigen). In Iran, this species has been reared from the phorid *Megaselia minuta* (Aldrich) (Ghahari & Gadallah, 2021).
Plant associations in Iran: *Triticum aestivum* (Poaceae) (Fischer et al., 2011; Ghotbi Ravandi et al., 2015; Khajeh et al., 2014; Lashkari Bod et al., 2011).

Dinotrema (Dinotrema) contracticorne (Fischer, 1974)
Catalogues with Iranian records: Farahani et al. (2016), Cortés et al. (2016), Yu et al. (2016), Samin, Coronado-Blanco, Kavallieratos et al. (2018), Samin, Coronado-Blanco, Fischer et al. (2018).
Distribution in Iran: Kerman (Ghotbi Ravandi et al., 2015).
Distribution in the Middle East: Iran.
Extralimital distribution: Austria, Czech Republic, Hungary, Russia.
Host records: Unknown.
Plant associations in Iran: *Medicago sativa* (Fabaceae) (Ghotbi Ravandi et al., 2015).

Dinotrema (Dinotrema) cratocera (Thomson, 1895)
Catalogues with Iranian records: Khajeh et al. (2014) as *Dinotrema cratocerum*, Gadallah et al. (2015), Farahani et al. (2016), Cortés et al. (2016), Yu et al. (2016), Samin, Coronado-Blanco, Kavallieratos et al. (2018), Samin, Coronado-Blanco, Fischer et al. (2018).
Distribution in Iran: Golestan (Ghahari, Fischer, Erdoğan et al., 2010), Kordestan (Samin, 2015).
Distribution in the Middle East: Iran.

Extralimital distribution: Austria, former Czechoslovakia, Hungary, Korea, Mongolia, Romania, Sweden.
Host records: Unknown.

Dinotrema (Dinotrema) cruciforme (Fischer, 1973)
Catalogues with Iranian records: No catalogue.
Distribution in Iran: West Azerbaijan (Sakenin et al., 2020).
Distribution in the Middle East: Iran.
Extralimital distribution: Austria, Denmark, Finland, Hungary, Korea, United Kingdom.
Host records: Unknown.

Dinotrema (Dinotrema) dimidiatum (Thomson, 1895)
Catalogues with Iranian records: No catalogue.
Distribution in Iran: West Azarbaijan (Gadallah et al., 2021).
Distribution in the Middle East: Iran (Gadallah et al., 2021), Turkey (Peris-Felipo, Belokobylskij, Jiménez-Peydró, 2014).
Extralimital distribution: Czech Republic, Denmark, Germany, Hungary, Kazakhstan, Kyrgyzstan, Moldova, Russia, Uzbekistan.
Host records: Recorded by Hedwig (1957) as being a parasitoid of the drosophilid *Scaptomyza pallida* Zetterstedt.

Dinotrema (Dinotrema) dimorpha (Fischer, 1976)
Catalogues with Iranian records: Cortés et al. (2016), Samin, Coronado-Blanco, Kavallieratos et al. (2018), Samin, Coronado-Blanco, Fischer et al. (2018).
Distribution in Iran: Hormozgan (Cortés et al., 2016).
Distribution in the Middle East: Iran.
Extralimital distribution: Armenia, Austria, Bosnia-Herzegovina, Bulgaria, Georgia, Germany, Hungary, Italy, Korea, North Macedonia, Romania, Slovakia, former Yugoslavia.
Host records: Unknown.

Dinotrema (Synaldis) distractum (Nees von Esenbeck, 1834)
Catalogues with Iranian records: Khajeh et al. (2014) as *Synaldis distracta*, Gadallah et al. (2015), Farahani et al. (2016), Cortés et al. (2016), Yu et al. (2016), Samin, Coronado-Blanco, Kavallieratos et al. (2018), Samin, Coronado-Blanco, Fischer et al. (2018).
Distribution in Iran: Isfahan (Ghahari et al., 2011b), Kerman (Ghotbi Ravandi et al., 2015; Safahani et al., 2017), East of Iran (no specific locality cited) (Yari et al., 2014), Iran (no specific locality cited) (Peris-Felipo, 2013).
Distribution in the Middle East: Iran (see references above), Israel—Palestine (Papp, 2012).
Extralimital distribution: Austria, Bulgaria, China, Croatia, Czech Republic, Finland, Germany, Greece, Hungary, Iceland, Ireland, Korea, Lithuania, Malaysia, Mongolia,

Poland, Portugal, Romania, Russia, Slovenia, Spain, Sweden, Switzerland, Tunisia, United Kingdom, Uzbekistan, former Yugoslavia.
Host records: Unknown.
Plant associations in Iran: *Medicago sativa* (Fabaceae) (Ghotbi Ravandi et al., 2015), *Triticum aestivum* (Poaceae) (Ghotbi Ravandi et al., 2015; Safahani et al., 2017), and *Anethum graveolens* (Apiaceae) (Safahani et al., 2017).

Dinotrema (Synaldis) exitiosae Fischer, 1976
Distribution in the Middle East: Turkey (Beyarslan & Inanç, 1992).
Extralimital distribution: United States of America.
Host records: Unknwon.

Dinotrema (Synaldis) glabripleura (Fischer, 1993)
Distribution in the Middle East: Israel—Palestine (Fischer, 1993; Papp, 2012).
Extralimital distribution: None.
Host records: Unknown.

Dinotrema (Dinotrema) intermissum (Fischer, 1974)
Catalogues with Iranian records: Khajeh et al. (2014), Gadallah et al. (2015), Farahani et al. (2016) as *Aspilota intermissa* Fischer, 1974, Cortés et al. (2016), Yu et al. (2016), Samin, Coronado-Blanco, Kavallieratos et al. (2018), Samin, Coronado-Blanco, Fischer et al. (2018).
Distribution in Iran: Guilan (Ghahari, 2018—around rice fields; Ghahari & Fischer, 2011c); Iran (no specific locality cited) (Peris-Felipo, 2013; Peris-Felipo, Belokobylskij, Jiménez-Peydró, 2014).
Distribution in the Middle East: Iran (see references above), Turkey (Fischer & Beyarslan, 2012; Yildirim et al., 2010).
Extralimital distribution: Austria.
Host records: Unknown.

Dinotrema (Synaldis) israelica (Fischer, 1993)
Distribution in the Middle East: Israel—Palestine (Fischer, 1993; Papp, 2012).
Extralimital distribution: None.
Host records: Unknown.

Dinotrema (Synaldis) jordanica (Fischer, 1993)
Distribution in the Middle East: Jordan (Fischer, 1993).
Extralimital distribution: None.
Host records: Unknown.

Dinotrema (Dinotrema) longicauda Tobias, 2003
Distribution in the Middle East: Turkey (Fischer & Beyarslan, 2012).
Extralimital distribution: Russia.
Host records: Unknown.

Dinotrema (*Dinotrema*) *macrocera* **(Thomson, 1895)**
Distribution in the Middle East: Turkey (Peris-Felipo, Belokobylskij, Jiménez-Peydró, 2014).
Extralimital distribution: Austria, former Czechoslovakia, Denmark, Georgia, Germany, Hungary, Mongolia, Romania, Serbia, Sweden.
Host records: Recorded by Hedwig (1958) as being a parasitoid of the drosophilid *Scaptomyza pallida* (Zetterstedt).

Dinotrema (*Synaldis*) *maxima* **(Fischer, 1962)**
Catalogues with Iranian records: Khajeh et al. (2014) as *Synaldis maxima*, Cortés et al. (2016).
Distribution in Iran: Ilam (Ghahari et al., 2011c), Kerman (Cortés et al., 2016).
Distribution in the Middle East: Iran.
Extralimital distribution: Palaearctic region.
Host records: Unknown.

Dinotrema (*Synaldis*) *megastigma* **(Fischer, 1967)**
Catalogues with Iranian records: Khajeh et al. (2014) as *Synaldis megastigma* Fischer, 1967, Gadallah et al. (2015) as *S. megastigma*, Farahani et al. (2016) as *S. megastigma*, Cortés et al. (2016) as *S. megastigma*, Yu et al. (2016) as *S. megastigma*, Samin, Coronado-Blanco, Kavallieratos et al. (2018), Samin, Coronado-Blanco, Fischer et al. (2018) both as *S. megastigma*.
Distribution in Iran: Golestan, Guilan (Ghahari, Fischer et al., 2011; Khajeh et al., 2014 both as *Synaldis megastigma* Fischer, 1967).
Distribution in the Middle East: Iran.
Extralimital distribution: Austria, Bulgaria, Greece, Hungary, North Macedonia, former Yugoslavia.
Host records: Unknown.

Dinotrema (*Dinotrema*) *naevium* **(Tobias, 1962)**
Catalogues with Iranian records: No catalogue.
Distribution in Iran: East Azerbaijan (Sakenin et al., 2020).
Distribution in the Middle East: Iran.
Extralimital distribution: Former Czechoslovakia, Hungary, Korea, Mongolia, Russia.
Host records: Unknown.

Dinotrema (*Dinotrema*) *nigricorne* **(Thomson, 1895)**
Distribution in the Middle East: Turkey (Peris-Felipo, Belokobylskij, Jiménez-Peydró, 2014).
Extralimital distribution: Austria, Bosnia-Herzegovina, former Czechoslovakia, Germany, Hungary, Korea, North Macedonia, Romania, Sweden, former Yugoslavia.
Host records: Unknown.

Dinotrema (*Dinotrema*) *oleraceum* **(Tobias, 1962)**
Catalogues with Iranian records: No catalogue.

Distribution in Iran: Kordestan (Samin, Beyarslan et al., 2020).
Distribution in the Middle East: Iran.
Extralimital distribution: Austria, Czech Republic, Hungary, Korea, Kosovo, Mongolia, Poland, Russia, Serbia, Slovakia.
Host records: Unknown.

Dinotrema (*Dinotrema*) *partimrufa* **Fischer, 2009**
Distribution in the Middle East: Turkey (Fischer, 2009; Peris-Felipo, Belokobylskij, Jiménez-Peydró, 2014).
Extralimital distribution: None.
Host records: Unknown.

Dinotrema (*Dinotrema*) *paucilia* **Papp, 2012**
Distribution in the Middle East: Israel−Palestine (Papp, 2012; Peris-Felipo, Belokobylskij, Jiménez-Peydró, 2014).
Extralimital distribution: None.
Host records: Unknown.

Dinotrema (*Dinotrema*) *perlustrandum* **(Fischer, 1973)**
Catalogues with Iranian records: Cortés et al. (2016).
Distribution in Iran: Sistan and Baluchestan (Cortés et al., 2016).
Distribution in the Middle East: Iran.
Extralimital distribution: Austria, Germany, Hungary, Mongolia.
Host records: Recorded by Fischer (1973) as being a parasitoid of unspecific phorid host.
Plant associations in Iran: *Medicago sativa* (Fabaceae) (Cortés et al., 2016).

Dinotrema (*Dinotrema*) *samsunense* **Fischer and Sullivan, 2014**
Distribution in the Middle East: Turkey (Fischer et al., 2014; Peris-Felipo, Belokobylskij, Jiménez-Peydró, 2014).
Extralimital distribution: None.
Host records: Unknown.

Dinotrema (*Dinotrema*) *significarium* **(Fischer, 1973)**
Catalogues with Iranian records: Khajeh et al. (2014), Gadallah et al. (2015), Farahani et al. (2016), Cortés et al. (2016), Yu et al. (2016), Samin, Coronado-Blanco, Kavallieratos et al. (2018), Samin, Coronado-Blanco, Fischer et al. (2018).
Distribution in Iran: Golestan (Samin, Ghahari et al., 2015), Sistan and Baluchestan (Ghahari, Fischer, Hedqvist et al., 2010), Iran (no specific locality cited) (Peris-Felipo, 2013; Peris-Felipo, Belokobylskij, Jiménez-Peydró, 2014).
Distribution in the Middle East: Iran.
Extralimital distribution: Armenia, Austria, Bulgaria, Denmark, Finland, Germany, Greece, Hungary, Korea, Romania, Slovakia, Spain, Tunisia.
Host records: Unknown.

***Dinotrema* (*Dinotrema*) *sinecarinum* (Fischer, 1993)**
Catalogues with Iranian records: Cortés et al. (2016).
Distribution in Iran: Kerman (Cortés et al., 2016).
Distribution in the Middle East: Iran (Cortés et al., 2016), Israel—Palestine (Fischer, 1993; Peris-Felipo, Belokobylskij, Jiménez-Peydró, 2014).
Extralimital distribution: None.
Host records: Unknown.

***Dinotrema* (*Synaldis*) *soederlundi* (Fischer, 2003)**
Distribution in the Middle East: Israel—Palestine (Fischer, 2003).
Extralimital distribution: None.
Host records: Unknown.

***Dinotrema* (*Prosapha*) *speculum* (Haliday, 1838)**
Catalogues with Iranian records: No catalogue.
Distribution in Iran: Kerman (Safahani et al., 2017).
Distribution in the Middle East: Iran.
Extralimital distribution: Austria, Belgium, Bulgaria, Czech Republic, France, Germany, Hungary, Ireland, Italy, Kazakhstan, Kyrgyzstan, Netherlands, Norway, Poland, Romania, Russia, Serbia, Slovenia, Spain, Sweden, United Kingdom.
Host records: Unknown.
Plant associations in Iran: *Medicago sativa* (Fabaceae), *Rubus* sp. (Rosaceae) (Safahani et al., 2017).

***Dinotrema* (*Dinotrema*) *tauricum* (Telenga, 1935)**
Catalogues with Iranian records: Cortés et al. (2016).
Distribution in Iran: Isfahan (Cortés et al., 2016).
Distribution in the Middle East: Iran.
Extralimital distribution: Austria, Belarus, Bulgaria, China, Finland, Germany, Hungary, Ireland, Korea, Netherlands, Norway, Russia, Switzerland, Ukraine, United Kingdom.
Host records: Unknown.

***Dinotrema* (*Synaldis*) *ultima* (Fischer, 1970)**
Catalogues with Iranian records: Farahani et al. (2016), Cortés et al. (2016), Yu et al. (2016), Samin, Coronado-Blanco, Kavallieratos et al. (2018), Samin, Coronado-Blanco, Fischer et al. (2018) (all as *Synaldis ultima* Fischer, 1970).
Distribution in Iran: Kerman (Ghotbi Ravandi et al., 2015 as *Synaldis ultima* Fischer, 1970).
Distribution in the Middle East: Iran.
Extralimital distribution: Austria, Czech Republic, Hungary, Russia, Spain.
Host records: Unknown.
Plant associations in Iran: *Triticum aestivum* (Poaceae) (Ghotbi Ravandi et al., 2015).

***Dinotrema* (*Dinotrema*) *varipes* (Tobias, 1962)**
Catalogues with Iranian records: No Catalogue.

Distribution in Iran: Lorestan (Samin, Beyarslan et al., 2020).
Distribution in the Middle East: Iran.
Extralimital distribution: Austria, Bulgaria, Croatia, Czech Republic, Finland, Georgia, Hungary, Korea, Mongolia, North Macedonia, Romania, Russia, Serbia.
Host records: Unknown.

Genus *Eudinostigma* Tobias, 1986
***Eudinostigma subpulvinatum* Fischer, 2009**
Distribution in the Middle East: Turkey (Fischer, 2009).
Extralimital distribution: None.
Host records: Unknown.

Genus *Grammospila* Foerster, 1863
***Grammospila rufiventris* (Nees von Esenbeck, 1812)**
Distribution in the Middle East: Turkey (Fischer & Beyraslan, 2012).
Extralimital distribution: Austria, Azores, Bulgaria, China, Czech Republic, France, Hungary, Ireland, Italy, Japan, Latvia, Lithuania, Mongolia, Netherlands, Poland, Portugal, Romania, Russia, Sweden, Switzerland, Ukraine, United Kingdom, former Yugoslavia.
Host records: Summarized by Yu et al. (2016) as being a parasitoid of several agromyzids of the genera *Agromyza* Fallén, *Amauromyza* Hendel, *Liriomyza* Mik, and *Phytomyza* Fallén.

Genus *Heterolexis* Foerster, 1863
***Heterolexis balteata* (Thomson, 1895)**
Distribution in the Middle East: Turkey (Fischer & Beyarslan, 2012).
Extralimital distribution: Austria, Bulgaria, Czech Republic, France, Germany, Hungary, Italy, Japan, Korea, Moldova, Netherlands, Norway, Poland, Russia, Serbia, Switzerland, United Kingdom.
Host records: Summarized by Yu et al. (2016) as being a parasitoid of the agromyzids *Agromyza* spp., *Cerodontha incisa* (Meigen), *C. pygmaea* (Meigen), *Liriomyza congesta* (Becker), *L. flaveola* (Fallén), *Phytomyza fallaciosa* Brischke, and *P. phellandrii* Hering.

***Heterolexis boscoli* Fischer, 1993**
Distribution in the Middle East: Turkey (Fischer & Beyarslan, 2012).
Extralimital distribution: Italy.
Host records: Unknown.

Genus *Idiasta* Foerster, 1863
***Idiasta* (*Idiasta*) *adanacola* Fischer and Beyarslan, 2012**
Distribution in the Middle East: Turkey (Fischer and Beyarslan, 2012).
Extralimital distribution: None.
Host records: Unknown.

Idiasta (Idiasta) argamani **Papp, 2012**

Distribution in the Middle East: Israel—Palestine (Papp, 2012).

Extralimital distribution: None.

Host records: Unknown.

Idiasta dichrocera **Königsmann, 1960**

Catalogues with Iranian records: No catalogue.

Distribution in Iran: Kerman (Safahani et al., 2017).

Distribution in the Middle East: Iran (Safahani et al., 2017), Israel—Palestine (Papp, 2012), Turkey (Fischer & Beyarslan, 2012).

Extralimital distribution: Austria, China, Czech Republic, Germany, Hungary, Kazakhstan, Korea, Russia, Sweden, Uzbekistan.

Host records: Unknown.

Plant associations in Iran: *Anethum graveolens* (Apiaceae), and *Medicago sativa* (Fabaceae) (Safahani et al., 2017).

Idiasta (Idiasta) picticornis **(Ruthe, 1854)**

Catalogues with Iranian records: Khajeh et al. (2014), Gadallah et al. (2015), Farahani et al. (2016), Cortés et al. (2016), Yu et al. (2016), Samin, Coronado-Blanco, Kavallieratos et al. (2018), Samin, Coronado-Blanco, Fischer et al. (2018).

Distribution in Iran: Sistan and Baluchestan (Khajeh et al., 2014; Sedighi et al., 2014).

Distribution in the Middle East: Iran (Khajeh et al., 2014; Sedighi et al., 2014), Israel—Palestine (Papp, 2012).

Extralimital distribution: China, former Czechoslovakia, Germany, Hungary, Korea, Serbia.

Host records: Unknown.

Idiasta (Idiasta) rugosipleurum **Fischer and Beyarslan, 2012**

Distribution in the Middle East: Turkey (Fischer and Beyarslan, 2012).

Extralimital distribution: None.

Host records: Unknown.

Idiasta (Idiasta) subannellata **(Thomson, 1895)**

Catalogues with Iranian records: No catalogue.

Distribution in Iran: West Azarbaijan (Samin, Sakenin Chelav et al., 2020).

Distribution in the Middle East: Iran (Samin, Sakenin Chelav et al., 2020), Turkey (Fischer & Beyarslan, 2012).

Extralimital distribution: Austria, China, Czech Republic, Finland, Hungary, Norway, Poland, Russia, Serbia, Sweden, Ukraine.

Host records: Unknown.

Genus *Orthostigma* **Ratzeburg, 1844**

Orthostigma (Orthostigma) beyarslani **Fischer, 1995**

Catalogues with Iranian records: Khajeh et al. (2014), Gadallah et al. (2015), Farahani et al. (2016), Cortés et al. (2016), Yu et al. (2016), Samin, Coronado-Blanco, Kavallieratos et al. (2018), Samin, Coronado-Blanco, Fischer et al. (2018).

Distribution in Iran: Fars (Fischer et al., 2011; Khajeh et al., 2014; Lashkari Bod et al., 2010, 2011), Kerman (Ghotbi Ravandi et al., 2015), Iran (no specific locality cited) (Peris-Felipo, 2013).

Distribution in the Middle East: Iran (see references above), Turkey (Fischer, 1995; Fischer & Beyarslan, 2012).

Extralimital distribution: Spain.

Host records: Unknown.

Plant associations in Iran: *Medicago sativa* (Fabaceae) (Ghotbi Ravandi et al., 2015; Khajeh et al., 2014; Lashkari Bod et al., 2011).

Orthostigma (Orthostigma) curtiradiale **Fischer, 1995**

Distribution in the Middle East: Turkey (Fischer, 1995).

Extralimital distribution: None.

Host records: Unknown.

Orthostigma (Orthostigma) impunctatum **Fischer, 1995**

Distribution in the Middle East: Turkey (Fischer, 1995).

Extralimital distribution: None.

Host records: Unknown.

Orthostigma (Orthostigma) laticeps **(Thomson, 1895)**

Catalogues with Iranian records: Khajeh et al. (2014), Gadallah et al. (2015), Farahani et al. (2016), Cortés et al. (2016), Yu et al. (2016), Samin, Coronado-Blanco, Kavallieratos et al. (2018), Samin, Coronado-Blanco, Fischer et al. (2018).

Distribution in Iran: Golestan (Ghahari & Fischer, 2011a), Isfahan (Ghahari et al., 2011b), Sistan and Baluchestan (Sedighi et al., 2014).

Distribution in the Middle East: Iran (Ghahari & Fischer, 2011b; Ghahari et al., 2011b; Sedighi et al., 2014), Turkey (Beyarslan & Inanç, 1992).

Extralimital distribution: Austria, Belgium, Bulgaria, China, former Czechoslovakia, Denmark, Germany, Greece, Hungary, Iceland, Italy, Korea, Netherlands, North Macedonia, Russia, Serbia, Spain, Sweden, Switzerland, Uzbekistan.

Host records: Unknown.

Orthostigma (Orthostigma) longicorne **Königsmann, 1969**

Catalogues with Iranian records: Cortés et al. (2016).

Distribution in Iran: Kerman (Cortés et al., 2016).
Distribution in the Middle East: Iran.
Extralimital distribution: Austria, China, Czech Republic, France, Germany, Hungary, Ireland, Italy, Korea, Latvia, Norway, Russia, Spain, Switzerland.
Host records: Unknown.

Orthostigma (Orthostigma) maculipes (Haliday, 1838)
Catalogues with Iranian records: Khajeh et al. (2014), Gadallah et al. (2015), Farahani et al. (2016), Cortés et al. (2016), Yu et al. (2016), Samin, Coronado-Blanco, Kavallieratos et al. (2018), Samin, Coronado-Blanco, Fischer et al. (2018).
Distribution in Iran: Isfahan (Ghahari et al., 2011b), Iran (no specific locality cited) (Peris-Felipo, 2013).
Distribution in the Middle East: Iran.
Extralimital distribution: Austria, Bosnia-Herzegovina, Bulgaria, Czech Republic, Faroe Islands, Germany, Greece, Hungary, Ireland, Montenegro, Netherlands, North Macedonia, Poland, Russia, Serbia, Spain, Sweden, Switzerland, United Kingdom.
Host records: Unknown.

Orthostigma (Orthostigma) mandibulare (Tobias, 1962)
Catalogues with Iranian records: No catalogue.
Distribution in Iran: Golestan, Mazandaran (Sakenin et al., 2020).
Distribution in the Middle East: Iran.
Extralimital distribution: Austria, Bulgaria, China, former Czechoslovakia, Germany, Hungary, Italy, Korea, Poland, Russia, Switzerland.
Host records: Unknown.

Orthostigma (Orthostigma) pseudolaticeps Königsmann, 1969
Distribution in the Middle East: Israel−Palestine (Papp, 2012), Turkey (Beyarslan & Inanç, 1992).
Extralimital distribution: Bulgaria, former Czechoslovakia, Germany, Hungary, North Macedonia, Spain, Switzerland, former Yugoslovakia.
Host records: Unknown.

Orthostigma (Orthostigma) pumilum (Nees von Esenbeck, 1834)
Catalogues with Iranian records: No catalogue.
Distribution in Iran: Chaharmahal and Bakhtiari (Sakenin et al., 2020).
Distribution in the Middle East: Iran.
Extralimital distribution: Austria, Bulgaria, China, Croatia, Czech Republic, Denmark, France, Germany, Hungary,

Iceland, Ireland, Italy, Lithuania, Mongolia, Montenegro, Netherlands, Poland, Portugal, Russia, Serbia, Slovakia, Spain, Switzerland, Ukraine, United Kingdom.
Host records: Summarized by Yu et al. (2016) as being a parasitoid of the agromyzid Phytomyza affinis Fallén; the cecidomyiid Rabdophaga salicis (Schrank); the muscid Musca domestica L.; and the phorids Megaselia giraudii (Egger), and M. rufipes (Meigen). In Iran, reared from Phytomyza sp. (Sakenin et al., 2020).

Orthostigma (Orthostigma) pusillum (Zettersdet, 1838)
Distribution in the Middle East: Turkey (Beyarslan & Inanç, 1992).
Extralimital distribution: Austria, Belgium, China, former Czechoslovakia, Hungary, Italy, Korea, Norway, Russia, Spain.
Host records: Unknown.

Orthostigma (Orthostigma) robusticeps Fischer, 1995
Distribution in the Middle East: Turkey (Fischer, 1995).
Extralimital distribution: None.
Host records: Unknown.

Genus Pentapleura Foerster, 1863
Pentapleura angustula (Haliday, 1838)
Catalogues with Iranian records: Samin, Coronado-Blanco, Kavallieratos et al. (2018).
Distribution in Iran: East Azarbaijan (Samin, Coronado-Blanco, Kavallieratos et al., 2018).
Distribution in the Middle East: Iran (Samin, Coronado-Blanco, Kavallieratos et al., 2018), Turkey (Fischer & Beyarslan, 2012).
Extralimital distribution: Austria, Bosnia-Herzegovina, Bulgaria, Czech Republic, Germany, Hungary, Iceland, Ireland, Lithuania, Mongolia, Netherlands, Poland, Russia, Sweden, Ukraine, United Kingdom.
Host records: Recorded by Hedwig (1958) as being a parasitoid of the drosophilid Scaptomyza pallida (Zetterstedt).

Pentapleura pumilio (Nees von Esenbeck, 1812)
Catalogues with Iranian records: This species is a new record for the fauna of Iran.
Distribution in Iran: Razavi Khorasan province, Fariman (Kalateh-Khosh), 2♀, 1♂, April 2011.
Distribution in the Middle East: Iran (new record).
Extralimital distribution: Afghanistan, Austria, Azores, Belarus, Bulgaria, Canada, Czech Republic, Faroe Islands, Germany, Hungary, Iceland, Ireland, Italy, Korea, Lithuania, Mongolia, Netherlands, Norway, Poland, Portugal, Russia, Spain, Sweden, Switzerland, United States of America, Ukraine, United Kingdom.

Host records: Summarized by Yu et al. (2016) as being a parasitoid of the drosophilid *Scaptomyza graminum* Fallén; and the sphaerocerid *Copromyza* sp. (Diptera).

Genus *Phaenocarpa* Foerster, 1863

Phaenocarpa (Phaenocarpa) bicolor (Foerster, 1863)
Catalogues with Iranian records: Cortés et al. (2016).
Distribution in Iran: Kerman (Cortés et al., 2016).
Distribution in the Middle East: Iran.
Extralimital distribution: Germany, Switzerland.
Host records: Unknown.
Plant associations in Iran: *Medicago sativa* (Fabaceae) (Cortés et al., 2016).

Phaenocarpa (Phaenocarpa) brevipalpis (Thomson, 1895)
Catalogues with Iranian records: Farahani et al. (2016), Cortés et al. (2016), Yu et al. (2016), Samin, Coronado-Blanco, Kavallieratos et al. (2018), Samin, Coronado-Blanco, Fischer et al. (2018).
Distribution in Iran: West Azarbaijan (Samin et al., 2014).
Distribution in the Middle East: Iran (Samin et al., 2014), Israel–Palestine (Papp, 2012).
Extralimital distribution: Greece, Hungary, Mongolia, Russia, Sweden.
Host records: Unknown.

Phaenocarpa (Phaenocarpa) canaliculata Stelfox, 1941
Catalogues with Iranian records: No catalogue.
Distribution in Iran: Guilan (Sakenin et al., 2020).
Distribution in the Middle East: Iran.
Extralimital distribution: Austria, Finland, Georgia, Germany, Hungary, Ireland, Italy, Kazakhstan, Netherlands, Switzerland, United Kingdom.
Host records: Recorded by van Achterberg (1981) as being a parasitoid of the fanniid *Fannia monilis* (Haliday).

Phaenocarpa (Phaenocarpa) carinthiaca Fischer, 1975
Catalogues with Iranian records: Cortés et al. (2016).
Distribution in Iran: Kerman (Cortés et al., 2016).
Distribution in the Middle East: Iran.
Extralimital distribution: Austria, China, Georgia, Germany, Spain.
Host records: Unknown.

Phaenocarpa (Phaenocarpa) conspurcator (Haliday, 1838)
Catalogues with Iranian records: No catalogue.
Distribution in Iran: Hamadan (Samin, Fischer et al., 2019).
Distribution in the Middle East: Iran.
Extralimital distribution: Austria, Belgium, Bulgaria, China, Czech Republic, Denmark, Faroe Islands, Finland, France, Germany, Hungary, Iceland, Ireland, Italy, Kazakhstan, Lithuania, Mongolia, Netherlands, Poland, Romania, Russia, Spain, Sweden, Switzerland, Turkmenistan, United Kingdom, Uzbekistan, former Yugoslavia.
Host records: Recorded by Woodward (1995) as being a parasitoid of the scatophagid *Scathophaga stercoraria* L.

Phaenocarpa (Phaenocarpa) picinervis (Haliday, 1838)
Catalogues with Iranian records: No catalogue.
Distribution in Iran: Razavi Khorasan (Gadallah et al., 2021).
Distribution in the Middle East: Iran (Gadallah et al., 2021), Turkey (Fischer & Beyarslan, 2012).
Extralimital distribution: Austria, Bulgaria, Czech Republic, Finland, Germany, Hungary, Ireland, Italy, Kazakhstan, Korea, Mongolia, Netherlands, Norway, Poland, Romania, Russia, Spain, Sweden, Switzerland, United States of America, Ukraine, United Kingdom, Uzbekistan.
Host records: Recorded by Fischer (1975) as being a parasitoid of the muscid *Hebecnema affinis* Malloch.

Phaenocarpa (Phaenocarpa) ruficeps (Nees von Esenbeck, 1811)
Catalogues with Iranian records: Khajeh et al. (2014), Gadallah et al. (2015), Farahani et al. (2016), Cortés et al. (2016), Yu et al. (2016), Samin, Coronado-Blanco, Kavallieratos et al. (2018), Samin, Coronado-Blanco, Fischer et al. (2018).
Distribution in Iran: Razavi Khorasan (Ghahari, Fischer et al., 2011; Ghahari & Beyarslan, 2017).
Distribution in the Middle East: Iran.
Extralimital distribution: Afrotropical, Nearctic, Oriental and Palaearctic regions [Adjacent countries to Iran: Armenia, Kazakhstan and Russia].
Host records: Summarized by Yu et al. (2016) as being a parasitoid of the anthomyiids *Botanophila fugax* (Meigen), *B. phrenione* (Séguy), *B. seneciella* (Meade), *Delia antiqua* (Meigen), *D. radicum* (L.), *Hylemya alcathoe* (Walker), *Pegomya bicolor* (Wiedemann), *P. hyoscyami* (Panzer), and *P. solennis* (Meigen); the lonchaeid *Lonchaea chorea* (Fabricius); and the piophilid *Piophila casei* (L.). In Iran, this species has been reared from the anthomyiid *Pegomya hyoscyami* (Panzer) in sugar-beet field (Ghahari & Beyarslan, 2017).

Phaenocarpa (Phaenocarpa) subruficeps Gurasashvili, 1983
Distribution in the Middle East: Turkey (Fischer & Beyarslan, 2012).
Extralimital distribution: Georgia, Hungary, Mongolia.
Host records: Unknown.

Phaenocarpa (Homophyla) pullata (Haliday, 1838)
Distribution in the Middle East: Turkey (Fischer & Beyarslan, 2012).

Extralimital distribution: Austria, Belgium, Bulgaria, Czech Republic, Finland, Germany, Hungary, Ireland, Kazakhstan, Moldova, Mongolia, Norway, Poland, Romania, Russia, Serbia, Sweden, Switzerland, Ukraine, United Kingdom.
Host records: Summarized by Yu et al. (2016) as being a parasitoid of the anthomyiid *Hylemya genitalis* (Schnabl).

Genus *Pseudopezomachus* Mantero, 1905

Pseudopezomachus cursitans (Ferrière, 1930)
Catalogues with Iranian records: Gadallah et al. (2015), Farahani et al. (2016), Cortés et al. (2016), Yu et al. (2016), Samin, Coronado-Blanco, Kavallieratos et al. (2018), Samin, Coronado-Blanco, Fischer et al. (2018).
Distribution in Iran: Golestan (Ghahari & Fischer, 2011b).
Distribution in the Middle East: Iran (Ghahari & Fischer, 2011b), Turkey (Fischer & Beyarslan, 2012).
Extralimital distribution: Greece, Montenegro.
Host records: Recorded by Griffiths (1967, 1968) as being a parasitoid of the agromyzids *Liriomyza* sp., *Phytomyza ferulae* Hering, and *P. obscura* Hendel.

Pseudopezomachus kasparyani Tobias, 1986
Distribution in the Middle East: Israel—Palestine (Belokobylskij & Kula, 2012).
Extralimital distribution: Azerbaijan, Ukraine.
Host records: Unknown.

Pseudopezomachus masii Nixon, 1940
Catalogues with Iranian records: Khajeh et al. (2014), Gadallah et al. (2015), Farahani et al. (2016), Cortés et al. (2016), Yu et al. (2016), Samin, Coronado-Blanco, Kavallieratos et al. (2018), Samin, Coronado-Blanco, Fischer et al. (2018).
Distribution in Iran: Ilam (Ghahari et al., 2011c).
Distribution in the Middle East: Egypt (Amer & Hegazi, 2014; Hegazi et al., 2014; Papp, 2012), Iran (Ghahari et al., 2011c), Israel—Palestine (Papp, 2012), Turkey (Fischer & Beyarslan, 2012; Papp, 2012).
Extralimital distribution: Greece, Libya, Romania.
Host records: Recorded by Hegazy et al. (2014) as being a parasitoid of the agromyzids *Liriomyza bryoniae* Kaltenbach, and *L. trifolii* (Burgess).

Genus *Tanycarpa* Foerster, 1863

Tanycarpa bicolor (Nees von Esenbeck, 1812)
Catalogues with Iranian records: No catalogue.
Distribution in Iran: Ardabil (Naderian et al., 2020).
Distribution in the Middle East: Iran (Naderian et al., 2020), Turkey (Fischer & Beyarslan, 2012).
Extralimital distribution: Austria, Bulgaria, China, Czech Republic, Germany, Hungary, Ireland, Netherlands, Poland, Russia, Spain, Ukraine, United Kingdom.

Host records: Summarized by Yu et al. (2016) as being a parasitoid of the drosophilids *Drosophila kuntzei* Duda, *D. phalerata* Meigen, and *Scaptomyza pallida* Meigen.

Tanycarpa rufinotata (Haliday, 1838)
Distribution in the Middle East: Turkey (Fischer & Beyarslan, 2012).
Extralimital distribution: Austria, China, Georgia, Germany, Hungary, Ireland, Italy, Poland, Russia, Sweden, Switzerland.
Host records: Unknown.

Genus *Trachyusa* Ruthe, 1854

Trachyusa aurora (Haliday, 1838)
Catalogues with Iranian records: This species is a new record for the fauna of Iran.
Distribution in Iran: Mazandaran province, Ramsar, near Dalkhani forest, 1♀, July 2018.
Distribution in the Middle East: Iran (new record).
Extralimital distribution: Austria, Belgium, Bulgaria, Finland, Germany, Hungary, Ireland, Italy, Korea, Lithuania, Netherlands, Norway, Poland, Romania, Russia, Serbia, Slovakia, Slovenia, Spain, Sweden, Switzerland, United Kingdom.
Host records: Unknown.

Tribe Dacnusini Foerster, 1863
Genus *Aristelix* Nixon, 1943

Aristelix persica Peris-Felipo, 2015
Catalogues with Iranian records: Farahani et al. (2016), Cortés et al. (2016), Yu et al. (2016), Samin, Coronado-Blanco, Kavallieratos et al. (2018), Samin, Coronado-Blanco, Fischer et al. (2018).
Distribution in Iran: Hormozgan (Peris-Felipo et al., 2015).
Distribution in the Middle East: Iran.
Extralimital distribution: None.
Host records: Unknown.

Genus *Chorebus* Haliday, 1833

Chorebus (*Chorebus*) *affinis* (Nees von Esenbeck, 1812)
Catalogues with Iranian records: Khajeh et al. (2014), Gadallah et al. (2015), Farahani et al. (2016), Cortés et al. (2016), Yu et al. (2016), Samin, Coronado-Blanco, Kavallieratos et al. (2018), Samin, Coronado-Blanco, Fischer et al. (2018).
Distribution in Iran: Fars (Fischer et al., 2011; Khajeh et al., 2014; Lashkari Bod et al., 2010, 2011), Razavi Khorasan (Ghahari, Fischer et al. 2011), Mazandaran (Ghahari, Fischer, Erdoğan et al., 2010 as *Chorebus longicornis* (Nees, 1811)), Semnan (Samin, 2015 as *C. longicornis*).
Distribution in the Middle East: Iran.

Extralimital distribution: Austria, Belgium, Czech Republic, Faroe Islands, France, Germany, Greece, Hungary, Iceland, Ireland, Italy, Korea, Mongolia, Netherlands, North Macedonia, Poland, Portugal, Russia, Serbia, Spain, Sweden, Ukraine, United Kingdom.
Host records: Erroneously recorded by Rondani (1876) as being a parasitoid of the hemipteran aphid *Rhopalosiphum nymphaeae* (L.).

Chorebus (*Stiphrocera*) *albipes* (Haliday, 1939)
Catalogues with Iranian records: No catalogue.
Distribution in Iran: Guilan (Ghahari et al., 2020).
Distribution in the Middle East: Iran.
Extralimital distribution: Azerbaijan, Bulgaria, Croatia, Denmark, Germany, Hungary, Ireland, Kazakhstan, Korea, Poland, Russia, Serbia, Sweden, Ukraine, United Kingdom, Uzbekistan.
Host records: Summarized by Yu et al. (2016) as being a parasitoid of the agromyzids *Aulagromyza populi* (Kaltenbach), *A. tremulae* (Hering), and *A. tridentata* (Loew). In Iran, this species has been reared from *A. populi* (Ghahari et al., 2020).

Chorebus (*Stiphrocera*) *anasellus* (Stelfox, 1951)
Catalogues with Iranian records: Cortés et al. (2016).
Distribution in Iran: Iran (no specific locality cited) (Cortés et al., 2016).
Distribution in the Middle East: Iran (Cortés et al., 2016), Turkey (Beyarslan & Inanç, 2001).
Extralimital distribution: Azerbaijan, Czech Republic, Ireland, Spain, Ukraine.
Host records: Recorded by Griffiths (1984) as a parasitoid of the agromyzid *Phytomyza plantaginis* Robineau-Desvoidy.

Chorebus (*Stiphrocera*) *anitus* (Nixon, 1943)
Distribution in the Middle East: Israel–Palestine (Papp, 2012).
Extralimital distribution: Hungary, Kazakhstan, Serbia, Sweden, Ukraine, United Kingdom.
Host records: Unknown.

Chorebus (*Stiphrocera*) *aphantus* (Marshall, 1896)
Catalogues with Iranian records: Gadallah et al. (2015), Farahani et al. (2016), Cortés et al. (2016), Yu et al. (2016), Samin, Coronado-Blanco, Kavallieratos et al. (2018), Samin, Coronado-Blanco, Fischer et al. (2018).
Distribution in Iran: Kermanshah (Hazini et al., 2015), Razavi Khorasan (Yari et al., 2016), East of Iran (no specific locality cited) (Yari et al., 2014).
Distribution in the Middle East: Iran (Hazini et al., 2015; Yari et al., 2014, 2016), Turkey (Beyarslan & Inanç, 2001).
Extralimital distribution: Austria, Azerbaijan, China, Denmark, Germany, Hungary, Iceland, Ireland, Poland, Russia, Spain, Sweden, Switzerland, United Kingdom.

Host records: Recorded by Griffiths (1967, 1968) as being a parasitoid of the agromyzids *Liriomyza flaveola* (Fallén), *Phytomyza horticola* Goureau, *P. milii* Kaltenbach, and *P. nigra* Meigen. In Iran, this species has been reared from the agromyzid *Phytomyza horticola* Goureau, 1851 (= *Chromatomyia horticola* (Goureau)) (Hazini et al., 2015).
Plant associations in Iran: *Medicago sativa* (Fabaceae) (Yari et al., 2016).

Chorebus (*Phaenolexis*) *ares* (Nixon, 1944)
Catalogues with Iranian records: Khajeh et al. (2014), Gadallah et al. (2015), Farahani et al. (2016), Cortés et al. (2016), Yu et al. (2016), Samin, Coronado-Blanco, Kavallieratos et al. (2018), Samin, Coronado-Blanco, Fischer et al. (2018).
Distribution in Iran: Sistan and Baluchestan (Yari et al., 2016), East of Iran (no specific locality cited) (Yari et al., 2014).
Distribution in the Middle East: Iran.
Extralimital distribution: Azerbaijan, Czech Republic, Hungary, Kazakhstan, Lithuania, Russia, Ukraine, United Kingdom.
Host records: Unknown.
Plant associations in Iran: *Medicago sativa* (Fabaceae) (Yari et al., 2016).

Chorebus (*Stiphrocera*) *asphodeli* Griffiths, 1968
Catalogues with Iranian records: Khajeh et al. (2014), Gadallah et al. (2015), Farahani et al. (2016), Cortés et al. (2016), Yu et al. (2016), Samin, Coronado-Blanco, Kavallieratos et al. (2018), Samin, Coronado-Blanco, Fischer et al. (2018).
Distribution in Iran: Ardabil (Rastegar et al., 2012), Qazvin (Ghahari & Fischer, 2011a).
Distribution in the Middle East: Iran.
Extralimital distribution: Croatia, Greece, North Macedonia, Spain, former Yugoslavia.
Host records: Recorded by Griffiths (1968) as being a parasitoid of the agromyzid *Liriomyza asphodeli* Spencer.

Chorebus (*Stiphrocera*) *avestus* (Nixon, 1944)
Catalogues with Iranian records: No catalogue.
Distribution in Iran: East Azarbaijan (Samin, Sakenin Chelav et al., 2020 as *C. avesta*).
Distribution in the Middle East: Iran (Samin, Sakenin Chelav et al., 2020), Turkey (Beyarslan & Inanç, 2001).
Extralimital distribution: Austria, Azerbaijan, Czech Republic, Germany, Hungary, Italy, Korea, Russia, Spain, Sweden, Ukraine, United Kingdom.
Host records: Summarized by Yu et al. (2016) as being a parasitoid of the agromyzids *Amauromyza labiatarum* (Hendel), *Galiomyza morio* (Brischke), and *Liriomyza eupatoriana* (Spencer).

Chorebus (Chorebus) axillaris Fischer, Lashkari Bod, Rakhshani and Talebi, 2011

Catalogues with Iranian records: Khajeh et al. (2014), Gadallah et al. (2015), Farahani et al. (2016), Cortés et al. (2016), Yu et al. (2016), Samin, Coronado-Blanco, Kavallieratos et al. (2018), Samin, Coronado-Blanco, Fischer et al. (2018).

Distribution in Iran: Fars (Fischer et al., 2011; Khajeh et al., 2014).

Distribution in the Middle East: Iran.

Extralimital distribution: None.

Host records: In Iran, recorded by Lotfalizadeh et al. (2015) as a parasitoid of the agromyzids *Liriomyza trifolii* (Burgess) and *Phytomyza horticola* Goureau.

Chorebus (Phaenolexis) brevicornis (Thomson, 1895)

Catalogues with Iranian records: No catalogue.

Distribution in Iran: West Azarbaijan (Ghahari et al., 2020).

Distribution in the Middle East: Iran.

Extralimital distribution: Austria, Azerbaijan, Czech Republic, Germany, Hungary, Ireland, Moldova, Russia, Serbia, Sweden, Ukraine, United Kingdom.

Host records: Recorded by Griffiths (1962) as being a parasitoid of the agromyzid *Melanagromyza aeneoventris* (Fallén). In Iran, this species has been reared from *M. aeneoventris* (Ghahari et al., 2020).

Chorebus (Chorebus) larides (Nixon, 1944)

Catalogues with Iranian records: No catalogue.

Distribution in Iran: Ardabil (Samin, Beyarslan et al., 2020).

Distribution in the Middle East: Iran.

Extralimital distribution: Hungary, Ireland, Malta, North Macedonia, Serbia, Spain, Sweden, Ukraine, United Kingdom.

Host records: Unknown.

Chorebus (Stiphrocera) baeticus Griffiths, 1967

Catalogues with Iranian records: Gadallah et al. (2015), Farahani et al. (2016), Cortés et al. (2016), Yu et al. (2016), Samin, Coronado-Blanco, Kavallieratos et al. (2018), Samin, Coronado-Blanco, Fischer et al. (2018).

Distribution in Iran: Golestan (Ghahari & Fischer, 2011b).

Distribution in the Middle East: Iran.

Extralimital distribution: Azerbaijan, Greece, Hungary, Serbia, Spain, Ukraine.

Host records: Recorded by Griffiths (1967) as being a parasitoid of the agromyzid *Agromyza beatica* Griffiths.

Chorebus (Phaenolexis) bathyzonus (Marshall, 1895)

Catalogues with Iranian records: Khajeh et al. (2014), Gadallah et al. (2015), Farahani et al. (2016), Cortés et al. (2016), Yu et al. (2016), Samin, Coronado-Blanco,

Kavallieratos et al. (2018), Samin, Coronado-Blanco, Fischer et al. (2018).

Distribution in Iran: Isfahan (Ghahari & Beyarslan, 2017), Kerman (Ghotbi Ravandi et al., 2015; Safahani et al., 2017), Sistan and Baluchestan (Sedighi et al., 2014; Yari et al., 2016).

Distribution in the Middle East: Iran.

Extralimital distribution: Austria, Azerbaijan, Bosnia-Herzegovina, Croatia, Czech Republic, Finland, France, Germany, Hungary, Ireland, Italy, Kazakhstan, Korea, Netherlands, North Macedonia, Poland, Portugal, Romania, Russia, Serbia, Spain, Sweden, Switzerland, Ukraine, United Kingdom.

Host records: Summarized by Yu et al. (2016) as being a parasitoid of the agromyzids *Ophiomyia heracleivora* Spencer, *O. renanunculicaulis* Hering, and *O. simplex* (Loew).

Plant associations in Iran: *Medicago sativa* (Fabaceae) (Ghotbi Ravandi et al., 2015; Safahani et al., 2017; Yari et al., 2016), *Mentha longifolia* (Lamiaceae), and *Triticum aestivum* (Poaceae) (Ghotbi Ravandi et al., 2015).

Chorebus (Phaenolyxis) caelebs (Nixon, 1944)

Distribution in the Middle East: Israel—Palestine (Papp, 2012).

Extralimital distribution: Hungary, United Kingdom.

Host records: Unknown.

Chorebus (Phaenolyxis) caesariatus Griffiths, 1967

Catalogues with Iranian records: Gadallah et al. (2015), Cortés et al. (2016), Yu et al. (2016), Samin, Coronado-Blanco, Kavallieratos et al. (2018), Samin, Coronado-Blanco, Fischer et al. (2018).

Distribution in Iran: Northern Khorasan (Yari et al., 2016), East of Iran (no specific locality cited) (Yari et al., 2014).

Distribution in the Middle East: Iran.

Extralimital distribution: Azerbaijan, Hungary, Italy, Ukraine (Yu et al., 2016), Sweden (Stigenberg & Peris-Felipo, 2019).

Host records: Recorded by Griffiths (1967) as being a parasitoid of the agromyzid *Ophiomyia* sp.

Plant associations in Iran: *Mentha pulegium* (Lamiaceae) (Yari et al., 2016).

Chorebus (Stiphrocera) calthae Griffiths, 1967

Catalogues with Iranian records: Yu et al. (2016), Samin, Coronado-Blanco, Kavallieratos et al. (2018), Samin, Coronado-Blanco, Fischer et al. (2018).

Distribution in Iran: East Azarbaijan (Lotfalizadeh et al., 2015).

Distribution in the Middle East: Iran.

Extralimital distribution: Denmark, Hungary, Montenegro, Poland, Russia, United Kingdom.

Host records: Recorded by Griffiths (1967) as being a parasitoid of the agromyzid *Phytomyza calthivora* Hendel.

In Iran, reared from the agromyzids *Liriomyza trifolii* (Burgess) and *Phytomyza horticola* Goureau.

Chorebus (Phaenolexis) compressiiventris (Telenga, 1935)

Catalogues with Iranian records: Khajeh et al. (2014), Gadallah et al. (2015), Farahani et al. (2016), Cortés et al. (2016), Yu et al. (2016), Samin, Coronado-Blanco, Kavallieratos et al. (2018), Samin, Coronado-Blanco, Fischer et al. (2018).
Distribution in Iran: Ardabil (Ghahari & Fischer, 2011a), Zanjan (Samin, 2015).
Distribution in the Middle East: Iran.
Extralimital distribution: Azerbaijan, Hungary, Mongolia, Ukraine.
Host records: Unknown.

Chorebus (Stiphocera) cubocephalus (Telenga, 1935)

Catalogues with Iranian records: Khajeh et al. (2014), Gadallah et al. (2015), Farahani et al. (2016), Cortés et al. (2016), Yu et al. (2016), Samin, Coronado-Blanco, Kavallieratos et al. (2018), Samin, Coronado-Blanco, Fischer et al. (2018).
Distribution in Iran: Northern Khorasan (Yari et al., 2016), Sistan and Baluchistan (Sedighi et al., 2014; Yari et al., 2016), East of Iran (no specific locality cited) (Yari et al., 2014).
Distribution in the Middle East: Iran (Sedighi et al., 2014; Yari et al., 2014, 2016), Israel–Palestine (Papp, 2012), Turkey (Beyarslan & Inanç, 2001).
Extralimital distribution: Austria, Azerbaijan, Germany, Hungary, Ireland, Italy, Kazakhstan, Korea, Poland, Portugal, Russia, Serbia, Spain, Sweden, Turkmenistan, United Kingdom, Uzbekistan.
Host records: Unknown.
Plant associations in Iran: *Medicago sativa* (Fabaceae) (Yari et al., 2016).

Chorebus (Stiphrocera) dagda (Nixon, 1943)

Catalogues with Iranian records: Samin, Coronado-Blanco, Fischer et al. (2018).
Distribution in Iran: Hamadan (Samin, Coronado-Blanco, Fischer et al., 2018).
Distribution in the Middle East: Iran.
Extralimital distribution: Germany, Greece, Hungary, Italy, Korea, Serbia, Spain, Ukraine, United Kingdom.
Host records: Recorded by Griffiths (1956, 1967) as being a parasitoid of the agromyzid *Phytomyza gentianae* Hendel.

Chorebus (Stiphrocera) diremtus (Nees von Esenbeck, 1834)

Catalogues with Iranian records: Khajeh et al. (2014), Gadallah et al. (2015), Farahani et al. (2016), Cortés et al. (2016), Yu et al. (2016), Samin, Coronado-Blanco,

Kavallieratos et al. (2018), Samin, Coronado-Blanco, Fischer et al. (2018).
Distribution in Iran: East Azarbaijan (Ghahari & Fischer, 2011a), Fars (Lashkari Bod et al., 2010), Golestan (Samin, Ghahari et al., 2015).
Distribution in the Middle East: Iran (Ghahari & Fischer, 2011a; Lashkari Bod et al., 2010; Samin, Ghahari et al., 2015), Turkey (Beyarslan & Inanç, 2001).
Extralimital distribution: Austria, Azerbaijan, Germany, Hungary, Ireland, Korea, Lithuania, Mongolia, Montenegro, Netherlands, Poland, Russia, Sweden, Switzerland, United Kingdom.
Host records: Recorded by Griffiths (1968) as being a parasitoid of the agromyzid *Cerodontha fulvipes* (Meigen).

Chorebus (Phaenolexis) femoratus (Tobias, 1962)

Catalogues with Iranian records: Khajeh et al. (2014), Gadallah et al. (2015), Farahani et al. (2016), Cortés et al. (2016), Yu et al. (2016), Samin, Coronado-Blanco, Kavallieratos et al. (2018), Samin, Coronado-Blanco, Fischer et al. (2018).
Distribution in Iran: Ilam (Ghahari et al., 2011c; Khajeh et al., 2014).
Distribution in the Middle East: Iran (Ghahari et al., 2011c; Khajeh et al., 2014), Turkey (Beyarslan & Inanç, 2001).
Extralimital distribution: Azerbaijan, Bosnia-Herzegovina, Croatia, Greece, Hungary, North Macedonia, Russia, former Yugoslavia.
Host records: Unknown.

Chorebus (Stiphrocera) flavipes (Goureau, 1851)

Catalogues with Iranian records: Khajeh et al. (2014), Gadallah et al. (2015), Farahani et al. (2016), Cortés et al. (2016), Yu et al. (2016), Samin, Coronado-Blanco, Kavallieratos et al. (2018), Samin, Coronado-Blanco, Fischer et al. (2018).
Distribution in Iran: Ardabil (Ghahari, Fischer et al. 2011), Chaharmahal and Bakhtiari (Samin, van Achterberg et al., 2015), Guilan (Ghahari & Fischer, 2011c), Kerman (Samin et al., 2011), Qazvin (Ghahari & Gadallah, 2021).
Distribution in the Middle East: Iran (see references above), Turkey (Yildirim et al., 2010).
Extralimital distribution: Denmark, France, Germany, Greece, Hungary, Ireland, Kazakhstan, Korea, Mongolia, North Macedonia, Poland, Russia, Serbia, Spain, Ukraine, United Kingdom, Uzbekistan.
Host records: Summarized by Yu et al. (2016) as being a parasitoid of the agromyzids *Agromyza nana* Meigen, *Cerodontha iraeos* (Robineau-Desvoidy), *Chromatomyia lonicerae* (Robineau-Desvoidy), and *Napomyza lateralis* (Fallén). In Iran, this species has been reared from the agromyzid *Phytomyza horticola* Goureau (Ghahari & Gadallah, 2021).

Chorebus (Chorebus) fordi (Nixon, 1954)

Catalogues with Iranian records: No catalogue.

Distribution in Iran: Ardabil (Samin, Beyarslan et al., 2020).

Distribution in the Middle East: Iran.

Extralimital distribution: Czech Republic, Germany, Hungary, Lithuania, Poland, Russia, Serbia, United Kingdom.

Host records: Recorded by Griffiths (1968) as being a parasitoid of the agromyzid *Cerodontha lateralis* (Macquart).

Chorebus (Stiphrocera) freya (Nixon, 1943)

Distribution in the Middle East: Israel−Palestine (Papp, 2012).

Extralimital distribution: Czech Republic, Hungary, North Macedonia, Poland, Serbia, Sweden.

Host records: Unknown.

Chorebus (Phaenolexis) fuscipennis (Nixon, 1937)

Catalogues with Iranian records: Khajeh et al. (2014), Gadallah et al. (2015), Farahani et al. (2016), Cortés et al. (2016), Yu et al. (2016), Samin, Coronado-Blanco, Kavallieratos et al. (2018), Samin, Coronado-Blanco, Fischer et al. (2018).

Distribution in Iran: Isfahan (Ghahari et al., 2011b; Khajeh et al., 2014).

Distribution in the Middle East: Iran.

Extralimital distribution: Austria, Azerbaijan, Croatia, Germany, Greece, Hungary, Italy, Korea, Lithuania, North Macedonia, Poland, Romania, Russia, Serbia, Spain, Sweden, Ukraine, United Kingdom.

Host records: Recorded by Griffiths (1962, 1967) as being a parasitoid of the agromyzids *Ophiomyia heringi* Stary and *O. labiatarum* Hering.

Chorebus (Phaenolexis) gedanensis (Ratzeburg, 1852)

Catalogues with Iranian records: Gadallah et al. (2015), Farahani et al. (2016), Cortés et al. (2016), Yu et al. (2016), Samin, Coronado-Blanco, Kavallieratos et al. (2018), Samin, Coronado-Blanco, Fischer et al. (2018).

Distribution in Iran: Razavi Khorasan (Samin et al., 2011; Yari et al., 2016), Qazvin (Samin, 2015; Yari et al., 2016).

Distribution in the Middle East: Iran.

Extralimital distribution: Bulgaria, Germany, Hungary, Italy, Kazakhstan, Mongolia, Netherlands, Poland, Russia, Sweden, United Kingdom, Uzbekistan.

Host records: Summarized by Yu et al. (2016) as being a parasitoid of the agromyzid *Hexomyza schineri* (Giraud).

Chorebus (Stiphrocera) geminus (Tobias, 1962)

Catalogues with Iranian records: No catalogue.

Distribution in Iran: Mazandaran (Sakenin et al., 2020).

Distribution in the Middle East: Iran.

Extralimital distribution: Azerbaijan, Hungary, Russia, Ukraine.

Host records: Unknown.

Chorebus (Chorebus) gnaphalii Griffiths, 1967

Catalogues with Iranian records: No catalogue.

Distribution in Iran: East Azarbaijan (Samin, Papp et al., 2018).

Distribution in the Middle East: Iran.

Extralimital distribution: Czech Republic, Germany, Greece, Hungary, former Yugoslavia.

Host records: Recorded by Griffiths (1967) as being a parasitoid of the agromyzid *Ophiomyia gnaphalii* Hering.

Chorebus (Chorebus) gracilipes (Thomson, 1895)

Catalogues with Iranian records: Khajeh et al. (2014), Gadallah et al. (2015), Farahani et al. (2016), Yu et al. (2016), Samin, Coronado-Blanco, Kavallieratos et al. (2018), Samin, Coronado-Blanco, Fischer et al. (2018).

Distribution in Iran: Fars (Samin, van Achterberg et al., 2015), Kordestan (Ghahari, Fischer, Hedqvist et al., 2010).

Distribution in the Middle East: Iran.

Extralimital distribution: Mongolia, Poland, Russia, Serbia, Sweden, Ukraine.

Host records: Recorded by Griffiths (1968) as being a parasitoid of the agromyzid *Cerodontha geniculata* (Fallén).

Chorebus (Chorebus) groschkei Griffiths, 1967

Catalogues with Iranian records: Khajeh et al. (2014), Gadallah et al. (2015), Farahani et al. (2016), Cortés et al. (2016), Yu et al. (2016), Samin, Coronado-Blanco, Kavallieratos et al. (2018), Samin, Coronado-Blanco, Fischer et al. (2018).

Distribution in Iran: Fars (Fischer et al., 2011; Lashkari Bod et al., 2010, 2011).

Distribution in the Middle East: Iran.

Extralimital distribution: Germany, Russia, Ukraine.

Host records: Recorded by Griffiths (1967) as being a parasitoid of the agromyzid *Agromyza prespana* Spencer.

Chorebus (Phaenolexis) heringianus Griffiths, 1967

Catalogues with Iranian records: Gadallah et al. (2015), Farahani et al. (2016), Cortés et al. (2016), Yu et al. (2016), Samin, Coronado-Blanco, Kavallieratos et al. (2018), Samin, Coronado-Blanco, Fischer et al. (2018).

Distribution in Iran: Golestan (Ghahari & Fischer, 2011b).

Distribution in the Middle East: Iran.

Extralimital distribution: Germany, Greece, Hungary, Russia, Serbia.

Host records: Recorded by Griffiths (1967) as being a parasitoid of the agromyzid *Ophiomyia thalictricaulis* Hering.

Chorebus (Phaenolexis) iridis Griffiths, 1968

Catalogues with Iranian records: Khajeh et al. (2014), Gadallah et al. (2015), Farahani et al. (2016), Cortés et al. (2016), Yu et al. (2016), Samin, Coronado-Blanco,

Kavallieratos et al. (2018), Samin, Coronado-Blanco, Fischer et al. (2018).

Distribution in Iran: Qazvin (Ghahari et al., 2011a).

Distribution in the Middle East: Iran (Ghahari et al., 2011a), Turkey (Beyarslan & Inanç, 2001).

Extralimital distribution: Azerbaijan, Greece, Hungary, Italy, Korea, Russia, Serbia.

Host record: Recorded by Griffiths (1968) as being a parasitoid of the agromyzid *Cerodontha iridis* (Hendel).

Chorebus (Stiphrocera) lar (Morley, 1924)

Catalogues with Iranian records: Khajeh et al. (2014), Gadallah et al. (2015), Farahani et al. (2016), Cortés et al. (2016), Yu et al. (2016), Samin, Coronado-Blanco, Kavallieratos et al. (2018), Samin, Coronado-Blanco, Fischer et al. (2018).

Distribution in Iran: Isfahan (Ghahari et al., 2011b); Kerman (Ghotbi Ravandi et al., 2015; Safahani et al., 2017), Northern Khorasan (Yari et al., 2016), Sistan and Baluchestan (Sedighi et al., 2014; Yari et al., 2016), East Iran (no specific locality cited) (Yari et al., 2014).

Distribution in the Middle East: Iran.

Extralimital distribution: Afghanistan, Austria, Azerbaijan, Croatia, Greece, Hungary, Korea, Moldova, North Macedonia, Poland, Russia, Serbia, Spain, Sweden, Ukraine, United Kingdom.

Host records: Recorded by Griffiths (1967) as being a parasitoid of the agromyzids *Agromyza johannae* de Meijere, and *Agromyza pulla* Meigen.

Plant associations in Iran: *Medicago sativa* (Fabaceae) (Ghotbi Ravandi et al., 2015; Safahani et al., 2017; Yari et al., 2016), *Mentha pulegium* (Lamiaceae), and *Triticum aestivum* (Poaceae) (Yari et al., 2016).

Chorebus (Phaenolexis) leptogaster (Haliday, 1839)

Catalogues with Iranian records: Khajeh et al. (2014), Gadallah et al. (2015), Farahani et al. (2016), Cortés et al. (2016), Yu et al. (2016), Samin, Coronado-Blanco, Kavallieratos et al. (2018), Samin, Coronado-Blanco, Fischer et al. (2018).

Distribution in Iran: Golestan (Ghahari, Fischer et al., 2011; Khajeh et al., 2014), Northern Khorasan (Yari et al., 2016), East of Iran (no specific locality cited) (Yari et al., 2014).

Distribution in the Middle East: Iran (Ghahari, Fischer et al. 2011; Khajae et al., 2014; Yari et al., 2014, 2016), Israel—Palestine (Papp, 2012).

Extralimital distribution: Afghanistan, Austria, Azerbaijan, Belgium, Finland, France, Greece, Hungary, Ireland, Italy, Kazakhstan, Korea, Montenegro, North Macedonia, Poland, Romania, Russia, Serbia, Spain, Sweden, Switzerland, Ukraine, United Kingdom.

Host records: Summarized by Yu et al. (2016) as being a parasitoid of the agromyzids *Chromatomyia syngenesiae* Hardy, *Napomyza lateralis* (Fallén), *Ophiomyia* spp., and *Phytomyza continua* Hendel.

Plant associations in Iran: *Medicago sativa* (Fabaceae) and *Mentha pulegium* (Lamiaceae) (Yari et al., 2016).

Chorebus (Chorebus) longiarticulis Fischer, Lashkari Bod, Rakhshani and Talebi, 2011

Catalogues with Iranian records: Khajeh et al. (2014), Gadallah et al. (2015), Farahani et al. (2016), Cortés et al. (2016), Yu et al. (2016), Samin, Coronado-Blanco, Kavallieratos et al. (2018), Samin, Coronado-Blanco, Fischer et al. (2018).

Distribution in Iran: Fars (Fischer et al., 2011), Kerman (Ghotbi Ravandi et al., 2015).

Distribution in the Middle East: Iran.

Extralimital distribution: None.

Host records: Unknown.

Plant associations in Iran: *Medicago sativa* (Fabaceae) (Ghotbi Ravandi et al., 2015).

Chorebus (Stiphrocera) merellus (Nixon, 1937)

Catalogues with Iranian records: Gadallah et al. (2015), Farahani et al. (2016), Cortés et al. (2016), Yu et al. (2016), Samin, Coronado-Blanco, Kavallieratos et al. (2018), Samin, Coronado-Blanco, Fischer et al. (2018).

Distribution in Iran: Kerman (Safahani et al., 2017), Northern Khorasan (Yari et al., 2016), East of Iran (no specific locality cited) (Yari et al., 2014).

Distribution in the Middle East: Iran (Safahani et al., 2017; Yari et al., 2014, 2016), Israel—Palestine (Papp, 2012).

Extralimital distribution: Austria, Germany, Hungary, Ireland, Korea, Poland, Russia, Sweden, Switzerland, Ukraine, United Kingdom.

Host records: Recorded by Griffiths (1968) as being a parasitoid of the agromyzids *Cerodontha caricicola* (Hering), *C. caricivora* (Groschke), *C. chaixiana* (Hering), *C. scirpi* (Karl), and *C. staryi* (Starý).

Plant associations in Iran: *Mentha pulegium* (Lamiaceae) (Yari et al., 2016), and *Medicago sativa* (Fabaceae) (Safahani et al., 2017).

Chorebus (Stiphocera) melanophytobiae Griffiths, 1968

Catalogues with Iranian records: Cortés et al. (2016).

Distribution in Iran: Fars, Kerman (Cortés et al., 2016).

Distribution in the Middle East: Iran.

Extralimital distribution: Azerbaijan, Germany, Hungary, Korea, Russia, Serbia, Spain, Ukraine, Uzbekistan.

Host records: Recorded by Griffiths (1968) as being a parasitoid of the agromyzid *Melanophytobia chamaebalani* Hering.

Chorebus (Stiphrocera) misellus (Marshall, 1895)

Catalogues with Iranian records: Khajeh et al. (2014), Gadallah et al. (2015), Farahani et al. (2016), Cortés et al.

(2016), Yu et al. (2016), Samin, Coronado-Blanco, Kavallieratos et al. (2018), Samin, Coronado-Blanco, Fischer et al. (2018).

Distribution in Iran: Hamadan (Ghahari & Beyarslan, 2017), Kerman (Ghotbi Ravandi et al., 2015), Semnan (Ghahari, Fischer, Hedqvist et al., 2010).

Distribution in the Middle East: Iran.

Extralimital distribution: Afghanistan, Austria, Azerbaijan, Czech Republic, Denmark, France, Germany, Greece, Hungary, Italy, Kazakhstan, Korea, Mongolia, Poland, Russia, Serbia, Spain, Ukraine, United Kingdom, Uzbekistan.

Host records: Summarized by Yu et al. (2016) as being a parasitoid of the agromyzids *Agromyza frontella* (Rondani), *A. nana* Meigen, *Chromatomyia syngenesia* Hardy, *L. balcania* Strobl, *L. centaureae* Hering, *L. congesta* (Becker), *L. pusilla* (Meigen), *L. trifolii* (Burgess), and *Phytomyza horticola* Goureau. In Iran, this species has been reared from *Liriomyza congesta* (Becker) (Ghahari & Beyarslan, 2017).

Plant associations in Iran: *Medicago sativa* (Fabaceae) (Ghahari & Beyarslan, 2017; Ghotbi Ravandi et al., 2015).

Chorebus (*Stiphrocera*) *mucronatus* (Telenga, 1935)

Catalogues with Iranian records: Khajeh et al. (2014), Gadallah et al. (2015), Farahani et al. (2016), Cortés et al. (2016), Yu et al. (2016), Samin, Coronado-Blanco, Kavallieratos et al. (2018), Samin, Coronado-Blanco, Fischer et al. (2018).

Distribution in Iran: Golestan (Samin, Ghahari et al., 2015), Ilam (Ghahari et al., 2011c), Kerman (Safahani et al., 2017), Mazandaran (Ghahari, Fischer, Hedqvist et al., 2010), Sistan and Baluchestan (Yari et al., 2016), East of Iran (no specific locality cited) (Yari et al., 2014).

Distribution in the Middle East: Iran.

Extralimital distribution: Azerbaijan, Germany, Greece, Hungary, Kazkhstan, Mongolia, North Macedonia, Poland, Russia, Serbia, Ukraine, Uzbekistan.

Host records: Recorded by Hochapfel (1937) as being a parasitoid of the agromyzid *Phytomyza rufipes* Meigen.

Plant associations in Iran: *Medicago sativa* (Fabaceae) (Safahani et al., 2017; Yari et al., 2016), and *Triticum aestivum* (Poaceae) (Yari et al., 2016).

Chorebus (*Stiphrocera*) *myles* (Nixon, 1943)

Distribution in the Middle East: Israel—Palestine (Papp, 2012).

Extralimital distribution: Azerbaijan, Germany, Serbia, Ukraine.

Host records: Unknown.

Chorebus (*Chorebus*) *nigridiremptus* Fischer, Lashkari Bod, Rakhshani and Talebi, 2011

Catalogues with Iranian records: Khajeh et al. (2014), Gadallah et al. (2015), Farahani et al. (2016), Cortés et al.

(2016), Yu et al. (2016), Samin, Coronado-Blanco, Kavallieratos et al. (2018), Samin, Coronado-Blanco, Fischer et al. (2018).

Distribution in Iran: Fars (Fischer et al., 2011; Khajeh et al., 2014).

Distribution in the Middle East: Iran.

Extralimital distribution: None.

Host records: Unknown.

Chorebus (*Chorebus*) *nigriscaposus* (Nixon, 1949)

Catalogues with Iranian records: Gadallah et al. (2015), Farahani et al. (2016), Cortés et al. (2016), Yu et al. (2016), Samin, Coronado-Blanco, Kavallieratos et al. (2018), Samin, Coronado-Blanco, Fischer et al. (2018).

Distribution in Iran: Sistan and Baluchestan (Yari et al., 2016), East of Iran (no specific locality cited) (Yari et al., 2014).

Distribution in the Middle East: Iran.

Extralimital distribution: Denmark, Hungary, Ireland, Poland, Spain.

Host records: Recorded by Griffiths (1968) as being a parasitoid of the agromyzids *Cerodontha calosoma* (Hendel), and *C. geniculata* (Fallén).

Plant associations in Iran: *Medicago sativa* (Fabaceae) (Yari et al., 2016).

Chorebus (*Chorebus*) *nixoni* Burghele, 1959

Catalogues with Iranian records: Khajeh et al. (2014), Gadallah et al. (2015), Farahani et al. (2016), Yu et al. (2016), Samin, Coronado-Blanco, Kavallieratos et al. (2018), Samin, Coronado-Blanco, Fischer et al. (2018).

Distribution in Iran: East Azarbaijan (Ghahari & Fischer, 2011a), Khuzestan (Samin, van Achterberg et al., 2015).

Distribution in the Middle East: Iran.

Extralimital distribution: Azerbaijan, Hungary, Mongolia, Romania, Russia, Spain, Ukraine.

Host records: Recorded by Burghele (1959) as being a parasitoid of the ephydrid *Hydrellia griseola* (Fallén).

Chorebus (*Phaenolexis*) *ornatus* (Telenga, 1935)

Catalogues with Iranian records: Khajeh et al. (2014), Gadallah et al. (2015), Farahani et al. (2016), Yu et al. (2016), Samin, Coronado-Blanco, Kavallieratos et al. (2018), Samin, Coronado-Blanco, Fischer et al. (2018).

Distribution in Iran: Qazvin (Ghahari et al., 2011a).

Distribution in the Middle East: Iran.

Extralimital distribution: Azerbaijan, Greece, Hungary, North Macedonia, Serbia, Ukraine.

Host records: Unknown.

Chorebus (*Stiphrocera*) *parvungula* (Thomson, 1895)

Catalogues with Iranian records: Gadallah et al. (2015), Farahani et al. (2016), Cortés et al. (2016), Yu et al. (2016),

Samin, Coronado-Blanco, Kavallieratos et al. (2018), Samin, Coronado-Blanco, Fischer et al. (2018).
Distribution in Iran: Kerman (Ghotbi Ravandi et al., 2015), Northern Khorasan (Yari et al., 2016), East of Iran (no specific locality cited) (Yari et al., 2014).
Distribution in the Middle East: Iran.
Extralimital distribution: Belgium, Germany, Hungary, Ireland, Kazakhstan, Netherlands, Spain, Sweden, Ukraine, United Kingdom.
Host records: Recorded by Griffiths (1967) as being a parasitoid of the agromyzid *Napomyza cichorii* Spencer, and *N. lateralis* (Fallén).
Plant associations in Iran: *Medicago sativa* (Fabaceae) (Ghotbi Ravandi et al., 2015; Yari et al., 2016), *Mentha pulegium* (Lamiaceae) (Yari et al., 2016).

Chorebus (Stiphrocera) perkinsi (Nixon, 1944)
Catalogues with Iranian records: No catalogue.
Distribution in Iran: Kordestan (Samin, Sakenin Chelav et al., 2020).
Distribution in the Middle East: Iran (Samin, Sakenin Chelav et al., 2020), Turkey (Beyarslan & Inanç, 2001).
Extralimital distribution: Austria, Azerbaijan, Germany, Hungary, Poland, Russia, Sweden, Switzerland, United Kingdom.
Host records: Recorded by Griffiths (1962) as being a parasitoid of the agromyzid *Agromyza albitarsis* Meigen. No yet recorded in Iran.

Chorebus (Phaenolexis) posticus (Haliday, 1839)
Catalogues with Iranian records: Khajeh et al. (2014), Gadallah et al. (2015), Farahani et al. (2016), Cortés et al. (2016), Yu et al. (2016), Samin, Coronado-Blanco, Kavallieratos et al. (2018), Samin, Coronado-Blanco, Fischer et al. (2018).
Distribution in Iran: Guilan (Ghahari and Beyarslan, 2019 as *Chorebus (Phaenolexis) gracilis* (Nees, 1834), Hamadan (Ghahari et al., 2009).
Distribution in the Middle East: Iran.
Extralimital distribution: Austria, Azerbaijan, Belgium, Canada, Croatia, Czech Republic, Finland, France, Germany, Hungary, Ireland, Italy, Kazakhstan, Mongolia, Netherlands, Norway, Poland, Russia, Serbia, Sweden, Switzerland, Ukraine, United Kingdom.
Host records: Summarized by Yu et al. (2016) as being a parasitoid of the psilid *Psila nigricornis* Meigen, and *P. rosae* (Fabricius).

Chorebus (Chorebus) properesam Fischer, Lashkari Bod, Rakhshani and Talebi, 2011
Catalogues with Iranian records: Khajeh et al. (2014), Gadallah et al. (2015), Farahani et al. (2016), Cortés et al. (2016), Yu et al. (2016), Samin, Coronado-Blanco, Kavallieratos et al. (2018), Samin, Coronado-Blanco, Fischer et al. (2018).

Distribution in Iran: Fars (Fischer et al., 2011).
Distribution in the Middle East: Iran.
Extralimital distribution: None.
Host records: Unknown.

Chorebus (Stiphrocera) pseudomisellus Griffiths, 1968
Catalogues with Iranian records: Gadallah et al. (2015), Farahani et al. (2016), Cortés et al. (2016), Yu et al. (2016), Samin, Coronado-Blanco, Kavallieratos et al. (2018), Samin, Coronado-Blanco, Fischer et al. (2018).
Distribution in Iran: Golestan (Ghahari & Fischer, 2011b).
Distribution in the Middle East: Iran (Ghahari & Fischer, 2011b), Israel–Palestine (Papp, 2012).
Extralimital distribution: Czech Republic, Germany, Greece, Hungary, Korea, Russia, Spain, Uzbekistan.
Host records: Recorded by Griffiths (1968) as being a parasitoid of the agromyzid *Liriomyza congesta* (Becker).

Chorebus (Chorebus) ruficollis (Stelfox, 1957)
Catalogues with Iranian records: Cortés et al. (2016).
Distribution in Iran: Kerman (Safahani et al., 2017), Northern Khorasan (Yari et al., 2016).
Distribution in the Middle East: Iran.
Extralimital distribution: Ireland, Mongolia, Romania, Spain.
Host records: Unknown.
Plant associations in Iran: *Mentha pulegium* (Lamiaceae) (Yari et al., 2016), and *Medicago sativa* (Fabaceae) (Safahani et al., 2017).

Chorebus (Chorebus) scabiosae Griffiths, 1967
Catalogues with Iranian records: Gadallah et al. (2015), Farahani et al. (2016), Cortés et al. (2016), Yu et al. (2016), Samin, Coronado-Blanco, Kavallieratos et al. (2018), Samin, Coronado-Blanco, Fischer et al. (2018).
Distribution in Iran: Northern Khorasan, Razavi Khorasan (Yari et al., 2016), East of Iran (no specific locality cited) (Yari et al., 2014).
Distribution in the Middle East: Iran.
Extralimital distribution: Germany, Greece, Hungary, Serbia, Spain, United Kingdom, Uzbekistan.
Host records: Recorded by Griffiths (1967) as being a parasitoid of the agromyzid *Chromatomyia scabiosae* (Hendel).
Plant associations in Iran: *Medicago sativa* (Fabaceae), and *Mentha pulegium* (Lamiaceae) (Yari et al., 2016).

Chorebus (Phaenolexis) senilis (Nees von Esenbeck, 1812)
Catalogues with Iranian records: No catalogue.
Distribution in Iran: East Azarbaijan (Samin, Beyarslan, Ranjith et al., 2020 as *Phaenolexis senilis* (Nees)).
Distribution in the Middle East: Iran.
Extralimital distribution: Austria, Azerbaijan, Belgium, Czech Republic, Faroe Islands, Finland, France, Germany,

Hungary, Ireland, Italy, Netherlands, Poland, Russia, Serbia, Spain, Sweden, Switzerland, Ukraine, United Kingdom.
Host records: Summarized by Yu et al. (2016) as being a parasitoid of the agromyzids *Melanagromyza aeneoventris* (Fallén), *Napomyza carotae* Spencer, *N. cichorii* Spencer, *N. lateralis* (Fallén), *N. scrophulariae* Spencer, and *Phytomyza albiceps* Meigen; the cecidomyiid *Mayetiola destructor* (Say); and the psilid *Psila rosae* (Fabricius).

Chorebus (Chorebus) solstitialis (Stelfox, 1951)
Catalogues with Iranian records: Gadallah et al. (2015), Farahani et al. (2016), Cortés et al. (2016), Yu et al. (2016), Samin, Coronado-Blanco, Kavallieratos et al. (2018), Samin, Coronado-Blanco, Fischer et al. (2018).
Distribution in Iran: Sistan and Baluchestan (Yari et al., 2016), East of Iran (no specific locality cited) (Yari et al., 2014).
Distribution in the Middle East: Iran.
Extralimital distribution: Azerbaijan, Bulgaria, Germany, Spain, Ukraine, United Kingdom.
Host records: Recorded by Griffiths (1967) as being a parasitoid of the agromyzid *Agromyza megalopsis* Hering.
Plant associations in Iran: *Medicago sativa* (Fabaceae) (Yari et al., 2016).

Chorebus (Striphrocera) spenceri Griffiths, 1964
Catalogues with Iranian records: Gadallah et al. (2015), Farahani et al. (2016), Cortés et al. (2016), Yu et al. (2016), Samin, Coronado-Blanco, Kavallieratos et al. (2018), Samin, Coronado-Blanco, Fischer et al. (2018).
Distribution in Iran: Kerman (Safahani et al., 2017), Northern Khorasan (Yari et al., 2016), East of Iran (no specific locality cited) (Yari et al., 2014).
Distribution in the Middle East: Iran.
Extralimital distribution: Hungary, Korea, Poland, Ukraine, United Kingdom.
Host records: Recorded by Griffiths (1964) as being a parasitoid of the agromyzid *Agromyza phragmatidis* Hendel.
Plant associations in Iran: *Medicago sativa* (Fabaceae) (Yari et al., 2016; Safahani et al., 2017).

Chorebus (Chorebus) stilifer Griffiths, 1968
Catalogues with Iranian records: Khajeh et al. (2014), Gadallah et al. (2015), Farahani et al. (2016), Cortés et al. (2016), Yu et al. (2016), Samin, Coronado-Blanco, Kavallieratos et al. (2018), Samin, Coronado-Blanco, Fischer et al. (2018).
Distribution in Iran: Fars (Fischer et al., 2011; Lashkari Bod et al., 2010, 2011), Razavi Khorasan (Yari et al., 2016), East of Iran (no specific locality cited) (Yari et al., 2014).
Distribution in the Middle East: Iran.

Extralimital distribution: None.
Host records: Recorded by Griffiths (1968) as being a parasitoid of the agromyzid *Cerodontha staryi* Starý.
Plant associations in Iran: *Medicago sativa* (Fabaceae) (Lashkari Bod et al., 2011; Yari et al., 2016).

Chorebus (Etriptes) subasper Griffiths, 1968
Distribution in the Middle East: Israel–Palestine (Papp, 2012).
Extralimital distribution: Hungary, Korea, Poland, Russia.
Host records: Recorded by Griffiths (1968) as being a parasitoid of the agromyzid *Cerodontha alpina* Nowakowski.

Chorebus (Phaenoloxis) tamsi (Nixon, 1944)
Catalogues with Iranian records: Khajeh et al. (2014), Gadallah et al. (2015), Farahani et al. (2016), Cortés et al. (2016), Yu et al. (2016), Samin, Coronado-Blanco, Kavallieratos et al. (2018), Samin, Coronado-Blanco, Fischer et al. (2018).
Distribution in Iran: Fars (Fischer et al., 2011; Lashkari Bod et al., 2010, 2011).
Distribution in the Middle East: Iran.
Extralimital distribution: Austria, Hungary, Russia, United Kingdom.
Host records: Unknown.
Plant associations in Iran: *Medicago sativa* (Fabaceae) (Lashkari Bod et al., 2011).

Chorebus (Stiphrocera) thecla (Nixon, 1943)
Distribution in the Middle East: Israel–Palestine (Papp, 2012).
Extralimital distribution: Germany, Hungary, United Kingdom.
Host records: Summarized by Yu et al. (2016) as being a parasitoid of the agromyzids *Liriomyza* sp., and *Phytomyza lithospermi* Nowakowski.

Chorebus (Chorebus) thusa (Nixon, 1937)
Catalogues with Iranian records: Cortés et al. (2016).
Distribution in Iran: Fars, Hormozgan (Cortés et al., 2016).
Distribution in the Middle East: Iran.
Extralimital distribution: Azerbaijan, Czech Republic, Germany, Ireland, Mongolia, Serbia, Spain, Sweden, Ukraine, United Kingdom.
Host records: Recorded by Nixon (1946) and Griffiths (1967) as being a parasitoid of the anthomyiid *Delia florilega* (Zetterstedt); and the agromyzid *Phytomyza rufipes* Meigen.

Chorebus (Stiphrocera) tumidus (Tobias, 1966)
Catalogues with Iranian records: Gadallah et al. (2015), Farahani et al. (2016), Cortés et al. (2016), Yu et al. (2016), Samin, Coronado-Blanco, Kavallieratos et al. (2018), Samin, Coronado-Blanco, Fischer et al. (2018).

Distribution in Iran: Golestan (Samin et al., 2011; Samin, Ghahari et al., 2015).

Distribution in the Middle East: Iran (Samin et al., 2011; Samin, Ghahari et al., 2015), Israel—Palestine (Papp, 2012).

Extralimital distribution: Afghanistan, Hungary, Korea, Mongolia, Serbia, Turkmenistan, Uzbekistan.

Host records: Unknown.

Chorebus (Chorebus) uliginosus (Haliday, 1839)

Catalogues with Iranian records: Khajeh et al. (2014), Gadallah et al. (2015), Farahani et al. (2016), Cortés et al. (2016), Yu et al. (2016), Samin, Coronado-Blanco, Kavallieratos et al. (2018), Samin, Coronado-Blanco, Fischer et al. (2018).

Distribution in Iran: Kermanshah (Hazini et al., 2015), Kuhgiloyeh and Boyerahmad (Samin, van Achterberg et al., 2015), Mazandaran (Ghahari, Fischer, Erdoğan et al., 2010).

Distribution in the Middle East: Iran.

Extralimital distribution: Belgium, Germany, Hungary, Ireland, Italy, Korea, Lithuania, Mongolia, Netherlands, Poland, Romania, Russia, Sweden, Ukraine, United Kingdom.

Host records: Summarized by Yu et al. (2016) as being a parasitoid of the agromyzids *Liriomyza strigata* (Meigen), and *Phytomyza horticola* Goureau; the ephydrids *Hydrellia griseola* (Fallén), and *H. nigripes* Robineau-Desvoidy. In Iran, this species has been reared from the agromyzid *Phytomyza horticola* Goureau on *Lactuca orientalis* and *Trifolium repens* (Hazini et al., 2015).

Chorebus (Stiphrocera) varunus (Nixon, 1945)

Catalogues with Iranian records: No catalogue.

Distribution in Iran: West Azarbaijan (Gadallah et al., 2021).

Distribution in the Middle East: Iran (Gadallah et al., 2021), Turkey (Beyarslan & Inanç, 2001).

Extralimital distribution: Azerbaijan, Hungary, Kazakhstan, Korea, Poland, Russia, Sweden, Ukraine, United Kingdom.

Host records: Recorded by Griffiths (1968) as being a parasitoid of the agromyzid *Metopomyza flavonotata* (Haliday).

Chorebus (Stiphrocera) venustus (Tobias, 1962)

Catalogues with Iranian records: Khajeh et al. (2014), Gadallah et al. (2015), Farahani et al. (2016), Cortés et al. (2016), Yu et al. (2016), Samin, Coronado-Blanco, Kavallieratos et al. (2018), Samin, Coronado-Blanco, Fischer et al. (2018).

Distribution in Iran: Isfahan (Ghahari et al., 2011c).

Distribution in the Middle East: Iran.

Extralimital distribution: Azerbaijan, Croatia, Germany, Greece, Hungary, Poland, Russia, Ukraine.

Host records: Recorded by Griffiths (1968) as being a parasitoid of the agromyzids *Liriomyza sonchi* Hendel and *L. soror* Hendel.

Chorebus (Chorebus) zarghanensis Fischer, Lashkari Bod, Rakhshani and Talebi, 2011

Catalogues with Iranian records: Khajeh et al. (2014), Gadallah et al. (2015), Farahani et al. (2016), Cortés et al. (2016), Yu et al. (2016), Samin, Coronado-Blanco, Kavallieratos et al. (2018), Samin, Coronado-Blanco, Fischer et al. (2018).

Distribution in Iran: Fars (Fischer et al., 2011).

Distribution in the Middle East: Iran.

Extralimital distribution: None.

Host records: Unknown.

Genus *Coelinidea* Viereck, 1913

Coelinidea albimana (Vollenhoven, 1873)

Distribution in the Middle East: Turkey (Beyarslan & Inanç, 2001).

Extralimital distribution: Azerbaijan, Hungary, Mongolia, Netherland, North Macedonia, Uzbekistan, former Yugoslavia.

Host records: Recorded by Rudow (1918) as being a parasitoid of the ulidiid *Seioptera vibrans* (L.).

Coelinidea elegans (Curtis, 1829)

Catalogues with Iranian records: Gadallah et al. (2015), Farahani et al. (2016), Cortés et al. (2016), Yu et al. (2016), Samin, Coronado-Blanco, Kavallieratos et al. (2018), Samin, Coronado-Blanco, Fischer et al. (2018).

Distribution in Iran: West Azarbaijan (Gadallah et al., 2015).

Distribution in the Middle East: Iran (Gadallah et al., 2015), Turkey (Beyarslan & Inanç, 2001).

Extralimital distribution: Azerbaijan, Belarus, Belgium, Bulgaria, Croatia, Czech Republic, France, Germany, Ireland, Italy, Korea, Lithuania, Mongolia, Netherlands, Norway, Poland, Romania, Russia, Serbia, Slovenia, Switzerland, Ukraine, United Kingdom.

Host records: Recorded by Rudow (1918) as being a parasitoid of the chloropid *Liapara lucens* Meigen.

Coelinidea gracilis (Curtis, 1829)

Catalogues with Iranian records: Khajeh et al. (2014), Gadallah et al. (2015), Farahani et al. (2016), Cortés et al. (2016), Yu et al. (2016), Samin, Coronado-Blanco, Kavallieratos et al. (2018), Samin, Coronado-Blanco, Fischer et al. (2018).

Distribution in Iran: Fars (Lashkari Bod et al., 2010, 2011 as *Lepton gracilis*; Fischer et al., 2011 as *L. gracilis*; Khajeh et al., 2014), Ilam (Ghahari et al., 2011c), East of Iran (no specific locality cited) (Yari et al., 2014).

Distribution in the Middle East: Iran.

Extralimital distribution: Azerbaijan, Belgium, former Czechoslovakia, France, Germany, Greece, Hungary, Ireland, Italy, Korea, Kazakhstan, Lithuania, Mongolia, Montenegro, Netherlands, North Macedonia, Norway, Poland, Russia, Serbia, Slovenia, Switzerland, United Kingdom, Uzbekistan (Yu et al., 2016), Sweden (Stigenberg & Peris-Felipo, 2019).

Host records: Erroneously recorded by Edelsten and Todd (1912) as being a parasitoid of the noctuid *Photedes extrema* (Hübner).

Coelinidea nigra (Nees von Esenbeck, 1811)

Catalogues with Iranian records: Samin, Coronado-Blanco, Fischer et al. (2018) as *Lepton nigrum* (Nees, 1812) under the subfamily Agathidinae Haliday, 1833.

Distribution in Iran: Fars (Samin, Coronado-Blanco, Fischer et al., 2018 as *Lepton nigrum* (Nees, 1812)), Markazi (Ghahari & Beyarslan, 2017).

Distribution in the Middle East: Iran.

Extralimital distribution: Austria, Azerbaijan, Belgium, Bulgaria, Croatia, Czech Republic, Finland, France, Germany, Hungary, India, Ireland, Italy, Lithuania, Moldova, Mongolia, Montenegro, Netherlands, Norway, Poland, Romania, Russia, Serbia, Spain, Sweden, Switzerland, Ukraine, United Kingdom, Uzbekistan, Zambia.

Host records: Summarized by Yu et al. (2016) as being a parasitoid of the chloropids *Chlorops pumilionis* (Bjerkander), *Chlorops strigulus* (Fabricius), and *Lasiosina herpini* (Guérin-Menéville). In Iran, this species has been reared from *Chlorops pumilionis* (Bjerkander) in a wheat field (Ghahari & Beyarslan, 2017).

Coelinidea vidua (Curtis, 1829)

Catalogues with Iranian records: No catalogue.

Distribution in Iran: Kuhgiloyeh and Boyerahmad (Ghahari & Beyarslan, 2019).

Distribution in the Middle East: Iran.

Extralimital distribution: Belgium, Finland, France, Germany, Hungary, Ireland, Italy, Korea, Mongolia, Poland, Russia, Serbia, Spain, Sweden, Switzerland, Ukraine, United Kingdom.

Host records: Summarized by Yu et al. (2016) as being a parasitoid of the chloropids *Chlorops pumilionis* (Bjerkander), *Meromyza pratorum* Meigen, and *M. saltatrix* (L.).

Genus *Coelinius* Nees von Esenbeck, 1819
Coelinius parvulus (Nees von Esenbeck, 1811)

Catalogues with Iranian records: No catalogue.

Distribution in Iran: Guilan (Samin, Beyarslan, Ranjith et al., 2020).

Distribution in the Middle East: Iran.

Extralimital distribution: Austria, Azerbaijan, Belgium, Bulgaria, Croatia, Czech Republic, Denmark, Finland, France, Georgia, Germany, Hungary, Ireland, Italy, Lithuania, Netherlands, Norway, Poland, Russia, Slovakia, Slovenia, Spain, Sweden, Switzerland, United Kingdom, former Yugoslavia.

Host records: Recorded (may be erroneously) by Rudow (1918) as being a parasitoid of the noctuid *Panolis flammea* (Denis and Schiffermüller).

Genus *Coloneura* Foerster, 1863
Coloneura arestor (Nixon, 1954)

Catalogues with Iranian records: Samin, Coronado-Blanco, Fischer et al. (2018).

Distribution in Iran: Zanjan (Samin, Coronado-Blanco, Fischer et al., 2018).

Distribution in the Middle East: Iran.

Extralimital distribution: Azerbaijan, Czech Republic, Greece, Hungary, Sweden, Uzbekistan.

Host records: Unknown.

Coloneura dice (Nixon, 1943)

Catalogues with Iranian records: Gadallah et al. (2015), Farahani et al. (2016), Cortés et al. (2016), Yu et al. (2016), Samin, Coronado-Blanco, Kavallieratos et al. (2018), Samin, Coronado-Blanco, Fischer et al. (2018).

Distribution in Iran: East of Iran (no specific locality cited) (Yari et al., 2014).

Distribution in the Middle East: Iran.

Extralimital distribution: Azerbaijan, former Czechoslovakia, Germany, Poland, Russia, Spain, Sweden, United Kingdom.

Host records: Recorded by Griffiths (1967, 1968) as being a parasitoid of the agromyzids *Liriomyza balcanica* Ströbl, *Phytomyza angelicivora* Hering and *P. silai* Hering.

Genus *Dacnusa* Haliday, 1833
Dacnusa (*Pachysema*) *abdita* (Haliday, 1839)

Catalogues with Iranian records: Farahani et al. (2016), Cortés et al. (2016), Yu et al. (2016), Samin, Coronado-Blanco, Kavallieratos et al. (2018), Samin, Coronado-Blanco, Fischer et al. (2018).

Distribution in Iran: Kerman (Ghotbi Ravandi et al., 2015).

Distribution in the Middle East: Iran.

Extralimital distribution: Bulgaria, France, Germany, Hungary, Ireland, Italy, Mongolia, Netherlands, Poland, Sweden, Switzerland, United Kingdom.

Host records: Summarized by Yu et al. (2016) as being a parasitoid of the agromyzids *Agromyza abiens* Zetterstedt, *A. anthracina* Meigen, *A. lithospermi* Spencer, *A. pseudoreptans* Nowakowski, *A. reptans* Fallén, *A. rufipes* Meigen, and *Aulagromyza fulvicornis* (Hendel).

Plant associations in Iran: *Triticum aestivum* (Poaceae) (Ghotbi Ravandi et al., 2015).

Dacnusa (Dacnusa) adducta (Haliday, 1839)

Catalogues with Iranian records: No catalogue.
Distribution in Iran: Ardabil (Samin, Beyarslan, Ranjith et al., 2020).
Distribution in the Middle East: Iran.
Extralimital distribution: Austria, Azerbaijan, Bulgaria, Czech Republic, Denmark, Germany, Hungary, Ireland, Italy, Korea, Poland, Russia, Serbia, Spain, Sweden, Switzerland, Ukraine, United Kingdom.
Host records: Summarized by Yu et al. (2016) as being a parasitoid of the agromyzids *Cerodontha pygmaea* (Meigen), and *Liriomyza flaveola* (Fallén).

Dacnusa (Pachysema) alpestris Griffiths, 1967

Catalogues with Iranian records: Khajeh et al. (2014), Gadallah et al. (2015), Farahani et al. (2016), Cortés et al. (2016), Yu et al. (2016), Samin, Coronado-Blanco, Kavallieratos et al. (2018), Samin, Coronado-Blanco, Fischer et al. (2018).
Distribution in Iran: Ilam (Ghahari et al., 2011c).
Distribution in the Middle East: Iran.
Extralimital distribution: Austria, Germany, Greece, Hungary, Poland, Russia.
Host records: Recorded by Griffiths (1967, 1984) as being a parasitoid of the agromyzids *Phytomyza alpina* Groschke, *P. marginella* Fallén, *P. senecionis* Kaltenbach, and *P. tussilaginis* Hendel.

Dacnusa (Pachysema) aquilegiae Marshall, 1896

Catalogues with Iranian records: Cortés et al. (2016).
Distribution in Iran: Fars, Hormozgan, Kerman (Cortés et al., 2016).
Distribution in the Middle East: Iran.
Extralimital distribution: Bosnia-Herzegovina, Germany, Hungary, Korea, Mongolia, Poland, Romania, Russia, Sweden, United Kingdom.
Host records: Recorded by Griffiths (1956, 1967) as being a parasitoid of the agromyzids *Phytomyza aconiti* Hendel, *P. actaeae* Hendel, *P. albimargo* Hering, *P. aquilegiae* Hardy, *P. heracleana* Hering, *P. medicaginis* Hering, *P. rydeni* Hering, and *P. thalictricola* Hendel.

Dacnusa (Dacnusa) areolaris (Nees von Esenbeck, 1811)

Catalogues with Iranian records: Cortés et al. (2016).
Distribution in Iran: Hormozgan, Kerman (Cortés et al., 2016).
Distribution in the Middle East: Iran.
Extralimital distribution: Austria, Azerbaijan, Belgium, Bosnia-Herzegovina, Bulgaria, Czech Republic, Faroe Islands, France, Germany, Hungary, Iceland, Ireland, Italy, Korea, Lithuania, Montenegro, Netherlands, New Zealand, Poland, Portugal, Romania, Russia, Serbia, Slovenia, Spain, Sweden, Switzerland, United Kingdom.

Host records: Summarized by Yu et al. (2016) as being a parasitoid of the agromyzids *Amauromyza labiatarum* (Hendel), *Chromatomyia asteris* (Hendel), *C. milii* (Kaltenbach), *C. nigra* (Meigen), *C. ramosa* (Hendel), and *C. sygenesiae* Hardy.

Dacnusa (Pachysema) aterrima Thomson, 1895

Catalogues with Iranian records: Gadallah et al. (2015), Farahani et al. (2016), Cortés et al. (2016), Yu et al. (2016), Samin, Coronado-Blanco, Kavallieratos et al. (2018), Samin, Coronado-Blanco, Fischer et al. (2018).
Distribution in Iran: East Azarbaijan (Gadallah et al., 2015).
Distribution in the Middle East: Iran.
Extralimital distribution: Azerbaijan, Hungary, Mongolia, Russia, Sweden, Switzerland, United Kingdom.
Host records: Unknown.

Dacnusa (Pachysema) clematidis Griffiths, 1967

Catalogues with Iranian records: Gadallah et al. (2015), Farahani et al. (2016), Cortés et al. (2016), Yu et al. (2016), Samin, Coronado-Blanco, Kavallieratos et al. (2018), Samin, Coronado-Blanco, Fischer et al. (2018).
Distribution in Iran: East of Iran (no specific locality cited) (Yari et al., 2014).
Distribution in the Middle East: Iran.
Extralimital distribution: Poland.
Host records: Recorded by Griffiths (1967) as being a parasitoid of the agromyzid *Phytomyza kaltenbachi atragenis* Hering.

Dacnusa (Dacnusa) confinis Ruthe, 1859

Catalogues with Iranian records: Khajeh et al. (2014), Gadallah et al. (2015), Farahani et al. (2016), Cortés et al. (2016), Yu et al. (2016), Samin, Coronado-Blanco, Kavallieratos et al. (2018), Samin, Coronado-Blanco, Fischer et al. (2018).
Distribution in Iran: Qazvin (Ghahari et al., 2011a).
Distribution in the Middle East: Iran.
Extralimital distribution: Azerbaijan, Bulgaria, Czech Republic, Denmark, Germany, Greece, Hungary, Iceland, Ireland, Netherlands, Spain, Sweden, United Kingdom.
Host records: Recorded by Griffiths (1967) and Vidal (1997) as being a parasitoid of the agromyzids *Chromatomyia sygenesiae* Hardy, *Liriomyza huidobrensis* (Blanchard), *Phytomyza glechomae* Kaltenbach, and *P. ranuculi* (Schrank).

Dacnusa (Pachysema) discolor (Foerster, 1863)

Catalogues with Iranian records: No catalogue.
Distribution in Iran: Ardabil (Samin, Beyarslan, Ranjith et al., 2020 as *Pachysema discolor*).
Distribution in the Middle East: Iran.

Extralimital distribution: Austria, Azerbaijan, former Czechoslovakia, Denmark, Germany, Hungary, Ireland, Italy, Norway, Poland, Russia, Serbia, Spain, Sweden, Switzerland, United Kingdom.
Host records: Summarized by Yu et al. (2016) as being a parasitoid of the agromyzids *Calycomyza humeralis* (von Roser), *Chromatomyia primulae* Robineau-Desvoidy, *Liriomyza strigata* (Meigen), *Phytomyza plantaginis* Goureau, and *P. sedicola* Hering. In Iran, this species has been reared from *Phytomyza plantaginis* (Samin, Beyarslan, Ranjith et al., 2020).

Dacnusa (Pachysema) evadne Nixon, 1937
Catalogues with Iranian records: Gadallah et al. (2015), Farahani et al. (2016), Cortés et al. (2016), Yu et al. (2016), Samin, Coronado-Blanco, Kavallieratos et al. (2018), Samin, Coronado-Blanco, Fischer et al. (2018).
Distribution in Iran: East of Iran (no specific locality cited) (Yari et al., 2014).
Distribution in the Middle East: Iran (Yari et al., 2014), Turkey (Beyarslan & Inanç, 2001).
Extralimital distribution: Azerbaijan, Bulgaria, Hungary, Ireland, Poland, Russia, Sweden, United Kingdom.
Host records: Recorded by Griffiths (1962, 1967) as being a parasitoid of the agromyzid *Agromyza potentillae* (Kaltenbach).

Dacnusa (Dacnusa) faeroeensis (Roman, 1917)
Catalogues with Iranian records: No catalogue.
Distribution in Iran: Guilan (Ghahari et al., 2020).
Distribution in the Middle East: Iran.
Extralimital distribution: Bulgaria, Czech Republic, Faroe Islands, Germany, Hungary, Iceland, Ireland, Lithuania, Netherlands, Norway, Poland, Russia, Serbia, Spain, Sweden, Switzerland, United Kingdom.
Host records: Summarized by Yu et al. (2016) as being a parasitoid of the agromyzids *Cerodontha* sp., and *Scaptomyza incana* Meigen.

Dacnusa (Dacnusa) gentianae Griffiths, 1967
Catalogues with Iranian records: Khajeh et al. (2014), Gadallah et al. (2015), Farahani et al. (2016), Cortés et al. (2016), Yu et al. (2016), Samin, Coronado-Blanco, Kavallieratos et al. (2018), Samin, Coronado-Blanco, Fischer et al. (2018).
Distribution in Iran: Isfahan (Ghahari et al., 2011b).
Distribution in the Middle East: Iran (Ghahari et al., 2011b), Turkey (Yildirim et al., 2010).
Extralimital distribution: Austria, Germany, Hungary, Poland, Serbia.
Host records: Recorded by Griffiths (1967) as being a parasitoid of the agromyzids *Chromatomyia gentianae* Hendel, *C. swertiae* (Hering), and *C. vernalis* (Groscke).

Dacnusa (Dacnusa) heringi Griffiths, 1967
Catalogues with Iranian records: Yu et al. (2016), Samin, Coronado-Blanco, Kavallieratos et al. (2018), Samin, Coronado-Blanco, Fischer et al. (2018).
Distribution in Iran: East Azerbaijan (Lotfalizadeh et al., 2015).
Distribution in the Middle East: Iran.
Extralimital distribution: Germany.
Host records: Recorded by Griffiths (1967) as being a parasitoid of the agromyzid *Phytomyza griffithsi* Spencer. In Iran, this species has been reared from *Liriomyza trifolii* (Burgess) and *Phytomyza horticola* Goureau (Lotfalizadeh et al., 2015).

Dacnusa (Aphanta) hospita (Foerster, 1863)
Catalogues with Iranian records: Khajeh et al. (2014), Gadallah et al. (2015), Farahani et al. (2016), Cortés et al. (2016), Yu et al. (2016), Samin, Coronado-Blanco, Kavallieratos et al. (2018), Samin, Coronado-Blanco, Fischer et al. (2018).
Distribution in Iran: Fars (Fischer et al., 2011; Khajeh et al., 2014; Lashkari Bod et al., 2010, 2011), Kermanshah (Hazini et al., 2015).
Distribution in the Middle East: Iran.
Extralimital distribution: Bulgaria, China, Denmark, Germany, Hungary, Ireland, Italy, Spain, United Kingdom.
Host records: Summarized by Yu et al. (2016) as being a parasitoid of the agromyzids *Liriomyza bryoniae* (Kaltenbach) and *L. huidobrensis* (Blanchard). In Iran, this species has been reared from the *Phytomyza horticola* Goureau on *Matthiola* sp. (Brassicaceae) (Hazini et al., 2015).
Plant associations in Iran: Wheat field (Lashkari Bod et al., 2011), and *Matthiola* sp. (Hazini et al., 2015).

Dacnusa (Dacnusa) laevipectus Thomson, 1895
Catalogues with Iranian records: No catalogue.
Distribution in Iran: Khuzestan (Samin, Beyarslan, Ranjith et al., 2020).
Distribution in the Middle East: Iran.
Extralimital distribution: Austria, Azerbaijan, Bosnia-Herzegovina, Bulgaria, China, Czech Republic, Faroe Islands, Germany, Hungary, Iceland, Ireland, Italy, Montenegro, Poland, Romania, Russia, Serbia, Spain, Sweden, Switzerland, United Kingdom.
Host records: Summarized by Yu et al. (2016) as being a parasitoid of the agromyzids *Amauromyza labiatarum* (Hendel), *A. verbasci* (Bouché), *Chromatomyia horticola* (Goureau), *C. syngenesiae* Hardy, and *Phytomyza* spp. In Iran, this species has been reared from *Phytomyza horticola* (Samin, Beyarslan, Ranjith et al., 2020).

Dacnusa (Pachysema) metula (Nixon, 1954)
Catalogues with Iranian records: Cortés et al. (2016).

Distribution in Iran: Hormozgan (Cortés et al., 2016).
Distribution in the Middle East: Iran.
Extralimital distribution: Azerbaijan, Hungary, Ireland, Mongolia, Russia, Serbia, Sweden, United Kingdom.
Host records: Recorded by Griffiths (1954, 1967) as being a parasitoid of the agromyzids *Chromatomyia ramose* (Hendel), *C. succisae* (Hering), and *Phytomyza* spp.

Dacnusa (Pachysema) monticola (Foerster, 1863)
Catalogues with Iranian records: Gadallah et al. (2015), Farahani et al. (2016), Cortés et al. (2016), Yu et al. (2016), Samin, Coronado-Blanco, Kavallieratos et al. (2018), Samin, Coronado-Blanco, Fischer et al. (2018).
Distribution in Iran: East of Iran (no specific locality cited) (Yari et al., 2014).
Distribution in the Middle East: Iran.
Extralimital distribution: Denmark, Germany, Hungary, Ireland, Mongolia, Romania, Sweden, Switzerland, Uzbekistan.
Host records: Recorded by Griffiths (1967) as being a parasitoid of the agromyzid *Phytomyza tenella* Meigen.

Dacnusa (Dacnusa) pubescens (Curtis, 1826)
Catalogues with Iranian records: Farahani et al. (2016), Cortés et al. (2016), Yu et al. (2016), Samin, Coronado-Blanco, Kavallieratos et al. (2018), Samin, Coronado-Blanco, Fischer et al. (2018).
Distribution in Iran: West Azarbaijan (Samin et al., 2014).
Distribution in the Middle East: Iran.
Extralimital distribution: Austria, Azerbaijan, Belgium, Bulgaria, Czech Republic, France, Germany, Greece, Hungary, Ireland, Italy, Lithuania, Netherlands, North Macedonia, Poland, Portugal, Russia, Serbia, Slovenia, Spain, Sweden, Switzerland, Ukraine, United Kingdom.
Host records: Summarized by Yu et al. (2016) as being a parasitoid of the agromyzids *Chromatomyia ramosa* (Hendel), *C. sygenesiae* Hardy, *Napomyza carotae* Spencer, *N. cichorii* Spencer, *N. lateralis* (Fallén), *Pegomya hyoscyami* (Panzer), *P. buhriella* Spencer, *P. conyzae* Hendel, *P. picridocecis* Hering, *P. robustella* Hendel, *P. rufipes*, and *P. wahlgreni* Rydén.

Dacnusa (Aphanta) sasakawai Takada, 1977
Catalogues with Iranian records: Gadallah et al. (2015), Farahani et al. (2016), Cortés et al. (2016), Yu et al. (2016), Samin, Coronado-Blanco, Kavallieratos et al. (2018), Samin, Coronado-Blanco, Fischer et al. (2018).
Distribution in Iran: East of Iran (no specific locality cited) (Yari et al., 2014, as *D. distracta* Tobias).
Distribution in the Middle East: Iran.
Extralimital distribution: Germany, Hungary, Italy, Japan, Korea, Mongolia, Russia.

Host records: Summarized by Yu et al. (2016) as being a parasitoid of the agromyzids *Liriomyza bryoniae* (Kaltenbach), and *Phytomyza horticola* Goureau.

Dacnusa (Pachysema) sibirica Telenga, 1935
Catalogues with Iranian records: Khajeh et al. (2014), Gadallah et al. (2015), Cortés et al. (2016), Farahani et al. (2016), Yu et al. (2016), Samin, Coronado-Blanco, Kavallieratos et al. (2018), Samin, Coronado-Blanco, Fischer et al. (2018).
Distribution in Iran: Ardabil (Fathi, 2011), Isfahan, Kerman (Ghahari & Gadallah, 2021), Sistan and Baluchestan (Sedighi et al., 2014), Tehran (Shojai et al., 2003, 2005).
Distribution in the Middle East: Iran.
Extralimital distribution: Armenia, Austria, Azerbaijan, Bulgaria, China, Denmark, Germany, Hungary, Ireland, Italy, Korea, Lithuania, Mongolia, Netherlands, Poland, Portugal, Russia, Spain, Sweden, United Kingdom, Uzbekistan, Vietnam, former Yugoslavia.
Host records: Summarized by Yu et al. (2016) as being a parasitoid of agromyzids of the genera *Chromatomyia*, *Liriomyza*, and *Phytomyza*. In Iran, this species has been reared from *Phytomyza horticola* Goureau (Fathi, 2011; Ghahari & Gadallah, 2021).

Genus *Epimicta* Foerster, 1863
Epimicta marginalis (Haliday, 1839)
Catalogues with Iranian records: Samin, Coronado-Blanco, Kavallieratos et al. (2018).
Distribution in Iran: Lorestan (Samin, Coronado-Blanco, Kavallieratos et al., 2018).
Distribution in the Middle East: Iran.
Extralimital distribution: Finland, France, Germany, Hungary, Moldova, Poland, Russia, Sweden, Switzerland, Ukraine, United Kingdom.
Host records: Recorded by van Achterberg et al. (2012) as being a parasitoid of the agromyzid *Phytobia carbonaria* (Zetterstedt).

Genus *Exotela* Foerster, 1863
Exotela gilvipes (Haliday, 1839)
Catalogues with Iranian records: Samin, Coronado-Blanco, Kavallieratos et al. (2018).
Distribution in Iran: Kermanshah (Samin, Coronado-Blanco, Kavallieratos et al., 2018).
Distribution in the Middle East: Iran.
Extralimital distribution: Austria, Azerbaijan, Belgium, Denmark, Finland, Hungary, Ireland, Italy, Netherlands, Poland, Russia, Slovakia, Sweden, Switzerland, United Kingdom.
Host records: Summarized by Yu et al. (2016) as being a parasitoid of the agromyzids *Chromatomyia milii* (Kaltenbach), and *Phytomyza ranunculi* (Schrank).

Exotela umbellina (Nixon 1954)
Catalogues with Iranian records: Samin, Coronado-Blanco, Fischer et al. (2018).
Distribution in Iran: Hamadan (Sakenin et al., 2018), Tehran (Ghahari & Beyarslan, 2019).
Distribution in the Middle East: Iran.
Extralimital distribution: Austria, Czech Republic, Germany, Hungary, Ireland, Poland, Portugal, Russia, Sweden, Switzerland, United Kingdom, former Yugoslavia.
Host records: Recorded by Griffiths (1956, 1967, 1984) as being a parasitoid of *Phytomyza* spp.

Genus *Polemochartus* Shulz, 1911
Polemochartus liparae (Giraud, 1863)
Catalogues with Iranian records: Yu et al. (2016), Samin, Coronado-Blanco, Kavallieratos et al. (2018), Samin, Coronado-Blanco, Fischer et al. (2018).
Distribution in Iran: West Azarbaijan (Karimpour, 2013).
Distribution in the Middle East: Iran.
Extralimital distribution: Austria, Azerbaijan, Belgium, former Czechoslovakia, Denmark, France, Germany, Hungary, Italy, Netherlands, Poland, Russia, Spain, Sweden, Switzerland, Tajikistan, Ukraine, United Kingdom.
Host records: Recorded by Fulmek (1968) and others as being a parasitoid of the chloropids *Lipara lucens* Meigen, and *L. similis* Schiner.

Genus *Protodacnusa* Griffiths, 1964
Protodacnusa aridula (Thomson, 1895)
Catalogues with Iranian records: Khajeh et al. (2014), Gadallah et al. (2015), Farahani et al. (2016), Cortés et al. (2016), Yu et al. (2016), Samin, Coronado-Blanco, Kavallieratos et al. (2018), Samin, Coronado-Blanco, Fischer et al. (2018).
Distribution in Iran: Fars (Ghahari, Fischer, Hedqvist et al., 2010).
Distribution in the Middle East: Iran.
Extralimital distribution: Former Czechoslovakia, Germany, Hungary, Korea, Mongolia, Spain, Sweden, Switzerland.
Host records: Recorded by Vidal (1995) as being a parasitoid of the agromyzid *Agromyza nigrella* (Rondani).

Protodacnusa litoralis Griffiths, 1964
Catalogues with Iranian records: Khajeh et al. (2014), Gadallah et al. (2015), Farahani et al. (2016), Cortés et al. (2016), Yu et al. (2016), Samin, Coronado-Blanco, Kavallieratos et al. (2018), Samin, Coronado-Blanco, Fischer et al. (2018).
Distribution in Iran: Qazvin (Ghahari et al., 2011a).
Distribution in the Middle East: Iran.

Extralimital distribution: Azerbaijan, Greece, Hungary, Ireland, Korea, Spain.
Host records: Unknown.

Protodacnusa ruthei Griffiths, 1964
Distribution in the Middle East: Turkey (Beyarslan & Inanç, 2001).
Extralimital distribution: Azerbaijan, Germany, Hungary, Mongolia.
Host records: Unknown.

Protodacnusa tristis (Nees von Esenbeck, 1834)
Catalogues with Iranian records: No catalogue.
Distribution in Iran: Ardabil (Samin, Fischer et al., 2019).
Distribution in the Middle East: Iran (Samin, Fischer et al., 2019), Turkey (Beyarslan & Inanç, 2001).
Extralimital distribution: Austria, Azerbaijan, Czech Republic, France, Germany, Hungary, Ireland, Kazakhstan, Norway, Poland, Russia, Serbia, Spain, Sweden, Switzerland, Tajikistan, Turkmenistan, Ukraine, United Kingdom, Uzbekistan.
Host records: Summarized by Yu et al. (2016) as being a parasitoid of the agromyzids *Agromyza nigrella* (Rondani) and *A. nigripes* Meigen; the chloropids *Chlorops pumilionis* Bjerkander, and *Oscinella frit* (L.).

Genus *Trachionus* Haliday, 1833
Trachionus (Planiricus) hians (Nees von Esenbeck, 1816)
Catalogues with Iranian records: No catalogue.
Distribution in Iran: Qazvin (Samin, Fischer et al., 2019 as *Planiricus hians* (Nees)).
Distribution in the Middle East: Iran.
Extralimital distribution: Austria, Azerbaijan, Belgium, Bulgaria, former Czechoslovakia, Denmark, Finland, France, Germany, Hungary, Ireland, Korea, Mongolia, Netherlands, Norway, Poland, Russia, Spain, Sweden, Switzerland, Ukraine, United Kingdom.
Host records: Recorded by Capek and Hofmann (1997) and van Achterberg et al. (2012) as being a parasitoid of the agromyzids *Cerodontha* sp. and *Phytobia cambii* (Hendel).

Trachionus (Trachionus) mandibularis (Nees von Esenbeck, 1816)
Distribution in the Middle East: Turkey (Yilmaz & Beyarslan, 2008).
Extralmilital distribution: Belgium, Bulgaria, Denmark, France, Germany, Hungary, Ireland, Japan, Korea, Moldova, Mongolia, Netherlands, Poland, Russia, Slovakia, Sweden, Switzerland, Ukraine, United Kingdom.
Host records: Summarized by Yu et al. (2016) as being parasitoid of the agromyzids *Cerodontha* sp., and *Phytobia cerasiferae* Kangas.

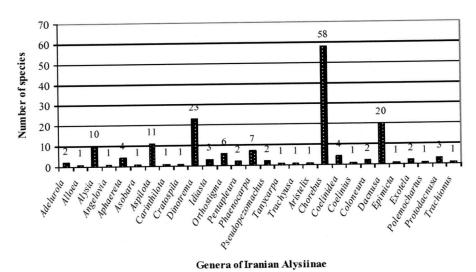

FIGURE 3.1 Number of Alysiinae species within genera known from Iran.

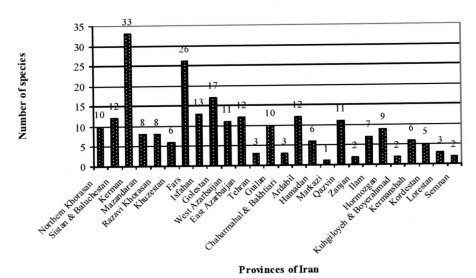

FIGURE 3.2 Number of reported species of Iranian Alysiinae by province.

Conclusion

Two hundred and thirteen species of the subfamily Alysiinae (8.7% of the world species) in 32 genera and two tribes (Alysiini Leach, 1815 and Dacnusini Foerster, 1863) have been reported from the Middle East countries. The genus *Chorebus* Haliday, 1833 is the most diverse, with 64 recorded species. In total, 171 species of Alysiinae in 28 genera have been recorded from Iran so far (7% of the world species). Among the genera, *Chorebus* (with 58 species) is more diverse than the others, followed by *Dinotrema* (with 23 species), and *Dacnusa* (with 20 species), respectively (Fig. 3.1). Eight species are so far only known from Iran (endemic or subendemic to Iran): *Aspilota alfalfae* Fischer et al., 2011, *Aspilota isfahanensis*

Peris-Felipo, 2016, *Aristelix persica* Peris-Felipo, 2015, *Chorebus axillaris* Fischer et al., 2011, *C. longiarticulis* Fischer et al., 2011, *C. nigridiremptus* Fischer et al., 2011, *C. properesam* Fischer et al., 2011 and *C. zarghanensis* Fischer et al., 2011. Iranian Alysiinae have been recorded from 26 provinces, which Kerman and Fars with 33 and 26 recorded species, respectively, have the highest number of species (Fig. 3.2). No record from some provinces especially Alborz, and also a few records from Guilan and Mazandaran provinces (with 10 and eight species, respectively) where comprise various agricultural crops and forest trees and shrubs, indicate that the faunistic surveys on Alysiinae were not conducted systematically in Iran. Host species have been detected for only nine Alysiinae species, which these hosts belong to three families of Diptera, Agromyzidae (three species, nine records), Anthomyiidae (one species), and Chloropidae (one species). *Phytomyza horticola* Goureau, 1851 (Agromyzidae) is the host of seven Alysiinae species in Iran. Comparison of the Alysiinae fauna of Iran with adjacent countries indicates the fauna of Russia with 969 recorded species in 62 genera (Belokobylskij & Lelej, 2019) is more diverse than Iran, followed by Azerbaijan (107 species), Turkey (61 species), Kazakhstan (54 species), Israel–Palestine (33 species), Armenia (26 species), Turkmenistan (13 species), Afghanistan (11 species), Egypt, Iraq, Jordan, Pakistan, and Yemen (each with one species). No species have been recorded from Cyprus, Bahrain, Kuwait, Lebanon, Oman, Qatar, Saudi Arabia, Syria, and United Arab Emirates (Yu et al., 2016). Russia might be expected to have the highest species diversity for Alysiinae because it is the largest country in the world. Furthermore, Russia has had a long history of taxonomic research on Braconidae to better describe its fauna. Additionally, among the 23 countries of the Middle East and adjacent to Iran, Russia shares 107 species with Iran, followed by Azerbaijan (54 species), Turkey (35 species), Israel–Palestine (33 species), Kazakhstan (21 species), Armenia (11 species), Afghanistan (eight species), Turkmenistan (four species), and Egypt, Iraq, Jordan, and Yemen (one species).

References

Amer, N. A., & Hagazi, E. M. (2014). Parasitoids of the leaf miners *Liriomyza* spp. (Diptera: Agromyzidae) attacking faba bean in Alexandria, Egypt. *Egyptian Journal of Biological Pest Control, 24*(2), 301–305.

Aubert, J. F. (1966). Liste d'identification No. 7 (Présentée par le Srvice d'Identification des Entomophages). *Entomophaga, 11*(1), 135–151.

Austin, M. D. (1934). Insect and allied fauna of cultivated mushroom. II. Laboratory investigations. *Annals of Applied Biology, 21*, 167–171.

Belokobylskij, S. A. (2005). On the systematic and distribution of three rare alysiine genera (Hymenoptera: Braconidae: Alysiinae). *Genus, 16*(3), 431–444.

Belokobylskij, S. A., & Kostromina, T. S. (2011). Two late-spring braconid genera of the subfamily Alysiinae (Hymenoptera: Braconidae) new for the fauna of Russia. *Zoosystematica Rossica, 20*(1), 85–95.

Belokobylskij, S. A., & Kula, R. R. (2012). Review of the brachypterous, macropterous, and apterous Braconidae of cyclostome lineage (Hymenoptera: Ichneumonoidea) from the Palearctic region. *Zootaxa, 3240*, 1–62.

Belokobylskij, S. A., & Tobias, V. I. (1997). On the braconid wasps of the subfamily Alysiinae (Hymenoptera, Braconidae) from Kuril Islands. *Far Eastern Entomologist, 47*, 1–17.

Belokobylskij, S. A., & Lelej, A. S. (2019). Annotated catalogue of the Hymenoptera of Russia. Volume II. Apocrita: Parasitica. In *Proceedings of the Zoological Institute of the Russian Academy of Sciences, Supplement No. 8* (p. 594).

Belshaw, R., Fitton, M., Hernion, E., Gimeno, C., & Quicke, D. L. J. (1998). A phylogenetic reconstruction of the Ichneumonoidea (Hymenoptera) based on the D2 variable region of the 28S ribosomal RNA. *Sytematic Entomology, 23*(2), 109–123.

Berry, J. A. (2007). Alysiinae (Insecta: Hymenoptera: Braconidae). *Fauna of New Zealand, 58*, 3–94.

Beyarslan, A., & Inanç, F. (1992). Turkish Alysiinae (Hymenoptera; Braconidae). *Türkiye Entomoloji Komgrei, II*, 661–670.

Beyarslan, A., & Inanç, F. (2001). Ein neuer Beitrag zur Kenntnis der Türkischen Dacnusini Foerster 1862 (Hymenoptera: Braconidae: Alysiinae). *Linzer Biologische Beiträge, 33*(1), 263–268.

Bignell, G. C. (1901). The Ichneumonidae (parasitic flies) of South Devon. Part II. Braconidae. *Transactions of the Devonshire Association for the Advancement of Science, 33*, 657–692.

Burghele, A. D. (1959). New Romanian species of Dacnusini (Hym. Braconidae) and some ecological observations upon them. *Entomologist's Monthly Magazine, 95*, 121–126.

Capek, M., & Hofmann, C. (1997). The Braconidae (Hymenoptera) in the collections of the Musée cantonal de Zoologie, Lausanne. *Litterae Zoologicae (Lausanne), 2*, 25–162.

Chen, X. X., & van Achterberg, C. (2019). Systematics, phylogeny, and evolution of braconid wasps: 30 years of progress. *Annual Review of Entomology, 64*, 1–24.

Cortés, E., Ameri, A., Rakhshani, E., & Peris-Felipo, F. J. (2016). Subfamily Alysiinae Leach, 1815 (Hymenoptera: Braconidae) from Iran: new records and updated list. *Journal of Insect Biodiversity and Systematics, 2*(4), 411–418.

Dowton, A., Austin, A. D., & Antolin, M. F. (1998). Evolutionary relationships among the microgastroid wasps (Hymenoptera: Braconidae): combined analysis of 16S and 28S rDNA genes. *Molecular Phylgenetics and Evolution, 10*(3), 129–150.

Edelsten, H. M., & Todd, R. G. (1912). Notes on the life-histories of *Tapinostola concolor* and *T. hellmanni*. *Entomologist, 45*, 285–287.

Farahani, S., Talebi, A. A., & Rakhshani, E. (2016). Iranian Braconidae (Insecta: Hymenoptera: Ichneumonoidea): diversity, distribution and host association. *Journal of Insect Biodiversity and Systematics, 2*(1), 1–92.

Fathi, S. A. A. (2011). Tritrophic interactions of nineteen canola cultivars - *Chromatomyia horticola* - parasitoids in Ardabil region. *Munis Entomology and Zoology, 6*(1), 449–454.

Fischer, M. (1966). Studien über Alysiinae (Hymenoptera, Braconidae). *Annalen des Naturhistorischen Museums in Wien, 69*, 177–205.

Fischer, M. (1973). *Aspilota*-wespen aus der weiteren Umgebung von Admont (Hym., Braconidae, Alysiinae). *Mitteilungen der Abteilung für Zoologie am Landesmuseum Joanneum, 2*, 137–167.

Fischer, M. (1975). Alysiinen-Wespen aus der Umgebung von Hüttenberg in Kärnten (Hymenoptera, Braconidae, Alysiinae). *Carinthia I, 2*, 303–342.

Fischer, M. (1993). Eine neue Studie über Buckelfliegen-Kieferwespen: *Synaldis* Foerster und *Dinotrema* Foerster (Hymenoptera, Braconidae, Alysiinae). *Linzer Biologiscje Beiträge, 25*(2), 565–592.

Fischer, M. (1995). Über die altweltlixhen *Orthostigma*-Arten und Ergänzungen zur *Aspilota*-Gattungs gruppe (Hymenoptera, Braconidae, Alysiinae). *Linzer Biologische Beiträge, 27*(2), 669–752.

Fischer, M. (2003). Ein Beitrag zur Kenntnis der gattungen *Synaldis* Foerster und *Adelphenaldis* Fischer, gen. nov. (Hymenoptera, Braconidae, Alysiinae). *Linzer Biologische Beiträge, 35*(1), 19–74.

Fischer, M. (2009). Neue Arten der Gattungen *Dinotrema* Foerster, 1862, *Aspilota* Foerster, 1862 and *Eudinostigma* Tobias, 1986 (Insecta: Hymenoptera: Braconidae: Alysiinae). *Annalen des Naturhistorischen Museums in Wien Serie B Botanik und Zoologie, 110,* 103−127.

Fischer, M., & Beyarslan, A. (2012). New species of *Synaldis* Foerster and *Idiasta* Foerster, and further records of Turkish Alysiini (Hymenoptera: Braconidae: Alysiinae). *Zoology in the Middle East, 55*(1), 55−64.

Fischer, M., Lashkari Bod, A., Rakhshani, E., & Talebi, A. A. (2011). Alysiinae from Iran (Insecta: Hymenoptera: Braconidae: Alysiinae). *Annalen des Naturhistorischen Museums in Wien B, 112,* 115−132.

Fischer, M., Sullivan, G. T., Karaca, I., & Ozmoun-Sullivan, S. K. (2014). A new species of the genus *Dinotrema* Foerster (Hymenoptera, Braconidae, Alysiinae) from Turkey. *Turkish Journal of Zoology, 38*(5), 651−654.

Gadallah, N. S., Ghahari, H., Fischer, M., & Peris-Felipo, F. J. (2015). An annotated catalogue of the Iranian Alysiinae (Hymenoptera: Braconidae). *Zootaxa, 3974*(1), 1−28.

Gadallah, N. S., Ghahari, H., & Quicke, D. L. J. (2021). Further addition to the braconid fauna of Iran (Hymenoptera: Braconidae). *Egyptian Journal of Biological Pest Control, 31,* 32. https://doi.org/10.1186/s41938-021-00376-8

Ghahari, H. (2018). Species diversity of the parasitoids in rice fields of northern Iran, especially parasitoids of rice stem borer. *Journal of Animal Environment, 9*(4), 289−298 (in Persian, English summary).

Ghahari, H., Fischer, M., Erdoğan, Ö.Ç., Beyarslan, A., Hedqvist, K. J., & Ostovan, H. (2009). Faunistic note on the Braconidae (Hymenoptera: Ichneumonoidea) in Iranian alfalfa fields and surrounding grasslands. *Entomofauna, 30*(24), 437−444.

Ghahari, H., Fischer, M., Erdoğan, Ö.Ç., Beyarslan, A., & Ostovan, H. (2010). A contribution to the braconid wasps (Hymenoptera: Braconidae) from the forests of northern Iran. *Linzer biologische Beiträge, 42*(1), 621−634.

Ghahari, H., Fischer, M., Hedqvist, K. J., Erdoğan, Ö.Ç., van Achterberg, C., & Beyarslan, A. (2010). Some new records of Braconidae (Hymenoptera) for Iran. *Linzer biologische Beiträge, 42*(2), 1395−1404.

Ghahari, H., & Fischer, M. (2011a). A contribution to the Braconidae (Hymenoptera: Ichneumonoidea) from north-western Iran. *Calodema, 134,* 1−6.

Ghahari, H., & Fischer, M. (2011b). A contribution to the Braconidae (Hymenoptera) from Golestan national Park, northern Iran. *Zeitschrift Arbeitsgemeinschaft Österreichischer Entomologen, 63,* 77−80.

Ghahari, H., & Fischer, M. (2011c). A study on the Braconidae (Hymenoptera: Ichneumonoidea) from some regions of northern Iran. *Entomofauna, 32*(8), 181−196.

Ghahari, H., Fischer, M., & Papp, J. (2011a). A study on the Braconidae (Hymenoptera: Ichneumonoidea) from Qazvin province, Iran. *Entomofauna, 32*(9), 197−208.

Ghahari, H., Fischer, M., & Papp, J. (2011b). A study on the braconid wasps (Hymenoptera: Braconidae) from Isfahan province, Iran. *Entomofauna, 32*(16), 261−272.

Ghahari, H., Fischer, M., & Papp, J. (2011c). A study on the Braconidae (Hymenoptera: Ichneumonoidea) from Ilam province, Iran. *Calodema, 160,* 1−5.

Ghahari, H., Fischer, M., Sakenin, H., & Imani, S. (2011). A contribution to the Agathidinae, Alysinae, Aphidiinae, Braconinae, Microgastrinae and Opiinae (Hymenoptera: Braconidae) from cotton fields and surrounding grasslands of Iran. *Linzer biologische Beiträge, 43*(2), 1269−1276.

Ghahari, H., & Beyarslan, A. (2017). A faunistic study on Braconidae (Hymenoptera: Ichneumonoidea) from Iran. *Natura Somogyiensis, 30,* 39−46.

Ghahari, H., & Beyarslan, A. (2019). A faunistic study on Braconidae (Hymenoptera: Ichneumonoidea) from Iran, and in Memoriam Dr. Jenő Papp (20 May 1933 − 11 December 2017). *Acta Biologica Turcica, 32*(4), 248−254.

Ghahari, H., Beyarslan, A., & Kavallieratos, N. G. (2020). New records of Braconidae (Hymenoptera: Ichneumonoidea) from Iran, and in Memoriam Dr. Maximilian Fischer (7 June 1929 − 15 June 2019). *Scientific Bulletin of Uzhhorod National University (Series: Biology), 48,* 48−55.

Ghahari, H., & Gadallah, N. S. (2021). Additional records to the braconid fauna (Hymenoptera: Ichneumonoidea) of Iran, with new host reports. *Entomological News* (in press).

Ghotbi Ravandi, S., Madjdzadeh, S. M., Peris-Felipo, F. J., Askari Hesni, M., & Rakhshani, E. (2015). A contribution to the Alysiinae of southeastern Iran, with description of the male of *Chorebus longiarticulis*. *Turkish Journal of Zoology, 39*(5), 836−841.

Gimeno, C., Belshaw, R., & Quicke, D. L. J. (1997). Phylogenetic relationships of the Alysiinae/Opiinae (Hymenoptera: Braconidae) and the utility of cytochorome b, 16S and 28S D2 rRNA. *Insect Molecular Biology, 6*(3), 273−284.

Griffiths, G. C. D. (1956). Host records of Dacnusini (Hym. Braconidae) from leaf-mining Diptera. *Entomologist's Monthly Magazine, 92,* 25−30.

Griffiths, G. C. D. (1962). The Agromyzidae (Diptera) of Woodwalton fen. *Entomologist's Monthly Magazine, 98,* 125−155.

Griffiths, G. C. D. (1964). The Alysiinae (Hym. Braconidae) parasites on the Agromyzidae (Diptera). I. General questions of taxonomy, biology and evolution. *Beiträge zur Entomologie, 14*(7−8), 823−914.

Griffiths, G. C. D. (1967). The Alysiinae (Hym. Braconidae) parasites of the Agromyzidae (Diptera). III. The parasites of *Paraphytomyza* Enderlein, *Phytagromyza* Hendel, and *Phytomyza* Fallén. *Beiträge zur Entomologie, 16*(7 & 8), 775−951.

Griffiths, G. C. D. (1968). The Alysiinae (Hym. Braconidae) parasites of the Agromyzidae. V. The parasites of *Liriomyza* Mik and certain small genera of Phytomyzinae. *Beiträge zur Entomologie, 18*(1 & 2), 5−62.

Griffiths, G. C. D. (1984). The Alysiinae (Hym. Braconidae) parasites of the Agromyzidae (Diptera). VII. Supplement. *Beiträge zur Entomologie, 34,* 343−362.

Győrfi, J. (1943). Beiträge zur Kenntnis der Wirte von Schlupwespen. *Zeitschrift für Angewandte Entomologie, 30,* 79−103.

Hazini, F., Zamani, A. A., Peris-Felipo, F. J., Yari, Z., & Rakhshani, E. (2015). Alysiinae (Hymenoptera: Braconidae) parasitoids of the pea leaf miner, *Chromatomyia horticola* (Goureau, 1851) (Diptera: Agromyzidae) in Kermanshah, Iran. *Journal of Crop Protection, 4*(1), 97−108.

Hedqvist, K. J. (1972). Two new species of *Aspilota* Först. (Hym., Ichneumonoidea, Braconidae, Alysiinae). *Entomologisk Tidschrift, 93*(4), 216−219.

Hedwig, K. (1958). Mitteleuropäische schlupwespen und ihre Wirte. *Nachrichten des Naturwissenschaftlichen Museums der Stadt Aschaffenburg, 58,* 21−37.

Hegazy, E., Karem, H., Amer, N., & Khafagi, W. (2014). *Pseudopezomachus masii* Nixon (Hymenoptera: Braconidae: Alysiinae) a newly recorded parasitoid on *Liriomyza* spp. in Egypt. *Egyptian Journal of Biological Pest Control, 24*(1), 163−167.

Hochapfel, H. (1937). Stäekeres Auftreten von *Phytomyza rufipes* in Schlesien. *Anzeiger für Schädlingskunde, 13*, 114−115.

Hussey, N. W. (1960). Biology of mushroom phorids. *Mushroom Science, 4*(1959), 260−270.

Karimpour, Y. (2013). Preliminary study on the biology of the common reed gall-forming fly, *Lipara lucens* (Dip.: Chloropidae), in Urmia region, Iran. *Journal of the Entomological Society of Iran, 33*(3), 11−24.

Khajeh, N., Yari, Z., Rakhshani, E., & Peris-Felipo, F. J. (2014). A regional checklist of Alysiinae (Hymenoptera: Braconidae) from Iran. *Journal of Crop Protection, 3*(4), 1−11.

Kolarov, J., & Bechev, D. (1995). Hymenopteren parasiten (Hymenoptera) auf Pilzmuchen (Mycetophiloidea: Diptera). *Acta Entomologica Bulgarica, 2*, 18−20.

Königsmann, E. (1959). Revision der paläarktischen Arten der Gattung *Dapsilarthra*. 1. Beitrag zur systematischen Bearbeitung der Alysiinae (Hymenoptera: Braconidae). *Beiträge zur Entomologie, 9*, 580−608.

Kostromina, T. S., Timokhov, A. V., & Belokobylskij, S. A. (2016). Braconid wasps of the subfamily Alysiinae (Hymenoptera: Braconidae) as endoparasitoids of *Selachops flavocinctus* Wahlberg, 1844 (Diptera: Agromyzidae) in the central Urals, Russia. *Zootaxa, 4200*(2), 305−319.

Kraaijeveld, A. R., & van der Wel, N. N. (1994). Geographic variation in reproductive success of the parasitoid *Asobara tabida* in larvae of several *Drosophila* species. *Ecological Entomology, 19*(3), 221−229.

Lashkari Bod, A., Rakhshani, E., Talebi, A. A., & Fischer, M. (2010). Identification and introduction of ten new records of Alysiinae (Hym.: Braconidae) from Iran. In *Proceedings of the 19th Iranian Plant Protection Congress, 31 July − 3 August 2010* (p. 160). Iranian Research Institute of Plant Protection.

Lashkari Bod, A., Rakhshani, E., Talebi, A. A., Lozan, A., & Zikic, V. (2011). A contribution to the knowledge of Braconidae (Hym., Ichneumonoidea) of Iran. *Biharean Biologist, 5*(2), 147−150.

Lotfalizadeh, H., Pourhaji, A., & Zargaran, M. R. (2015). Hymenopterous parasitoids (Hymenoptera: Braconidae, Eulophidae, Pteromalidae) of the alfalfa leafminers in Iran and their diversity. *Far Eastern Entomologist, 288*, 1−24.

Naderian, H., Penteado-Dias, A. M., Sakenin Chelav, H., & Samin, N. (2020). A faunistic study on Braconidae and Ichneumonidae (Hymenoptera, Ichneumonoidea) of Iran. *Calodema, 844*, 1−9.

Nixon, G. E. J. (1946). A revision of the European Dacnusini (Hym., Braconidae, Dacnusinae). *Entomologist's Monthly Magazine, 82*, 279−300.

Papp, J. (2012). Contribution to the braconid fauna of Israel (Hymenoptera: Braconidae), 3. *Israel Journal of Entomology, 41−42*, 165−219.

Peris-Felipo, F. J. (2013). Aspilota-*group in Natural Parks of Valencia and European Dinotrema revision* (Ph.D. thesis) (p. 518). Facultat de Ciencies Biologiques.

Peris-Felipo, F. J., & Jiménez-Peydró, R. (2011). Biodiversity within the subfamily Alysiinae (Hymenoptera, Braconidae) in the Natural Park Penas de Aya (Spain). *Revista de Entomologia, 55*(3), 406−410.

Peris-Felipo, F. J., Belokobylskij, S. A., Falcó-Garí, J. V., & Jiménez-Peydró, R. (2014). *Aspilota-*group (Hymenoptera: Braconidae: Alysiinae) diversity in Mediterranean natural Parks of Spain. *Biodiversity Data Journal, 2*, e1112.

Peris-Felipo, F. J., Belokobylskij, S. A., & Jiménez-Peydró, R. (2014). Revision of the western Palaearctic species of the genus *Dinotrema* Foerster, 1862 (Hymenoptera, Braconidae, Alysiinae). *Zootaxa, 3885*(1), 1−483.

Peris-Felipo, F. J., Ameri, A., Talebi, A. A., & Belokobylskij, S. A. (2015). Review of the genus *Aristelix* Nixon, 1943 (Hymenoptera, Braconidae, Alysiinae), with description of a new species from Iran and clarification of the status of *Antrusa chrysogastra* (Tobias, 1986). *Journal of Hymenoptera Research, 45*, 97−111.

Peris-Felipo, F. J., Yari, Z., Rakhshani, E., & Belokobylskij, S. A. (2016). *Aspilota isfahanensis*, a new species of the genus *Aspilota* Foerster, 1863 from Iran (Hymenoptera, Braconidae, Alysiinae). *ZooKeys, 582*, 121−127.

Peris-Felipo, F. J., Yari, Z., van Achterberg, C., Rakhshani, E., & Belokobylskij, S. A. (2016). Review of species of the genus *Adelurola* Strand, 1928, with a key to species (Hymenoptera, Braconidae, Alysiinae). *ZooKeys, 566*, 13−30.

Peris-Felipo, F. J., Ameri, A., Rakhshani, E., & Belokobylskij, S. A. (2016). The genus *Aspilota* Foerster (Hymenoptera: Braconidae: Alysiinae) in western Asia. *Journal of Insect Biodiversity and Systematics, 2*(2), 259−283.

Peris-Felipo, F. J., van Achterberg, C., & Belokobylskij, S. A. (2019). Revision of the Afrotropical *Asobara* Foerster, 1863 (Hymenoptera: Braconidae: Alysiinae), with the description of twenty five new species. *European Journal of Taxonomy, 557*, 1−146.

Peris-Felipo, F. J., & Belokobylskij, S. A. (2020). Species of the subgenus *Synaldis* Foerster, 1863 (Hymenoptera: Braconidae: Alysiinae: *Dinotrema* Foerster, 1863) in Papua New Guinea: descriptions of three new species and a key to the Australasian taxa, pp. 191−208. In T. Robillard, F. Legendre, C. Villemant, & M. Leponce (Eds.), *Insects of Mount Wilhelm, Papua New Guinea* (vol. 2, p. 573). Paris: Muséum national d'Histoire naturelle.

Quicke, D. L. J., & van Achterberg, C. (1990). Phylogeny of the subfamilies of the family Braconidae. *Zoologica Scripta, 21*, 403−416.

Rastegar, J., Sakenin, H., Khodaparast, S., & Havaskary, M. (2012). On a collection of Braconidae (Hymenoptera) from East Azarbaijan and vicinity, Iran. *Calodema, 226*, 1−4.

Rondani, C. (1876). Repertorio degli insetti parasitie e delle Loro Vittime. *Bollettino della Societa Entomologica Italiana, 8*, 54−70.

Rudow, F. (1918). Braconiden und ihre Wirte. *Entomologische Zeitschrift, 32*, 4, 7−8, 11−12, 15−16.

Safahani, S., Iranmanesh, M., Madjdzadeh, S. M., & Peris-Felipo, F. J. (2017). Contribution to the knowledge of Alysiinae (Hymenoptera: Braconidae) of Kerman province, with three new records for Iran. *Journal of Insect Biodiversity and Systematics, 3*(3), 265−271.

Sakenin, H., Coronado-Blanco, M., Samin, N., & Fischer, M. (2018). New records of Braconidae (Hymenoptera) from Iran. *Far Eastern Entomologist, 362*, 13−16.

Sakenin, H., Samin, N., Beyarslan, A., Coronado-Blanco, J. M., Navaeian, M., Fischer, M., & Hosseini Boldaji, S. A. (2020). A faunistic study on braconid wasps (Hymenoptera: Braconidae) from Iran. *Boletin de la SAE, 30*, 96−102.

Samin, N. (2015). A faunistic study on the Braconidae of Iran (Hymenoptera: Ichneumonoidea). *Arquivos Entomoloxicos, 13*, 339−345.

Samin, N., Sakenin, H., Imani, S., & Shojai, M. (2011). A study on the Braconidae (Hymenoptera) of Khorasan province and vicinity, Northeastern Iran. *Phegea, 39*(4), 137—143.

Samin, N., Ghahari, H., Gadallah, N. S., & Davidian, E. (2014). A study on the Braconidae (Hymenoptera: Ichneumonoidea) from West Azarbaijan province, Northern Iran. *Linzer Biologische Beiträge, 46*(2), 1447—1478.

Samin, N., Ghahari, H., Gadallah, N. S., & Monaem, R. (2015). A study on the braconid wasps (Hymenoptera: Ichneumonoidea: Braconidae) from Golestan province, northern Iran. *Linzer biologische Beiträge, 47*(1), 731—739.

Samin, N., van Achterberg, C., & Ghahari, H. (2015). A faunistic study of Braconidae (Hymenoptera: Ichneumonoidea) from southern Iran. *Linzer biologische Beiträge, 47*(2), 1801—1809.

Samin, N., Coronado-Blanco, J. M., Kavallieratos, N. G., Fischer, M., & Sakenin, H. (2018). Recent findings on Braconidae (Hymenoptera: Ichneumonoidea) of Iran with an updated checklist. *Acta Biologica Turcica, 31*(4), 160—173.

Samin, N., Coronado-Blanco, J. M., Fischer, M., van Achterberg, C., Sakenin, H., & Davidian, E. (2018). Updated checklist of Iranian Braconidae (Hymenoptera: Ichneumonoidea) with twenty-three new records. *Natura Somogyiensis, 32*, 21—36.

Samin, N., Papp, J., & Coronado-Blanco, J. M. (2018). A faunistic study on braconid wasps (Hymenoptera: Ichneumonoidea: Braconidae) of Iran. *Scientific Bulletin of the Uzhgorod University (Series: Biology), 45*, 15—19.

Samin, N., Coronado-Blanco, J. M., Hosseini, A., Fischer, M., & Sakenin Chelav, H. (2019). A faunistic study on the braconid wasps (Hymenoptera: Braconidae) of Iran. *Natura Somogyiensis, 33*, 75—80.

Samin, N., Fischer, M., Sakenin, H., Coronado-Blanco, J. M., & Tabari, M. (2019). A faunistic study on Agathidinae, Alysiinae, Doryctinae, Helconinae, Microgastrinae, and Rogadinae (Hymenoptera: Braconidae), with eight new country records. *Calodema, 734*, 1—7.

Samin, N., Beyarslan, A., Coronado-Blanco, J. M., & Navaeian, M. (2020). A contribution to the braconid wasps (Hymenoptera: Braconidae) from Iran. *Natura Somogyiensis, 35*, 25—28.

Samin, N., Sakenin Chelav, H., Ahmad, Z., Penteado-Dias, A. M., & Samiuddin, A. (2020). A faunistic study on the family Braconidae (Hymenoptera: Ichneumonoidea) from Iran. *Scientific Bulletin of Uzhhorod National University (Series: Biology), 48*, 14—19.

Samin, N., Beyarslan, A., Ranjith, A. P., Ahmad, Z., Sakenin Chelav, H., & Hosseini Boldaji, S. A. (2020c). A faunistic study on Braconidae (Hymenoptera: Ichneumonoidea) from Ardebil and East Azarbayjan provinces, Northwestern Iran. *Egyptian Journal of Plant Protection Research Institute, 3*(4), 955—963.

Sedighi, S., Madjdzadeh, S. M., & Rakhshani, E. (2014). A survey on Alysiinae (Hymenoptera: Braconidae) associated with alfalfa in central part of Sistan & Baluchistan province. In *Proceedings of the third Integrated pest management Conference (IPMC), 21—22 January 2014, Kerman* (p. 506).

Shenefelt, R. D. (1974). *Braconidae 7: Alysiinae. Hymenopterorum catalogus. Nova Edito, Pars II* (pp. 937—1113).

Shi, M., Chen, X. X., & van Achterberg, C. (2005). Molecular phylogeny of the Aphidiinae (Hymenoptera: Braconidae) based on the DNA sequences of the 16S rRNA and ATPase 6 genes. *European Journal of Entomology, 102*(2), 133—138.

Shojai, M., Ostovan, H., Zamanizadeh, H., Labbafi, Y., Nasrollahi, A., Ghasemzadeh, M., & Rajabi, M. Z. (2003). The management on the intercropping of cucumber and tomato, with the implementation of non-chemical and reasonably control of pests and diseases for organic crop production in greenhouse. *Journal of the Agricultural Sciences, 9*(2), 1—39 (in Persian, English summary).

Shojai, M., Ostovan, H., Darvish, F., Tirgari, S., Labbafyi, Y., & Rajabi, M. Z. (2005). Technology of biological control and pollination of Iranian cucumber cultivar in protected cultivation and organic production crop. *Journal of the Agricultural Sciences, 11*(1), 69—104 (in Persian, English summary).

Stigenberg, J., & Peris-Felipo, F. J. (2019). Contribution to the knowledge of Swedish Dacnusini (Hymenoptera, Braconidae: Alysiinae): checklist and seven new species records. *Journal of Insect Biodiversity and Systematics, 5*(3), 221—230.

Tobias, V. I., & Jakimavicius, I. G. K. (1986). Alysiinae and Opiinae, pp. 7-231. In G. S. Medvedev (Ed.), *Opredelitel Nasekomych Evrospeiskoi Tsasti SSSR. 3, Peredpontdakrylye. 4. Opr. Faune USSR* (vol. 147, p. 308). Section 3, Part 5.

van Achterberg, C. (1981). The genera of *Aspilota*-group and some descriptions of fungicolous Alysiini from The Netherlands (Hymenoptera: Braconidae: Alysiinae). *Zoologische Verhandelingen, 247*, 1—88.

van Achterberg, C. (1993). Illustrated key to the subfamilies of the Braconidae (Hymenoptera: Ichneumonoidea). *Zoologische Verhandelingen, 283*, 189.

van Achterberg, C., Gumez, J.-L., Martinez, M., & Rasplus, J.-Y. (2012). *Orientopius* Fischer (Hymenoptera, Braconidae, Opiinae) new for Europe, with first notes on its biology and description of a new species. *Journal of Hymenoptera Research, 28*, 123—134.

Vidal, S. (1997). Determination list of entomophagous insects. Nr. 13. *IOBC-WPRS Bulletin, 20*(2), 1—8.

Wharton, R. A. (1984). Biology of the Alysiini (Hymenoptera: Braconidae), parasitoids of the cyclorrhaphous Diptera. Texas Experimental Agricultural Station. *Technical Monograph, 11*, 1—39.

Wharton, R. A. (1986). The braconid genus *Alysia* (Hymenoptera): the description of the subgenera and a revision of the subgenus *Alysia*. *Systematic Entomology, 11*, 453—504.

Woodward, J. (1995). *Phaenocarpa conspurcator* (Haliday) (Hymenoptera: Braconidae) reared from *Scathophaga stercoraria* L. (Diptera: Scathophagidae). *Entomologist, 114*(1), 62—69.

Yari, Z., Khajeh, N., Rahmani, Z., Rakhshani, E., & Peris-Felipo, F. J. (2014). A faunistic study on Alysiinae (Hym.: Braconidae) in Eastern part of Iran. In *Proceedings of the 21st Iranian plant protection Congress, 23—26 August 2014* (p. 749). Urmia University.

Yari, Z., Cortés, E., Peris-Felipo, F. J., & Rakhshani, E. (2016). A faunistic survey on the genus *Chorebus* Haliday (Hymenoptera: Braconidae, Alysiinae, Dacnusini) in Eastern Iran. *Journal of Insect Biodiversity and Systematics, 2*(3), 355—366.

Yildirim, E. M., Civelek, H. S., Cikman, E., Dursun, O., & Eski, A. (2010). Contributions to the Turkish Braconidae (Hymenoptera) fauna with seven new records. *Turkiye Entomoloji Dergisi, 3491*, 29—35.

Yilmaz, T., & Beyarslan, A. (2008). The first record of *Trachionus mandibularis* (Nees, 1816) (Hymenoptera: Braconidae: Alysiinae) in Turkey. *Linzer Biologische Beiträge, 40*(2), 1363—1366.

Yu, D. S., van Achterberg, C., & Horstmann, K. (2016). *Taxapad 2016, Ichneumonoidea 2015, Database on flash-drive. Ottawa, Ontario, Canada.*

Zaldívar-Riverón, A., Areekul, B., Shaw, M. R., & Quicke, D. L. J. (2004). Comparative morphology of the venom apparatus in the braconid wasp subfamily Rogadinae (Insecta, Hymenoptera, Braconidae) and related taxa. *Zoologica Scripta, 33*(3), 223−237.

Zaldívar-Riverón, A., Mori, M., & Quicke, D. L. J. (2006). Systematics of the cyclostomes subfamilies of braconid parasitic wasps (Hymenoptera, Ichneumonoidea): a simultaneous molecular and morphological Bayesian approach. *Molecular Phylogenetics and Evolution, 38*(1), 130−145.

Zhu, J., van Achterberg, C., & Chen, X.-X. (2017). *An illustrated key to the genera and subgenera of the Alysiini (Hymenoptera, Braconidae, Alysiinae), with three genera new for China.*

Aphidius sp. (Aphidiinae), ♀, lateral habitus. *Photo prepared by S.R. Shaw.*

Pauesia sp. (Aphidiinae), ♀, lateral habitus. *Photo prepared by S.R. Shaw.*

Chapter 4

Subfamily Aphidiinae Haliday, 1833

Neveen Samy Gadallah[1], Nickolas G. Kavallieratos[2], Hassan Ghahari[3] and Scott Richard Shaw[4]

[1]Entomology Department, Faculty of Science, Cairo University, Giza, Egypt; [2]Department of Crop Science, Agricultural University of Athens, Athens, Greece; [3]Department of Plant Protection, Yadegar-e Imam Khomeini (RAH) Shahre Rey Branch, Islamic Azad University, Tehran, Iran; [4]UW Insect Museum, Department of Ecosystem Science and Management, University of Wyoming, Laramie, WY, United States

Introduction

Aphidiinae Haliday, 1833 is a cosmopolitan subfamily of the family Braconidae that includes about 657 valid species classified into 63 genera and three tribes (Aphidiini Haliday, 1833; Ephedrini Mackauer, 1961; Praini Mackauer, 1961) (Belshaw & Quicke, 1997; Shi & Chen, 2005; Yu et al., 2016). Aphidiinae are mostly solitary, koinobiont endoparasitoids of nymphal, and adult aphids living exposed on plant surfaces (Shaw & Huddleston, 1991; van Achterberg, 1993). On the other hand, a few genera (i.e., *Aclitus, Lysiphlebus, Paralipsis*) attack root aphids (Shaw & Huddleston, 1991, Starý et al., 1998). They play important roles as biocontrol agents of numerous species of pest aphids (Kocić et al., 2020; Shaw & Huddleston, 1991; Tomanović et al., 2018) in different plant ecosystems (Hagvar & Hofsvang, 1991; Hughes, 1989; Kavallieratos et al., 2005, 2008, 2010, 2013, 2016). Several aphidiine species have been successfully used in biological control programs in different parts of the world (Benelli et al., 2014; Boivin et al., 2012; Bolkmans & Tetteroo, 2002; Desneux & Ramirez-Romero, 2009; González et al., 1992; Grasswitz & Reese, 1998; Hofsvang & Hagvar, 1975; Latham & Milles, 2012; Messing & Klungness, 2002; Petrović et al., 2019; Pons et al., 2011; Starý, 1970, 1975, 2002; Summers, 1998; Takada, 1998; van Lenteren, 2012; van Schelt et al., 1990; Wang & Messing, 2006).

Aphidiinae have been treated as either cyclostomes or noncyclostomes by different authors (e.g., Belshaw et al., 1998; Quicke & van Achterberg, 1990; Shi et al., 2005; van Achterberg & Quicke, 1992); however, on the basis of phylogenetic and molecular analyses, they are now considered as a sister group of cyclostomes (Chen & van Achterberg, 2019; Zaldívar-Riverón et al., 2006). From a taxonomic point of view, although formerly treated as a valid family (Aphidiidae), the aphidiines have been classified as a separate subfamily of Braconidae by most authors for the last 30 years (Chen & van Achterberg, 2019). A recent molecular phylogenetic study by Quicke et al. (2020) proposes that the Aphidiinae, together with the Mesostoinae and Maxfisheriinae, comprise a lineage that is the sister-group to the cyclostomes.

Members of the subfamily Aphidiinae are diagnosed by their small and weakly sclerotized bodies, with rather reduced forewing venation, except in *Ephedrus* and *Toxares*, in which the wing venation remains complete (Rakhshani et al., 2019; Shaw & Huddleston, 1991; van Achterberg, 1993). The presence of a flexible suture between the second and third metasomal tergites is unique among the Braconidae, and their parasitism of aphids is unique among all the Ichneumonoidea, making them easy to diagnose from any other Braconidae or Ichneumonidae if the host is known (Shaw & Huddleston, 1991). Most species pupate inside the dried and mummified host remains, which are thus naturally preserved, and such "aphid mummies" should be archived along with the associated wasp, whenever aphidiines are reared from their host aphids.

Checklists of Iranian Aphidiinae. In the checklist of economically important insects of Iranian agricultural products, Farahbakhsh (1961) listed three aphidiine species: *Lysiphlebus* sp., *Praon palitans* Muesebeck, 1956, and *Trioxys utilis* Muesebeck, 1956. Modarres Awal (1997) treated Aphidiinae as a separate family (Aphidiidae) and listed 15 species in six genera. However, he reported *Praon exsoletum palitans* Muesebeck, 1956 under the family Braconidae. Later, Modarres Awal (2012), still keeping the family status, listed 61 species in 13 genera. Nevertheless, he placed *Binodoxys angelicae* (Haliday, 1833) and 10 species of the genus *Praon* under the family Braconidae. Fallahzadeh and Saghaei (2010) listed 55 aphidiine species in 12 genera as members of Iranian Braconidae. In another catalog of aphid parasitoids and their associated aphid hosts of Iran, Barahoei et al. (2014) listed 78 aphidiine species in 17 genera. Yu et al. (2016) recorded 67 aphidiine species in 11 genera occurring in Iran. In a recent review of the family Braconidae of Iran, Farahani et al. (2016) covered 73 aphidiine species belonging to 15 genera. Updated checklists of Iranian Braconidae have been published by Samin, Coronado-Blanco, Kavallieratos et al. (2018) and Samin, Coronado-Blanco, Fischer et al. (2018) who listed 67 and 68 species, respectively, in 15 genera of the Aphidiinae. In a more recent study, the Aphidiinae of the Middle East and Africa is reviewed by Rakhshani et al. (2019), who listed and provided keys for 108 aphidiine species in 18 genera, of which 76 species have been reported from the fauna of Iran. A recent background information is provided by Rakhshani and Starý (2021) on Aphidiinae of Iran, comprising a rich database of trophic associations based on a thorough review of all taxonomic studies both on primary and secondary aphid parasitoids. In the present chapter, the subfamily Aphidiinae in Iran comprises 92 known species classified into 18 genera and three tribes (Aphidiini, Ephedrini, and Praini). *Ephedrus lacertosus* (Haliday) is recorded for the first time for the Iranian fauna. The genus *Toxares* is represented by a single unidentified species. In the current study, we follow Yu et al. (2016) and Rakhshani et al. (2019) for the classification of Iranian Aphidiinae and Yu et al. (2016) for the global distribution of species. We also supplement Aphidiinae distribution with recent references.

Key to genera of the subfamily Aphidiinae of the Middle East (modified from Rakhshani et al., 2019)

1. Forewing with eight closed cells 2
— Forewing with less than eight closed cells 3
2. Antenna 18-segmented; ovipositor sheath short, triangular in shape .... *Toxares* Haliday
— Antenna 13-segmented; ovipositor sheath elongate .... *Ephedrus* Haliday
3. Notauli complete; vein RS+M of forewing present, sometimes partly or completely pigmented 4
— Notauli absent or incomplete; vein RS+M of forewing absent 5
4. Antenna 13-14-segmented; propodeum areolated; ovipositor sheath densely setose at distal half .... *Areopraon* Mackauer
— Antenna 15-22-segmented; propodeum smooth; ovipositor sheath sparsely setose .... *Praon* Haliday
5. Vein 3RSb of forewing long, reaching vein R1 at wing margin; petiole 0.8× as long as wide at spiracles; ovipositor sheath spike-shaped; eyes reduced in size .... *Aclitus* Foerster
— Vein 3RSb of forewing short, not reaching vein R1; petiole longer than width at spiracles; ovipositor sheath other than spike-shaped; eyes normal 6
6. Last metasomal sternite with a pair of prongs 7
— Last metasomal sternite without prongs 9
7. Petiole with two pairs of tubercles, the anterior pair bearing the spiracles .... *Binodoxys* Mackauer
— Petiole with a single pair of spiracular tubercles 8
8. Apical portion of prongs differentiated, bearing several stout basally dilated bristles .... *Betuloxys* Mackauer
— Apical portion of prongs tubular, with one to two bristles of various shapes .... *Trioxys* Haliday
9. Propodeum reticulated or irregularly areolated; petiole short, subquadrate; antenna moniliform ..... *Monoctonia* Starý
— Propodeum smooth or regularly areolate; petiole elongate; antenna filiform 10
10. Ovipositor sheath ploughshare shaped .... *Monoctonus* Haliday
— Ovipositor sheath of various shapes 11
11. Veins r & RS of forewing extending over the tip of the R1 vein, reaching wing margin; dorsal surface of petiole with a pair of strong carinae diverging backward; ovipositor sheath elongated cup-shape, strongly curved downward .... *Lipolexis* Foerster
— Veins r & Rs of forewing reaching end of vein R1; dorsal surface of petiole smooth; ovipositor sheath variable, upwardly curved 12
12. Dorsal side of propodeum carinated, with complete or incomplete central areola; ovipositor sheath truncated at apex 13
— Dorsal side of propodeum smooth, or with two divergent carinae at lower part; ovipositor sheath pointed apically 16
13. Propodeum with a narrow central areola 14

— Propodeum with wide central pentagonal areola, or only with anterolateral central carina (in the form of an inverted T) 15

14. Veins r-m and M+m-cu of forewing absent .... *Diaeretiella* Starý

— Veins r-m and M+m-cu of forewing present, sometimes vein M+m-cu reduced anteriorly .... *Aphidius* Nees

15. Ovipositor sheath stout, subquadrate; notauli absent; antenna with 15—16 segments .... *Diaeretus* Foerster

— Ovipositor sheath elongated, different shaped; notauli developed anteriorly; antenna with more than 17 segments .... *Pauesia* Quilis

16. Veins M+m-cu and r of forewing absent .... *Adialytus* Foerster

— Veins M+m-cu and r of forewing present, vein M+m-cu incomplete .... *Lysiphlebus* Foerster

List of species of the subfamily Aphidiinae recorded in the Middle East

Subfamily Aphidiinae Haliday, 1833

Tribe Aphidiini Haliday, 1833

Genus *Aclitus* Foerster, 1863

Aclitus obscuripennis Foerster, 1863

Catalogs with Iranian records: Rakhshani et al. (2019).

Distribution in Iran: Mazandaran (Farahani et al., 2017).

Distribution in the Middle East: Iran.

Extralimital distribution: Czech Republic, France, Germany, Hungary, Poland, Serbia.

Host records: Summarized by Yu et al. (2016) as being a parasitoid of the aphid species *Anoecia corni* (Fabricius), *A. vagans* (Koch), *Tuberculatus annulatus* (Hartig), and *T. neglectus* (Krzwiec).

Genus *Adialytus* Foerster, 1863

Adialytus ambiguus (Haliday, 1834)

Catalogs with Iranian records: Fallahzadeh and Saghaei (2010), Barahoei et al. (2014), Yu et al. (2016), Farahani et al. (2016), Samin, Coronado-Blanco, Kavallieratos et al. (2018), Samin, Coronado-Blanco, Fischer et al. (2018), Rakhshani et al. (2019).

Distribution in Iran: East Azarbaijan (Starý, 1979; Starý et al., 2000; Modarres Awal, 2012 as *Lysiphlebus arvicola* Starý), Fars (Barahoei et al., 2013; Rakhshani, Tomanović et al., 2008, Rakhshani, Starý et al., 2012; Taheri & Rakhshani, 2013), Isfahan, Kerman (Asadizade et al., 2014 as *Adialytus arvicola*; Barahoei et al., 2013; Rakhshani, Starý et al., 2012), Kermanshah (Nazari et al., 2012; Rakhshani, Starý et al., 2012 as *Adialytus arvicola*), Kordestan (Nazari et al., 2012), Mazandaran (Modarres Awal, 1997, 2012 as *Lysiphlebus ambiguus* Haliday), Northern Khorasan (Kazemzadeh et al., 2010; Modarres Awal, 2012; Rakhshani, Starý et al., 2012, Rakhshani, Kazemzadeh et al., 2012), Razavi Khorasan (Modarres Awal, 2012; Rakhshani, Starý et al., 2012), Tehran (Modarres Awal, 2012 as *Lysiphlebus ambiguus*), Iran (no specific locality cited) (Arockia Lenin, 2015).

Distribution in the Middle East: Egypt (Starý, 1976, 1981), Iran (see references above), Iraq (Al-Azawi, 1970; Starý, 1969; Starý & Kaddou, 1971), Israel—Palestine (Starý, 1981), Turkey (Starý, 1981).

Extralimital distribution: Widely distributed in the Nearctic, Oriental and Palaearctic regions [Adjacent countries to Iran: Azerbaijan, Kazakhstan, Pakistan, Russia, Turkmenistan].

Host records: Summarized by Yu et al. (2016) as being a parasitoid of the following aphid species: *Acyrthosiphon pisum* Harris, *Aphis* spp., *Atheroides serrulatus* Haliday, *Aulacorthum* sp., *Brachycaudus cardui* L., *B. persicae* Passerini, *B. prunicola* Kaltenbach, *B. tragopogonis* (Kaltenbach), *Brachyunguis atraphaxidis* (Nevsky), *B. harmalae* Das, *Brevicoryne brassicae* (L.), *Capitophorus hippophaes* Walker, *Chaetosiphella berlesei* (del Guercio), *Chromaphis juglandicola* (Kaltenbach), *Dysaphis devecta* (Walker), *Dysaphis ranunculi* (Kaltenbach), *Hydaphias hofmanni* Börner, *Laingia psammae* Theobald, *Macrosiphoniella pulvera* (Walker), *M. tapuskae* (Hottes and Frison), *Macrosiphum* sp., *Melanaphis donacis* (Passerini), *M. sacchari* (Zehntner), *Metopeurum fuscoviride* Stroyan, *Protaphis* sp., *Rhopalosiphum maidis* (Fitch), *R. padi* (L.), *Rungsia* sp., *Semiaphis dauci* (Fabricius), *Sipha elegans* del Gurcio, *S. glyceriae* (Kaltenbach), *Sitobiun miscanthi* (Takahashi), *Staticobium* sp., *Toxoptera aurantii* (Boyer de Fonscolombe), and *Xerophilaphis* sp.

Parasitoid—aphid—plant associations in Iran: Hemiptera: Aphididae: *Sipha elegans* del Guercio on *Hordeum vulgare*, and *Triticum aestivum* (Barahoei et al., 2013; Rakhshani, Tomanović et al., 2008, Rakhshani, Starý et al., 2012; Taheri & Rakhshani, 2013), *Sipha flava* (Forbes) on *Agropyron repens* (Nazari et al., 2012; Rakhshani, Starý et al., 2012), *Sipha maydis* Passerini on *Agropyrum* sp., *Avena fatua*, *Bromus tectorum*, *Cynodon dactylon*, and *Sorghum halepense* (Barahoei et al., 2013; Nazari et al., 2012; Rakhshani, Starý et al., 2012, Rakhshani, Kazemzadeh et al., 2012; Starý, 1979; Starý et al., 2000), *Sipha* sp. (Starý, 1979).

Adialytus salicaphis (Fitch, 1855)
Catalogs with Iranian records: Fallahzadeh and Saghaei (2010), Barahoei et al. (2014), Yu et al. (2016), Farahani et al. (2016), Samin, Coronado-Blanco, Kavallieratos et al. (2018), Samin, Coronado-Blanco, Fischer et al. (2018), Rakhshani et al. (2019).

Distribution in Iran: Alborz (Heidari et al., 2004; Rakhshani, Talebi, Starý et al., 2007; Rakhshani, Starý et al., 2012; Starý, 1979; Starý et al., 2000), Fars (Rakhshani, Talebi, Starý et al., 2007; Rakhshani, Starý et al., 2012; Taheri & Rakhshani, 2013), Golestan (Ghahari et al., 2011; Barahoei, Sargazi et al., 2012), Guilan (Rakhshani, Talebi, Starý et al., 2007; Starý, 1979; Talebi et al., 2009; Yaghubi, 1997; Yaghubi & Sahragard, 1998), Isfahan (Barahoei et al., 2013; Rakhshani, Talebi, Starý et al., 2007), Kerman (Barahoei, Madjdzadeh et al., 2012, 2013; Rakhshani, Starý et al., 2012), Kermanshah (Nazari et al., 2012), Markazi (Alikhani et al., 2013; Barahoei et al., 2013; Rakhshani, Talebi, Starý et al., 2007; Rakhshani, Starý et al., 2012), Kordestan (Nazari et al., 2012; Rakhshani, Talebi, Starý et al., 2007, Rakhshani, Starý et al., 2012), Mazandaran (Ghahari et al., 2010; Rakhshani, Talebi, Starý et al., 2007), Northern Khorasan (Kazemzadeh et al., 2010; Rakhshani, Starý et al., 2012; Rakhshani, Kazemzadeh et al., 2012; Starý, 1979; Starý et al., 2000), Qazvin (Rakhshani, Talebi, Starý et al., 2007; Talebi et al., 2009), Sistan and Baluchestan, Tehran (Rakhshani, Talebi, Starý et al., 2007; Rakhshani, Starý et al., 2012; Talebi et al., 2009), Zanjan (Rakhshani, Talebi, Starý et al., 2007), Iran (no specific locality cited) (Arockia Lenin, 2015; Modarres Awal, 2012; Rahimi et al., 2010; Rakhshani, Talebi, Manzari et al., 2006).

Distribution in the Middle East: Iran (see references above), Iraq (Starý & Kaddou, 1971), Turkey (Çetin Erdoğan et al., 2008; Ölmez & Ulusoy, 2003).

Extralimital distribution: Nearctic, Neotropical, Oriental and Palaearctic regions [Adjacent countries to Iran: Kazakhstan, Pakistan, and Russia].

Host records: Summarized by Yu et al. (2016) as being a parasitoid of *Aphis craccivora* Koch, *A. fabae* Scopoli, *A. farinosa* Gmelin, *A. illinoisensis* Shimer, *A. spiraecola* Patch, *Chaitophorus* spp., *Cryptomyzus ribis* (L.), *Periphyllus* sp., and *Uroleucon ambrosiae* (Thomas).

Parasitoid−aphid−plant associations in Iran: Hemiptera: Aphididae: *Chaitophorus euphraticus* Hodjat on *Populus euphratica* (Rakhshani, Talebi, Starý et al., 2007; Rakhshani, Starý et al., 2012; Starý, 1979; Starý et al., 2000; Talebi et al., 2009), *Chaitophorus leucomelas* Koch on *Populus nigra*, and *Populus* sp. (Barahoei et al., 2013; Heidari et al., 2004; Rakhshani, Talebi, Starý et al., 2007; Rakhshani, Starý et al., 2012; Taheri & Rakhshani, 2013), *Chaitophorus pakistanicus* Hille Ris Lambers on *Salix alba* (Alikhani et al., 2013), *C. populeti* (Panzer) on *Populus alba*, and *Populus nigra* (Barahoei et al., 2013; Rakhshani, Talebi, Starý et al., 2007; Rakhshani, Starý et al., 2012), *Chaitophorus populialbae* (Boyer de Fonscolombe) on *Popolus alba* (Barahoei et al., 2013; Rakhshani, Talebi, Starý et al., 2007; Rakhshani, Starý et al., 2012; Starý et al., 2000; Yaghubi, 1997), *Chaitophorus remaudierei* Pintera on *Salix alba*, and *S. nubica* (Nazari et al., 2012; Rahimi et al., 2010; Rakhshani, Talebi, Starý et al., 2007; Rakhshani, Starý et al., 2012; Starý, 1979; Starý et al., 2000; Talebi et al., 2009), *Chaitophorus salijaponicus* Essig and Kuwana on *S. alba*, and *S. nigra* (Barahoei et al., 2013; Nazari et al., 2012; Rahimi et al., 2010; Rakhshani, Talebi, Starý et al., 2007; Rakhshani, Kazemzadeh et al., 2012; Starý et al., 2000; Taheri & Rakhshani, 2013), *Chaitophorus salijaponicus niger* Mordvilko on *S. alba*, and *Salix* sp. (Rakhshani, Starý et al., 2012; Starý, 1979), *Chaitophorus truncatus* (Hausmann) on *Salix purpurea* (Rakhshani, Talebi, Starý et al., 2007; Talebi et al., 2009), *Chaitophorus vitellinae* (Schrank) on *Salix alba* (Alikhani et al., 2013; Barahoei et al., 2013; Rakhshani, Talebi, Starý et al., 2007; Rakhshani, Starý et al., 2012; Talebi et al., 2009), *Chaitophorus* sp. on *Populus alba*, *P. nigra*, and *Salix aegyptiaca* (Barahoei, Madjdzadeh et al., 2012; Barahoei et al., 2013; Rakhshani, Talebi, Manzari et al., 2006; Rakhshani, Starý et al., 2012; Rakhshani, Kazemzadeh et al., 2012; Starý, 1979; Starý et al., 2000), aphids on *Populus euphratica* and willow (Modarres Awal, 2012), *Sipha maydis* Passerini on *Cynodon dactylum* (Barahoei, Madjdzadeh et al., 2012).

Adialytus thelaxis (Starý, 1961)
Catalogs with Iranian records: Fallahzadeh and Saghaei (2010), Barahoei et al. (2014), Yu et al. (2016), Farahani et al. (2016), Samin, Coronado-Blanco, Kavallieratos et al. (2018), Samin, Coronado-Blanco, Fischer et al. (2018), Rakhshani et al. (2019).

Distribution in Iran: Golestan (Barahoei, Sargazi et al., 2012; Rakhshani, Starý et al., 2012; Starý, 1979; Starý et al., 2000), Guilan (Modarres Awal, 2012; Rakhshani, Starý et al., 2012), Kermanshah (Nazari et al., 2012; Starý, 1979; Starý et al., 2000), Mazandaran (Babaee et al., 2000; Modarres Awal, 2012; Starý, 1979; Starý et al., 2000), Razavi Khorasan, Tehran (Modarres Awal, 2012).

Distribution in the Middle East: Iran (see references above), Iraq (Starý, 1969, 1976; Starý & Kaddou, 1971), Israel−Palestine (Mescheloff & Rosen, 1990a), Turkey (Starý, 1976).

Extralimital distribution: Andorra, Bulgaria, Czech Republic, Finland, France, Germany, Hungary, Italy, Moldova, Montenegro, Poland, Russia, Serbia, Slovakia, Spain.

Host records: Summarized by Yu et al. (2016) as being a parasitoid of *Thelaxes dryophila* (Schrank) and *T. suberi* (del Guercio).

Parasitoid—aphid—plant associations in Iran: Hemiptera: Aphididae: *Thelaxes suberi* (del Guercio) on *Quercus castanifolia*, *Q. persica*, and *Quercus* sp. (Babaee et al., 2000; Nazari et al., 2012; Rakhshani, Starý et al., 2012; Starý, 1979; Starý et al., 2000).

Adialytus veronicaecola Starý, 1968

Catalogs with Iranian records: Barahoei et al. (2014), Yu et al. (2016), Farahani et al. (2016), Samin, Coronado-Blanco, Kavallieratos et al. (2018), Samin, Coronado-Blanco, Fischer et al. (2018), Rakhshani et al. (2019).

Distribution in Iran: Isfahan (Rakhshani, Starý et al., 2012).

Distribution in the Middle East: Iran (Rakhshani, Starý et al., 2012), Turkey (Çetin Erdoğan & Akar, 2018).

Extralimital distribution: Bulgaria, Kazakhstan.

Host records: Recorded by Rakhshani et al. (2012) as being a parasitoid of *Aphis craccivora* Koch and *A. gossypii* Glover.

Parasitoid—aphid—plant associations in Iran: Hemiptera: Aphididae: *Aphis craccivora* Koch on *Phaseolus vulgaris*, *A. gossypii* Glover on *Cucurbita pepo*, *Aphis* sp. on *Rubia tinctorum* (Rakhshani, Starý et al., 2012).

Genus Aphidius Nees von Esenbeck, 1819

Aphidius absinthii Marshall, 1896

Catalogs with Iranian records: Modarres Awal (2012), Barahoei et al. (2014), Farahani et al. (2016), Rakhshani et al. (2019).

Distribution in Iran: Ardabil, Fars (Modarres Awal, 2012), Guilan (Modarres Awal, 2012; Rakhshani, Talebi, Starý et al., 2008; Starý et al., 2000; Talebi et al., 2009; Yaghubi, 1997; Yaghubi & Sahragard, 1998), Mazandaran (Modarres Awal, 2012; Starý, 1979; Starý et al., 2000), Northern Khorasan, Kuhgiloyeh and Boyerahmad (Starý, 1979; Starý et al., 2000), Tehran (Starý et al., 2000).

Distribution in the Middle East: Iran (see references above), Israel—Palestine (Rakhshani et al., 2019).

Extralimital distribution: Holarctic, Indo-Malayan.

Host records: Summarized by Yu et al. (2016) as being a parasitoid of the following aphid species: *Aphis glycines* Matsumura, *Brachycaudus persicae* Passerini, *Capitophorus* sp., *Lipaphis erysimi* (Kaltenbach), *Macrosiphoniella* spp., *Macrosiphum* sp., *Melanaphis sacchari* (Zehntner), *Myzus persicae* (Sulzer), *Paczoskia major* Börner, *Phalangomyzus* sp., *Rhopalosiphum padi* (L.), *Sitobion avenae* Fabricius, *Staticobium limonii* (Contarini), *Titanosiphon artemisiae* (Koch), *T. bellicosum* Nevsky, *T. dracunculi* Nevsky, and *Uroleucon inulae* (Ferrari).

Parasitoid—aphid—plant associations in Iran: Hemiptera: Aphididae: *Macrosiphoniella abrotani* (Walker) on *Artemisia annua* L. (Rakhshani, Talebi, Starý et al., 2008; Rakhshani et al., 2011; Starý et al., 2000; Talebi et al., 2009), *Macrosiphoniella artemisiae* (Boyer de Fonscolombe) on *Artemisia absinthium* L. (Rakhshani, Talebi, Starý et al., 2008; Rakhshani et al., 2011; Talebi et al., 2009), *Macrosiphoniella helichrysi* Remaudière on *Helichrysum* sp. (Rakhshani et al., 2011; Starý et al., 2000), *Macrosiphoniella* nr. *macrura* Hille Ris Lambers (Rakhshani et al., 2011), *Macrosiphoniella oblonga* (Mordvilko) on *Artemisia annua* L. (Rakhshani, Talebi, Starý et al., 2008; Rakhshani et al., 2011; Starý et al., 2000; Talebi et al., 2009; Yaghubi, 1997), *Macrosiphoniella pulvera* (Walker) on *Artemisia absinthium* L. (Rakhshani et al., 2011; Starý, 1979; Starý et al., 2000), *Macrosiphoniella riedeli* Szelegiewicz on *Centaurea* sp. (Rakhshani et al., 2011), *Macrosiphoniella tuberculata* (Nevsky) (Modarres Awal, 2012; Rakhshani et al., 2011; Starý, 1979; Starý et al., 2000), *Macrosiphoniella* sp. (Rakhshani et al., 2011; Starý et al., 2000), *Titanosiphon dracunculi* Nevsky on *Artemisia dracunculus* L. (Modarres Awal, 2012).

Comments: This species was formerly synonymized with *Aphidius asteris* Haliday (1834) (Yu et al., 2016); according to Rakhshani et al. (2019), it is considered as a valid species.

Aphidius aquilus Mackauer, 1961

Catalogs with Iranian records: No catalog.

Distribution in Iran: Guilan (Samin, Beyarslan, Coronado-Blanco et al., 2020).

Distribution in the Middle East: Iran (Samin, Beyarslan, Coronado-Blanco et al., 2020).

Extralimital distribution: Bulgaria, Czech Republic, Finland, France, Georgia, Germany, Hungary, Italy, Japan, Latvia, Lithuania, Mongolia, Netherlands, Poland, Russia, Serbia, Slovakia, Spain, United Kingdom.

Host records: Summarized by Yu et al. (2016) as being a parasitoid of the following aphid species: *Aphis pomi* De Geer, *Betulaphis brevipilosa* Börner, *B. quadrituberculata* (Kaltenbach), *Calaphis betulicola* (Kaltenbach), *C. flava* Mordvilko, *Callipterinella calliptera* (von Heyden), *Cavariella aegopodi* (Scopoli), *Euceraphis punctipennis* (Zetterstedt), *Glyphina betulae* (L.), *Monaphis antennata* (Kaltenbach), and *Panaphis* sp.

Parasitoid-aphid-plant associations in Iran: Aphidiinae: *Betulaphis quadrituberculata* (Kaltenbach), plant is unknown (Samin, Beyarslan, Coronado-Blanco et al., 2020).

Aphidius arvensis (Starý, 1960)

Catalogs with Iranian records: Fallahzadeh and Saghaei (2010 as *Lysaphidius arvensis* Starý, 1960), Barahoei et al.

(2014), Yu et al. (2016), Farahani et al. (2016), Samin, Coronado-Blanco, Kavallieratos et al. (2018), Samin, Coronado-Blanco, Fischer et al. (2018), Rakhshani et al. (2019).

Distribution in Iran: Fars (Barahoei et al., 2013; Taheri & Rakhshani, 2013), Golestan (Barahoei, Sargazi et al., 2012), Hamadan (Nazari et al., 2012; Talebi et al., 2009), Kordestan (Nazari et al., 2012), Tehran (Modarres Awal, 2012; Starý, 1979; Starý et al., 2000; all as *Lysaphidius arvensis* Starý, 1960).

Distribution in the Middle East: Iran (see references above), Iraq (Rakhshani et al., 2019).

Extralimital distribution: Bulgaria, Czech Republic, France, Germany, Hungary, Moldova, Montenegro, Poland, Russia, Slovakia, Serbia, Spain.

Host records: Summarized by Yu et al. (2016) as being a parasitoid of *Coloradoa* spp.

Parasitoid−aphid−plant associations in Iran: Hemiptera: Aphididae: *Coloradoa achilleae* Hille Ris Lambers on *Achillea millefolium* L. (Barahoei et al., 2013; Nazari et al., 2012; Starý, 1979; Starý et al., 2000; Taheri & Rakhshani, 2013; Talebi et al., 2009), and *Coloradoa santolinae* Hille Ris Lambers on *Achillea millefolium* (Starý, 1979; Starý et al., 2000).

Aphidius artemisicola Tizado and Núñez-Perez, 1994

Catalogs with Iranian records: No catalog.

Distribution in Iran: Golestan (Barahoei, Sargazi et al., 2012).

Distribution in the Middle East: Iran.

Extralimital distribution: Andorra, Montenegro, Russia, Serbia, Spain.

Host records: Unknown.

Aphidius asteris Haliday, 1834

Catalogs with Iranian records: Fallahzadeh and Saghaei (2010), Barahoei et al. (2014), Yu et al. (2016), Farahani et al. (2016), Samin, Coronado-Blanco, Kavallieratos et al. (2018), Samin, Coronado-Blanco, Fischer et al. (2018), Rakhshani et al. (2019).

Distribution in Iran: Ardabil, Fars (Modarres Awal, 2012), Golestan (Barahoei, Sargazi et al., 2012), Guilan (Modarres Awal, 2012; Rakhshani, Talebi, Starý et al., 2008; Starý et al., 2000; Talebi et al., 2009; Yaghubi, 1997), Kuhgiloyeh and Boyerahmad, Northern Khorasan (Starý, 1979; Starý et al., 2000), Mazandaran (Modarres Awal, 2012; Starý, 1979; Starý et al., 2000), Tehran (Starý et al., 2000).

Distribution in the Middle East: Egypt (Abu El-Ghiet et al., 2014), Iran (see references above), Iraq (Starý, 1976), Israel−Palestine (Mescheloff & Rosen, 1990b), Turkey (Aslan et al., 2004; Çetin Erdoğan et al., 2008, 2010).

Extralimital distribution: Holarctic, Indo-Malayan.

Host records: Summarized by Yu et al. (2016) as being a parasitoid of the following aphid species: *Aphis glycines* Matsumura, *Brachycaudus persicae* Passerini, *Capitophorus* sp., *Lipaphis erysimi* (Kaltenbach), *Macrosiphoniella* spp., *Macrosiphum* sp., *Melanaphis sacchari* (Zehntner), *Myzus persicae* (Sulzer), *Paczoskia major* Börner, *Phalangomyzus* sp., *Rhopalosiphum padi* (L.), *Sitobion avenae* Fabricius, *Staticobium limonii* (Contarini), *Titanosiphon artemisiae* (Koch), *T. bellicosum* Nevsky, *T. dracunculi* Nevsky, and *Uroleucon inulae* (Ferrari).

Parasitoid−aphid−plant associations in Iran: Hemiptera: Aphididae: *Macrosiphoniella* spp., plant is unknown (Barahoei et al., 2014).

Aphidius avenae Haliday, 1834

Catalogs with Iranian records: Barahoei et al. (2014), Farahani et al. (2016), Rakhshani et al. (2019).

Distribution in Iran: Fars (Taheri & Rakhshani, 2013), Hormozgan (Ameri et al., 2019); Hamadan province, Shirinsu, 3♀, 1♂, April 2014, ex. *Sitobion aveanae* (Fabricius) (Hemiptera: Aphididae) on *Avena sativa* (Poaceae); Kordestan province, Bijar, 2♀, 2♂, June 2016, ex. *Acyrthosiphon pisum* (Harris) (Hemiptera: Aphididae).

Distribution in the Middle East: Iran (Ameri et al., 2019; Taheri & Rakhshani, 2013), Lebanon (Tremblay et al., 1985), Turkey (Çetin Erdoğan et al., 2008, 2010; Akar & Çetin Erdoğan, 2017), Saudi Arabia (Rakhshani et al., 2019).

Extralimital distribution: Indo-Malayan, Nearctic, Palaearctic regions. Introduced into Brazil, Burundi, Chile and United States of America [Adjacent countries to Iran: Russia].

Host records: Summarized by Yu et al. (2016) as being a parasitoid of the following aphid species: *Acyrthosiphon auctum* (Walker), *A. caraganae* (Cholodkovsky), *A. cypharissiae* (Koch), *A. malvae* (Mosley), *A. nigripes* Hille Ris Lambers, *A. pisum* (Harris), *Amphorophora rubi* (Kaltenbach), *Aphis gossypii* Glover, *A. spiraephaga* Müller, *A. ulmariae* Schrank, *Aulacorthum solani* (Kaltenbach), *Brachycaudus cardui* (L.), *B. helichrysis* (Kaltenbach), *B. lychidis* (L.), *B. tragopogonis* (Kaltenbach), *Brevicoryne brassicae* (L.), *Capitophorus carduina* (Walker), *Chaetosiphon glabrum* David, Rajasingh and Narayanan, *Diuraphis noxia* (Mordvilko), *Dysaphis crithmi* (Buckton), *D. plantaginea* (Passerini), *D. ranunculi* (Kaltenbach), *Haydaphis foeniculi* Passerini, *Hyalopterus pruni* (Geoffroy), *Hyperomyzus lactucae* (L.), *Longicaudus trirhodus* (Walker), *Macrosiphum* spp., *Metopolophium dirhodum* (Walker), *M. festucae* (Theobald), *Microlophium carnosum* (Buckton), *Myzaphis turanica* Nevsky, *Myzus certus* (Walker), *M. persicae* (Sulzer), *Nasonovia ribisnigri* (Mosley), *Phorodon humuli* (Schrank), *Prociphilus xylostei*

(De Geer), *Rhopalomyzus lonicerae* (Siebold), *Rhopalosi-phum nymphaeae* (L.), *R. padi* (L.), *Schizaphis graminum* (Rondani), *S. longicaudata* Hille Ris Lambers, *Sitobion akebiae* (Shinji), *S. avenae* (Fabricius), *S. fragariae* (Walker), *S. ibarae* (Matsumura), *S. rosaeiforme* Das, *Toxoptera aurantii* Boyer de Fonscolombe, *Uroleucon sonchi* (L.), and *Wahlgreniella ossiannilssoni* Hille Ris Lambers. Parasitoid—aphid—plant associations in Iran: Hemiptera: Aphididae: *Acyrthosiphon pisum* (Harris) (present work) plant is unknown, *Myzus persicae* (Sulzer) on *Capsicum annum*, and *Prunus persica*, *Rhopalosiphum maidis* (Fitch) on *Hordeum vulgare* (Taheri & Rakhshani, 2013), *Sitobion aveanae* (Fabricius) on *Avena sativa* (present work).

Aphidius banksae Kittel, 2016

Distribution in the Middle East: Israel—Palestine (Tomanović et al., 2018), Turkey (Chen et al., 1991 as *A. staryi*). Extralimital distribution: Belgium, France, Greece, Montenegro, Serbia, Slovenia, United Kingdom.
Host records: Recorded by Chen et al. (1991) as being a parasitoid of *Acyrthosiphon pisum* Harris.

Aphidius colemani Viereck, 1912

Catalogs with Iranian records: Fallahzadeh and Saghaei (2010), Barahoei et al. (2014), Yu et al. (2016), Samin, Coronado-Blanco, Kavallieratos et al. (2018), Samin, Coronado-Blanco, Fischer et al. (2018), Rakhshani et al. (2019).
Distribution in Iran: Alborz (Rakhshani, Talebi, Manzari et al., 2006; Rakhshani, Talebi, Starý et al., 2008; Talebi et al., 2009), Fars (Barahoei et al., 2013; Rakhshani, Talebi, Starý et al., 2008; Rakhshani, 2012; Starý et al., 2000; Taheri & Rakhshani, 2013), Golestan (Barahoei, Sargazi et al., 2012; Rakhshani, Tomanović et al., 2008; Ghahari & Fischer, 2011), Hamadan (Rajabi Mazhar et al., 2010), Isfahan (Barahoei et al., 2013; Rakhshani, Talebi, Starý et al., 2008; Rakhshani, Tomanović et al., 2008; Rakhshani et al., 2010; Talebi et al., 2009), Kerman (Takallozadeh, 2003; Barahoei, Madjdzadeh et al., 2012; Barahoei et al., 2013 as *Aphidius* cf. *colemani*), Kermanshah (Bagheri-Matin et al., 2008; Nazari et al., 2012; Rakhshani, Tomanović et al., 2008), Khuzestan (Mossadegh et al., 2010, 2011; Nazari et al., 2012; Salehipour et al., 2010; Talebi et al., 2009), Kuhgiloyeh and Boyerahmad (Barahoei et al., 2013), Kordestan (Nazari et al., 2012; Rakhshani, Talebi, Starý et al., 2008; Rakhshani, Tomanović et al., 2008), Markazi (Alikhani et al., 2013), Northern Khorasan (Kazemzadeh et al., 2010; Rakhshani, Tomanović et al., 2008; Rakhshani, Kazemzadeh et al., 2012), Razavi Khorasan (Rakhshani, Talebi, Starý et al., 2008; Darsouei et al., 2011b), Qazvin (Modarres Awal, 2012; Starý, 1979), Qom (Rakhshani, Tomanović et al., 2008), Sistan and Baluchestan

(Bandani, 1992; Bandani et al., 1993; Modarres Awal, 1997, 2012; Modarres Najafabadi & Gholamian, 2007; Rakhshani et al., 2005a, 2006a, 2008a, b; Starý et al., 2000; Talebi et al., 2009), Tehran (Rakhshani, 2012; Rakhshani, Talebi, Kavallieratos et al., 2005; Rakhshani, Talebi, Starý et al., 2008; Rakhshani, Tomanović et al., 2008; Talebi et al., 2009), Iran (no specific locality cited) (Tazerouni et al., 2011b).
Distribution in the Middle East: Egypt (Abdel-Rahman, 2005; Abdel-Rahman et al., 2000; Abu El-Ghiet et al., 2014; El-Gantiry et al., 2012; El-Heneidy & Abdel-Samad, 2001; El-Heneidy et al., 2001, 2002, 2003, 2004; Gadallah et al., 2017; Ghanim & El-Adl, 1983; Salman, 2006; Sobhy et al., 2004; Starý, 1976), Iran (see references above), Iraq (Al-Azawi, 1970; Starý, 1976; Starý & Kaddour, 1971), Israel—Palestine (Avidov & Harpaz, 1969; Mescheloff & Rosen, 1990b; Rosen, 1967, 1969; Samara & Qubbaj, 2012), Jordan (Irshaid & Hasan, 2011; Hasan, 2016), Lebanon (Abou-Fakhr, 1982; Abou-Fakhr & Kawar, 1998; Starý, 1976; Tremblay et al., 1985), Syria (Starý, 1976), Turkey (Akar & Çetin Erdoğan, 2017; Çetin Erdoğan et al., 2008; Ölmez & Ulusoy, 2003; Starý, 1976; Tozlu et al., 2002), United Arab Emirates (Starý et al., 2013), Yemen (Starý & Erdelen, 1982; Starý et al., 2013).
Extralimital distribution: Nearly cosmopolitan species; introduced into Australia (Victoria), Brazil, Czech Republic, Tonga, United States of America.
Host records: Summarized by Yu et al. (2016) as being a parasitoid of the following aphid species: *Acyrthosiphon gossypii* Mordvilko, *A. pisum* (Harris), *Aphis* spp., *Aulacorthum solani* (Kaltenbach), *Brachycaudus* spp., *Brevicoryne brassicae* (L.), *Capitophorus elaeagni* (del Guercio), *C. mitegoni* Eastop, *Cavariella aegopodii* (Scopoli), *Chaitophorus* sp., *Corylobium avellanae* (Schrank), *Diuraphis noxia* (Modvilko), *Dysaphis apiifolia* (Theobald), *D. tulipae* (Boyer de Fonscolombe), *Eucarazzia elegans* (Ferrari), *Hayhurstia atriplicis* (L.), *Hyadaphis coriandri* (Das), *H. foeniculi* Passerini, *Hyalopterus pruni* (Geoffroy), *Hyperomyzus carduellinus* (Theobald), *H. lactucae* (L.), *Hysteroneura setariae* (Thomas), *Impatientinum asiaticum* Nevsky, *Ipuka dispersum* (van der Goot), *Lipaphis erysimi* (Kaltenbach), *Macrosiphoniella sanborni* (Gillette), *Macrosiphum euphorbiae* (Thomas), *M. rosae* (L.), *Melanaphis donacis* (Psserini), *Metopolophium dirhodum* (Walker), *M. festucae* (Theobold), *Microparsus vignaphilus* (Blanchard), *Myzus cerasi* (Fabricius), *M. ornatus* Laing, *M. persicae* (Sulzer), *Nasonovia ribisnigri* (Mosley), *Pentalonia nigronervosa* Coquerel, *Phorodon humuli* (Schrank), *Pterocomma populeum* (Kaltenbach), *Rhopalosiphum maidis* (Fitch), *R. nymphaeae* (L.), *R. padi* (L.), *R. rufiabdominalis* (Sasaki), *Schizaphis eastopi* van Harten and Ilharco, *S.*

graminum (Rondani), *S. rotundiventris* (Signoret), *Sitobion avenae* (Fabricius), *S. fragariae* (Walker), *Thelaxes suberi* (del Guercio), *Toxoptera aurantii* Boyer de Fonscolombe, *T. citricida* Kirkaldy, *Uroleucon aeneum* Hille Ris Lambers, *U. ambrosiae* (Thomas), *U. bereticum* (Blanchard and Everard Eel), and *U. erigeronense* (Thomas).

Parasitoid—aphid—plant associations in Iran: Hemiptera: Aphididae: *Amegosiphon platicaudum* (Narzikulov) on *Berberis thunbergii* (Nazari et al., 2012), *Aphis craccivora* Koch on *Alhagi maurorum, Calendula officinalis, Capsella bursa-pastoris, Francocuria* sp., *Gerbera jamesonii, Hedera helix, Hibiscus esculentus, Medicago sativa, Polygonum* sp., *Robinia pseudoacacia, Rumex* sp., *Solanum nigrum*, and *Vicia faba* (Barahoei, Madjdzadeh et al., 2012; Barahoei et al., 2013; Rakhshani, Talebi, Kavallieratos et al., 2005; Rakhshani, Talebi, Manzari et al., 2006; Rakhshani, Talebi, Starý et al., 2008; Rakhshani et al., 2010; Taheri & Rakhshani, 2013; Takallozadeh 2003; Talebi et al., 2009), *Aphis acetosae* L. on *Rumex* sp., *Aphis affinis* del Guercio on *Mentha longifolia* (Barahoei et al., 2013 as *Aphidius* cf. *colemani*), *Aphis fabae* Scopoli on *Abelmoschus esculentes, Althaea* sp., *Callendula officinalis, Centaurea iberica, Chenopodium album, Chenopodium* sp., *Citrus aurantiifolia, Rumex* sp., *Solanum lycopersicum*, and *Solanum nigrum* (Barahoei, Madjdzadeh et al., 2012; Barahoei et al., 2013; Rakhshani, Talebi, Starý et al., 2008; Rakhshani, Kazemzadeh et al., 2012; Taheri & Rakhshani, 2013; Talebi et al., 2009), *Aphis gossypii* Glover on *Catalpa bignonioides, Crysanthemum morifolium, Cucumis sativus, Cucurbita pepo, Evonymus* sp., *Hibiscus esculentus, Hibiscus seriacus, Malva neglecta, Rumex* sp., *Salsola* sp., *Solanum alatum*, and *Whitania* sp. (Attarzadeh et al., 2019; Barahoei, Madjdzadeh et al., 2012; Barahoei et al., 2013; Nazari et al., 2012; Rakhshani, Talebi, Starý et al., 2008; Talebi et al., 2009; Zamani et al., 2006, 2007, 2012), *Aphis intybi* Koch on *Calendula officinalis, Aphis nerii* Boyer de Fonscolombe on *Nerium oleander* (Barahoei et al., 2013; Rakhshani, Talebi, Starý et al., 2008; Taheri & Rakhshani, 2013; Talebi et al., 2009), *Aphis plantaginis* Goeze on *Plantago lanceolata* (Barahoei et al., 2013 as *Aphidius* cf. *colemani*), *Aphis punicae* Passerini on *Punica granatum* (Barahoei et al., 2013; Rakhshani, Talebi, Starý et al., 2008; Rakhshani, Kazemzadeh et al., 2012; Taheri & Rakhshani, 2013), *Aphis rumicis* L. on *Rumex acetosella* (Rakhshani, Kazemzadeh et al., 2012), *Aphis solanella* Theobald on *Solanum alatum, Aphis* sp. on *Aster* sp., *Chrysanthemum* sp., and *Solanum* sp. (Barahoei, Madjdzadeh et al., 2012; Mossadegh et al., 2011), *Aphis* sp. on *Plantago lanceolata*, and *Solanum* sp., *Brachycaudus tragopogonis* (Kaltenbach) on *Tragopogon graminifolius* (Barahoei,

Madjdzadeh et al., 2012), *Aphis terricola* Rondani on *Calendula officinalis* (Rajabi Mazhar et al., 2010), *Aphis umbrella* Börner on *Malva neglecta* (Nazari et al., 2012; Rakhshani, Talebi, Starý et al., 2008; Talebi et al., 2009), *Brachycaudus amygdalinus* (Schouteden) on *Prunus persica* (Barahoei et al., 2013; Rakhshani, 2012), *Brachycaudus cardui* (L.) on *Carduus australis*, and *Cirsium arvense* (Barahoei et al., 2013; Rakhshani, Talebi, Starý et al., 2008; Taheri & Rakhshani, 2013; Talebi et al., 2009), *Brachycaudus helichrysi* (Kaltenbach) on *Calendula* sp., and *Calendula officinalis* (Barahoei, Madjdzadeh et al., 2012; Barahoei et al., 2013 as *Aphidius* cf. *colemani*; Rakhshani, Talebi, Starý et al., 2008; Talebi et al., 2009), *Brachycaudus tragopogonis* (Kaltenbach) on *Tragopogon graminifolius* (Barahoei, Madjdzadeh et al., 2012), *Capitophorus elaeagni* (Del Guercio) on *Eleagnus angustifolia* (Rakhshani, Talebi, Starý et al., 2008; Talebi et al., 2009), *Dysaphis pulverina* ssp. *iranica* Stroyan on *Plantago lanceolata, Dysaphis radicola* (Mordvilko) on *Rumex* sp. (Barahoei et al., 2013 as *Aphidius* cf. *colemani*), *Diuraphis noxia* (Kurdjumov) on *Saccharum* sp. (Bandani, 1992; Bandani et al., 1993, Starý et al., 2000), *Hayhurstia atriplicis* (L.) on *Chenopodium album* (Rakhshani, Kazemzadeh et al., 2012), *Hyalopterus amygdali* (Blanchard) on *Prunus dulcis* (Taheri & Rakhshani, 2013), *Hyalopterus pruni* Geoffroy on *Phragmites* sp. (Starý, 1979), *Macrosiphum euphorbiae* (Thomas) on *Calendula officinalis* (Barahoei et al., 2013; Taheri & Rakhshani, 2013), *Macrosiphum rosae* (L.) on *Rosa hybrida* (Taheri & Rakhshani, 2013), *Metopolophium dirhodum* (Walker) on *Rosa damscena*, and *Triticum aestivum* (Alikhani et al., 2013; Barahoei et al., 2013), *Myzus persicae* (Sulzer) on *Althaea* sp., *Capsicum annuum, Convolvulus arvensis, Cucurbita pepo, Datura stramonium, Helianthus annuus, Lycopersicum esculentum, Malva neglecta, Prunus padus, Prunus persica, Sisymbrium* sp., and *Spergula arvensis* (Barahoei et al., 2012a, 2013; Mossadegh et al., 2011; Rakhshani, Talebi, Starý et al., 2008; Taheri & Rakhshani, 2013; Talebi et al., 2009; Rakhshani, 2012), *Phorodon humuli* (Schrank) on *Prunus persica* (Rakhshani, 2012; Rakhshani, Talebi, Starý et al., 2008), *Rhopalosiphum maidis* (Fitch) on *Sorghum bicolor* (Barahoei, Madjdzadeh et al., 2012; Barahoei et al., 2013 as *Aphidius* cf. *colemani*), *Rhopalosiphum padi* (L.) on *Hordeum vulgare, Triticum aestivum*, and *Zea mays* (Barahoei et al., 2013; Nazari et al., 2012; Rakhshani, Talebi, Starý et al., 2008; Rakhshani, Tomanović et al., 2008), *Schizaphis graminum* (Rondani) on *Triticum aestivum* (Bandani et al., 1992, 1993; Barahoei et al., 2013; Modarres Najafabadi & Gholamian, 2007; Rakhshani, Talebi, Starý et al., 2008; Rakhshani, Tomanović et al., 2008; Starý et al., 2000), *Sitobion avenae* (Fabricius) on

Hordeum vulgare, Triticum aestivum, Triticum vulgare, and *Zea mays* (Bandani, 1992; Bandani et al., 1993; Barahoei et al., 2013; Rakhshani, Talebi, Starý et al., 2008; Rakhshani, Tomanović et al., 2008; Starý et al., 2000), *Uroleucon sonchi* (Linnaeus) on *Sonchus uleraceus, Uroleucon* sp. on *Artemisia biennis* (Barahoei, Madjdzadeh et al., 2012).

Aphidius eadyi Starý, Gonzalez and Hall, 1980

Catalogs with Iranian records: Fallahzadeh and Saghaei (2010), Fallahzadeh and Saghaei (2010), Barahoei et al. (2014), Yu et al. (2016), Farahani et al. (2016), Samin, Coronado-Blanco, Kavallieratos et al. (2018), Samin, Coronado-Blanco, Fischer et al. (2018), Rakhshani et al. (2019).

Distribution in Iran: Alborz (González et al., 1978; Rakhshani, Talebi, Manzari et al., 2006; Rakhshani, Talebi et al., 2008; Starý, 1979; Starý et al., 2000), Ardabil (González et al., 1978; Starý, 1979; Starý et al., 2000), East Azarbaijan (Ghahari et al., 2011), Hamadan (Nazari et al., 2012; Rakhshani, Talebi, Manzari et al., 2006; Rakhshani, Talebi et al., 2008), Isfahan (Barahoei et al., 2013; Rakhshani, Talebi, Manzari et al., 2006; Rakhshani, Talebi et al., 2008; Rakhshani et al., 2010), Kermanshah (Nazari et al., 2012; Rakhshani, Talebi, Manzari et al., 2006), Kordestan, Qazvin, Sistan and Baluchestan, Zanjan (Rakhshani, Talebi, Manzari et al., 2006), Markazi (Alikhani et al., 2013; Rakhshani, Talebi, Manzari et al., 2006; Rakhshani, Talebi et al., 2008), Tehran (Modarres Awal, 2012; Rakhshani, Talebi, Starý et al., 2008).

Distribution in the Middle East: Iran (see references above), Turkey (Akar & Çetin Erdoğan, 2017; Çetin Erdoğan et al., 2008; Ölmez and Ulusoy, 2003).

Extralimital distribution: Algeria, Andorra, Bulgaria, Czech Republic, Finland, France, Georgia, Greece, Italy, Kazakhstan, Lithuania, Moldova, Morocco, New Zealand, Russia, Serbia, Slovakia, Slovenia, Spain, Switzerland, United Kingdom, Uzbekistan (Yu et al., 2016). Introduced into Burundi, New Zealand (Autrique et al., 1989; Cameron & Walker, 1989).

Host records: Summarized by Yu et al. (2016) as being a parasitoid of *Acyrthosiphon caraganae* (Cholodkovsky), *A. pisum* Harris, *A. pisum ononis* Koch, *Aphis loti* Kaltenbach, and *Sitobion avenae* (Fabricius).

Parasitoid–aphid–plant associations in Iran: Hemiptera: Aphididae: *Acyrthosiphon pisum* (Harris) on *Medicago sativa, Vicia pannonica,* and *Vicia villosa* (Alikhani et al., 2013; Barahoei et al., 2013; González et al., 1978; Modarres Awal, 2012; Nazari et al., 2012; Rakhshani, Talebi, Manzari et al., 2006; Rakhshani, Talebi et al., 2008; Rakhshani et al., 2010; Starý, 1979; Starý et al., 1980, 2000), *Acrythosiphon* sp. on cahalzion plant (Starý et al., 2000).

Aphidius eglanteriae Haliday, 1834

Distribution in the Middle East: Turkey (Barjadze et al., 2010).

Extralimital distribution: Czech Republic, France, Germany, Hungary, India, Italy, Lithuania, Moldova, Montenegro, Poland, Russia, Serbia, United Kingdom.

Host records: Summarized by Yu et al. (2016) as being a parasitoid of the aphids *Capitophorus formasartemisiae* L., *Chaetosiphon fragaefolii* (Cockerell), *C. totrarhodum* (Walker), *Longicaudus trirhodus* (Walker), *Macrosiphoniella pseudoartemisiae* Shinji, *Myzaphis rosarum* (Kaltenbach), and *Myzus sorbi* Bhattacharya and Chakrabarti.

Aphidius ervi Haliday, 1834

Catalogs with Iranian records: Fallahzadeh and Saghaei (2010), Barahoei et al. (2014), Yu et al. (2016), Farahani et al. (2016), Samin, Coronado-Blanco, Kavallieratos et al. (2018), Samin, Coronado-Blanco, Fischer et al. (2018), Rakhshani et al. (2019).

Distribution in Iran: Alborz (González et al., 1978; Monajemi & Esmaili 1981; Rakhshani, Talebi, Manzari et al., 2006; Rakhshani, Talebi et al., 2008; Shojai, 1998; Starý, 1979; Starý et al., 2000), Ardabil (Razmgou et al., 2002), Fars (Barahoei et al., 2013; Keyhanian et al., 2005; Rakhshani, Talebi, Starý et al., 2008; Rakhshani, Tomanović et al., 2008; Taheri & Rakhshani, 2013), Golestan (Darvish-Mojeni & Bayat-Asadi, 1995; Ghahari, 2019; Keyhanian et al., 2005; Modarres Awal, 1997, 2012; Rakhshani, Talebi, Starý et al., 2008; Rakhshani, Tomanović et al., 2008; Barahoei, Sargazi et al., 2012; Starý et al., 2000), Guilan (Rakhshani, Talebi, Starý et al., 2008; Talebi et al., 2009), Hamadan (Ghahari, 2019; Nazari et al., 2012; Rakhshani, Talebi, Manzari et al., 2006; Rakhshani, Talebi et al., 2008), Hormozgan (Ameri et al., 2019), Isfahan (Barahoei et al., 2013; Mehrparvar et al., 2016; Rakhshani, Talebi, Manzari et al., 2006; Rakhshani, Talebi et al., 2008; Rakhshani et al., 2010), Kerman (Asadizade et al., 2014; Barahoei et al., 2013; Ghotbi Ravandi et al., 2017; González et al., 1978; Modarres Awal, 2012; Starý et al., 2000), Kermanshah (Nazari et al., 2012), Khuzestan (Keyhanian et al., 2005; Khajehzadeh et al., 2006), Kordestan (Mansour Ghazi et al., 2008; Nazari et al., 2012; Rakhshani, Talebi, Starý et al., 2008), Markazi (Alikhani et al., 2013; Keyhanian et al., 2005; Rakhshani, Tomanović et al., 2008), Mazandaran (Keyhanian et al., 2005; Modarres Awal, 1997; Rakhshani, Kazemzadeh et al., 2012), Northern Khorasan (Kazemzadeh et al., 2010; Rakhshani, Kazemzadeh et al., 2012), Qazvin, Zanjan (Rakhshani, Talebi, Manzari et al., 2006), Razavi Khorasan (Darsouei et al., 2011b), Tehran (Modarres Awal, 1997, 2012; Rakhshani, Talebi, Starý et al., 2008; Rasulian, 1985, 1989; Shahrokhi et al., 2004; Starý et al., 2000; Talebi et al., 2009), West Azarbaijan (Ghahari, 2019; Keyhanian et al., 2005), Iran (no specific locality cited) (Khanjani, 2006a).

Distribution in the Middle East: Cyprus (Beyarslan et al., 2017), Egypt (Gadallah et al., 2017), Iran (see references above), Iraq (Al-Azawi, 1970; Starý, 1976), Israel—Palestine (Avidov & Harpaz, 1969; Bodenheimer & Swirski, 1957; Mescheloff & Rosen, 1990b), Lebanon (Abou-Fakhr, 1982; Abou-Fakhr & Kawar, 1998; Mackauer & Starý, 1967; Starý, 1976; Tremblay et al., 1985), Saudi Arabia, Turkey, United Arab Emirates, Yemen (Yu et al., 2016).

Extralimital distribution: Cosmopolitan species (Gadallah et al., 2017; Yu et al., 2016).

Host records: Summarized by Yu et al. (2016) as being a parasitoid of the following aphid species: *Acyrthosiphon* spp., *Amphorophora rubi* (Kaltenbach), *Aphis* spp., *Aspidophorodon longicauda* (Richards), *Aulacorthum solani* (Kaltenbach), *Brachycaudus cardui* L., *B. helichrysi* (Kaltenbach), *B. schwartzi* (Börner), *Brevicoryne brassicae* (L.), *Capitophorus* sp., *Cinara juniperi* (De Geer), *Corylobium avellanae* (Schrank), *Delphiniobium junackianum* (Karsch), *Diuraphis noxia* (Mordvilko), *Dysaphis plantaginea* (Passerini), *Eriosoma ulmi* (L.), *Hayhurstia atriplicis* (L.), *Hyalopterus humilis* (Walker), *H. pruni* (Geoffroy), *Hyperomyzus lactucae* (L.), *H. nigricornis* (Knowlton), *Illinoia liriodendri* (Monell), *Lambersius* sp., *Macrosiphoniella absinthii* (L.), *Macrosiphum* spp., *Megoura viciae* Buckton, *Metopolophium dirhodum* (Walker), *Megoura festucae* (Theobald), *Microlophium carnosum* (Buckton), *Myzus certus* (Walker), *M. persicae* (Sulzer), *M. varians* Davidson, *Nasonovia ribisnigri* (Mosley), *Ovatus crataegarius* (Walker), *Phorodon humuli* (Schrank), *Pleotrichophorus pycnorhysus* Knowlton and Smith, *Rhodobium porosum* (Sanderson), *Rhopalosiphum maidis* (Fitch), *R. padi* (L.), *Schizaphis graminum* (Rondani), *Sitobion avenae* (Fabricius), *S. fragariae* (Walker), *Thereoaphis luteola* (Boerner), *T. trifolii* (Monell), *Uroleucon aeneum* Hille Ris Lambers, *U. ambrosiae* (Thomas), *U. picridis* (Fabricius), *U. sonchi* (L.), and *Wahlgreniella ossinnilssoni* Hille Ris Lambers.

Parasitoid—aphid—plant associations in Iran: Hemiptera: Aphididae: *Acyrthosiphon gossypii* Mordvilko on *Sophora alopecuroides* (Barahoei et al., 2013), *Acyrthosiphon kondoi* Shinji on *Medicago sativa* (Barahoei et al., 2013; González et al., 1978; Nazari et al., 2012; Rakhshani, Talebi, Manzari et al., 2006; Rakhshani, Talebi, Starý et al., 2008; Rakhshani, Tomanović et al., 2008; Starý et al., 2000), *Acyrthosiphon pisum* (Harris) on *Medicago sativa*, *Sophora alopecuroides*, *Vicia pannonica*, and *V. villosa* (Alikhani et al., 2013; Barahoei et al., 2013; González et al., 1978; Modarres Awal, 1997, 2012; Monajemi & Esmaili, 1981; Nazari et al., 2012; Rasulian, 1985; 1989; Rakhshani, Talebi, Manzari et al., 2006; Rakhshani, Talebi, Starý et al., 2008; Rakhshani, Kazemzadeh et al., 2012;

Shojai, 1998; Starý, 1979; Starý & González, 1978; Starý et al., 2000; Taheri & Rakhshani, 2013; Talebi et al., 2009), *Acyrthosiphon* sp. on *Medicago sativa* (Rakhshani et al., 2010), *Aphis gossypii* Glover (Razmgou et al., 2002), *Brevicoryne brassicae* (L.) (Khajehzadeh et al., 2006), *Diuraphis noxia* (Kurdjumov) (Khanjani, 2006a), *Macrosiphum euphorbiae* (Thomas) on *Solanum melongena* (Rakhshani, Talebi, Starý et al., 2008), *Metopolophium dirhodum* (Walker) (Shahrokhi et al., 2004), *Microlophium carnosum* (Buckton) on *Urtica dioica* (Rakhshani, Talebi, Starý et al., 2008; Talebi et al., 2009), *Myzus persicae* (Sulzer) on *Nicotiana tabacum*, and *Orobanche cilicica* (Barahoei et al., 2013; Rakhshani, Talebi, Starý et al., 2008), *Rhopalosiphum padi* (L.) on *Triticum aestivum* (Taheri & Rakhshani, 2013), *Sitobion avenae* (Fabricius) on *Gastridium phleoides*, *Triticum aestivum*, and *Zea mays* (Barahoei et al., 2013; Bazyar et al., 2011; Darvish-Mojeni & Bayat-Asadi, 1995; Modarres Awal, 1997, 2012; Rakhshani, Talebi, Starý et al., 2008), *Wahlgreniella nervata* (Gillette) on *Rosa damascena* (Barahoei et al., 2013), *Macrosiphum rosae* (L.) on *Dipsacus fullonum* (Mehrparvar et al., 2016).

Aphidius funebris Mackauer, 1961

Catalogs with Iranian records: Fallahzadeh and Saghaei (2010), Barahoei et al. (2014), Yu et al. (2016), Farahani et al. (2016), Samin, Coronado-Blanco, Kavallieratos et al. (2018), Samin, Coronado-Blanco, Fischer et al. (2018), Rakhshani et al. (2019).

Distribution in Iran: Fars (Barahoei et al., 2013; Taheri & Rakhshani, 2013), Guilan (Rakhshani, Talebi, Starý et al., 2008; Talebi et al., 2009), Isfahan (Barahoei et al., 2013; Rakhshani, Talebi, Starý et al., 2008; Talebi et al., 2009), Kerman (Barahoei, Madjdzadeh et al., 2012; Barahoei et al., 2013), Kermanshah (Nazari et al., 2012), Markazi (Alikhani et al., 2013), Qazvin (Rakhshani, Talebi, Starý et al., 2008), Tehran (Modarres Awal, 2012; Rakhshani, Talebi, Starý et al., 2008; Starý, 1979; Starý et al., 2000).

Distribution in the Middle East: Iran (see references above), Iraq (Al-Azawi, 1970; Starý, 1976; Starý & Kaddou, 1971), Turkey (Akar & Çetin Erdoğan, 2017; Aslan et al., 2004; Çetin Erdoğan et al., 2008; Düzgüneş et al., 1982; Güz & Kilinçer, 2005).

Extralimital distribution: Algeria, Andorra, Bulgaria, China, Czech Republic, Finland, France, Georgia, Germany, Greece, Hungary, Italy, Japan, Kazakhstan, Moldova, Montenegro, Morocco, Poland, Portugal, Russia, Serbia, Slovakia, Spain, Tajikistan, Turkmenistan, Ukraine, United Kingdom, Uzbekistan, former Yugoslavia.

Host records: Summarized by Yu et al. (2016) as being a parasitoid of the aphids *Acyrthosiphon pisum* Harris, *Aphis fabae* Scopoli, *A. farinose* Gmelin, *Aulacorthum solanae*

(Kaltenbach), *Capitophorus carduina* (Walker), *Linosiphon* sp., *Macrosiphum euphorbiae* (Thomas), *Nasonovia pilosellae* (Börner), *Paczoskia major* Börner, *Uroleucon* spp., and *Uromelan* sp.

Parasitoid—aphid—plant associations in Iran: Hemiptera: Aphididae: *Uroleucon acroptilidis* Kadyrbekov, Renxin and Shao on *Acroptilon repens* (Barahoei et al., 2013), *Uroleucon chondrillae* (Nevsky) on *Chondrilla juncea* (Rakhshani, Talebi, Starý et al., 2008; Starý, 1979; Starý et al., 2000), *Uroleucon cichorii* (Koch) on *Tragopogon graminifolius* (Rakhshani, Talebi, Starý et al., 2008), *Uroleucon compositae* (Theobald) on *Carthamus oxyacanthus*, *C. tinctorius*, and *Cirsium* sp. (Barahoei et al., 2013; Nazari et al., 2012; Rakhshani, Talebi, Starý et al., 2008; Talebi et al., 2009), *Uroleucon erigeronense* (Thomas) on *Conyza canadensis* (Rakhshani, Talebi, Starý et al., 2008; Talebi et al., 2009), *Uroleucon jaceae* (L.) on *Acroptilon repens*, *Carduus onopordiodes*, *Centauria depressa*, and *C. solstitialis* (Barahoei et al., 2013; Modarres Awal, 2012; Nazari et al., 2012; Rakhshani, Talebi, Starý et al., 2008; Taheri & Rakhshani, 2013; Talebi et al., 2009), *Uroleucon sonchi* (L.) on *Sonchus asper* (Barahoei et al., 2013; Rakhshani, Talebi, Starý et al., 2008; Taheri & Rakhshani, 2013), *Uroleucon* sp. on *Acroptilon repens*, *Catharanthus* sp., and *Centaurea iberica* (Alikhani et al., 2013; Barahoei, Madjdzadeh et al., 2012; Barahoei et al., 2013).

Aphidius hieraciorum Starý, 1962
Catalogs with Iranian records: Barahoei et al. (2014), Yu et al. (2016), Farahani et al. (2016), Samin, Coronado-Blanco, Kavallieratos et al. (2018), Samin, Coronado-Blanco, Fischer et al. (2018), Rakhshani et al. (2019).
Distribution in Iran: Khuzestan (Mossadegh et al., 2010, 2011; Nazari et al., 2012; Salehipour et al., 2010).
Distribution in the Middle East: Iran.
Extralimital distribution: Czech Republic, France, Germany, Hungary, Moldova, Montenegro, Netherlands, Poland, Russia, Serbia, Slovakia, Spain, Ukraine.
Host records: Summarized by Yu et al. (2016) as being a parasitoid of the aphids *Aphis schneideri* (Börner), *Aphis triglochinis* Theobald, *Cryptomyzus ribis* (L.), *Nasonovia brachycyclica* Holman, *N. brevipes* (Börner), *N. compositellae* Theobald, *N. dasyphylli* Stroyan, *N. ribisnigri* (Mosley), and *Trama troglodytes* von Heyden.
Parasitoid—aphid—plant associations in Iran: Hemiptera: Aphididae: *Nasonovia ribisnigri* (Mosley) on *Lactuca sativa* (Nazari et al., 2012), *Nasonovia* sp. on *Lactua sativa* (Mossadegh et al., 2011).

Aphidius hortensis Marshall, 1896
Catalogs with Iranian records: No catalog.
Distribution in Iran: Golestan (Ghahari et al. 2020).

Distribution in the Middle East: Iran.
Extralimital distribution: Bulgaria, Canada, Czech Republic, Estonia, Ethiopia, Finland, France, Georgia, Germany, Hungary, India, Italy, Moldova, Netherlands, Poland, Serbia, Slovakia, Spain, United States of America, Ukraine, United Kingdom.
Host records: Summarized by Yu et al. (2016) as being a parasitoid of the aphids *Diuraphis noxia* Kurdjumov, *Drepanosiphum platanoides* (Schrank), *Liosomaphis berberidis* (Kaltenbach), *L. himalayensis* Basu, *Macrosiphum stellariae* Theobald, and *Schizaphis graminum* (Rondani).
Parasitoid—aphid—plant associations: An unknown aphid species (Hemiptera: Aphididae) around wheat fields (Ghahari et al., 2020).

Aphidius iranicus Rakhshani and Starý, 2007
Catalogs with Iranian records: Fallahzadeh and Saghaei (2010), Barahoei et al. (2014), Yu et al. (2016), Farahani et al. (2016), Samin, Coronado-Blanco, Kavallieratos et al. (2018), Samin, Coronado-Blanco, Fischer et al. (2018), Rakhshani et al. (2019).
Distribution in Iran: Guilan (Talebi et al., 2009; Tomanović et al., 2007).
Distribution in the Middle East: Iran.
Extralimital distribution: None.
Host records: Recorded by Tomanović et al. (2007) as a parasitoid of the aphid *Titanosiphon bellinosiphon* Nevsky.
Parasitoid—aphid—plant associations in Iran: Hemiptera: Aphididae: *Titanosiphon bellicosum* Nevsky, 1928 on *Atrtemisia absinthium* (Talebi et al., 2009; Tomanović et al., 2007).

Aphidius matricariae Haliday, 1834
Catalogs with Iranian records: Fallahzadeh and Saghaei (2010), Barahoei et al. (2014), Yu et al. (2016), Farahani et al. (2016), Samin, Coronado-Blanco, Kavallieratos et al. (2018), Samin, Coronado-Blanco, Fischer et al. (2018), Rakhshani et al. (2019).
Distribution in Iran: Alborz (Rakhshani, Talebi, Starý et al., 2008; Starý, 1979; Starý et al., 2000; Talebi et al., 2009), East Azarbaijan (Modarres Awal, 2012; Rakhshani, Talebi, Starý et al., 2008; Starý, 1979; Starý et al., 2000), Fars (Barahoei et al., 2013; González et al., 1992; Radjabi, 1989; Modarres Awal, 1997, 2012; Rakhshani, 2012; Rakhshani, Talebi, Starý et al., 2008; Starý et al., 2000; Taheri & Rakhshani, 2013), Golestan (Barahoei, Sargazi et al., 2012; Ghahari, 2019; Ghahari et al., 2010; Rakhshani, 2012; Sakenin et al., 2012), Guilan (Modarres Awal, 2012; Mokhtari et al., 2000; Shojai et al., 1999; Starý et al., 2000; Talebi et al., 2009; Yaghubi, 1997; Yaghubi & Sahragard, 1998), Hamadan (Nazari et al., 2012; Rakhshani, Talebi, Starý et al., 2008; Rakhshani, Tomanović et al., 2008), Hormozgan (Ameri et al., 2019), Isfahan (Barahoei et al., 2013; Ghahari,

2019; Modarres Awal, 1997, 2012; Radjabi, 1989; Rakhshani, Talebi, Starý et al., 2008; Rakhshani et al., 2010), Kerman (Barahoei, Madjdzadeh et al., 2012; Barahoei et al., 2013; Ghotbi Ravandi et al., 2017), Kermanshah (Bagheri-Matin et al., 2008; Nazari et al., 2012; Rakhshani, 2012; Rakhshani, Tomanović et al., 2008), Khuzestan (Farsi et al., 2012, 2019; Modarres Awal, 2012; Mossadegh et al., 2010, 2011, 2017; Rakhshani, Talebi, Starý et al., 2008; Rakhshani, Tomanović et al., 2008; Rezaei et al., 2006; Sabbaghan & Soleymannejadian, 2007; Salehipour et al., 2010; Talebi et al., 2009), Kordestan (Rakhshani, 2012), Kuhgiloyeh and Boyerahmad (Barahoei et al., 2013; Sabbaghan & Soleymannejadian, 2007), Markazi (Alikhani et al., 2013; Barahoei et al., 2013; Modarres Awal, 1997, 2012; Radjabi, 1989; Rakhshani, Tomanović et al., 2008), Mazandaran (Modarres Awal, 2012; Shojai et al., 1999; Starý et al., 2000), Northern Khorasan (Kazemzadeh et al., 2010; Rakhshani, Kazemzadeh et al., 2012), Qom (Barahoei et al., 2013; Rakhshani, Talebi, Starý et al., 2008; Rakhshani, Tomanović et al., 2008), Razavi Khorasan (Darsouei et al., 2011a,b; Jafari-Ahmadabadi and Modarres Awal, 2012; Jafari-Ahmadabadi et al., 2011; Rakhshani, 2012), Sistan and Baluchestan (Bandani, 1992; Bandani et al., 1993; Fakhireh et al., 2016; Modarres Awal, 1997, 2012; Modarres Najafabadi & Gholamian, 2007; Rakhshani, Talebi, Starý et al., 2008; Rakhshani, Tomanović et al., 2008; Starý et al., 2000; Talebi et al., 2009), Tehran (Modarres Awal, 1997, 2012; Radjabi, 1989; Rakhshani, 2012; Rakhshani, Talebi, Starý et al., 2008; Rakhshani, Tomanović et al., 2008; Shahrokhi et al., 2004; Shojai et al., 2005; Starý, 1979; Starý et al., 2000; Talebi et al., 2009), West Azarbaijan (Samin et al., 2014), Iran (no specific locality cited) (Shojai et al., 2003; Tazerouni et al., 2011b).

Distribution in the Middle East: Cyprus (Beyarslan et al., 2017), Egypt (Abdel-Samad & Ahmad, 2009; Abu El-Ghiet et al., 2014; El-Heneidy, 1994; El-Heneidy & Abdel-Samad, 2001; El-Heneidy et al., 2001, 2002, 2003, 2004; Abdel-Rahman, 2005; Gadallah et al., 2017; Ibrahim, 1990a,b; Mahmoud et al., 2009), Iran (see references above), Iraq (Starý, 1976; Starý & Kaddou, 1971), Israel–Palestine (Avidov & Harpaz, 1969; Mescheloff & Rosen, 1990b; Rosen, 1967, 1969), Lebanon (Abou-Fakhr, 1982; Abou-Fakhr & Kawar, 1998; Tremblay et al., 1985), Turkey (Akar & Çetin Erdoğan, 2017; Aslan et al., 2004; Barjadze et al., 2010; Ölmez & Ulusoy, 2003; Starý, 1976), United Arab Emirates, Yemen (Starý et al., 2013).

Extralimital distribution: Worldwide in distribution (except Australasian region); introduced into Burundi and United States of America.

Host records: Summarized by Yu et al. (2016) as being a parasitoid of the aphids *Acyrthosiphon ghanii* Eastop, *A.*

lambersi Leclant and Remaudière, *A. primulae* Theobald, *Anoecia* sp., *Aphis* spp., *Aphthargelia symphoricarpi* (Thomas), *Aulacorthum solani* Kaltenbach, *Brachycaudus* spp., *Brevicoryne brassicae* L., *Calaphis betulaecolens* (Fitch), *Capitophorus* spp., *Cavariella aegopodii* (Scopoli), *Chaitophorus kapuri* Hille Ris Lambers, *Coloradoa rufomaculata* (Wilson and H.F.), *C. tanacetina* (Walker), *Corylobium avellanae* (Schrank), *Dysaphis* spp., *Eucarazzia elegans* (Ferrari), *Galiobium* sp., *Hayhurstia atriplicis* (L.), *Hyalopterus amygdali* (Blanchard), *H. pruni* (Geoffroy), *Hydaphis hofmanni* Boerner, *Hyperomyzus* sp., *Israelaphis lambersi* Illharco, *Linosiphon asperulophagum* Wimshrust, *L. galii* Mamontova, *Linosiphon galiophagus* (Wimshrust), *Liosomaphis berberidis* (Kaltenbach), *Lipaphis erysimi* (Kaltenbach), *L. lepidii* (Nevsky), *Macrosiphoniella formosartemisiae* Takahashi, *M. pseudoartemisiae* Shinji, *M. sanborni* (Gillette), *Microsiphum euphorbiae* (Thomas), *M. rosae* (L.), *Metopolophium dirhodum* (Walker), *Microlophium carnosum* (Buckton), *Myzus* spp., *Nasonovia aquilegiae* (Essig), *N. ribisnigri* (Mosley), *Neomyzus circumflexus* (Buckton), *Ovatus crataegarius* (Walker), *O. insitus* (Walker), *O. mentharius* (van der Goot), *Paramyzus heracleid* Börner, *Phorodon cannabis* Passerini, *P. humuli* (Schrank), *Rhopalomyzus* sp., *Rhopalosiphum* spp., *Schizaphis garminum* (Rondani), *S. rosazebedoi* (Illharco), *Shinjia orientalis* (Mordvilko), *Sipha burakowskii* (Holman and Szelegiewicz), *Sitobion avenae* (Fabricius), *S. fragariae* (Walker), *Toxoptera aurantii* Boyer de Fonscolombe, *Tubaphis ranunculinus* (Walker), and *Uroleucon* spp.

Parasitoid—aphid—plant associations in Iran: Hemiptera: Aphididae: *Acyrthosiphon gossypii* Mordvilko on *Malva neglecta* (Taheri & Rakhshani, 2013), *Amegosiphon platicaudum* (Narzikulov) on *Berberis thunbergii* (Nazari et al., 2012), *Aphis affinis* del Guercio on *Mentha longifolia* (Barahoei et al., 2013; Nazari et al., 2012; Rakhshani, Kazemzadeh et al., 2012), *Aphis craccivora* Koch on *Capsella bursa-pastoris*, *Centaura* sp., *Gundelia tournefortii*, *Medicago polymorpha*, *Medicago sativa*, *Melilotus officinalis*, *Robinia pseudoacacia*, and *Solanum tuberosum* (Barahoei et al., 2013; Mossadegh et al., 2011; Nazari et al., 2012; Rakhshani, Talebi, Starý et al., 2008; Rakhshani et al., 2010; Taheri & Rakhshani, 2013; Talebi et al., 2009), *Aphis crepidis* (Börner) on *Crepis* sp. (Starý, 1979; Starý et al., 2000), *Aphis dlabolai* Holman on *Euphorbia* sp. (Barahoei et al., 2013), *Aphis euphorbiae* Kaltenbach on *Euphorbia* sp., *E. cyparisias*, and *E. sequierana* (Barahoei et al., 2013; Mossadegh et al., 2011; Rakhshani, Talebi, Starý et al., 2008; Talebi et al., 2009), *Aphis fabae* Scopoli on *Althaea* sp., *Ammi majus*, *Chenopodium album*, *Marrubium* sp., *Mentha longifolia*, *Picnomon acarna*, *Plantago lanceolata*, *Rumex acetosa*, *R. crispus*, *Solanum lycopersicus*, *Spinacia oleratia*, and *Triticum aestivum* (Barahoei et al., 2013; Mossadegh et al., 2011; Nazari et al., 2012; Rakhshani, Talebi, Starý et al., 2008; Taheri & Rakhshani, 2013; Tahriri et al., 2007, 2010; Talebi et al., 2009), *Aphis fabae cirsiiacanthoides* Scopoli on *Cirsium arvens*, *Aphis gossypii* Glover on *Abelmoschus esculentus*, *Catalpa bignonioides*, *Cestrum nocturnum*, *Citrus* sp., *Convulvolus* sp., *Cucumis melo*, *Cucurbita pepo*, *Ranunculus* sp., *Solanum melongena*, *Tecoma stans*, and *Veronica* sp. (Barahoei et al., 2013; Mossadegh et al., 2011; Nazari et al., 2012; Rakhshani, Talebi, Starý et al., 2008; Starý, 1979; Starý et al., 2000, 2007, 2012; Talebi et al., 2009; Zamani et al., 2006), *Aphis intybi* Koch on *Calendula officinalis* (Barahoei et al., 2013; Taheri & Rakhshani, 2013), *Aphis nasturtii* Kaltenbach on *Plantago lanceolata* (Nazari et al., 2012), *Aphis plantaginis* Goeze on *Plantago lanceolata* (Barahoei et al., 2013; Radjabi, 1989), *Aphis pomi* De Geer on *Malus domestica*, and *Pyrus malus* (Barahoei et al., 2013; Modarres Awal, 2012; Rakhshani, 2012; Rakhshani, Talebi, Starý et al., 2008), *Aphis punicae* Passerini on *Punica granatum* var. *sativa* (Alikhani et al., 2013; Barahoei et al., 2013; Nazari et al., 2012), *Aphis solanella* Theobald on *Capsella bursa-pastoris*, and *Solanum melongena* (Barahoei et al., 2013), *Aphis spiraecola* Patch on *Crataegus monogyna*, and *Malus domestica* (Rakhshani, 2012), *Aphis umbrella* Börner on *Capitophorus hippophaes*, *Crepis* sp., and *Malva neglecta* (Nazari et al., 2012; Rakhshani, Talebi, Starý et al., 2008; Schlinger & Mackauer, 1963; Starý et al., 2000; Talebi et al., 2009), *Aphis* sp. on *Ammi majus*, *Anagallis arvensis*, *Aster* sp., *Callendula officinalis*, *Centura* sp., *Chrysanthemum* sp., *Duranta plumieri*, *Hibiscus esculentus*, *Lamium* sp., *Malva* sp., *Mentha pulegium*, *Rumex acetosa*, *Silybum marianum*, *Tagetes* sp., and *Urtica dioica* (Mossadegh et al., 2011), *Brachycaudus amygdalinus* (Schouteden) on *Prunus dulcis* (Nazari et al., 2012; Rakhshani, 2012), *Brachycaudus cardui* (L.) on *Prunus* sp., and *Prunus padus* (Rakhshani, 2012; Rakhshani, Talebi, Starý et al., 2008), *Brachycaudus divaricatae* Shaposhnikov on *Prunus domestica* (Jafari-Ahmadabadi & Modarres Awal, 2012; Jafari-Ahmadabadi et al., 2011), *Brachycaudus helichrysi* (Kaltenbach) on *Calendula officinalis*, *Calendula* sp., *Carthamus oxyacanthus*, *Chrysanthemum* sp., *Prunus armeniaca*, *P. persica*, *Zinia elegans* (Barahoei et al., 2013; Jafari-Ahmadabadi & Modarres Awal, 2012; Jafari-Ahmadabadi et al., 2011; Modarres Awal, 2012; Mokhtari et al., 2000; Mossadegh et al., 2011; Nazari et al., 2012; Rakhshani, 2012; Rakhshani, Talebi, Starý et al., 2008; Rakhshani, Kazemzadeh et al., 2012; Starý et al., 2000; Talebi et al., 2009; Yaghubi, 1997), *Brachycaudus persicae* (Passerini) on *Prunus persica* (Rakhshani, 2012), *Brachycaudus tragopogonis* (Kaltenbach) on *Tragopogon graminifolius*, and *Scrozonera isphahanica* (Alikhani et al., 2013; Barahoei et al., 2013), *Capitophorus*

hippophaes Walker (1852) (Modarres Awal, 2012), *Capitophorus similis* van der Goot on *Eleagnus angustifolia* (Rakhshani, Talebi, Starý et al., 2008; Talebi et al., 2009), *Brachycaudus* sp. on Brassicaceae (Starý et al., 2000), *Diuraphis noxia* (Kurdjumov) (Bandani et al., 1992; González et al., 1992; Modarres Awal, 1997, 2012; Starý et al., 2000; Zareh et al., 1995), *Dysaphis devecta* (Walker) (Modarres Awal, 1997, 2012; Radjabi, 1989; Starý et al., 2000), *Dysaphis plantaginea* (Passerini) on *Malus* sp., *M. domestica*, and *Prunus communis* (Modarres Awal, 1997, 2012; Mossadegh et al., 2011; Radjabi, 1989; Rakhshani, 2012; Starý et al., 2000), *Dysaphis pyri* (Boyer de Fonscolombe) (Modarres Awal, 1997, 2012; Radjabi, 1989; Starý et al., 2000), *Eucarazzia elegans* (Ferrari) on *Salvia officinalis* (Barahoei et al., 2013; Rakhshani, Talebi, Starý et al., 2008; Talebi et al., 2009), *Hyalopterus amygdali* (Blanchard) on *Prunus amygdalinus*, and *P. dulcis* (Jafari-Ahmadabadi & Modarres Awal, 2012; Jafari-Ahmadabadi et al., 2011; Nazari et al., 2012), *Hyalopterus pruni* (Geoffroy) on *Prunus armeniaca* (Modarres Awal, 2012; Mokhtari et al., 2000; Starý et al., 2000), *Hyperomyzus lactucae* (L.) on *Antirrhinum majus*, *Coropsis delphinifolia*, and *Silybum marianum* (Mossadegh et al., 2011), *Lipaphis erysmi* (Kaltenbach) on *Brassica kaber*, and *B. oleracea* (Mossadegh et al., 2011), *Lipaphis lepidii* Nevsky on *Brassica kaber*, *Convolvulus arvensis*, *Lepidium draba*, and *L. latifolia* (Mossadegh et al., 2011; Starý, 1979; Starý et al., 2000), *Lipaphis pseudobrassicae* (Davis) on *Brassica napus* (Barahoei et al., 2013), *Lipaphis* sp. on *Raphanus sativus* (Mossadegh et al., 2011), *Metopolophium dirhodum* (Walker) on *Triticum aestivum* (Bandani et al., 1992; Barahoei et al., 2013; Nazari et al., 2012; Rakhshani, Talebi, Starý et al., 2008; Rakhshani, Tomanović et al., 2008; Starý et al., 2000; Taheri & Rakhshani, 2013), *Myzus ascalonicus* Doncaster on *Brassica rapa* (Barahoei et al., 2013), *Myzus beybienkoi* (Narzykulov) on *Fraxinus oxycarpa* (Starý, 1979; Starý et al., 2000), *Myzus cerasi* (Fabricius) on *Prunus cerasus*, and *P. spinosa* (Modarres Awal, 2012; Mokhtari et al., 2000; Rakhshani, 2012; Rakhshani, Talebi, Starý et al., 2008; Starý et al., 2000), *Myzus certus* (Walker) on *Catharanthus* sp. (Barahoei et al., 2013), *Myzus ornatus* Laing (Starý et al., 2000), *Myzus persicae* (Sulzer) on *Achillea* sp., *Althaea* sp., *Amaranthus* sp., *Antirrhinum majus*, *Beta maritima*, *Brassica kaber*, *B. napus*, *Callendula officinalis*, *Capsicum annuum*, *Cardaria draba*, *Caryophylla* sp., *Caucalis* sp., *Celosia cristata*, *Centura* sp., *Chrysanthemum* sp., *Convolvulus arvensis*, *C.* sp., *Cucurbita pepo*, *Dodona viscosa*, *Eruca* sp., *Euphorbia helioscopia*, *Euphorbia* sp., *Halianthus annus*, *Lactuca sativa*, *Lamium* sp., *Lepidium draba*, *Malva neglecta*, *M. sylvestris*, *Malva* sp., *Medicago orbicularis*, *Mentha longifolia*, *Nicotiana tabacum*, *Orobanche cilicica*, *Papavum dubium*, *Prunus armeniaca*, *P. dulcis*, *P. padus*, *P. persica*,

Raphanus sativus, *Silybum marianum*, *Sinapis arvenses*, *Sisymbrium* sp., *Solanum melongena*, *Spinacia oleracea*, *Tagetes* sp., and *Tropaeolum* sp. (Barahoei et al., 2012a, 2013; Jafari-Ahmadabadi & Modarres Awal, 2012; Jafari-Ahmadabadi et al., 2011; Modarres Awal, 1997, 2012; Mossadegh et al., 2011; Nazari et al., 2012; Radjabi, 1989; Rakhshani, 2012; Rakhshani, Talebi, Starý et al., 2008; Schlinger & Mackauer, 1963; Starý, 1979; Starý et al., 2000; Taheri & Rakhshani, 2013; Talebi et al., 2009; Tazerouni et al., 2017, 2018; Yaghubi, 1997), *Nasonovia ribisnigri* Mosely on lettuce plant (Farsi et al., 2012, 2019), *Myzus* sp. (Shojai et al., 1999), *Ovatus crataegarius* (Walker) on *Cydonia oblonga* (Alikhani et al., 2013), *Ovatus insitus* (Walker) on *Cydonia vulgaris* (Rakhshani, Talebi, Starý et al., 2008; Rakhshani, 2012), *Phorodon humuli* (Schrank) on *Prunus persicae* (Bandani et al., 1992; Modarres Awal, 2012; Mokhtari et al., 2000; Schlinger & Mackauer, 1963; Starý et al., 2000), *Rhopalomyzus* sp. on *Fraxinus* sp. (Starý, 1979), *Rhopalosiphum maidis* (Fitch) on *Hordeum marinum*, *H. spontaneum*, *Setaria glauca*, and *Sorghum bicolor* (Bandani et al., 1992, 1993; Barahoei, Madjdzadeh et al., 2012; Barahoei et al., 2013; Nazari et al., 2012; Starý et al., 2000), *Rhopalosiphum nymphaeae* (L.) on *Prunus* sp. (Modarres Awal, 2012; Mokhtari et al., 2000; Starý et al., 2000), *Rhopalosiphum padi* (L.) on *Bromus tectorum*, *Bromus* sp., *Cydonia oblonga*, *Hordeum murinum*, *H. vulgare*, *Triticum aestivum*, and *Zea mays* (Bagheri-Matin et al., 2008; Bandani et al., 1992, 1993; Barahoei et al., 2013; Mossadegh et al., 2011; Nazari et al., 2012; Rakhshani, 2012; Rakhshani, Talebi, Starý et al., 2008; Rakhshani, Tomanović et al., 2008; Shahrokhi et al., 2004; Starý et al., 2000; Taheri & Rakhshani, 2013), *Schizaphis graminum* (Rondani) on *Bromus tectorum*, *Hordeum marinum*, *H. tectorum*, *H. vulgare*, and *Triticum aestivum* (Bagheri-Matin et al., 2008; Bandani et al., 1992, 1993; Barahoei et al., 2013; Modarres Najafabadi & Gholamian, 2007; Nazari et al., 2012; Rakhshani, Talebi, Starý et al., 2008; Rakhshani, Tomanović et al., 2008; Shahrokhi et al., 2004; Starý et al., 2000), *Sitobion avenae* (Fabricius) on *Triticum aestivum*, and *Gastridium phleoides* (Bandani, 1992; Bandani et al., 1992, 1993; Barahoei et al., 2013; Mossadegh et al., 2011; Rakhshani, Talebi, Starý et al., 2008; Rakhshani, Tomanović et al., 2008; Shahrokhi et al., 2004; Starý et al., 2000), *Sitobion fragariae* (Walker) on *Malva* sp. (Barahoei et al., 2013), *Wahlgreniella nervata* (Gillette) on *Rosa damascena* (Barahoei et al., 2013). Comments: Based on Barahoei et al. (2014), the record of *Acyrthosiphon gossypii* Mordvilko as a host of *A. matricariae* (Taheri & Rakhshani, 2013) is doubtful.

Aphidius microlophii Pennachio and Tremblay, 1987
Distribution in the Middle East: Turkey (Akar & Çetin Erdoğan, 2017).

Extralimital distribution: Bulgaria, Czech Republic, France, Italy, Montenegro, Russia, Serbia, Slovenia, United Kingdom.
Host records: Summarized by Yu et al. (2016) as being a parasitoid of the aphids *Microlophium carnosum* (Buckton), and *Wahlgreniella ossiannilssoni* Hille Ris Lambers.

Aphidius myzocallidis Mescheloff and Rosen, 1990
Distribution in the Middle East: Israel−Palestine (Mescheloff & Rosen, 1990b).
Extralimital distribution: None.
Host records: Recorded by Mescheloff and Rosen (1990b) as being a parasitoid of the aphid *Myzocallis glandulosa* Hille Ris Lambers.

Aphidius (*Euaphidius*) *cingulatus* Ruthe, 1859
Catalogs with Iranian records: Fallahzadeh and Saghaei (2010), Barahoei et al. (2014) as *Euaphidius cingulatus* Ruthe, 1859, Yu et al. (2016), Farahani et al. (2016), Samin, Coronado-Blanco, Kavallieratos et al. (2018), Samin, Coronado-Blanco, Fischer et al. (2018), Rakhshani et al. (2019).
Distribution in Iran: Alborz (Rakhshani, Talebi, Starý et al., 2007; Rakhshani, Talebi, Starý et al., 2008; Starý et al., 2000; Starý, 1979 as *Euaphidius cingulatus*), Golestan (Barahoei, Sargazi et al., 2012), Guilan (Rakhshani, Talebi, Starý et al., 2007; Rakhshani, Talebi, Starý et al., 2008; Talebi et al., 2009 as *Euaphidius cingulatus* (Ruthe)), Mazandaran (Babaee et al., 2004), Northern Khorasan (Kazemzadeh et al., 2010; Rakhshani, Kazemzadeh et al., 2012), Razavi Khorasan (Modarres Awal, 2012; Rakhshani, Talebi, Starý et al., 2007; Starý, 1979; Starý et al., 2000), Tehran (Modarres Awal, 2012), Iran (no specific locality cited) (Rakhshani, Talebi, Manzari et al., 2006; Rahimi et al., 2010).
Distribution in the Middle East: Iran (see references above), Turkey (Rakhshani et al., 2019).
Extralimital distribution: Indo-Malayan, Nearctic, Neotropical, Oceanic, and Palaearctic regions [Adjacent countries to Iran: Kazakhstan and Russia].
Host records: Summarized by Yu et al. (2016) as being a parasitoid of the aphids *Acyrthosiphon pisum* Harris, *Aphis glycines* Matsumara, *Brevicoryne brassicae* (L.), *Chaitophorus populicola* Thomas, *Cinara pini* (L.), *Lachnus* sp., *Metopeurum fuscoviride* Stroyan, *Pterocomma bicolor* (Oesthund), *P. pilosum* Buckton, *P. populeum* (Kaltenbach), *P. populifoliae* (Fitch), *P. pseudopopuleum* Palmer, *P. rufipes* (Hartig), *Pterocomma salicicola* Thomas, *Pterocomma salicis* (L.), *P. sanguiceps* Richards, *P. smithiae* Boerner, and *Schizolachnus pineti* (Fabricius).
Parasitoid−aphid−plant associations in Iran: Hemiptera: Aphididae: *Pterocomma pilosum* Buckton on *Salix alba*, *Salix australiator*, and *Salix* sp. (Rahimi et al., 2010; Rakhshani, Talebi, Starý et al., 2007; Rakhshani, Talebi,

Starý et al., 2008; Rakhshani, Kazemzadeh et al., 2012; Starý, 1979; Starý et al., 2000; Talebi et al., 2009), *Pterocomma populeum* (Kaltenbach) on *Populus alba* (Babaee et al., 2004; Rakhshani, Talebi, Starý et al., 2007; Rakhshani, Talebi, Starý et al., 2008; Rakhshani, Kazemzadeh et al., 2012; Starý, 1979), *Pterocomma* sp. on *Populus alba* (Rakhshani, Talebi, Manzari et al., 2006; Starý, 1979; Starý et al., 2000).

Aphidius persicus Rakhshani and Starý, 2006

Catalogs with Iranian records: Fallahzadeh and Saghaei (2010), Barahoei et al. (2014), Yu et al. (2016), Farahani et al. (2016), Samin, Coronado-Blanco, Kavallieratos et al. (2018), Samin, Coronado-Blanco, Fischer et al. (2018), Rakhshani et al. (2019).

Distribution in Iran: Alborz (Rakhshani, Talebi, Starý et al., 2006; Rakhshani, Talebi, Starý et al., 2008), Fars (Barahoei et al., 2013; Taheri & Rakhshani, 2013), Hamadan (Nazari et al., 2012; Rakhshani, Talebi, Starý et al., 2006; Rakhshani, Talebi, Starý et al., 2008), Isfahan (Barahoei et al., 2013), Kerman (Barahoei, Madjdzadeh et al., 2012; Barahoei et al., 2013), Khuzestan (Mossadegh et al., 2010, 2011; Salehipour et al., 2010), Markazi (Alikhani et al., 2013), Northern Khorasan (Kazemzadeh et al., 2010; Rakhshani, Kazemzadeh et al., 2012), Tehran, Zanjan (Rakhshani, Talebi, Starý et al., 2006; Rakhshani, Talebi, Starý et al., 2008).

Distribution in the Middle East: Iran (see references above), Iraq (Rakhshani et al., 2006).

Extralimital distribution: None.

Host records: Summarized by Yu et al. (2016) as being a parasitoid of the aphids *Uroleucon chondrillae* (Nevsky), *U. jaceae* (L.), and *U. sonchi* (L.).

Parasitoid–aphid–plant associations in Iran: Hemiptera: Aphididae: *Macrosiphoniella* sp. on *Artemisia biennis*, *Uroleucon bielawskii* (Szelegiewicz) on *Lactusa* sp., *Uroleucon carthami* Hille Ris Lambers on *Carthamus oxyacantha* (Barahoei et al., 2013), *Uroleucon chondrillae* (Nevsky) on *Chondrilla juncea* (Alikhani et al., 2013; Barahoei et al., 2013; Nazari et al., 2012; Rakhshani, Talebi, Starý et al., 2008; Taheri & Rakhshani, 2013), *Uroleucon compositae* (Theobald) on *Carthamus oxyacantha*, *C. tinctorius*, and *Lactuca* sp., *Uroleucon ochropus* (Hille Ris Lambers) on *Launaea acanthodes* (Barahoei et al., 2013), *Uroleucon jaceae* (L.) on *Centaurea virgata sequarrosa* (Alikhani et al., 2013), *Uroleucon sonchi* (L.) on *Lactuca serriola*, *Launaea acanthodes*, *Sonchus asper*, *Sonchus* sp., *S. oleraceus*, and *Tragopogon pratensis* (Barahoei, Madjdzadeh et al., 2012; Barahoei et al., 2013; Mossadegh et al., 2011; Rakhshani, Talebi, Starý et al., 2008), *Uroleucon* sp. on *Launaea acanthodes*, *L. arborescens*, and *Picnomon acarna* (Rakhshani et al., 2012b; Barahoei et al., 2012a, 2013).

Aphidius phalangomyzi Starý, 1963

Catalogs with Iranian records: No catalog.

Distribution in Iran: Golestan (Barahoei, Sargazi et al., 2012).

Distribution in the Middle East: Iran.

Extralimital distribution: Czech Republic, France, Germany, Hungary, Moldova, Russia, Serbia, Slovakia.

Host records: Reported by Yu et al. (2016) as being a parasitoid of the aphid *Microsiphoniella oblonga* (Mordvilko).

Aphidius platensis Brethes, 1913

Catalogs with Iranian records: Yu et al. (2016), Farahani et al. (2016), Samin, Coronado-Blanco, Kavallieratos et al. (2018), Samin, Coronado-Blanco, Fischer et al. (2018), Rakhshani et al. (2019).

Distribution in Iran: Alborz (Rakhshani, Talebi, Manzari et al., 2006; Rakhshani, Talebi, Starý et al., 2008; Talebi et al., 2009), Fars (Barahoei et al., 2013; Rakhshani, 2012; Rakhshani, Talebi, Starý et al., 2008; Starý et al., 2000; Taheri & Rakhshani, 2013), Golestan (Rakhshani, Tomanović et al., 2008), Ilam (Rakhshani, 2012), Hormozgan (Ameri et al., 2019), Isfahan (Rakhshani, Talebi, Starý et al., 2008; Rakhshani, Tomanović et al., 2008; Talebi et al., 2009), Kerman (Tomanović et al., 2014), Kermanshah (Rakhshani, Talebi, Starý et al., 2008; Rakhshani, Tomanović et al., 2008), Khuzestan (Mossadegh et al., 2011, 2017; Talebi et al., 2009), Kordestan (Rakhshani, Talebi, Starý et al., 2008; Rakhshani, Tomanović et al., 2008), Markazi (Khaki et al., 2020), Northern Khorasan (Rakhshani, Tomanović et al., 2008; Rakhshani, Kazemzadeh et al., 2012), Kuhgiloyeh and Boyerahmad (Barahoei et al., 2013), Qazvin (Starý, 1979), Qom (Rakhshani, Tomanović et al., 2008), Razavi Khorasan (Rakhshani, Talebi, Starý et al., 2008), Sistan and Baluchestan (Rakhshani et al., 2005a; 2006a; Rakhshani, Talebi, Starý et al., 2008; Rakhshani, Tomanović et al., 2008; Starý et al., 2000; Talebi et al., 2009; Tomanović et al., 2014), Tehran (Rakhshani, Talebi, Kavallieratos et al., 2005; Rakhshani, Talebi, Starý et al., 2008; Rakhshani, Tomanović et al., 2008; Talebi et al., 2009; Rakhshani, 2012; Tomanović et al., 2014), Iran (no specific locality cited) (Arockia Lenin, 2015).

Distribution in the Middle East: Iran (see references above), Saudi Arabia (Rakhshani et al., 2019).

Extralimital distribution: Argentina, Australia, Brazil, Chile, France, India, Kenya, Uruguay.

Host records: Summarized by Yu et al. (2016) as being a parasitoid of the aphids *Aphis* spp., *Brachycaudus aegyptiacus* (Hall), *B. helichrysis* (Kaltenbach), *B. persicae* Passerini, *B. schwartzi* (Börner), *Brevicoryne brassicae* (L.), *Capitophorus elaeagni* (del Guercio), *Hyadaphis foeniculi* Passerini, *Macrosiphum euphorbiae* (Thomas), *M.*

rosae (L.), *Myzus ornatus* Laing, *M. persicae* (Sulzer), *Pterocomma populeum* (Kaltenbach), *Rhopalosiphum maidis* (Fitch), *R. nymphaeae* (L.), *Schizaphis graminum* (Rondani), *Sitobion avenae* (Fabricius), *Toxoptera aurantii* Boyer de Fonscolombe, and *T. citricida* (Kirkaldy).

Parasitoid—aphid—plant associations in Iran: Hemiptera: Aphididae: *Aphis gossypii* Glover on *Malva neglecta* (Nazari et al., 2012; Rakhshani, Talebi, Starý et al., 2008; Talebi et al., 2009; Tomanović et al., 2014), *Aphis intybi* Koch and *A. nerii* Boyer de Fonscolombe on *Nerium oleander* (Barahoei et al., 2013; Rakhshani, Talebi, Starý et al., 2008; Taheri & Rakhshani, 2013; Talebi et al., 2009; Tomanović et al., 2014), *Brachycaudus helichrysi* (Kaltenbach) on *Calendula officinalis* (Rakhshani, Talebi, Starý et al., 2008; Talebi et al., 2009; Tomanović et al., 2014), *Brachycaudus tragopogonis* (Kaltenbach) on *Tragopogon graminifolius* (Tomanović et al., 2014), *Capitophorus elaeagni* (Del Guercio) on *Elaeagnus angustifolia* (Rakhshani, Talebi, Starý et al., 2008; Talebi et al., 2009; Tomanović et al., 2014), *Myzus persicae* (Sulzer) on *Althaea* sp. (Mossadegh et al., 2011; Rakhshani, 2012; Rakhshani, Talebi, Starý et al., 2008; Taheri & Rakhshani, 2013; Talebi et al., 2009; Tomanović et al., 2014), *M. persicae* (Khaki et al., 2020).

Comments: *Pachyneuron aphidis* (Bouché) (Hymenoptera: Pteromalidae) is hyperparasitoid of *A. platensis* (Khaki et al., 2020).

Aphidius popovi Starý, 1978

Catalogs with Iranian records: Fallahzadeh and Saghaei (2010), Barahoei et al. (2014), Yu et al. (2016), Farahani et al. (2016), Samin, Coronado-Blanco, Kavallieratos et al. (2018), Samin, Coronado-Blanco, Fischer et al. (2018), Rakhshani et al. (2019).

Distribution in Iran: Alborz (Rakhshani, Talebi, Starý et al., 2008; Talebi et al., 2009), Kerman (Barahoei, Madjdzadeh et al., 2012; Barahoei et al., 2013), Kermanshah (Nazari et al., 2012), Northern Khorasan (Kazemzadeh et al., 2010; Rakhshani, Kazemzadeh et al., 2012), Tehran (Khayat-Zadeh et al., 2006, on *Rosa damascena*; Rakhshani, Talebi, Starý et al., 2008; Talebi et al., 2009).

Distribution in the Middle East: Iran.

Extralimital distribution: Uzbekistan.

Host records: Recorded by Starý and González (1978) as being a parasitoid of *Amphorophora catharinae* (Nevsky).

Parasitoid—aphid—plant associations in Iran: Hemiptera: Aphididae: *Amphorophora catharinae* (Nevsky) on *Rosa damascena*, *R. hybrida*, and *Rosa* sp. (Barahoei et al., 2013; Nazari et al., 2012; Rakhshani, Talebi, Starý et al., 2008; Rakhshani, Kazemzadeh et al., 2012; Talebi et al., 2009),

Metopolophium dirhodum (Walker) on *Rosa beggeriana* (Barahoei, Madjdzadeh et al., 2012; Barahoei et al., 2013).

Aphidius rhopalosiphi De Stefani-Perez, 1902

Catalogs with Iranian records: Fallahzadeh and Saghaei (2010), Barahoei et al. (2014), Yu et al. (2016), Farahani et al. (2016), Samin, Coronado-Blanco, Kavallieratos et al. (2018), Samin, Coronado-Blanco, Fischer et al. (2018), Rakhshani et al. (2019).

Distribution in Iran: Fars (Alichi et al., 2008; Barahoei et al., 2013; Rakhshani, Talebi, Starý et al., 2008; Rakhshani, Tomanović et al., 2008; Starý et al., 2000; Taheri & Rakhshani, 2013), Golestan (Barahoei, Sargazi et al., 2012; Rakhshani, Tomanović et al., 2008; Starý et al., 2000), Hamadan (Nazari et al., 2012; Rakhshani, Talebi, Starý et al., 2008; Rakhshani, Tomanović et al., 2008), Hormozgan (Ameri et al., 2019), Isfahan, Markazi, Qom (Barahoei et al., 2013; Rakhshani, Talebi, Starý et al., 2008; Rakhshani, Tomanović et al., 2008), Kermanshah (Bagheri-Matin et al., 2008; Nazari et al., 2012; Rakhshani, Talebi, Starý et al., 2008; Rakhshani, Tomanović et al., 2008), Khuzestan (Modarres Awal, 1997, 2012; Mossadegh et al., 2010, 2011; Rakhshani, Talebi, Starý et al., 2008; Rakhshani, Tomanović et al., 2008; Rezaei et al., 2006, 2010; Salehipour et al., 2010), Kordestan (Mansour Ghazi et al., 2008; Rakhshani, Talebi, Starý et al., 2008), Markazi (Alikhani et al., 2013), Mazandaran (Modarres Awal, 2012), Razavi Khorasan (Darsouei et al., 2011b; Modarres Awal, 2012; Rakhshani, Talebi, Starý et al., 2008; Rakhshani, Tomanović et al., 2008), Sistan and Baluchestan (Rakhshani, Talebi, Starý et al., 2008), Tehran (Rakhshani, Talebi, Starý et al., 2008; Rakhshani, Tomanović et al., 2008), Iran (no specific locality cited) (Tazerouni et al., 2011b).

Distribution in the Middle East: Egypt (Gadallah et al., 2017; Ghanim & El-Adl, 1983; El-Serafy, 1999), Iran (see references above), Israel—Palestine (Starý, 1981), Turkey (Uysal et al., 2004).

Extralimital distribution: Andorra, Argentina, Belgium, Brazil, Bulgaria, Chile, China, Czech Republic, Denmark, Finland, France, Germany, Greece, Hungary, India, Ireland, Italy, North Macedonia, Morocco, Netherlands, New Zealand, Norway, Pakistan, Poland, Portugal, Russia, Serbia, Slovakia, Slovenia, Spain, Sweden, Switzerland, Ukraine, United Kingdom, United States of America, Uzbekistan.

Host records: Summarized by Yu et al. (2016) as being a parasitoid of the following aphid species: *Diuraphis noxia* (Mordvilko), *Hyalopteroides humilis* (Walker), *Israelaphis lambersi* Illharco, *Macrosiphum dryopteridis* (Holman), *M. equiseti* (Holman), *Metopolophium albidum* Hille Ris

Lambers, *M. dirhodum* (Walker), *Myzus persicae* (Sulzer), *Rhopalosiphum maidis* (Fitch), *R. nymphaeae* (L.), *R. padi* (L.), *Schizaphis graminum* (Rondani), *Sitobion avenae* (Fabricius), *S. fragariae* (Walker), and *Uroleucon* sp.

Parasitoid—aphid—plant associations in Iran: Hemiptera: Aphididae: *Diuraphis noxia* (Kurdjumov) on *Hordeum* sp. (Starý et al., 2000), *Metopolophium dirhodum* (Walker) on *Hordeum spontaneum, H. vulgare*, and *Triticum aestivum* (Alichi et al., 2007, 2008; Alikhani et al., 2013; Barahoei et al., 2013; Nazari et al., 2012; Rakhshani, Talebi, Starý et al., 2008; Rakhshani, Tomanović et al., 2008), *Rhopalosiphum maidis* (Fitch) on *Hordeum spontaneum* (Nazari et al., 2012), *Rhopalosiphum padi* (L.) on *Hordeum* sp., and *Triticum aestivum* (Alichi et al., 2008; Bagheri-Matin et al., 2008; Barahoei et al., 2013; Nazari et al., 2012; Rakhshani, Talebi, Starý et al., 2008; Rakhshani, Tomanović et al., 2008; Starý et al., 2000; Taheri & Rakhshani, 2013), *Schizaphis graminum* (Rondani) on *Triticum aestivum* (Alikhani et al., 2013; Bagheri-Matin et al., 2008; Barahoei et al., 2013; Nazari et al., 2012; Rakhshani, Talebi, Starý et al., 2008; Rakhshani, Tomanović et al., 2008), *Sitobion avenae* (Fabricius) on *Avena fatua, Gastridium phleoides, Hordeum spontaneum*, and *Triticum aestivum* (Alikhani et al., 2013; Bagheri-Matin et al., 2008; Barahoei et al., 2013; Darvish-Mojeni, 1994; Mansour Ghazi et al., 2008; Mossadegh et al., 2011; Nazari et al., 2012; Rakhshani, Talebi, Starý et al., 2008; Rakhshani, Tomanović et al., 2008; Starý et al., 2000).

Aphidius rosae Haliday, 1833

Catalogs with Iranian records: Fallahzadeh and Saghaei (2010), Barahoei et al. (2014), Yu et al. (2016), Farahani et al. (2016), Samin, Coronado-Blanco, Kavallieratos et al. (2018), Samin, Coronado-Blanco, Fischer et al. (2018), Rakhshani et al. (2019).

Distribution in Iran: Alborz (Rakhshani, Talebi, Starý et al., 2008), Golestan (Barahoei, Sargazi et al., 2012), Hamadan (Nazari et al., 2012; Rajabi Mazhar et al., 2010; Talebi et al., 2009), Isfahan (Barahoei et al., 2013; Mehrparvar & Hatami, 2003; Mehrparvar et al., 2005, 2016; Modarres Awal, 2012; Rakhshani, Talebi, Starý et al., 2008), Kerman (Barahoei et al., 2013), Khuzestan (Mossadegh et al., 2010, 2011; Salehipour et al., 2010), Markazi (Alikhani et al., 2013), Tehran (Rakhshani, Talebi, Starý et al., 2008; Talebi et al., 2009).

Distribution in the Middle East: Iran (see references above), Iraq (Starý, 1976; Starý & Kaddou, 1971), Israel—Palestine (Mescheloff & Rosen, 1990b), Turkey (Düzgüneş et al., 1982; Uysal et al., 2004).

Extralimital distribution: Andorra, Belarus, Belgium, Brazil, Bulgaria, Canada, China, Croatia, Czech Republic, Denmark, Finland, France, Georgia, Germany, Greece, Hungary, Iceland, India, Ireland, Italy, Latvia, Lithuania, Moldova, Montenegro, Netherlands, Norway, Pakistan, Poland, Portugal, Romania, Russia, Serbia, Slovakia, Spain, Switzerland, Ukraine, United Kingdom, United States of America, Uzbekistan, Venezuela.

Host records: Summarized by Yu et al. (2016) as being a parasitoid of the following aphid species: *Acyrthosiphon pisum* (Harris), *Aphis aceris* L., *A. fabae* Scopoli, *A. pomi* De Geer, *A. urticata* Gmelin, *Brachycaudus cardui* (L.), *B. helichrysis* (Kaltenbach), *B. tragopogonis* (Kaltenbach), *Chaetosiphon tetrarhodum* (Walker), *Drepanosiphum aceris* Koch, *D. platanoides* (Schrank), *Hyalopterus pruni* (Geoffroy), *Illinoia liriodendri* (Monell), *Lipaphis erysimi* (Kaltenbach), *Macrosiphonella ludoviciana* (Oestlund), *Macrosiphum* spp., *Melomys rubicola* (Thomas), *M. dirhodum* (Walker), *M. montanum* Hille Ris Lambers, *Myzus persicae* (Sulzer), *Schizaphis graminum* (Rondani), *Sitobion fragariae* (Walker), *Uroleucon ambrosiae* (Thomas), *U. rudbeckiae* (Fitch), and *U. sonchi* (L.).

Parasitoid—aphid—plant associations in Iran: Hemiptera: Aphididae: *Macrosiphum rosae* (L.) on *Rosa canina, R. damascene, R. hybrida*, and *Rosa* sp. (Alikhani et al., 2013; Barahoei et al., 2013; Mehrparvar & Hatami, 2003; Mehrparvar et al., 2005; 2016; Modarres Awal, 2012; Mossadegh et al., 2011; Nazari et al., 2012; Rajabi Mazhar et al., 2010; Rakhshani, Talebi, Starý et al., 2008; Talebi et al., 2009).

Aphidius salicis Haliday, 1834

Catalogs with Iranian records: Fallahzadeh & Saghaei (2010), Barahoei et al. (2014), Yu et al. (2016), Farahani et al. (2016), Samin, Coronado-Blanco, Kavallieratos et al. (2018), Samin, Coronado-Blanco, Fischer et al. (2018), Rakhshani et al. (2019).

Distribution in Iran: Alborz (Starý, 1979; Starý et al., 2000), Fars (Barahoei et al., 2013; Taheri & Rakhshani, 2013), Kermanshah (Nazari et al., 2012 as *Aphidius* cf. *salicis* Haliday, 1834), Kuhgiloyeh and Boyerahmad (Barahoei et al., 2013), Tehran (Modarres Awal, 2012; Rakhshani, Talebi, Starý et al., 2007; Rakhshani, Talebi, Starý et al., 2008; Starý, 1979; Starý et al., 2000; Talebi et al., 2009), Iran (no specific locality cited) (Rahimi et al., 2010; Rakhshani, Talebi, Manzari et al., 2006).

Distribution in the Middle East: Iran (see references above), Israel—Palestine (Mescheloff & Rosen, 1990b), Turkey (Düzgüneş et al., 1982; Uysal et al., 2004).

Extralimital distribution: Indo-Malayan, Nearctic, Neotropical, Oceanic, and Palaearctic regions.
Host records: Summarized by Yu et al. (2016) as being a parasitoid of the following aphis species: *Aphis* spp., *Cavariella* spp., *Chaitophorus saliciti* (Schrank), *C. salijaponicus* Essig and Kowana, *Dysaphis crataegi* (Kaltenbach), *Hayhurstia foeniculi* Passerini, *Macrosiphum rosae* (L.), *M. rosaeiformes* Das, *Semiaphis dauci* (Fabricius), *S. heraclei* (Takahashi), and *Tricaudatus polygoni* (Narzikulov).
Parasitoid−aphid−plant associations in Iran: Hemiptera: Aphididae: *Cavariella aegopodii* (Scopoli) on *Salix* sp., *S. alba*, and *S. australior* (Barahoei et al., 2013; Rakhshani, Talebi, Starý et al., 2007; Rakhshani, Talebi, Starý et al., 2008; Rahimi et al., 2010; Starý, 1979; Starý et al., 2000; Taheri & Rakhshani, 2013; Talebi et al., 2009), *Cavariella aquatica* (Gillette and Bragg) on *Salix* sp. (Starý, 1979; Starý et al., 2000), *Cavariella aspidaphoides* Hille Ris Lambers on *Salix australior* (Starý, 1979; Starý et al., 2000), *Cavariella theobaldi* (Gillette and Bragg) on *Salix* sp. (Starý, 1979; Starý et al., 2000), *Capitophorus elaeagni* (del Guercio) on *Elaeagnus angustifolia* (Nazari et al., 2012), *Hyadaphis foeniculi* (Passerini) on *Ammi majus* (Rakhshani, Talebi, Starý et al., 2008; Talebi et al., 2009).

Aphidius (*Euaphidius*) *setiger* (Mackauer, 1961)
Catalogs with Iranian records: Fallahzadeh and Saghaei (2010), Barahoei et al. (2014) as *Euaphidius setiger* (Mackauer, 1961), Yu et al. (2016), Farahani et al. (2016), Samin, Coronado-Blanco, Kavallieratos et al. (2018), Samin, Coronado-Blanco, Fischer et al. (2018), Rakhshani et al. (2019).
Distribution in Iran: Lorestan (Modarres Awal, 2012; Nazari et al., 2012; Starý, 1979 as *Euaphidius setiger*; Starý et al., 2000; Rakhshani, Talebi, Starý et al., 2008).
Distribution in the Middle East: Iran (see references above), Turkey (Starý, 1976).
Extralimital distribution: Austria, Bulgaria, Canada, Czech Republic, Ethiopia, Finland, France, Georgia, Germany, Greece, Hungary, India, Italy, Lithuania, Moldova, Netherlands, Poland, Serbia, Slovakia, Spain, Switzerland, Ukraine, United Kingdom United States of America.
Host records: Summarized by Yu et al. (2016) as being a parasitoid of the following aphid species: *Aphis fabae* Scopoli, *Diuraphis noxia* (Mordvilko), *Metopolophium dirhodum* (Walker), *Myzocallis carpini* (Koch), *Periphyllus acericola* (Walker), *P. aceris* (L.), *P. bulgaricus* Tashev, *P. californiensis* (Shinji), *P. coracinus* (Koch), *P. hirticornis* (Walker), *P. lyropictus* (Kessler), *P. testudinaceus* (Fernie), *Rhopalosiphum padi* (L.), and *Schizaphis graminum* (Rondani).

Parasitoid−aphid−plant associations in Iran: Hemiptera: Aphididae: *Periphyllus testudinaceus* (Frenie) on *Acer compestre* (Rakhshani, Talebi, Starý et al., 2008), *Periphyllus* sp. on *Acer cinerascens* (Nazari et al., 2012; Starý, 1979; Starý et al., 2000).

Aphidius smithi Sharma and Subba Rao, 1959
Catalogs with Iranian records: Fallahzadeh and Saghaei (2010), Barahoei et al. (2014), Yu et al. (2016), Farahani et al. (2016), Samin, Coronado-Blanco, Kavallieratos et al. (2018), Samin, Coronado-Blanco, Fischer et al. (2018), Rakhshani et al. (2019).
Distribution in Iran: Alborz (González et al., 1978; Rakhshani, Talebi, Manzari et al., 2006; Rakhshani, Talebi, Starý et al., 2008; Starý, 1979; Starý et al., 2000), Fars (Rakhshani, Talebi, Starý et al., 2008), Golestan (Barahoei, Sargazi et al., 2012), Guilan (Modarres Awal, 2012; Rakhshani, Talebi, Starý et al., 2008; Starý et al., 2000; Yaghubi, 1997; Yaghubi & Sahragard, 1998), Hamadan, Kermanshah, Kordestan (Nazari et al., 2012; Rakhshani, Talebi, Manzari et al., 2006; Rakhshani, Talebi, Starý et al., 2008), Hormozgan (Ameri et al., 2019), Isfahan (Barahoei et al., 2013; Rakhshani, Talebi, Manzari et al., 2006; Rakhshani, Talebi, Starý et al., 2008; Rakhshani et al., 2010), Kerman (Asadizade et al., 2014; Barahoei, Madjdzadeh et al., 2012; Ghotbi Ravandi et al., 2017), Markazi (Alikhani et al., 2013), Northern Khorasan (Kazemzadeh et al., 2010; Rakhshani, Kazemzadeh et al., 2012), Sistan and Baluchestan (Rakhshani, Talebi, Manzari et al., 2006; Rakhshani, Talebi, Starý et al., 2008), Tehran (Modarres Awal, 2012).
Distribution in the Middle East: Cyprus (Beyarslan et al., 2017), Iran (see references above), Iraq (El-Azawi, 1970), Israel−Palestine (Mescheloff & Rosen, 1990b), Lebanon (Tremblay et al., 1985), Turkey (Akar & Çetin Erdoğan, 2017).
Extralimital distribution: Afghanistan, Algeria, Andorra, Argentina, Australia, Belgium, Brazil, Bulgaria, Canada, Chile, China, Croatia, Czech Republic, Denmark, Finland, France, Georgia, Germany, Greece, Hungary, India, Ireland, Italy, Japan, Korea, Lithuania, Mexico, Moldova, Morocco, Netherlands, New Zealand, Norway, Pakistan, Poland, Portugal (Madeira Islands), Russia, Serbia, Slovakia, Spain, Switzerland, Tajikistan, Ukraine, United Kingdom, United States of America, Uzbekistan.
Host records: Summarized by Yu et al. (2016) as being a parasitoid of the following aphid species: *Acyrthosiphon* spp., *Amphorophora rubi* (Kaltenbach), *Aphis* spp., *Aspidophorodon longicauda* (Richards), *Aulacorthum solani* Kaltenbach, *Brachycaudus cardui* (L.), *B. helichrysi* (Kaltenbach), *B. schwartzi* (Boerner), *Brevicoryne brassicae* (L.), *Capiphorus* sp., *Cinara juniperi* (De Geer), *Corylobium avellanae* (Schrank), *Delphiniobium junackianum*

(Karsch), *Diuraphis noxia* (Mordvilko), *Dysaphis plantaginea* (Passerini), *Hayhurstia atriplicis* (L.), *Hyalopteroides humilis* (Walker), *Hyalopterus pruni* (Geoffroy), *Hyperomyzus lactucae* (L.), *H. nigricornis* (Knowlton), *Illinoia liriodendri* (Monell), *Lambersius* sp., *Macrosiphoniella absinthii* (L.), *Macrosiphum* spp., *Megoura viciae* Buckton, *Metopolophium dirhodum* (Walker), *M. festucae* (Theobald), *Microlophium carnosum* (Buckton), *Myzus certus* (Walker), *M. persicae* (Sulzer), *M. varians* Davidson, *Nasonovia ribisnigri* (Mosley), *Ovatus crataegarius* (Walker), *Phorodon humuli* (Schrank), *Rhopalosiphum maidis* (Fitch), *R. padi* (L.), *Schizaphis graminum* (Rondani), *Sitobion avenae* (Fabricius), *S. fragariae* (Walker), *Therioaphis luteola* (Boerner), *T. trifolii* (Monell), *Uroleucon aeneum* (Hille Ris Lambers), *U. picridis* (Fabricius), and *Wahlgreniella ossinnilssoni* Hille Ris Lambers.

Parasitoid—aphid—plant associations in Iran: Hemiptera: Aphididae: *Acyrthosiphon kondoi* Shinji on *Medicago sativa* (González et al., 1978; Rakhshani, Talebi, Manzari et al., 2006; Starý et al., 2000), *Acyrthosiphon pisum* (Harris) on *Lathyrus* sp., *Medicago sativa*, *M.* sp., *Trifolium repens*, *Vicia pannonica*, *V. villosa*, and *Vicia* sp. (Alkhani et al., 2013; Barahoei et al., 2013; González et al., 1978; Modarres Awal, 2012; Nazari et al., 2012; Rakhshani, Talebi, Manzari et al., 2006; Rakhshani, Talebi, Starý et al., 2008; 2010; Rakhshani, Kazemzadeh et al., 2012; Starý, 1979; Starý et al., 2000; Yaghubi, 1997), *Aphis fabae* Scopoli on *Raphanus sativus* (Barahoei, Madjdzadeh et al., 2012), *Nearctaphis bakeri* (Cowen) on *Trifolium campestre* (Barahoei et al., 2013), *Sitobion avenae* (Fabricius) on *Triticum aestivum* (Rakhshani, Talebi, Starý et al., 2008; Rakhshani, Tomanović et al., 2008).

Comments: Record of *Aphis fabae* Scopoli as the host of *Aphidius smithi* by Barahoei et al. (2013) is doubtful (Barahoei et al., 2014).

Aphidius sonchi Marshall, 1896

Catalogs with Iranian records: No catalog.
Distribution in Iran: Khuzestan (Ghahari et al., 2020).
Distribution in the Middle East: Egypt (Hassan, 1957; Risbec, 1960; Starý, 1976), Iraq, Turkey (Aeschlimann & Vitou, 1985), Iran (Ghahari et al., 2020), Israel—Palestine (Starý, 1976).
Extralimital distribution: Algeria, Andorra, Australia, China, Czech Republic, Faeroe Islands, Finland, France, Greece, Hungary, India, Italy, Japan, Kazakhstan, Lithuania, Moldova, Montenegro, Netherlands, New Zealand, Poland, Serbia, Slovenia, Spain, United Kingdom.
Host records: Sammurized by Yu et al. (2016) as being a parasitoid of the following aphid species: *Acyrthosiphon cyparissiae* (Koch), *A. lactucae* (Passerini), *Aphis punicae* Passerini, *Aulacorthum solani* (Kaltenbach), *Cryptomyzus ribis* (L.), *Hyalopterus carduellinus* (Theobald), *H.*

lactucae (L.), *H. picridis* (Börner), *Rhopalosiphum maidis* (Fitch), and *Uroleucon sonchi* (L.).
Parasitoid—aphid—plant associations in Iran: *Hyperomyzus lactucae* (Linnaeus) (Hemiptera: Aphididae) on *Lactuca sativa* L. (Asteraceae) (Ghahari et al., 2020).

Aphidius stigmaticus Rakhshani and Tomanović, 2011

Catalogs with Iranian records: Barahoei et al. (2014), Yu et al. (2016), Farahani et al. (2016), Samin, Coronado-Blanco, Kavallieratos et al. (2018), Samin, Coronado-Blanco, Fischer et al. (2018), Rakhshani et al. (2019).
Distribution in Iran: Hamadan (Rakhshani et al., 2011).
Distribution in the Middle East: Iran.
Extralimital distribution: None.
Host records: Recorded by Rakhshani et al. (2011) as a parasitoid of the aphid *Macrosiphoniella tanacetaria* Kaltenbach.
Parasitoid—aphid—plant associations in Iran: Hemiptera: Aphididae: *Macrosiphoniella tanacetaria* Kaltenbach on *Tanacetum polycephalum* (Rakhshani et al., 2011).

Aphidius tanacetarius Mackauer, 1962

Catalogs with Iranian records: Barahoei et al. (2014).
Distribution in Iran: Hamadan (Modarres Awal, 2012; Rajabi-Mazhar et al., 2008, 2010).
Distribution in the Middle East: Iran.
Extralimital distribution: Czech Republic, France, Germany, Hungary, Moldova, Poland, Romania, Russia, Serbia, Slovakia, United Kingdom.
Host records: Summarized by Yu et al. (2016) as being a parasitoid of the following aphid species: *Aphis rumicis* L., *Metopeurum fuscoviride* Stroyan, and *Microsiphum millefolii* Wahlgren.
Parasitoid—aphid—plant associations in Iran: Hemiptera: Aphididae: *Macrosiphoniella tuberculata* (Nevsky, 1928) (Rajabi Mazhar et al., 2010).

Aphidius transcaspicus Telenga, 1958

Catalogs with Iranian records: Barahoei et al. (2014), Yu et al. (2016), Farahani et al. (2016), Samin, Coronado-Blanco, Kavallieratos et al. (2018), Samin, Coronado-Blanco, Fischer et al. (2018), Rakhshani et al. (2019).
Distribution in Iran: East Azarbaijan (Mosavi et al., 2012; Rastegar et al., 2012), Fars (Barahoei et al., 2013; Rakhshani, 2012; Starý et al., 2000; Taheri & Rakhshani, 2013), Golestan (Barahoei, Sargazi et al., 2012), Guilan (Modarres Awal, 2012; Mokhtari et al., 2000; Starý et al., 2000), Hamadan (Rakhshani, 2012), Ilam (Nazari et al., 2012; Rakhshani, 2012), Isfahan (Barahoei et al., 2013; Rakhshani, 2012; Rakhshani, Talebi, Starý et al., 2008), Kerman (Barahoei et al., 2013), Kermanshah

(Nazari et al., 2012), Khuzestan (Mossadegh et al., 2010, 2011, 2017; Salehipour et al., 2010), Markazi (Alikhani et al., 2013), Mazandaran (Sakenin et al., 2012), Northern Khorasan (Kazemzadeh et al., 2010; Rakhshani, 2012; Rakhshani, Kazemzadeh et al., 2012), Qazvin (Starý et al., 2000), Razavi Khorasan (Jafari-Ahmadabadi et al., 2011; Jafari-Ahmadabadi & Modarres Awal, 2012), Sistan and Baluchestan (New record), Tehran (Rakhshani, Talebi, Starý et al., 2008; Rakhshani, 2012), West Azarbaijan (Samin et al., 2014).

Distribution in the Middle East: Cyprus (Lozier et al., 2008, 2009), Egypt (Gadallah et al., 2017; Lozier et al., 2008, 2009), Iran (see references above), Iraq (Al-Azawi, 1970; Starý, 1969; Starý & Kaddou, 1971; Tomanović et al., 2012), Israel–Palestine (Lozier et al., 2008; 2009; Tomanović et al., 2012), Lebanon (Bartlett et al., 1978), Turkey (Aslan et al., 2004; Çetin Erdoğan et al., 2008; 2010; Lozier et al., 2008; 2009; Ölmez & Ulusoy, 2003; Tomanović et al., 2012), Saudi Arabia (Rakhshani et al., 2019).

Extralimital distribution: Algeria, Bulgaria, China, Czech Republic, France, Georgia, Greece, Hungary, India, Italy, Japan, Madeira Islands, Morocco, Pakistan, Russia, Spain, Tajikistan, Tunisia, Turkmenistan, Uzbekistan.

Host records: Summarized by Yu et al. (2016) as being a parasitoid of the following aphid species: *Amphorophora rubi* (Kaltenbach), *Aphis craccivora, A. fabae* Scopoli, *A. gossypii* Glover, *A. nasturtii, Brachycaudus amygdalinus, B. cardui* (L.), *Hyalopterus amygdali, H. pruni* (Geoffroy), *Melanaphis donacis, Phorodon humuli* (Schrank), *Rhopalosiphum maidis* (Fitch), and *R. nymphaeae* (L.).

Parasitoid–aphid–plant associations in Iran: Hemiptera: Aphididae: *Hyalopterus amygdali* (Blanchard) on *Phragmites* sp., *Prunus amygdalinus, P. armeniaca, P. dulcis,* and *P. persica* (Barahoei et al., 2013; Jafari-Ahmadabadi et al., 2011; Jafari-Ahmadabadi & Modarres Awal, 2012; Mossadegh et al., 2011; Rakhshani, 2012; Rakhshani, Talebi, Starý et al., 2008; Taheri & Rakhshani, 2013), *Hyalopterus pruni* (Geoffroy) on *Phragmites australis, Phragmites* sp., *Prunus armeniaca, P. domestica, P. dulcis,* and *P. persica* (Alikhana et al., 2013; Barahoei et al., 2013; Jafari-Ahmadabadi & Modarres Awal, 2012; Jafari-Ahmadabadi et al., 2011; Modarres Awal, 2012; Mokhtari et al., 2000; Nazari et al., 2012; Rakhshani, Talebi, Starý et al., 2008; Rakhshani, Kazemzadeh et al., 2012; Rakhshani, 2012; Starý et al., 2000; Taheri & Rakhshani, 2013; Tomanović et al., 2012), *Phorodon humuli* (Schrank) on *Prunus* sp. (Modarres Awal, 2012; Mokhtari et al., 2000; Starý et al., 2000), *Rhopalosiphum nymphaeae* (L.) on *Prunus* sp. (Mokhtari et al., 2000; Starý et al., 2000).

Comments: *Aphidius magdae* Mescheloff & Rosen, 1990 which is distributed in Israel–Palestine, and Japan is synonym to *Aphidius transcaspicus.*

Aphidius urticae Haliday, 1834
Catalogs with Iranian records: Fallahzadeh & Saghaei (2010), Barahoei et al. (2014), Yu et al. (2016), Farahani et al. (2016), Samin, Coronado-Blanco, Kavallieratos et al. (2018), Samin, Coronado-Blanco, Fischer et al. (2018), Rakhshani et al. (2019).

Distribution in Iran: Alborz (Shojai, 1998), Golestan (Barahoei, Sargazi et al., 2012), Guilan (Rakhshani, Talebi, Starý et al., 2008; Talebi et al., 2009), Kerman (Barahoei et al., 2013), Mazandaran (Starý et al., 2000), Tehran (Modarres Awal, 1997, 2012).

Distribution in the Middle East: Iran (see references above), Israel–Palestine (Mescheloff & Rosen, 1990b), Turkey (Akar & Çetin Erdoğan, 2017; Starý, 1976; Tomanović et al., 2008).

Extralimital distribution: Afghanistan, Andorra, Belgium, Bulgaria, Canada, China, Czech Republic, Finland, France, Georgia, Germany, Greece, Hungary, India, Italy, Japan, Korea, Latvia, Lithuania, Moldova, Montenegro, Morocco, Netherlands, New Zealand, Poland, Portugal, Romania, Russia, Serbia, Slovakia, Slovenia, Spain, Sweden, Turkmenistan, Ukraine, United Kingdom, United States of America, Uzbekistan.

Host records: Summarized by Yu et al. (2016) as being a parasitoid of the following aphid species: *Acyrthosiphon* spp., *Amphorophora ampullata, A. amurensis, A. gei, A. rubi, Aphis* spp., *Aulacorthum majanthami, A. solani* Kaltenbach, *Brachycolus cucubali, Hyadaphis foeniculi* Passerini, *Impatientinum asiaticum, Macrosiphum* spp., *Masanaphis* sp., *Metopholophium dirhodum* (Buckton), *Microsiphum carnosum* (Buckton), *Myzus persicae* (Sulzer), *M. varians* Davidson, *Neoacyrthosiphon holsti, Prociphalus xylostei, Rhopalomyzus lonicerae, Schizaphis scirpi, Sitobion avenae* (Fabricius), *S. dryopteridis, S. fragariae* (Walker), *Toxoptera aurantii* Boyer de Fonscolombe, *Uroleucon cichorii, U. jaceae, U. sonchi* (L.), *Wahlgreniella arbuti,* and *W. ossiannilssoni* Hille Ris Lambers.

Parasitoid–aphid–plant associations in Iran: Hemiptera: Aphididae: *Acyrthosiphon pisum* (Harris) on *Onobrychis altissima* (Modarres Awal, 1997, 2012; Rakhshani, Talebi, Starý et al., 2008; Shojai, 1998), *Acyrthosiphon gossypii* Mordvilko on *Zygophyllum fabago* (Barahoei et al., 2013), *Acyrthosiphon* sp. (Starý & González, 1978), *Microlophium carnosum* (Buckton) on *Urtica dioica,* and *Urtica* sp. (Rakhshani, Talebi, Starý et al., 2008; Starý et al., 2000; Talebi et al., 2009), *Amphorophora rubi* (Kaltenbach) (Modarres Awal, 2012).

Aphidius uzbekistanicus Luzhetzki, 1960

Catalogs with Iranian records: Fallahzadeh and Saghaei (2010), Barahoei et al. (2014), Yu et al. (2016), Farahani et al. (2016), Samin, Coronado-Blanco, Kavallieratos et al. (2018), Samin, Coronado-Blanco, Fischer et al. (2018), Rakhshani et al. (2019).

Distribution in Iran: Alborz (Shojai, 1998; Starý, 1979; Starý et al., 2000), East Azarbaijan (Ghahari et al., 2011), Fars (Barahoei et al., 2013; Rakhshani, Talebi, Starý et al., 2008; Rakhshani, Tomanović et al., 2008; Starý et al., 2000), Golestan (Barahoei, Sargazi et al., 2012; Darvish-Mojeni & Bayat-Asadi, 1995; Modarres Awal, 1997, 2012; Rakhshani, Tomanović et al., 2008; Starý et al., 2000), Hamadan, Khuzestan, Qom (Rakhshani, Tomanović et al., 2008), Hormozgan (Ameri et al., 2019), Isfahan (Barahoei et al., 2013; Rakhshani, Talebi, Starý et al., 2008; Rakhshani, Tomanović et al., 2008), Kerman (Asadizade et al., 2014; Ghotbi Ravandi et al., 2017), Kermanshah (Nazari et al., 2012), Kordestan (Mansour Ghazi et al., 2008; Rakhshani, Talebi, Starý et al., 2008; Rakhshani, Tomanović et al., 2008), Markazi (Alikhani et al., 2013; Rakhshani, Talebi, Starý et al., 2008; Rakhshani, Tomanović et al., 2008), Mazandaran (Modarres Awal, 1997, 2012; Starý et al., 2000), Northern Khorasan (Kazemzadeh et al., 2010; Rakhshani, Tomanović et al., 2008; Rakhshani, Kazemzadeh et al., 2012), Razavi Khorasan (Darsouei et al., 2011b; Rakhshani, Tomanović et al., 2008; Starý et al., 2000), Sistan and Baluchestan (Bandani, 1992; Bandani et al., 1993; Modarres Awal, 1997, 2012; Modarres Najafabadi & Gholamian, 2007; Rakhshani, Talebi, Starý et al., 2008; Rakhshani, Tomanović et al., 2008; Starý et al., 2000), Tehran (Modarres Awal, 1997, 2012; Rakhshani, Talebi, Starý et al., 2008; Rakhshani, Tomanović et al., 2008; Shahrokhi et al., 2004; Starý et al., 2000), Iran (no specific locality cited) (Khanjani, 2006a; Tazerouni et al., 2011b).

Distribution in the Middle East: Egypt (El-Serafy, 1999; Gadallah et al., 2017; Ibrahim, 1990a), Iran (see references above), Israel—Palestine (Mescheloff & Rosen, 1990b), Turkey (Çetin Erdoğan et al., 2008; Ölmez & Ulusoy, 2003; Özder & Toroz, 1999; Uysal et al., 2004).

Extralimital distribution: Andorra, Argentina, Belgium, Brazil, Bulgaria, Chile, China, Czech Republic, Denmark, Finland, France, Germany, Greece, Hungary, India, Italy, Japan, Montenegro, Morocco, Netherlands, Norway, Pakistan, Poland, Portugal, Romania, Serbia, Slovakia, Slovenia, Spain, Sweden, Tajikistan, Ukraine, United Kingdom, United States of America, Uzbekistan.

Host records: Summarized by Yu et al. (2016) as being a parasitoid of the following aphid species: *Acyrthosiphon pisum* Harris, *Aphis fabae* Scopoli, *Diuraphis noxia* (Mordvilko), *Impatientinum asiaticum*, *Mealaphis donacis*, *Metopolophium dirhodum* (Walker), *Metopolophium festucae*, *Myzus mumecola*, *M. persicae* (Sulzer), *Rhopalosiphum maidis* (Fitch), *R. padi* (L.), *Schizaphis graminum* (Rondani), *S. pyri*, *Shinjia orientalis*, *Sipha maydis*, and *Sitobion* spp.

Parasitoid—aphid—plant associations in Iran: Hemiptera: Aphididae: *Diuraphis noxia* (Kurdjumov) on cereals (Bandani et al., 1992, 1993; Khanjani, 2006a; Modarres Awal, 1997, 2012; Starý et al., 2000; Zareh et al., 1995), *Metopolophium dirhodum* (Walker) on *Triticum aestivum* (Barahoei et al., 2013; Rakhshani, Talebi, Starý et al., 2008; Rakhshani, Tomanović et al., 2008), *Rhopalosiphum maidis* (Fitch) (Bandani et al., 1992, 1993; Starý et al., 2000), *Rhopalosiphum padi* (L.) on *Triticum aestivum* (Bandani et al., 1992, 1993; Nazari et al., 2012; Rakhshani, Talebi, Starý et al., 2008; Rakhshani, Tomanović et al., 2008; Starý, 1979; Starý et al., 2000), *Schizaphis graminum* (Rondani) on *Avena fatua*, *Hordeum vulgare*, and *Triticum aestivum* (Alikhani et al., 2013; Bandani et al., 1992; Barahoei et al., 2013; Modarres Najafabadi & Gholamian, 2007; Rakhshani, Talebi, Starý et al., 2008; Rakhshani, Tomanović et al., 2008; Modarres Awal, 2012; Starý et al., 2000), *Sipha maydis* Passerini (Bandani et al., 1992, 1993; Starý et al., 2000), *Sitobion avenae* (Fabricius) on *Dactylis glomerata*, and *Triticum aestivum* (Alikhani et al., 2013; Barahoei et al., 2013; Bandani et al., 1992, 1993; Darvish-Mojeni & Bayat-Asadi, 1995; Modarres Awal, 1997, 2012; Shojai, 1998; Shahrokhi et al., 2004; Starý, 1979; Starý et al., 2000; Rakhshani, Talebi, Starý et al., 2008; Rakhshani, Tomanović et al., 2008; Rakhshani, Kazemzadeh et al., 2012).

Aphidius sp.

Distribution in Iran: Alborz, Zanjan (Shojai, 1998), Khorasan (Starý, 1979), Markazi (Radjabi, 1989), Iran (no specific locality cited) (Khanjani et al., 2006b).

Parasitoids-aphid-plant assocaitions in Iran: Hemiptera: Aphididae: *Amphorophora catharinae* (Nevsky) on *Rosa* sp. (Starý, 1979), *Aphis craccivora* Koch on *Cousinia* sp. (Starý, 1979), *Aphis loti* Kaltenbach on *Lotus gebelia* (Starý, 1979), *Brachycaudus cardui* (L.) (Khanjani et al., 2006b; Radjabi, 1989), *Brevicoryne brassicae* (L.) on *Brassica oleracea* (Khanjani et al., 2006b; Starý et al., 2000), *Chionaspis asiatica* Archangelskaya, *Pterochloroides persicae* (Cholodkovsky) (Radjabi, 1989; Shojai, 1998), *Diuraphis noxia* (Kurdjumov) (Ahmadi, 2000; Starý et al., 2000), *Myzus persicae* (Sulzer) on *Bellis* sp. (Radjabi, 1989; Starý et al., 2000).

Genus *Betuloxys* Mackauer, 1960

Betuloxys hortorum (Starý, 1960)

Catalogs with Iranian records: Fallahzadeh and Saghaei (2010), Barahoei et al. (2014), Yu et al. (2016), Farahani et al. (2016), Samin, Coronado-Blanco, Kavallieratos et al. (2018), Samin, Coronado-Blanco, Fischer et al. (2018), Rakhshani et al. (2019).

Distribution in Iran: Iran (no specific locality cited) (Mackauer & Starý, 1967; Modarres Awal, 2012 as *Trioxys hortorum*; Starý, 1979; Starý et al., 2000).

Distribution in the Middle East: Iran.

Extralimital distribution: Czech Republic, Georgia, Germany, Hungary, India, Moldova, Poland, Russia, former Yugoslavia.

Host records: Summarized by Yu et al. (2016) as being a parasitoid of the following aphid species: *Myzocallis carpini* (Koch), *M. coryli* (Goetze), *Tinocallis nevskyi* Remaudière Quednau and Heie, *T. platani* (Kaltenbach), and *T. saltans* (Nevsky).

Parasitoid–aphid–plant associations in Iran: Hemiptera: Aphididae: *Tinocallis saltans* (Nevsky) (Mackauer & Starý, 1967; Starý et al., 2000).

Genus *Binodoxys* Mackauer, 1960

Binodoxys acalephae (Marshall, 1896)

Catalogs with Iranian records: Fallahzadeh and Saghaei (2010), Barahoei et al. (2014), Yu et al. (2016), Farahani et al. (2016), Samin, Coronado-Blanco, Kavallieratos et al. (2018), Samin, Coronado-Blanco, Fischer et al. (2018), Rakhshani et al. (2019).

Distribution in Iran: Alborz, Sistan and Baluchestan (Rakhshani, Talebi, Kavallieratos et al., 2005a), Golestan (Barahoei, Sargazi et al., 2012; Ghahari, 2019; Rakhshani, 2012), Guilan, Qazvin (Talebi et al., 2009), Hamadan (Ghahari, 2019; Nazari et al., 2012; Rakhshani, 2012; Talebi et al., 2009), Isfahan (Barahoei et al., 2013), Kerman (Barahoei, Madjdzadeh et al., 2012; Barahoei et al., 2013), Kermanshah (Starý, 1979; Starý et al., 2000; Talebi et al., 2009), Khuzestan (Mossadegh et al., 2010, 2011, 2017; Nazari et al., 2012; Salehipour et al., 2010), Kordestan (Nazari et al., 2012), Markazi (Alikhani et al., 2013), Mazandaran (Ghahari et al., 2011), Northern Khorasan (Rakhshani, Kazemzadeh et al., 2012), Razavi Khorasan (Jafari-Ahmadabadi et al., 2011; Jafari-Ahmadabadi & Modarres Awal, 2012), Tehran (Rakhshani, Talebi, Kavallieratos et al., 2005; Starý, 1979; Starý et al., 2000; Talebi et al., 2009), West Azarbaijan (Modarres Awal, 2012; Starý, 1979; Starý et al., 2000).

Distribution in the Middle East: Egypt (Abdel-Rahman et al., 2000; El-Heneidy, 1994; El-Heneidy & Abdel-Samad, 2001; El-Heneidy et al., 2001, 2002, 2003, 2004), Iran (see references above), Iraq (Starý, 1976; Starý & Kaddou, 1971), Turkey (Akar & Çetin Erdoğan, 2017; Güz & Kilinçer, 2005; Ölmez & Ulusoy, 2003; Starý, 1976; Uysal et al., 2004), United Arab Emirates, Yemen (Starý et al., 2013).

Extralimital distribution: Andorra, Bulgaria, Canada, China, Czech Republic, Finland, France, Georgia, Germany, Greece, Hungary, India, Italy, Kazakhstan, Moldova, Mongolia, Montenegro, Morocco, Netherlands, Poland, Portugal, Russia, Serbia, Spain, Tajikistan, United Kingdom, Uzbekistan, former Yugoslavia.

Host records: Summarized by Yu et al. (2016) as being a parasitoid of the following aphid species: *Auraphis* sp., *Aphis* spp., *Brachycaudus amygadalinus*, *B. cardui* (L.), *B. helichrysi* (Kaltenbach), *Hysteroneura setariae*, *Macrosiphum* sp., *Phorodon* sp., *Rhopalosiphum padi* (L.), *Shinjia orientalis*, *Toxoptera odinae*, *Uhlmannia singularis*, and *Uroleucon sonchi* (L.).

Parasitoid–aphid–plant associations in Iran: Hemiptera: Aphididae: *Aphis affinis* del Guercio on *Mentha latifolia*, and *M. logifolium* (Barahoei et al., 2013; Talebi et al., 2009), *Aphis craccivora* Koch on *Alhagi maurorum*, *Gerbera jamesonii*, *Glycerrhiza glabra*, *Matricaria chamomella*, *Medicago sativa*, *Partulaca oleracea*, *Robinia pseudoacacia*, *Rumex scutatus*, *Sophora alopecuroides*, and *Vicia faba* (Alikhani et al., 2013; Barahoei et al., 2013; Nazari et al., 2012; Rakhshani, Talebi, Kavallieratos et al., 2005; Rakhshani, Kazemzadeh et al., 2012; Starý, 1979; Starý et al., 2000; Talebi et al., 2009), *Aphis euphorbiae* Kaltenbach on *Euphorbia* sp. (Talebi et al., 2009; Nazari et al., 2012), *Aphis fabae* Scopoli on *Beta vulgaris*, *Phaseolus vulgaris*, *Solanum nigrum*, and *Vicia fabae* (Barahoei et al., 2013; Nazari et al., 2012; Rakhshani, Kazemzadeh et al., 2012; Talebi et al., 2009), *Aphis gossypii* Glover on *Cucumis sativus* (Barahoei et al., 2013), *Aphis idaei* van der Goot on *Rubus idaeus* (Barahoei et al., 2014), *Aphis nerii* Boyer de Fonscolombe on *Nerium oleander* (Talebi et al., 2009), *Aphis pomi* de Geer on *Malus domestica* (Rakhshani, 2012), *Aphis spiraecola* Patch on *Cydonia oblonga* (Rakhshani, 2012), *Aphis umbrella* (Börner) on *Malva sylvestris* (Mackauer & Starý, 1967; Nazari et al., 2012; Starý et al., 2000; Talebi et al., 2009), *Aphis urticata* Gmelin on *Urtica dioica* (Talebi et al., 2009), *Aphis* sp. on *Euphorbia* sp., *Euphorbia* cf. *esula*, and *Vicia villosa* (Mossadegh et al., 2011; Starý, 1979), *Aphis* (*Protaphis*) sp. on *Tragopogon graminifolius* (Barahoei, Madjdzadeh et al., 2012; Barahoei et al., 2013), *Brachycaudus helichrysi* (Kaltenbach) on *Prunus domestica* (Jafari-Ahmadabadi et al., 2011; Jafari-Ahmadabadi & Modarres Awal, 2012).

Binodoxys angelicae (Haliday, 1833)

Catalogs with Iranian records: Fallahzadeh and Saghaei (2010), Barahoei et al. (2014), Yu et al. (2016), Farahani et al. (2016), Samin, Coronado-Blanco, Kavallieratos et al. (2018), Samin, Coronado-Blanco, Fischer et al. (2018), Rakhshani et al. (2019).

Distribution in Iran: Alborz (Rakhshani, 2012; Talebi et al., 2009), Fars (Barahoei et al., 2013; Taheri & Rakhshani, 2013; Talebi et al., 2009), Golestan (Barahoei, Sargazi et al., 2012; Ghahari & Fischer 2011; Rakhshani, 2012), Guilan (Modarres Awal, 2012; Starý et al., 2000;

Yaghubi, 1997; Yaghubi & Sahragard, 1998), Hormozgan (Ameri et al., 2019), Isfahan (Barahoei et al., 2013), Kerman (Barahoei, Madjdzadeh et al., 2012; Barahoei et al., 2013), Kermanshah (Nazari et al., 2012), Khuzestan (Modarres Awal, 2012; Moodi & Mossadegh 2006; Mossadegh et al., 2010, 2011, 2017; Nazari et al., 2012; Salehipour et al., 2010; Talebi et al., 2009), Markazi (Alikhani et al., 2013; Modarres Awal, 1997, 2012, both as *Trioxys angelicae* (Haliday); Radjabi, 1989; Starý et al., 2000), Razavi Khorasan (Darsouei et al., 2011a,b; Farokhzadeh et al., 2014; Jafari-Ahmadabadi et al., 2011; Jafari-Ahmadabadi & Modarres Awal, 2012), Tehran (Rakhshani, Talebi, Kavallieratos et al., 2005; Talebi et al., 2009).

Distribution in the Middle East: Egypt (Gadallah et al., 2017; Mackaeur & Starý, 1967; Ragab, 1996; Starý, 1976), Iran (see references above), Iraq (Al-Azawi, 1970; Starý, 1976; Starý & Kaddou, 1971), Israel—Palestine (Avidov & Harpaz, 1969; Mescheloff & Rosen, 1993; Rosen, 1967, 1969; Starý, 1976), Lebanon (Abu-Fakhr & Kawar, 1998; Hussein & Kawar, 1984; Mackauer & Starý, 1967; Starý, 1976; Tremblay et al., 1985), Turkey (Akar & Çetin Erdoğan, 2017; Aslan et al., 2004; Çetin Erdoğan et al., 2008; Güz & Kilinçer, 2005; Ölmez & Ulusoy, 2003; Uysal et al., 2004), United Arab Emirates (Starý et al., 2013).

Extralimital distribution: Algeria, Andorra, Austria, Azerbaijan, Belgium, Bulgaria, China, Czech Republic, Finland, France, Georgia, Germany, Greece, Hungary, India, Ireland, Italy, Kazakhstan, Latvia, Lithuania, Moldova, Monaco, Montenegro, Morocco, Netherlands, Pakistan, Poland, Portugal, Romania, Russia, Serbia, Slovakia, Slovenia, Spain, Switzerland, Tajikistan, Tunisia, United Kingdom, Uzbekistan.

Host records: Summarized by Yu et al. (2016) as being a parasitoid of the following aphid species: *Acyrthosiphon* spp., *Amphorophora rubi*, *Aphis* spp., *Brachycaudus cardui* (L.), *B. helichrysi* (Kaltenbach), *B. persicae* (L.), *B. prunicola*, *Brevicoryne brassicae* (L.), *Capitophorus elaeagni*, *Ceruraphis ribis*, *Dysaphis* spp., *Greenidea ficicola*, *Hyalopterus pruni*, *Lipaphis lepidii*, *Macrosiphum euphorbiae* (Thomas), *Mariaella lambersi*, *Melanaphis pyraria*, *Myzocallis coryli*, *Myzus persicae* (Sulzer), *Ovatus insitus* (Walker), *Rhopalosiphum maidis* (Fitch), *R. padi* (L.), *Sitobion avenae* (Fabricus), *Toxoptera aurantii* Boyer de Fonscolombe, and *Uhlmannia singularis* (Börner).

Parasitoid—aphid—plant associations in Iran: Hemiptera: Aphididae: *Aphis acetosae* L. on *Rumex* sp. (Barahoei et al., 2013), *Aphis affinis* del Guercio on *Mentha longifolia*, and *M. piperita* (Barahoei et al., 2013; Mossadegh et al., 2011; Taheri & Rakhshani, 2013), *Aphis craccivora* Koch on *Alhagi maurorum*, *Medicago polymorpha*, *M. officinalis*, *M. sativa*, *Rumex* sp., *Solanum nigrum*, and

Vicia faba (Barahoei et al., 2013; Mossadegh et al., 2011; Nazari et al., 2012; Rakhshani, Talebi, Kavallieratos et al., 2005), *Aphis euphorbiae* on *Euphorbia* sp., *Aphis fabae* Scopoli on *Carduus* sp., *C. getulus*, *Chenopodium album*, *Chrysanthemum pyrethrum*, *Cestrum parqui*, *Marrubium* sp., *Rumex* sp., *Solanum nigrum*, and *Vicia faba* (Alikhani et al., 2013; Barahoei, Madjdzadeh et al., 2012; Barahoei et al., 2013; Modarres Awal, 2012; Mossadegh et al., 2011; Nazari et al., 2012; Taheri & Rakhshani, 2013; Talebi et al., 2009), *Aphis fabae solanella* Theobald on *Solanum alatum*, and *S. nigrum* (Alikhani et al., 2013; Barahoei, Madjdzadeh et al., 2012), *Aphis farinosa* Gmelin on *Salix aegyptiaca* (Moodi & Mossadegh 2006; Starý et al., 2000; Yaghubi, 1997), *Aphis gossypii* Glover on *Althaea rosea*, *Catalpa bignonoides*, *Cestrum nocturnum*, *C. pyrethrum*, *Cucurbita pepo*, *Dahlia excelsa*, *Duranta plumieri*, *Gossypium herbaceum*, *Hibiscus rosa-sinensis*, *H. syriacus*, *Hibiscus* sp., *Lycopersicum esculentum*, *Malva parviflora*, *M. piperita*, *Prosopis juliflora*, *Solanum nigrum*, and *Tecoma stans* (Barahoei et al., 2013; Mossadegh et al., 2011; Nazari et al., 2012; Starý et al., 2000; Yaghubi, 1997), *Aphis hederae* Kaltenbach on *Hedera helix* (Starý et al., 2000; Yaghubi, 1997), *Aphis idaei* van der Goot on *Rubus persicus* (Barahoei et al., 2013; Taheri & Rakhshani, 2013), *Aphis nerii* Boyer de Fonscolombe on *Nerium oleander* (Talebi et al., 2009), *Aphis pomi* de Geer on *Malus domistica* (Modarres Awal, 1997, 2012; Radjabi, 1989; Rakhshani, 2012; Starý et al., 2000), *Aphis polygonata* Rusanova on *Polygonum aviculare* (Mossadegh et al., 2011), *Aphis punicae* Passerini (Farokhzadeh et al., 2014), *Aphis rumicis* L. on *Rumex crispus* (Talebi et al., 2009), *Aphis solanella* Theobald (Barahoei, Madjdzadeh et al., 2012; Barahoei et al., 2013), *Aphis spiraecola* Patch on *Crataegus monogyna* (Rakhshani, 2012), *Aphis umbrella* (Börner) on *Malva neglecta*, *M. parviflora*, and *M. sylvestris* (Mackauer & Starý, 1967; Mossadegh et al., 2011; Nazari et al., 2012; Starý et al., 2000; Talebi et al., 2009), *Aphis* sp. on *Althaea rosea*, *Capsella bursa-pastoris*, *Cosmos* sp., *Duranta plumieri*, *Hibiscus rosa-sinensis*, *Lamium* sp., *Lantana camera*, *Malva* sp., *M. piperita*, *M. pulegium*, *Rumex acetosa*, *Scrophularia* sp., and *Tagetes* sp. (Mossadegh et al., 2011), *Brachycaudus helichrysi* (Kaltenbach) on *Duranta plumieri*, and on *Prunus domestica* (Jafari-Ahmadabadi et al., 2011; Jafari-Ahmadabadi & Modarres Awal, 2012; Mossadegh et al., 2011), *Hypeomyzus* sp. on *Silybum marianum* (Mossadegh et al., 2011), *Lipaphis lepidii* Nevsky on *Cardaria draba* (Talebi et al., 2009), *Myzus persicae* (Sulzer) on *Beta vulgaris*, *Callendula officinalis*, *C. pyrethrum*, *Emex spinosus*, *Malva parviflora*, *Papaver dubium*, *Silybum marianum*, and *Tagetes* sp. (Mossadegh et al., 2011), *Myzus* sp. on *Papaver rhoeas* (Mossadegh et al., 2011).

Binodoxys brevicornis (Haliday, 1833)

Catalogs with Iranian records: Fallahzadeh and Saghaei (2010), Barahoei et al. (2014), Yu et al. (2016), Farahani et al. (2016), Samin, Coronado-Blanco, Kavallieratos et al. (2018), Samin, Coronado-Blanco, Fischer et al. (2018), Rakhshani et al. (2019).

Distribution in Iran: Guilan (Yaghubi, 1997; Yaghubi & Sahragard, 1998; Starý et al., 2000; Modarres Awal, 2012 as *Troxys brevicornis* (Haliday)), Markazi (Alikhani et al., 2013; Barahoei et al., 2013; Rakhshani, Talebi, Starý et al., 2007; Talebi et al., 2009), Iran (no specific locality cited) (Modarres Awal, 2012; Rahimi et al., 2010; Rakhshani, Talebi, Manzari et al., 2006).

Distribution in the Middle East: Iran (see references above), Israel–Palestine (Mescheloff & Rosen, 1993; Starý, 1976), Turkey (Güz & Kilinçer, 2005; Starý, 1976, 1990).

Extralimital distribution: Algeria, Andorra, Argentina, Austria, Brazil, Bulgaria, Czech Republic, Finland, France, Germany, Hungary, India, Ireland, Italy, Moldova, Montenegro, Poland, Romania, Russia, Serbia, Slovakia, Spain, Sweden, Switzerland, United Kingdom, Uzbekistan, Venezuela, Vietnam.

Host records: Summarized by Yu et al. (2016) as being a parasitoid of the following aphid species: *Brachycorynella asparagi* (Mordvilko), *Cavariella aegopodii* (Scopoli), *Chaitophorus pakistanicus* Hille Ris Lambers, *Dysaphis anthrisci* Börner, *Haydaphis bupleuri* Börner, *H. coriandri* (Das), *H. foeniculi* Passerini, *Macrosiphum* sp., *Myzus cerasi* (Fabricius), *M. persicae* (Sulzer), *Semiaphis dauci* (Fabricius), *Staegeriella necopinata* (Börner), and *Uhlmannia singularis* (Börner).

Parasitoid–aphid–palnt associations in Iran: Hemiptera: Aphididae: *Cavariella aegopodii* (Scopoli) on *Salix alba* (Alikhani et al., 2013; Barahoei et al., 2013; Rahimi et al., 2010; Rakhshani, Talebi, Starý et al., 2007; Talebi et al., 2009), *Hyadaphis* sp. on *Turgenia* sp. (Starý et al., 2000; Yaghubi, 1997).

Binodoxys centaureae (Haliday, 1833)

Catalogs with Iranian records: No catalog.

Distribution in Iran: Northern Khorasan (Naderian et al., 2020).

Distribution in the Middle East: Cyprus (Fulmek, 1968; Wilkinson, 1926), Iran (Naderian et al., 2020).

Extralimital distribution: Algeria, Andorra, Austria, China, Czech Republic, Finland, France, Georgia, Germany, Greece, Hungary, India, Ireland, Italy, Japan, Korea, Moldova, Montenegro, Poland, Romania, Russia, Serbia, Slovakia, Slovenia, Spain, Sweden, Switzerland, Tajikistan, United Kingdom, Uzbekistan.

Host records: Summarized by Yu et al. (2016) as being a parasitoid of the aphids *Amphorophora rubi* (Kaltenbach), *Aphis craccae* L., *A. fabae* Scopoli, *A. pomi* De Geer, *A. rumicis* L., *Brachycaudus helichrysi* (Kaltenbach), *Macrosiphoniella artemisiae* (Fonscolombe), *M. millefolii* (De Geer), *Macrosiphum rosaeibarae* Tao, *Microlophium cornosum* (Buckton), *Rhopalosiphum padi* (L.), *Sitobion ibarae* (Matsumura), *S. miscanthi* (Takahashi), and *Uroleucon* spp.

Parasitoid–aphid–plant associations in Iran: *Uroleucon cichori* (Koch) (Hemiptera: Aphididae) on *Centaurea* sp. (Naderian et al., 2020)

Binodoxys heraclei (Haliday, 1833)

Catalogs with Iranian records: Fallahzadeh and Saghaei (2010), Barahoei et al. (2014), Yu et al. (2016), Farahani et al. (2016), Samin, Coronado-Blanco, Kavallieratos et al. (2018), Samin, Coronado-Blanco, Fischer et al. (2018), Rakhshani et al. (2019).

Distribution in Iran: Kermanshah (Nazari et al., 2012), Razavi Khorasan (Rakhshani, Talebi, Starý et al., 2007; Talebi et al., 2009), Iran (no specific locality cited) (Modarres Awal, 2012; Rakhshani, Talebi, Manzari et al., 2006).

Distribution in the Middle East: Iran (see references above), Turkey (Rakhshani et al., 2019).

Extralimital distribution: Andorra, Belgium, Czech Republic, France, Georgia, Germany, Hungary, Ireland, Italy, Montenegro, Nepal, Netherlands, Poland, Serbia, Slovenia, Spain, Tajikistan, United Kingdom.

Host records: Summarized by Yu et al. (2016) as being a parasitoid of the following aphid species: *Aphis fabae* Scopoli, *A. rumicis* L., *Capitophorus hippophaes* Walker, *Cavariella aegopodii* (Scopoli), *C. aquatica* (Gillette and Bragg), *C. archangelicae* (Scopoli), *C. theobaldi* (Gillette and Bragg), *Cryptomyzus galeopsidis* Kaltenbach, *Pterocomma salicis* (L.), *Semiaphis* sp., and *Uroleucon sonchi* (L.).

Parasitoid–aphid–plant associations in Iran: Hemiptera: Aphididae: *Cavariella aspidaphoides* Hille Ris Lambers on *Salix babylonica*, *S. fragilis*, and *Salix* sp. (Nazari et al., 2012; Rakhshani, Talebi, Starý et al., 2007; Talebi et al., 2009).

Binodoxys sp.

Distribution in Iran: Kermanshah (Bagheri-Matin et al., 2008).

Parasitoid–aphid–plant associations in Iran: Hemiptera: Aphididae: *Metopolophium dirhodum* (Walker) on *Rosa canina* (Barahoei et al., 2013), *Sitobion avenae* on *Triticum aestivum* (Fabricius) (Bagheri-Matin et al., 2008).

Genus *Diaeretiella* Starý, 1960

Diaeretiella rapae (M'Intosh, 1855)

Catalogs with Iranian records: Fallahzadeh and Saghaei (2010), Fallahzadeh and Saghaei (2010), Barahoei et al. (2014), Yu et al. (2016), Farahani et al. (2016), Samin, Coronado-Blanco, Kavallieratos et al. (2018), Samin, Coronado-Blanco, Fischer et al. (2018), Rakhshani et al. (2019).

Distribution in Iran: Alborz (Malkeshi et al., 2004; Starý, 1979; Starý et al., 2000; Talebi et al., 2009), Ardabil (Lotfalizadeh 2002a,b), Fars (Ahmadi & Sarafrazi, 1993; Barahoei et al., 2013; González et al., 1992; Keyhanian et al., 2005; Khazduzi Nejad Jamali et al., 2012; Modarres Awal, 2012; Rakhshani, Tomanović et al., 2008; Starý et al., 2000; Taheri & Rakhshani, 2013), Golestan (Barahoei, Sargazi et al., 2012; Ghahari & Fischer 2011; Keyhanian et al., 2005; Malkeshi et al., 2004, 2010; Rakhshani, Tomanović et al., 2008), Guilan (Sakenin et al., 2012; Shojai et al., 1999; Starý et al., 2000; Yaghubi & Sahragard, 1998), Hamadan (Starý et al., 2000; Rakhshani, Tomanović et al., 2008), Hormozgan (Ameri et al., 2019), Isfahan (Barahoei et al., 2013; Ghahari, 2019; Rakhshani, Tomanović et al., 2008; Rakhshani et al., 2010; Talebi et al., 2009), Kerman (Barahoei, Madjdzadeh et al., 2012, Barahoei et al., 2013; Ghotbi Ravandi et al., 2017, in alfalfa and wheat fields), Kermanshah (Bagheri-Matin et al., 2008; Nazari et al., 2012; Rakhshani, Tomanović et al., 2008), Khuzestan (Farsi et al., 2009, 2012; Keyhanian et al., 2005; Khajehzadeh et al., 2006; Malkeshi et al., 2004, 2010; Modarres Awal, 1997, 2012; Mossadegh & Kocheili, 2003; Mossadegh et al., 2010, 2011, 2017; Rezaei et al., 2006; Salehipour et al., 2010), Kordestan (Mansour Ghazi et al., 2008; Rakhshani, Tomanović et al., 2008), Lorestan (Malkeshi et al., 2004), Markazi (Alikhani et al., 2013; Barahoei et al., 2013; Keyhanian et al., 2005; Rakhshani, Tomanović et al., 2008), Mazandaran (Ghahari et al., 2010; Keyhanian et al., 2005; Sakenin et al., 2012; Shojai et al., 1999; Starý et al., 2000), Northern Khorasan (Kazemzadeh et al., 2010; Rakhshani, Kazemzadeh et al., 2012; Starý et al., 2000), Qazvin (Davatchi & Shojai, 1968; Shojai, 1998; Starý et al., 2000), Qom (Barahoei et al., 2013; Rakhshani, Tomanović et al., 2008), Razavi Khorasan (Darsouei et al., 2011a,b; Jafari-Ahmadabadi et al., 2011; Jafari-Ahmadabadi & Modarres Awal, 2012; Rakhshani, 2012; Rakhshani, Tomanović et al., 2008), Semnan (Starý et al., 2000), Sistan and Baluchestan (Bandani, 1992; Bandani et al., 1993; Fakhireh et al., 2016; Modarres Awal, 2012; Modarres Najafabadi et al., 2005; Modarres Najafabadi & Gholamian, 2007; Rakhshani, Talebi, Manzari et al., 2006; Rakhshani, Tomanović et al., 2008; Starý et al., 2000; Talebi et al., 2009), Tehran (Hodjat & Moradeshaghi, 1988; Kazemi et al., 2020; Malkeshi et al., 2010; Modarres Awal, 2012; Rakhshani, Tomanović et al., 2008; Shahrokhi et al., 2004;

Shojai et al., 2005; Starý et al., 2000; Talebi et al., 2009), West Azarbaijan (Keyhanian et al., 2005; Malkeshi et al., 2004, 2010; Modarres Awal, 2012; Samin et al., 2014), Zanjan (Modarres Awal, 2012), Iran (no specific locality cited) (Khanjani, 2006a,b; Shojai et al., 2003; Tazerouni et al., 2011b).

Distribution in the Middle East: Cyprus, Egypt, Iran, Iraq, Israel—Palestine, Jordan, Lebanon, Saudi Arabia, Syria, Turkey, Yemen (Yu et al., 2016).

Extralimital distribution: Afghanistan, Algeria, Andorra, Argentina, Australia, Austria, Azerbaijan, Belgium, Brazil, Bulgaria, Canada, Cape Verde, Chile, China, Costa Rica, Croatia, Cuba, Czech Republic, Finland, France, Georgia, Germany, Greece, Hungary, India, Ireland, Italy, Japan, Kazakhstan, Kenya, Korea, Kyrgyzstan, Latvia, Libya, Lithuania, North Macedonia, Mexico, Moldova, Mongolia, Montenegro, Morocco, Netherlands, New Zealand, Norway, Pakistan, Peru, Poland, Portugal, Puerto Rico, Romania, Russia, Serbia, Slovakia, Slovenia, South Africa, Spain, Sri Lanka, Tajikistan, Ukraine, United Kingdom, United States of America, Uruguay, Uzbekistan, Venezuela.

Host records: Summarized by Yu et al. (2016) as being a parasitoid of the following aphid species: *Acyrthosiphon* sp., *Anuraphis* sp., *Aphis* spp., *Aulacorthum solani* (Kaltenbach), *Brachycaudus amygdalinus* (Schouteden), *B. cardui* (L.), *B. helichrysi* Kaltenbach, *B. persicae* Passerini, *B. rumexicolens* (Patch), *B. tragopogonis* (Kaltenbach), *Brachycolus cucubali* (Passerini), *Brachycorynella asparagi* (Mordvilko), *Brachyunguis lycii* L., *Braggia* sp., *Brevicoryne barbareae* Nevsky, *B. brassicae* (L.), *Calaphis betulaecolens* (Fitch), *Capitophorus inulae* (Passerini), *Cavariella salicicola* (Matsumura), *Diuraphis bromicola* (Hille Ris Lambers), *D. frequens* (Walker), *D. muehlei* (Boerner), *D. noxia* (Mordvilko), *Dysaphis devecta* (walker), *D. plantaginea* (Passerini), *D. tulipae* (Boyer de Fonscolombe), *Ericaphis leclanti* Remaudière, *Glyphina betulae* (L.), *Greenidea ficicola* Takahashi, *Hayhurstia atriplicis* (L.), *Hyadaphis coriandri* (Das), *H. foeniculi* Passerini, *Hyalopterus amygdali* Blanchard, *H. pruni* (Geoffroy), *Hyperomyzus lactucae* (L.), *Hysteroneura setariae* (Thomas), *Illinoia leriodendri* (Monell), *Lipaphis* sp., *Macrosiphoniella absinthii* (L.), *M. pseudoartemisiae* Shinji, *M. sanborni* (Gillette), *Macrosiphum euphorbiae* (Thomas), *M. rosaeiforme* (Atwal and Dingra), *Mariaella lambersi* Szelegiewicz, *Melanaphis donacis* (Passerini), *M. sacchari* (Zehntner), *Metopholophium dirhodum* (Walker), *Myzus* spp., *Nasonovia ribisnigri* (Mosley), *Phorodon humuli* (Schrank), *Pleotrichophorus pycnorhysus* Knowlton and Remaudière, *Pseudobrevicoryne leclanti* Petrović and Remaudière, *Pterochloroides persicae* (Cholodkovsky), *Rhopalosiphum insertum* (Walker), *R. maidis* (Fitch), *R. padi* (L.), *Saltusaphis scripus* (Theobald),

Schizaphis graminum (Rondani), *S. longicaudata* Hille Ris Lambers, *Sitobion avenae* (Fabricius), *S. fragariae* (Walker), *Toxoptera aurantii* (Boyer de Fonscolombe), *Uroleucon ivae* Robinson, *U. sonchi* (L.), and *Xerobion eriosomatinum* Nevsky.

Parasitoid—aphid—plant associations in Iran: Hemiptera: Aphididae: *Amegosiphon platicaudum* (Narzikulov) on *Berberis thunbergii* (Nazari et al., 2012), *Aphis craccivora* Koch on *Atriplex* sp., *Cardaria draba*, and *Glycyrrhiza glabra* (Barahoei et al., 2013; Nazari et al., 2012; Rakhshani, Talebi, Manzari et al., 2006; Rakhshani et al., 2010; Taheri & Rakhshani, 2013), *Aphis fabae* Scopoli on *Capsella bursa-pastoris*, *Lactuca sativa*, *Phaseolus vulgaris*, and *Spinacia oleracea* (Alikhani et al., 2013; Barahoei et al., 2013), *Aphis gossypii* Glover on *Cucumis sativus*, *Hibiscus rosa-sinensis*, *Prosopis juliflora*, and *Salsola* sp. (Barahoei et al., 2013; Khanjani, 2006b; Mossadegh et al., 2011), *Aphis punicae* Passerini on *Punica granatum* (Nazari et al., 2012), *Aphis solanella* Theobald on *Solanum melongena*, *Aphis umbrella* (Börner) on *Malva sylvestris* (Barahoei et al., 2013), *Aphis* sp. on *Ammi majus* (Mossadegh et a., 2011), *Brachycaudus amygdalinus* (Schouteden) on *Prunus persica* (Starý, 1979; Starý et al., 2000), *Brachycaudus cardui* (L.) on *Prunus armeniaca* (Modarres Awal, 2012; Starý, 1979; Starý et al., 2000), *Brevicoryne brassicae* (L.) on *Brassica gongiloides*, *B. juncea*, *B. napus*, *B. napus* var. *oleifera*, *B. oleracea*, *B. rapa*, *Brassica* sp., *Crambe orientalis*, *Descurainia sophia*, *Fortuyia bungei*, *Isatis* sp., *Raphanus* sp., and *Sisymbrium irio* (Alikhani et al., 2013; Barahoei, Madjdzadeh et al., 2012; Barahoei et al., 2013; Davatchi & Shojai, 1968; Farsi et al., 2009; Fathipour et al., 2006; Khajehzadeh et al., 2006; Khanjani, 2006b; Khazduzi Nejad Jamali et al., 2012; Lotfalizadeh, 2002a; Malkeshi et al., 2004, 2010; Modarres Awal, 1997; 2012; Modarres Najafabadi et al., 2005; Mossadegh et al., 2011; Nazari et al., 2012; Rakhshani, Kazemzadeh et al., 2012; Sanati et al., 2012; Shojai, 1998; Starý, 1979; Starý et al., 2000; Taheri & Rakhshani, 2013; Talebi et al., 2009), *Brevicoryne* sp. (Starý et al., 2000), *Diuraphis noxia* (Kurdjumov) on *Brumus tectorum*, *Hordeum* sp., and *Triticum aestivum* (Ahmadi, 2000; Ahmadi & Sarafrazi, 1993; Alikhani et al., 2013; Bandani, 1992; Bandani et al., 1993; Barahoei et al., 2013; González et al., 1992; Modarres Awal, 1997, 2012; Nazari et al., 2012; Rakhshani, Tomanović et al., 2008; Rakhshani, Kazemzadeh et al., 2012; Starý et al., 2000; Tazerouni et al., 2011a; Zareh et al., 1995), *Dysaphis devecta* (Walker) on *Malus orientalis* (Alikhani et al., 2013), - *Hayhurstia atriplicis* (L.) on *Atriplex* sp., and *Chenopodium album* (Modarres Awal, 2012; Nazari et al., 2012;

Rakhshani, Kazemzadeh et al., 2012; Starý, 1979; Starý et al., 2000), *Hyalopterus pruni* (Geoffroy) on *Prunus armeniaca*, and *P. domestica* (Jafari-Ahmadabadi et al., 2011; Jafari-Ahmadabadi & Modarres Awal, 2012), *Lipaphis erysimi* (Kaltenbach) on *Brassica chinensis*, *B. gongiloides*, *B. kaber*, *B. napus*, *B. oleracea*, *B. rapa*, *Eruca* sp., *Lipaphis* sp., *Raphanus sativus*, and *Thlaspi* sp. (Farsi et al., 2009; Malkeshi et al., 2010; Modarres Awal, 2012; Mossadegh et al., 2011; Starý, 1979; Starý et al., 2000), *Lipaphis lepidii* Nevsky on *Cardaria draba*, *Lipidium draba*, and *L. latifolium* (Mossadegh et al., 2011; Nazari et al., 2012; Starý et al., 2000; Talebi et al., 2009), *Lipaphis pseudobrassicae* (Davis) on *Brassica napus* (Barahoei et al., 2013), *Mariaella lambersi* Szelegiewicz on *Myricaria germanica* (Starý, 1979; Starý et al., 2000), *Metopolophium dirhodum* (Walker) on *Triticum aestivum* (Rakhshani, Tomanović et al., 2008), *Myzus beybienkoi* (Narzykulov) on *Fraxinus oxycarpa* (Starý, 1979; Starý et al., 2000), *Myzus* sp. (Shojai et al., 1999), *Myzus persicae* (Sulzer) on *Achillea* sp., *Alhagi maurorum*, *Althaea rosea*, *Ammi majus*, *Brassica oleracea*, *B. napus*, *Capsella bursa-pastoris*, *Convolvulus arvensis*, *Dianthus* sp., *Eruca* sp., *Glycyrrhiza glabra*, *Lepidium draba*, *Prunus armeniaca*, *P. persica*, *Raphanus sativus*, *Silybum marianum*, *Solanum melongena*, *Spinacia oleracea*, and *Tragopogon graminifolius* (Barahoei et al., 2013; Farsi et al., 2009; Hodjat & Moradeshaghi, 1988; Jafari-Ahmadabadi & Modarres Awal, 2012; Jafari-Ahmadabadi et al., 2011; Modarres Awal, 1997, 2012; Mossadegh et al., 2011; Nazari et al., 2012; Rakhshani, 2012; Starý et al., 2000; Starý, 1979; Taheri & Rakhshani, 2013), *Nasonovia ribisnigri* Mosely (Farsi et al., 2012), *Rhopalosiphum maidis* (Fitch) on *Hordeum* sp., and *Triticum* sp. (Bandani, 1992; Bandani et al., 1993; Barahoei et al., 2013; Starý, 1979; Starý et al., 2000), *Rhopalosiphum padi* (L.) on *Hordeum vulgare*, *Triticum aestivum*, and *Triticum* sp. (Bandani, 1992; Bandani et al., 1993; Barahoei et al., 2013; Mossadegh et al., 2011; Rakhshani, Tomanović et al., 2008; Shahrokhi et al., 2004; Starý et al., 2000), *Saltusaphis scirpus* Theobald on *Carex* sp. (Starý et al., 2000), *Schizaphis graminum* (Rondani) on *Brumus* sp., and *Triticum aestivum* (Bandani, 1992; Bandani et al., 1993; Barahoei et al., 2013; Kazemi et al., 2020; Modarres Najafabadi & Gholamian, 2007; Rakhshani, Tomanović et al., 2008; Starý, 1979; Shahrokhi et al., 2004; Starý et al., 2000), *Sitobion avenae* (Fabricius) on *Gastridium phleoides*, and *Triticum aestivum* (Bandani, 1992; Bandani et al., 1993; Barahoei et al., 2013; Mossadegh et al., 2011; Rakhshani, Tomanović et al., 2008; Rakhshani, Kazemzadeh et al., 2012; Starý et al., 2000).

Comments: Various biological characteristics and parameters of *Diaeretiella rapae* have been studied by Tazerouni et al. (2011a, 2012a,b,c, 2013) and Elahii et al. (2017).

Genus *Diaeretus* Foerster, 1863

Diaeretus leucopterus (Haliday, 1834)

Catalogs with Iranian records: No catalog.

Distribution in Iran: West Azarbaijan (Samin, Beyarslan, Coronado-Blanco et al., 2020).

Distribution in the Middle East: Iran (Samin, Beyarslan, Coronado-Blanco et al., 2020), Israel—Palestine (Mescheloff & Rosen, 1990b; Starý, 1976).

Extralimital distribution: Andorra, Bulgaria, China, Czech Republic, France, Germany, Greece, Hungary, India, Italy, Japan, Korea, Moldova, Poland, Russia, Serbia, Slovakia, Spain, Tajikistan, Thailand, Ukraine, United Kingdom, Uzbekistan.

Host records: Summarized by Yu et al. (2016) as being a parasitoid of the following aphid species: *Brachycaudus cardui* (L.), *Cinara piniformosana* (Takahashi), *Eulachnus agilis* (Kaltenbach), *E. mediterraneus* Binazzi, *E. rileyi* (Williams), *E. thunbergii* Wilson, *E. tuberculostemmata* (Theobald), *Mindarus abietinus* Koch, *Schizaphis graminum* (Rondani), and *Schizolachnus pineti* (Fabricius).

Aphid—plant—host associations in Iran: Aphididae: *Schizaphis graminum* (Rondani) on *Hordeum vulgare* (Samin, Beyarslan, Coronado-Blanco et al., 2020).

Genus *Lipolexis* Foerster, 1863

Lipolexis gracilis Foerster, 1863

Catalogs with Iranian records: Barahoei et al. (2014), Yu et al. (2016), Farahani et al. (2016), Samin, Coronado-Blanco, Kavallieratos et al. (2018), Samin, Coronado-Blanco, Fischer et al. (2018), Rakhshani et al. (2019).

Distribution in Iran: Markazi (Alikhani et al., 2013; Barahoei et al., 2013), Northern Khorasan (Rakhshani, Kazemzadeh et al., 2012), Iran (no specific locality cited) (Modarres Awal, 2012; Rakhshani, Talebi, Manzari et al., 2008).

Distribution in the Middle East: Iran (see references above), Lebanon (Abou-Fakhr, 1982; Abou-Fakhr & Kawar, 1998; Tremblay et al., 1985), Turkey (Aslan & Karaca, 2005; Tomanović et al., 2008).

Extralimital distribution: Indo-Malayan and Palaearctic regions; introduced into Burundi [Adjacent countries to Iran: Kazakhstan, Pakistan, and Russia].

Host records: Summarized by Yu et al. (2016) as being a parasitoid of the following aphid species: *Acyrthosiphon loti* (Theobald), *A. rubi* Narzikulov, *Anoeca corni* (Fabricius), *Aphis* spp., *Brachycaudus* spp., *Capitophorus pakansus* Hottes and Frison, *Dysaphis apiifolia* (Theobald), *D. plantaginea* (Passerini), *D. pyri* (Boyer de Fonscolombe), *H. pruni* (Geoffroy), *Liosomaphis himalayensis* (Narzikolov), *Lipaphis erysimi* (Kaltenbach), *Macchiatiella* sp., *Melanaphis sacchari* (Zehntner), *Metopeurum fuscoviride* Stroyan, *Myzocallis coryli* (Goeze), *Myzus cerasi* (Fabricius), *M. lythri* (Schrank), *M. persicae* (Sulzer), *M. varians* Davidson, *Rhopalosiphum maidis* (Fitch), *R. padi* (L.), *Semiaphis heraclei* (Takahashi), *Therioaphis trifolii* (Monell), *Toxoptera aurantii* Boyer de Fonscolombe, *T. citricida* (Kirkaldy), *T. odinae* (Goot), and *Toxopterina* sp.

Parasitoid—aphid—plant associations in Iran: Hemiptera: Aphididae: *Aphis salviae* Walker on *Salvia nemorosa* (Alikhani et al., 2013; Barahoei et al., 2013), *Brachycaudus amygdalinus* (Schouteden) on *Amygdalis scoparia* (Rakhshani, Kazemzadeh et al., 2012).

Comments: A taxonomic revision of the *Lipolexis gracilis* Foerster—species group has been studied by Poodineh Moghaddam et al. (2012).

Genus *Lysiphlebus* Foerster, 1863

Lysiphlebus cardui (Marshall, 1896)

Catalogs with Iranian records: Barahoei et al. (2014), Rakhshani et al. (2019).

Distribution in Iran: Markazi (Alikhani et al., 2013).

Distribution in the Middle East: Iran.

Extralimital distribution: Algeria (Laamari et al., 2012).

Host records: Summarized by Yu et al. (2016) as being a parasitoid of the following aphid species: *Acyrthosiphon cyparissiae* (Koch), *Aphis* spp., *Aulacorthum solani* (Kaltenbach), *Brachycaudus cardui* (L.), *B. tragopogonis* (Kaltenbach), *Cavariella archangelicae* (Scopoli), *Hyalopterus pruni* (Geoffroy), *Myzus persicae* (Sulzer), and *Uroleucon cirsii* (L.).

Parasitoid—aphid—plant associations in Iran: Hemiptera: Aphididae: *Aphis fabae* Scopoli (Alikhani et al., 2013).

Comments: This species was treated as a synonym with *Lysiphlebus fabarum* (Marshall, 1896) (Yu et al., 2016); however, Rakhshani et al. (2019) treated both as valid species on the basis of several morphological characters.

Lysiphlebus confusus Tremblay and Eady, 1978

Catalogs with Iranian records: Fallahzadeh and Saghaei (2010), Barahoei et al. (2014), Yu et al. (2016), Farahani et al. (2016), Samin, Coronado-Blanco, Kavallieratos et al. (2018), Samin, Coronado-Blanco, Fischer et al. (2018), Rakhshani et al. (2019).

Distribution in Iran: Ardabil (Talebi et al., 2009), Fars (Barahoei et al., 2013; Taheri & Rakhshani, 2013), Guilan (Modarres Awal, 2012; Mokhtari et al., 2000; Rahimi et al., 2012; Starý et al., 2000; Talebi et al., 2009; Yaghubi, 1997; Yaghubi & Sahragard, 1998), Hamadan (Nazari et al., 2012; Rajabi Mazhar et al., 2010; Rakhshani, Talebi, Starý et al., 2007; Talebi et al., 2009), Kerman (Barahoei, Madjdzadeh et al., 2012; Barahoei et al., 2013), Kermanshah (Nazari et al., 2012), Khuzestan (Modarres Awal 2012; Moodi & Mossadegh 2006; Mossadegh et al., 2017; Rahimi et al., 2012), Markazi (Alikhani et al., 2013), Mazandaran (Aghajanzadeh et al., 1995; Starý et al., 2000), Northern Khorasan (Kazemzadeh et al., 2010; Rakhshani, Kazemzadeh et al., 2012; Starý, 1979; Starý et al., 2000), Razavi Khorasan (Ghahari et al., 2011; Starý, 1979; Starý et al., 2000), Sistan and Baluchestan (Rakhshani, Talebi, Kavallieratos et al., 2005), Tehran (Rakhshani, Talebi, Kavallieratos et al., 2005; Starý, 1979; Starý et al., 2000; Talebi et al., 2009), Iran (no specific locality cited) (Rakhshani, Talebi, Manzari et al., 2006).

Distribution in the Middle East: Egypt (Mackauer & Starý, 1967; Starý, 1976), Iran (see references above), Iraq (Al-Azawi, 1970; Starý & Kaddou, 1971), Israel–Palestine (Avidov & Harpaz, 1969; Mescheloff & Rosen, 1990a; Rosen, 1967, 1969; Starý, 1976), Lebanon (Abou-Fakhr, 1982; Abou-Fakhr & Kawar, 1998; Hussein & Kawar, 1984; Tremblay et al., 1985), Turkey (Akar & Çetin Erdoğan, 2017; Aslan et al., 2004; Güz & Kilinçer, 2005; Starý, 1976; Ölmez & Ulusoy, 2003), United Arab Emirates (Starý et al., 2013).

Extralimital distribution: Andorra, China, Czech Republic, Germany, Greece, India, Italy, Lithuania, Moldova, Montenegro, Norway, Poland, Portugal, Russia, Serbia, Spain, former Yugoslavia.

Host records: Summarized by Yu et al. (2016) as being a parasitoid of the following aphid species: *Acyrthosiphon* sp., *Aphis* spp., *Brachycaudus amygdalinus* (Schouteden), *B. cardui* L., *B. helichrysi* (Kaltenbach), *Callipterinella calliptera* (Hartig), *Chaitophorus salijaponicus* Essig and Kuwana, *Dysaphis apiifolia* (Theobald), *D. crataegi* (Kaltenbach), *D. foeniculus* (Theobald and F.V.), *D. pyri* (Boyer de Fonscolombe), *Ephedraphis ephedrae* (Nevsky), *Hyadaphis foeniculi* Passerini, *Macrosiphoniella tanacetaria* Kaltenbach, *M. tapuskae* (Hottes and Frison), *Melanaphis sacchari* (Zehntner), *Myzus dycei* Carver, *M. persicae* (Sulzer), *Semiaphis dauci* (Fabricius), *Sipha elegans* del Guercio, *S. maydis* Passerini, *Sitobion miscanthi* (Takahashi), and *Toxoptera aurantii* Boyer de Fonscolombe.

Parasitoid–aphid–plant associations in Iran: Hemiptera: Aphididae: *Acyrthosiphon* sp. on *Cousinia* sp. (Starý, 1979; Starý et al., 2000), *Aphis affinis* del Guercio on *Mentha longifolia*, and *Marrubium* sp. (Taheri & Rakhshani, 2013), *Aphis craccivora* Koch on *Alhagimaurorum* sp., *Astragalus* sp., *Glycyrrhiza glabra*, *Kochia scoparia*, *Medicago sativa*, *Robinia pseudoacacia*, and *Zygophyllum* sp. (Barahoei et al., 2013; Nazari et al., 2012; Rakhshani, Talebi, Kavallieratos et al., 2005; Rakhshani, Talebi, Manzari et al., 2006; Starý, 1979; Starý et al., 2000; Taheri & Rakhshani, 2013; Talebi et al., 2009), *Aphis fabae* Scolopli on *Carduus arabicus*, *Rumex cripus*, *Vicia faba* (Alikhani et al., 2013; Moodi & Mossadegh 2006; Nazari et al., 2012; Talebi et al., 2009), *Aphis farinosa* Gmelin on *Salix aegyptiaca*, and *S. alba* (Nazari et al., 2012; Rakhshani, Talebi, Starý et al., 2007; Starý et al., 2000; Talebi et al., 2009; Yaghubi, 1997), *Aphis gossypii* Glover on *Cucumis sativus* (Nazari et al., 2012), *Aphis idaei* van der Goot on *Rubus persicus* (Barahoei et al., 2013; Taheri & Rakhshani, 2013), *Aphis intybi* Koch on *Calendula officinalis*, and *Cichorium intybus* (Barahoei et al., 2013; Taheri & Rakhshani, 2013), *Aphis urticata* Gmelin on *Urtica dioica* (Rajabi Mazhar et al., 2010), *Aphis verbasci* Schrank on *Verbascum songaricum*, *V. thapsus*, and *Verbascum* sp. (Barahoei et al., 2013; Rakhshani, Kazemzadeh et al., 2012; Starý et al., 2000; Talebi et al., 2009; Yaghubi, 1997), *Aphis* sp. on *Cousinia* sp., *Mentha* sp., and *Urtica* sp. (Starý, 1979; Starý et al., 2000), *Aulacorthum* sp. (Starý, 1979), *Brachycaudus cardui* (L.) on *Echinops* sp. (Starý et al., 2000; Yaghubi, 1997), *Brachycaudus helichrysi* (Kaltenbach) on *Chrysanthemum morifolium* (Nazari et al., 2012), *Brachyunguis skafi* Remaudière and Talhouk on *Astragalus* sp. (Starý, 1979; Starý et al., 2000), *Ephedraphis ephedrae* (Nevsky) on *Ephedra major* (Starý, 1979; Starý et al., 2000), *Hyperomyzus lactucae* (L.) on *Lactuca* sp. (Barahoei et al., 2013), *Macrosiphoniella tapuskae* Hottes and Frison on *Achillea* sp. (Rakhshani et al., 2011; Starý, 1979; Starý et al., 2000), *Phorodon humuli* (Schrank) and *Rhopalosiphum nymphaeae* (L.) (Mokhtari et al., 2000), *Aphis* (*Protaphis*) sp. on *Cousinia* sp. (Starý, 1979; Starý et al., 2000), *Toxoptera aurantii* (Boyer de Fonscolombe) (Aghajanzadeh et al., 1995, 1997; Starý et al., 2000), *Uroleucon* sp. on *Picnomon acarna* (Barahoei, Madjdzadeh et al., 2012; Barahoei et al., 2013).

Lysiphlebus desertorum Starý, 1965

Catalogs with Iranian records: Fallahzadeh and Saghaei (2010), Barahoei et al. (2014), Yu et al. (2016), Farahani et al. (2016), Samin, Coronado-Blanco, Kavallieratos et al. (2018), Samin, Coronado-Blanco, Fischer et al. (2018), Rakhshani et al. (2019).

Distribution in Iran: Ardabil (Tomanović et al., 2018), Guilan (Modarres Awal, 2012); Yaghubi & Sahragard, 1998, Kerman (Barahoei et al., 2013), Kermanshah (Talebi et al., 2009), Kordestan (Nazari et al., 2012; Tomanović et al., 2018), Northern Khorasan (Modarres Awal, 2012; Rakhshani, Kazemzadeh et al., 2012; Starý, 1979; Starý et al., 2000), Tehran (Starý, 1979).

Distribution in the Middle East: Iran.

Extralimital distribution: China, Hungary, Moldova, Uzbekistan.

Host records: Recorded by Starý (1979) as a parasitoid of the aphid *Xerobion cinae* (Nevsky) (= *Aphis cinae*).

Parasitoid—aphid—plant associations in Iran: Hemiptera: Aphididae: *Aphis craccivora* Koch on *Peganum harmala* (Barahoei et al., 2013), *Xerobion cinae* (Nevsky) on *Artemisia biennis* and *A. herba-alba* (Asteraceae) (Starý, 1979; Starý et al., 2000); *Aphis terricola* Rondani on *Cirsium hygrophilum* (Barahoei et al., 2013), *Aphis* sp. on *Artemisia herba-alba* (Rakhshani, Kazemzadeh et al., 2012), *Protaphis* sp. on *Achellia millefolium* (Nazari et al., 2012; Talebi et al., 2009), *Protaphis* sp. on *Artemisia* sp. and *Achellia millefolium* (Tomanović et al., 2018).

Lysiphlebus fabarum (Marshall, 1896)

Catalogs with Iranian records: Fallahzadeh and Saghaei (2010), Barahoei et al. (2014), Yu et al. (2016), Farahani et al. (2016), Samin, Coronado-Blanco, Kavallieratos et al. (2018), Samin, Coronado-Blanco, Fischer et al. (2018), Rakhshani et al. (2019).

Distribution in Iran: Alborz (Rakhshani, Talebi, Kavallieratos et al., 2005; Rakhshani, Talebi, Manzari et al., 2006; Starý, 1979; Starý et al., 2000; Talebi et al., 2009), Ardabil (Lotfalizadeh, 2002b), East Azarbaijan (Modarres Awal, 1997, 2012; Starý, 1979; Starý et al., 2000), Fars (Barahoei et al., 2013; Davatchi & Shojai, 1968; Modarres Awal, 1997, 2012; Radjabi, 1989; Shojai, 1998; Starý, 1979; Starý et al., 2000; Taheri & Rakhshani, 2013), Golestan (Afshari et al., 2006; Barahoei, Sargazi et al., 2012; Ghahari & Fischer 2011; Rakhshani, 2012; Starý, 1979; Starý et al., 2000), Guilan (Modarres Awal, 1997, 2012; Mokhtari et al., 2000; Rahimi et al., 2012; Starý et al., 2000; Talebi et al., 2009; Yaghubi, 1997; Yaghubi & Sahragard, 1998), Hamadan (Nazari et al., 2012; Rajabi Mazhar et al., 2010; Rakhshani, Talebi, Manzari et al., 2006; Talebi et al., 2009), Hormozgan (Ameri et al., 2019), Isfahan (Barahoei et al., 2013; Nematollahi & Bagheri, 2018; Rakhshani, 2012; Rakhshani, Talebi, Manzari et al., 2006; Rakhshani, Tomanović et al., 2008; Rakhshani et al., 2010; Starý et al., 2000; Talebi et al., 2009), Kerman (Barahoei et al., 2011; Barahoei, Madjdzadeh et al., 2012; Barahoei et al., 2013; Modarres Awal, 2012; Ghotbi Ravandi et al., 2017), Kermanshah (Bagheri-Matin et al., 2008; Nazari et al., 2012; Rakhshani, 2012; Rakhshani, Talebi, Manzari et al., 2006; Talebi et al., 2009), Khuzestan (Baroon et al., 2008; Modarres Awal, 2012; Mohammadi et al., 2019; Moodi & Mossadegh 2006; Mossadegh & Kocheili, 2003; Mossadegh et al., 2010, 2011, 2017; Nazari et al., 2012; Salehipour et al., 2010), Kordestan (Nazari et al., 2012; Rakhshani, Talebi, Manzari et al., 2006), Markazi (Alikhani et al., 2013; Barahoei et al., 2013; Talebi et al., 2009), Mazandaran (Aghajanzadeh et al., 1995;

Modarres Awal, 1997, 2012; Starý et al., 2000), Northern Khorasan (Kazemzadeh et al., 2010; Malkeshi 1997; Malkeshi et al., 1998; Rakhshani, Kazemzadeh et al., 2012), Qazvin (Rakhshani, Talebi, Manzari et al., 2006; Talebi et al., 2009), Qom (Barahoei et al., 2013), Razavi Khorasan (Darsouei et al., 2011a,b; Farokhzadeh et al., 2014; Jafari-Ahmadabadi et al., 2011; Jafari-Ahmadabadi & Modarres Awal, 2012; Jalali Moghadam Ziabari et al., 2014; Modarres Awal, 1997, 2012; Rakhshani, 2012; Starý, 1979; Starý et al., 2000), Sistan and Baluchestan (Fakhireh et al., 2016; Rakhshani, Talebi, Kavallieratos et al., 2005; Rakhshani, Talebi, Manzari et al., 2006; Rakhshani, Tomanović et al., 2008; Talebi et al., 2009), Tehran (Modarres Awal, 1997, 2012; Monajemi & Esmaili, 1981; Radjabi, 1989; Rakhshani, 2012; Rakhshani, Talebi, Kavallieratos et al., 2005; Rakhshani, Talebi, Manzari et al., 2006; Shojai et al., 2005; Starý, 1979; Starý et al., 2000; Talebi et al., 2009), West Azarbaijan (Samin et al., 2014; Starý, 1979; Starý et al., 2000), Zanjan (Rakhshani, Talebi, Manzari et al., 2006; Rasekh et al., 2012), Iran (no specific locality cited) (Khanjani, 2006a,b; Shojai et al., 2003).

Distribution in the Middle East: Egypt, Iran, Iraq, Israel—Palestine, Lebanon, Syria, Turkey, United Arab Emirates (Rakhshani et al., 2019; Yu et al., 2016).

Extralimital distribution: Australasian, Indo-Malayan, Oceanic, and Palaearctic regions [Adjacent countries to Iran: Afghanistan, Azerbaijan, Kazakhstan, Pakistan, and Russia].

Host records: Summarized by Yu et al. (2016) as being a parasitoid of the following aphid species: *Acyrthosiphon* ssp., *Amphorophora catharinae* (Nevsky), *Anoecia corni* (Fabricius), *Aphis* spp., *Aulacorthum solani* (Kaltenbach), *Brachycaudus* spp., *Brachyunguis atraphaxidis* (Nevsky), *B. calligoni* (Nevsky), *B. tamaricis* (Lichtensteini), *B. tamaricophilus* (Nevsky), *Capitophorus carduina* (Walker), *C. eleagni* (Del Guercio), *C. inulae* (Passerini), *Cavariella aegopodii* (Scopoli), *C. aquatica* (Gillette and Bragg), *C. archangelicae* (Scopoli), *C. pastinacae* (L.), *Chaitophorus salijaponicus* Essig and Kuwana, *Chomaphis* sp., *Coloradoa* sp., *Cryptomyzus ribis* L., *Drepanosiphum platanoides* (Schrank), *Dysaphis apiifolia* (Theobald), *D. crataegi* (Kaltenbach), *D. emicis* (Mimeur), *D. lappae* (Koch), *D. plantaginea* (Passerini), *Hayhurstia atriplicis* (L.), *Hyadaphis foeniculi* Passerini, *Hyalopterus pruni* (Geoffroy), *Hyperomyzus lactucae* (L.), *H. picridis* (Börner), *Liosomaphis berberidis* (Kaltenbach), *Lipaphis lepidii* (Nevsky), *Macchiatiella rhamni* (Boyer de Fonscolombe), *Macrosiphoniella sanborni* (Gillette), *M. subterranea* (Koch), *Melanaphis donacis* (Passerini), *Metopeurum fuscoviride* Stroyan, *Microsiphum millefolii* Wahlgren, *Myzus* spp., *Ovatus crataegarius* (Walker), *O. insitus* (Walker), *Paczoskia major* Börner, *Pemphigus immunis* Buckton, *Periphyllus aceris* (L.), *Phorodon humuli* (Schrank), *Protaphis* sp., *Rhopalosiphum maidis* (Fitch), *R. nymphaeae* (L.), *R. padi* (L.), *Saltusaphis scripus* (Theobald), *Schizaphis rufula* (Walker), *Semiaphis dauci* (Fabricius), *Sipha maydis* Passerini, *Sitobion aveanae* (Fabricius), *S. fragariae* (Walker), *Toxoptera aurantii* Boyer de Fonscolombe, *Trama troglodytes* von Heyden, *Uroleucon cirsii* (L.), *U. hypochoeridis* Hille Ris Lambers, *U. jaceae* (L.), and *U. sonchi* (L.).

Parasitoid—aphid—plant associations in Iran: Hemiptera: Aphididae: *Acyrthosiphon bidentis* Eastop on *Papaver* sp. (Starý, 1979; Starý et al., 2000), *Aphis davletshinae* Hille Ris Lambers on *Malva parviflora*, *Acyrthosiphon gossypii* Mordvilko on *Sophora alopecuroides* (Barahoei et al., 2013), *Acyrthosiphon lactucae* (Passerini) on *Lactuca scariola* (Starý, 1979; Starý et al., 2000), *Aphis acetosae* L. on *Rumex* sp. (Barahoei et al., 2013; Rajabi Mazhar et al., 2010; Starý et al., 2000; Yaghubi, 1997), *Aphis affinis* del Guercio on *Ligustrum vulgare*, *Mentha aquatica*, *Mentha longifolia*, and *Mentha* sp. (Barahoei et al., 2013; Nazari et al., 2012; Rakhshani, Kazemzadeh et al., 2012; Starý et al., 2000; Taheri & Rakhshani, 2013; Talebi et al., 2009; Tomanović et al., 2012; Yaghubi, 1997), *Aphis alexandrae* (Nevsky) on *Carthamus oxyacantha* (Rakhshani, Kazemzadeh et al., 2012), *Aphis anthemidis* Börner on *Anthemis nobilis* (Nazari et al., 2012; Talebi et al., 2009), *Aphis craccivora* Koch on *Achillea* sp., *Alhagi maurorum*, *Alhagi* sp., *Alcea rosea*, *Amaranthus blitoides*, *A. caudatus*, *A. retroflexus*, *Amaranthus* sp., *Anagallis arvensis*, *Astragalus gossypinus*, *Astragalus* sp., *Capsella bursa-pastoris*, *Cardaria draba*, *Cardius onopordiodes*, *Celosia cristata*, *Chenopodium* sp., *C. album*, *Citrus aurantiifolia*, *Elaeagnus angustifoliae*, *Erodium deserti*, *Gerbera jamesonii*, *Glycerrhiza asprima*, *G. glabra*, *Gundelia tournefortii*, *Hibiscus esculentus*, *Kochia scoparia*, *Malva althea*, *Medicago sativa*, *Melilothus officinalis*, *Onobrychis sativa*, *Onobrychis* sp., *Papaver* sp., *Peganum harmala*, *Phaseolus vulgaris*, *Portulaca oleracea*, *Robinia pseudoacacia*, *Rumex* sp., *Sinapes arvensis*, *Solanum nigrum*, *Sophora alopecuroides*, *S. mollis*, *S. pachycarpa*, *Sophora* sp., *Onobrychis altissima*, *Astragalus brevidens*, *A. podocarypus*, *Tragopogon graminifolius*, *Trigonella foenumgraecum*, *Triplospemum discifernus*, *Vicia saliva*, and *V. villosa* (Alikhani et al., 2013; Barahoei et al., 2011; Barahoei, Madjdzadeh et al., 2012; Barahoei et al., 2013; Modarres Awal, 1997, 2012; Monajemi & Esmaili, 1981; Mossadegh et al., 2011; Nazari et al., 2012; Rajabi Mazhar et al., 2010; Rakhshani, Talebi, Kavallieratos et al., 2005; Rakhshani, Talebi, Manzari et al., 2006; Rakhshani et al., 2010; Rakhshani, Kazemzadeh et al., 2012; Starý, 1979; Starý et al., 2000; Taheri & Rakhshani, 2013; Talebi et al., 2009; Yaghubi, 1997), *Aphis davletshinae* Hille Ris Lambers on *Althaea* sp. (Starý, 1979), *Aphis epilobii* Kaltenbach on *Epilobium angustifolium* (Nazari et al., 2012), *Aphis evonymi* Fabricius on *Arctium lappa* (Talebi et al., 2009), *Aphis euphorbicola* Rezwani and Lampel on *Euphorbia aucheri* (Barahoei et al., 2013; Rakhshani, Kazemzadeh et al., 2012), *Aphis fabae* Scopoli on *Althaea* sp., *Amaranthus retroflexus*, *Anthemis* sp., *Arctium lappa*, *Beta vulgaris*, *Brassica rapa*, *Carduus onopordiodes*, *Centaurea iberica*, *Centaurea* sp., *Chenopodium album*, *Cirsium hygrophilum*, *C. arvense*, *C. vulgare*, *Cousinia* sp., *Citrus aurantiifolia*, *Faba vulgaris*, *Glycyrrhiza glabra*, *Hibiscus trionum*, *Lawsonia inermis*, *Ligustrum vulgare*, *Phaseolus vulgaris*, *Phaseolus* sp., *Raphanus sativus*, *Rosa damascene*, *Rumex acetosa*, *R. acetosella*, *R. crispus*, *Rumex* sp., *Solanum nigrum*, *Salsola canescens*, *Spinacia aleracea*, *Sinapis arvensis*, and *Vicia faba* (Bagheri-Matin et al., 2009; Barahoei et al., 2011; Barahoei, Madjdzadeh et al., 2012; Barahoei et al., 2013; Baroon et al., 2008; Davatchi & Shojai, 1968; Mahmoudi et al., 2012; Modarres Awal, 1997, 2012; Mohammadi & Rasekh, 2018; Moodi & Mossadegh 2006; Nazari et al., 2012; Radjabi, 1989; Rakhshani, Kazemzadeh et al., 2012; Rasekh et al., 2010, 2012; Shahbaz Gahroee et al., 2018; Shojai, 1998; Starý et al., 2000; Taheri & Rakhshani, 2013; Talebi et al., 2009; Yaghubi, 1997), *Aphis fabae* Scopoli on *Amaranthus* sp., *Cirsium arvense*, and *Solanum nigrum*, *Aphis fabae cirsiiacanthoidis* on *Cirsium arvense*, *Aphis fabae solanella* on *Solanum nigrum*, *Aphis pomi* de Geer on *Malus domestica*, *Aphis* sp. on *Consinia cylindracea*, *Borago officialis* and *Picnomon acarna* (Alikhani et al., 2013; Barahoei, Madjdzadeh et al., 2012), *Aphis gerardianae* Mordvilko on *Euphorbia aellinae* (Rakhshani, Kazemzadeh et al., 2012), *Aphis gossypii* Glover on *Calendula persica*, *Chrysanthenum* sp., *Cucumis sativa*, *Cucurbita pepo*, *Hibiscus* sp., *Gossypium hirsutum*, *Gossypium* sp., *Lepidium* sp., *Malva neglecta*, *Merabilis jalapa*, *Nonea* sp., *Picnomon acarna*, *Pyrus communis*, *Ranunculus* sp., *Rosa damascena*, *Solanum melongena*, *Salsola* sp., *Veronica persica*, and *Zinnia elegans* (Afshari et al., 2006; Barahoei et al., 2011; Barahoei, Madjdzadeh et al., 2012; Barahoei et al., 2013; Khanjani, 2006a,b; Modarres Awal, 2012; Mohammadi et al., 2019; Mossadegh et al., 2011; Nazari et al., 2012; Rakhshani, Kazemzadeh et al., 2012; Starý, 1979; Starý et al., 2000; Taheri & Rakhshani, 2013; Talebi et al., 2009; Yaghubi, 1997), *Aphis idaei* van der Goot on *Rubus idaeus*, and *R. persicus* (Barahoei et al., 2013; Taheri & Rakhshani, 2013; Talebi et al., 2009), *Aphis intybi* Koch on *Cichorium intybus* (Rajabi Mazhar et al., 2010; Rakhshani, Kazemzadeh et al., 2012; Starý et al., 2000; Taheri & Rakhshani, 2013; Talebi et al., 2009; Yaghubi, 1997), *Aphis nasturtii* Kaltenbach on *Marrubium* sp., *Plantago lanceolate*, and *Veronica anagallis* (Barahoei et al., 2013; Nazari et al., 2012; Starý, 1979; Starý et al., 2000), *Aphis nerii* Boyer de Fonscolombe on *Nerium oleander* (Barahoei et al., 2013; Starý et al., 2000; Talebi et al., 2009; Yaghubi, 1997), *Aphis origani* Passerini on *Mentha longifolia*; *Aphis plantaginis* Goeze on *Plantago lanceolata* (Barahoei et al., 2013), *Aphis pomi* De Geer on *Malus domestica*, and

M. pumila (Malkeshi, 1997; Radjabi, 1989; Rakhshani, 2012; Starý et al., 2000), *Aphis punicae* (Passerini) on *Punica grannatum* (Barahoei et al., 2013; Farokhzadeh et al., 2014; Jalali Moghadam Ziabari et al., 2014; Nazari et al., 2012; Taheri & Rakhshani, 2013; Talebi et al., 2009), *Aphis ruborum* Blackman on *Rubus* sp., and *R. idaeus* (Starý, 1979; Starý et al., 2000; Yaghubi, 1997), *Aphis rumicis* L. on *Rumex crispus*, and *Rumex* sp. (Modarres Awal, 2012; Nazari et al., 2012; Starý et al., 2000; Talebi et al., 2009; Yaghubi, 1997), *Aphis salviae* Walker on *Salvia* sp. (Starý, 1979; Starý et al., 2000), *Aphis solanella* Theobald on *Capsella bursa-pastoris*, *Ctenopodium album*, and *Beta vulgaris* (Barahoei et al., 2013; Nazari et al., 2012), *Aphis spiraecola* Patch on *Chaenomeles japonica*, *Cydonia oblonga*, *Crataega monogyna*, and *Pyracantha coccinea* (Rakhshani, 2012; Starý et al., 2000; Yaghubi, 1997), *Aphis taraxacicola* (Börner) on *Taraxacum seriacum* (Starý, 1979; Starý et al., 2000), *Aphis terricola* Rondani on *Anthemis arvensis*, *Aphis umbrella* (Börner) on *Althaea* sp., *Althaea rosea*, *Malva althea*, *M. neglecta*, and *M. sylvestris* (Alikhani et al., 2013; Barahoei et al., 2013; Modarres Awal, 2012; Nazari et al., 2012; Starý et al., 2000; Talebi et al., 2009), *Aphis urticata* Gmelin on *Urtica* sp. (Starý et al., 2000; Yaghubi, 1997), *Aphis* sp. on *Astragalus* sp. aff. *mollis*, *Buxus hyrcanus*, *Centaurea europaea*, *Centaurea* sp., *Euphorbia aucheri*, *E. cyparissias*, *Ficus* sp., *Galium* sp., *Hypericum* sp., *Lythrum* sp., *Mangifera indica*, *Marrubium* sp., *Potentella* sp., *Rumex acetosa*, *Rubus idaeus*, *Rumex* sp., and *Salsola canescens*, on *Trifolium* sp., and *Urtica* sp. (Barahoei et al., 2013; Nazari et al., 2012; Starý, 1979; Starý et al., 2000; Yaghubi, 1997), *Aphis (Protaphis)* sp. on *Alhagi maurorum*, *Angelica* sp., and *Picnomon acarna*, *Brachycaudus amygdalinus* (Schouteden) on *Prunus dulcis* (Nazari et al., 2012; Rakhshani, 2012), *Brachycaudus cardui* (L.) on *Borago officinalis*, *Carduus arabicus*, *C. onopordiodes*, *Centaurea* sp., *Cirsium arvense*, *Cirsium* sp., *Borago officialis*, *Echinops* sp., and *Onopordon acanthium* (Alikhani et al., 2013; Barahoei et al., 2013; Modarres Awal, 2012; Nazari et al., 2012; Rakhshani, Kazemzadeh et al., 2012; Starý et al., 2000; Taheri & Rakhshani, 2013; Talebi et al., 2009; Yaghubi, 1997), *Brachycaudus helichrysi* (Kaltenbach) on *Calendula officinalis*, *Calendula* sp., *Prunus* sp., *Centaurea* sp., *Chrysanthemum morifolium domestica*, *Prunus domestica*, and *Tanacetum vulgare* (Barahoei et al., 2013; Jafari-Ahmadabadi et al., 2011; Jafari-Ahmadabadi & Modarres Awal, 2012; Nazari et al., 2012; Rakhshani, 2012; Starý et al., 2000; Taheri & Rakhshani, 2013; Yaghubi, 1997), *Brachycaudus persicae* (Passerini) on *Prunus persica* (Rakhshani, 2012), *Brachycaudus tragopogonis* (Kaltenbach) on *Scrozonera isphahanica*, and *Tragopogon graminifolius* (Barahoei, Madjdzadeh et al., 2012; Barahoei et al., 2013), *Brachycaudus tragopogonis setosus* Hille Ris Lambers on *Scorzonera* sp. (Alikhani et al., 2013; Barahoei, Madjdzadeh et al., 2012;

Barahoei et al., 2013; Rakhshani, Kazemzadeh et al., 2012; Starý et al., 2000), *Brachycaudus* sp. on *Cirsium* sp., *Euphorbia amygdaloides*, and *Prunus* sp. (Barahoei et al., 2013), *Brachyunguis harmalae* Das on *Peganum harmala* (Talebi et al., 2009; Rakhshani, Kazemzadeh et al., 2012), *Brachyunguis zygophylli* (Nevsky) on *Zygophyllum fabago* (Barahoei et al., 2011; Barahoei, Madjdzadeh et al., 2012; Barahoei et al., 2013), *Brachyunguis* sp. on *Cynanchum acutum* (Barahoei et al., 2013), *Coloradoa* sp. on *Artemisia herba-alba* (Starý, 1979; Starý et al., 2000), *Dysaphis lappae* Koach (Starý, 1979), *Dysaphis plantaginea* (Passerini) on *Malus domestica*, and *Malus* sp. (Barahoei et al., 2013; Rakhshani, 2012; Starý et al., 2000), *Dysaphis radicola* (Mordvilko) on *Rheum palmatum*, *Hyadaphis coriandri* (Passerini) on *Coriandrum sativum*, *Lipaphis erysimi* (Kaltenbach) on *Capsella bursa-pasturis*, *Lipaphis fritzmuelleri* Börner on *Sisymbrium irio*; *Myzus beybienkoi* (Narzikulov) on *Fraxinus oxycarpa*, *Protaphis elongata* (Nevsky) on *Artemisia absinthium*, *Trigonella foenum-graesum* (Talebi et al., 2009), *Dysaphis* sp. on *Lappa* sp., *Plantago* sp., *Pyrus* sp., and *Rumex acetosa* (Starý et al., 2000), *Hayhurstia atriplicis* (L.) on *Chenopodium album* (Nazari et al., 2012), *Hyalopterus amygdali* (Blanchard) on *Prunus persica* (Rakhshani, 2012), *Hyalopterus pruni* (Geoffroy) on *Prunus armeniaca*, and *P. domestica* (Jafari-Ahmadabadi et al., 2011; Jafari-Ahmadabadi & Modarres Awal, 2012), *Lipaphis lepidii* Nevsky on *Cardaria draba*, and *Lepidium* sp. (Starý, 1979; Starý et al., 2000; Talebi et al., 2009), *Macrosiphoniella papilata* Holman on *Artemisia biennis* (Barahoei et al., 2013), *Macrosiphoniella sanborni* (Gillette) on *Chrysanthemum* sp. (Starý et al., 2000; Yaghubi, 1997), *Macrosiphum rosae* (L.) on *Rosa damascena* (Taheri & Rakhshani, 2013), *Melanaphis* sp. on *Sorghum halepense* (Barahoei, Madjdzadeh et al., 2012; Barahoei et al., 2013), *Melanaphis sacchari* (Zehntner) (Barahoei et al., 2011), *Metopolophium dirhodum* (Walker) on *Rosa damascena* (Barahoei et al., 2013), *Myzus persicae* (Sulzer) on *Lycopersicum esculentum*, *Malva neglecta*, *Melilotus* sp., and *Trigonella foenum-graesum* (Barahoei et al., 2013; Talebi et al., 2009), *Phorodon humuli* (Schrank) on *Prunus* sp. (Mokhtari et al., 2000; Starý et al., 2000), *Aphis (Protaphis) terricola* Walker on *Anthemis arvensis* (Barahoei et al., 2013), *Aphis (Protaphis)* sp. on *Alhagi maurorum*, and *Picnomon acarna* (Barahoei et al., 2013; Starý, 1979), *Rhopalosiphum nymphaeae* (L.) on *Prunus* sp. (Mokhtari et al., 2000; Starý et al., 2000), *Rhopalosiphum padi* (L.) on *Triticum aestivum* (Nazari et al., 2012; Taheri & Rakhshani, 2013; Rakhshani, Tomanović et al., 2008), *Saltusaphis scirpus* Theobald on *Carex* sp. (Starý, 1979; Starý et al., 2000; Tomanović et al., 2012), *Sitobion avenae* (Fabricius) (Bagheri-Matin et al., 2008), *Toxoptera aurantii* (Boyer de Fonscolombe) (Aghajanzadeh et al., 1995; Starý et al., 2000), *Uroleucon compositae* (Theobald) on *Carthamus oxyacantha*, *Uroleucon jacaea* (Linnaeus) on *Centaurea iberica* (Barahoei, Madjdzadeh et al., 2012), *Uroleucon* sp. on *Conyza* sp. (Starý et al., 2000; Yaghuri, 1997).

Lysiphlebus fritzmuelleri Mackauer, 1960
Distribution in the Middle East: Turkey (Akar & Çetin Erdoğan, 2017).
Extralimital distribution: Bulgaria, Czech Republic, Finland, Germany, Hungary, Kazakhstan, Moldova, Montenegro, Poland, Romania, Russia, Serbia, Slovakia, Ukraine.
Host records: Summarized by Yu et al. (2016) as being a parasitoid of *Aphis craccae* L., *A. fabae* Scopoli, and *Dysaphis apiifolia* (Theobald).

Lysiphlebus marismotui Mescheloff and Rosen, 1989
Distribution in the Middle East: Israel—Palestine (Mescheloff & Rosen, 1989, 1990a).
Extralimital distribution: None.
Host records: Recorded by Mescheloff and Rosen (1990a) as being a parasitoid of the aphid *Aphis craccivora* Koch.
Comments: According to Rakhshani et al. (2019), *Lysiphlebus marismortui* is a synonym of *Lysiphlebus confusus* on the basis of morphological characters.

Lysiphlebus testaceipes (Cresson, 1880)
Catalogs with Iranian records: Fallahzadeh and Saghaei (2010), Barahoei et al. (2014), Yu et al. (2016), Farahani et al. (2016), Samin, Coronado-Blanco, Kavallieratos et al. (2018), Samin, Coronado-Blanco, Fischer et al. (2018), Rakhshani et al. (2019).
Distribution in Iran: Fars, Khuzestan (Alesafoor & Mossadegh, 2015), Hamadan (Rajabi Mazhar et al., 2010), Razavi Khorasan (Darsouei et al., 2011b), Tehran (Rakhshani, Talebi, Kavallieratos et al., 2005).
Distribution in the Middle East: Egypt (Rakhshani et al., 2019), Iran (see references above), Turkey (Yoldaş et al., 2011; Uysal et al., 2004).
Extralimital distribution: Algeria, Argentina, Australia, Belgium, Brazil, Canada, Chile, China, Costa Rica, Croatia, Cuba, Czech Republic, Dominica, France, Greece, Haiti, India, Italy, Korea, Mexico, Montenegro, Pakistan, Peru, Portugal, Puerto Rico, Serbia, South Africa, Spain, Trinidad and Tobago, United Kingdom, United States of America, Uzbekistan.
Host records: Summarized by Yu et al. (2016) as being a parasitoid of the aphids *Acyrthosiphon lactucae* (Passerini), *A. macrosiphum* (Wilson and H.F.), *A. pisum* (Harris), *Anoecia corni* (Fabricius), *Aphis* spp., *Aphthargelia symphoricarpi* (Thomas), *Boernerina variabilis* Richards, *Brachycaudus cardui* (L.), *B. helichrysi* (Kaltenbach),

B. salicinae Börner, *B. schwartzi* (Börner), *B. tragopogonis* (Kaltenbach), *Brachycorynella asparagi* (Mordvilko), *Brachyunguis bonnevillensis* Knowlton, *B. tamaricis* (Lichtenstein), *B. tetrapteralis* (Cockerell and T.D.A.), *Braggia eriogoni* (Cowen and J.H.), *B. urovaneta* Hottes, *Brevicoryne brassicae* (L.), *Capitophorus elaeagni* (Del Guercio), *Capitophorus inulae* (Passerini), *C. xanthii* (Oestlund), *Carolinaia cyperi* Ainslie, *Cavariella aegopodii* (Scopoli), *C. pastinacea* (L.), *Cedoaphis* sp., *Ceruraphis viburnicola* (Gillette), *Chaetosiphon fragaefolii* (Cockerell), *Chaitophorus populicola* Thomas, *C. viminalis* Monell, *Cinara chinokiana* Hottes, *Cryptomyzus ribis* (L.), *Diuraphis noxia* (Mordvilko), *Dysaphis apiifolia* (Theobald), *D. plantaginea* (Passerini), *D. pyri* (Boyer de Fonscolombe), *D. tulipae* (Boyer de Fonscolombe), *Ericaphis* sp., *Eriosoma americanum* (Riley), *E. lanigerum* (Hausmann), *Hayhurstia atriplicis* (L.), *Haydaphis foeniculi* Passerini, *Hyalopterus pruni* (Geoffroy), *Hyperomyzus lactucae* (L.), *H. nigricornis* (Knowlton), *Hysteroneura setariae* (Thomas), *Illinoia liriodendra* (Monell), *Kakimia* sp., *Letigerina orizabaensis* Remaudière, *Lipaphis erysimi* (Kaltenbach), *Macchiatiella rhamni* (Boyer de Fonscolombe), *Macrosiphum* spp., *Melanaphis bambusae* (Fullaway), *M. sacchari* (Zehntner), *Metopolophium dirhodum* (Walker), *Myzus* spp., *Nasonovia houghtonensis* (Troop), *Nearctaphis* spp., *Neonasonovia* sp., *Neotoxoptera formosana* (Takahashi), *Obtusicauda* sp., *Ovatus crataegarius* (Walker), *Pentalonia nigronervosa* Coquerel, *Phorodon humuli* (Schrank), *Pleotrichophorus oestlundi* (Knowlton), *Pseudoepameibaphis tridentatae* (Wilson and H.F.), *Pterocomma* sp., *Rhopalosiphoninus latysiphon* (Davidson), *Rhopalosiphum* spp., *Schizaphis graminum* (Rondani), *Sipha flava* (Forbes), *Siphonatrophia cupressi* (Swain), *Sitobion* spp., *Tinocallis kahawaluohalani* (Kirkady), *Toxoptera aurantii* (Boyer de Fonscolombe), *T. criticida* (Kirkaldy), *Tuberculatus columbiae* Richards, *T. eggleri* Börner, *Uroleucon* spp., and *Utamphorophora commelinensis* (Smith).
Parasitoid—aphid—plant associations in Iran: Hemiptera: Aphididae: *Aphis craccivora* Koch on *Medicago sativa*, and *Robinia pseudoacacia* (Rakhshani, Talebi, Kavallieratos et al., 2005), *Coloradoa achilleae* Hille Ris Lambers (Rajabi Mazhar et al., 2010), *Aphis nerii* Boyer de Fonscolombe (Alesafoor & Mossadegh, 2015).

Lysiphlebus volkli Tomanović and Kavallieratos, 2018
Catalogs with Iranian records: Rakhshani et al. (2019).
Distribution in Iran: Ardabil, Hamadan, Isfahan (Tomanović et al., 2018).
Distribution in the Middle East: Iran.

Extralimital distribution: Lithuania, Serbia (Tomanović et al., 2018).

Host records: Unknown.

Parasitoid—aphid—plant associations in Iran: Hemiptera: Aphididae: *Brachycaudus tragopogonis* (Kaltenbach, 1843) on *Tragopogon pratensis* (Asteraceae), *Aphis verbasci* Schrank, 1801 on *Verbascum* sp. (Scrophulariaceae) (Tomanović et al., 2018).

Lysiphlebus sp.

Distribution in Iran: Khuzestan (Mossadegh & Kocheili, 2003; Salehipour et al., 2010), Iran (no specific locality cited) (Farahbakhsh, 1961).

Parasitoid—aphid—plant associations in Iran: *Aphis* sp. (Farahbakhsh, 1961), *Aphis crassivora* on *Medicago polymorpha*, *Medigaco sativa*, *Melilotus officinalis*, *Portulaca oleracea*, *Vicia faba*, *V. sativa*, *V. villosa*, and *Xanthium stramarium*, *Aphis fabae* on *Carduus getulus*, and *Vicia faba*, *Aphis frangulae* Kaltenbach on *Viola tricolor*, *Anchosa rosae*, *A. strigosa*, *Cestrum nocturnum*, *Gossypium hebaceum*, *Hibiscus esculentus*, *Malva parviflora*, *Prosopis juliflora*, and *Zinnia elegans*, *Aphis gossypii* on *Anchosa rosea*, and *Dahlia excelsa*, *Aphis umbrella* on *Malva nigriflora*, and *Malva* sp., *Aphis polygonata* (Nevsky) on *Polygonum aviculare*, *Aphis* sp. on *Atriplex* sp., *Carthamus oxyacanthus*, *C. tinctorius*, *Centaurea* sp., *Citrus* sp., *Dahlia excels*, *Hibiscus rosa-sinensis*, *Malva* sp., *Malva piperita*, *Rumex acetosa*, *Silybum marianum*, and *S. nigrum*, *Brachyunguis harmalae* Das on *Peganum harmala*, *Lipaphis lepidii* (Nevsky) on *Lepidium draba*, *Myzus persicae* on *Ammi majus*, *Atriplex* sp., *Beta maritima*, *Calendula officinalis*, *Raphanus sativus*, *Silubum marianum*, and *Solanum tuberosum*.

Genus *Monoctonia* Starý, 1962

Monoctonia pistaciaecola Starý, 1962

Catalogs with Iranian records: Rakhshani et al. (2019).

Distribution in Iran: Fars (Kargarian et al., 2016).

Distribution in the Middle East: Iran (Kargarian et al., 2016), Iraq (Starý, 1976, 1981; Starý & Kaddou, 1971), Israel—Palestine (Wool & Berstein, 1991).

Extralimital distribution: Czech Republic, Georgia, Italy, Spain, Tajikistan, Ukraine.

Host records: Summarized by Yu et al. (2016) as being a parasitoid of the following aphid species: *Aploneura lentisci* Passerini, *Forda marginata* Koch, *Geoica utricularia* (Passerine), *Pemphigus bursarius* (L.), *P. populinigrae* (Schrank), *P. spyrothecae* Passerini, and *Symnthurodes betae* (Westwood).

Parasitoid—aphid—plant associations in Iran: In wheat field (*Triticum vulgaris*) (Kargarian et al., 2016).

Monoctonia vesicarii Tremblay, 1991

Catalogs with Iranian records: Barahoei et al. (2014), Yu et al. (2016), Farahani et al. (2016), Samin, Coronado-Blanco, Kavallieratos et al. (2018), Samin, Coronado-Blanco, Fischer et al. (2018), Rakhshani et al. (2019).

Distribution in Iran: Ardabil (Ghafouri Moghaddam et al., 2012).

Distribution in the Middle East: Iran.

Extralimital distribution: Czech Republic, Italy, Japan, Spain.

Host records: Summarized by Yu et al. (2016) as being a parasitoid of the following aphid species: *Pemphigus bursarius* (L.), *P. matsumurai* Monzen, *P. populinigrae* (Schrank), *P. spyrothecae* Passerini, and *P. vesicarii* Passerini.

Parasitoid—aphid—plant associations in Iran: Hemiptera: Aphididae: *Pemphigus spirothecae* Passerini on *Populus nigra* (Ghafouri Moghaddam et al., 2012).

Genus *Monoctonus* Haliday, 1833

Monoctonus (Monoctonus) crepidis (Haliday, 1834)

Distribution in the Middle East: Turkey (Tomanović et al., 2008).

Extralimital distribution: Belgium, Canada, Czech Republic, Finland, France, Germany, Hungary, India, Latvia, Moldova, Montenegro, Netherlands, Norway, Poland, Russia, Serbia, Slovakia, Slovenia, Spain, Sweden, United Kingdom, United States of America.

Host records: Summarized by Yu et al. (2016) as being a parasitoid of the following aphid species: *Aphis fabae* Scopoli, *Hyperomyzus* spp., *Liosomaphis atra* Hille Ris Lambers, *Myzus ligustri* (Mosley), *Nasonovia compositellae* Theobald, *N. pilosellae* (Börner), and *N. ribisnigri* (Mosley).

Monoctonus (Monoctonus) mali van Achterberg, 1989

Catalogs with Iranian records: Barahoei et al. (2014), Yu et al. (2016), Farahani et al. (2016), Samin, Coronado-Blanco, Kavallieratos et al. (2018), Samin, Coronado-Blanco, Fischer et al. (2018), Rakhshani et al. (2019).

Distribution in Iran: Guilan (Sakenin et al., 2012), West Azarbaijan (Mosavi et al., 2012; Rastegar et al., 2012; Samin et al., 2014).

Distribution in the Middle East: Iran (Mosavi et al., 2012; Rastegar et al., 2012; Sakenin et al., 2012; Samin et al., 2014), Turkey (Ölmez & Ulusoy, 2003).

Extralimital distribution: Bulgaria, Czech Republic, Montenegro, Netherlands, Serbia, Slovakia, Spain.

Host records: Summarized by Yu et al. (2016) as being a parasitoid of the following aphid species: *Aphis fabae* Scopoli, *A. spiraephaga* Patch, *Dysaphis reaumuri* (Mordvilko), *Ovatus insitus* (Walker), and *Rhopalosiphum insertum* (Walker).

Monoctonus sp.

Distribution in Iran: Markazi, Tehran (Radjabi, 1989).

Parasitoid—aphid—plant associations in Iran: Hemiptera: Aphididae: *Dysaphis plantaginea* (Passerini), *D. pyri* (Boyer de Fonscolombe) (Radjabi, 1989).

Genus *Pauesia* Quilis, 1931

Pauesia abietis (Marshall, 1896)

Distribution in the Middle East: Turkey (Starý, 1976).

Extralimital distribution: Bulgaria, China, Czech Republic, Finland, France, Germany, Hungary, Italy, Japan, Kenya, Korea, Moldova, Poland, Russia, Spain, Sweden, Ukraine, United Kingdom.

Host records: Summarized by Yu et al. (2016) as being a parasitoid of the following aphid species: *Aphis craccivora* Koch, *Cinara cuneomaculata* (del Guercio), *C. laricis* (Hartig), *C. maritimae* (Dufour), *C. pilicornis* (Hartig), *C. pinea* (Mordvilko), *C. piniformosana* (Takahashi), and *Schizolachnus pineti* (Fabricius).

Pauesia akamtsucola Takada, 1968

Distribution in the Middle East: Turkey (Davidian, 2005).

Extralimital distribution: Japan, Moldova, Russia, Ukraine.

Host records: Summarized by Yu et al. (2016) as being a parasitoid of *Cinara maritimae* (Dufour), and *C. pinea* (Mordvilko).

Pauesia anatolica Michelene, Assael & Mendel, 2005

Distribution in the Middle East: Israel—Palestine (imported), Turkey (Michelena et al., 2015).

Extralimital distribution: None.

Host records: recorded by Michelene et al. (2005) as being a parasitoid of the aphid *Cinara cedri* Mimeur.

Pauesia antennata (Mukerji, 1950)

Catalogs with Iranian records: Fallahzadeh and Saghaei (2010), Barahoei et al. (2014), Yu et al. (2016), Farahani et al. (2016), Samin, Coronado-Blanco, Kavallieratos et al. (2018), Samin, Coronado-Blanco, Fischer et al. (2018), Rakhshani et al. (2019).

Distribution in Iran: Ardabil, Tehran (Rakhshani, Talebi, Starý et al., 2005; Rakhshani, 2012), Hamadan (Rakhshani, Talebi, Starý et al., 2005), Kerman (Barahoei et al., 2013; Rakhshani, 2012), Kermanshah (Nazari et al., 2012;

Rakhshani, 2012), Northern Khorasan (Kazemzadeh et al., 2010; Rakhshani, 2012; Rakhshani et al., 2012b), Sistan and Baluchestan (Rakhshani, Talebi, Starý et al., 2005; Rakhshani, 2012; Talebi et al., 2009), Iran (no specific locality cited) (Modarres Awal, 2012).

Distribution in the Middle East: Iran (see references above), Iraq (Starý, 1976), Yemen (Rakhshani et al., 2019).

Extralimital distribution: India, Pakistan, Tajikistan, Uzbekistan.

Host records: Summarized by Yu et al. (2016) as being a parasitoid of *Cinara cupressi* Buckton, *Pterochloroides persicae* (Cholodkowski), and *Sipha maydis* Passerini.

Parasitoid—aphid—plant associations in Iran: Hemiptera: Aphididae: *Pterochloroides persicae* (Cholodkovsky) on *Amygdalus arabica*, *Prunus armeniaca*, *P. dulcis*, *P. padus*, and *P. persica* (Barahoei et al., 2013; Mackauer & Starý, 1967; Nazari et al., 2012; Rakhshani, 2012; Rakhshani, Talebi, Starý et al., 2005; Rakhshani, Kazemzadeh et al., 2012; Starý et al., 2000; Talebi et al., 2009).

Pauesia hazratbalensis Bhagat, 1981

Catalogs with Iranian records: Fallahzadeh and Saghaei (2010), Barahoei et al. (2014), Yu et al. (2016), Farahani et al. (2016), Samin, Coronado-Blanco, Kavallieratos et al. (2018), Samin, Coronado-Blanco, Fischer et al. (2018), Rakhshani et al. (2019).

Distribution in Iran: Kerman (Rakhshani et al., 2017), Northern Khorasan (Kazemzadeh et al., 2010; Rakhshani, Kazemzadeh et al., 2012; Rakhshani et al., 2017), Razavi Khorasan (Heidari Latibari et al., 2020), Tehran (Starý et al., 2005; Rakhshani et al., 2017).

Distribution in the Middle East: Iran.

Extralimital distribution: India, Kyrgyzstan.

Host records: Summarized by Yu et al. (2016) as a parasitoid of *Cinara tujafilina* (Del Gurcio), and *Lachnus* sp. Additionally, Heidari Latibari et al. (2020) reported the pine aphid, *Cinara pinihabitans* (Mordvilko) (Hemiptera, Lachnidae) as a new host record for *Pauesia hazratbalensis*.

Parasitoid—aphid—plant associations in Iran: Hemiptera: Aphididae: *Cinara tujafilina* (del Guercio) on *Thuja orientalis* (Rakhshani, Kazemzadeh et al., 2012; Rakhshani et al., 2017; Starý et al., 2005), *Cinara pinihabitans* (Mordvilko) on *Pinus mugo* (Pinaceae) (Heidari Latibari et al., 2020).

Pauesia picta (Haliday, 1834)

Distribution in the Middle East: Turkey (Aslan et al., 2004).

Extralimital distribution: Andorra, Bulgaria, Czech Republic, Finland, France, Germany, Hungary, India, Italy, Madeira Islands, Moldova, Netherlands, Poland, Russia, Serbia, Slovakia, Spain, Sweden, Switzerland, Ukraine, United Kingdom.

Host records: Summarized by Yu et al. (2016) as being a parasitoid of *Cinara* spp., *Eulachnus thunbergii* Wilson, and *Nasonovia snigri* (Mosley).

Pauesia pini (Haliday, 1834)

Distribution in the Middle East: Israel—Palestine (Starý, 1976).

Extralimital distribution: Andorra, China, Czech Republic, Finland, France, Germany, Hungary, India, Italy, Japan, Korea, Latvia, Moldova, Mongolia, Poland, Romania, Russia, Serbia, Slovakia, Slovenia, Spain, Sweden, United Kingdom.

Host records: Summarized by Yu et al. (2016) as being a parasitoid of *Aphis frangulae* Kaltenbach, *A. pomi* De Geer, *Cinara* spp., *Macchiatiella rhamni* (Boyer de Fonscolombe), and *Macrosiphum rosae* (L.).

Pauesia silana Tremblay, 1969

Distribution in the Middle East: Israel—Palestine (Mescheloff & Rosen, 1990b).

Extralimital distribution: Bulgaria, France, Greece, Italy, Malta, Russia, Spain.

Host records: Summarized by Yu et al. (2016) as being a parasitoid of several aphid species of the genus *Cinara* Curtis.

Pauesia unilachni (Gahan, 1926)

Distribution in the Middle East: Turkey (Rakhshani et al., 2019).

Extralimital distribution: Andorra, Belarus, Bulgaria, Burundi, China, Czech Republic, Finland, France, Germany, Hungary, India, Italy, Japan, Korea, Latvia, Lithuania, Moldova, Netherlands, Poland, Russia, Serbia, Slovakia, Spain, Sweden.

Host records: Summarized by Yu et al. (2016) as being a parasitoid of the following aphid species: *Cinara* spp., *Cranaphis formosanus* (Takahashi), *Eulachnus agilis* (Kaltenbach), *E. thunbergia* Wilson, *Mindarus abietinus* Koch, *Schizolachnus obscurus* Börner, and *S. pineti* (Fabricius).

Genus Tanytrichophorus Mackauer, 1961

Tanytrichophorus petiolaris Mackauer, 1961

Catalogs with Iranian records: Fallahzadeh and Saghaei (2010), Barahoei et al. (2014), Yu et al. (2016), Farahani et al. (2016), Samin, Coronado-Blanco, Kavallieratos et al. (2018), Samin, Coronado-Blanco, Fischer et al. (2018), Rakhshani et al. (2019).

Distribution in Iran: Tehran (Starý et al., 2000), Iran (no specific locality cited) (Modarres Awal, 2012).

Distribution in the Middle East: Iran.

Extralimital distribution: None.

Host recocords: Recorded by Mackauer (1961) and Starý et al. (2000) as a parasitoid of *Brachycaudus persicae* (Passerini).

Parasitoid—aphid—plant associations in Iran: Hemiptera: Aphididae: *Brachycaudus persicae* (Passerini) on *Prunus* sp. (Mackauer, 1961; Starý et al., 2000).

Genus Trioxys Haliday, 1833

Trioxys asiaticus Telenga, 1953

Catalogs with Iranian records: Fallahzadeh and Saghaei (2010), Barahoei et al. (2014), Yu et al. (2016), Farahani et al. (2016), Samin, Coronado-Blanco, Kavallieratos et al. (2018), Samin, Coronado-Blanco, Fischer et al. (2018), Rakhshani et al. (2019).

Distribution in Iran: Alborz (Starý et al., 2000), Kerman (Asadizade et al., 2014; Barahoei et al., 2013; Talebi et al., 2009), Markazi (Alikhani et al., 2013), Tehran (Talebi et al., 2009), Iran (no specific locality cited) (Modarres Awal, 2012).

Distribution in the Middle East: Iran.

Extralimital distribution: Armenia, China, Kazakhstan, Korea, Mongolia, Russia, Tajikistan, Turkmenistan, Uzbekistan.

Host records: Summarized by Yu et al. (2016) as being a parasitoid of the aphids *Acyrthosiphon gossypii* Mordvilko, and *Aphis craccivora* Koch.

Parasitoid—aphid—plant associations in Iran: Hemiptera: Aphididae: *Acyrthosiphon gossypii* Mordvilko on *Sophora alopecuroides* (Alikhani et al., 2013; Barahoei et al., 2013; Mackauer, 1960; Modarres Awal, 2012; Rakhshani, Kazemzadeh et al., 2008; Starý et al., 2000; Talebi et al., 2009), *Acyrthosiphon pisum* (Harris) on *Glycyrrhiza glabra* (Barahoei et al., 2013).

Trioxys auctus (Haliday, 1833)

Catalogs with Iranian records: Samin, Coronado-Blanco, Fischer et al. (2018) as *Aphidius auctus*.

Distribution in Iran: Mazadaran (Samin, Coronado-Blanco, Fischer et al., 2018 as *Aphidius auctus*).

Distribution in the Middle East: Iran (Samin, Coronado-Blanco, Fischer et al., 2018).

Extralimital distribution: Canada, Czech Republic, Finland, France, Germany, Hungary, India, Ireland, Italy, Japan, Lithuania, Netherlands, Norway, Poland, Russia, Serbia, Slovenia, Sweden, United Kingdom, Uzbekistan.

Host records: Summarized by Yu et al. (2016) as being a parasitoid of the following aphid species: *Acyrthosiphon gossypii* Mordvilko, *Aphis fabae evonymi* Fabricius, *A. frangulae* Kaltenbach, *A. oxyacanthae* (Schrank), *Myzus persicae* (Sulzer), *Rhopalosiphum insertum* (Walker), *R. nymphaeae* (L.), *R. padi* (L.), *Schizaphis scripi* (Passerini), and *Sitobion avenae* (Fabricius).

Trioxys betulae Marshall (1896)

Catalogs with Iranian records: No catalog.

Distribution in Iran: Guilan (Sakenin et al., 2020).

Distribution in the Middle East: Iran (Sakenin et al., 2020). Extralimital distribution: Austria, Bulgaria, Canada, China, Czech Republic, Denmark, Finland, France, Germany, Kazakhstan, Montenegro, Netherlands, Poland, Russia, Serbia, Slovakia, Switzerland, Ukraine, United Kingdom, United States of America.

Host records: Summarized by Yu et al. (2016) as being a parasitoid of the aphids *Calaphis betulaecolens* (Fitch), *C. flava* Mordvilko, *Clethrobius comes* (Walker), *Eucallipterus tiliae* (L.), *Euceraphis betulae* (Koch), and *Symydobius oblongus* (von Heyden).

Parasitoid–aphid–plant associations in Iran: Unknown Aphididae on *Alnus subcordata* Mey (Betulaceae) (Sakenin et al., 2020).

Comments: Host of *Trioxys betulae* was determined as *Eucallipterus* sp. (Hemiptera: Aphididae) (H. Sakenin, unpublished data).

Trioxys cirsii (Curtis, 1831)
Catalogs with Iranian records: Fallahzadeh and Saghaei (2010), Barahoei et al. (2014), Yu et al. (2016), Farahani et al. (2016), Samin, Coronado-Blanco, Kavallieratos et al. (2018), Samin, Coronado-Blanco, Fischer et al. (2018), Rakhshani et al. (2019).

Distribution in Iran: Mazandaran (Babaee et al., 2000; Modarres Awal, 2012; Starý et al., 2000).

Distribution in the Middle East: Iran.

Extralimital distribution: Australia, Czech Republic, Finland, Germany, Hungary, Italy, Latvia, Moldova, Montenegro, Netherlands, Poland, Romania, Russia, Sebia, Slovakia, Spain, Tajikistan, United Kingdom, United States of America.

Host records: Summarized by Yu et al. (2016) as being a parasitoid of the aphids *Aphis fabae* Scopoli, *A. nasturtii* (Kaltenbach), *A. pomi* De Geer, *Drepanosiphum oregenense* Granovsky, *D. platanoidis* (Schrank), *Periphyllus aceris* (L.), *Rhopalomyzus lonicerae* (Siebold), *Tuberculatus annulatus* (Hartig), and *Uroleucon cirsii* (L.).

Parasitoid–aphid–plant associations in Iran: Hemiptera: Aphididae: *Drepanosiphum platanoidis* (Schrank) on *Acer* sp. (Babaee et al., 2000; Modarres Awal, 2012; Rakhshani, Kazemzadeh et al., 2008; Starý et al., 2000).

Trioxys complanatus (Quilis, 1931)
Catalogs with Iranian records: Fallahzadeh and Saghaei (2010), Barahoei et al. (2014), Yu et al. (2016), Farahani et al. (2016), Samin, Coronado-Blanco, Kavallieratos et al. (2018), Samin, Coronado-Blanco, Fischer et al. (2018), Rakhshani et al. (2019).

Distribution in Iran: Alborz (Monajemi & Esmaili, 1981 as *Praon utilis*; Rakhshani, Talebi, Manzari et al., 2006), Ardabil, Markazi, Qazvin, Semnan, Zanjan (Rakhshani, Talebi, Manzari et al., 2006), East Azarbaijan (Modarres Awal, 2012), Hamadan, Kermanshah, Kordestan (Nazari et al., 2012; Rakhshani, Talebi, Manzari et al., 2006), Hormozgan (Ameri et al., 2019), Isfahan (Barahoei et al., 2013; Rakhshani, Talebi, Manzari et al., 2006; Rakhshani et al., 2010), Kerman (Barahoei et al., 2013; Hadadian et al., 2016; Ghotbi Ravandi et al., 2017), Markazi (Alikhani et al., 2013), Northern Khorasan (Kazemzadeh et al., 2010; Rakhshani, Kazemzadeh et al., 2012), Sistan and Baluchestan (Fakhireh et al., 2016; Rakhshani, Talebi, Manzari et al., 2006), Tehran (Hadadian et al., 2016; Rakhshani, Talebi, Manzari et al., 2006; Starý et al., 2000), West Azarbaijan (Starý, 1979; Starý et al., 2000), Iran (no specific locality cited) (Farahbakhsh, 1961; Modarres Awal, 1997, 2012).

Distribution in the Middle East: Iran (see references above), Iraq (Al-Azawi, 1970; Starý, 1976; van den Bosch, 1957), Israel–Palestine (Mescheloff & Rosen, 1993; Starý, 1976; van den Bosch, 1957), Lebanon (Tremblay et al., 1985), Turkey (Düzgüneş et al., 1982; Starý, 1976; Uysal et al., 2004), United Arab Emirates (Starý et al., 2013), Yemen (Starý & Erdelen, 1982; Starý et al., 2013).

Extralimital distribution: Cosmopolitan species.

Host records: Summarized by Yu et al. (2016) as being a parasitoid of the following aphid species: *Acyrthosiphon pisum* (Harris), *Pleotrichophorus elongatus* (Knowlton), *P. oestlundi* (Knowlton), *Pterocallis maculatus* (von Heyden), *Therioaphis* spp., and *Uroleucon escalantii* (Knowlton).

Parasitoid–aphid–plant associations in Iran: Hemiptera: Aphididae: *Therioaphis khayami* Remaudière on *Astragalus* sp. (Starý, 1979; Starý et al., 2000), *Therioaphis* ? *riehmi* (Börner) (Starý, 1979; Starý et al., 2000; van den Bosch, 1957), *Therioaphis trifolii* (Monell) (= *Therioaphis maculata* (Buckton)) on *Medicago sativa* (Alikhani et al., 2013; Barahoei et al., 2013; Farahbakhsh, 1961; Modarres Awal, 1997, 2012; Monajemi & Esmaili, 1981; Nazari et al., 2012; Rakhshani, Talebi, Manzari et al., 2006; Rakhshani, Kazemzadeh et al., 2008; Rakhshani et al., 2010; Rakhshani, Kazemzadeh et al., 2012; Starý, 1979; Starý et al., 2000; van den Bosch, 1957), *Therioaphis* spp. (Starý, 1979).

Trioxys curvicaudus Mackauer, 1967
Catalogs with Iranian records: Barahoei et al. (2014), Farahani et al. (2016), Rakhshani et al. (2019).

Distribution in Iran: Tehran (Rakhshani, Kazemzadeh et al., 2008 on *Eucalipterus tilliae* (L.) (Myrtaceae)), Iran (no specific locality cited) (Modarres Awal, 2012).

Distribution in the Middle East: Iran (Modarres Awal, 2012; Rakhshani, Kazemzadeh et al., 2008).

Extralimital distribution: Bulgaria, Czech Republic, Finland, France, Germany, Greece, Italy, Korea, Lithuania, Montenegro, Russia, Serbia, Slovakia, Spain, Ukraine, United Kingdom, United States of America (introduced).

Host records: Summarized by Yu et al. (2016) as being a parasitoid of the following aphid species: *Eucallipterus tiliae* (L.), *Hoplocallis picta* (Ferrari), *Mesocallis sawashibae* (Matsumura), *Myzocallis carpini* Koch, *M. coryli* (Goeze), *M. komareki* (Pašek), *M. walshii* (Monell), *Tinocallis platani* (Kaltenbach), *T. zelkawae* (Takahashi), *Tuberculatus* sp., and *T. annulatus* (Hartig).

Trioxys metacarpalis Rakhshani and Starý, 2012

Catalogs with Iranian records: Barahoei et al. (2014), Farahani et al. (2016), Rakhshani et al. (2019).

Distribution in Iran: Northern Khorasan (Kazemzadeh et al., 2010 as *Trioxys parauctus* Starý, 1960; Rakhshani, Kazemzadeh et al., 2012).

Distribution in the Middle East: Iran.

Extralimital distribution: None.

Host records: Recorded by Barahoei et al. (2014) as a parasitoid of the aphid *Chaitaphis tenuicauda* Nevsky.

Parasitoid–aphid–plant associations in Iran: Hemiptera: Aphididae: *Chaitaphis tenuicauda* Nevsky on *Kochia scoparia* (Rakhshani, Kazemzadeh et al., 2012).

Comments: This species was originally recorded as *Trioxys parauctus* Starý, 1960 from Iran (Rakhshani, Kazemzadeh et al., 2008 but later was described as a new species (Barahoei et al., 2014).

Trioxys moshei Mescheloff and Rosen, 1993

Distribution in the Middle East: Israel–Palestine (Mescheloff & Rosen, 1993).

Extralimital distribution: None.

Host records: Recorded by Mescheloff and Rosen (1993) as being a parasitoid of the aphid *Hoplocallis picta* (Ferrari).

Trioxys pallidus (Haliday, 1833)

Catalogs with Iranian records: Fallahzadeh and Saghaei (2010), Barahoei et al. (2014), Yu et al. (2016), Farahani et al. (2016), Samin, Coronado-Blanco, Kavallieratos et al. (2018), Samin, Coronado-Blanco, Fischer et al. (2018), Rakhshani et al. (2019).

Distribution in Iran: Alborz (Talebi et al., 2009), Chaharmahal and Bakhtiari (Khajehali & Poorjavad, 2015), Fars (Barahoei et al., 2013; Taheri & Rakhshani, 2013), Golestan (Ghahari et al., 2010; Ghahari & Fischer, 2011; Talebi et al., 2009), Isfahan, Kuhgiloyeh and Boyerahmad (Barahoei et al., 2013; Talebi et al., 2009), Kerman (Asadizade et al., 2014; Barahoei, Madjdzadeh et al., 2012; Barahoei et al., 2013), Kermanshah (Nazari et al., 2012), Markazi (Alikhani et al., 2013; Khaki et al., 2020), Mazandaran (Babaee et al., 2000; Modarres Awal, 2012; Starý et al., 2000), Northern Khorasan (Kazemzadeh et al., 2010), Qazvin (Modarres Awal, 2012; Mohammadbeigi, 2000; Starý et al., 2000), Tehran

(Modarres Awal, 2012; Rakhshani et al., 2002; Rakhshani, Talebi, Kavallieratos et al., 2004; Rakhshani, Talebi, Sadeghi et al., 2004; Starý et al., 2000; Talebi et al., 2009), Iran (no specific locality cited) (Modarres Awal, 1997).

Distribution in the Middle East: Egypt (Yu et al., 2016), Iran (see references above), Iraq (Starý & Kaddou, 1971; Starý, 1976), Israel–Palestine (Mescheloff & Rosen, 1993), Turkey (Aslan et al., 2004; Ölmez & Ulusoy, 2003; Starý, 1976; Uysal et al., 2004).

Extralimital distribution: Andorra, Bulgaria, China, China, Czech Republic, Finland, France, Georgia, Germany, Greece, Hungary, India, Italy, Latvia, Lithuania, Moldova, Montenegro, Morocco, Netherlands, Poland, Russia, Serbia, Spain, Sweden, Tajikistan, United Kingdom, United States of America, Uzbekistan, former Yugoslavia.

Host records: Summarized by Yu et al. (2016) as being a parasitoid of the following aphid species: *Chromaphis juglandicola* (Kaltenbach), *Eucallipterus tiliae* (L.), *Hoplocallis picta* (Ferrari), *Monellia caryella* (Fitch), *Monelliopsis caryae* (Monell), *Myzocallis carpini* Koch, *M. castanicola* Backer, *M. coryli* (Goeze), *M. komareki* (Pašek), *M. multisetis* Boudreaux and Tissot, *M. schreiberi* Hille Ris Lambers and Stroyan, *M. walshii* (Monell), *Panaphis juglandis* (Goeze), *Pterocallis alni* (De Geer), *Thelaxes suberi* (Del Guercio), *Therioaphis trifolii* (Monell), *Tinocallis platani* (Kaltenbach), *T. saltans* (Nevsky), *Tuberculatus albosiphonatus* Hille Ris Lambers, *T. annulatus* (Hartig), *T. eggleri* Börner, *T. moerickei* Hille Ris Lambers, *Tuberculoides* sp., and *Uroleucon sonchi* (L.).

Parasitoid–aphid–plant associations in Iran: Hemiptera: Aphididae: *Chromaphis juglandicola* (Kaltenbach) on *Juglans regia*, and *Juglans* sp. (Alikhani et al., 2013; Barahoei, Madjdzadeh et al., 2012; Barahoei et al., 2013; Khaki et al., 2020; Modarres Awal, 1997, 2012; Mohammadbeigi, 2000; Nazari et al., 2012; Rakhshani et al., 2002, Rakhshani, Talebi, Kavallieratos et al., 2004; Rakhshani, Talebi, Sadeghi et al., 2004; Rakhshani, Kazemzadeh et al., 2008; Rakhshani, Kazemzadeh et al., 2012; Starý, 1979; Starý et al., 2000; Taheri & Rakhshani, 2013; Talebi et al., 2009; van den Bosch et al., 1970; van den Bosch & Messenger, 1973), *Hoplocallis pictus* (Ferrari) on *Quercus ilex* (Rakhshani, Kazemzadeh et al., 2008), *Pterocallis alni* (De Geer) on *Alnus* sp. (Babaee et al., 2000; Modarres Awal, 2012; Starý et al., 2000), *Tinocallis nevskyi* Remaudière, Quednau & Heie (1988) (Khajehali & Poorjavad, 2015), *Tinocallis* sp. on *Ulmus* sp. (Alikhani et al., 2013).

Trioxys pannonicus Starý, 1960

Catalogs with Iranian records: Fallahzadeh & Saghaei (2010), Barahoei et al. (2014), Yu et al. (2016), Farahani et al. (2016),

Samin, Coronado-Blanco, Kavallieratos et al. (2018), Samin, Coronado-Blanco, Fischer et al. (2018), Rakhshani et al. (2019).

Distribution in Iran: Guilan (Talebi et al., 2009), Kermanshah (Starý, 1979; Starý et al., 2000), Sistan and Baluchestan (Barahoei et al., 2014), Iran (no specific locality cited) (Modarres Awal, 2012).

Distribution in the Middle East: Iran.

Extralimital distribution: Finland, Germany, Hungary, India, Italy, Kazakhstan, Moldova, Mongolia, Montenegro, Pakistan, Poland, Russia, Serbia, Slovakia, Spain.

Host records: Summarized by Yu et al. (2016) as being a parasitoid of *Macrosiphoniella* sp., *M. tuberculata* (Nevsky), *Titanosiphon artemisiae* (Koch), and *T. bellicosum* Nevsky.

Parasitoid—aphid—plant associations in Iran: Hemiptera: Aphididae: *Macrosiphoniella tuberculata* (Nevsky) on *Picnomon acarna* (Starý, 1979; Starý et al., 2000; Rakhshani et al., 2011), *Macrosiphoniella* sp. on *Artemisia absinthium* (Rakhshani, Kazemzadeh et al., 2008; Talebi et al., 2009), *Titanosiphon neoartemisiae* (Takahashi) on *Artemisia* sp. (Barahoei et al., 2014).

Trioxys pappi Takada, 1979

Catalogs with Iranian records: Barahoei et al. (2014) as *Trioxys persicus* Davidian, 2005, Yu et al. (2016) as *Trioxys persicus*, Farahani et al. (2016), Samin, Coronado-Blanco, Kavallieratos et al. (2018), Samin, Coronado-Blanco, Fischer et al. (2018) as *Trioxys persicus*, Rakhshani et al. (2019).

Distribution in Iran: Iran (no specific locality cited) (Davidian, 2005).

Distribution in the Middle East: Iran (Davidian, 2005).

Extralimital distribution: Mongolia.

Host records: Unknown.

Comments: *Trioxys pappi* was reported as *Trioxys persicus*, but the examination of type specimen indicated that this was misidentification, and proved to be *Trioxys pappi* (Farahani et al., 2016).

Trioxys quercicola Starý, 1969

Distribution in the Middle East: Iraq (Starý, 1969, 1976; Starý & Kaddou, 1971).

Extralimital distribution: None.

Host records: Recorded by Starý (1976) as being a parasitoid of the aphid *Thelaxes suberi* (del Guercio).

Trioxys tanaceticola Starý, 1971

Catalogs with Iranian records: Fallahzadeh and Saghaei (2010), Barahoei et al. (2014), Yu et al. (2016), Farahani et al. (2016), Samin, Coronado-Blanco, Kavallieratos et al. (2018), Samin, Coronado-Blanco, Fischer et al. (2018), Rakhshani et al. (2019).

Distribution in Iran: Guilan (Talebi et al., 2009), Mazandaran (Starý, 1979; Starý et al., 2000).

Distribution in the Middle East: Iran.

Extralimital distribution: France, Kazakhstan, Russia, Ukraine.

Host records: Summarized by Yu et al. (2016) as being a parasitoid of *Coloradoa heinzi* Börner, and *Metopeurum fuscoviride* Storyan.

Parasitoid—aphid—plant associations in Iran: Hemiptera: Aphididae: *Coloradoa heinzei* Börner on *Artemisia absinthium* (Starý, 1979; Starý et al., 2000), *Coloradoa absinthii* (Lichtenstein) on *Artemisia absinthium* (Talebi et al., 2009), *Coloradoa* sp. on *Artemisia absinthium* L. (Barahoei et al., 2014), *Titanosiphum bellicosum* Nevsky on *Artemisia absinthium* (Talebi et al., 2009), *Titanosiphon bellicosum* Nevsky (=*Titanosiphon neoartemisiae* Nevsky) on *Artemisia absinthium* (Rakhshani, Kazemzadeh et al., 2008).

Trioxys sp.

Distribution in Iran: Markazi, Tehran (Radjabi, 1989).

Parasitoid—aphid—plant associations in Iran: Hemiptera: Aphididae: *Aphis epilobii* Kaltenbach on *Epilobium* sp. (Yaghubi, 1997), *Aphis gossypii* Glover (Starý et al., 2000), *Aphis spiraecola* Patch on *Chaenomeles japonica* (Starý et al., 2000; Yaghubi, 1997), *Brachycaudus amygdalinus* (Schouteden) (Radjabi, 1989), *Chromaphis juglandicola* (Kaltenbach) on *Juglans regia* (Starý et al., 2000; Yaghubi, 1997), *Coloradoa* sp. on *Artemisia herba-alba* (Starý, 1979; Starý et al., 2000), *Myzocallis persicus* Quednau, and Remaudière on *Quercus persica* (Starý, 1979; Starý et al., 2000).

Tribe Ephedrini Mackauer, 1961

Genus *Ephedrus* Haliday, 1833

Ephedrus cerasicola Starý, 1962

Catalogs with Iranian records: Fallahzadeh and Saghaei (2010), Barahoei et al. (2014), Yu et al. (2016), Farahani et al. (2016), Samin, Coronado-Blanco, Kavallieratos et al. (2018), Samin, Coronado-Blanco, Fischer et al. (2018), Rakhshani et al. (2019).

Distribution in Iran: Guilan (Modarres Awal, 2012; Mokhtari et al., 2000; Rakhshani, 2012; Starý et al., 2000), Isfahan (Barahoei et al., 2013; Rakhshani, 2012), Kordestan, Razavi Khorasan (Rakhshani, 2012).

Distribution in the Middle East: Iran (see references above), Turkey (Starý & Stechmann, 1990).

Extralimital distribution: Bulgaria, Czech Republic, Finland, France, Germany, Hungary, Italy, Lithuania, Moldova, Montenegro, Netherlands, New Zealand, Norway, Poland, Russia, Serbia, Slovakia, Slovenia, Spain, Sweden, United Kingdom, United States of America.

Host records: Summarized by Yu et al. (2016) as being a parasitoid of the following aphid species: *Brachycaudus helichrysi* (Kaltenbach), *B. prunicola* (Kaltenbach), *Capitophorus inulae* (Passerini), *Cryptomyzus* sp., *C. galeopsidis* Kaltenbach, *Hyperomyzus lactucae* (L.), *H. rhinanthi* (Schouteden), *Myzus cerasi* (Fabricius), *M. ligustri* (Mosley), *M. persicae* (Sulzer), *M. varians* Davidson, *Nasonovia ribisnigri* (Mosley), *Ovatus crataegarius* (Walker), *Phorodon humuli* (Scharnk), and *Rhopalosiphum nymphaeae* (L.).

Parasitoid—aphid—plant associations in Iran: Hemiptera: Aphididae: *Dysaphis plantaginea* (Passerini) on *Malus domestica*, and *Pyrus malus* (Barahoei et al., 2013; Rakhshani, 2012), *Myzus cerasi* (Fabricius) on *Prunus cerasus* (Rakhshani, 2012), *Myzus persicae* (Sulzer) on *Prunus persica* (Rakhshani, 2012), *Phorodon humuli* (Schrank) on *Prunus persica* (Modarres Awal, 2012; Mokhtari et al., 2000; Starý et al., 2000), *Rhopalosiphum nymphaeae* (L.) on *Prunus* sp. (Modarres Awal, 2012; Mokhtari et al., 2000; Starý et al., 2000).

Ephedrus chaitophori Gärdenfors, 1986

Catalogs with Iranian records: Fallahzadeh and Saghaei (2010), Barahoei et al. (2014), Yu et al. (2016), Farahani et al. (2016), Samin, Coronado-Blanco, Kavallieratos et al. (2018), Samin, Coronado-Blanco, Fischer et al. (2018), Rakhshani et al. (2019).

Distribution in Iran: Tehran (Rakhshani, Talebi, Starý et al., 2007), Iran (no specific locality cited) (Modarres Awal, 2012; Rakhshani, Talebi, Manzari et al., 2006).

Distribution in the Middle East: Iran.

Extralimital distribution: Czech Republic, Finland, Germany, Serbia, Spain, Sweden, United States of America.

Host records: Summarized by Yu et al. (2016) as being a parasitoid of the following aphid species: *Chaitophorus leucomelas* Koch, *C. nigricantis* Pintera, *C. populeti* (Panzer), *C. populialbae* (Boyer de Fonscolombe), *C. populifolii* (Essig), *C. salijaponicus* Essig and Kuwana, and *C. tremulae* Koch.

Parasitoid—aphid—plant associations in Iran: Hemiptera: Aphididae: *Chaitophorus populeti* (Panzer) on *Populus nigra* (Rakhshani, Talebi, Starý et al., 2007).

Ephedrus helleni Mackauer, 1968

Catalogs with Iranian records: Fallahzadeh and Saghaei (2010), Barahoei et al. (2014), Yu et al. (2016), Farahani et al. (2016), Samin, Coronado-Blanco, Kavallieratos et al. (2018), Samin, Coronado-Blanco, Fischer et al. (2018), Rakhshani et al. (2019).

Distribution in Iran: Golestan (Barahoei, Sargazi et al., 2012), Tehran (Barahoei et al., 2014; Modarres Awal, 2012 as *Ephedrus salicicola* Takada, 1968; Rakhshani, Talebi, Starý

et al., 2007; Starý, 1979; Starý et al., 2000; Talebi et al., 2009), Iran (no specific locality cited) (Modarres Awal, 2012; Rahimi et al., 2010; Rakhshani, Talebi, Manzari et al., 2006).

Distribution in the Middle East: Iran.

Extralimital distribution: Czech Republic, Finland, France, India, Japan, Montenegro, Russia, Serbia, Sweden, United Kingdom.

Host records: Summarized by Yu et al. (2016) as being a parasitoid of *Cavariella aegopodii* (Scopoli), *C. aquatica* (Gillette and Bragg), *C. archangelicae* (Scopoli), *C. pastinacae* (L.), *C. salicicola* (Matsumura), *C. theobaldi* (Gillette and Bragg), and *Eumyzus* sp.

Parasitoid—aphid—plant associations in Iran: Hemiptera: Aphididae: *Cavariella aquatica* (Gillette and Bragg) on *Salix alba* (Rahimi et al., 2010; Rakhshani, Talebi, Starý et al., 2007; Starý, 1979; Starý et al., 2000; Talebi et al., 2009), *Hayhurstia atriplicis* (L.) on *Atriplex* sp. and on *Chenopodium album* L. (Barahoei et al., 2014).

Ephedrus lacertosus (Haliday, 1833)

Catalogs with Iranian records: This species is a new record for the fauna of Iran.

Distribution in Iran: Mazandaran province, Tonekabon (Jangal-e 3000), 2♀, 1♂, July 2015, ex *Aphis* sp. (Hemiptera: Aphididae) on *Urtica dioica* (Urticaceae).

Distribution in the Middle East: Cyprus (Beyarslan et al., 2017), Iran (new record), Turkey (Tomanović et al., 2008).

Extralimital distribution: Indo-Malayan, Nearctic, Neotropical, and Palaearctic regions [Adjacent countries to Iran: Kazakhstan and Russia].

Host records: Summarized by Yu et al. (2016) as being a parasitoid of the following aphid species: *Acyrthosiphon malvae* (Mosley), *A. pisum* (Harris), *Amphorophora agathonica* (Hottes), *A. idaei* (Börner), *A. rubi* (Kaltenbach), *Aphis fabae* (Scopoli), *A. oxyacanthae* (Schrank), *A. pomi* De Geer, *A. rumicis* L., *Aspidophorodon* sp., *Aulacorthum majanthemi* Müller, *Brachysiphoniella montana* (van der Goot), *Cinara puniperi* (De Geer), *C. mordvilkoi* (Pašek), *Cryptomyzus ribis* (L.), *Dysaphis crataegi* (Kaltenbach), *D. plantaginea* (Passerini), *D. pyri* (Boyer de Fonscolombe), *Ericaphis gentneri* Mason, *Hyalopteroides humilis* (Walker), *Hyalopterus pruni* (Geoffroy), *Impatientinum balswardiae* (Takahashi), *Macromyzus woodwardiae* (Takahashi), *Macrosiphoniella absinthii* (L.), *M. usquertensis* Hille Ras Lambers, *Macrosiphum* spp., *Masonaphis maxima* (Mason), *Metopolophium carnosum* Buckton, *Myzus cerasi* (Fabricius), *Rhopalosiphoninus* sp., *R. padi* (L.), *Sitobion avenae* (Fabricius), *S. equiseti* Holman, *S. miscanthi* (Takahashi), *Takecallis affinis* Ghosh, *Tetraneura ulmi* (L.), *Uroleucon gravicorne* (Patch), *U. jaceae* (L.), and *U. minatii* Das.

Ephedrus (Ephedrus) laevicollis (Thomson, 1895)

Catalogs with Iranian records: Fallahzadeh and Saghaei (2010), Yu et al. (2016), Samin, Coronado-Blanco, Kavallieratos et al. (2018), Samin, Coronado-Blanco, Fischer et al. (2018).

Distribution in Iran: Iran (no specific locality cited) (Starý, 1979).

Distribution in the Middle East: Iran.

Extralimital distribution: Austria, Bulgaria, Czech Republic, Finland, France, Germany, Hungary, India, Ireland, Kazakhstan, Lithuania, Moldova, Mongolia, Montenegro, Netherlands, Poland, Serbia, Slovakia, Spain, Sweden, Switzerland, United Kingdom.

Host records: Summarized by Yu et al. (2016) as being a parasitoid of the following aphid species: *Brachycaudus helichrysi* (Kaltenbach), *Cavariella* sp., *C. aegopodii* (Scopoli), *C. aquatica* (Gillette and Bragg), *C. panstinacae* (L.), *Chaetosiphon fragaefolii* (Cockerell), *C. tetrarhodum* (Walker), *Longicaudus trirhodus* (Walker), *Macrosiphum rosae* (L.), *Myzaphis* sp., *M. bucktoni* Jacob, *M. rosarum* (Kaltenbach), *Pentatrichopus* sp., and *Uroleucon sonchi* (L.).

Ephedrus nacheri Quilis, 1934

Catalogs with Iranian records: No catalog.

Distribution in Iran: Mazandaran (Sakenin et al., 2020).

Distribution in the Middle East: Iran (Sakenin et al., 2020), Turkey (Rakhshani et al., 2019).

Extralimital distribution: Bulgaria, China, Czech Republic, Finland, France, Georgia, Germany, Greece, Hungary, India, Italy, Japan, Korea, Moldova, Nepal, Netherlands, Poland, Russia, Serbia, Slovakia, Spain, Sweden.

Host records: Summarized by Yu et al. (2016) as being a parasitoid of the following aphid species: *Acyrthosiphon kondoi* Shinji, *A. pisum* (Harris), *Amphicercidus japonicus* (Hori), *Aphis* spp., *Brachycaudus* sp., *Brevicoryne brassicae* (L.), *Cavariella salicicola* (Matsumura), *Coloradoa rufomaculata* (Wilson and H.F.), *Cryptosiphum artemisiae* Buckton, *Hayhurstia atriplicis* (L.), *Hyadaphis foeniculi* (Passerini), *Hyalopterus pruni* (Geoffroy), *Hyperomyzus lactucae* (L.), *H. rhinanthi* (Schouteden), *Lipaphis erysimi* (Kaltenbach), *Macrosiphoniella sanborni* (Gillette), *Melanaphis arundinariae* (Takahashi), *Myzus persicae* (Sulzer), *Pleotrichophorus glandulosus* (Kaltenbach), *Prociphilus konoi* Hori, *Rhopalosiphoninus deutzifoliae* Shinji, *R.. padi* (L.), *R. rufiabdominalis* (Sasaki), *Schizaphis graminum* (Rondani), *Semiaphis anthrisci* (Kaltenbach), *Sitobion akebiae* (Shinji), *S. avenae* (Fabricius), *S. ibarae* (Matsumura), *Staegeriella necopinata* (Börner), and *Trichosiphonaphis lonicerae* (Uye).

Parasitoid—aphid—plant associations in Iran: Hemiptera: Aphididae: *Hayhurstia atriplicis* (L.), plant is unknown (Sakenin et al., 2020).

Ephedrus niger Gautier, Bonnamour and Gaumont, 1929

Catalogs with Iranian records: Fallahzadeh and Saghaei (2010), Barahoei et al. (2014), Yu et al. (2016), Farahani et al. (2016), Samin, Coronado-Blanco, Kavallieratos et al. (2018), Samin, Coronado-Blanco, Fischer et al. (2018), Rakhshani et al. (2019).

Distribution in Iran: Fars (Taheri & Rakhshani, 2013), Golestan (Barahoei, Sargazi et al., 2012), Guilan (Modarres Awal, 2012; Starý et al., 2000; Talebi et al., 2009; Yaghubi, 1997; Yaghubi & Sahragard, 1998), Isfahan (Barahoei et al., 2013; Talebi et al., 2009), Kerman (Barahoei, Madjdzadeh et al., 2012; Barahoei et al., 2013), Kermanshah (Nazari et al., 2012), Markazi (Alikhani et al., 2013), Mazandaran (Modarres Awal, 2012; Starý, 1979; Starý et al., 2000), Northern Khorasan (Kazemzadeh et al., 2010; Rakhshani, Kazemzadeh et al., 2012; Starý, 1979; Starý et al., 2000), Tehran (Modarres Awal, 2012; Starý, 1979; Starý et al., 2000; Talebi et al., 2009).

Distribution in the Middle East: Iran (see references above), Israel—Palestine (Mescheloff & Rosen, 1988).

Extralimital distribution: Algeria, Andorra, Azerbaijan, Bulgaria, China, Czech Republic, Finland, France, Georgia, Germany, Greece, Hungary, India, Italy, Korea, Moldova, Monaco, Mongolia, Montenegro, Netherlands, Poland, Portugal, Russia, Serbia, Spain, Sweden, Taiwan, Tajikistan, former Yugoslavia.

Host records: Summarized by Yu et al. (2016) as being a parasitoid of the following aphid species: *Acrthosiphon pisum* (Harris), *Aphis fabae* (Scopoli), *A. sambuci* L., *Dysaphis* sp., *Macrosiphoniella* spp., *Macrosiphum rosae* (L.), *Megoura viciae* Buckton, *Metopeurum fuscoviride* (L.), *Microsiphoniella artemisiae* (Boyer de Fonscolombe), *Titanosiphon dracunculi* Nevsky, *Uroleucon* spp., and *Uromelan* sp.

Parasitoid—aphid—plant associations in Iran: Hemiptera: Aphididae: *Macrosiphoniella abrotani* (Walker) on *Artemisia annua* (Mackauer & Starý, 1967; Rakhshani et al., 2011; Starý et al., 2000), *Macrosiphoniella sanborni* (Gillette) on *Chrysanthemum morifolium* (Alikhani et al., 2013; Nazari et al., 2012; Rakhshani et al., 2011), *Macrosiphoniella pulvara* Walker (Rakhshani et al., 2011; Starý, 1979), *Macrosiphoniella* sp. on *Artemisia absinthium*, *Artemisia biennis*, and *Artemisia* sp. (Barahoei et al., 2013; Rakhshani et al., 2011; Starý, 1979; Starý et al., 2000; Yaghubi, 1997), *Uroleucon cichorii* (Koch) on *Sonchus* sp. (Barahoei, Madjdzadeh et al., 2012; Barahoei et al., 2013;

Starý et al., 2000; Yaghubi, 1997), *Uroleucon acroptilidis* Kadyrbekov, Renxin and Shao on *Acroptilon repens*; *Uroleucon bielawskii* (Szelegiewicz) on *Lactuca* sp.; *Uroleucon cichorii* (Koch) on *Cichorium intybus* (Barahoei, Madjdzadeh et al., 2012; Barahoei et al., 2013), *Uroleucon compositae* (Theobald) on *Carthamus oxyacantha*, *Cirsium* sp., and *Lactuca* sp. (Barahoei et al., 2013; Nazari et al., 2012), *Uroleucon erigeronense* (Thomas) on *Conyza canadensis*, and *Conyza* sp. (Nazari et al., 2012; Starý et al., 2000; Talebi et al., 2009), *Uroleucon chondrillae* on *Chondrilla juncea* (Alikhani et al., 2013), *Uroleucon jaceae* (L.) on *Acroptilon repens*, *Centaurea hyrcanica*, *C. iberica*, and *C. solstitialis* (Barahoei, Madjdzadeh et al., 2012; Barahoei et al., 2013; Nazari et al., 2012; Rakhshani, Kazemzadeh et al., 2012; Starý, 1979; Starý et al., 2000; Talebi et al., 2009), *Uroleucon jacaea aeneum* Hille Ris Lambers on *Acroptilon repens*, *Uroleucon sonchi* (L.) on *Launaea acanthodes*, *Lactuca serriola*, *Sonchus* sp., *Sonchus asper*, and *Sonchus oleraceus* (Barahoei, Madjdzadeh et al., 2012; Barahoei et al., 2013; Nazari et al., 2012; Rakhshani, Kazemzadeh et al., 2012; Taheri & Rakhshani, 2013; Talebi et al., 2009), *Uroleucon* sp. on *Acroptilon repens*, *Artemisia biennis*, *Cantharanthus* sp., and *Mindium laevigatum* (Barahoei, Madjdzadeh et al., 2012; Barahoei et al., 2013; Starý, 1979; Starý et al., 2000).

Ephedrus persicae Froggatt, 1904

Catalogs with Iranian records: Fallahzadeh and Saghaei (2010), Barahoei et al. (2014), Yu et al. (2016), Farahani et al. (2016), Samin, Coronado-Blanco, Kavallieratos et al. (2018), Samin, Coronado-Blanco, Fischer et al. (2018), Rakhshani et al. (2019).

Distribution in Iran: Alborz (Rakhshani, 2012; Starý, 1979; Starý et al., 2000), Fars (Barahoei et al., 2013; Modarres Awal, 2012; Rakhshani, 2012; Starý, 1979; Starý et al., 2000; Taheri & Rakhshani, 2013), Golestan (Barahoei, Sargazi et al., 2012; Ghahari et al., 2011), Guilan (Modarres Awal, 2012; Starý et al., 2000; Yaghubi & Sahragard, 1998), Hormozgan (Ameri et al., 2019), Isfahan (Barahoei et al., 2013; Rakhshani, 2012), Kerman (Barahoei, Madjdzadeh et al., 2012; Barahoei et al., 2013), Kermanshah (Nazari et al., 2012; Rakhshani, 2012; Rakhshani, Tomanović et al., 2008), Khuzestan (Mossadegh et al., 2010, 2011; Salehipour et al., 2010), Kordestan, Qazvin (Rakhshani, 2012), Markazi (Alikhani et al., 2013; Radjabi, 1989; Rakhshani, Tomanović et al., 2008), Northern Khorasan (Malkeshi, 1997; Malkeshi et al., 1998; Modarres Awal, 2012; Rakhshani, 2012; Rakhshani, Kazemzadeh et al., 2012; Starý et al., 2000), Qom (Rakhshani, Tomanović et al., 2008), Razavi Khorasan (Darsouei et al., 2011a,b; Farokhzadeh et al., 2014; Jafari-Ahmadabadi et al., 2011; Jafari-Ahmadabadi & Modarres Awal, 2012; Jalali Moghadam Ziabari et al.,

2014; Modarres Awal, 2012; Rakhshani, 2012), Sistan and Baluchestan (Bandani, 1992; Bandani et al., 1993; Fakhireh et al., 2016; Malkeshi et al., 2004, 2010; Modarres Awal, 1997, 2012; Modarres Najafabadi & Gholamian, 2007; Rakhshani, Talebi, Kavallieratos et al., 2005; Rakhshani, Talebi, Manzari et al., 2006; Rakhshani, Tomanović et al., 2008; Starý et al., 2000; Talebi et al., 2009), Tehran (Mackauer, 1963; Modarres Awal, 2012; Rakhshani, 2012; Rakhshani, Talebi, Kavallieratos et al., 2005; Starý, 1979; Starý et al., 2000; Talebi et al., 2009), Iran (no specific locality cited) (Bogdanovi et al., 2009; Khanjani, 2006b; Žikić et al., 2009).

Distribution in the Middle East: Cyprus, Egypt (Abu El-Ghiet et al., 2014; El-Heneidy & Abdel-Samad, 2001; El-Heneidy et al., 2001, 2002, 2003), Iran (see references above), Iraq (Al-Azawi, 1970; Starý, 1976), Israel—Palestine (Avidov & Harpaz, 1969; Bodenheimer & Swirski, 1957; Mescheloff & Rosen, 1988; Rosen, 1967, 1969; Starý, 1976), Jordan (Hasan, 2016), Lebanon (Abou-Fakhr, 1982; Hussein & Kawar, 1982; Mackauer & Starý, 1967; Starý, 1976; Talhouk, 1961; Tremblay et al., 1985), Syria (Starý, 1976), Turkey (Akar & Çetin Erdoğan, 2017; Aslan & Karaca, 2005; Aslan et al., 2004; Düzgüneş et al., 1982), United Arab Emirates, Yemen (Starý et al., 2013).

Extralimital distribution: Algeria, Andorra, Argentina, Australia, Belgium, Brazil, Bulgaria, Canada, Chile, China, Czech Republic, Finland, France, Georgia, Germany, Greece, Hungary, India, Ireland, Italy, Japan, Kazakhstan, Korea, Kyrgyzstan, Latvia, Libya, Lithuania, Madagascar, Moldova, Mongolia, Montenegro, Morocco, Netherlands, Norway, Pakistan, Poland, Portugal, Romania, Russia, Serbia, Slovakia, South Africa, Spain, Sweden, Switzerland, Tajikistan, United Kingdom, United States of America, Uzbekistan.

Host records: Summarized by Yu et al. (2016) as being a parasitoid of the following aphid species: *Acyrthosiphon malvae* (Mosley), *A. pisum* (Harris), *Allocotaphis quaestionis* (Börner), *Aphis* spp., *Brachycaudus* spp., *Brachyunguis tamaricis* (Lichtenstein), *B. tamaricophilus* (Nevsky), *Brevicoryne brassicae* (L.), *Capitophorus* spp., *Chaitophorus leucomelas* Koch, *C. populeti* (Panzer), *C. truncatus* (Haussmann), *Diuraphis calamagrostis* (Ossiannilsson), *D. noxia* (Mordvilko), *Dysaphis* spp., *Eulachnus tuberculostemmata* (Theobald), *Hayhurstia atriplicis* (L.), *Hyadaphis coriandri* (Das), *H. foeniculi* Passerini, *H. paserinii* (Del Guercio), *Hyalopterus amygdali* (Blanchard), *H. pruni* (Geoffroy), *Hyperomyzus lactucae* (L.), *Lipaphis erysimi* (Kaltenbach), *Macrosiphum euphorbiae* (Thomas), *Melanaphis donacis* (Passerini), *M. pyraria* (Passerini), *Metopolophium dirhodum* (Walker), *Myzus* spp., *Nasonovia brachycyclica* Holman, *N. ribisnigri* (Mosley), *Ovatus malisuctus* (Matsumura), *Phorodon humuli* (Schrank), *Rhopalomyzus lonicerae* (Siebold), *Rhopalosiphum* spp., *Roepkea marchali* (Börner), *Schizaphis graminum* (Rondani), *Sipha maydis* Passerini, *Sitobion akebiae* (Shinji), *Sitobion avenae* (Fabricius), *Tetraneura* sp., *Tinocallis ulmiparvifoliae* Matsumura, *Tinocallis viridis* (Takahashi), *Toxoptera aurantii* (Boyer de Fonscolombe), *Trichosiphonaphis lonicerae* (Uye), *Tuberocephalus momonis* (Matsumura), *Uroleucon* sp., and *Vesiculaphis caricis* (Fullaway).

Parasitoid—aphid—plant associations in Iran: Hemiptera: Aphididae: *Aphis affinis* del Guercio on *Mentha aquatica*, and on *Mentha longifolia* (Barahoei et al., 2013; Taheri & Rakhshani, 2013; Talebi et al., 2009), *Aphis craccivora* Koch on *Alhagi maurorum*, *Citrus* spp., *Medicago sativa*, *Polygonum* sp., *Prunus padus*, *Pyrus communis*, *Robinia pseudoacacia*, *Solanum dulcamara*, and *Sophora alopecuroides* (Barahoei, Madjdzadeh et al., 2012; Barahoei et al., 2013; Mackauer, 1963; Nazari et al., 2012; Rakhshani, Talebi, Kavallieratos et al., 2005; Rakhshani, Talebi, Manzari et al., 2006; Starý, 1979; Starý et al., 2000; Žikić et al., 2009), *Aphis fabae* Scopoli on *Vicia* sp., and *Vicia faba* (Barahoei et al., 2013; Starý, 1979; Taheri & Rakhshani, 2013; Žikić et al., 2009), *Aphis gossypii* Glover on *Lycopersicum esculentum*, and *Salsola* sp. (Barahoei et al., 2013), *Aphis nerii* Boyer de Fonscolombe on *Nerium oleander* (Talebi et al., 2009), *Aphis plantaginis* Goeze on *Rumex* sp. (Barahoei et al., 2013; Radjabi, 1989), *Aphis pomi* De Geer on *Malus pumila* (Malkeshi, 1997; Modarres Awal, 2012; Starý et al., 2000), *Aphis punicae* Passerini (Farokhzadeh et al., 2014; Jalali Moghadam Ziabari et al., 2014; Mackauer, 1963; Starý et al., 2000), *Aphis* sp. on *Beta maritima* (Mossadegh et al., 2011), *Brachycaudus amygdalinus* (Schouteden) on *Amygdalus arabica*, *Prunus amygdalinus*, and *Prunus dulcis* (Alikahni et al., 2013; Jafari-Ahmadabadi et al., 2011; Jafari-Ahmadabadi & Modarres Awal, 2012; Nazari et al., 2012; Rakhshani, 2012; Rakhshani, Kazemzadeh et al., 2012; Talebi et al., 2009; Žikić et al., 2009), *Brachycaudus cardui* (L.) (Žikić et al., 2009), *Brachycaudus helichrysi* (Kaltenbach) on *Amygdalus* sp., *Prunus domestica*, *Prunus dulcis*, and *Prunus persica* (Alikhani et al., 2013; Barahoei et al., 2013; Nazari et al., 2012; Rakhshani, 2012; Starý, 1979; Starý et al., 2000; Žikić et al., 2009), *Brachyunguis tamaricis* (Lichtenstein) on *Tamarix* sp. (Starý, 1979; Starý et al., 2000), *Brevicoryne brassicae* (L.) (Khanjani, 2006b; Malkeshi et al., 2004, 2010), *Diuraphis noxia* (Kurdjumov) (Bandani, 1992; Bandani et al., 1993; Starý et al., 2000), *Dysaphis crataegi* (Kaltenbach) on *Crataegus monogyna* (Starý, 1979; Starý et al., 2000), *Dysaphis devecta* (Walker) on *Malus orientalis* (Alikhani et al., 2013; Žikić et al., 2009), *Dysaphis foeniculus* Theobald on *Foeniculum vulgare* (Talebi et al., 2009), *Dysaphis plantaginea* (Passerini) on *Malus domestica*, *Malus pumila*, and *Pyrus malus*

(Barahoei et al., 2013; Malkeshi, 1997; Nazari et al., 2012; Rakhshani, 2012; Starý et al., 2000), *Dysaphis pyri* (Boyer de Fonscolombe) on *Pyrus communis* (Mackauer, 1963; Malkeshi, 1997; Radjabi, 1989; Rakhshani, 2012; Starý et al., 2000; Žikić et al., 2009), *Dysaphis reaumuri* (Mordvilko) on *Pyrus communis* (Barahoei et al., 2013; Rakhshani, 2012), *Hayhurstia atriplicis* (L.) on *Cichorium* sp. (Barahoei et al., 2013), *Hyadaphis coriandri* Das on *Coriandrum sativum* (Talebi et al., 2009), *Hyalopterus amygdali* (Blanchard) on *Prunus amygdalinus*, and *Prunus dulcis* (Barahoei et al., 2013; Jafari-Ahmadabadi et al., 2011; Jafari-Ahmadabadi & Modarres Awal, 2012; Rakhshani, 2012), *Hyadaphis coriandri* Das on *Coriandrum* sp. (Barahoei et al., 2014), *Macrosiphum* sp. on *Malva* sp. (Starý et al., 2000), *Metopolophium dirhodum* (Walker) on *Triticum aestivum* (Bandani, 1992; Bandani et al., 1993; Rakhshani, Tomanović et al., 2008; Starý et al., 2000), *Myzus cerasi* (Fabricius) on *Prunus incana* (Rakhshani, 2012; Žikić et al., 2009), *Myzus persicae* (Sulzer) on *Althaea* sp., *Cucumis sativus*, *Hibiscus rosasinensis*, *Malva neglecta*, *Prunus dulcis*, *Prunus persica*, and *Sysimbrium* sp. (Barahoei et al., 2013; Nazari et al., 2012; Rakhshani, 2012; Talebi et al., 2009; Žikić et al., 2009), *Phorodon humuli* (Schrank) on *Prunus dulcis* (Barahoei et al., 2013; Rakhshani, 2012), *Rhopalosiphum maidis* (Fitch) on *Hordeum vulgare*, *Hordeum* sp., *Setaria glauca*, and *Triticum* sp. (Bandani et al., 1992, 1993; Barahoei, Madjdzadeh et al., 2012; Barahoei et al., 2013; Gärdenfors 1986; Mossadegh et al., 2011; Starý, 1979; Starý et al., 2000), *Rhopalosiphum padi* (L.) on *Triticum aestivum* (Bandani, 1992; Bandani et al., 1993; Gärdenfors 1986; Rakhshani, Tomanović et al., 2008; Starý, 1979; Starý et al., 2000), *Schizaphis graminum* (Rondani) on *Hordeum* sp., and *Triticum aestivun* (Alikhani et al., 2013; Bandani, 1992; Bandani et al., 1993; Bogdanovi et al., 2009; Modarres Najafabadi & Gholamian, 2007; Rakhshani, Tomanović et al., 2008; Rakhshani, Kazemzadeh et al., 2012; Starý et al., 2000; Žikić et al., 2009), *Sipha maydis* Passerini (Bandani, 1992; Bandani et al., 1993; Starý et al., 2000), *Sitobion avenae* (Fabricius) on *Triticum aestivum* (Bandani, 1992; Bandani et al., 1993; Barahoei et al., 2013; Starý et al., 2000), *Toxoptera aurantii* Boyer de Fonscolombe (Mackauer, 1963).

Ephedrus plagiator (Nees von Esenbeck, 1811)

Catalogs with Iranian records: Fallahzadeh and Saghaei (2010), Barahoei et al. (2014), Yu et al. (2016), Farahani et al. (2016), Samin, Coronado-Blanco, Kavallieratos et al. (2018), Samin, Coronado-Blanco, Fischer et al. (2018), Rakhshani et al. (2019).

Distribution in Iran: Alborz (Rakhshani, 2012), East Azarbaijan (Starý, 1979; Starý et al., 2000), Fars

(Modarres Awal, 2012), Golestan (Barahoei, Sargazi et al., 2012; Darvish-Mojeni & Bayat-Asadi, 1995; Modarres Awal, 1997; 2012; Starý et al., 2000), Guilan (Modarres Awal, 2012; Mokhtari et al., 2000; Starý et al., 2000; Yaghubi, 1997; Yaghubi & Sahragard, 1998), Hormozgan (Ameri et al., 2019), Isfahan (Barahoei et al., 2013; Mehrparvar et al., 2016; Rakhshani, Tomanović et al., 2008), Kerman (Asadizade et al., 2014), Khuzestan (Nazari et al., 2012; Rakhshani, Tomanović et al., 2008), Mazandaran (Modarres Awal, 1997, 2012), Razavi Khorasan (Darsouei et al., 2011b; Rakhshani, 2012), Tehran (Modarres Awal, 2012; Rakhshani, Tomanović et al., 2008), Iran (no specific locality cited) (Khanjani, 2006a).

Distribution in the Middle East: Egypt (Abdel-Rahman, 2005; Abu El-Ghiet et al., 2014), Iran (see references above), Iraq (Starý, 1976), Turkey (Akar & Çetin Erdoğan, 2017; Çetin Erdoğan et al., 2008; Düzgüneş et al., 1982; Starý, 1976; Uysal et al., 2004), Lebanon (Rakhshani et al., 2019).

Extralimital distribution: Andorra, Australia, Austria, Belgium, Brazil, Bulgaria, Chile, China, Czech Republic, Denmark, Estonia, Finland, France, Georgia, Germany, Greece, Hungary, Iceland, India, Ireland, Italy, Japan, Kazakhstan, Korea, Latvia, Lithuania, Moldova, Mongolia, Montenegro, Netherlands, New Zealand, Norway, Pakistan, Poland, Portugal, Romania, Russia, Serbia, Slovakia, Slovenia, Spain, Sweden, Switzerland, Tajikistan, Ukraine, United Kingdom, Uzbekistan.

Host records: Summarized by Yu et al. (2016) as being a parasitoid of the following aphid species: *Acyrthosiphon* spp., *Amphicercidus japonicus* (Hori), *Anoecia corni* (Fabricius), *Anuraphis farfarae* (Koch), *Aphis* spp., *Aulacorthum circumflexum* (Buckton), *Aulacorthum magnoliae* (Essid and Kuwana), *Aulacorthum muradachi* (Shinji), *Brachycaudus* spp., *Brachycorynella asparagi* (Mordvilko), *Brachysiphoniella montana* (van der Goot), *Capitophorus inulae* (Passerini), *Cavariella aegopodi* (Scopoli), *Cavariella araliae* Takahashi, *Ceratovacuna silvestrii* Takahashi, *Ceruraphis eriophori* (Walker), *Corylobium avellanae* (Schrank), *Cryptomyzus galeopsidis* Kaltenbach, *Cryptomyzus ribis* (L.), *Cryptosiphum artemisiae* Buckton, *Diuraphis calamagrostis* (Ossiannilsson), *Diuraphis noxia* (Mordvilko), *Dysaphis* spp., *Elatobium abietinum* Walker, *Eriosoma ulmi* (L.), *Hayhurstia atriplicis* (L.), *Hyadaphis foeniculi* Passerini, *Hyalopteroides humilis* (Walker), *Hyalopterus pruni* (Geoffroy), *Hyperomyzus* spp., *Hysteroneura setariae* (Thomas), *Impatientinum asiaticum* Nevsky, *Impatientinum balsamines* (Kaltenbach), *Liosomaphis berberidis* (Kaltenbach), *Lipaphis erysimi* (Kaltenbach), *Macromyzus wooddwardiae* (Takahashi), *Macrosiphoniella absinthii* (L.), *Macrosiphum* spp., *Melanaphis bambusae* (Fullaway), *Melanaphis sacchari* (Zehntner), *Metopolophium dirhodum* (Walker), *Metopolophium festucae* (Theobald), *Myzaphis rosarum* (Kaltenbach), *Myzocallis coryli* (Goeze), *Myzus* spp., *Nasonovia ribisnigri* (Mosley), *Neorhopalomyzus*

lonicericola Takahashi, *Ovatus crataegarius* (Walker), *Pachypappa tremulae* (L.), *Parachaitophorus spiraea* (Takahashi), *Phorodon humuli* (Schrank), *Pleotrichophorus duponti* Hille Ris Lambers, *Prociphilus bumeliae* (Schrank), *Prociphilus fraxini* (Hartig), *Prociphilus konoi* Hori, *Rhopalomyzus lonicerae* (Siebold), *Rhopalosiphoninus deutzifoliae* Shinji, *Rhopalosiphum* spp., *Sappaphis* sp., *Schizaphis* spp., *Sinomegoura citricola* (van der Goot), *Sipha maydis* Passerini, *Sitobion* spp., *Staticobium limonii* (Contarini), *Tetraneura ulmi* (L.), *Toxoptera aurantii* (Boyer de Fonscolombe), *Toxoptera odinae* (van der Goot), and *Uroleucon* spp.

Parasitoid—aphid—plant associations in Iran: Hemiptera: Aphididae: *Brachycaudus cardui* (L.) on *Onopordon acanthium* (Starý et al., 2000; Yaghubi, 1997), *Brachycaudus helichrysi* (Kaltenbach) on *Prunus persica* (Rakhshani, 2012), *Brachycaudus* sp. on *Onopordon* sp. (Starý et al., 2000), *Dysaphis crataegi* (Kaltenbach) on *Daucus* sp. (Starý, 1979; Starý et al., 2000), *Dysaphis pyri* (Boyer de Fonscolombe) on *Pyrus communis* (Rakhshani, 2012), *Dysaphis* sp. (Starý, 1979), *Myzus cerasi* (Fabricius) on *Prunus cerasus* (Mokhtari et al., 2000; Starý et al., 2000), *Diuraphis noxia* (Kurdjumov) (Khanjani, 2006a), *Rhopalosiphum padi* (L.) on *Triticum aestivum* (Barahoei et al., 2013; Rakhshani, Tomanović et al., 2008), *Schizaphis graminum* (Rondani) on *Triticum aesticum* L. (Rakhshani, Tomanović et al., 2008), *Sitobion avenae* (Fabricius) on *Triticum aestivum* (Darvish-Mojeni & Bayat-Asadi, 1995; Modarres Awal, 1997, 2012; Nazari et al., 2012; Starý et al., 2000; Rakhshani, Tomanović et al., 2008), *Macrosiphum rosae* (L.) (Mehrparvar et al., 2016).

Ephedrus validus (Haliday, 1833)
Distribution in the Middle East: Cyprus (Fulmek, 1968; Wilkinson, 1926).

Extralimital distribution: Austria, Bulgaria, Czech Republic, Denmark, Finland, France, Germany, Hungary, Ireland, Italy, Korea, Moldova, Montenegro, Netherlands, Norway, Poland, Romania, Russia, Sweden, Switzerland, Tajikistan, United Kingdom.

Host records: Summarized by Yu et al. (2016) as being a parasitoid of the aphids *Anuraphis farfarae* (Koch), *Aphis fabae* Scopoli, *Aphis glycines* Matsumura, *Apis pomi* De Geer, *Aphis rumicis* L., *Brachycaudus helichrysi* Kaltenbach, *Dysaphis plantaginea* (Passerini), *Dysaphis pyri* (Boyer de Fonscolombe), *Myzus cerasi* (fabricius), and *Thecabius affinis* (Kaltenbach).

Ephedrus sp.
Distribution in Iran: Khuzestan (Mossadegh & Kocheili, 2003).
Parasitoid—aphis—plant associations in Iran: *Aphis craccivora* Koch on *Cousinia* sp., *Myzus persicae* (Sulzer) on *Isatis* sp. (Starý, 1979; Starý et al., 2000).

Genus *Toxares* Haliday, 1840

Toxares deltiger (Haliday, 1833)

Distribution in the Middle East: Turkey (Tomanović et al., 2008).

Extralimital distribution: Belgium, Bulgaria, Czech Republic, Finland, France, Germany, Hungary, India, Ireland, Italy, Lithuania, Moldova, Netherlands, Norway, Poland, Russia, Slovakia, Spain, Sweden, United Kingdom, United States of America, former Yugoslavia.

Host records: Summarized by Yu et al. (2016) as being a parasitoid of the following aphid species: *Acyrthosiphon caraganae* (Cholodkovsky), *A. pisum* (Harris), *Amphorophora rubi* (Kaltenbach), *Aphis craccivora* Koch, *Aphis nerii* Boyer de Fonscolombe, *Aphis spiraecola* Patch, *Betacallis* sp., *Brachycaudus helichrysi* (Kaltenbach), *Capitophorus hippophaes* Walker, *Cryptomyzus ribis* (L.), *Eumyzus* sp., *Dysaphis plantaginea* (Passerini), *Hyperomyzus picridis* (Börner), *Impatientinum balsamines* (Kaltenbach), *Macrosiphoniella* sp., *Macrosiphum euphorbiae* (Thomas), *Metopolophium dirhodum* (Walker), *Myzus certus* (Walker), *Myzus ornatus* Laing, *Myzus persicae* (Sulzer), *Ovatus* sp., *Prociphilus* sp., *Rhopalosiphum nymphaea* (L.), *Schizaphis rotundiventris* (Signoret), *Sitobion adianti* (Oestlund), and *Sitobion avenae* (Fabricius).

Toxares sp.

Distribution in Iran: Fars (Ahmadi & Sarafrazi, 1993; Starý et al., 2000).

Host records in Iran: Hemiptera: Aphididae: *Diuraphis noxia* (Kurdjumov) (Ahmadi & Sarafrazi, 1993; Starý et al., 2000).

Comments: The association of *Toxares* sp. with *Diuraphis noxia* (Kurdjumov) has been considered doubtful by Starý et al. (2000).

Tribe Praini Mackauer, 1916

Genus *Areopraon* Mackauer, 1959

Areopraon lepelleyi (Waterson, 1926)

Catalogs with Iranian records: Barahoei et al. (2014), Yu et al. (2016), Farahani et al. (2016), Samin, Coronado-Blanco, Kavallieratos et al. (2018), Samin, Coronado-Blanco, Fischer et al. (2018), Rakhshani et al. (2019).

Distribution in Iran: Northern Khorasan (Kazemzadeh et al., 2009, 2010; Rakhshani, Kazemzadeh et al., 2012).

Distribution in the Middle East: Iran.

Extralimital distribution: Czech Republic, Georgia, Germany, Hungary, India, Italy, Poland, Russia, Sweden, United Kingdom.

Host records: Summarized by Yu et al. (2016) as being a parasitoid of *Eriosoma lanigerum* (Hausmann), *E. lanuginosum* (Hartig), *E. patchiae* (Börner and Blunk), *E. phaenax* (Mordvilko), *Mindarus abietinus* Koch, and *Schizoneurella indica* Hille Ris Lambers.

Parasitoid−aphid−plant associations in Iran: Hemiptera: Aphididae: *Eriosoma lanuginosum* (Hartig) on *Ulmus carpinifolia* var. *umbraculifera* (Kazemzadeh et al., 2009; Rakhshani, Kazemzadeh et al., 2012).

Genus *Praon* Haliday, 1833

Praon abjectum (Haliday, 1833)

Catalogs with Iranian records: Fallahzadeh and Saghaei (2010), Barahoei et al. (2014), Yu et al. (2016), Farahani et al. (2016), Samin, Coronado-Blanco, Kavallieratos et al. (2018), Samin, Coronado-Blanco, Fischer et al. (2018), Rakhshani et al. (2019).

Distribution in Iran: Tehran (Modarres Awal, 2012; Rakhshani, 2012; Rakhshani, Talebi, Manzari et al., 2007).

Distribution in the Middle East: Iran (Modarres Awal, 2012; Rakhshani, Talebi, Manzari et al., 2007; Rakhshani, Starý et al., 2012; Rakhshani, Kazemzadeh et al., 2012), Iraq (Starý, 1976; Starý & Kaddou, 1971), Turkey (Akar & Çetin Erdoğan, 2017; Tomanovic et al., 2008).

Extralimital distribution: Andorra, Austria, Belgium, Bosnia-Herzegovina, Bulgaria, China, Czech Republic, Denmark, Finland, France, Germany, Greece, Hungary, Iceland, India, Ireland, Italy, Moldova, Montenegro, Norway, Poland, Russia, Serbia, Slovakia, Slovenia, Spain, Sweden, Switzerland, Tajikistan, United Kingdom, Uzbekistan.

Host records: Summarized by Yu et al. (2016) as being a parasitoid of the following aphid species: *Acyrthosiphon scariolae* Nevsky, *Aphis* spp., *Aulacorthum solani* (Kaltenbach), *Brachycaudus cardui* (L.), *Brachycaudus helichrysi* (Kaltenbach), *Cryptomyzus ribis* (L.), *Dysaphis anthrisci* Boerner, *Dysaphis sorbi* (Kaltenbach), *Hyadaphis coriandri* (Das), *Hyalopterus pruni* (Geoffroy), *Hyperomyzus lactucae* (L.), *Liosomaphis berberidis* (Kaltenbach), *Longicaudus trirhodus* (Walker), *Macrosiphum stellariae* Theobald, *Megoura viciae* Buckton, *Melanaphis donacis* (Passerini), *Myzaphis rosarum* (Kaltenbach), *Myzus persicae* (Sulzer), *Nasonovia compositella* Theobald, *Nasonovia ribisnigri* (Mosley), *Rhopalosiphum nymphaeae* (L.), *Rhopalosiphum padi* (L.), *Semiaphis anthrisci* (Kaltenbach), *Sitobion avenae* (Fabricius), and *Symnthurodes betae* (Westwood).

Parasitoid−aphid−plant associations in Iran: Hemiptera: Aphididae: *Brachycaudus cardui* (L.) on *Prunus padus* (Modarres Awal, 2012; Rakhshani, 2012; Rakhshani, Talebi, Manzari et al., 2007).

Praon athenaeum Kavallieratos and Lykouressis, 2000

Distribution in the Middle East: Turkey (Akar & Çetin Erdoğan, 2017).

Extralimital distribution: Greece.

Host records: recorded by Kavallieratos and Lykouressis (2000) as being a parasitoid of the aphid *Hyperomyzus lactucae* (L.).

Praon barbatum Mackauer, 1967

Catalogs with Iranian records: Fallahzadeh and Saghaei (2010), Barahoei et al. (2014), Yu et al. (2016), Farahani et al. (2016), Samin, Coronado-Blanco, Kavallieratos et al. (2018), Samin, Coronado-Blanco, Fischer et al. (2018), Rakhshani et al. (2019).

Distribution in Iran: Alborz (Starý, 1979; Starý et al., 2000), Ardabil (González et al., 1978; Modarres Awal, 2012; Starý & González, 1978; Starý, 1979; Starý et al., 2000), Fars (Taheri & Rakhshani, 2013), Golestan (Barahoei, Sargazi et al., 2012), Isfahan (Barahoei et al., 2013; Rakhshani et al., 2010), Kerman (Asadizade et al., 2014), Kermanshah (Nazari et al., 2012), Markazi (Alikhani et al., 2013; Modarres Awal, 2012), Tehran (Modarres Awal, 2012; Rakhshani, Talebi, Manzari et al., 2007).

Distribution in the Middle East: Cyprus (Aeschlimann, 1981; Kavallieratos et al., 2004), Iran (see references above), Lebanon (Tremblay et al., 1985), Saudi Arabia (Rakhshani et al., 2019).

Extralimital distribution: Afghanistan, Andorra, Austria, China, Czech Republic, Finland, France, Germany, Greece, Italy, Japan, Moldova, Mongolia, Montenegro, Morocco, Poland, Serbia, Slovakia, Slovenia, Spain, Switzerland, Tajikistan, Turkmenistan, Uzbekistan.

Host records: Summarized by Yu et al. (2016) as being a parasitoid of *Acyrthosiphon gossypii* Mordvilko, *Acyrthosiphon kondoi* Shinji, and *Acyrthosiphon pisum* (Harris).

Parasitoid–aphid–plant associations in Iran: Hemiptera: Aphididae: *Acyrthosiphon pisum* (Harris) on *Medicago sativa*, *Vicia pannonica*, and *Vicia villosa* (Alikhani et al., 2013; Barahoei et al., 2013; González et al., 1978; Modarres Awal, 2012; Nazari et al., 2012; Rakhshani, Talebi, Manzari et al., 2007; Starý & González, 1978; Starý, 1979; Starý et al., 2000; 2010; Taheri & Rakhshani, 2013), *Acyrthosiphon* sp. on *Medicago sativa* (Starý et al., 2000), *Nearctaphis bakeri* (Cowen) on *Trifolium campestre* (Barahoei et al., 2013).

Praon bicolor Mackauer, 1959

Catalogs with Iranian records: Farahani et al. (2016), Rakhshani et al. (2019).

Distribution in Iran: Alborz (Farahani et al., 2015).

Distribution in the Middle East: Iran.

Extralimital distribution: Czech Republic, France, Germany, Hungary, Italy, Moldova, Poland, Russia, Serbia, Slovakia, Spain, United Kingdom.

Host records: Summarized by Yu et al. (2016) as being a parasitoid of the aphids *Eulachnus agilis* (Kaltenbach), *Eulachnus rileyi* (Williams), *Metopolophium dirhodum* (Walker), *Schizolachnus obscurus* Börner, *Schizolachnus pineti* (Fabricius), and *Semiaphis dauci* (Fabricius).

Praon dorsale (Haliday, 1833)

Catalogs with Iranian records: Barahoei et al. (2014).
Distribution in Iran: Guilan (Starý et al., 2000).

Distribution in the Middle East: Iran (Starý et al., 2000), Iraq (Starý, 1976), Israel–Palestine (Mescheloff & Rosen, 1989), Turkey (Aeschlimann & Vitou, 1985).

Extralimital distribution: Andorra, Austria, Bulgaria, China, Czech Republic, Finland, France, Germany, Greece, Hungary, India, Ireland, Italy, Japan, Korea, Moldova, Morocco, Netherlands, Poland, Portugal, Russia, Serbia, Spain, Sweden, Switzerland, Taiwan, Tajikistan, Turkmenistan, Ukraine, United Kingdom, Uzbekistan, former Yugoslavia.

Host records: Summarized by Yu et al. (2016) as being a parasitoid of the following aphid species: *Acyrthosiphon gossypii* Mordvilko, *A. kondoi* Shinji, *A. pisum* (Harris), *Amphicercidus tuberculatus* David, Narayanan and Rajisingh, *Amphorophora rubi* (Kaltenbach), *Aphis craccivora* Koch, *Corylobium avellanae* (Schrank), *Hyperomyzus lactucae* (L.), *Indomegoura indica* (van der Goot), *Macrosiphoniella sanborni* (Gillette), *Macrosiphum euphorbiae* (Thomas), *M. funestum* (Macchiati), *Megoura viciae* Buckton, *Microlophium carnosum* (Buckton), *Paczoskia* sp., *Sitobion ptericolens* Patch, *Staticobium limonii* (Contarini), and *U.* spp.

Parasitoid–aphid–plant associations in Iran: Hemiptera: Aphididae: *Uroleucon jaceae* (L.) on *Centaurea hyrcanica* (Starý et al., 2000).

Praon exsoletum (Nees von Esenbeck, 1811)

Catalogs with Iranian records: Fallahzadeh and Saghaei (2010), Barahoei et al. (2014), Yu et al. (2016), Farahani et al. (2016), Samin, Coronado-Blanco, Kavallieratos et al. (2018), Samin, Coronado-Blanco, Fischer et al. (2018), Rakhshani et al. (2019).

Distribution in Iran: Alborz (Monajemi & Esmaili, 1981 as *Praon palitans* Muesebeck, 1956; Rakhshani, Talebi, Manzari et al., 2006; Rakhshani, Talebi, Manzari et al., 2007), Ardabil (Modarres Awal, 2012), Isfahan (Rakhshani et al., 2010; Barahoei et al., 2013; Hadadian et al., 2016), Hamadan, Kordestan (Rakhshani, Talebi, Manzari et al., 2006; Rakhshani, Talebi, Manzari et al., 2007; Modarres Awal, 2012; Nazari et al., 2012), Kerman (Ghotbi Ravandi et al., 2017), Kermanshah (Rakhshani, Talebi, Manzari et al., 2006; Modarres Awal, 2012; Nazari et al., 2012), Khuzestan (Mossadegh & Kocheili, 2003 as *Praon palitans*), Markazi (Alikhani et al., 2013), Northern Khorasan (Kazemzadeh et al., 2010; Rakhshani, Kazemzadeh et al., 2012), Qazvin, Semnan (Rakhshani, Talebi, Manzari et al., 2006), Sistan and Baluchestan (Rakhshani, Talebi, Manzari et al., 2006, 2007), Tehran (Modarres Awal, 2012; Rasulian, 1985, 1989; Starý et al., 2000; Rakhshani, Talebi, Manzari et al., 2006; Rakhshani, Talebi, Manzari et al., 2007), Zanjan (Rakhshani, Talebi, Manzari et al., 2006; Rakhshani, Talebi, Manzari et al., 2007), Iran (no specific locality cited) (Farahbakhsh, 1961 as *Praon palitans*; Modarres Awal, 1997, 2012 as *Praon exsoletum palitans*).

Distribution in the Middle East: Cyprus, Egypt (Starý, 1976), Iran (see references above), Iraq (Starý, 1976; Starý & Kaddou, 1971), Israel–Palestine (Avidov & Harpaz, 1969; Mescheloff & Rosen, 1988; Starý, 1976), Lebanon (van den Bosch, 1957; Starý, 1976; Tremblay et al., 1985), Saudi Arabia (Rakhshani et al., 2019), Turkey (Akar & Çetin Erdoğan, 2017; Çetin Erdoğan et al., 2008; Starý, 1976; Uysal et al., 2004), Yemen (Starý & Erdelen, 1982; Starý et al., 2013).

Extralimital distribution: Andorra, Belgium, Bulgaria, Canada, China, Czech Republic, Finland, France, Germany, Greece, Hungary, Italy, Mexico, Moldova, Montenegro, Poland, Serbia, Slovakia, Spain, Tajikistan, United Kingdom, United States of America, Uzbekistan.

Host records: Summarized by Yu et al. (2016) as being a parasitoid of the following aphid species: *Acyrthosiphon pisum* (Harris), *Amphorophora rubi* (Kaltenbach), *Aphis medicaginis* Koch, *Eriosoma lanuginosum* (Hartig), *Eriosoma ulmi* (L.), *Hyalopterus pruni* (Geoffroy), *Macrosiphum rosae* (L.), *Macrosiphum stellariae* Theobald, *Metopolophium dirhodum* (Walker), *Therioaphis ononidis* (Kaltenbach), *Therioaphis riehmi* (Börner), and *Therioaphis trifolii* (Monell).

Parasitoid–aphid–plant associations in Iran: Hemiptera: Aphididae: *Acyrthosiphon pisum* (Harris) (Modarres Awal, 1997, 2012; Monajemi & Esmaili, 1981; Starý et al., 2000), *Therioaphis riehmi* (Börner) (Starý et al., 2000; van den Bosch, 1957), *Therioaphis trifolii* (Monell) on *Medicago sativa* (Alikhani et al., 2013; Barahoei et al., 2013; Farahbakhsh, 1961; Modarres Awal, 1997; 2012; Monajemi & Esmaili, 1981; Nazari et al., 2012; Rakhshani, Talebi, Manzari et al., 2006; Rakhshani et al., 2010; Rakhshani, Kazemzadeh et al., 2012; Rasulian, 1985; 1989; Starý, 1979; Starý et al., 2000), *Therioaphis trifolii maculata* (Buckton) on *Medicago sativa* (Modarres Awal, 2012; Rakhshani, Talebi, Manzari et al., 2007).

Praon flavinode (Haliday, 1833)

Catalogs with Iranian records: Fallahzadeh & Saghaei (2010), Barahoei et al. (2014) as *Praon absinthii*, and *Praon flavinode*, Yu et al. (2016), Farahani et al. (2016) as *Praon absinthii*, and *Praon flavinode*, Samin, Coronado-Blanco, Kavallieratos et al. (2018), Samin, Coronado-Blanco, Fischer et al. (2018), Rakhshani et al. (2019).

Distribution in Iran: Ardabil (Modarres Awal, 2012 as *P. absinthii*), Kerman (Barahoei et al., 2010; Barahoei, Madjdzadeh et al., 2012; Barahoei et al., 2013), Guilan (Starý, 1979; Starý et al., 2000, both as *Praon absinthii*), Tehran (Rakhshani, Talebi, Manzari et al., 2007; Modarres Awal, 2012, both as *P. absinthii*), Zanjan (Rakhshani, Talebi, Manzari et al., 2007; Talebi et al., 2009; Modarres Awal, 2012, all as *P. absinthii*).

Distribution in the Middle East: Egypt (Hassan, 1957; Risbec, 1960), Iran (see references above), Iraq (Starý, 1969, 1976; Starý & Kaddou, 1971).

Extralimital distribution: Austria, Bulgaria, China, Croatia, Czech Republic, Denmark, Finland, France, Georgia, Germany, Hungary, India, Italy, Japan, Latvia, Moldova, Netherlands, Poland, Romania, Russia, Serbia, Slovakia, Spain, Switzerland, Ukraine, United Kingdom, Uzbekistan.

Host records: Summarized by Yu et al. (2016) as being a parasitoid of the following aphid species: *Aphis medicaginis* Koch, *Aphis pomi* De Geer, *Betulaphis quadrituberculata* (Kaltenbach), *Calaphis flava* Mordvilko, *Callipterinella tuberculata* (von Heyden), *Corylobium avellanae* (Schrank), *Elatobium abietinum* Walker, *Eucallipterus tiliae* (L.), *Euceraphis betulae* (Koch), *Euceraphis punctipennis* (Zetterstedt), *Hyalopterus pruni* (Geoffroy), *Macrosiphoniella* spp., *Microlophium carnosum* (Buckton), *Myzocallis carpini* (Koch), *Myzocallis castanicola* Baker, *Myzocallis coryli* (Goeze), *Panaphis juglandis* (Goeze), *Phyllaphis fagi* (L.), *Pleotrichophorus* sp., *Rhopalosiphum maidis* (Fitch), *Sitobion avenae* (Fabricius), *Tinocallis platani* (Kaltenbach), *Titanosiphon artemisiae* (Koch), *Tuberculatus albosiphonatus* Hille Ris Lambers, *Tuberculatus annulatus* (Hartig), *Tuberculatus moerickei* Hille Ris Lambers, *Tuberculoides* sp., *Uroleucon cichorii* (Koch), and *Uroleucon jaceae* (L.).

Parasitoid–aphid–plant associations in Iran: Hemiptera: Aphididae: *Macrosiphoniella abrotani* (Walker) on *Artemisia annua*, *M. artemisiae* (Boyer de Fonscolombe) on *Artemisia absinthium* (Rakhshani, Talebi, Starý et al., 2008; Rakhshani, Tomanović et al., 2008; Rakhshani, Kazemzadeh et al., 2008; Rakhshani, Talebi, Manzari et al., 2008; Rakhshani et al., 2011), *Macrosiphoniella absinthii* (L.) on *Artemisia absinthium* (Modarres Awal, 2012; Rakhshani et al., 2011; Starý, 1979; Starý et al., 2000), *Macrosiphoniella oblonga* (Mordvilko) on *Artemisia annua* (Modarres Awal, 2012; Rakhshani, Talebi, Manzari et al., 2007; Rakhshani, Talebi et al., 2008; Rakhshani, Tomanović et al., 2008; Rakhshani, Kazemzadeh et al., 2008; Rakhshani, Talebi, Manzari et al., 2008; Rakhshani et al., 2011; Talebi et al., 2009), *Macrosiphoniella sanborni* (Gillette) on *Chrysanthemum morifolium* (Modarres Awal, 2012; Rakhshani, Talebi, Manzari et al., 2007; Rakhshani et al., 2011), *Tinocallis nevskyi* Remaudiere, Qucdnau, and Heie on *Ulmus campestris* (Barahoei et al., 2010; Barahoei, Madjdzadeh et al., 2012; Barahoei, Sargazi et al., 2012; Barahoei et al., 2013).

Praon gallicum Starý, 1971

Catalogs with Iranian records: Barahoei et al. (2014), Farahani et al. (2016), Rakhshani et al. (2019).

Distribution in Iran: Kermanshah (Bagheri-Matin et al., 2008; 2010; Nazari et al., 2012), Khuzestan (Modarres Awal, 2012; Mossadegh et al., 2017; Rezaei et al., 2006, 2010), Razavi Khorasan (Darsouei et al., 2011b).

Distribution in the Middle East: Egypt (Rakhshani et al., 2019), Iran (see references above).

Extralimital distribution: Andorra, Belgium, Brazil, Canada, Chile, China, Czech Republic, France, Germany, Greece, Hungary, Netherlands, Norway, Poland, Russia, Serbia, Slovakia, Sweden, United States of America, Uzbekistan.

Host records: Summarized by Yu et al. (2016) as being a parasitoid of the following aphid species: *Acyrthosiphon kondoi* Shinji, *Acyrthosiphon pisum* (Harris), *Aphis helianthi* Monell, *Diuraphis noxia* (Mordvilko), *Metopolophium dirhodum* (Walker), *Pleotrichophorus glandulosus* (Kaltenbach), *Rhopalosiphum maidis* (Fitch), *Rhopalosiphum padi* (L.), *Schizaphis graminum* (Rondani), *Sitobion avenae* (Fabricius), and *Sitobion fragariae* (Walker).

Parasitoid—aphid—plant associations in Iran: Hemiptera: Aphididae: *Rhopalosiphum padi* (L.) on *Triticum aestivum* (Bagheri-Matin et al., 2008, 2010; Nazari et al., 2012).

Praon longicorne Marshall, 1896

Catalogs with Iranian records: No catalog.

Distribution in Iran: Ardabil (Samin, Beyarslan, Ranjith et al., 2020).

Distribution in the Middle East: Iran (Samin, Beyarslan, Ranjith et al., 2020), Turkey (Tomanović et al., 2008).

Extralimital distribution: Bulgaria, Czech Republic, Finland, France, Germany, Hungary, India, Italy, Moldova, Montenegro, Netherlands, Poland, Russia, Serbia, Slovakia, Spain, Tajikistan, United Kingdom.

Host records: Summarized by Yu et al. (2016) as being a parasitoid of *Acyrthosiphon chelidonii* (Kaltenbach), *A. malvae* (Mosley), *A. pisum* (Harris), *Amphorophora ampullata* Buckton, *Amphorophora rubi* (Kaltenbach), *Aulacorthum solani* (Kaltenbach), *Corylobium avellana* L., *Hyadaphis foeniculi* (Passerini), *Impatientinum asiaticum* Necksy, *Impatientinum balsamines* (Kaltenbach), *Macrosiphum* spp., and *Microlophium carnosum* (Buckton).

Parasitoid—aphid—plant associations in Iran: Hemiptera: Aphididae: *Macrosiphum euphorbiae* (Thomas) on *Euphorbia pulcherrima* (Samin, Beyarslan, Ranjith et al., 2020).

Praon necans Mackauer, 1959

Catalogs with Iranian records: Barahoei et al. (2014), Yu et al. (2016), Farahani et al. (2016), Samin, Coronado-Blanco, Kavallieratos et al. (2018), Samin, Coronado-Blanco, Fischer et al. (2018), Rakhshani et al. (2019).

Distribution in Iran: Golestan (Barahoei, Sargazi et al., 2012), Hormozgan (Ameri et al., 2019), Kermanshah (Nazari et al., 2012), Khuzestan (Mossadegh et al., 2010, 2011, 2017; Salehipour et al., 2010).

Distribution in the Middle East: Egypt (Abdel-Rahman, 2005; Abdel-Rahman et al., 2000; Abu El-Ghiet et al., 2014; El-heneidy & Abdel-Samad, 2001; El-Heneidy et al., 2001, 2002, 2003), Iran (see references above), Iraq (Starý, 1976).

Extralimital distribution: China, Czech Republic, Finland, France, Germany, Hungary, India, Italy, Moldova, Netherlands, Pakistan, Poland, Portugal, Russia, Serbia, Slovakia, Slovenia, Ukraine, United Kingdom.

Host records: Summarized by Yu et al. (2016) as being a parasitoid of *Rhopalosiphoninus calthae* (Koch), *Rhopalosiphum nymphaeae* (L.), *Rhopalosiphum padi* (L.), *Rhopalosiphum rufulum* (Richards), *Schizaphis* sp., and *Staticobium limonii* (Contarini).

Parasitoid—aphid—plant associations in Iran: Hemiptera: Aphididae: *Aphis craccivora* Koch on *Robinia pseudacacia* (Nazari et al., 2012), *Aphis gossypii* Glover on *Prosopis juliflora*, *Aphis nerii* Boyer de Fonscolombe on *Nerium oleander* (Mossadegh et al., 2011), *Aphis punicae* Passerini on *Punica granatum*, *Macrosiphoniella sanborni* (Gillette) on *Chrysanthemum morifolium* (Nazari et al., 2012), *Aphis* sp. on *Anagallis arvensis*, *Scrophularia* sp., and *Spergullaria* sp. (Mossadegh et al., 2011), *Myzus persicae* (Sulzer) on *Euphorbia helioscopia* (Mossadegh et al., 2011), *Rhopalosiphum nymphaeae* (L.) on *Ranunculus repens* (Nazari et al., 2012), *Rhopalosiphum padi* (L.) on *Hordeum vulgare*, *Triticum aestivum*, and *Zea mays* (Mossadegh et al., 2011; Nazari et al., 2012), *Sitobion avenae* (Fabricius) on *Triticum aestivum* (Mossadegh et al., 2011).

Praon nonveilleri Tomanović & Kavallieratos, 2003

Distribution in the Middle East: Turkey (Akar & Çetin Erdoğan, 2017).

Extralimital distribution: Montenegro, Serbia.

Host records: Recorded by Tomanović et al. (2003) as being a parasitoid of the aphid *Uroleucon inulicola* (Hille Ris lambers).

Praon orpheusi Kavallieratos, Athanassiou & Tomanović, 2003

Catalogs with Iranian records: Fallahzadeh and Saghaei (2010), Barahoei et al. (2014), Yu et al. (2016), Farahani et al. (2016), Samin, Coronado-Blanco, Kavallieratos et al. (2018), Samin, Coronado-Blanco, Fischer et al. (2018), Rakhshani et al. (2019).

Distribution in Iran: Tehran (Modarres Awal, 2012; Rakhshani, Talebi, Manzari et al., 2007; Talebi et al., 2009).

Distribution in the Middle East: Iran.

Extralimital distribution: Bulgaria, Greece.
Host records: Summarized by Yu et al. (2016) as being a parasitoid of the aphids *Hyperomyzus lactucae* (L.), and *Macrosiphum rosae* (L.).
Parasitoid−aphid−plant associations in Iran: Hemiptera: Aphididae: *Hyperomyzus lactucae* (L.) on *Lactuca oleracea* (Modarres Awal, 2012; Rakhshani, Talebi, Manzari et al., 2007; Talebi et al., 2009).

Praon pubescens Starý, 1961
Catalogs with Iranian records: Barahoei et al. (2014), Farahani et al. (2016), Rakhshani et al. (2019).
Distribution in Iran: Khuzestan (Nazari et al., 2012).
Distribution in the Middle East: Iran (Nazari et al., 2012), Turkey (Tomanović et al., 2008).
Extralimital distribution: Bulgaria, Czech Republic, France, Hungary, India, Montenegro, Netherlands, Poland, Serbia, Slovakia.
Host records: Summarized by Yu et al. (2016) as being a parasitoid of the aphids *Metopolophium dirhodum* (Walker), *Nasonovia brachycyclica* Holman, *Nasonovia compositellae* Theobald, *Nasonovia ribisnigri* (Mosley), and *Uroleucon* sp.
Parasitoid−aphid−plant associations in Iran: Hemiptera: Aphididae: *Nasonovia ribisnigri* (Mosley) on *Lactuca sativa* (Nazari et al., 2012).

Praon rosaecola Starý, 1961
Catalogs with Iranian records: Fallahzadeh and Saghaei (2010), Barahoei et al. (2014), Yu et al. (2016), Farahani et al. (2016), Samin, Coronado-Blanco, Kavallieratos et al. (2018), Samin, Coronado-Blanco, Fischer et al. (2018), Rakhshani et al. (2019).
Distribution in Iran: Golestan (Barahoei, Sargazi et al., 2012), Guilan (Rakhshani et al., 2007a), Isfahan (Barahoei et al., 2013), Kerman (Barahoei et al., 2010; 2012a,b, 2013), Mazandaran (Modarres Awal, 2012), Northern Khorasan (Kazemzadeh et al., 2010; Rakhshani et al., 2012b), Tehran (Rakhshani, Talebi, Manzari et al., 2007; Talebi et al., 2009).
Distribution in the Middle East: Iran (see references above), Israel−Palestine (Mescheloff & Rosen, 1989).
Extralimital distribution: Bulgaria, Czech Republic, France, Hungary, Moldova, Poland, Serbia, Spain.
Host records: Summarized by Yu et al. (2016) as a parasitoid of *Aphis fabae* (Scopoli), *Chaetosiphon tetrarhodum* (Walker), *Macrosiphum rosae* (L.), and *Sitobion fragariae* (Walker).
Parasitoid−aphid−plant associations in Iran: Hemiptera: Aphididae: *Macrosiphum rosae* (L.) on *Rosa damascena*, and *Rosa* sp. (Barahoei et al., 2010; Barahoei, Madjdzadeh et al., 2012; Barahoei et al., 2013; Modarres Awal, 2012; Rakhshani, Talebi, Manzari et al., 2007; Rakhshani, Kazemzadeh et al., 2012; Talebi et al., 2009),

Wahlgreniella nervata (Gillette) on *Rosa damascena* (Barahoei et al., 2013).

Praon unitum Mescheloff and Rosen, 1988
Catalogs with Iranian records: Barahoei et al. (2014), Yu et al. (2016), Farahani et al. (2016), Samin, Coronado-Blanco, Kavallieratos et al. (2018), Samin, Coronado-Blanco, Fischer et al. (2018), Rakhshani et al. (2019).
Distribution in Iran: Kerman (Barahoei et al., 2010; Barahoei, Madjdzadeh et al., 2012; Barahoei, Sargazi et al., 2012; Barahoei et al., 2013).
Distribution in the Middle East: Iran (Barahoei et al., 2010; Barahoei, Madjdzadeh et al., 2012; Barahoei, Sargazi et al., 2012; Barahoei et al., 2013), Israel−Palestine (Mescheloff & Rosen, 1989).
Extralimital distribution: None.
Host records: Recorded by Mescheloff and Rosen (1989) as a parasitoid of the aphids *Macrosiphoniella sanborni* (Gillette), and *Uroleucon sonchi* (L.).
Parasitoid−aphid−plant associations in Iran: Hemiptera: Aphididae: *Macrosiphoniella sanborni* (Gillette) (Rakhshani et al., 2011), *Sipha maydis* Passerini on *Cynodon dactylon*, *Uroleucon acroptilidis* Kadyrbekov, Renxin and Shao on *Acroptilon repens* (Barahoei et al., 2010; Barahoei, Madjdzadeh et al., 2012; Barahoei, Sargazi et al., 2012; Barahoei et al., 2013), *Uroleucon jaceae* (L.) on *Acroptilon repens*, *Uroleucon sonchi* (L.) on *Sonchus oleraceus* (Barahoei et al., 2013).

Praon volucre (Haliday, 1833)
Catalogs with Iranian records: Fallahzadeh and Saghaei (2010), Barahoei et al. (2014), Yu et al. (2016), Farahani et al. (2016), Samin, Coronado-Blanco, Kavallieratos et al. (2018), Samin, Coronado-Blanco, Fischer et al. (2018), Rakhshani et al. (2019).
Distribution in Iran: Alborz (Rakhshani, 2012; Rakhshani, Talebi, Manzari et al., 2007; Starý, 1979; Starý et al., 2000; Talebi et al., 2009), Fars (Alichi et al., 2008; Barahoei et al., 2013; Modarres Awal, 2012; Rakhshani, Talebi, Manzari et al., 2007; Rakhshani, Tomanović et al., 2008; Rakhshani, 2012; Taheri & Rakhshani, 2013), Golestan (Barahoei, Sargazi et al., 2012; Ghahari & Fischer 2011; Ghahari et al., 2010; Rakhshani, Talebi, Manzari et al., 2007; Rakhshani, Tomanović et al., 2008), Hamadan (Modarres Awal, 2012; Nazari et al., 2012; Rakhshani, 2012; Rakhshani, Talebi, Manzari et al., 2007; Rakhshani, Tomanović et al., 2008), Hormozgan (Ameri et al., 2019), Isfahan (Barahoei et al., 2013; Mehrparvar & Hatami, 2003; Mehrparvar et al., 2016; Modarres Awal, 2012; Rakhshani, Talebi, Manzari et al., 2007; Rakhshani, Tomanović et al., 2008; Rakhshani et al., 2010; Talebi et al., 2009), Kerman (Asadizade et al., 2014; Barahoei et al., 2010; Barahoei, Madjdzadeh et al., 2012; Barahoei et al., 2013; Ghotbi Ravandi et al., 2017, in alfalfa

and wheat fields), Kermanshah (Modarres Awal, 2012; Nazari et al., 2012; Rakhshani, 2012; Rakhshani, Tomanović et al., 2008), Khuzestan (Farsi et al., 2012; Modarres Awal, 2012; Nazari et al., 2012; Rakhshani, Tomanović et al., 2008), Kordestan (Modarres Awal, 2012; Rakhshani, 2012; Rakhshani, Talebi, Manzari et al., 2007; Rakhshani, Tomanović et al., 2008), Markazi (Alikhani et al., 2013; Rakhshani, Tomanović et al., 2008; Modarres Awal, 2012), Qom (Modarres Awal, 2012; Rakhshani, Tomanović et al., 2008), Mazandaran (Sakenin et al., 2012), Northern Khorasan (Kazemzadeh et al., 2010; Malkeshi 1997; Malkeshi et al., 1998; Rakhshani, 2012; Rakhshani, Kazemzadeh et al., 2012), Qazvin (Rakhshani, Talebi, Manzari et al., 2007), Razavi Khorasan (Darsouei et al., 2011a,b; Jafari-Ahmadabadi et al., 2011; Jafari-Ahmadabadi & Modarres Awal, 2012; Modarres Awal, 2012; Rakhshani, 2012; Rakhshani, Tomanović et al., 2008), Sistan and Baluchestan (Modarres Awal, 2012; Rakhshani, 2012; Rakhshani, Talebi, Manzari et al., 2006; Rakhshani, Talebi, Manzari et al., 2007; Rakhshani, Tomanović et al., 2008; Talebi et al., 2009), Tehran (Modarres Awal, 2012; Rakhshani, 2012; Rakhshani, Talebi, Kavallieratos et al., 2005; Rakhshani, Talebi, Manzari et al., 2007; Rakhshani, Tomanović et al., 2008; Shahrokhi et al., 2004; Starý, 1979; Starý et al., 2000), West Azarbaijan (Modarres Awal, 2012; Samin et al., 2014; Starý et al., 2000), Zanjan (Rakhshani, Talebi, Manzari et al., 2006), Iran (no specific locality cited) (Tazerouni et al., 2011b).

Distribution in the Middle East: Egypt (Abu El-Ghiet et al., 2014; Gadallah et al., 2017; Hafez, 1994), Iran (see references above), Iraq (Starý, 1976; Starý & Kaddou, 1971), Israel—Palestine (Avidov & Harpaz, 1969; Mescehloff & Rosen, 1988; Rosen, 1967, 1969; Starý, 1976), Lebanon (Abou- Fakhr, 1982; Abou-Fakhr & Kawar, 1998; Hussein & Kawar, 1984; Tremblay et al., 1985), Saudi Arabia (Rakhshani et al., 2019), Turkey (Akar & Çetin Erdoğan, 2017; Aslan & Karaca, 2005; Çetin Erdoğan et al., 2008; Düzgüneş et al., 1982; Güz & Kilinçer, 2005; Yoldaş et al., 2011).

Extralimital distribution: Algeria, Andorra, Argentina, Austria, Azerbaijan, Belgium, Bosnia-Herzegovina, Brazil, Bulgaria, Chile, China, Czech Republic, Denmark, Finland, France, Georgia, Germany, Greece, Hungary, Iceland, India, Ireland, Italy, Japan, Kazakhstan, Korea, Kyrgyzstan, Lithuania, Macedonia, Moldova, Mongolia, Montenegro, Morocco, Netherlands, Norway, Pakistan, Poland, Portugal, Romania, Russia, Serbia, Slovakia, Slovenia, Spain, Sweden, Switzerland, Tajikistan, Ukraine, United Kingdom, Uzbekistan.

Host records: Summarized by Yu et al. (2016) as being a parasitoid of the following aphid species: *Acyrthosiphon* spp., *Amphorophora amurensis* (Mordvilko), *A. catherinae* (Nevsky), *Amphorophora rubi* (Kaltenbach), *Aphis* spp., *Aulacorthum circumflexum* (Buckton), *Aulacorthum magnoliae* Essig and Kuwana, *Aulacorthum solani* (Kaltenbach), *Aulacorthum syringae* (Matsumura), *Brachycaudus* spp., *Brachycorynella asparagi* (Mordvilko), *Capitophorus elaeagni* (Del Guercio), *Chaetosiphon tetrarhodum* (Walker), *Chromaphis juglandicola* (Kaltenbach), *Cinara pinea* (Mordvilko), *Corylobium avellanae* (Schrank), *Cryptomyzus ribis* (L.), *Diuraphis calamagrostis* (Ossiannilsson), *Diuraphis noxia* (Mordvilko), *Dysaphis crataegi* (Kaltenbach), *Dysaphis plantaginea* (Passerini), *Eucarazzia elegans* (Ferrari), *Euceraphis punctipennis* (Zetterstedt), *Eulachnus agilis* (Kaltenbach), *Hyadaphis foeniculi* (Passerini), *Hyalopterus amygdali* (Blanchard), *Hyalopterus pruni* (Geoffroy), *Hyperomyzus lactucae* (L.), *Hyperomyzus picridis* (Börner), *Impatientinum asiaticum* Nevsky, *Impatientinum balsamines* (Kaltenbach), *Linosiphon galiophagus* (Wimshrust), *Liosomaphis atra* Hille Ris Lambers, *Lipaphis erysimi* (Kaltenbach), *Lipaphis fritzmuelleri* Börner, *Macrosiphoniella riedeli* Szelegiewics, *Macrosiphum* spp., *Megoura viciae* Buckton, *Melanaphis donacis* (Passerini), *Metopeurum fuscoviride* Stroyan, *Metopolophium albidum* Hille Ris Lambers, *Metopolophium dirhosum* (Walker), *Metopolophium festucae cerealium* (Stroyan), *Metopolophium montanum* Hille Ris Lambers, *Microlophium carnosum* (Buckton), *Myzaphis turanica* Nevsky, *Myzocallis castanicola* Baker, *Myzocallis coryli* (Goeze), *Myzus* spp., *Nasonovia ribisnigri* (Mosley), *Ovatus crataegarius* (Walker), *Phorodon humuli* (Schrank), *Rhodobium porosum* (Saderson), *Rhopalomyzus lonicerae* (Siebold), *Rhopalosiphoninus calthae* (Koch), *Rhopalosiphum maidis* (Fitch), *Rhopalosiphum padi* (L.), *Schizaphis graminum* (Rondani), *Schizolachnus pineti* (Fabricius), *Sitobion* spp., *Toxoptera aurantii* (Boyer de Fonscolombe), *Uroleucon* spp., and *Wahlgreniella ossiannilssoni* Hille Ris Lambers.

Parasitoid—aphid—plant associations in Iran: Hemiptera: Aphididae: *Acyrthosiphon gossypii* Mordrilko on *Sophora alopecuroides* (Alikhani et al., 2013), *Acyrthosiphon lactucae* (Passerini) on *Lactuca serriola* (Rakhshani, Talebi, Manzari et al., 2007), *Acyrthosiphon pisum* (Harris) on *Medicago sativa*, *Robinia pseudoacacia*, and *Vicia villosa* (Rakhshani, Talebi, Manzari et al., 2006; Rakhshani, Talebi, Manzari et al., 2007; Rakhshani et al., 2010), *Amphorophora catharinae* (Nevsky) on *Rosa damascena*, and *Rosa* sp. (Barahoei et al., 2010, 2013; Barahoei, Madjdzadeh et al., 2012; Rakhshani, Kazemzadeh et al., 2012), *Aphis affinis* del Guercio on *Mentha longifolia* (Rakhshani, Kazemzadeh et al., 2012), *Aphis craccivora* Koch on *Cardaria draba*, *Medicago sativa*, and *Portulaca oleracea* (Alikhani et al., 2013; Barahoei et al., 2010, 2013; Barahoei, Madjdzadeh et al., 2012; Rakhshani, Talebi, Kavallieratos et al., 2005; Rakhshani, Talebi, Manzari et al., 2006; Rakhshani, Talebi, Manzari et al., 2007; Rakshani et al., 2010;

Rakhshani, Kazemzadeh et al., 2012), *Aphis dlabolai* Holman on *Euphorbia* sp. (Barahoei et al., 2013), *Aphis fabae* Scopoli on *Arctium lappa, Chinopodium album, Cirsium arvense, Cirsium* sp., *Mentha longifolia, Solanum nigrum*, and *Vicia fabae* (Barahoei et al., 2013; Nazari et al., 2012; Rakhshani, Talebi, Manzari et al., 2007; Starý, 1979; Starý et al., 2000; Taheri & Rakhshani, 2013), *Aphis fabae circiiacanthoides* Scopoli on *Circium arvensis* (Rakhshani, Talebi, Manzari et al., 2007), *Aphis fabae solanella* on *Solnaum* sp., *Solanum nigrum*, and *Solanum lycopersicum* (Alikhani et al., 2013; Barahoei, Madjdzadeh et al., 2012; Rakhshani, Talebi, Manzari et al., 2007), *Aphis gossypii* Glover on *Solanum melongena* (Barahoei et al., 2013), *Aphis pomi* de Geer on *Malus domestica* (Modarres Awal, 2012; Rakhshani, 2012), *Aphis solanella* Theobald on *Solanum* sp. (Barahoei et al., 2010, 2013; Barahoei, Madjdzadeh et al., 2012; Rakhshani, Talebi, Manzari et al., 2007; Talebi et al., 2009), *Aphis urticata* Gmelin on *Urtica dioica* (Rakhshani, Kazemzadeh et al., 2012), *Brachycaudus amygdalinus* (Schouteden) on *Amygdalus arabica*, and *Prunus dulcis* (Rakhshani, 2012; Rakhshani, Talebi, Manzari et al., 2007; Talebi et al., 2009), *Brachycaudus cardui* (L.) on *Prunus persicae*, and *Prunus* sp. (Rakhshani, 2012), *Brachycaudus helichrysi* (Kaltenbach) on *Calendula officinalis, Prunus armeniaca, Prunus domestica*, and *Prunus persica* (Barahoei et al., 2013; Jafari-Ahmadabadi et al., 2011; Jafari-Ahmadabadi & Modarres Awal, 2012; Rakhshani, 2012; Rakhshani, Talebi, Manzari et al., 2007; Talebi et al., 2009), *Brachycaudus persicae* (Passerini) on *Prunus persica* (Rakhshani, 2012), *Diuraphis noxia* (Kurdjumov) on *Triticum aestivum* (Barahoei et al., 2013; Modarres Awal, 2012; Rakhshani, Talebi, Manzari et al., 2007; Rakhshani, Tomanović et al., 2008; Starý et al., 2000; Zareh et al., 1995), *Dysaphis pyri* (Boyer de Fonscolombe) on *Pyrus communis* (Rakhshani, 2012), *Hyalopterus amygdali* (Blanchard) on *Prunus amygdalinus*, and *Prunus dulcis* (Nazari et al., 2012; Rakhshani, 2012; Rakhshani, Talebi, Manzari et al., 2007; Taheri & Rakhshani, 2013), *Hyalopterus pruni* (Geoffrey) on *Prunus armeniaca, Prunus padus*, and *Prunus persicae* (Alikhani et al., 2013; Barahoei et al., 2013; Jafari-Ahmadabadi et al., 2011; Jafari-Ahmadabadi & Modarres Awal, 2012; Rakhshani, 2012; Rakhshani, Talebi, Manzari et al., 2007), *Hyperomyzus lactucae* (L.) on *Lactuca oleracea*, and *Lactuca serriola* (Rakhshani, Talebi, Manzari et al., 2007; Talebi et al., 2009), *Macrosiphum rosae* (L.) on *Rosa canina, Rosa damascena*, and *Rosa hybrida* (Alikhani et al., 2013; Barahoei et al., 2010; 2012a; 2013; Mehrparvar & Hatami, 2003; Mehrparvar et al., 2016; Nazari et al., 2012; Rakhshani, Talebi, Manzari et al., 2007; Talebi et al., 2009), *Macrosiphum* sp. on *Symphytum* sp. (Starý et al., 2000), *Macrosiphoniella* sp. on *Chrysanthemum*

morifolium (Rakhshani et al., 2011), *Metopolophium dirhodum* (Walker) on *Triticum aestivum* (Alichi et al., 2007, 2008; Barahoei et al., 2013; Rakhshani, Tomanović et al., 2008; Shahrokhi et al., 2004; Taheri & Rakhshani, 2013), *Myzus beybienkoi* (Narzykulov) on *Fraxinus oxycarpa* (Starý et al., 2000), *Myzus persicae* (Sulzer) on *Althaea* sp., *Capsicum annuum, Caryophila* sp., *Euphorbia helioscopia, Mentha longifolia, Nicotiana tabacum, Nicotiana* sp., *Prunus padus*, and *Prunus persica* (Barahoei et al., 2013; Jafari-Ahmadabadi et al., 2011; Jafari-Ahmadabadi & Modarres Awal, 2012; Nazari et al., 2012; Rakhshani, 2012; Rakhshani, Talebi, Manzari et al., 2007; Starý, 1979; Starý et al., 2000; Tazerouni et al., 2017, 2018), *Nasonovia ribisnigri* Mosely (Farsi et al., 2012), *Phorodon humuli* (Schrank) on *Prunus persica* (Rakhshani, Talebi, Manzari et al., 2007), *Rhopalosiphum maidis* Fitch on *Setaria glauca, Triticum aestivum*, and *Zea mays* (Alikhani et al., 2013; Barahoei et al., 2010, 2013; Barahoei, Madjdzadeh et al., 2012; Nazari et al., 2012; Rakhshani, Talebi, Manzari et al., 2007; Rakhshani, Tomanović et al., 2008), *Rhopalosiphum padi* (L.) on *Triticum aestivum*, and *Hordeum vulgare* (Alichi et al., 2008; Barahoei et al., 2013; Modarres Awal, 2012; Rakhshani, Talebi, Manzari et al., 2007; Rakhshani, Tomanović et al., 2008), *Schizaphis graminum* (Rondani) on *Bromus tectorum*, and *Triticum aestivum* (Barahoei et al., 2013; Nazari et al., 2012; Rakhshani, Talebi, Manzari et al., 2007; Rakhshani, Tomanović et al., 2008; Shahrokhi et al., 2004), *Sitobion avenae* (Fabricius) on *Triticum aestivum* (Barahoei et al., 2013; Farhad et al., 2011, 2012; Rakhshani, Talebi, Manzari et al., 2007; Rakhshani, Tomanović et al., 2008; Shahrokhi et al., 2004), *Therioaphis trifolii* (Monell) on *Medicago sativa* (Rakhshani et al., 2010), *Uroleucon cichorii* (Koch) on *Cichorium intybus*, and *Lactuca oleracea* (Rakhshani, Talebi, Manzari et al., 2007), *Uroleucon compositae* (Theobald) on *Carthamus tinctorius* (Barahoei et al., 2013; Rakhshani, Talebi, Manzari et al., 2007), *Uroleucon jaceae* (L.) on *Lactuca oleracea* (Rakhshani, Talebi, Manzari et al., 2007), *Uroleucon sonchi* (L.) on *Lactuca serriola, Sonchus asper, Sonchus oleraceus*, and *Sonchus arvensis* (Barahoei et al., 2010, 2013; Barahoei, Madjdzadeh et al., 2012; Nazari et al., 2012; Rakhshani, Talebi, Manzari et al., 2007; Rakhshani, Kazemzadeh et al., 2012; Talebi et al., 2009), *Uroleucon* sp. (Rakhshani, Talebi, Manzari et al., 2007), *Neomyzus circumflexus* Buckton (Modarres Awal, 2012).

Praon yomenae Takada, 1968
Catalogs with Iranian records: Fallahzadeh and Saghaei (2010), Barahoei et al. (2014), Yu et al. (2016), Farahani et al. (2016), Samin, Coronado-Blanco, Kavallieratos et al. (2018), Samin, Coronado-Blanco, Fischer et al. (2018), Rakhshani et al. (2019).

Distribution in Iran: Alborz (Rakhshani, Talebi, Manzari et al., 2007; Talebi et al., 2009), Golestan (Barahoei, Sargazi et al., 2012), Hamadan (Nazari et al., 2012), Isfahan (Barahoei et al., 2013; Modarres Awal, 2012; Rakhshani, Talebi, Manzari et al., 2007; Talebi et al., 2009), Kerman (Barahoei et al., 2010, 2013; Barahoei, Madjdzadeh et al., 2012), Markazi (Alikhani et al., 2013), Northern Khorasan (Rakhshani, Kazemzadeh et al., 2012), Tehran, Zanjan (Modarres Awal, 2012; Rakhshani, Talebi, Manzari et al., 2007).

Distribution in the Middle East: Iran (see references above), Israel–Palestine (Mescheloff & Rosen, 1988), Turkey (Akar & Çetin Erdoğan, 2017; Çetin Erdoğan et al., 2008; Kavallieratos et al., 2004; Tomanović et al., 2003; Uysal et al., 2004), Yemen (Starý et al., 2013), Saudi Arabia (Rakhshani et al., 2019).

Extralimital distribution: Algeria, Bulgaria, Czech Republic, France, Greece, Italy, Japan, Korea, Malta, Montenegro, Poland, Russia, Serbia, Slovakia, Slovenia, Spain, United Kingdom.

Host records: Summarized by Yu et al. (2016) as being a parasitoid of the aphids *Hyalopterus pruni* (Geoffroy), *Hyperomyzus lactucae* (L.), *Macrosiphoniella yomenae* (Shinji), *Staticobium limonii* (Contarini), and *Uroleucon* spp.

Parasitoid–aphid–plant associations in Iran: Hemiptera: Aphididae: *Acyrthosiphon lactucae* (Pass.) on *Sonchus oleraceus* (Barahoei et al., 2010, Barahoei, Madjdzadeh et al., 2012), *Uroleucon acroptilidis* Kadyrbekov, Renxin and Shao on *Acroptilon repens* (Barahoei et al., 2010, 2013; Barahoei, Madjdzadeh et al., 2012; Barahoei, Sargazi et al., 2012), *Uroleucon carthami* on *Carthamus oxyacantha* (Hille Ris Lambers) (Barahoei et al., 2013), *Uroleucon cichorii* (Koch) on *Cichorium intybus* (Barahoei et al., 2010, 2013; Barahoei, Madjdzadeh et al., 2012; Barahoei, Sargazi et al., 2012), *Uroleucon chondrillae* (Nevsky) on *Chondrilla juncea* (Alikhani et al., 2013; Nazari et al., 2012; Rakhshani, Talebi, Manzari et al., 2007), *Uroleucon compositae* (Theobald) on *Carthamus tinctorius* (Alikhani et al., 2013; Rakhshani, Talebi, Manzari et al., 2007), *Uroleucon jaceae* (L.) on *Acroptilon repens*, *Centaurea depressa*, and *Centaurea iberica* (Barahoei et al., 2010, 2013; Barahoei, Madjdzadeh et al., 2012; Rakhshani, Talebi, Manzari et al., 2007), *Uroleucon sonchi* (L.) on *Sonchus* sp., *Sonchus asper*, and *S. oleraceus* (Barahoei et al., 2010, 2013; Barahoei, Madjdzadeh et al., 2012; Modarres Awal, 2012; Rakhshani, Talebi, Manzari et al., 2007; Rakhshani, Kazemzadeh et al., 2012; Talebi et al., 2009), *Uroleucon tortuosissimae* Rezwani and Lampel on *Launaea acanthodes* (Barahoei et al., 2013), *Uroleucon* sp. (Rakhshani, Talebi, Manzari et al., 2007), *Uroleucon* sp. on *Acroptilon repens* and *Picnomon acarna* (Barahoei, Madjdzadeh et al., 2012).

Praon spp.

Distribution in Iran: Fars (Ahmadi & Sarafrazi, 1993), Tehran (Starý, 1979; Starý et al., 2000).

Parasitoid–aphid–plant associations in Iran: Hemiptera: Aphididae: *Diuraphis noxia* (Kurdjumov) (Ahmadi & Sarafrazi, 1993; Starý et al., 2000), *Dysaphis plantagenia* (Passerini) on *Malus pumila* (Starý et al., 2000), *Liosomaphis berberidis* (Kaltenbach) on *Berberis* sp. (Starý, 1979), *Macrosiphum* sp. on *Symphytum* sp. (Starý, 1979; Starý et al., 2000), *Macrosiphoniella sanborni* (Gillette) on *Chrysanthemum* sp. (Starý, 1979; Starý et al., 2000), *Myzus persicae* (Sulzer) on *Bellis* sp. (Starý et al., 2000).

Species excluded from the fauna of Iran

Aphidius nigripes Ashmead, 1901

Catalogs with Iranian records: No catalog.

Distribution in Iran: Iran (no specific locality cited) (Khanjani, 2006b, as the parasitoid of *Myzus persicae* (Sulzer) (Hemiptera: Aphididae)).

Extralimital distribution: Canada, Mexico, United States of America.

Host records: Summarized by Yu et al. (2016) as being a parasitoid of the aphids *Acyrthosiphon churchillense* Robinson, *Acyrthosiphon pisum* (Harris), *Aphis nasturtii* (Kaltenbach), *Aphis viburniphila* Patch, *Aulacorthum solani* (Kaltenbach), *Capitophorus* sp., *Illinoia* sp., *I. liriodendri* (Monell), *Macrosiphoniella tapuskae* (Hottes and Frison), *Macrosiphum albifrons* Essig, *Macrosiphum creelii* Davis, *Macrosiphum euphorbiae* (Thomas), *Macrosiphum pallidum* (Oestlund), *Macrosiphum rosae* (L.), *Myzus persicae* (Sulzer), *Nasonovia borealis* Heie, *Nasonovia houghtonensis* (Troop), *Pleotrichophorus pseudopatonkus* Corpuz-Raros and Cook, *Pterocomma smithiae* (Monell), *Sitobion avenae* (Fabricius), *Sitobion clydesmithi* Robinson, *Uroleucon ambrosiae* (Thomas), and *Uroleucon erigeronense* (Thomas).

Comments: *Aphidius nigripes* is distribued in Nearctic; therefore, its presence in Iran is questionable. For this reason, we exclude it from the fauna of Iran.

Conclusion

A 111 species of the subfamily Aphidiinae (16.9% of the world species) in 18 genera and three tribes (Aphidiini Haliday, 1833, Ephedrini Mackauer, 1961, and Praini Mackaeur, 1961) have been reported from the Middle East countries. The genus *Aphidius* Nees von Esenbeck, 1819 is the most diverse, with 33 recorded species. The Iranian fauna is found to the richest with Aphidiinae, including 92 known species (14% of the world species) classified into 18 genera, where the genus *Aphidius* is the more diverse among all other aphidiine genera, with 30 recorded species, followed by *Praon* (15 species), *Trioxys* (11 species), *Ephedrus* (nine species), *Lysiphlebus* (six species),

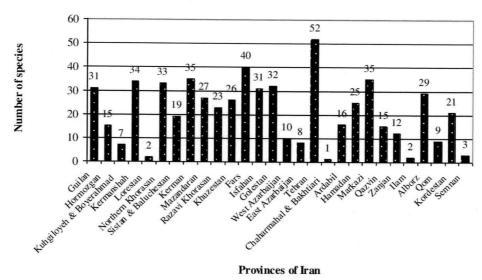

FIGURE 4.1   Number of reported species of Iranian Aphidiinae by province.

Binodoxys (five species), *Adialytus* (four species), *Monoctonia*, *Pauesia* (each with two species), *Aclitus*, *Betuloxys*, *Diaeretiella*, *Diaeretus*, *Lipolexis*, *Monoctonus*, *Tanytrichophorus*, and *Areopraon* (each with one species); *Toxares* with an unknown species. Among the 28 provinces, where the Iranian Aphidiinae have been recorded, Tehran exhibits the highest recorded diversity by holding 52 species (Fig. 4.1). Most of the Iranian aphidiine species are abundant in various regions and ecosystems. Four species, *Aphidius iranicus* Rakhshani and Starý, 2007, *Aphidius stigmaticus* Rakhshani and Tomanović, 2011, *Tanytrichophorus petiolaris* Mackauer, 1961, and *Trioxys metacarpalis* Rakhshani and Starý, 2012 (Aphidiini), are known only from Iran (endemic to Iran). Six species, *Aphidius* (*Aphidius*) *matricariae* Haliday, 1834, *Aphidius* (*Aphidius*) *colemani* Viereck, 1912, *Diaeretiella rapae* (Mc'Intosh, 1855), *Ephedrus persicae* Froggatt, 1904, *Lysiphlebus fabarum* (Marshall, 1896), and *Praon volucre* (Haliday, 1833), have wider distributions and host ranges than other Iranian aphidiines. Our study clearly shows that numerous aphids are attacked by aphidiine wasps in various Iranian agroecosystems. Thus, it should be emphasized that aphidiine parasitoids can serve as efficient natural biocontrol agents against these important agricultural pests. The knowledge of bitrophic (aphid—parasitoid) associations is an important basic tool for the design of focused biological control strategies since it enhances the potential long-term establishment of Aphidiinae in given regions. Although the aphidiine fauna of Iran is rich, some issues should be noted. The exact localities are unknown for two species, *Betuloxys hortorum* (Starý, 1960), and *Trioxys pappi* Takada, 1979, while the aphid host associations of 18 aphidiines are currently unknown. Therefore, further faunistic surveys are necessary to confirm the local distribution of these species in Iran, as well as to reveal missing host range patterns. Comparison of the Aphidiinae fauna of Iran with adjacent countries indicates the fauna of Russia with 151 recorded species in 24 genera (Belokobylskij & Lelej, 2019) is more diverse than Iran; followed by Turkey (58 species), Israel—Palestine (39 species), Pakistan (38 species), Iraq (32 species), Kazakhstan (31 species), Egypt (23 species), Lebanon (17 species), Cyprus (11 species), Yemen (10 species), Saudi Arabia and United Arab Emirates (both with nine species), Turkmenistan (eight species), Afghanistan and Azerbaijan (both with seven species), Syria (four species), Jordan (three species), and Armenia (two species). No species have been recorded from Bahrain, Kuwait, Oman, and Qatar (Yu et al., 2016). Additionally, among the 23 countries of the Middle East and adjacent to Iran, Russia shares 53 species with Iran, followed by Israel—Palestine (32 species), Iraq and Turkey (both with 31 species), Egypt (23 species), Kazakhstan (21 species), Pakistan (20 species), Lebanon (16 species), Cyprus (10 species), Yemen (10 species), Saudi Arabia (nine species), Turkmenistan (seven species), Afghanistan (five species), Syria (four species), Azerbaijan and Jordan (both with three species), and Armenia (one species). There is no aphidiine species shared between Iran and five other countries (Bahrain, Kuwait, Oman, Qatar, and United Arab Emirates).

References

Abdel-Rahman, M. A. A. (2005). The relative abundance and species composition of hymenopterous parasitoids attacking cereal aphids (Homoptera: Aphididae) infesting wheat plants at Upper Egypt. *Egyptian Journal of Agricultural Research, 83*(2), 633—645.

Abdel-Rahman, M. A. A., Nasser, M. A. K., & Ali, A. M. (2000). Incidence of hymenopterous parasitoids attacking cereal aphids in wheat fields in Upper Egypt. *Assiut Journal of Agricultural Sciences, 31*(2), 317—328.

Abdel-Samad, S. S., & Ahmad, M. A. (2009). Population fluctuations of *Aphis crassivora* and *Liriomyza trifolii* endoparasitoids on faba bean varieties. In S. Kumari, B. Bayaa, K. Makkouk, M. El-Ahmed Haidar, A. Dawabah, A. Shehab, & Y. Abu-Jawdeh (Eds.), *10th Arab Congress of Plant Protection: 131*. Beirut: Arab Society for Plant protection.

Abou-Fakhr, I. (1982). *Aphid endoparasitoid complex on certain vegetables ornamentals and weeds in Lebanon* (Ph.D. thesis). American University of Beirut.

Abou-Fakhr, E. M., & Kawar, N. S. (1998). Complex of endoparasitoids of aphids (Hom.: Aphididae) on vegetables and other plants. *Entomologicheskoe Obozrenie, 77*, 753—763.

Abu El-Ghiet, U. M., Edmardash, Y. A., & Gadallah, N. S. (2014). Braconidae diversity (Hymenoptera: Ichneumonoidea) in alfalfa fields *Medicago sativa* L. of some Western Desert Oases. *Journal of Crop Protection, 3*(4), 543—556.

Aeschlimann, J. P. (1981). Occurrence and natural enemies of *Theriophilus trifolii* Morell and *Acyrthosiphon pisum* Harris (Homoptera, Aphididae) on lucerne in the Mediterranean region. *Acta Oecologica Oecoliga Applicata, 2*(1), 3—11.

Aeschlimann, J. P., & Vitou, J. (1985). Aphids (Homoptera, Aphididae) and their natural enemies occurring on *Sonchus* spp. in the Mediterranean region. *Acta Oecologica Oecoliga Applicata, 6*, 69—76.

Afshari, A., Soleymannejadian, E., Bayat Asadi, H., & Shishehbor, P. (2006). Population fluctuation of cotton aphid, *Aphis gossypii* (Hom.: Aphididae), and its natural enemies on cotton, under two sprayed and unsprayed conditions. *Entomology and Phytopathology, 73*(2), 39—60 (in Persian, English summary).

Aghajanzadeh, S., Rasulian, G., Rezwani, A., & Esmaili, M. (1995). Identification of the aphids attacking citrus trees in West-Mazandaran and their population dynamics. In *Proceedings of the 12th Iranian plant protection congress, 2—7 September 1995* (p. 208). Karaj: Karaj Junior College of Agriculture.

Aghajanzadeh, S., Rasoulian, G., Rezwani, N., & Esmaili, M. (1997). Study on faunistic aspects of citrus aphids in West-Mazandaran. *Applied Entomology and Phytopathology, 65*(1), 62—78, 15—17.

Ahmadi, R. (2000). Natural enemies of Russian wheat aphid, *Diuraphis noxia* in Hamadan province. In *Proceedings of the 14th Iranian plant protection congress, 5—8 September 2000* (p. 226). Isfahan University of Technology.

Ahmadi, A. A., & Sarafrazi, A. M. (1993). Distribution and natural enemies of Russian wheat aphid, *Duraphis noxia*, in the Fars province. In *Proceedings of the 11th Iranian plant protection congress, 28 August — 2 September 1993* (p. 1). University of Guilan.

Akar, S., & Çetin Erdoğan, Ö. (2017). Contributions to the Aphidiinae (Hymenoptera: Braconidae) fauna of Turkey, with new records. *Trakya University Journal of Natural Sciences, 18*(2), 89—96.

Al-Azawi, A. F. (1970). Some aphid parasites from central and south Iraq with notes on their occurrence. *Bulletin of the Iraq Natural History Museum, 4*, 27—31.

Alesafoor, M., & Mossadegh, M. S. (2015). Natural enemies of oleander aphid, *Aphis nerii* B. de F. (Hom., Aphididae), effective factors in attraction of coccinellids and population fluctuation of oleander aphid in Shiraz and Ahvaz (Iran). *Entomofauna, 36*(42), 549—560.

Alichi, M., Shishehbor, P., Mossadegh, M. S., & Soleiman Nejadian, E. (2007). The effect of different temperatures on biology and life tables of *Aphidius rhopalosiphi* and *Praon volucre*, parasitizing *Metopophilium dirhodum* under laboratory condition. *Scientific Journal of Agriculture, 29*(4), 99—109 (in Persian, English summary).

Alichi, M., Shishehbor, P., Mossadegh, M. S., & Soleiman Nejadian, E. (2008). Species structure and distribution of wheat aphids and their parasitoids in Shiraz and study on the seasonal population dynamics of dominant species. *Journal of Science and Technology of Agriculture and Natural Resources, 12*, 287—295 (in Persian, English summary).

Alikhani, M., Rezwani, A., Starý, P., Kavallieratos, N. G., & Rakhshani, E. (2013). Aphid parasitoids (Hymenoptera: Braconidae: Aphidiinae) in cultivated and non-cultivated areas of Markazi Province, Iran. *Biologia, 68*, 966—973.

Ameri, A., Talebi, A. A., Rakhshani, E., & Ebrahimi, E. (2019). A survey on the Aphidiinae (Hym., Braconidae) of Hormozgan province, South of Iran. *Journal of Insect Biodiversity and Systematics, 4*(4), 227—239.

Arockia Lenin, E. (2015). Intergeneric affinity of sixty-two Aphidiine parasitoids (Hymenoptera: Braconidae: Aphidiinae) from different geographical regions. *Journal of Entomology and Zoology Studies, 3*(6), 405—410.

Asadizade, A., Mahdiyan, K., Talebi, A. A., & Esfandiarpour, I. (2014). Faunistic survey of parasitoid wasps subfamily Aphidiinae from Baft region, Kerman province. In *Proceedings of the 3rd Integrated Pest Management Conference (IPMC), 21—22 January 2014, Kerman* (p. 630).

Aslam, M. M., Uygun, N., & Starý, P. (2004). A survey of aphid parasitoides in Kahramanmaras, Turkey (Hymenoptera: Braconidae, Aphidiinae). *Phytoparasitica, 32*(3), 255—263.

Aslan, B., & Karaca, I. (2005). Fruit tree Aphids and their natural enemies in Isparta region Turkey. *Journal of Pest Science, 78*, 227—229.

Attarzadeh, M., Rajabpour, A., Farkhari, M., & Rasekh, A. (2019). Interactions between *Orius albidipennis* and *Aphidius colemani* (Hymenoptera: Braconidae) for the control of *Aphis gossypii* on greenhouse cucumber. *Journal of Crop Protection, 8*(1), 21—31.

Autrique, A., Starý, P., & Ntahimpera, L. (1989). Biological control of pest aphids by hymenopterous parasitoids in Burundi. *FAO Plant Protection Bulletin, 37*, 1—74.

Avidov, Z., & Harpaz, L. (1969). *Plant pests of Israel*. Jerusalem: Israel Universities Press. x + 549 pp.

Babaee, M. R., Sahragard, A., & Rezwani, A. (2000). Three species of parasitoids (Aphidiidae) on forest trees aphids in Mazandaran and a new method for determining percent parasitism. In *Proceedings of the 14th Iranian plant protection congress, 5—8 September 2000* (p. 128). Isfahan University of Technology.

Babaee, M. R., Mohammadi, M., Barimani, H., & Rezvani, A. (2004). Biological study of *Pterocomma populeum* (Kaltenbach) on poplar trees in Mazandaran province. In *Proceedings of the 16th Iranian plant protection congress, 28 August — 1 September 2004* (p. 413). University of Tabriz.

Bagheri-Matin, S., Shahrokhi, S., & Starý, P. (2008). Introduction of wheat aphids and their parasitoids (Hym.: Aphidiinae and Aphelinidae) in Kermanshah province, Iran. In *Proceedings of the 18th Iranian plant protection congress, 24—27 August 2008* (p. 84). University of Bu-Ali Sina Hamedan.

Bagheri-Matin, S., Sahragard, A., & Rasoolian, G. (2009). Some biological parameters of *Lysiphlebus fabarum* (Hymenoptera: Aphidiidae) a parasitoid of *Aphis fabae* (Homoptera: Aphidiidae) under laboratory conditions. *Munis Entomology & Zoology, 4*(1), 193–200.

Bagheri-Matin, S., Shahrokhi, S., & Starý, P. (2010). Report of *Praon gallicum* (Hymenoptera: Braconidae: Aphidiinae) from Iran. *Applied Entomology and Phytopathology, 78*(1), 33–34.

Bandani, A. R. (1992). *Investigation on the fauna of cereal aphids (wheat and barely) and their parasitoids in Sistan region*. M. Sc. thesis of. Entomology, University of Tehran, 129 pp. (in Persian, English summary).

Bandani, A. R., Rasoulian, G., Kharazi Pakdel, A., Esmaili, M., & Azmayesh Fard, P. (1993). Cereal aphids and their hymenopterous parasites in Sistan province. In *Proceedings of the 11th Iranian plant protection congress, 28 August – 2 September 1993* (p. 6). University of Guilan.

Barahoei, H., Madjdzadeh, S. M., Mehrparvar, M., & Starý, P. (2010). A study of *Praon* Haliday (Hymenoptera: Braconidae: Aphidiinae) in south-east Iran with two new records. *Acta Entomologica Serbica, 15*, 107–120.

Barahoei, H., Madjdzadeh, S. M., & Mehrparvar, M. (2011). Morphometric differentiation of five biotypes of *Lysiphlebus fabarum* (Marshall) (Hymenoptera: Braconidae: Aphidiinae) in Iran. *Zootaxa, 2745*, 43–52.

Barahoei, H., Madjdzadeh, S. M., & Mehrparvar, M. (2012). Aphid parasitoids (Hymenoptera: Braconidae: Aphidiinae) and their tritrophic relationships in Kerman province, Southeastern Iran. *Iranian Journal of Animal Biosystematics, 8*(1), 1–14.

Barahoei, H., Sargazi, A., & Rakhshani, E. (2012). Biosystematic study of the aphid parasitoids (Hym., Braconidae, Aphidiinae) in Golestan province. In *Proceedings of the 20th Iranian plant protection congress, 26–29 August 2012* (p. 100). University of Shiraz.

Barahoei, H., Rakhshani, E., Madjdzadeh, S. M., Alipour, A., Taheri, S., Nader, E., Mitrovski Bogdanović, A., Petrović-Obradović, O., Starý, P., Kavallieratos, N. G., & Tomanović, Ž. (2013). Aphid parasitoid species (Hymenoptera: Braconidae: Aphidiinae) of central submountains of Iran. *North-Western Journal of Zoology, 9*, 70–93.

Barahoei, H., Rakhshani, E., Nader, E., Starý, P., Kavallieratos, N. G., Tomanović, Ž., & Mehrparvar, M. (2014). Checklist of Aphidiinae parasitoids (Hymenoptera: Braconidae) and their host aphid associations in Iran. *Journal of Crop Protection, 3*(2), 199–232.

Barjadze, S., Karaca, I., Yasar, B., & Gratiashvili, N. (2010). New evidence of parasitoids of pest aphids on roses and grapevine in Turkey (Hem.: Aphididae; Hym.: Braconidae: Aphidiinae). *Journal of Entomology and Acarology Research, 42*(3), 143–145.

Baroon, N., Kocheili, F., & Mosadegh, M. S. (2008). The population dynamics and parasitism rate of black bean aphid *Aphis fabae* Scopoli (Hom.: Aphididae) by parasitoid wasp *Lysiphlebus fabarum* Marshall (Hym.: Braconidae) on broad bean Barket and Shakh Bozzi varieties in Ahwaz, southern Iran. In *Proceedings of the 18th Iranian plant protection congress, 24–27 August 2008* (p. 39). University of Bu-Ali Sina Hamedan.

Bartlett, B. R., Clausen, C. P., DeBach, P., Goeden, R. D., Legner, E. F., McMurtry, J., & Oatman, E. R. (1978). Introduced parasites and predators of arthropod pests and weeds. A world review. In *Agriculture Handbook* (vol. 480). Agricultural Research Service. United States Department of Agriculture, 544 pp.

Bazyar, M., Hodjat, M., & Alichi, M. (2011). The functional response of *Aphidius ervi* (Haliday) (Hym.: Braconidae, Aphidiinae) to different densities of *Sitobion avenae* (Fabricius) (Hom.: Aphididae) on two wheat cultivars. *Iran Agricultural Research, 30*(1 & 2), 61–72.

Belokobylskij, S. A., & Lelej, A. S. (2019). Annotated catalogue of the Hymenoptera of Russia. In , *Apocrita: Parasitica: Vol. II. Proceedings of the Zoological Institute of the Russian Academy of Sciences*. Supplement No. 8, 594 pp.

Belsahw, R., & Quicke, D. L. J. (1997). A molecular phylogeny of Aphidiinae (Hymenoptera: Braconidae). *Molecular Phylogenetics and Evolution, 7*(3), 281–293.

Belshaw, R., Fitton, M., Hernion, E., Gimeno, C., & Quicke, D. L. J. (1998). A phylogenetic reconstruction of the Ichneumonoidea (Hymenoptera) based on the D2 variable region of the 28S ribosomal RNA. *Systematic Entomology, 23*(2), 109–123.

Benelli, G., Messing, R. H., Wright, M. G., Giunti, G., Kavallieratos, N. G., & Canale, A. (2014). Cues triggering mating and host-seeking behavior in the aphid parasitoid *Aphidius colemani* (Hymenoptera: Braconidae: Aphidiinae): implications for biological control. *Journal of Economic Entomology, 107*, 2005–2022.

Beyarslan, A., Gözüaçik, C., Güllü, M., & Konuksal, A. (2017). Taxonomical investigation on Braconidae (Hymenoptera: Ichneumonoidea) fauna in northern Cyprus, with twenty six new records for the country. *Journal of Insect Biodiversity and Systematics, 3*(4), 319–334.

Bodenheimer, F. S., & Swirski, E. (1957). *The Aphidoidea of the Middle East*. Jerusalem: Weisman Science Press, 378 pp.

Bogdanović, A. M., Ivanović, A., Tomanović, Ž., Žikić, V., Starý, P., & Kavallieratos, N. G. (2009). Sexual dimorphism in *Ephedrus persicae* (Hymenoptera: Braconidae: Aphidiinae): intraspecific variation in size and shape. *The Canadian Entomologist, 141*, 550–560.

Boivin, G., Hance, T., & Brodeur, J. (2012). Aphid parasitoids in biological control. *Canadian Journal of Plant Science, 92*, 1–12.

Bolckmans, K., & Tetteroo, A. (2002). Biological pest control in eggplants in The Netherlands. *IOBC-WPRS Bulletin, 25*, 25–28.

Cameron, P. J., & Walker, G. P. (1989). Release and establishment of *Aphidius* spp. (Hymenoptera: Aphidiinae), parasitoids of the pea aphid and blue green aphid in New Zealand. *New Zealand Journal of Agricultural Research, 32*(2), 281–290.

Çetin Erdoğan, Ö, & Akar, S. (2018). First record of the species *Adialytus veronicaecola* (Starý, 1978) (Hymenoptera: Braconidae: Aphidiinae) from the West Palaearctic Region. *Bitki KorumaBülteni / Plant Protection Bulletin, 58*(4), 227–230.

Çetin Erdoğan, Ö., Tomanović, Z., & Beyarslan, A. (2008). New aphid parasitoids (Hymenoptera: Braconidae: Aphidiinae) in the region of Marmara, Turkey. *Acta Entomologica Serbica, 13*(1–2), 85–88.

Çetin Erdoğan, Ö., Tomanović, Z., & Beyarslan, A. (2010). New distributional records on the subfamily Aphidiinae (Hymenoptera: Braconidae) in black Sea region, Turkey. *Linzer Biologische Beiträge, 42*(1), 613–616.

Chen, J. H., González, D., & Luhman, J. (1991). A new species of *Aphidius* (Hymenoptera) attacking the pea aphid, *Acyrthosiphon pisum*. *Entomophaga, 35*(4), 509–514.

Chen, X. X., & van Achterberg, C. (2019). Systematics, phylogeny, and evolution of braconid wasps: 30 years of progress. *Annual Review of Entomology, 64*, 1–24.

Darvish-Mojeni, T. (1994). An introduction to one parasitoid wasp species and on new hyperparasitoid wasp genus for the fauna of Iran. *Journal of Entomological Society of Iran, 14*, 79.

Darvish-Mojeni, T., & Bayat-Asadi, H. (1995). Identification of natural enemies of wheat green aphid, *Sitobion avenae*. In *Gorgan and Dasht. Proceedings of the 12th Iranian plant protection congress, 2–7 September 1995* (p. 21). Karaj: Karaj Junior College of Agriculture.

Davatchi, A., & Shojai, M. (1968). *Entomophagous hymenopterous wasps of Iran* (Vol. 107). University of Tehran, Faculty of Agriculture, 88 pp.

Davidian, E. M. (2005). A review of species of the subgenus *Trioxys* s. str., genus *Trioxys* Haliday (Hymenoptera: Aphidiidae) of Russia and adjacent countries. *Entomological Review, 84*(3), 579–609.

Darsouei, R., Karimi, J., & Modarres Awal, M. (2011a). A molecular approach for study the aphids of poem fruit orchards and their parasitoids in Mashhad. In *Proceedings of the 2nd Iranian pest Management Conference, 14–15 September 2011, Kerman* (p. 65).

Darsouei, R., Karimi, J., & Modarres Awal, M. (2011b). Parasitic wasps as natural enemies of aphid populations in the Mashhad region of Iran: new data from DNA barcodes and SEM. *Archives of Biological Science Belgrade, 63*(4), 1225–1234.

Desneux, N., & Ramirez-Romero, R. (2009). Plant characteristics mediated by growing conditions can impact parasitoid's ability to attack host aphids in winter canola. *Journal of Pest Science, 82*, 335–342.

Düzgüneş, Z., Toros, S., Kilincer, N., & Kovanci, B. (1982). *The parasites and the predators of Aphidoidea in Ankara*. Turkish Ministry of Agriculture Ankara, 251 pp.

Elahii, A., Shirvani, A., & Rashki, M. (2017). Effect of different canola genotypes on some biological parameters of the parasitoid wasp *Diaeretiella rapae* (Hymenoptera: Braconidae). *Biological Control of Pests and Plant Diseases, 6*(2), 155–164 (in Persian, English summary).

El-Gantiry, A. M., El-Heneidy, A. H., Mousa, S. F., & Adly, D. (2012). *Aphis illinoisensis* Shimer (Hemiptera: Aphididae) a recent invasive aphid species in Egypt. *Egyptian Journal of Biological Pest Control, 22*(2), 225–226.

El-Heneidy, A. H. (1994). Efficacy of aphidophagous insects against aphids at wheat fields in Egypt, a 5- year evaluation. *Egyptian Journal of Biological Pest Control, 4*(2), 113–123.

El-Heneidy, A. H., & Abdel-Samad, S. S. (2001). Tritrophic interaction among Egyptian wheat plant, cereal aphids and natural enemies. *Egyptian Journal of Biological Pest Control, 11*(2), 119–125.

El-Heneidy, A. H., González, D., Starý, P., Adly, D., & El-Khawas, M. A. (2001). A survey of primary and secondary parasitoid species of cereal aphids on wheat in Egypt. *Egyptian Journal of Biological Pest Control, 11*(2), 193–194.

El-Heneidy, A. H., González, D., Starý, P., & Adly, D. (2002). Significance of hyperparasitization of primary cereal aphid parasitoids in Egypt "Hymenoptera Parasitic". *Egyptian Journal of Biological Pest Control, 12*(2), 109–114.

El-Heneidy, A. H., Agamy, E. A., El-Husseini, M. M., & Adly, D. (2003). Seasonal occurrence of the aphid parasitoid *Aphidius matricariae* Hal. (Hymenoptera: Aphidiidae) in Egyptian wheat fields. *Agricultural Research Journal Suez Canal University, 2*(1), 103–108.

El-Heneidy, A. H., Sobhy, H. M., Abd-El-Wahed, S. M. N., & Mikhail, W. Z. A. (2004). Biological aspects and life table analysis of cereal aphid species and their parasitoid *Aphidius colemani* Viereck (Hymenoptera: Aphidiidae). *Egyptian Journal of Biological Pest Control, 14*(1), 43–51.

El-Serafy, H. A. (1999). Population density of cereal aphids' parasitoids and their role in suppressing cereal aphids on wheat plantations at Mansoura district. *Archiv für Phytopathologie und Pflanzenschutz, 32*(3), 257–264.

Fakhireh, F., Barahoei, H., & Madjdzadeh, S. M. (2016). Fauna of Aphidiinae (Hym., Braconidae) in sistan region, north of Sistan-Baluchestan province. In *Proceedings of the 3rd National Conference of biological control in agriculture and natural resources* (p. 39). Ferdowsi University of Mashhad.

Fallahzadeh, M., & Saghaei, N. (2010). Checklist of Braconidae (Insecta: Hymenoptera) from Iran. *Munis Entomology and Zoology, 5*(1), 170–186.

Farahani, S., Talebi, A. A., & Barahoei, H. (2015). Occurrence of the rare aphid parasitoid *Praon bicolor* Mackauer, 1959 (Hymenoptera, Braconidae, Aphidiinae) in central Asia. *Journal of Insect Biodiversity and Systematics, 1*(1), 11–15.

Farahani, S., Talebi, A. A., & Rakhshani, E. (2016). Iranian Braconidae (Insecta: Hymenoptera: Ichneumonoidea): diversity, distribution and host association. *Journal of Insect Biodiversity* and Systematics, 2(1), 1–92.

Farahani, S., Talebi, A. A., Starý, P., & Rakhshani, E. (2017). Occurrence of the rare root aphid parasitoid, *Aclitus obscuripennis* (Hymenoptera: Braconidae: Aphidiinae) in Iran. *Biologia, 72*(12), 1494–1498.

Farahbakhsh, G. (1961). *A checklist of economically important insects and other enemies of plants and agricultural products of Iran* (vol. 1, p. 153). Tehran: Ministry of Agriculture Department of Plant Protection.

Farhad, A., Talebi, A. A., & Fathipour, Y. (2011). Foraging behavior of *Praon volucre* (Hymenoptera: Braconidae) a parasitoid of *Sitobion avenae* (Hemiptera: Aphididae) on wheat. *Psyche*. https://doi.org/10.1155/2011/868546

Farhad, A., Talebi, A. A., & Fathipour, Y. (2012). Thermal requirements of *Sitobion avenae* (Hem.: Aphididae) and its parasitoid, *Praon volucre* (Hym.: Braconidae). *Iranian Journal of Plant Protection Science, 43*(1), 143–154.

Farokhzadeh, H., Moravvej, G., Modarres Awal, M., & Karimi, J. (2014). Molecular and morphological identification of hymenopteran parasitoids from the pomegranate aphid, *Aphis punicae* in Razavi Khorasan province, Iran. *Turkiye Entomoloji Dergisi, 38*(3), 291–306.

Farsi, A., Kocheili, F., Soleymannejadian, E., & Khajehzadeh, Y. (2009). Population dynamics of canola aphids and their dominant natural enemies in Ahvaz. *Journal of Plant Protection, 32*(2), 55–66 (in Persian, English summary).

Farsi, A., Kocheili, F., Mossadegh, M., Rasekh, A., & Bagheri, S. (2012). Natural enemies of aphid lettuce *Nasonovia ribisnigri* Mosely and the dominant species in Khuzestan province. In *Proceedings of the 20th Iranian plant pprotection congress, 26–29 August 2012* (p. 148). University of Shiraz.

Farsi, A., Kocheili, F., Mossadegh, M., & Rasekh, A. (2019). Temperature-dependent life table parameters of *Aphidius matricariae* (Hym.: Braconidae), an important parasitoid of the currant lettuce aphid, *Nasonovia ribisnigri* (Hemiptera: Aphididae). *Journal of Entomological Society of Iran, 38*(4), 365–375.

Fathipour, Y., Hosseini, A., Talebi, A. A., & Moharramipour, S. (2006). Functional response and mutual interference of *Diaeretiella rapae* (Hymenoptera: Aphidiidae) on *Brevicoryne brassicae* (Homoptera: Aphididae). *Entomologica Fennica, 17*, 90–97.

Fulmek, L. (1968). Parasitinsekten der insektengallen Europas. *Beiträge zur Entomologie, 18*(7/8), 719–952.

Gadallah, N. S., El-Heneidy, A. H., Mahmoud, S. M., & Kavallieratos, N. G. (2017). Identification key, diversity and host associations of parasitoids (Hymenoptera: Braconidae: Aphidiinae) of aphids attacking cereal crops in Egypt. *Zootaxa, 4312*(1), 143–154.

Gärdenfors, U. (1986). Taxonomic and biological revision of Palearctic *Ephedrus* Haliday (Hymenoptera: Braconidae: Aphidiinae). *Entomologia Scandinavica (Supplement), 27*, 1–95.

Ghafouri Moghaddam, M., Rakhshani, E., Starý, P., Tomanović, Ž., & Kavallieratos, N. G. (2012). Occurrence of *Monoctonia vesicarii* Tremblay (Hym., Braconidae, Aphidiinae), a very rare parasitoid of the gall forming aphids, *Pemphigus* spp. (Hem., Eriosomatidae) in Iran. In *Proceedings of the 20th Iranian plant protection congress, 26–29 August 2012* (p. 212). University of Shiraz.

Ghahari, H. (2019). Study on the natural enemies of agricultural pests in some sugar beet fields, Iran. *Journal of Sugar Beet, 35*(1), 91–102 (in Persian, English summary).

Ghahari, H., & Fischer, M. (2011). A contribution to the Braconidae (Hymenoptera) from Golestan National Park, northern Iran. *Zeitschrift Arbeitsgemeinschaft Österreichischer Entomologen, 63*, 77–80.

Ghahari, H., Fischer, M., Çetin Erdoğan, Ö., Beyarslan, A., & Ostovan, H. (2010). A contribution to the braconid wasps (Hymenoptera: Braconidae) from the forests of northern Iran. *Linzer Biologische Beitrage, 42*(1), 621–634.

Ghahari, H., Fischer, M., Sakenin, H., & Imani, S. (2011). A contribution to the Agathidinae, Alysiinae, Aphidiinae, Braconinae, Microgastrinae and Opiinae (Hymenoptera: Braconidae) from cotton fields and surrounding grasslands of Iran. *Linzer Biologische Beiträge, 43*(2), 1269–1276.

Ghahari, H., Beyarslan, A., & Kavallieratos, N. G. (2020). New records of Braconidae (Hymenoptera: Ichneumonoidea) from Iran, and in Memoriam Dr. Maximilian Fischer (7 June 1929 – 15 June 2019). *Scientific Bulletin of Uzhhorod National University (Series: Biology), 48*, 48–55.

Ghanim, A., & El-Adl, M. (1983). Aphids infesting wheat and the effect of their predators in suppressing their populations in fields at Mansoura district Egypt. *Journal of Agriculctural Science, 8*(4), 958–968.

Ghotbi Ravandi, S., Askari Hesni, M., & Madjdzadeh, S. M. (2017). Species diversity and distribution pattern of Aphidiinae (Hym.: Braconidae) in Kerman province, Iran. *Journal of Crop Protection, 6*(2), 245–257.

González, D., White, W., Hall, J., & Dickson, R. C. (1978). Geographical distribution of Aphidiidae (Hym.) imported to California for biological control of *Acyrthosiphon kondoi* and *Acyrthosiphon pisum* (Hom.: Aphididae). *Entomophaga, 23*, 239–248.

González, D., Gilstrap, F., McKinnon, L., Zhang, J., Zareh, N., Zhang, G., Starý, P., Woolley, J. B., & Wang, P. (1992). Foreign exploration for natural enemies of Russian wheat aphid in Iran and in the Kunlun, Tian Shan, and Altai mountain valleys of the Peoples Republic of China. In , *Vol. 142. Proceedings of the 5th Russian Wheat Aphid Conference, for Worth 1992 (Texas)* (pp. 197–209). Publication Great Plains Agriculture Council.

Grasswitz, T. R., & Reese, B. (1998). Biology and host selection behaviour of the aphid hyperparasitoid *Alloxysta victrix* in association with the primary parasitoid *Aphidius colemani* and the host aphid *Myzus persicae. Biological Control, 43*, 261–271.

Güz, N., & Kilinçer, N. (2005). Aphid parasitoids (Hymenoptera: Braconidae: Aphidiinae) on weeds from Ankara Turkey. *Phytoparasitica, 33*(4), 359–366.

Hadadian, M., Zamani, A. A., Rakhshani, E., & Marefat, A. (2016). Investigation of various geographic population of two competitor species of aphid parasitoid *Praon exsoletum* (Nees, 1811) and *Trioxys complanatus* Quilis, 1931 (Braconidae: Aphidiinae), speciality parasitoids of *Therioaphis trifolii* maculate. In *Proceedings of the 22nd Iranian plant protection congress, College of agriculture and natural resources, 27–30 August 2016* (p. 540). Karaj: University of Tehran.

Hafez, A. A. (1994). Increasing the role of biocontrol agents against cereal aphids infesting wheat in Qalubia-Egypt. *Egyptian Journal of Biological Pest Control, 4*(2), 57–71.

Hagvar, E. B., & Hofsang, T. (1991). Aphid parasitoids (Hymenoptera, Aphidiinae): biology, host selection and use in biological control. *Biocontrol News and Information, 12*(1), 13–41.

Hasan, H. S. (2016). Survey of aphid species and associated parasitoids in Al-Homra Jordan. *Journal of Entomology and Zoology Studies, 4*(5), 1–4.

Hassan, M. S. (1957). Studies on the damage and control of *Aphis maidis* Fitch, in Egypt (Hemiptera: Homoptera: Aphididae). *Bulletin of the Entomological Society of Egypt, 41*, 213–230.

Heidari, S., Fathipour, Y., Sadeghi, S. E., Moharramipour, S., & Talebi, A. A. (2004). Spatial distribution pattern of *Chaitophorus leucomelas* Koch and its relation to spatial distribution of natural enemies, under monoculture and poplar-alfalfa agroforestry systems in Karaj. In *Proceedings of the 16th Iranian plant protection congress, 28 August – 1 September 2004* (p. 358). University of Tabriz.

Heidari Latibari, M., Moravej, G., Ghafouri Moghaddam, M., Barahoei, H., & Hanley, G. A. (2020). The novel host associations for the aphid parasitoid, *Pauesia hazratbalensis* (Hymenoptera: Braconidae: Aphidiinae). *Oriental Insects, 54*, 88–95.

Hodjat, S. H., & Moradeshaghi, M. J. (1988). Citrus aphids of Iran. *Journal of Plant Protection, 31*, 1–40.

Hofsvang, T., & Hagvar, E. B. (1975). Duration of development and longevity in *Aphidius ervi* and *Aphidius platensis* (Hym.: Aphidiinae), two parasites of *Myzus persicae* (Hom.: Aphididae). *Entomophaga, 20*(1), 11–22.

Hughes, R. D. (1989). Biological control in the open field. In A. K. Minks, & P. Harrewijn (Eds.), *Aphids. Their biology, natural enemies and control* (vol. C, pp. 167–198). Amsterdam: Elsevier. xvi + 312 pp.

Hussein, M. K., & Kawar, N. S. (1984). A study of aphids and their natural enemies in Southern Lebanon. *Arab Journal of Plant Protection, 2*, 17–82 (in Arabic).

Ibrahim, A. M. A. (1990a). Corn leaf aphid *Rhopalosiphum maidis* (F) (Hom.: Aphididae) on wheat and associated primary parasitoids and hyperparasitoids. *Bulletin of the Entolomological Society of Egypt, 69*, 149–157.

Ibrahim, A. M. A. (1990b). Population dynamics of bird-cherry aphid *Rhopalosiphum padi* L. (Hom.: Aphididae) and its primary parasitoids and hyperparasitoids association on wheat in Egypt. *Bulletin of the Entolomological Society of Egypt, 69*, 137–147.

Irshaid, L. A., & Hasan, H. S. (2011). Bioresidual effect of two insecticides on melon aphid *Aphis gossypii* Glover (Homoptera: Aphididae) and its parasitoid *Aphidius colemani* Viereck (Hymenoptera: Brachonidae). *American-Eurasian Journal of Agricultural & Environmental Sciences, 11*(2), 228–236.

Jafari-Ahmadabadi, N., Karimi, J., Modarres Awal, M., & Rakhshani, E. (2011). Morphological and molecular methods in identification of *Aphidius transcaspicus* Telenga (Hym: Braconidae: Aphidiinae) parasitoid of *Hyalopterus* spp. (Hom: Aphididae) with additional data on Aphidiinae phylogeny. *Journal of the Entomological Research Society, 13*(2), 91–103.

Jafari-Ahmadabadi, N., & Modarres Awal, M. (2012). Aphid parasitoids associations on stone fruit trees in Khorasan-e-Razavi province (Iran) (Hymenoptera: Braconidae: Aphidiinae). *Munis Entomology & Zoology, 7*(1), 418–423.

Jalali Moghadam Ziabari, N., Hosseini, M., & Olyie Torshiz, A. (2014). Introduction of natural enemies of *Aphis punicae* Passerini (Hemiptera: Aphididae) in pomegranate orchards in Kashmar. In *Proceedings of the 3rd Integrated Pest Management Conference (IPMC), 21–22 January 2014, Kerman* (p. 555).

Kargarian, F., Hesami, S., & Rakhshani, E. (2016). First report of *Monoctonia pistaciaecola* (Hymenoptera: Braconidae) from Iran. *Journal of Entomological Research, 8*(3), 263–267 (in Persian, English summary).

Kavallieratos, N. G., & Lykouressis, D. P. (2000). Two new species of *Praon* Haliday (Hymenoptera: Aphidiinae) from Greece. *Entomologia Hellenica, 13*, 5–12.

Kavallieratos, N. G., Tomanović, Z., Starý, P., Athanassiou, C. G., Sarlis, G. P., Petrović, O., Niketić, M., & Veroniki, M. A. (2004). A survey of aphid parasitoids (Hymenoptera: Braconidae: Aphidiinae) of Southeastern Europe and their aphid-plant associations. *Applied Entomology and Zoology, 39*(3), 527–563.

Kavallieratos, N. G., Tomanović, Ž., Athanassiou, C. G., Starý, P., Žikić, V., Sarlis, G. P., & Fasseas, C. (2005). Aphid parasitoids (Hymenoptera: Braconidae: Aphidiinae) infesting cotton, citrus, tobacco and cereal crops in southeastern Europe: aphid-plant associations and keys. *The Canadian Entomologist, 137*, 516–531.

Kavallieratos, N. G., Tomanović, Ž., Starý, P., & Mitrovski Bogdanović, A. (2008). Parasitoids (Hymenoptera: Braconidae: Aphidiinae) attacking aphids feeding on Prunoideae and Maloideae crops in Southeast Europe: aphidiine-aphid-plant associations and key. *Zootaxa, 1793*, 47–64.

Kavallieratos, N. G., Tomanović, Ž., Starý, P., Žikić, V., & Petrović Obradović, O. (2010). Parasitoids (Hymenoptera: Braconidae: Aphidiinae) attacking aphids feeding on Solanaceae and Cucurbitaceae crops in southeastern Europe: aphidiine-aphid-plant associations and key. *Annals of the Entomological Society of America, 103*, 153–164.

Kavallieratos, N. G., Tomanović, Ž., Petrović, A., Janković, M., Starý, P., Yovkova, M., & Athanassiou, C. G. (2013). Review and key for the identification of parasitoids (Hymenoptera: Braconidae: Aphidiinae) of aphids infesting herbaceous and shrubby ornamental plants in southeastern Europe. *Annals of the Entomological Society of America, 106*(3), 294–309.

Kavallieratos, N. G., Tomanović, Ž., Petrović, A., Kocić, K., Janković, M., & Starý, P. (2016). Parasitoids (Hymenoptera: Braconidae: Aphidiinae) of aphids feeding on ornamental trees in southeastern Europe: key for identification and tritrophic associations. *Annals of the Entomological Society of America, 109*, 473–487.

Kazemi, M., Talebi, A. A., Tazerouni, Z., Fathipour, Y., Rezaei, M., & Mehrabadi, M. (2020). Thermal requirements of parasitoid wasp, *Diaretiella rapae* (Hym.: Braconidae) reared on *Schizaphis graminum* (Hem.: Aphididae) under laboratory conditions. *Plant Pest Research, 10*(2), 1–13.

Kazemzadeh, S., Rakhshani, E., Tomanović, Ž., Starý, P., & Petrović, A. (2009). *Areopraon lepelleyi* (Waterston) (Hymenoptera: Braconidae: Aphidiinae), a parasitoid of Eriosomatinae (Hemiptera: Aphidoidea: Pemphigidae) new to Iran. *Acta Entomologica Serbica, 14*, 55–63.

Kazemzadeh, S., Rakhshani, E., & Rezwani, A. (2010). Aphid parasitoids (Hymenoptera, Braconidae, Aphidiinae) fauna of Northern Khorasan province and the first report of one species from Iran. In *Proceedings of the 19th Iranian plant protection congress, 31 July – 3 August 2010* (p. 164). Iranian Research Institute of Plant Protection.

Keyhanian, A. A., Taghizadeh, M., Taghadosi, M. V., & Khajehzadeh, Y. (2005). A faunistic study on insect pests and its natural enemies in canola fields at different regions of Iran. *Journal of Pajouhesh and Sazandegi, 68*, 2–8 (in Persian, English summary).

Khajehali, J., & Poorjavad, N. (2015). Identification and seasonal fluctuations of natural enemies of Elm aphid, *Tinocallis nevskyi* (Hem.: Aphididae) in Shahrekord. *Plant Pests Research, 5*(2), 1–12 (in Persian, English summary).

Khajehzadeh, Y., Kazemzadeh, H., & Melkeshi, H. (2006). The population dynamics of the natural enemies of cabbage aphid (*Brevicoryne brassicae*) and their efficiency in the rape seed fields of Ahvaz region. In *Proceedings of the 17th Iranian plant protection congress, Campus of agriculture and natural resources, 2–5 September 2006* (p. 57). Karaj: University of Tehran.

Khaki, F., Nazari, A., Madadi, H., & Rafie Karahrodi, Z. (2020). The first report of two aphids parasitoid *Aphidius platensis, Trioxys pallidus* (Hym., Braconidae) and a hyperparasitoid *Pachyneuron aphidis* (Hym., Pteromalidae) from Markazi province. *Journal of Entomological Research, 12*(3), 227–238 (in Persian, English summary).

Khanjani, M. (2006a). *Field crop pests in Iran* (3rd ed., p. 719). Bu-Ali Sina University Publication [in Persian].

Khanjani, M. (2006b). *Vegetable pests in Iran* (2nd ed., p. 467). Bu-Ali Sina University Publication [in Persian].

Khayat-Zadeh, B., Sadeghi, S. E., Ostovan, H., Shojai, M. M., Moharami-Poor, S., & Rakhshani, E. (2006). *Aphidius popovi* a new record from Iran. *Iranian Journal of Forest and Range Protection Research, 3*(2), 207–208 (in Persian, English summary).

Khazduzi Nejad Jamali, E., Fallahzadeh, M., & Dousti, A. (2012). Study of the population dynamics of cabbage aphid, *Brevicoryne brassicae* (L.) and identification the natural enemies in canola farms in the North of Fars province. In *Proceedings of the 20th Iranian plant protection congress, 26–29 August 2012* (p. 645). University of Shiraz.

Kocić, K., Petrović, A., Čkrkić, J., Kavallieratos, N. G., Rakhshani, E., Arnó, J., Aparicio, Y., Hebert, P. D. N., & Tomanović, Ž. (2020). Resolving the taxonomic status of potential biocontrol agents belonging to the neglected genus *Lipolexis* Förster (Hymenoptera, Braconidae, Aphidiinae) with descriptions of six new species. *Insects, 11*, 667.

Laamari, M., Chaouche, S. T., Halimi, C. W., Benferhat, S., Abbes, S. B., Khenissa, N., & Starý, P. (2012). A review of aphid parasitoids and their associations in Algeria (Hymenoptera: Braconidae: Aphidiinae; Hemiptera: Aphidoidea). *African Entomology, 20*(1), 161–176.

Latham, D. R., & Mills, N. J. (2012). Host instar preference and functional response of *Aphidius transcaspicus*, a parasitoid of mealy aphids (*Hyalopterus* species). *BioControl, 57*, 603–610.

Lotfalizadeh, H. (2002a). Parasitoids of cabbage aphid, *Brevicoryne brassicae* (L.) (Hom.: Aphididae) in Moghan region. *Agricultural Science, 12*(1), 15–25 (in Persian, English summary).

Lotfalizadeh, H. (2002b). Natural enemies of cotton aphids in Moghan region. In *Proceedings of the 15th Iranian plant protection congress, 7–11 September 2002* (p. 36). Razi University of Kermanshah.

Lozier, D. J., Roderick, K. G., & Mills, J. N. (2008). Evolutionarily significant units in natural enemies: Identifying regional populations of *Aphidius transcaspicus* (Hymenoptera: Braconidae) for use in biological control of mealy plum aphid. *Biological Control, 46*, 532–541.

Lozier, D. J., Roderick, K. G., & Mills, J. N. (2009). Molecular markers reveal strong geographic, but not host associated, genetic differentiation in *Aphidius transcaspicus*, a parasitoid of the aphid genus *Hyalopterus*. *Bulletin of Entomological Research, 99*(1), 83–96.

Mackauer, M. (1960). Zur Kenntnis der wearktischen Arten der gattung *Lysiphlebus* Foerster (Hymenoptera: Braconidae, Aphidiinae). *Bollettino del Laboratorio di Entomologia Agraria Filippo Silvestri, 18*, 230−256.

Mackauer, M. (1961). Neue europaeische Blattlatus-Schlupfwespen (Hymenoptera: Aphidiidae). *Bollettino del Laboratorio di Entomologia Agraria Filippo Silvestri, 19*, 270−290.

Mackauer, M. (1963). A re-examination of C. F. Baker's collection of aphid parasites (Hymenoptera: Aphidiiae). *The Canadian Entomologist, 95*, 921−935.

Mackauer, M., & Starý, P. (1967). Hymenoptera. Ichneumonoidea. World Aphidiidae. In V. Delluchi, & G. Remaudière (Eds.), *Index of entomophagous insects*. Le Francois, paris (pp. 1−195).

Mahmoud, S. M., El-Heneidy, A. H., Gadallah, N. S., & Ahmad, R. S. (2009). Survey and abundance of common ichneumonoid parasitoid species in Suez Canal Region, Egypt. *Egyptian Journal of Biological Pest Control, 19*(2), 185−190.

Mahmoudi, M., Sahragard, A., & Jalali, J. (2012). Aggregative behavior of the parasitoid, *Lysiphlebus fabarum* Marshall (Hymenoptera: Aphidiidae) to black bean aphid, *Aphis fabae* Scopoli (Hemiptera: Aphididae), spatial distribution pattern under laboratory conditions. In *Proceedings of the 20th Iranian plant protection congress, 26−29 August 2012* (p. 55). University of Shiraz.

Malkeshi, H. (1997). *Identification of important natural enemies of pome fruit trees aphids in Bojnurd region and investigation on their interaction with Aphis pomi De Geer* (M. Sc thesis) (p. 356). Tarbiat Modarres University College of Agriculture.

Malkeshi, H., Rezwani, A., & Talebi, A. A. (1998). Identification of important natural enemies of pome fruit trees aphids in Bojnurd. In *Proceedings of the 13th Iranian plant protection congress, 23−27 August 1998* (p. 163). Karaj: Karaj Junior College of Agriculture.

Malkeshi, S. H., Ghilasian, A., Ranji, H., Ghadiri Rad, S., Modarres Najafabadi, S., Pirhadi, A., & Khajezadeh, Y. (2004). An investigation of natural enemies of the cabbage aphid, *Brevicoryne brassicae* L., in canola farms. In *Proceedings of the 16th Iranian plant protection congress, 28 August − 1 September 2004* (p. 48). University of Tabriz.

Malkeshi, S. H., Khajezadeh, Y., Ghadiri Rad, S., Ranji, H., & Modarres Najafabadi, S. (2010). Survey of biodiversity, frequently and population fluctuation of dominant natural enemies of cabbage aphid. In *Proceedings of the 19th Iranian plant protection congress, 31 July − 3 August 2010* (p. 626). Iranian Research Institute of Plant Protection.

Mansour Ghazi, M., Shahrokhi, S. H., & Khanizad, A. (2008). Investigation of wheat aphid parasitoid and population dynamics of dominant species in the Kurdistan province. In *Proceedings of the 18th Iranian plant protection congress, 24−27 August 2008* (p. 73). University of Bu-Ali Sina Hamedan.

Mehrparvar, M., & Hatami, B. (2003). Report of two parasitoid wasps on *Macrosiphum rosae* L. *Newsletter of Entomological Society of Iran (Khabarnameh), 19*, 1 (in Persian).

Mehrparvar, M., Hatami, B., & Starý, P. (2005). Report of *Aphidius rosae* (Hym.: Braconidae), a parasitoid of rose aphid, *Macrosiphum rosae* (Hom.: Aphididae) from Iran. *Journal of Entomological Society of Iran, 25*(1), 63−64.

Mehrparvar, M., Mansouri, S. M., & Hatami, B. (2016). Some bioecological aspects of the rose aphid, *Macrosiphum rosae* (Hemiptera: Aphididae) and its natural enemies. *Acta Universitatis Sapientiae, Agriculture and Environment, 8*, 74−88.

Mescheloff, E., & Rosen, D. (1988). Biosystematic studies on the Aphidiidae of Israel (Hymenoptera: Ichneumonoidea), 1. Introduction and key to genera. *Israel Journal of Entomology, 22*, 61−73.

Mescheloff, E., & Rosen, D. (1989). Biosystematic studies on the Aphidiidae of Israel (Hymenoptera: Ichneumonidae). 2. The genera *Ephedrus* and *Praon*. *Israel Journal of Entomology, 23*, 75−100.

Mescheloff, E., & Rosen, D. (1990a). Biosystematic studies on the Aphidiidae of Israel (Hymenoptera: Ichneumonoidea): 3. The genera *Adialytus* and *Lysiphlebus*. *Israel Journal of Entomology, 24*, 35−50.

Mescheloff, E., & Rosen, D. (1990b). Biosystematic studies on the Aphidiinae of Israel (Hymenoptera: Ichneumonoidea), 3. The genera *Pauesia, Diaeretus, Aphidius* and *Diaeretiella*. *Israel Journal of Entomology, 24*, 51−91.

Mescheloff, E., & Rosen, D. (1993). Biosystematic studies on the Aphidiinae of Israel (Hymenoptera: Ichneumonoidea), 5. The genera *Trioxys* and *Binodoxys*. *Israel Journal of Entomology, 27*, 31−47.

Messing, R. H., & Klungness, L. M. (2002). A two-year survey of the melon aphid, *Aphis gossypii* Glover on crop plants in Hawaii. *Proceedings of the Hawaiian Entomological Society, 35*, 101−111.

Michelena, J. M., Assael, F., & Mendel, Z. (2005). Description of *Pauesia (Pauesia) anatolica* (Hymenoptera: Braconidae, Aphidiinae) sp. nov., a parasitoid of the cedar aphid *Cinara cedri*. *Phytoparasitica, 33*(5), 499−505.

Modarres Awal, M. (1997). Aphidiidae (Hymenoptera), p. 263; Braconidae (Hymenoptera). In M. Modarres Awal (Ed.), *List of agricultural pests and their natural enemies in Iran* (2nd ed., pp. 265−267). Ferdowsi University of Mashhad Press, 429 pp.

Modarres Awal, M. (2012). Aphidiidae (Hymenoptera), pp. 477-480; Braconidae (Hymenoptera). In M. Modarres Awal (Ed.), *List of agricultural pests and their natural enemies in Iran* (3rd ed., pp. 483−486). Ferdowsi University of Mashhad Press, 759 pp.

Modarres Najafabadi, S. S., Akbari Moghaddam, H., & Gholamian, G. (2005). Study on population dynamic of *Brevicoryne brassicae* and identification of its natural enemies in Sistan region. *Journal of Science and Technology of Agriculture and Natural Resources, 8*(4), 175−184 (in Persian, English summary).

Modarres Najafabadi, S. S., & Gholamian, G. (2007). Seasonal population dynamics of greenbug, *Schizaphis graminum* Rondani (Hemiptera: Aphididae) and introducing of its natural enemies in Sistan region. *Journal of Science and Technology of Agriculture and Natural Resources, 10*(4), 367−379 (in Persian, English summary).

Mohammadbeigi, A. (2000). Natural enemies of the walnut aphids in Qazvin province. In *Proceedings of the 14th Iranian plant protection congress, 5−8 September 2000* (p. 273). Isfahan University of Technology.

Mohammadi, Z., & Rasekh, A. (2018). Comparison of defensive behaviors in second and fourth instars of *Aphis fabae* (Aphididae), exposed to the parasitoid wasp, *Lysiphlebus fabarum*. In *Proceedings of the 23rd Iranian plant protection congress, 27−30 August 2018* (p. 1228). Gorgan University of Agricultural Sciences and Natural Resources.

Mohammadi, Z., Rasekh, A., Esfandiari, M., Michaud, J. P., & Kocheili, F. (2019). Effects of single and simultaneous application of the parasitoid, *Lysiphlebus fabarum* and the ladybird beetle, *Hippodamia variegata* on control of *Aphis gossypii* on cucumber. *Plant Protection (Scientific Journal of Agriculture), 42*(1), 1−17 (in Persian, English summary).

Mokhtari, A., Sahragard, A., Rezwani, A., & Salehi, L. (2000). Study on five parasitoid species of the stone fruit aphids in Gilan province. In *Proceedings of the 14th Iranian plant protection congress, 5—8 September 2000* (p. 278). Isfahan University of Technology.

Monajemi, N., & Esmaili, M. (1981). Population dynamics of alfalfa aphids and their natural controlling factors, in Karadj. *Journal of Entomological Society of Iran, 6*(1 & 2), 41—63.

Moodi, S., & Mossadegh, M. S. (2006). Natural enemies of *Aphis fabae* on *Solanum nigrum* plant in Khuzestan province. In *Proceedings of the 17th Iranian plant protection congress, Campus of agriculture and natural resources, 2—5 September 2006* (p. 56). Karaj: University of Tehran.

Mosavi, H., Havaskary, M., Monem, R., Khodaparast, S., & Sakenin, H. (2012). On a collection of Braconidae (Hymenoptera) from East Azarbaijan and vicinity, Iran. *Calodema, 226*, 1—4.

Mossadegh, M. S., & Kocheili, F. (2003). *A semi descriptive checklist of identified species of arthropods (agricultural, medical, ...) and other pests from Khuzestan.* Iran: Shahid Chamran University Press, 475 pp. (in Persian).

Mossadegh, M. S., Starý, P., & Salehipour, H. (2010). Aphid parasitoids in a dry lowland of Khuzestan, Iran (Hymenoptera, Braconidae, Aphidiinae). In *Proceedings of the IX European Congress of Entomology, 22-27 August 2010, Budapest* (p. 122).

Mossadegh, M. S., Starý, P., & Salehipour, H. (2011). Aphid parasitoids in a dry lowland area of Khuzestan, Iran (Hymenoptera, Braconidae, Aphidiinae). *Asian Journal of Biological Science, 4*, 175—181.

Mossadegh, M. S., Stary, P., Tamoli Torfi, E., Abolfarsi, R., Bahrami, R., Mohseni, L., Shahini, A., Seifollahi, F., Soheilyfar, P., Ravan, B., & Alaghemand, F. (2017). Aphid parasitoids (Hem.: Aphididae, Hym.: Braconidae, Aphidiinae) in Khuzestan, southwest Iran. In *Proceedings of the 2nd Iranian International congress of entomology* (p. 131). University of Tehran, College of Agriculture and Natural Resources.

Naderian, H., Penteado-Dias, A. M., Sakenin Chelav, H., & Samin, N. (2020). A faunistic study on Braconidae and Ichneumonidae (Hymenoptera, Ichneumonoidea) of Iran. *Calodema, 844*, 1—9.

Nazari, Y., Zamani, A. A., Masoumi, S. M., Rakhshani, E., Petrović-Obradović, O., Tomanović, S., Starý, P., & Tomanović, Ž. (2012). Diversity and host associations of aphid parasitoids (Hymenoptera: Braconidae: Aphidiinae) in the farmlands of western Iran. *Acta Entomologica Musei Nationalis Pragae, 52*, 559—584.

Nematollahi, M. R., & Bagheri, M. R. (2018). Distribution and population density of safflower pests and their natural enemies in Isfahan province, Iran. *Applied Researches in Plant Protection, 7*(3), 91—101 (in Persian, English summary).

Ölmez, S., & Ulusoy, M. R. (2003). A survey of aphid parasitoids (Hymenoptera: Braconidae: Aphidiinae) in Diyarbakir, Turkey. *Phytoparasitica, 31*(5), 524—528.

Özder, N., & Toros, S. (1999). Investigations on the natural enemies of aphid species damaging to wheat plants in Tekirdag province. In *Proceedings of the 4th Turkish National congress of biological control* (pp. 501—512). Adana: Çukurova Üniversitesi.

Petrović, A., Mitrović, M., Ghaliow, M. E., Ivanović, A., Kavallieratos, N. G., Starý, P., & Tomanović, Ž. (2019). Resolving the taxonomic status of biocontrol agents belonging to the *Aphidius eadyi* species group (Hymenoptera: Braconidae: Aphidiinae): an integrative approach. *Bulletin of Entomological Research, 109*, 342—355.

Pons, X., Lumbierres, B., Antoni, R., & Starý, P. (2011). Parasitoid complex of alfalfa aphids in an IPM intensive crop system in northern Catalonia. *Journal of Pest Science, 84*, 437—456.

Poodineh Moghaddam, M., Rakhshani, E., Ravan, S., Barahoei, H., & Starý, P. (2012). A taxonomic revision of the *Lipolexis gracilis* Förster - species group (Hym., Braconidae, Aphidiinae). In *Proceedings of the 20th Iranian plant protection congress* (p. 218). University of Shiraz.

Quicke, D. L. J., & van Achterberg, C. (1990). Phylogeny of the subfamilies of the family Braconidae (Hymenoptera: Ichneumonoidea). *Zoologische Verhandelingen, 258*, 1—95.

Quicke, D. L. J., Ward, D. F., Belokobylskij, S. A., & Butcher, B. A. (2020). *Zealastoa* Quicke and Ward, gen. nov., a new basal cyclostome braconid wasp (Hymenoptera: Braconidae) from New Zealand. *Austral Entomology, 59*, 455—466.

Radjabi, G. (1989). Insects attacking rosaceous fruit trees in Iran. In *Homoptera, Third volume*. Ministry of Agriculture, Plant Pests and Diseases Research Institute, 256 pp. (in Persian).

Ragab, M. E. (1996). Biology and efficiency of *Trioxys angelicae* Hal. (Hymenoptera: Aphidiidae) a newly recorded parasitoid of *Aphis craccivora* Koch (Homoptera: Aphididae) in Egypt. *Egyptian Journal of Biological Pest Control, 6*(1), 7—11.

Rahimi, S., Sadeghi, S. E., Rakhshani, E., Moharramipour, S., Shojai, M., & Zeinali, S. (2010). Parasitoids wasps reported on willow aphids in Iran. In *Proceedings of the 7th International congress of Hymenopterists, 20-26 June 2010, Köszeg, Hungary* (pp. 110—111).

Rahimi, R., Hosseini, R., Hajizadeh, J., & Sohani, M. M. (2012). Evaluation of COI and ITS2 sequences ability to distinguish *Lysiphlebus fabarum* and *Lysiphlebus confusus* (Hymenoptera: Aphidiidae). *Journal of Entomological Society of Iran, 31*(2), 53—66.

Rajabi-Mazhar, N., Rakhshani, E., Tomanović, Ž., & Starý, P. (2008). *Aphidius tanacetarius* Mackauer (Hym., Braconidae, Aphidiinae) a newly detected aphid parasitoid from Iran. In *Proceedings of the 18th Iranian plant protection congress* (p. 78). University of Bu-Ali Sina Hamedan.

Rajabi Mazhar, N. A., Rezvani, A., Rakhshani, E., & Yarmand, H. (2010). Survey of medicinal plants aphids and it's natural enemies in Hamadan province of Iran. *Iranian Journal of Forest and Range Protection Research, 7*(2), 115—127 (in Persian, English summary).

Rakhshani, E. (2012). Aphid parasitoids (Hym., Braconidae, Aphidiinae) associated with pome and stone fruit trees in Iran. *Journal of Crop Protection, 1*(2), 81—95.

Rakhshani, E., & Starý, P. (2021). Aphid parasitoids: Aphidiinae (Hym., Braconidae). In J. Karimi, & H. Madadi (Eds.), *Biological control of insect and mite pests in Iran. A review from fundamental and applied aspects* (pp. 333—399). Springer, 621 pp.

Rakhshani, E., Talebi, A. A., Sadeghi, S. E., & Fathipour, Y. (2002). Host stage preference, juvenile mortality and functional response of parasitoid wasps, *Trioxys pallidus* (Haliday) (Hom., Aphidiidae). In *Proceedings of the 15th Iranian plant protection congress* (p. 95). Razi University of Kermanshah.

Rakhshani, E., Talebi, A. A., Kavallieratos, N., & Fathipour, Y. (2004). Host stage preference, juvenile mortality and functional response of *Trioxys pallidus* (Hymenoptera: Aphidiinae). *Biologia, 59*, 197—203.

Rakhshani, E., Talebi, A. A., Sadeghi, S. E., Kavallieratos, N. G., & Rashed, A. (2004). Seasonal parasitism and Hyperparasitism of walnut aphid, *Chromaphis juglandicola* (Kaltenbach) (Hom., Aphididae) in Tehran province. *Journal of Entomological Society of Iran, 23*(2), 1−11.

Rakhshani, E., Talebi, A. A., Kavallieratos, N. G., Rezwani, A., Manzari, S., & Tomanović, Ž. (2005). Parasitoid complex (Hymenoptera, Braconidae, Aphidiinae) of *Aphis craccivora* Koch (Hemiptera: Aphidoidea) in Iran. *Journal of Pest Science, 78,* 193−198.

Rakhshani, E., Talebi, A. A., Starý, P., Manzari, S., & Rezwani, A. (2005). Re-description and biocontrol information of *Pauesia antennata* (Mukerji) (Hym., Braconidae, Aphidiinae), parasitoid of *Pterochloroides persicae* (Chol.) (Hom., Aphidoidea, Lachnidae). *Journal of the Entomological Research Society, 7*(3), 59−69.

Rakhshani, E., Talebi, A. A., Manzari, S., Rezwani, A., & Rakhshani, H. (2006). An investigation on alfalfa aphids and their parasitoids in different parts of Iran, with a key to the parasitoids (Hemiptera: Aphididae; Hymenoptera: Braconidae: Aphidiinae). *Journal of Entomological Society of Iran, 25*(2), 1−14.

Rakhshani, E., Talebi, A. A., Starý, P., Tomanović, Ž., Manzari, S., Kavallieratos, N. G., & Cetković, A. (2006). A new species of *Aphidius* Nees, 1818 (Hymenoptera, Braconidae, Aphidiinae) attacking *Uroleucon* aphids (Homoptera, Aphididae) from Iran and Iraq. *Journal of Natural History, 40*(32−34), 1923−1929.

Rakhshani, E., Talebi, A. A., Manzari, S., Rezwani, A., Starý, P., & Tomanović, Ž. (2006). Aphid parasitoid (Hym.: Braconidae: Aphidiinae) associations on willows and poplars in Iran. In *Proceedings of the 17th Iranian plant protection congress, Campus of agriculture and natural resources* (p. 27). Karaj: University of Tehran.

Rakhshani, E., Talebi, A. A., Manzari, S., Tomanović, Ž., Starý, P., & Rezwani, A. (2007). Preliminary taxonomic study of the genus *Praon* (Hymenoptera: Braconidae: Aphidiinae) and its host associations in Iran. *Journal of Entomological Society of Iran, 26*(2), 19−34.

Rakhshani, E., Talebi, A. A., Starý, P., Tomanović, Ž., & Manzari, S. (2007). Aphid-parasitoid (Hymenoptera: Braconidae: Aphidiinae) associations on willows and poplars in Iran. *Acta Zoologica Academiae Scientiarum Hungricae, 53*(3), 281−292.

Rakhshani, E., Talebi, A. A., Starý, P., Tomanović, Ž., Manzari, S., & Kavallieratos, N. G. (2008). A review of *Aphidius* Nees (Hymenoptera, Braconidae, Aphidiinae) in Iran: host associations, distribution and taxonomic notes. *Zootaxa, 1767,* 37−54.

Rakhshani, E., Tomanović, Ž., Starý, P., Talebi, A. A., Kavallieratos, N. G., Zamani, A. A., & Stanković, S. (2008). Distribution and diversity of wheat aphid parasitoids (Hymenoptera: Braconidae: Aphidiinae) in Iran. *European Journal of Entomology, 105,* 863−870.

Rakhshani, E., Kazemzadeh, S., Talebi, A. A., Starý, P., Manzari, S., Rezwani, A., & Asadi, G. (2008). Preliminary taxonomic study of genus *Trioxys* Haliday (Hym., Braconidae, Aphidiinae) and its host associations in Iran. In *Proceedings of the 18th Iranian plant protection congress* (p. 97). University of Bu-Ali Sina Hamedan.

Rakhshani, E., Talebi, A. A., Manzari, S., Starý, P., Tomanović, Ž., & Arjmandi, A. A. (2008d). *Lipolexis gracilis* Förster (Hymenoptera, Braconidae, Aphidiinae), an aphid parasitoid newly determined in Iran and Turkey. In *Proceedings of the 18th Iranian plant protection congress* (p. 98). University of Bu-Ali Sina Hamedan.

Rakhshani, H., Ebadi, R., Hatami, B., Rakhshani, E., & Gharali, B. (2010). A survey of alfalfa aphids and their natural enemies in Isfahan, Iran, and the effect of alfalfa strip-harvesting on their populations. *Journal of Entomological Society of Iran, 30*(1), 13−28.

Rakhshani, E., Tomanović, Ž., Starý, P., Kavallieratos, N. G., Ilić, M., Stanković, S., & Rajabi-Mazhar, N. (2011). Aphidiinae parasitoids (Hymenoptera: Braconidae) of *Macrosiphoniella* aphids (Hemiptera: Aphididae) in the western Palaearctic region. *Journal of Natural History, 45,* 41−42.

Rakhshani, E., Starý, P., & Tomanović, Ž. (2012). Species of *Adialytus* Förster, 1862 (Hymenoptera, Braconidae, Aphidiinae) in Iran: taxonomic notes and tritrophic associations. *ZooKeys, 221,* 81−95.

Rakhshani, E., Kazemzadeh, S., Starý, P., Barahoei, H., Kavallieratos, N. G., Ćetković, A., Popović, A., Bodlah, I., & Tomanović, Ž. (2012). Parasitoids (Hymenoptera: Braconidae: Aphidiinae) of northeastern Iran: aphidiine-aphid-plant associations, key and description of a new species. *Journal of Insect Science, 12*(143), 1−26.

Rakhshani, E., Starý, P., & Davidian, D. (2017). A taxonomic review of the subgenus, *Pauesiella* Sedlag and Starý, 1980 (Hym.: Braconidae, Aphidiinae). *Plant Pest Research, 7*(3), 53−66.

Rakhshani, E., Barahoei, H., Ahmad, Z., Starý, P., Ghafouri-Moghaddam, M., Mehrparvar, M., Kavallieratos, N. G., Čkrkić, J., & Tomanović, Ž. (2019). Review of Aphidiinae parasitoids (Hymenoptera: Braconidae) of the Middle East and north Africa: key to species and host associations. *European Journal of Taxonomy, 552,* 1−132.

Rasekh, A., Kharazi-Pakdel, A., Allahyari, H., & Michaud, J. P. (2010). A survey on foraging behavior of *Lysiphlebus fabarum* (Marshall), a thelytokous parasitoid of *Aphis fabae* Scopoli, and the effect of hungry on this behavior. In *Proceedings of the 19th Iranian plant protection congress* (p. 34). Iranian Research Institute of Plant Protection.

Rasekh, A., Kharazi-Pakdel, A., Michaud, J. P., Allahyari, H., & Rakhshani, E. (2012). Report of a thelytokous population of *Lysiphlebus fabarum* (Hym.: Aphidiidae) from Iran. *Journal of Entomological Society of Iran, 30*(2), 83−84.

Rastegar, J., Sakenin, H., Khodaparast, S., & Havaskary, M. (2012). On a collection of Braconidae (Hymenoptera) from East Azarbaijan and vicinity, Iran. *Calodema, 226,* 1−4.

Rasulian, G. (1985). *Investigation of the biology and population fluctuations of important alfalfa aphids in Karaj* (Ph.D dissertation). University of Tehran, 126 pp. (in Persian, English summary).

Rasulian, G. (1989). Evaluation of some natural enemies of alfa aphids. In *Proceedings of the 9th Iranian plant protection congress* (p. 16). Ferdowsi University of Mashhad.

Razmgou, J., Hajizade, J., & Asadi, A. (2002). Some of important natural enemies of cotton aphid in Moghan. In *Proceedings of the 15th Iranian plant protection congress* (p. 36). Razi University of Kermanshah.

Rezaei, N., Mosaddegh, M. S., & Hodjat, H. (2006). Aphids and their natural enemies in wheat and barley fields in Khuzestan. *Scientific Journal of Agriculture, 29*(2), 127−137.

Rezaei, N., Mossadegh, M. S., & Hodjat, S. H. (2010). Wheat aphids and their natural enemies population dynamics with particular reference to dominant species in Ahvaz, Mollasani, and Safiabad regions in Khuzestan. In *Proceedings of the 19th Iranian plant protection congress* (p. 159). Iranian Research Institute of Plant Protection.

Risbec, J. (1960). Les parasites des insects d'importance en Afrique Tropicale et à Madagascar. *Agronomie Tropicale, 15*, 624–656.

Rosen, D. (1967). The hymenopterous parasites and hyperparasites of aphids on citrus in Israel. *Annals of the Entomological Society of America, 60*, 394–399.

Rosen, D. (1969). The parasites of coccids, aphids and aleyrodids on citrus in Israel: some zoogeographical considerations. *Israel Journal of Entomology, 4*, 45–53.

Sabbaghan, K., & Soleymannejadian, E. (2007). Seasonal population fluctuation of *Myzus persicae* (Hom.: Aphididae) and its parasitoid, *Aphidius matricariae* (Hymenoptera: Braconidae) on potato in Behbahan and Yassuj regions. *The Scientific Journal of Agriculture, 29*(4), 153–162 (in Persian, English summary).

Sakenin, H., Naderian, H., Samin, N., Rastegar, J., Tabari, M., & Papp, J. (2012). On a collection of Braconidae (Hymenoptera) from northern Iran. *Linzer Biologische Beiträge, 44/2*, 1319–1330.

Sakenin, H., Samin, N., Beyarslan, A., Coronado-Blanco, J. M., Navaeian, M., Fischer, M., & Hosseini Boldaji, S. A. (2020). A faunistic study on braconid wasps (Hymenoptera: Braconidae) from Iran. *Boletin de la Sociedad Andaluza de Entomologia, 30*, 96–102.

Salehipour, H., Mossadegh, M. S., & Starý, P. (2010). Aphid parasitoids, individual patterns and combination of aphid parasitoid species at two different sampling periods in Khuzestan, southwest Iran. In *Proceedings of the 19th Iranian plant protection congress* (p. 157). Iranian Research Institute of Plant Protection.

Salman, F. A. A. (2006). Incidence of cereal aphids and seasonal abundance of their parasitoids in wheat fields in Sohag (Upper Egypt). *Assiut Journal of Agricultural Science, 37*, 211–220.

Samara, R. Y., & Qubbaj, T. A. (2012). Preliminary study of some natural enemies of the Tulkarm-Northern West-Bank and their aphid-plant associations. *International Journal of Agronomy and Plant Protection, 3*(4), 123–127.

Samin, N., Ghahari, H., Gadallah, N. S., & Davidian, E. (2014). A study on the Braconidae (Hymenoptera: Ichneumonoidea) from West Azarbaijan province, Northwestern Iran. *Linzer Biologische Beiträge, 46/2*, 1447–1478.

Samin, N., Coronado-Blanco, J. M., Kavallieratos, N. G., Fischer, M., & Sakenin, H. (2018). Recent Findings on Braconidae (Hymenoptera: Ichneumonoidea) of Iran with an updated checklist. *Acta Biologica Turcica, 31*(4), 160–173.

Samin, N., Coronado-Blanco, J. M., Fischer, M., van Achterberg, C., Sakenin, H., & Davidian, E. (2018). Updated checklist of Iranian Braconidae (Hymenoptera: Ichneumonoidea) with twenty-three new records. *Natura Somogyiensis, 32*, 21–36.

Samin, N., Beyarslan, A., Coronado-Blanco, J. M., & Navaeian, M. (2020). A contribution to the braconid wasps (Hymenoptera: Braconidae) from Iran. *Natura Somogyiensis, 35*, 25–28.

Samin, N., Beyarslan, A., Ranjith, A. P., Ahmad, Z., Sakenin Chelav, H., & Hosseini Boldaji, S. A. (2020). A faunistic study on Braconidae (Hymenoptera: Ichneumonoidea) from Ardebil and east Azarbayjan provinces, Northwestern Iran. *Egyptian Journal of Plant Protection Research Institute, 3*(4), 955–963.

Sanati, S., Takalloozadeh, H. M., & Ahmadi, K. (2012). *Brevicoryne brassicae* Linnaeus (Aphididae) different stages preference by *Diaeretiella rapa* McIntosch (Hym: Aphidiidae). In *Proceedings of the 20th Iranian plant protection congress* (p. 835). University of Shiraz.

Schlinger, E. I., & Mackauer, M. (1963). Identify, distribution and hosts of *Aphidius matricariae* Haliday, an important parasite of the green peach aphid, *Mysus persicae* (Hymenoptera: Aphidiidae - Homoptera: Aphidoidea). *Annals of the Entomological Society of America, 65*, 648–653.

Shahbaz Gahroee, M. R., Rasekh, A., & Michaud, J. P. (2018). Life table parameters of the black bean aphid, *Aphis fabae* Scopoli, change in response to maternal parasitism. In *Proceedings of the 23rd Iranian plant protection congress* (p. 1015). Gorgan University of Agricultural Sciences and Natural Resources.

Shahrokhi, S., Shojai, M., Rezwani, A., & Ostovan, H. (2004). Introduction of wheat aphids and their parasitoids (Hym., Aphidiidae) in Varamin region of Iran. In *Proceedings of the 23rd Iranian plant protection congress* (p. 52). Gorgan University of Agricultural Sciences and Natural Resources.

Shaw, M. R., & Huddleston, T. (1991). Classification and biology of braconid wasps (Hymenoptera: Braconidae). *Handbooks for the Identification of British Insects, 7*, 1–126.

Shi, M., & Chen, X. X. (2005). Molecular phylogeny of the Aphidiinae (Hymenoptera: Braconidae) based on DNA sequences of 16S rRNA, 18S rDNA and ATPase 6 genes. *European Journal of Entomology, 102*(2), 133–138.

Shi, M., Chen, X. X., & van Achterberg, C. (2005). Phylogenetic relationships among the Braconidae (Hymenoptera: Ichneumonoidea) inferred from partial 16S rDNA, 28S rDNA D2, 18S rDNA gene sequences and morphological characters. *Molecular Phylogenetics and Evolution, 37*(1), 104–116.

Shojai, M. (1998). *Entomology (Ethology, social life and natural enemies) (Biological control)*. Tehran University Publication, 550 pp. (in Persian).

Shojai, M., Gamintchi, A., Mesbah, M., Sadeghi, H., Abbasipour, H., Labafi, Y., & Nasrollahi, A. (1999). A research in bioecology on *Myzus* aphids and effective factors for their biocontrol in the tobacco fields *Myzus* spp. (Hom.: Aphididae). *Journal of the Agricultural Sciences, 5*, 5–35 (in Persian, English summary).

Shojai, M., Ostovan, H., Zamanizadeh, H., Labbafi, Y., Nasrollahi, A., Ghasemzadeh, M., & Rajabi, M. Z. (2003). The management on the intercropping of cucumber and tomato, with the implementation of non-chemical and reasonably control of pests, and diseases for organic crop production in greenhouse. *Journal of the Agricultural Sciences, 9*(2), 1–39 (in Persian, English summary).

Shojai, M., Ostovan, H., Darvish, F., Tirgari, S., Labbafyi, Y., & Rajabi, M. Z. (2005). Technology of biological control and pollination of Iranian cucumber cultivar in protected cultivation and organic production crop. *Journal of the Agricultural Sciences, 11*(1), 69–104 (in Persian, English summary).

Sobhy, H. M., El-Heneidy, A. H., Abd-El-Wahed, S. M. N., & Mikhail, W. Z. A. (2004). Seasonal occurrence of aphid parasitoids *Aphidius colemani* Viereck (Hymenoptera: Aphidiidae) in the Middle Delta Egypt. *Egyptian Journal of Biological Pest Control, 14,* 213–216.

Starý, P. (1969). Aphid-ant-parasite relationship in Iraq. *Insect Sociaux, 16*(4), 269–277.

Starý, P. (1970). *Biology of aphid parasites (Hymenoptera; Aphidiidae) with respect to integrated control. The Hagne, The Netherlands,* 643 pp.

Starý, P. (1975). *Aphidius colemani* Viereck: its taxonomy, distribution, and host range. *Acta Entomologica Bohemoslov, 72,* 56–163.

Starý, P. (1976). *Aphid parasites (Hymenoptera: Aphidiidae) of the Mediterranean area.* Prague: Dr. W. Junk b.v., The Hague Academia, 95 pp.

Starý, P. (1979). *Aphid parasites (Hymenoptera, Aphidiidae) of the central Asian area.* Praha: Academia, 127 pp.

Starý, P. (1981). Bisosystematic synopsis of parasitoids on cereal aphids in the western Palaearctic (Hymenoptera, Aphidiidae, Homoptera, Aphididae). *Acta Entomologica Bohemoslovaca, 78,* 382–396.

Starý, P. (1990). *Trioxys brevicornis,* a new parasitoid and potential biocontrol agent of the asparagus aphid, *Brachycorynella asparagi* (Hymenoptera, Aphidiidae; Homoptera, Aphidoidea). *Acta Entomologica Behemoslovaca, 87*(2), 87–96.

Starý, P. (2002). Field establishment of *Aphidius colemani* Viereck (Hymenoptera: Braconidae: Aphidiinae) in the Czech Republic. *Journal of Applied Entomology, 126,* 405–408.

Starý, P., & Erdelen, C. (1982). Aphid parasitoids (Hym.: Aphidiidae, Aphelinidae) from the Yemen Arab Republic. *Entomophaga, 27,* 105–108.

Starý, P., & González, D. (1978). Parasitoid spectrum of *Acyrthosiphon* aphids in central Asia (Hymenoptera: Aphidiidae). *Scandinavian Entomology, 9,* 140–145.

Starý, P., & Kaddou, I. K. (1971). Fauna and distribution of aphid parasites (Hym. Aphidiidae) in Iraq. *Acta Faunistica Entomologica Musei Nationalis Prague, 14,* 179–198.

Starý, P., & Stechmann, D. H. (1990). *Ephedrus cerasicola* Starý, a new biocontrol agent of the banana aphid, *Pentalonia nigronervosa* Coq. (Hym., Aphidiidae; Hom., Aphididae). *Journal of Applied Entomology, 109,* 457–462.

Starý, P., González, D., & Hall, J. (1980). *Aphidius eadyi* n. sp. (Hymenoptera: Aphidiidae), a widely distributed parasitoid of pea aphid *Acyrthosiphon pisum* (Harris) in the Palearctic. *Scandinavian Entomology, 11,* 473–480.

Starý, P., Tomanović, Z., & Petrović, O. (1998). A new parasitoid of root-feeding aphids from the Balkan Mountains (Hymenoptera, Braconidae, Aphidiinae). *Mitteilungen aus dem Museum für Naturkunde in Berlin Deutsche Entomologische Zeitschrift, 45*(2), 175–179.

Starý, P., Remaudière, G., González, D., & Shahrokhi, S. (2000). A review and host associations of aphid parasitoids (Hymenoptera: Braconidae: Aphidiinae) of Iran. *Parasitica (Gembloux), 56,* 15–41.

Starý, P., Rakhshani, E., & Talebi, A. A. (2005). Parasitoids of aphid pests on conifers and their state as biocontrol agents in the Middle East to Central Asia on the world background (Hym., Braconidae,
Aphidiinae; Hom., Aphididae). *Egyptian Journal of Biological Pest Control, 15*(2), 147–151.

Starý, P., Rakhshani, E., Tomanović, Ž., Kavallieratos, N. G., & Havelka, J. (2013). Order Hymenoptera, family Braconidae - Aphidiinae aphid parasitoids of the Arabian Peninnsula. In A. van Harten (Ed.), *Arthropod fauna of UAE* (Vol. 5, pp. 407–425). Al Amal Printing Press.

Abu Dhabi Summers, C. G. (1998). Integrated pest management in forage alfalfa. *Integrated Pest Management Reviews, 3,* 127–154.

Taheri, S., & Rakhshani, E. (2013). Identification of aphid parasitoids (Hym., Braconidae, Aphidiinae) and determination of their host relationships in Southern Zagros. *Journal of Plant Protection, 27*(1), 85–95 (in Persian, English summary).

Tahriri, S., Talebi, A. A., Fathipour, Y., & Zamani, A. A. (2007). Host stage preference, functional response and mutual interference of *Aphidius matricariae* (Hym.: Braconidae: Aphidiinae) on *Aphis fabae* (Hom.: Aphididae). *Entomological Science, 10,* 323–331.

Tahriri, S., Talebi, A. A., Fathipour, Y., & Zamani, A. A. (2010). Life history and demographic parameters of *Aphis fabae* (Hemiptera: Aphididae) and its parasitoid, *Aphidius matricariae* (Hymenoptera: Aphidiidae) on four sugar beet cultivars. *Acta Entomologica Serbica, 15*(1), 61–73.

Takada, H. (1998). A review of *Aphidius colemani* (Hymenoptera: Braconidae: Aphidiinae) and closely related species indigenous to Japan. *Applied Entomology and Zoology, 33,* 59–66.

Takallozadeh, H. M. (2003). *Biology, population dynamics and seasonal parasitism of Aphis craccivora and study on interaction samong plant hosts, aphid and major parasitoid in Kerman* (Ph. D thesis of Entomology). Tarbiat Modarres University, 177 pp.

Talebi, A. A., Rakhshani, E., Fathipour, Y., Starý, P., Tomanović, Ž., & Rajabi-Mazhar, N. (2009). Aphids and their parasitoids (Hym., Braconidae: Aphidiinae) associated with medicinal plants in Iran. *American-Eurasian Journal of Sustainable Agriculture, 3*(2), 205–219.

Talhouk, A. S. (1961). Records of entomophagous insects from Lebanon. *Entomophaga, 6,* 207–209.

Tazerouni, Z., Talebi, A. A., & Rakhshani, E. (2011a). The foraging behavior of *Diaeretiella rapae* (Hymenoptera: Braconidae) on *Diuraphis noxia* (Hemiptera: Aphididae). *Archives of Biological Sciences, 63*(1), 225–234.

Tazerouni, Z., Talebi, A. A., & Rakhshani, E. (2011b). Biological control of wheat aphids in Iran. *Proceedings of the Biological Control Development Congress in Iran,* 64–74.

Tazerouni, Z., Talebi, A. A., & Rakhshani, E. (2012a). Comparison of development and demographic parameters of *Diuraphis noxia* (Hem., Aphididae) and its parasitoid, *Diaeretiella rapae* (Hym., Braconidae: Aphidiinae). *Archives of Phytopathology and Plant Protection, 45*(8), 886–897.

Tazerouni, Z., Talebi, A. A., & Rakhshani, E. (2012b). Temperature-dependent functional response of *Diaeretiella rapae* (Hymenoptera: Braconidae), a parasitoid of *Diuraphis noxia* (Hemiptera: Aphididae). *Journal of the Entomological Research Society, 14*(1), 31–40.

Tazerouni, Z., Talebi, A. A., & Rakhshani, E. (2012c). Effect of temperature on biological characteristics and population growth parameters of *Diaeretiella rapae*, parasitoid of Russian wheat aphid, *Diuraphis noxia*. *Iranian Journal of Plant Protection Science, 43*(1), 83−95.

Tazerouni, Z., Talebi, A. A., & Rakhshani, E. (2013). Temperature thresholds and thermal requirements for development of Iranian *Diuraphis noxia* population (Hemiptera: Aphididae) on wheat. *Zoology and Ecology, 23*(4), 323−329.

Tazerouni, Z., Talebi, A. A., Fathipour, Y., & Soufbaf, M. (2017). Interspecific interaction between *Aphidius matricariae* and *Praon volucre* (Hym., Braconidae) the parasitoids of *Myzus persicae* (Hem., Aphididae). In *Proceedings of the 2nd Iranian International congress of entomology* (p. 148). University of Tehran, College of Agriculture and Natural Resources.

Tazerouni, Z., Talebi, A. A., & Fathipour, Y. (2018). Biological characteristics of *Aphidius matricariae* and *Praon volucre* on *Myzus persicae* influenced by interspecific interaction between parasitoid wasps. In *Proceedings of the 23rd plant protection congress* (p. 1027). Gorgan University of Agricultural Sciences and Natural Resources.

Tomanović, Ž., Kavallieratos, N. G., Athanassiou, C. G., & Stanisavljević, L. Z. (2003). A review of the West Palaearctic aphidiines (Hymenoptera, Braconidae: Aphidiinae) parasitic on *Uroleucon* spp., with the description of a new species. *Annales de la Société Entomologique de France (N.S.), 39*(4), 343−353.

Tomanović, Ž., Rakhshani, E., Starý, P., Kavallieratos, N. G., Stanisavljević, L.Ž., Žikić, V., & Athanassiou, C. G. (2007). Phylogenetic relationships between the genera *Aphidius* Nees and *Lysaphidus* (Hymenoptera: Braconidae: Aphidiinae) with description of *Aphidius iranicus* sp. nov. *The Canadian Entomologist, 139*, 297−307.

Tomanović, Ž., Beyerslan, A., Çetin Erdoğan, Ö., & Žikić, Ž. (2008). New records of aphid parasitoids (Hymenoptera: Braconidae: Aphidiinae) from Turkey. *Periodicum Biologorum, 110*(4), 335−338.

Tomanović, Ž., Starý, P., Kavallieratos, N. G., Gagić, V., Plećaš, M., Janković, M., Rakhshani, E., Ćetković, A., & Petrović, A. (2012). Aphid parasitoids (Hymenoptera: Braconidae: Aphidiinae) in wetland habitats in western Palaearctic: key and associated aphid parasitoid guilds. *Annales de la Société Entomologique de France (N.S.), 48*(1−2), 189−198.

Tomanović, Z., Petrović, A., Mitrović, M., Kavallieratos, N. G., Starý, P., Rakhshanipour, M., Popović, A., Shushuk, A. H., & Ivanović, A. (2014). Molecular and morphological variability within the *Aphidius colemani* group with redescription of *Aphidius platensis* Brethes (Hymenoptera: Braconidae: Aphidiinae). *Bulletin of Entomological Research, 104*(5), 552−565.

Tomanović, Ž., Mitrović, M., Petrović, A., Kavallieratos, N. G., Zikić, V., Ivanović, A., Rakhshani, E., Starý, P., & Vorburger, C. (2018). Revision of the European *Lysiphlebus* species (Hymenoptera: Braconidae: Aphidiinae) on the basis of COI and 28SD2 molecular markers and morphology. *Arthropod Systematics and Phylogeny, 76*(2), 179−213.

Tremblay, E., & Pennacchio, F. (1985). Taxonomic status of some species of the genus *Praon* Haliday (Hymenoptera: Braconidae: Aphidiinae). *Bollettino del Laboratorio di Entomologia Agraria Filippo Silvestri, 42*, 143−147.

Tremblay, E., Kawar, N., & Barbagallo, S. (1985). Aphids (Homoptera: Aphidoidea) and aphidiines (Hymenoptera: Braconidae) of Lebanon. *Bollittino del Laboratorio di Entomologia Agraria 'Filippo Silvestri' Portici, 42*, 19−32.

Tozlu, G., Gültekin, L., Hayat, R., & Güçlü, Ş. (2002). Studies on the natural enemies of cabbage pests in Erzurum. In H. Özbek, Ş. Güçlü, & R. Hayat (Eds.), *Proceedings of the 5th Turkish National Congress of biological control* (pp. 227−235). Erzurum: Ataturk University.

Uysal, M., Starý, P., Sahbaz, A., & Özsemerci, F. (2004). A review of aphid parasitoids (Hymenoptera: Braconidae: Aphidiinae) of Turkey. *Egyptian Journal of Biological Pest Control, 12*(2), 355−370.

van Achterberg, C. (1993). Illustrated key to the subfamilies of the Braconidae (Hymenoptera: Ichneumonoidea). *Zoologische Verhandelingen, 283*, 198 pp.

van Achterberg, C., & Quicke, D. L. J. (1992). Phylogeny of the subfamilies of the family Braconidae: a reassessment assessed. *Cladistics, 8*, 237−264.

van den Bosch, R. (1957). The spotted alfalfa and its parasites in Mediterranean region, Middle East and East Africa. *Journal of Economic Entomology, 50*, 352−356.

van den Bosch, R., & Messenger, P. S. (1973). *Biological control*. New York and London: Intext Educational Publishers, 108 pp.

van den Bosch, R., Frazer, B. D., Davis, C. S., Messenger, P. S., & Hom, R. (1970). *Trioxys pallidus* - an effective new walnut aphid parasite from Iran. *California Agriculture, 24*(11), 8−10.

van Lenteren, J. C. (2012). The state of commercial augmentative biological control: plenty of natural enemies, but a frustrating lack of uptake. *BioControl, 57*, 1−20.

van Schelt, J., Douma, J. B., & Raveneberg, W. J. (1990). Recent developments in the control of aphids in sweet pepper and cucumber. *IOBC-WPRS Bulletin, 13*, 190−193.

Wang, X. G., & Messing, R. H. (2006). Potential host range of the newly introduced aphid parasitoid *Aphidius transcaspicus* (Hymenoptera: Braconidae) in Hawaii. In , *38. Proceedings of the Hawaiian Entomological Society* (pp. 81−86).

Wilkinson, D. S. (1926). Entomological notes. *Cyprus Agricultural Journal, 21*, 47−48.

Wool, D., & Burstein, M. (1991). Parasitoids of the gall-forming aphid *Smynthurodes betae* [Aphidoidea: Fordinae] in Israel. *BioControl, 36*(4), 531−538.

Yaghubi, P. (1997). *Faunal study of Aphidiinae in Gilan province* (M.Sc thesis) (p. 105). Faculty of Science, University of Tehran.

Yaghubi, P., & Sahragard, A. (1998). Introduction of Aphidiidae (parasitoid of aphids) in Guilan province. In *Proceedings of the 13th Iranian plant protection congress* (p. 257). Karaj Junior College of Agriculture.

Yoldaş, Z., Güncan, A., & Koçlu, T. (2011). Seasonal occurrence of aphids and their natural enemies in Satsuma Mandarin orchards in Izmir Turkey. *Turkish Journal of Entomology, 35*(1), 59−74.

Yu, D. S., van Achterberg, C., & Horstmann, K. (2016). *Taxapad 2016, Ichneumonoidea 2015, database on flash-drive. Ottawa, Ontario, Canada.*

Zaldívar-Riverón, A., Mori, M., & Quicke, D. L. J. (2006). Systematics of the cyclostome subfamilies of braconid parasitic wasps (Hymenoptera: Ichneumonoidea): a simultaneous molecular and morphological Baysian approach. *Molecular Phylogenetics and Evolution, 38*(1), 130–145.

Zamani, A. A., Talebi, A. A., Fathipour, Y., & Baniameri, V. (2006). Temperature-dependent functional response of two aphid parasitoids, *Aphidius colemani* and *Aphidius matricariae* (Hymenoptera: Aphidiidae), on the cotton aphid. *Journal of Pest Science, 79*, 183–188.

Zamani, A. A., Talebi, A. A., Fathipour, Y., & Baniameri, V. (2007). Effect of temperature on life history of *Aphidius colemani* and *Aphidius matricariae* (Hymenoptera: Braconidae), two parasitoids of *Aphis gossypii* and *Myzus persicae* (Homoptera: Aphididae). *Environmental Entomology, 36*(2), 263–271.

Zamani, A. A., Haghani, M., & Kheradmand, K. (2012). Effect of temperature on reproductive parameters of *Aphidius colemani* and *Aphidius matricariae* (Hymenoptera: Braconidae) on *Aphis gossypii* (Hemiptera: Aphididae) in laboratory conditions. *Journal of Crop Protection, 1*(1), 35–40.

Zareh, N., Gonzalez, D., Ahmadi, A., Esmaili, M., Maleki-Milani, H., Vafabakhsh, J., Gilstrap, F., Starý, P., Woolley, J. B., & Thompson, F. C. (1995). A search for the Russian wheat aphid, *Diuraphis noxia* and its natural enemies in Iran. In *Proceedings of the 12th Iranian plant protection congress* (p. 12). Karaj: Karaj Junior College of Agriculture.

Žikić, Ž., Tomanović, Ž., Ivanović, A., Kavallieratos, N. G., Starý, P., Stanisavljević, L. Z., & Rakhshani, E. (2009). Morphological characterization of *Ephedrus persicae* biotypes (Hymenoptera: Braconidae: Aphidiinae) in the Palaearctic. *Annals of the Entomological Society of America, 102*(1), 1–11.

Schizoprymnus sp. (Brachistinae), ♀, lateral habitus. *Photo prepared by S.R. Shaw.*

Blacus ruficornis (Nees, 1811) (Brachistinae), ♀, lateral habitus. *Photo prepared by S.R. Shaw.*

Chapter 5

Subfamily Brachistinae Foerster, 1863

Hassan Ghahari[1], Donald L.J. Quicke[2], Neveen Samy Gadallah[3] and Scott Richard Shaw[4]

[1]Department of Plant Protection, Yadegar-e Imam Khomeini (RAH) Shahre Rey Branch, Islamic Azad University, Tehran, Iran; [2]Integrative Ecology Laboratory, Department of Biology, Faculty of Science, Chulalongkorn University, Pathumwan, Bangkok, Thailand; [3]Entomology Department, Faculty of Science, Cairo University, Giza, Egypt; [4]UW Insect Museum, Department of Ecosystem Science and Management, University of Wyoming, Laramie, WY, United States

Introduction

The Brachistinae, up until recently regarded as a tribe of Helconinae (e.g., Čapek, 1970; Mason, 1974), is a large subfamily of small to large-sized parasitoid wasps with about 794 described species worldwide (Sharanowski et al., 2011; Yu et al., 2016). There are 41 brachistine genera classified into five tribes: Blacini Foerster, 1863, Brachistini Foerster, 1863, Brulleiini van Achterberg, 1983, Diospilini Foerster, 1863, and Tainitermini van Achterberg, 2001 (Chen & van Achterberg, 2019; Sharanowski et al., 2011, 2014). As far as known, brachistines are small koinobiont endoparasitoids of coleopteran larvae, especially of the families Brentidae (Apioninae), Anobiidae, Cerambycidae, Chrysomelidae, Curculionidae, and Mordellidae (Yu et al., 2016). Hosts also include some destructive agricultural pests (Belokobylskij, 1998; Belokobylskij et al., 2004; Sharanowski et al., 2014). A few species of them have been recorded in association with other insect taxa including Diptera, Hymenoptera, and Lepidoptera (Shaw & Huddleston, 1991; Tobias, 1986; Yu et al., 2016) though many of these are dubious and all refer to substrate rearings.

Recognition of Brachistinae is difficult. Some have two submarginal cells in fore wing, others only one; the metasoma is generally short and broad (Shaw & Huddleston, 1991). Some genera (as *Triaspis* and *Schizoprymnus*) have the first tergite forming an unarticulated carapace (either with two transverse furrows in the former genus or none in the later) concealing beneath the rest of metasomal tergites; and the ovipositor is distinctly long. Both resemble chelonines but differ in wing venation and ovipositor length (Shaw & Huddleston, 1991). *Foerteria* and *Polydegmon* also have the three-segmented carapace, but in these there is a flexible articulation between the first tergite and the second (van Achterberg, 1990).

The tribe Blacini Foerster, 1863 was critically revised by van Achterberg (1976) who included it in the subfamily Helconinae. He later treated it as a distinct subfamily (Blacinae), reporting a total of 165 species, of which 53 were newly described, and the status of several genera and subgenera were also adjusted (van Achterberg, 1988). Based on molecular studies, Sharanowski et al. (2011) included the tribe Blacini within the subfamily Brachistinae, together with the tribes Brachistini, Brulleiini, and Diospilini.

The tribe Tainitermini was first described and illustrated under the subfamily Euphorinae by van Achterberg & Shaw (2001), with the single genus *Tainiterma*. A phylogenetic study based on a molecular study was conducted by Stigenberg et al. (2015). In their study, Stigenberg et al. (2015) treated the Tainitermini as "unplaced" based on insufficient data; however, they excluded it from Euphorinae, and noted a possible sister-group relationship with *Diospilus*. Chen and van Achterberg (2019) later reclassified Tainitermini into the subfamily Brachistinae based on the work of Stigenberg et al. (2015). However, that action contradicts Stigenberg's conclusion that *Tainiterma* should remain "unplaced" because of incomplete data and too few outgroup taxa in the analysis to obtain a more conclusive placement.

Checklists of Regional Brachistinae Modarres Awal (1997, 2012) represented one species, *Triaspis thoracica* (Curtis, 1860). Fallahzadeh and Saghaei (2010) listed two species in one genus (without precise localities). Gadallah and Ghahari (2013) cataloged 17 species in four genera (Brachistinae without Blacini). Farahani et al. (2016) listed

33 species in five genera [Brachistinae (18 species in four genera) + Blacinae (15 species in single genus)]. Yu et al. (2016) represented 21 species in six genera of Brachistinae, and 15 species in single genus of Blacinae. Samin, Coronado-Blanco, Kavallieratos et al. (2018) listed 22 species in five genera of Brachistinae and 15 species in one genus of Blacinae; Samin, Coronado-Blanco, Fischer et al. (2018) represented 24 species in six genera of Brachistinae and 15 species in one genus of Blacinae; Gadallah and Ghahari (2019) cataloged 18 species in the genus *Blacus* (tribe Blacini). The present checklist includes 100 species in the Middle Eastern countries classified into ten genera and three tribes (Blacini, Brachistini, and Diospilini). *Eubazus lepidus* is here recorded for the first time for the Iranian fauna. Here we follow the classification used by Sharanowski et al. (2011, 2014), and Chen and van Achterberg (2019). General distribution of species is based on Yu et al. (2016) or upon others (see additional references given below).

List of species of the subfamily Brachistinae recorded in the Middle East

Subfamily Brachistinae Foerster, 1863

Tribe Blacini Foerster, 1863

Genus *Blacus* Nees von Esenbeck, 1819

Blacus (Tarpheion) achterbergi Haeselbarth, 1976
Catalogs with Iranian records: Farahani et al. (2016), Samin, Coronado-Blanco, Kavallieratos et al. (2018), Samin, Coronado-Blanco, Fischer et al. (2018), Gadallah and Ghahari (2019).
Distribution in Iran: Mazandaran (Farahani et al., 2013).
Distribution in the Middle East: Iran.
Extralimital distribution: Austria, Czech Republic, Finland, Hungary, Italy, Korea, Moldova, Poland, Russia, Switzerland, Ukraine.
Host records: Unknown.

Blacus (Ganychorus) armatulus Ruthe, 1861
Catalogs with Iranian records: Farahani et al. (2016), Yu et al. (2016), Samin, Coronado-Blanco, Kavallieratos et al. (2018), Samin, Coronado-Blanco, Fischer et al. (2018), Gadallah and Ghahari (2019).
Distribution in Iran: Guilan (Ghahari, Fischer, & Tobias, 2012; Farahani et al., 2013), Mazandaran, Qazvin (Farahani et al., 2013), Semnan (Samin, Fischer et al., 2015).
Distribution in the Middle East: Iran (see references above), Turkey (Çetin Erdoğan & Beyarslan, 2005; Güçlü, 2011).

Extralimital distribution: Albania, Austria, Azerbaijan, Canada, former Czechoslovakia, Denmark, Finland, France, Georgia, Germany, Hungary, Italy, Kazakhstan, Moldova, Poland, Portugal, Romania, Russia, Serbia, Sweden, Switzerland, Ukraine, United Kingdom, United States of America.
Host records: Summarized by Yu et al. (2016) as being a parasitoid of the curculionids *Barynotus* sp., and *Gymnetron campanulae* Schönherr.

Blacus (Blacus) bovistae Haeselbarth, 1973
Catalogs with Iranian records: Farahani et al. (2016), Yu et al. (2016), Samin, Coronado-Blanco, Kavallieratos et al. (2018), Samin, Coronado-Blanco, Fischer et al. (2018), Gadallah and Ghahari (2019).
Distribution in Iran: Kermanshah (Ghahari & Fischer, 2012).
Distribution in the Middle East: Iran (Ghahari & Fischer, 2012), Turkey (Güçlü, 2011).
Extralimital distribution: Austria, Bulgaria, former Czechoslovakia, France, Greece, Hungary, Italy, Switzerland, Tunisia, Ukraine, United Kingdom.
Host records: Unknown.

Blacus (Ganychorus) capeki Haeselbarth, 1973
Distribution in the Middle East: Turkey (Çetin Erdoğan, 2010).
Extralimital distribution: Austria, former Czechoslovakia, Hungary, Korea, Moldova, Romania, Serbia, Slovakia, Slovenia.
Host records: Unknown.

Blacus (Ganychorus) conformis Wesmael, 1835
Catalogs with Iranian records: No catalog.
Distribution in Iran: Guilan (Sakenin et al., 2020 under Blacinae).
Distribution in the Middle East: Iran (Sakenin et al., 2020), Turkey (Haeselbarth, 1973).
Extralimital distribution: Austria, Belgium, France, Germany, Hungary, Italy, Moldova, Switzerland.
Host records: Summarized by Yu et al. (2016) as being a parasitoid of the curculionids *Gymnetron* sp., and *Rynchaenus* sp.

Blacus (Ganychorus) diversicornis (Nees von Esenbeck, 1834)
Catalogs with Iranian records: Farahani et al. (2016), Yu et al. (2016), Samin, Coronado-Blanco, Kavallieratos et al. (2018), Samin, Coronado-Blanco, Fischer et al. (2018), Gadallah and Ghahari (2019).

Distribution in Iran: Lorestan (Ghahari, Fischer, Papp et al., 2012).

Distribution in the Middle East: Iran (Ghahari, Fischer, Papp et al., 2012), Turkey (Çetin Erdoğan & Beyarslan, 2005; Güçlü, 2011).

Extralimital distribution: Albania, Austria, Azerbaijan, Belgium, Bulgaria, Czech Republic, Denmark, Finland, France, Georgia, Germany, Hungary, Italy, Kazakhstan, Korea, Latvia, Lithuania, Moldova, Mongolia, Netherlands, Poland, Romania, Russia, Slovenia, Spain, Sweden, Switzerland, Ukraine, United Kingdom, former Yugoslavia.

Host records: Recorded by Donisthorpe (1927) in a nest of *Formica glebaria* Nylander situated under a stone in England. Not yet recorded in Iran.

Blacus (Blacus) errans (Nees von Esenbeck, 1811)

Catalogs with Iranian records: Farahani et al. (2016), Yu et al. (2016), Samin, Coronado-Blanco, Kavallieratos et al. (2018), Samin, Coronado-Blanco, Fischer et al. (2018), Gadallah and Ghahari (2019).

Distribution in Iran: Lorestan (Ghahari, Fischer, Papp et al., 2012).

Distribution in the Middle East: Iran (Ghahari, Fischer, Papp et al., 2012), Turkey (Çetin Erdoğan & Beyarslan, 2005; Güçlü, 2011).

Extralimital distribution: Armenia, Austria, Czech Republic, Denmark, Finland, France, Germany, Hungary, Italy, Latvia, Lithuania, North Macedonia, Moldova, Netherlands, Poland, Russia, Slovakia, Sweden, Switzerland, Turkmenistan, United Kingdom, Uzbekistan, former Yugoslavia.

Host records: Summarized by Yu et al. (2016) as being a parasitoid of cerambycids *Exocentrus adspersus* Mulsant, and *Pogonocherus hispidus* (L.); the curculionids *Polygraphus grandiclava* (Thomson), *Scolytus mali* (Bechstein), and *Scolytus rugulosus* (Müller); and the melyrid beetle *Dasytes plumbeus* (Müller). It was also recorded by Telenga (1935) as being a parasitoid of the cecidomyiid *Rabdophaga saliciperda* (Dufour).

Blacus (Blacus) exilis (Nees von Esenbeck, 1811)

Catalogs with Iranian records: Farahani et al. (2016), Yu et al. (2016), Samin, Coronado-Blanco, Kavallieratos et al. (2018), Samin, Coronado-Blanco, Fischer et al. (2018), Gadallah and Ghahari (2019).

Distribution in Iran: Alborz, Mazandaran (Farahani et al., 2013), Kermanshah (Ghahari & Fischer, 2012).

Distribution in the Middle East: Iran (Farahani et al., 2013; Ghahari & Fischer, 2012), Israel−Palestine (Papp, 2012), Turkey (Güçlü, 2011).

Extralimital distribution: Afghanistan, Austria, Azerbaijan, Belgium, Bulgaria, Canada, Czech Republic, Denmark, Finland, France, Georgia, Germany, Greece, Hungary, Italy, Kazakhstan, Korea, Latvia, Lithuania, Moldova,

Mongolia, Morocco, Netherlands, Norway, Poland, Portugal, Russia, Serbia, Slovakia, Spain, Sweden, Switzerland, Turkmenistan, United Kingdom, United States of America, Uzbekistan, former Yugoslavia.

Host records: Summarized by Yu et al. (2016) as being a parasitoid of the curculionids *Magdalis armigera* (Geoffroy), and *Pityokteines vorontzovi* (Jacobson); and the ptenid *Ernobius nigrinus* (Sturm).

Blacus (Blacus) filicornis Haeselbarth, 1973

Catalogs with Iranian records: Farahani et al. (2016), Yu et al. (2016), Samin, Coronado-Blanco, Kavallieratos et al. (2018), Samin, Coronado-Blanco, Fischer et al. (2018), Gadallah and Ghahari (2019).

Distribution in Iran: Guilan (Ghahari, Fischer, & Tobias, 2012).

Distribution in the Middle East: Iran (Ghahari, Fischer, & Tobias, 2012), Turkey (Çetin Erdoğan, 2010; Güçlü, 2011).

Extralimital distribution: Austria, former Czechoslovakia, Denmark, Finland, France, Germany, Greece, Hungary, Italy, Kazakhstan, Latvia, Mongolia, Poland, Romania, Russia, Sweden, Switzerland, United Kingdom.

Host records: Unknown.

Blacus (Blacus) forticornis Haeselbarth, 1973

Catalogs with Iranian records: No catalog.

Distribution in Iran: West Azarbaijan (Samin et al., 2019).

Distribution in the Middle East: Iran (Samin et al., 2019), Turkey (Çetin Erdoğan & Beyarslan, 2015).

Extralimital distribution: Czech Republic, Denmark, Hungary, Poland, Russia, United Kingdom.

Host records: Unknown.

Blacus (Blacus) hastatus Haliday, 1835

Catalogs with Iranian records: Farahani et al. (2016), Yu et al. (2016), Samin, Coronado-Blanco, Kavallieratos et al. (2018), Samin, Coronado-Blanco, Fischer et al. (2018), Gadallah and Ghahari (2019).

Distribution in Iran: Lorestan (Ghahari, Fischer, Papp et al., 2012), Semnan (Naderian et al., 2012).

Distribution in the Middle East: Iran (Ghahari, Fischer, Papp et al., 2012; Naderian et al., 2012), Turkey (Çetin Erdoğan & Beyarslan, 2005; Güçlü, 2011).

Extralimital distribution: Austria, Croatia, Czech Republic, Denmark, Finland, Germany, Hungary, Italy, Lithuania, Netherlands, Poland, Russia, Serbia, Slovakia, Sweden, Switzerland, Ukraine, United Kingdom.

Host records: Summarized by Yu et al. (2016) as being a parasitoid of the nitidulid *Meligethes* sp.; and the curculionid *Stereonychus fraxini* (DeGeer).

Blacus (Blacus) humilis (Nees von Esenbeck, 1811)

Catalogs with Iranian records: Farahani et al. (2016), Yu et al. (2016), Samin, Coronado-Blanco, Kavallieratos et al.

(2018), Samin, Coronado-Blanco, Fischer et al. (2018), Gadallah and Ghahari (2019).

Distribution in Iran: Guilan (Farahani et al., 2013), Kermanshah (Ghahari & Fischer, 2012).

Distribution in the Middle East: Cyprus (Haeselbarth, 1973), Iran (Ghahari & Fischer, 2012; Farahani et al., 2013), Turkey (Güçlü, 2011).

Extralimital distribution: Austria, Belgium, Bulgaria, Canada, Croatia, Czech Republic, Denmark, Finland, France, Germany, Hungary, India, Italy, Japan, Korea, Latvia, Lithuania, Mongolia, Netherlands, Poland, Portugal, Russia, Slovenia, Sweden, Switzerland, Ukraine, United Kingdom, United States of America, former Yugoslavia.

Host records: Summarized by Yu et al. (2016) as being a parasitoid of the cryptophagid *Cryptophagus lycoperdi* (Scopoli); the curculionids *Cionus thapsus* (Fabricius), *Gymnetron campanulae* Schönherr, *Tomicus piniperda* (L.); and the ptinid *Stegobium paniceum* (L.). It was also reported by Rondani (1876) as being a parasitoid of the noctuid moth *Mamestra brassicae* (L.).

Blacus (*Blacus*) instabilis Ruthe, 1861
Catalogs with Iranian records: No catalog.

Distribution in Iran: Golestan (Ghahari et al., 2020).

Distribution in the Middle East: Iran (Ghahari et al., 2020), Turkey (Haeselbarth, 1975).

Extralimital distribution: Austria, Azerbaijan, Bulgaria, Czech Republic, Denmark, Finland, Germany, Hungary, India, Ireland, Italy, Korea, Latvia, Lithuania, Moldova, Mongolia, Norway, Poland, Russia, Slovakia, Sweden, Switzerland, Turkmenistan, Ukraine, United Kingdom.

Host records: Unknown.

Blacus (*Blacus*) interstitialis Ruthe, 1861
Catalogs with Iranian records: Farahani et al. (2016), Yu et al. (2016), Samin, Coronado-Blanco, Kavallieratos et al. (2018), Samin, Coronado-Blanco, Fischer et al. (2018), Gadallah and Ghahari (2019).

Distribution in Iran: Guilan (Ghahari, Fischer, & Tobias, 2012; Farahani et al., 2013).

Distribution in the Middle East: Iran (Ghahari, Fischer, & Tobias, 2012; Farahani et al., 2013), Turkey (Çetin Edoğan & Beyarslan, 2005; Güçlü, 2011).

Extralimital distribution: Afghanistan, Austria, Czech Republic, Denmark, Finland, Germany, Hungary, Morocco, Netherlands, Poland, Slovakia, Sweden, Ukraine, United Kingdom.

Host records: Recorded by Fischer (1963) as being a parasitoid of the chloropid *Oscinella frit* (L.).

Blacus (*Blacus*) leptostigma Ruthe, 1861
Distribution in the Middle East: Turkey (Çetin Erdoğan & Beyarslan, 2005).

Extralimital distribution: Czech Republic, Germany, Hungary, Ireland.

Host records: Unknown.

Blacus (*Blacus*) longipennis (Gravenhorst, 1807)
Catalogs with Iranian records: Gadallah and Ghahari (2019).

Distribution in Iran: West Azarbaijan (Ghahari, 2016).

Distribution in the Middle East: Iran.

Extralimital distribution: Austria, Bulgaria, Czech Republic, Denmark, Finland, Germany, Hungary, Kazakhstan, Latvia, Netherlands, Norway, Poland, Russia, Slovakia, Sweden, Switzerland, United Kingdom.

Host records: Summarized by Yu et al. (2016) as being a parasitoid of the ptelinid *Anobium* sp.; and the cerambycid *Molorchus umbellatarum* (Schreber).

Blacus (*Ganychorus*) maculipes Wesmael, 1835
Catalogs with Iranian records: Farahani et al. (2016), Yu et al. (2016), Samin, Coronado-Blanco, Kavallieratos et al. (2018), Samin, Coronado-Blanco, Fischer et al. (2018), Gadallah and Ghahari (2019).

Distribution in Iran: Kermanshah (Ghahari & Fischer, 2012).

Distribution in the Middle East: Iran (Ghahari & Fischer, 2012), Turkey (Çetin Erdoğan & Beyarslan, 2005; Çetin Erdoğan, 2010; Güçlü, 2011).

Extralimital distribution: Austria, Belgium, Bulgaria, Czech Republic, Denmark, Estonia, Finland, France, Germany, Hungary, Italy, Latvia, Lithuania, North Macedonia, Netherlands, Poland, Romania, Russia, Serbia, Slovakia, Sweden, Switzerland, United Kingdom.

Host records: Summarized by Yu et al. (2016) as being a parasitoid of the attelabid *Apoderus coryli* (L.), and the bucculatricid beetle *Bucculatrix nigricomella* (Zeller).

Blacus (*Ganychorus*) madli Haeselbarth, 1992
Distribution in the Middle East: Turkey (Haeselbarth, 1992).

Extralimital distribution: None.

Host records: Unknown.

Blacus (*Hysterobolus*) mamillanus Ruthe, 1861
Distribution in the Middle East: Cyprus (Hellén, 1958).

Extralimital distribution: Croatia, former Czechoslovakia, Denmark, Finland, Germany, Hungary, Italy, Mongolia, Netherlands, Poland, Russia, Sweden, Switzerland, United Kingdom, former Yugoslavia.

Host records: Recorded by Donisthorpe (1927) as being a parasitoid of the formicid *Lasius niger* (L.).

Blacus (*Hysterobolus*) nixoni Haeselbarth, 1973
Catalogs with Iranian records: Farahani et al. (2016), Yu et al. (2016), Samin, Coronado-Blanco, Kavallieratos et al. (2018), Samin, Coronado-Blanco, Fischer et al. (2018), Gadallah and Ghahari (2019).

Distribution in Iran: Guilan, Mazandaran (Farahani et al., 2013).

Distribution in the Middle East: Cyprus (Haeselbarth, 1973, 1975), Iran (Farahani et al., 2013), Turkey (Çetin Erdoğan & Beyarslan, 2005).

Extralimital distribution: Bulgaria, Korea, Netherlands, Portugal, Russia, Spain.

Host records: Unknown.

Blacus (Blacus) paganus Haliday, 1835

Catalogs with Iranian records: Farahani et al. (2016), Yu et al. (2016), Samin, Coronado-Blanco, Kavallieratos et al. (2018), Samin, Coronado-Blanco, Fischer et al. (2018), Gadallah and Ghahari (2019).

Distribution in Iran: Lorestan (Ghahari, Fischer, Papp et al., 2012), West Azarbaijan (Rastegar et al., 2012).

Distribution in the Middle East: Iran (Ghahari, Fischer, Papp et al., 2012; Rastegar et al., 2012), Turkey (Güçlü, 2011).

Extralimital distribution: Austria, Belgium, Bulgaria, Canada, Denmark, Finland, France, Georgia, Germany, Hungary, Ireland, Italy, Latvia, Moldova, Mongolia, Netherlands, Norway, Poland, Russia, Serbia, Slovakia, Slovenia, Sweden, Switzerland, United States of America.

Host records: Recorded by Čapek and Hofmann (1997) as being a parasitoid of the cryptophagid Antherophagus sp.

Blacus (Blacus) rufescens Ruthe, 1861

Catalogs with Iranian records: Gadallah and Ghahari (2019).

Distribution in Iran: Ardabil (Gadallah & Ghahari, 2019).

Distribution in the Middle East: Iran (Gadallah & Ghahari, 2019), Turkey (Çetin Erdoğan & Beyarslan, 2005).

Extralimital distribution: Austria, former Czechoslovakia, Denmark, Finland, Germany, Hungary, Moldova, Netherlands, Poland, Romania, Sweden, former Yugoslavia.

Host records: Unknown.

Blacus (Ganychorus) ruficornis (Nees von Esenbeck, 1811)

Catalogs with Iranian records: Farahani et al. (2016), Yu et al. (2016), Samin, Coronado-Blanco, Kavallieratos et al. (2018), Samin, Coronado-Blanco, Fischer et al. (2018), Gadallah and Ghahari (2019).

Distribution in Iran: Alborz, Guilan, Qazvin (Farahani et al., 2013), Mazandaran (Farahani et al., 2013; Kian et al., 2020).

Distribution in the Middle East: Cyprus (Haeselbarth, 1975), Iran (see references above), Turkey (Beyarslan et al., 2006).

Extralimital distribution: Afghanistan, Albania, Austria, Azerbaijan, Belgium, Bulgaria, Canada, Czech Republic, Denmark, Estonia, Finland, France, Germany, Greece, Hungary, Italy, Japan, Kazakhstan, Korea, Latvia, Lithuania, North Macedonia, Mexico, Moldova, Nepal, Netherlands, Norway, Poland, Portugal, Romania, Russia, Serbia, Slovakia, Slovenia, Spain, Sweden, Switzerland, Tunisia, United Kingdom, United States of America, Uzbekistan, former Yugoslavia.

Host records: Recorded by Tobias (1976) as being a parasitoid of the curculionid Stereonychus fraxini (DeGeer); and the staphylinid Tachyphorus obtusus (L.).

Blacus (Blacus) stelfoxi Haeselbarth, 1973

Catalogs with Iranian records: Farahani et al. (2016), Yu et al. (2016), Samin, Coronado-Blanco, Kavallieratos et al. (2018), Samin, Coronado-Blanco, Fischer et al. (2018), Gadallah and Ghahari (2019).

Distribution in Iran: Kermanshah (Ghahari & Fischer, 2012).

Distribution in the Middle East: Iran (Ghahari & Fischer, 2012), Turkey (Çetin Erdoğan & Beyarslan, 2005; Güçlü, 2011).

Extralimital distribution: Austria, Bulgaria, Finland, France, Georgia, Germany, Greece, Hungary, Mongolia, Netherlands, Romania, Russia, Sweden, Switzerland (Yu et al., 2016), England (Broad et al., 2016).

Host records: Unknown.

Blacus tripudians Haliday, 1835

Catalogs with Iranian records: No catalog.

Distribution in Iran: Ardabil (Samin, Beyarslan et al., 2020).

Distribution in the Middle East: Iran.

Extralimital distribution: Austria, Azerbaijan, Belgium, Bulgaria, Croatia, Czech Republic, Denmark, France, Georgia, Germany, Greece, Hungary, Ireland, Netherlands, Poland, Romania, Serbia, Sweden, Switzerland, Tunisia, United Kingdom.

Host records: Unknown.

Blacus sp.

Distribution in Iran: Fars (Lashkari Bod et al., 2010, 2011).

Tribe Brachistini Foerster, 1863

Genus Chelostes van Achterberg, 1990

Chelostes robustus van Achterberg, 1990

Distribution in the Middle East: Turkey (van Achterberg, 1990).

Extralimital distribution: None.

Host records: Unknown.

Chelostes subrobustus Yilmaz and Beyarslan, 2009

Distribution in the Middle East: Turkey (Yilmaz & Beyarslan, 2009).

Extralimital distribution: None.

Host records: Unknown.

Genus *Eubazus* Nees von Esenbeck, 1812

Eubazus (*Aliolus*) *lepidus* (Haliday, 1835)

Catalogs with Iranian records: This species is a new record for the fauna of Iran.

Distribution in Iran: East Azarbaijan province, Arasbaran forest, 1♀, 1♂, September 2014.

Distribution in the Middle East: Iran (new record).

Extralimital distribution: Azerbaijan, former Czechoslovakia, Finland, Georgia, Germany, Hungary, Italy, Lithuania, Netherlands, Norway, Poland, Russia, Sweden, Switzerland, United Kingdom.

Host records: Summarized by Yu et al. (2016) as being a parasitoid of the curculionids *Hylobius* sp., and *Pissodes harcyniae* (Herbst).

Eubazus (*Brachistes*) *cingulatus* (Szépligeti, 1896)

Catalogs with Iranian records: No catalog.

Distribution in Iran: Razavi Khorasan (Sakenin et al., 2020).

Distribution in the Middle East: Iran (Sakenin et al., 2020), Turkey (Güçlü & Özbek, 2011).

Extralimital distribution: Armenia, Azerbaijan, Georgia, Greece, Hungary, Moldova, Russia.

Host records: Unknown.

Eubazus (*Brachistes*) *claviventris* (Reinhard, 1867)

Distribution in the Middle East: Israel−Palestine (Papp, 2012).

Extralimital distribution: Bulgaria, former Czechoslovakia, Germany, Hungary, Italy, Moldova, Poland, Russia.

Host records: Recorded by Tobias (1976, 1986) as being a parasitoid of the curculionid *Pityogenes bidentatus* (Herbst).

Eubazus (*Brachistes*) *fasciatus* (Nees von Esenbeck, 1816)

Catalogs with Iranian records: No catalog.

Distribution in Iran: Golestan (Ghahari et al., 2019 as *Eubazus fuscipalpis* (Wesmael)).

Distribution in the Middle East: Iran (Ghahari et al., 2019), Turkey (Güçlü & Özbek, 2011).

Extralimital distribution: Belgium, Finland, France, Germany, Hungary, Ireland, Italy, Japan, Kazakhstan, Moldova, Montenegro, Poland, Romania, Russia, Slovakia, Switzerland, Turkmenistan, United Kingdom.

Host records: Unknown.

Eubazus (*Brachistes*) *fuscipes* (Herrich-Schäffer, 1838)

Catalogs with Iranian records: No catalog.

Distribution in Iran: Semnan (Ghahari et al., 2019).

Distribution in the Middle East: Iran (Ghahari et al., 2019), Turkey (Güçlü & Özbek, 2011).

Extralimital distribution: Azerbaijan, Finland, Georgia, Germany, Greece, North Macedonia, Moldova, Poland, Romania, Russia, Serbia, Turkmenistan.

Host records: Unknown.

Eubazus (*Brachistes*) *gallicus* (Reinhard, 1867)

Catalogs with Iranian records: No catalog.

Distribution in Iran: Hamadan (Sakenin et al., 2020).

Distribution in the Middle East: Iran (Sakenin et al., 2020), Israel−Palestine (Papp, 2012).

Extralimital distribution: Azerbaijan, Bulgaria, Croatia, France, Germany, Greece, Hungary, Italy, Kazakhstan, Romania, Serbia, Slovakia.

Host records: Recorded by Lukas (1981) as being a parasitoid of the curculionid *Magdalis ruficornis* Porta.

Eubazus (*Brachistes*) *gigas* (Fahringer, 1925)

Distribution in the Middle East: Turkey (Atay et al., 2019).

Extralimital distribution: Hungary, Montenegro, Romania, Serbia, Slovakia.

Host records: Unknown.

Eubazus (*Brachistes*) *minutus* (Ratzeburg, 1848)

Catalogs with Iranian records: Gadallah and Ghahari (2013), Farahani et al. (2016), Yu et al. (2016), Samin, Coronado-Blanco, Kavallieratos et al. (2018), Samin, Coronado-Blanco, Fischer et al. (2018).

Distribution in Iran: East Azarbaijan (Gadallah and Ghahari, 2013).

Distribution in the Middle East: Iran (Gadallah & Ghahari, 2013), Israel−Palestine (Papp, 2012).

Extralimital distribution: Azerbaijan, Bulgaria, China, former Czechoslovakia, Finland, Germany, Greece, Hungary, Italy, Norway, Poland, Switzerland, Ukraine, United Kingdom.

Host records: Summarized by Yu et al. (2016) as being a parasitoid of the brentid *Protapion apricans* (Herbst); and the curculionids *Rhyncaenus alni* (L.), *Rhynchaenus fagi* (L.), and *Rhynchaenus quercus* (L.).

Eubazus (*Brachistes*) *nigricoxis* (Wesmael, 1835)

Catalogs with Iranian records: Samin, Coronado-Blanco, Kavallieratos et al. (2018).

Distribution in Iran: Lorestan (Samin, Coronado-Blanco, Kavallieratos et al., 2018).

Distribution in the Middle East: Iran (Samin, Coronado-Blanco, Kavallieratos et al., 2018), Turkey (Güçlü & Özbek, 2011).

Extralimital distribution: Armenia, Belgium, Finland, France, Georgia, Germany, Greece, Hungary, Italy, Kazakhstan, Lithuania, Moldova, Poland, Romania, Russia, Sweden, Switzerland, Ukraine.

Host records: Unknown.

Eubazus (Brachistes) rufithorax (Abdinbekova, 1969)

Distribution in the Middle East: Israel−Palestine (Papp, 2012), Turkey (Güçlü & Özbek, 2011).

Extralimital distribution: Azerbaijan (Abdinbekova, 1969 as *Calyptus rufithorax*).

Host records: Unknown.

Eubazus (Eubazus) flavipes (Haliday, 1835)

Catalogs with Iranian records: No catalog.

Distribution in Iran: Guilan (Ghahari et al., 2019).

Distribution in the Middle East: Iran (Ghahari et al., 2019), Turkey (Güçlü & Özbek, 2011).

Extralimital distribution: Armenia, Austria, Bulgaria, France, Germany, Hungary, Ireland, Italy, Korea, Lithuania, Moldova, Netherlands, Poland, Russia, Slovakia, Switzerland, United Kingdom.

Host records: Recorded by Tobias (1986) and Čapek and Hofmann (1997) as being a parasitoid of the cerambycid *Exocentrus punctipennis* Mulsant and Guillebeu.

Eubazus (Eubazus) pallipes Nees von Esenbeck, 1812

Catalogs with Iranian records: Samin, Coronado-Blanco, Fischer et al. (2018).

Distribution in Iran: Kordestan, West Azarbaijan (Sakenin et al., 2018).

Distribution in the Middle East: Iran (Sakenin et al., 2018), Turkey (Güçlü & Özbek, 2011).

Extralimital distribution: Albania, Azerbaijan, Belgium, Czech Republic, France, Germany, Greece, Hungary, Ireland, Moldova, Poland, Romania, Russia, Serbia, Switzerland, Ukraine, United Kingdom, United States of America.

Host records: Summarized by Yu et al. (2016) as being a parasitoid of the bostrichids *Lyctus brunneus* (Stephens), *Lyctus carbonarius* (Waltl), and *Lyctus linearis* (Goeze); the cerambycid *Acanthocinus griseus* (Fabricius); the chrysomelid *Psylliodes chrysocephala* L.; and the curculionids *Anthonomus pomorum* (L.), *Ceutorhynchus pleurostigma* (Stephens), *Ceutorhynchus rapae* (Gyllenhal), *Gymnetron antirrhini* (Paykull), *Gymnetron tetrum* (Fabricius), and *Pissodes harcyniae* (Herbst).

Eubazus (Brachistes) parvulus (Reinhard, 1867)

Catalogs with Iranian records: No catalog.

Distribution in Iran: Kordestan (Ghahari et al., 2020).

Distribution in the Middle East: Iran.

Extralimital distribution: Azerbaijan, Czech Republic, Germany, Greece, Hungary, Kazakhstan, North Macedonia, Moldova, Poland, Romania, Slovakia, former Yugoslavia.

Host records: Unknown.

Eubzus (Brachistes) ruficoxis (Wesmael, 1835)

Catalogs with Iranian records: No catalog.

Distribution in Iran: Guilan (Ghahari & Gadallah, 2022).

Distribution in the Middle East: Iran (Ghahari & Gadallah, 2022), Turkey (Güçlü, 2011).

Extralimital distribution: Belgium, Finland, France, Germany, Greece, Hungary, Italy, Japan, Kazakhstan, Mongolia, Netherlands, Poland, Romania, Russia, Slovakia, Switzerland, Ukraine, United Kingdom.

Host records: Summarized by Yu et al. (2016) as being a parasitoid of the curculionid *Pteleobius vittatus* (Fabricius); and the rhynchitids *Bytiscus betulae* (L.), *Bytiscus populi* (L.), and *Bytiscus venustus* Sharp. In Iran, this species has been reared from the rhynchitid *Bytiscus populi* (L.) (Ghahari & Gadallah, 2022).

Eubazus (Allodorus) semirugosus (Nees von Esenbeck, 1816)

Catalogs with Iranian records: No catalog.

Distribution in Iran: Guilan, Mazandaran (Gadallah et al., 2021), Hamadan (Ghahari & Gadallah, 2022).

Distribution in the Middle East: Iran (Gadallah et al., 2021; Ghahari & Gadallah, 2022), Turkey (Güçlü & Özbek, 2011).

Extralimital distribution: Armenia, Austria, Azerbaijan, Belarus, Belgium, Bosnia-Herzegovina, Bulgaria, Canada (introduced), China, Czech Republic, Finland, France, Georgia, Germany, Hungary, Italy, Latvia, Moldova, Mongolia, Netherlands, Norway, Poland, Romania, Russia, Slovakia, Sweden, Switzerland, United Kingdom.

Host records: Summarized by Yu et al. (2016) as being a parasitoid of the buprestid *Chrysobothris solieri* Gory and Laporte; the cerambycids *Acanthocinus reticulatus* (Razoumowsky), and *Pogonocherus fasciculatus* (DeGeer); the curculionids *Dendroctonus micans* (Kugelann), *Hylurgops glabratus* Zetterstedt, *Ips sexdendatus* (Boerner and I.C.H.), *Magdalis armigera* (Geoffroy), *Magdalis frontalis* (Gyllenhal), *Onthotomicus proximus* Eichhoff, *Orchestes quercus* (L.), *Pissodes* spp., *Pityophthorus lichtensteini* (Ratzeburg), and *Tomicus piniperla* (L.); the ptinid *Ernobius nigrinus* (Sturm); the pyralid *Dioryctria abietella* (Denis and Schiffermüller); the tortricids *Rhyacionia buoliana* (Denis and Schiffermüller), and *Tischeria ekebladella* (Denis and Schiffermüller); and the xiphydriid *Xyphydria longicollis* (Geoffroy). In Iran, this species has been reared from the buprestid *Chrysobothris solieri* Gory and Laporte (Ghahari & Gadallah, 2022).

Eubazus (Brachistes) aydae Beyarslan, 2011

Distribution in the Middle East: Turkey (Beyarslan, 2011).

Extralimital distribution: None.

Host records: Unknown.

Eubazus (Brachistes) tibialis (Haliday, 1835)

Catalogs with Iranian records: Gadallah and Ghahari (2013), Farahani et al. (2016), Yu et al. (2016), Samin,

Coronado-Blanco, Kavallieratos et al. (2018), Samin, Coronado-Blanco, Fischer et al. (2018).
Distribution in Iran: Razavi Khorasan (Samin et al., 2011), Semnan (Samin, 2015).
Distribution in the Middle East: Iran (Samin, 2015; Samin et al., 2011), Turkey (Güçlü & Özbek, 2011).
Extralimital distribution: Belgium, Bulgaria, Croatia, Finland, France, Germany, Hungary, Ireland, Italy, Latvia, Lithuania, Moldova, Mongolia, Montenegro, Netherlands, Norway, Poland, Russia, Serbia, Slovakia, Sweden, Switzerland, Ukraine, United Kingdom.
Host records: Unknown.

Genus *Foersteria* Szépligeti, 1896
Foersteria longicauda van Achterberg, 1990

Catalogs with Iranian records: Gadallah and Ghahari (2013), Farahani et al. (2016), Yu et al. (2016), Samin, Coronado-Blanco, Kavallieratos et al. (2018), Samin, Coronado-Blanco, Fischer et al. (2018).
Distribution in Iran: Kuhgiloyeh and Boyerahmad (Samin, van Achterberg et al., 2015), West Azarbaijan (Ghahari & Fischer, 2011a).
Distribution in the Middle East: Iran (Ghahari & Fischer, 2011a; Samin, van Achterberg et al., 2015), Turkey (Güçlü & Özbek, 2011).
Extralimital distribution: None.
Host records: Unknown.

Foersteria polonoca Fahringer, 1934

Distribution in the Middle East: Turkey (Beyarslan & Deveci, 2019).
Extralimital distribution: Hungary, Poland, Slovakia.
Host records: Unknown.

Genus *Polydegmon* Foerster, 1863
Polydegmon foveolatus (Herrich-Schäffer, 1838)

Catalogs with Iranian records: No catalog.
Distribution in Iran: Guilan (Ghahari et al., 2019).
Distribution in the Middle East: Iran (Ghahari et al., 2019), Turkey (Güçlü & Özbek, 2011; van Achterberg, 1990).
Extralimital distribution: Bosnia-Herzegovina, Croatia, former Czechoslovakia, Germany, Greece, Hungary, Kazakhstan, Moldova, Russia, Ukraine, Uzbekistan, former Yugoslavia.
Host records: Unknown.

Polydegmon sinuatus Foerster, 1863

Catalogs with Iranian records: No catalog.
Distribution in Iran: Hamadan (Gadallah et al., 2018).
Distribution in the Middle East: Iran (Gadallah et al., 2018), Turkey (Güçlü & Özbek, 2011; van Achterberg, 1990).
Extralimital distribution: Armenia, Austria, Azerbaijan, Bulgaria, China, Georgia, Germany, Hungary, Kazakhstan,

Moldova, Montenegro, Russia, Serbia, Slovakia, Uzbekistan.
Host records: Unknown.

Genus *Schizoprymnus* Foerster, 1863
Schizoprymnus (*Schizoprymnus*) *ambiguus* (Nees von Esenbeck, 1816)

Catalogs with Iranian records: No catalog.
Distribution in Iran: Razavi Khorasan (Samin, Sakenin Chelav et al., 2020).
Distribution in the Middle East: Iran (Samin, Sakenin Chelav et al., 2020), Turkey (Güçlü & Özbek, 2011).
Extralimital distribution: Azerbaijan, Belgium, Croatia, Finland, France, Germany, Greece, Hungary, Ireland, Italy, Kazakhstan, Latvia, Lithuania, North Macedonia, Moldova, Netherlands, Poland, Russia, Serbia, Slovakia, Sweden, Tajikistan, Ukraine, United Kingdom.
Host records: Summarized by Yu et al. (2016) as being a parasitoid of the curculionid *Ceutorhynchus maculaalba* Germar; and the pyralid *Myelois circumvoluta* (Fourcroy).

Schizoprymnus (*Schizoprymnus*) *angustatus* (Herrich-Schäffer, 1838)

Catalogs with Iranian records: Gadallah and Ghahari (2013), Farahani et al. (2016), Yu et al. (2016), Samin, Coronado-Blanco, Kavallieratos et al. (2018), Samin, Coronado-Blanco, Fischer et al. (2018).
Distribution in Iran: Golestan (Samin et al., 2011), Semnan (Samin, 2015).
Distribution in the Middle East: Iran (Samin, 2015; Samin et al., 2011), Turkey (Güçlü & Özbek, 2011).
Extralimital distribution: Azerbaijan, former Czechoslovakia, Finland, Germany, Hungary, Italy, Kazakhstan, Lithuania, North Macedonia, Moldova, Mongolia, Russia, Spain, Sweden, Switzerland, Ukraine, former Yugoslavia.
Host records: Unknown.

Schizoprymnus (*Schizoprymnus*) *azerbajdzhanicus* (Abdinbekova, 1967)

Catalogs with Iranian records: No catalog.
Distribution in Iran: Alborz, Guilan, Tehran (Talebi et al., 2018), Mazandaran (Kian et al., 2020).
Distribution in the Middle East: Iran (Kian et al., 2020; Talebi et al., 2018), Turkey (Güçlü & Özbek, 2011).
Extralimital distribution: Azerbaijan, Bosnia-Herzegovina, Bulgaria, Hungary, Korea, North Macedonia, Russia, Spain, former Yugoslavia.
Host records: Unknown.

Schizoprymnus (*Schizoprymnus*) *bidentulus* (Szépligeti, 1901)

Catalogs with Iranian records: No catalog.
Distribution in Iran: Mazandaran, Qazvin (Talebi et al., 2018), Southern Khorasan (Ghahari, 2020).

Distribution in the Middle East: Iran (Ghahari, 2020; Talebi et al., 2018), Turkey (Güçlü & Özbek, 2011).
Extralimital distribution: Armenia, Bulgaria, former Czechoslovakia, Hungary, Kazakhstan, Serbia, Spain.
Host records: Unknown.

Schizoprymnus (*Schizoprymnus*) *brevicornis* (Herrich-Schäffer, 1838)

Catalogs with Iranian records: Samin, Coronado-Blanco, Fischer et al. (2018).
Distribution in Iran: Golestan, Mazandaran (Samin, Coronado-Blanco, Fischer et al., 2018).
Distribution in the Middle East: Iran (Sakenin et al., 2018), Turkey (Güçlü & Özbek, 2011).
Extralimital distribution: Former Czechoslovakia, Germany, Hungary, Italy, Kazakhstan, Moldova, Poland, Spain, Ukraine.
Host records: Recorded by Güçlü and Özbek (2011) as being a parasitoid of the curculionid *Gymnetron* sp.

Schizoprymnus (*Schizoprymnus*) *crassiceps* (Thomson, 1892)

Catalogs with Iranian records: No catalog.
Distribution in Iran: Alborz, Mazandaran (Talebi et al., 2018).
Distribution in the Middle East: Iran (Talebi et al., 2018), Turkey (Güçlü & Özbek, 2011).
Extralimital distribution: Azerbaijan, Bulgaria, Croatia, former Czechoslovakia, Germany, Hungary, Kazakhstan, Mongolia, Norway, Russia, Serbia, Sweden.
Host records: Unknown.

Schizoprymnus (*Schizoprymnus*) *elongatus* (Szépligeti, 1898)

Catalogs with Iranian records: Gadallah and Ghahari (2013), Farahani et al. (2016), Yu et al. (2016), Samin, Coronado-Blanco, Kavallieratos et al. (2018), Samin, Coronado-Blanco, Fischer et al. (2018).
Distribution in Iran: Ardabil (Ghahari & Fischer, 2011a), Golestan (Samin, Ghahari et al., 2015a), Hamadan (Samin et al., 2019).
Distribution in the Middle East: Iran (see references above), Turkey (Güçlü & Özbek, 2011).
Extralimital distribution: Afghanistan, Armenia, Azerbaijan, Belarus, Czech Republic, Georgia, Hungary, Kazakhstan, Lithuania, Moldova, Mongolia.
Host records: Unknown.

Schizoprymnus (*Schizoprymnus*) *erzurumus* Belokobylskij, Güclü and Özbek, 2004

Distribution in the Middle East: Turkey (Belokobylskij et al., 2004; Güçlü et al., 2011).
Extralimital distribution: None.
Host records: Unknown.

Schizoprymnus (*Schizoprymnus*) *excisus* (Šnoflák, 1953)

Catalogs with Iranian records: Gadallah and Ghahari (2013), Farahani et al. (2016), Yu et al. (2016), Samin, Coronado-Blanco, Kavallieratos et al. (2018), Samin, Coronado-Blanco, Fischer et al. (2018).
Distribution in Iran: Ardabil (Samin et al., 2019), Golestan (Samin, Ghahari et al., 2015), Guilan (Ghahari & Fischer, 2011b; Sakenin et al., 2012), Kerman (Ghahari et al., 2010).
Distribution in the Middle East: Iran (see references above), Turkey (Güçlü & Özbek, 2011).
Extralimital distribution: Armenia, Czech Republic, Hungary, Kazakhstan, Mongolia, Russia, Serbia, Ukraine.
Host records: Recorded by Güçlü and Özbek (2011) as a parasitoid of the curculionid *Curculio rubido* (Gyllenhal).

Schizoprymnus (*Schizoprymnus*) *hilaris* (Herrich-Schäffer, 1838)

Catalogs with Iranian records: No catalog.
Distribution in Iran: Alborz, Guilan, Mazandaran (Talebi et al., 2018).
Distribution in the Middle East: Iran (Talebi et al., 2018), Israel—Palestine (Papp, 2012).
Extralimital distribution: Croatia, former Czechoslovakia, Finland, France, Germany, Hungary, Kazakhstan, Moldova, Russia, Sweden, Switzerland, former Yugoslavia.
Host records: Recorded by Fahringer (1934) and Telenga (1941) as being a parasitoid of the chrysomelid *Bruchidius varius* (Olivier).

Schizoprymnus (*Schizoprymnus*) *longiseta* (Herrich-Schäffer, 1838)

Distribution in the Middle East: Turkey (Güçlü & Özbek, 2011).
Extralimital distribution: Austria, Germany, Italy, Spain.
Host records: Recorded by Priore and Tremblay (1987) as being a parasitoid of the curculionid *Curculio elephas* (Gyllenhal).

Schizoprymnus (*Schizoprymnus*) *luteipalpis* (Šnoflák, 1953)

Distribution in the Middle East: Turkey (Güçlü & Özbek, 2011).
Extralimital distribution: Czech Republic, Hungary.
Host records: Unknown.

Schizoprymnus (*Schizoprymnus*) *nigripes* (Thomson, 1892)

Catalogs with Iranian records: Gadallah and Ghahari (2013), Farahani et al. (2016), Yu et al. (2016), Samin, Coronado-Blanco, Kavallieratos et al. (2018), Samin, Coronado-Blanco, Fischer et al. (2018).

Distribution in Iran: Golestan (Samin, 2015), Sistan and Baluchestan (Samin et al., 2011).

Distribution in the Middle East: Iran (Samin, 2015; Samin et al., 2011), Turkey (Güçlü & Özbek, 2011).

Extralimital distribution: Croatia, former Czechoslovakia, Hungary, Kazakhstan, Korea, Moldova, Mongolia, Russia, Sweden, Switzerland.

Host records: Recorded by Güçlü & Özbek (2011) as being a parasitoid of the curculionid *Acentrus histrio* (Boheman).

Schizoprymnus (*Schizoprymnus*) *obscurus* (Nees von Esenbeck, 1816)

Catalogs with Iranian records: Fallahzadeh and Saghaei (2010), Gadallah and Ghahari (2013), Farahani et al. (2016), Yu et al. (2016), Samin, Coronado-Blanco, Kavallieratos et al. (2018), Samin, Coronado-Blanco, Fischer et al. (2018).

Distribution in Iran: Guilan (Ghahari, Fischer, & Tobias, 2012), Isfahan (Ghahari et al., 2011), Iran (no specific locality cited) (Hellén, 1958; Papp, 2011−12; Telenga, 1941).

Distribution in the Middle East: Iran (see references above), Israel−Palestine (Papp, 2012), Turkey (Güçlü & Özbek, 2011).

Extralimital distribution: Austria, Belgium, Bulgaria, Czech Republic, Denmark, Finland, France, Georgia, Germany, Greece, Hungary, Italy, Kazakhstan, Lithuania, Moldova, Mongolia, Morocco, Netherlands, Norway, Poland, Russia, Slovenia, Spain, Sweden, Switzerland, Tunisia, Turkmenistan, Ukraine, United Kingdom, former Yugoslavia.

Host records: Summarized by Yu et al. (2016) as being a parasitoid of the brentid *Protapion apicrans* (Herbst); the curculionids *Apion semivittatum* Gyllenhal, *Ceutorhynchus picitarsis* Gyllenhal, *Ceutorhynchus pleurostigma* (Stephens), *Ceutorhynchus sulcicollis* (Paykull), *Gymnetron antirrhini* (Paykull), and *Gymnetron coliinum* (Gyllenhal); and the mordellid *Mordellistena parvula* (Gyllenhal).

Schizoprymnus (*Schizoprymnus*) *opacus* (Thomson, 1892)

Distribution in the Middle East: Turkey (Güçlü & Özbek, 2011).

Extralimital distribution: Armenia, Azerbaijan, Bulgaria, former Czechoslovakia, Finland, Georgia, Germany, Greece, Hungary, Kazakhstan, Korea, North Macedonia, Moldova, Mongolia, Russia, Serbia, Spain, Sweden, Switzerland, Ukraine, United Kingdom.

Host records: Unknown.

Schizoprymnus (*Schizoprymnus*) *ozlemae* Beyarslan, 1988

Distribution in the Middle East: Turkey (Beyarslan, 1988).

Extralimital distribution: None.

Host records: Unknown.

Schizoprymnus (*Schizoprymnus*) *pallidipennis* (Herrich-Schäffer, 1838)

Catalogs with Iranian records: Gadallah and Ghahari (2013), Farahani et al. (2016), Yu et al. (2016), Samin, Coronado-Blanco, Kavallieratos et al. (2018), Samin, Coronado-Blanco, Fischer et al. (2018).

Distribution in Iran: Isfahan (Ghahari & Beyarslan, 2017), Mazandaran (Ghahari & Fischer, 2011b; Sakenin et al., 2012).

Distribution in the Middle East: Iran (see references above), Turkey (Güçlü & Özbek, 2011).

Extralimital distribution: Armenia, China, Croatia, former Czechoslovakia, Germany, Hungary, Kazakhstan, North Macedonia, Mongolia, Russia, Serbia, Spain, Switzerland, Tajikistan.

Host records: Recorded by Papp (1998) as being a parasitoid of mordellid *Mordellistena parvula* Gyllenhal. In Iran, this species has been reared from the mordellid *Mordellistena* (*Mordellistena*) *parvula* Gyllenhal (Ghahari & Beyarslan, 2017).

Schizoprymnus (*Schizoprymnus*) *parvus* (Thomson, 1892)

Catalogs with Iranian records: Gadallah and Ghahari (2013), Farahani et al. (2016), Yu et al. (2016), Samin, Coronado-Blanco, Kavallieratos et al. (2018), Samin, Coronado-Blanco, Fischer et al. (2018).

Distribution in Iran: Ardabil, West Azarbaijan (Rastegar et al., 2012), Chaharmahal and Bakhtiari (Samin, van Achterberg et al., 2015), Isfahan (Ghahari et al., 2011), Mazandaran (Ghahari, 2017; Talebi et al., 2018), Razavi Khorasan (Samin et al., 2011).

Distribution in the Middle East: Iran (see references above), Turkey (Güçlü & Özbek, 2011).

Extralimital distribution: Croatia, former Czechoslovakia, Finland, Germany, Greece, Hungary, Kazakhstan, North Macedonia, Mongolia, Netherlands, Sweden, former Yugoslavia.

Host records: Unknown.

Schizoprymnus (*Schizoprymnus*) *pullatus* (Dahlbom, 1833)

Catalogs with Iranian records: Fallahzadeh and Saghaei (2010), Gadallah and Ghahari (2013), Farahani et al. (2016), Yu et al. (2016), Samin, Coronado-Blanco, Kavallieratos et al. (2018), Samin, Coronado-Blanco, Fischer et al. (2018).

Distribution in Iran: Khuzestan (Samin et al., 2019), Qazvin (Talebi et al., 2018), Iran (no specific locality cited) (Szépligeti, 1898, as *S. globosus*; Tobias, 1986).

Distribution in the Middle East: Iran (see references above), Turkey (Güçlü & Özbek, 2011).
Extralimital distribution: Bulgaria, Croatia, former Czechoslovakia, Georgia, Germany, Hungary, Kazakhstan, Moldova, Poland, Russia, Sweden, Ukraine, Uzbekistan.
Host records: Recorded by Papp (1997) as being a parasitoid of the mordellid *Mordella leucaspis bicoloripilosa* Horak.

Schizoprymnus (Schizoprymnus) tantalus Papp, 1981
Catalogs with Iranian records: No catalog.
Distribution in Iran: Alborz, Guilan, Qazvin (Talebi et al., 2018).
Distribution in the Middle East: Iran (Talebi et al., 2018), Israel−Palestine (Papp, 2012), Turkey (Güçlü & Özbek, 2011; Papp, 1981).
Extralimital distribution: Bosnia-Herzegovina, Croatia, Greece, Hungary, North Macedonia, Moldova, Serbia, Spain.
Host records: Recorded by Güçlü and Özbek (2011) as being a parasitoid of the curculionid *Glocianus* sp.

Schizoprymnus (Schizoprymnus) telengai Tobias, 1976
Catalogs with Iranian records: No catalog.
Distribution in Iran: Alborz (Talebi et al., 2018), West Azarbaijan (Shahand & Karimpour, 2017).
Distribution in the Middle East: Iran (Shahand & Karimpour, 2017; Talebi et al., 2018), Turkey (Güçlü & Özbek, 2011).
Extralimital distribution: Former Czechoslovakia, Kazakhstan, Russia.
Host records: Recorded by Tobias (1976, 1986) and Güçlü and Özbek (2011) as being a parasitoid of the nanophyid *Titanomalia komaroffi* (Faust). In Iran, this species has been reared from the curculionid *Lixus (Dilixellus) fasciculatus* Boheman (Shahand & Karimpour, 2017).

Schizoprymnus (Schizoprymnus) terebralis (Šnoflák, 1953)
Catalogs with Iranian records: Gadallah and Ghahari (2013), Farahani et al. (2016), Yu et al. (2016), Samin, Coronado-Blanco, Kavallieratos et al. (2018), Samin, Coronado-Blanco, Fischer et al. (2018).
Distribution in Iran: East Azarbaijan (Sakenin et al., 2009), Fars (Samin et al., 2015b), Isfahan (Ghahari et al., 2010), Mazandaran (Kian et al., 2020; Talebi et al., 2018), Tehran (Talebi et al., 2018).
Distribution in the Middle East: Iran (see references above), Turkey (Güçlü & Özbek, 2011).
Extralimital distribution: Armenia, Azerbaijan, Bulgaria, Czech Republic, Greece, Hungary, Kazakhstan, North Macedonia, Moldova, Mongolia, Russia, Serbia, Ukraine.
Host records: Recorded by Papp (1998) as being a parasitoid of the mordellid *Mordellistena parvula* (Gyllenhal).

Genus Triaspis Haliday, 1835
Triaspis armeniaca Tobias, 1976
Catalogs with Iranian records: No catalog.
Distribution in Iran: Mazandaran (Pirouzeh et al., 2016).
Distribution in the Middle East: Iran (Pirouzeh et al., 2016), Turkey (Koldaş et al., 2018).
Extralimital distribution: Armenia (Tobias, 1976, 1986).
Host records: Unknown.

Triaspis caucasica Abdinbekova, 1969
Catalogs with Iranian records: Samin, Coronado-Blanco, Fischer et al. (2018).
Distribution in Iran: Semnan (Samin, Coronado-Blanco, Fischer et al., 2018).
Distribution in the Middle East: Iran (Sakenin et al., 2018), Turkey (Güçlü & Özbek, 2011).
Extralimital distribution: Azerbaijan (Abdinbekova, 1969).
Host records: Unknown.

Triaspis caudata (Nees von Esenbeck, 1816)
Catalogs with Iranian records: Samin et al. (2018b as *Sigalphus caudatus* Nees, 1816).
Distribution in Iran: Guilan, Mazandaran, Qazvin (Pirouzeh et al., 2016), Hamadan (Samin, Coronado-Blanco, Fischer et al., 2018).
Distribution in the Middle East: Iran (Pirouzeh et al., 2016; Samin, Coronado-Blanco, Fischer et al., 2018), Turkey (Koldaş et al., 2018).
Extralimital distribution: Armenia, Azerbaijan, Belgium, Bosnia-Herzegovina, Bulgaria, Croatia, former Czechoslovakia, Denmark, Finland, France, Germany, Greece, Hungary, Ireland, Italy, Kazakhstan, Latvia, Lithuania, Moldova, Netherlands, Poland, Russia, Serbia, Spain, Sweden, United Kingdom.
Host records: Summarized by Yu et al. (2016) as being a parasitoid of a wide range of insect hosts belonging to the orders Coleoptera: *Apion aestivum* Germar, *Apion asiimile* Kirby, *Apion trifoli* (L.), and *Protapion aprican* (Herbst) (Apionidae), *Byctiscus betulae* (L.), and *Bytiscus populi* (L.) (Attelabidae), *Bruchidius fasciatus* (Olivier) (Chrysomelidae), *Orchestes fagi* (L.), *Orchestes quercus* (L.), *Pissodes notatus* Duftschmidt, and *Thamnurgus euphorbiae* Kuester (Curculionidae); and *Ochina ptinoides* (Marsham) (Ptinidae); Diptera: *Chlorops pumilionis* (Bjerkander), *Oscinella frit* (L.), and *Oscinella vastator* (Curtis) (Chloropidae); and Lepidoptera: *Coleophora laricella* (Hübner) (Coleophoridae), *Lathronympha strigana* (Fabricius) (Tortricidae), and *Tischeria ekebladella* (Bjerkander) (Tischeriidae).

Triaspis collaris (Thomson, 1874)
Distribution in the Middle East: Israel−Palestine (Papp, 2012).

Extralimital distribution: France, Germany, Sweden, United Kingdom.

Host records: Unknown.

Triaspis complanellae (Hartig, 1847)

Catalogs with Iranian records: Gadallah and Ghahari (2013), Farahani et al. (2016), Yu et al. (2016), Samin, Coronado-Blanco, Kavallieratos et al. (2018), Samin, Coronado-Blanco, Fischer et al. (2018).

Distribution in Iran: Guilan (Ghahari, Fischer, & Tobias, 2012, under *Triaspis flavipes* (Ivanov); Pirouzeh et al., 2016), Mazandaran (Pirouzeh et al., 2016).

Distribution in the Middle East: Iran (see references above), Syria (Čapek & Hofmann, 1997).

Extralimital distribution: Belgium, former Czechoslovakia, France, Germany, Hungary, Italy, Kazakhstan, Korea, Mongolia, Russia, Tunisia, Ukraine.

Host records: Summarized by Yu et al. (2016) as being a parasitoid of the curculionid *Gymnetron antirhini* (Paykull); and the tischeriid moth *Tischeria ekebladella* (Bjerkander).

Triaspis facialis (Ratzeburg, 1852)

Distribution in the Middle East: Israel—Palestine (Papp, 2012).

Extralimital distribution: Bulgaria, Croatia, Germany, Hungary, Italy.

Host records: Summarized by Yu et al. (2016) as being a parasitoid of the chrysomelid *Bruchus lentis* Frölich, and the ptinid *Ochina ptinoides* (Marsham).

Triaspis flavipalpis (Wesmael, 1835)

Distribution in the Middle East: Israel—Palestine (Halperin, 1986).

Extralimital distribution: Belgium, former Czechoslovakia, France, Germany, Hungary, Italy, North Macedonia, Netherlands, Poland, Portugal, Ukraine, former Yugoslavia.

Host records: Summarized by Yu et al. (2016) as being a parasitoid of cerambycid *Niphona picticornis* Mulsant; the following curculionids: *Gymnetron antirrhini* (Paykull), *Gymnetron netum* Schönherr, *Gymnetron tetrum* (Fabricius), *Scolytus rugulosus* (Müller), and *Thamnurgus euphorbiae* Küster; and the ptinid *Mesocoelopus collaris* Mulsant and Rey.

Triaspis floricola (Wesmael, 1835)

Catalogs with Iranian records: Gadallah and Ghahari (2013), Farahani et al. (2016), Yu et al. (2016), Samin, Coronado-Blanco, Kavallieratos et al. (2018), Samin, Coronado-Blanco, Fischer et al. (2018).

Distribution in Iran: Isfahan (Ghahari et al., 2011).

Distribution in the Middle East: Iran (Ghahari et al., 2011), Israel—Palestine (Halperin, 1986; Papp, 2012).

Extralimital distribution: Algeria, Belgium, Czech Republic, France, Germany, Greece, Hungary, Ireland, Italy, Malta, Netherlands, Poland, Russia, Slovakia, Spain, Switzerland, United Kingdom.

Host records: Summarized by Yu et al. (2016) as being a parasitoid of the apionids *Apion lemoroi* (Brisout), *Apion longirostre* Olivier, *Apion loti* Kirby, *Apion rufirostre* (Fabricius), *Apion trifoli* (L.), and *Protapion apricans* (Herbst); the bostrichids *Sinoxylon sexdendatum* Olivier, and *Xylopertha retusa* (Olivier); and the curculionids *Ceutorhynchus pleurostigma* Stephens, *Ceutorhynchus sulcatus* Brisout, *Ceutorhynchus sulcicollis* (Paykull). It was also reported by Telenga (1941) as being a parasitoid of the tenthredinid *Caliroa cerasi* L.

Triaspis glaberrima Šnoflák, 1953

Distribution in the Middle East: Turkey (Güçlü & Özbek, 2011).

Extralimital distribution: Former Czechoslovakia, Hungary, Poland, Serbia.

Host records: Unknown.

Triaspis lugubris Šnoflák, 1953

Catalogs with Iranian records: Gadallah and Ghahari (2013), Farahani et al. (2016), Yu et al. (2016), Samin, Coronado-Blanco, Kavallieratos et al. (2018), Samin, Coronado-Blanco, Fischer et al. (2018).

Distribution in Iran: Kuhgiloyeh and Boyerahmad (Samin, van Achterberg et al., 2015), Mazandaran (Ghahari & Fischer, 2011b; Sakenin et al., 2012).

Distribution in the Middle East: Iran.

Extralimital distribution: Czech Republic, Hungary, Kazakhstan, Korea, Mongolia, Russia.

Host records: Recorded by Ku et al. (2001) as being a parasitoid of the curculionid *Rynchaenus sanguinipes* (Roelof).

Triaspis luteipes (Thomson, 1874)

Catalogs with Iranian records: No catalog.

Distribution in Iran: Fars (Samin et al., 2019).

Distribution in the Middle East: Iran.

Extralimital distribution: Bulgaria, Finland, France, Germany, Greece, Hungary, Italy, Poland, Romania, Spain, Sweden, Switzerland, Ukraine, United Kingdom.

Host records: Summarized by Yu et al. (2016) as being a solitary endoparasitoid of the anobiid *Anobium rufipes* Fabricius; the chrysomelids *Bruchus affinis* Frölich, *Bruchus rufimanus* Boheman, *Timarcha tenebricosa* (Fabricius), and *Timarcha tenebricosa normandiana* Bechyne; the curculionid *Hylesinus fraxini* (Panzer); and the ptinid *Ochina ptinoides* (Marsham); the choreutids *Choreutis nemorana* (Hübner); and the tortricids *Lobesia littoralis* (Westwood and Humphreys), and *Ryacionia buoliana* (Denis and Schiffermüller).

Triaspis obscurella (Nees von Esenbeck, 1816)

Catalogs with Iranian records: Gadallah and Ghahari (2013), Farahani et al. (2016), Yu et al. (2016), Samin, Coronado-Blanco, Kavallieratos et al. (2018), Samin, Coronado-Blanco, Fischer et al. (2018).

Distribution in Iran: East Azarbaijan (Rastegar et al., 2012), Guilan, Mazandaran (Pirouzeh et al., 2016), Isfahan (Ghahari et al., 2011), Mazandaran (Sakenin et al., 2020).

Distribution in the Middle East: Iran (see references above), Israel—Palestine (Halperin, 1986; Papp, 2012), Turkey (Güçlü & Özbek, 2011).

Extralimital distribution: Azerbaijan, Bulgaria, Croatia, Czech Republic, France, Germany, Greece, Hungary, Italy, Kazakhstan, Lithuania, North Macedonia, Malta, Moldova, Mongolia, Netherlands, Poland, Russia, Serbia, Slovakia, Spain, Switzerland, Ukraine, United Kingdom, Uzbekistan.

Host records: Summarized by Yu et al. (2016) as being a parasitoid of several coleopteran species of the families Apionidae, Chrysomelidae, Curculionidae, and Ptinidae.

Triaspis pallipes (Nees von Esenbeck, 1816)

Catalogs with Iranian records: Gadallah and Ghahari (2013), Farahani et al. (2016), Yu et al. (2016), Samin, Coronado-Blanco, Kavallieratos et al. (2018), Samin, Coronado-Blanco, Fischer et al. (2018).

Distribution in Iran: Ardabil (Gadallah & Ghahari, 2013), Guilan (Pirouzeh et al., 2016), Mazandaran (Kian et al., 2020; Pirouzeh et al., 2016).

Distribution in the Middle East: Iran (see references above), Israel—Palestine (Papp, 2012), Turkey (Koldaş et al., 2018).

Extralimital distribution: Azerbaijan, Bulgaria, China, Czech Republic, France, Germany, Hungary, Ireland, Italy, Kazakhstan, Latvia, Moldova, Norway, Poland, Russia, Spain, Switzerland, Ukraine, United Kingdom.

Host records: Summarized by Yu et al. (2016) as being a parasitoid of the apionid *Apion aeneum* (Fabricius); the chrysomelids *Brichidius seminaries* (L.) and *Bruchus* spp.; and the curculionids *Anthonomus pomorum* (L.), *Ceutorhynchus* spp., *Gymnetron* spp., and *Rynchaenus* spp.

Triaspis sulcata (Szépligeti, 1901)

Distribution in the Middle East: Turkey (Koldaş et al., 2018).

Extralimital distribution: Greece, Hungary, Italy.

Host records: Unknown.

Triaspis thoracica (Curtis, 1860)

Catalogs with Iranian records: Farahani et al. (2016).

Distribution in Iran: Guilan (Pirouzeh et al., 2016), Mazandaran, Tehran (Shahhosseini & Kamali, 1989; Modarres Awal, 1997, 2012, as *Triaspis thoracicus*).

Distribution in the Middle East: Cyprus (Morice, 1928), Iran (see references above), Israel—Palestine (Bodenheimer, 1930; Halperin, 1986; Papp, 2012), Turkey (Koldaş et al., 2018).

Extralimital distribution: Argentina, Austria, Azerbaijan, Bulgaria, Croatia, former Czechoslovakia, France, Georgia, Greece, Hungary, Italy, Mexico, Moldova, Portugal, Romania, Russia, Spain, Switzerland, Tunisia, Ukraine, United Kingdom, Uruguay, former Yugoslavia.

Host records: Summarized by Yu et al. (2016) as being a parasitoid of several coleopterous species belonging to the families Bostrichidae, Cerambycidae, Chrysomelidae, and Curculionidae. In Iran, this species has been reared from the chrysomelid *Bruchus rufimanus* Boheman (Modarres Awal, 1997, 2012; Shahhosseini & Kamali, 1989).

Triaspis xylophagi Fischer, 1966

Distribution in the Middle East: Turkey (Koldaş et al., 2018).

Extralimital distribution: Algeria (Fischer, 1966).

Host records: Unknown.

Tribe Diospilini Foerster, 1863

Genus *Aspicolpus* Wesmael, 1838

Aspicolpus carinator (Nees von Esenbeck, 1812)

Catalogs with Iranian records: Gadallah et al. (2016), Farahani et al. (2016), Yu et al. (2016), Samin, Coronado-Blanco, Kavallieratos et al. (2018), Samin, Coronado-Blanco, Fischer et al. (2018).

Distribution in Iran: Ardabil (Gadallah et al., 2016).

Distribution in the Middle East: Iran.

Extralimital distribution: Austria, Belgium, Finland, France, Germany, Greece, Hungary, Italy, Latvia, Moldova, Poland, Russia, Slovakia, Sweden, Switzerland, Ukraine, Uzbekistan.

Host records: Summarized by Yu et al. (2016) as being a parasitoid of the bostrichid *Xylopertha retusa* (Olivier); the buprestid *Anthaxia aurulenta* (Fabricius); the cerambycids *Callidium aeneum* (DeGeer), *Callidium violaceum* (L.), *Clytus arietis* (L.), *Clytus tropicus* (Panzer), *Exocentrus punctipennis* Mulsant and Guillebeu, *Leiopus nebulosus* (L.), *Obrium cantharinum* (L.), *Oplosia cinerea* (Mulsant), *Phymatodes testaceus* (L.), *Plagionotus arcuatus* (L.), *Saperda punctata* (L.), *Xylotrechus arvicola* (Olivier), *Xylotrechus capricornis* Chevrolat, and *Xylotrechus rusticus* (L.); and the curculionid *Scolytus intricatus* (Ratzeburg).

Aspicolpus sibiricus (Fahringer, 1934)

Aspicolpus borealis (Thomson, 1892) (Synonymized by van Achterberg, 2014).

Catalogs with Iranian records: Fallahzadeh and Saghaei (2010) as *Aspicolpus borealis*, Gadallah et al. (2016),

Farahani et al. (2016) as *Aspicolpus borealis*, Yu et al. (2016), Samin, Coronado-Blanco, Kavallieratos et al. (2018), Samin, Coronado-Blanco, Fischer et al. (2018).

Distribution in Iran: Golestan (Ghahari et al., 2010), Tehran (Hedwig, 1957 as *Helcon borealis*).

Distribution in the Middle East: Iran.

Extralimital distribution: Hungary, Russia, Sweden, Switzerland.

Host records: Recorded by Hedqvist (1967, 1998) as being a parasitoid of the cerambycid *Chlorophorus herbstii* (Brahm).

Genus *Diospilus* Haliday, 1833

Diospilus (Diospilus) angorensis Beyarslan, 2014

Distribution in the Middle East: Turkey (Beyarslan & Coban, 2014).

Extralimital distribution: None.

Host records: Unknown.

Diospilus (Diospilus) belokobylskiji Beyarslan, 2008

Distribution in the Middle East: Turkey (Beyarslan et al., 2008).

Extralimital distribution: None.

Host records: Unknown.

Diospilus (Diospilus) capito (Nees von Esenbeck, 1834)

Catalogs with Iranian records: Gadallah et al. (2016), Farahani et al. (2016), Yu et al. (2016), Samin, Coronado-Blanco, Kavallieratos et al. (2018), Samin, Coronado-Blanco, Fischer et al. (2018).

Distribution in Iran: Chaharmahal and Bakhtiari (Gadallah et al., 2016), Kuhgiloyeh and Boyerahmad (Ghahari & Gadallah, 2022).

Distribution in the Middle East: Iran (Gadallah et al., 2016; Ghahari & Gadallah, 2022), Israel—Palestine (Belokobylskij et al., 1997; Papp, 1970, 2012), Turkey (Beyarslan et al., 2008).

Extralimital distribution: Armenia, Austria, Azerbaijan, Belarus, Belgium, Bulgaria, Croatia, former Czechoslovakia, Denmark, Estonia, Finland, France, Germany, Greece, Hungary, Ireland, Italy, Kazakhstan, Latvia, Lithuania, Moldova, Mongolia, Morocco, Netherlands, Norway, Poland, Russia, Serbia, Sweden, Switzerland, Ukraine, United Kingdom.

Host records: Summarized by Yu et al. (2016) as being a parasitoid of the chrysomelid *Psylliodes chrysocephala* L.; the curculionids *Ceutorhynchus* spp., *Gymnetron antirrhinin* (Paykull), *Gymnetron tetrum* (Fabricius), *Hypurus bertrandi* (Perris), and *Magdalis ruficornis* (L.); the nitidulids *Brassicogethes aeneus* (Fabricius), and *Meligethes viridiscens* (Fabricius); the ptinid *Anobium punctatum* De Geer; and the rhynchitid *Byctiscus betulae* (L.). In Iran, this species has been reared from the curculionid *Ceutorhynchus assimilis* (Paykull) (Ghahari & Gadallah, 2022).

Diospilus (Diospilus) dispar (Nees von Esenbeck, 1811)

Distribution in the Middle East: Israel—Palestine (Kugler, 1966).

Extralimital distribution: Argentina, Austria, Belarus, Bosnia-Herzegovina, Bulgaria, Finland, France, Germany, Hungary, Italy, Korea, Moldova, Mongolia, Poland, Sweden, Switzerland, United Kingdom, former Yugoslavia.

Host records: Summarized by Yu et al. (2016) as being a parasitoid of the cerambycid *Molorchus umbellatarum* (Schreber); the chrysomelid *Paropsis atomaria* Olivier; the elaterid *Agriotes lineatus* L.; the following ptinids: *Dorcatoma chrysomelina* (Sturm), *Dercatoma dresdensis* Herbst, *Dorcatoma serra* Sturm, and *Dorcotoma setosella* Mulsant and Rey; and the tenebrionid *Diaperis boleti* (L.).

Diospilus (Diospilus) inflexus Reinhard, 1862

Distribution in the Middle East: Turkey (Beyarslan et al., 2013).

Extralimital distribution: France, Germany, Hungary, Poland, Russia, Ukraine.

Host records: Unknown.

Diospilus melanoscelus (Nees von Esenbeck, 1834)

Catalogs with Iranian records: No catalog.

Distribution in Iran: West Azarbaijan (Naderian et al., 2020).

Distribution in the Middle East: Iran (Naderian et al., 2020), Turkey (Beyarslan et al., 2008).

Extralimital distribution: Belarus, Germany, Korea, Russia, Slovakia, Sweden.

Host records: Summarized by Yu et al. (2016) as being a parasitoid of the ptinids *Dorcatoma chrysomelina* Sturm, and *Dorcatoma dresdensis* Herbst. Not yet recorded in Iran.

Diospilus (Diospilus) morosus Reinhard, 1862

Catalogs with Iranian records: No catalog.

Distribution in Iran: Ardabil (Gadallah et al., 2018).

Distribution in the Middle East: Iran (Gadallah et al., 2018), Turkey (Beyarslan et al., 2008).

Extralimital distribution: Czech Republic, Finland, France, Germany, Greece, Hungary, Italy, Kazakhstan, Lithuania, Moldova, Poland, Serbia, Sweden, Switzerland, Ukraine, United Kingdom.

Host records in Iran: Summarized by Yu et al. (2016) as being a parasitoid of the chrysomelids *Phyllotreta memorum* L., and *Psylliodes chrysocephala* L.; the curculionid *Ceutorhynchus assimilis* (Paykull); and the erotylid *Dacne bipustulata* (Thunberg). In Iran, this species has been reared from the curculionid *Ceutorhynchus assimilis* Paykull (Gadallah et al., 2018).

Diospilus (Diospilus) nigricornis (Wesmael, 1835)

Catalogs with Iranian records: Gadallah et al. (2016), Farahani et al. (2016), Yu et al. (2016), Samin, Coronado-Blanco, Kavallieratos et al. (2018), Samin, Coronado-Blanco, Fischer et al. (2018).

Distribution in Iran: Khuzestan (Ghahari et al., 2010), Golestan (Samin, Ghahari et al., 2015).

Distribution in the Middle East: Iran (Ghahari et al., 2010; Samin, Ghahari et al., 2015), Turkey (Beyarslan et al., 2008).

Extralimital distribution: Austria, Azerbaijan, Belarus, Belgium, Bulgaria, former Czechoslovakia, Finland, France, Georgia, Germany, Greece, Hungary, Italy, Korea, Latvia, Lithuania, Moldova, Netherlands, Norway, Poland, Russia, Slovenia, Sweden, Switzerland, Ukraine, United Kingdom, former Yugoslavia.

Host records: Summarized by Yu et al. (2016) as being a parasitoid of the anobiid *Xestobium plumbeum* (Illiger); the attelabid *Byctiscus populi* (L.); the cerambycid *Pogonocherus hispidus* (L.); and the curculionids *Ceutorhynchus pleurostigma* Stephens, *Ceutorhynchus sulcicollis* (Paykull), and *Tachyerges salicis* (L.). It was also reported by Kopelke (1994) as being a parasitoid of the tenthredinids *Pontania proxima* (Lepeletier) and *Pontania vesicator* (Bremi).

Diospilus (Diospilus) oleraceus Haliday, 1833

Catalogs with Iranian records: No catalog.

Distribution in Iran: Golestan (Ghahari & Gadallah, 2022).

Distribution in the Middle East: Iran.

Extralimital distribution: Belarus, Belgium, China, Czech Republic, Finland, France, Germany, Greece, Hungary, Ireland, Italy, Kazakhstan, Korea, Lithuania, North Macedonia, Moldova, Netherlands, Poland, Romania, Russia, Sweden, Switzerland, Ukraine, United Kingdom, former Yugoslavia.

Host records: Summarized by Yu et al. (2016) as being a parasitoid of the chrysomelid *Psylliodes chrysocephala* (L.); the curculionids *Ceutorhynchus assimilis* (Paykull), *Ceutorhynchus leprieuri* Brisout, *Ceutorhynchus obstrictus* (Marsham), *Ceutorhynchus pallidactylus* (Marsham), *Ceutorhynchus pleurostigma* (Marsham), *Ceutorhynchus rapae* Gyllenhal, and *Scolytus multistriatus* (Paykull); and the nitidulid *Meligethes aeneus* (Fabricius). In Iran, this species has been reared from the curculionid *Scolytus multistriatus* (Paykull) (Ghahari & Gadallah, 2022).

Diospilus (Diospilus) productus Marshall, 1894

Catalogs with Iranian records: Gadallah et al. (2016), Farahani et al. (2016), Yu et al. (2016), Samin, Coronado-Blanco, Kavallieratos et al. (2018), Samin, Coronado-Blanco, Fischer et al. (2018).

Distribution in Iran: Kordestan (Gadallah et al., 2016).

Distribution in the Middle East: Iran (Gadallah et al., 2016), Israel−Palestine (Papp, 2012), Turkey (Beyarslan et al., 2008).

Extralimital distribution: Armenia, Greece, Hungary, Italy, Switzerland, Ukraine, United Kingdom.

Host records: Unknown.

Genus *Taphaeus* Wesmael, 1835

Taphaeus hiator (Thunberg, 1822)

Catalogs with Iranian records: Gadallah et al. (2016), Farahani et al. (2016), Yu et al. (2016), Samin, Coronado-Blanco, Kavallieratos et al. (2018), Samin, Coronado-Blanco, Fischer et al. (2018).

Distribution in Iran: Chaharmahal and Bakhtiari (Ghahari et al., 2010; Samin, 2015), East Azarbaijan (Sakenin et al., 2009), Mazandaran (Ghahari, 2017—around rice fields).

Distribution in the Middle East: Iran (see references above), Turkey (Beyarslan et al., 2008).

Extralimital distribution: Belgium, Bulgaria, Canada (introduced), former Czechoslovakia, Finland, France, Germany, Greece, Hungary, Ireland, Italy, Kazakhstan, Lithuania, Moldova, Mongolia, Netherlands, Norway, Poland, Russia, Sweden, Switzerland, United Kingdom, United States of America.

Host records: Summarized by Yu et al. (2016) as being a parasitoid of the curculionid *Polydrusus impressifrons* (Gyllenhal); the melandryid *Orchesia micans* (Panzer); and the depressariid moth *Agonopterix subpropinquella* (Stainton).

Taphaeus rufocephalus (Telenga, 1950)

Distribution in the Middle East: Turkey (Beyarslan et al., 2008).

Extralimital distribution: Germany, Kazakhstan, Russia.

Host records: Unknown.

Conclusion

A hundred species of the subfamily Brachistinae in 10 genera and three tribes (Blacini Foerster, 1863, Brachistinae Foerster, 1863; Diospilinae Foerster, 1863) have been reported from five of the Middle East countries (Cyprus, Iran, Israel−Palestine, Syria, and Turkey). The genus *Blacus* is the most diverse, with 25 recorded species. The fauna of Iranian Brachistinae comprises 73 species in nine genera (9.1% of the world species), of which *Blacus* and *Schizoprymnus* are the most diverse genera, with 21 and 17 species, respectively; followed by *Eubazus* (13 species), *Triaspis* (10 species), *Diospilus* (six species), *Aspicolpus* and *Polydegmon* (both with two species), and *Foersteria* and *Taphaeus* (both with one species). Iranian Brachistinae

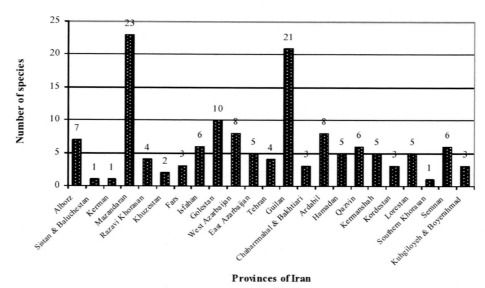

FIGURE 5.1   Number of reported species of Iranian Brachistinae by province.

species have been recorded from 23 provinces, which Mazandaran and Guilan with 23 and 21 species, respectively, have the highest number of species (Fig. 5.1). In Iran, host species have been recorded for four brachistine species. These hosts belong to three families of Coleoptera (Chrysomelidae: one species; Curculionidae: three species; Mordellidae: one species). Among the countries of the Middle East and those adjacent to Iran, Russia and Turkey with 54 and 52 species, respectively, share the largest number of species with Iran; followed by Kazakhstan (37 species), Azerbaijan (27 species), Armenia (15 species), Israel—Palestine (12 species), Turkmenistan (six species), Afghanistan and Cyprus (both with four species), and Syria (one species). There is no brachistine species shared between Iran and 13 other Middle Eastern countries.

References

Abdinbekova, A. A. (1969). On the knowledge about braconids (Hymenoptera, Braconidae) in the lenkoran zone of Azerbaijan. *Doklady Akademii Nauk Zaerbaidzhanskoi SSR, 25*(9), 59—66.

Atay, C., Çetin Erdoğan, Ö., & Beyarslan, A. (2019). The Braconidae (Hymenoptera-Apocrita) of Gala Lake and the surrounding area. *Turkish Journal of Zoology, 43*, 131—141.

Belokobylskij, S. A. (1998). Subfamily Brachistinae (Calyptinae), pp. 440—489. In , *Key to the insects of Russian Far East, Vladivostok, Dal'nauka: Vol. IV. Neuropteroidea, Mecoptera, Hymenoptera. Part 3*, 708 pp.

Belokobylskij, S. A., Lobodenko, & Yu, S. (1997). Brief review of Palaearctic species of the genus *Diospilus* (Hymenoptera, Braconidae) with description of four new species. *Zoologicheskii Zhurnal, 76*(8), 915—924.

Belokobylskij, S. A., Güçlü, C., & Özbek, H. (2004). A new species of the genus *Schizoprymnus* Förster (Hymenoptera, Braconidae, Brachistinae) from Turkey. *Zoosystematica Rossica, 12*, 245—248.

Beyarslan, A. (1988). Zwei neue arten der Familie Braconidae (Hymenoptera) aus der Türkei. *Zeitschrift der Arabeitsgemeinschaft Österreichischer Entomologen, 39*(3/4), 71—76.

Beyarslan, A. (2011). *Eubazus (Brachistes) aydae* sp. nov. from Turkey (Hymenoptera: Braconidae: Brachistinae). *Journal of the Entomological Research Society, 13*(1), 107—111.

Beyarslan, A., & Coban, E. (2014). Checklist of Turkish Helconinae with a new species (Hymenoptera, Braconidae). *Turkish Journal of Zoology, 38*(1), 89—95.

Beyarslan, A., & Deveci, R. (2019). Taxonomic studies on the Brachistini (Hymenoptera, Braconidae, Brachistinae) fauna of the Eastern Anatolia region (Bingöl, Bitlis, Muş, and Van) from Turkey. *Journal of Science and Technology, 9*(2), 63—66.

Beyarslan, A., Yurktan, M., Çetin Erdoğan, O., & Aydoğdu, M. (2006). A study on Braconidae and Ichneumonidae from Ganos Mountain (Thrace Region, Turkey) (Hymenoptera, Braconidae, Ichneumonidae). *Linzer Biologische Beiträge, 38/1*, 409—422.

Beyarslan, A., Çetin Erdoğan, Ö., & Aydoğdu, M. (2008). *Diospilus belokobylskiji* Beyarslan sp. nov., with new records of Diospilini (Hymenoptera: Braconidae: Helconini) from Turkey. *Entomological News, 119*(4), 403—410.

Beyarslan, A., Gozuacik, C., & Inanç, O. (2013). A contribution of the subfamilies Helconinae, Homolobinae, Macrocentrinae, Meteorinae, and Orgilinae (Hymenoptera: Braconidae) of Southern Anatolia with new records from other parts of Turkey. *Turkish Journal of Zoology, 37*(4), 501—505.

Broad, G. R., Shaw, M. R., & Godfray, H. C. J. (2016). Checklist of British and Irish Hymenoptera-Braconidae. *Biodiversity Data Journal, 4*, e8151.

Čapek, M. (1970). The taxonomy of braconid larvae (Braconidae, Hym.). *Vedecke Prace, 12*, 243—279.

Čapek, M., & Hofmann, C. (1997). The Braconidae (Hymenoptera).in the collections of the Musée contonal Zoologie, Laussane. *Litterae Zoologicae (Laussane), 2*, 25—162.

Çetin Erdoğan, Ö. (2010). Further addition to the Blacinae (Hymenoptera: Braconidae) of Turkey and Coleoptera species as hosts. *Egyptian Journal of Biological Pest Control, 20*(2), 105—109.

Çetin Erdoğan, Ö., & Beyarslan, A. (2005). Contributions to the Blacinae fauna of Turkey (Hymenoptera, Braconidae). *Entomophaga, 26*(1), 1−8.

Çetin Erdoğan, Ö., & Beyarslan, A. (2015). First record of *Blacus (Blacus) forticornis* Haeselbarth, 1973 (Hymenoptera: Braconidae: Blacinae) from Turkey. *Turkish Journal of Zoology, 39*(5), 965−966.

Chen, X. X., & van Achterberg, C. (2019). Systematics, phylogeny, and evolution of braconid wasps: 30 years of progress. *Annual Review of Entomology, 64*, 1−24.

Donisthorpe, H. (1927). *The guests of the British ants*. London, 224 pp. [Ichneumonoidea in pp. 85−91].

Fahringer, J. (1934). *Opuscula braconologica Band 3. Palaearktischen region. Liefrung 5-8. Opuscula braconologica* (pp. 321−594). Wien: Fritz Wagner.

Fallahzadeh, M., & Saghaei, N. (2010). Checklist of Braconidae (Insecta: Hymenoptera) from Iran. *Munis Entomology & Zoology, 5*(1), 170−186.

Farahani, S., Talebi, A. A., Rakhshani, E., & van Achterberg, C. (2013). Study of the genus *Blacus* Nees (Hymenoptera: Braconidae: Blacinae) in Iran, with three new records. *Ukrainska Entomofaunistyka, 4*(2), 3−11.

Farahani, S., Talebi, A. A., & Rakhshani, E. (2016). Iranian Braconidae (Insecta: Hymenoptera: Ichneumonoidea): diversity, distribution and host association. *Journal of Insect Biodiversity and Systematics, 2*(1), 1−92.

Fischer, M. (1963). Neue Zuchtergebnisse von Braconiden (Hymenoptera). *Zeitschrift für Angewandte Zoologie, 50*, 195−214.

Fischer, M. (1966). Zei neue aus Buprestiden gezüchtete Braconiden (Hymenoptera). *Entomophaga, 11*(4), 341−346.

Gadallah, N. S., & Ghahari, H. (2013). An annotated catalogue of the Iranian Agathidinae and Brachistinae (Hymenoptera: Braconidae). *Linzer biologische Beiträge, 45*(2), 1873−1901.

Gadallah, N. S., & Ghahari, H. (2019). An updated checklist of Iranian Cardiochilinae, Ryssalinae and Blacini (Hymenoptera: Ichneumonoidea: Braconidae). *Oriental Insects, 54*(2), 143−161.

Gadallah, N. S., Ghahari, H., & van Achterberg, C. (2016). An annotated catalogue of the Iranian Euphorinae, Gnamptodontinae, Helconinae, Hormiinae and Rhysipolinae (Hymenoptera: Braconidae). *Zootaxa, 4072*(1), 1−38.

Gadallah, N. S., Ghahari, H., Papp, J., & Beyarslan, A. (2018). New records of Braconidae (Hymenoptera) from Iran. *Wuyi Science Journal, 34*, 43−48.

Gadallah, N. S., Ghahari, H., & Quicke, D. L. J. (2021). Further addition to the braconid fauna of Iran (Hymenoptera: Braconidae). *Egyptian Journal of Biological Pest Control, 31*, 32. https://doi.org/10.1186/s41938-021-00376-8

Ghahari, H. (2016). Five new records of Iranian Braconidae (Hymenoptera: Ichnemonoidea) for Iran and annotated catalogue of the subfamily Homolobinae. *Wuyi Science Journal, 32*, 35−43.

Ghahari, H. (2017). Species diversity of Ichneumonoidea (Hymenoptera) from rice fields of Mazandaran province, northern Iran. *Journal of Animal Environment, 9*(3), 371−378 (in Persian, English summary).

Ghahari, H., & Beyarslan, A. (2017). A faunistic study on Braconidae (Hymenoptera: Ichneumonoidea) from Iran. *Natura Somogyiensis, 30*, 39−46.

Ghahari, H., & Fischer, M. (2011a). A contribution to the Braconidae (Hymenoptera: Ichneumonoidea) from north-western Iran. *Calodema, 134*, 1−6.

Ghahari, H., & Fischer, M. (2011b). A study on the Braconidae (Hymenoptera: Ichneumonoidea) from some regions of northern Iran. *Entomofauna, 32*(8), 181−196.

Ghahari, H., & Fischer, M. (2012). A study on the Braconidae (Hymenoptera: Ichneumonoidea) from some regions of northern Iran. *Entomofauna, 32*, 181−196.

Ghahari, H., & Gadallah, N. S. (2022). *Additional records to the braconid fauna (Hymenoptera: Ichneumonoidea) of Iran, with new host reports.* Entomological News (in press).

Ghahari, H., Fischer, M., Hedqvist, K. J., Çetin Erdoğan, Ö., van Achterberg, C., & Beyarslan, A. (2010). Some new records of Braconidae (Hymenoptera) for Iran. *Linzer biologische Beiträge, 42*(2), 1395−1404.

Ghahari, H., Fischer, M., & Papp, J. (2011). A study on the braconid wasps (Hymenoptera: Braconidae) from Isfahan province, Iran. *Entomofauna, 32*(16), 261−272.

Ghahari, H., Fischer, M., Papp, J., & Tobias, V. (2012). A contribution to the knowledge of braconids (Hymenoptera: Braconidae) from Lorestan province Iran. *Entomofauna, 33*(7), 65−72.

Ghahari, H., Fischer, M., & Tobias, V. (2012). A study on the Braconidae (Hymenoptera: Ichneumonoidea) from Guilan province, Iran. *Entomofauna, 33*(22), 317−324.

Ghahari, H., Fischer, M., Beyarslan, A., Navaeian, M., & Hosseini Boldaji, S. A. (2019). New records of Brachistinae, Braconinae, Cheloninae, and Microgastrinae (Hymenoptera: Braconidae) from Iran. *Wuyi Science Journal, 35*(2), 135−141.

Ghahari, H., Beyarslan, A., & Kavallieratos, N. G. (2020). New records of Braconidae (Hymenoptera: Ichneumonoidea) from Iran, and in Memoriam Dr. Maximilian Fischer (7 June 1929 − 15 June 2019). *Scientific Bulletin of Uzhhorod National University (Series: Biology), 48*, 48−55.

Güçlü, C. (2011). A contribution to the knowledge of Blacinae (Hymenoptera, Braconidae) from Turkey. *Scientific Research and Essays, 6*(3), 575−579.

Güçlü, C., & Özbek, H. (2011). A contribution to the knowledge of the subfamily Brachistinae (Hymenoptera: Braconidae) of Turkey. *Journal of the Entomological Research Society, 13*(3), 15−26.

Haeselbarth, E. (1973). Die *Blacus*-Arten Europas und Zentral-Asiens (Hymenoptera: Braconidae). *Veröffentlichungen der Zoologischen Stäatssammlung, 16*, 69−170.

Haeselbarth, E. (1975). Faunistische und taxonomische Notizen zur europäischen *Blacus*-Arten (Hymenoptera, Braconidae). *Nachrichtenblatt der Bayerischen Entomologe, 24*, 28−31.

Haeselbarth, E. (1992). *Blacus (Ganychorus) madli* sp. n., eine neue Braconidae aus der Türkei (Insecta, Hymenoptera, Braconidae). *Nachrichtenblatt der Bayerischen Entomologen, 41*(3), 95−98.

Halperin, J. (1986). Braconidae (Hymenoptera) associated with forest and ornamental trees and shrubs in Israel. *Phytoparasitica, 14*(2), 119−135.

Hedqvist, K. J. (1967). Notes on helconini (Ichneumonoidea, Braconidae, Helconinae) Part I. *Entomologisk Tidskrift, 88*, 133−143.

Hedqvist, K. J. (1998). Bark beetle enemies in Sweden 2. Braconidae (Hymenoptera). *Entomologica Scandinavica Supplement, 52*, 1−87.

Hedwig, K. (1957). Ichneumoniden und Braconiden aus den Iran 1954 (Hymenoptera). *Jahresheft des Vereins für Vaterlaendische Naturkunde, 112*(1), 103−117.

Hellén, W. (1958). Zur Kenntnis der Braconiden (Hym.) Finnlands. II. Subfamilia Helconinae (part.). *Fauna Fennica, 4*, 3−37.

Kian, N., Goldasteh, S., & Farahani, S. (2020). A survey on abundance and species diversity of braconid wasps in forest of Mazandaran province. *Journal of Entomological Research, 12*(1), 61−69 (in Persian, English summary).

Koldaş, T., Çetin Erdoğan, Ö., & Beyarslan, A. (2018). Taxonomic and faunistic data on the genus *Triaspis* Haliday, 1855 (Hymenoptera: Braconidae: Brachistinae) from Turkey. *International Journal of Biological and Ecological Engineering, 12*(11), 424−429.

Kopelke, J. P. (1994). Der Schmarotzerkomplex (Brutparasiten und Parasitoide) der gallenbilden *Pontania* Arten (Insecta: Hymenoptera: Tenthredinidae). *Senckenbergiana Biologica, 73*(1−2), 83−133.

Ku, D. S., Belokobylskij, S. A., & Cha, J. Y. (2001). Hymenoptera (Braconidae). Economic insects of Korea 16. *Insecta Koreana. Supplement, 23*, 283.

Kugler, J. (1966). A list of parasites of lepidoptera from Israel. *Israel Journal of Entomology, 1*, 75−88.

Lashkari Bod, A., Rakhshani, E., Talebi, A. A., & Lozan, A. (2010). Introduction of twelve newly recorded species of Braconidae (Hymenoptera) from Iran. In *Proceedings of the 19th Iranian plant protection congress, 31 July − 3 August 2010* (p. 161). Iranian Research Institute of Plant Protection.

Lashkari Bod, A., Rakhshani, E., Talebi, A. A., Lozan, A., & Žikič, V. (2011). A contribution to the knowledge of Braconidae (Hym., Ichneumonoidea) of Iran. *Biharean Biologist, 5*(2), 147−150.

Lukas, J. (1981). Some new information on the braconids (Hym.-Braconidae) in the southern region of the central part of the River Vah Valley. *Biologia (Bratislava), 36*(8), 649−657.

Mason, W. R. M. (1974). A generic synopsis of Brachistini (Hymenoptera: Braconidae) and recognition of the name *Charmon* Haliday. *Proceedings of the Entomological Society of Washington, 76*(3), 235−246.

Modarres Awal, M. (1997). Family Braconidae (Hymenoptera), pp. 265−267. In M. Modarres Awal (Ed.), *List of agricultural pests and their natural enemies in Iran* (2nd ed., p. 429). Ferdowsi University of Mashhad Press.

Modarres Awal, M. (2012). Family Braconidae (Hymenoptera). In M. Modarres Awal (Ed.), *List of agricultural pests and their natural enemies in Iran* (3rd ed., pp. 483−486). Ferdowsi University of Mashhad Press, 759 pp.

Morice, H. M. (1928). Entomological notes. *Cyprus Agricultural Journal, 23*, 32−33.

Naderian, H., Ghahari, H., & Asgari, S. (2012). Species diversity of natural enemies in corn fields and surrounding grasslands of Semnan province, Iran. *Calodema, 217*, 1−8.

Naderian, H., Penteado-Dias, A. M., Sakenin Chelav, H., & Samin, N. (2020). A faunistic study on Braconidae and Ichneumonidae (Hymenoptera, Ichneumonoidea) of Iran. *Calodema, 844*, 1−9.

Papp, J. (1970). A contribution to the braconid fauna of Israel (Hymenoptera). *Israel Journal of Entomology, 5*, 63−76.

Papp, J. (1981). New species of Braconidae from Hortobágy National Park, Hungary (Hymenoptera). *Acta Zoologica Hungarica, 27*, 369−379.

Papp, J. (1997). Revision of the *Chelonus* species described by A.G. Dahlbom (Hymenoptera, Braconidae: Cheloninae). *Acta Zoologica Academiae Scientiarum Hungaricae, 43*(1), 1−19.

Papp, J. (1998). Contribution to the braconid fauna of Hungary. XIII. Calyptinae-2., Helconinae (Hymenoptera, Braconidae). *Folia Entomologica Hungarica, 59*, 163−184.

Papp, J. (2011−2012). A contribution to the braconid fauna of Israel (Hymenoptera: Braconidae), 3. *Israel Journal of Entomology, 41−42*(2011−2012), 165−219.

Pirouzeh, F. Z., Talebi, A. A., & Farahani, S. (2016). Study on the genus *Triaspis* Haliday, 1835 (Hymenoptera: Braconidae, Brachistinae) in northern Iran, with two new records. *Journal of Insect Biodiversity and Systematics, 2*(4), 395−403.

Priore, R., & Tremblay, E. (1987). Contributo alla revision delle species italiane dei generei *Triaspis* Haliday e *Schizoprymnus* Foerster (Hymenoptera Braconidae). *Bollitino del Laboratorio di Entomologia Agraria Filippo Silvestri, 44*, 47−61.

Rastegar, J., Sakenin, H., Khodaparast, S., & Havaskary, M. (2012). On a collection of Braconidae (Hymenoptera) from East Azarbaijan and vicinity, Iran. *Calodema, 226*, 1−4.

Rondani, C. (1876). Repertorio degli insetti parassiti e delle Loro Vittime. *Bolletino della Societa Entomologica Italiana, 8*, 54−70.

Sakenin, H., Ghahari, H., Lehr, P. A., Ostovan, H., & Havaskary, M. (2009). A contribution to the robber flies (Diptera: Asilidae) from Arasbaran and vicinity, and northwestern Iran. In *Proceedings of the 10th Arab congress of plant protection, 26−30 October 2009, Beirut, Lebanon, p. 42*.

Sakenin, H., Naderian, H., Samin, N., Rastegar, J., Tabari, M., & Papp, J. (2012). On a collection of Braconidae (Hymenoptera) from northern Iran. *Linzer biologische Beiträge, 44*(2), 1319−1330.

Sakenin, H., Coronado-Blanco, M., Samin, N., & Fischer, M. (2018). New records of Braconidae (Hymenoptera) from Iran. *Far Eastern Entomologist, 362*, 13−16.

Sakenin, H., Samin, N., Beyarslan, A., Coronado-Blanco, J. M., Navaeian, M., Fischer, M., & Hosseini Boldaji, S. A. (2020). A faunistic study on braconid wasps (Hymenoptera: Braconidae) from Iran. *Boletin de la Sociedad Andaluza de Entomologia, 30*, 96−102.

Samin, N. (2015). A faunistic study on the Braconidae of Iran (Hymenoptera: Ichneumonoidea). *Arquivos Entomoloxicos, 13*, 339−345.

Samin, N., Sakemin, H., Imani, S., & Shojai, M. (2011). A study on the Braconidae (Hymenoptera) of Khorasan province and vicinity, Northeastern Iran. *Phagea, 39*(4), 137−143.

Samin, N., Ghahari, H., Gadallah, N. S., & Monaem, R. (2015a). A study on the braconid wasps (Hymenoptera: Ichneumonoidea: Braconidae) from Golestan province, northern Iran. *Linzer biologische Beiträge, 47*(1), 731−739.

Samin, N., van Achterberg, C., & Ghahari, H. (2015b). A faunistic study of Braconidae (Hymenoptera: Ichneumonoidea) from southern Iran. *Linzer biologische Beiträge, 47*(2), 1801−1809.

Samin, N., Fischer, M., & Ghahari, H. (2015c). A contribution to the study on the fauna of Braconidae (Hymenoptera, Ichneumonoidea) from the province of Semnan, Iran. *Arquivos Entomoloxicos, 13*, 429−433.

Samin, N., Coronado-Blanco, J. M., Kavallieratos, N. G., Fischer, M., & Sakenin, H. (2018a). Recent findings on Braconidae (Hymenoptera: Ichneumonoidea) of Iran with an updated checklist. *Acta Biologica Turcica, 31*(4), 160−173.

Samin, N., Coronado-Blanco, J. M., Fischer, M., van Achterberg, C., Sakenin, H., & Davidian, E. (2018b). Updated checklist of Iranian Braconidae (Hymenoptera: Ichneumonoidea) with twenty-three new records. *Natura Somogyiensis, 32*, 21−36.

Samin, N., Coronado-Blanco, J. M., Hosseini, A., Fischer, M., & Sakenin Chelav, H. (2019). A faunistic study on the braconid wasps (Hymenoptera: Braconidae) of Iran. *Natura Somogyiensis, 33*, 75−80.

Samin, N., Sakenin Chelav, H., Ahmad, Z., Penteado-Dias, A. M., & Samiuddin, A. (2020a). A faunistic study on the family Braconidae (Hymenoptera: Ichneumonoidea) from Iran. *Scientific Bulletin of Uzhhorod National University (Series: Biology), 48*, 14−19.

Samin, N., Beyarslan, A., Ranjith, A. P., Ahmad, Z., Sakenin Chelav, H., & Hosseini Boldaji, S. A. (2020b). A faunistic study on Braconidae (Hymenoptera: Ichneumonoidea) from Ardebil and East Azarbayjan provinces, Northwestern Iran. *Egyptian Journal of Plant Protection Research Institute, 3*(4), 955–963.

Shahhosseini, M. J., & Kamali, K. (1989). A checklist of insects mites and rodents affecting stored products in Iran. *Journal of Entomological Society of Iran, 5,* 1–47.

Shahand, S., & Karimpour, Y. (2017). Biology of mugwort weevil, *Lixus fasciculatus* (Col.: Curculionidae) on *Artemisia vulgaris* (Asteraceae) in Urmia region. *Biocontrol in Plant Protection, 5*(1), 45–57.

Sharanowski, B. J., Dowling, A. P. G., & Sharkey, M. J. (2011). Molecular phylogenetics of Braconidae (Hymenoptera: Ichneumonoidea), based on multiple nuclear genes, and implications for classification. *Systematic Entomology, 36*(3), 549–572.

Sharanowski, B. J., Zhang, Y. M., & Wanigasekara, R. W. M. U. M. (2014). Annotated checklist of Braconidae (Hymenoptera) in the Canadian Prairies Ecozone. In D. J. Giberson, & H. A. Cárcamo (Eds.), *Biodiversity and systematics Part 2: Vol. 4. Arthropods of Canadian grasslands* (pp. 399–425). Biological Survey of Canada, 479 pp.

Shaw, M. R., & Huddleston, T. (1991). Classification and biology of braconid wasps (Hymenoptera: Braconidae). *Handbooks for the Identification of British Insects, 7,* 1–126.

Stigenberg, J., Boring, C. A., & Ronquist, F. (2015). Phylogeny of the parasitic wasp subfamily Euphorinae (Braconidae) and evolution of its host preferences. *Systematic Entomology, 40,* 570–591.

Szépligeti, G. (1898). Beitrag zur Kenntnis der ungarischen Braconiden Braconiden, 3. Teil. *Természetrajzi Füzetek, 21,* 381–396.

Talebi, A. A., Farahani, S., & Pirouzeh, F. Z. (2018). Six new record species of the genus *Schizoprymnus* Förster, 1862 (Hymenoptera: Braconidae, Brachistinae) from Iran. *Entomofauna, 39*(1), 325–333.

Telenga, N. A. (1935). Neue und weniger bekannte palaearktische Braconiden (Hym.). *Arbeiten über Physiologische und Angewandte Entomologie, 2,* 271–275.

Telenga, N. A. (1941). Family Braconidae, subfamily Braconinae (continuation) and Sigalphinae. Fauna USSR. *Hymenoptera, 5*(3), 466.

Tobias, V. I. (1976). *Braconids of the Caucasus (Hymenoptera, Braconidae). Opred. Faune SSSR* (Vol. 110). Leningrad: Nauka Press, 286 pp.

Tobias, V. I. (1986). Subfam. Euphorinae. In G. S. Medvedev (Ed.), *Opredelitel Nasekomych Evrospeiskoi Tsasti SSSR 3, Peredpontdatokrylye 4. Opr. Faune SSSR* (Vol. 145, pp. 1–501) (Keys to the insects of the European part of USSR. Hymenoptera).

van Achterberg, C. (1976). A revision of the tribus Blacini (Hymenoptera, Braconidae, Helconinae). *Tijdschrift Voor Entomologie, 118*(7), 159–322.

van Achterberg, C. (1988). Revision of the subfamily Blacinae Foerster (Hymenoptera: Braconidae). *Zoologische Verhandlingen, 249,* 1–324.

van Achterberg, C. (1990). Revision of the genera *Foersteria* Szépligeti and *Polydegmon* Foerster (Hymenoptera: Braconidae) with the description of a new genus. *Zoologische verhandelingen, 275,* 1–32.

van Achterberg, C. (1993). Illustrated key to the subfamilies of the Braconidae (Hymenoptera: Ichneumonoidea). *Zoologische Verhandelingen, 238,* 1–189.

van Achterberg, C. (2014). Notes on the checklist of Braconidae (Hymenoptera) from Switzerland. *Mitteilungen der Schweizerischen Entomologischen Gesellschaft (Bulletin de la Société Entomologique Suisse), 87,* 191–213.

van Achterberg, C., & Shaw, S. R. (2001). *Tainiterma,* a new genus of the subfamily of Euphorinae (Hymenoptera: Braconidae) from Vietnam and China. *Zoologische Mededelingen Leiden, 75*(3), 69–78.

Yilmaz, T., & Beyarslan, A. (2009). A new species of *Chelostes* van Achterberg, 1990 (Hymenoptera: Braconidae: Brachistinae) from Turkey. *Biologia (Bratislava), 64*(2), 340–342.

Yu, D. S., van Achterberg, C., & Horstmann, K. (2016). *Taxapad 2016, Ichneumonoidea 2015, database on flash-drive.* Nepean, Ontario, Canada.

Coeloides scolyticida Wesmael, 1838 (Braconinae), ♀, lateral habitus. *Photo prepared by S.R. Shaw.*

Cyanopterus sp. (Braconinae), ♀, lateral habitus. *Photo prepared by S.R. Shaw.*

Chapter 6

Subfamily Braconinae Nees von Esenbeck, 1811

Donald L.J. Quicke[1], Neveen Samy Gadallah[2], Hassan Ghahari[3] and Scott Richard Shaw[4]

[1]Integrative Ecology Laboratory, Department of Biology, Faculty of Science, Chulalongkorn University, Pathumwan, Bangkok, Thailand; [2]Entomology Department, Faculty of Science, Cairo University, Giza, Egypt; [3]Department of Plant Protection, Yadegar-e Imam Khomeini (RAH) Shahre Rey Branch, Islamic Azad University, Tehran, Iran; [4]UW Insect Museum, Department of Ecosystem Science and Management, University of Wyoming, Laramie, WY, United States

Introduction

Braconinae is one of the largest subfamilies in the family Braconidae (Hymenoptera: Ichneumonoidea) (Quicke, 1987). Braconines are one of the most diverse lineages of any insects, constituting more than 3054 described species in 190 genera (Yu et al., 2016) and 12 tribes (Chen & van Achterberg, 2019; Yu et al., 2016). Most braconine species are indigenous to tropical and subtropical areas of the Old World, and there are comparatively fewer genera and species in temperate regions (Mason, 1978; Quicke, 1987; Yu et al., 2016). Members of the subfamily Braconinae are often brightly colored species, with black, red, orange, yellow, and/or white patterns. The Middle Eastern species are small to large-sized insects, characterized by their concave labrum, absence of an epicnemial carina, absence of an occipital carina (although it can be weakly developed in the putatively basal *Argamania*), and females having moderately long to very long ovipositors (Sharkey, 1993).

Most of the Braconinae are idiobiont ectoparasitoids of concealed or semiconcealed holometabolous larvae of numerous xylophagous and stem-boring moth caterpillars or beetle larvae. Several species are known to parasitize concealed fly larvae (Diptera), predominantly gall midges (Cecidomyiidae) and fruit flies (Tephritidae) (Loni et al., 2016), as well as some sawfly larvae (Hymenoptera, Tenthredinidae). Exceptionally, some extralimital genera are known to be gregarious endoparasitoids of exposed butterfly pupae (Quicke, 2015; Quicke & van Achterberg, 1990; Shaw & Huddleston, 1991; van Achterberg, 1984, 1988, 1993). Accordingly, the Braconinae represent a large and powerful biological asset (e.g., Baird, 1958; Lewis et al., 1990; Quicke, 1983, 1987; Thomson, 1953) against various pests from the holometabolous insect orders (Coleoptera, Diptera, Hymenoptera, and Lepidoptera). Several species attack stored products and field crops pests (Quicke, 2015).

Important early work on Braconinae was done by Szépligeti (1904) who constructed a key to 32 of genera known at that time. Shenefelt's 18th part of Catalogus Hymenopterorum (1978) has greatly facilitated many studies, as it covered systematic as well as biological data for all braconine taxa up to that date. Quicke (1987) provided a key for 123 of the Old World genera, and five new genera were also described in that paper. Additional taxonomic studies were also provided by Mason (1978), van Achterberg (1980, 1983); and Quicke (1981, 1982, 1983a,b,c, 1985a,b,c).

Braconines are diagnosed by the absence of an occipital carina (at most represented by an angulation of the head capsule); by having the clypeus ventrally depressed and forming the dorsal part of hypoclypeal depression; by having the prepectal carina entirely absent; by the mesopleuron being usually flat, without a wide depression; by the fore wing having vein 1cu-a more or less opposing the 1M vein (except *Argamania*); by having the hind wing with the vein 1-M at least 1.5× the vein M+CU length (except *Argamania*), somewhat broad basally, and by the hind wing subbasal cell being small (at most as long as one third basal cell) (Shaw & Huddleston, 1991; van Achterberg, 1993).

Several tribe level taxa have been created in the past based almost entirely on the possession of one or two character states without reference to phylogenetics, and largely based on the Russian fauna (Quicke 2015). Van Achterberg (1989) effectively synonymized all members of a large clade largely characterized by possession of a ventrally elongate scapus within the Aphrastobraconini, a group that had previously been recognised only on the basis of the derived wing venation of the type genus and relatives. It should be noted that scapus shape is somewhat homeoplastic also in the Doryctinae and therefore probably adaptive.

Checklists of Regional Braconinae. Farahbakhsh (1961): two species in one genus (*Bracon* Fabricius, 1804); Modarres Awal (1997, 2012): six and eight species in four genera, respectively; Fallahzadeh and Saghaei (2010) listed 33 species in seven genera (without precise localities); Gadallah and Ghahari (2015) and Farahani et al. (2016): both with 115 species and subspecies in 11 genera; Yu et al. (2016): 114 species in 10 genera; Samin, Coronado-Blanco, Kavallieratos et al. (2018), 120 species in 10 genera;

Samin, Coronado-Blanco, Fischer et al. (2018), 122 species in 10 genera. The present checklist includes 347 species in 35 genera and four tribes (Aphrastobraconini, Argamaniini, Braconini, and Coeloidini) in the Middle Eastern countries. Three new records, *Atanycolus genalis* (Thomson), *Bracon curticaudis* Szépligeti, and *Bracon tenuicornis* Wesmael, are here added for the Iranian fauna. Here we follow Chen and van Achterberg for the tribal classification, and Yu et al. (2016) for the general distribution of species, as well as other recent papers in some cases.

This chapter is dedicated to the late Jenő Papp (May 20, 1933−December 11, 2017; see Ghahari and Beyarslan, 2019) *for his invaluable cooperation with H. Ghahari by identifying many specimens of Iranian* Braconidae.

Key to the genera of the subfamily Braconinae in the Middle East (modified from Quicke, 1987—based on females)

1. Metasomal spiracles in the epipleura; head, mesosoma, and metasoma completely granulate; blackish but with distinctly bronze metallic sheen; hind wing vein M+CU approximately 0.5 × length of 1-M [Body length less than 5 mm] .... *Argamania* Enderlein
− Metasomal spiracles in the tergum; head, mesosoma, and metasoma not completely granulate, usually shiny; variously colored, may be black but never with metallic sheen; hind wing vein M+CU less than 0.5 × length of 1-M 2
2. Fore tibia with two well-developed apical spurs [large colourful wasps, body length >1.5 cm] .... *Rhammura* Enderlein
− Fore tibia with a single apical spur; outer mid and hind tibial spurs more or less setose 3
3. First, and usually second and third basal flagellomeres conspicuously expanded at apex, especially ventrally .... *Coeloides* Wesmael
− Basal flagellomeres normal, not expanded 4
4. First three metasomal tergites forming a carapace, completely hiding the following tergites, sutures between them largely indistinct; posterior margin of carapace produced into a pair of submedian spines .... *Physaraia* Shenefelt
− First and second metasomal tergites movably joined, not fused laterally, not separated by a suture 5
5. First subdiscal cell of fore wing oval, formed in part of rather thickened veins, and in particular vein CU1b usually conspicuously widened anteriorly; ovipositor without a dorsal nodus .... *Megalommum* Szépligeti
− First subdiscal cell of fore wing not or only slightly ovoid, or if rather petiolate and formed of rather thickened veins with CU1b marginally thicker than 3-CU1, then ovipositor is more than 1.5× as long as body and/or has a preapical dorsal nodus 6
6. Ovipositor with three to four narrowly separated distinct arched zones distally .... *Zaglyptogastra* Ashmead
− Ovipositor without such arched zones 7
7. Fifth metasomal tergite with well-developed, spaced, denticulate projections posteriorly 8
− Fifth metasomal tergite not denticulate posteriorly, although sometimes emarginate medially 9
8. Scapus large, emarginate on both apico-laterally and apico-medially; second submarginal cell of fore wing expanded distally; vein 3-SR sigmoid; second metasomal tergite with well-marked antero-lateral areas, with only a small, usually striated midbasal area; fourth and fifth metasomal tergites often rather raised around midline .... *Rhytimorpha* Szépligeti
− Scapus small, rather narrow, oblique at apex, not emarginate apico-medially; second submarginal cell of fore wing long, more or less parallel-sided; vein 3-SR of fore wing not sigmoid; second metasomal tergite with a medio-basal triangular area that is usually produced posteriorly into a longitudinal median carina, and with a small antero-lateral areas that give rise to a pair of posteriorly converging carinae, bordered by lateral grooves .... *Soter* Saussure
9. Last antennal flagellomere blunt, more or less conspicuously compressed laterally; vein 1- SR+M of fore wing straight or slightly curved toward wing margin after arising from 1-M; face without a vertical lamella, not conspicuously produced into a plate-like suture ventrally 10
− Last antennal flagellomere not blunt, and not compressed laterally, or if blunt and rather compressed, then either vein 1-SR+M of fore wing distinctly bent or curved away from anterior wing margin after arising from vein 1-M, or face has vertical lamella and is produced ventrally into a plate-like suture 11

10. Scapus longer ventrally than dorsally, more or less cylindrical; labio-maxillary complex not elongate; hamuli at apex of vein C+SC+R of hind wing normal, robust .... *Rhadinobracon* Szépligeti
— Scapus shorter ventrally than dorsally, subglobose; labio-maxillary complex elongate; hamuli at apex of vein C+SC+R of hind wing blunt and peg-like .... *Glyptomorpha* Holmgren
11. Scapus subcylindrical, with a very large apico-medial ledge, sharply and concavely narrowed at base; pedicellus always long and conspicuously protruding medially and petiolate when seen in dorsal view 12
— Scapus variable, sometimes with a weak to moderate apico-medial ledge, but never conspicuously protruding medially and petiolate, occasionally rather cylindrical and marginally narrowing basally when seen in dorsal view 14
12. Apical ledge of scapus comprising two distinct regions, a large ventral to lateral crescent and a smaller one sided at the apex of the medial prominence; face usually with a conspicuous depression above clypeus, bordered dorsally by lamelliform, horizontal ridge, then this ridge is not distinctly produced anteriorly .... *Odontoscapus* Kriechbaumer
— Apical ledge of scapus composed of a single large ventral to lateral crescent; face without a lamella, and not distinctly compressed above clypeus 13
13. Third metasomal tergite usually distinctly striated or rugose, and always with a subposterior transverse crenulate groove .... *Monilobracon* Quicke
— Third metasomal tergite largely smooth and shiny, without transverse subposterior groove .... *Atanycolus* Foerster
14. Face with a well-developed anterior cariniform protruberance(s) below antennal sockets 15
— Face without cariniform protruberances below antennal sockets, rarely with a protruberance at the level of antennal sockets, or with a more or less gently rounded raised medial area, or mid-longitudinal ridge 16
15. Lower part of face not forming a horizontal ledge or spoon-shaped structure, instead, the middle and upper part of face forming a W-shaped lamella, that is well-separated from clypeal carina; first metasomal tergite without dorsal carinae, or rarely are represented by a pair of weak ridges at the extreme anterior part of a raised median area .... *Zanzopsis* van Achterberg
— Lower part of face more or less including dorsal margin of clypeus, produced medially to form a semicircular ledge or a narrowed spoon-shaped structure at base, that is more or less simple, with a weak mid-longitudinal ridge, or with a well-developed curving spine dorsally .... *Plaxopsis* Szépligeti
16. Tarsal claws with distinctly pointed or at least angular basal lobe, rarely rather square and lamelliform, or with the basal lobe produced into a distinct tooth 17
— Tarsal claws with rounded basal lobe, or if protrudes, then not pointed or sharply angled distally, not formed into a large protruding, rather square lamella 24
17. Vein SR1 of fore wing reaching wing margin less than 0.65 of the way from pterostigmal apex to wing tip; labio-maxillary complex not elongate; suture between second and third metasomal tergites with well-develop crenulations .... *Vipiomorpha* Tobias
— Vein SR1 of fore wing reaching wing margin more than 0.7 the way from pterostigmal apex to wing tip or intermediate, then labio-maxillary complex is elongate, and/or suture of second metasomal tergite not crenulate 18
18. First metasomal tergite more than 2.6× as long as median width, lateral areas largely membranous, spiracles located within the unsclerotized region; hind tibia rather swollen posteriorly, with a lateral longitudinal groove, rather densely setose; metasoma smooth and shiny, second suture smooth .... *Amyosoma* Viereck
— First metasomal tergite less than 2.5× as long as median width, with well-developed and often crenulate lateral areas, sometimes with dorso-lateral carinae; second tergite less triangular, not largely membranous antero-laterally; spiracles placed within the sclerotized area of the tergum; hind tibia not swollen and grooved 19
19. Hind tibia strongly flattened and with well-developed longitudinal lateral groove 20
— Hind tibia simple, not compressed or with lateral groove 21
20. Hind femur with a conspicuous tooth near mid-ventral margin; scapus simple, shorter ventrally than dorsally in lateral aspect .... *Braconella* Szépligeti
— Hind femur without ventral tooth; scapus distinctly elongate ventrally .... *Doggerella* Quicke, Mahmood & Papp
21. Face forming a Y-shaped projection between antennal sockets .... *Ceratobracon* Telenga
— Face without a Y-shaped projection between antennal sockets 22
22. Hind telotarsus (excluding the claw) at least 0.9× as long as hind basitarsus; claws with very large, ventrally rather flattened basal lobe; median area of metanotum with a complete longitudinal carina .... *Baryproctus* Ashmead
— Hind telotarsus (excluding claw) less than 0.8× as long as hind basitarsus; basal lobe of claws smaller, rounded ventrally; median area of metanotum at most forming a point or short anterior carina, always smooth medio-posteriorly 23
23. Vein 3-SR of fore wing less than 1.5× as long as vein r (usually less than 1.2); vein 2-SR+M of fore wing relatively long; vein r distinctly sinuate .... *Habrobracon* Ashmead
— Vein 3-SR of fore wing more than 1.6× as long as vein r (usually more than 1.9); vein 2-SR+M of fore wing shorter; vein r nearly straight .... *Bracon* Fabricius

24. Vein 1-SR+M of fore wing straight or slightly curved toward anterior margin of wing after arising from 1-SR; vein C+SC+R forming an angle of 50° with vein 1-SR; vein r of fore wing at least 0.69× as long as vein m-cu 25

— Vein 1-SR+M of fore wing usually distinctly curved after arising from vein 1-SR, either evenly or sharply, or if vein 1-SR+M straight or slightly curved anteriorly, then vein 1-SR forming an angle of 55° with vein C+SC+R, and vein r of fore wing less than 0.65× as long as m-cu 27

25. Propodeum almost completely coarsely foveate to foveate-rugose, sometimes also with a mid-longitudinal carina .... *Odesia* Cameron

— Propodeum almost completely smooth and shiny, except for some punctures at base of setae when present; without a median longitudinal carina 26

26. Third and fourth tarsal articles strongly produced ventrolaterally and fifth tarsal article strongly arched and widening from a narrow base; vein 2-SC+R of hind wing distinctly transverse; scutellar sulcus smooth .... *Bathyaulax* Szépligeti

— Third and fourth tarsal articles not produced ventrolaterally and fifth article not strongly arched; vein 2-SC+R of hind wing interstitial or distinctly longitudinal; scutellar suture at least with weak crenulations .... *Stenobracon* Szépligeti

27. Marginal cell of fore wing short, vein SR1 reaching wing margin less than 0.58 of the way between pterostigmal apex and wing tip; second submarginal cell of fore wing distinctly expanded distally; scapus small, shorter ventrally than dorsally, not emarginate apico-laterally; vein 1-SR+M of fore wing straight or slightly curved toward anterior wing margin after arising from vein 1-M; labio-maxillary complex very long 28

— Marginal cell of fore wing long, vein SR1 reaching wing margin more than 0.6 of the way between pterostigmal apex and wing tip, or if marginal cell shorter, then scapus distinctly longer ventrally than dorsally and usually emarginate apico-laterally; labio-maxillary complex not elongate, and/or second submarginal cell of fore wing not expanded distally 29

28. Clypeal guard setae fused apically to form a pair pencil-hair brush like structures .... *Vipio* Latreille

— Clypeal guard setae separated, not fused apically .... *Liomorpha* Szépligeti

29. Dorsal valves of ovipositor smooth, lower valves smooth or with very weak serrations only near extreme apex; ovipositor sheaths shorter than body 30

— Dorsal valves of ovipositor at least with a sharp preapical angle, usually with a well-developed nodus or notch 31

30. Second metasomal tergite with a large smooth mid-basal triangular area reaching more than half way along the tergite .... *Bracomorpha* Papp

— Second metasomal tergite without a mid-basal area or at most with small striated or rugose area which does not reach 0.25× tergite length .... *Iphiaulax* Foerster

31. Marginal cell of fore wing long, vein SR1 reaching wing margin more than 0.8 way between pterostigmal apex and wing tip; second metasomal tergite usually with a well-developed and clearly distinct mid-basal area 32

— Marginal cell of fore wing short, vein SR1 reaching wing margin less than 0.7 way between pterostigmal apex and wing tip; second metasomal tergite without a mid-basal triangular area, with a pair of posteriorly diverging grooves surrounding a large antero-lateral areas; scapus subglobose, shorter ventrally than dorsally, or at most only slightly longer than dorsally, truncate or weakly emarginate apico-laterally .... *Pseudovipio* Szépligeti

32. Scapus small, shorter ventrally than dorsally, or rarely about as long as ventrally as dorsally, not emarginate apico-laterally .... *Bracon* (part)

— Scapus usually much longer ventrally than dorsally in lateral aspect, or if only slightly longer ventrally than dorsally, then strongly emarginate apico-laterally and often distinctly emarginate apico-medially 33

33. Ovipositor sheaths more than 1.2× as long as fore wing; median area of first metasomal tergite either conspicuously striated or rugose, or with distinct median longitudinal dorsal carina; third and fourth tergites conspicuously striated, rugose or foveate; metasoma more or less long, narrow and parallel-sided 34

— Ovipositor sheaths less than 1.1× as long as fore wing, or if longer, then median area of second tergite, and third and fourth tergites are virtually entirely smooth and shiny; metasoma not especially long and parallel-sided, often either widely oval in dorsal view or conspicuously compressed laterally 36

34. Face coarsely reticulate to foveate; medioposterior margin of propodeum with simple or with only a pair of weak carinae 35

— Face smooth and shiny, or with fine coriaceous sculpture but still shiny; medioposterior margin of propodeum with several short carinae demarcating a line of pits .... *Monilobracon* Quicke

35. Notauli well-developed; median area of first metasomal tergite and most of second and third tergites with a reticulate pattern composed of triangular and kite-shaped area bordered by ridges; metasomal third tergite usually with a middle longitudinal, striate band which may be elevated .... *Bacuma* Cameron

—Notauli absent or virtually so; metasoma largely longitudinally striate except for midbasal triangular area of tergite two; metasoma third tergite usually without a distinctly differentiated, longitudinal, band .... *Merinotus* Szépligeti

36. Third and/or fourth metasomal tergites with a well-developed (usually) crenulate transverse subposterior groove, rest of tergite usually coarsely sculptured .... *Cyanopterus* Haliday
— Third and fourth metasomal tergites without a transverse crenulate subposterior groove, rest of these tergites largely smooth and shiny 37

37. Third metasomal tergite more than 2.75× as wide as its minimum length; hind wing with setosity not or hardly reduced near the base; anterior 0.6 of subbasal cell and base of discal and subdiscal cell more or less densely setose, occasionally glabrous along the flexure line running in front of vein 1-1A .... *Campyloneurus* Szépligeti
— Third metasomal tergite less than 2.7× as wide as its minimal length; hind wing largely glabrous basally, at most with a narrow setose band along the anterior edge of the subbasal cell and with a large glabrous patch at base of discal and subdiscal cell extending along the whole length of vein cu-a .... *Tsavobracon* Quicke

List of species of the subfamily Braconinae recorded in the Middle East

Subfamily Braconinae Nees von Esenbeck, 1811

Tribe Aphrastobraconini Ashmead, 1900

Genus *Atanycolus* Foerster, 1863

Atanycolus denigrator (Linnaeus, 1758)

Catalogs with Iranian records: No catalog.
Distribution in Iran: Golestan (Ghahari et al., 2019).
Distribution in the Middle East: Iran (Ghahari et al., 2019), Israel–Palestine (Papp, 1989).
Extralimital distribution: Austria, Bulgaria, China, Croatia, Finland, France, Germany, Greece, Hungary, Italy, Kazakhstan, Korea, Mongolia, Niger, Norway, Poland, Portugal, Russia, Serbia, Slovakia, Sweden, Switzerland, United Kingdom.
Host records: Summarized by Yu et al. (2016) as being a parasitoid of the buprestids *Chrysobothris chrysostigma* (L.), *Chrysobothris solieri* Gory and Laporte, *Lampra rutilans* (Fabricius), and *Poecilonota variolosa* (Paykull); the cerambycids *Acanthocinus aedilis* (L.), *Acanthocinus griseus* (Fabricius), *Arhopalus syriacus* (Reitter), *Monochamus galloprovincialis* (Olivier), *Monochamus sutor* (L.), *Rhagium inquisitor* (L.), *Rhagium mordax* (DeGeer), *Saperda populnea* (L.), *Tetropium castaneum* (L.), *Tetropium fuscum* (Fabricius), and *Tetropium gabrieli* Weise; and the curculionid *Ips sexdentatus* (Boerner & I.C.H.).

Atanycolus fulviceps (Kriechbaumer, 1898)

Catalogs with Iranian records: No catalog.
Distribution in Iran: West Azarbaijan (Gadallah et al., 2021).
Distribution in the Middle East: Iran (Gadallah et al., 2021), Turkey (Beyarslan 1999, 2014; Fahringer, 1926).
Extralimital distribution: Former Czechoslovakia, Germany, Hungary, Ukraine.
Host records: Summarized by Yu et al. (2016) as being a parasitoid of the buprestids *Lampra mirifica* (Mulsant), and *Lampra rutilans* (Fabricius).

Atanycolus genalis (Thomson, 1892)

Catalogs with Iranian records: This species is a new record for the fauna of Iran.
Distribution in Iran: Semnan province, Shahrud, Jangal-e Abr, 2♀, June 2011.
Distribution in the Middle East: Iran (new record), Turkey (Beyarslan, 2016).
Extralimital distribution: Germany, Hungary, Italy, Kazakhstan, Mongolia, Poland, Sweden.
Host records: Summarized by Yu et al. (2016) as being a parasitoid of the bupresids *Lampra miridica* (Mulsant), *Melanophila cyanea* (Fabricius), and *Melanophila guttulata* (Olivier); the cerambycids *Acanthocinus aedilis* (L.), *Arhopalus rusticus* (L.), *Asemum striatum* (L.), *Callidium rufipennis* (Motschulsky), *Cerambyx scopolii* Fuessly, *Leioderus kollari* Redtenbacher, *Monochamus galloprovincialis* (Olivier), *Phymatodes pusillus* (Fabricius), *Rhagium bifasciatum* Fabricius, *Rhagium inquisitor* (L.), *Stenostola ferrea* (Schrank), and *Tetropium fuscum* (Fabricius); the curculionids *Hylesinus crenatus* (Fabricius), *Ips subelongatus* Motschulsky, and *Tomicus piniperda* (L.); and the sesiid moths *Synanthedon flaviventris* (Staudinger), and *Synanthedon vespiformis* (L.).

Atanycolus initiator (Fabricius, 1793)

Catalogs with Iranian records: No catalog.
Distribution in Iran: Ardabil (Samin, Beyarslan, Ranjith et al., 2020).
Distribution in the Middle East: Iran (Samin, Beyarslan, Ranjith et al., 2020), Turkey (Beyarslan, 1999, 2014).
Extralimital distribution: Austria, Azerbaijan, Belgium, China, Croatia, Czech Republic, Finland, France, Germany, Hungary, Italy, Japan, Latvia, Lithuania, Mongolia, Norway, Poland, Russia, Slovakia, Spain, Sweden, Switzerland, Turkmenistan, Ukraine, United Kingdom.
Host records: Summarized by Yu et al. (2016) as being a parasitoid of the buprestids *Chrysobothris solieri* Groy and Laporte, *Lampra mirifica* (Mulsant), *Melanophila cyanea* (Fabricius), and *Melanophila guttulata* (Orlinski); the following cerambycids: *Acanthocinus aedilis* (L.), *Acanthocinus griseus* (Fabricius), *Acanthocinus reticulatus* (Razoumowsky), *Arhopalus coreanus* (Sharp), *Arhopalus*

rusticus (L.), *Arhopalus syriacus* (Reitter), *Asenum striatum* (L.), *Callidium abdominale* (Olivier), *Callidium coriaceum* Paykull, *Callidium rufipennis* (Motschulsky), *Cerambyx scopolii* Fuessly, *Clytus* sp., *Leioderus kollari* Redtenbacher, *Monochamus galloprovincialis* (Olivier), *Monochamus sutor* (L.), *Phymatodes pusillus* (Fabricius), *Rhagium bifasciatum* Fabricius, *Rhagium inquisitor* (L.), *Saperda punctata* (L.), *Stenostola ferrea* (Schrank), *Tetropium castaneum* (L.), *Tetropium fuscum* (Fabricius), *Tetropium gabrieli* Weise, and *Tetropium gracilicorne* Reitter; the curculionids *Blastophagus piniperda* (L.), *Ips subelongatus* Motschulsky, *Ips topographus* (L.), *Niphades variegatus* J. Faust, *Pissodes notatus* (Fabricius), *Pissodes obscurus* Roelofs, and *Scolytus* sp.; the hymenopteran argids *Arge cyaneocrocea* (Foerster), and *Arge pagana* (Panzer); and the lepidopteran sesiids *Sesia apiformis* Clerck, *Synanthedon flaviventris* (Staudinger), and *Synanthedon vespiformis* (L.). In Iran, this species has been reared from the buprestid *Chrysobothris affinis* (Fabricius) (Samin, Beyarslan, Ranjith et al., 2020).

Atanycolus ivanowi (Kokujev, 1898)

Catalogs with Iranian records: Gadallah and Ghahari (2015), Farahani et al. (2016), Yu et al. (2016), Samin, Coronado-Blanco, Kavallieratos et al. (2018), Samin, Coronado-Blanco, Fischer et al. (2018).

Distribution in Iran: Alborz (Davatchi and Shojai, 1969; Shojai, 1968, 1998; Zargar, Talebi, Hajiqanbar, Farahani, & Ameri, 2014), Fars (Lashkari Bod et al., 2011 as *A. sculpturatus* Thomson), Golestan (Ghahari & Fischer, 2011c), Guilan (Zargar, Talebi, Hajiqanbar, Farahani, & Ameri, 2014), Ilam (Ghahari, Fischer, & Papp, 2011c as *A. sculpturatus*), Kerman (Iranmanesh et al., 2018), Kordestan, Marzaki, Semnan (Dastgheyb Beheshti, 1980; Davatchi & Shojai, 1969; Modarres Awal, 1997; Radjabi, 1976, 1991; Shojai, 1968 all as *A. sculpturatus*), Qazvin (Ghahari & Beyarslan, 2017; Zargar, Talebi, Hajiqanbar, Farahani, & Ameri, 2014), Tehran (Dastgheyb Beheshti, 1980; Davatchi & Shojai, 1969; Modarres Awal, 1997; 2012 as *A. sculpturatus*; Radjabi, 1976, 1991; Shojai, 1968; Zargar, Talebi, Hajiqanbar, Farahani, & Ameri, 2014), Iran (no specific locality cited) (Behdad, 1991 as *A. sculpturatus*; Shestakov, 1926).

Distribution in the Middle East: Iran (see references above), Turkey (Beyarslan, 2014; Beyarslan & Çetin Erdoğan, 2012; Bolu et al., 2009).

Extralimital distribution: Armenia, Austria, Azerbaijan, China, Croatia, Czech Republic, Finland, France, Germany, Greece, Hungary, Italy, Japan, Kazakhstan, Portugal, Russia, Serbia, Slovakia, Switzerland, Tajikistan, Turkmenistan, Ukraine, Uzbekistan (Yu et al., 2016), South Korea (Samartsev & Ku, 2021).

Host records: Summarized by Yu et al. (2016) as being a parasitoid of buprestids *Anthaxia aurulenta* (Fabricius), *Chrysobothris solieri* Gory and Laporte, *Lampra mirifica* (Mulsant), *Sphenoptera tappesi* Marseul, and *Trachypterus picta* (Pallas); and the cerambycids *Arhopalus syriacus*

(Reitter), *Stictoleptura rubra* (L.), *Monochamus galloprovincialis* (Olivier), *Osphrateria coerulescens inaurata* (Holzschuh), and *Tetropium fuscum* (Fabricius). In Iran, this species has been reared from the curculionid *Scolytus rugulosus* (Müller) (Behdad, 1991; Modarres Awal, 1997, 2012; Davatchi & Shojai, 1969; Shojai, 1998 as *Ruguloscolytus mediterraneus* Eggers), and the buprestids *Sphenoptera servistana* Obenberger (Behdad, 1991; Modarres Awal, 1997, 2012 both as *Sphenoptera kambyses*), *Chrysobothris affinis* (Fabricius), *Sphenoptera davatchii* Descarpentries (Modarres Awal, 1997, 2012), *Sphenoptera tappesi* Marseul on a peach tree (Ghahari & Fischer, 2011c), and *Trachypteris picta* (Pallas) (Ghahari & Beyarslan, 2017).

Plant associations in Iran: Apple orchards (Lashkari Bod et al., 2011), *Medicago sativa* L. (Fabaceae) (Iranmanesh et al., 2018).

Atanycolus neesii (Marshall, 1897)

Distribution in the Middle East: Israel—Palestine (Papp, 2015), Turkey (Beyarslan, 1999, 2014; Fahringer, 1922, 1926).

Extralimital distribution: Austria, Czech Republic, France, Germany, Hungary, Italy, Kazakhstan, Liechtenstein, Russia, Slovakia, Spain, Switzerland, Ukraine.

Host records: Summarized by Yu et al. (2016) as being a parasitoid of the following buprestids: *Agrilus biguttatus* Fabricius, *Agrilus pannonicus* (Piller and Mitterpacher), *Agrilus viridis* (L.), *Lampra decipiens* Mannerheim, *Lampra rutilans* (Fabricius), *Melanophila cyanea* (Fabricius), and *Poecilonota variolosa* (Paykull); the following cerambycids: *Acanthocinus aedilis* (L.), *Arhopalus rusticus* (L.), *Mesosa nebulosa* (Fabricius), *Molorchus minor* (L.), *Plagionotus arcuatus* (L.), *Pyrrhidium sanguineum* L., *Rhagium inquisitor* (L.), *Rutpela maculata* (Poda), *Saperda carcharias* (L.), *Tetropium fuscum* (Fabricius), and *Tetropium gabrielli* Weise; the curculionid *Cryptorhynchus lapathi* (L.); and the sesiids *Sesia apiformis* (Clerck), and *Synanthedon vespiformis* (L.).

Atanycolus sp.

Distribution in Iran: Ilam (Jozeyan et al., 2017).

Host records in Iran: Oak wood borer beetles (Jozeyan et al., 2017).

Genus *Bacuma* Cameron, 1906

Bacuma maculipennis (Szépligeti, 1914)

Distribution in the Middle East: Egypt (Fahringer, 1926).

Extralimital distribution: Congo, Ethiopia, Kenya, Somalia, Sudan, Tanzania.

Host records: Unknown.

Genus *Bathyaulax* Szépligeti, 1906

Bathyaulax fortisulcatus (Strand, 1912)

Distribution in the Middle East: Syria (Quicke & Koch, 1990; Strand, 1912).

Extralimital distribution: Kenya.

Host records: Unknown.

***Bathyaulax fritzeni* Kaartinen and Quicke, 2007**
Distribution in the Middle East: Saudi Arabia (Kaartinen & Quicke, 2007).
Extralimital distribution: None.
Host records: Unknown.

***Bathyaulax juhai* Kaartinen and Quicke, 2007**
Distribution in the Middle East: Saudi Arabia (Kaartinen & Quicke, 2007).
Extralimital distribution: None.
Host records: Unknown.

***Bathyaulax kersteni* (Gerstaeker, 1870)**
Distribution in the Middle East: Egypt, Yemen (including Socotra) (Brues, 1926; Fahringer, 1935; Kirby, 1903; Kohl, 1907).
Extralimital distribution: Angola, Democratic Republic of Congo, Eritrea, Ethiopia, Kenya, Mozambique, Namibia, Somalia, South Africa, Tanzania.
Host records: Unknown.

***Bathyaulax kossui* Kaartinen and Quicke, 2007**
Distribution in the Middle East: Yemen (Kaartinen & Quicke, 2007).
Extralimital distribution: None.
Host records: Unknown.

***Bathyaulax marjae* Kaartinen and Quicke, 2007**
Distribution in the Middle East: Yemen (Kaartinen & Quicke, 2007).
Extralimital distribution: None.
Host records: Unknown.

***Bathyaulax ollilae* Kaartinen and Quicke, 2007**
Distribution in the Middle East: Saudi Arabia (Kaaertinen & Quicke, 2007).
Extralimital distribution: None.
Host records: Unknown.

***Bathyaulax ruber* (Bingham, 1902)**
Distribution in the Middle East: Saudi Arabia (Kaartinen & Quicke, 2007).
Extralimital distribution: Ivory Coast, Kenya, Malawi, Sudan, Uganda, Zimbabwe.
Host records: Unknown.

***Bathyaulax somaliensis* (Szépligeti, 1914)**
Distribution in the Middle East: Egypt (Fahringer, 1926).
Extralimital distribution: Somalia.
Host records: Unknown.

***Bathyaulax striolatus* (Szépligeti, 1914)**
Distribution in the Middle East: Egypt (Fahringer, 1926), Saudi Arabia (Quicke, 1985a).
Extralimital distribution: Morocco, Sierra Leone, Sudan.
Host records: Unknown.

***Bathyaulax syraensis* (Strand, 1912)**
Distribution in the Middle East: Syria (Quicke & Koch, 1990; Strand, 1912).
Extralimital distribution: None.
Host records: Unknown.

***Bathyaulax varipennis* (Szépligeti, 1914)**
Distribution in the Middle East: Egypt (Fahringer, 1926).
Extralimital distribution: Somalia.
Host records: Unknown.

Genus *Campyloneurus* Szépligeti, 1900

***Campyloneurus manni* Fahringer, 1928**
Distribution in the Middle East: Syria (Fahringer, 1928).
Extralimital distribution: None.
Host records: Unknown.

Genus *Glyptomorpha* Holmgren, 1868

***Glyptomorpha* (*Glyptomorpha*) *discolor* Thunberg, 1822**
Catalogs with Iranian records: Gadallah and Ghahari (2015), Farahani et al. (2016), Yu et al. (2016), Samin, Coronado-Blanco, Kavallieratos et al. (2018), Samin, Coronado-Blanco, Fischer et al. (2018).
Distribution in Iran: Golestan (Ghahari, Fischer, Erdoğan, Tabari et al., 2009).
Distribution in the Middle East: Iran (Ghahari, Fischer, Erdoğan, Tabari et al., 2009), Turkey (Beyarslan, 1991, 1999, 2014; Beyarslan & Çetin Erdoğan, 2012; Beyarslan & Inanç, 1994; Beyarslan et al., 2006).
Extralimital distribution: Azerbaijan, France, Georgia, Kazakhstan, Morocco, Russia, Turkmenistan, Ukraine.
Host records: Unknown.

***Glyptomorpha* (*Glyptomorpha*) *dispar* Tobias, 1986**
Distribution in the Middle East: Turkey (Beyarslan, 2016).
Extralimital distribution: Greece, Russia, Ukraine.
Host records: Unknown.

***Glyptomorpha* (*Glyptomorpha*) *elector* (Kokujev, 1898)**
Catalogs with Iranian records: No catalog.
Distribution in Iran: Qazvin (Ghahari & Beyarslan, 2019).
Distribution in the Middle East: Iran (Ghahari & Beyarslan, 2019), Turkey (Beyarslan, 2014; Kohl, 1905).
Extralimital distribution: China, Hungary, Kazakhstan, Mongolia, Turkmenistan, Uzbekistan.
Host records: Unknown.

***Glyptomorpha* (*Glyptomorpha*) *exsculpta* Shestakov, 1926**
Catalogs with Iranian records: Samin, Coronado-Blanco, Kavallieratos et al. (2018).
Distribution in Iran: West Azarbaijan (Samin, Coronado-Blanco, Kavallieratos et al., 2018).

Distribution in the Middle East: Iran (Samin, Coronado-Blanco, Kavallieratos et al., 2018), Turkey (Beyarslan, 2014; Čapek & Hofmann, 1997).
Extralimital distribution: Kazakhstan, Turkmenistan.
Host records: Unknown.

Glyptomorpha (Glyptomorpha) gracilis (Szépligeti, 1901)

Distribution in the Middle East: Israel—Palestine (Papp, 2015), Turkey (Beyarslan, 1999, 2014; Beyarslan et al., 2006).
Extralimital distribution: Greece, Hungary, Italy, North Macedonia, Russia, Slovakia, Spain, Sweden, Switzerland, Uzbekistan, former Yugoslavia.
Host records: Unknown.

Glyptomorpha (Glyptomorpha) irreptor (Klug, 1817)

Distribution in the Middle East: Egypt (Fahringer, 1926, 1928), Israel—Palestine (Papp, 1989, 2015), Saudi Arabia (Brues, 1926; Fahringer, 1926, 1928), Yemen (Kohl, 1907).
Extralimital distribution: Croatia, Italy, former Yugoslavia.
Host records: Unknown.

Glyptomorpha (Glyptomorpha) kasparyani Tobias, 1976

Catalogs with Iranian records: Gadallah and Ghahari (2015), Farahani et al. (2016), Yu et al. (2016), Samin, Coronado-Blanco, Kavallieratos et al. (2018), Samin, Coronado-Blanco, Fischer et al. (2018).
Distribution in Iran: Tehran (Zargar, Talebi, Hajiqanbar, Farahani, & Ameri, 2014).
Distribution in the Middle East: Iran (Zargar, Talebi, Hajiqanbar, Farahani, & Ameri, 2014), Turkey (Beyarslan, 1999, 2014; Beyarslan et al., 2006).
Extralimital distribution: Armenia, Azerbaijan.
Host records: Unknown.

Glyptomorpha (Glyptomorpha) nachitshevanica Tobias, 1976

Catalogs with Iranian records: Gadallah and Ghahari (2015), Farahani et al. (2016), Yu et al. (2016), Samin, Coronado-Blanco, Kavallieratos et al. (2018), Samin, Coronado-Blanco, Fischer et al. (2018).
Distribution in Iran: Fars (Lashkari Bod et al., 2010, 2011).
Distribution in the Middle East: Iran.
Extralimital distribution: Azerbaijan.
Host records: Unknown.

Glyptomorpha (Glyptomorpha) pectoralis (Brullé, 1832)

Catalogs with Iranian records: Fallahzadeh and Saghaei (2010), Gadallah and Ghahari (2015), Farahani et al. (2016), Yu et al. (2016), Samin, Coronado-Blanco, Kavallieratos et al. (2018), Samin, Coronado-Blanco, Fischer et al. (2018).

Distribution in Iran: Guilan, Hormozgan, Qazvin (Zargar, Talebi, Hajiqanbar, Farahani, & Ameri, 2014), Hamadan (Rajabi Mazhar et al., 2019), Kermanshah (Ghahari & Fischer, 2012), Mazandaran (Ghahari, 2019a), Iran (no specific locality cited) (Shestakov, 1926 as *Glyptomorpha elongata* Shestakov; Telenga, 1936; Watanbe, 1937 as *Glyptomorpha elongata* Shestakov; Tobias, 1976; Čapek & Hofmann, 1997).
Distribution in the Middle East: Cyprus (Beyarslan et al., 2017), Egypt, Iran, Israel—Palestine, Turkey (Yu et al., 2016).
Extralimital distribution: Afghanistan, Algeria, Austria, Azerbaijan, China, Croatia, Czech Republic, France, Georgia, Greece, Hungary, India, Italy, Kazakhstan, Libya, Malaysia, Moldova, Mongolia, Morocco, Mozambique, Pakistan, Poland, Russia, Serbia, Slovakia, South Africa, Spain, Tajikistan, Tunisia, Turkmenistan, Ukraine, United Kingdom, Uzbekistan.
Host records: Summarized by Yu et al. (2016) as being a parasitoid of the buprestids *Chrysobothris affinis* (Fabricius), *Sphenoptera gossypii* Cotes, and *Sphenoptera laticollis* Théry; and the cerambycid *Plagionotus arcuatus* (L.). In Iran, this species has been reared from *Plagionotus arcuatus* L. (Ghahari & Fischer, 2012).

Glyptomorpha (Glyptomorpha) sicula (Marshall, 1888)

Catalogs with Iranian records: No catalog.
Distribution in Iran: East Azarbaijan (Ghahari et al., 2019).
Distribution in the Middle East: Iran (Ghahari et al., 2019), Israel—Palestine (Papp, 2015), Turkey (Beyarslan, 1991, 1999, 2014; Beyarslan & Inanç, 1994).
Extralimital distribution: Croatia, former Czechoslovakia, France, Greece, Hungary, Italy, Romania, Spain.
Host records: Unknown.

Glyptomorpha (Glyptomorpha) turcomanica Fahringer, 1928

Distribution in the Middle East: Cyprus (Papp, 1998), Israel—Palestine (Papp, 2012).
Extralimital distribution: Kazakhstan, Turkmenistan.
Host records: Unknown.

Glyptomorpha (Teraturus) semenowi (Kokujev, 1898)

Distribution in the Middle East: Israel—Palestine (Papp, 2015).
Extralimital distribution: Kazakhstan, Mongolia, Tajikistan, Turkmenistan.
Host records: Unknown.

Glyptomorpha (Zanporia) aegyptiaca Sarhan and Quicke, 1989

Distribution in the Middle East: Egypt, Saudi Arabia, Yemen (Sarhan & Quicke, 1989).
Extralimital distribution: Pakistan.
Host records: Unknown.

Genus *Iphiaulax* Foerster, 1863

Iphiaulax (Euglyptobracon) impeditor (Kokujev, 1898)

Catalogs with Iranian records: Fallahzadeh and Saghaei (2010), Gadallah and Ghahari (2015), Farahani et al. (2016), Yu et al. (2016), Samin, Coronado-Blanco, Kavallieratos et al. (2018), Samin, Coronado-Blanco, Fischer et al. (2018).
Distribution in Iran: Kerman (Iranmanesh et al., 2018), Iran (no specific locality cited) (Fahringer, 1926).
Distribution in the Middle East: Iran (Fahringer, 1926; Iranmanesh et al., 2018), Israel–Palestine (Papp, 2015), Turkey (Beyarslan, 1999, 2014; Beyarslan & Inanç, 1994).
Extralimital distribution: Azerbaijan, China, Georgia, Kazakhstan, Lithuania, Moldova, Russia, Slovakia.
Host records: Unknown.
Plant associations in Iran: *Medicago sativa* L. (Fabaceae) (Iranmanesh et al., 2018).

Iphiaulax (Euglyptobracon) tauricus Shestakov, 1927

Catalogs with Iranian records: Samin, Coronado-Blanco, Fischer et al. (2018).
Distribution in Iran: West Azarbaijan (Sakenin et al., 2018).
Distribution in the Middle East: Iran (Sakenin et al., 2018), Israel–Palestine (Papp, 1989, 2012, 2015), Turkey (Beyarslan, 1999, 2014; Beyarslan & Inanç, 1994; Beyarslan et al., 2014; Papp, 2001).
Extralimital distribution: Armenia, Azerbaijan, Italy, Kazakhstan, Malta, Ukraine.
Host records: Unknown.

Iphiaulax (Euglyptobracon) umbraculator (Nees von Esenbeck, 1834)

Catalogs with Iranian records: No catalog.
Distribution in Iran: Hormozgan (Ameri et al., 2016 as *Pseudovipio umbraculator* (Nees, 1834)).
Distribution in the Middle East: Cyprus (Fahringer, 1926; Papp, 1998), Iran (Ameri et al., 2016), Israel–Palestine (Papp, 1989, 2012, 2015), Turkey (Beyarslan, 1999, 2014, 2016; Beyarslan & Inanç, 1994).
Extralimital distribution: Azerbaijan, Bosnia-Herzegovina, Croatia, Czech Republic, France, Georgia, Germany, Greece, Hungary, Italy, Kazakhstan, Lithuania, Moldova, Poland, Russia, Serbia, Slovakia, Uzbekistan.
Host records: Unknown.

Iphiaulax (Iphiaulax) agnatus Kohl, 1906

Distribution in the Middle East: Egypt (Fahringer, 1935), Israel–Palestine (Halperin, 1986), Saudi Arabia (Brues, 1926; Fahringer, 1926, 1935), Yemen (Kohl, 1907).
Extralimital distribution: Niger, Sudan.
Host records: Recorded by Halperin (1986) as being a parasitoid of the buprestid *Chalcogenia theryi* Abeille de Perrin.

Iphiaulax (Iphiaulax) ardens (Walker, 1871)

Distribution in the Middle East: Egypt (Brues, 1926; Fahringer, 1928; Shenefelt, 1978).

Extralimital distribution: Djibouti.
Host records: Unknown.

Iphiaulax (Iphiaulax) bohemani (Holmgren, 1868)

Distribution in the Middle East: Egypt (Fahringer, 1926).
Extralimital distribution: Ethiopia, Namibia, Somalia.
Host records: Unknown.

Iphiaulax (Iphiaulax) bohemani habesiensis Szépligeti, 1913

Distribution in the Middle East: Egypt (Fahringer, 1926).
Extralimital distribution: Ethiopia, Somalia.
Host records: Unknown.

Iphiaulax (Iphiaulax) concolor (Walker, 1871)

Distribution in the Middle East: Egypt (Brues, 1926; Fahringer, 1935; Shenefelt, 1978; Walker, 1871).
Extralimital distribution: Sudan.
Host records: Unknown.

Iphiaulax (Iphiaulax) congruus (Walker, 1871)

Distribution in the Middle East: Egypt (Brues, 1926; Fahringer, 1935; Walker, 1871).
Extralimital distribution: None.
Host records: Unknown.

Iphiaulax (Iphiaulax) determinatus (Walker, 1871)

Distribution in the Middle East: Egypt (Brues, 1926; Fahringer, 1935; Walker, 1871).
Extralimital distribution: Sudan.
Host records: Unknown.

Iphiaulax (Iphiaulax) ehrenbergi Strand, 1912

Distribution in the Middle East: Syria (Quicke & Koch, 1990; Strand, 1912).
Extralimital distribution: None.
Host records: Unknown.

Iphiaulax (Iphiaulax) fastidiator (Fabricius, 1781)

Catalogs with Iranian records: No catalog.
Distribution in Iran: Guilan (Ghahari et al., 2019).
Distribution in the Middle East: Cyprus (Beyarslan et al., 2017), Egypt (Fahringer, 1926; Walker, 1871), Iran (Ghahari et al., 2019), Turkey (Beyarslan, 2014; Beyarslan et al., 2014).
Extralimital distribution: Cameroon, Democratic Republic of Congo, Equatorial Guinea, Eritrea, France, Germany, Ghana, Guinea, Italy, Kenya, Madagascar, Mozambique, Namibia, Senegal, Somalia, South Africa, Sudan, Tanzania, Togo, Tunisia, Uganda.
Host records: Summarized by Yu et al. (2016) as being a parasitoid of the cerambycid *Saperda populnea* (L.).

Iphiaulax (Iphiaulax) gracilites Shenefelt, 1978

Distribution in the Middle East: Egypt (Fahringer, 1926), Israel–Palestine (Papp, 1989).

Extralimital distribution: Eritrea, Ethiopia, Tanzania.
Host records: Unknown.

Iphiaulax (Iphiaulax) hians Pérez, 1907
Catalogs with Iranian records: Fallahzadeh and Saghaei (2010), Gadallah and Ghahari (2015), Farahani et al. (2016), Yu et al. (2016), Samin, Coronado-Blanco, Kavallieratos et al. (2018), Samin, Coronado-Blanco, Fischer et al. (2018).
Distribution in Iran: Iran (no specific locality cited) (Papp, 2015; Telenga, 1936).
Distribution in the Middle East: Iran (Papp, 2015; Telenga, 1936), Israel−Palestine (Papp, 2015), Oman (Shenefelt, 1978).
Extralimital distribution: None.
Host records: Unknown.

Iphiaulax (Iphiaulax) impostor (Scopoli, 1763)
Catalogs with Iranian records: Gadallah and Ghahari (2015), Farahani et al. (2016), Yu et al. (2016), Samin, Coronado-Blanco, Kavallieratos et al. (2018), Samin, Coronado-Blanco, Fischer et al. (2018).
Distribution in Iran: East Azarbaijan (Ghahari, Fischer, Erdoğan, Tabari et al., 2009), Fars, Mazandaran (Sakenin et al., 2021—around cotton fields), Guilan (Ghahari, 2018; Sakenin et al., 2008—around rice fields).
Distribution in the Middle East: Cyprus (Yu et al., 2016), Iran (see references above), Israel−Palestine (Papp, 2015), Turkey (Beyarslan, 1999, 2014; Beyarslan et al., 2002).
Extralimital distribution: Europe, Afrotropical, Oriental, Palaearctic [Adjacent countries to Iran: Armenia, Azerbaijan, Kazakhstan, Russia, Turkmenistan].
Host records: Summarized by Yu et al. (2016) as being a parasitoid of the buprestids Anthaxia inornata (Randall), and Trachypteris picta (Pallas); the following cerambycids: Acanthocinus aedilis (L.), Acanthoderes varius Küster, Aegomorphus clavipes (Schrank), Apriona germari Hope, Icosium tomentosum Lucas, Leiopus nebulosus (L.), Monochamus galloprovincialis (Olivier), Monochamus sutor (L.), Obera linearis (L.), Purpuricenus budensis (Goeze), and Rhagium inquisitor (L.). It was also recorded by Minamikawa (1955) as being a parasitoid of the cossid moth Xyleutes persona (Le Guillou). In Iran, this species has been reared from the cerambycid Saperda populnea (L.) (Sakenin et al., 2008).

Iphiaulax (Iphiaulax) rufosignatus (Kokujev, 1898)
Distribution in the Middle East: Cyprus (Papp, 1990; Szépligeti, 1906), Israel−Palestine (Papp, 1989, 2012, 2015), Turkey (Szépligeti, 1906).
Extralimital distribution: Greece, Ukraine.
Host records: Unknown.

Iphiaulax (Iphiaulax) iranicus Quicke, 1985
Catalogs with Iranian records: Fallahzadeh and Saghaei (2010), Gadallah and Ghahari (2015), Farahani et al. (2016), Yu et al. (2016), Samin, Coronado-Blanco, Kavallieratos

et al. (2018), Samin, Coronado-Blanco, Fischer et al. (2018) [all as Iphiaulax mirabilis (Hedwig, 1957)].
Distribution in Iran: Sistan & Baluchestan (Hedwig, 1957 as Rhytimorpha mirabilis Hedwig).
Distribution in the Middle East: Iran (Hedwig, 1957 as Rhytimorpha mirabilis Hedwig; Quicke, 1985).
Extralimital distribution: None.
Host records: Unknown.

Iphiaulax (Iphiaulax) jacobsoni Shestakov, 1927
Catalogs with Iranian records: No catalog.
Distribution in Iran: Ardabil (Samin et al., 2019).
Distribution in the Middle East: Cyprus (Beyarslan et al., 2017), Iran (Samin et al., 2019), Israel−Palestine (Papp, 2015), Turkey (Beyarslan, 1999, 2014; Beyarslan et al., 2002).
Extralimital distribution: Malta, Uzbekistan.
Host records: Unknown.

Iphiaulax (Iphiaulax) jakowlewi (Kokujev, 1898)
Catalogs with Iranian records: No catalog.
Distribution in Iran: East Azarbaijan (Ghahari & Gadallah, 2019).
Distribution in the Middle East: Iran (Ghahari & Gadallah, 2019), Turkey (Beyarslan, 2014; Beyarslan et al., 2014).
Extralimital distribution: China, Kazakhstan, Mongolia.
Host records: Unknown.

Iphiaulax (Iphiaulax) mactator (Klug, 1817)
Catalogs with Iranian records: Fallahzadeh and Saghaei (2010), Gadallah and Ghahari (2015), Farahani et al. (2016), Yu et al. (2016), Samin, Coronado-Blanco, Kavallieratos et al. (2018), Samin, Coronado-Blanco, Fischer et al. (2018).
Distribution in Iran: Iran (no specific locality cited) (Tobias, 1976, 1986).
Distribution in the Middle East: Iran (Tobias, 1976, 1986), Israel−Palestine (Papp, 2015), Syria (Szépligeti, 1906), Turkey (Beyarslan, 2014; Beyarslan & Çetin Erdoğan, 2012).
Extralimital distribution: Azerbaijan, Bosnia-Herzegovina, China, Croatia, Germany, Greece, Hungary, Italy, Kazakhstan, Mongolia, Poland, Romania, Russia, Serbia, Slovakia, Spain, Switzerland, Ukraine (Yu et al., 2016), South Korea (Samartsev & Ku, 2021).
Host records: Recorded by Telenga (1936) and Tobias (1986) as being a parasitoid of the cerambycid beetle Acanthocinus aedilis (L.).

Iphiaulax (Iphiaulax) mactator pictus Kawall, 1856
Distribution in the Middle East: Israel−Palestine (Papp, 2015), Syria (Szépligeti, 1906).
Extralimital distribution: Hungary, Russia, Serbia.
Host records: Unknown.

Iphiaulax (Iphiaulax) melanarius (Walker, 1871)
Distribution in the Middle East: Egypt (Brues, 1926).

Extralimital distribution: Eritrea, Sudan.
Host records: Unknown.

Iphiaulax (Iphiaulax) perezi (Fahringer, 1926)

Catalogs with Iranian records: Fallahzadeh and Saghaei (2010), Gadallah and Ghahari (2015), Farahani et al. (2016), Yu et al. (2016), Samin, Coronado-Blanco, Kavallieratos et al. (2018), Samin, Coronado-Blanco, Fischer et al. (2018).
Distribution in Iran: Iran (no specific locality cited) (Papp, 2015; Telenga, 1936).
Distribution in the Middle East: Iran (Papp, 2015; Telenga, 1936), Israel—Palestine (Papp, 2015), Oman (Shenefelt, 1978).
Extralimital distribution: None.
Host records: Unknown.

Iphiaulax (Iphiaulax) potanini (Kokujev, 1898)

Catalogs with Iranian records: No catalog.
Distribution in Iran: Chaharmahal & Bakhtiari (Ghahari & Beyarslan, 2019; Gadallah et al., 2018).
Distribution in the Middle East: Iran (Ghahari & Beyarslan, 2019; Gadallah et al., 2018), Turkey (Beyarslan, 2014; Beyarslan & Çetin Erdoğan, 2012).
Extralimital distribution: Mongolia, Tajikistan, Turkmenistan.
Host records: Recorded by Tobias (1971, 1986) as being a parasitoid of the cerambycid *Cleroclytus banghaasi* (Reitter).

Iphiaulax (Iphiaulax) sculpturalis (Walker, 1871)

Distribution in the Middle East: Egypt (Fahringer, 1935), Saudi Arabia (Brues, 1926).
Extralimital distribution: Sudan.
Host records: Unknown.

Iphiaulax sp.

Distribution in Iran: Mazandaran (Amooghli-Tabari & Ghahari, 2021).
Host records in Iran: *Chilo suppressalis* Walker (Lepidoptera: Crambidae) (Amooghli-Tabari & Ghahari, 2021).

Genus *Megalommum* Szépligeti, 1900

Megalommum antefurcalis (Szépligeti, 1915)

Distribution in the Middle East: Israel—Palestine (Papp, 2015).
Extralimital distribution: Cameroon, Cape Verde Islands, Croatia, Democratic republic of Congo, Mozambique.
Host records: Unknown.

Megalommum fasciatipenne (Ashmead, 1900)

Distribution in the Middle East: Egypt (Fahringer, 1926).
Extralimital distribution: Cameroon, Ivory Coast, Liberia, Morocco, Senegal, Togo.

Host records: Unknown.

Megalommum jacobsoni (Tobias, 1968)

Distribution in the Middle East: Israel—Palestine (Papp, 2015).
Extralimital distribution: Cape Verde Islands, Croatia, France, Kazakhstan, Morocco, Senegal, Spain, Turkmenistan.
Host records: Unknown.

Megalommum pistacivorae van Achterberg and Mehrnejad, 2011

Catalogs with Iranian records: Gadallah and Ghahari (2015), Farahani et al. (2016), Yu et al. (2016), Samin, Coronado-Blanco, Kavallieratos et al. (2018), Samin, Coronado-Blanco, Fischer et al. (2018).
Distribution in Iran: Kerman (van Achterberg & Mehrnejad, 2011).
Distribution in the Middle East: Iran (van Achterberg & Mehrnejad, 2011).
Extralimital distribution: None.
Host records: In Iran, this species was recorded by van Achterberg and Mehrnejad (2011) as being a parasitoid of the cerambycid *Calchaenesthes pistacivora* Holzschuh.

Megalommum xanthoceps (Fahringer, 1928)

Distribution in the Middle East: Yemen (Quicke et al., 2000).
Extralimital distribution: Ghana, Ivory Coast, Kenya, Uganada.
Host records: Unknown.

Genus *Monilobracon* Quicke, 1984

Monilobracon tessmanni (Szépligeti, 1914)

Distribution in the Middle East: Egypt (Fahringer, 1926).
Extralimital distribution: Algeria, Cameroon, Democratic Republic of Congo, Equatorial Guinea, Guinea, Morocco, Senegal, Sierra Leone, Senegal, Tanzania, Togo.
Host records: Unknown.

Genus *Plaxopsis* Szépligeti, 1905

Plaxopsis abyssinica Szépligeti, 1913

Distribution in the Middle East: Egypt (Fahringer, 1926; Shenefelt, 1978).
Extralimital distribution: Ethiopia, Uganda.
Host records: Unknown.

Plaxopsis schroederi Szépligeti, 1914

Distribution in the Middle East: Egypt (Fahringer, 1926).
Extralimital distribution: Cameroon, Sudan, Togo.
Host records: Unknown.

Genus *Liomorpha* Szépligeti, 1914

Liomorpha nigrirostris Szépligeti, 1914

Distribution in the Middle East: Israel—Palestine (Papp, 1989).
Extralimital distribution: Morocco, Tunisia.
Host records: Unknown.

Genus *Odesia* Cameron, 1906

Odesia pulchripes Szépligeti, 1914

Distribution in the Middle East: Egypt (Fahringer, 1926, 1928).
Extralimital distribution: Ethiopia, South Africa, Sudan, Tanzania.
Host records: Unknown.

Genus *Odontoscapus* Kriechbaumer, 1894

Odontoscapus kriechbaumeri Kohl, 1906

Distribution in the Middle East: Saudi Arabia (Brues, 1926), Yemen (Fahringer, 1928; Kohl, 1907).
Extralimital distribution: None.
Host records: Unknown.

Genus *Pseudovipio* Szépligeti, 1896

Pseudovipio baeticus (Spinola, 1843)

Distribution in the Middle East: Israel—Palestine (Papp, 2015), Turkey (Beyarslan, 2014; Beyarslan et al., 2006).
Extralimital distribution: Greece, Morocco, Portugal, Spain.
Host records: Recorded by Beyarslan et al. (2006) as being a parasitoid of the noctuid *Sesamia nonagrioides* (Lefèbvre).

Pseudovipio barchanicus (Telenga, 1936)

Catalogs with Iranian records: Samin, Coronado-Blanco, Kavallieratos et al. (2018).
Distribution in Iran: Hamadan (Gadallah et al., 2018), West Azarbaijan (Samin, Coronado-Blanco, Kavallieratos et al., 2018).
Distribution in the Middle East: Iran (Gadallah et al., 2018; Samin et al., 2018a), Turkey (Beyarslan, 2014; Beyarslan et al., 2014).
Extralimital distribution: Kazakhstan, Turkmenistan.
Host records: Unknown.

Pseudovipio castrator (Fabricius, 1798)

Catalogs with Iranian records: Gadallah and Ghahari (2015), Farahani et al. (2016), Yu et al. (2016), Samin, Coronado-Blanco, Kavallieratos et al. (2018), Samin, Coronado-Blanco, Fischer et al. (2018).
Distribution in Iran: East Azarbaijan (Ghahari & Gadallah, 2022), Hormozgan, Qazvin (Ameri et al., 2016; Zargar, Talebi, Hajiqanbar, Farahani, & Ameri, 2014), Kerman

(Iranmanesh et al., 2018), Lorestan (Ghahari, Fischer, Papp, & Tobias, 2012), Mazandaran (Ghahari, 2019a), West Azarbaijan (Samin et al., 2014).
Distribution in the Middle East: Cyprus (Papp, 1998), Egypt (Fahringer, 1928), Iran (see references above), Israel—Palestine (Papp, 1989, 2012, 2015), Syria (Čapek & Hofmann, 1997), Turkey (Yu et al., 2016).
Extralimital distribution: Albania, Algeria, Azerbaijan, Croatia, Czech Republic, France, Georgia, Germany, Greece, Hungary, Italy, Kazakhstan, Moldova, Mongolia, Montenegro, Romania, Russia, Serbia, Slovakia, Spain, Sudan, Switzerland, Ukraine.
Host records: Summarized by Yu et al. (2016) as being a parasitoid of the buprestid *Chrysobothris affinis* (Fabricius); the cerambycid *Plagionotus arcuatus* (L.); the curculionid *Lixus juncii* Boheman; and the noctuid *Gortyna xanthenes* Germar. In Iran, this species has been reared from the buprestid *Chrysobothris affinis* (Fabricius) (Ghahari & Gadallah, 2022) and the cerambycid *Plagionotus arcuatus* L. (Ghahari, Fischer, Papp, & Tobias, 2012).
Plant associations in Iran: *Medicago sativa* L. (Fabaceae) (Iranmanesh et al., 2018).

Pseudovipio gorgoneus (Marshall, 1897)

Distribution in the Middle East: Israel—Palestine (Papp, 2015), Turkey (Beyarslan, 1999, 2014; Beyarslan & Inanç, 1994).
Extralimital distribution: Greece, Spain.
Host records: Unknown.

Pseudovipio guttiventris (Thomson, 1892)

Catalogs with Iranian records: No catalog.
Distribution in Iran: Lorestan (Ghahari & Gadallah, 2019).
Distribution in the Middle East: Iran (Ghahari & Gadallah, 2019), Turkey (Beyarslan, 1999, 2014; Beyarslan & Inanç, 1994).
Extralimital distribution: Czech Republic, Finland, Germany, Hungary, Kazakhstan, Poland, Sweden, Ukraine, United Kingdom.
Host records: Summarized by Yu et al. (2016) as being a parasitoid of the cerambycids *Oberea erythrocephala* Schrank, and *Xylotrechus rusticus* (L.); and the curculionid *Pilopedon plagiatum* (Schaller).

Pseudovipio inscriptor (Nees von Esenbeck, 1834)

Catalogs with Iranian records: Gadallah and Ghahari (2015), Farahani et al. (2016), Yu et al. (2016), Samin, Coronado-Blanco, Kavallieratos et al. (2018), Samin, Coronado-Blanco, Fischer et al. (2018).
Distribution in Iran: Guilan (Ghahari, Fischer, & Tobias, 2012), Hamadan (Rajabi Mazhar et al., 2019), Hormozgan (Ameri et al., 2016; Zargar, Talebi, Hajiqanbar, Farahani, & Ameri, 2014), Ilam (Ghahari, Fischer, & Papp, 2011c),

Qazvin (Zargar, Talebi, Hajiqanbar, Farahani, & Ameri, 2014).

Distribution in the Middle East: Cyprus (Papp, 1998), Iran (see references above), Israel—Palestine (Papp, 1989, 2015), Turkey (Yu et al., 2016).

Extralimital distribution: Azerbaijan, Croatia, France, Georgia, Germany, Greece, Hungary, Italy, Kazakhstan, Moldova, Mongolia, Romania, Russia, Serbia, Slovakia, Spain, Ukraine, Uzbekistan.

Host records: Recorded by Tobias (1986), Čapek and Hofmann (1997), and Ghahari, Fischer, & Tobias (2012) as being a parasitoid of the crambid *Ostrinia nubilalis* Hübner.

Pseudovipio insubricus (Fahringer, 1926)

Catalogs with Iranian records: No catalog.

Distribution in Iran: Hormozgan (Ameri et al., 2016).

Distribution in the Middle East: Cyprus (Fahringer, 1926; Papp, 1998; Szépligeti, 1906), Iran (Ameri et al., 2016).

Extralimital distribution: None.

Host records: Unknown.

Pseudovipio kirgisorum (Shestakov, 1932)

Distribution in the Middle East: Israel—Palestine (Papp, 2012).

Extralimital distributed: Kazakhstan.

Host records: Unknown.

Pseudovipio kirmanensis (Kokujev, 1907)

Catalogs with Iranian records: Fallahzadeh and Saghaei (2010), Gadallah and Ghahari (2015), Farahani et al. (2016), Yu et al. (2016), Samin, Coronado-Blanco, Kavallieratos et al. (2018), Samin, Coronado-Blanco, Fischer et al. (2018).

Distribution in Iran: Sistan & Baluchestan (Hedwig, 1957), Iran (no specific locality cited) (Fahringer, 1926; Kokujev, 1907 as *Vipio kermanensis* Kokujev; Papp, 2015; Telenga, 1936).

Distribution in the Middle East: Iran (see references above), Israel—Palestine (Papp, 2015).

Extralimital distribution: Afghanistan, Turkmenistan.

Host records: Unknown.

Pseudovipio minutus (Telenga, 1936)

Catalogs with Iranian records: No catalog.

Distribution in Iran: Hamadan (Sakenin et al., 2020).

Distribution in the Middle East: Iran (Sakenin et al., 2020), Turkey (Beyarslan, 1999, 2014; Beyarslan & Inanç, 1994).

Extralimital distribution: Kazakhstan, Russia, Ukraine.

Host records: Unknown.

Pseudovipio nigrirostris (Kokujev, 1907)

Catalogs with Iranian records: Fallahzadeh and Saghaei (2010), Gadallah and Ghahari (2015), Farahani et al.

(2016), Yu et al. (2016), Samin, Coronado-Blanco, Kavallieratos et al. (2018), Samin, Coronado-Blanco, Fischer et al. (2018).

Distribution in Iran: Iran (no specific locality cited) (Fahringer, 1926; Kokujev, 1907 as *Vipio nigrirostris* Kokujev).

Distribution in the Middle East: Iran.

Extralimital distribution: None.

Host records: Unknown.

Pseudovipio schaeuffelei (Hedwig, 1957)

Catalogs with Iranian records: Fallahzadeh and Saghaei (2010), Gadallah and Ghahari (2015), Farahani et al. (2016), Yu et al. (2016), Samin, Coronado-Blanco, Kavallieratos et al. (2018), Samin, Coronado-Blanco, Fischer et al. (2018).

Distribution in Iran: Southern Khorasan (Hedwig, 1957 as *Bracon laetus* var. *schäuffelei* Hedwig), Iran (no specific locality cited) (Papp, 2012 as *Bracon laetus*; van Achterberg, 1980).

Distribution in the Middle East: Iran.

Extralimital distribution: None.

Host records: Unknown.

Pseudovipio tataricus (Kokujev, 1898)

Catalogs with Iranian records: Fallahzadeh and Saghaei (2010), Gadallah and Ghahari (2015), Farahani et al. (2016), Yu et al. (2016), Samin, Coronado-Blanco, Kavallieratos et al. (2018), Samin, Coronado-Blanco, Fischer et al. (2018).

Distribution in Iran: Kerman (Iranmanesh et al., 2018 on *Medicago sativa*—Fabaceae), Khuzestan (Ameri et al., 2020), Iran (no specific locality cited) (Shenefelt, 1978; Telenga, 1936).

Distribution in the Middle East: Cyprus (Papp, 1990), Iran (Ameri et al., 2020; Iranmanesh et al., 2018; Shenefelt, 1978; Telenga, 1937), Israel—Palestine (Papp, 1989, 2015), Turkey (Beyarslan, 2014; Beyarslan & Çetin Erdoğan, 2012).

Extralimital distribution: Armenia, Azerbaijan, China, Kazakhstan, Moldova, Mongolia, Russia, Tajikistan, Turkmenistan, Uzbekistan.

Host records: Summarized by Yu et al. (2016) as being a parasitoid of the curculionid *Lixus incanescens* Boheman; and the gelechiid moth *Amblypalpis tamaricella* Danilevsky.

Genus *Rhadinobracon* Szépligeti 1906

Rhadinobracon zarudnyi (Telenga, 1936)

Catalogs with Iranian records: Fallahzadeh and Saghaei (2010), Gadallah and Ghahari (2015), Farahani et al. (2016), Yu et al. (2016), Samin, Coronado-Blanco, Kavallieratos et al. (2018), Samin, Coronado-Blanco, Fischer et al. (2018).

Distribution in Iran: Kerman (Iranmanesh et al., 2018), Sistan & Baluchestan (Hedwig, 1957 as *Pseudovipio nigrocephalus* Hedwig), Iran (no specific locality cited) (Papp, 2015; Shenefelt, 1978 as *Heliobracon zarudnyi* Telenga; Telenga, 1936).
Distribution in the Middle East: Iran (see references above), Israel—Palestine (Papp, 2012, 2015).
Extralimital distribution: None.
Host records: Unknown.

Genus *Rhammura* Enderlein, 1905
Rhammura clavata Szépligeti, 1911
Distribution in the Middle East: Egypt (Fahringer, 1926; Shenefelt, 1978).
Extralimital distribution: Democratic Republic of Congo, Sudan, Uganda.
Host records: Unknown.

Rhammura longiseta (Szépligeti, 1905)
Distribution in the Middle East: Egypt (Fahringer, 1926; Shenefelt, 1978).
Extralimital distribution: Cameroon, Democratic Republic of Congo, Equatorial Guinea, Gabon, Ghana, Morocco, Senegal, Sierra Leone, Sudan, Togo, Uganda.
Host records: Unknown.

Genus *Rhytimorpha* Szépligeti, 1901
Rhytimorpha coccinea Szépligeti, 1901
Distribution in the Middle East: Egypt (Fahringer, 1926), Israel—Palestine (Papp, 1989), Saudi Arabia (Quicke et al., 2018).
Extralimital distribution: Democratic Republic of Congo, Namibia, South Africa, Sudan, Tanzania.
Host records: Unknown.

Rhytimorpha pappi Quicke and Butcher, 2018
Distribution in the Middle East: Israel—Palestine (Quicke et al., 2018).
Extralimital distribution: None.
Host records: Unknown.

Genus *Soter* Saussure, 1892
Soter abyssinica (Szépligeti, 1913)
Distribution in the Middle East: Egypt (Fahringer, 1926), Israel—Palestine (Papp, 2015).
Extralimital distribution: Ethiopia.
Host records: Unknown.

Genus *Stenobracon* Szépligeti, 1901
Stenobracon deesae (Cameron, 1902)
Catalogs with Iranian records: No catalog.
Distribution in Iran: Mazandaran (Amooghli-Tabari & Ghahari, 2021).

Distribution in the Middle East: Iran.
Extralimital distribution: Bangladesh, China, India, Madagascar, Malaysia, Mauritius, Pakistan, Sudan.
Host records: Summarized by Yu et al. (2016) as being a parasitoid of the crambids *Bissectia steniellus* (Hampson), *Chilo* spp., *Scirpophaga excerptalis* (Walker), *Scirpophaga incertulas* (Walker), and *Scirpophaga nivella* (Fabricius); the noctuid *Sesamia inferens* (Walker); and the pyralids *Emmalocera depresella* (Swinhoe), and *Rhaphimetopus oblutella* (Zeller). In Iran, this species has been reared from *Chilo suppressalis* Walker (Lepidoptera: Crambidae) (Amooghli-Tabari & Ghahari, 2021).

Stenobracon nicevillei (Bingham, 1901)
Catalogs with Iranian records: No catalog.
Distribution in Iran: Sistan & Baluchestan (Gadallah et al., 2021).
Distribution in the Middle East: Iran.
Extralimital distribution: Bangladesh, China, India, Malaysia, Nepal, Pakistan, Philippines, Sri Lanka, Vietnam (Chisti & Quicke, 1996; Yu et al., 2016).
Host records: Summarized by Yu et al. (2016) as being a parasitoid of the following crambids: *Bissetia steniellus* (Hampson), *Chilo* spp., *Glaucocharis reniella* Wang and Sung, *Scirpophaga incertulas* (Walker), *Scirpophaga innotata* (Walker), and *Scirpophaga nivella* (Fabricius). No yet recorded in Iran.

Stenobracon unifasciatus (Brullé, 1846)
Distribution in the Middle East: Saudi Arabia, Yemen (van Achterberg & Polaszek, 1996).
Extralimital distribution: Algeria, Democratic Republic of Congo, Ivory Coast, Kenya, Libya, Madagascar, Malawi, Mali, Morocco, Mozambique, Niger, Nigeria, Senegal, Sierra Leone, South Africa, Sudan, Tanzania, Uganda, Zimbabwe.
Host records: Summarized by Yu et al. (2016) as being a parasitoid of the crambids *Chilo diffusilineus* de Joanis, *Chilo partellus* (Swinhoe), and *Chilo sacchariphagus* (Bojer); the noctuids *Busseola fusca* (Fuller), *Sesamia calamistis* hampson, *Sesamia cretica* Lederer, and *Sesamia nonagrioides* (Lefèbver); and the pyralid *Eldana saccharina* Walker.

Genus *Tsavobracon* Quicke, 1985
Tsavobracon atripennis (Szépligeti, 1906)
Distribution in the Middle East: Egypt (Fahringer, 1926).
Extralimital distribution: Sudan, Somalia, Tanzania.
Host records: Unknown.

Genus *Vipio* Latreille, 1804
Vipio abdelkader Schmiedeknecht, 1896

Distribution in the Middle East: Israel—Palestine (Papp, 1989, 2015).

Extralimital distribution: Algeria, Hungary, Libya, Tunisia.

Host records: Unknown.

Vipio alpi Beyarslan, 2002

Distribution in the Middle East: Turkey (Beyarslan, 2002, 2014, 2016; Beyarslan et al., 2008).

Extralimital distribution: None.

Host records: Unknown.

Vipio appellator (Nees von Esenbeck, 1834)

Catalogs with Iranian records: Farahani et al. (2016), Samin, Coronado-Blanco, Kavallieratos et al. (2018), Samin, Coronado-Blanco, Fischer et al. (2018).

Distribution in Iran: Lorestan (Sakenin et al., 2018), Tehran (Farahani et al., 2016).

Distribution in the Middle East: Iran (Farahani et al., 2016: Sakenin et al., 2018), Turkey (Beyarslan, 1991, 1999, 2014; Beyarslan et al., 2008; Kohl, 1905).

Extralimital distribution: Azerbaijan, China, Croatia, Czech Republic, Finland, France, Georgia, Germany, Hungary, Italy, Kazakhstan, Latvia, Lithuania, Mongolia, Montenegro, Netherlands, Poland, Russia, Serbia, Slovakia, Spain, Sweden, Switzerland, Tajikistan, Ukraine, Uzbekistan.

Host records: Summarized by Yu et al. (2016) as being a parasitoid of the buprestid *Chrysobothris affinis* (Fabricius); and the curculionid *Chromoderus fasciatus* (Müller).

Vipio cinctellus Brullé, 1832

Distribution in the Middle East: Israel—Palestine (Papp, 2015).

Extralimital distribution: Greece.

Host records: None.

Vipio humerator (Costa, 1885)

Catalogs with Iranian records: Gadallah and Ghahari (2015), Farahani et al. (2016), Yu et al. (2016), Samin, Coronado-Blanco, Kavallieratos et al. (2018), Samin, Coronado-Blanco, Fischer et al. (2018).

Distribution in Iran: Golestan (Ghahari, Gadallah et al., 2009), Guilan (Ghahari et al., 2019), Ilam (Ghahari, Fischer, & Papp, 2011c), Qazvin (Ghahari & Beyarslan, 2017).

Distribution in the Middle East: Iran (see references above), Israel—Palestine (Papp, 2015), Turkey (Yu et al., 2016).

Extralimital distribution: Albania, Algeria, Azerbaijan, Bulgaria, Croatia, France, Georgia, Greece, Hungary, Italy, Moldova, Romania, Russia, Serbia, Spain, Ukraine.

Host records: Unknown.

Vipio illusor (Klug, 1817)

Catalogs with Iranian records: Gadallah and Ghahari (2015), Farahani et al. (2016), Yu et al. (2016), Samin,

Coronado-Blanco, Kavallieratos et al. (2018), Samin, Coronado-Blanco, Fischer et al. (2018).

Distribution in Iran: Fars (Lashkari Bod et al., 2010, 2011).

Distribution in the Middle East: Iran (Lashkari Bod et al., 2010, 2011), Israel—Palestine (Papp, 2015), Turkey (Beyarslan, 1999, 2014; Beyarslan & Inanç, 1994; Beyarslan et al., 2008, 2014).

Extralimital distribution: Austria, Azerbaijan, Croatia, France, Greece, Hungary, Italy, Kazakhstan, North Macedonia, Moldova, Romania, Russia, Spain, Tunisia, Ukraine, former Yugoslavia.

Host records: Unknown.

Plant associations in Iran: Peach (Lashkari Bod et al., 2010, 2011).

Vipio indecisus (Walker, 1871)

Distribution in the Middle East: Egypt (Brues, 1926; Fahringer, 1928; Wlaker, 1871).

Extralimital distribution: None.

Host records: Unknown.

Vipio insectator Kokujev, 1898

Distribution in the Middle East: Turkey (Beyarslan, 1999, 2014; Beyarslan & Inanç, 1994; Beyarslan et al., 2008).

Extralimital distribution: Czech Republic, France, Hungary, Kazakhstan, Moldova, Russia, Slovakia, Ukraine.

Host records: Unknown.

Vipio intermedius Szépligeti, 1896

Catalogs with Iranian records: Gadallah and Ghahari (2015), Farahani et al. (2016), Yu et al. (2016), Samin, Coronado-Blanco, Kavallieratos et al. (2018), Samin, Coronado-Blanco, Fischer et al. (2018).

Distribution in Iran: East Azarbaijan, West Azarbaijan (Rastegar et al., 2012), Lorestan (Ghahari, Fischer, Papp, & Tobias, 2012), Semnan (Ghahari & Gadallah, 2015).

Distribution in the Middle East: Egypt (Fahringer, 1925, 1928), Iran (Ghahari & Gadallah, 2015; Ghahari, Fischer, Papp, & Tobias, 2012; Rastegar et al., 2012), Israel—Palestine (Papp, 2015), Turkey (Beyarslan, 2014).

Extralimital distribution: Albania, Algeria, Armenia, Azerbaijan, Bulgaria, China, Croatia, Georgia, Hungary, Italy, Kazakhstan, Moldova, Mongolia, Morocco, Romania, Russia, Slovakia, Spain, Tajikistan, Turkmenistan, Ukraine, Uzbekistan, former Yugoslavia.

Host records: Unknown.

Vipio lalapasaensis (Beyarslan, 1992)

Distribution in the Middle East: Turkey (Beyarslan, 1992, 1999, 2014; Beyarslan et al., 2008).

Extralimital distribution: None.

Host records: Unknown.

Vipio longicauda (Boheman, 1853)

Catalogs with Iranian records: Gadallah and Ghahari (2015), Farahani et al. (2016), Yu et al. (2016), Samin, Coronado-Blanco, Kavallieratos et al. (2018), Samin, Coronado-Blanco, Fischer et al. (2018).

Distribution in Iran: Guilan (Ghahari, Fischer, & Tobias, 2012 as *Vipio nominator* (Fabricius, 1787); Ghahari and Sakenin, 2018), Kermanshah (Ghahari & Gadallah, 2022).

Distribution in the Middle East: Cyprus (Papp, 1998), Iran (Ghahari, Fischer, & Tobias, 2012; Ghahari & Gadallah, 2015; Ghahari & Sakenin, 2018), Israel—Palestine (Papp, 2012), Syria (Fahringer, 1925), Turkey (Beyarslan, 1991, 1999, 2014: Beyarslan & Inanç, 1994; Beyarslan et al., 2005, 2008).

Extralimital distribution: Albania, Algeria, Azerbaijan, Belgium, Bosnia-Herzegovina, Croatia, Czech Republic, France, Georgia, Germany, Greece, Hungary, Italy, Kazakhstan, North Macedonia, Moldova, Mongolia, Montenegro, Netherlands, Poland, Russia, Serbia, Slovakia, Spain, Sweden, Switzerland, Tajikistan, Ukraine, United Kingdom.

Host records: Summarized by Yu et al. (2016) as being a parasitoid of the cerambycids *Acanthocinus griseus* (Fabricius), and *Molorchus minor* (L.); and the lymexylids *Hylecoetus dermestoides* (L.), and *Lymexylon navale* L. In Iran, this species has been reared from the cerambycid *Acanthocerus elegans* Ganglbauer (Ghahari & Gadallah, 2022).

Vipio mlokossewiczi Kokujev, 1898

Catalogs with Iranian records: Fallahzadeh and Saghaei (2010), Gadallah and Ghahari (2015), Farahani et al. (2016), Yu et al. (2016), Samin, Coronado-Blanco, Kavallieratos et al. (2018), Samin, Coronado-Blanco, Fischer et al. (2018).

Distribution in Iran: Alborz, Qazvin, Tehran (Zargar, Talebi, Hajiqanbar, Farahani, & Ameri, 2014), Mazandaran (Ghahari, Fischer, Erdoğan, Tabari et al., 2009 as *Isomecus mlokossewiczi* Kokujev), Semnan (Ghahari & Gadallah, 2015), Iran (no specific locality cited) (Papp, 2012; Telenga, 1936).

Distribution in the Middle East: Cyprus (Papp, 1998), Iran (see references above), Israel—Palestine (Papp, 1970, 1989, 2012, 2015), Turkey (Beyarslan, 2014; Beyarslan et al., 2008, 2014).

Extralimital distribution: Afghanistan, Azerbaijan, Georgia, Romania, Tajikistan, Turkmenistan, Uzbekistan.

Host records: Unknown.

Vipio nomioides Shestakov, 1926

Catalogs with Iranian records: Fallahzadeh and Saghaei (2010), Gadallah and Ghahari (2015), Farahani et al. (2016), Yu et al. (2016), Samin, Coronado-Blanco, Kavallieratos et al. (2018), Samin, Coronado-Blanco, Fischer et al. (2018).

Distribution in Iran: Alborz (Zargar, Talebi, Hajiqanbar, Farahani, & Ameri, 2014), Iran (no specific locality cited) (Shestakov, 1926; Telenga, 1936; Tobias, 1976, 1986).

Distribution in the Middle East: Iran (see references above), Turkey (Čapek & Hofmann, 1997; Beyarslan, 2014; Beyarslan et al., 2008).

Extralimital distribution: Azerbaijan, former Czechoslovakia.

Host records: Unknown.

Vipio sareptanus Kawall, 1865

Catalogs with Iranian records: Gadallah and Ghahari (2015), Farahani et al. (2016), Yu et al. (2016), Samin, Coronado-Blanco, Kavallieratos et al. (2018), Samin, Coronado-Blanco, Fischer et al. (2018).

Distribution in Iran: Kermanshah (Ghahari & Fischer, 2012).

Distribution in the Middle East: Iran.

Extralimital distribution: China, Kazakhstan, Korea, Latvia, Mongolia, Russia, Ukraine.

Host records: Unknown.

Vipio shestakovi Telenga, 1936

Catalogs with Iranian records: No catalog.

Distribution in Iran: Chaharmahal & Bakhtiari (Ghahari & Beyarslan, 2019).

Distribution in the Middle East: Iran (Ghahari & Beyarslan, 2019), Turkey (Beyarslan, 2014; Beyarslan & Çetin Erdoğan, 2012).

Extralimital distribution: Turkmenistan.

Host records: Unknown.

Vipio simulator Kokujev, 1898

Catalogs with Iranian records: No catalog.

Distribution in Iran: West Azarbaijan (Samin, Beyarslan, Coronado-Blanco et al., 2020).

Distribution in the Middle East: Iran (Samin, Beyarslan, Coronado-Blanco et al., 2020), Turkey (Beyarslan, 1991, 1999, 2014; Beyarslan et al., 2008).

Extralimital distribution: Former Czechoslovakia, Hungary, Kazakhstan, Russia.

Host records: Unknown.

Vipio spilogaster (Walker, 1871)

Distribution in the Middle East: Egypt (Brues, 1926; Fahringer, 1928; Walker, 1871).

Extralimital distribution: None.

Host records: Unknown.

Vipio striolatus Telenga, 1936

Catalogs with Iranian records: Gadallah and Ghahari (2015), Farahani et al. (2016), Yu et al. (2016), Samin,

Coronado-Blanco, Kavallieratos et al. (2018), Samin, Coronado-Blanco, Fischer et al. (2018).
Distribution in Iran: Tehran (Zargar, Talebi, Hajiqanbar, Farahani, & Ameri, 2014).
Distribution in the Middle East: Iran (Zargar, Talebi, Hajiqanbar, Farahani, & Ameri, 2014), Israel—Palestine (Papp, 2015).
Extralimital distribution: Azerbaijan, Croatia, Morocco, Tajikistan, Turkmenistan, Uzbekistan.
Host records: Unknown.

Vipio tentator (Rossi, 1790)
Catalogs with Iranian records: Gadallah and Ghahari (2015), Farahani et al. (2016), Yu et al. (2016), Samin, Coronado-Blanco, Kavallieratos et al. (2018), Samin, Coronado-Blanco, Fischer et al. (2018).
Distribution in Iran: Kermanshah (Ghahari & Fischer, 2012), Northern Khorasan (Sakenin et al., 2021—around cotton fields).
Distribution in the Middle East: Iran (Ghahari & Fischer, 2012; Sakenin et al., 2021), Israel—Palestine (Papp, 1970, 1989, 2012, 2015), Syria (Čapek & Hofmann, 1997), Turkey (Beyarslan, 1991, 1999, 2014, 2015; Beyarslan & Çetin Erdoğan, 2012; Beyarslan & Inanç, 1994; Beyarslan et al., 2006, 2008, 2014).
Extralimital distribution: Algeria, Armenia, Austria, Azerbaijan, Croatia, Czech Republic, France, Georgia, Germany, Greece, Hungary, Italy, Kazakhstan, North Macedonia, Moldova, Romania, Russia, Serbia, Slovakia, Spain, Tajikistan, Tunisia, Ukraine.
Host records: Recorded by Tobias (1986) as being a parasitoid of the buprestid *Agrilus cyanescens* Ratzeburg.

Vipio terrefactor (Villers, 1789)
Catalogs with Iranian records: Fallahzadeh and Saghaei (2010), Gadallah and Ghahari (2015), Farahani et al. (2016), Yu et al. (2016), Samin, Coronado-Blanco, Kavallieratos et al. (2018), Samin, Coronado-Blanco, Fischer et al. (2018).
Distribution in Iran: Iran (no specific locality cited) (Szépligeti, 1901 as *Vipio persica* Szépligeti, 1901; Maidl, 1923; Telenga, 1936; Tobias, 1976, 1986 as *Zavipio terrefactor* Villers).
Distribution in the Middle East: Iran (see references above), Turkey (Beyarslan, 1991, 1999, 2008, 2014).
Extralimital distribution: Armenia, Austria, Azerbaijan, Bosnia-Herzegovina, Croatia, Czech Republic, France, Germany, Greece, Hungary, Italy, Kazakhstan, Latvia, North Macedonia, Moldova, Romania, Russia, Serbia, Slovakia, Spain, Ukraine, United Kingdom.
Host records: recorded by Volovnik (1994) as being a parasitoid of the curculionid *Cyphocleonus achates* Fahraeus.
Comments: *Vipio terrefactor persica* Szépligeti, 1901 was recorded from Iran (no specific locality cited) by Szépligeti (1901 as *Vipio persica* Szépligeti) and Telenga (1936).

Vipio walkeri (Dalla Torre, 1898)
Distribution in the Middle East: Egypt (Brues, 1926; Fahringer, 1928; Walker, 1871).
Extralimital distribution: None.
Host records: Unknown.

Vipio xanthurus (Fahringer, 1926)
Catalogs with Iranian records: Fallahzadeh and Saghaei (2010), Gadallah and Ghahari (2015), Farahani et al. (2016), Yu et al. (2016), Samin, Coronado-Blanco, Kavallieratos et al. (2018), Samin, Coronado-Blanco, Fischer et al. (2018).
Distribution in Iran: Iran (no specific locality cited) (Fahringer, 1926).
Distribution in the Middle East: Iran (Fahringer, 1926).
Extralimital distribution: None.
Host records: Unknown.

Genus *Vipiomorpha* Tobias, 1962
Vipiomorpha fischeri Beyarslan, 1992
Distribution in the Middle East: Lebanon (Beyarslan, 1992), Turkey (Beyarslan, 1992, 1999).
Extralimital distribution: None.
Host records: Unknown.

Genus *Zaglyptogastra* Ashmead, 1900
Zaglyptogastra brevicaudis (Szépligeti, 1914)
Distribution in the Middle East: Egypt (Fahringer, 1926).
Extralimital distribution: Ethiopia, Somalia.
Host records: Unknown.

Zaglyptogastra seminigra (Szépligeti, 1906)
Distribution in the Middle East: Egypt (Fahringer, 1926).
Extralimital distribution: Djibouti, Kenya, Somalia, Tanzania.
Host records: Unknown.

Genus *Zanzopsis* van Achterberg, 1983
Zanzopsis buettneri (Szépligeti, 1914)
Distribution in the Middle East: Egypt (Fahringer, 1926).
Extralimital distribution: Equatorial Guinea, Togo, Uganda.
Host records: Unknown.

Tribe Argamaniini van Achterberg, 1991
Genus *Argamania* Papp, 1989
Argamania aereus Papp, 1989
Distribution in the Middle East: Israel—Palestine (van Achterberg, 1991; Papp, 1989, 2012).
Extralimital distribution: Algeria, Tunisia.
Host records: Unknown.

Tribe Braconini Nees von Esenbeck, 1811
Genus *Amyosoma* Viereck, 1913
Amyosoma chinense (Szépligeti, 1902)

Catalogs with Iranian records: No catalog.

Distribution in Iran: Mazandaran (Amooghli-Tabari & Ghahari, 2021; Ghahari & Sakenin, 2018), Northern Khorasan (Rahmani et al., 2019).

Distribution in the Middle East: Iran (Rahmani et al., 2019), Oman (van Achterberg & Polaszek, 1996).

Extralimital distribution: Bangladesh, Barbados (introduced), Borneo, China, Grenada (introduced), India, Indonesia (introduced in some parts), Japan, Korea, Madagascar, Malaysia, Mauricius, Mexico, Nepal, Pakistan, Papua New Guinea (introduced), Philippines, Sri Lanka, St. Vincent (introduced), Thailand, Trinidad and Tobago (introduced), United States of America (introduced in some states), Vietnam.

Host records: Summarized by Yu et al. (2016) as being a parasitoid of the crambids *Chilo* spp., *Diatraea grandiosella* Dyar, and *Diatraea lineolate* (Walker); the gelechiid *Pictinophora gossypiella* (Saunders); the geometrid *Hyposidra aquilaria* (Walker); the noctuids *Sesamia calamistis* Hampson, and *Sesamia inferens* (Walker); the pyralids *Euzophera perticella* Ragonot, and *Thylacoptila paurosema* Meyrick; and the pyraustid *Scirpophaga incertulus* (Walker). In Iran, this species has been reared from the crambid *Chilo suppressalis* (Walker) (Amooghli-Tabari & Ghahari, 2021; Ghahari & Sakenin, 2018).

Plant associations in Iran: *Medicago sativa* L. (Fabaceae) (Rahmani et al., 2019).

Genus *Baryproctus* Ashmead, 1900

Baryproctus barypus (Marshall, 1897)

Catalogs with Iranian records: No catalog.

Distribution in Iran: Golestan (Sakenin et al., 2021—around cotton fields).

Distribution in the Middle East: Iran (Sakenin et al., 2021), Israel—Palestine (Papp, 2015), Turkey (Beyarslan, 1999, 2014, 2016; Beyarslan & Erdoğan, 2012; Beyarslan et al., 2002).

Extralimital distribution: Azerbaijan, France, Germany, Greece, Hungary, Kazakhstan, Moldova, Montenegro, Russia, Slovakia, Spain, Sweden, Switzerland, Ukraine, United Kingdom.

Host records: Summarized by Yu et al. (2016) as being a parasitoid of the chloropids *Lipara orientalis* Martshuk, *Lipara rufitarsis* Loew, *Lipara similis* Schiner, and *Platycephala planifrons* (Fabricius).

Baryproctus turanicus Telenga, 1936

Distribution in the Middle East: Israel—Palestine (Papp, 2015).

Extralimital distribution: Mongolia, Uzbekistan.

Host records: Unknown.

Baryproctus zarudnianus Telenga, 1936

Catalogs with Iranian records: Fallahzadeh and Saghaei (2010), Gadallah and Ghahari (2015), Farahani et al. (2016), Yu et al. (2016), Samin, Coronado-Blanco, Kavallieratos et al. (2018), Samin, Coronado-Blanco, Fischer et al. (2018).

Distribution in Iran: Iran (no specific locality cited) (Shenefelt, 1978; Telenga, 1936).

Distribution in the Middle East: Iran (Shenefelt, 1978; Telenga, 1936), Israel—Palestine (Papp, 1970).

Extralimital distribution: None.

Host records: Unknown.

Genus *Bracomorpha* Papp, 1971

Bracomorpha rector (Thunberg, 1822)

Distribution in the Middle East: Turkey (Beyarslan, 1999).

Extralimital distribution: Croatia, Czech Republic, Estonia, Finland, Georgia, Germany, Hungary, Italy, Norway, Russia, Sweden, Switzerland, Tunisia.

Host records: Summarized by Yu et al. (2016) as being a parasitoid of the cerambycid *Rhamnusium bicolor* (Schrank); the curculionids *Pissodes notatus* (Fabricius), and *Scolytus scolytus* (Fabricius); and the tortricid *Cydia pactolana* (Zeller).

Genus *Bracon* Fabricius, 1804

Bracon (*Asiabracon*) *quadrimaculatus* Telenga, 1936

Catalogs with Iranian records: Gadallah and Ghahari (2015), Farahani et al. (2016), Yu et al. (2016), Samin, Coronado-Blanco, Kavallieratos et al. (2018), Samin, Coronado-Blanco, Fischer et al. (2018).

Distribution in Iran: Hormozgan (Ameri et al., 2014), Isfahan, Razavi Khorasan, Sistan & Baluchestan (Rahmani et al., 2017), Khuzestan (Zargar, Samartsev et al., 2020 as *B.* (*Asiabracon*) *quardrimaculatus* Telenga, 1936), Mazandaran (Zargar et al., 2015).

Distribution in the Middle East: Iran (see references above), Israel—Palestine (Papp, 2012, 2015), Turkey (Beyarslan, 1988, 1999, 2014; Beyarslan & Çetin Erdoğan, 2010; Beyarslan et al., 2014).

Extralimital distribution: Afghanistan, Azerbaijan, Greece, Turkmenistan.

Host records: Recorded by Tobias (1986) as being a parasitoid of the noctuids *Helicoverpa armigera* (Hübner) and *Helicoverpa zea* (Boddie).

Plant associations in Iran: *Medicago sativa* L. (Fabaceae), *Cyperus globosus* All. (Cyperaceae) (Rahmani et al., 2017).

Bracon (*Bracon*) *alutaceus* Szépligeti, 1901

Catalogs with Iranian records: No catalog.

Distribution in Iran: Khuzestan (Zargar, Samartsev et al., 2020), Northern Khorasan (Rahmani et al., 2017).

Distribution in the Middle East: Cyprus (Papp, 1998, 2008), Iran (Rahmani et al., 2017; Zargar, Samartsev et al., 2020). Extralimital distribution: Austria, Bulgaria, Czech Republic, Denmark, Germany, Hungary, Kazakhstan, Lithuania, Montenegro, Poland, Russia, Serbia, Slovakia, Spain, Sweden, Tajikistan, Ukraine, United Kingdom.
Host records: Unknown.

Bracon (Bracon) andriescui Papp, 1992
Distribution in the Middle East: Israel—Palestine (Papp, 2012).
Extralimital distribution: Greece, Hungary, Korea.
Host records: Unknown.

Bracon (Bracon) bachtiae Beyarslan, 2012
Distribution in the Middle East: Turkey (Beyarslan, 2014; Beyarslan & Çetin Erdoğan, 2012).
Extralimital distribution: None.
Host records: Unknown.

Bracon (Bracon) bilecikator Beyarslan, 1996
Distribution in the Middle East: Turkey (Beyarslan, 1996).
Extralimital distribution: None.
Host records: Unknown.

Bracon (Bracon) bipustulatus Szépligeti, 1913
Distribution in the Middle East: Egypt (Fahringer, 1927).
Extralimital distribution: Democratic Republic of Congo, Ethiopia, Tanzania.
Host records: Unknown.

Bracon (Bracon) cakili Beyarslan, 1996
Distribution in the Middle East: Turkey (Beyarslan, 1996).
Extralimital distribution: None.
Host records: Unknown.

Bracon (Bracon) celer Szépligeti, 1913
Distribution in the Middle East: Israel—Palestine (Kuslitzky & Argov, 2013).
Extralimital distribution: Cape Verde Islands, Eritrea, Italy, Kenya, South Africa.
Host records: Summarized by Yu et al. (2016) as being a parasitoid of the tephritids *Bactrocera oleae* (Rossi), and *Trirhithrum nigrum* (Graham).

Bracon (Bracon) chagrinicus Beyarslan, 2002
Distribution in the Middle East: Turkey (Beyarslan, 2002, 2014, 2016; Beyarslan & Çetin Erdoğan, 2005, 2010).
Extralimital distribution: None.
Host records: Unknown.

Bracon (Bracon) chivensis Telenga, 1936
Catalogs with Iranian records: Fallahzadeh and Saghaei (2010), Gadallah and Ghahari (2015), Farahani et al. (2016), Yu et al. (2016), Samin, Coronado-Blanco, Kavallieratos et al. (2018), Samin, Coronado-Blanco, Fischer et al. (2018).

Distribution in Iran: Iran (no specific locality cited) (Čapek & Hofmann, 1997).
Distribution in the Middle East: Iran (Čapek & Hofmann, 1997).
Extralimital distribution: Hungary, Turkmenistan, Uzbekistan.
Host records: Unknown.

Bracon (Bracon) cisellatus Papp, 1998
Catalogs with Iranian records: No catalog.
Distribution in Iran: Khuzestan (Zargar, Samartsev et al., 2020).
Distribution in the Middle East: Iran.
Extralimiat distribution: Korean Peninsula (Papp, 1998).
Host records: Unknown.

Bracon (Bracon) curticornis (Brues, 1924)
Distribution in the Middle East: Turkey (Beyarslan, 2014; Beyarslan & Četin Erdoğan, 2012).
Extralimital distribution: South Africa.
Host records: Unknown.

Bracon (Bracon) depressiusculus Szépligeti, 1904
Distribution in the Middle East: Turkey (Beyarslan, 2014; Beyarslan & Četin Erdoğan, 2012).
Extralimital distribution: Armenia, Azerbaijan, Hungary, Kazakhstan, Lithuania, Romania, Russia, Slovakia, Sweden, Switzerland, United Kingdom.
Host records: Unknown.

Bracon (Bracon) extasus Papp, 1990
Distribution in the Middle East: Israel—Palestine (Papp, 2015).
Extralimital distribution; China, Tunisia.
Host records: Unknown.

Bracon (Bracon) fulvipes Nees von Esenbeck, 1834
Catalogs with Iranian records: Gadallah and Ghahari (2015), Farahani et al. (2016), Yu et al. (2016), Samin, Coronado-Blanco, Kavallieratos et al. (2018), Samin, Coronado-Blanco, Fischer et al. (2018).
Distribution in Iran: East Azarbaijan (Ghahari, Fischer, Erdoğan, Tabari et al., 2009), Guilan (Ghahari & Fischer, 2011b), Isfahan (Ghahari & Beyarslan, 2017), Khuzestan (Zargar, Samartsev et al., 2020), Mazandaran (Ghahari, Fischer, Erdoğan et al., 2010; Zargar et al., 2015), Northern Khorasan (Rahmani et al., 2017), Semnan (Ghahari & Gadallah, 2015), Iran (no specific locality cited) (Papp, 2012 as *B. fulvipes* var. *carinatus*).
Distribution in the Middle East: Iran (see references above), Israel—Palestine (Papp, 2015), Turkey (Beyarslan, 1987, 1999, 2014).
Extralimital distribution: Afghanistan, Austria, Azerbaijan, Belgium, Croatia, Czech Republic, Finland, France, Germany, Greece, Hungary, Ireland, Italy, Kazakhstan, Korea, Lithuania, Moldova, Mongolia, Netherlands, Poland,

Russia, Serbia, Slovenia, Spain, Sweden, Switzerland, Tunisia, Ukraine, United Kingdom, Uzbekistan, former Yugoslavia.

Host records: Summarized by Yu et al. (2016) as being a parasitoid of various insect species of the orders Coleoptera (Apionidae, Curculionidae), Diptera (Cecidomyiidae, Lonchaeidae), Hymenoptera (Cynipidae, Eurytomidae, Tenthredinidae), and Lepidoptera (Coleophoridae, Sesiidae). In Iran, this species has been reared from the curculionid *Mononychus punctumalbum* (Herbst) (Ghahari & Beyarslan, 2017).

Plant associations in Iran: *Cortaderia selloana* (Schult. and Schult.) (Poaceae) (Rahmani et al., 2017).

Bracon (Bracon) furthi Papp, 2015
Distribution in the Middle East: Israel—Palestine (Papp, 2015).

Extralimital distribution: None.

Host records: Unknown.

Bracon (Bracon) heberola Papp, 2012
Distribution in the Middle East: Israel—Palestine (Papp, 2012).

Extralimital distribution: None.

Host records: Unknown.

Bracon (Bracon) intercessor Nees von Esenbeck, 1834
Catalogs with Iranian records: Gadallah and Ghahari (2015), Farahani et al. (2016), Yu et al. (2016), Samin, Coronado-Blanco, Kavallieratos et al. (2018), Samin, Coronado-Blanco, Fischer et al. (2018).

Distribution in Iran: Chaharmahal & Bakhtiari, Fars, Southern Khorasan (Ghahari & Beyarslan, 2017), Hamadan (Rajabi Mazhar et al., 2019), Hormozgan (Ameri et al., 2014), Kerman (Iranmanesh et al., 2018 on *Medicago sativa*—Fabaceae; Rahmani et al., 2017), Kermanshah, Northern Khorasan, Sistan & Baluchestan (Rahmani et al., 2017), Khuzestan (Zargar, Samartsev et al., 2020), Lorestan (Ghahari, Fischer, Papp, & Tobias, 2012), Razavi Khorasan (Abedi et al., 2015; Fathi et al., 2016; Ghahari, Gadallah et al., 2009; Mahmoudi et al., 2013; Rahmani et al., 2017), Semnan (Naderian et al., 2012; Samin, Fischer, & Ghahari, 2015), Tehran (Abbasipour et al., 2012; Ghahari, Fischer et al., 2011; Zargar et al., 2015).

Distribution in the Middle East: Cyprus, Iran, Israel—Palestine, Syria, Turkey (Yu et al., 2016).

Extralimital distribution: Europe, Palaearctic [Adjacent countries to Iran: Afghanistan, Kazakhstan, Russia, Turkmenistan].

Host records: Summarized by Yu et al. (2016) as being a parasitoid of various insects pests belonging to the orders Coleoptera (Apionidae, Attelabidae, Cerambycidae, Curculionidae), Diptera (Agromyzidae), Hymenoptera (Eurytomidae, Tenthredinidae), and Lepidoptera (Argyresthiidae, Coleophoridae, Elachistidae, Gelechiidae,

Plutellidae, Sesiidae, Tortricidae). In Iran, this species has been reared from the gelechiid moth *Scrobipalpa ocellatella* Boyd (Abbasipour et al., 2012; Mahmoudi et al., 2013), the curculionids *Sibinia femoralis* Germar (Abedi et al., 2015; Ghahari & Beyarslan, 2017), *Lixus incanescens* Boheman (Abedi et al., 2015; Fathi et al., 2016; Ghahari & Beyarslan, 2017), and *Archarius crux* (Fabricius) (Ghahari & Beyarslan, 2017).

Plant associations in Iran: *Beta vulgaris* L. (Amaranthaceae) (Abedi et al., 2015), *Medicago sativa* L. (Fabaceae) (Abedi et al., 2015; Rahmani et al., 2017), *Amaranthus* sp. (Amaranthaceae), *Apium* sp. (Apiaceae), *Convolvulus arvensis* L. (Convolvulaceae), *Mentha pulegium* L. (Lamiaceae) (Lamiaceae), *Sisymbrium irio* L. (Brassicaceae), *Lactuca serriola* L. (Asteraceae) (Rahmani et al., 2017).

Bracon (Bracon) intercessor laetus (Wesmael, 1838)
Catalogs with Iranian records: Farahani et al. (2016).

Distribution in Iran: Kerman (Iranmanesh et al., 2018), Mazandaran (Zargar et al., 2015), Iran (no specific locality cited) (Hedwig, 1957; Papp, 2012).

Distribution in the Middle East: Cyprus, Iran, Israel—Palestine, Turkey (Yu et al., 2016).

Extralimital distribution: Belgium, Bosnia-Herzegovina, Bulgaria, Croatia, Czech Republic, France, Georgia, Germany, Greece, Hungary, Italy, Kazakhstan, Korea, Portugal, Romania, Slovenia, Spain, United Kingdom, Uzbekistan, former Yugoslavia.

Host records: Summarized by Yu et al. (2016) as being a parasitoid of the curculionids *Anthonomus pomorum* (L.), and *Magdalis nitida* (Gyllenhal); the Argyresthiid moth *Argyresthia chrysidella* Peyerimhoff; and the plutellid *Prays citri* Millière.

Bracon (Bracon) israelicus Papp, 2015
Distribution in the Middle East: Israel—Palestine (Papp, 2015).

Extralimital distribution: None.

Host records: Unknown.

Bracon (Bracon) kozak Telenga, 1936
Catalogs with Iranian records: Farahani et al. (2016), Yu et al. (2016), Samin, Coronado-Blanco, Kavallieratos et al. (2018), Samin, Coronado-Blanco, Fischer et al. (2018).

Distribution in Iran: Hormozgan (Ameri et al., 2014), Khuzestan (Zargar, Samartsev et al., 2020 as *B. (B.) ovoides* Telenga, 1936, and as *B. (B.) kozak* Telenga, 1936), Razavi Khorasan (Fathi et al., 2016).

Distribution in the Middle East: Iran (Ameri et al., 2014; Fathi et al., 2016; Zargar, Samartsev et al., 2020), Israel—Palestine (Papp, 2012, 2015), Turkey (Beyarslan, 2014, 2016; Beyarslan & Çetin Erdoğan, 2005, 2010).

Extralimital distribution: Azerbaijan, Kazakhstan, Malta, Moldova, Russia, Turkmenistan, Uzbekistan.

Host records: In Iran, this species has been reared from the curculionid *Lixus incanescens* Boheman (Fathi et al., 2016).

Bracon (*Bracon*) *kuslitzkyi* Tobias, 1986
Distribution in the Middle East: Turkey (Beyarslan, 2014; Beyarslan & Četin Erdoğan, 2012).
Extralimital distribution: Moldova.
Host records: Unknown.

Bracon (*Bracon*) *lefroyi* (Dudgeon & Gough, 1914)
Catalogs with Iranian records: Fallahzadeh and Saghaei (2010), Gadallah and Ghahari (2015), Farahani et al. (2016), Yu et al. (2016), Samin, Coronado-Blanco, Kavallieratos et al. (2018), Samin, Coronado-Blanco, Fischer et al. (2018).
Distribution in Iran: Fars (Hussain et al., 1976), Iran (no specific locality cited) (Modarres Awal, 1997, 2012).
Distribution in the Middle East: Iran.
Extralimital distribution: India.
Host records: Summarized by Yu et al. (2016) as being a parasitoid of the crambid *Crocidolomia pavonana* (Fabricius); the gelechiid *Pectinophora gossypiella* (Saunders); the noctuids *Adisura atkinsoni* Moore, and *Helicoverpa armigera* (Hübner); the nolid *Earias insulana* (Boisduval); and the pyralid *Etiella zinckenella* (Treitschke). In Iran, this species has been reared from the nolid *Earias insulana* (Boisduval) (Modarres Awal, 1997, 2012).

Bracon (*Bracon*) *leptus* Marshall, 1897
Catalogs with Iranian records: Gadallah and Ghahari (2015), Farahani et al. (2016), Yu et al. (2016), Samin, Coronado-Blanco, Kavallieratos et al. (2018), Samin, Coronado-Blanco, Fischer et al. (2018).
Distribution in Iran: Golestan (Ghahari & Fischer, 2011c), Guilan (Ghahari, Fischer, Erdoğan, Tabari et al., 2009; Ghahari, Fischer, & Tobias, 2012), Hormozgan (Ameri et al., 2014), Isfahan (Rahmani et al., 2017), Kerman (Iranmanesh et al., 2018), Northern Khorasan (Ghahari, Fischer, Erdoğan, Beyarslan, & Havaskary, 2009), Qazvin (Ghahari, Fischer, & Papp, 2011b).
Distribution in the Middle East: Cyprus (Papp, 1998), Iran (see references above), Israel—Palestine (Papp, 2012, 2015), Turkey (Beyarslan, 2014, 2016; Beyarslan & Četin Erdoğan, 2005, 2010; Beyarslan et al., 2002).
Extralimital distribution: Austria, Czech Republic, Georgia, Germany, Greece, Hungary, Italy, Kazakhstan, Mongolia, Romania, Russia, Serbia, Spain, Ukraine, Uzbekistan.
Host records: Recorded by Beyarslan et al. (2005) as being a parasitoid of the gelechiid *Metzneria lapella* (L.).
Plant associations in Iran: *Medicago sativa* L. (Fabaceae) (Ghahari, Fischer, Erdoğan, Beyarslan, & Havaskary, 2009; Iranmanesh et al., 2018), *Mentha longifolia* L. (Lamiaceae) (Iranmanesh et al., 2018).

Bracon (*Barcon*) *longicollis* (Wesmael, 1838)
Catalogs with Iranian records: Fallahzadeh and Saghaei (2010), Gadallah and Ghahari (2015), Farahani et al. (2016), Yu et al. (2016), Samin, Coronado-Blanco, Kavallieratos et al. (2018), Samin, Coronado-Blanco, Fischer et al. (2018).
Distribution in Iran: Isfahan, Kerman, Northern Khorasan, Sistan & Baluchestan (Rahmani et al., 2017), Hamadan (Rajabi Mazhar et al., 2019), Khuzestan (Zargar, Samartsev et al., 2020), Iran (no specific locality cited) (Beyarslan et al., 2005; Papp, 2012, 2015; Tobias, 1976, 1986, 1995).
Distribution in the Middle East: Cyprus (Papp, 1998), Iran (see references above), Israel—Palestine (Papp, 2015), Turkey (Beyarslan, 2014, 2016; Beyarslan & Četin Erdoğan, 2005; Beyarslan et al., 2002).
Extralimital distribution: Afghanistan, Austria, Azerbaijan, Belgium, Bulgaria, Croatia, Finland, France, Georgia, Germany, Greece, Hungary, Italy, Kazakhstan, Korea, Lithuania, Moldova, Mongolia, Netherlands, Poland, Serbia, Slovenia, Spain, Sweden, Switzerland, Ukraine, United Kingdom, former Yugoslavia.
Host records: Recorded by Čapek and Hofmann (1997) as being a parasitoid of the chlropid fly *Chlorops pumilionis* (Bjerkander).
Plant associations in Iran: *Chenopodium album* L. (Amaranthaceae), *Cortaderia selloana* (Schult. and Schult.) (Poaceae), *Medicago sativa* L. (Fabaceae), *Mentha pulegium* L. (Lamiaceae), *Plantago major* L. (Plantaginaceae) (Rahmani et al., 2017).

Bracon (*Bracon*) *luteator* Spinola, 1808
Catalogs with Iranian records: Fallahzadeh and Saghaei (2010), Gadallah and Ghahari (2015), Farahani et al. (2016), Yu et al. (2016), Samin, Coronado-Blanco, Kavallieratos et al. (2018), Samin, Coronado-Blanco, Fischer et al. (2018).
Distribution in Iran: Chaharmahal & Bakhtiari East Azarbaijan (Lotfalizadeh & Gharali, 2014) Guilan (Ghahari & Gadallah, 2022), Ilam (Gharali, 2004; Lotfalizadeh & Gharali, 2014; Modarres Awal, 2012), Kerman (Iranmanesh et al., 2018), Kermanshah (Ghahari & Fischer, 2012), Khuzestan (Zargar, Samartsev et al., 2020), Kuhgiloyeh & Boyerahmad (Khatima & Reza, 2015; Saeidi, 2013, 2015), Mazandaran (Ghahari, 2019a), Northern Khorasan, Razavi Khorasan (Rahmani et al., 2017), Iran (no specific locality cited) (Beyarslan et al., 2005).
Distribution in the Middle East: Cyprus (Papp, 1998), Iran (see references above), Israel—Palestine (Papp, 1970, 1989, 2012, 2015), Syria (Basheer et al., 2014), Turkey (Belshaw et al., 2001; Beyarslan, 1986, 1999, 2014, 2016; Beyarslan & Četin Erdoğan, 2005; Beyarslan et al., 2002; Kohl, 1905).
Extralimital distribution: Afghanistan, Albania, Austria, Azerbaijan, Bulgaria, Croatia, former Czechoslovakia,

France, Germany, Georgia, Greece, Hungary, Italy, Kazakhstan, Latvia, North Macedonia, Moldova, Mongolia, Russia, Slovenia, Slovenia, Spain, Sweden, Switzerland, Tajikistan, Tunisia, Turkmenistan, Ukraine, United Kingdom, Uzbekistan, former Yugoslavia.

Host records: Summarized by Yu et al. (2016) as being a parasitoid of the gelchiids *Metzneria aestivella* (Zeller), and *Metzneria lapella* (L.); the tephritids *Acanthiophilus helianthin* (Rossi), and *Urophora solstitialis* L. In Iran, this species has been reared from the tephritids *Acanthiophilus helianthi* (Rossi) (Gharali, 2004; Modarres Awal, 2012; Saeidi, 2013, 2015), *Urophora* sp. (Ghahari & Fischer, 2012), and *Urophora solstitialis* (L.) (Ghahari & Gadallah, 2022).

Plant associations in Iran: *Carthamus tinctorius* L. (Asteraceae) (Khatima & Reza, 2015; Lotfalizadeh & Gharali, 2014), *Medicago sativa* L. (Fabaceae), *Beta vulgaris* L. (Amaranthaceae), *Cynodon dactylon* (L.) (Poaceae), *Mentha pulegium* L. (Lamiaceae) (Rahmani et al., 2017).

Bracon (Bracon) luteator filicauda Costa, 1838
Distribution in the Middle East: Cyprus (Papp, 1998).
Extralimital distribution: Italy.
Host records: Unknown.

Bracon (Bracon) mariae Dalla Torre, 1898
Catalogs with Iranian records: Gadallah and Ghahari (2015), Farahani et al. (2016), Yu et al. (2016), Samin, Coronado-Blanco, Kavallieratos et al. (2018), Samin, Coronado-Blanco, Fischer et al. (2018).
Distribution in Iran: Guilan (Ghahari, Fischer, & Tobias, 2012), Semnan (Ghahari & Gadallah, 2015; Samin, Fischer, & Ghahari, 2015), Iran (no specific locality cited) (Tobias, 1961, 1976).
Distribution in the Middle East: Iran (Ghahari & Gadallah, 2015; Ghahari, Fischer, & Tobias, 2012; Samin, Fischer, & Ghahari, 2015; Tobias, 1961, 1976), Israel−Palestine (Papp, 1989, 2015), Turkey (Beyarslan, 1999; 2014; Beyarslan & Çetin Erdoğan, 2005; 2010; Beyarslan et al., 2014).
Extralimital distribution: Armenia, Azerbaijan, Croatia, former Czechoslovakia, Germany, Hungary, Italy, Kazakhstan, Moldova, Russia, Slovenia, Spain, Sweden, Switzerland, Tunisia, Ukraine, former Yugoslavia.
Host records: Recorded by Vidal (1993) in association with the tachinid fly *Phasia costalis* (Malloch).

Bracon (Bracon) murgabensis Tobias, 1957
Catalogs with Iranian records: No catalog.
Distribution in Iran: Alborz, Markazi, Tehran (Shimi et al., 1995 as *B. myrgabansis* Tobias [misspelling]), Kerman, Northern Khorasan (Rahmani et al., 2017).
Distribution in the Middle East: Iran (Rahmani et al., 2017; Shimi et al., 1995), Turkey (Beyarslan, 2014, 2016; Beyarslan & Çetin Erdoğan, 2012).

Extralimital distribution: Azerbaijan, Kazakhstan, Malta, Moldova.
Host records: Recorded by Papp (2015) as being a parasitoid of the coleophorid *Coleophora festivella* Toll; and by Tobias (1986) as being a parasitoid of the curculionid *Smicronyx tataricus* Faust. In Iran, this species has been reared from *Smicronyx robustus* Faust (Shimi et al., 1995). Plant associations in Iran: *Medicago sativa* L. (Fabaceae) (Rahmani et al., 2017).

Bracon (Bracon) necator (Fabricius, 1777)
Catalogs with Iranian records: Samin, Coronado-Blanco, Fischer et al. (2018).
Distribution in Iran: West Azarbaijan (Sakenin et al., 2018).
Distribution in the Middle East: Iran (Sakenin et al., 2018), Turkey (Yilmaz et al., 2010).
Extralimital distribution: France, Germany, Italy, Latvia, Poland, United Kingdom.
Host records: Unknown.

Bracon (Bracon) nigratus Wesmael, 1838
Catalogs with Iranian records: Gadallah and Ghahari (2015), Farahani et al. (2016), Yu et al. (2016), Samin, Coronado-Blanco, Kavallieratos et al. (2018), Samin, Coronado-Blanco, Fischer et al. (2018).
Distribution in Iran: Hormozgan (Ameri et al., 2014), Lorestan (Ghahari, Fischer, Papp, & Tobias, 2012).
Distribution in the Middle East: Iran (Ghahari, Fischer, Papp, & Tobias, 2012; Ameri et al., 2014), Turkey (Beyarslan, 2014, 2016; Beyarslan & Çetin Erdoğan, 2005, 2010; Beyarslan et al., 2014; Papp, 2012).
Extralimital distribution: Albania, Austria, Belgium, Bulgaria, Denmark, France, Georgia, Germany, Hungary, Ireland, Italy, Lithuania, Moldova, Poland, Romania, Russia, Serbia, Slovakia, Slovenia, Spain, Sweden, Switzerland, Ukraine, United Kingdom.
Host records: Summarized by Yu et al. (2016) as being a parasitoid of the coleophorid *Coleophora mellifolii* Zeller; the tortricid *Cydia compositella* (Fabricius); and the zygaenid *Zygaena minos* (Denis and Schiffermüller). It was also recorded by Čapek and Hofmann (1997) as being a parasitoid of the tephritid *Chaetostomella cylindrica* (Robineau-Desvoidy). In Iran, this species has been reared from *Zygaena loti* (Denis and Schiffermüller) (Ghahari, Fischer, Papp, & Tobias, 2012).

Bracon (Bracon) obscuricornis Szépligeti, 1896
Distribution in the Middle East: Israel−Palestine (Papp, 2015), Turkey (Kohl, 1905).
Extralimital distribution: Croatia, Greece, Hungary, Italy, North Macedonia, Slovenia, Ukraine, former Yugoslavia.
Host records: Unknown.

Bracon (Bracon) pectoralis (Wesmael, 1838)
Catalogs with Iranian records: Gadallah and Ghahari (2015), Farahani et al. (2016), Yu et al. (2016).

Distribution in Iran: Ardabil (Ghahari & Fischer, 2011b), Golestan (Ghahari, Fischer, Erdoğan, Beyarslan, & Havaskary, 2009, in alfalfa fields; Ghahari, Fischer, Erdoğan et al., 2010; Ghahari & Gadallah, 2022), Guilan, Qazvin (Zargar et al., 2015), Hormozgan (Ameri et al., 2015), Isfahan (Ghahari, Fischer, & Papp, 2011a), Kerman (Rahmani et al., 2017; Iranmanesh et al., 2018 on *Medicago sativa*—Fabaceae), Khuzestan (Zargar, Samartsev et al., 2020), Northern Khorasan (Rahmani et al., 2017), Qazvin (Ghahari & Gadallah, 2022).

Distribution in the Middle East: Cyprus (Papp, 1998, 2012), Egypt (Papp, 2012), Iran (see references above), Israel−Palestine (Papp, 2015), Jordan (Papp, 2012), Syria (Yu et al., 2016), Turkey (Beyarslan, 1987, 2014, 2016; Beyarslan & Çetin Erdoğan, 2005, 2010; Beyarslan et al., 2002, 2014).

Extralimital distribution: Europe, Palaearctic, Puerto Rico (introduced), United States of America (introduced) [Adjacent countries to Iran: Afghanistan, Azerbaijan, Kazakhstan, Russia, Turkmenistan].

Host records: Summarized by Yu et al. (2016) as being a parasitoid of the alucitids *Alucita hexadactyla* L., and *Alucita huebneri* Wallengren; the gelechiid *Sitotroga cerealella* (Olivier); the pyralid *Etiella zenkenella* (Treitschke); and the tortricid *Aethes francillana* (Fabricius). It was also recorded by Vidal (1993) as being a parasitoid of the curculionid *Larinus flavescens* Germar, and by Halil (2006) as being a parasitoid of the rhynchitid *Tatianaerhynchites aequatus* (L.). Aubert (1966) recorded it as being a parasitoid of the tiphritid *Xyphosia miliaria* Schrank. In Iran, this species has been reared from the curculionid *Larinus flavescens* Germar, and the rhynchitid *Tatianaerhynchites aequatus* (L.) (Ghahari & Gadallah, 2022).

Plant associations in Iran: *Medicago sativa* L. (Fabaceae), *Mentha pulegium* L. (Lamiaceae), *Convolvulus arvensis* L. (Convolvulaceae), *Cynodon dactylon* (L.) (Poaceae) (Rahmani et al., 2017).

Bracon (*Bracon*) *pectoralis fumigatus* Széplgeti, 1901
Distribution in the Middle East: Cyprus (Papp, 1998, 2012), Egypt (Papp, 2012), Israel−Palestine (Papp, 2015).
Extralimital distribution: Albania, Algeria, Bulgaria, Germany, Greece, Hungary, Italy, Romania, Russia, Slovenia, Spain, Turkmenistan, United Kingdom, former Yugoslavia.
Host records: Unknown.

Bracon (*Bracon*) *querceus* Tobia, 1986
Distribution in the Middle East: Turkey (Beyarslan, 2016).
Extralimital distribution: Russia.
Host records: Unknown.

Bracon (*Bracon*) *robustus* Hedwig, 1961
Catalogs with Iranian records: Gadallah and Ghahari (2015), Farahani et al. (2016), Yu et al. (2016), Samin,

Coronado-Blanco, Kavallieratos et al. (2018), Samin, Coronado-Blanco, Fischer et al. (2018).
Distribution in Iran: Golestan (Samin, Ghahari et al., 2015), Razavi Khorasan (Samin et al., 2011).
Distribution in the Middle East: Iran (Samin et al., 2011; Samin, Ghahari et al., 2015).
Extralimital distribution: Afghanistan.
Host records: Unknown.

Bracon (*Bracon*) *scabriusculus* Dalla Torre, 1898
Catalogs with Iranian records: Fallahzadeh and Saghaei (2010), Gadallah and Ghahari (2015), Farahani et al. (2016), Yu et al. (2016), Samin, Coronado-Blanco, Kavallieratos et al. (2018), Samin, Coronado-Blanco, Fischer et al. (2018).
Distribution in Iran: Iran (no specific locality cited) (Telenga, 1936).
Distribution in the Middle East: Iran (Telenga, 1936).
Extralimital distribution: Croatia, Germany, Hungary, Italy, Kazakhstan, Romania, Slovenia, Spain, former Yugoslavia.
Host records: Unknown.

Bracon (*Bracon*) *schmidti* Kokujev, 1912
Catalogs with Iranian records: Fallahzadeh and Saghaei (2010), Gadallah and Ghahari (2015), Farahani et al. (2016), Yu et al. (2016), Samin, Coronado-Blanco, Kavallieratos et al. (2018), Samin, Coronado-Blanco, Fischer et al. (2018).
Distribution in Iran: Iran (no specific locality cited) (Papp, 2012; Shenefelt, 1978; Telenga, 1936; Tobias, 1976, 1986).
Distribution in the Middle East: Cyprus (Papp, 1998), Iran (Papp, 2012; Shenefelt, 1978; Telenga, 1936; Tobias, 1976, 1986), Israel−Palestine (Papp, 2012), Turkey (Beyarslan & Çetin Erdoğan, 2010; Beyarslan et al., 2014).
Extralimital distribution: Azerbaijan, Hungary, Uzbekistan.
Host records: Unknown.

Bracon (*Bracon*) *scutellaris* (Wesmael, 1838)
Distribution in the Middle East: Turkey (Papp, 2012).
Extralimital distribution: Belgium, Denmark, Finland, France, Germany, Hungary, Ireland, Italy, Korea, Netherlands, Poland, Romania, Sweden, United Kingdom.
Host records: Summarized by Yu et al. (2016) as being a parasitoid of the anobiid *Anobium abietis* (Fabricius); and the tenthredinids *Euura mucronata* (Hartig), *Euura pedunculi* (Hartig), and *Euura viminalis* (L.).

Bracon (*Bracon*) *selviae* Beyarslan, 2016
Distribution in the Middle East: Turkey (Beyarslan, 2016).
Extralimital distribution: None.
Host records: Unknown.

Bracon (*Bracon*) *speerschneideri* Schmiedeknecht, 1897
Catalogs with Iranian records: No catalog.
Distribution in Iran: Kerman (Iranmanesh et al., 2018).

Distribution in the Middle East: Iran.

Extralimital distribution: Austria, France, Germany, Italy, Norway, Sweden, Switzerland, United Kingdom.

Host records: Recorded by Papp (1999) as being a parasitoid of the argyresthiid weevil *Argyresthia thuiella* (Packard); and the yponomeutid moth *Ocnerostoma* sp.

Plant associations in Iran: *Medicago sativa* L. (Fabaceae) (Iranmanesh et al., 2018).

Bracon (Bracon) subrugosus Szépligeti, 1901

Catalogs with Iranian records: Gadallah and Ghahari (2015), Farahani et al. (2016), Yu et al. (2016), Samin, Coronado-Blanco, Kavallieratos et al. (2018), Samin, Coronado-Blanco, Fischer et al. (2018).

Distribution in Iran: Kerman (Iranmanesh et al., 2018), Khuzestan (Zargar, Samartsev et al., 2020), Lorestan (Ghahari, Fischer, Papp, & Tobias, 2012 as *Bracon (Bracon) subglaber* Szépligeti, 1904), Iran (no specific locality cited) (Papp, 2012).

Distribution in the Middle East: Iran (Ghahari, Fischer, Papp, & Tobias, 2012; Iranmanesh et al., 2018; Papp, 2012; Zargar, Samartsev et al., 2020), Israel—Palestine (Papp, 2015), Turkey (Beyarslan, 2014, 2016; Beyarslan & Çetin Erdoğan, 2005, 2010; Beyarslan et al., 2002; Papp, 2008).

Extralimital distribution: Austria, Azerbaijan, Bulgaria, Croatia, former Czechoslovakia, France, Germany, Hungary, Italy, Kazakhstan, Moldova, Netherlands, Norway, Poland, Romania, Russia, Slovakia, Slovenia, Sweden, Switzerland, Ukraine, United Kingdom, Uzbekistan, former Yugoslavia.

Host records: Summarized by Yu et al. (2016) as being a parasitoid of various hosts belonging to the orders Coleoptera (Curculionidae), Diptera (Tachinidae, Tephritidae), and Lepidoptera (Gelechiidae, Psychidae, Sesiidae, Tortricidae). In Iran, this species has been recorded as a parasitoid of the curculionid *Ceutorhynchus* sp. (Ghahari, Fischer, Papp, & Tobias, 2012).

Plant assocaitions in Iran: *Medicago sativa* L. (Fabaceae) (Iranmanesh et al., 2018).

Bracon (Bracon) tenuicornis (Wesmael, 1838)

Catalogs with Iranian records: This species is a new record for the fauna of Iran.

Distribution in Iran: Golestan province, Golestan National Park, 2♀, June 2016.

Distribution in the Middle East: Iran (new record), Turkey (Beyarslan, 2014, 2016; Beyarslan & Çetin Erdoğan, 2010, 2012).

Extralimital distribution: Armenia, Belgium, Bulgaria, France, Georgia, Germany, Hungary, Italy, Kazakhstan, Moldova, Netherlands, Norway, Russia, Sweden, United Kingdom.

Host records: Recorded by Tobias (1986) as being a parasitoid of the curculionid *Phloeotribus scarabaeoides* (Bernard).

Bracon (Bracon) trucidator Marshall, 1888

Catalogs with Iranian records: Gadallah and Ghahari (2015), Farahani et al. (2016), Yu et al. (2016), Samin, Coronado-Blanco, Kavallieratos et al. (2018), Samin, Coronado-Blanco, Fischer et al. (2018).

Distribution in Iran: Ardabil (Ghahari & Fischer, 2011b), Hamadan (Samin, 2015), Hormozgan (Ameri et al., 2015), Kerman (Rahmani et al., 2017; Iranmanesh et al., 2018), Northern Khorasan (Rahmani et al., 2017), Kermanshah (Ghahari, Fischer, Hedqvist et al., 2010; Rahmani et al., 2017), Mazandaran (Ghahari, 2017—around rice fields), Qazvin (Ghahari, Fischer, & Papp, 2011b), Iran (no specific locality cited) (Papp, 2015).

Distribution in the Middle East: Iran (see references above), Israel—Palestine (Papp, 2015), Turkey (Beyarslan, 1987, 2014, 2015; Beyarslan & Çetin Erdoğan, 2010, 2012; Beyarslan et al., 2002).

Extralimital distribution: Albania, Algeria, Armenia, Austria, Azerbaijan, Croatia, Czech Republic, France, Georgia, Germany, Greece, Hungary, Italy, Kazakhstan, Moldova, Romania, Russia, Slovenia, Spain, Sweden, Switzerland, Tajikistan, Tunisia, Ukraine, former Yugoslavia.

Host records: Summarized by Yu et al. (2016) as being a parasitoid of the tephritid *Urophora solstitialis* (L.); the gelechiid *Metzneria lapella* (L.); and the pyralid *Homoeosoma electellum* (Hulst).

Plant associations in Iran: *Amaranthus* sp. (Amaranthaceae), *Mentha pulegium* L. (Lamiaceae) (Rahmani et al., 2017), *Medicago sativa* L. (Fabaceae) (Iranmanesh et al., 2018; Rahmani et al., 2017).

Bracon (Bracon) variegator Spinola, 1808

Catalogs with Iranian records: Gadallah and Ghahari (2015), Farahani et al. (2016), Yu et al. (2016), Samin, Coronado-Blanco, Kavallieratos et al. (2018), Samin, Coronado-Blanco, Fischer et al. (2018).

Distribution in Iran: Hormozgan (Ameri et al., 2014), Kermanshah (Ghahari & Fischer, 2012), Khuzestan (Zargar, Samartsev et al., 2020), Qazvin (Zargar et al., 2015), Semnan (Ghahari & Gadallah, 2015), Tehran (Ghahari & Gadallah, 2022).

Distribution in the Middle East: Cyprus, Iran, Israel—Palestine, Lebanon, Turkey (Yu et al., 2016).

Extralimital distribution: Europe, Oceanic, Oriental, Palaearctic, New Zeland (introduced), United States of America (introduced) [Adjacent countries to Iran: Afghanistan, Armenia, Kazakhstan, Russia, Turkmenistan].

Host records: Summarized by Yu et al. (2016) as being a parasitoid of a wide range of lepidopteran pests of the families Coleophoridae, Depressariidae, Gelechiidae, Gracillariidae, Nolidae, Oecophoridae, Pyralidae, Tortricidae, and Yponomeutidae. It was also recorded by Tobias (1986) as being a parasitoid of the ptinid weevil *Ernobium abietis* (Fabricius). In Iran, this species has been reared from the gelechiid *Anarsia lineatella* Zeller (Ghahari & Fischer, 2012; Ghahari & Gadallah, 2022); and the tortricid *Pandemis cerasana* (Hübner) (Ghahari & Gadallah, 2022).

Bracon (*Cyanopterobracon*) *armeniacus* Telenga, 1936

Distribution in the Middle East: Cyprus (Papp, 1998), Turkey (Beyarslan, 2014, 2016; Beyarslan & Četin Erdoğan, 2012).
Extralimital distribution: Armenia.
Host records: Unknown.

Bracon (*Cyanopterobracon*) *fallax* Szépligeti, 1901

Catalogs with Iranian records: Samin, Coronado-Blanco, Kavallieratos et al. (2018).
Distribution in Iran: Chaharmahal & Bakhtiari (Samin, Coronado-Blanco, Kavallieratos et al., 2018).
Distribution in the Middle East: Cyprus (Papp, 1998, 2008), Iran (Samin, Coronado-Blanco, Kavallieratos et al., 2018), Turkey (Beyarslan, 1986; 1999, 2014, 2016; Beyarslan & Četin Erdoğan, 2010; Beyarslan et al., 2014; Papp, 2008).
Extralimital distribution: Georgia, Hungary, Italy, Kazakhstan, North Macedonia, Moldova, Morocco, Romania, Russia, Slovakia, Tajikistan, Ukraine, former Yugoslavia.
Host records: Unknown.

Bracon (*Cyanopterobracon*) *illyricus* Marshall, 1888

Catalogs with Iranian records: Gadallah and Ghahari (2015), Farahani et al. (2016), Yu et al. (2016), Samin, Coronado-Blanco, Kavallieratos et al. (2018), Samin, Coronado-Blanco, Fischer et al. (2018).
Distribution in Iran: Guilan (Ghahari, Fischer, & Tobias, 2012), Isfahan (Ghahari, Fischer, & Papp, 2011a), Iran (no specific locality cited) (Papp, 2015).
Distribution in the Middle East: Cyprus (Papp, 1998), Iran (Ghahari, Fischer, & Papp, 2011a; Ghahari, Fischer, & Tobias, 2012; Papp, 2015), Israel−Palestine (Belshaw et al., 2001; Papp, 2015), Turkey (Beyarslan, 1986, 1999, 2014; Beyarslan & Četin Erdoğan, 2005, 2010).
Extralimital distribution: Albania, Algeria, Armenia, Austria, Azerbaijan, Croatia, Georgia, Greece, Hungary, Italy, North Macedonia, Moldova, Spain, Switzerland, Tunisia, Ukraine, Uzbekistan, former Yugoslavia.
Host records: Recorded by Papp (1997) as being a parasitoid of the depressarriid *Agonopterix ferulae* (Zeller). It was also summarized by Yu et al. (2016) as being a parasitoid of the curculionids *Lachnaeus horridus* (Reitter), and *Larinus turbinatus* Gyllenhal. In Iran, this species has been reared from *Larinus turbinatus* Gyllenhal (Ghahari, Fischer, & Tobias, 2012).

Bracon (*Cyanopterobracon*) *sabulosus* Szépligeti, 1896

Catalogs with Iranian records: Fallahzadeh and Saghaei (2010), Gadallah and Ghahari (2015), Farahani et al. (2016), Yu et al. (2016), Samin, Coronado-Blanco, Kavallieratos et al. (2018), Samin, Coronado-Blanco, Fischer et al. (2018).
Distribution in Iran: East Azarbaijan (Ghahari, Fischer, Erdoğan, Tabari et al., 2009), Golestan (Ghahari & Fischer, 2011b), Iran (no specific locality cited) (Shenefelt, 1978; Telenga, 1936; Tobias, 1976, 1986).
Distribution in the Middle East: Cyprus (Papp, 1998), Iran (Ghahari & Fischer, 2011b; Ghahari, Fischer, Erdoğan, Tabari et al., 2009; Shenefelt, 1978; Telenga, 1936; Tobias, 1976, 1986), Jordan (Papp, 2008), Turkey (Beyarslan, 2014; Beyarslan & Četin Erdoğan, 2010; Beyarslan et al., 2014; Papp, 2008).
Extralimital distribution: Azerbaijan, Bulgaria, Greece, Hungary, Kazakhstan, North Macedonia, Moldova, Romania, Russia, Serbia and Montenegro, Tajikistan, Turkmenistan, Ukraine, Uzbekistan.
Host records: Unknown.

Bracon (*Cyanopterobracon*) *spectabilis* (Telenga, 1936)

Catalogs with Iranian records: No catalog.
Distribution in Iran: Mazandaran (Ghahari, 2019a).
Distribution in the Middle East: Iran (Ghahari, 2019a), Turkey (Beyarslan, 1986, 1999, 2014; Beyarslan & Četin Erdoğan, 2010).
Extralimital distribution: Georgia, Kazakhstan, Tajikistan.
Host records: Unknown.

Bracon (*Cyanopterobracon*) *urinator* (Fabricius, 1798)

Catalogs with Iranian records: Fallahzadeh and Saghaei (2010), Gadallah and Ghahari (2015), Farahani et al. (2016), Yu et al. (2016), Samin, Coronado-Blanco, Kavallieratos et al. (2018), Samin, Coronado-Blanco, Fischer et al. (2018).
Distribution in Iran: Golestan (Ghahari & Fischer, 2011c), Kerman (Iranmanesh et al., 2018), Khuzestan (Zargar, Samartsev et al., 2020 as *B.* (*Rostrobracon*) *urinator* (Fabricius, 1798)), Mazandaran (Ghahari, Fischer, Erdoğan et al., 2010; Ghahari & Fischer, 2011b), Iran (no specific locality cited) (Beyarslan et al., 2005; Shenefelt, 1978; Telenga, 1936; Tobias, 1995).
Distribution in the Middle East: Cyprus, Egypt, Iran, Israel−Palestine, Saudi Arabia, Syria, Turkey (Yu et al., 2016).
Extralimital distribution: Europe, Afrotropical, Oriental, Palaearctic [Adjacent countries to Iran: Afghanistan, Azerbaijan, Kazakhstan, Russia, Turkmenistan].

Host records: Summarized by Yu et al. (2016) as being a parasitoid of the curculionids *Larinus* spp., *Lixus obesus* Petri, and *Rhinocyllus conicus* Frölich; the lonchaeid fly *Protearomyia nigra* (Meigen); and the tephritid *Tephritis pulchra* (Loew).

Plant associations in Iran: *Hordeum vulgare* L. (Poaceae) (Iranmanesh et al., 2018).

Bracon (*Glabrobracon*) abbreviator Nees von Esenbeck, 1834

Catalogs with Iranian records: Gadallah and Ghahari (2015), Farahani et al. (2016), Yu et al. (2016), Samin, Coronado-Blanco, Kavallieratos et al. (2018), Samin, Coronado-Blanco, Fischer et al. (2018).

Distribution in Iran: Qazvin (Ghahari, Fischer, & Papp, 2011b), Semnan (Ghahari & Gadallah, 2015), Iran (no specific locality cited) (Tobias, 1961, 1976).

Distribution in the Middle East: Cyprus (Papp, 1998), Iran, Israel−Palestine (Papp, 2015), Jordan (Papp, 2008), Turkey (Beyarslan, 1986; 1999, 2014; Beyarslan & Četin Erdoğan, 2012).

Extralimital distribution: Europe, and Palaearctic region.

Host records: Summarized by Yu et al. (2016) as being a parasitoid of the curculionid *Anthonomus pomorum* (L.); the chloropid fly *Liapara lucens* Meigen; the cephid sawfly *Cephus pygmeus* L.; and the cynipid *Diplopis rosae* (L.). In addition to several lepidopteran pests including: the coleophorid *Coleophora follicularis* (Vallot); the elachistid *Elachista gangabella* Zeller; the gracillariid *Gracillaria syringella* (Fabricius); the noctuid *Oria musculosa* (Hübner); and the tortricids *Cochylis posterana* Zeller, *Leguminivora glycinivorella* Matsumura, *Pandemis heparana* (Denis and Schiffermüller), and *Rhyacionia buoliana* (Denis and Schiffermüller).

Bracon (*Glabrobracon*) admotus Papp, 2000

Distribution in the Middle East: Turkey (Beyarslan, 2014; Beyarslan & Četin Erdoğan, 2012).

Extralimital distribution: Bulgaria, Hungary.

Host records: Recorded by Papp (2000) as being a parasitoid of the rhynchitid *Byctiscus betulae* (L.).

Bracon (*Glabrobracon*) ahngeri Telenga, 1936

Catalogs with Iranian records: Farahani et al. (2016), Yu et al. (2016), Samin, Coronado-Blanco, Kavallieratos et al. (2018), Samin, Coronado-Blanco, Fischer et al. (2018).

Distribution in Iran: Hormozgan (Ameri et al., 2015).

Distribution in the Middle East: Cyprus (Beyarslan et al., 2017), Iran (Ameri et al., 2015), Turkey (Beyarslan, 2014, 2016; Beyarslan & Četin Erdoğan, 2012; Beyarslan et al., 2014).

Extralimital distribution: China, Korea, Russia.

Host records: Summarized by Yu et al. (2016) as being a parasitoid of several host species of the orders Coleoptera (Curculionidae), Diptera (Chloropidae), Hymenoptera (Cephidae, Cynipidae), and Lepidoptera (Coleophoridae, Elachistidae, Gracillariidae, Noctuidae, Tortricidae).

Bracon (*Glabrobracon*) angustiventris Tobias, 1957

Catalogs with Iranian records: Gadallah and Ghahari (2015), Farahani et al. (2016), Yu et al. (2016), Samin, Coronado-Blanco, Kavallieratos et al. (2018), Samin, Coronado-Blanco, Fischer et al. (2018).

Distribution in Iran: Kermanshah (Ghahari & Fischer, 2012).

Distribution in the Middle East: Iran (Ghahari & Fischer, 2012), Turkey (Beyarslan, 1986, 1999, 2014, 2016; Beyarslan & Četin Erdoğan, 2010, 2012).

Extralimital distribution: Armenia, Azerbaijan, Kazakhstan, Moldova, Turkmenistan.

Host records: Unknown.

Bracon (*Glabrobracon*) arcuatus Thomson, 1892

Catalogs with Iranian records: No catalog.

Distribution in Iran: Khuzestan (Zargar, Samartsev et al., 2020 as *Bracon* (*Bracon*) *arcuatus* Thomson, 1892), Semnan (Ghahari & Gadallah, 2019).

Distribution in the Middle East: Cyprus (Papp, 1998), Iran (Ghahari & Gadallah, 2019; Zargar, Samartsev et al., 2020), Turkey (Beyarslan, 2014).

Extralimital distribution: Austria, Bulgaria, China, Croatia, Czech Republic, Germany, Greece, Hungary, North Macedonia, Mongolia, Serbia, Slovenia, Sweden, United Kingdom.

Host records: Unknown.

Barcon (*Glabrobracon*) baseflavus Beyarslan, 2002

Distribution in the Middle East: Turkey (Beyarslan, 2002, 2014; Beyarslan & Četin Erdoğan, 2010).

Extralimital distribution: None.

Host records: Unknown.

Bracon (*Glabrobracon*) batis Papp, 1981

Catalogs with Iranian records: No catalog.

Distribution in Iran: Khuzestan (Zargar, Samartsev et al., 2020 as *Bracon* (*Bracon*) *batis* Papp, 1981).

Distribution in the Middle East: Iran.

Extralimital distribution: Hungary.

Host records: Unknown.

Bracon (*Glabrobracon*) brevicalcaratus Tobias, 1957

Catalogs with Iranian records: No catalog.

Distribution in Iran: Ardabil (Samin et al., 2019).

Distribution in the Middle East: Iran (Samin et al., 2019), Turkey (Beyarslan, 2014, 2016; Beyarslan & Četin Erdoğan, 2010).

Extralimital distribution: Hungary, Kazakhstan, Turkmenistan.

Host records: Unknown.

Bracon (*Glabrobracon*) *caudatus* Ratzeburg, 1848

Catalogs with Iranian records: No catalog.

Distribution in Iran: East Azarbaijan (Sakenin et al., 2020).

Distribution in the Middle East: Iran (Sakenin et al., 2020), Turkey (Beyarslan, 2014; Beyarslan & Çetin Erdoğan, 2010, 2012).

Extralimital distribution: Former Czechoslovakia, Germany, Greece, Hungary, Italy, Poland, Russia, Slovenia, Switzerland, United Kingdom, former Yugoslavia.

Host records: Summarized by Yu et al. (2016) as being a parasitoid of the curculionids *Curculio villosus* Fabricius, and *Hylesinus fraxini* (Panzer); the agromyzid fly *Napomyza lateralis* (Fallén); the cynipids *Andricus terminalis* (Hartig), and *Biorhiza terminalis* (Hartig); and the tortricid *Pammene gallicolana* (Lienig and Zeller).

Bracon (*Glabrobracon*) *caudiger* Nees von Esenbeck, 1834

Catalogs with Iranian records: No catalog.

Distribution in Iran: Hamadan (Sakenin et al., 2020).

Distribution in the Middle East: Iran (Sakenin et al., 2020), Israel−Palestine (Papp, 1970), Turkey (Beyarslan, 1999, 2014; Beyarslan & Çetin Erdoğan, 2010; Papp, 2012).

Extralimital distribution: Austria, Belgium, Croatia, former Czechoslovakia, France, Germany, Greece, Hungary, Italy, Poland, Sweden, Ukraine, former Yugoslavia.

Host records: Summarized by Yu et al. (2016) as being a parasitoid of the coleophorid *Coleophora obscenella* Herrich-Schäffer; the tortricids *Aphelia amplana* Hübner, *Cydia splendana* (Hübner), and *Cydia strobilella* (L.).

Bracon (*Glabrobracon*) *chrysostigma* Greese, 1928

Catalogs with Iranian records: Gadallah and Ghahari (2015), Farahani et al. (2016), Yu et al. (2016), Samin, Coronado-Blanco, Kavallieratos et al. (2018), Samin, Coronado-Blanco, Fischer et al. (2018).

Distribution in Iran: Golestan (Ghahari & Fischer, 2011c), Hamadan (Rajabi Mazhar et al., 2019), Northern Khorasan (Rahmani et al., 2017).

Distribution in the Middle East: Iran (Rahmani et al., 2017), Israel−Palestine (Papp, 1989, 2012), Turkey (Beyarslan & Çetin Erdoğan, 2010; Beyarslan, 2014).

Extralimital distribution: Austria, France, Greece, Hungary, North Macedonia, Moldova, Romania, Russia, Ukraine, former Yugoslavia.

Host records: Recorded by Fulmek (1968) as being a parasitoid of the tephritid *Acanthophilus helianthi* Rossi. It was also recorded by Volovnik (1994) as being a parasitoid of the curculionid *Lixus canescens* Steven.

Bracon (*Glabrobracon*) *claripennis* Thomson, 1892

Distribution in the Middle East: Cyprus (Papp, 2000).

Extralimital distribution: Austria, Denmark, Finland, Germany, Greece, Hungary, Russia, Sweden, United Kingdom, former Yugoslavia.

Host records: Summarized by Yu et al. (2016) as being a parasitoid of the coleophorid *Coleophora serratella* (L.), and the tortricid *Cydia nigricana* (Fabricius).

Bracon (*Glabrobracon*) *colpophorus* (Wesmael, 1838)

Catalogs with Iranian records: No catalog.

Distribution in Iran: Ardabil (Samin, Beyarslan, Coronado-Blanco et al., 2020).

Distribution in the Middle East: Iran (Samin, Beyarslan, Coronado-Blanco et al., 2020), Turkey (Beyarslan, 2016).

Extralimital distribution: Austria, Belgium, Bulgaria, Czech Republic, France, Germany, Hungary, Italy, Kazakhstan, Netherlands, Poland, Romania, Russia, Serbia, Spain, Switzerland, Ukraine, United Kingdom.

Host records: Summarized by Yu et al. (2016) as being a parasitoid of the tenthredinids *Hoplocampa fulvicornis* (Panzer), and *Hoplocampa minuta* Christoph; the apionid *Apion craccae* (L.); the chrysomelid *Bruchus spartii* (Eric); and the curculionid *Curculio villosus* Fabricius.

Bracon (*Glabrobracon*) *conjugellae* Bengtsson, 1924

Distribution in the Middle East: Israel−Palestine (Papp, 2015).

Extralimital distribution: Former Czechoslovakia, Germany, Hungary, Netherlands, Sweden, Switzerland.

Host records: Recorded by Fahringer (1928) as being a parasitoid of the yponomeutid *Argyresthia conjugella* Zeller.

Bracon (*Glabrobracon*) *crassiceps* Thomson, 1892

Catalogs with Iranian records: No catalog.

Distribution in Iran: Kermanshah (Sakenin et al., 2020).

Distribution in the Middle East: Iran.

Extralimital distribution: Austria, Croatia, Finland, Germany, Hungary, Norway, Russia, Sweden, Switzerland, former Yugoslavia.

Host records: Recorded by Čapek and Hofmann (1997) as being a parasitoid of the tortricid *Archips xylosteana* (L.).

Bracon (*Glabrobracon*) *curticaudis* Szépligeti, 1901

Catalogs with Iranian records: This species is a new record for the fauna of Iran.

Distribution in Iran: East Azarbaijan province, Arasbaran forest, 1♀, August 2007.

Distribution in the Middle East: Iran (new record), Turkey (Papp, 2008).

Extralimital distribution: Austria, Azerbaijan, Bulgaria, Finland, Germany, Hungary, Kazakhstan, Poland, Russia, Sweden, Switzerland, United Kingdom.

Host records: Unknown.

Bracon (*Glabrobracon*) *debitor* Papp, 1971

Catalogs with Iranian records: No catalog.

Distribution in Iran: Khuzestan (Zargar, Samartsev et al., 2020 as *B.* (*B.*) *debitor* Papp, 1971).

Distribution in the Middle East: Iran.
Extralimital distribution: Mongolia.
Host records: Unknown.

Bracon (*Glabrobracon*) *delusor* Spinola, 1808
Catalogs with Iranian records: No catalog.
Distribution in Iran: Kermanshah (Ghahari et al., 2020).
Distribution in the Middle East: Iran.
Extralimital distribution: Austria, Azerbaijan, France, Germany, Hungary, Italy, Poland, Russia, Slovenia, former Yugoslavia.
Host records: Recorded by Györfi (1956) as being a parasitoid of the tortricid *Cydia strobilella* (L.).

Bracon (*Bracon*) *delusorius* Telenga, 1936
Distribution in the Middle East: Turkey (Beyarslan, 1986, 1999, 2014; Beyarslan & Çetin Erdoğan, 2010).
Extralimital distribution: Turkmenistan.
Host records: Unknown.

Bracon (*Glabrobracon*) *densipilosus* Tobias, 1957
Catalogs with Iranian records: Samin, Coronado-Blanco, Fischer et al. (2018).
Distribution in Iran: Fars (Sakenin et al., 2018).
Distribution in the Middle East: Iran (Sakenin et al., 2018), Turkey (Beyarslna, 2014, 2016; Beyarslan & Çetin Erdoğan, 2010, 2012).
Extralimital distribution: Tajikistan.
Host records: Unknown.

Bracon (*Glabrobracon*) *dersimensis* Beyarslan, 2012
Distribution in the Middle East: Turkey (Beyarslan, 2014; Beyarslan & Erdoğan, 2012).
Extralimital distribution: None.
Host records: Unknown.

Bracon (*Glabrobracon*) *dichromus* (Wesmael, 1838)
Catalogs with Iranian records: Gadallah and Ghahari (2015), Farahani et al. (2016), Yu et al. (2016), Samin, Coronado-Blanco, Kavallieratos et al. (2018), Samin, Coronado-Blanco, Fischer et al. (2018).
Distribution in Iran: East Azarbaijan (Rastegar et al., 2012 as *Bracon variator maculiger* (Wesmael, 1838)), Isfahan (Ghahari, Fischer, & Papp, 2011a as *B. variator maculiger*), Iran (no specific locality cited) (Yu et al., 2012 as *B. variator maculiger*).
Distribution in the Middle East: Cyprus (Papp, 1998, 2012), Iran (Ghahari, Fischer, & Papp, 2011a; Rastegar et al., 2012), Israel—Palestine (Papp, 2012, 2015), Jordan, Syria (Papp, 2012), Turkey (Beyarslan, 1986, 1987, 1999, 2014; Kohl, 1905; Papp, 2012).
Extralimital distribution: Algeria, Armenia, Austria, Azerbaijan, Belgium, Bulgaria, Croatia, France, Georgia,

Germany, Greece, Hungary, Italy, Kazakhstan, Latvia, Lithuania, North Macedonia, Mongolia, Poland, Portugal, Romania, Russia, Serbia, Slovakia, Spain, Sweden, Tunisia, Turkmenistan, Ukraine, United Kingdom, Uzbekistan.
Host records: Summarized by Yu et al. (2016) as being a parasitoid of the buprestid *Coraebus florentinus* (Herbst); the curculionids *Ceutorhynchus assimilis* (Paykull), *Ceutorhynchus punctiger* Gyllenhal, *Gymnetron campanulae* Schoenherr, and *Rhinusa asellus* (Gravenhorst); the tenthredinid *Hoplocampa minuta* Christoph; the gelechiid *Platyedra subcinerea* (Haworth); the pyralid *Myelois circumvoluta* (Furcroy); the noctuid *Gortyna xanthenes* Germar; and the tortricid *Grapholita funebrana* (Treitschke).
Comments: Farahani et al. (2016) considered *Bracon* (*Glabrobracon*) *dichromus* as incorrect report for the fauna of Iran which was recorded by Yu et al. (2012). It was recorded from Iran as *Bracon variator maculiger* by Ghahari, Fischer, & Papp (2011a), and Rastegar et al. (2012).

Bracon (*Glabrobracon*) *dilatus* Papp, 1999
Catalogs with Iranian records: Fallahzadeh and Saghaei (2010), Gadallah and Ghahari (2015), Farahani et al. (2016), Yu et al. (2016), Samin, Coronado-Blanco, Kavallieratos et al. (2018), Samin, Coronado-Blanco, Fischer et al. (2018).
Distribution in Iran: Mazandaran (Papp, 1999).
Distribution in the Middle East: Iran (Papp, 1999), Iraq (Papp, 1999).
Extralimital distribution: None.
Host records: Unknown.

Bracon (*Glabrobracon*) *discoideus* (Wesmael, 1838)
Catalogs with Iranian records: No catalog.
Distribution in Iran: Hamadan (Gadallah et al., 2018; Ghahari & Beyarslan, 2019).
Distribution in the Middle East: Egypt (Fahringer, 1928), Iran (Ghahari & Beyarslan, 2019; Gadallah et al., 2018), Turkey (Beyarslan & Çetin Erdoğan, 2010; Beyarslan, 1986, 1999, 2014).
Extralimital distribution: Armenia, Austria, Belgium, Bulgaria, France, Germany, Hungary, Ireland, Italy, Kazakhstan, Kyrgyzstan, Moldova, Netherlands, Poland, Russia, Serbia, Slovakia, Slovenia, Sweden, Switzerland, Ukraine, United Kingdom.
Host records: Summarized by Yu et al. (2016) as being a parasitoid of several insect pests belonging to the orders Coleoptera (Attelabidae, Cerambycidae, Curculionidae, Rhynchitidae), Diptera (Chloropidae), Hymenoptera (Cynipidae, Tenthredinidae), and Lepidoptera (Geometridae, Tortricidae). In Iran, this species has been reared from the curculionid *Anthonomus pomorum* (L.) (Ghahari & Beyarslan, 2019; Gadallah et al., 2018).

Bracon (*Glabrobracon*) *dolichurus* Marshall, 1897

Catalogs with Iranian records: Gadallah and Ghahari (2015), Farahani et al. (2016), Yu et al. (2016), Samin, Coronado-Blanco, Kavallieratos et al. (2018), Samin, Coronado-Blanco, Fischer et al. (2018).

Distribution in Iran: Kermanshah (Ghahari & Fischer, 2012).

Distribution in the Middle East: Iran (Ghahari & Fischer, 2012), Israel—Palestine (Papp, 2015), Turkey (Beyarslan, 2014, 2016; Beyarslan & Çetin Erdoğan, 2005, 2010; Beyarslan et al., 2002, 2014).

Extralimital distribution: Former Czechoslovakia, France, Hungary, Romania, Russia, Spain.

Host records: Unknown.

Bracon (*Glabrobracon*) *ductor* Telenga, 1936

Distribution in the Middle East: Israel—Palestine (Papp, 2015).

Extralimital distribution: Bulgaria, Hungary, Mongolia, Montenegro, Russia.

Host records: Unknown.

Bracon (*Glabrobracon*) *epitriptus* Marshall, 1885

Catalogs with Iranian records: Gadallah and Ghahari (2015), Farahani et al. (2016), Yu et al. (2016), Samin, Coronado-Blanco, Kavallieratos et al. (2018), Samin, Coronado-Blanco, Fischer et al. (2018).

Distribution in Iran: Chaharmahal & Bakhtiari, Kuhgiloyeh & Boyerahmad (Samin, van Achterberg et al., 2015 as *Bracon* (*Orthobracon*) *epitriptus*), East Azarbaijan (Rastegar et al., 2012), Fars (Lashkari Bod et al., 2010, 2011), Hormozgan (Ameri et al., 2014 as *B.* (*Orthobracon*) *epitriptus*), Isfahan (Ghahari, Fischer, & Papp, 2011a), Khuzestan (Zargar, Samartsev et al., 2020 as *B.* (*B.*) *epitriptus* Marshall, 1885; Ghahari & Gadallah, 2022), Mazandaran (Zargar et al., 2015; Ghahari, 2017—around rice fields), Razavi Khorasan (Samin et al., 2011 as *B.* (*Orthobracon*) *epitriptus*).

Distribution in the Middle East: Iran (see references above), Turkey (Beyarslan, 2014, 2016; Beyarslan & Çetin Erdoğan, 2005, 2010).

Extralimital distribution: Armenia, Austria, Azerbaijan, Belarus, China, Croatia, Czech Republic, Georgia, Germany, Greece, Hungary, Italy, Kazakhstan, Korea, Lithuania, North Macedonia, Moldova, Mongolia, Netherlands, Poland, Romania, Russia, Serbia, Slovenia, Switzerland, Ukraine, United Kingdom.

Host records: Summarized by Yu et al. (2016) as being a parasitoid of the curculionids *Cryptorhynchus lapathi* (L.), and *Hylobius transversovittatus* (Goeze); the agromyzid *Agromyza flaviceps* Fallén; the cecedomyiid *Iteomyia capreae* (Winnertz); and the tenthredinid *Fenusa pusilla* (Lepeletier). In Iran, this species has been reared from the agromyzid *Agromyza* sp. (Ghahari & Gadallah, 2022).

Bracon (*Glabrobracon*) *exhilarator* Nees von Esenbeck, 1834

Catalogs with Iranian records: Gadallah and Ghahari (2015), Farahani et al. (2016), Yu et al. (2016), Samin, Coronado-Blanco, Kavallieratos et al. (2018), Samin, Coronado-Blanco, Fischer et al. (2018).

Distribution in Iran: Hormozgan (Ameri et al., 2014), Kermanshah (Ghahari & Fischer, 2012).

Distribution in the Middle East: Cyprus (Papp, 1998), Iran (Ameri et al., 2014; Ghahari & Fischer, 2012), Israel—Palestine (Papp, 2015), Turkey (Beyarslan, 1999, 2014, 2016; Beyarslan & Çetin Erdoğan, 2005, 2010, 2012).

Extralimital distribution: Europe, Palaearctic [Adjacent countries to Iran: Kazakhstan, Russia].

Hosts records: Summarized by Yu et al. (2016) as being a parasitoid of the curculionid *Apion hookeri* Kirby; the scatophagids *Nanna armillata* (Zetterstedt), and *Nanna flavipes* (Fallén); the tephritids *Plioreocepta poeciloptera* (Schrank), and *Urophora stylata* (Fabricius); and the tortricid *Acleris rhombana* (Denis and Schiffermüller). In Iran, this species has been reared from the curculionid *Apion* sp. (Ghahari & Fischer, 2012).

Bracon (*Glabrobracon*) *fadiche* Beyarslan, 1996

Distribution in the Middle East: Turkey (Beyarslan, 1996, 2014, 2016; Beyarslan & Çetin Erdoğan, 2005; Beyarslan et al., 2002, 2010, 2014).

Extralimital distribution: None.

Host records: Unknown.

Bracon (*Glabrobracon*) *filicornis* Thomson, 1892

Catalogs with Iranian records: No catalog.

Distribution in Iran: Fars (Ghahari & Gadallah, 2019).

Distribution in the Middle East: Iran (Ghahari & Gadallah, 2019), Turkey (Beyarslan, 2014; Beyarslan et al., 2014).

Extralimital distribution: Austria, Sweden.

Host records: Unknown.

Bracon (*Glabrobracon*) *flamargo* Papp, 2011

Catalogs with Iranian records: No catalog.

Distribution in Iran: Khuzestan (Zargar, Samartsev et al., 2020 as *B.* (*B.*) *flamargo* Papp, 2011).

Distribution in the Middle East: Iran.

Extralimital distribution: Bulgaria.

Host records: Unknown.

Bracon (*Glabrobracon*) *flavipalpis* Thomson, 1892

Distribution in the Middle East: Turkey (Beyarslan, 2014; Beyarslan & Çetin Erdoğan, 2010, 2012).

Extralimital distribution: Austria, Sweden.

Host records: Unknown.

Bracon (*Glabrobracon*) frater Tobias, 1957
Catalogs with Iranian records: Samin, Coronado-Blanco, Kavallieratos et al. (2018).
Distribution in Iran: Chaharmahal & Bakhtiari (Samin, Coronado-Blanco, Kavallieratos et al., 2018).
Distribution in the Middle East: Iran (Samin, Coronado-Blanco, Kavallieratos et al., 2018), Turkey (Beyarslan, 2014, 2016; Beyarslan & Çetin Erdoğan, 2010; Beyarslan et al., 2014).
Extralimital distribution: Turkmenistan.
Host records: Unknown.

Bracon (*Glabrobracon*) fumatus Szépligeti, 1901
Catalogs with Iranian records: No catalog.
Distribution in Iran: Isfahan (Sakenin et al., 2020).
Distribution in the Middle East: Cyprus (Papp, 2008), Iran (Sakenin et al., 2020), Israel–Palestine (Papp, 2012, 2015), Turkey (Beyarslan, 2014; Beyarslan & Çetin Erdoğan, 2012; Papp, 2008).
Extralimital distribution: Austria, Croatia, France, Greece, Hungary, Kazakhstan, Moldova, Montenegro, Serbia, Spain, Switzerland, Tunisia, Ukraine.
Host records: Unknown.

Bracon (*Glabrobracon*) fuscicoxis (Wesmael, 1838)
Distribution in the Middle East: Turkey (Papp, 2012).
Extralimital distribution: Austria, Belgium, Croatia, former Czechoslovakia, Denmark, France, Germany, Hungary, Ireland, Italy, Korea, Netherlands, Poland, Romania, Russia, Sweden, Ukraine, United Kingdom.
Host records: Recorded by Papp (2012) as being a parasitoid of the chrysomelid *Prasocuris phellandrii* (L.).

Bracon (*Glabrobracon*) gusaricus Telenga, 1933
Catalogs with Iranian records: Gadallah and Ghahari (2015), Farahani et al. (2016), Yu et al. (2016), Samin, Coronado-Blanco, Kavallieratos et al. (2018), Samin, Coronado-Blanco, Fischer et al. (2018).
Distribution in Iran: Hormozgan (Ameri et al., 2014), Kerman (Iranmanesh et al., 2018; Rahmani et al., 2017), Khuzestan (Zargar, Samartsev et al., 2020 as *B.* (*B.*) *gusaricus* Telenga, 1933), Lorestan (Ghahari, Fischer, Papp, & Tobias, 2012), Northern Khorasan, Sistan & Baluchestan (Rahmani et al., 2017).
Distribution in the Middle East: Cyprus (Beyarslan et al., 2017), Iran (see references above), Israel–Palestine (Papp, 2012, 2015), Turkey (Beyarslan, 2014, 2016; Beyarslan & Çetin Erdoğan, 2005, 2010, 2012; Beyarslan et al., 2014).
Extralimital distribution: Afghanistan, Azerbaijan, Bulgaria, Georgia, Russia, Switzerland, Turkmenistan, Uzbekistan.
Host records: Unknown.

Plant associations in Iran: *Elymus repens* (L.) (Poaceae), *Beta vulgaris* L. (Amaranthaceae), *Chenopodium album* L. (Amaranthaceae), *Cortaderia selloana* (Schult. and Schult.) (Poaceae), *Cynodon dactylon* L. (Poaceae), *Cyperus globosus* All. (Cyperaceae), *Mentha pulegium* L. (Lamiaceae), *Plantago major* L. (Plantaginaceae), *Triticum dicoccum* Schrank (Poaceae) (Rahmani et al., 2017), *Medicago sativa* L. (Fabaceae) (Rahmani et al., 2017; Iranmanesh et al., 2018), *Rubus* sp. (Rosaceae), and *Triticum aestivum* L. (Poaceae) (Iranmanesh et al., 2018).

Bracon (*Glabrobracon*) helleni Telenga, 1936
Catalogs with Iranian records: Gadallah and Ghahari (2015), Farahani et al. (2016), Yu et al. (2016), Samin, Coronado-Blanco, Kavallieratos et al. (2018), Samin, Coronado-Blanco, Fischer et al. (2018).
Distribution in Iran: Golestan (Samin, Ghahari et al., 2015), West Azarbaijan (Ghahari & Fischer, 2011a).
Distribution in the Middle East: Cyprus (Papp, 1998), Iran (Ghahari & Fischer, 2011a; Samin, Ghahari et al., 2015), Israel–Palestine (Papp, 1989, 2012), Turkey (Beyarslan, 2014; Beyarslan & Çetin Erdoğan, 2010; Beyarslan et al., 2014).
Extralimital distribution: Hungary, Kazakhstan, Russia, Switzerland.
Host records: Unknown.

Bracon (*Glabrobracon*) hemiflavus Szépligeti, 1901
Catalogs with Iranian records: Gadallah and Ghahari (2015), Farahani et al. (2016), Yu et al. (2016), Samin, Coronado-Blanco, Kavallieratos et al. (2018), Samin, Coronado-Blanco, Fischer et al. (2018).
Distribution in Iran: Khuzestan (Zargar, Samartsev et al., 2020 as *B.* (*B.*) *hemiflavus* Szépligeti, 1901), Iran (no specific locality cited) (Papp, 2008).
Distribution in the Middle East: Cyprus (Papp, 1998, 2008), Iran (Papp, 2008; Zargar, Samartsev et al., 2020), Israel–Palestine (Papp, 1989, 2008, 2015), Syria (Papp, 2008), Turkey (Beyarslan, 2014; Beyarslan & Çetin Erdoğan, 2010; Beyarslan et al., 2002; Papp, 2008).
Extralimital distribution: Azerbaijan, Bulgaria, Croatia, France, Greece, Hungary, Italy, Kazakhstan, North Macedonia, Poland, Romania, Russia, Slovakia, Spain, Turkmenistan, Uzbekistan, former Yugoslavia.
Host records: Summarized by Yu et al. (2016) as being a parasitoid of the curculionids *Larinus flavescens* Germar, and *Rhinocyllus conicus* Frölich; and the noctuid *Gortyna xanthenes* Germar.

Bracon (*Glabrobracon*) immutator Nees von Esenbeck, 1834
Catalogs with Iranian records: Gadallah and Ghahari (2015), Farahani et al. (2016), Yu et al. (2016), Samin, Coronado-Blanco, Kavallieratos et al. (2018), Samin, Coronado-Blanco, Fischer et al. (2018).

Distribution in Iran: Hormozgan (Ameri et al., 2014), Mazandaran (Ghahari, 2019a).

Distribution in the Middle East: Cyprus (Papp, 1998), Iran (Ameri et al., 2014; Ghahari et al., 2019), Turkey (Beyarslan, 2014; Beyarslan & Četin Erdoğan, 2005, 2010).

Extralimital distribution: Austria, Belgium, Bosnia-Herzegovina, Bulgaria, Croatia, Denmark, Finland, France, Germany, Hungary, Italy, Kazakhstan, Korea, Latvia, North Macedonia, Moldova, Mongolia, Netherlands, Norway, Poland, Russia, Serbia, Slovakia, Slovenia, Sweden, Switzerland, Ukraine, United Kingdom.

Host records: Summarized by Yu et al. (2016) as being a parasitoid of the curculionids *Anthonomus rubi* Herbst, *Ceutorhynchus assimilis* (Paykull), *Cryptorhynchus lapathi* (L.), *Curculio villosus* (Fabricius), and *Pissodes* sp.; the tiphritid *Urophora stylata* (Fabricius); the cynipids *Biorhiza pallida* L., and *Biorhiza terminalis* (Hartig); the tenthredid *Pontania* spp.; and the tortricid *Pammene aurana* (Fabricius).

Bracon (Glabrobracon) jaroslavensis Telenga, 1936

Catalogs with Iranian records: No catalog.

Distribution in Iran: Ardabil (Samin, Papp, & Coronado-Blanco, 2018).

Distribution in the Middle East: Iran (Samin, Papp, & Coronado-Blanco, 2018), Turkey (Beyarslan, 2014; Beyarslan & Četin Erdoğan, 2010, 2012).

Extralimital distribution: Azerbaijan, Greece, Hungary, Russia, Switzerland.

Host records: Unknown.

Bracon (Glabrobracon) jenoi Beyarslan, 2010

Distribution in the Middle East: Turkey (Beyarslan, 2010, 2014).

Extralimital distribution: None.

Host records: Unknown.

Bracon (Glabrobracon) karakumicus Tobias, 1967

Distribution in the Middle East: Israel—Palestine (Papp, 2012), Turkey (Beyarslan, 2016).

Extralimital distribution: Turkmenistan.

Host records: Unknown.

Bracon (Glabrobracon) kirgisorum Telenga, 1936

Catalogs with Iranian records: Gadallah and Ghahari (2015), Farahani et al. (2016), Yu et al. (2016), Samin, Coronado-Blanco, Kavallieratos et al. (2018), Samin, Coronado-Blanco, Fischer et al. (2018).

Distribution in Iran: Golestan (Ghahari & Fischer, 2011b; Sakenin et al., 2012), Northern Iran (Sakenin et al., 2011), Iran (no specific locality cited) (Papp, 2015).

Distribution in the Middle East: Iran (see references above), Israel—Palestine (Papp, 2015), Turkey (Beyarslan, 2014, 2016; Beyarslan & Četin Erdoğan, 2010, 2012; Çikman et al., 2006).

Extralimital distribution: Kyrgyzstan, Lithuania, Moldova.

Host records: Recorded by Çikman et al. (2006) as being a parasitoid of the agromyzid *Phytomyza orobanchia* Kaltenbach.

Bracon (Glabrobracon) lividus Telenga, 1936

Catalogs with Iranian records: Gadallah and Ghahari (2015), Farahani et al. (2016), Yu et al. (2016), Samin, Coronado-Blanco, Kavallieratos et al. (2018), Samin, Coronado-Blanco, Fischer et al. (2018).

Distribution in Iran: Ardabil (Ghahari & Fischer, 2011b), East Azerbaijan (Ghahari & Fischer, 2011a), Golestan (Ghahari & Fischer, 2011c), Khuzestan (Samin, van Achterberg et al., 2015).

Distribution in the Middle East: Cyprus (Beyarslan et al., 2017), Iran (Ghahari & Fischer, 2011a,b,c; Samin, van Achterberg et al., 2015), Israel—Palestine (Papp, 2012), Turkey (Beyarslan, 1999, 2014, 2016; Beyarslan & Četin Erdoğan, 2005, 2010; Beyarslan et al., 2002, 2014).

Extralimital distribution: Armenia, Germany, Greece, Hungary, Russia.

Host records: Recorded by Papp (1991) as being a parasitoid of the tenthredinid *Pontania vesicator* (Bremi).

Bracon (Glabrobracon) longiantennatus Tobias, 1957

Distribution in the Middle East: Turkey (Beyarslan, 2016).

Extralimital distribution: Kazakhstan.

Host records: Unknown.

Bracon (Glabrobracon) longulus Thomson, 1892

Catalogs with Iranian records: No catalog.

Distribution in Iran: West Azerbaijan (Ghahari et al., 2019).

Distribution in the Middle East: Cyprus (Papp, 1998), Iran (Ghahari et al., 2019), Israel—Palestine (Papp, 2015), Turkey (Beyarslan, 2014; Beyarslan & Çetin Erdoğan, 2010; Papp, 2000).

Extralimital distribution: Austria, Bulgaria, Croatia, Czech Republic, Denmark, Finland, France, Germany, Greece, Hungary, Italy, North Macedonia, Norway, Slovenia, Spain, Sweden, former Yugoslavia.

Host records: Recorded by Hellén (1957) as being a parasitoid of the agromyzid *Melanagromyza aenea* (Meigen).

Bracon (Glabrobracon) malatyensis Beyarslan, 2009

Distribution in the Middle East: Turkey (Beyarslan, 2009, 2014; Beyarslan et al., 2014).

Extralimital distribution: None.

Host records: Unknown.

Bracon (Glabrobracon) marshalli Szépligeti, 1901

Catalogs with Iranian records: No catalog.

Distribution in Iran: Northern Khorasan (Rahmani et al., 2017).

Distribution in the Middle East: Cyprus (Papp, 2008), Iran (Rahmani et al., 2017), Israel—Palestine (Papp, 2012).

Extralimital distribution: Azerbaijan, Bulgaria, Croatia, Denmark, Finland, Germany, Greece, Hungary, Italy, Kazakhstan, Mongolia, Montenegro, Norway, Portugal, Romania, Russia, Slovenia, Spain, Sweden, Switzerland, Tajikistan, United Kingdom.

Host records: Summarized by Yu et al. (2016) as being a parasitoid of the epermeniid *Phaulernis fulviguttella* (Zeller); and the pyralid *Homoeosoma sinuella* (Fabricius).

Bracon (Glabrobracon) minutator (Fabricius, 1798)

Catalogs with Iranian records: Fallahzadeh and Saghaei (2010), Gadallah and Ghahari (2015), Farahani et al. (2016 as *B. (Bracon) minutator*), Yu et al. (2016), Samin, Coronado-Blanco, Kavallieratos et al. (2018), Samin, Coronado-Blanco, Fischer et al. (2018).

Distribution in Iran: Hormozgan (Ameri et al., 2014), Isfahan (Ghahari, Fischer, & Papp, 2011a), Kermanshah (Ghahari & Fischer, 2012), Khuzestan (Zargar, Samartsev et al., 2020 as *B. (B.) minutator* (Fabricius, 1798), Iran (no specific locality cited) (Tobias, 1961, 1976).

Distribution in the Middle East: Cyprus, Iran, Israel—Palestine, Jordan, Turkey (Yu et al., 2016).

Extralimital distribution: Europe, Palaearctic [Adjacent countries to Iran: Azerbaijan, Kazakhstan, Russia, Turkmenistan].

Host records: Summarized by Yu et al. (2016) as being a parasitoid of several insect pests belonging to the orders: Coleoptera (Curculionidae), Diptera (Chloropidae, Tephritidae), Hymenoptera (Cephidae), and Lepidoptera (Gelechiidae, Gracillariidae, Noctuidae, Psychidae, Sesiidae, Tortricidae). In Iran, this species has been reared from the curculionid *Anthonomus pomorum* (L.) (Ghahari & Fischer, 2012).

Bracon (Glabrobracon) negativus Tobias, 1957

Catalogs with Iranian records: Samin, Coronado-Blanco, Fischer et al. (2018).

Distribution in Iran: Semnan (Sakenin et al., 2018).

Distribution in the Middle East: Iran (Sakenin et al., 2018), Turkey (Beyarslan, 2014, 2016; Beyarslan & Çetin Erdoğan, 2010).

Extralimital distribution: Turkmenistan.

Host records: Unknown.

Bracon (Glabrobracon) nigricollis (Wesmael, 1838)

Distribution in the Middle East: Cyprus (Papp, 1998, 2012).

Extralimital distribution: Belgium, France, Georgia, Germany, Hungary, Italy, Netherlands, Romania, Switzerland.

Host records: Summarized by Yu et al. (2016) as being a parasitoid of the prodoxid *Lampronia fuscatella* (Tengström), and the tortricid *Cydia pactolana* (Zeller).

Bracon (Glabrobracon) nigripilosus Tobias, 1957

Catalogs with Iranian records: Gadallah and Ghahari (2015), Farahani et al. (2016), Yu et al. (2016), Samin, Coronado-Blanco, Kavallieratos et al. (2018), Samin, Coronado-Blanco, Fischer et al. (2018).

Distribution in Iran: Kermanshah (Ghahari & Fischer, 2012).

Distribution in the Middle East: Iran (Ghahari & Fischer, 2012), Turkey (Beyarslan, 2014; Beyarslan & Çetin Erdoğan, 2005, 2010).

Extralimital distribution: Russia.

Host records: Unknown.

Bracon (Glabrobracon) nigriventris (Wesmael, 1838)

Catalogs with Iranian records: Gadallah and Ghahari (2015), Farahani et al. (2016), Yu et al. (2016), Samin, Coronado-Blanco, Kavallieratos et al. (2018), Samin, Coronado-Blanco, Fischer et al. (2018).

Distribution in Iran: Khuzestan (Zargar, Samartsev et al., 2020 as *B. (B.) nigriventris* Wesmael, 1838), Lorestan (Ghahari, Fischer, Papp, & Tobias, 2012).

Distribution in the Middle East: Cyprus (Papp, 1998), Iran (Ghahari, Fischer, Papp, & Tobias, 2012; Zargar, Samartsev et al., 2020), Turkey (Beyarslan, 2014; Beyarslan & Çetin Erdoğan, 2005, 2010; Beyarslan et al., 2002; Papp, 2012).

Extralimital distribution: Albania, Armenia, Austria, Azerbaijan, Belgium, Bulgaria, Croatia, Czech Republic, Denmark, France, Germany, Hungary, Italy, Kazakhstan, Latvia, North Macedonia, Moldova, Mongolia, Poland, Romania, Russia, Serbia, Slovakia, Sweden, Switzerland, Ukraine, United Kingdom.

Host records: Summarized by Yu et al. (2016) as being a parasitoid of several coleopteran host species belonging to the families Cerambycidae and Curculionidae.

Bracon (Glabrobracon) nigriventris indubius Szépligeti, 1901

Catalogs with Iranian records: No catalog.

Distribution in Iran: Khuzestan (Zargar, Samartsev et al., 2020 as *B. (B.) indubius* Szépligeti, 1901).

Distribution in the Middle East: Cyprus (Papp, 1998), Iran (Zargar, Samartsev et al., 2020).

Extralimital distribution: Hungary, Kazakhstan, North Macedonia, Romania, Russia, Ukraine, former Yugoslavia.

Host records: Unknown.

Bracon (*Glabrobracon*) *novus* Szépligeti, 1901

Catalogs with Iranian records: No catalog.
Distribution in Iran: Hamadan (Rajabi Mazhar et al., 2019), West Azarbaijan (Sakenin et al., 2020).
Distribution in the Middle East: Iran.
Extralimital distribution: Austria, Czech Republic, Germany, Hungary, Italy, Poland, Romania, Slovakia, Slovenia, Switzerland.
Host records: Unknown.

Bracon (*Glabrobracon*) *obscurator* Nees von Esenbeck, 1811

Catalogs with Iranian records: Gadallah and Ghahari (2015), Farahani et al. (2016), Yu et al. (2016), Samin, Coronado-Blanco, Kavallieratos et al. (2018), Samin, Coronado-Blanco, Fischer et al. (2018).
Distribution in Iran: Hormozgan (Ameri et al., 2015), Qazvin (Ghahari, Fischer, & Papp, 2011b).
Distribution in the Middle East: Cyprus (Papp, 1998), Iran (Ghahari, Fischer, & Papp, 2011b; Ameri et al., 2015), Israel—Palestine (Papp, 1989, 2012, 2015), Turkey (Beyarslan, 1986, 1999, 2014, 2016; Beyarslan & Çetin Erdoğan, 2005, 2010; Beyarslan et al., 2002, 2014).
Extralimital distribution: Austria, Azerbaijan, Belgium, Bosnia-Herzegovina, Bulgaria, China, Croatia, former Czechoslovakia, Denmark, Finland, France, Germany, Greece, Hungary, Italy, Kazakhstan, Korea, North Macedonia, Moldova, Mongolia, Netherlands, Norway, Poland, Romania, Russia, Serbia, Slovenia, Spain, Sweden, Switzerland, Tajikistan, Tunisia, United Kingdom.
Host records: Summarized by Yu et al. (2016) as being a parasitoid of several insect pests of the orders: Coleoptera (Buprestidae, Curculionidae), Diptera (Syrphidae, Tachinidae, Tephritidae), Hymenoptera (Cephidae), and Lepidoptera (Coleophoridae, Epermeniidae, Pyralidae, Tineidae, Tortricidae).

Bracon (*Glabrobracon*) *otiosus* Marshall, 1885

Catalogs with Iranian records: No catalog.
Distribution in Iran: Hamadan (Ghahari & Beyarslan, 2019).
Distribution in the Middle East: Cyprus (Beyarslan et al., 2017), Iran (Ghahari & Beyarslan, 2019), Israel—Palestine (Papp, 2015), Turkey (Beyarslan, 1986, 1999, 2014, 2016; Beyarslan & Çetin Erdoğan, 2005; 2010; Beyarslan et al., 2002, 2014).
Extralimital distribution: Austria, Finland, Georgia, Greece, Hungary, Korea, Lithuania, Netherlands, Norway, Poland, Romania, Russia, Sweden, Switzerland, Ukraine, United Kingdom.
Host records: Summarized by Yu et al. (2016) as being a parasitoid of the curculionid *Anthonomus pomorum* (L.); the elachistid *Blastodacna atra* (Haworth); the tenthredinid

Euura pumilio (Konow); and the tortricid *Pseudargyrotoza conwagana* (Fabricius).

Bracon (*Glabrobracon*) *pachyceri* Quintaret, 1912

Distribution in the Middle East: Turkey (Papp, 2000).
Extralimital distribution: Bulgaria, Czech Republic, France, Hungary, Sweden.
Host records: Recorded by Fahringer (1928) as being a parasitoid of the curculionid *Pachycerus varius* (Herbst).

Bracon (*Glabrobracon*) *pallicarpus* Thomson, 1892

Catalogs with Iranian records: No catalog.
Distribution in Iran: Khuzestan (Zargar, Samartsev et al., 2020 as *B.* (*B.*) *pallicarpus* Thomson, 1892), Northern Khurasan (Samin, Beyarslan, Coronado-Blanco et al., 2020).
Distribution in the Middle East: Iran.
Extralimital distribution: Austria, Finland, Germany, Hungary, Italy, Kazakhstan, Serbia, Spain, Sweden, United Kingdom.
Host records: Unknown.

Bracon (*Glabrobracon*) *pauris* Beyarslan, 1996

Distribution in the Middle East: Israel—Palestine (Papp, 2015), Turkey (Beyarslan, 1996, 2014; Beyarslan & Çetin Erdoğan, 2005, 2010; Beyarslan et al., 2002; Papp, 2000).
Extralimital distribution: Bulgaria, Greece, Hungary, Spain.
Host records: Unknown.

Bracon (*Glabrobracon*) *parvicornis* Thomson, 1892

Catalogs with Iranian records: No catalog.
Distribution in Iran: Isfahan (Rahmani et al., 2017).
Distribution in the Middle East: Iran (Rahmani et al., 2017), Turkey (Beyarslan, 1987, 2014, 2016; Beyarslan & Çetin Erdoğan, 2010; Beyarslan et al., 2014).
Extralimital distribution: Austria, Czech Republic, Germany, Greece, Hungary, Italy, Kazakhstan, North Macedonia, Mongolia, Poland, Russia, Slovakia, Sweden, Ukraine, former Yugoslavia.
Host records: Unknown.

Bracon (*Glabrobracon*) *parvulus* (Wesmael, 1838)

Catalogs with Iranian records: Gadallah and Ghahari (2015), Farahani et al. (2016), Yu et al. (2016), Samin, Coronado-Blanco, Kavallieratos et al. (2018), Samin, Coronado-Blanco, Fischer et al. (2018).
Distribution in Iran: Golestan (Ghahari & Fischer, 2011c), Mazandaran (Ghahari, 2019a; Zargar, Talebi, Hajiqanbar, & Papp, 2014, 2015), Northern Khorasan (Sakenin et al., 2021 as *Bracon* (*Glabrobracon*) *fumipennis* (Thomson, 1892)—around cotton fields).
Distribution in the Middle East: Cyprus (Beyarslan et al., 2017), Iran (see references above), Israel—Palestine (Papp,

2012), Turkey (Beyarslan, 1986, 1999, 2014, 2016; Beyarslan & Četin Erdoğan, 2005, 2010; Beyarslan et al., 2002, 2014).

Extralimital distribution: Europe, Oriental, Palaearctic [Adjacent countries to Iran: Armenia, Azerbaijan, Russia].

Host records: Summarized by Yu et al. (2016) as being a parasitoid of the agromyzid *Amouromyza flavifrons* (Meigen); the cecidomyiid *Clinodiplosis cilicrus* (Kieffer); and the tephritids *Chaetostomella cylindrica* (Robineau-Desvoidy), *Tephritis bardanae* (Schrank), *Tephritis conura* (Loew), *Tiphritis pulchra* (Loew), and *Trupanea stellata* (Fuesslin).

Bracon (*Glabrobracon*) peroculatus (Wesmael, 1838)
Catalogs with Iranian records: No catalog.
Distribution in Iran: Razavi Khorasan, Semnan (Samin, Beyarslan, Coronado-Blanco et al., 2020).
Distribution in the Middle East: Iran.
Extralimital distribution: Austria, Belgium, Finland, France, Germany, Hungary, Italy, North Macedonia, Russia, former Yugoslavia.
Host records: Recorded by Shenefelt (1978) as being a parasitoid of the tortricid *Cnephasia chrysantheana* (Duponchel).

Bracon (*Glabrobracon*) persiangulfensis Ameri, Beyarslan & Talebi, 2013
Catalogs with Iranian records: Gadallah and Ghahari (2015), Farahani et al. (2016), Yu et al. (2016), Samin, Coronado-Blanco, Kavallieratos et al. (2018), Samin, Coronado-Blanco, Fischer et al. (2018).
Distribution in Iran: Hormozgan (Ameri et al., 2014).
Distribution in the Middle East: Iran.
Extralimital distribution: None.
Host records: Unknown.

Bracon (*Glabrobracon*) picticornis (Wesmael, 1838)
Catalogs with Iranian records: Gadallah and Ghahari (2015), Farahani et al. (2016), Yu et al. (2016), Samin, Coronado-Blanco, Kavallieratos et al. (2018), Samin, Coronado-Blanco, Fischer et al. (2018).
Distribution in Iran: Hormozgan (Ameri et al., 2015), Kerman (Iranmanesh et al., 2018; Rahmani et al., 2017), Kermanshah (Ghahari & Fischer, 2012; Rahmani et al., 2017), Khuzestan (Zargar, Samartsev et al., 2020 as *B.* (*B.*) *picticornis* Wesmael, 1838), Mazandaran (Zargar et al., 2015), Northern Khorasan (Rahmani et al., 2017), Iran (no specific locality cited) (Papp, 2012).
Distribution in the Middle East: Iran (see references above), Turkey (Beyarslan, 1986, 1999, 2014, 2016; Beyarslan & Četin Erdoğan, 2005, 2010; Beyarslan et al., 2002, 2014).
Extralimital distribution: Europe, Oriental, Palaearctic [Adjacent countries to Iran: Afghanistan, Armenia, Azerbaijan, Kazakhstan, Russia].
Host records: Summarized by Yu et al. (2016) as being a parasitoid of the cerambycid *Plagionotus arcuatus* (L.); the

curculionid *Archarius salicivorus* (Paykull); the agromyzid *Malanagromyza albocilia* Hendel; the cecidomyiids *Rabdophaga* spp.; the tephritid *Euphranta connexa* (Fabricius); the tenthredinids *Euura mucronate* (Hartig), *Nematus* spp., and *Pontania* spp.; and the tortricids *Adoxophyes orana* (Fischer), *Cochylis pallidana* Zeller, and *hedya* spp. In Iran, this species has been reared from the tenthredinid *Pontania vesicator* (Bremi) (Ghahari & Fischer, 2012).
Plant associations in Iran: *Amaranthus* sp. (Amaranthaceae), *Beta vulgaris* (Amaranthaceae), *Chenopodium album* L. (Amaranthaceae), *Cynodon dactylon* (L.) (Poaceae), *Cyperus globosus* All. (Cyperaceae), *Lactuca serriola* L. (Asteraceae), *Mentha pulegium* L. (Lamiaceae), *Sisymbrium irio* L. (Brassicaceae) (Rahmani et al., 2017), *Medicago sativa* L. (Fabaceae) (Iranmanesh et al., 2018; Rahmani et al., 2017).

Bracon (*Glabrobracon*) pineti Thomson, 1892
Catalogs with Iranian records: Gadallah and Ghahari (2015), Farahani et al. (2016), Yu et al. (2016), Samin, Coronado-Blanco, Kavallieratos et al. (2018), Samin, Coronado-Blanco, Fischer et al. (2018).
Distribution in Iran: Lorestan (Ghahari, Fischer, Papp, & Tobias, 2012).
Distribution in the Middle East: Iran (Ghahari, Fischer, Papp, & Tobias, 2012), Israel–Palestine (Papp, 2015), Turkey (Beyarslan, 1986, 1999, 2014, 2016; Beyarslan & Četin Erdoğan, 2005, 2010; Beyarslan et al., 2014).
Extralimital distribution: Armenia, Austria, Azerbaijan, Croatia, Finland, France, Georgia, Germany, Hungary, Kazakhstan, Lithuania, Moldova, Norway, Poland, Russia, Slovakia, Slovenia, Sweden, Switzerland, former Yugoslavia.
Host records: Summarized by Yu et al. (2016) as being a parasitoid of the anobiid *Anobium abietis* (Fabricius); the ptinids *Ernobius abietinus* (Gyllenhal), and *Ernobius nigrinus* (Sturm); the geometrid *Thera variegata* (Denis and Schiffermüller); and the tortricids *Cydia strobilella* (L.), and *Retinia perangustata* (Snellen).

Bracon (*Glabrobracon*) planinotus Tobias, 1957
Catalogs with Iranian records: Gadallah and Ghahari (2015), Farahani et al. (2016), Yu et al. (2016), Samin, Coronado-Blanco, Kavallieratos et al. (2018), Samin, Coronado-Blanco, Fischer et al. (2018).
Distribution in Iran: Golestan (Samin, Ghahari et al., 2015), Guilan (Ghahari, Fischer, & Tobias, 2012), Kermanshah (Ghahari, Fischer, Hedqvist et al., 2010).
Distribution in the Middle East: Iran (Ghahari, Fischer, Hedqvist et al., 2010; Ghahari, Fischer, & Tobias, 2012; Samin, Ghahari et al., 2015), Turkey (Beyarslan 2014, 2016; Beyarslan & Četin Erdoğan, 2005, 2010).
Extralimital distribution: Kazakhstan, Russia, Ukraine.
Host records: Unknown.

Bracon (*Glabrobracon*) *popovi* Telenga, 1936

Catalogs with Iranian records: Gadallah and Ghahari (2015), Farahani et al. (2016), Yu et al. (2016), Samin, Coronado-Blanco, Kavallieratos et al. (2018), Samin, Coronado-Blanco, Fischer et al. (2018).

Distribution in Iran: Mazandaran (Ghahari, Fischer et al., 2011).

Distribution in the Middle East: Cyprus (Papp, 1998), Iran (Ghahari, Fischer et al., 2011), Israel–Palestine (Papp, 2015), Turkey (Beyarslan, 1986, 1987, 1999, 2014, 2016; Beyarslan & Çetin Erdoğan, 2010; Beyarslan et al., 2002, 2005).

Extralimital distribution: Armenia, Azerbaijan, Greece, Kazakhstan, Moldova, Mongolia, Russia, Turkmenistan.

Host records: Unknown.

Bracon (*Glabrobracon*) *praecox* (Wesmael, 1838)

Catalogs with Iranian records: No catalog.

Distribution in Iran: Mazandaran (Ghahari & Sakenin, 2018).

Distribution in the Middle East: Cyprus (Papp, 1998, 2012), Iran (Ghahari & Sakenin, 2018), Iraq (Papp, 2012), Israel–Palestine (Papp, 2012, 2015), Jordan (Papp, 2012), Turkey (Beyarslan, 2014; Beyarslan & Çetin Erdoğan, 2012; Papp, 2012).

Extralimital distribution: Algeria, Austria, Azerbaijan, Belarus, Belgium, Bulgaria, Croatia, Denmark, Finland, France, Germany, Greece, Hungary, Italy, Japan, Kazakhstan, North Macedonia, Mongolia, Poland, Romania, Russia, Spain, Sweden, Switzerland, Turkmenistan, Ukraine, United Kingdom, Uzbekistan, former Yugoslavia.

Host records: Summarized by Yu et al. (2016) as being a parasitoid of the curculionids *Gymnetron campanulae* Schönherr, and *Pissodes validirostris* Gyllenhal; the cynipid *Biorhiza pallida* L.; the tenthredinid *Hoplocampa brevis* (Klug); the tephritidid *Tephritis leotodontis* (DeGeer); the coleophorid *Coleophora gallipennella* (Hübner); and the tortricid *Gravitarmata margatana* (von Heinemann).

Bracon (*Glabrobracon*) *propebella* Papp, 2012

Distribution in the Middle East: Israel–Palestine (Papp, 2012).

Extralimital distribution: None.

Host records: Unknown.

Bracon (*Glabrobracon*) *pulcher* Bengtsson, 1924

Distribution in the Middle East: Israel–Palestine (Papp, 2012, 2015).

Extralimital distribution: Austria, Denmark, Germany, Hungary, Poland, Sweden.

Host records: Summarized by Yu et al. (2016) as being a parasitoid of the gelechiid *Metzneria lappella* Zeller; and the yponomeutid *Argyresthia conjugella* Zeller.

Bracon (*Glabrobracon*) *punicus* Schmiedecknecht, 1900

Distribution in the Middle East: Egypt (Papp, 1999).

Extralimital distribution: Spain, Tunisia.

Host records: Unknown.

Bracon (*Glabrobracon*) *surucicus* Beyarslan, 2002

Distribution in the Middle East: Turkey (Beyarslan, 2002, 2014; Beyarslan & Çetin Erdoğan, 2010).

Extralimital distribution: None.

Host records: Unknown.

Bracon (*Glabrobracon*) *tekkensis* Telenga, 1936

Catalogs with Iranian records: Gadallah and Ghahari (2015), Farahani et al. (2016), Yu et al. (2016), Samin, Coronado-Blanco, Kavallieratos et al. (2018), Samin, Coronado-Blanco, Fischer et al. (2018).

Distribution in Iran: Isfahan (Ghahari, Fischer, & Papp, 2011a), Iran (no specific locality cited) (Papp, 2015).

Distribution in the Middle East: Iran (Ghahari, Fischer, & Papp, 2011a; Papp, 2015), Israel–Palestine (Papp, 2015), Turkey (Beyarslan, 2014, 2016; Beyarslan & Çetin Erdoğan, 2010; Beyarslan et al., 2014).

Extralimital distribution: Greece, Hungary, North Macedonia, Turkmenistan, former Yugoslavia.

Host records: Unknown.

Bracon (*Glabrobracon*) *terebella* (Wesmael, 1838)

Catalogs with Iranian records: No catalog.

Distribution in Iran: Mazandaran (Ghahari, 2019a).

Distribution in the Middle East: Cyprus (Papp, 1998), Iran (Ghahari, 2019a), Israel–Palestine (Papp, 2015), Syria (Miller et al., 1992), Turkey (Beyarslan, 1986, 1999, 2014, 2016; Beyarslan & Çetin Erdoğan, 2010).

Extralimital distribution: Algeria, Armenia, Azerbaijan, Belgium, Canada (introduced), Czech Republic, Denmark, Finland, France, Germany, Hungary, Italy, Korea, Lithuania, Moldova, Poland, Romania, Russia, Slovakia, Slovenia, Sweden, Switzerland, United States, Ukraine, United Kingdom, former Yugoslavia.

Host records: Summarized by Yu et al. (2016) as being a parasitoid of the curculionids *Cionus scrophulariae* (L.), and *Gymnetron camapanulae* Schönherr; the cephids *Cephus cinctus* Norton, *Cephus pygmeus* L., and *Trachelus tabidus* (Fabricius).

Bracon (*Glabrobracon*) *titubans* (Wesmael, 1838)

Catalogs with Iranian records: No catalog.

Distribution in Iran: Kuhgiloyeh & Boyerahmad (Ghahari & Beyarslan, 2019), Qazvin (Gadallah et al., 2018).

Distribution in the Middle East: Iran (Gadallah et al., 2018; Ghahari & Beyarslan, 2019), Turkey (Beyarslan, 2014; Beyarslan & Çetin Erdoğan, 2012).

Extralimital distribution: Afghanistan, Armenia, Austria, Belgium, Bulgaria, Croatia, Finland, France, Georgia, Hungary, Germany, Korea, Mongolia, Netherlands, Poland, Romania, Russia, Slovakia, Slovenia, Sweden, Switzerland, Ukraine, United Kingdom, former Yugoslavia.

Host records: Recorded by Cox (1994) as being a parasitoid of the chrysomelids *Gastrophyza polygoni* (L.), *Gastrophyza viridula* (DeGeer), and *Plagidera versicolora* Laicharting.

Bracon (*Glabrobracon*) *triangularis* (Nees von Esenbeck, 1834)

Catalogs with Iranian records: Gadallah and Ghahari (2015), Farahani et al. (2016), Yu et al. (2016), Samin, Coronado-Blanco, Kavallieratos et al. (2018), Samin, Coronado-Blanco, Fischer et al. (2018).

Distribution in Iran: Kermanshah (Ghahari & Fischer, 2012).

Distribution in the Middle East: Iran (Ghahari & Fischer, 2012), Turkey (Beyarslan, 2014; Beyarslan & Çetin Erdoğan, 2010, 2012; Beyarslan et al., 2005, 2014).

Extralimital distribution: Austria, Bulgaria, Czech Republic, Finland, Germany, Hungary, Ireland, Italy, Norway, Poland, Slovenia, Switzerland, United Kingdom, former Yugoslavia.

Host records: Summarized by Yu et al. (2016) as being a parasitoid of the sesiids *Paranthrene tabaniformis* (Rottemburg), *Pennisetia hylaeiformis* (Laspeyres), *Synanthedon myopaeformis* (Borkhausen), and *Synanthedon tipuliformis* (Clerck).

Bracon (*Glabrobracon*) *tschitscherini* Kokujev, 1904

Catalogs with Iranian records: Gadallah and Ghahari (2015), Farahani et al. (2016), Yu et al. (2016), Samin, Coronado-Blanco, Kavallieratos et al. (2018), Samin, Coronado-Blanco, Fischer et al. (2018).

Distribution in Iran: Golestan (Ghahari & Fischer, 2011b), Hamadan (Rajabi Mazhar et al., 2019), Kordestan (Samin, 2015), West Azarbaijan (Ghahari & Fischer, 2011a).

Distribution in the Middle East: Cyprus (Papp, 1998), Iran (see references above), Israel—Palestine (Papp, 1989, 2012, 2015), Turkey (Beyarslan, 1986, 1999, 2014; Beyarslan & Çetin Erdoğan, 2010).

Extralimital distribution: Azerbaijan, Greece, Hungary, Italy, Kazakhstan, Kyrgyzstan, Russia, Tajikistan, Turkmenistan, Uzbekistan.

Host records: Recorded by Gultekin et al. (2008) as being a parasitoid of the curculionid *Larinus filiformis* Petri.

Bracon (*Glabrobracon*) *variator* Nees von Esenbeck, 1811

Catalogs with Iranian records: Fallahzadeh and Saghaei (2010), Gadallah and Ghahari (2015), Farahani et al. (2016), Yu et al. (2016), Samin, Coronado-Blanco,

Kavallieratos et al. (2018), Samin, Coronado-Blanco, Fischer et al. (2018).

Distribution in Iran: Alborz, Qazvin, Tehran (Zargar et al., 2015), Guilan (Ghahari & Fischer, 2011b), Hamadan (Rajabi Mazhar et al., 2019), Hormozgan (Ameri et al., 2014), Kermanshah (Valipour et al., 2017), Khuzestan (Zargar, Samartsev et al., 2020 as *B. (B.) variator* Nees, 1811), Northern Khorasan (Rahmani et al., 2017), West Azarbaijan (Samin et al., 2014), Iran (no specific locality cited) (Beyarslan et al., 2005; Čapek & Hofmann, 1997; Papp, 1966, 1967, 2012).

Distribution in the Middle East: Cyprus, Iran, Israel—Palestine, Jordan, Syria, Turkey (Yu et al., 2016).

Extralimital distribution: Europe, Palaearctic [Adjacent countries to Iran: Armenia, Azerbaijan, Kazakhstan, Russia, Turkmenistan].

Host records: Summarized by Yu et al. (2016) as being a parasitoid of several insect pests belonging to the orders Coleoptera (Anobiidae, Chrysomelidae, Curculionidae), Diptera (Anthomyiidae, Cecidomyiidae, Tephritidae), Hymenoptera (Tenthredinidae), and Lepidoptera (Carposinidae, Coleophoridae, Gelechiidae, Geometridae, Gracillaridae, Lycaenidae, Noctuidae, Nymphalidae, Pyralidae, Sesiidae, Tortricidae). In Iran, this species has been reared from the tortricid *Cydia johanssoni* Aarvik and Karsholt (Valipour et al., 2017).

Plant associations in Iran: *Medicago sativa* L. (Fabaceae) (Rahmani et al., 2017).

Bracon (*Glabrobracon*) *variator bipartitus* (Wesmael, 1838)

Catalogs with Iranian records: Gadallah and Ghahari (2015), Farahani et al. (2016).

Distribution in Iran: Alborz, Qazvin, Tehran (Zargar et al., 2015), Hamadan (Rajabi Mazhar et al., 2019 as *Bracon (Glabrobracon) bipartitus* Wesmael, 1838), Kerman, Northern Khorasan (Rahmani et al., 2017), Khuzestan (Zargar, Samartsev et al., 2020 as *B. (B.) bipartitus*), Mazandaran (Ghahari, 2017—around rice fields), Iran (no specific locality cited) (Papp, 2012 as *Bracon bipartitus*).

Distribution in the Middle East: Cyprus, Iran, Israel—Palestine, Jordan, Syria, Turkey (Yu et al., 2016).

Extralimital distribution: Algeria, Belgium, Bulgaria, Croatia, Finland, France, Germany, Greece, Hungary, Italy, North Macedonia, Mongolia, Netherlands, Romania, Russia, Spain, Turkmenistan, Ukraine, United Kingdom, former Yugoslavia.

Host records: Unknown

Plant associations in Iran: *Amaranthus* sp. (Amaranthaceae), *Cortaderia selloana* (Schult. And Schult.) (Poaceae), *Cynodon dactylon* (L.) (Poaceae), *Medicago sativa* L. (Fabaceae), *Mentha pulegium* L. (Lamiaceae), and *Phragmites australis* (Cav.) (Poaceae) (Rahmani et al., 2017).

Bracon (Glabrobracon) variator collinus Szépligeti, 1896
Distribution in the Middle East: Turkey (Papp, 2012).
Extralimital distribution: Austria, Bosnia-Herzegovina, Croatia, Denmark, Greece, Hungary, Italy, Romania, Slovakia, Switzerland, Ukraine, former Yugoslavia.
Host records: Unknown.

Bracon (Glabrobracon) kotulai Niezabitowski, 1910
Distribution in the Middle East: Israel—Palestine (Papp, 2015).
Extralimital distribution: Austria, Germany, Poland.
Host records: Unknown.

Bracon (Glabrobracon) zonites Marshall, 1897
Distribution in the Middle East: Turkey (Beyarslan, 2014; Beyarslan & Četin Erdoğan, 2012).
Extralimital distribution: China, Italy, Switzerland.
Host records: Unknown.

Bracon (Lucobracon) achterbergi Beyarslan, 2010
Distribution in the Middle East: Turkey (Beyarslan, 2014; Beyarslan & Četin Erdoğan, 2010).
Extralimital distribution: None.
Host records: Unknown.

Bracon (Lucobracon) apricus Schmiedeknecht, 1897
Catalogs with Iranian records: Gadallah and Ghahari (2015), Farahani et al. (2016), Yu et al. (2016), Samin, Coronado-Blanco, Kavallieratos et al. (2018), Samin, Coronado-Blanco, Fischer et al. (2018).
Distribution in Iran: Semnan (Ghahari, Fischer et al., 2011).
Distribution in the Middle East: Iran.
Extralimital distribution: Germany, Greece, Italy, Slovenia, former Yugoslavia.
Host records: Unknown.

Bracon (Lucobracon) attilae Papp, 2011
Distribution in the Middle East: Turkey (Papp, 2012).
Extralimital distribution: None.
Host records: Unknown.

Bracon (Lucobracon) biroicus Papp, 1990
Catalogs with Iranian records: No catalog.
Distribution in Iran: Khuzestan (Zargar, Samartsev et al., 2020 as B. (Bracon) biroicus Papp, 1990).
Distribution in the Middle East: Iran (Zargar, Samartsev et al., 2020), Israel—Palestine (Papp, 1990).
Extralimital distribution: Greece.
Host records: Unknown.

Bracon (Lucobracon) brachycerus Thomson, 1892
Catalogs with Iranian records: No catalog.
Distribution in Iran: Khuzestan (Zargar, Samartsev et al., 2020 as B. (Bracon) brachycerus Thomson, 1892).
Distribution in the Middle East: Iran (Zargar, Samartsev et al., 2020), Turkey (Beyarslan, 2014, 2016; Beyarslan & Četin Erdoğan, 2012).
Extralimital distribution: Austria, former Czechoslovakia, Finland, Germany, Hungary, Norway, Romania, Russia, Sweden.
Host records: Recorded by Fuchs (1912) and Telenga (1936) as being a parasitoid of the curculionid *Hylobius abietis* (L.).

Bracon (Lucobracon) brachypterus Tobias, 1959
Distribution in the Middle East: Turkey (Beyarslan, 2014; Beyarslan & Četin Erdoğan, 2012).
Extralimital distribution: Azerbaijan, Kazakhstan, Mongolia.
Host records: Unknown.

Bracon (Lucobracon) breviradius Beyarslan, 2011
Distribution in the Middle East: Turkey (Beyarslan, 2011, 2014; Beyarslan & Četin Erdoğan, 2014).
Extralimital distribution: None.
Host records: Unknown.

Bracon (Lucobracon) brevitemporis Tobias, 1959
Catalogs with Iranian records: Gadallah and Ghahari (2015), Farahani et al. (2016), Yu et al. (2016), Samin, Coronado-Blanco, Kavallieratos et al. (2018), Samin, Coronado-Blanco, Fischer et al. (2018).
Distribution in Iran: Lorestan (Ghahari, Fischer, Papp, & Tobias, 2012), West Azarbaijan (Rastegar et al., 2012).
Distribution in the Middle East: Iran (Ghahari, Fischer, Papp, & Tobias, 2012; Rastegar et al., 2012), Turkey (Beyarslan, 2005, 2014, 2016; Beyarslan & Četin Erdoğan, 2010).
Extralimital distribution: Kazakhstan.
Host records: Unknown.

Bracon (Lucobracon) brunescens Fahringer, 1928
Distribution in the Middle East: Cyprus (Papp, 1998, 1999).
Extralimital distribution: Germany, Spain.
Host records: Unknown.

Bracon (Lucobracon) byurakanicus Tobias, 1976
Catalogs with Iranian records: No catalog.
Distribution in Iran: Kermanshah (Rahmani et al., 2017).

Distribution in the Middle East: Iran (Rahmani et al., 2017), Turkey (Beyarslan, 2014; Beyarslan & Çetin Erdoğan, 2012).
Extralimital distribution: Armenia.
Host records: Unknown.
Plant associations in Iran: *Amaranthus* sp. (Amaranthaceae), *Medicago sativa* L. (Fabaceae), *Mentha pulegium* L. (Lamiaceae), and *Trifolium repens* L. (Fabaceae) (Rahmani et al., 2017).

Bracon (*Lucobracon*) concavus Tobias, 1957
Catalogs with Iranian records: No catalog.
Distribution in Iran: Hamadan (Rajabi Mazhar et al., 2019).
Distribution in the Middle East: Iran (Rajabi Mazhar et al., 2019), Turkey (Beyarslan, 2014; Beyarslan & Çetin Erdoğan, 2010, 2012).
Extralimital distribution: Turkmenistan.
Host records: Unknown.

Bracon (*Lucobracon*) crassungula Thomson, 1892
Catalogs with Iranian records: Samin, Coronado-Blanco, Kavallieratos et al. (2018).
Distribution in Iran: Razavi Khorasan (Samin, Coronado-Blanco, Kavallieratos et al., 2018).
Distribution in the Middle East: Iran (Samin, Coronado-Blanco, Kavallieratos et al., 2018), Turkey (Beyarslan, 2014; Beyarslan et al., 2008).
Extralimital distribution: Austria, Hungary, Italy, Sweden, Switzerland.
Host records: Unknown.

Bracon (*Lucobracon*) erraticus (Wesmael, 1838)
Catalogs with Iranian records: Gadallah and Ghahari (2015 as *Bracon praetermissus* Marshall, 1885), Farahani et al. (2016 as *Bracon praetermissus* Marshall, 1885), Samin, Coronado-Blanco, Kavallieratos et al. (2018), Samin, Coronado-Blanco, Fischer et al. (2018).
Distribution in Iran: Ardabil (Ghahari & Gadallah, 2022), Fars (Lashkari Bod et al., 2010, 2011 as *Bracon praetermissus* Marshall, 1885), Guilan (Ghahari, Fischer, & Tobias, 2012 as *Bracon praetermissus*), Kerman (Iranmanesh et al., 2018), Khuzestan (Zargar, Samartsev et al., 2020 as *B.* (*B.*) *erraticus* Wesmael, 1838).
Distribution in the Middle East: Cyprus, Iran, Israel—Palestine, Syria, Turkey, United Arab Emirates (Yu et al., 2016).
Extralimital distribution: Europe, Afrotropical, Palaearctic [Adjacent countries to Iran: Azerbaijan, Kazakhstan, Russia, Turkmenistan].
Host records: Summarized by Yu et al. (2016) as being a parasitoid several insects pests of the orders Coleoptera (Chrysomelidae, Curculionidae), Diptera (Tephritidae), Hymenoptera (Cephidae, Cynipidae, Eurytomidae, Tenthredinidae), and Lepidoptera (Gelechiidae, Nepticulidea, Sesiidae, Tortricidae). In Iran, this species has been reared from the cynipid gall wasp *Biorhiza pallida* L. (Ghahari,

Fischer, & Tobias, 2012) and the chrysomelid *Gastrophysa viridula viridula* (DeGeer) (Ghahari & Gadallah, 2022).
Plant associations in Iran: *Medicago sativa* L. (Fabaceae) (Iranmanesh et al., 2018).

Bracon (*Lucobracon*) femoralis (Brullé, 1832)
Catalogs with Iranian records: Gadallah and Ghahari (2015), Farahani et al. (2016), Yu et al. (2016), Samin, Coronado-Blanco, Kavallieratos et al. (2018), Samin, Coronado-Blanco, Fischer et al. (2018).
Distribution in Iran: Alborz (Zargar et al., 2015), Guilan, Northern Khorasan (Rahmani et al., 2017), Khuzestan (Zargar, Samartsev et al., 2020 as *B.* (*B.*) *femoralis* (Brullé, 1832)), Mazandaran (Ghahari, Fischer et al., 2011).
Distribution in the Middle East: Cyprus (Papp, 1998), Iran (Ghahari, Fischer et al., 2011; Zargar et al., 2015, Zargar, Samartsev et al., 2020), Israel—Palestine (Papp, 1989, 2015; Szépligeti, 1901), Syria (Fahringer, 1928), Turkey (Beyarslan, 2014; Beyarslan & Çetin Erdoğan, 2010, 2012).
Extralimital distribution: Algeria, Azerbaijan, Bulgaria, Georgia, Gibraltar, Greece, Hungary, North Macedonia, Montenegro, Spain, Switzerland, Tunisia, Turkmenistan, Uzbekistan.
Host records: Unknown.

Bracon (*Lucobracon*) filizae Beyarslan, 2002
Distribution in the Middle East: Turkey (Beyarslan, 2002, 2014; Beyarslan & Çetin Erdoğan, 2010).
Extralimital distribution: None.
Host records: Unknown.

Bracon (*Lucobracon*) flagellaris Thomson, 1892
Catalogs with Iranian records: Samin, Coronado-Blanco, Kavallieratos et al. (2018).
Distribution in Iran: Chaharmahal & Bakhtiari (Samin, Coronado-Blanco, Kavallieratos et al., 2018).
Distribution in the Middle East: Iran (Samin, Coronado-Blanco, Kavallieratos et al., 2018), Turkey (Beyarslan, 2014; Beyarslan & Çetin Erdoğan, 2010; Beyarslan et al., 2008; Gultekin, 2006).
Extralimital distribution: Germany, Hungary, Moldova, Poland, Romania, Russia, Slovenia, Sweden, Ukraine.
Host records: Recorded by Gultekin (2006) as being a parasitoid of the curculionid *Larinus onopordi* (Fabricius).

Bracon (*Lucobracon*) fortipes (Wesmael, 1838)
Catalogs with Iranian records: Gadallah and Ghahari (2015), Farahani et al. (2016), Yu et al. (2016), Samin, Coronado-Blanco, Kavallieratos et al. (2018), Samin, Coronado-Blanco, Fischer et al. (2018).
Distribution in Iran: Alborz, Qazvin (Zargar et al., 2015), Hormozgan (Ameri et al., 2014), Kermanshah (Ghahari & Fischer, 2012), Iran (no specific locality cited) (Papp, 2012).
Distribution in the Middle East: Cyprus (Papp, 1998, 2012), Iran (Ameri et al., 2014; Zargar et al., 2015),

Israel—Palestine (Papp, 1989, 2012), Turkey (Beyarslan, 2014, 2016; Beyarslan & Četin Erdoğan, 2010, 2012; Beyarslan et al., 2005, 2014; Papp, 2012; Tamer, 1995).
Extralimital distribution: Algeria, Armenia, Austria, Azerbaijan, Belgium, Bulgaria, Croatia, France, Germany, Greece, Hungary, Italy, Kazakhstan, Korea, Mongolia, Poland, Portugal, Russia, Serbia, Slovenia, Spain, Sweden, Switzerland, Tajikistan, Tunisia, Turkmenistan, Ukraine, Uzbekistan.
Host records: Summarized by Yu et al. (2016) as being a parasitoid of the cerambycid *Plagionotus floralis* (Pallas); the curculionid *Lixus anguinus* (L.); the noctuid *Gortyna xanthenes* Germar; and the sesiid *Bembecia scopigera* (Scopoli).

Bracon (Lucobracon) freidbergi Papp, 2015
Distribution in the Middle East: Israel—Palestine (Papp, 2015).
Extralimital distribution: None.
Host records: Unknown.

Bracon (Lucobracon) fumarius Szépligeti, 1901
Catalogs with Iranian records: No catalog.
Distribution in Iran: Golestan, Northern Khorasan (Sakenin et al., 2021—around cotton fields), Mazandaran (Ghahari, 2019a).
Distribution in the Middle East: Cyprus (Papp, 1998), Iran (Ghahari, 2019a; Sakenin et al., 2021).
Extralimital distribution: Ausria, Bulgaria, Croatia, France, Germany, Greece, Hungary, Italy, Romania.
Host records: Unknown.

Bracon (Lucobracon) fumigidus Szépligeti, 1901
Catalogs with Iranian records: No catalog.
Distribution in Iran: Guilan (Ghahari & Sakenin, 2018), Hamadan (Rajabi Mazhar et al., 2019), Khuzestan (Zargar, Samartsev et al., 2020 as *B. (B.) fumigidus* Szépligeti, 1901).
Distribution in the Middle East: Iran (Ghahari and Sakenin, 2018; Rajabi Mazhar et al., 2019; Zargar, Samartsev et al., 2020), Turkey (Beyarslan, 1987, 2014, 2016, Beyarslan & Četin Erdoğan, 2010; Papp, 2005).
Extralimital distribution: Armenia, Austria, Bulgaria, Germany, Greece, Hungary, Italy, Mongolia, Montenegro, Romania, Russia, Serbia.
Host records: Unknown.

Bracon (Lucobracon) grandiceps Thomson, 1892
Catalogs with Iranian records: Gadallah and Ghahari (2015), Farahani et al. (2016), Yu et al. (2016), Samin, Coronado-Blanco, Kavallieratos et al. (2018), Samin, Coronado-Blanco, Fischer et al. (2018).
Distribution in Iran: Lorestan (Ghahari, Fischer, Papp, & Tobias, 2012), Semnan (Naderian et al., 2012).

Distribution in the Middle East: Iran (Ghahari, Fischer, Papp, & Tobias, 2012; Naderian et al., 2012), Turkey (Beyarslan, 1999, 2014; Beyarslan & Četin Erdoğan, 2010; Beyarslan et al., 2005, 2008; Gultekin & Güclü, 1997).
Extralimital distribution: Austria, Croatia, Finland, France, Germany, Greece, Hungary, Italy, North Macedonia, Mongolia, Montenegro, Norway, Poland, Russia, Serbia, Slovenia, Sweden.
Host records: Recorded by Gultekin and Guçlu (1997) as being a parasitoid of the sesiid *Bembecia scopigera* (Scopoli).

Bracon (Lucobracon) guttiger (Wesmael, 1838)
Catalogs with Iranian records: Gadallah and Ghahari (2015), Farahani et al. (2016), Yu et al. (2016), Samin, Coronado-Blanco, Kavallieratos et al. (2018), Samin, Coronado-Blanco, Fischer et al. (2018).
Distribution in Iran: Hamadan, Zanjan (Ghahari, Fischer, Hedqvist et al., 2010), Kerman (Iranmanesh et al., 2018), Qazvin (Samin, 2015).
Distribution in the Middle East: Iran (Ghahari, Fischer, Hedqvist et al., 2010; Iranmanesh et al., 2018; Samin, 2015), Turkey (Beyarslan, 2014, 2016; Beyarslan et al., 2014; Papp, 2012).
Extralimital distribution: Armenia, Austria, Belgium, former Czechoslovakia, Denmark, Finland, France, Germany, Hungary, Italy, Latvia, Lithuania, Moldova, Mongolia, Netherlands, Poland, Russia, Slovenia, Sweden, Switzerland, United Kingdom.
Host records: Summarized by Yu et al. (2016) as being a parasitoid of the chrysomelids *Gastrophyza viridula* (DeGeer), *Hydrothassa marginella* (L.), *Phaedon cochleariae* (Fabricius), and *Phyllotreta nemorum* L.; the coleophorids *Coleophora laricella* (Hübner), and *Coleophora lutioennella* (Zeller).

Bracon (Lucobracon) humidus Tobias, 1976
Catalogs with Iranian records: Gadallah and Ghahari (2015), Farahani et al. (2016), Yu et al. (2016), Samin, Coronado-Blanco, Kavallieratos et al. (2018), Samin, Coronado-Blanco, Fischer et al. (2018).
Distribution in Iran: Guilan (Ghahari, Fischer, & Tobias, 2012).
Distribution in the Middle East: Iran (Ghahari, Fischer, & Tobias, 2012), Turkey (Beyarslan, 2014; Beyarslan & Četin Erdoğan, 2010; Beyarslan et al., 2005, 2008, 2014).
Extralimital distribution: Georgia, Russia.
Host records: Unknown.

Bracon (Lucobracon) hungaricus (Szépligeti, 1896)
Catalogs with Iranian records: No catalog.
Distribution in Iran: Lorestan (Samni et al., 2018c).
Distribution in the Middle East: Iran (Samin, Papp, & Coronado-Blanco, 2018), Turkey (Beyarslan, 1986, 1987,

1999, 2014, 2016; Beyarslan & Četin Erdoğan, 2010; Beyarslan et al., 2002, 2005, 2008).

Extralimital distribution: Azerbaijan, Bulgaria, Croatia, Czech Republic, former Czechoslovakia, France, Georgia, Hungary, Kazakhstan, Moldova, Mongolia, Montenegro, Romania, Russia, Serbia, Slovakia, Ukraine, former Yugoslavia.

Host records: Recorded by Györfi (1959) as being a parasitoid of the cerambycid *Clytus arietis* (L.).

Bracon (Lucobracon) hylobii Ratzeburg, 1848
Catalogs with Iranian records: No catalog.
Distribution in Iran: Guilan (Sakenin et al., 2020).
Distribution in the Middle East: Iran (Sakenin et al., 2020), Turkey (Beyarslan, 2014, 2016; Beyarslan & Četin Erdoğan, 2012).
Extralimital distribution: Austria, Czech Republic, Denmark, Germany, Hungary, Ireland, Italy, Latvia, Lithuania, Norway, Poland, Russia, Slovenia, Sweden, Switzerland, United Kingdom, former Yugoslavia.
Host records: Summarized by Yu et al. (2016) as being a parasitoid of the following curculionids: *Hylobius abietis* L., *Hylobius piceus* (DeGeer), *Pissodes harcyniae* (Herbst), *Pissodes notatus* (Fabricius), and *Pissodes piceae* (Illiger).

Bracon (Lucobracon) infernalis Telenga, 1936
Distribution in the Middle East: Cyprus (Papp, 1998), Turkey (Beyarslan, 2014, 2016; Beyarslan & Četin Erdoğan, 2010).
Extralimital distribution: Armenia, China, Greece, Kazakhstan, Russia.
Host records: Unknown.

Bracon (Lucobracon) irkutensis Telenga, 1936
Distribution in the Middle East: Turkey (Beyarslan, 2016).
Extralimital distribution: Korea, Russia.
Host records: Unknown.

Bracon (Lucobracon) isiklericus Beyarslan, 2002
Distribution in the Middle East: Turkey (Beyarslan, 2002, 2014).
Extralimital distribution: None.
Host records: Unknown.

Bracon (Lucobracon) iskilipus Beyarslan and Tobias, 2008
Catalogs with Iranian records: No catalog.
Distribution in Iran: Hamadan (Rajabi Mazhar et al., 2019).
Distribution in the Middle East: Iran (Rajabi Mazhar et al., 2019), Turkey (Beyarslan, 2014; Beyarsln & Erdoğan, 2010; Beyarslan & Tobias, 2008).
Extralimital distribution: None.
Host records: Unknown.

Bracon (Lucobracon) jakuticus Tobias, 1961
Distribution in the Middle East: Turkey (Beyarslan, 2014; Beyarslan & Četin Erdoğan, 2010; Beyarslan et al., 2008).
Extralimital distribution: Russia.
Host records: Unknown.

Bracon (Lucobracon) kuzguni Beyarslan, 2011
Distribution in the Middle East: Turkey (Beyarslan, 2011, 2014; Beyarslan & Četin Erdoğan, 2012).
Extralimital distribution: None.
Host records: Unknown.

Bracon (Lucobracon) larvicida (Wesmael, 1838)
Catalogs with Iranian records: Gadallah and Ghahari (2015), Farahani et al. (2016), Yu et al. (2016), Samin, Coronado-Blanco, Kavallieratos et al. (2018), Samin, Coronado-Blanco, Fischer et al. (2018).
Distribution in Iran: Golestan (Ghahari & Fischer, 2011c), Lorestan (Ghahari, Fischer, Papp, & Tobias, 2012), Semnan (Samin, Fischer, & Ghahari, 2015).
Distribution in the Middle East: Iran (Ghahari & Fischer, 2011c; Ghahari, Fischer, Papp, & Tobias, 2012), Turkey (Beyarslan, 1986, 1987, 1999, 2014, 2016; Beyarslan & Četin Erdoğan, 2010; Beyarslan et al., 2002, 2005, 2008).
Extralimital distribution: Armenia, Austria, Azerbaijan, Belgium, Bulgaria, Croatia, Czech Republic, France, Germany, Greece, Hungary, Italy, Kazakhstan, Moldova, Mongolia, Poland, Romania, Russia, Serbia, Slovenia, Sweden, Ukraine, United Kingdom.
Host records: Unknown.

Bracon (Lucobracon) meyeri Telenga, 1936
Catalogs with Iranian records: Gadallah and Ghahari (2015), Farahani et al. (2016), Yu et al. (2016), Samin, Coronado-Blanco, Kavallieratos et al. (2018), Samin, Coronado-Blanco, Fischer et al. (2018).
Distribution in Iran: Golestan (Samin, Ghahari et al., 2015), Guilan (Ghahari, Fischer, & Tobias, 2012), Razavi Khorasan (Samin et al., 2011).
Distribution in the Middle East: Iran (Ghahari, Fischer, & Tobias, 2012; Samin et al., 2011; Samin, Ghahari et al., 2015), Turkey (Beyarslan, 2014; Beyarslan & Četin Erdoğan, 2010; Beyarslan et al., 2005).
Extralimital distribution: Kazakhstan, Moldova, Mongolia, Russia, Uzbekistan.
Host records: Unknown.

Bracon (Lucobracon) mirus Szépligeti, 1901
Catalogs with Iranian records: No catalog.
Distribution in Iran: Qazvin (Gadallah et al., 2018).
Distribution in the Middle East: Iran (Gadallah et al., 2018), Turkey (Beyarslan, 2014; Beyarslan & Četin Erdoğan, 2010, 2012).

Extralimital distribution: China, Czech Republic, Georgia, Hungary, Kazakhstan, Mongolia, Netherlands, Sweden.
Host records: Recorded by Hedqvist (1973) as being a parasitoid of the ptinid *Xyletinus hanseni* Jansson.

Bracon (*Lucobracon*) *moczari* Papp, 1969
Catalogs with Iranian records: No catalog.
Distribution in Iran: Hamadan (Rajabi Mazhar et al., 2019).
Distribution in the Middle East: Iran (Rajabi Mazhar et al., 2019), Turkey (Beyarslan, 2014; Beyarslan & Çetin Erdoğan, 2010, 2012).
Extralimital distribution: Greece, Hungary, Russia.
Host records: Unknown.

Bracon (*Lucobracon*) *ochraceus* Szépligeti, 1896
Catalogs with Iranian records: No catalog.
Distribution in Iran: Qazvin (Sakenin et al., 2020).
Distribution in the Middle East: Iran.
Extralimital distribution: Austria, Azebaijan, Croatia, Hungary, Italy, Spain, former Yugoslavia.
Host records: Unknown.

Bracon (*Lucobracon*) *pliginskii* Telenga, 1936
Distribution in the Middle East: Israel–Palestine (Papp, 2015), Turkey (Beyarslan, 2014; Beyarslan & Çetin Erdoğan, 2012; Beyarslan et al., 2014).
Extralimital distribution: Armenia, Austria, Croatia, Georgia, Germany, Hungary, North Macedonia, Moldova, Ukraine, former Yugoslavia.
Host records: Recorded by Papp (1974) as being a parasitoid of the curculionid *Baris morio* Portevin.

Bracon (*Lucobracon*) *punctifer* Thomson, 1892
Catalogs with Iranian records: No catalog.
Distribution in Iran: Razavi Khorasan (Sakenin et al., 2018).
Distribution in the Middle East: Iran (Sakenin et al., 2018), Turkey (Beyarslan, 2014; Beyarslan & Çetin Erdoğan, 2010, 2012; Beyarslan et al., 2014).
Extralimital distribution: Finland, Germany, Hungary, Mongolia, Norway, Russia, Sweden.
Host records: Unknown.

Bracon (*Lucobracon*) *punctithorax* Tobias, 1959
Catalogs with Iranian records: Gadallah and Ghahari (2015), Farahani et al. (2016), Yu et al. (2016), Samin, Coronado-Blanco, Kavallieratos et al. (2018), Samin, Coronado-Blanco, Fischer et al. (2018).
Distribution in Iran: Kermanshah (Ghahari & Fischer, 2012).
Distribution in the Middle East: Iran (Ghahari & Fischer, 2012), Turkey (Beyarslan, 2014; Beyarslan & Erdoğan, 2010; Beyarslan et al., 2005).
Extralimital distribution: Kazakhstan.
Host records: Unknown.

Bracon (*Lucobracon*) *radiatus* Tobias, 1957
Catalogs with Iranian records: No catalog.
Distribution in Iran: Kordestan (Ghahari & Beyarslan, 2019).
Distribution in the Middle East: Iran (Ghahari & Beyarslan, 2019), Turkey (Beyarslan, 2014; Beyarslan & Çetin Erdoğan, 2012).
Extralimital distribution: Russia, Tajikistan.
Host records: Unknown.

Bracon (*Lucobracon*) *roberti* (Wesmael, 1838)
Catalogs with Iranian records: No catalog.
Distribution in Iran: Isfahan, Kerman, Northern Khorasan (Rahmani et al., 2017).
Distribution in the Middle East: Iran (Rahmani et al., 2017), Turkey (Beyarslan, 1987, 2014; Beyarslan & Çetin Erdoğan, 2010).
Extralimital distribution: Austria, Belgium, Bulgaria, France, Germany, Hungary, Ireland, Italy, Romania, Ukraine, United Kingdom.
Host records: Summarized by Yu et al. (2016) as being a parasitoid of the sesiid *Synanthedon andrenaeformis* (Laspeyres); and the tortricid *Cydia strobilella* (L.).
Plant associations in Iran: *Medicago sativa* L. (Fabaceae) and *Cynodon dactylon* L. (Poaceae) (Rahmani et al., 2017).

Bracon (*Lucobracon*) *santaecrucis* Schmiedeknecht, 1897
Catalogs with Iranian records: No catalog.
Distribution in Iran: Mazandaran (Samni et al., 2018c).
Distribution in the Middle East: Iran.
Extralimital distribution: Algeria, Bosnia-Herzegovina, Croatia, Greece, Spain, Tunisia.
Host records: Unknown.

Bracon (*Lucobracon*) *semifusus* Papp, 1965
Distribution in the Middle East: Turkey (Beyarslan & Çetin Erdoğan, 2012; Beyarslan, 2014).
Extralimital distribution: Hungary.
Host records: Unknown.

Bracon (*Lucobracon*) *shestakoviellus* Tobias, 1957
Catalogs with Iranian records: No catalog.
Distribution in Iran: East Azarbaijan (Sakenin et al., 2020).
Distribution in the Middle East: Iran (Sakenin et al., 2020), Turkey (Beyarslan, 2016; Beyarslan & Çetin Erdoğan, 2010; Beyarslan et al., 2008).
Extralimital distribution: Kazakhstan.
Host records: Unknown.

Bracon (*Lucobracon*) *sphaerocephalus* Szépligeti, 1901
Catalogs with Iranian records: Gadallah and Ghahari (2015), Farahani et al. (2016), Yu et al. (2016), Samin,

Coronado-Blanco, Kavallieratos et al. (2018), Samin, Coronado-Blanco, Fischer et al. (2018).
Distribution in Iran: Iran (no specific locality cited) (Papp, 2005), Hamadan (Ghahari et al., 2019).
Distribution in the Middle East: Iran (Ghahari et al., 2019; Papp, 2005), Turkey (Beyarslan, 1987, 2014; Beyarslan & Çetin Erdoğan, 2010; Papp, 2005).
Extralimital distribution: Austria, Bosnia-Herzegovina, Bulgaria, former Czechoslovakia, France, Germany, Greece, Hungary, Italy, North Macedonia, Mongolia, Montenegro, Russia, Serbia, Slovenia, Switzerland, United Kingdom.
Host records: Summarized by Yu et al. (2016) as being a parasitoid of the curculionids *Cleopis piger* (Scopoli), and *Pachycerus varius* Schönherr; and the cosmopterigid *Eteobalea serratella* (Treitschke).

Bracon (Lucobracon) subhylobii Tobias, 1986
Distribution in the Middle East: Turkey (Beyarslan, 2014; Beyarslan & Çetin Erdoğan, 2012).
Extralimital distribution: Moldova.
Host records: Unknown.

Bracon (Lucobracon) suchorukovi Telenga, 1936
Catalogs with Iranian records: Gadallah and Ghahari (2015), Farahani et al. (2016), Yu et al. (2016), Samin, Coronado-Blanco, Kavallieratos et al. (2018), Samin, Coronado-Blanco, Fischer et al. (2018).
Distribution in Iran: Guilan (Ghahari, Fischer, & Tobias, 2012; Sakenin et al., 2012), Semnan (Samin, Fischer, & Ghahari, 2015).
Distribution in the Middle East: Iran (Ghahari, Fischer, & Tobias, 2012; Sakenin et al., 2012), Turkey (Beyarslan & Çetin Erdoğan, 2010; Beyarslan, 1987, 2014; Beyarslan et al., 2005, 2008, 2014).
Extralimital distribution: Armenia, Austria, Azerbaijan, Kazakhstan, Lithuania, Moldova, Mongolia, Russia, Ukraine.
Host records: Unknown.

Bracon (Lucobracon) superciliosus (Wesmael, 1838)
Catalogs with Iranian records: Gadallah and Ghahari (2015), Yu et al. (2016).
Distribution in Iran: Qazvin (Ghahari, Fischer, & Papp, 2011b as *Bracon (Lucobracon) erraticus* var. *superciliosus*).
Distribution in the Middle East: Cyprus (Papp, 1998), Iran (Ghahari, Fischer, & Papp, 2011b), Israel−Palestine (Papp, 2015), Turkey (Papp, 2012).
Extralimital distribution: Armenia, Austria, Belgium, Bulgaria, Denmark, Germany, Greece, North Macedonia, Mongolia, Romania, Slovakia, Sweden, Switzerland, United Kingdom.

Host records: Unknown.

Bracon (Lucobracon) talyschicus Tobias, 1976
Distribution in the Middle East: Turkey (Beyarslan, 2014; Beyarslan & Erdoğan, 2012).
Extralimital distribution: Azerbaijan.
Host records: Unknown.

Bracon (Lucobracon) thuringiacus Schmiedeknecht, 1897
Catalogs with Iranian records: Gadallah and Ghahari (2015), Farahani et al. (2016), Yu et al. (2016), Samin, Coronado-Blanco, Kavallieratos et al. (2018), Samin, Coronado-Blanco, Fischer et al. (2018).
Distribution in Iran: Lorestan (Ghahari, Fischer, Papp, & Tobias, 2012), Semnan (Samin, Fischer, & Ghahari, 2015), West Azarbaijan (Rastegar et al., 2012).
Distribution in the Middle East: Iran (Ghahari, Fischer, Papp, & Tobias, 2012; Rastegar et al., 2012; Samin, Fischer, & Ghahari, 2015), Turkey (Beyarslan & Çetin Erdoğan, 2010; Beyarslan et al., 2005).
Extralimital distribution: Germany, Italy, former Yugoslavia.
Host records: Unknown.

Bracon (Ophthalmobracon) nocturnus (Tobias, 1962)
Catalogs with Iranian records: No catalog.
Distribution in Iran: Kerman (Iranmanesh et al., 2018).
Distribution in the Middle East: Iran.
Extralimital distribution: Turkmenistan.
Host records: Unknown.
Plant associationss in Iran: *Medicago sativa* L. (Fabaceae) and *Triticum* sp. (Poaceae) (Iranmanesh et al., 2018).

Bracon (Ophthalmobracon) ophtalmicus Telenga, 1933
Catalogs with Iranian records: Farahani et al. (2016), Yu et al. (2016), Samin, Coronado-Blanco, Kavallieratos et al. (2018), Samin, Coronado-Blanco, Fischer et al. (2018).
Distribution in Iran: Hormozgan (Ameri et al., 2015), Kerman (Iranmanesh et al., 2018).
Distribution in the Middle East: Iran (Ameri et al., 2015; Iranmanesh et al., 2018), Israel−Palestine (Halperin, 1986; Papp, 2012, 2015).
Extralimital distribution: Armenia, Azerbaijan, Kyrgyzstan, Mongolia, Russia, Tajikistan, Turkmenistan, Uzbekistan.
Host records: Summarized by Yu et al. (2016) as being a parasitoid of the gelechiids *Amblypalpis olivierella* Ragonot, *Amblypalpis tamaricella* Danilevsky, *Pexicopia malvella* (Huebner), and *Recurvaria pistaciicola* (Danilevsky).
Plant associations in Iran: *Medicago sativa* L. (Fabaceae) (Iranmanesh et al., 2018).

Bracon (Bracon) longigenis Tobias, 1957

Distribution in the Middle East: Cyprus (Beyarslan et al., 2017), Israel—Palestine (Papp, 2015), Turkey (Beyarslan, 2014, 2016; Beyarslan et al., 2014).
Extralimital distribution: Russia.
Host records: Unknown.

Bracon (Osculobracon) bilgini Beyarslan, 2002

Catalogs with Iranian records: No catalog.
Distribution in Iran: Kerman (Rahmani et al., 2017).
Distribution in the Middle East: Iran (Rahmani et al., 2017), Turkey (Beyarslan, 2002, 2014, 2016; Beyarslan & Çetin Erdoğan, 2010; Beyarslan et al., 2005, 2008).
Extralimital distribution: Russia.
Host records: Unknown.
Plant associationss in Iran: *Cynodon dactylon* (L.) (Poaceae) (Rahmani et al., 2017).

Bracon (Osculobracon) cingulator Szépligeti, 1901

Catalogs with Iranian records: No catalog.
Distribution in Iran: Chaharmahal & Bakhtiari (Ghahari & Gadallah, 2019).
Distribution in the Middle East: Iran (Ghahari & Gadallah, 2019), Turkey (Beyarslan, 2014; Beyarslan & Çetin Erdoğan, 2010; Beyarslan et al., 2008).
Extralimital distribution: Austria, Bulgaria, Croatia, Czech Republic, Germany, Greece, Hungary, Italy, Korea, Moldova, Netherlands, Norway, Poland, Romania, Russia, Slovakia, Tunisia, Turkmenistan.
Host records: Unknown.

Bracon (Osculobracon) ciscaucasicus Telenga, 1936

Catalogs with Iranian records: Gadallah and Ghahari (2015), Farahani et al. (2016), Yu et al. (2016), Samin, Coronado-Blanco, Kavallieratos et al. (2018), Samin, Coronado-Blanco, Fischer et al. (2018).
Distribution in Iran: Ardabil (Rastegar et al., 2012), Lorestan (Ghahari, Fischer, Papp, & Tobias, 2012).
Distribution in the Middle East: Iran (Ghahari, Fischer, Papp, & Tobias, 2012; Rastegar et al., 2012), Turkey (Beyarslan, 2014; Beyarslan & Çetin Erdoğan, 2005, 2010; Beyarslan et al., 2008).
Extralimital distribution: Azerbaijan, Hungary, Kazakhstan, Kyrgyzstan, Mongolia, Russia, Ukraine.
Host records: Unknown.

Bracon (Osculobracon) erzurumiensis Beyarslan, 2002

Catalogs with Iranian records: Farahani et al. (2016), Yu et al. (2016), Samin, Coronado-Blanco, Kavallieratos et al. (2018), Samin, Coronado-Blanco, Fischer et al. (2018).
Distribution in Iran: Hormozgan (Ameri et al., 2015), Isfahan, Kerman, Kermanshah, Northern Khorasan (Rahmani et al., 2017).
Distribution in the Middle East: Iran (Ameri et al., 2015; Rahmani et al., 2017), Turkey (Beyarslan, 2002, 2014, 2016; Beyarslan & Çetin Erdoğan, 2010; Beyarslan et al., 2014).
Extralimital distribution: None.
Host records: Unknown.
Plant associationss in Iran: *Medicago sativa* L. (Fabaceae) and *Trifolium repens* L. (Fabaceae) (Rahmani et al., 2017).

Bracon (Osculobracon) osculator Nees von Esenbeck, 1811

Catalogs with Iranian records: Gadallah and Ghahari (2015), Farahani et al. (2016), Yu et al. (2016), Samin, Coronado-Blanco, Kavallieratos et al. (2018), Samin, Coronado-Blanco, Fischer et al. (2018).
Distribution in Iran: Ardabil (Ghahari & Fischer, 2011b), East Azarbaijan (Ghahari & Fischer, 2011a), Fars (Samin, van Achterberg et al., 2015), Hormozgan (Ameri et al., 2015), Hamadan (Rajabi Mazhar et al., 2019 as *Bracon (Osculobracon) minutus* Szépligeti, 1901), Kermanshah (Rahmani et al., 2017 as *Bracon (Osculobracon) minutus*), Khuzestan (Zargar, Samartsev et al., 2020 as *B. (Osculobracon) minutus*), Mazandaran (Ghahari, 2017—around rice fields), Qazvin (Zargar et al., 2015), Iran (no specific locality cited) (Papp, 2012 as *Bracon osculator* var. *temporalis* Telenga, 1936).
Distribution in the Middle East: Cyprus, Iran, Iraq, Israel—Palestine, Turkey (Yu et al., 2016).
Extralimital distribution: Afghanistan, Armenia, Austria, Azerbaijan, Belgium, Bosnia-Herzegovina, Bulgaria, Croatia, Czech Republic, Denmark, Finland, France, Georgia, Germany, Greece, Hungary, Ireland, Italy, Kazakhstan, Korea, Latvia, Liechtenstein, Lithuania, North Macedonia, Moldova, Mongolia, Montenegro, Netherlands, New Zealand (introduced), Norway, Poland, Portugal, Romania, Russia, Serbia, Slovakia, Slovenia, Spain, Sweden, Switzerland, Turkmenistan, Ukraine, United Kingdom.
Host records: Summarized by Yu et al. (2016) as being a parasitoid of several lepidopteran pests of the families Choreutidae, Coleophoridae, Cosmopteridae, Elachistidae, Gelechiidae, Gracillariidae, Nepticulidae, and Tortricidae.

Bracon (Osculobracon) pelliger rumezensis Samartsev and Zargar, 2020

Catalogs with Iranian records: No catalog.
Distribution in Iran: Khuzestan (Zargar, Samartsev et al., 2020).
Distribution in the Middle East: Iran.
Extralimital distribution: None.
Host records: Unknown.

Bracon (Palpibracon) atrator Nees von Esenbeck, 1834

Catalogs with Iranian records: Gadallah and Ghahari (2015), Farahani et al. (2016), Yu et al. (2016), Samin, Coronado-Blanco, Kavallieratos et al. (2018), Samin, Coronado-Blanco, Fischer et al. (2018).

Distribution in Iran: Guilan (Ghahari, Fischer, & Tobias, 2012), Hormozgan (Ameri et al., 2015), West Azarbaijan (Samin et al., 2014).

Distribution in the Middle East: Cyprus (Papp, 1998, 2012), Iran (Ameri et al., 2015; Ghahari, Fischer, & Tobias, 2012; Samin et al., 2014), Israel−Palestine (Papp, 1989, 2012, 2015), Turkey (Beyarslan, 1986, 1987, 1999, 2014, 2016; Beyarslan & Çetin Erdoğan, 2005, 2010; Beyarslan et al., 2008).

Extralimital distribution: Austria, Azerbaijan, Belarus, Belgium, Croatia, Czech Republic, Finland, France, Georgia, Germany, Greece, Hungary, Ireland, Italy, Kazakhstan, Latvia, Lithuania, North Macedonia, Moldova, Mongolia, Netherlands, Poland, Russia, Serbia, Spain, Sweden, Switzerland, Tunisia, United Kingdom.

Host records: Summarized by Yu et al. (2016) as being a parasitoid of the apionid *Apion buddenbergi* Bedel; the curculionids *Gymnetron antirrhini* (Paykull), *Gymnetron campanulae* Schönherr, and *Gymnetron villosum* Gyllenhal; the agromyzid *Amauromyza flavifrons* (Meigen); the tephritids *Tephritis conura* (Loew), *Tephritis neesii* (Meigen), and *Tephritis separate* Rondani; and the coleophorid *Coleophora coronillae* Zeller.

Bracon (Palpibracon) delibator Haliday, 1833

Catalogs with Iranian records: Gadallah and Ghahari (2015), Farahani et al. (2016 under *Bracon (Glabrobracon) delibator* Haliday, 1833), Yu et al. (2016), Samin, Coronado-Blanco, Kavallieratos et al. (2018), Samin, Coronado-Blanco, Fischer et al. (2018).

Distribution in Iran: Alborz (Zargar et al., 2015), Fars, Razavi Khorasan (Ghahari & Beyarslan, 2017), Guilan (Rahmani et al., 2017), Ilam (Ghahari, Fischer, & Papp, 2011c), Mazandaran (Ghahari, 2019a), Semnan (Ghahari & Gadallah, 2015 as *Bracon (Palpibracon) delibator*).

Distribution in the Middle East: Cyprus, Iran, Israel−Palestine, Syria, Turkey (Yu et al., 2016).

Extralimital distribution: Europe, Palaearctic [Adjacent countries to Iran: Azerbaijan, Kazakhstan, Russia].

Host records: Summarized by Yu et al. (2016) as being a parasitoid of the curculionids *Gymnetron campanulae* Schönherr, and *Hylobius piceus* (DeGeer); the phalacrid *Olibrus aeneus* (Fabricius); the tephritid *Urophora cuspidata* (Meigen); and the tortricid *Cydia strobilella* (L.). In Iran, this species has been reared from *Urophora* sp. (Ghahari & Beyarslan, 2017).

Bracon (Palpibracon) mongolicus Telenga, 1936

Distribution in the Middle East: Turkey (Beyarslan & Çetin Erdoğan, 2010; Beyarslan et al., 2008).

Extralimital distribution: Kazakhstan, Mongolia, Russia, Tajikistan, Ukraine.

Host records: Unknown.

Bracon (Pigeria) piger (Wesmael, 1838)

Catalogs with Iranian records: Gadallah and Ghahari (2015), Farahani et al. (2016), Yu et al. (2016), Samin, Coronado-Blanco, Kavallieratos et al. (2018), Samin, Coronado-Blanco, Fischer et al. (2018).

Distribution in Iran: Chaharmahal & Bakhtiari, Fars (Samin, van Achterberg et al., 2015), Guilan (Ghahari et al., 2012; Zargar et al., 2015), Hormozgan (Ameri et al., 2014), Kerman (Iranmanesh et al., 2018), Khuzestan (Zargar, Samartsev et al., 2020), Mazandaran (Ghahari & Fischer, 2011b; Sakenin et al., 2012), Northern Khorasan, Sistan & Baluchestan (Rahmani et al., 2017).

Distribution in the Middle East: Cyprus (Papp, 1998), Iran (see references above), Israel−Palestine (Papp, 2012, 2015), Lebanon (Papp, 2012), Turkey (Beyarslan, 1986, 1987, 1999, 2014, 2016).

Extralimital distribution: Armenia, Austria, Azerbaijan, Belgium, Bulgaria, Croatia, former Czechoslovakia, France, Georgia, Germany, Greece, Hungary, Italy, Kazakhstan, North Macedonia, Malta, Montenegro, Netherlands, Poland, Puerto Rico (introduced), Romania, Russia, Serbia, Spain, Switzerland, Tunisia, Turkmenistan, United Kingdom, United States of America (introduced).

Host records: Summarized by Yu et al. (2016) as being a parasitoid of the curculionids *Magdalis rufa* Germar, and *Pissodes validirostris* Gyllenhal; the tenthredinid *Hoplocampa brevis* (Klug); the noctuid *Heliothis peltigera* (Denis and Schiffermüller); the pyralid *Etiella zinckenella* (Treitschke); and the tortricids *Cnephania longana* (Howarth), and *Cydia nigricana* (Fabricius).

Plant associations in Iran: *Medicago sativa* (Fabaceae) (Iranmanesh et al., 2018; Rahmani et al., 2017) and *Mentha pulegium* (Lamiaceae) (Rahmani et al., 2017).

Bracon sp.

Distribution in Iran: Bushehr (Fasihi et al., 2017; Sohrabi et al., 2014), Kordestan (Kamangar et al., 2017), Mazandaran (Amooghli-Tabari & Ghahari, 2021; Kiadaliri et al., 2005), Qazvin (Shojai, 1998), Sistan & Baluchestan (Afrouzian et al., 2001), West Azarbaijan (Akbarzadeh Shoukat, 2012).

Host records in Iran: *Phyllonorycter* sp. (Lepidoptera: Gracillariidae) (Shojai, 1998), *Caryedon palestinicus* Southgate, 1979 (Coleoptera: Chrysomelidae: Bruchinae) (Afrouzian et al., 2001), *Erannis defoliaria* (Clerck, 1759) (Lepidoptera: Geometridae) (Kiadaliri et al., 2005), *Lobesia botrana* Denis and Schiffermüller (Lepidoptera: Tortricidae) (Akbarzadeh Shoukat, 2012), *Tuta absoluta* (Meyrick) (Lepidoptera: Gelechiidae) (Fasihi et al., 2017; Sohrabi et al., 2014), and *Tortrix viridana* Linnaeus (Lepidoptera: Tortricidae) (Kamangar et al., 2017), *Chilo suppressalis* Walker (Lepidoptera: Crambidae) (Amooghli-Tabari & Ghahari, 2021).

Genus *Braconella* Szépligeti, 1906

Braconella fuscipennis Szépligeti, 1913
Distribution in the Middle East: Egypt (Fahringer, 1927).
Extralimital distribution: Ethiopia.
Host records: Unknown.

Genus *Ceratobracon* Telenga, 1936

Ceratobracon adaniensis Beyarslan, 1987
Distribution in the Middle East: Turkey (Beyarslan, 1987, 1999).
Extralimital distribution: None.
Host records: Unknown.

Ceratobracon stschegolevi (Telenga, 1933)
Catalogs with Iranian records: No catalog.
Distribution in Iran: Khuzestan (Zargar, Talebi, & Farahani, 2020), Mazandaran (Ghahari, 2019a).
Distribution in the Middle East: Iran (Ghahari, 2019a; Zargar, Talebi, & Farahani, 2020), Turkey (Beyarslan, 2014, 2016; Beyarslan & Çetin Erdoğan, 2010; Beyarslan et al., 2002, 2005, 2008, 2014).
Extralimital distribution: Austria, Azerbaijan, Georgia, Germany, Hungary, North Macedonia, Moldova, Mongolia, Poland, Russia, Uzbekistan, former Yugoslavia.
Host records: Recorded by Telenga (1933) as being a parasitoid of cephid *Cephus pygmeus* L.

Genus *Cyanopterus* Haliday, 1835

Cyanopterus (*Cyanopterus*) *flavator* (Fabricius, 1793)
Catalogs with Iranian records: No catalog.
Distribution in Iran: Kuhgiloyeh & Boyerahmad (Ghahari et al., 2019).
Distribution in the Middle East: Cyprus (Papp, 1998), Iran (Ghahari et al., 2019), Israel—Palestine (Halperin, 1986; Papp, 2009), Syria (Szépligeti, 1901).
Extralimital distribution: Algeria, Croatia, Czech Republic, Finland, France, Germany, Greece, Hungary, Italy, Japan, Kazakhstan, Korea, Latvia, Morocco, Netherlands, Poland, Portugal, Romania, Russia, Spain, Sweden, Switzerland, Tunisia, Ukraine, United Kingdom, former Yugoslavia.
Host records: Summarized by Yu et al. (2016) as being a parasitoid of the cerambycids *Acanthocinus griseus* (Fabricius), *Acanthoderes clavipes* Lecomte, *Icosum tometosum atticum* Lucas, *Monochamus galloprovincialis* (Olivier), *Monochamus sartor* (Fabricius), *Morinus asper* Sulzer, *Phymatodes testaceus* (L.), *Pogonocherus fasciculatus* (DeGeer), *Pogonocherus hispidus* (L.), *Rhagium inquisitor* (L.), and *Saperda scalaris* (L.). It was also recorded by Fulmek (1968) as being a parasitoid of the tephritid *Sphenella marginata* (Fallén).

Cyanopterus (*Ipobracon*) *amorosus* (Kohl, 1906)
Distribution in the Middle East: Yemen (including Socotra) (Brues, 1926; Fahringer, 1926, 1927; Kohl, 1907).

Extralimital distribution: None.
Host records: Unknown.

Cyanopterus (*Ipobracon*) *extricator* (Nees von Esenbeck, 1834)
Catalogs with Iranian records: No catalog.
Distribution in Iran: Ardabil (Ghahari et al., 2019).
Distribution in the Middle East: Iran.
Extralimital distribution: Algeria, former Czechoslovakia, Finland, France, Georgia, Germany, Greece, Italy, Mongolia, Poland, Russia, Spain, Sweden, Switzerland, Ukraine.
Host records: Summarized by Yu et al. (2016) as being a parasitoid the cerambycids *Pogonocherus fasciculatus* (DeGeer), *Pogonocherus perroudi* Mulsant, and *Ropalopus macropus* (Germar); the curculionids *Onthotomicus suturalis* (Gyllenhal), and *Pissodes castaneus* (DeGeer); and the sesiid *Paranthrene tabaniformis* (Rottemburg).

Genus *Habrobracon* Ashmead, 1895
Remarks: Based on Quicke (1987, 2015) and Belshaw et al. (2001), we treat *Habrobracon* Ashmead, 1895 as a separate genus from typical *Bracon*, under which for many years was synonymized.

Habrobracon brevicornis (Wesmael, 1838)
Catalogs with Iranian records: Gadallah and Ghahari (2015 as *Habrobracon hebetor*), Farahani et al. (2016 as *H. hebetor*), Yu et al. (2016), Samin, Coronado-Blanco, Kavallieratos et al. (2018), Samin, Coronado-Blanco, Fischer et al. (2018) as *Bracon brevicornis*).
Distribution in Iran: Alborz, Kerman, Khuzestan (Farahbakhsh, 1961; Modarres Awal, 1997, 2012 as *Bracon brevicornis*), Bushehr (Karampour & Fasihi, 2004), East Azarbaijan (Lotfalizadeh & Gharali, 2014), Ilam (Gharali, 2004; Modarres Awal, 2012 as *Bracon brevicornis*; Lotfalizadeh & Gharali, 2014), Semnan (Dezianian and Jalali, 2004), Tehran (Modarres Awal, 1997, 2012 as *Bracon brevicornis*), Northern Iran (Shojai et al., 1997), Iran (no specific locality cited) (Behdad, 1991 as *Bracon brevicornis*; Khanjani, 2006a as *Microbracon brevicornis* (Wesmael)).
Distribution in the Middle East: Cyprus, Egypt, Iran, Israel—Palestine, Syria, Turkey (Yu et al., 2016).
Extralimital distribution: Afrotropical, Oriental, Neotropical, Oceanic, Palaearctic; Canada (introduced), Mexico (introduced), United States of America (introduced) [Adjacent countries to Iran: Armenia, Azerbaijan, Russia, Turkmenistan].
Host records: Summarized by Yu et al. (2016) as being a parasitoid of several lepidopteran pests of the families Austostichidae, Crambidae, Cryptophasidae, Depressariidae, Erebidae, Gelechiidae, Noctuidae, Nolidae, Plutellidae, Pyralidae, Tineidae, Tortricidae, and Yponomeutidae. In Iran, this species has been reared from the tephritids

Acanthiophilus helianthi (Rossi), *Chaetorellia carthami* Stackelberg, *Terellia luteola* (Wiedemann), and *Urophora mauritanica* Macquart (Gharali, 2004); the batracherid *Batrachedra amydraula* Meyrick (Behdad, 1991; Karampour & Fasihi, 2004; Modarres Awal, 1997; 2012); the gelechiid *Phthorimaea operculella* (Zeller) (Dezianian & Jalali, 2004; Dezianian & Quicke, 2006); the nolid *Earias insulana* Boisduval (Farahbakhsh, 1961; Khanjani, 2006a,b; Modarres Awal, 1997, 2012); and the noctuids *Helicoverpa armigera* (Hübner), and *Spodoptera exigua* (Hübner) (Farahbakhsh, 1961; Modarres Awal, 1997, 2012).

Plant associationss in Iran: *Carthamus tinctorius* (Asteraceae) (Lotfalizadeh & Gharali, 2014).

Comments: *Habrobracon brevicornis* was formerly synonymized with *Habrobracon hebetor*, however, based on a molecular study carried out by Kittel & Maeto (2019), it is now treated as a valid species.

Habrobracon breviradiatus Tobias, 1957

Distribution in the Middle East: Turkey (Beyarslan, 2014; Beyarslan & Četin Erdoğan, 2012).

Extralimital distribution: Mongolia, Turkmenistan.

Host records: Unknown.

Habrobracon concolorans (Marshall, 1900)

Catalogs with Iranian records: Fallahzadeh and Saghaei (2010 under *B.* (*Habrobracon*) *nigricans* Szépligeti, 1901), Gadallah and Ghahari (2015), Farahani et al. (2016), Yu et al. (2016), Samin, Coronado-Blanco, Kavallieratos et al. (2018), Samin, Coronado-Blanco, Fischer et al. (2018).

Distribution in Iran: Fars (Lashkari Bod et al., 2011 as *Habrobracon nigricans*), Isfahan, Northern Khorasan, Razavi Khorasan, Sistan & Baluchestan (Rahmani et al., 2017), Kerman (Iranmanesh et al., 2018), Iran (no specific locality cited) (Szépligeti, 1901; Haeselbarth, 1983 as *Habrobracon nigricans*).

Distribution in the Middle East: Cyprus (Beyarslan et al., 2017), Iran (see references above), Israel—Palestine (Papp, 2015), Jordan (Al-Jboory et al., 2012), Turkey (Beyarslan, 1999, 2014, 2016; Beyarslan & Četin Erdoğan, 2010; Beyarslan et al., 2002, 2005, 2008, 2014).

Extralimital distribution: Armenia, Azerbaijan, Bulgaria, China, Croatia, former Czechoslovakia, Denmark, France, Georgia, Greece, Hungary, Italy, Kazakhstan, Kyrgyzstan, Lithuania, Moldova, Mongolia, Poland, Romania, Russia, Sweden, Tunisia, Turkmenistan.

Host records: Summarized by Yu et al. (2016) as being a parasitoid of the crambid *Loxostege sticticalis* (L.); the gelechiids *Pexicopia malvella* (Hübner), and *Tuta absoluta* (Meyrick); the ptinid *Ernobius nigrinus* (Sturm), the pyralid *Etiella zinckenella* (Treitschke); and the tortricids *Cnephasia sedana* (Constant), and *Cydia strobilella* (L.).

Plant associations in Iran: *Amaranthus* sp. (Amaranthaceae), *Convolvulus arvensis* L., *Medicago sativa* L., *Cyperus difformis* L., *Cyperus globosus* All. (Cyperaceae), *Lactuca serriola* L. (Asteraceae), and *Triticum dicoccum* Schrank (Rahmani et al., 2017).

Habrobracon crassicornis (Thomson, 1894)

Catalogs with Iranian records: Gadallah and Ghahari (2015), Farahani et al. (2016), Yu et al. (2016), Samin, Coronado-Blanco, Kavallieratos et al. (2018), Samin, Coronado-Blanco, Fischer et al. (2018).

Distribution in Iran: Ardabil (Ghahari et al., 2019 as *H. Flavosignatus* (Tobias, 1957)), Ilam (Ghahari, Fischer, & Papp, 2011c), Khuzestan (Zargar et al., 2019), Kordestan (Samin, Papp, & Coronado-Blanco, 2018).

Distribution in the Middle East: Cyprus (Papp, 1998), Iran (Samin, Papp, & Coronado-Blanco, 2018; Zargar et al., 2019), Israel—Palestine (Papp, 2012, 2015), Jordan (Papp, 2008), Turkey (Beyarslan, 2014; Beyarslan & Četin Erdoğan, 2010; Beyarslan et al., 2002, 2008; Papp, 2008).

Extralimital distribution: Armenia, Bulgaria, Denmark, France, Greece, Hungary, Italy, North Macedonia, Romania, Spain, Sweden, Tunisia, Turkmenistan, United Kingdom, former Yugoslavia.

Host records: Summarized by Yu et al. (2016) as being a parasitoid of the gelechiids *Anacampsis populella* (Clerck), and *Scrobipalpa acuminatella* (Sircom); the plutellid *Prays oleae* Bernard; the pyralids *Assara temerella* (Zincken), and *Ephestia kuehniella* Zeller; and the tortricid *Sparganothis pilleriana* (Denis and Schiffermüller).

Plant associations in Iran: Citrus plant (Zargar et al., 2019).

Habrobracon didemie (Beyarslan, 2002)

Catalogs with Iranian records: Farahani et al. (2016), Yu et al. (2016), Samin, Coronado-Blanco, Kavallieratos et al. (2018), Samin, Coronado-Blanco, Fischer et al. (2018).

Distribution in Iran: Hormozgan (Ameri et al., 2015).

Distribution in the Middle East: Iran (Ameri et al., 2015), Turkey (Beyarslan, 2002, 2014; Beyarslan & Četin Erdoğan, 2010).

Extralimital distribution: None.

Host records: Unknown.

Habrobracon excisus Tobias, 1957

Catalogs with Iranian records: Fallahzadeh and Saghaei (2010), Gadallah and Ghahari (2015 as *Habrobracon excisus* Tobias, 1957), Farahani et al. (2016), Yu et al. (2016), Samin, Coronado-Blanco, Kavallieratos et al. (2018), Samin, Coronado-Blanco, Fischer et al. (2018).

Distribution in Iran: Iran (no specific locality cited) (Haeselbarth, 1983).

Distribution in the Middle East: Iran.

Extralimital distribution: Kazakhstan, Mongolia, Uzbekistan.

Host records: Recorded by Haeselbarth (1983) as being a parasitoid of the deprassariid *Ethmia lybiella* (Ragonot).

Habrobracon gelechiae (Ashmead, 1889)

Distribution in the Middle East: Cyprus (introduced) (Bartlett et al., 1978), Israel—Palestine (Coll et al., 2000).

Extralimital distribution: Afghanistan, Australia (introduced), Bermuda (introduced), Canada, Chile (introduced), Cuba, France (introduced), India (introduced), Malta (introduced), Mexico (introduced), South Africa (introduced), United States of America.

Host records: Summarized by Yu et al. (2016) as being a parasitoid of several lepidopteran species belonging to the families Choreutidae, Coleophoridae, Depressariidae, Gelechiidae, Geometridae, Heliodinidae, Hesperiidae, Lasiocampidae, Lyonetiidae, Noctuidae, Plutellidae, Pterophoridae, Pyralidae, and Tortricidae.

Habrobracon hebetor (Say, 1836)

Catalogs with Iranian records: Fallahzadeh and Saghaei (2010), Gadallah and Ghahari (2015), Farahani et al. (2016), Yu et al. (2016), Samin, Coronado-Blanco, Kavallieratos et al. (2018), Samin, Coronado-Blanco, Fischer et al. (2018) as *Bracon hebetor*).

Distribution in Iran: Alborz (Shojai, 1998), Ardabil (Najafi Navaei et al., 2004; Khanjani, 2006a), Chaharmahal & Bakhtiari (Ghahari, 2019b), East Azarbaijan (Shahhosseini & Kamali, 1989; Ghahari, Fischer et al., 2011), Fars (Lashkari Bod et al., 2011), Golestan (Shojai, 1998; Mojeni et al., 2005; Eyidozehi et al., 2013; Afshari et al., 2014), Guilan (Rahmani et al., 2017), Hormozgan (Ameri et al., 2014 as *Bracon (Habrobracon) hebetor*), Isfahan (Afiunizadeh Isfahani & Karimzadeh Isfahani, 2010; Afiunizadeh Isfahani et al., 2010; Bagheri & Nasr Isfahani, 2010 as *Bracon hebetor* Say; Bagheri & Nematollahi, 2006; Haghshenas & Esfandiari, 2016; Nematollahi & Bagheri, 2018 as *Bracon hebetor*; Nobakht et al., 2015; Rahmani et al., 2017; Sobhani et al., 2012), Kermanshah (Noori, 1994; Rahmani et al., 2017), Khuzestan (Siahpoush et al., 1993; Modarres Awal, 1997, 2012 as *Microbracon hebetor* Say; Habibpour et al., 2002; Shahabi & Rajabpour, 2015; Zargar et al., 2019), Markazi (Kishani Farahani et al., 2008; Modarres Awal, 2012 as *Microbracon hebetor*), Northern Khorasan (Ghahari & Gadallah, 2022; Rahmani et al., 2017), Mazandaran (Ghahari & Gadallah, 2022), Sistan & Baluchestan (Rahmani et al., 2017), Kuhgiloyeh & Boyerahmad (Khatima & Reza, 2015; Saeidi, 2013, 2015), Razavi Khorasan (Akbari-Asl et al., 2009), Qom (Goldansaz et al., 2011; Kishani Farahani et al., 2008; Modarres Awal, 2012 as *Microbracon hebetor*), Semnan (Dezianian & Quicke, 2006; Ghahari & Gadallah, 2015; Modarres Awal, 2012 as *Microbracon hebetor*), Tehran

(Farahbakhsh, 1961; Golizadeh, 2008; Kishani Farahani et al., 2008, 2010; Modarres Awal, 1997, 2012 as *Microbracon hebetor*; Shahhosseini & Kamali, 1989), West Azarbaijan (Adldoost, 2010; Adldoost & Shayesteh, 2010; Akbarzadeh Shoukat, 2012; Akbarzadeh Shoukat et al., 2008; Modarres Awal, 1997, 2012 as *Microbracon hebetor*), Northern Iran (Shojai et al., 1997), generally distributed in Iran (Shojai et al., 1995), Iran (no specific locality cited) (Behdad, 1991 as *Microbracon hebetor*; Haeselbarth, 1983).

Distribution in the Middle East: Cyprus, Egypt, Iran, Iraq, Israel—Palestine, Saudi Arabia, Turkey (Yu et al., 2016).

Extralimital distribution: Nearly cosmopolitan. Introduced to Burkina Faso, China, Fiji, India, Mali, Niger.

Host records: Summarized by Yu et al. (2016) as being a parasitoid of wide range of insect pests belonging to the orders Coleoptera (Chrysomelidae, Curculionidae), Diptera (Tephritidae), Hymenoptera (Braconidae, Cynipidae), and Lepidoptera (Blastobasidae, Depressariidae, Gelechiidae, Noctuidae, Nolidae, Oecophoridae, Papilionidae, Pieridae, Pyralidae, Sesiidae, Tineidae, Tortricidae, Yponomeutidae). In Iran, reared from the batrachedrid *Batrachedra amydraula* (Meyrick); the crambid *Antigastra catalaunalis* (Duponchel); the gelechiids *Phthorimaea operculella* Zeller (Ghahari & Gadallah, 2022; Shahabi & Rajabpour, 2015), and *Tuta absoluta* (Meyrick) (Haghshenas & Esfandiari, 2016); the noctuids *Spodoptera exigua* (Hübner) (Modarres Awal, 1997, 2012), *Helicoverpa armigera* (Hübner) (Adldoost & Shayesteh, 2010; Afshari et al., 2014; Faal-Mohamad Ali et al., 2010; Modarres Awal, 1997; Mojeni et al., 2005; Shojai, 1998; Vaez & Piurgoli, 2016), *Sesamia cretica* Lederer (Khanjani, 2006a; Shojai et al., 1995), and *Sesamia nonagrioides botanephaga* Lefebvre (Khanjani, 2006a); the pyralids *Ephestia* spp. (Farahbakhsh, 1961; Modarres Awal, 1997, 2012; Shahhosseini & Kamali, 1989), *Ephestia kuehniella* (Zeller) (Akbari-Asl et al., 2009; Vaez & Piurgoli, 2016), *Plodia interpunctella* (Hübner) (Akbari-Asl et al., 2009; Bagheri Zonuz, 1973; Farahbakhsh, 1961; Modarres Awal, 1997, 2012; Shahhosseini & Kamali, 1989; Shojai, 1998), *Homoeosoma nebulella* Denis and Schiffermüller (Shojai, 1998), *Ectomyelois ceratoniae* Zeller (Goldansaz et al., 2011; Kishani Farahani et al., 2008, 2010; Nobakht et al., 2015; Sobhani et al., 2012), and *Galleria mellonella* (L.) (Dweck et al., 2010; Foruzan et al., 2010; Shahhosseini & Kamali, 1989; Vaez & Piurgoli, 2016); the plutellid *Plutella xylostella* (L.) (Afiunizadeh Isfahani & Karimzadeh Isfahani, 2010; Afiunizadeh Isfahani et al., 2010; Ghahari & Gadallah, 2022; Golizadeh, 2008)); the noctuids *Heliothis* sp. (=*Chloridea* sp.) (Noori, 1988, 1994), *Heliothis viriplaca* (Hufnagel) (Adldoost, 2010), and *Leucania loreyi* (Duponchel) (Siahpoush et al., 1993; Modarres Awal, 1997, 2012 as *Mythimna loreyi* (Duponchel)); the tortricid *Lobesia botrana* Denis and

Schiffermüller (Akbarzadeh Shoukat, 2012; Akbarzadeh Shoukat et al., 2008); the curculionid *Lixus incanescens* Boheman; and the tephritid *Acanthiophilus helianthi* (Rossi) (Bagheri & Nematollahi, 2006; Nematollahi & Bagheri, 2018; Ghahari, 2019b; Saeidi 2013, 2015).

Plant associations in Iran: Corn field (Najafi Navaei et al., 2004), *Carthamus tinctorius* L. (Asteraceae) (Khatima & Reza, 2015), *Capsella bursapastoris* L. (Brassicaceae), *Medicago sativa* L. (Fabaceae), *Mentha pulegium* L. (Lamiaceae), *Sisymbrium irio* L. (Brassicaceae) (Rahmani et al., 2017), and palm orchards (Zargar et al., 2019).

Note: For many years, the separate species *Habrobracon brevicornis* was treated as a synonym of *H. hebetor* (see Kittel & Maeto, 2019).

Habrobracon iranicus Fischer, 1972

Catalogs with Iranian records: Fallahzadeh and Saghaei (2010 under *Bracon (Habrobracon) iranicus*), Gadallah and Ghahari (2015), Farahani et al. (2016), Yu et al. (2016 under *Bracon iranicus*), Samin, Coronado-Blanco, Kavallieratos et al. (2018), Samin, Coronado-Blanco, Fischer et al. (2018).

Distribution in Iran: Alborz (Shojai, 1998), Khuzestan (Zargar et al., 2019), Razavi Khorasan, West Azarbaijan (Shojai et al., 2000, 2002), Tehran (Modarres Awal, 1997, 2012; Shojai, 1998), Iran (no specific locality cited) (Fischer, 1972; Shenefelt, 1978).

Distribution in the Middle East: Iran.

Extralimital distribution: None.

Host records: In Iran, this species has been reared from gelechiid *Phthorimaea operculella* Zeller (Modarres Awal, 2012); the noctuid *Heliothis viriplaca* (Hufnagel) (Modarres Awal, 1997, 2012; Shojai, 1998); the pyralid *Ectomyelois ceratoniae* (Zeller); the tortricids *Cydia pomonella* L. (Modarres Awal, 1997, 2012; Shojai, 1998; Shojai et al., 2000, 2002), and *Lobesia botrana* (Denis and Schiffermüller); and the yponomeutid *Yponomeuta malinellus* Zeller.

Plant associations in Iran: Citrus orchards (Zargar et al., 2019).

Habrobracon kitcheneri (Dudgeon & Gough, 1914)

Distribution in the Middle East: Egypt (Ballou, 1918; Brues, 1926; Dudgeon & Gough, 1914; Fahringer, 1928; Shenefelt, 1978; Willcocks, 1914, 1916), Israel−Palestine (Bodenheimer, 1930; Halperin, 1986; Kugler, 1966).

Extralimital distribution: China, India.

Host records: Summarized by Yu et al. (2016) as being a parasitoid of the gelechiid *Pectinophora gossypiella* (Saunders); the lasiocampid *Eriogaster philippsi* Bartel; the nolid *Earias insulana* (Boisduval); and the pyralids *Cadra cautella* (Walker), and *Pterothrixidia rufella* (Duponchel).

Habrobracon kopetdagi Tobias, 1957

Catalogs with Iranian records: No catalog.

Distribution in Iran: Khuzestan (Zargar et al., 2019).

Distribution in the Middle East: Iran (Zargar et al., 2019), Turkey (Beyarslan, 2014; Beyarslan & Çetin Erdoğan, 2010, 2012).

Extralimital distribution: Azerbaijan, Georgia, Kazakhstan, Moldova, Mongolia, Tajikistan, Turkmenistan.

Host records: Recorded by Tobias (1986) as being a parasitoid of the gelechiid *Anarsia linealtella* Zeller; and the buprestid *Anthaxia plaviltschikovi* Obenb.

Plant associations in Iran: Olive orchards (Zargar et al., 2019).

Habrobracon lissothorax Tobias, 1967

Catalogs with Iranian records: No catalog.

Distribution in Iran: Kerman (Iranmanesh et al., 2018 in stem galls of *Tamarix* sp.—Tamaricaceae).

Distribution in the Middle East: Iran.

Extralimital distribution: Turkmenistan.

Host records: Unknown.

Plant association in Iran: *Tamarix* sp. (Tamaricaceae) (Iranmanesh et al., 2018).

Habrobracon notatus Szépligeti, 1914

Distribution in the Middle East: Egypt (Fahringer, 1928).

Extralimital distribution: Equatorial Guinea, Morocco, Senegal.

Host records: Unknown.

Habrobracon nygmiae Telenga, 1936

Catalogs with Iranian records: No catalog.

Distribution in Iran: West Azarbaijan (Ghahari & Gadallah, 2022).

Distribution in the Middle East: Iran (Ghahari & Gadallah, 2022), Turkey (Beyarslan, 2014; Beyarslan & Çetin Erdoğan, 2010; Beyarslan et al., 2008).

Extralimital distribution: Armenia, Germany, Moldova, Russia, Ukraine.

Host records: Recorded by Tobias (1976, 1986) as being a parasitoid of the erebid *Euproctis chrysorrhoae* (L.), and the lasiocampid *Malacosoma neustria* (L.). In Iran, this species has been reared from *Malacosoma neustria* (L.) (Ghahari & Gadallah, 2022).

Habrobracon pillerianae Fischer, 1980

Distribution in the Middle East: Turkey (Beyarslan, 2014; Fischer, 1980).

Extralimital distribution: None.

Host records: Recorded by Fischer (1980) as being a parasitoid of the tortricid *Sparganothis pilleriana* (Denis and Schiffermüller).

Habrobracon ponticus (Tobias, 1986)
Catalogs with Iranian records: No catalog.
Distribution in Iran: Khuzestan (Zargar et al., 2019).
Distribution in the Middle East: Iran (Zargar et al., 2019), Turkey (Beyarslan, 2014; Beyarslan & Çetin Erdoğan, 2010, 2012).
Extralimital distribution: Greece, Hungary, Moldova, Russia, Ukraine (Yu et al., 2016 as *Bracon* (*Habrobracon*) *ponticus* Tobias, 1986).
Host records: Unknown.

Habrobracon radialis Telenga, 1936
Catalogs with Iranian records: Fallahzadeh and Saghaei (2010), Gadallah and Ghahari (2015), Farahani et al. (2016), Yu et al. (2016), Samin, Coronado-Blanco, Kavallieratos et al. (2018), Samin, Coronado-Blanco, Fischer et al. (2018).
Distribution in Iran: Golestan (Ghahari & Fischer, 2011b; Sakenin et al., 2012), Guilan (Ghahari, Fischer, & Tobias, 2012), Semnan (Dezianian & Quicke, 2006 as *Bracon* (*Habrobracon*) aff. *radialis*; Modarres Awal, 2012 as *Bracon radialis*; Samin, Fischer, & Ghahari, 2015).
Distribution in the Middle East: Iran (see references above), Turkey (Beyarslan, 2014; Beyarslan et al., 2005; Beyarslan & Çetin Erdoğan, 2010).
Extralimital distribution: Hungary, Italy, Kazakhstan, Macedonia, Mongolia, Russia, Tunisia, Turkmenistan, Uzbekistan, former Yugoslavia.
Host records: Recorded by Tobias (1986) as being a parasitoid of the plutellid *Plutella xylostella* (L.). In Iran, reared from the gelechiid *Phthorimaea operculella* Zeller (Dezianian & Quicke, 2006; Modarres Awal, 2012).

Habrobracon simonovi Kokujev, 1914
Catalogs with Iranian records: No catalog.
Distribution in Iran: Markazi (Ghahari and Gadallah, 2019).
Distribution in the Middle East: Cyprus (Papp, 1998), Iran (Ghahari & Gadallah, 2019), Israel–Palestine (Papp, 2015), Turkey (Beyarslan, 2014, 2016; Beyarslan & Çetin Erdoğan, 2010, 2012).
Extralimital distribution: Azerbaijan, China, Korea, Mongolia, Turkmenistan, Uzbekistan.
Host records: Summarized by Yu et al. (2016) as being a parasitoid of the crambid *Ostrinia nubilalis* (Hübner); the noctuids *Helicoverpa armigera* (Hübner), *Helicoverpa zea* (Boddie), and *Spodoptera exigua* (Hübner); and the nolid *Earias roseipes* Filipjev.

Habrobracon stabilis (Wesmael, 1838)
Catalogs with Iranian records: Gadallah and Ghahari (2015), Farahani et al. (2016), Yu et al. (2016), Samin, Coronado-Blanco, Kavallieratos et al. (2018), Samin, Coronado-Blanco, Fischer et al. (2018) as *Bracon stabilis* (Wesmael, 1838)).
Distribution in Iran: Hamadan (Rajabi Mazhar et al., 2019), Khuzestan (Zargar et al., 2019), Qazvin (Ghahari, Fischer, & Papp, 2011b).
Distribution in the Middle East: Cyprus, Iran, Israel–Palestine, Turkey (Yu et al., 2016).
Extralimital distribution: Europe, Nearctic, Oriental, Palaearctic, United States of America (introduced) [Adjacent countries to Iran: Armenia, Azerbaijan, Kazakhstan, Russia].
Host records: Summarized by Yu et al. (2016) as being a parasitoid of several insect pests belonging to the orders Coleoptera (Curculionidae), Diptera (Tephritidae), and Lepidoptera (Coleophoridae, Depressariidae, Erebidae, Gelechiidae, Pterolonchaidae, Pyralidae, Tortricidae, Yponomeutidae).
Plant associations in Iran: Olive orchards (Zargar et al., 2019).

Habrobracon telengai Muljarskaya, 1955
Catalogs with Iranian records: Gadallah and Ghahari (2015), Farahani et al. (2016), Yu et al. (2016), Samin, Coronado-Blanco, Kavallieratos et al. (2018), Samin, Coronado-Blanco, Fischer et al. (2018).
Distribution in Iran: Hormozgan (Ameri et al., 2014 as *Bracon* (*Habrobracon*) *telengai*), Kerman (Mehrnejad, 2010).
Distribution in the Middle East: Cyprus (Papp, 1998), Iran (Ameri et al., 2014; Mehrnejad, 2010), Israel–Palestine (Papp, 1989, 2012, 2015), Turkey (Beyarslan, 2014; Beyarslan & Çetin Erdoğan, 2010; Beyarslan et al., 2002, 2014).
Extralimital distribution: Armenia, Azerbaijan, Georgia, Germany, Kazakhstan, Moldova, Russia, Tajikistan, Turkmenistan, Uzbekistan.
Host records: Summarized by Yu et al. (2016) as being a parasitoid of the buprestids *Anthaxia plaviltschikovi* Obemb., and *Sphenoptera kaznakovi* Jakovlov; the curculionids *Chaetoptelius vestitus* (Mulsant and Rey), and *Scolytus rugulosus* (Müller); the gelechiid *Pexicopia malvella* (Hübner); and the tortricids *Gypsonoma minutana* (Hübner), and *Spilonota ocellana* (Denis and Schiffermüller). In Iran, this species has been recorded by Mehrnejad (2020) as being a parasitoid of the pyralid *Arimania komaroffi* Ragonot attacking pistachio fruits.

Habrobracon viktorovi Tobias, 1961
Catalogs with Iranian records: No catalog.
Distribution in Iran: East Azarbaijan (Samin et al., 2019).
Distribution in the Middle East: Cyprus (Papp, 1998), Iran (Samin et al., 2019), Turkey (Beyarslan, 2014; Beyarslan & Çetin Erdoğan, 2010; Beyarslan et al., 2008).

Extralimital distribution: Fomer Czechoslovakia, Greece, Hungary, Korea, Russia.
Host records: Unkown.

Habrobracon sp.
Distribution in Iran: Alborz (Davatchi & Shojai, 1969; Shojai, 1998).
Host records: *Heliothis dipsacea* (Linnaeus) (Lepidoptera: Noctuidae) on *Pisum sativum* (Fabaceae) (Davatchi & Shojai, 1969; Shojai, 1998).

Genus *Doggerella* Quicke, Mahmood & Papp, 2011
Doggerella sp.
Distribution in Iran: Hormozgan (Ameri et al., 2014 as *Bracon* (*Habrobracon*) *nigricans* (Szépligeti, 1904) —misidentification).
Comments: Samartsev (2016) stated that *B. nigricans* which was recorded by Ameri et al. (2014), is a misidentification of *Doggerella* sp.
The genus *Doggerella* comprises 15 species worldwide which are distributed in the Afrotropical region (Angola, Erritrea, Ivory Coast, Madagascar, Namibia, South Africa, Tanzania, Zimbabwe) (14 species) (Mahmood et al., 2011) and Yu et al. (2016), as well as Palaearctic (Russian Far East) (one species) (Samatsev, 2016).

Genus *Physaraia* Shenefelt, 1978
Physaraia furcata (Guérin-Méneville, 1848)
Distribution in the Middle East: Egypt (Fahringer, 1934).
Extralimital distribution: Democratic Republic of Congo, Equatorial Guinea, Ethiopia, Kenya, Libya, Madagascar, Malawi, Mozambique, Sudan, Tanzania, Togo, Uganda, Zimbabwe.
Host records: Unknown.

Tribe Coeloidini Tobias, 1957
Genus *Coeloides* Wesmael, 1838
Coeloides abdominalis (Zetterstedt, 1838)
Catalogs with Iranian records: Samin, Coronado-Blanco, Fischer et al. (2018).
Distribution in Iran: Golestan (Sakenin et al., 2008 as *Coeloides brunneri* Viereck, 1911—misidentification), Lorestan (Sakenin et al., 2018).
Distribution in the Middle East: Iran (Sakenin et al., 2018), Turkey (Schimitscheck, 1941; Beyarslan, 2014, 2016).
Extralimital distribution: Armenia, Austria, China, Czech Republic, Denmark, Finland, France, Germany, Hungary, Japan, Korea, Norway, Poland, Romania, Russia, Slovakia, Spain, Sweden, Switzerland, United Kingdom.

Host records: Summarized by Yu et al. (2016) as being a parasitoid of the buprestid *Melanophila cyanea* (Fabricius); the cerambycid *Acanthocinus aedilis* (L.); the curculionids *Hylesinus fraxini* (Panzer), *Hylurgops palliates* (Gyllenhal), *Ips* spp., *Onthotomicus proximus* Echhoff, *Phloeotribus scarabaeoides* (Bernard), *Pissodes* spp., *Tomicus pilifer* Wood and Bright, and *Tomicus piniperda* (L.); the chloropid *Lipara lucens* Meigen; and the tortricid *Epinotia pygmaeana* (Hübner). In Iran, was recorded as a larval parasitoid of *Agapanthia violacea* (Fabricius) (Coleoptera: Cerambycidae) (Sakenin et al., 2008).

Coeloides bostrichorum Giraud, 1872
Catalogs with Iranian records: Gadallah and Ghahari (2015), Farahani et al. (2016), Yu et al. (2016), Samin, Coronado-Blanco, Kavallieratos et al. (2018), Samin, Coronado-Blanco, Fischer et al. (2018).
Distribution in Iran: Ardabil (Ghahari & Fischer, 2011b).
Distribution in the Middle East: Iran.
Extralimital distribution: Austria, Bulgaria, China, Czech Republic, Finland, France, Germany, Hungary, Italy, Japan, Mongolia, Poland, Russia, Slovakia, Sweden, Switzerland.
Host records: Summarized by Yu et al. (2016) as being a parasitoid of the curculionids of the genera *Blastophagus* Eichhoff, *Carphoborus* Eichhoff, *Ips* De Geer, *Onthotomicus* Ferrari, *Pissodes* Germar, *Pityogenes* Bedel, *Pityokteines* Germar, and *Scolytus* Geoffroy. In Iran, it has been reared from the curculionid *Ips typographus* (Linnaeus) (Ghahari & Fischer, 2011b).

Coeloides filiformis Ratzeburg, 1852
Catalogs with Iranian records: None.
Distribution in Iran: Fars (Ghahari et al., 2020).
Distribution in the Middle East: Egypt (Fahringer, 1928), Iran (Ghahari et al., 2020).
Extralimital distribution: Austria, Czech Republic, Finland, France, Germany, Hungary, Italy, Lithuania, Netherlands, Norway, Poland, Romania, Russia, Serbia, Slovakia, Spain, Sweden, Switzerland, Ukraine, United Kingdom.
Host records: Summarized by Yu et al. 2016 as being a parasitoid of the curculionids *Hylesinus crenatus* (Fabricius), *Hylesinus fraxini* (Panzer), *Hylesinus varius* (Fabricius), *Phloeotribus scarabaeoides* (Bernard), and *Pissodes harcyniae* (Herbst).

Coeloides foersteri Haeselbarth, 1967
Distribution in the Middle East: Turkey (Beyarslan, 2015; Beyarslan & Çetin Erdoğan, 2012).

Extralimital distribution: Austria, Czech Republic, Finland, France, Georgia, Germany, Poland, Slovakia, Switzerland. Host records: Summarized by Yu et al. (2016) as being a parasitoid of the curculionids *Ips typographus* (L.), *Pissodes harcyniae* (Fabricius), *Pissodes notatus* (Fabricius), *Pissodes piceae* (Illiger), and *Pissodes piniphilus* (Herbst).

Coeloides rossicus (Kokujev, 1902)

Catalogs with Iranian records: Gadallah and Ghahari (2015), Farahani et al. (2016), Yu et al. (2016), Samin, Coronado-Blanco, Kavallieratos et al. (2018), Samin, Coronado-Blanco, Fischer et al. (2018).
Distribution in Iran: Golestan (Samin, Ghahari et al., 2015), Razavi Khorasan (Samin et al., 2011).
Distribution in the Middle East: Iran.
Extralimital distribution: Afghanistan, Belgium, Canada, Czech Republic, Finland, Germany, Hungary, Kazakhstan, Lithuania, Norway, Poland, Russia, Sweden, Ukraine, United States of America.
Host records: Summarized by Yu et al. (2016) as being a parasitoid of the buprestid *Agrilus anxius* Gory; the xiphydriids *Xiphydria camelus* (L.), *Xiphydria maculata* Say, *Xiphydria prolongata* (Geoffroy), and *Xiphydria tibialis* Say.

Coeloides sordidator (Ratzeburg, 1844)

Distribution in the Middle East: Turkey (Beyarslan, 2016).
Extralimital distribution: Austria, Croatia, former Czechoslovakia, France, Germany, Greece, Hungary, Italy, Latvia, Lithuania, Poland, Portugal, Russia, Serbia, Sweden, Switzerland, United Kingdom.
Host records: Summarized by Yu et al. (2016) as being a parasitoid of the buprestid *Melanophila cyanea* (Fabricius); the cerambycid *Monochamus galloprovincialis* (Olivier); and the following curculionids: *Carphoborus minimus* (Fabricius), *Ips typographus* (L.), *Orthotomicus proximus* (Eichhoff), *Pissodes castaneus* (DeGeer), *Pissodes notatus* (Fabricius), *Pissodes piniphilus* (Herbst), *Pityogenes quadridens* Hartig, *Tomicus minor* (Hartig), and *Tomicus piniperda* (L.).

Genera and species names excluded from the fauna of Iran

Coeloides brunneri Viereck, 1911

Records from Iran: Golestan (Sakenin et al., 2008).
General distribution: Nearctic (Canada, United States of America).
Host records reported in Iran: *Agapanthia violacea* (Fabricius) (Coleoptera: Cerambycidae) (Sakenin et al., 2008).
Comments: Reexamination of specimens proved Sakenin et al. (2008) identification of *Coeloides brunneri* was a misidentification of *Coeloides abdominalis* (Zetterstedt) and we therefore exclude *C. brunneri* from the fauna of Iran and record the data under *C. abdominalis*.

Habrobracon johnson Nomen Nudum

Comments: Hashemi Aghajari et al. (1995) recorded *Habrobracon johnson* from East Azarbaijan as the parasitoid of *Heliothis peltigera* (Denis and Schiffermüller) and *Heliothis viviplaca* (Hufnagel) [under *Chloridea* Duncan and Westwood = *Heliothis* Ochsenheimer] (Lepidoptera: Noctuidae). *Habrobracon johnson* which is represented in Modarres Awal (1997, 2012) is an invalid taxon, and we exclude it from the fauna of Iran. The name refers to *Habrobracon* Johnson, 1895 which is regarded as a generic synonym of *Habrobracon* Ashmead, 1900 which is the accepted name (https://www.gbif.org/species/7799773/metrics: accessed August 14, 2020).

Shelfordia capensis (Cameron, 1904)

Comments: The synonymization of the Oriental genus *Sigalphogastra* Cameron, 1903 with the Oriental genus *Shelfordia* Cameron, 1903, by Quicke (1981) led to the unintended record of the genus *Shelfordia* from Africa and the Palaearctic. The type specimen of *S. capensis* was studied by the Philippine entomologist Clare Baltazar who transferred it from *Merinotus* Szépligeti to *Sigalphogastra* Cameron and so the subsequent generic synonymization wrongly records *Shelfordia* from Africa and adjacent areas. Since *S. capensis* had been transferred to *Merinotus* Szépligeti (superficially similar to *Shelfordia*), we consider it most likely that this species represents *Merinotus*.

Conclusion

In total, 347 valid species of the subfamily Braconinae in 35 genera and four tribes (Aphrastobraconini, Argamaniini, Braconini, and Coeloidini) have been reported from most of the Middle East countries, with the exception of Bahrain, Kuwait, and Qatar. It was found that Turkey is the richest country, with 217 species. The most species rich genus in the Braconinae of the Middle East is *Bracon* (with 191 species), followed by *Iphiaulax* (with 23 species), *Vipio* (with 22 species), then *Habrobracon* (with 21 species). The subfamily Braconinae of Iran comprises 212 species in 15 genera (6.9% of the world species), of which the genus *Bracon* with 134 species is more diverse than the others, followed by *Habrobracon* (with 16 species), *Vipio* (with 14 species), *Iphiaulax* (12 species), *Pseudovipio* (10 species), *Glyptomorpha* (seven species), *Atanycolus* (five species), *Coeloides* (four species), *Baryproctus*, *Cyanopterus*, and *Stenobracon* (each with two species), and *Amyosoma*, *Ceratobracon*, *Megalommum*, and *Rhadinobracon* (each with one species). Seven species are so far only known from Iran (endemic or subendemic to Iran), *Bracon persiangulfensis* Ameri, Beyarslan and Talebi, 2013, *Habrobracon iranicus* Fischer, 1972, *Iphiaulax iranicus* Quicke, 1985, *Megalommum pistacivora* van Achterberg and Mehrnejad, 2011, *Pseudovipio nigrirostris* (Kokujev,

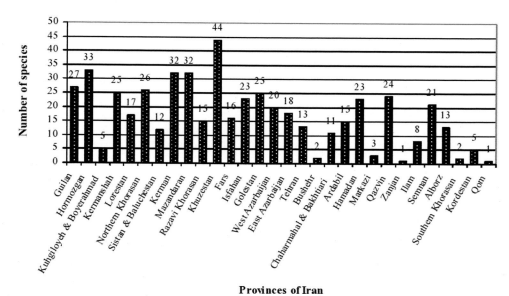

Provinces of Iran

FIGURE 6.1 Number of reported species of Iranian Braconinae by province.

1907), *Pseudovipio schaeuffelei* (Hedwig, 1957), and *Vipio xanthurus* (Fahringer, 1926). Iranian Braconinae species have been recorded from 30 provinces (all provinces except Yazd), of which Khuzestan with 44 records has the highest number of species, followed by Hormozgan, Kerman and Mazandaran with 33, 32, and 32 species, respectively (Fig. 6.1). Host species have been discovered for 36 Braconinae species. These hosts belong to four orders, Coleoptera (Brentidae: one unknown species; Buprestidae: nine species; Cerambycidae: seven species; Curculionidae: 12 species), Diptera (Agromyzidae: one species; Tephritidae: four species), Hymenoptera (Argidae: one species; Cynipidae: one species; Tenthredinidae: one species) and Lepidoptera (Batrachedridae: one species; Crambidae: two species; Gelechiidae: three species; Geometridae: one species; Gracillariidae: one unknown species; Lasiocampidae: one species; Noctuidae: 10 species; Plutellidae: one species; Pyralidae: four species; Sesiidae: two species; Tortricidae: five species; Yponomeutidae: one species; Zygaenidae: one species). Comparison of the Braconinae fauna of Iran with the Middle East and adjacent countries indicates the fauna of Russia with 232 species in 15 genera (Belokobylskij & Lelej, 2019) is more diverse than Iran, followed by Turkey (with 219 species), Kazakhstan (121 species), Israel–Palestine (with 127 species), Azerbaijan (89 species), Turkmenistan (71 species), Cyprus (78 species), Armenia (58 species), Egypt (50 species), Afghanistan (23 species), Syria (23 species), Pakistan (11 species), Saudi Arabia (14 species), Jordan and Yemen (both with 10 species), Iraq (four species), Lebanon and Oman (three species), and United Arab Emirates (one species). No species have been recorded from Bahrain, Kuwait and Qatar (Yu et al., 2016). Among the Middle East countries and adjacent to Iran, Turkey shares 153 species with Iran, followed by Russia (115 species), Kazakhstan (94 species), Israel–Palestine (84 species), Azerbaijan (77 species), Cyprus (67 species), Turkmenistan (54 species), Armenia (41 species), Syria (18 species) Afghanistan (16 species), Jordan (10 species), Egypt (eight species), Pakistan (four species), Iraq (three species), Lebanon (two species), and Oman (one species).

References

Abbasipour, H., Mahmoudvand, M., Basij, M., & Lozan, A. (2012). First report of the parasitoid wasps, *Microchelonus subcontractus* and *Bracon intercessor* (Hym.: Braconidae), from Iran. *Journal of Entomological Society of Iran, 32*(1), 89–92.

Abedi, A. A., Fathi, S. A. A., & Nouri-Ganbalani, G. (2015). Effect of strip cropping of sugar beet-alfalfa on population density of the sugar beet weevil, *Lixus incanescens* (Col.: Curculionidae) and species diversity of its natural enemies. *Journal of Entomological Society of Iran, 34*(4), 1–14 (in Persian, English summary).

Adldoost, H. (2010). Study on the larval parasitism of *Heliothis viriplaca* Huf. In the western Azarbaijan province of Iran. In *Proceedings of the 19th Iranian Plant Protection Congress, 31 July – 3 August 2010* (p. 127). Iranian Research Institute of Plant Protection.

Adldoost, H., & Shayesteh, N. (2010). Study on the larval parasitism of *Heliothis viriplaca* Huf in the western Azarbaijan provience of Iran. In *Proceedings of the IX European Congress of Entomology, 22–27 August 2010, Budapest* (p. 139).

Afiunizadeh Isfahani, M., & Karimzadeh Isfahani, J. (2010). Larval and pupal parasitoids of *Plutella xylostella* (Lep.: Plutellidae) in Isfahan province, Iran. *Plant Protection Journal, 2*(2), 79–97 (in Persian, English summary).

Afiunizadeh Isfahani, M., Karimzadeh, J., Broad, G., Shojai, M., Emami, M. S., Lotfalizadeh, H., Papp, J., LaSalle, J., Whietfield, J. B., van Achterberg, C., & Shaw, M. R. (2010). Larval and pupal parasitoids of *Plutella xylostella* in Isfahan province. In *Proceedings of the 19th Iranian Plant Protection Congress, 31 July − 3 August 2010* (p. 115). Iranian Research Institute of Plant Protection.

Afrouzian, M., Bagheri-Zenous, E., & Shojai, M. (2001). Biology of seed beetle, *Caryedon palestinicus* Southgate at south Baluchestan region, Iran. *Journal of Agricultural Sciences, 7*(3), 59−70 (in Persian, English summary).

Afshari, A., Yazdanian, M., Shabanipour, M., & Ghadiri-Rad, S. (2014). Natural parasitism of tomato fruitworm, (*Helicoverpa armigera* Hübner) in tomato fields of Golestan province, northern Iran. In *Proceedings of the 22nd Iranian Plant Protection Congress, 27−30 August 2016, College of Agriculture and Natural Resources* (p. 503). Karaj: University of Tehran.

Akbari-Asl, M. H., Talebi, A. A., Kamali, H., & Kazemi, S. (2009). Stored product pests and their parasitoid wasps in Mashhad, Iran. *Advances in Environmental Biology, 3*(3), 239−243.

Akbarzadeh Shoukat, G. (2012). Larval parasitoids of *Lobesia botrana* (Denis and Schiffermüller, 1775) (Lepidoptera: Tortricidae) in Orumieh vineyards. *Journal of Agricultural Science and Technology A, 14*, 267−274 (in Persian, English summary).

Akbarzadeh Shoukat, G., Ebrahimi, E., & Masnadi Yazdinejad, A. (2008). Larval parasitoid of *Lobesia botrana* Denis and Schiff. (Lepidoptera: Tortricidae) on grape in Ourmieh, Iran. In *Proceedings of the 18th Iranian Plant Protection Congress, 24−27 August 2008* (p. 20). University of Bu-Ali Sina Hamedan.

Al-Jboory, I. J., Katbeh-Bader, A., & Shakir, A. (2012). First observation and identification of some natural enemies collected from heavily infested tomato by *Tuta absoluta* (Meyrick) (Lepidoptera: Gelechiidae), in Jordan. *World Applied Siences Journal, 17*(5), 589−592.

Ameri, A., Talebi, A. A., Beyarslan, A., Kamali, K., & Rakhshani, E. (2014). Study of the genus *Bracon* Fabricius, 1804 (Hymenoptera: Braconidae) of southern Iran with description of a new species. *Zootaxa, 3754*(4), 353−380.

Ameri, A., Talebi, A. A., Rakhshani, E., Beyarslan, A., & Kamali, K. (2015). Additional evidence and new records of the genus *Bracon* Fabricius, 1804 (Hymenoptera: Braconidae) in southern Iran. *Turkish Journal of Zoology, 39*, 1110−1120.

Ameri, A., Talebi, A. A., Rakhshani, E., Ebtahi, Y., & Askari, M. (2016). A faunistic survey on the genus *Pseudovipio* Szépligeti, 1896 (Hym., Braconidae, Braconinae) in Hormozgan province. In *Proceedings of the 22nd Iranian Plant Protection Congress, 27−30 August 2016, College of Agriculture and Natural Resources* (p. 480). Karaj: University of Tehran.

Ameri, A., Ebrahimi, E., & Talebi, A. A. (2020). Additions to the fauna of Braconidae (Hym., Ichneumonoidea) of Iran based on specimens housed in Hayk Mirzayans insect Museum with six new records for Iran. *Journal of Insect Biodiversity and Systematics, 6*(4), 353−364.

Amooghli-Tabari, M., & Ghahari, H. (2021). A study on the transitional population of overwintering larvae of *Chilo suppressalis* Walker (Lepidoptera: Crambidae) with percent rice infestation in main crop, and their natural enemies in Mazandaran, Iran. *Experimental Animal Biology, 9*(4), 39−54 (in Persian, English summary).

Aubert, J. F. (1966). In: Liste d'indentification No. 6 (presentée par le service d'identification des Entomophages). *Entomophaga, 11*(1), 115−134.

Bagheri Zenuz, E. (1973). *Plodia interpunctella* (Lep., Phycitidae) et ses ennemis naturels. *Journal of Entomological Society of Iran, 1*(1), 23−41.

Bagheri, M. R., & Nematollahi, M. R. (2006). Biology and damage rate of safflower shoot fly *Acanthiophilus helianthi* in Esfahan province. In *Proceedings of the 17th Iranian Plant Protection Congress, 2−5 September 2006, Campus of Agriculture and Natural Resources* (p. 268). Karaj: University of Tehran.

Bagheri, M. R., & Nasr Isfahani, M. (2010). The fauna of harmful and beneficial arthropods of medicinal and range plants in Isfahan. *Journal of Entomological Research, 3*(2), 119−132.

Baird, A. B. (1958). Biological control of insects and plant pests in Canada. In E. C. Baecker (Ed.), *Proceedings of the 10th International Congress of Entomologist, 17−25 August 1956, Montreal* (vol. 4, pp. 483−485).

Ballou, H. A. (1918). The pink bollworm (*Gelechia gossypiella*) in Egypt. *Journal of Economic Entomology, 11*, 236−245.

Bartlett, B. R., Clausen, C. P., DeBach, P., Goeden, R. D., Legner, E. F., McMurtry, J., & Oatman, E. R. (1978). Introduced parasites and predators of arthropod pests and weeds. A wold review. In *, 480. Agriculture Handbook*. Agricultural Research Service, United States Department of Agriculture, 544 pp.

Basheer, A., Asslan, L., & Abdalrazak, F. (2014). Survey of the parasitoids of the tephritid flies of the safflower *Carthamus tinctorius* (Asteracea) in Damascus, Syria. *Egyptian Journal of Biological Pest Control, 24*(1), 169−172.

Behdad, E. (1991). *Pests of fruit crops in Iran*. Isfahan: Neshat Publication, 882 pp. (in Persian).

Belokobylskij, S. A., & Lelej, A. S. (2019). Annotated catalogue of the Hymenoptera of Russia. In *, Apocrita: Parasitica: Vol. II. Proceedings of the Zoological Institute of the Russian Academy of Sciences, Supplement No. 8*, 594 pp.

Belshaw, R., Lopez-Vaamonde, C., Degerli, N., & Quicke, D. L. J. (2001). Paraphyletic taxa and taxonomic chaining: evaluating the classification of braconine wasps (Hymenoptera: Braconidae) using 28S D2-3 r-DNAsequences and morphological characters. *Biological Journal of the Linnean Society, 73*, 411−424.

Beyarslan, A. (1986). Untersuchungen über die im Mediteranischen Gebiet der Türkei festgestellen *Bracon* Fabricius-Arten (Hymenoptera: Braconidae; Braconinae) II. *Ulusal Biyoloji Kongresi, VIII*, 387−402.

Beyarslan, A. (1987). Untersuchungen über fauna von Braconinae (Hymenopetra: Braconidae) im Thrakien Gebiet. *Türkei Entomoloji Kongresi, I*, 595−604.

Beyarslan, A. (1988). Zwei neue Arten der familie Braconidae (Hymenoptera) aus der Türkei. *Zeitschrift der Arbeitsgemeinschaft Österreichischer Entomologen, 39*(3/4), 71−76.

Beyarslan, A. (1991). Die Arten der Tribus Vipionini Telenga aus der Turkei (Hymenoptera: Braconidae: Braconinae). *Linzer biologische Beiträge, 23*(2), 495−519.

Beyarslan, A. (1992). *Isomecus lalapasaensis* sp. nov. und *Vipiomorpha fischeri* sp. nov., zwei neue Arten der Tribu Vipionini (Hymenoptera, Braconidae, Braconinae). *Entomofauna, 13*(15), 253−259.

Beyarslan, A. (1996). Vier neue Arten der Tribus Braconini aus der Türkei (Hymenoptera, Braconidae, Braconinae). *Entomofauna, 17*(21), 345−352.

Beyarslan, A. (1999). Liste der Braconinae-Arten der Mittelmeer- und Marmara Region der Turkei (Hymenoptera, Braconidae). *Entomofauna, 20*(5), 93−118.

Beyarslan, A. (2002). Four new species of the genus *Bracon* (Hymenoptera: Braconidae: Braconinae) from Turkey. *Biologia (Bratislava), 57*(2), 139−146.

Beyarslan, A. (2009). A new species *Bracon (Orthobracon) malatyensis* sp. n. from Eastern Anatolia (Hymenoptera, Braconidae, Braconinae). *Journal of the Entomological Research Society, 11*(3), 31−36.

Beyarslan, A. (2010). *Bracon (Glabrobracon) jenoi* sp. n. (Hymenoptera: Braconidae: Braconinae) from Turkey. *Biologia, 65/1*, 110−112.

Beyarslan, A. (2011). Two new species, *Bracon (Lucobracon) kuzguni* sp. n. and *Bracon (Lucobracon) breviradius* sp. n., from Turkey (Hymenoptera: Braconidae: Braconinae). *Turrkish Journal of Zoology, 35*(4), 503−508.

Beyarslan, A. (2014). Checklist of Braconinae species of Turkey (Hymenoptera: Braconidae). *Zootaxa, 3790*(2), 201−242.

Beyarslan, A. (2015). A faunal study of the subfamily Doryctinae in Turkey (Hymenoptera: Braconidae). *Turkish Journal of Zoology, 39*(1), 126−143.

Beyarslan, A. (2016). Taxonomic investigations of the Braconinae fauna (Hymenoptera: Braconidae) in north-eastern Anatolian region, Turkey, with the description of a new species. *Zootaxa, 4079*(1), 1−33.

Beyarslan, A., & Inanç, F. (1994). In *Taxonomische Untersuchungen über die Braconinae-Fauna der Marmara Region (Hymenoptera: Braconidae). 1. Anon. Turkiye. 3. Biyolojik Mucadele Kongresi Bildirileri 25-28 Ocak 1994.*

Beyarslan, A., & Çetin Erdoğan, Ö. (2010). New data on the zoogeography and taxonomy of the east Black Sea region species of Braconinae (Hymenoptera: Braconidae) in Turkey. *Journal of the Entomological Research Society, 12*(2), 51−56.

Beyarslan, A., & Çetin Erdoğan, Ö. (2012). The Braconinae (Hymenoptera: Braconidae) of Turkey, with new locality records and descriptions of two new species of *Bracon* Fabricius, 1804. *Zootaxa, 3343*, 45−56.

Beyarslan, A., & Tobias, V. I. (2008). *Bracon (Lucobracon) iskilipus* sp. n. (Hymenoptera: Braconidae: Braconinae) from the central Black Sea region of Turkey. *Biologia (Bratislava), 63/4*, 550−552.

Beyarslan, A., Aydoğdu, M., & Erdoğan, Ö.Ç. (2006). A survey of Turkish *Glyptomorpha* (Hymenoptera, Braconidae, Braconinae) fauna, with redescription of *G. baetica* from a new host. *Biologia (Bratislava), 61*(2), 139−143.

Beyarslan, A., Aydoğdu, M., & Çetin Erdoğan, Ö. (2008). The subfamily Braconinae in northern Turkey, with new records of *Bracon* species from the Western Palaearctic (Hymenoptera: Braconidae). *Linzer biologische Beiträge, 40/2*, 1341−1361.

Beyarslan, A., Erdoğan, Ö.Ç., & Aydoğdu, M. (2005). A survey of Braconinae (Hymenoptera, Braconidae) of Turkish Western Black Sea region. *Linzer biologische Beiträge, 37*(1), 195−213.

Beyarslan, A., Gözüaçik, C., & Inanç, O. (2014). First research on Braconinae fauna of South-eastern Anatolia region with new localities of Turkey (Hymenoptera: Braconidae). *Entomofauna, 35*(10), 177−201.

Beyarslan, A., Gözüaçik, C., Güllü, M., & Konuksal, A. (2017). Taxonomical investigation on Braconidae (Hymenoptera: Ichneumonoidea) fauna in northern Cyprus, with twenty six new records for the country. *Journal of Insect Biodiversity and Systematics, 3*(4), 319−334.

Beyarslan, A., Inanç, F., Çetin Erdoğan, Ö., & Aydoğdu, M. (2002). Braconiden von den Tuerkischen Inseln Imbros und Tenedos (Hymenoptera, Braconidae: Agathidinae, Braconinae, Cheloninae, Microgastrinae). *Entomofauna, 23*(15), 173−188.

Bodenheimer, F. S. (1930). Die Schädlingsfauna Palästinas. *Monographien zur Angewandten Entomologie, 10*, 438 pp.

Bolu, H., Beyarslan, A., Yildirim, H., & Aktürk, Z. (2009). Two new host records of *Atanycolus ivanowi* (Kokujev, 1898) (Hymenoptera: Braconidae) from Turkey. *Turkiye Entomoloji Dergisi, 33*(4), 279−287.

Brues, C. I. (1926). Studies on the Ethiopian Braconidae with a catalogue of the African species. *Proceedings of the American Academy of Arts and Sciences, 6*, 206−436.

Čapek, M., & Hofmann, C. (1997). The Braconidae (Hymenoptera) in the collections of the Musee cantonal de Zoologie, Lausanne. *Litterae Zoologicae (Lausanne), 2*, 25−163.

Chen, X. X., & van Achterberg, C. (2019). Systematics, phylogeny, and evolution of braconid wasps: 30 years of progress. *Annual Review of Entomology, 64*, 1−24.

Chishti, M. J. K., & Quicke, D. L. J. (1996). A revision of the Indo-Australian species of *Stenobracon* (Hymenoptera: Braconidae) parasitoids of lepidopterous stem-borers of graminaceous crops. *Bulletin of the Entomological Research, 86*, 227−245.

Çikman, E., Beyarslan, A., & Civelek, H. S. (2006). Parasitoids of leafminers (Diptera: Agromyzidae) from southeast Turkey with 3 new records. *Turkish Journal of Zoology, 30*, 167−173.

Coll, M., Gavish, S., & Dori, L. (2000). Population biology of potato tuber moth, *Phthorimaea operculella* (Lepidoptera: Gelechiidae), in two potato cropping systems in Israel. *Bulletin of the Entomological Research, 90*(4), 309−315.

Cox, M. L. (1994). The Hymenoptera and Diptera parasitoids of Chrysomelidae, pp. 419−467. In P. H. Jolivet, M. L. Cox, & E. Petitpierre (Eds.), *Novel Aspects of the Biology of Chrysomelidae*. The Netherlands: Kluwer Academic Publishers, 582 pp.

Dastgheyb Beheshti, N. (1980). *Insect pests of cold region fruit trees of Esfahan*. Esfahan: Plant Pests and Diseases Research Institute, 145 pp.

Davatchi, A., & Shojai, M. (1969). *Les hymenopteres entomophages de l'Iran (etudes faunestiques)*. Tehran University, Faculty of Agriculture, 88 pp.

Dezianian, A., & Jalali, A. (2004). Investigation on the distribution and seasonal fluctuation and important natural enemies of potato tuber moth *Phthorimaea operculella* (Zeller) in Shahrood region. In *Proceedings of the 16th Iranian Plant Protection Congress, 28 August − 1 September 2004* (p. 17). University of Tabriz.

Dezianian, A., & Quicke, D. (2006). Introduction of potato tuber moth parasite wasp, *Bracon (Habrobracon)* aff. *radialis* Telenga from Iran. In *Proceedings of the 17th Iranian Plant Protection Congress, 2−5 September 2006, Campus of Agriculture and Natural Resources* (p. 65). Karaj: University of Tehran.

Dudgeon, G. C., & Gough, L. H. (1914). *Rogas kitcheneri* n. sp., a new braconid destructive to the Egyptian cotton bollworm. *Bulletin de la Société Entomologique d'Egypte, 1912*, 140−141.

Dweck, H. K., Svensson, G. P., Gündüz, E. A., & Anderbrant, O. (2010). Kaironomal response of the parasitoid, *Bracon hebetor* Say, to the male-produced sex pheromone of its host, the greater waxmoth, *Galleria mellonella* (L.). *Journal of Chemical Ecology, 36*, 171−178.

Eyidozehi, K., Khormali, S., Ravan, S., & Barahoei, H. (2013). Introduction of seven wasps' parasitoid species associated with stored food product pests in Golestan province. *Plant Pests Research, 3*(1), 69−72.

Faal-Mohamad Ali, H., Seraj, A. A., Talebi-Jahromi, K., Shishebor, P., & Mosadegh, M. S. (2010). Investigation sublethal effect of conventional pesticides of tomato fields on life cycle parameters of *Habrobracon hebetor* Say (Hymenoptera: Braconidae) in adult stage. In

Proceedings of the 19th Iranian Plant Protection Congress, 31 July — 3 August 2010 (p. 264). Tehran: Iranian Research Institute of Plant Protection.

Fahringer, J. (1922). Hymenopterologische Ergebnisse einer wissenschaftlichen Studienreise nach der Türkei und Kleinasien (mit Ausschluβ des Amanusgebirges). *Archiv für naturgeschichte, A88*(9), 149–222.

Fahringer, J. (1925). *Opuscula braconologica. Band 1. Palaearktische region. Lieferung 1. Opuscula braconologica. Fritz Wagner, Wien* (pp. 1–60).

Fahringer, J. (1926). *Opuscula braconologica. Band 1. Palaearktischen region. Lieferung 2–3. Opuscula braconologica, Fritz Wagner, Wien* (pp. 61–220).

Fahringer, J. (1927). *Opuscula braconologica. Band 1. Palaearktischen region. Lieferung 4-6. Opuscula braconologica. Fritz Wagner, Wien* (pp. 221–432).

Fahringer, J. (1928). *Opuscula braconologica. Band 1. Palaearctischen Region. Lieferung 7–9. Opuscula braconologica* (pp. 433–606).

Fahringer, J. (1934). *Opuscula braconologica. Band 3. Palaearktischen region. Lieferung 5-8. Opuscula braconologica. Fritz Wagner, Wien* (pp. 321–594).

Fahringer, J. (1935). *Opuscula braconologica. Band 2. Aethiopische region. Lieferung 6-8. Opuscula braconologica. Fritz Wagner, Wien* (pp. 385–635).

Fallahzadeh, M., & Saghaei, N. (2010). Checklist of Braconidae (Insecta: Hymenoptera) from Iran. *Munis Entomology & Zoology, 5*(1), 170–186.

Farahani, S., Talebi, A. A., & Rakhshani, E. (2016). Iranian Braconidae (Insecta: Hymenoptera: Ichneumonoidea), diversity, distribution and host association. *Journal of Insect Biodiversity and Systematics, 2*(1), 1–92.

Farahbakhsh, G. (1961). Family Braconidae (Hymenoptera), p. 124. In G. Farahbakhsh (Ed.), *A checklist of economically important insects and other enemies of plants and agricultural products in Iran*. The Ministry of Agriculture, Department Plant Protection, 153 pp.

Fasihi, M. T., Farrokhi, S., Sarafrazi, A., Sohrabi, F., Heidari, A., & Salehi, M. (2017). Natural enemies of tomato leafminer, *Tuta absoluta* (Meyrick) (Lep.: Gelechiidae) in Busher province. In *Proceedings of the 8th National Conference of Biological Control in Agriculture and Natural Resources* (p. 35).

Fathi, S. A. A., Abedi, A., & Lotfalizadeh, H. A. (2016). Effects of sugar beet field margin vegetation on population density of the sugar beet weevil, *Lixus incanescens* Boheman and the percentage of its larval parasitism. *Journal of Applied Researches in Plant Protection, 4*(2), 201–214 (in Persian, English summary).

Fischer, M. (1972). Eine neue *Habrobracon*-Art aus dem Iran (Hymenoptera, Braconidae). *Entomophaga, 17*(1), 89–91.

Fischer, M. (1980). Fünf neue Raupenwespen (Hymenoptera, Braconidae). *Frustula Entomologica, 1*(1978), 147–160.

Fouruzan, M., Sahragard, A., & Amir Maafi, M. (2010). Age-specific two sex life table of the parasitoid wasp, *Habrobracon hebetor* Say (Hym.: Braconidae) reared on *Galleria mellonella*. In *Proceedings of the 19th Iranian Plant Protection Congress, 31 July — 3 August 2010* (p. 62). Tehran: Iranian Research Institute of Plant Protection.

Fuchs, G. (1912). Generationsfragen bei Rüsselkafern. *Naturwissenschaftliche Zeitschrift für Forst- und Landwirtschaft, 10*, 43–54.

Fulmek, L. (1968). Parasitinsekten der Insektengallen Europas. *Beiträge zur Entomologie, 18*(7/8), 719–952.

Gadallah, N. S., & Ghahari, H. (2015). An annotated catalogue of the Iranian Braconinae (Hymenoptera: Braconidae). *Entomofauna, 36*, 121–176.

Gadallah, N. S., Ghahari, H., Papp, J., & Beyarslan, A. (2018). New records of Braconidae (Hymenoptera) from Iran. *Wuyi Science Journal, 34*, 43–48.

Gadallah, N. S., Ghahari, H., & Quicke, D. L. J. (2021). Further addition to the braconid fauna of Iran (Hymenoptera: Braconidae). *Egyptian Journal of Biological Pest Control, 31*, 32. https://doi.org/10.1186/s41938-021-00376-8

Ghahari, H. (2017). Species diversity of Ichneumonoidea (Hymenoptera) from rice fields of Mazandaran province, northern Iran. *Journal of Animal Environment, 9*(3), 371–378 (in Persian, English summary).

Ghahari, H. (2018). Species diversity of the parasitoids in rice fields of northern Iran, especially parasitoids of rice stem borer. *Journal of Animal Environment, 9*(4), 289–298 (in Persian, English summary).

Ghahari, H. (2019a). Faunistic survey of parasitoid wasps (Hymenoptera) in forest areas of Mazandaran province, northern Iran. *Iranian Journal of Forest, 11*(1), 61–79 (in Persian, English summary).

Ghahari, H. (2019b). Study on the natural enemies of agricultural pests in some sugar beet fields, Iran. *Journal of Sugar Beet, 35*(1), 91–102 (in Persian, English summary).

Ghahari, H., & Sakenin, H. (2018). Species diversity of Chalcidoidea and Ichneumonoidea (Hymenoptera) in some paddy fields and surrounding grasslands of Mazandaran and Guilan provinces, northern Iran. *Applied Plant Protection, 7*(1), 11–19 (in Persian, English summary).

Ghahari, H., & Gadallah, N. S. (2022). Additional records to the braconid fauna (Hymenoptera: Ichneumonoidea) of Iran, with new host reports. *Entomological News* (in press).

Ghahari, H., Fischer, M., Erdoğan, Ö.Ç., Tabari, M., Ostovan, H., & Beyarslan, A. (2009). A contribution to Braconidae (Hymenoptera) from rice fields and surrounding grasslands of northern Iran. *Munis Entomology and Zoology, 4*(2), 432–435.

Ghahari, H., Fischer, M., Erdoğan, Ö.Ç., Beyarslan, A., Hedqvist, K. J., & Ostovan, H. (2009). Faunistic note on the Braconidae (Hymenoptera: Ichneumonoidea) in Iranian alfalfa fields and surrounding grasslands. *Entomofauna, 30*, 437–444.

Ghahari, H., Gadallah, N. S., Erdoğan, Ö.Ç., Hedqvist, K. J., Fischer, F., Beyarslan, A., & Ostovan, H. (2009). Faunistic note on the Braconidae (Hymenoptera: Ichneumonoidea) in Iranian cotton fields and surrounding grasslands. *Egyptian Journal of Biological Pest Control, 19*(2), 115–118.

Ghahari, H., Fischer, M., Erdoğan, Ö.Ç., Beyarslan, A., & Ostovan, H. (2010). A contribution to the braconid wasps (Hymenoptera: Braconidae) from the forests of northern Iran. *Linzer biologische Beiträge, 42*(1), 621–634.

Ghahari, H., Fischer, M., Hedqvist, K. J., Erdoğan, Ö.Ç., van Achterberg, C., & Beyarslan, A. (2010). Some new records of Braconidae (Hymenoptera) for Iran. *Linzer biologische Beiträge, 42*(2), 1395–1404.

Ghahari, H., & Fischer, M. (2011a). A contribution to the Braconidae (Hymenoptera: Ichneumonoidea) from north-western Iran. *Calodema, 134*, 1–6.

Ghahari, H., & Fischer, M. (2011b). A study on the Braconidae (Hymenoptera: Ichneumonoidea) from some regions of northern Iran. *Entomofauna, 32*, 181–196.

Ghahari, H., & Fischer, M. (2011c). A contribution to the Braconidae (Hymenoptera) from Golestan National Park, northern Iran. *Zeitschrift Arbeitsgemeinschaft Österreichischer Entomologen, 63*, 77–80.

Ghahari, H., Fischer, M., Sakenin, H., & Imani, S. (2011). A contribution to the Agathidinae, Alysiinae, Aphidiinae, Braconinae, Microgastrinae and Opiinae (Hymenoptera: Braconidae) from cotton fields and surrounding grasslands of Iran. *Linzer biologische Beiträge, 43*(2), 1269−1276.

Ghahari, H., Fischer, M., & Papp, J. (2011). A study on the braconid wasps (Hymenoptera: Braconidae) from Isfahan province, Iran. *Entomofauna, 32*, 261−272.

Ghahari, H., Fischer, M., & Papp, J. (2011). A study on the Braconidae (Hymenoptera: Ichneumonoidea) from Qazvin province, Iran. *Entomofauna, 32*, 197−208.

Ghahari, H., Fischer, M., & Papp, J. (2011). A study on the Braconidae (Hymenoptera: Ichneumonoidea) from Ilam province, Iran. *Calodema, 160*, 1−5.

Ghahari, H., & Fischer, M. (2012). A faunistic survey on the braconid wasps (Hymenoptera: Braconidae) from Kermanshah province, Iran. *Entomofauna, 33*, 305−312.

Ghahari, H., Fischer, M., & Tobias, V. (2012). A study on the Braconidae (Hymenoptera: Ichneumonoidea) from Guilan province, Iran. *Entomofauna, 33*, 317−324.

Ghahari, H., Fischer, M., Papp, J., & Tobias, V. (2012). A contribution to the knowledge of braconids (Hymenoptera: Braconidae) from Lorestan province Iran. *Entomofauna, 33*, 65−72.

Ghahari, H., & Gadallah, N. S. (2015). A faunistic study on the Braconidae (Hymenoptera) from some regions of Semnan, Iran. *Entomofauna, 36*(11), 177−184.

Ghahari, H., & Beyarslan, A. (2017). A faunistic study on Braconidae (Hymenoptera: Ichneumonoidea) from Iran. *Natura Somogyiensis, 30*, 39−46.

Ghahari, H., & Beyarslan, A. (2019). A faunistic study on Braconidae (Hymenoptera: Ichneumonoidea) from Iran, and in Memoriam Dr. Jenő Papp (20 May 1933 − 11 December 2017). *Acta Biologica Turcica, 32*(4), 248−254.

Ghahari, H., & Gadallah, N. S. (2019). New records of braconid wasps (Hymenoptera: Ichneumonoidea: Braconidae) from Iran. *Entomological News, 129*(3), 384−392.

Ghahari, H., Fischer, M., Beyarslan, A., Navaeian, M., & Hosseini Boldaji, S. A. (2019). New records of Brachistinae, Braconinae, Cheloninae and Microgastrinae (Hymenoptera: Braconidae) from Iran. *Wuyi Science Journal, 35*(2), 135−141.

Ghahari, H., Beyarslan, A., & Kavallieratos, N. G. (2020). New records of Braconidae (Hymenoptera: Ichneumonoidea) from Iran, and in Memoriam Dr. Maximilian Fischer (7 June 1929 − 15 June 2019). In *Scientific Bulletin of Uzhhorod National University (Series: Biology), 48* pp. 48−55).

Gharali, B. (2004). Study of natural enemies of safflower shoot flies in Ilam province. In *Proceedings of the 16th Iranian Plant Protection Congress, 28 August − 1 September 2004* (p. 54). University of Tabriz.

Goldansaz, S. H., Kishani, H., Talaei, L., Poorjavad, N., & Sobhani, M. (2011). Biological control of carob moth, past, present and future. In *Proceedings of the Biological Control Development Congress in Iran, 27-28 July 2011* (pp. 55−63). Iranian Research Institute of Plant Protection (in Persian, English summary).

Golizadeh, A. (2008). Thermal requirements and population dynamics of diamondback moth, *Plutella xylostella* (Lep., Plutellidae) (Ph. D. thesis). In *Tehran region*. Tehran, Iran: Tarbiat Modares University (in Persian).

Gultekin, L. (2006). Seasonal occurrence and biology of globe thistle capitulum weevil *Larinus anopordi* (F.) (Coleoptera: Curculionidae) in northeastern Turkey. *Munis Entomology & Zoology, 1*(2), 191−198.

Gultekin, L., & Guclu, S. (1997). Biologia of *Bembecia scopigera* (Scopoli) (Lep.: Sesiidae) pest of sainfain in Erzurum. *Bitki Koruma Bulteni, 37*(3−4), 101−110.

Gultekin, L., Cristofaro, M., Tronci, C., & Smith, L. (2008). Natural history studies for the preliminary evaluation of *Larinus filiformis* (Coleoptera: Curculionidae) as a prospective biological control agent of yellow starthistle. *Enviromental Entomology, 37*(5), 11851−21199.

Györfi, J. (1956). Nadelholzzaofen und Nadelhozamenschädlinge und ihre parasiten. *Acta Agronomica Hungarica, 6*, 321−373.

Györfi, J. (1959). Beiträge zur Kenntnis der Wirte verschiedener Braconiden-Arten (Hymenoptera, Braconidae). *Acta Zoologica Hungarica, 5*, 49−65.

Haeselbarth, E. (1983). Determination list of entomophagous insects. Nr. 9. Bulletin. *Section Regionale Ouest Palaearctique, Organisation Internationale de Lutte Biologiqu, 6*(1), 1−49.

Haghshenas, A. R., & Esfandiari, H. (2016). Study of natural enemies of *Tuta obsoluta* (Lepidoptera, Gelechiidae) in Esfahan province. In *Proceedings of the 3rd Iranian National Conference of Agriculture and Natural Resources* (p. 41). Ferdowsi University of Mashhad.

Halil, B. (2006). A new host (*Tatianaerhynchites aequatus* (L.) Coleoptera: Rhynchitidae) recorded from *Bracon pectoralis* Wesmael, *Baryscapus britchidi* (Erdos), *Eupelmus urozonus* Dalman and *Exopristud trigonomerus* (Masi) from Turkey. *Journal of the Entomological Research Society, 8*(3), 51−62.

Halperin, J. (1986). Braconidae (Hymenoptera) associated with forest and ornamental trees and shrubs in Israel. *Phytoparasitica, 14*(2), 119−135.

Hashemi Aghajari, M. H., Hasani, M. H., & Tebyani, M. (1995). Some biological and ecological aspects of *Chloridea viviplaca* and *Ch. peltigera* in Maragheh and Hashtrood. In *Proceedings of the 12th Iranian Plant Protection Congress, 2−7 September 1995* (p. 297). Karaj: Karaj Junior College of Agriculture.

Hedqvist, K. J. (1973). A new species of the genus *Bracon* F. From north Sweden (Hym. Ichneumonoidea, Braconidae). *Entomologisk Tidskrift, 94*, 89−90.

Hedwig, von K. (1957). Ichneumoniden und Braconiden aus Iran 1954 (Hymenoptera). *Jahresheft des Vereins für Vaterlaendische Naturkunde Vürttemberg, 112*(1), 103−117.

Hellén, W. (1957). Zur kenntnis der Braconidae: Cyclostomi Finnland. *Notulae Entomologicae, 37*(2), 33−52.

Hussain, M., Askari, A., & Asadi, G. (1976). A study of *Bracon lefroyi* (Hymenoptera: Braconidae) from Iran. *Entomological News, 87*, 299−302.

Iranmanesh, M., Changizi, M., Madjdzadeh, S. M., & Samartsev, K. (2018). A faunistic survey of the Braconinae (Hymenoptera: Braconidae) of Kerman province, Iran. *Journal of Entomological Society of Iran, 38*(2), 235−246.

Jozeyan, A., Vafaei Shoushtari, R., & Askari, H. (2017). The survey of oak wood borer beetles and natural enemeis in the forest of Ilam province. *Iranian Journal of Forest and Range Protection Research, 14*(2), 107−121.

Kaartinen, R., & Quicke, D. L. J. (2007). A revision of the parasitic wasp genus *Bathyaulax* Szépligeti (Hymenoptera: Braconidae: Braconinae) from Africa and the Arabian Peninsula. *Journal of Natural History, 41*(1−4), 125−212.

Kamangar, S., Lotfalizadeh, H., Mohammadi-Khoramabadi, A., & Seyedi-Sahebari, F. (2017). Parasitoids of *Tortrix viridana* L. in Kurdistan province. *Iranian Journal of Forest and Range Protection Research, 15*(2), 176−186 (in Persian, English summary).

Karampour, F., & Fasihi, M. (2004). Collection, identification and study of the natural enemies of *Batrachedra amydraula* Meyr. In *Proceedings of the 16th Iranian Plant Protection Congress, 28 August − 1 September 2004* (p. 49). University of Tabriz.

Khanjani, M. (2006a). *Field crop pests in Iran* (3rd ed.). Bu-Ali Sina University, 719 pp. (in Persian).

Khanjani, M. (2006b). *Vegetable pests in Iran* (2nd ed.). Bu-Ali Sina University, 467 pp. (in Persian).

Khatima, A., & Reza, M. (2015). An overview on pest insect fauna of safflower fields in Iran. *African Journal of Insects, 2*(1), 34−38.

Kiadaliri, H., Ostovan, H., Abaei, M., & Ahangaran, Y. (2005). Investigation on the behaviour treat of leaf feeder moth (*Erannis defoliaria* Clerck) and natural enemies in forests of the in west of Mazandaran province. *Journal of Agricultural Science, 11*(1), 145−159 (in Persian, English summary).

Kirby, W. F. (1903). *The Ichneumons, wasps and bees of Sokotra and Abd-el-Kuri.* London (pp. 235−260).

Kishani Farahani, H., Goldansaz, H., & Sabahi, G. (2008). Study on the larval parasitoids of carob moth, *Ectomyelois ceratoniae* Zeller (Lep.: Pyralidae) in Saveh, Qom, and Varamin. In *Proceedings of the 18th Iranian Plant Protection Congress, 24−27 August 2008* (p. 72). University of Bu-Ali Sina Hamedan.

Kishani Farahani, H., Goldansaz, H., Sabahi, G., & Shakeri, M. (2010). Study on the larval parasitoids of carob moth *Ectomyelois ceratoniae* Zeller (Lepidoptera: Pyralidae) in Varamin, Qom and Saveh. *Iranian Journal of Plant Protection Science, 41*(2), 337−344 (in Persian, English summary).

Kittel, R. N., & Maeto, K. (2019). Revalidation of *Habrobracon brevicornis* stat. rest. (Hymenoptera: Braconidae) based on the CO1, 16S, and 28S gene fragments. *Journal of Economic Entomology, 112*, 906−911.

Kohl, F. F. (1905). Hymenoptera. In A. Penther, & E. Zederbauer (Eds.), *Annalen des Naturhistorischen Museum in Wien: Vol. 20. Ergebnisse einer naturwissenschatlischen Reise zur Erdschias-Dagh (Kleinasien)* (pp. 220−246).

Kohl, F. F. (1907). Zoologische Ergebnisse der Expidition der Keiserlichen Akademie der Wissenschaften nach südarabien und Sokotra in Jahre 1898-1899. *Denkschriften der Akademie der Wissenschaften Mathematisch-naturwissenschaftliche Wien, 71*, 169−301.

Kokujev, N. R. (1907). Sur quelques espèces de Braconides de collections du Musée zoologique de l'Academie imperial des Sciences. *Annales du Musée Zoologique. Academie Imperiale des Sciences, St. Petersburg, 10*(1908), 244−250.

Kugler, J. (1966). A list of parasites of Lepidoptera from Israel. *Israel Journal of Entomology, 1*, 75−88.

Kuslitzky, W., & Argov, Y. (2013). A new record of *Bracon celer* (Hymenoptera: Braconidae), a parasitoid of the olive fruit fly in Israel. *Israel Journal of Entomology, 43*, 91−93.

Lashkari Bod, A., Rakhshani, E., Talebi, A. A., & Lozan, A. (2010). Introduction of twelve newly recorded species of Braconidae (Hymenoptera) from Iran. In *Proceedings of the 19th Iranian Plant Protection Congress, 31 July − 3 August 2010* (p. 161). Iranian Research Institute of Plant Protection.

Lashkari Bod, A., Rakhshani, E., Talebi, A. A., Lozan, A., & Žikič, V. (2011). A contribution to the knowledge of Braconidae (Hym., Ichneumonoidea) of Iran. *Biharean Biologist, 5*(2), 147−150.

Lewis, W. J., Vet, L. E. M., Tumlinson, J. H., van Lateren, J. C., & Papaj, D. R. (1990). Variations in parasitoid foraging behaviour: essential element of a sound biological control theory. *Environmental Entomology, 19*, 1183−1193.

Loni, A., Samartsev, K. G., Scaramozzino, R. L., Belokobylskij, S. A., & Lucchi, A. (2016). Braconinae parasitoids (Hymenoptera, Braconidae) emerged from larvae of *Lobesia botrana* (Denis and Schiffermüller) (Lepidoptera, Tortricidae) feeding on *Daphne gnidium* L. *ZooKeys, 587*, 125−150.

Lotfalizadeh, H., & Gharali, B. (2014). Hymenopterous parasitoids of safflower seed pests in Iran. *Applied Entomology and Phytopathology, 82*, 1−11.

Mahmood, K., Papp, J., & Quicke, D. L. J. (2011). A new Afrotropical genus *Doggerella* gen. nov. of braconine wasp (Hymenoptera: Braconidae) with twelve new species. *Zootaxa, 2927*, 1−37.

Mahmoudi, J., Askarianzadeh, Karimi, A. J., & Abbasipour, H. (2013). Introduction of two parasittoids of braconid wasps on the sugar beet moth, *Scrobipalpa ocellatella* Boyd. (Lep.: Gellechiidae) from Khorasan-e-Razavi province, Iran. *Journal of Sugar Beet, 28*(2), 103−106 (in Persian, English summary).

Maidl, F. (1923). Beiträge zur Hymenopteren fauna Dalmatiens, Montenegros und Albaniens. III. Teil: Braconidae, Aphidiidae und Serphidae by Dr. *Josef Fahringer (Wien) Annalen Bd, 36*.

Mason, W. R. M. (1978). A synopsis of the Nearctic Braconini, with revisions of the Nearctic species of *Coeloides* and *Myosoma* (Hymenopetra: Braconidae). *The Canadian Entomologist, 110*, 721−768.

Mehrnejad, M. R. (2010). The parasitoids of the pistachio fruit hull borer moth, *Arimania komaroffi*. *Journal of Applied Entomology and Phytopathology, 78*(1), 129−130 (in Persian, English summary).

Mehrnejad, M. R. (2020). Arthropod pests of pistachios, their natural enemies and management. *Plant Protection Science, 56*(4), 231−260.

Miller, R. H., El Masri, S., & Al Jundi, K. (1992). Incidence of wheat stem sawflies and their natural enemies on wheat and barley in northern Syria. *Arab Journal of Plant Protection, 10*(1), 25−30.

Minamikawa, J. (1955). On the hosts of three parasitic Hymenoptera in Japan. *Kontyu, 22*, 57−88.

Modarres Awal, M. (1997). Family Braconidae (Hymenoptera), pp. 265-267. In M. Modarres Awal (Ed.), *List of agricultural pests and their natural enemies in Iran* (2nd ed.). Ferdowsi University of Mashhad Press, 429 pp.

Modarres Awal, M. (2012). Family Braconidae (Hymenoptera), pp. 483-486. In M. Modarres Awal (Ed.), *List of agricultural pests and their natural enemies in Iran* (3rd ed.). Ferdowsi University of Mashhad Press, 759 pp.

Mojeni, T. D., Bayat-Asadi, H., Noori Ghanbalani, G., & Shojai, M. (2005). Study on bioregional aspects of bollworm, *Helicoverpa armigera* (Hüb.) (Lepidoptera: Noctuidae), in the cotton fields of Golestan province. *Journal of Agricultural Sciences, 11*(2), 97−115 (in Persian, English summary).

Naderian, H., Ghahari, H., & Asgari, S. (2012). Species diversity of natural enemies in corn fields and surrounding grasslands of Semnan province, Iran. *Calodema, 217*, 1−8.

Najafi Navaei, I., Taghizadeh, M., Havanmoghadam, H., Asadi, A., Osko, T., & Attaean, M. R. (2004). Study on the efficiency of *Trichogramma pintoi* and *Habrobracon hebetor* in corn fields ob Moghan. In *Proceedings of the 3rd National Conference on the Development in the Application of Biological Products and optimum utilization of Chemical fertilizers and Chemical Pesticides in Agriculture* (p. 402).

Nematollahi, M. R., & Bagheri, M. R. (2018). Distribution and population density of safflower pests and their natural enemies in Isfahan province, Iran. *Applied Researches in Plant Protection, 7*(3), 91−101 (in Persian, English summary).

Nobakht, Z., Karimzadeh, J., Shakaram, J., & Jafari, S. (2015). Identification of parasitoids of *Apomyelois ceratoniae* (Zeller) (Lepidoptera, Pyralidae) on pomegranate in Isfahan province. *Journal of Entomology and Zoology Studies, 3*(1), 287−289.

Noori, P. (1994). The parasitism trend of the wasp *Habrobracon hebetor* Say on *Chloridea* spp. in chickpea fields of Kermanshah province. *Applied Entomology and Phytopathology, 61*(1−2), 22−30.

Papp, J. (1966). Faunistic catalogue and ethological data of the *Bracon* Fabr. Species in the Carpathian Basin (Hym., Braconidae)-1. (Cat. Hym. XIX). *Folia Hungarica, 19,* 177−202.

Papp, J. (1967). Faunistic catalogue and ethological data of the *Bracon* Fabr. Species in the Carpathian Basin (Hym., Braconidae)-2. (Cat. Hym. XXI). *Folia Hungarica, 20*(25), 589−603.

Papp, J. (1970). A contribution to the braconid fauna of Israel (Hymenoptera). *Israel Journal of Entomology, 5,* 63−76.

Papp, J. (1974). Zur kenntnis der *Bracon*-arten Österreichs (Hymenoptera, Braconidae). *Annales des Naturhistorischen Museums in Wien, 87,* 415−435.

Papp, J. (1989). A contribution to the braconid fauna of Israel (Hymenoptera), 2. *Israel Journal of Entomology, 22,* 45−59.

Papp, J. (1990). Braconidae (Hymenoptera) from Greece, 3. *Annales Musei Goulandris, 8,* 269−290.

Papp, J. (1991). First outline of the braconid fauna of southern Transdanubia, Hungary (Hymenoptera: Braconidae), 4. Braconinae and Exothecinae. *A Janus Pannonius Múzeum Évkönyve (Pécs), 35*(1990), 71−76.

Papp, J. (1997). Taxonomic revision of seven European species of the genus *Bracon* Fabricius (Hymenoptera: Braconidae). *Folia Entomologica Hungarica, 58,* 115−135.

Papp, J. (1998). Contributions to the braconid fauna of Cyprus (Hymenoptera, Braconidae: Braconinae). *Entomofauna, 19*(14), 241−251.

Papp, J. (1999). *Bracon (Glabrobracon) dilatus* sp. n. from Iran and Iraq with taxonomical remarks on several related species (Hymenoptera: Braconidae, Braconinae). *Folia Entomologica Hungarica, 60,* 269−282.

Papp, J. (2000). First synopsis of the species of *obscurator* species-group, genus *Bracon*, subgenus *Habrobracon* (Hymenoptera: Braconidae, Braconinae). *Annales Historico-Naturales Musei Nationalis Hungarici, 92,* 229−244.

Papp, J. (2001). Taxonomic studies on six genera of Braconini (Insecta: Hymenoptera: Braconidae). *Reichenbachia, 34*(2), 167−174.

Papp, J. (2005). A revision of the bracon (leucobracon) species described by szépligeti from the western Palaearctic Region (Hymenoptera: Braconidae, Braconinae). *Annales Historica Naturale Musei Nationalis Hungarici, 97,* 197−224.

Papp, J. (2008). A revision of the *Bracon* (subgenera *Bracon* s.str., *Cyanopterobracon, Glabrobracon, Lucobracon, Osculobracon* subgen. N., *Pigeria*) species described by Szépligeti from the western palaearctic region (Hymenoptera: Braconidae, Braconinae). *Linzer biologische Beiträge, 40*(1), 1741−1837.

Papp, J. (2009). Contribution to the braconid fauna of the former Yugoslavia, V. Ten subfamilies (Hymenoptera: Braconidae). *Entomofauna, 30*(1), 1−35.

Papp, J. (2012). A contribution to the braconid fauna of Israel (Hymenoptera: Braconidae), 3. *Israel Journal of Entomology, 41−42,* 165−219.

Papp, J. (2012). A revision of the *Bracon* Fabricius species in Wesmael's collection deposited in Brussels (Hymenoptera: Braconidae: Braconinae). *European Journal of Taxonomy, 21,* 1−154.

Papp, J. (2015). A contribution to the braconid fauna of Israel (Hymenoptera: Braconidae) 4. Braconinae. *Entomofauna, 36,* 1−32.

Quicke, D. L. J. (1981). A reclassification of some Oriental and Ethiopean species of Braconinae (Hymenoptera: Braconidae). *Oriental Insects, 14,* 493−498.

Quicke, D. L. J. (1982). Hamuli number in the Braconinae (Hymenoptera: Braconidae), an inter- and intraspecific, size dependent, taxonomic character. *Oriental Insects, 15,* 235−240.

Quicke, D. L. J. (1983a). Some new host records for genera and species off Braconinae (Hym., Braconidae) including economically important species. *Entomologist's Monthly Magazine, 119,* 91−93.

Quicke, D. L. J. (1983b). Reclassification of twenty species of tropical, Old world Braconinae described by Cameron, Strand and Szépligeti (Hymenoptera: Braconidae). *Entomologist's Monthly Magazine, 119,* 81−83.

Quicke, D. L. J. (1983c). The Afrotropical genus *Archibracon* Saussure (Hymenoptera: Braconidae: Braconinae), characteristics and new generic synonymy. *Entomologist's Monthly Magazine, 119,* 147−150.

Quicke, D. L. J. (1983d). Some new hosts records for genera and species of Braconinae (Hym., Braconidae) including economically important species. *Entomol. Month. Magaz., 119,* 91−93.

Quicke, D. L. J. (1985a). Further reclassification of Afrotropical and Indo-Australian Braconinae (Hymenoptera: Braconidae). *Oriental Insects, 18,* 339−353.

Quicke, D. L. J. (1985b). Two new genera of Braconinae (Insecta, Hymenoptera) from the Afrotropical region. *Zoologica Scripta, 14,* 117−122.

Quicke, D. L. J. (1985c). Reclassification of Indo-Australian and Afrotropical Braconinae (Hym., Braconidae). *Entomologist's Monthly Magazine, 121,* 215−216.

Quicke, D. L. J. (1985d). Reclassification of three species of Iranian Braconinae (Hymenoptera) described by Hedwig and Telenga. *Stuttgarter Beiträge zur Naturkunde Serie A (Biologie) Nr, 382,* 1−6.

Quicke, D. L. J. (1987). The Old world genera of braconine wasps (Hymenoptera: Braconidae). *Journal of Natural History, 21,* 43−157.

Quicke, D. L. J., & Koch, F. (1990). Die Braconinae-Typen der Beiden bedeutendsten Hymenopteran sammlingen der DDR (Hymenoptera). *Deutsche Entomologische Zeitschrift, 37,* 213−227.

Quicke, D. L. J., & van Achterberg, C. (1990). Phylogeny of the subfamilies of the family Braconidae. *Zoologica Scripta, 21,* 403−416.

Quicke, D. L. J., Brandt, A. P., & Falco, J. V. (2000). Revision of the Afrotropical species of *Curriea* Ashmead (Hymenoptera: Braconidae: Braconinae): a genus with diverse ovipositor morphology. *African Entomology, 8*(1), 109−139.

Quicke, D. L. J. (2015). *The braconid and ichneumonid parasitoid wasps: Biology, systematics, evolution and ecology.* Hoboken, NJ, USA: John Wiley and Sons, 704 pp.

Quicke, D. L. G., Koch, F., Broad, G. R., Bennett, A. M. R., van Noort, S., Hebert, P. D. N., & Butcher, B. A. (2018). New species of *Rhytimorpha* Szépligeti (Hymenoptera: Braconidae: Braconinae) from Israel. *Zoology in the Middle East, 64*(3), 1−9.

Radjabi, G. (1976). *Xylophagous insects of rosaceous fruit trees in Iran.* Tehran, Iran: Plant pests and diseases research institute, 241 pp. (in Persian).

Radjabi, G. (1991). *Insects attacking Rosaceous fruit trees in Iran* (Vol. I). Tehran: Coleoptera. Plant pests and deases research Institute, 221 pp. (in Persian).

Rahmani, Z., Rakhshani, E., Samartsev, K. G., & Mokhtari, A. (2017). A survey of the genera *Bracon* Fabricius, 1804 and *Habrobracon* Ashmead, 1895 (Hymenoptera, Braconidae, Braconinae) in Iran. *Turkish Journal of Zoology, 41,* 821−840.

Rahmani, Z., Samartsev, K. G., Ghafouri Moghaddam, M., & Rakhshani, E. (2019). Occurrence of an uncommon genus, *Amyosoma* Viereck, 1913 (Hymenoptera, Braconidae, Braconinae) in Iran. *Oriental Insects, 54*(3), 327–334.

Rajabi Mazhar, A., Samartsev, K., Goldasteh, S., & Farahani, S. (2019). Additional evidence and new records of the subfamily Braconinae (Hymenoptera: Braconidae) in Western Iran. *Journal of Insect Biodiversity and Systematics, 5*(2), 87–94.

Rastegar, J., Sakenin, H., Khodaparast, S., & Havaskary, M. (2012). On a collection of Braconidae (Hymenoptera) from East Azarbaijan and vicinity, Iran. *Calodema, 226*, 1–4.

Saeidi, K. (2013). Hymenopterous pupal parasitoids of *Acanthiophilus helianthi* Rossi (Diptera: Tephritidae) in Kohgiloyeh safflower farms. In *Proceedings of the 2nd Global Conference on Entomology, 8-12 November 2013, Kuching, Sarawak, Malaysia* (p. 28).

Saeidi, K. (2015). Pupal hymenopterous parasitoids of *Acanthiophilus helianthi* (Diptera: Tephritidae) in Kohgiluyeh safflower farms. *Plant Protection Journal, 7*, 47–57 (in Persian, English summary).

Sakenin, H., Eslami, B., Samin, N., Imani, S., Shirdel, F., & Havaskary, M. (2008). A contribution to the most important trees and shrubs as the hosts of wood-boring beetles in different regions of Iran and identification of many natural enemies. *Plant and Ecosystem, 16*, 27–46 (in Persian, English abstract).

Sakenin, H., Fischer, M., Samin, N., Imani, S., Papp, J., Ghahari, H., & Rastegar, J. (2011). A faunistic survey on the Braconidae wasps (Hymenoptera: Braconidae) from northern Iran. In *Proceedings of Global Conference on Entomology, Chiang Mai, Thailand* (p. 123).

Sakenin, H., Naderian, H., Samin, N., Rastegar, J., Tabari, M., & Papp, J. (2012). On a collection of Braconidae (Hymenoptera) from northern Iran. *Linzer biologische Beiträge, 44*(2), 1319–1330.

Sakenin, H., Coronado-Blanco, M., Samin, N., & Fischer, M. (2018). New records of Braconidae (Hymenoptera) from Iran. *Far Eastern Entomologist, 362*, 13–16.

Sakenin, H., Samin, N., Beyarslan, A., Coronado-Blanco, J. M., Navaeian, M., Fischer, M., & Hosseini Boldaji, S. A. (2020). A faunistic study on braconid wasps (Hymenoptera: Braconidae) from Iran. *Boletin de la Sociedad Andaluza de Entomologia, 30*, 96–102.

Sakenin, H., Ghahari, H., & Navaeian, M. (2021). A study on the predator and parasitoid insects in some cotton fields of Iran. *Journal of Animal Environment, 13*(1), 397–406 (in Persian, English summary).

Samartsev, K. G. (2016). A new subgenus of the genus *Doggerella* Quicke, Mahmood et Papp, 2011 (Hymenoptera: Braconidae: Braconinae) from Russian Far East. *Euroasian Entomological Journal, 15*(1), 123–128.

Samartsev, K. G., & Ku, D. S. (2021). New records of Braconinae (Hymenoptera, Braconidae) from South Korea. *Journal of Hymenoptera Research, 83*, 21–72.

Samin, N. (2015). A faunistic study on the Braconidae of Iran (Hymenoptera: Ichneumonoidea). *Arquivos Entomoloxicos, 13*, 339–345.

Samin, N., Sakemin, H., Imani, S., & Shojai, M. (2011). A study on the Braconidae (Hymenoptera) of Khorasan province and vicinity, Northeastern Iran. *Phagea, 39*(4), 137–143.

Samin, N., Ghahari, H., Gadallah, N. S., & Davidian, E. (2014). A study on the Braconidae (Hymenoptera: Ichneumonoidea) from West Azarbaijan province, Northern Iran. *Linzer biologische Beiträge, 46*(2), 1447–1478.

Samin, N., Ghahari, H., Gadallah, N. S., & Monaem, R. (2015). A study on the braconid wasps (Hymenoptera: Ichneumonoidea: Braconidae) from Golestan province, northern Iran. *Linzer biologische Beiträge, 47*(1), 731–739.

Samin, N., van Achterberg, C., & Ghahari, H. (2015). A faunistic study of Braconidae (Hymenoptera: Ichneumonoidea) from southern Iran. *Linzer biologische Beiträge, 47*(2), 1801–1809.

Samin, N., Fischer, M., & Ghahari, H. (2015). A contribution to the study on the fauna of Braconidae (Hymenoptera, Ichneumonoidea) from the province of Semnan, Iran. *Arquivos Entomoloxicos, 13*, 429–433.

Samin, N., Coronado-Blanco, J. M., Kavallieratos, N. G., Fischer, M., & Sakenin, H. (2018). Recent findings on Braconidae (Hymenoptera: Ichneumonoidea) of Iran with an updated checklist. *Acta Biologica Turcica, 31*(4), 160–173.

Samin, N., Coronado-Blanco, J. M., Fischer, M., van Achterberg, C., Sakenin, H., & Davidian, E. (2018). Updated checklist of Iranian Braconidae (Hymenoptera: Ichneumonoidea) with twenty-three new records. *Natura Somogyiensis, 32*, 21–36.

Samin, N., Papp, J., & Coronado-Blanco, J. M. (2018). A faunistic study on braconid wasps (Hymenoptera: Ichneumonoidea: Braconidae) of Iran. *Scientific Bulletin of the Uzhgorod University (Series Biology), 45*, 15–19.

Samin, N., Coronado-Blanco, J. M., Hosseini, A., Fischer, M., & Sakenin Chelav, H. (2019). A faunistic study on the braconid wasps (Hymenoptera: Braconidae) of Iran. *Natura Somogyiensis, 33*, 75–80.

Samin, N., Beyarslan, A., Coronado-Blanco, J. M., Navaeian, M., & Kavallieratos, N. G. (2020). A contribution to the braconid wasps (Hymenoptera: Braconidae) from Iran. *Natura Somogyiensis, 35*, 25–28.

Samin, N., Beyarslan, A., Ranjith, A. P., Ahmad, Z., Sakenin Chelav, H., & Hosseini Boldaji, S. A. (2020). A faunistic study on Braconidae (Hymenoptera: Ichneumonoidea) from Ardebil and east Azarbayjan provinces, Northwestern Iran. *Egyptian Journal of Plant Protection Research Institute, 3*(4), 955–963.

Sarhan, A. A., & Quicke, D. L. J. (1989). A new subgenus and species of *Glyptomorpha* (Hymenoptera: Braconidae) from Arabia, Egypt, Pakistan, and Yemen, with a reappraisal of the status of *Teraturus*. *Systematic Entomology, 14*, 403–409.

Schimitschek, E. (1941). Beiträge zur Forestentomologie der Türkei. III. Die Massenvermhrung des *Ips sexdendatus* Börner in Gebiete der orientalischen Fichte. II. Teil. *Zeitschrift für Angewandte Entomologie, 27*, 84–113.

Shahabi, M., & Rajabpour, A. (2015). Activity evaluation of Bracon hebetor Say (Hym.: Braconidae) against Phthorimea operculella Zeller on two potato varieties in north of Khouzestan province, Iran. In *Proceedings of the 1st Iranian International Congress of Entomology* (pp. 192–196).

Shahhosseini, M. J., & Kamali, K. (1989). A checklist of insects, mites and rodents affecting stored products in Iran. *Journal of Entomological Society of Iran, 5*, 1–47.

Sharkey, M. J. (1993). Family Braconidae, pp. 362–394. In H. Goulet, & J. Huber (Eds.), *Hymenoptera of the World: An identification guide to families*. Agriculture Canada Research Branch Monograph, No. 1894E, 668 pp.

Shaw, M. R., & Huddleston, T. (1991). Classification and biology of braconid wasps (Hymenoptera: Braconidae). *Royal Entomological Society of London, 7*(11), 1–125.

Shenefelt, R. D. (1978). Hymenopterum Catalogus (nov. Ed.). Pars 15- Braconidae 10. *Junk, S- Gravenhage, 10*, 1459–1638.

Shestakov, A. (1926). Species palaearcticae novae Braconidarum subfamiliae Braconinarum (Hymenoptera). *Entomologicheskoye Obozreniye, 19*, 208–212.

Shimi, P., Bayat-Asadi, H., Reza Panah, M. R., & Koliaii, R. (1995). A study of *Smycronyx robustus* Faust (Curculionidae) as a biological control agent of eastern dodder (*Cuscuta monogyna* Vahl.) in Iran. *Journal of Agricultural Science, 1*(2), 43−51 (in Persian, English summary).

Shojai, M. (1968). Resultats de l'étude faunestiques des hyménoptères parasites (Terebrants) en Iran et l'importance de leur utilization dans la lutte biologique. In *Proceedings of the 1ˢᵗ Iranian Plant Protection Congress, 14−19 September 1968* (pp. 25−35). Karaj: Karaj Junior College of Agriculture.

Shojai, M. (1998). *Entomology (Ethology, social life and natural enemies) (Biological control)* (3rd ed., Vol. III). Tehran University Publications, 550 pp (in Persian).

Shojai, M., Abbaspour, H., Nasrollahi, A., & Labbafi, Y. (1995). Technology and biocenotic aspects of integrated biocontrol of corn stem borer: *Sesamia cretica* Led. (Lep.: Noctuidae). *Journal of Agricultural Science, 1*(2), 5−32 (in Persian, English summary).

Shojai, M., Nasrollahi, A., Labafi, Y., Azma, M., Amiri, B., Baiat, H., Daniali, M., & Maghsodi, A. (1997). The biocenotic aspects of the Iranian subspecies corn stem borer, *Ostrinia nubilalis persica* (Lepidoptera: Pyraustididae) and its role in the increasing efficiency of *Trichogramma* wasps in IPM program in corn field of northern Iran. *Journal of Agricultural Sciences, 3*, 5−48 (in Persian, English summary).

Shojai, M., Esmaili, M., Ostovan, H., Khodaman, A., Daniali, M., Hosseini, M., Assadi, Y., Sadighfar, M., Korosh-Najad, A., Nasrollahi, A., Labbafi, Y., Azma, M., Ghavam, F., & Honarbakhsh, S. (2000). Integrated pest management of codling moth and other important pests of *Pomoidea* fruit trees. *Journal of Agricultural Sciences, 6*(2), 15−45 (in Persian, English summary).

Shojai, M., Ostovan, H., Hosseini, M., Sadighfar, M., Khodaman, A., Labbafi, Y., Nasrollahi, A., Ghavam, F., & Honarbakhsh, S. (2002). Biocenotic potentials of apple orchards IPM in organic crop production programme. *Journal of Agricultural Sciences, 8*(1), 1−27 (in Persian, English summary).

Siahpoush, A., Azimi, A., Rabee, R., & Mozaffari, M. (1993). Introduction of three species of *Mythimna* (Lepidoptera) at Khuzestan corn fields. In *Proceedings XI Plant protection Congress of Iran* (p. 94).

Sobhani, M., Goldansaz, S. H., & Hatami, B. (2012). Study of larval parasitoids of carob moth *Ectomyelois ceratoniae* (Lep.: Pyralidae) in Kashan region. In *Proceedings of the 20th Iranian Plant Protection Congress, 26−29 August 2012* (p. 83). University of Shiraz.

Sohrabi, F., Lotfalizadeh, H., & Salehipour, H. (2014). Report of two larval parasitoids of *Tuta absoluta* (Meyreck) (Lepidoptera: Gelechidae) from Iran. In *Proceedings of the 21st Iranian Plant Protection Congress, 23−26 August 2014* (p. 726). Urmia University.

Strand, E. (1912). Über exotische Schlupfwespen. *Archiv für Naturgeschichte, (A), 78*(6), 24−75.

Szépligeti, G. (1901). Tropische Cenocolioniden und Braconiden aus der Sammlung des Ungarischen National-Museums. *Természetrajzi Füsetek, 24*, 354−402.

Szépligeti, G. (1904). Hymenoptera, Fam. Braconidae. In P. Wytsman (Ed.), *1902−32. Genera Insectorum* (Vol. 22, pp. 1−253).

Szépligeti, G. (1906). Braconiden aus der Sammlung des ungarischen National-Museums, 1. *Annales Historico-Naturales Musei Nationalis Hungarici, 4*, 547−618.

Tamer, A. (1995). Investigations on natural enemies and biological control possibilities of *Bembecia scopigera* (Scopoli) (Lepidoptera: Sesiidae). *BCPC Symposium Proceedings, 63*, 87−90.

Telenga, N. A. (1933). Einige neue Braconiden-Arten aus USSR (Hymenoptera). *Konowia, 12*(3−4), 242−244.

Telenga, N. A. (1936). Braconidae Pt.1. Hymenoptera fauna USSR. *Hymenoptera, 5*(2), 402 pp.

Thompson, W. R. (1953). *A catalogue of the parasites and predators of insect pests. Section 2. Part 2, Hosts of the Hymenoptera (Agaonidae to Braconidae)*. Ottawa: Commonwealth Institute of Biological Control, 190 pp.

Tobias, V. I. (1961). New subgenera and species of the genus *Bracon* F. (Hymenoptera, Braconidae). *Entomologischeskoe obozrenie, 40*, 659−668.

Tobias, V. I. (1971). Review of the Braconidae (Hymenoptera) of the USSR. *Trudy Vsesoyuznogo Entomologischeskogo Obshchestva, 54*, 156−268.

Tobias, V. I. (1976). Braconids of the Caucasus (Hymenoptera, Braconidae). *Opred. Faune SSSR, 110*, 286 pp. Nauke Press. Leningrad (in Russian).

Tobias, V. I. (1986). Gnaptodontinae, Braconinae, Telengainae, pp. 85−149. In G. S. Medvedev (Ed.), *Opredelitel nasekomyeh Evropeiskoi Tsasti SSSR 3, Peredpontdatokrylye 4. Opr. Faune SSSR. Nauka, Leningrad* (vol. 145, pp. 1−501) (in Russian).

Tobias, V. I. (1995). Subfamily Braconinae, pp. 156-255. In G. S. Medvedev (Ed.), *Hymenoptera, Part 4: Vol. 3. Keys to the insects of the European part of the USSR*. New Delhi: Oxonian Press, 900 pp.

Vaez, N., & Piurgoli, Z. (2016). Functional response of *Habrobracon hebetor* (Say) (Hym.: Braconidae) on *H. armigera, Anagasta kuhniella* and *Galleria mellonella* larvae. In *Proceedings of the 22ⁿᵈ Iranian Plant Protection Congress, 27−30 August 2016, College of Agriculture and Natural Resources* (p. 530). Karaj: University of Tehran.

Valipour, J., Vahedi, H. A., & Zamani, A. A. (2017). An outline on biology and behavior of *Bracon variator* Nees, 1812 (Hym.: Braconidae), an ectoparasitoid of *Cydia johanssoni* Aarvik and Karsholt, 1993 (Lep.: Tortricidae) from Iran. *Biharian Biologist, 11*(1), 15−19.

van Achterberg, C. (1980). Three new palaearctic genera of Braconidae (Hymenoptera). *Entomologische Berichten, Amesterdam,, 40*, 72−80.

van Achterberg, C. (1983). Six new genera of Braconinae from the Afrotropical region (Hymenoptera: Braconidae). *Tijdschrift voor Entomologie, 126*, 175−202.

van Achterberg, C. (1984). Essay on the phylogeny of the Braconidae (Hymenoptera: Ichneumonoidea). *Entomologisk Tidskirift, 105*, 41−58.

van Achterberg, C. (1988). Parallelisms in the Braconidae (Hymenoptera) with special reference to the biology, pp. 85−115. In V. K. Gupta (Ed.), *Advances in Parasitic Hymenoptera Research*. Leiden/New York: E.J. Brill, 546 pp.

van Achterberg, C. (1989). *Pheloura* gen. nov., a neotropical genus with an extremely long pseudo-ovipositor (Hymenoptera: Braconidae). *Entomologische Berichten, Amsterdam, 49*, 105−108.

van Achterberg, C. (1991). The taxonomic position of the genus *Argamania* Papp (Hymenoptera: Braconidae). *Zoologische Mededelingen, 65*(14), 203−207.

van Achterberg, C. (1993). Illustrated key to the subfamilies of the Braconidae (Hymenoptera: Ichneumonoidea). *Zoologische Verhandelingen*, 1−189.

van Achterberg, C., & Polaszek, A. (1996). The parasites of cereal stem borers (Lepidoptera: Cossidea, Crambidae, Noctuidae, Pyralidae) in Africa belonging to the family Braconidae (Hymenoptera: Ichneumonoidea). *Zoologische Verhandelingen, 304*, 1−123.

van Achterberg, C., & Mehrnejad, M. (2011). A new species of *Megalommum* Szépligeti (Hymenoptera, Braconidae, Braconinae), a parasitoid of the pistachio longhorn beetle (*Calchaenesthes pistacivora* Holzschuh, Coleoptera, Cerambycidae) in Iran. *ZooKeys, 112*, 21−38.

Vidal, S. (1993). Determination list of entomophagous insects. Nr. 12. *IOBC-WPRS Bulletin, 16*(3), 1−9.

Volovnik, S. V. (1994). On parasites and predators of Cleoninae weevils (Col. Curculionidae) in Ukrainian steppe. *Anzeiger für Schaedlingskunde Pflanzenschutz Umweltschutz, 67*(4), 77—79.

Walker, F. (1871). *A list of Hymenoptera collected by J.K. Lord Esq. in Egypt, in the neighbourhood of the Red Sea and in Arabia*. London, 59 pp.

Watanabe, C. (1937). On some species of Braconidae from Manchoukuo (Contributions to the knowledge of the Braconid fauna of Manchukuo, 1). *Insecta Matsumarana, 12*(1), 39—44.

Willcocks, F. C. (1914). Note préliminaire sur *Bracon* sp. Insecte parasite du ver de la cottonier (*Earias insulana* Boisd.). *Bulletin de la Société Entomologique d'Egypte, 6*(1913), 56—67.

Willcocks, F. C. (1916). *The insect and related pests of Egypt. The insect and related pests injurious to the cotton plant. PartI. The pink boll worm. Sultanic Agriculture Society, Cairo*, 339 pp.

Yilmaz, T., Aydoğdu, M., & Beyarslan, A. (2010). The distribution of Euphorine wasps (Hymenoptera: Braconidae), with phytogeographical notes. *Turkish Journal of Zoology, 3491*, 181—194.

Yu, D. S., van Achterberg, C., & Horstmann, K. (2012). *Taxapad 2005, Ichneumonoidea 2005, Database on flash-drive*. Nepean, Ontario, Canada.

Yu, D. S., van Achterberg, C., & Horstmann, K. (2016). *Taxapad 2016, Ichneumonoidea 2015, Database on flash-drive*. Nepean, Ontario, Canada.

Zargar, M., Talebi, A. A., Hajiqanbar, H. R., & Papp, J. (2014). First record of *Bracon (Glabrobracon) parvulus* (Hymenoptera: Braconidae: Braconinae) from Iran. In *Proceedings of the 21st Iranian Plant Protection Congress, 23—26 August 2014* (p. 491). Urmia University.

Zargar, M., Talebi, A. A., Hajiqanbar, H. R., Farahani, S., & Ameri, A. (2014). A contribution to the knowledge of Braconinae (Hymenoptera: Braconidae) in some parts of northern and southern Iran. *Journal of Crop Protection, 3*(2), 233—243.

Zargar, M., Talebi, A. A., Hajiqanbar, H. R., & Papp, J. (2015). A study on the genus *Bracon* Fabricius (Hymenoptera: Braconidae) in north central Iran with four new records for Iranian fauna. *Entomofauna, 36*, 425—440.

Zargar, M., Talebi, A. A., & Farahani, S. (2019). Faunistic study of the genus *Habrobracon* Ashmead (Hymenoptera: Braconidae) from Iran. *Journal of Insect Biodiversity and Systematics, 5*(3), 159—169.

Zargar, M.,K., Talebi, A. A., & Farahani, S. (2020). Occurrence of *Ceratobracon stschegolevi* (Telenga, 1936) (Hymenotera: Braconidae: Braconinae), a rare genus and species new for Iran. *Biharean Biologist, 14*(1), 1—4.

Zargar, M., Samartsev, K., Talebi, A. A., & Farahani, S. (2020). Study of the genus *Bracon* Fabricius, 1804 (Hymenoptera: Braconidae) from Iran: new subspecies, new records and an updated checklist. *Zootaxa, 4758*(2), 201—230.

Cardiochiles sp. (Cardiochilinae), ♀, lateral habitus. *Photo prepared by S.R. Shaw.*

Cardiochiles sp. (Cardiochilinae), ♀, lateral habitus. *Photo prepared by S.R. Shaw.*

Chapter 7

Subfamily Cardiochilinae Ashmead, 1900

Hassan Ghahari[1], Ilgoo Kang[2], Neveen Samy Gadallah[3], Michael J. Sharkey[4] and Scott Richard Shaw[5]

[1]Department of Plant Protection, Yadegar-e Imam Khomeini (RAH) Shahre Rey Branch, Islamic Azad University, Tehran, Iran; [2]Department of Entomology, Louisiana State University Agricultural Center, Baton Rouge, LA, United States; [3]Entomology Department, Faculty of Science, Cairo University, Giza, Egypt; [4]UW Insect Museum, Department of Entomology, University of Kentucky, Lexington, KY, United States; [5]UW Insect Museum, Department of Ecosystem Science and Management, University of Wyoming, Laramie, WY, United States

Introduction

Cardiochilinae Ashmead, 1900 is a small cosmopolitan braconid subfamily (Chen et al., 2004; Mercado & Wharton, 2003; Quicke & van Achterberg, 1990; Whitfield & Dangerfield, 1997; Whitfield & Mason, 1994; Yu et al., 2016). Members of the subfamily Cardiochilinae tend to be most species rich in arid and tropical areas (Dangerfield et al., 1999). The subfamily is considered to be a member of the microgastroid complex based on morphological data (Mason, 1981; Quicke & van Achterberg, 1990; Wharton et al., 1992), as well as molecular data (Dowton, 1999; Dowton et al., 1998; Sharanowski et al., 2011). More than 220 species in 17 genera have been described (Yu et al., 2016), with seven species and *Orientocardiochiles* Kang and Long, 2020 just recently described (Dabek et al., 2020; Edmardash et al., 2018; Long et al., 2019; Kang, Long et al., 2020; Kang, Shaw, & Lord, 2020). Fourteen genera have fewer than 15 species, and most of these genera (12 genera) are restricted to one or a few biogeographical regions. The two largest genera *Cardiochiles* Nees, 1819 and *Schoenlandella* Cameron, 1905 are cosmopolitan, and together they comprise about 55% of the total number of described species of Cardiochilinae (Dangerfield et al., 1999; Yu et al., 2016). In the Palaearctic region, 49 species in seven genera have been recorded showing the second highest species-level diversity and the third highest genus-level diversity among the biogeographic realms (Dangerfield et al., 1999; Edmardash et al., 2018; Kang et al., 2021; Yu et al., 2016).

Diagnoses of the Cardiochilinae have been published by van Achterberg (1993), Dangerfield and Austin (1995), Dangerfield et al. (1999); and Mercado and Wharton (2003). The following diagnostic characters are based on study of Cardiochilinae female specimens: body size small to large (3.5–14 mm), usually medium in size, robust, variable in coloration (black; pale orange; pale yellow; and/ or whitish pale); eyes globular or setaceous; clypeus sometimes with two tubercles; mandible usually bidentate; maxillary palpus 5- or 6-segmented; labial palpus 4-segmented; mouth parts (length of galea and glossa) short to extremely elongate; occipital carina usually absent (if present, restricted to malar area); notauli present, usually complete and meeting before scutellar sulcus, smooth to crenulate; scutellar sulcus present, usually with at least one crenula; axilla reduced to a vertical carinate lobe; propodeum usually rugose and with a medium-sized areola; epicnemial carina usually absent; mesopleuron usually smooth, precoxal sulcus usually present; fore wing entirely hyaline or infuscate, or with one or two black bands, (RS+M) always present, 1r usually absent, 3r vein usually absent (spectrally present in most members of *Schoenlandella*), RS smoothly curved or sharply angled; hindwing 2-1A variable; tarsal claw usually pectinate; T1 usually not elongate, with an inverted Y-shaped suture, lateral suture of T1 usually incomplete; T2 usually smooth (a lens-shaped area usually present in *Austerocardiochiles* Dangerfield et al., 1999 and a plateau-like projection present in *Orientocardiochiles*); hypopygium apically sharp or obtuse (members of *Hymenicis* Dangerfield et al., 1999 possess apically membranous hypopygia); hypopygium ventrally usually without median longitudinal suture (present in members of *Austerocardiochiles, Cardiochiles*, and a few members of *Heteropteron* Brullé, 1846 and *Schoenlandella*); and ovipositor short to elongate, straight to sharply downcurved; and ovipositor sheath variable in length and shape, usually setose. All known members of Cardiochilinae are koinobiont endoparasitoids of lepidopteran larvae. The following are known host families: Apatelodidae; Cosmopterigidae; Crambidae; Gelechiidae; Pyralidae; Noctuidae; Tortricidae; Uraniidae (Dangerfield et al., 1999; Huddleston & Walker, 1988). Among the lepidopteran families, some members of Crambidae, Pyralidae, and Noctuidae are serious agricultural pest insects according to Huddleston and Walker (1988) and Shaw and Huddleston (1991). The most well-known cardiochiline species is the biological control agent *Toxoneuron nigriceps* (Viereck, 1912), which mainly utilizes caterpillars of *Heliothis virescens* (F.) (tobacco budworm). Its toxins and polydnavirus have been widely investigated. The members of *T. nigriceps* have been introduced to other continents from the Nearctic region. Based on previous biological investigations, a few other cardiochiline species have biological control potential (Dangerfield & Austin, 1995; Marsh, 1986), for example, *Schoenlandella diaphaniae* (Marsh, 1986) and *Schoenlandella montserratensis* Kang, 2021 (Kang et al., 2021) collected in Montserrat possibly attack *Diaphania hyalinata* (L.) (melonworm moth).

Checklists of Regional Cardiochilinae. Fallahzadeh and Saghaei (2010) listed four species in two genera (without precise localities).

Yu et al. (2016) and Farahani et al. (2016) represented five and seven species, respectively, in two genera. Samin, Coronado-Blanco, Kavallieratos et al. (2018) and Samin, Coronado-Blanco, Fischer et al. (2018) listed five species in two genera, and finally, Gadallah and Ghahari (2019) cataloged eight species in three genera. The present checklist includes 22 species in five genera, *Bohayella* Belokobylskij, *Cardiochiles* Nees, *Pseudocardiochilus* Hedwig, *Retusigaster* Dangerfield, Austin & Whitfield, and *Schoenlandella* Cameron in the Middle East. We follow the classification of Dangerfield et al. (1999), Sharanowski et al. (2011), and van Achterberg et al. (2017).

Key to genera of the subfamily Cardiochilinae in the Middle East (modified from Dangerfield et al., 1999)

1. Ovipositor and sheaths very short, less than 0.2 × as long as hind tibia, stout, strongly curved downwardly; hypopygium hardly seen behind apex of metasoma 2
— Ovipositor and sheaths much longer, 0.3—1.8 × as long as hind tibia, not stout, straight to gently curved downward at apex; hypopygium pointed or truncate at apex, if truncate, then moderately long and visible behind apex of metasoma 4
2. Tarsal claws simple; notauli smooth; apex of hind tibia variously produced, from slight pointed protrusion to long cup-shaped apex .... *Pseudcardiochilus* Hedwig
— Tarsal claws pectinate; notauli variable; hind tibia variously produce, but never with cup-shaped apex 3
3. First metasomal tergum moderately short, less than 2.0× as long as its apical width; clypeus more than 2.5× as wide as long … .... *Retusigaster* Dangerfield, Austin & Whitfield
— First metasomal tergite moderately long, more than 3.0× as longa its apical width; clypeus less than 2.5× as broad as long … .... *Bohayella* Belokobylskij
4. Spectral node of vein 3r of fore wing virtually always present, if apparently absent, then mouth parts elongate and deeply to moderately bilobed at apex; galea mostly long, narrow and blade-like .... *Schoenlandella* Cameron
— Spectral node of vein 3r of fore wing absent; glossa variable; galea otherwise, usually short, if moderately long, then broad .... *Cardiochiles* Nees von Esenbeck

List of species of the subfamily Cardiochilinae recorded in the Middle East

Subfamily Cardiochilinae Ashmead, 1900

Genus *Bohayella* Belokobylskij, 1987

Bohayella temporalis (Fischer, 1958)

Distribution in the Middle East: Egypt (Edmardash et al., 2018; Fischer, 1958).
Extralimital distribution: None.
Host records: Unknown.

Genus *Cardiochiles* Nees von Esenbeck, 1819

Cardiochiles fallax Kokujev, 1895

Catalogs with Iranian records: Farahani et al. (2016), Yu et al. (2016), Samin, Coronado-Blanco, Kavallieratos

et al. (2018), Samin, Coronado-Blanco, Fischer et al. (2018), Gadallah and Ghahari (2019).
Distribution in Iran: Fars (Taghizadeh et al., 2013), Tehran (Farahani et al., 2015).
Distribution in the Middle East: Iran (Farahani et al., 2015; Taghizadeh et al., 2013), Turkey (Inanç, 2002).
Extralimital distribution: Armenia, Azerbaijan, Georgia, Kazakhstan, Romania, Russia, Tajikistan, Ukraine.
Host records: Summarized by Yu et al. (2016) as being a parasitoid of the pyralids *Etiella zinckenella* (Treitschke), and *Sciota marmorata* (Alphérky). In Iran, this species has been reared from *Etiella zinckenella* (Treitschke) (Taghizadeh et al., 2013).

Cardiochiles fumatus Telenga, 1949

Catalogs with Iranian records: Farahani et al. (2016), Gadallah and Ghahari (2019).
Distribution in Iran: Alborz, Tehran (Farahani et al., 2015).

Distribution in the Middle East: Iran.
Extralimital distribution: Afghanistan, Azerbaijan, Kazakhstan, Tajikistan, Uzbekistan.
Host records: Unknown.

Cardiochiles kasachstanicus Tobias and Alexeev, 1977

Distribution in the Middle East: Turkey (Inanç, 2002).
Extralimital distribution: Kazakhstan, Ukraine.
Host records: Unknown.

Cardiochiles priesneri Fischer, 1958

Distribution in the Middle East: Egypt (Edmardash et al., 2018; Fischer, 1958).
Extralimital distribution: None.
Host records: Unknown.

Cardiochiles pseudofallax Telenga, 1955

Catalogs with Iranian records: No catalog.
Distribution in Iran: East Azarbaijan (Ameri et al., 2020), Khuzestan (Ameri et al., 2019).
Distribution in the Middle East: Egypt (Edmardash et al., 2018), Iran (Ameri et al., 2019, 2020), Turkey (Inanç & Beyarslan, 1994).
Extralimital distribution: Afghanistan, Azerbaijan, Turkmenistan.
Host records: Unknown.

Cardiochiles saltator (Fabricius, 1781)

Catalogs with Iranian records: Fallahzadeh and Saghaei (2010), Farahani et al. (2016), Yu et al. (2016), Samin, Coronado-Blanco, Kavallieratos et al. (2018), Samin, Coronado-Blanco, Fischer et al. (2018), Gadallah and Ghahari (2019).
Distribution in Iran: Golestan (Telenga, 1955, as *Cardiochiles brachialis* Rondani (under the subfamily Microgastrinae)), Isfahan (Ghahari et al., 2011), Iran (no specific locality cited) (Belokobylskij, 1998; Shenefelt, 1973, as *C. brachialis*).
Distribution in the Middle East: Iran (see references above), Israel−Palestine (Papp, 2012), Turkey (Čapek & Hofmann, 1997; Fahringer, 1922; Inanç, 2002; Inanç & Beyarslan, 1994).
Extralimital distribution: Albania, Austria, Belgium, Bulgaria, Croatia, Czech Republic, Finland, France, Germany, Greece, Hungary, India, Italy, Kazakhstan, Korea, Latvia, North Macedonia, Moldova, Morocco, Netherlands, Poland, Romania, Russia, Serbia, Spain, Sweden, Switzerland, Turkmenistan, Ukraine, United Kingdom.
Host records: Recorded by Čapek and Hofmann (1997) as a parasitoid of the crambid *Loxostege sticticalis* (L.); and the pyralid, *Etiella zinketiella* (Treitschke).

Cardiochiles shestakovi Telenga, 1949

Catalogs with Iranian records: Farahani et al. (2016), Gadallah and Ghahari (2019).
Distribution in Iran: Tehran (Farahani et al., 2015), Yazd (Goldansaz et al., 1996; Shamszadeh, 1998).
Distribution in the Middle East: Iran (see references above).
Extralimital distribution: Tajikistan, Turkmenistan, Uzbekistan.
Host records: In Iran, this species has been reared from the pyralids *Achroia grisella* (Fabricius) (Goldansaz et al., 1996), and *Epischidia caesariella* (Hampson) (Shamszadeh, 1998).

Cardiochiles tibialis Hedwig, 1957

Catalogs with Iranian records: Fallahzadeh and Saghaei (2010), Farahani et al. (2016), Yu et al. (2016), Samin, Coronado-Blanco, Kavallieratos et al. (2018), Samin, Coronado-Blanco, Fischer et al. (2018), Gadallah and Ghahari (2019).
Distribution in Iran: Sistan and Baluchestan (Hedwig, 1957).
Distribution in the Middle East: Iran.
Extralimital distribution: None.
Host records: Unknown.

Cardiochiles triplus Shenefelt, 1973

Catalogs with Iran records: Farahani et al. (2016), Yu et al. (2016), Samin, Coronado-Blanco, Kavallieratos et al. (2018), Samin, Coronado-Blanco, Fischer et al. (2018), Gadallah and Ghahari (2019).
Distribution in Iran: Sistan and Baluchestan (Hedwig, 1957, under *Cardiochiles minutus* Hedwig, 1957), Iran (no specific locality cited) (Shenefelt, 1973, under the subfamily Microgastrinae).
Distribution in the Middle East: Iran.
Extralimital distribution: None.
Host records: Unknown.

Cardiochiles weidholzi Fischer, 1958

Distribution in the Middle East: Egypt (Edmardash et al., 2018; Fischer, 1958).
Extralimital distribution: None.
Host records: Unknown.

Genus Pseudcardiochilus Hedwig, 1957
Pseudcardiochilus abnormipes Hedwig, 1957

Catalogs with Iranian records: Fallahzadeh and Saghaei (2010), Farahani et al. (2016), Yu et al. (2016), Samin, Coronado-Blanco, Kavallieratos et al. (2018), Samin, Coronado-Blanco, Fischer et al. (2018), Gadallah and Ghahari (2019).
Distribution in Iran: Sistan and Baluchestan (Hedwig, 1957; van Achterberg, 1980).

Distribution in the Middle East: Iran.
Extralimital distribution: None.
Host records: Unknown.

Genus *Retusigaster* Dangerfield, Austin & Whitfield, 1999

Retusigaster eremita (Kokujev, 1904)

Distribution in the Middle East: Turkey (Erdoğan, 2015; Telenga, 1955; Tobias, 1995).
Extralimital distribution: Kazakhstan, Mongolia, Turkmenistan.
Host records: Unknown.

Genus *Schoenlandella* Cameron, 1905

According to Mercado and Wharton (2003) and Papp (2014), the taxonomic status of *Schoenlandella* is uncertain. Kang and Sharkey (Pers. Comm. in 2021) indicated the possibility of synonymization based on the phylogeny by Murphy et al. (2008). However, we tentatively treat *Schoenlandella* as a valid genus in this chapter following Dangerfield et al. (1999), Edmardash et al., 2018; and Gadallah and Ghahari (2019). The taxonomic status of the genus will be tested based on molecular data (Kang et al., in prep).

Schoenlandella acrenulata (Fischer, 1958)

Distribution in the Middle East: Egypt (Edmardash et al., 2018; Fischer, 1958 as *Cardiochiles acrenulatus*).
Extralimital distribution: None.
Host records: Unknown.

Schoenlandella angustigena Kang, 2021

Catalogs with Iranian records: No catalog.
Distribution in Iran: Hormozgan (Kang et al., 2021).
Distribution in the Middle East: Iran.
Extralimital distribution: None.
Host records: Unknown.

Schoenlandella deserta (Telenga, 1955)

Catalogs with Iranian records: Gadallah and Ghahari (2019).
Distribution in Iran: Guilan (Gadallah & Ghahari, 2019).
Distribution in the Middle East: Egypt (Edmardash et al., 2018), Iran (Gadallah & Ghahari, 2019).
Extralimital distribution: Azerbaijan, Morocco, Spain, Turkmenistan, Uzbekistan (Yu et al., 2016 as *Cardiochiles (Schoenlandella) desertus* Telenga, 1955).
Host records: Unknown.

Schoenlandella glaber (Fischer, 1958)

Distribution in the Middle East: Egypt (Fischer, 1958 as *Cardiochiles glaber*; Edmardash et al., 2018).
Extralimital distribution: None.
Host records: Unknown.

Schoenlandella latigena Kang, 2021

Catalogs with Iranian records: No catalog.
Distribution in Iran: Hormozgan (Kang et al., 2021).
Distribution in the Middle East: Iran.
Extralimital distribution: None.
Host records: Unknown.

Schoenlandella maculata (Fischer, 1958)

Distribution in the Middle East: Egypt (Edmardash et al., 2018; Fischer, 1958 as *Cardiochiles maculatus*).
Extralimital distribution: None.
Host records: Unknown.

Schoenlandella obscuriceps (Fischer, 1958)

Distribution in the Middle East: Egypt (Edmardash et al., 2018; Fischer, 1958 as *Cardiochiles obscuriceps*).
Extralimital distribution: None.
Host records: Unknown.

Schoenlandella pseudoglabra Edmardash, Gadallah and Sharkey, 2018

Distribution in the Middle East: Egypt (Edmardash et al., 2018).
Extralimital distribution: None.
Host records: Unknown.

Scheonlandella variegata (Szépligeti, 1913)

Distribution in the Middle East: Egypt (Papp, 2014 as *Cardiochiles variegatus*; Edmardash et al., 2018).
Extralimital distribution: Ethiopia, Gambia, Niger, Nigeria, Senegal, Tanzania.
Host records: Summarized by Yu et al. (2016) as being a parasitoid of the noctuids *Helicoverpa armigera* (Hübner), and *Heliocheilus albipunctella* (de Joannis).

Conclusion

Twenty-two valid species of the subfamily Cardiochilinae in five genera have been reported from four of the Middle East countries (Egypt, Iran, Israel—Palestine, and Turkey), of which Egypt and Iran, each with 11 species, are the richest. The known Cardiochilinae of Iran (with 11 species) represents 5% of the world species in three genera *Cardiochiles* (seven species), *Pseudcardiochilus* (one species), and *Schoenlandella* (three species). Five species, *Cardiochiles tibialis* Hedwig, 1957, *Cardiochiles triplus* Shenefelt, 1973, *Pseudcardiochilus abnormipes* Hedwig, 1957, *Schoenlandella angustigena* Kang, 2021, and *Schoenlandella latigena* Kang, 2021, are so far only known from Iran (endemic to Iran). Iranian Cardiochilinae have been recorded from 11 provinces, Sistan and Baluchestan, and Tehran (both with three species), Hormozgan (two species), Alborz, East Azarbaijan, Fars, Golestan, Guilan, Isfahan, Khuzestan, and Yazd (each with one species).

Regarding the host information of the Iranian cardiochiline wasps, three pyralid species (Lepidoptera) are known as the hosts of two cardiochiline species. Caterpillars of *Etiella zinckenella* (Treitschke) are attacked by *Cardiochiles fallax*. Individuals of *Cardiochiles shestakovi* utilize caterpillars of *Achroia grisella* (Fabricus) and *Epischedia caesariella* (Hampson) as their hosts (Yu et al., 2016). Comparison of the Cardiochilinae fauna of Iran with the Middle East and adjacent countries to Iran indicates the fauna of Turkmenistan (20 species) is more diverse than Iran, followed by Egypt (11 species), Kazakhstan (10 species), Russia (seven species), Turkey (five species), Azerbaijan (four species), Afghanistan, Pakistan (each with two species), and Armenia and Israel—Palestine (each with one species). No cardiochiline species have been recorded from Bahrain, Cyprus, Iraq, Jordan, Kuwait, Lebanon, Oman, Qatar, Saudi Arabia, Syria, United Arab Emirates, and Yemen. Additionally, among the adjacent countries, Azerbaijan and Turkmenistan share the highest number of species with Iran (both with four species), followed by Kazakhstan and Turkey (both with three species), Afghanistan, Egypt, Russia (each with two species), and Armenia and Israel—Palestine (both with one species). *Bohayella* species have been found in Egypt and India, and *Asiacardiochiles* has been collected from Oman, so it seems possible that species of either or both genera might eventually be discovered from Iran.

References

Ameri, A., Ebrahimi, E., Talebi, A. A., & Beyarslan, A. (2019). First record of *Cardiochiles pseudofallax* Telenga, 1955 (Braconidae: Cardiochilinae) from Iran. In *Proceedings of the 3rd International Iranian Congress of Entomology, 17—19 August 2019* (p. 145). University of Tabriz.

Ameri, A., Ebrahimi, E., & Talebi, A. A. (2020). Additions to the fauna of Braconidae (Hym., Ichneumonoidea) of Iran based on specimens housed in Hayk Mirzayans insect Museum with six new records for Iran. *Journal of Insect Biodiversity and Systematics, 6*(4), 353—364.

Belokobylskij, S. A. (1998). Braconidae: Cardiochilinae. In P. A. Ler (Ed.), *Key to the insects of Russian far east* (pp. 543—546). Vladivostok: Dal'nauka, 706 pp. (in Russian).

Čapek, M., & Hofmann, C. (1997). The Braconidae (Hymenoptera) in the collections of the Musée contonal Zoologie, Laussane. *Litterae Zoologicae, 2*, 25—162.

Chen, X., Whitfield, J. B., & He, J. (2004). Revision of the subfamily Cardiochilinae (Hymenoptera: Braconidae) on China. I. The genera *Austerocardiochiles* Dangerfield, Austin and Whitfield and *Psilomiscus* Enderlein. *Proceedings of the Entomological Society of Washington, 106*(1), 33—51.

Dabek, E. Z., Whitfield, J. B., Hallwachs, W., & Janzen, D. H. (2020). Two new reared species of *Heteropteron* Brullé (Hymenoptera, Braconidae, Cardiochilinae) from northwest Costa Rica, with the first definitive host records for the genus. *Journal of Hymenoptera Research, 77*, 151—165.

Dangerfield, P. C., & Austin, A. D. (1995). Revision of the Australian species of Cardiochilinae (Hymenoptera: Braconidae). *Invertebrate Taxonomy, 9*(3), 387—445.

Dangerfield, P. C., Austin, A. D., & Whitfield, J. B. (1999). Systematics of the world genera of Cardiochilinae. *Invertebrate Taxonomy, 13*, 917—976.

Dowton, M. (1999). Relationships among the cyclostome braconid (Hymenoptera: Braconidae) subfamilies inferred from a mitochondrial tRNA gene rearrangement. *Molecular Phylogenetics and Evolution, 11*(2), 283—287.

Dowton, M., Austin, A. D., & Antolin, M. F. (1998). Evolutionary relationships among Braconidae (Hymenoptera: Ichneumonoidea) inferred from partial 16S rDNA gene sequences. *Insect Molecular Biology, 7*, 129—150.

Edmardash, Y. A., Gadallah, N. S., & Sharkey, M. J. (2018). Revision of the subfamily Cardiochilinae Ashmead, 1900 (Hymenoptera: Braconidae) in Egypt, with new records and a new species. *Journal of Natural History, 52*(5—6), 269—297.

Erdoğan, Ö. C. (2015). First record of the genus *Retusigaster* Dangerfield, Austin & Whitfield, 1999 (Hymenoptera: Braconidae: Cardiochilinae) from the west Palaearctic region. *Biharean Biologist, 9*(2), 160—161.

Fahringer, J. (1922). Hymenopterlogische Ergebnisse einer Wissenschaftlichen Studienreise nach der Türkei und Kleinasien (mit Aussschluβ des Amanusgebirges). *Archiv für Naturgeschichte, A88*(9), 149—222.

Fallahzadeh, M., & Saghaei, N. (2010). Checklist of Braconidae (Insecta: Hymenoptera) from Iran. *Munis Entomology & Zoology, 5*(1), 170—186.

Farahani, S., Talebi, A. A., & Rakhshani, E. (2015). Study of the genus *Cardiochiles* (Hymenoptera: Braconidae: Cardiochilinae) in Tehran and Alborz provinces, with one new record from Iran. In *Proceedings of the 1st Iranian International Congress of Entomology, 29—31 August 2015* (p. 199). Iranian Research Institute of Plant Protection.

Farahani, S., Talebi, A. A., & Rakhshani, E. (2016). Iranian Braconidae (Insecta: Hymenoptera: Ichneumonoidea): diversity, distribution and host association. *Journal of Insect Biodiversity and Systematics, 2*(1), 1—92.

Fischer, M. (1958). Neue *Cardiochiles*-Arten aus Aegypten. *Polskie Pismo Entomologiczne, 28*, 13—33.

Gadallah, N. S., & Ghahari, H. (2019). An updated checklist of Iranian Cardiochilinae, Ryssalinae and Blacini (Hymenoptera: Ichneumonoidea: Braconidae). *Oriental Insects, 54*(2), 143—161.

Ghahari, H., Fischer, M., & Papp, J. (2011). A study on the braconid wasps (Hymenoptera: Braconidae) from Isfahan province, Iran. *Entomofauna, 32*, 261—272.

Goldansaz, S. H., Esmaili, M., & Ebadi, R. (1996). Lesser wax moth, *Achroia grisella* F., and its parasitic wasps. In *Proceedings of XX International Congress of Entomology, 25—31 August 1996, Firenze, Italy* (p. 663).

Hedwig, K. (1957). Ichneumoniden und Braconiden aus den Iran 1954 (Hymenoptera). *Jahresheft des Vereins für Vaterlaendische Naturkunde, 112*(1), 103—117.

Huddleston, T., & Walker, A. K. (1988). *Cardiochiles* (Hymenoptera: Braconidae), a parasitoid of lepidopterous larvae, in the Sahel of Africa, with a review of the biology and host relationships of the genus. *Bulletin of Entomological Research, 78*, 435—461.

Inanç, F. (2002). Untersuchungen über Cardiochilinae Fauna der Türkei (Hymenoptera, Braconidae). *Entomofauna, 23*(10), 121—124.

Inanç, F., & Beyarslan, A. (1994). New records Cardiochilinae (Hymenoptera, Braconidae) fauna of Turkey. *Ulusal Biyoloji Kongresi, XII*, 243—247.

Kang, I., Long, K. D., Sharkey, M. J., Whitfield, J. B., & Lord, N. P. (2020). *Orientocardiochiles*, a new genus of Cardiochilinae (Hymenoptera: Braconidae), with descriptions of two new species from Malaysia and Vietnam. *ZooKeys, 971*, 1−15.

Kang, I., Shaw, S. R., & Lord, N. (2020). Two new species and distribution records for the genus *Bohayella* Belokobylskij, 1987 from Costa Rica (Hymenoptera: Braconidae: Cardiochilinae). *ZooKeys, 996*, 93−105.

Kang, I., Ameri, A., & Sharkey, M. J. (2021). Revision of the Iranian *Schoenlandella* Cameron, 1905 (Hymenoptera, Braconidae, Cardiochilinae) with descriptions of two new species from Hormozgan province. *Deutsche Entomologische Zeitschrift, 68*(2), 262−268.

Kang, I., Sharkey, M. J., & Diaz, R. (2021). Revision of the genus *Schoenlandella* (Hymenoptera, Braconidae, Cardiochilinae) in the New World, with a potential biological control agent for a lepidopteran pest of bitter gourd (*Momordica charantia* L.). *Journal of Hymenoptera Research, 86*, 47−61.

Long, K. D., Oanh, N. T., van Dzuong, N., & Hoa, D. T. (2019). Two new species of the genus *Austerocardiochiles* Dangerfield, Austin and Whitfield, 1999 (Hymenoptera: Braconidae: Cardiochilinae) from Vietnam. *Zootaxa, 4657*(3), 587−595.

Marsh, P. M. (1986). A new species of *Cardiochiles* (Hymenoptera: Braconidae) introduced into Florida to control *Diaphania* spp. (Lepidoptera: Pyralidae). *Proceedings of the Entomological Society of Washington, 88*(1), 131−133.

Mason, W. R. M. (1981). The phylogenetic nature of *Apanteles* Foerster (Hymenoptera: Braconidae): Phylogeny and reclassification of Microgastrinae. *Memoirs of the Entomological Society of Canada, 115*, 1−147.

Mercado, I., & Wharton, R. A. (2003). Mexican Cardiochilinae genera (Hymenoptera: Braconidae), including a preliminary assessment of species-groups in *Toxoneuron* Say and *Retusigaster* Dangerfield, Austin and Whitfield. *Journal of Natural History, 37*, 845−902.

Murphy, N., Banks, J. C., Whitfield, J. B., & Austin, A. D. (2008). Phylogeny of the parasitic microgastroid subfamilies (Hymenoptera: Braconidae) based on sequence data from seven genes, with an improved time estimate of the origin of the lineage. *Molecular Phylogenetics and Evolution, 47*(1), 378−395.

Papp, J. (2012). A contribution to the Braconid fauna of Israel (Hymenoptera: Braconidae), 3. *Israel Journal of Entomology, 41−42*, 165−219.

Papp, J. (2014). A revisional study on Szépligeti's cardiochiline type specimens deposited in the Hungarian Natural History Museum, Budapest (Hymenoptera, Braconidae: Cardiochilinae). *Annales Historico-Naturales Musei Nationalis Hungarici, 106*, 169−214.

Quicke, D. L. J., & van Achterberg, C. (1990). Phylogeny of the subfamilies of the family Braconidae (Hymenoptera: Ichneumonoidea). *Zoologische Verhandelingen, 258*, 1−180.

Samin, N., Coronado-Blanco, J. M., Kavallieratos, N. G., Fischer, M., & Sakenin, H. (2018). Recent findings on Braconidae (Hymenoptera: Ichneumonoidea) of Iran with an updated checklist. *Acta Biologica Turcica, 31*(4), 160−173.

Samin, N., Coronado-Blanco, J. M., Fischer, M., van Achterberg, C., Sakenin, H., & Davidian, E. (2018). Updated checklist of Iranian Braconidae (Hymenoptera: Ichneumonoidea) with twenty-three new records. *Natura Somogyiensis, 32*, 21−36.

Shamszadeh, M. (1998). Introducing of *Cardiochiles shestakovi* (Hym.: Braconidae) parasitoid of *Proceratia caesariella* larva in Yazd province. *Pajohesh and Sazandegi, 39*, 47−49 (in Persian, English summary).

Sharanowski, B. J., Dowling, A. P. G., & Sharkey, M. J. (2011). Molecular phylogenetics of Braconidae (Hymenoptera: Ichneumonoidea), based on multiple nuclear genes, and implications for classification. *Systematic Entomology, 36*, 549−571.

Shaw, M. R., & Huddleston, T. (1991). Classification and biology of braconid wasps (Hymenoptera: Braconidae). *Handbook for identification of British Insects, 7*(11), 1−126.

Shenefelt, R. D. (1973). *Braconidae 5. Microgasterinae and Ichneutinae. Hymenopterorum Catalogus (nova editio). Pars 9* (pp. 669−812).

Taghizadeh, R., Talebi, A. A., Fathipour, Y., & Lozan, A. I. (2013). *Cardiochiles fallax* (Hym.: Braconidae), a new species record for Iran. *Journal of Entomological Society of Iran, 33*(2), 83−85.

Telenga, N. A. (1955). *Braconidae, subfamily Microgasterinae, subfamily Agathinae. Fauna USSR, Hymenoptera* (p. 311). The Smithsonian Institution and The National Science Foundation Publishing.

Tobias, V. I. (1995). *Keys to insects of the European part of the USSR.* Volume III. Hymenoptera, Part IV. Braconidae. New Delhi: chief. Amerind Publishing Co.

van Achterberg, C. (1980). Notes on some species of Braconidae (Hymenoptera) described by Hedwig from Iran and Afghanistan. *Entomologische Berichten, Amsterdam, 40*, 25−31.

van Achterberg, C. (1993). Illustrated key to the subfamilies of the Braconidae (Hymenoptera: Ichneumonoidea). *Zoologische Verhandelingen, 238*, 1−189.

van Achterberg, C., Taeger, A., Blank, S. M., Zwakhals, K., Viitassari, M., Yu, D. S. K., & de Jong, Y. (2017). Fauna Europaea: Hymenoptera - Symphyta and Ichneumonoidea. *Biodiversity Data Journal, 5*, e14650.

Wharton, R. A., Shaw, S. R., Sharkey, M. J., Whal, D. B., Wooley, J. B., Marsh, P. M., & Johnson, W. (1992). Phylogeny of the subfamilies of the family Braconidae (Hymenoptera: Ichneumonoidea): A reassessment. *Cladistics, 8*(3), 199−235.

Whitfield, J. B., & Mason, W. R. M. (1994). Mendesellinae, a new subfamily of braconid wasps (Hymenoptera, Braconidae) with a review of relationships within the microgastroid assemblage. *Systematic Entomology, 19*(1), 61−76.

Whitfield, J. B., & Dangerfield, P. C. (1997). Subfamily Cardiochilinae. In R. A. Wharton, P. M. Marsh, & M. J. Sharkey (Eds.), *Manual of the New World Genera of the Family Braconidae (Hymenoptera)* (Vol. 1, pp. 177−183). International Society of Hymenopterists. Special Publication, No., 439 pp.

Yu, D. S., van Achterberg, C., & Horstmann, K. (2016). *Taxapad 2016, Ichneumonoidea 2015, Database on flash-drive.* Nepean, Ontario, Canada.

Cenocoelius analis (Nees, 1834) (Cenocoeliinae), ♀, lateral habitus. *Photo prepared by S.R. Shaw.*

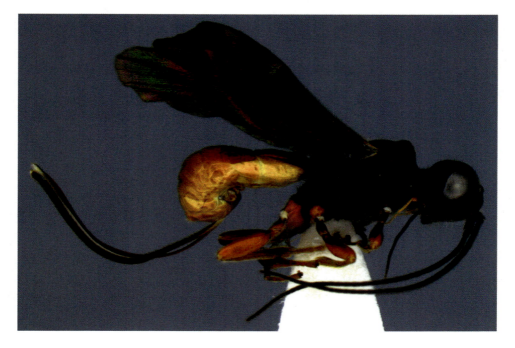

Cenocoelius sp. (Cenocoeliinae), ♀, lateral habitus. *Photo prepared by S.R. Shaw.*

Chapter 8

Subfamily Cenocoeliinae Szépligeti, 1901

Scott Richard Shaw[1], Neveen Samy Gadallah[2], Hassan Ghahari[3] and Michael J. Sharkey[4]

[1]UW Insect Museum, Department of Ecosystem Science and Management, University of Wyoming, Laramie, WY, United States; [2]Entomology Department, Faculty of Science, Cairo University, Giza, Egypt; [3]Department of Plant Protection, Yadegar-e Imam Khomeini (RAH) Shahre Rey Branch, Islamic Azad University, Tehran, Iran; [4]UW Insect Museum, Department of Entomology, University of Kentucky, Lexington, KY, United States

Introduction

The subfamily Cenocoeliinae is a rather small subfamily (Chen & van Achterberg, 2019; van Achterberg, 1984), comprising 91 species classified into six genera and one tribe (Cenocoeliini) worldwide (Yu et al., 2016). They are mainly encountered in tropical and subtropical areas and are particularly well represented in the New World (S. Shaw, 1995; S. Shaw & Huddleston, 1991; van der Ent & S. Shaw, 1998, 1999). Species of the subfamily Cenocoeliinae are specialized solitary koinobiont endoparasitoids of saproxylic beetle larvae most commonly on members of the families Cerambycidae and sometimes Curculionidae (Scolytinae) or Buprestidae (Saffer, 1982; van Achterberg, 1984). The hosts are killed after the parasitoid has prepared for pupation and completed their feeding externally (M. Shaw & Huddleston, 1991). A rather broad host range was described by Saffer (1982), in which some cenocoeliine species were reported to parasitize phytophagous beetle larvae of several families such as Cerambycidae, Buprestidae, Scolytidae, and Curculionidae within herbaceous stems or fruits. Van der Ent and S. Shaw (1998) studied the diversity of cenocoeliines along latitudinal and altitudinal gradients and found 57 morphospecies (mostly undescribed species of *Capitonius* and *Cenocoelius*) mostly occurring at low altitudes (below 500 m el.) and never encountered above 1600 m in Costa Rica. The fauna of Costa Rican Cenocoeliine is a least five times more speciose than that of Canada and the United States combined. While most temperate region Cenocoeliinae are associated with wood—boring and bark-boring beetle larvae in dead or dying trees, van der Ent and S. Shaw (1999) described a *Capitonius* species from Costa Rica that attacks cerambycid larvae in stem galls and leaf petioles of epiphytic Solanaceae species. Given the broad diversity of tropical lowland species, and noting that few cenocoeliine occur at high elevations where potential hosts such as cerambycids and scolytine curculionids are diverse, they hypothesized that neotropical Cenocoeliinae will be found to have a much wider variety of host habitat adaptations than their temperate zone counterparts (van der Ent & S. Shaw, 1998, 1999). Neotropical cenocoeliines are also characterized by commonly exhibiting orange and contrasting black aposematic color patterns (Mora & Hanson, 2019; van der Ent & S. Shaw, 1998, 1999).

Because of their resemblance in biology and larval structures to the Helconinae, the cenocoeliines were included in the subfamily Helconinae by Čapek (1970). However, cenocoeliines were later reclassified as a separate subfamily because of the unique high insertion of the metasoma on the propodeum far above hind coxa (M. Shaw & Huddleston, 1991; van Achterberg, 1984, 1993; Wharton et al., 1992). The only other hymenopterans to display a similar characteristic are the Evanioidea (ensign wasps and relatives), as well as a few doryctine braconid genera. Cenocoeliine species are diagnosed by the absence of a transverse scutellar depression; the more or less developed postpectal carina; and the metasoma without carapace, inserted high on the propodeum (Chen & van Achterberg, 2019; van Achterberg, 1984, 1993). Molecular studies carried out by Sharanowski et al. (2011), suggest a basal position of cenocoeliines within the euphoroid complex, which accords with earlier morphological assessments by Wharton et al. (1992).

Checklists of Regional Cenocoeliinae. Samin et al. (2018) recorded two species of Cenocoeliinae from Iran for the first time. The present checklist comprises two species, *Cenocoelius analis* (Nees, 1834) and *Lestricus secalis* (Linnaeus, 1758) in the Middle East. Here we follow Yu et al. (2016) for the global distribution of species.

Braconidae of the Middle East (Hymenoptera). https://doi.org/10.1016/B978-0-323-96099-1.00008-X

Key to genera of the subfamily Cenocoeliinae of the Middle East (modified from van Achterberg, 1994)

1. Second metasomal tergite entirely smooth; propleuron weakly convex to slightly concave, without transverse ridge .... *Lestricus* Reinhard
— Second metasomal tergite sculptured basally, rarely entirely smooth; propleuron with strong scaly, transverse ridge, with rather concave area behind .... *Cenocoelius* Westwood

List of species of the subfamily Cenocoeliinae recorded in the Middle East

Subfamily Cenocoeliinae Szépligeti, 1901

Genus *Cenocoelius* Westwood, 1840

Cenocoelius analis (Nees von Esenbeck, 1834)

Catalogs with Iranian records: Samin et al. (2018).
Distribution in Iran: Isfahan (Samin et al., 2018).
Distribution in the Middle East: Iran.
Extralimital distribution: Azerbaijan, Belgium, Bulgaria, Czech Republic, Finland, France, Germany, Hungary, Lithuania, Moldova, Netherland, Poland, Romania, Russia, Slovakia, Sweden, United Kingdom.
Host records: Summarized by Yu et al. (2016) as being a parasitoid of the buprestid *Agrilus convexicollis* Redtenbacher; the following cerambycids: *Molorchus umbellatarum* (Schreber), *Pogonocherus fasciculatus* (DeGeer), *Pogonocherus hispidulus* (L.), *Tetrops praeustus* (L.), and *Tetrops starki* Chevrolat; and the following curculionids *Magdalis armigera* (Geoffroy), *Magdalis ruficornis* (L.), and *Scolytus rugulosus* (Müller).

Genus *Lestricus* Reinhard, 1865

Lestricus secalis (Linnaeus, 1758)

Catalogs with Iranian records: Samin et al. (2018).
Distribution in Iran: Kordestan (Samin et al., 2018).
Distribution in the Middle East: Iran.
Extralimital distribution: Bulgaria, Czech Republic, Finland, France, Germany, Hungary, Italy, Lithuania, Netherlands, Norway, Poland, Romania, Russia, Spain, Sweden, Switzerland, United Kingdom.
Host records: Summarized by Yu et al. (2016) as being a parasitoid of the following cerambycids: *Brachyclytus singularis* Kraatz, *Leiopus nebulosus* (L.), *Monochamus sutor* (L.), *Pogonocherus decoratus* Fairnaire, *Pogonocherus euginiae* Ganglbauer, *Pogonocherus fasciculatus* (DeGeer), *Pogonocherus hispidulus* (Piller and Mitterpacher), *Pogonocherus hispidus* (L.), and *Pogonocherus sturanii* Sama and Schurmann; and the following curculionids: *Ips typographus* (L.), *Magdalis nitida* (Gyllenhal), *Magdalis rufa* Germar, *Magdalis ruficornis* (L.), *Magdalis violacea* (L.), and *Pissodes validirostris* (Sahlberg).

Conclusion

Two species of the subfamily Cenocoeliinae in two genera, *Cenocoelius* and *Lestricus*, have been reported from only one of the Middle East countries, Iran (2.2% of the world species). Iranian Cenocoeliinae species have been recorded from two provinces (Isfahan and Kordestan). To date, no host records have been reported for these parasitoids in Iran. Among the Middle East and adjacent countries of Iran, the subfamily Cenocoeliinae has been recorded only from Russia (six species) and Azerbaijan (one species), which Russia and Azerbaijan share two and one species with Iran, respectively. No species have been recorded from the other 21 countries of the Middle East and adjacent to Iran.

References

Čapek, M. (1970). A new classification of the Braconidae (Hym.) based on the cephalic structures of the final instar larvae and biological evidence. *The Canadian Entomologist, 102*, 846–875.

Chen, X. X., & van Achterberg, C. (2019). Systematics, phylogeny, and evolution of braconid wasps: 30 years of progress. *Annual Review of Entomology, 64*, 1–24.

Mora, R., & Hanson, P. E. (2019). Widespread occurrence of black-orange-black color pattern in Hymenoptera. *Journal of Insect Science, 19*(2), 13, 1–12.

Saffer, B. (1982). A systematic revision of the genus *Cenocoelius* (Hym., Braconidae) in North America, including Mexico. *Polskie Pismo, 52*, 73–167.

Samin, N., Coronado-Blanco, J. M., Fischer, M., van Achterberg, C., Sakenin, H., & Davidian, E. (2018). Updated checklist of Iranian Braconidae (Hymenoptera: Ichneumonoidea) with twenty-three new records. *Natura Somogyiensis, 32*, 21–36.

Sharanowski, B. J., Dowling, A. P. G., & Sharkey, M. J. (2011). Molecular phylogenetics of Braconidae (Hymenoptera: Ichneumonoidea), based on multiple nuclear genes, and implications for classification. *Systematic Entomology, 36*(3), 549–572.

Shaw, S. R. (1995). Chapter 12.2, Braconidae. In P. Hanson, & I. D. Gauld (Eds.), *The Hymenoptera of Costa Rica* (pp. 431–463). Oxford University Press, 893 pp.

Shaw, M. R., & Huddleston, T. (1991). *Classification and biology of braconid wasps (Hymenoptera: Braconidae)* (Vol. 7, pp. 1–125). Royal Entomological Society of London.

van Achterberg, C. (1984). Essay on the phylogeny of Braconidae (Hymenoptera, Ichneumonoidea). *Entomologisk Tidskrift, 105*, 41–58.

van Achterberg, C. (1993). Illustrated key to the subfamilies of the Braconidae (Hymenoptera: Ichneumonoidea). *Zoologische Verhandelingen, 283,* 1−189.

van Achterberg, C. (1994). Generic revision of the subfamily Cenocoeliinae Szépligeti (Hymenoptera: Braconidae). *Zoologische Verhandlingen, 292,* 1−52.

van der Ent, L.-J., & Shaw, S. R. (1998). Species richness of Costa Rican Cenocoeliini (Hymentoptera: Braconidae): A latitudinal and altitudinal search for anomalous diversity. *Journal of Hymenoptera Research, 7*(1), 15−24.

van der Ent, L.-J., & Shaw, S. R. (1999). A new species of *Capitonius* (Hymenoptera: Braconidae) from Costa Rica with rearing records. *Pan-Pacific Entomologist, 75*(2), 112−120.

Wharton, R., Shaw, S. R., Sharkey, M. J., Wahl, D. B., Wooley, J. B., Whitfield, J. B., Marsh, P. M., & Johnson, J. B. (1992). Phylogeny of the subfamilies of the family Braconidae: A reassessment. *Cladistics, 8,* 199−235.

Yu, D. S., van Achterberg, C., & Horstmann, K. (2016). *Taxapad 2016, Ichneumonoidea 2015, Database on flash-drive.* Nepean, Ontario, Canada.

Charmon cruentatus Haliday, 1833 (Charmontinae), ♀, lateral habitus. *Photo prepared by S.R. Shaw.*

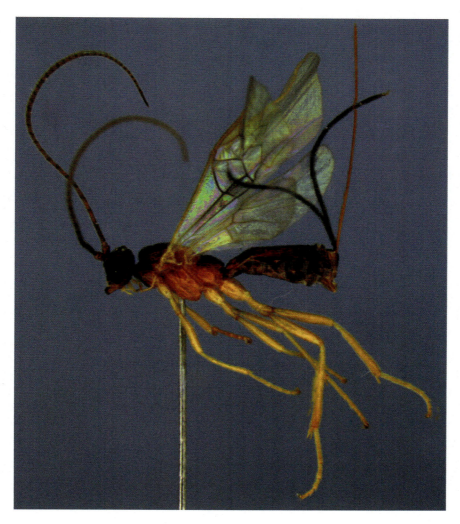

Charmon extensor (Linnaeus, 1758) (Charmontinae), ♀, lateral habitus. *Photo prepared by S.R. Shaw.*

Chapter 9

Subfamily Charmontinae van Achterberg, 1979

Neveen Smay Gadallah[1], Hassan Ghahari[2], Michael J. Sharkey[3] and Scott Richard Shaw[4]

[1]*Entomology Department, Faculty of Science, Cairo University, Giza, Egypt;* [2]*Department of Plant Protection, Yadegar-e Imam Khomeini (RAH) Shahre Rey Branch, Islamic Azad University, Tehran, Iran;* [3]*UW Insect Museum, Department of Entomology, University of Kentucky, Lexington, KY, United States;* [4]*UW Insect Museum, Department of Ecosystem Science and Management, University of Wyoming, Laramie, WY, United States*

Introduction

Charmontinae van Achterberg (1979) comprises a small subfamily of Braconidae that are distributed in almost all parts of the world (Yu et al., 2016). Currently, charmontines comprise 10 species classified into three genera: *Charmon* Haliday, 1833, *Charmontina* van Achterberg, 1979 and the fossil genus *Palaeocharmon* Belokobylskij, Nel, Waller and De Plöeg, 2010. The genera are arranged into two tribes: Charmontini van Achterberg and Palaeocharmontini Belokobylskij, Nel, Waller, and De Plöeg (based on fossils records only) (Chen & van Achterberg, 2019; Rousse, 2013; Sabahatullah et al., 2014; Yu et al., 2016). The majority of the species belong to the genus *Charmon*, with eight species (80% of the total number of species) (Rousse, 2013; Sabahatullah et al., 2014; Yu et al., 2016). Charmontines are koinobiont endoparasitoids of the concealed larvae of at least 16 lepidopteran families (Chen & van Achterberg, 2019; M. Shaw & Huddleston, 1991; Yu et al., 2016). The genus *Charmon* was placed in Orgilini by Mason (1974). In 1979, van Achterberg included it in the tribe Charmontini (in his new subfamily Homolobinae), but it was later upgraded to a separate subfamily, Charmontinae (Quicke & van Achterberg, 1990).

Members of the subfamily Charmontinae are easily diagnosed by the following combination of characters: slender bodies with a very long, longitudinally ridged ovipositor; occipital carina present; r-m of forewing absent, forewing with only two submarginal cells; and hind wing with anal cross-vein (Rousse, 2013; Salim et al., 2016; M. Shaw & Huddleston, 1991; van Achterberg, 1979). The longitudinally ridged ovipositor, together with molecular studies, indicates a phylogenetic affinity with the macrocentroid subcomplex of the helconoid subcomplex (Quicke & van Achterberg, 1990; Sharanowski et al., 2011; Shi et al., 2005).

Checklists of Regional Charmontinae. Yu et al. (2016), Samin, Coronado-Blanco, Kavallieratos et al. (2018), Samin, Coronado-Blanco, Fischer et al. (2018) and Gadallah et al. (2019) all listed two species in one genus, *Charmon* Haliday, 1833. The present Middle Eastern checklist includes two species in one genus as well.

List of species of the subfamily Charmontinae recorded in the Middle East

Subfamily Charmontinae van Achterberg, 1979

Tribe Charmontini van Achterberg, 1979

Genus *Charmon* Haliday, 1833

Charmon cruentatus Haliday, 1833

Catalogs with Iranian records: Yu et al. (2016), Samin, Coronado-Blanco, Kavallieratos et al. (2018), Samin, Coronado-Blanco, Fischer et al. (2018), Gadallah et al. (2019).

Distribution in Iran: Kordestan (Samin et al., 2016).

Distribution in the Middle East: Iran.

Extralimital distribution: Austria, Belgium, Bulgaria, Canada, China, Czech Republic, Denmark, France, Germany, Hungary, Ireland, Italy, Mexico, Mongolia, Netherlands, Norway, Poland, Russia, Slovakia, South Africa, South Korea, Sweden, Switzerland, United Kingdom, United States of America (introduced).

Host records: Summarized by Yu et al. (2016) as being a parasitoid of the depressariid *Agonopterix nervosa*

(Haworth); the following gelechiids: *Gelechia hippophaella* (Schrank), and *Hypatima rhomboidella* (L.), and the following tortricids: *Acleris variana* (Ferland), *Ancylis comptana* (Frölich), *Choristoneura fumiferana* (Clemens), *Choristoneura rosaceana* (Harris), *Cydia pomonella* (L.), *Epinotia lindana* (Fernald), *Grapholita molesta* (Busck), and *Spilonota ocellana* (Denis and Schiffermüller).

Charmon extensor (Linnaeus, 1758)

Catalogs with Iranian records: Yu et al. (2016), Samin, Coronado-Blanco, Kavallieratos et al. (2018), Samin, Coronado-Blanco, Fischer et al. (2018), Gadallah et al. (2019).

Distribution in Iran: Fars (Samin et al., 2016), Razavi Khorasan (Ghahari, 2020), West Azarbaijan (Masnady-Yazdinejad, 2010).

Distribution in the Middle East: Cyprus (Čapek & Hofmann, 1997), Iran (see references above), Turkey (Beyarslan & Aydoğdu, 2013, 2014).

Extralimital distribution: Austria, Azerbaijan, Belgium, Bulgaria, Canada, China, Congo, Croatia, Czech Republic, Finland, France, Germany, Hungary, India, Ireland, Italy, Japan, Latvia, Lithuania, Mexico, Moldova, Mongolia, Netherlands, Norway, Poland, Portugal, Romania, Russia, Slovakia, South Korea, Spain, Sweden, Switzerland, United States of America (introduced).

Host records: Summarized by Yu et al. (2016) as being a parasitoid of a wide range of lepidopteran host insects of the families Argyresthiidae, Coleophoridae, Depressariidae, Gelechiidae, Geometridae, Nepticulidae, Noctuidae, Oecophoridae, Pyralidae, Tineidae, Tischeriidae, Tortricidae, Ypsolophidae, and Zygaenidae.

Conclusion

Two species of the subfamily Charmontinae in one genus have been reported from only three of the Middle Eastern countries, Cyprus, Iran (being the richest in the Middle Eastern countries), and Turkey. Iranian Charmontinae comprises two species in one genus *Charmon* Haliday, 1833, which have been recorded from four provinces: Fars, Kordestan, Razavi Khorasan, and West Azarbaijan. No host records have been established for these parasitoids so far in Iran. Among the Middle East and adjacent countries of Iran, the subfamily Charmontinae has been recorded only from Russia (three species), Azerbaijan, Cyprus, and Turkey (each with one species). Russia shares two species with Iran, and one species for three other countries. No species have been recorded from the other 18 countries of the Middle East and adjacent to Iran.

References

Beyarslan, A., & Aydoğdu, M. (2013). Additions to the rare species of Braconidae fauna (Hymenoptera: Braconidae) from Turkey. *Munis Entomology & Zoology, 8*(1), 369−374.

Beyarslan, A., & Aydoğdu, M. (2014). Additions to the rare species of Braconidae fauna (Hymenoptera: Braconidae) from Turkey. *Munis Entomology & Zoology, 9*(1), 103−108.

Čapek, M., & Hofmann, C. (1997). The Braconidae (Hymenoptera) in the collections of the Musée cantonal de Zoologie, Lausanne. *Litterae Zoologicae Lausanne, 2*, 25−162.

Chen, X. X., & van Achterberg, C. (2019). Systematics, phylogeny, and evolution of braconid wasps: 30 years of progress. *Annual Review of Entomology, 64*, 1−24.

Gadallah, N. S., Ghahari, H., & Kavallieratos, N. G. (2019). An annotated catalogue of the Iranian Charmontinae, Ichneutinae, Macrocentrinae and Orgilinae (Hymenoptera: Braconidae). *Journal of the Entomological Research Society, 21*(3), 333−354.

Ghahari, H. (2020). A study on the fauna of predator and parasitoid arthropods in saffron fields (*Crocus sativus* L.). *Journal of Saffron Research, 7*(2), 203−215 (in Persian, English summary).

Masnady-Yazdinejad, A. (2010). The first braconid species record of subfamily Charmontinae from Iran: *Charmon extensor* (L.) (Hymenoptera: Braconidae: Charmontinae). In *Proceedings of the 7th International Congress of Hymenopterists, 20−26 June, 2010* (pp. 96−97). Hungary: Köszeg.

Mason, W. R. M. (1974). A generic synopsis of Brachistini (Hymenoptera: Braconidae) and recognition of the name *Charmon* Haliday. *Proceedings of the Entomological Society of Washington, 76*(3), 235−246.

Quicke, D. L. J., & van Achterberg, C. (1990). Phylogeny of the subfamilies of the family Braconidae (Hymenoptera: Ichneumonoidea). *Zoologische Verhandelingen, 258*, 1−95.

Rousse, P. (2013). *Charmon ramagei* sp. nov., a new Charmontinae (Hymenoptera: Braconidae) from reunion, with a synopsis of world species. *Zootaxa, 3626*(4), 583−588.

Sabahatullah, M., Mashwani, M. A., Tahira, Q. A., & Inayatullah, M. (2014). New record of the subfamily Charmontinae (Braconidae: Hymenoptera) in Pakistan with the description of a new species. *Pakistan Journal of Agricultural Research, 27*(4), 296−302.

Samin, N., van Achterberg, C., & Çetin Erdoğan, Ö. (2016). A faunistic study on some subfamilies of Braconidae (Hymenoptera: Ichneumonoidea) from Iran. *Arquivos Entomolóxicos, 15*, 153−161.

Samin, N., Coronado-Blanco, J. M., Kavallieratos, N. G., Fischer, M., & Sakenin, H. (2018). Recent findings on Braconidae (Hymenoptera: Ichneumonoidea) of Iran with an updated checklist. *Acta Biologica Turcica, 31*(4), 160−173.

Samin, N., Coronado-Blanco, J. M., Fischer, M., van Achterberg, C., Sakenin, H., & Davidian, E. (2018). Updated checklist of Iranian Braconidae (Hymenoptera: Ichneumonoidea) with twenty-three new records. *Natura Somogyiensis, 32*, 21−36.

Sharanowski, B. J., Dowling, A. P. G., & Sharkey, M. J. (2011). Molecular phylogenetics of Braconidae (Hymenoptera: Ichneumonoidea), based on multiple nuclear genes, and implications for classification. *Systematic Entomology, 36*(3), 549−572.

Shaw, M. R., & Huddleston, T. (1991). Classification and biology of braconid wasps (Hymenoptera: Braconidae). *Handbook for the Identification of British Insects, 7*(11), 1−126.

Shi, M. M., Chen, X. X., & van Achterberg, C. (2005). Phylogenetic relationships among Braconidae (Hymenoptera: Ichneumonoidea) inferred from partial 16S rDNA, 28S rDNA, 18S rDNA gene sequences and morphological characters. *Molecular Phylogenetics and Evolution, 37*(1), 104–116.

van Achterberg, C. (1979). A revision of the subfamily Zelinae auct. (Hymenoptera: Braconidae). *Tijdschrift voor Entomologie, 122*, 241–479.

Yu, D. S., van Achterberg, C., & Horstmann, K. (2016). *Taxapad 2016, Ichneumonoidea 2015, Database on flash-drive*. Nepean, Ontario, Canada.

Chelonus subcontractus Abdinbekova, 1971 (Cheloninae), ♀, lateral habitus. *Photo prepared by S.R. Shaw.*

Ascogaster grahami Huddleston, 1984 (Cheloninae), ♀, lateral habitus. *Photo prepared by S.R. Shaw.*

Chapter 10

Subfamily Cheloninae Foerster, 1863

Hassan Ghahari[1], Neveen Samy Gadallah[2], Rebecca N. Kittel[3], Scott Richard Shaw[4] and Donald L.J. Quicke[5]

[1]Department of Plant Protection, Yadegar-e Imam Khomeini (RAH) Shahre Rey Branch, Islamic Azad University, Tehran, Iran; [2]Entomology Department, Faculty of Science, Cairo University, Giza, Egypt; [3]Laboratory of Insect Biodiversity and Ecosystems Science, Graduate School of Agricultural Science, Kobe University, Kobe, Japan; [4]UW Insect Museum, Department of Ecosystem Science and Management, University of Wyoming, Laramie, WY, United States; [5]Integrative Ecology Laboratory, Department of Biology, Faculty of Science, Chulalongkorn University, Pathumwan, Bangkok, Thailand

Introduction

The cosmopolitan subfamily Cheloninae Foerster, 1863, is one of the large lineages of the family Braconidae with 1523 described species worldwide classified into 23 genera and four tribes (Adeliini Vierek, 1918; Chelonini Foerster, 1863; Odontosphaeropygini Zettel, 1990a,b; and Phanerotomini Baker, 1926) (Chen & van Achterberg, 2019). Some chelonine genera, such as *Chelonus* Panzer, 1806, *Ascogaster* Wesmael, 1835, and *Leptodrepana* Shaw, 1983, are distributed in both temperate and tropical areas (Braet et al., 2012; Dadelahi et al., 2018; Huddleston, 1984; Shaw, 1983, 1997). Zettel (1990) proposed four tribes (Chelonini; Odontosphaeropygini; Pseudophanerotomini Zettel, 1990; and Phanerotomini), whereas van Achterberg (1990) proposed only two tribes (Chelonini and Phanerotomini). Both authors treated Adeliinae as a separate subfamily. A phylogenetic study carried out by Kittel et al. (2016), based on both morphological and molecular data, supported van Achterberg's classification (1990) recognizing the two tribes but adeliine genera were additionally considered as a tribe of the Cheloninae (Belsahw et al., 1998; Kittel et al., 2016). Subsequent revisionary studies have followed this placement for Adeliinae (Shimbori et al., 2017). Many phylogenetic studies indicate that the subfamily Cheloninae is part of the microgastroid lineage (Murphy et al., 2008; Quicke & van Achterberg, 1990; Sharanowski et al., 2011; Shi et al., 2005; van Achterberg, 1984).

Members of this subfamily are generally black, yellow, or orange in color, and they vary in size from 1.8 to 11 mm. Chelonines are easily recognized from most other braconid subfamilies by their rigid, sculptured metasomal carapace that is formed by the fusion of the first three tergites, with the remaining parts of metasoma usually being concealed ventrally (Duarenko, 1974). Characteristics of the chelonine carapace, such as size, shape, microsculpture types, and color patterns, vary greatly both between genera and species and such characters provide rich data for taxonomic assessments (Dadelahi et al., 2018). Additional useful characters for Cheloninae are the presence of a complete postpectal carina in front of the mid coxae, and the fore wing having three submarginal cells, although the first submarginal cell may be fused with the first discal cell (van Achterberg, 1990, 1993; Walker & Huddleston, 1987). Chelonines are considered to be mostly solitary koinobiont egg-larval endoparasitoids of concealed Lepidoptera, especially Tortricoidea and Pyralidoidea (Yu et al., 2016) or Nepticulidae (adeliines; Whitfield & Wagner, 1991; Shimbori et al., 2019). Chelonines play an important role as regulatory agents for phytophagous insect population dynamics, particularly the economically important insect pests that mostly belong to the families Noctuidae, Geometridae, Tortricidae, Pyralidae, and Gelechiidae (Kaeslin et al., 2005; LaSalle & Gauld, 1993; Shaw & Huddleston, 1991; van Achterberg, 1990). The biology of this group has been reviewed by Shaw and Huddleston (1991).

Shaw (1983) revised the Nearctic species of *Ascogaster* and proposed a new genus, *Leptodrepana* Shaw, 1983, based on two new species and two others formerly placed in *Ascogaster*. Subsequent authors studying Palaearctic *Ascogaster* treated *Leptodrepana* as a junior synonym (Huddleston, 1984; Tang & Marsh, 1994); however, Kittel et al. (2016) provided molecular evidence for the distinctness of the New World *Leptodrepana* clade and Dadelahi et al. (2018) treated *Leptodrepana* as a valid genus with at least 24 species just in Costa Rica. The issue of whether or not any valid species of *Leptodrepana* occur in the Old World has not been fully resolved, and it remains possible that some may discovered in the Palaearctic region.

Braconidae of the Middle East (Hymenoptera). https://doi.org/10.1016/B978-0-323-96099-1.00001-7

Checklists of Regional Cheloninae. Farahbakhsh (1961): one species in one genus (*Chelonus* Panzer) plus an unknown species of *Phanerotoma* Wesmael, 1838; Modarres Awal (1997): six species in two genera plus one unknown species of the genus *Chelonella* Szépligeti, 1908; Fallahzadeh and Saghaei (2010): 24 species in three genera (without precise localities) (*Ascogaster* Wesmael, *Chelonus*, *Phanerotoma*); Modarres Awal (2012): five species in two genera; Gadallah and Ghahari (2013): 48 species in four genera; Farahani et al. (2016) and Yu et al. (2016): both with 65 species in six genera. The present checklist includes 219 species in five genera and three tribes (Adeliini, Chelonini, Phanerotomini) in the Middle East, of which two species, *Chelonus longiventris* (Tobias, 1964), and *Chelonus risorius* Reinhard, 1867, are newly recorded for the Iranian fauna. Here we follow the classification of Chen and van Achterberg (2019); and the global distribution of the different species is based mainly on Yu et al. (2016), supplemented with additional more recent references.

Key to genera of the subfamily Cheloninae of the Middle East (modified from van Achterberg, 1990; Edmardash & Gadallah, 2019)

1. Metasoma without distinct transverse sutures; body usually dark brown to black 2
— Metasoma with two distinct sutures; body usually yellowish to brownish 4
2. Second submarginal cell of fore wing small, sometimes open apically; notauli generally absent or very weak ....
.... *Adelius* Haliday
— Second submarginal cell of fore wing larger, and closed; notauli usually well-developed 3
3. Vein 1-SR+M of fore wing present; male carapace without apical aperture; vein r of fore wing usually arises far distad the middle of pterostigma .... *Ascogaster* Wesmael
— Vein 1-SR+M of fore wing absent; male carapace with or without apical aperture; vein r of fore wing arises near middle of pterostigma .... *Chelonus* Panzer
4. Vein CU1b of fore wing present, resulting in a closed subdiscal cell; free margin of clypeus usually with two to three teeth; antenna 23-segmented in both sexes .... *Phanerotoma* Wesmael
— Vein CU1b of fore wing absent, resulting in an open subdiscal cell; free margin of clypeus usually without teeth; antenna 24-60-segmented .... *Phanerotomella* Szépligeti

List of species of the subfamily Cheloninae recorded in the Middle East

Subfamily Cheloninae Foerster, 1863

Tribe Adeliini Viereck, 1918

Genus *Adelius* Haliday, 1833

Adelius (Adelius) aridus (Tobias, 1967)

Catalogs with Iranian records: No catalog.
Distribution in Iran: Sistan and Baluchestan (Derafshan et al., 2017).
Distribution in the Middle East: Iran.
Extralimital distribution: Turkmenistan.
Host records: In Iran, this species has been reared from an unknown nepticulid leaf miner (Lepidoptera: Nepticulidae) (Derafshan et al., 2017).
Plant associations in Iran: *Tamarix stricta* Boissier, and *Tamarix aphylla* (L.) (Tamaricaceae) (Derafshan et al., 2017).

Adelius (Adelius) erythronotus (Foerster, 1851)

Catalogs with Iranian records: Yu et al. (2016), Samin, Coronado-Blanco, Kavallieratos et al. (2018), Samin, Coronado-Blanco, Fischer et al. (2018).
Distribution in Iran: Razavi Khorasan (Samin et al., 2016).

Distribution in the Middle East: Iran (Samin et al., 2016), Turkey (Beyarslan & Aydoğdu, 2013, 2014).
Extralimital distribution: Azerbaijan, Bosnia-Herzegovina, Bulgaria, former Czechoslovakia, France, Georgia, Germany, Hungary, Ireland, Korea, Moldova, Poland, Romania, Russia, Turkmenistan, Ukraine, United Kingdom, Uzbekistan, former Yugoslavia.
Host records: Summarized by Yu et al. (2016) as being a parasitoid of the nepticulids of the genera *Ectoedemia* Busck, *Stigmella* Schrank, and *Trifurcula bupleurella* (Chretein).

Adelius (Adelius) subfasciatus Haliday, 1833

Catalogs with Iranian records: Yu et al. (2016), Samin, Coronado-Blanco, Kavallieratos et al. (2018), Samin, Coronado-Blanco, Fischer et al. (2018).
Distribution in Iran: Hamadan (Samin et al., 2016).
Distribution in the Middle East: Iran (Samin et al., 2016), Turkey (Beyarslan & Aydoğdu, 2013, 2014; Beyarslan et al., 2006).
Extralimital distribution: Belgium, Bulgaria, Croatia, Czech Republic, Finland, France, Georgia, Germany, Hungary, Italy, Kazakhstan, Latvia, Lithuania, Malta, Moldova, Netherlands, Poland, Romania, Russia, Serbia, Slovakia, Spain, Sweden, Switzerland, Ukraine, United Kingdom.

Host records: Summarized by Yu et al. (2016) as being a parasitoid of several species of the dipteran family Cecidomyiidae; and from the lepidopteran families Coleophoridae, Gracillariidae, Lyonetiidae, Nepticulidae, Tischeriidae, and Tortricidae.

Tribe Chelonini Foerster, 1863

Genus *Ascogaster* Wesmael, 1835

Palaearctic species of *Ascogaster* have been revised by Huddleston (1984) and the Chinese species by Tang and Marsh (1994).

Ascogaster (Ascogaster) abdominator (Dahlbom, 1833)

Catalogs with Iranian records: No catalog.
Distribution in Iran: Guilan (Gadallah et al., 2018).
Distribution in the Middle East: Iran (Gadallah et al., 2018), Turkey (Aydoğdu & Beyarslan, 2007, 2012).
Extralimital distribution: Austria, Azerbaijan, Belgium, Bulgaria, Czech Republic, Finland, France, Georgia, Germany, Hungary, Ireland, Italy, Lithuania, North Macedonia, Moldova, Netherlands, Norway, Poland, Romania, Russia, Slovenia, Sweden, Switzerland, United Kingdom, former Yugoslavia.
Host records: Summarized by Yu et al. (2016) as being a parasitoid of the crambid *Scoparia basistrigalis* Knaggs; the tortricid *Epinotia cruciana* (L.); and the yponomeutid *Roeslerstammia erxlebella* (Fabricius).

Ascogaster (Ascogaster) annularis (Nees von Esenbeck, 1816)

Catalogs with Iranian records: Farahani et al. (2016), Yu et al. (2016), Samin, Coronado-Blanco, Kavallieratos et al. (2018), Samin, Coronado-Blanco, Fischer et al. (2018).
Distribution in Iran: Alborz, Qazvin (Farahani et al., 2013, 2014), Hamadan (Rajabi Mazhar et al., 2018).
Distribution in the Middle East: Iran (see references above), Israel−Palestine (Halperin, 1986; Papp, 2012), Turkey (Aydoğdu & Beyarslan, 2007, 2012; Beyarslan, 2021; Beyarslan et al., 2002).
Extralimital distribution: Armenia, Austria, Bulgaria, Finland, France, Georgia, Germany, Hungary, Italy, Kazakhstan, Korea, Latvia, Lithuania, Moldova, Netherlands, Norway, Poland, Romania, Russia, Slovakia, Spain, Sweden, Switzerland, Ukraine, United Kingdom.
Host records: Summarized by Yu et al. (2016) as being a parasitoid of the agonoxenid *Chrysoclista linneella* (Clerck); the coleophorid *Coleophora lutipennella* (Zeller); the gelechiids *Batia lambdella* (Donovan), *Dichomeris* sp., *Parachronistis albiceps* (Zeller), *Recurvaria leucatella* (Clerck), *Recurvaria nanella* (Denis and Schiffermüller), and *Stenolechia gemmella* (L.); the psychid *Narycia*

duplicella (Goeze); the tortricids *Pandemis heparana* (Denis and Schiffermüller), and *Spilonota ocellana* (Denis and Schiffermüller); and the yponomeutids *Yponomeuta malinella* Zeller, and *Yponomeuta padella* (Linnaeus).

Ascogaster (Ascogaster) armata Wesmael, 1835

Catalogs with Iranian records: Samin, Coronado-Blanco, Kavallieratos et al. (2018).
Distribution in Iran: West Azarbaijan (Samin, Coronado-Blanco, Kavallieratos et al., 2018).
Distribution in the Middle East: Iran (Samin, Coronado-Blanco, Kavallieratos et al., 2018), Turkey (Aydoğdu & Beyarslan, 2012).
Extralimital distribution: Austria, Belgium, Bulgaria, former Czechoslovakia, France, Germany, Hungary, Italy, Lithuania, Moldova, Netherlands, Poland, Romania, Russia, Switzerland, United Kingdom.
Host records: Recorded by Huddleston (1984) and Čapek and Hofmann (1997) as being a parasitoid of the coleophorid *Coleophora violacea* (Ström).

Ascogaster (Ascogaster) bicarinata (Herrich-Schäffer, 1838)

Catalogs with Iranian records: Fallahzadeh and Saghaei (2010), Gadallah and Ghahari (2013), Farahani et al. (2016), Yu et al. (2016), Samin, Coronado-Blanco, Kavallieratos et al. (2018), Samin, Coronado-Blanco, Fischer et al. (2018).
Distribution in Iran: Isfahan (Ghahari, Fischer, & Papp, 2011), Semnan (Ghahari & Gadallah, 2015), Iran (no specific locality cited) (Telenga, 1941; Tobias, 1976, 1986).
Distribution in the Middle East: Iran (see references above), Turkey (Beyarslan, 2021; Lozan, 2005).
Extralimital distribution: Azerbaijan, Croatia, Georgia, Germany, Greece, Hungary, Italy, North Macedonia, Moldova, Romania, Russia, Serbia, Slovakia, Spain, Ukraine, United Kingdom.
Host records: Recorded by Aydoğdu and Beyarslan (2012) and Beyarslan (2021) as being a parasitoid of the tortricid *Archips rosana* (L.).

Ascogaster (Ascogaster) bidentula Wesmael, 1835

Catalogs with Iranian records: Samin, Coronado-Blanco, Fischer et al. (2018).
Distribution in Iran: Ilam (Sakenin et al., 2018).
Distribution in the Middle East: Iran (Sakenin et al., 2018), Turkey (Beyarslan, 2021; Beyarslan et al., 2002).
Extralimital distribution: Austria, Belgium, Bulgaria, China, Czech Republic, Finland, Germany, Hungary, Ireland, Italy, Japan, Korea, Latvia, Lithuania, Moldova, Netherlands, Poland, Slovakia, Russia, Sweden, Switzerland, United Kingdom.

Host records: Summarized by Yu et al. (2016) as being a parasitoid of the geometrid *Eupithecia venosata* (Fabricius); and the following tortricids: *Archips rosana* (L.), *Epinotia cruciana* (L.), *Gypsonoma sociana* (Haworth), *Notocelia roborana* (Denis and Schiffermüller), *Pandemis* sp., and *Rhopobota myrtillana* (Humphryes and Westwood).

Ascogaster (Ascogaster) bimaris Tobias, 1986
Catalogs with Iranian records: Farahani et al. (2016), Yu et al. (2016), Samin, Coronado-Blanco, Kavallieratos et al. (2018), Samin, Coronado-Blanco, Fischer et al. (2018).
Distribution in Iran: Guilan (Farahani et al., 2014).
Distribution in the Middle East: Iran.
Extralimital distribution: Azerbaijan, Georgia, Russia.
Host records: Unknown.

Ascogaster (Ascogaster) brevicornis Wesmael, 1835
Catalogs with Iranian records: No catalog.
Distribution in Iran: East Azarbaijan (Ghahari & Gadallah, 2019).
Distribution in the Middle East: Iran.
Extralimital distribution: Austria, Belgium, former Czechoslovakia, France, Germany, Hungary, Ireland, Netherlands, Switzerland, United Kingdom.
Host records: Unknown.

Ascogaster (Ascogaster) canifrons Wesmael, 1835
Distribution in the Middle East: Turkey (Beyarslan, 2021).
Extralimital distribution: Austria, Belgium, Finland, France, Germany, Hungary, Ireland, Kazakhstan, Mongolia, Netherlands, Poland, Russia, Slovakia, Spain, Sweden, Switzerland, United Kingdom, Uzbekistan.
Host records: Summarized by Yu et al. (2016) as being a parasitoid of the geometrid *Eupithecia pyreneata* Mabille; as well as the following tortricids: *Cydia pomonella* (L.), *Cydia splendana* (Hübner), *Eupoecilia angustana* (Hübner), *Gypsonoma dealbana* (Frölich), and *Rhopobota naevana* (Hübner).

Ascogaster (Ascogaster) caucasica Kokujev, 1895
Catalogs with Iranian records: Fallahzadeh and Saghaei (2010), Gadallah and Ghahari (2013), Farahani et al. (2016), Yu et al. (2016), Samin, Coronado-Blanco, Kavallieratos et al. (2018), Samin, Coronado-Blanco, Fischer et al. (2018).
Distribution in Iran: Khuzestan (Susa = Shush) (Huddleston, 1984).
Distribution in the Middle East: Cyprus, Iran (Huddleston, 1984), Turkey (Beyarslan & Inanç, 1992).
Extralimital distribution: Bulgaria, Croatia, former Czechoslovakia, Georgia, Greece, Hungary, North Macedonia, Romania, Spain.
Host records: Unknown.

Ascogaster (Ascogaster) dentifer Tobias, 1976
Distribution in the Middle East: Turkey (Aydoğdu & Beyarslan, 2007, 2012).
Extralimital distribution: Armenia, Austria, former Czechoslovakia, Georgia, Germany, Hungary, Italy, Switzerland.
Host records: Unknown.

Ascogaster (Ascogaster) dispar Fahringer, 1934
Catalogs with Iranian records: Gadallah and Ghahari (2013), Farahani et al. (2016), Yu et al. (2016), Samin, Coronado-Blanco, Kavallieratos et al. (2018), Samin, Coronado-Blanco, Fischer et al. (2018).
Distribution in Iran: Iran (no specific locality cited) (Huddleston, 1984).
Distribution in the Middle East: Iran (Huddleston, 1984), Turkey (Beyarslan & Inanç, 1992; Lozan, 2005).
Extralimital distribution: Austria, former Czechoslovakia, Germany, Hungary, Kazakhstan, Mongolia, Netherlands, Romania, Russia, Switzerland, United Kingdom.
Host records: Summarized by Yu et al. (2016) as being a parasitoid of the tortricids *Cydia delineana* (Walker), and *Endothenia gentianaeana* (Hübner).

Ascogaster (Ascogaster) disparilis Tobias, 1986
Catalogs with Iranian records: Gadallah and Ghahari (2013), Farahani et al. (2016), Yu et al. (2016), Samin, Coronado-Blanco, Kavallieratos et al. (2018), Samin, Coronado-Blanco, Fischer et al. (2018).
Distribution in Iran: Guilan (Farahani et al., 2014).
Distribution in the Middle East: Iran (Farahani et al., 2014), Turkey (Aydoğdu & Beyarslan, 2006, 2012).
Extralimital distribution: Russia.
Host records: Unknown.

Ascogaster (Ascogaster) excavata Telenga, 1941
Catalogs with Iranian records: Farahani et al. (2016), Yu et al. (2016), Samin, Coronado-Blanco, Kavallieratos et al. (2018), Samin, Coronado-Blanco, Fischer et al. (2018).
Distribution in Iran: Alborz (Farahani et al., 2013, 2014).
Distribution in the Middle East: Iran.
Extralimital distribution: Kazakhstan, Russia, Switzerland.
Host records: Unknown.

Ascogaster (Ascogaster) excisa (Herrich-Schäffer, 1838)
Catalogs with Iranian records: Samin, Coronado-Blanco, Kavallieratos et al. (2018).
Distribution in Iran: Ardabil (Samin, Coronado-Blanco, Kavallieratos et al., 2018), Southern Khorasan (Ghahari, 2020).
Distribution in the Middle East: Egypt (Edmardash et al., 2011; Edmardash & Gadallah, 2019), Iran (Ghahari, 2020;

Samin, Coronado-Blanco, Kavallieratos et al., 2018), Turkey (Aydoğdu & Beyarslan, 2007, 2012).

Extralimital distribution: Azerbaijan, Bulgaria, Croatia, France, Germany, Hungary, Italy, Kazakhstan, Moldova, Spain, Switzerland.

Host records: Unknown.

Ascogaster (Ascogaster) gonocephala Wesmael, 1835

Catalogs with Iranian records: Samin, Coronado-Blanco, Fischer et al. (2018).

Distribution in Iran: Semnan (Sakenin et al., 2018).

Distribution in the Middle East: Iran (Sakenin et al., 2018), Turkey (Aydoğdu & Beyarslan, 2007, 2012; Gultekin & Güçlü, 1997).

Extralimital distribution: Belgium, France, Hungary, Moldova, Serbia, Slovakia, Switzerland, United Kingdom.

Host records: Recorded by Gultekin and Güçlü (1997) as being a parasitoid of the sesiid *Bembecia scopigera* (Scopoli).

Ascogaster (Ascogaster) grahami Huddleston, 1984

Catalogs with Iranian records: Farahani et al. (2016), Yu et al. (2016), Samin, Coronado-Blanco, Kavallieratos et al. (2018), Samin, Coronado-Blanco, Fischer et al. (2018).

Distribution in Iran: Guilan, Mazandaran (Farahani et al., 2013, 2014), Hamadan (Rajabi Mazhar et al., 2018).

Distribution in the Middle East: Iran (see references above), Israel—Palestine (Halperin, 1986), Turkey (Aydoğdu & Beyarslan, 2007, 2012).

Extralimital distribution: Austria, China, former Czechoslovakia, France, Germany, Hungary, Italy, Korea, Netherlands, Russia, Sweden, Switzerland, United Kingdom.

Host records: Recorded by Huddleston (1984) and Halperin (1986) as being a parasitoid of the cosmopetrigid *Sorhagenia lophyrella* (Douglas); the oecophorid *Borkhausenia einsleri* Amsel; and the tineid *Infurcitinea argentimaculella* (Stainton).

Ascogaster (Ascogaster) kasparyani Tobias, 1976

Catalogs with Iranian records: Farahani et al. (2016), Yu et al. (2016), Samin, Coronado-Blanco, Kavallieratos et al. (2018), Samin, Coronado-Blanco, Fischer et al. (2018).

Distribution in Iran: Alborz, Qazvin, Tehran (Farahani et al., 2013, 2014).

Distribution in the Middle East: Iran (Farahani et al., 2013, 2014), Turkey (Aydoğdu & Beyarslan, 2012).

Extralimital distribution: Georgia, Greece.

Host records: Unknown.

Ascogaster (Ascogaster) klugii (Nees von Esenbeck, 1816)

Catalogs with Iranian records: Farahani et al. (2016), Yu et al. (2016), Samin, Coronado-Blanco, Kavallieratos et al. (2018), Samin, Coronado-Blanco, Fischer et al. (2018).

Distribution in Iran: Mazandaran (Farahani et al., 2014).

Distribution in the Middle East: Iran (Farahani et al., 2014), Turkey (Aydoğdu & Beyarslan, 2012).

Extralimital distribution: Austria, Azerbaijan, Belgium, Bulgaria, Czech Republic, Finland, France, Georgia, Germany, Hungary, Italy, Japan, Korea, Latvia, Lithuania, Moldova, Netherlands, Norway, Poland, Russia, Serbia, Slovakia, Sweden, Switzerland.

Host records: Summarized by Yu et al. (2016) as being a parasitoid of the curculionid *Dryocoetes villosus* (Fabricius); and the oecophorids *Denisia stipella* (L.), and *Pseudatemelia subochreella* (Doubleday).

Ascogaster (Ascogaster) magnidentis Tobias, 1986

Distribution in the Middle East: Turkey (Beyarslan, 2021).

Extralimital distribution: Russia, Spain.

Host records: Recorded by Tobias (1986) as being a parasitoid of the tortricid *Cydia milleniana* (Adamczewski).

Ascogaster (Ascogaster) quadridentata Wesmael, 1835

Catalogs with Iranian records: Gadallah and Ghahari (2013), Farahani et al. (2016), Yu et al. (2016), Samin, Coronado-Blanco, Kavallieratos et al. (2018), Samin, Coronado-Blanco, Fischer et al. (2018).

Distribution in Iran: Alborz, Guilan (Farahani et al., 2013, 2014), East Azarbaijan (Ranjbar Aghdam & Fathipour, 2010), Khuzestan (Samin, van Achterberg, & Ghahari, 2015), Kordestan (Ghahari, Fischer, Çetin Erdoğan et al., 2010), Mazandaran (Farahani et al., 2013, 2014; Ghahari, 2017—around rice fields; Kian et al., 2020), West Azarbaijan (Akbarzadeh Shoukat et al., 2015).

Distribution in the Middle East: Cyprus (Aydoğdu & Beyarslan, 2007), Egypt (Edmardash et al., 2011; Edmardash & Gadallah, 2019), Iran (see references above), Syria (Bashir et al., 2010), Turkey (Aydoğdu & Beyarslan, 2007).

Extralimital distribution: Australia (introduced), Nearctic (introduced to Canada and United States of America) (Shaw, 1983), Neotropical (introduced to Argentina and Peru), Oceanic (introduced to New Zealand), Oriental (introduced to Pakistan), Palaearctic [Adjacent countries to Iran: Armenia, Kazakhstan, Russia, Turkmenistan].

Host records: Summarized by Yu et al. (2016) as being a parasitoid of several host species of the orders Coleoptera (Curculionidae), Diptera (Anthomyiidae, Clusiidae), Hymenoptera (Cynipidae, Tenthredinidae), and Lepidoptera (Elachistidae, Gelechiidae, Geometridae, Limacodidae, Pyralidae, Tortricidae, Yponomeutidae). In Iran, this species has been reared from the tortricids *Cydia pomonella* (L.) (Ranjbar Aghdam & Fathipour, 2010), and *Lobesia botrana* (Denis and Schiffermüller) (Akbarzadeh Shoukat et al., 2015).

Ascogaster (Ascogaster) rufipes (Latreille, 1809)

Catalogs with Iranian records: No catalog.

Distribution in Iran: Lorestan (Gadallah et al., 2018).

Distribution in the Middle East: Iran (Gadallah et al., 2018), Turkey (Beyarslan & Inanç, 1992).

Extralimital distribution: Algeria, Austria, Azerbaijan, Belarus, Belgium, Bulgaria, China, Croatia, Czech Republic, Finland, France, Germany, Hungary, Ireland, Italy, Japan, Korea, Kyrgyzstan, Latvia, Moldova, Mongolia, Netherlands, Norway, Poland, Russia, Slovakia, Slovenia, Spain, Sweden, Switzerland, Ukraine, United Kingdom, Uzbekistan.

Host records: Summarized by Yu et al. (2016) as being a parasitoid of the curculionid *Anthonomus pomorum* (L.); the tenthredinid *Blennocampa pusilla* (Klug); the coleophorids *Coleophora discordella* Zeller, and *Coleophora gryphipennella* (Hübner); the depressariids *Agonopterix conterminella* (Zeller), and *Ethmia quadrillella* (Goeze); the erebid *Coscinia cribraria* (L.); the nolid *Earias clorana* (L.); the tortricids *Acleris hastiana* (L.), *Archips oporana* (L.), *Archips rosana* (L.), *Argyroploce lediana* (L.), *Croesia holmiana* (L.), *Cydia pomonella* (L.), *Grapholita funebrana* (Treitschke), *Dichelia histrionana* (Frölich), *Hedya nubiferana* Haworth, *Lobesia botrana* (Denis and Schiffermüller), *Notocelia uddmanniana* (L.), *Rhopobota ustomaculana* (Curtis), and *Selania leplastriana* (Curtis); and the yponomeutids *Yponomeuta evonymella* (L.), and *Yponomeuta padella* (L.).

Ascogaster (Ascogaster) scabricula (Dahlbom, 1833)

Distribution in the Middle East: Turkey (Beyarslan, 2021).

Extralimital distribution: Austria, Belgium, China, former Czechoslovakia, Finland, France, Germany, Greece, Hungary, Norway, Romania, Sweden, Switzerland.

Host records: Recorded by Zettel (1987) as being a parasitoid of the tortricid *Cydia zebeana* (Ratzeburg).

Ascogaster (Ascogaster) similis (Nees von Esenbeck, 1816)

Catalogs with Iranian records: No catalog.

Distribution in Iran: East Azarbaijan (Ghahari & Gadallah, 2021).

Distribution in the Middle East: Iran (Ghahari & Gadallah, 2021), Turkey (Beyarslan & Şahin, 2018).

Extralimital distribution: Belgium, Czech Republic, France, Germany, Hungary, Italy, Latvia, Lithuania, Netherlands, Poland, Russia, Slovenia, Spain, United Kingdom, former Yugoslavia.

Host records: Summarized by Yu et al. (2016) as being a parasitoid of the gelechiid *Teleiodes saltuum* (Zeller); the tortricids *Cydia splendana* (Hübner), and *Spilonota ocellana* (Denis and Schiffermüller); and the yponomeutid *Yponomeuta padella* (L.). In Iran, this species has been reared from *Yponomeuta padella* (Ghahari & Gadallah, 2021).

Ascogaster (Ascogaster) varipes Wesmael, 1835

Catalogs with Iranian records: Farahani et al. (2016), Yu et al. (2016), Samin, Coronado-Blanco, Kavallieratos et al. (2018), Samin, Coronado-Blanco, Fischer et al. (2018).

Distribution in Iran: Alborz, Guilan, Mazandaran (Farahani et al., 2014).

Distribution in the Middle East: Iran (Farahani et al., 2014), Turkey (Aydoğdu & Beyarslan, 2012; Beyarslan, 2021).

Extralimital distribution: Albania, Armenia, Austria, Azerbaijan, Belgium, China, Finland, France, Germany, Greece, Hungary, Ireland, Italy, Kazakhstan, Korea, Latvia, Lithuania, Moldova, Mongolia, Netherlands, Norway, Poland, Romania, Russia, Serbia, Slovakia, Slovenia, Spain, Sweden, Switzerland, Ukraine, United Kingdom.

Host records: Summarized by Yu et al. (2016) as being a parasitoid of the curculionid *Magdalis violacea* (L.); the coleophorid *Coleophora binderella* (Kollar); the gelechiid *Gelechia rhombella* (Denis and Schiffermüller); and the tortricids *Cydia* spp., *Endothenia gentianaeana* (Hübner), and *Epinotia cruciana* (L.).

Genus *Chelonus* Panzer 1806
Chelonus ahngeri Tobias, 1966

Distribution in the Middle East: Turkey (Beyarslan, 2021).

Extralimital distribution: Turkmenistan.

Host records: Unknown.

Chelonus albor (Tobias, 1994)

Distribution in the Middle East: Turkey (Papp, 2014a).

Extralimital distribution: Russia.

Host records: Unknown.

Chelonus andrievskii Tobias, 1972

Distribution in the Middle East: Turkey (Tobias, 2011).

Extralimital distribution: Armenia.

Host records: Unknown.

Chelonus annulatus (Nees von Esenbeck, 1816)

Catalogs with Iranian records: Fallahzadeh & Saghaei (2010), Gadallah & Ghahari (2013), Farahani et al. (2016), Yu et al. (2016), Samin, Coronado-Blanco, Kavallieratos et al. (2018), Samin, Coronado-Blanco, Fischer et al. (2018).

Distribution in Iran: Guilan (Ghahari, Fischer, & Tobias, 2012), Iran (no specific locality cited) (Telenga, 1941).

Distribution in the Middle East: Iran (Telenga, 1941; Ghahari, Fischer, & Tobias, 2012), Turkey (Aydoğdu & Beyarslan, 2012; Beyarslan, 2021).

Extralimital distribution: Belarus, Bulgaria, China, Croatia, Finland, France, Georgia, Germany, Hungary, Italy,

Kazakhstan, Kyrgyzstan, Lithuania, Moldova, Mongolia, Norway, Poland, Russia, Slovakia, Spain, Sweden, Switzerland, Ukraine, United Kingdom.

Host records: Summarized by Yu et al. (2016) as being a parasitoid of the following tephritids: *Chaetostomella cylindrica* (Robineau-Desvoidy), *Terellia longicauda* (Meigen), and *Urophora eriolepidis* Loew; the agonoxenid *Chrysoclista lineella* (Clerck); the gelechiids *Parachronistis albiceps* (Zeller), *Recurvaria nanella* (Denis and Schiffermüller), and *Stenolechia gemmella* (L.); the oecophorid *Batia lambdella* (Donovan); the psychid *Narycia duplicella* (Goeze); the tortricid *Dichrorampha petiverella* (L.); and the yponomeutids *Yponomeuta malinellus* (Zeller) and *Yponomeuta padella* (L.).

Chelonus annulipes Wesmael, 1835

Catalogs with Iranian records: Fallahzadeh and Saghaei (2010), Gadallah and Ghahari (2013), Farahani et al. (2016), Yu et al. (2016), Samin, Coronado-Blanco, Kavallieratos et al. (2018), Samin, Coronado-Blanco, Fischer et al. (2018).

Distribution in Iran: Alborz, Guilan, Qazvin, Tehran (Farahani et al., 2013), Hamadan (Rajabi Mazhar et al., 2018), Kerman (Madjdzadeh et al., 2021), Mazandaran (Kian et al., 2020), Semnan (Ghahari & Gadallah 2015), Iran (no specific locality cited) (Telenga, 1941; Tobias, 1976, 1986).

Distribution in the Middle East: Iran (see references above), Turkey (Aydoğdu & Beyarslan, 2007; Beyarslan, 2021).

Extralimital distribution: Nearctic (introduced), Oriental, Palaearctic [Adjacent countries to Iran: Afghanistan, Armenia, Azerbaijan, Kazakhstan, Russia, Turkmenistan].

Host records: Summarized by Yu et al. (2016) as being a parasitoid of the crambids *Loxostege sticticalis* (L.), and *Ostrinia nubilalis* (Hübner); the erebid *Eublemma pannonica* (Freyer); the noctuids *Agrotis segetum* (Denis and Schiffermüller), *Heliothis viriplaca* (Hufnagel), and *Spodoptera exigua* (Hübner); the pyralid *Ephestia kuehniella* Zeller; and the tortricid *Cydia pomonella* (L.).

Chelonus areolatus Cameron, 1906

Catalogs with Iranian records: Fallahzadeh and Saghaei (2010), Gadallah and Ghahari (2013), Farahani et al. (2016), Yu et al. (2016 as *Microchelonus* (*M.*) *areolatus* (Cameron, 1906)), Samin, Coronado-Blanco, Kavallieratos et al. (2018), Samin, Coronado-Blanco, Fischer et al. (2018).

Distribution in Iran: Iran (no specific locality cited) (Papp, 1996a; Shenefelt, 1973).

Distribution in the Middle East: Iran.

Extralimital distribution: Pakistan, Turkmenistan.

Host records: Unknown.

Chelonus argamani Papp, 2012

Distribution in the Middle East: Israel—Palestine (Papp, 2012).

Extralimital distribution: None.

Host records: Unknown.

Chelonus armeniacus Tobias, 1976

Catalogs with Iranian records: Farahani et al. (2016), Yu et al. (2016), Samin, Coronado-Blanco, Kavallieratos et al. (2018), Samin, Coronado-Blanco, Fischer et al. (2018).

Distribution in Iran: Alborz, Qazvin, Tehran (Farahani et al., 2013), Hamadan (Rajabi Mazhar et al., 2018).

Distribution in the Middle East: Iran.

Extralimital distribution: Armenia.

Host records: Unknown.

Chelonus arnoldi (Tobias, 1964)

Distribution in the Middle East: Turkey (Aydoğdu & Beyarslan, 2006).

Extralimital distribution: Hungary, Kazakhstan, Romania, Russia.

Host records: Unknown.

Chelonus artus (Tobias, 1986)

Distribution in the Middle East: Turkey (Papp, 2014a; Beyarslan, 2021 as *Microchelonus* (*Microchelonus*) *artus*).

Extralimital distribution: Armenia, Bulgaria, Moldova, Spain.

Host records: Unknown.

Chelonus atripes Thomson, 1874

Distribution in the Middle East: Turkey (Beyarslan, 1995; Aydoğdu & Beyarslan, 2006; Beyarslan, 2021 as *Microchelonus* (*Microchelonus*) *atripes*).

Extralimital distribution: Albania, Bulgaria, China, Croatia, Finland, Germany, Hungary, Kazakhstan, Kyrgyzstan, Lithuania, North Macedonia, Mongolia, Romania, Russia, Serbia, Slovakia, Sweden, Uzbekistan.

Host records: Summarized by Yu et al. (2016) as being a parasitoid of the coleophorids *Coleophora alticolella* Zeller, and *Coleophora glaucicolella* Wood.

Chelonus atrotibia (Papp, 2012)

Distribution in the Middle East: Israel—Palestine (Papp, 2012).

Extralimital distribution: None.

Host records: Unknown.

Chelonus balkanicus (Tobias, 2003)

Distribution in the Middle East: Israel—Palestine (Papp, 2012, 2014a).

Extralimital distribution: Croatia, Hungary.
Host records: Unknown.

Chelonus basalis Curtis, 1837

Catalogs with Iranian records: Fallahzadeh and Saghaei (2010), Gadallah and Ghahari (2013), Farahani et al. (2016), Yu et al. (2016 as *Microchelonus basalis* (Curtis, 1837)), Samin, Coronado-Blanco, Kavallieratos et al. (2018), Samin, Coronado-Blanco, Fischer et al. (2018).
Distribution in Iran: Iran (no specific locality cited) (Telenga, 1941).
Distribution in the Middle East: Egypt (Edmardash et al., 2011; Edmardash & Gadallah, 2019), Iran (Telenga, 1941), Israel−Palestine (Papp, 1970).
Extralimital distribution: Austria, Croatia, Czech Republic, Denmark, Finland, Germany, Hungary, Kazakhstan, Poland, Russia, Sweden, United Kingdom.
Host records: Unknown.

Chelonus beyarslani Aydoğdu, 2008

Distribution in the Middle East: Turkey (Aydoğdu, 2008).
Extralimital distribution: None.
Host records: Unknown.

Chelonus bidens Tobias, 1972

Catalogs with Iranian records: Gadallah and Ghahari (2013), Farahani et al. (2016), Yu et al. (2016), Samin, Coronado-Blanco, Kavallieratos et al. (2018), Samin, Coronado-Blanco, Fischer et al. (2018).
Distribution in Iran: Golestan (Ghahari, Fischer, Çetin Erdoğan et al., 2010), Kordestan (Samin, 2015), Lorestan (Ghahari, Fischer, Papp, & Tobias, 2012), Mazandaran (Ghahari & Fischer, 2011b; Ghahari, Fischer, Çetin Erdoğan et al., 2010; Sakenin et al., 2012).
Distribution in the Middle East: Iran (see references above), Turkey (Aydoğdu & Beyarslan, 2002; Beyarslan, 2021; Beyarslan et al., 2002).
Extralimital distribution: Former Czechoslovakia, Hungary, Kazakhstan, Russia, Ukraine.
Host records: Unknown.

Chelonus bimaculatus Szépligeti, 1896

Catalogs with Iranian records: No catalog.
Distribution in Iran: Ilam (Gadallah et al., 2018).
Distribution in the Middle East: Iran (Gadallah et al., 2018), Turkey (Aydoğdu & Beyarslan, 2011).
Extralimital distribution: Albania, Bulgaria, China, Croatia, former Czechoslovakia, Germany, Hungary, Kazakhstan, Mongolia, Russia, Serbia, Switzerland, Ukraine, Uzbekistan.
Host records: Summarized by Yu et al. (2016) as being a parasitoid of the tortricids *Gypsonoma dealbana* (Frölich), and *Gypsonoma incarnana* (Haworth).

Chelonus blackburni Cameron, 1886

Distribution in the Middle East: Egypt (Edmardash et al., 2011; Edmardash & Gadallah, 2019).
Extralimital distribution: Costa Rica, Hondouras, India, Mexico, Midway Islands, Pakistan, United States of America.
Host records: Summarized by Yu et al. (2016) as being a parasitoid of the acrolepiid *Acrolepiopsis assectella* (Zeller); the agonoxenid *Asymphorodes dimorpha* (Busck); the batrachedrid *Batrachedra cuniculator* Busck; the following crambids: *Hellula undalis* (Fabricus), *Lineodes ochrae* Walsington, *Omphisa anastomosalis* (Guenée), and *Spoladea recurvalis* (Fabricius); the following gelechiids: *Keifera lycopersicella* (Walsingham), *Pectinophora gossypiella* (Saunders), and *Phthorimaea operculella* (Zeller); the noctuid *Helicoverpa armigera* (Hübner); the nolid *Earias vittella* Fabricius; the plutellids *Plutella capparidis* Swezey, and *Plutella xylostella* L.; and the following pyralids: *Corcyra cephalonica* (Stainton), *Genophantis leahi* Swezey, and *Unadella humeralis* (Butler).

Chelonus bonellii (Nees von Esenbeck, 1816)

Catalogs with Iranian records: Farahani et al. (2016), Yu et al. (2016), Samin, Coronado-Blanco, Kavallieratos et al. (2018), Samin, Coronado-Blanco, Fischer et al. (2018).
Distribution in Iran: Tehran (Farahani et al., 2013).
Distribution in the Middle East: Iran.
Extralimital distribution: Albania, Azerbaijan, Bulgaria, China, Croatia, Germany, Italy, Kyrgyzstan, North Macedonia, Russia.
Host records: Unknown.

Chelonus brevimetacarpus (Tobias, 1995)

Distribution in the Middle East: Turkey (Papp, 2014a).
Extralimital distribution: Hungary, Russia.
Host records: Unknown.

Chelonus brevis Tobias, 1976

Distribution in the Middle East: Israel−Palestine (Papp, 2012).
Extralimital distribution: Hungary, Italy, Russia.
Host records: Unknown.

Chelonus breviventris Thomson, 1874

Catalogs with Iranian records: Fallahzadeh and Saghaei (2010), Gadallah and Ghahari (2013), Farahani et al. (2016), Yu et al. (2016), Samin, Coronado-Blanco, Kavallieratos et al. (2018), Samin, Coronado-Blanco, Fischer et al. (2018).
Distribution in Iran: Iran (no specific locality cited) (Papp, 1997).
Distribution in the Middle East: Iran, Jordan, Turkey (Papp, 1997).

Extralimital distribution: Austria, China, Croatia, Finland, Greece, Hungary, Sweden, former Yugoslavia.
Host records: Unknown.

Chelonus calcaratus (Tobias, 1989)
Distribution in the Middle East: Syria, Turkey (Papp, 2014a).
Extralimital distribution: China, North Macedonia, Mongolia.
Host records: Unknown.

Chelonus canescens Wesmael, 1835
Catalogs with Iranian records: Gadallah and Ghahari (2013), Farahani et al. (2016), Yu et al. (2016), Samin, Coronado-Blanco, Kavallieratos et al. (2018), Samin, Coronado-Blanco, Fischer et al. (2018).
Distribution in Iran: East Azarbaijan (Ghahari, Fischer, Çetin Erdoğan, Tabari et al., 2009), Guilan, Mazandaran (Farahani et al., 2013), Hamadan (Rajabi Mazhar et al., 2018).
Distribution in the Middle East: Iran (see references above), Turkey (Aydoğdu & Beyarslan, 2002, 2007, 2011; Beyarslan et al., 2002).
Extralimital distribution: Belarus, Belgium, Czech Republic, France, Germany, Hungary, Italy, Mongolia, Poland, Romania, Russia, Spain, Sweden, Switzerland, United Kingdom.
Host records: Recorded by Papp (1995, 1996b) as being a parasitoid of the tortricid *Cnephasia pasinana* (Hübner).

Chelonus capsa Tobias, 1972
Distribution in the Middle East: Israel−Palestine (Papp, 2012), Turkey (Aydoğdu & Beyarslan, 2011).
Extralimital distribution: China, Hungary, Mongolia, Russia.
Host records: Unknown.

Chelonus caradrinae Kokujev, 1914
Catalogs with Iranian records: No catalog.
Distribution in Iran: Hormozgan (Ameri et al., 2018), West Azarbaijan (Ghahari et al., 2020).
Distribution in the Middle East: Iran (Ameri et al., 2018), Turkey (Aydoğdu & Beyarslan, 2007).
Extralimital distribution: Azerbaijan, Croatia, Greece, Hungary, Italy, Kazakhstan, Moldova, Mongolia, Romania, Russia, Serbia, Slovakia, Turkmenistan, Ukraine, Uzbekistan.
Host records: Recorded by Kokujev (1914) and Ghahari et al. (2020) as being a parasitoid of the noctuid *Spodoptera exigua* (Hübner).

Chelonus carbonator Marshall, 1885
Catalogs with Iranian records: Gadallah and Ghahari (2013), Farahani et al. (2016), Yu et al. (2016), Samin,

Coronado-Blanco, Kavallieratos et al. (2018), Samin, Coronado-Blanco, Fischer et al. (2018).
Distribution in Iran: Isfahan (Ghahari, Fischer, & Papp, 2011).
Distribution in the Middle East: Iran.
Extralimital distribution: China, Croatia, India, Kazakhstan, Mongolia, Russia, Tajikistan.
Host records: Summarized by Yu et al. (2016) as being a parasitoid of the noctuid *Spodoptera litura* (Fabricius); and the tortricid *Cydia janthinana* (Duponchel).

Chelonus cesa Koçak and Kemal, 2013
Catalogs with Iranian records: Gadallah and Ghahari (2013 as *Chelonus asiaticus* Telenga, 1941), Farahani et al. (2016 as *Chelonus asiaticus*), Samin, Coronado-Blanco, Kavallieratos et al. (2018), Samin, Coronado-Blanco, Fischer et al. (2018).
Distribution in Iran: Golestan (Ghahari, Fischer, Çetin Erdoğan, Tabari et al., 2009 as *Chelonus asiaticus*), Hormozgan (Ameri et al., 2018 as *C. asiaticus*), Kermanshah (Ghahari & Fischer, 2012 as *Chelonus asiaticus*).
Distribution in the Middle East: Iran (see references above), Israel−Palestine (Papp, 2012), Turkey (Aydoğdu & Beyarslan, 2002, 2007; Beyarslan, 2021).
Extralimital distribution: Armenia, China, Czech Republic, Greece, Hungary, Kazakhstan, Lithuania, Moldova, Mongolia, Russia, Slovakia, Spain, Uzbekistan.
Host records: Unknown.

Chelonus chasanicus (Tobias, 2000)
Distribution in the Middle East: Jordan (Papp, 2014a).
Extralimital distribution: Russia.
Host records: Unknown.

Chelonus chetini Bayarslan and Şahan, 2019
Distribution in the Middle East: Turkey (Beyarslan & Şahan, 2019).
Extralimital distribution: None.
Host records: Unknown.

Chelonus circumfossa (Tobias, 2002)
Distribution in the Middle East: Turkey (Papp, 2014a).
Extralimital distribution: Croatia, Russia.
Host records: Unknown.

Chelonus cisapicalis (Tobias, 1989)
Catalogs with Iranian records: Gadallah and Ghahari (2013), Farahani et al. (2016), Yu et al. (2016 as *Microchelonus cisapicalis* (Tobias, 1989)), Samin, Coronado-Blanco, Kavallieratos et al. (2018), Samin, Coronado-Blanco, Fischer et al. (2018).
Distribution in Iran: Guilan (Ghahari, Fischer, & Tobias, 2012 as *Microchelonus cisapicalis* Tobias, 1989).

Distribution in the Middle East: Iran.
Extralimital distribution: Mongolia, Russia.
Host records: Unknown.

Chelonus continens (Tobias, 1989)
Distribution in the Middle East: Turkey (Papp, 2014a).
Extralimital distribution: Hungary, North Macedonia, Mongolia, Russia, former Yugoslavia.
Host records: Unknown.

Chelonus contractus (Nees von Esenbeck, 1816)
Catalogs with Iranian records: Fallahzadeh and Saghaei (2010), Gadallah and Ghahari (2013), Farahani et al. (2016), Yu et al. (2016), Samin, Coronado-Blanco, Kavallieratos et al. (2018), Samin, Coronado-Blanco, Fischer et al. (2018).
Distribution in Iran: Fars (Lashkari Bod et al., 2011a,b), Guilan (Papp, 2014a as *Microchelonus* (*Microchelonus*) *contractus* (Nees, 1816)), Tehran (Farahbakhsh, 1961 as *Chelonus contractor*; Modarres Awal, 1997, 2012).
Distribution in the Middle East: Iran (see references above), Syria (Papp, 2014a), Turkey (Lozan, 2005).
Extralimital distribution: Albania, Armenia, Azerbaijan, Belgium, Bosnia-Herzegovina, Bulgaria, China, Croatia, Czech Republic, Finland, France, Germany, Greece, Hungary, Italy, Kazakhstan, Korea, Kyrgyzstan, Lithuania, North Macedonia, Mongolia, Netherlands, Norway, Poland, Romania, Russia, Serbia, Slovakia, Spain, Sweden, Switzerland, Turkmenistan, Ukraine, United Kingdom.
Host records: Summarized by Yu et al. (2016) as being a parasitoid of the choreutids *Anthophila fabriciana* (L.) and *Prochoreutis myllerana* (Fabricius); the coleophorid *Coleophora hungariae* Gozmany; the cosmopterigid *Stagmatophora extremella* Wocke; the gelechiids *Mirificarma mulinella* (Zeller) and *Phthorimaea operculella* (Zeller); the plutellid *Plutella xylostella* (L.); the tischeriid *Emmetia szoecsi* (Kasy); the tortricids *Cydia nigricana* (Fabricius) and *Rhyacionia buoliana* (Denis and Schiffermüller); and the yponomeutid *Argyresthia pygmaeela* (Denis and Schiffermüller). In Iran, this species has been reared from the gelechiid *Scorbipalpa ocellatella* (Boyd) (Behdad, 1991; Farahbakhsh, 1961; Modarres Awal, 1997, 2012).

Chelonus corvulus Marshall, 1885
Catalogs with Iranian records: No catalog.
Distribution in Iran: Mazandaran (Ghahari & Beyarslan, 2019).
Distribution in the Middle East: Iran (Ghahari & Beyarslan, 2019), Israel–Palestine (Bodenheimer, 1930; Halperin, 1986; Kugler, 1966), Turkey (Aydoğdu & Beyarslan, 2011; Beyarslan, 2021; Kohl, 1905; Lozan, 2005).
Extralimital distribution: Azerbaijan, Belgium, China, Croatia, Czech Republic, Finland, France, Germany, Greece, Hungary, India, Italy, Kazakhstan, Latvia, Lithuania, North Macedonia, Moldova, Mongolia, Netherlands, Poland, Romania, Russia, Serbia, Slovakia, Slovenia, Spain, Switzerland, Tajikistan, United Kingdom.
Host records: In Iran, this species has been reared from the noctuid *Spodoptera exigua* (Hübner) (Ghahari & Beyarslan, 2019).

Chelonus curvimaculatus Cameron, 1906
Distribution in the Middle East: Egypt (Edmardash et al., 2011; Edmardash & Gadallah, 2019), Israel–Palestine (Papp, 2012).
Extralimital distribution: Australia, Botswana, Democratic Republic of Congo, Ethiopia, India, Kenya, Madagascar, Malawi, Mauritius, Namibia, Senegal, Somalia, South Africa, Sudan, Tanzania, Uganda, Zambia, Zimbabwe.
Host records: Summarized by Yu et al. (2016) as being a parasitoid of the curculionid *Piezotrachelus varium* (Walker); as well as the crambid *Epascestria frustalis* (Zeller); the following gelechiids: *Pectinophora gossypiella* (Saunders), *Phthorimaea operculella* (Zeller), *Platyedra cunctatrix* Meyrick, *Scrobipalpa aptatella* (Walker), and *Scrobipalpa ergasima* (Meyrick); the following noctuids: *Busseola fusca* (Fuller), *Helicoverpa armigera* (Hübner), *Helicoverpa assulta* (Guenée), *Sesamia calamistis* Hampson, *Spodoptera exigua* (Hübner), *Trichoplusia ni* (Hübner), and *Trichoplusia oricalcea* (Fabricius); the nolid *Earias insulana* (Boisduval); the plutellid *Plutella xylostella* (L.); and the following pyralids: *Chilo orichalcociliellus*, *Chilo partellus* (Swinhoe), *Chilo zacconius* Bleszyński, and *Phycita diaphana* (Staudinger).

Chelonus cylindrus (Klug, 1816)
Distribution in the Middle East: Turkey (Papp, 1997).
Extralimital distribution: Belarus, China, Denmark, Finland, Germany, Hungary, Kazakhstan, Moldova, Mongolia, Netherlands, Romania, Russia, Spain, Sweden, Switzerland, United Kingdom.
Host records: Unknown.

Chelonus cyprensis (Tobias, 2001)
Distribution in the Middle East: Cyprus (Tobias, 2001).
Extralimital distribution: None.
Host records: Unknown.

Chelonus cypri (Tobias, 2001)
Distribution in the Middle East: Cyprus (Tobias, 2001).
Extralimital distribution: None.
Host records: Unknown.

Chelonus cyprianus Fahringer, 1937
Distribution in the Middle East: Cyprus (Fahringer, 1937).
Extralimital distribution: None.
Host records: Unknown.

Chelonus dauricus Telenga, 1941

Catalogs with Iranian records: Samin, Coronado-Blanco, Fischer et al. (2018).
Distribution in Iran: East Azarbaijan (Sakenin et al., 2018).
Distribution in the Middle East: Iran (Sakenin et al., 2018), Turkey (Aydoğdu & Beyarslan, 2011; Beyarslan, 2021; Lozan, 2005).
Extralimital distribution: China, former Czechoslovakia, Mongolia, Montenegro, Russia, Serbia.
Host records: Unknown.

Chelonus depressus Thomson, 1874

Catalogs with Iranian records: Fallahzadeh and Saghaei (2010), Gadallah and Ghahari (2013), Farahani et al. (2016), Yu et al. (2016 as *Microchelonus depressus* (Thomson, 1874)), Samin, Coronado-Blanco, Kavallieratos et al. (2018), Samin, Coronado-Blanco, Fischer et al. (2018).
Distribution in Iran: Tehran (Davatchi & Shojai, 1969), Iran (no specific locality cited) (Jafaripour, 1969).
Distribution in the Middle East: Iran (Davatchi & Shojai, 1969; Jafaripour, 1969), Turkey (Aydoğdu & Beyarslan, 2006).
Extralimital distribution: Finland, Sweden.
Host records: Recorded by Izhevskiy (1985) as being a parasitoid of the gelechiid *Phthorimaea operculella* (Zeller).

Chelonus devius (Tobias, 1964)

Catalogs with Iranian records: No catalog.
Distribution in Iran: Golestan (Ghahari et al., 2019).
Distribution in the Middle East: Iran (Ghahari et al., 2019), Turkey (Aydoğdu & Beyarslan, 2006).
Extralimital distribution: China, Croatia, Greece, Kazakhstan, Kyrgyzstan, Serbia, Uzbekistan.
Host records: Unknown.

Chelonus elaeaphilus Silvestri, 1908

Distribution in the Middle East: Egypt (Edmardash & Gadallah, 2019), Israel−Palestine (Halperin, 1986; Papp, 2012).
Extralimital distribution: Hungary, Italy, Kazakhstan, Mongolia, Portugal, Serbia, Spain, Tajikistan, Tunisia, Uzbekistan.
Host records: Summarized by Yu et al. (2016) as being a parasitoid of the coleophorid *Coleophora hemerobiella* (Scopoli); the epermeniid *Ochromolopis staintonellus* (Millière); the plutellid *Prays oleae* Bernard; and the pyralid *Merulempista turturella* (Zeller).

Chelonus elongatus Szépligeti, 1898

Catalogs with Iranian records: Farahani et al. (2016), Yu et al. (2016), Samin, Coronado-Blanco, Kavallieratos et al. (2018), Samin, Coronado-Blanco, Fischer et al. (2018).

Distribution in Iran: Alborz, Guilan (Farahani et al., 2013), Hamadan (Rajabi Mazhar et al., 2018).
Distribution in the Middle East: Iran (Farahani et al., 2013; Rajabi Mazhar et al., 2018), Turkey (Aydoğdu & Beyarslan, 2007).
Extralimital distribution: China, Finland, Germany, Hungary, Poland, Serbia, Switzerland.
Host records: Unknown.

Chelonus erdosi (Tobias, 2001)

Distribution in the Middle East: Cyprus, Syria (Tobias, 2001), Israel−Palestine (Papp, 2012; Tobias, 2001).
Extralimital distribution: None.
Host records: Unknown.

Chelonus erosus Herrich-Schäffer, 1838

Catalogs with Iranian records: No catalog.
Distribution in Iran: Kermanshah (Samin et al., 2020 as *Microchelonus erosus*).
Distribution in the Middle East: Iran.
Extralimital distribution: Albania, Alegria, Austria, Azerbaijan, Belarus, Bulgaria, Finland, Georgia, Germany, Hungary, Kazakhstan, North Macedonia, Mongolia, Montenegro, Romania, Russia, Serbia.
Host records: Unknown.

Chelonus erythrogaster Lucas, 1849

Catalogs with Iranian records: Gadallah and Ghahari (2013), Farahani et al. (2016), Yu et al. (2016 as *Microchelonus erythrogaster* (Lucas, 1846)), Samin, Coronado-Blanco, Kavallieratos et al. (2018), Samin, Coronado-Blanco, Fischer et al. (2018).
Distribution in Iran: Fars (Lashkari Bod et al., 2011a,b), Hamadan (Rajabi Mazhar et al., 2018).
Distribution in the Middle East: Iran.
Extralimital distribution: Algeria, Croatia, Italy, Russia, Tunisia.
Host records: Unknown.

Chelonus exilis Marshall, 1885

Catalogs with Iranian records: No catalog.
Distribution in Iran: Mazandaran (Ghahari et al., 2020).
Distribution in the Middle East: Iran (Ghahari et al., 2020), Turkey (Aydoğdu & Beyarslan, 2006; Beyarslan, 1995).
Extralimital distribution: Armenia, Azerbaijan, Bulgaria, Croatia, former Czechoslovakia, France, Germany, Hungary, Italy, Kazakhstan, Latvia, Lithuania, Moldova, Mongolia, Netherlands, Poland, Russia, Serbia, Slovenia, Spain, Sweden, Switzerland, United Kingdom, Uzbekistan.
Host records: Summarized by Yu et al. (2016) as being a parasitoid of the cosmopterigids *Cosmopterix lienigiella* Zeller, and *Cosmopterix scribaiella* Zeller; the elachistid *Elachista gangabella* Zeller; and the pyralid *Myelois circumvoluta* (Furcroy).

Chelonus fenestratus (Nees von Esenbeck, 1816)

Catalogs with Iranian records: No catalog.

Distribution in Iran: Northern Khorasan (Samin et al., 2020 as *Microchelonus fenestralis*).

Distribution in the Middle East: Iran (Samin et al., 2020), Turkey (Aydoğdu & Beyarslan, 2006).

Extralimtal distribution: Azerbaijan, Belgium, Czech Republic, Finland, France, Germany, Hungary, Ireland, Italy, Kazakhstan, Korea, Moldova, Mongolia, Poland, Romania, Russia, Slovakia, Slovenia, Ukraine, United Kingdom, Uzbekistan, former Yugoslavia.

Host records: Summarized by Yu et al. (2016) as being a parasitoid of the tortricids *Dichrorampha alpinana* (Treitschke), *Dichrorampha petiverella* (L.), *Epiblema foenella* (L.), and *Rhyiacionia buoliana* (Denis and Schiffermüller).

Chelonus ferganicus (Tobias, 2001)

Distribution in the Middle East: Israel—Palestine (Papp, 2012).

Extralimital distribution: Uzbekistan.

Host records: Unknown.

Chelonus flavipalpis Szépligeti, 1896

Catalogs with Iranian records: Farahani et al. (2016), Yu et al. (2016 as *Microchelonus flavipalpis* (Szépligeti, 1896)), Samin, Coronado-Blanco, Kavallieratos et al. (2018), Samin, Coronado-Blanco, Fischer et al. (2018).

Distribution in Iran: Guilan (Farahani et al., 2013).

Distribution in the Middle East: Iran (Farahani et al., 2013), Turkey (Aydoğdu & Beyarslan, 2006).

Extralimital distribution: Georgia, Hungary, Moldova, Mongolia, Russia, Ukraine.

Host records: Summarized by Yu et al. (2016) as being a parasitoid of the elachistid *Haplochois theae* (Kusnezoc); the tineid *Kermania pistaciella* Amsel; and the tortricid *Sparganothis pilleriana* (Denis and Schiffermüller).

Chelonus flavonaevulus Abdinbekova, 1971

Catalogs with Iranian records: Farahani et al. (2016), Yu et al. (2016 as *Microchelonus flavonaevulus* (Abdinbekova, 1971)), Samin, Coronado-Blanco, Kavallieratos et al. (2018), Samin, Coronado-Blanco, Fischer et al. (2018).

Distribution in Iran: Iran (no specific locality cited) (Tobias, 2010; Papp, 2014a both as *Microchelonus (Microchelonus) flavonaevulus*).

Distribution in the Middle East: Iran (Papp, 2014a; Tobias, 2010), Turkey (Aydoğdu & Beyarslan, 2006; Beyarslan, 1995).

Extralimital distribution: Azerbaijan, Bulgaria, Greece, Hungary, North Macedonia, Moldova, Russia, former Yugoslavia.

Host records: Unknown.

Chelonus flavoscaposus (Tobias, 2001)

Distribution in the Middle East: Israel—Palestine (Papp, 2012).

Extralimital distribution: Italy.

Host records: Unknown.

Chelonus gracilis (Lozan & Tobias, 2006)

Distribution in the Middle East: Turkey (Papp, 2014a).

Extralimital distribution: Czech Republic, Greece, Macedonia, former Yugoslavia.

Host records: Unknown.

Chelonus halperini (Papp, 2012)

Distribution in the Middle East: Israel—Palestine (Papp, 2012).

Extralimital distribution: None.

Host records: Unknown.

Chelonus ibericus (Tobias, 2001)

Distribution in the Middle East: Turkey (Papp, 2014a).

Extralimital distribution: Czech Republic, Spain.

Host records: Unknown.

Chelonus inanitus (Linnaeus, 1767)

Catalogs with Iranian records: Fallahzadeh and Saghaei (2010), Gadallah and Ghahari (2013), Farahani et al. (2016), Yu et al. (2016), Samin, Coronado-Blanco, Kavallieratos et al. (2018), Samin, Coronado-Blanco, Fischer et al. (2018).

Distribution in Iran: Alborz (Davatchi & Shojai, 1969; Kheyri, 1977; Shojai, 1968, 1998), Isfahan (Bagheri & Basiri, 2004), Kermanshah (Ghahari & Fischer, 2012), Mazandaran (Ghahari, Fischer, Çetin Erdoğan, Tabari et al., 2009), Tehran (Modarres Awal, 1997, 2012), West Azerbaijan (Alizzadeh & Javan Moghaddam, 2004; Khanjani, 2006a; Modarres Awal, 2012), Iran (no specific locality cited) (Aubert, 1966; Shenefelt, 1973; Khanjani, 2006b).

Distribution in the Middle East: Cyprus (Yu et al., 2016), Egypt (Edmardash et al., 2011; Edmardash & Gadallah, 2019), Iran (see references above), Israel—Palestine (Halperin, 1986), Turkey (Aydoğdu & Beyarslan, 2011).

Extralimital distribution: Nearctic (introduced), Palaearctic [Adjacent countries to Iran: Armenia, Russia].

Host records: Summarized by Yu et al. (2016) as being a parasitoid of the crambids *Haritalodes derogata* (Fabricius), and *Ostrinia nubilalis* (Hübner); the noctuids *Leucania loreyi* (Duponchel), *Longalatedes elymi* (Treitschke), *Mesoligia literosa* (Haworth), *Mythimna unipuncta* (Haworth), *Oligia strigilis* (L.), *Peridroma saucia*

(Hübner), *Spodoptera exigua* (Hübner), and *Spodoptera littoralis* (Boisduval); the pyralid *Etiella zinckenella* (Treitschke); and the tortricids *Aethes francillana* (Fabricius), and *Eucosma aemulana* (Schlaeger). In Iran, this species has been reared from the gelechiid *Phthorimaea operculella* (Zeller) (Khanjani, 2006b), and the noctuids *Spodoptera exigua* (Hübner) (Davatchi & Shojai, 1969; Khanjani, 2006a,b; Kheyri, 1977; Modarres Awal, 1997, 2012), and *Agrotis segetum* (Denis & Schiffermüller, 1755) (Alizzadeh & Javan Moghaddam, 2004; Khanjani, 2006a,b; Modarres Awal, 2012).

Chelonus incisus (Tobias, 1986)

Catalogs with Iranian records: Gadallah and Ghahari (2013), Farahani et al. (2016), Yu et al. (2016 as *Microchelonus incisus* (Tobias, 1986)), Samin, Coronado-Blanco, Kavallieratos et al. (2018), Samin, Coronado-Blanco, Fischer et al. (2018).
Distribution in Iran: Guilan (Ghahari, Fischer, & Tobias, 2012 as *Microchelonus incises*), Semnan (Samin, Fischer, & Ghahari, 2015).
Distribution in the Middle East: Iran (Ghahari, Fischer, & Tobias, 2012; Samin, Fischer, & Ghahari, 2015), Syria (Papp, 2014a).
Extralimital distribution: Hungary, North Macedonia, Mongolia, Netherlands, Russia, former Yugoslavia.
Host records: Unknown.

Chelonus insidiator (Tobias, 1989)

Distribution in the Middle East: Egypt (Edmardash & Gadallah, 2019; Papp, 2014a).
Extralimital distribution: Hungary, Mongolia.
Host records: Unknown.

Chelonus iranicus Tobias, 1972

Catalogs with Iranian records: Fallahzadeh and Saghaei (2010), Gadallah and Ghahari (2013), Farahani et al. (2016), Yu et al. (2016), Samin, Coronado-Blanco, Kavallieratos et al. (2018), Samin, Coronado-Blanco, Fischer et al. (2018)—all as both *Chelonus (Chelonus) iranicus* and *Microchelonus (Microchelonus) iranicus* Tobias, 2001 [Synonym].
Distribution in Iran: Alborz (Papp, 2014a as *Microchelonus (Microchelonus) iranicus* Tobias, 2001), East Azarbaijan (Tabriz, Bagh-e Shomal: Holotype) (Tobias, 1972), Iran (no specific locality cited) (Tobias, 2001; Papp, 2014b as *Microchelonus iranicus*).
Distribution in the Middle East: Iran.
Extralimital distribution: None.
Host records: Unknown.

Chelonus jacobsoni Tobias, 1986

Distribution in the Middle East: Turkey (Beyarslan, 2021).
Extralimital distribution: Finland, Hungary, Romania, Russia, Switzerland.
Host records: Unknown.

Chelonus jordanicus (Tobias, 2001)

Distribution in the Middle East: Jordan (Tobias, 2001).
Extralimital distribution: None.
Host records: Unknown.

Chelonus karadagensis (Tobias, 2001)

Distribution in the Middle East: Turkey (Papp, 2014a).
Extralimital distribution: Croatia, Ukraine.
Host records: Unknown.

Chelonus kermakiae (Tobias, 2001)

Catalogs with Iranian records: Fallahzadeh and Saghaei (2010), Gadallah and Ghahari (2013), Farahani et al. (2016), Yu et al. (2016 as *Microchelonus kermakiae* Tobias, 2001), Samin, Coronado-Blanco, Kavallieratos et al. (2018), Samin, Coronado-Blanco, Fischer et al. (2018).
Distribution in Iran: Kerman (Mehrnejad, 2009; Mehrnejad & Basirat, 2009; van Achterberg & Mehrnejad, 2002).
Distribution in the Middle East: Iran.
Extralimital distribution: Kyrgyzstan.
Host records: Recorded by Tobias (2001) as being a parasitoid of the gelechiid *Schneideraria pistaciicola* (Danilevski). In Iran, this species has been reared from the teneid *Kermania pistaciella* Amsel (Mehrnejad, 2009, 2020; Mehrnejad & Basirat, 2009; van Achterberg & Mehrnejad, 2002).
Comments: Davatchi and Shojai (1969), Modarres Awal (1997) and Shojai (1998) recorded *Chelonella* sp. from Alborz, Kerman, and Tehran as the parasitoid of *Kermania pistaciella* Amsel (Lepidoptera: Tineidae) and *Scrobipalpa ocellatella* Boyd (Lepidoptera: Gelechiidae). Regarding to its host, it could be *C. (Microchelonus) kermakiae*.

Chelonus kiritshenkoi Tobias, 1976

Distribution in the Middle East: Turkey (Beyarslan, 2021 as *Microchelonus (Microchelonus) kiritshenkoi*).
Extralimital distribution: Azerbaijan, Bulgaria, North Macedonia, Russia, former Yugoslavia.
Host records: Unknown.

Chelonus kopetdagicus (Tobias, 1966)

Catalogs with Iranian records: Farahani et al. (2016).
Distribution in Iran: Hormozgan (Ameri et al., 2018), Kordestan (Ghahari et al., 2019 as *Chelonus caucasicus* (Abdinbekova, 1969)), Sistan and Baluchestan (Khajeh et al., 2014).
Distribution in the Middle East: Iran (see references above), Israel—Palestine (Papp, 2012), Turkey (Aydoğdu & Beyarslan, 2006; Beyarslan, 1995; Lozan, 2005; Beyarslan, 2021 as *Microchelonus (Microchelonus) kopetdagicus*).
Extralimital distribution: Azerbaijan, Croatia, Greece, Hungary, Kazakhstan, North Macedonia, Moldova, Montenegro, Russia, Serbia, Spain, Switzerland, Turkmenistan (Yu et al., 2016 as *Microchelonus kopetdagicus* Tobias, 1966).
Host records: Unknown.

Chelonus kotenkoi (Tobias, 1992)
Distribution in the Middle East: Turkey (Papp, 2014a).
Extralimital distribution: North Macedonia, Russia, former Yugoslavia.
Host records: Unknown.

Chelonus kryzhanovskii Tobias, 1966
Distribution in the Middle East: Israel—Palestine (Papp, 2012).
Extralimital distribution: China, Hungary, Spain, Turkmenistan.
Host records: Unknown.

Chelonus larsi Kittel, 2016
Distribution in the Middle East: Turkey (Papp, 2014a).
Extralimital distribution: Austria, Bulgaria, Finland, France, Hungary, Lithuania, Portugal, Russia, Spain.
Host records: Unknown.

Chelonus latifossa (Tobias, 1990)
Distribution in the Middle East: Syria, Turkey (Papp, 2014a).
Extralimital distribution: Bulgaria, Mongolia, Spain.
Host records: Unknown.

Chelonus latrunculus Marshall, 1885
Catalogs with Iranian records: No catalog.
Distribution in Iran: East Azarbaijan (Gadallah et al., 2021).
Distribution in the Middle East: Iran (Gadallah et al., 2021), Turkey (Aydoğdu & Beyarslan, 2006).
Extralimital distribution: Armenia, Azerbaijan, Czech Republic, Finland, Germany, Hungary, Italy, Moldova, Poland, Russia, Slovakia, Sweden, Switzerland, Tajikistan, Ukraine, United Kingdom.
Host records: Unknown.

Chelonus lissogaster Tobias, 1972
Catalogs with Iranian records: Samin, Coronado-Blanco, Kavallieratos et al. (2018).
Distribution in Iran: Kordestan (Samin, Coronado-Blanco, Kavallieratos et al., 2018).
Distribution in the Middle East: Iran (Samin, Coronado-Blanco, Kavallieratos et al., 2018), Turkey (Aydoğdu & Beyarslan, 2011).
Extralimital distribution: China, Kazakhstan, Norway, Ukraine.
Host records: Unknown.

Chelonus longiventris (Tobias, 1964)
Catalogs with Iranian records: This species is a new record for the fauna of Iran.
Distribution in Iran: Guilan province, Talesh, Khalifeh Sara, 2♀, August 2015.

Distribution in the Middle East: Iran (new record), Turkey (Tezcan et al., 2006; Beyarslan, 2021 as Microchelonus (Microchelonus) longiventris).
Extralimital distribution: Azerbaijan, Germany, Hungary, Kazakhstan, Lithuania, Moldova, Russia, Serbia.
Host records: Unknown.

Chelonus luteipalpis (Tobias, 1994)
Distribution in the Middle East: Turkey (Papp, 2014a).
Extralimital distribution: Russia.
Host records: Unknown.

Chelonus luzhetzkji (Tobias, 1966)
Catalogs with Iranian records: No catalog.
Distribution in Iran: Mazandaran (Samin et al., 2020 as Microchelonus luzhetzkji).
Distribution in the Middle East: Iran (Samin et al., 2020), Turkey (Beyarslan, 2021 as Microchelonus (Microchelonus) luzhetzkii).
Extralimital distribution: Armenia, Hungary, Kazakhstan, Mongolia, Romania, Tajikistan, Turkmenistan, Uzbekistan.
Host records: Recorded by Tobias (1966) as being a parasitoid of the yponomeutids Yponomeuta malinellus (Zeller), and Yponomeuta padella (L.).

Chelonus magnifissuralis Abdinbekova, 1971
Distribution in the Middle East: Turkey (Aydoğdu & Beyarslan, 2006; Beyarslan, 1995).
Extralimital distribution: Azerbaijan, Spain.
Host records: Unknown.

Chelonus medus Telenga, 1941
Catalogs with Iranian records: Fallahzadeh and Saghaei (2010), Gadallah and Ghahari (2013), Farahani et al. (2016), Yu et al. (2016), Samin, Coronado-Blanco, Kavallieratos et al. (2018), Samin, Coronado-Blanco, Fischer et al. (2018).
Distribution in Iran: Sistan and Baluchestan (Telenga, 1941).
Distribution in the Middle East: Iran.
Extralimital distribution: Turkmenistan.
Host records: Unknown.

Chelonus microphtalmus Wesmael, 1838
Catalogs with Iranian records: Farahani et al. (2016).
Distribution in Iran: Sistan and Baluchestan (Khajeh et al., 2014).
Distribution in the Middle East: Iran (Khajeh et al., 2014), Turkey (Aydoğdu & Beyarslan, 2006; Beyarslan, 1995; Lozan, 2005; Papp, 2014a).
Extralimital distribution: Albania, Belgium, Bulgaria, Croatia, Czech Republic, Denmark, Finland, France, Germany, Greece, Hungary, Italy, Kazakhstan, Korea, Kyrgyzstan, North Macedonia, Mongolia, Poland,

Romania, Russia, Serbia, Spain, Sweden, Switzerland, Tunisia, Turkmenistan, Ukraine, Uzbekistan (Yu et al., 2016 as *Microchelonus microphalmus* (Wesmael, 1838)). Host records: Recorded by Fulmek (1968) as being a parasitoid of the tephritid *Myopites inulaedyssentericae* Blot. It was also recorded by Meyer (1934) and Tobias (1986) as being a parasitoid of the coleophorid *Coleophora hemerobiella* (Scopoli).

Chelonus microsomus Tobias, 1964

Catalogs with Iranian records: Gadallah and Ghahari (2013), Farahani et al. (2016), Yu et al. (2016), Samin, Coronado-Blanco, Kavallieratos et al. (2018), Samin, Coronado-Blanco, Fischer et al. (2018).
Distribution in Iran: Golestan (Ghahari & Fischer, 2011b; Sakenin et al., 2012), Hamadan (Rajabi Mazhar et al., 2018), Kuhgiloyeh and Boyerahmad (Samin, van Achterberg, & Ghahari, 2015), Mazandaran (Ghahari, Fischer, Çetin Erdoğan et al., 2010; Ghahari, 2017—around rice fields), North of Iran (no specific locality cited) (Sakenin et al., 2011).
Distribution in the Middle East: Iran (see references above), Turkey (Aydoğdu & Beyarslan, 2002; Beyarslan et al., 2002).
Extralimital distribution: Kazakhstan, Uzbekistan.
Host records: Unknown.

Chelonus milkoi (Tobias, 2003)

Catalogs with Iranian records: Farahani et al. (2016), Yu et al. (2016 as *Microchelonus milkoi* Tobias, 2003), Samin, Coronado-Blanco, Kavallieratos et al. (2018), Samin, Coronado-Blanco, Fischer et al. (2018).
Distribution in Iran: Guilan (Papp, 2014a as *Microchelonus (Microchelonus) milkoi* Tobias, 2003).
Distribution in the Middle East: Iran.
Extralimital distribution: China, Kazakhstan, Kyrgyzstan.
Host records: Unknown.

Chelonus minifossa (Tobias, 1986)

Distribution in the Middle East: Turkey (Papp, 2014a; Beyarslan, 2021 as *Microchelonus (Microchelonus) minifossa*).
Extralimital distribution: Czech Republic, Denmark, Hungary, North Macedonia, Moldova, Slovakia, former Yugoslavia.
Host records: Unknown.

Chelonus mirandus Tobias, 1964

Catalogs with Iranian records: No catalog.
Distribution in Iran: Zanjan (Samin et al., 2019).
Distribution in the Middle East: Cyprus (Beyarslan et al., 2017), Iran (Samin et al., 2019).
Extralimital distribution: Hungary, Kazakhstan, Moldova, Mongolia, Poland, Slovakia, Spain.

Host records: Unknown.

Chelonus moczari (Papp, 2014)

Catalogs with Iranian records: Farahani et al. (2016), Yu et al. (2016), Samin, Coronado-Blanco, Kavallieratos et al. (2018), Samin, Coronado-Blanco, Fischer et al. (2018).
Distribution in Iran: Alborz (Papp, 2014a as *Microchelonus (M.) moczari* Papp, 2014).
Distribution in the Middle East: Iran.
Extralimital distribution: None.
Host records: Unknown.

Chelonus mongolicus (Telenga, 1941)

Catalogs with Iranian records: Gadallah and Ghahari (2013), Farahani et al. (2016), Yu et al. (2016 as *Microchelonus mongolicus* Telenga, 1941), Samin, Coronado-Blanco, Kavallieratos et al. (2018), Samin, Coronado-Blanco, Fischer et al. (2018).
Distribution in Iran: Guilan (Ghahari, Fischer, & Tobias, 2012 as *Microchelonus mongolicus* Telenga, 1941).
Distribution in the Middle East: Iran.
Extralimital distribution: Mongolia, Russia.
Host records: Unknown.

Chelonus monticola (Tobias, 2003)

Distribution in the Middle East: Turkey (Lozan, 2005).
Extralimital distribution: Kazakhstan.
Host records: Unknown.

Chelonus nachitshevanicus Abdinbekova, 1971

Distribution in the Middle East: Syria, Turkey (Papp, 2014a).
Extralimital distribution: Algeria, Armenia, Austria, Azerbaijan.
Host records: Unknown.

Chelonus nigritibialis Abdinbekova, 1971

Catalogs with Iranian records: No catalog.
Distribution in Iran: Hormozgan (Ameri et al., 2018).
Distribution in the Middle East: Iran (Ameri et al., 2018), Israel—Palestine (Papp, 2012), Jordan, Syria (Papp, 2014a), Turkey (Beyarslan, 1995, 2021 as *Microchelonus (Microchelonus) nigritibialis*; Aydoğdu & Beyarslan, 2006; Papp, 2014a).
Extralimital distribution: Algeria, Azerbaijan, Bulgaria, France, Hungary, Moldova, Netherlands, Spain, Tunisia (Yu et al., 2016 as *Microchelonus (M.) nigritibialis* (Abdinbekova, 1971)).
Host records: Unknown.

Chelonus nigritulus Dahlbom, 1833

Distribution in the Middle East: Turkey (Papp, 2014a).
Extralimital distribution: Bulgaria, Finland, Germany, Hungary, Sweden.

Host records: Unknown.

Chelonus obscuratus Herrich-Schäffer, 1838
Catalogs with Iranian records: No catalog.
Distribution in Iran: Hormozgan (Ameri et al., 2018).
Distribution in the Middle East: Egypt (Edmardash et al., 2011; Edmardash & Gadallah, 2019), Iran (Ameri et al., 2018), Israel–Palestine (Papp, 2012), Turkey (Aydoğdu & Beyarslan, 2007, 2011; Sertkaya & Beyarslan, 2004).
Extralimital distribution: Bulgaria, China, Czech Republic, Finland, France, Germany, Greece, Hungary, Italy, Kazakhstan, Moldova, Mongolia, Poland, Russia, Slovakia, Spain, Sweden, Switzerland, Tunisia, Ukraine, United Kingdom.
Host records: Summarized by Yu et al. (2016) as being a parasitoid of the noctuid *Spodoptera exigua* (Hübner); and the tortricid *Rhyacionia buoliana* (Denis and Schiffermüller).

Chelonus ocellatus Alexeev, 1971
Catalogs with Iranian records: Gadallah and Ghahari (2013), Farahani et al. (2016), Yu et al. (2016), Samin, Coronado-Blanco, Kavallieratos et al. (2018), Samin, Coronado-Blanco, Fischer et al. (2018).
Distribution in Iran: Kermenshah (Ghahari & Fischer, 2012), Mazandaran (Ghahari, Fischer, Çetin Erdoğan et al., 2010; Ghahari & Fischer, 2011b; Sakenin et al., 2012), Qazvin (Samin, 2015).
Distribution in the Middle East: Iran (see references above), Turkey (Aydoğdu & Beyarslan, 2002).
Extralimital distribution: Hungary, Slovakia, Turkmenistan, Ukraine.
Host records: Unknown.

Chelonus oculator (Fabricius, 1775)
Catalogs with Iranian records: Fallahzadeh and Saghaei (2010), Gadallah and Ghahari (2013), Farahani et al. (2016), Yu et al. (2016), Samin, Coronado-Blanco, Kavallieratos et al. (2018), Samin, Coronado-Blanco, Fischer et al. (2018).
Distribution in Iran: Alborz, Qazvin, Tehran (Farahani et al., 2013), Hormozgan (Ameri et al., 2018), Kerman (Madjdzadeh et al., 2021), Lorestan (Ghahari, Fischer, Papp, & Tobias, 2012), Mazandaran (Ghahari, 2017—around rice fields), Iran (no specific locality cited) (Telenga, 1941; Tobias, 1976, 1986), Hamadan (Rajabi Mazhar et al., 2018).
Distribution in the Middle East: Cyprus (Beyarslan et al., 2017), Egypt (Edmardash et al., 2011; Edmardash & Gadallah, 2019), Iran (see references above), Israel–Palestine (Papp, 2012), Turkey (Beyarslan et al., 2002; 2006; Beyarslan, 2021).
Extralimital distribution: Widely distributed in the Palaearctic region [Adjacent countries to Iran: Afghanistan, Azerbaijan, Kazakhstan, Russia, Turkmenistan].

Host records: Summarized by Yu et al. (2016) as being a parasitoid of the coleophorid *Coleophora anatipennella* (Hübner); the crambids *Loxostege sticticalis* (L.) and *Ostrinia nubilalis* (Hübner); the noctuids *Agrotis segetum* (Denis and Schiffermüller), *Helicoverpa armigera* (Hübner), *Heliothis peltigera* (Denis and Schiffermüller), *Heliothis viriplaca* (Hufnagel), *Leucania loreyi* (Duponchel), *Longalatedes elymi* (Treitschke), *Spodoptera exigua* (Hübner), and *Spodoptera littoralis* (Boisduval); the pyralids *Ephestia kuehniella* Zeller, and *Etiella zinckenella* (Treitschke); and the tortricids *Apotomis turbidana* (Hübner), *Archips rosana* (L.), and *Zeiraphera isertana* (Fabricius).

Chelonus olgae Kokujev, 1895
Catalogs with Iranian records: No catalog.
Distribution in Iran: Hamadan (Rajabi Mazhar et al., 2018).
Distribution in the Middle East: Iran (Rajabi Mazhar et al., 2018), Turkey (Beyarslan et al., 2002; Aydoğdu & Beyarslan, 2007).
Extralimital distribution: Georgia, Germany, Hungary, Kazakhstan, Spain, Uzbekistan.
Host records: Unknown.

Chelonus pannonicus Szépligeti, 1896
Catalogs with Iranian records: No catalog.
Distribution in Iran: Fars (Samin et al., 2019).
Distribution in the Middle East: Iran.
Extralimital distribution: China, Croatia, Czech Republic, Greece, Hungary, Mongolia, Ukraine, former Yugoslavia.
Host records: Recorded by Papp (1996b, 2003) as being a parasitoid of the erebid *Eublemma pannonica* Freyer.

Chelonus pectinophorae Cushman, 1931
Catalogs with Iranian records: Farahani et al. (2016).
Distribution in Iran: Sistan and Baluchestan (Khajeh et al., 2014).
Distribution in the Middle East: Iran.
Extralimital distribution: China, Costa Rica, Japan, Korea, Mexico, Mongolia, Russia (Yu et al., 2016 as *Microchelonus pectinophorae* (Cushman, 1931)).
Host records: Summarized by Yu et al. (2016) as being a parasitoid of the gelechiids *Dichomeris acuminata* Staudinger, and *Pectinophora gossyiella* (Saunders); the nolid *Earias cupreoviridis* (Walker); and the tortricids *Leguminivora glycinivorella* (Matsumura), and *Tetramoera schistaceana* (Snellen).

Chelonus pectoralis Tobias, 1976
Distribution in the Middle East: Turkey (Beyarslan, 2021 as *Microchelonus* (*Microchelonus*) *pectoralis*).
Extralimital distribution: Azerbaijan, Mongolia.
Host records: Unknown.

Chelonus phalloniae (Telenga, 1941)
Distribution in the Middle East: Cyprus (Beyarslan et al., 2017).

Extralimital distribution: Kazakhstan.
Host records: Unknown.

Chelonus phthorimaea Gahan, 1917
Distribution in the Middle East: Yemen (Kroschel, 1995; Kroschel & Koch, 1994).
Extralimital distribution: Australia (introduced), Bermuda (introduced), Chile (introduced), Mexico, South Africa (introduced), United States of America.
Host records: Summarized by Yu et al. (2016) as being a parasitoid of the following gelechiids: *Athrips rancidella* (Herrich-Schäffer), *Keiferia lycopersicella* (Walsingham), *Phtorimaea oprculella* (Zeller), *Tildenia glochinella* (Zeller), *Tildenia gudmannella* Walsingham, and *Tildenia inconspicuella* (Murtfeldt); and the tortricid *Ancylis comptana* (Frölich).

Chelonus planiventria Tobias, 1960
Distribution in the Middle East: Turkey (Beyarslan, 2021).
Extralimital distribution: Hungary, Tajikistan, Uzbekistan.
Host records: Unknown.

Chelonus processiventris Tobias, 1964
Distribution in the Middle East: Turkey (Aydoğdu & Beyarslan, 2007).
Extralimital distribution: Kazakhstan.
Host records: Unknown.

Chelonus productus Herrich-Schäffer, 1838
Catalogs with Iranian records: Gadallah and Ghahari (2013), Farahani et al. (2016), Yu et al. (2016), Samin, Coronado-Blanco, Kavallieratos et al. (2018), Samin, Coronado-Blanco, Fischer et al. (2018).
Distribution in Iran: Isfahan (Ghahari, Fischer, & Papp, 2011).
Distribution in the Middle East: Cyprus (Papp, 2003), Iran (Ghahari, Fischer, & Papp, 2011), Israel—Palestine (Papp, 2012).
Extralimital distribution: Germany, Greece, Hungary, Mongolia, Romania.
Host records: Unknown.

Chelonus przewalskii (Tobias, 2001)
Distribution in the Middle East: Turkey (Papp, 2014a).
Extralimital distribution: Kyrgyzstan.
Host records: Unknown.

Chelonus pusilloides Tobias, 1972
Distribution in the Middle East: Israel—Palestine (Papp, 2012).
Extralimital distribution: Kazakhstan, Mongolia.
Host records: Unknown.

Chelonus retusus (Nees von Esenbeck, 1816)
Catalogs with Iranian records: No catalog.
Distribution in Iran: Golestan (Gadallah et al., 2021).
Distribution in the Middle East: Iran.
Extralimital distribution: Armenia, Azerbaijan, Bulgaria, Czech Republic, Finland, France, Germany, Hungary, Italy, Moldova, Mongolia, Montenegro, Romania, Russia, Serbia, Slovakia, Sweden, Switzerland.
Host records: Recorded by Fahringer (1934) as being a parasitoid of the alucitid *Alucita hexadactyla* L.

Chelonus ripaeus (Tobias, 1986)
Distribution in the Middle East: Turkey (Papp, 2014a).
Extralimital distribution: Croatia, Denmark, France, Germany, North Macedonia, Netherlands, Russia, Spain, former Yugoslavia.
Host records: Unknown.

Chelonus risorius Reinhard, 1867
Catalogs with Iranian records: This species is a new record for the fauna of Iran.
Distribution in Iran: Mazandaran province, Ramsar, Dalkhani forest, 2♂, July 2016.
Distribution in the Middle East: Iran (new record), Turkey (Aydoğdu & Beyarslan, 2006; Beyarslan, 1995).
Extralimital distribution: Armenia, Croatia, Czech Republic, Finland, Germany, Hungary, Italy, Kazakhstan, Kyrgyzstan, Moldova, Mongolia, Russia, Slovakia, United Kingdom.
Host records: recorded by Papp (2004) as being a parasitoid of the cynipids *Biorhiza pallida* L., and *Biorhiza terminalis* (Hartig).

Chelonus rostratus (Tobias, 1966)
Catalogs with Iranian records: No catalog.
Distribution in Iran: Hormozgan (Ameri et al., 2018).
Distribution in the Middle East: Iran (Ameri et al., 2018), Turkey (Beyarslan, 1995, 2021 as *Microchelonus (Microchelonus) rostratus*); Lozan, 2005; Aydoğdu & Beyarslan, 2006; Papp, 2014a).
Extralimital distribution: Armenia, Azerbaijan, Bulgaria, Czech Republic, Denmark, Hungary, North Macedonia, Moldova, Russia, Slovakia, Spain, Turkmenistan, Ukraine, former Yugoslavia (Yu et al., 2016 as *Microchelonus (Microchelonus) rostratus* (Tobias)).
Host records: Unknown.

Chelonus rugicollis Thomson, 1870
Distribution in the Middle East: Turkey (Papp, 2014a; Beyarslan, 2021 as *Microchelonus (Microchelonus) rugicollis*), Yemen (Papp, 2014a).

Extralimital distribution: Albania, Azerbaijan, Hungary, Italy, Norway, Slovakia, Spain, Sweden, former Yugoslavia. Host records: Recorded by Györfi (1959) as being a parasitoid of *Alucita hexadactyla* L. (Lepidoptera: Alucitidae).

Chelonus rugilobus (Tobias, 1986)
Distribution in the Middle East: Israel−Palestine (Papp, 2012).
Extralimital distribution: Moldova.
Host records: Unknown.

Chelonus scabrator (Fabricius, 1793)
Catalogs with Iranian records: Fallahzadeh and Saghaei (2010), Gadallah and Ghahari (2013), Farahani et al. (2016), Yu et al. (2016), Samin, Coronado-Blanco, Kavallieratos et al. (2018), Samin, Coronado-Blanco, Fischer et al. (2018).
Distribution in Iran: Fars (Ghahari & Beyarslan, 2017), Guilan (Ghahari, Fischer, Çetin Erdoğan, Tabari et al., 2009), Hormozgan (Ameri et al., 2018), Kuhgiloyeh and Boyerahmad (Ghahari & Beyarslan, 2017), Sistan and Baluchestan (Khajeh et al., 2014), Iran (no specific locality cited) (Shenefelt, 1973; Telenga, 1941).
Distribution in the Middle East: Iran (see references above), Turkey (Aydoğdu & Beyarslan, 2002, 2007, 2011; Beyarslan et al., 2002, 2006; Lozan, 2005).
Extralimital distribution: Armenia, China, Kazakhstan, Mongolia, Russia, Ukraine.
Host records: Summarized by Yu et al. (2016) as being a parasitoid of the crambid *Loxostege sticticalis* (L.); the noctuid *Oligia strigilis* (L.); and the tortricid *Rhopobota stagnana* (Denis & Schiffermüller).

Chelonus scabrosus Szépligeti, 1896
Catalogs with Iranian records: No catalog.
Distribution in Iran: Kermanshah (Samin et al., 2020 as *Microchelonus scabrosus*).
Distribution in the Middle East: Iran (Samin et al., 2020), Turkey (Lozan, 2005).
Extralimital distribution: Algeria, Bulgaria, former Czechoslovakia, Greece, Hungary, Kyrgyzstan, Moldova, Poland, Romania, Russia, Serbia, Slovakia, Spain, Uzbekistan.
Host records: Unknown.

Chelonus setaceus Papp, 1993
Catalogs with Iranian records: Fallahzadeh and Saghaei (2010), Gadallah and Ghahari (2013), Farahani et al. (2016), Yu et al. (2016), Samin, Coronado-Blanco, Kavallieratos et al. (2018), Samin, Coronado-Blanco, Fischer et al. (2018).
Distribution in Iran: Hormozgan (Ameri et al., 2018), Tehran (Farahani et al., 2013; Papp, 1993).
Distribution in the Middle East: Iran.
Extralimital distribution: None.
Host records: Unknown.

Chelonus silvestrii (Papp, 1999)
Distribution in the Middle East: Israel−Palestine, Yemen (Papp, 1999).
Extralimital distribution: Italy, Spain.
Host records: Recorded by Papp (1999) as being a parasitoid of the plutellid *Prays oleae* Bernard.

Chelonus smirnovi (Telenga, 1953)
Catalogs with Iranian records: Fallahzadeh and Saghaei (2010), Gadallah and Ghahari (2013), Farahani et al. (2016), Yu et al. (2016), Samin, Coronado-Blanco, Kavallieratos et al. (2018), Samin, Coronado-Blanco, Fischer et al. (2018).
Distribution in Iran: Lorestan (Ghahari, Fischer, Papp, & Tobias, 2012), Iran (no specific locality cited) (Tobias, 1976).
Distribution in the Middle East: Iran (Tobias, 1976; Ghahari, Fischer, Papp, & Tobias, 2012), Turkey (Aydoğdu & Beyarslan, 2002, 2007).
Extralimital distribution: Mongolia, Russia, Turkmenistan.
Host records: Recorded by Telenga (1953) and Tobias (1976) as being a parasitoid of the pyralid *Etiella zinckenella* (Treitschke).

Chelonus subarcuatilis (Tobias, 1986)
Catalogs with Iranian records: No catalog.
Distribution in Iran: Razavi Khorasan, Semnan (Gadallah et al., 2021).
Distribution in the Middle East: Iran (Gadallah et al., 2021), Turkey (Aydoğdu & Beyarslan, 2006).
Extralimital distribution: Armenia, Bulgaria, Hungary, Kazakhstan, Moldova, Spain, Turkmenistan, Uzbekistan.
Host records: Unknown.

Chelonus subcontractus Abdinbekova, 1971
Catalogs with Iranian records: Gadallah and Ghahari (2013), Farahani et al. (2016), Yu et al. (2016 as *Microchelonus subcontractus* (Abdinbekova, 1971)), Samin, Coronado-Blanco, Kavallieratos et al. (2018), Samin, Coronado-Blanco, Fischer et al. (2018).
Distribution in Iran: Razavi Khorasan (Mahmoudi et al., 2013), Tehran (Abbasipour et al., 2012 as *Microchelonus subcontractus* (Abdinbekova, 1971)).
Distribution in the Middle East: Iran (Abbasipour et al., 2012; Mahmoudi et al., 2013), Turkey (Aydoğdu & Beyarslan, 2006; Beyarslan, 1995).
Extralimital distribution: Azerbaijan, Georgia, Greece, Hungary, Kazakhstan, Lithuania, Moldova, Mongolia, Poland, Romania, Russia, Serbia, Slovakia, Ukraine.
Host records: Summarized by Yu et al. (2016) as being a parasitoid of the cosmopterigid *Stagmatophora extremella* Wocke; the elachistid *Elachista* sp.; and the gelechiid *Phthorimaea operculella* (Zeller). In Iran, this species has been reared from the gelechiid *Scrobipalpa ocellatella* Boyd (Abbasipour et al., 2012; Mahmoudi et al., 2013).

Chelonus submuticus Wesmael, 1835
Catalogs with Iranian records: No catalog.
Distribution in Iran: Lorestan (Ghahari & Gadallah, 2019).
Distribution in the Middle East: Iran (Ghahari & Gadallah, 2019), Israel—Palestine (Gerling, 1969; Halperin, 1986), Turkey (Aydoğdu & Beyarslan, 2011).
Extralimital distribution: Albania, Azerbaijan, Belgium, Bulgaria, Croatia, Finland, France, Georgia, Hungary, Italy, Kazakhstan, North Macedonia, Moldova, Mongolia, Poland, Romania, Russia, Serbia, Slovakia, Sweden, Switzerland, Ukraine, United Kingdom, Uzbekistan.
Host records: Summarized by Yu et al. (2016) as being a parasitoid of the noctuid *Spodoptera littoralis* (Boisduval); and the pyralid *Homoeosoma nebulella* (Denis and Schiffermüller).

Chelonus subpamiricus Farahani and van Achterberg, 2018
Catalogs with Iranian records: No catalog.
Distribution in Iran: Hamadan (Rajabi Mazhar et al., 2018).
Distribution in the Middle East: Iran.
Extralimital distribution: None.
Host records: Unknown.

Chelonus subpusillus (Tobias, 1997)
Catalogs with Iranian records: Farahani et al. (2016), Yu et al. (2016 as *Microchelonus subpusillus* Tobias, 1997), Samin, Coronado-Blanco, Kavallieratos et al. (2018), Samin, Coronado-Blanco, Fischer et al. (2018).
Distribution in Iran: Fars (Papp, 2014a as *Microchelonus (M.) subpusillus* Tobias, 1997).
Distribution in the Middle East: Iran, Turkey (Papp, 2014a).
Extralimital distribution: Kazakhstan, Romania, Tajikistan.
Host records: Unknown.

Chelonus subseticornis Tobias, 1971
Catalogs with Iranian records: No catalog.
Distribution in Iran: Lorestan (Ghahari et al., 2019).
Distribution in the Middle East: Iran.
Extralimital distribution: Azerbaijan, Bulgaria, China, Greece, Hungary, Kazakhstan, Moldova, Mongolia, Russia, Serbia, Slovakia.
Host records: Unknown.

Chelonus subsulcatus Herrich-Schäffer, 1838
Distribution in the Middle East: Turkey (Papp, 2014a).
Extralimital distribution: Germany, Hungary, Macedonia, Sweden, former Yugoslavia.
Host records: Unknown.

Chelonus sulcatus Jurine, 1807
Catalogs with Iranian records: No catalog.
Distribution in Iran: Hormozgan (Ameri et al., 2018).

Distribution in the Middle East: Egypt (Edmardash et al., 2011; Edmardash & Gadallah, 2019), Iran (Ameri et al., 2018), Israel—Palestine (Kugler, 1966), Turkey (Beyarslan, 1995, 2021 as *Microchelonus (Microchelonus) sulcatus*).
Extralimital distribution: Widely distributed in the Palaearctic region [Adjacent countries to Iran: Armenia, Azerbaijan, Kazakhstan, Russia] (Yu et al., 2016 as *Microchelonus (M.) sulcatus* (Jurine)).
Host records: Summarized by Yu et al. (2016) as being a parasitoid of the mycetophilid *Noeompheria striata* Meigen; the carposinid *Carposina sasakii* Matsumara; the coleophorids *Augasma aeratella* (Zeller), and *Coleophora tadzhikiella* Danilevsky; the gelechiids *Pectinophora gossypiella* (Saunders), *Pexicopia malvella* (Hübner), and *Scrobipalpa ocellatella* (Boyd); the plutellid *Prays citri* Millière; and the tortricid *Rhyacionia buoliana* (Denis and Schiffermüller).

Chelonus szepligetii Dalla Torre, 1898
Catalogs with Iranian records: Gadallah and Ghahari (2013), Farahani et al. (2016), Yu et al. (2016), Samin, Coronado-Blanco, Kavallieratos et al. (2018), Samin, Coronado-Blanco, Fischer et al. (2018).
Distribution in Iran: Chaharmahal and Bakhtiari (Samin, van Achterberg, & Ghahari, 2015), Razavi Khorasan (Samin et al., 2011), Kermanshah (Ghahari & Fischer, 2012), Qazvin (Farahani et al., 2013).
Distribution in the Middle East: Iran (see references above), Israel—Palestine (Papp, 2012), Turkey (Aydoğdu & Beyarslan, 2002, 2007).
Extralimital distribution: Azerbaijan, Croatia, Hungary, former Yugoslavia.
Host records: Unknown.

Chelonus talizkii (Tobias, 1986)
Distribution in the Middle East: Turkey (Papp, 2014a).
Extralimital distribution: Bulgaria, Hungary, North Macedonia, Moldova, former Yugoslavia.
Host records: Unknown.

Chelonus telengai (Abdinbekova, 1965)
Catalogs with Iranian records: Fallahzadeh and Saghaei (2010), Gadallah and Ghahari (2013), Farahani et al. (2016), Yu et al. (2016 as *Microchelonus telengai* (Abdinbekova, 1965)), Samin, Coronado-Blanco, Kavallieratos et al. (2018), Samin, Coronado-Blanco, Fischer et al. (2018).
Distribution in Iran: Alborz, Qazvin, Tehran (Farahani et al., 2013), Iran (no specific locality cited) (Abdinbekova, 1965; Tobias, 1976, 1986, 2001).
Distribution in the Middle East: Iran (see references above), Israel—Palestine (Papp, 2012).
Extralimital distribution: Armenia, Azerbaijan.
Host records: Unknown.

Chelonus temporalis (Tobias, 1986)
Distribution in the Middle East: Turkey (Lozan, 2005).
Extralimital distribution: Austria, China, Czech Republic, Kyrgyzstan, Russia, Ukraine.
Host records: Unknown.

Chelonus turcius (Tobias, 2008)
Distribution in the Middle East: Turkey (Tobias, 2008).
Extralimital distribution: None.
Host records: Unknown.

Chelonus versatilis (Wilkinson, 1932)
Distribution in the Middle East: Egypt (Edmardash & Gadallah, 2019).
Extralimital distribution: Somalia, Sudan.
Host records: Summarized by Yu et al. (2016) as being a parasitoid of the gelechiid *Pectinophora gossypiella* (Saunders); and the noctuids *Helicoverpa armigera* (Hübner), and *Helicoverpa zea* (Boddie).

Chelonus vescus (Kokujev, 1899)
Catalogs with Iranian records: No catalog.
Distribution in Iran: Guilan (Ghahari et al., 2020).
Distribution in the Middle East: Iran (Ghahari et al., 2020), Turkey (Beyarslan, 2021 as *Microchelonus (Microchelonus) vescus*).
Extralimital distribution: Armenia, Azerbaijan, Hungary, Kazakhstan, North Macedonia, Russia, former Yugoslavia.
Host records: Unknown.

Chelonus wesmaelii Curtis, 1837
Catalogs with Iranian records: Samin, Coronado-Blanco, Fischer et al. (2018).
Distribution in Iran: Hamadan (Sakenin et al., 2018).
Distribution in the Middle East: Iran (Sakenin et al., 2018), Turkey (Aydoğdu & Beyarslan, 2011).
Extralimital distribution: Germany, Hungary, Poland, Russia, Slovakia, United Kingdom.
Host records: Recorded by Shenefelt (1973) as being a parasitoid of the curculionid *Scolytus scolytus* (Fabricius).

Chelonus xanthoscaposus (Tobias, 2001)
Distribution in the Middle East: Syria (Tobias, 2001).
Extralimital distribution: None.
Host records: Unknown.

Chelonus xenia (Tobias, 2000)
Distribution in the Middle East: Turkey (Papp, 2014a).
Extralimital distribution: Greece, Hungary, North Macedonia, Mongolia, Russia, former Yugoslavia.
Host records: Unknown.

Chelonus (Parachelonus) pellucens (Nees von Esenbeck, 1816)
Catalogs with Iranian records: Gadallah and Ghahari (2013 as *Chelonus (Parachelonus) pellucens* (Nees, 1816)),
Farahani et al. (2016), Yu et al. (2016 as *Microchelonus pellucens* (Nees, 1816)), Samin, Coronado-Blanco, Kavallieratos et al. (2018), Samin, Coronado-Blanco, Fischer et al. (2018).
Distribution in Iran: Fars (Samin, van Achterberg, & Ghahari, 2015 as *Chelonus (Parachelonus) pellucens*), Guilan (Ghahari & Fischer, 2011b; Sakenin et al., 2012, both as *C. varimaculatus* Tobias, 1986), Lorestan (Ghahari, Fischer, Papp, & Tobias, 2012 as *C. varimaculatus*), West Azarbaijan (Ghahari & Fischer, 2011a as *C. (Parachelonus) pellucens*), North of Iran (no specific locality cited) (Sakenin et al., 2011 as *C. varimaculatus*).
Distribution in the Middle East: Iran (see references above), Israel−Palestine (Papp, 2012), Turkey (Aydoğdu & Beyarslan, 2002; Beyarslan, 2021; Beyarslan et al., 2002; Doganlar, 1982).
Extralimital distribution: Armenia, Azerbaijan, Bosnia-Herzegovina, Bulgaria, Croatia, former Czechoslovakia, France, Germany, Hungary, Italy, Kazakhstan, Kyrgyzstan, Lithuania, North Macedonia, Moldova, Mongolia, Poland, Russia, Serbia, Spain, Switzerland, Ukraine, United Kingdom, Uzbekistan.
Host records: Summarized by Yu et al. (2016) as being a parasitoid of the sesiids *Bembecia ichneumoniformis* (Denis and Schiffermüller), *Bembecia scopigera* (Scopoli), *Chamaesphecia euceraeformis* (Ochsenheimer), *Chamaesphecia hungarica* (Tomala), and *Chamaesphecia tenthrediniformis* (Denis & Schiffermüller).

Chelonus (Parachelonus) starki (Telenga, 1953)
Distribution in the Middle East: Turkey (Beyarslan, 2021 as *Microchelonus (Parachelonus) starki*).
Extralimital distribution: Bulgaria, Czech Republic, Hungary, Kazakhstan, Moldova, Mongolia, Russia.
Host records: Summarized by Yu et al. (2016) as being a parasitoid of the following sesiids: *Chamaesphecia bibioniformis* (Esper), *Paranthrene tabaniformis* (Rottemburg), *Synanthedon formicaeformis* (Esper), *Synanthedon myopaeformis* (Borkhauser), and *Synanthedon tipuliformis* (Clerck); as well as the tortricid *Cydia pomonella* (L.).

Chelonus (Stylochelonus) elachistae (Tobias, 1995)
Distribution in the Middle East: Turkey (Tobias, 1995).
Extralimital distribution: Czech Republic, Georgia, Hungary, Poland.
Host records: Unknown.

Chelonus (Stylochelonus) mucronatus Thomson, 1874
Catalogs with Iranian records: Gadallah and Ghahari (2013), Farahani et al. (2016), Yu et al. (2016 as *Microchelonus (Stylochelonus) mucronatus* (Thomson, 1874)), Samin, Coronado-Blanco, Kavallieratos et al. (2018), Samin, Coronado-Blanco, Fischer et al. (2018).
Distribution in Iran: Ilam (Ghahari, Fischer, Hedqvist et al., 2010 as *Chelonus (Microchelonus) mucronatus*

(Thomson, 1874)), Kordestan (Samin, 2015 as *C. (M.)* *mucronatus*).
Distribution in the Middle East: Iran (Ghahari, Fischer, Hedqvist et al., 2010; Samin, 2015), Turkey (Aydoğdu & Beyarslan, 2002).
Extralimital distribution: China, Hungary, Russia, Sweden, Ukraine, Uzbekistan.
Host records: Unknown.

Chelonus (Stylochelonus) pusillus (Szépligeti, 1908)
Catalogs with Iranian records: No catalog.
Distribution in Iran: Mazandaran (Ghahari et al., 2019).
Distribution in the Middle East: Cyprus (Beyarslan et al., 2017), Iran (Ghahari et al., 2019).
Extralimital distribution: Azerbaijan, Croatia, Czech Republic, Denmark, Finland, Georgia, Hungary, Italy, Lithuania, Moldova, Mongolia, Poland, Russia, Serbia, Spain, Ukraine.
Host records: Recorded by Papp (1996) as being a parasitoid of the elachistid *Elachista* sp.

Chelonus sp.
Distribution in Iran: Qazvin (Shojai, 1998), Qom (Goldansaz et al., 2011), Tehran (Kishani Farahani et al., 2010).
Host records in Iran: *Kermania pistaciella* Amsel (Lepidoptera: Tineidae) (Shojai, 1998), and *Ectomyelois ceratoniae* (Zeller, 1839) (Lepidoptera: Pyralidae) (Goldansaz et al., 2011; Kishani Farahani et al., 2010).

Tribe Phanerotomini Baker, 1926
Genus *Phanerotoma* Wesmael, 1838
Phanerotoma acara van Achterberg, 1990
Distribution in the Middle East: Egypt (Edmardash & Gadallah, 2019).
Extralimital distribution: Spain.
Host records: Unknown.

Phanerotoma acuminata Szépligeti, 1908
Catalogs with Iranian records: Gadallah and Ghahari (2013), Farahani et al. (2016), Yu et al. (2016), Samin, Coronado-Blanco, Kavallieratos et al. (2018), Samin, Coronado-Blanco, Fischer et al. (2018).
Distribution in Iran: Guilan (Farahani, Talebi, & Rakhshani, 2012).
Distribution in the Middle East: Iran (Farahani, Talebi, & Rakhshani, 2012), Turkey (Beyarslan et al., 2002).
Extralimital distribution: Armenia, Azerbaijan, Canada, Europe, Georgia, Korea, Mongolia, Russia, United States of America, Ukraine.
Host records: Recorded by Papp (1996) as being a parasitoid of the pyralid *Hypsopygia costalis* (Fabricius).

Phanerotoma angusticrus van Achterberg, 2021
Distribution in the Middle East: United Arab Emirates, Yemen (van Achterberg, 2021).
Extralimital distribution: None.
Host records: Unknown.

Phanerotoma arabica Ghramh, 2011
Distribution in the Middle East: Saudi Arabia (Ghramh, 2011).
Extralimital distribution: None.
Host records: Unknown.

Phanerotoma artocornuta van Achterberg, 2021
Distribution in the Middle East: United Arab Emirates, Yemen (van Achterberg, 2021).
Extralimital distribution: None.
Host records: Unknown.

Phanerotoma aspidiota van Achterberg, 2021
Distribution in the Middle East: Yemen (van Achterberg, 2021).
Extralimital distribution: None.
Host records: Unknown.

Phanerotoma atra Šnoflak, 1951
Catalogs with Iranian records: No catalog.
Distribution in Iran: Mazandaran (Gadallah et al., 2021).
Distribution in the Middle East: Iran (Gadallah et al., 2021), Turkey (Beyarslan et al., 2002).
Extralimital distribution: Armenia, Austria, Azerbaijan, former Czechoslovakia, France, Greece, Hungary, North Macedonia, Moldova, Russia, Serbia, Spain, Switzerland.
Host records: Unknown.

Phanerotoma bilinea Lyle, 1924
Catalogs with Iranian records: No catalog.
Distribution in Iran: Fars (Ghahari et al., 2019).
Distribution in the Middle East: Iran (Ghahari et al., 2019), Yemen (van Achterberg, 2021).
Extralimital distribution: Austria, Azerbaijan, Belgium, Czech Republic, France, Germany, Greece, Hungary, Japan, Korea, Moldova, Netherlands, Poland, Romania, Russia, Serbia, Slovakia, Spain, Switzerland, Ukraine, United Kingdom.
Host records: Summarized by Yu et al. (2016) as being a parasitoid of the praydid *Prays citri* (Millière); and the tortricid *Argyrotaenia ljungiana* (Thunberg).

Phanerotoma brunneivena van Achterberg, 2021
Distribution in the Middle East: United Arab Emirates, Yemen (van Achterberg, 2021).

Extralimital distribution: None.
Host records: Unknown.

Phanerotoma capeki van Achterberg, 1990

Distribution in the Middle East: Turkey (van Achterberg, 1990).
Extralimital distribution: Former Czechoslovakia, Hungary, Switzerland.
Host records: Recorded by van Achterberg (1990) as being a parasitoid of *Teleiodes* sp. (Lepidoptera: Gelechiidae).

Phanerotoma caudatoides van Achterberg, 2021

Distribution in the Middle East: Yemen (van Achterberg, 2021).
Extralimital distribution: None.
Host records: Unknown.

Phanerotoma dentata (Panzer, 1805)

Catalogs with Iranian records: No catalog.
Distribution in Iran: Guilan (Ghahari & Gadallah, 2021), Hamadan (Gadallah et al., 2021).
Distribution in the Middle East: Egypt (Edmardash et al., 2011; Edmardash & Gadallah, 2019), Iran (Gadallah et al., 2021; Ghahari & Gadallah, 2021).
Extralimital distribution: Algeria, Argentina, Armenia, Austria, Azerbaijan, Belgium, Croatia, Finland, France, Germany, Greece, Hungary, Italy, Kazakhstan, Kenya, Korea, Latvia, Lithuania, Moldova, Netherlands, Norway, Poland, Romania, Russia, Slovakia, Slovenia, Spain, Sweden, Switzerland, Ukraine, United Kingdom, Uzbekistan, former Yugoslavia.
Host records: Summarized by Yu et al. (2016) as being a parasitoid of the gelechiid *Gelechia turpella* (Denis and Schiffermüller); the erebid *Lyamantria dispar* (L.); the praydid *Prays citri* Millière; the pyralids *Acrobasis advenella* Zincken, *Acrobasis consociella* (Hübner), *Acrobasis sodalella* Zeller, *Acrobasis tumidana* (Denis and Schiffermüller), and *Etiella zinckenella* (Treitscke); the sesiid *Synanthedon andrenaeformis* (Laspeyres); and the following tortricids: *Cydia pomonella* (L.), *Grapholita funebrana* (Treitschke), *Pammene amygdalana* (Duponchel), *Pammene populana* (Fabricius), *Pammene gallicolana* (Lienig and Zeller), *Pammene regiana* (Zeller), *Phalonidia curvistrigana* (Stainton), *Rhopobota ustomaculana* (Curtis), *Spilonota ocellana* (Denis and Schiffermüller), and *Zeiraphera isertana* (Fabricius). In Iran, this species has been reared from the erebid *Lymantria dispar* (L.) (Ghahari & Gadallah, 2021), and the tortricid *Cydia pomonella* (L.) (Gadallah et al., 2021).

Phanerotoma diversa (Walker, 1874)

Distribution in the Middle East: Egypt (Edmardash & Gadallah, 2019).
Extralimital distribution: Canada, former Czechoslovakia, Hungary, Italy, Japan, Korea, Mongolia, Russia, Spain, United States of America.

Host records: Recorded by Tobias (1971) as being a parasitoid of the pyralid *Acrobasis cymindella* (Ragonot).

Phanerotoma ejuncida van Achterberg, 2021

Distribution in the Middle East: United Arab Emirates (van Achterberg, 2021).
Extralimital distribution: None.
Host records: Unknown.

Phanerotoma elbaiensis Edmardash, Abdel Dayem and Gadallah, 2011

Distribution in the Middle East: Egypt (Edmardash et al., 2011; Edmardash & Gadallah, 2019).
Extralimital distribution: None.
Host records: Unknown.

Phanerotoma flavivena Edmardash and Gadallah, 2019

Distribution in the Middle East: Egypt (Edmardash and Gadallah, 2019), Yemen (Edmardash and Gadallah, 2019; van Achterberg, 2021).
Extralimital distribution: None.
Host records: Unknown.

Phanerotoma fracta Kokujev, 1903

Catalogs with Iranian records: Fallahzadeh and Saghaei (2010), Gadallah and Ghahari (2013), Farahani et al. (2016), Yu et al. (2016), Samin, Coronado-Blanco, Kavallieratos et al. (2018), Samin, Coronado-Blanco, Fischer et al. (2018).
Distribution in Iran: Sistan and Baluchestan (Hedwig, 1957), Iran (no specific locality cited) (Shenefelt, 1973; van Achterberg, 1990).
Distribution in the Middle East: Egypt (Edmardash & Gadallah, 2019), Iran (see references above), Israel—Palestine (Papp, 2012).
Extralimital distribution: Austria, Croatia, France, Greece, Hungary, Italy, Kazakhstan, Korea, Kyrgyzstan, Mongolia, Romania, Russia, Slovakia, Spain.
Host records: Recorded by Shenefelt (1973) as being a parasitoid of the pyralid *Etiella zinckenella* (Treitschke).

Phanerotoma glabritemporalis van Achterberg, 2021

Distribution in the Middle East: United Arab Emirates, Yemen (van Achterberg, 2021).
Extralimital distribution: None.
Host records: Unknown.

Phanerotoma graciloides van Achterberg, 1990

Distribution in the Middle East: Egypt (Edmardash & Gadallah, 2019), Saudi Arabia (van Achterberg, 1990), United Arab Emirates, Yemen (van Achterberg, 2021).
Extralimital distribution: None.
Host records: Unknown.

Phanerotoma granulata van Achterberg, 2021
Distribution in the Middle East: United Arab Emirates, Yemen (van Achterberg, 2021).
Extralimital distribution: None.
Host records: Unknown.

Phanerotoma hellyeri van Achterberg, 2021
Distribution in the Middle East: United Arab Emirates, Yemen (van AChterberg, 2021).
Extralimital distribution: None.
Host records: Unknown.

Phanerotoma hendecasisella Cameron, 1905
Distribution in the Middle East: Egypt (Edmardash et al., 2011; Edmardash & Gadallah, 2019).
Extralimital distribution: Bangladesh, India, Myanmar, Sri Lanka.
Host records: Summarized by Yu et al. (2016) as being a parasitoid of the following crambids: *Chilo suppressalis* (Walker), *Conogethes punctideralis* (Guenée), *Diaphania caesalis* (Walker), *Diaphania indica* (Saunders), *Diaphania pyloalis* (Hampson), *Haritalodes derogata* (Fabricius), *Hendecasis duplifascialis* (Hampson), *Maruca vitrata* (Fabricius), *Nephopterix rhodobasalis* Hampson, *Notarcha obrinusalis* (Walker), *Notarcha quaternalis* (Zeller), *Pilocrocis milvinalis* (Swinhoe), and *Pleuroptya balteata* (Fabricius); the erbid *Autoba silicule* Swinhoe; the eucosmid *Pammene theristis* Meyrick; the gelechiid *Dichomeris eridantis* (Meyrick); the nolid *Earias insulana* (Boisduval); and the pyralids *Hyapargyria metalliferella* (Ragonot), and *Hypsipyla robusta* Moore.

Phanerotoma intermedia van Achterberg, 1990
Catalogs with Iranian records: No catalog.
Distribution in Iran: Iran (no specific locality cited) (Aydoğdu & Beyarslan, 2009).
Distribution in the Middle East: Iran (Aydoğdu & Beyarslan, 2009), Israel–Palestine (Papp, 2012), Turkey (van Achterberg, 1990).
Extralimital distribution: None.
Host records: Unknown.

Phanerotoma katkowi Kokujev, 1900
Catalogs with Iranian records: Gadallah and Ghahari (2013), Farahani et al. (2016), Yu et al. (2016), Samin, Coronado-Blanco, Kavallieratos et al. (2018), Samin, Coronado-Blanco, Fischer et al. (2018).
Distribution in Iran: Fars (Al-e Mansour & Moustafavi, 1993; Modarres Awal, 1997 as *Phanerotoma kaktay* Kokujev).

Distribution in the Middle East: Iran (Al-e Mansour & Moustafavi, 1993), Israel–Palestine (Brues, 1926).
Extralimital distribution: Albania, Kazakhstan, Mongolia, Russia.
Host records: Unknown.

Phanerotoma kozlovi Shestakov, 1930
Catalogs with Iranian records: Gadallah and Ghahari (2013), Farahani et al. (2016), Yu et al. (2016), Samin, Coronado-Blanco, Kavallieratos et al. (2018), Samin, Coronado-Blanco, Fischer et al. (2018).
Distribution in Iran: Hormozgan (Ameri et al., 2012).
Distribution in the Middle East: Iran.
Extralimital distribution: Afghanistan, Azerbaijan, China, Kazakhstan, Mongolia, Turkmenistan.
Host records: Unknown.

Phanerotoma latifemorata van Achterberg, 2021
Distribution in the Middle East: Yemen (van Achterberg, 2021).
Extralimital distribution: None.
Host records: Unknown.

Phanerotoma lepta van Achterberg, 2021
Distribution in the Middle East: United Arab Emirates, Yemen (van Achterberg, 2021).
Extralimital distribution: None.
Host records: Unknown.

Phanerotoma leucobasis Kriechbaumer, 1894
Catalogs with Iranian records: Fallahzadeh and Saghaei (2010), Gadallah and Ghahari (2013), Farahani et al. (2016), Yu et al. (2016), Samin, Coronado-Blanco, Kavallieratos et al. (2018), Samin, Coronado-Blanco, Fischer et al. (2018).
Distribution in Iran: Isfahan (Goldansaz et al., 2011; Sobhani et al., 2012), Kerman (Madjdzadeh et al., 2021), Sistan and Baluchestan (Hedwig, 1957 as *Phanerotoma hispanica* var. *desertorum* Hedwig, 1957; van Achterberg 1990 as *Phanerotoma ocularis* Kohl, 1906), Iran (no specific locality cited) (Gharib, 1968 as *P. ocularis*; Behdad, 1991; Modarres Awal, 1997, 2012 as *P. ocularis*, in date palm planting areas).
Distribution in the Middle East: Cyprus, Turkey, Saudi Arabia (van Achterberg, 1990), Egypt (Edmardash et al., 2011; Edmardash & Gadallah, 2019), Iran (see references above), Israel–Palestine (van Achterberg, 1990; Papp, 2012), United Arab Emirates (van Achterberg, 2021), Yemen (including Socotra) (van Achterberg, 1990, 2021).

Extralimital distribution: Algeria, Belgium, Benin, Burkina Faso, Cameroon, Cape Verde Islands, Comoros, Croatia, Democratic Republic of Congo, Ethiopia, France, Greece, Hungary, Italy, Kenya, Korea, Libya, Madagascar, Morocco, Niger, Nigeria, Serbia, Somalia, Spain, Switzerland, Tanzania, Togo, Tunisia (introduced), United States of America (introduced), United Kingdom, Uzbekistan.

Host records: Summarized by Yu et al. (2016) as being a parasitoid of the crambids *Desmia incomposita* (Bethune—Baker), and *Maruca vitrata* (Fabricius); the gelechiid *Pectinophora gossypiella* (Saunders); the noctuid *Busseola fusca* (Fuller); the nolid *Earias biplaga* Walker; the praydid *Prays citri* Millière; and the pyralids *Amyeloides transitella* (Walker), *Apomyeloides decolor* (Zeller), *Cadra calidella* (Guenée), *Ectomyelois ceratoniae* (Zeller), *Edulicodes hylobatis* Ghesqiuère, *Etiella zinckenella* (Treitschke), and *Gallaria mellonella* (L.). In Iran, this species has been reared from the batrachedrid *Batrachedra amydraula* Meyrick (Behdad, 1991; Gharib, 1968; Modarres Awal, 1997, 2012), and the pyralids *Plodia interpunctella* (Hübner) (Modarres Awal, 1997, 2012), and *Ectomyelois ceratoniae* Zeller (Goldansaz et al., 2011; Sobhani et al., 2012).

Phanerotoma longiradialis van Achterberg, 1990
Distribution in the Middle East: Iraq (van Achterberg, 1990).
Extralimital distribution: None.
Host records: Unknown.

Phanerotoma longivena van Achterberg, 2021
Distribution in the Middle East: Yemen (van Achterberg, 2021).
Extralimital distribution: None.
Host records: Unknown.

Phanerotoma masiana Fahringer, 1934
Catalogs with Iranian records: Fallahzadeh and Saghaei (2010), Gadallah and Ghahari (2013), Farahani et al. (2016), Yu et al. (2016).
Distribution in Iran: Fars (Al-e Mansour & Mostafavi, 1993; Modarres Awal, 1997, 2012).
Distribution in the Middle East: Egypt (Edmardash et al., 2011; Edmardash & Gadallah, 2019), Iran (see references above), Saudi Arabia (van Achterberg, 1990), United Arab Emirates, Yemen (van Achterberg, 2021).
Extralimital distribution: Libya.
Host records: Unknown.

Phanerotoma mesocellata van Achterberg, 2021
Distribution in the Middle East: United Arab Emirates, Yemen (van Achterberg, 2021).
Extralimital distribution: None.
Host records: Unknown.

Phanerotoma microdonta van Achterberg, 2021
Distribution in the Middle East: United Arab Emirates, Yemen (van Achterberg, 2021).
Extralimital distribution: None.
Host records: Unknown.

Phanerotoma micrommata van Achterberg, 2021
Distribution in the Middle East: United Arab Emirates (van Achterberg, 2021).
Extralimital distribution: None.
Host records: Unknown.

Phanerotoma minuta Kokujev, 1903
Catalogs with Iranian records: Gadallah and Ghahari (2013), Farahani et al. (2016), Yu et al. (2016), Samin, Coronado-Blanco, Kavallieratos et al. (2018), Samin, Coronado-Blanco, Fischer et al. (2018).
Distribution in Iran: Sistan and Baluchestan (Farahani, Talebi, Rashkani et al., 2012).
Distribution in the Middle East: Egypt (Edmardash & Gadallah, 2019), Iran (Farahani, Talebi, Rashkani et al., 2012).
Extralimital distribution: Azerbaijan, China, Kazakhstan, Mongolia, Turkmenistan.
Host records: Unknown.

Phanerotoma ocularis Kohl, 1906
Distribution in the Middle East: Egypt (Edmardash & Gadallah, 2019), United Arab Emirates, Yemen (including Socotra) (van Achterberg, 2021).
Extralimital distribution: Non-Arabian reports in literature (Yu et al., 2016) refer mostly to *P. leucobasis* Kriechbaumer.
Host records: see comment above.

Phanerotoma parva Kokujev 1903
Catalogs with Iranian records: Fallahzadeh and Saghaei (2010), Gadallah and Ghahari (2013), Farahani et al. (2016), Yu et al. (2016), Samin, Coronado-Blanco, Kavallieratos et al. (2018), Samin, Coronado-Blanco, Fischer et al. (2018).
Distribution in Iran: Fars (Al-e Mansour & Mostafavi, 1993; Modarres Awal, 1997, 2012), Mazandaran (Ghahari, 2017—around rice fields).

Distribution in the Middle East: Cyprus (van Achterberg, 1990), Iran (see references above), Israel–Palestine (Papp, 2012).
Extralimital distribution: Armenia, Greece, Kazakhstan, Kyrgyzstan, Libya, Mongolia, Turkmenistan.
Host records: Unknown.

Phanerotoma permixtellae Fischer, 1968
Catalogs with Iranian records: Gadallah and Ghahari (2013), Farahani et al. (2016), Yu et al. (2016), Samin, Coronado-Blanco, Kavallieratos et al. (2018), Samin, Coronado-Blanco, Fischer et al. (2018).
Distribution in Iran: Golestan (Samin, Ghahari et al., 2015), Hormozgan (Ameri et al., 2012), Kerman (Madjdzadeh et al., 2021), West Azarbaijan (Ghahari & Fischer, 2011a).
Distribution in the Middle East: Egypt (Edmardash & Gadallah, 2019), Iran (see references above), Israel–Palestine (Papp, 2012), Syria (van Achterberg, 1990), Yemen (van Achterberg, 2021).
Extralimital distribution: Greece.
Host records: Recorded by Fischer (1968) and van Achterberg (1990) as being a parasitoid of the depressariid *Cacochroa permixtella* (Herrich-Schäffer). In Turkey, this species has been recorded by Aydoğdu and Beyarslan (2009) as being a parasitoid of gelechiid *Recurvaria pistaciicola* (Danilevsky).

Phanerotoma planifrons (Nees von Esenbeck, 1816)
Catalogs with Iranian records: No catalog.
Distribution in Iran: Markazi (Ghahari et al., 2020).
Distribution in the Middle East: Iran (Ghahari et al., 2020), Turkey (Beyarslan et al., 2002).
Extralimital distribution: Afghanistan, Alegria, Azerbaijan, Belgium, Bulgaria, China, former Czechoslovakia, France, Georgia, Germany, Hungary, Italy, Japan, Kazakhstan, Korea, Moldova, Mongolia, Morocco, Portugal, Russia, Serbia, Slovenia, Spain, Switzerland, Tunisia, Ukraine, United Kingdom, United States of America, Uzbekistan.
Hosts records: Summarized by Yu et al. (2016) as being a parasitoid of the crambids *Diaphania indica* (Saunders), *Diaphania pyloalis* Walker, *Haritalodes derogata* (Fabricius), and *Glyphodes pyloalis* Walker; the lasiocampid *Dendrolimus punctatus* Walker; the pyralids *Assara terebrella* (Zincken), *Conobathra repandana* (Fabricius), *Ephestia elutella* (Hübner), *Etiella zinckenella* (Treitscke), and *Phycita diaphana* (Staudinger); and the tortricids *Cydia delineana* (Walker), *Cydia strobilella* (L.), *Cydia zebeana* (Ratzeburg), *Grapholita molesta* (Busck), and *Leguminivora glycinivorella* (Matsumura). In Iran, this species has been reared from the tortricid *Grapholita funebrana* (Treitschke) (Ghahari et al., 2020).

Phanerotoma ponti Edmardash, Abdel Dayem and Gadallah, 2011
Distribution in the Middle East: Egypt (Edmardash et al., 2011; Edmardash & Gadallah, 2019).
Extralimital distribution: None.
Host records: Unknown.

Phanerotoma puchneriana Zettel, 1992
Distribution in the Middle East: Turkey (Zettel, 1992).
Extralimital distribution: Hungary.
Host records: Unknown.

Phanerotoma robusta Zettel, 1988
Distribution in the Middle East: Egypt, Kuwait (Edmardsah & Gadallah, 2019), United Arab Emirates (van Achterberg, 2021).
Extralimital distribution: None.
Host records: Unknown.
Comments: Zettel (1988) reported the presence of this species in Dasmat (Saudi Arabia), however this location is found in Kuwait, not in Saudi Arabia (Edmardash & Gadallah, 2019).

Phanerotoma rufescens (Latreille, 1809)
Catalogs with Iranian records: Gadallah and Ghahari (2013), Farahani et al. (2016), Yu et al. (2016), Samin, Coronado-Blanco, Kavallieratos et al. (2018), Samin, Coronado-Blanco, Fischer et al. (2018).
Distribution in Iran: Hormozgan (Ameri et al., 2012).
Distribution in the Middle East: Egypt (Edmardash et al., 2011; Edmardash & Gadallah, 2019), Iran (Ameri et al., 2012).
Extralimital distribution: Austria, Belgium, former Czechoslovakia, France, Germany, Hungary, Netherlands, Poland, Romania, Russia, Switzerland, United Kingdom.
Host records: Unknown.

Phanerotoma sculptilis van Achterberg, 2021
Distribution in the Middle East: United Arab Emirates, Yemen (van Achterberg, 2021).
Extralimital distribution: None.
Host records: Unknown.

Phanerotoma signifera van Achterberg, 2021
Distribution in the Middle East: Yemen (van Achterberg, 2021).
Extralimital distribution: None.
Host records: Unknown.

Phanerotoma sinaitica Edmardash and Gadallah, 2019
Distribution in the Middle East: Egypt (Edmardash & Gadallah, 2019).
Extralimital distribution: None.
Host records: Unknown.

Phanerotoma soror van Achterberg, 1990

Distribution in the Middle East: Egypt (Edmardash & Gadallah, 2019), Israel−Palestine (Papp, 2012).
Extralimital distribution: France, Greece, Italy, Russia, Spain, Switzerland.
Host records: Unknown.

Phanerotoma spuriserrata van Achterberg, 2021

Distribution in the Middle East: Yemen (van Achterberg, 2021).
Extralimital distribution: None.
Host records: Unknown.

Phanerotoma stenochora van Achterberg, 2021

Distribution in the Middle East: United Arab Emirates (van Achterberg, 2021).
Extralimital distribution: None.
Host records: Unknown.

Phanerotoma syleptae Zettel, 1990

Catalogs with Iranian records: Fallahzadeh and Saghaei (2010), Gadallah and Ghahari (2013), Farahani et al. (2016), Yu et al. (2016), Samin, Coronado-Blanco, Kavallieratos et al. (2018), Samin, Coronado-Blanco, Fischer et al. (2018).
Distribution in Iran: Tehran (Zettel, 1990a), Iran (no specific locality cited) (Zettel, 1990b).
Distribution in the Middle East: Iran.
Extralimital distribution: India, Indonesia, Myanmar, Sri Lanka, Thailand.
Host records: Summarized by Yu et al. (2016) as being a parasitoid of the crambids *Diaphania indica* (Saunders), *Haritalodes derogata* (Fabricius), *Hendecasis duplifascialis* (Hampson), and *Palpita machaeralis* (Walker).

Phanerotoma tritoma (Marshall, 1898)

Catalogs with Iranian records: No catalog.
Distribution in Iran: Kordestan (Ghahari & Gadallah, 2021).
Distribution in the Middle East: Cyprus (van Achterberg, 1990), Iran (Ghahari & Gadallah, 2021).
Extralimital distribution: Austria, Bulgaria, former Czechoslovakia, France, Greece, Hungary, Ireland, Italy, Korea, Netherlands, Poland, Romania, Russia, Spain, Switzerland, United Kingdom.
Host records: Summarized by Yu et al. (2016) as being a parasitoid of the coleophorid *Coleophora lutipennella* (Zeller); the gelechiid *Gelechia senticetella* (Staudinger); and the tortricids *Grapholita delineana* (Walker), and *Pammene regiana* (Zeller). In Iran, this species has been reared from the tortricid *Cydia pomonella* (L.) (Ghahari & Gadallah, 2021).

Phanerotoma vanharteni van Achterberg, 2021

Distribution in the Middle East: United Arab Emirates, Yemen (van Achterberg, 2021).
Extralimital distribution: None.
Host records: Unknown.

Phanerotoma sp.

Distribution in Iran: Khuzestan (Farahbakhsh, 1961).
Host records in Iran: *Antigastra catalaunalis* (Duponchel) (Lepidoptera: Crambidae) (Farahbakhsh, 1961).

Genus *Phanerotomella* Szépligeti, 1900

Phanerotomella (Phanerotomella) bisulcata (Herrich-Schäffer, 1838)

Catalogs with Iranian records: No catalog.
Distribution in Iran: Ardabil (Ghahari et al., 2019).
Distribution in the Middle East: Iran (Ghahari et al., 2019), Israel−Palestine (Papp, 2012).
Extralimital distribution: Austria, former Czechoslovakia, Germany, Greece, Hungary, Italy, Korea, Moldova, Russia, Serbia, Spain.
Host records: Unknown.

Phanerotomella (Phanerotomella) rufa (Marshall, 1898)

Catalogs with Iranian records: Gadallah and Ghahari (2013), Farahani et al. (2016), Yu et al. (2016), Samin, Coronado-Blanco, Kavallieratos et al. (2018), Samin, Coronado-Blanco, Fischer et al. (2018).
Distribution in Iran: Alborz, Tehran (Farahani, Talebi, & Rakhshani, 2012), Hormozgan (Ameri et al., 2012), Kerman (Asadizadeh et al., 2014).
Distribution in the Middle East: Cyprus (Zettel, 1989), Iran (see references above), Israel−Palestine (Papp, 2012), Turkey (Beyarslan & Inanç, 1992; Beyarslan et al., 2002; van Achterberg, 1990; Zettel, 1989).
Extralimital distribution: Austria, Bulgaria, China, Croatia, Greece, Hungary, Italy, Moldova, Romania, Serbia, Spain, Switzerland.
Host records: Unknown.

Phanerotomella yemenitica van Achterberg, 2021

Distribution in the Middle East: Yemen (van Achterberg, 2021).
Extralimital distribution: None.
Host records: Unknown.

Conclusion

A total of 219 species of the subfamily Cheloninae (14.3% of the world species) in five genera and three tribes (Adeliini, Cheloninae, Phanerotomini) have been reported from the Middle East countries. The genus *Chelonus* is the

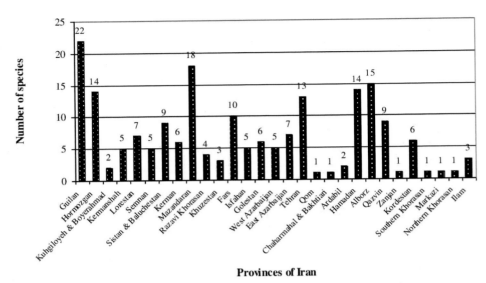

FIGURE 10.1 Number of reported species of Iranian Cheloninae by province.

most diverse, with 137 recorded species. Cheloninae of Iran comprises 113 species (7.4% of the world species) in five genera, *Chelonus* (71 species), *Ascogaster* (20 species), *Phanerotoma* (17 species), *Adelius* (three species), and *Phanerotomella* (two species) and three tribes. Four species, *Chelonus iranicus* Tobias, 1972, *Chelonus setaceus* Papp, 1993, *Chelonus moczari* (Papp, 2014), and *Chelonus subpamiricus* Farahani and van Achterberg, 2018 are so far only known from Iran (endemic to subendemic to Iran). Distributions of the Iranian Cheloninae have been recorded from 29 provinces which among them, Guilan with 22 species has the highest number of species; followed by Mazandaran (18 species), Alborz (15 species), Hamadan, and Hormozgan (both with 14 species) (Fig. 10.1). Host species have been reported for eight Cheloninae species, which these hosts belong to seven families of Lepidoptera, Cosmopteritidae (one species), Erebidae (one species), Gelechiidae (two species), Nepticulidae (one species), Noctuidae (three species), Pyralidae (two species), Tineidae (one species), Tortricidae (five species), and Yponomeutidae (one species). Comparison of the Cheloninae fauna of Iran with the Middle East and adjacent countries to Iran indicates the fauna of Russia with 344 species in five genera (Belokobylskij & Lelej (2019) is more diverse than Iran, followed by Turkey (118 species), Kazakhstan (113 species), Turkmenistan (59 species), Azerbaijan (54 species), Israel—Palestine (40 species), Armenia (31 species), Egypt (30 species), Yemen (28 species), United Arab Emirates (19 species), Cyprus (16 species), Afghanistan and Syria (both with 10 species), Pakistan (six species), Jordan and Saudi Arabia (each with four species), and Iraq (one species). No species have been recorded from Bahrain, Lebanon, Oman, and Qatar (Yu et al., 2016). Russia is indicated as having the most diverse fauna, which probably should be expected because it is the largest country, but

the greater number of reported species undoubtedly also reflects a history of more extensive taxonomic research on Cheloninae in that country. The much lower number of species recorded from some other countries compared to Iran probably not only reflects a smaller land area but also fewer comprehensive surveys in those countries. Additionally, among the countries of the Middle East and adjacent to Iran, Russia shares 72 species with Iran, followed by Turkey (67 species), Kazakhstan (49 species), Azerbaijan (41 species), Armenia and Israel—Palestine (both with 24 species), Turkmenistan (19 species), Egypt (14 species), Cyprus (11 species), Syria (five species), Afghanistan and Yemen (each with four species), Jordan, Pakistan, Saudi Arabia, and United Arab Emirates (each with two species).

References

Abbasipour, H., Mahmoudvand, M., Basij, M., & Lozan, A. (2012). First report of the parasitoid wasps, *Microchelonus subcontractus* and *Bracon intercessor* (Hym.: Braconidae), from Iran. *Journal of Entomological Society of Iran, 32*(1), 89—92.

Abdinbekova, A. A. (1965). New species and forms of insects (fam. Ichneumonidae and Braconidae) from Kuba-Khachmask zone in Azerbaidjan. *Doklady Akademii Nauk Azerbaidzhanskoi SSR, 21*(4), 61—64 (in Russian).

Akbarzadeh Shoukat, G., Safaralizadeh, M., Ranjbar Aghdam, H., & Aramideh, S. (2015). Study on the parasitoid wasps belonging to the superfamily Ichneumonoidea on grape berry moth, *Lobesia botrana* (Lep., Tortricidae) in Urmia vineyards. In *Proceedings of the 1st Iranian International Congress of Entomology, 29.31 August 2015* (p. 115). Iranian Research Institute of Plant Protection.

Al-e Mansour, H., & Mostafavi, M. S. (1993). The first record of Braconidae bees on forest and range vegetations in Fars province. In *Proceedings of the 11th Iranian Plant Protection Congress, 28 August — 2 September 1993* (p. 236). Rasht: University of Guilan.

Alizadeh, S., & Javan Moghaddam, H. (2004). Introduction of some natural enemies of common cutworm (*Agrotis segetum* Schiff) in Miyandoab. In *Proceedings of the 16th Iranian Plant Protection congress, 28 August — 1 September 2004* (p. 45). Tabriz: University of Tabriz.

Ameri, A., Talebi, A. A., Kamali, K., & Rakhshani, E. (2012). Study of the tribe Phanerotomini Baker (Hymenoptera: Braconidae) in Hormozgan province of Iran with two new records. *Biosystema, 6*(2), 31—38.

Ameri, A., Talebi, A. A., & Ebrahimi, E. (2018). A taxonomic study on genus *Chelonus* Panzer, 1806 (Hymenoptera: Braconidae: Cheloninae) from Hormozgan province with five new records for Iranian fauna. In *Proceedings of the 23rd Iranian Plant Protection Congress, 27—30 August 2018* (p. 962). Gorgan University of Agricultural Sciences and Natural Resources.

Asadizadeh, A., Mahriyan, K., Talebi, A. A., & Esfandiarpour, I. (2014). Faunistic survey of parasitoid wasps family of Braconidae from Anar Region, Kerman province. In *Proceedings of the 3rd Integrated Pest Management Conference (IPMC), 21—22 January 2014* (p. 629). Kerman: Shahid Bahonar University of Kerman.

Aubert, J. F. (1966). Liste d'identification No. 6 (présentée par le service d'identification des Entomophages). *Entomophaga, 11*(1), 115—134.

Aydoğdu, M. (2008). A new species of the genus *Chelonus* Panzer, 1806 (Hymenoptera: Braconidae: Cheloninae) from western Anatolia (Turkey). *Biologia Bratislava, 63*(2), 245—248.

Aydoğdu, M., & Beyarslan, A. (2002). *Chelonus* Jurine (Hymenoptera: Braconidae: Cheloninae) species of the Marmara region. *Turkish Journal of Zoology, 26*(1), 1—13.

Aydoğdu, M., & Beyarslan, A. (2006). Microchelonus Szépligeti 1908 (Hymenoptera: Braconidae: Cheloninae) species from the Marmara, Western and Black Sea regions of Turkey. *Linzer biologische Beiträge, 38*(1), 397—407.

Aydoğdu, M., & Beyarslan, A. (2007). Parasitoid species of the genera *Ascogaster* and *Chelonus* (Hymenoptera: Braconidae: Cheloninae) from the Marmara, western and middle black Sea regions of Turkey. *Acta Entomologica Slovenica, 15*(1), 75—90.

Aydoğdu, M., & Beyarslan, A. (2009). A review of the tribe Phanerotomini (Hymenoptera, Braconidae, Cheloninae) in Turkey, with a new host record for *Phanerotoma (Bracotritoma) permixtellae*. *Biologia, 64*(4), 748—756.

Aydoğdu, M., & Beyarslan, A. (2011). Additional notes on *Chelonus* Panzer, 1806 fauna of Turkey with new records (Hymenoptera, Braconidae, Cheloninae). *Journal of the Entomological Research Society, 13*(2), 75—81.

Aydoğdu, M., & Beyarslan, A. (2012). A review of the genus *Ascogaster* Wesmael, 1835 (Hymenoptera, Braconidae, Cheloninae) in Turkey, with a new host record for *Ascogaster bicarinata* (Herrich-Schäffer, 1838). *North-Western Journal of Zoology, 8*(1), 31—40.

Bagheri, M. R., & Basiri, G. H. (2004). Preliminary studies of field biology of sorghum stem borers, *Sesamia cretica*. In Ardestan area. *Proceedings of the 16th Iranian Plant Protection Congress, 28 August — 1 September 2004* (p. 258). University of Tabriz.

Bashir, A., Aslan, L., & Al-Haj, S. (2010). Survey of parasitoids of codling moth *Cydia pomonella* L. in Eramo region in Lattakia Governorate (Syria). *Arab Journal of Plant Protection, 28*(1), 91—95 (in Arabic, English summary).

Behdad, E. (1991). *Pests of fruit crops in Iran* (p. 822). Isfahan Neshat Publication (in Persian).

Belokobylskij, S. A., & Lelej, A. S. (2019). Annotated catalogue of the Hymenoptera of Russia. Volume II. Apocrita: Parasitica. In *Proceedings of the Zoological Institute of the Russian Academy of Sciences.* Suppl. 8, 594 pp.

Belshaw, R., Fitton, M., Herniou, E., Gimeno, C., & Quicke, D. L. J. (1998). A phylogenetic reconstruction of the Ichneumonoidea (Hymenoptera) based on the D2 variable region of 28S ribosomal RNA. *Systematic Entomology, 23*(2), 109—123.

Beyarslan, A. (1995). Für die fauna der Turkei neu destgestelte Arten der Gattung *Microchelonus* Szépligeti, 1908 (Hymenoptera, Braconidae, Cheloninae). *Entomofauna, 16*(6), 121—133.

Beyarslan, A. (2021). Taxonomic studies on Cheloninae (Hymenoptera: Braconidae) fauna of Northeatern Anatolia region (Ardahan, Erzurum, Iğdir and Kars) of Turkey. *Natura Somogyiensis, 36*, 111—124.

Beyarslan, A., & Inanç, F. (1992). Taxonomic investigations on the species of Cheloninae (Hymenoptera: Braconidae). *X Ulusal Biyoloji Kongresi*, 141—151.

Beyarslan, A., & Aydoğdu, M. (2013). Additions to the rare species of Braconidae fauna (Hymenoptera: Braconidae) from Turkey. *Munis Entomology & Zoology, 8*(1), 369—374.

Beyarslan, A., & Aydoğdu, M. (2014). Additions to the rare species of Braconidae fauna (Hymenoptera: Braconidae) from Turkey. *Munis Entomology & Zoology, 9*(1), 103—108.

Beyarslan, A., & Şahin, F. (2018). Taxonomic investigations on Braconidae (Hymenoptera) fauna of Bitlis Nemrut Caldera and its around. *Munis Entomology & Zoology, 13*(1), 292—308.

Beyarslan, A., & Şahan, Y. B. (2019). A new species of Chelonini (Hymenoptera, Braconidae) from pistachio twig borer moth (*Kermania pistaciella* Amsel (Lep.: Tineidae) in Gaziantep. *Turkish Journal of Zoology, 43*, 388—392.

Beyarslan, A., Inanç, F., Çetin Erdoğan, Ö., & Aydoğdu, M. (2002). Braconidae species of Turkish Aegean region (Hymenoptera). In G. Melika, & C. Thuroczy (Eds.), *Parasitic wasps: Evolution, systematics, biodiversity and biological control. International Symposium "Parasitic Hymenoptera: taxonomy and biological control" (14—17 May, 2001, Koszek, Hungary).*

Beyarslan, A., Yurtcan, M., Çetin Erdoğan, Ö., & Aydoğdu, M. (2006). A study on Braconidae and Ichneumonidae from Ganos mountain (Thrace region, Turkey) (Hymenoptera, Braconidae, Ichneumonoidea). *Linzer biologische Beiträge, 38*(1), 409—422.

Beyarslan, A., Arisoy, M., & Çolak, R. D. (2018). Records of some Braconidae species (Hymenoptera, Ichneumonoidea) from Gürün (Sivas) and Ilgaz Montain (Kastamonu). *Acta Biologica Turcica, 31*(4), 178—192.

Bodenheimer, F. S. (1930). Die Schädlingsfauna Palästinas. *Monographien zur Angewandten Entomologie, 10*, 438.

Braet, Y., Rousse, P., & Sharkey, M. (2012). New data on African Cheloninae (Hymenoptera, Braconidae) show a strong biogeographic signal for taxa with spined propodea. *Zootaxa, 3385*, 1—32.

Brues, C. I. (1926). Studies on the Ethiopian Braconidae with a catalogue of the African species. *Proceedings of the American Academy of Arts and Sciences, 6*, 206—436.

Čapek, M., & Hofmann, C. (1997). The Braconidae (Hymenoptera) in the collections of the Musée Cantonal de Zoologie, Lausanne. *Litterae Zoologicae, 2*, 25—162.

Chen, X. X., & van Achterberg, C. (2019). Systematics, phylogeny, and evolution of braconid wasps: 30 years of progress. *Annual Review of Entomology, 64*, 1—24.

Dadelahi, S., Shaw, S. R., Aguirre, H., & Almeida, L. F. de (2018). A taxonomic study of Costa Rican *Leptodrepana* with descriptions of twenty-four new species (Hymenoptera: Braconidae: Cheloninae). *ZooKeys, 750,* 59—130.

Davatchi, A., & Shojai, M. (1969). Les hymenopteres entomophages de l'Iran (etudes faunestiques). *Tehran University, Faculty of Agriculture, 107,* 1—88.

Derafshan, H. A., Rakhshani, E., Farahani, S., & Peris-Felipo, F. J. (2017). *Adelius aridus* (Tobias, 1967). (Hym., Braconidae, Cheloninae) associated with a Tamarix leafminer (Lepidoptera: Nepticulidae), new for Iran. *Journal of Insect Biodiversity and Systematics, 3*(3), 229—238.

Doganlar, M. (1982). Hymenopterous parasites of some lepidopterous pests in eastern Anatolia. *Türkiye bitki koruma dergisi, 6,* 197—205.

Dudarenko, G. P. (1974). Formation of the abdominal carapace in braconids (Hymenoptera, Braconidae) and some aspects of the classification of the family. *Entomological Review, 53,* 80—90.

Edmardash, Y. A., Abdel-Dayem, M. S., & Gadallah, N. S. (2011). The subfamily Cheloninae (Hymenoptera: Braconidae) from Egypt, with the description of two new species. *ZooKeys, 115,* 85—102.

Edmardash, Y. A., & Gadallah, N. S. (2019). Revision of the subfamily Cheloninae (Hymenoptera: Braconidae) from Egypt, with description of two new species. *Annales Zoologici, 69*(2), 339—380.

Fahringer, J. (1934). *Opuscula braconologica. Band 3. Plaearktischen Region. Lieferung 5-8. Opuscula braconologica 321-594.* Wien: Fritz Wagner.

Fahringer, J. (1937). *Opuscula braconologica. Band 4. Palaearktische region. Lieferung 4—6* (pp. 257—520). Opuscula braconologica.

Fallahzadeh, M., & Saghaei, N. (2010). Checklist of Braconidae (Insecta: Hymenoptera) from Iran. *Munis Entomology & Zoology, 5*(1), 170—186.

Farahani, S., Talebi, A. A., Rashkani, E., Hajiqanbar, H., & van Achterberg, C. (2012). First report of parasitoid wasp, *Phanerotoma minuta* Kokujev (Hym., Braconidae, Cheloninae) from Iran. In *Proceedings of the 20th Iranian Plant Protection Congress, 26—29 August 2012* (p. 117). Shiraz: University of Shiraz.

Farahani, S., Talebi, A. A., & Rakhshani, E. (2012). New records of *Phanerotomella rufa* (Marshall, 1898) and *Phanerotoma (Phanerotoma) acuminata* Szépligeti, 1908 (Hymenoptera: Braconidae) from northern Iran. *Biharean Biologist, 6*(1), 61—64.

Farahani, S., Talebi, A. A., & Rakhshani, E. (2013). A contribution to the tribe Chelonini Foerster (Hymenoptera: Braconidae: Cheloninae) of northern Iran, with first records for eight species and an updated check list of Iranian species. *Zoosystematics and Evolution, 89*(2), 227—238.

Farahani, S., Talebi, A. A., & Rakhshani, E. (2016). Iranian Braconidae (Insecta: Hymenoptera: Ichneumonoidea): diversity, distribution, and host association. *Journal of Insect Biodiversity and Systematics, 2*(1), 1—92.

Farahani, S., Talebi, A. A., van Achterberg, C., & Rakhshani, E. (2014). A review of species of the genus *Ascogaster* Wesmael (Hymenoptera: Braconidae: Cheloninae) from Iran. *Far Eastern Entomologist, 275,* 1—12.

Farahbakhsh, G. (1961). Family Braconidae (Hymenoptera), p. 124. In G. Farahbakhsh (Ed.), *Vol. 1. A checklist of economically important insects and other enemies of plants and agricultural products in Iran* (pp. 1—153). The Ministry of Agriculture, Department Plant Protection.

Fischer, M. (1968). Über gezüchtete Raupenwespen (Hymenoptera, Braconidae). *Pflanzenschutzberichte, 37*(7/8/9), 97—140.

Fulmek, L. (1968). Parasitinsekten der insektengallen Europas. *Beiträge zur Entomologie, 18*(7 and 8), 719—952.

Gadallah, N. S., & Ghahari, H. (2013). An annotated catalogue of the Iranian Cheloninae (Hymenoptera: Braconidae). *Linzer biologische Beiträge, 45*(2), 1921—1943.

Gadallah, N. S., Ghahari, H., Papp, J., & Beyarslan, A. (2018). New records of Braconidae (Hymenoptera) from Iran. *Wuyi Science Journal, 34,* 43—48.

Gadallah, N. S., Ghahari, H., & Quicke, D. L. J. (2021). Further addition to the braconid fauna of Iran (Hymenoptera: Braconidae). *Egyptian Journal of Biological Pest Control, 31,* 32. https://doi.org/10.1186/s41938-021-00376-8

Gerling, D. (1969). The parasites of *Spodoptera littoralis* Bois. (Lep., Noctuidae) egg and larvae in Israel. *Israel Journal of Entomology, 4,* 73—81.

Ghahari, H. (2017). Species diversity of Ichneumonoidea (Hymenoptera) from rice fields of Mazandaran province, northern Iran. *Journal of Animal Environment, 9*(3), 371—378 (in Persian, English summary).

Ghahari, H. (2020). A study on the fauna of predator and parasitoid arthropods in saffron fields (*Crocus sativus* L.). *Journal of Saffron Research, 7*(2), 203—215 (in Persian, English summary).

Ghahari, H., Fischer, M., Çetin Erdoğan, Ö., Tabari, M., Ostovan, H., & Beyarslan, A. (2009). A contribution to Braconidae (Hymenoptera) from rice fields and surrounding grasslands of northern Iran. *Munis Entomology & Zoology, 4*(2), 432—435.

Ghahari, H., Fischer, M., Çetin Erdoğan, Ö., Beyarslan, A., & Ostovan, H. (2010). A contribution to the braconid wasps (Hymenoptera: Braconidae) from forests of northern Iran. *Linzer biologische Beiträge, 42*(1), 621—634.

Ghahari, H., Fischer, M., Hedqvist, K. J., Çetin Erdoğan, Ö., van Achterberg, C., & Beyarslan, A. (2010). Some new records of Braconidae (Hymenoptera) for Iran. *Linzer biologische Beiträge, 42*(2), 1395—1404.

Ghahari, H., & Fischer, M. (2011a). A contribution to the Braconidae (Hymenoptera: Ichneumonoidea) from north-western Iran. *Calodema, 134,* 1—6.

Ghahari, H., & Fischer, M. (2011b). A study on the Braconidae (Hymenoptera: Ichneumonoidea) from some regions of northern Iran. *Entomofauna, 32*(8), 181—196.

Ghahari, H., & Fischer, M. (2012). A faunistic survey on the braconid wasps (Hymenoptera: Braconidae) from Kermanshah province, Iran. *Entomofauna, 33,* 305—312.

Ghahari, H., Fischer, M., & Papp, J. (2011). A study on the braconid wasps (Hymenoptera: Braconidae) from Isfahan province, Iran. *Entomofauna, 32*(16), 261—272.

Ghahari, H., Fischer, M., & Tobias, V. (2012). A study on the Braconidae (Hymenoptera: Ichneumonoidea) from Guilan province, Iran. *Entomofauna, 33,* 317—324.

Ghahari, H., Fischer, M., Papp, J., & Tobias, V. (2012). A contribution to the knowledge of braconids (Hymenoptera: Braconidae) from Lorestan province Iran. *Entomofauna, 33,* 65—72.

Ghahari, H., & Gadallah, N. S. (2015). A faunistic study on the Braconidae (Hymenoptera) from some regions of Semnan, Iran. *Entomofauna, 36*(11), 177—184.

Ghahari, H., & Beyarslan, A. (2017). A faunistic study on Braconidae (Hymenoptera: Ichneumonoidea) from Iran. *Natura Somogyiensis, 30,* 39—46.

Ghahari, H., & Beyarslan, A. (2019). A faunistic study on Braconidae (Hymenoptera: Ichneumonoidea) from Iran, and in Memoriam Dr. Jenő Papp (20 May 1933 − 11 December 2017). *Acta Biologica Turcica, 32*(4), 248−254.

Ghahari, H., & Gadallah, N. S. (2019). New records of braconid wasps (Hymenoptera: Ichneumonoidea: Braconidae) from Iran. *Entomological News, 129*(3), 384−394.

Ghahari, H., & Gadallah, N. S. (2022). Additional records to the braconid fauna (Hymenoptera: Ichneumonoidea) of Iran, with new host reports. *Entomological News* (in press).

Ghahari, H., Fischer, M., Beyarslan, A., Navaeian, M., & Hosseini Boldaji, S. A. (2019). New records of Brachistinae, Braconinae, Cheloninae and Microgastrinae (Hymenoptera: Braconidae) from Iran. *Wuyi Science Journal, 35*(2), 135−141.

Ghahari, H., Beyarslan, A., & Kavallieratos, N. G. (2020). New records of Braconidae (Hymenoptera: Ichneumonoidea) from Iran, and in Memoriam Dr. Maximilian Fischer (7 June 1929 − 15 June 2019). *Scientific Bulletin of Uzhhorod National University (Series: Biology), 48*, 48−55.

Gharib, A. (1968). *Batrachedra amydraula* Meyr (superfamily: Gelechoidea) (Momphidae (Cosmopterygidae)). *Entomologie et Phytopatologie Appliqees, 27*, 63−67.

Ghramh, H. A. (2011). Description of a new species of the genus *Phanerotoma* Wesmael (Hymenoptera: Braconidae) and a key to the Saudi Arabian species. *African Journal of Biotechnology, 10*(65), 14649−14651.

Goldansaz, S. H., Kishani, H., Talaei, L., Poorjavad, N., & Sobhani, M. (2011). Biological control of carob moth, past, present and future. In *Proceedings of the biological Control development congress in Iran, 27−28 july 2011* (pp. 55−63). Tehran: Iranian Research Institute of Plant Protection (in Persian, English summary).

Gultekin, L., & Güçlü, S. (1997). Bioecology of *Bembecia scopigera* (Scopoli) (Lep.: Sesiidae) pest of sainfoin in Erzurum. *Bitki Koruma Bulteni, 37*(3−4), 101−110.

Györfi, J. (1959). Beiträge zur Kenntnis der Wirte verschiedener Braconiden-Arten (Hymenoptera, Braconidae). *Acta Zoologica Hungarica, 5*, 49−65.

Halperin, J. (1986). Braconidae (Hymenoptera) associated with forest and ornamental trees and shrubs in Israel. *Phytoparasitica, 14*(2), 119−135.

Hedwig, K. (1957). Ichneumoniden und Braconiden aus den Iran 1954 (Hymenoptera). *Jahresheft des Vereins für Vaterlaendische Naturkunde, 112*(1), 103−117.

Huddleston, T. (1984). The Palaeartic species of *Ascogaster* (Hymenoptera: Braconidae). *Bulletin of the British Museum (Natural History), Entomology series, 49*(5), 341−392.

Izhevskiy, S. S. (1985). Review of the parasites of potato tuber moth *Phthorimaea operculella* (Lepidoptera: Gelechiidae). *Entomolo-gicheskoye Obozreniye, 64*(5), 516−524.

Jafaripour, M. (1969). New pistachio wood borer (*Kermania pistaciella* Amsel.). In *Proceedings of the 2nd Iranian Plant Protection Congress* (pp. 124−131). University of Tehran.

Kaeslin, M., Wehrle, I., Gossliknaus-Bürgin, C., Wyler, T., Guggisberg, U., Schittny, J. C., & Lanzrein, B. (2005). Stage-dependent strategies of hosts invasion in the egg-larval parasitoid *Chelonus inanitus. Journal of Insect Physiology, 51*, 287−296.

Khajeh, N., Rakhshani, E., Barahoei, H., & Arjomandinejad, A. (2014). Taxonomic study of the genus *Chelonus* Panzer (Hymenoptera: Braconidae: Cheloninae) in Sistan region. In *Proceedings of the 21st Iranian Plant Protection Congress, 23−26 August 2014* (p. 739). Urmia University.

Khanjani, M. (2006a). *Field crop pests in Iran* (Third edition). Bu-Ali Sina University, 719 pp. (in Persian).

Khanjani, M. (2006b). *Vegetable pests in Iran* (Second edition). Bu-Ali Sina University, 467 pp. (in Persian).

Kheyri, M. (1977). The necessary of integrated control application against beet armyworm. *Entomologie and Phytopathologie Applique, 45*, 5−28.

Kian, N., Goldasteh, S., & Farahani, S. (2020). A survey on abundance and species diversity of Braconid wasps in forest of Mazandaran province. *Journal of Entomological Research, 12*(1), 61−69 (in Persian, English summary).

Kishani Farahani, H., Goldansaz, H., Sabahi, G., & Shakeri, M. (2010). Study on the larval parasitoids of carob moth *Ectomyelois ceratoniae* Zeller (Lepidoptera: Pyralidae) in Varamin, Qom and Saveh. *Iranian Journal of Plant Protection Science, 41*(2), 337−344 (in Persian, English summary).

Kittel, R. N., Austin, A. D., & Klopfstein, S. (2016). Molecular and morphological phylogenetics of chelonine parasitoid wasps (Hymenoptera: Braconidae), with a critical assessment of divergence time estimations. *Molecular Phylogenetics and Evolution, 101*, 224−241.

Kohl, F. F. (1905). Hymenoptera. In A. Penther, & E. Zederbauer (Eds.), *Ergebnisse einer naturwissenschatlischen Reise zur Erdschias-Dagh (Kleinasien). Annalen des Naturhistorischen Museum in Wien* (Vol. 20, pp. 220−246).

Kokujev, N. R. (1914). Hymenoptera parasitic nove fauna d'Entomologie. *Revue Russe d'Entomologie, 13*, 513−514.

Kroschel, J. (1995). Integrated pest management in potato production in the Republic of Yemen with special reference to the integrated biological control of the potato tuber moth (Phthorimaea operculella Zeller). *Tropical Agriculture (Weikersheim), 8*, 1−232.

Kroschel, J., & Koch, W. (1994). Studies on the population dynamics of the potato tuber moth (Phthorimaea operculella Zell. (Lep., Gelechiidae)) in the Republic of Yemen. *Journal of Applied Entomology, 118*(4-5), 327−341.

Kugler, J. (1966). A list of parasites of Lepidoptera from Israel. *Israel Journal of Entomology, 1*, 75−88.

Lashkari Bod, A., Rakhshani, E., Talebi, A. A., Lozan, A., & Žikič, V. (2011a). New records of Cheloninae [Förster, 1862] and Braconinae [Nees, 1811] [Insecta: Hymenoptera: Braconidae] from Iran. *Check List, 7*(5), 632−634.

Lashkari Bod, A., Rakhshani, E., Talebi, A. A., Lozan, A., & Zikic, V. (2011b). A contribution to the knowledge of Braconidae (Hym., Ichneumonoidea) of Iran. *Biharean Biologist, 5*(2), 147−150.

LaSalle, J., & Gauld, I. D. (1993). *Hymenoptera and biodiversity.* England: CAB International, 384 pp.

Lozan, A. (2005). Cheloninae wasps (Hymenoptera: Braconidae, Cheloninae) from uplands of Turkey. *Entomologist's Monthly Magazine, 141*(1694−1696), 151−159.

Madjdzadeh, S. M., Parrezaali, M., Dolati, S., & Ghassemi-Khademi, T. (2021). New data on the braconid wasps (Hymenoptera: Braconidae: Cheloninae, Opiinae, Rogadinae) of South-eastern Iran. *Faunistic Entomology, 74*, 1−7.

Mahmoudi, J., Askarianzadeh, Karimi, A. J., & Abbasipour, H. (2013). Introduction of two parasitoids of braconid wasps on the sugar beet moth, *Scrobipalpa ocellatella* Boyd. (Lep.: Gellechiidae) from Khorasan-e-Razavi province, Iran. *Journal of Sugar Beet, 28*(2), 103−106 (in Persian, English summary).

Mehrnejad, M. R. (2009). Primary and secondary parasitoids of *Kermania pistaciella* Amsel (Lepidoptera: Tineidae: Hieroxestinae). II. *Applied Entomology and Phytopathology, 76*(2), 135−136.

Mehrnejad, M. R. (2020). Arthropod pests of pistachios, their natural enemies and management. *Plant Protection Science, 56*(4), 231−260.

Mehrnejad, M. R., & Basirat, M. (2009). Parasitoid complex of the pistachio twig borer moth *Kermania pistaciella* Amsel (Lepidoptera: Tineidae), in Iran. *Biocontrol Science and Technology, 5*, 499−510.

Meyer, N. F. (1934). Schlipfwespen die in Russland in den letzten Jahren aus schädlingen gezogen sind. *Zeitschrift für Angewandte Entomologie, 20*, 611−618.

Modarres Awal, M. (1997). Family Braconidae (Hymenoptera), pp. 265−267. In M. Modarres Awal (Ed.), *List of agricultural pests and their natural enemies in Iran* (2nd ed., p. 429). Ferdowsi University of Mashhad Press.

Modarres Awal, M. (2012). Family Braconidae (Hymenoptera), pp. 483−486. In M. Modarres Awal (Ed.), *List of agricultural pests and their natural enemies in Iran* (3rd ed., p. 759). Ferdowsi University of Mashhad Press.

Murphy, N., Banks, J. C., Whitfield, J. B., & Austin, A. D. (2008). Phylogeny of the parasitic microgastroid subfamilies (Hymenoptera: Braconidae) based on sequence data from seven genes, with an improved time estimate of the origin of the lineage. *Molecular Phylogenetics and Evolution, 47*(1), 378−395.

Papp, J. (1970). A contribution to the braconid fauna of Israel (Hymenoptera). *Israel Journal of Entomology, 5*, 63−76.

Papp, J. (1993). New braconid wasps (Hymenoptera, Braconidae) in the Hungarian Natural History Museum, 4. *Annales Historico-Naturales Musei Nationalis Hungarici, 84*, 155−180.

Papp, J. (1995). Revision of C. Wesmael's *Chelonus* species (Hymenoptera Braconidae Cheloninae). *Bulletin Institut Royal des Sciences Naturelles de Belgique, 65*, 113−134.

Papp, J. (1996a). On the taxonomy of *Microchelonus cycloporus* (Franz 1930) (Insecta: Hymenoptera: Braconidae: Cheloninae). *Senckenbergiana Biologica, 75*(1−2), 203−206.

Papp, J. (1996b). Contribution to the braconid fauna of Hungary, XI. Cheloninae and Sigalphinae (Hymenoptera: Braconidae). *Folia Entomologica Hungarica, 57*, 131−156.

Papp, J. (1997). Revision of the *Chelonus* s.str. species described by C.G. Thomson (Hymenoptera: Braconidae, Cheloninae). *Stobaeana (Lund), 10*, 1−21.

Papp, J. (1999). Redescription of F. Silvestri's two chelonine species (Hymenoptera, Braconidae, Cheloninae). *Bolletino del Laboratorio di Entomologia Agraria 'Filippo Silvestri' Portici, 55*, 15−26.

Papp, J. (2003). Revision of the Szépligeti's *Chelonus* s. str. species described from Hungary (Hymenoptera: Braconidae: Cheloninae). *Annales Historico-Naturales Musei Nationalis Hungarici, 95*, 113−133.

Papp, J. (2004). A revision of Szépligeti's *Microchelonus* species described from Hungary (Hymenoptera: Braconidae: Cheloninae). *Annales Historico-Naturales Musei Nationalis Hungarici, 96*, 225−259.

Papp, J. (2012). A contribution to the braconid fauna of Israel (Hymenoptera: Braconidae), 3. *Israel Journal of Entomology, 41−42*, 165−219.

Papp, J. (2014a). Faunistic contributions to the *Microchelonus* Szépligeti species of the Palaearctic region, with descriptions of two new species (Hymenoptera: Braconidae: Cheloninae). *Acta Zoologica Academiae Scientiarum Hungaricae, 60*(4), 325−358.

Papp, J. (2014b). *Microchelonus deplanus* sp. n. from Canada and checklists of the nearctic and Palaearctic species of the genus *Microchelonus* Szépligeti, 1908 (Hymenoptera, Braconidae: Cheloninae). *Natura Somogyiensis, 25*, 115−140.

Quicke, D. L. J., & van Achterberg, C. (1990). Phylogeny of the subfamilies of the family Braconidae (Hymenoptera: Ichneumonoidea). *Zoologische Verhandelingen, 258*(1), 1−95.

Rajabi Mazhar, A., Goldasteh, S., Farahani, S., & van Achterberg, C. (2018). New species and new record of Chelonini (Hymenoptera: Braconidae) from western Iran. *Turkish Journal of Zoology, 42*, 416−421.

Ranjbar Aghdam, H., & Fathipour, Y. (2010). Fist report of parasitoid wasps, *Ascogaster quadridentata* and *Bassus rufipes* (Hym.: Braconidae) on codling moth (Lep.: Tortricidae) larvae from Iran. *Journal of Entomological Society of Iran, 30*(1), 55−58.

Sakenin, H., Fischer, M., Samin, N., Imani, S., Papp, J., Ghahari, H., & Rastegar, J. (2011). A faunistic survey on the Braconidae wasps (Hymenoptera: Braconidae) from northern Iran. In *Proceedings of global Conference on Entomology, March 5−9, 2011* (p. 123). Thailand: Chiang Mai.

Sakenin, H., Naderian, H., Samin, N., Rastegar, J., Tabari, M., & Papp, J. (2012). On a collection of Braconidae (Hymenoptera) from northern Iran. *Linzer biologische Beiträge, 44*(2), 1319−1330.

Sakenin, H., Coronado-Blanco, M., Samin, N., & Fischer, M. (2018). New records of Braconidae (Hymenoptera) from Iran. *Far Eastern Entomologist, 362*, 13−16.

Samin, N. (2015). A faunistic study on the Braconidae of Iran (Hymenoptera: Ichneumonoidea). *Arquivos Entomoloxicos, 13*, 339−345.

Samin, N., Sakemin, H., Imani, S., & Shojai, M. (2011). A study on the Braconidae (Hymenoptera) of Khorasan province and vicinity, Northeastern Iran. *Phagea, 39*(4), 137−143.

Samin, N., Ghahari, H., Gadallah, N. S., & Monaem, R. (2015). A study on the braconid wasps (Hymenoptera: Ichneumonoidea: Braconidae) from Golestan province, northern Iran. *Linzer biologische Beiträge, 47*(1), 731−739.

Samin, N., van Achterberg, C., & Ghahari, H. (2015). A faunistic study of Braconidae (Hymenoptera: Ichneumonoidea) from southern Iran. *Linzer biologische Beiträge, 47*(2), 1801−1809.

Samin, N., Fischer, M., & Ghahari, H. (2015). A contribution to the study on the fauna of Braconidae (Hymenoptera, Ichneumonoidea) from the province of Semnan, Iran. *Arquivos Entomoloxicos, 13*, 429−433.

Samin, N., van Achterberg, C., & Çetin Erdoğan, Ö. (2016). A faunistic study on some subfamilies of Braconidae (Hymenoptera: Ichneumonoidea) from Iran. *Arquivos Entomoloxicos, 15*, 153−161.

Samin, N., Coronado-Blanco, J. M., Kavallieratos, N. G., Fischer, M., & Sakenin, H. (2018). Recent findings on Braconidae (Hymenoptera: Ichneumonoidea) of Iran with an updated checklist. *Acta Biologica Turcica, 31*(4), 160−173.

Samin, N., Coronado-Blanco, J. M., Fischer, M., van Achterberg, C., Sakenin, H., & Davidian, E. (2018b). Updated checklist of Iranian Braconidae (Hymenoptera: Ichneumonoidea) with twenty-three new records. *Natura Somogyiensis, 32*, 21−36.

Samin, N., Coronado-Blanco, J. M., Hosseini, A., Fischer, M., & Sakenin Chelav, H. (2019). A faunistic study on the braconid wasps (Hymenoptera: Braconidae) of Iran. *Natura Somogyiensis, 33*, 75−80.

Samin, N., Sakenin Chelav, H., Ahmad, Z., Penteado-Dias, A. M., & Samiuddin, A. (2020). A faunistic study on the family Braconidae (Hymenoptera: Ichneumonoidea) from Iran. *Scientific Bulletin of Uzhhorod National University (Series: Biology), 48*, 14–19.

Sertkaya, E., Beyarslan, A., & Kornosor, S. (2004). Egg and larval parasitoids of the beet armyworm *Spodoptera exigua* on maize in Turkey. *Phytoparasitica, 32*(3), 305–312.

Sharanowski, J. B., Dowling, A. P. G., & Sharkey, M. J. (2011). Molecular phylogenetics of Braconidae (Hymenoptera: Ichneumonoidea), based on multiple nuclear genes, and amplications for classification. *Systematic Entomology, 36*, 549–572.

Shaw, M. R., & Huddleston, T. (1991). Classification and biology of braconid wasps (Hymenoptera: Braconidae). *Handbooks for the Identification of British Insects, 7*(11), 1–126.

Shaw, S. R. (1983). A taxonomic study of Nearctic *Ascogaster* and a description of a new genus *Leptodrepana* (Hymenoptera: Braconidae). *Entomography, 2*, 1–54.

Shaw, S. R. (1997). Subfamily Cheloninae, pp.192–201. In R. A. Wharton, P. M. Marsh, & M. J. Sharkey (Eds.), *Manual of the New World Genera of the Family Braconidae (Hymenoptera)* (p. 439). Washington D.C: Special Publication of the International Society of Hymenopterists.

Shenefelt, R. D. (1973). Braconidae 6. Cheloninae. *Hymenopterorum catalogus* (nova ed., pp. 813–936). Pars 10.

Shi, M., Chen, X. X., & van Achterberg, C. (2005). Phylogenetic relationships among the Braconidae (Hymenoptera: Ichneumonoidea) inferred from partial 16S rDNA D2, 18S rDNA gene sequences and morphological characters. *Molecular Phylogenetics and Evolution, 37*, 104–116.

Shimbori, E. M., Bortoni, M. A., Shaw, S. R., Soussa-Gessner, C., da, S., Cerântola, P., da, C. M., & Penteado-Diaz, A. M. (2019). Revision of the new world genera *Adelius* Haliday and *Paradelius* DeSaeger (Hymenoptera: Braconidae: Cheloninae: Adeliini). *Zootaxa, 4571*(2), 151–200.

Shojai, M. (1968). Resultats de létude faunestiques des Hyménopteres parasites (Terebrants) en Iran et l'importance de leur utilization des la lute biologique. In *Proceedings of the 1st Iranian Plant Protection Congress* (pp. 25–35). University of Tehran.

Shojai, M. (1998). *Entomology (ethology, social life and natural enemies) (biological control)* (3rd ed., Vol. III). Tehran University Publications, 550 pp. (in Persian).

Sobhani, M., Goldansaz, S. H., & Hatami, B. (2012). Study of larval parasitoids of carob moth *Ectomyelois ceratoniae* (Lep.: Pyralidae) in Kashan region. In *Proceedings of the 20th Iranian Plant Protection Congress, 26–29 August 2012* (p. 83). Shiraz: University of Shiraz.

Tang, Y., & Marsh, P. M. (1994). A taxonomic study of the genus *Ascogaster* in China (Hymenoptera: Braconidae: Cheloninae). *Journal of Hymenoptera Research, 3*, 279–302.

Telenga, N. A. (1941). Family Braconidae, subfamily Braconinae (continuation) and Sigalphinae. Fauna USSR. *Hymenoptera, 5*(3), 466.

Telenga, N. A. (1953). Contributions to systematics of *Chelonus* Jur. And *Chelonella* Szép. (Hymenoptera, Braconidae). *Zoologicheskii Zhurnal, 32*(6), 1175–1177.

Tezcan, S., Yildirim, E., Anlaş, S., & beyaz, G. (2006). Hymenoptera fauna of oregano (Lamiaceae) in Manisa province of Turkey. *Ege Universitesi Ziraat Fakultesi Dergisi, 43*(1), 55–62.

Tobias, V. I. (1966). New species and genus of the braconids (Hymenoptera: Braconidae) from Turkmenia and adjacent territories. *Trudy Zoologicheskogo Instituta, Linengrad, 37*, 111–131.

Tobias, V. I. (1972). To the knowledge of the genus *Chelonus* s. str. (Hymenoptera, Braconidae) of the fauna of the USSR and adjacent countries-. In O. A. Scarlato (Ed.), *Insects of Mongolia* (Vol. 1, pp. 585–612). Leningrad: Nauka.

Tobias, V. I. (1976). Opred. Faune SSSR. *Braconids of the Caucasus (Hymenoptera, Braconidae)* (Vol. 110). Leningrad: Nauka Press, 286 pp. (in Russian).

Tobias, V. I. (1986). Acaeliinae, Cardiochilinae, Microgastrinae, Miracinae. Supplement. In G. S. Medvedev (Ed.), *Opredelitel Nasekomych Evrospeiskoi Tsasti SSSR 3, Peredpontdatokrylye 4* (Vol. 145, pp. 336–501). Opr. Faune SSSR, 501 pp. (Keys to the insects of the European part of USSR. Hymenoptera).

Tobias, V. I. (1995). New subgenus and species of the genus *Microchelonus* (Hymenoptera, Braconidae) with some comments on synonymy. *Entomological Review, 75*(7), 158–170.

Tobias, V. I. (2001). Species of the genus *Microchelonus* Szépl. (Hymenoptera, Braconidae) with yellow abdominal spots and pale coloration of the body from the western Palaearctic region. *Entomologicheskoye Obozreniye, 80*, 137–179.

Tobias, V. I. (2008). Palaearctic species of *Microchelonus retusus* group (Hymenoptera, Braconidae, Cheloninae). *Entomological Review, 88*(9), 1171–1191.

Tobias, V. I. (2010). Palaearctic species of the genus *Microchelonus* Szépligeti (Hymenoptera: Braconidae, Cheloninae): key to species. *Proceedings of the Russian Entomological Society, 81*(1), 1–354 (in Russian).

Tobias, V. I. (2011). Genus *Chelonus* Jurine (Hymenoptera, Braconidae, Cheloninae) in the fauna of Russia and adjacent territories. Ch. Olgae Kok. Species group. *Entomological Review, 91*(8), 1011–1030.

van Achterberg, C. (1984). Essay on the phylogeny of Braconidae (Hymenoptera: Ichneumonoidea). *Entomologisk Tidskrift, 105*, 41–58.

van Achterberg, C. (1990). Revision of the western Palaearctic Phanerotomini (Hymenoptera: Braconidae). *Zoologische Verhandelingen, 255*, 1–106.

van Achterberg, C. (1993). Illustrated key to the subfamilies of the Braconidae (Hymenoptera, Ichneumonoidea). *Zoologische Verhandelingen, 283*, 1–189.

van Achterberg, C. (2021). The tribe Phanerotomini (Hymenoptera, Braconidae, Cheloninae) of the Arabian Peninsula, with especial reference to the United Arab Emirates and Yemen. *ZooKeys, 1014*, 1–118.

van Achterberg, C., & Mehrnejad, M. R. (2002). The braconid parasitoids (Hymenoptera: Braconidae) of *Kermania pistaciella* Amsel (Lepidoptera: Tineidae: Hieroxestinae) in Iran. *Zoologische Mededelingen, 76*(2), 27–39.

Walker, A. K., & Huddleston, T. (1987). New Zealand chelonine wasps (Hymenoptera). *Journal of Natural History, 21*(2), 339–361.

Whitfield, J. B., & Wagner, D. L. (1991). Annotated key to the genera of Braconidae (Hymenoptera) attacking leafmining Lepidoptera in Holarctic region. *Journal of Natural History, 25*, 733–754.

Yu, D. S., van Achterberg, C., & Horstmann, K. (2016). *Taxapad 2016, Ichneumonoidea 2015, Database on flash-drive*. Nepean, Ontario, Canada.

Zettel, H. (1987). Beitrag zur kenntnis der Siagalphinen und Cheloninen Fauna in Österreich (Hym.: Braconidae). *Linzer biologische Beiträge, 19*, 359–376.

Zettel, H. (1988). Eine neue *Phanerotoma*-Art aus Saudi Arabien (Hym.: Braconidae, Cheloninae). *Linzer biologische Beiträge, 20,* 199−201.

Zettel, H. (1989). Die Gattung *Phanerotomella* Szépligeti (Hymenoptera: Braconidae, Cheloninae). *Linzer biologische Beiträge, 21*(1), 15−142.

Zettel, H. (1990a). Die *Phanerotoma*-Arten des indischen Subkontinentes (Insecta, Hymenoptera, Braconidae: Cheloninae). *Reichenbachia Staatliches Museum für Tierkunde Dresden, 27,* 147−158.

Zettel, H. (1990b). Beschreibung von vier neuen *Phanerotoma*-Arten aus Ostasien mit einem Bestimmungschlüssel zu den Arten der orientalischen region (Hymenoptera: Braconidae, Cheloninae). *Zeitschrift der Arbeitsgmeinschaft österreichischer Entomologen, 42*(3 & 4), 110−120.

Zettel, H. (1992). *Phanerotoma puchneriana* sp. n. aus Ungarn und der Türkei (Hymenoptera: Braconidae: Cheloninae). *Acta Zoologica Hungarica, 38*(1−2), 145−147.

Dirrhope sp. (Dirrhopinae), ♀, lateral habitus. *Photo prepared by S.R. Shaw.*

Dirrhope sp. (Dirrhopinae), ♀, lateral habitus. *Photo prepared by S.R. Shaw.*

Chapter 11

Subfamily Dirrhopinae van Achterberg, 1984

Neveen Samy Gadallah[1], Hassan Ghahari[2], Scott Richard Shaw[3] and Donald L.J. Quicke[4]

[1]Entomology Department, Faculty of Science, Cairo University, Giza, Egypt; [2]Department of Plant Protection, Yadegar-e Imam Khomeini (RAH) Shahre Rey Branch, Islamic Azad University, Tehran, Iran; [3]UW Insect Museum, Department of Ecosystem Science and Management, University of Wyoming, Laramie, WY, United States; [4]Integrative Ecology Laboratory, Department of Biology, Faculty of Science, Chulalongkorn University, Pathumwan, Bangkok, Thailand

Introduction

The Dirrhopinae is a small, cosmopolitan subfamily in the family Braconidae (Quicke & van Achterberg, 1990). It is represented by only a single, rare genus *Dirrhope* Foerster (Belokobylskij et al., 2003; Quicke & van Achterberg, 1990; van Achterberg, 1984). It currently comprises five species (Yu et al., 2016). *Dirrhope* was traditionally included in the subfamily Microgastrinae (e.g., Marsh, 1979; Musesbeck, 1935; Tobias 1967) or within Adeliinae (Čapek, 1970; Shenefelt, 1973; Telenga, 1955). It was excluded from Microgastrini (Microgastrinae *sensu* Mason) by Nixon (1965). His scheme was accepted by Mason (1981). The subfamily Dirrhopinae was first erected by van Achterberg (1984) including only the genus *Dirrhope*, based on its peculiar autapomorphy (the flattened first metasomal tergite with the spiracle positioned behind the middle of the tergum length). Van Achterberg placed Dirrhopinae as a sister group to Ichneutinae + (Miracinae + Adeliini), remote from Microgastrinae and other higher-level taxa constituting the microgastroid group of subfamilies (*sensu* Whitfield & Mason, 1994).

The genus *Dirrhope* was subsequently placed within Microgastrinae by some workers (e.g., Belokobylskij, 1989; Tobias, 1995), who elected not to follow van Achterberg's scheme (1984). More cladistic analyses using morphological data were carried out by several authors (Quicke & van Achterberg, 1990; Wharton et al., 1992; Whitfield and Mason, 1994) who corroborated placement of the Dirrhopinae as a part of the microgastroid clade. However, to date no molecular data are known (Belokobylskij et al., 2003; Chen & van Achterberg, 2019).

As far as is known, the Dirrhopinae are thought to be koinobiont endoparasitoids of the leaf-mining Nepticulidae larvae (van Achterberg, 1984; Whitfield & Wagner, 1991). One host record, *Ectoedemia phyloeophaga* Busck, has been established for the North American species *D. americana* Muesebeck (Muesebeck, 1935). However, other *Dirrhope* species are predicted to be leaf-miner parasitoids as well, because of their small size and short ovipositor. This is consistent with the hosts of other small members of the microgastroid group such as adeliines and some microgastrines, which also parasitize leaf miners (Belokobylskij et al., 2003).

In Iran, which is the only Middle Eastern country having Dirrhopinae species, the genus *Dirrhope* is represented by a single species, *D. rufa* Foerster (Samin et al., 2020).

List of species of the subfamily Dirrhopinae recorded in the Middle East

Subfamily Dirrhopinae van Achterberg, 1984

Genus *Dirrhope* Foerster, 1851

Dirrhope rufa Foerster, 1851

Catalogs with Iranian records: No catalog.
Distribution in Iran: East Azarbaijan (Samin et al., 2020).
Distribution in the Middle East: Iran.
Extralimital distribution: Germany, Hungary, Japan, Korea, Moldova, Romania, Russia.
Host records: Unknown.

Braconidae of the Middle East (Hymenoptera). https://doi.org/10.1016/B978-0-323-96099-1.00015-7

Conclusion

The fauna of the Middle East Dirrhopinae is found to comprise a single species, *Dirrhope rufa*, recorded only from Iran (East Azarbaijan province). No host species has been so far recorded for this parasitoid in Iran or elsewhere (Yu et al., 2016). Among the 23 countries of the Middle East and adjacent to Iran, the subfamily Dirrhopinae has been recorded from only Russia with three species in one genus (Belokobylskij & Lelej, 2019), of which just one of them is known to be shared with Iran.

References

Belokobylskij, S. A. (1989). Eastern Palaearctic braconid species of the genera *Dirrhope and Mirax* (Hymenoptera, Braconidae, Miracinae). *Vestnik Zoologii, 4*, 34−46.

Belokobylskij, S. A., & Lelej, A. S. (2019). Annotated catalogue of the Hymenoptera of Russia. Volume II. Apocrita: Parasitica. In *Proceedings of the Zoological Institute of the Russian Academy of Sciences*. Supplement No. 8, 594 pp.

Belokobylskij, S. A., Iqbal, M., & Austin, A. D. (2003). First record of the subfamily Dirrhopinae (Hymenoptera: Braconidae) from the Australian region, with a discussion of relationships and biology. *Australian Journal of Entomology, 42*, 260−265.

Čapek, M. (1970). A new classification of the Braconidae (Hymenoptera) based on the cephalic structures of the final instar larvae and biological evidence. *The Canadian Entomologist, 102*, 846−865.

Chen, X. X., & van Achterberg, C. (2019). Systematics, phylogeny, and evolution of braconid wasps: 30 years of progress. *Annual Review of Entomology, 64*, 1−24.

Marsh, P. M. (1979). Family Braconidae, pp. 144−313. In K. V. Krombein, P. D. Hurd, D. R. Smith, & B. D. Burks (Eds.), *Catalog of Hymenoptera North of Mexico* (p. 1198). Washington, USA: Smithsonian Institution Press.

Mason, W.R.M. (1981). The phylogenetic nature of *Apanteles* Foerster (Hymenoptera: Braconidae): A phylogeny and reclassification of Microgastrinae. *Memoirs of the Entomological Society of Canada, 115*, 147 pp.

Muesebeck, C. F. W. (1935). On two little known genera of Braconidae (Hymenoptera). *Proceedings of the Entomological Society of Washington, 37*, 173−177.

Nixon, G. E. J. (1965). A reclassification of the tribe Microgasterini (Hymenoptera: Braconidae). *Bulletin of the British Museum (Natural History) Entomology Supplement, 2*, 1−284.

Quicke, D. L. J., & van Achterberg, C. (1990). Phylogeny of the subfamilies of the family Braconidae (Hymenoptera: Ichneumonoidea). *Zoologische Verhandelingen, 258*, 1−95.

Samin, N., Sakenin Chelav, H., Ahmad, Z., Penteado-Dias, A. M., & Samiuddin, A. (2020). A faunistic study on the family Braconidae (Hymenoptera: Ichneumonoidea) from Iran. *Scientific Bulletin of Uzhhorod National University (Series: Biology), 48*, 14−19.

Shenefelt, R. D. (1973). Pars 9. Braconidae 5, Microgasterinae and Ichneutinae. In J. van der Vecht, & R. D. Shenefelt (Eds.), *Hymenopterorum catalogus* (pp. 669−805). The Hague, Netherlands: Dr. W. Junk, 136 pp.

Telenga, N. A. (1955). *Fauna of the USSR. Hymenoptera, Vol. 5, No. 4. Braconidae: Microgasterinae and Agathidinae*. Moscow, USSR: Zoologicheskii Institut Akademii Nauk SSSR, 908 pp.

Tobias, V. I. (1967). A review of the classification, phylogeny and evolution of the family Braconidae (Hym.). *Entomological Review, 46*, 387−399.

Tobias, V. I. (1995). *Key to the insects of the European part of the USSR. III, Hymenoptera* (Vol. 4). Leningrad, USSR: Akademia Nauka, 883 pp.

van Achterberg, C. (1984). Essay on the phylogeny of Braconidae (Hymenoptera: Ichneumonoidea). *Entomologisk Tidskrift, 105*, 41−58.

Wharton, R. A., Shaw, S. R., Sharkey, M. J., Whal, D. B., Wooley, J. B., Whitfield, J. B., Marsh, P. M., & Johnson, J. W. (1992). Phylogeny of the subfamilies of the family Braconidae (Hymenoptera: Ichneumonoidea): a reassessment. *Cladistics, 8*, 199−235.

Whitfield, J. B., & Mason, W. R. M. (1994). Mendesellinae, a new subfamily of braconid wasps (Hymenoptera: Braconidae) with a review of relationships within the microgasteroid assemblage. *Systematic Entomology, 19*, 61−76.

Whitfield, J. B., & Wagner, D. L. (1991). Annotated key to the genera of Braconidae (Hymenoptera) attacking leaf-mining Lepidoptera in the Holarctic region. *Journal of Natural History, 25*, 733−754.

Yu, D. S., van Achterberg, C., & Horstmann, K. (2016). *Taxapad 2016, Ichneumonoidea 2015, Database on flash-drive*. Nepean, Ontario, Canada.

Dendrosoter protuberans (Nees, 1834) (left), ♂, *Ecphylus silesiacus* (Ratzeburg, 1848) (right), ♀, (Doryctinae), lateral habitus. *Photos prepared by S.R. Shaw.*

Dendrosoter protuberans (Nees, 1834) (Doryctinae), ♀, lateral habitus. *Photo prepared by S.R. Shaw.*

Chapter 12

Subfamily Doryctinae Foerster, 1863

Hassan Ghahari[1], Sergey A. Belokobylskij[2], Neveen Samy Gadallah[3], Donald L.J. Quicke[4] and Scott Richard Shaw[5]

[1]Department of Plant Protection, Yadegar-e Imam Khomeini (RAH) Shahre Rey Branch, Islamic Azad University, Tehran, Iran; [2]Zoological Institute Russian Academy of Sciences, Universitetskaya nab. 1, St. Petersburg, Russia; [3]Entomology Department, Faculty of Science, Cairo University, Giza, Egypt; [4]Integrative Ecology Laboratory, Department of Biology, Faculty of Science, Chulalongkorn University, Pathumwan, Bangkok, Thailand; [5]UW Insect Museum, Department of Ecosystem Science and Management, University of Wyoming, Laramie, WY, United States

Introduction

The subfamily Doryctinae Foerster is one of the richest, most speciose, and diverse lineages of Braconidae, currently comprising about 2000 described species classified into almost 200 genera and 15 tribes (Chen & van Achterberg, 2019; Martínez et al., 2016; Yu et al., 2016). Some genera are highly speciose and widely distributed, most notably *Spathius* Nees (with over 450 species), *Heterospilus* Haliday (with at least 418 species), *Rhaconotus* Ruthe s.l. (132 species), *Doryctes* Haliday (89 species), and *Ecphylus* Foerster (59 species) (Jasso-Martínez et al., 2019; Yu et al., 2016; Zaldívar-Riverón & Belokobylskij, 2009; Zaldívar-Riverón et al., 2018). Just from Costa Rica alone, a country only one-third the size of Iran, there are 280 named species of *Heterospilus* (Marsh et al., 2013). Morphological definition of the entire subfamily Doryctinae is challenging (Chen & van Achterberg, 2019). Relying upon morphological characters alone is not fully supportive of all the subdivisions within the subfamily because of the presence of many homoplasious characters (Belokobylskij et al., 2004). The phylogeny of Doryctinae was recently extensively revised based on molecular data (Zaldívar-Riverón et al., 2006, 2008).

Doryctines, especially the large-bodied species, are most easily distinguished by the presence of a row of stout spines or distinct pegs on the fore tibia, and often the middle tibiae as well. However, these spines may be tiny and difficult to see in small-bodied doryctines, and they may be reduced or absent in some cases. Therefore, the following characteristics may be useful for identifying the subfamily Doryctinae, especially the small-bodied species. The hind coxa is often angulated and has baso-ventral tubercle; the propleuron possess a distinct dorso-posterior flange; usually both the epicnemial and occipital carinae are present; the ovipositor is strongly sclerotized and distinctly blackened apically, and the dorsal valve of ovipositor nearly always has a double nodus subapically (Marsh, 2002; Quicke et al., 1993; van Achterberg, 1993).

Species of Doryctinae are idiobiont ectoparasitoids of mostly concealed or semiconcealed larvae of bark and wood–boring and xylophagous beetles (Quicke, 2015; Shaw & Huddleston, 1991). A few taxa are found attacking wood boring or leaf-mining Lepidoptera and sawflies (van Achterberg, 1993), some other genera are associated with termites (Blattodea: Isoptera) but their biologies are unknown. One species from Trinidad described in the Doryctinae is an endoparasitoid of Embioptera (Shaw & Edgerly, 1985) but the subfamily placement of this taxon has not yet been confirmed with molecular data. Recently, several tropical genera have been discovered to be gall inducers (phytophages), whereas others are suspected to be predators of gall-formers (Zaldívar-Riverón et al., 2007, 2014). Some species were used in biological control programs, either through release or serendipitous utilization of available harmful hosts (Quicke, 2015).

Checklists of Regional Doryctinae. Modarres Awal (1997, 2012) listed four species in three genera, *Ecphylus* Foerster, 1863 (one species), *Hecabalodes* Wilkinson, 1929 (two species), and *Spathius* Nees, 1819 (one species). Farahani et al. (2016) and Yu et al. (2016) represented 31 and 35 species, respectively, in 13 genera. Gadallah and Ghahari (2017) cataloged 40 species in 15 genera. Samin, Coronado-Blanco, Kavallieratos et al. (2018), Samin, Coronado-Blanco, Fischer et al. (2018) listed 41 and 39 species, respectively, in 16 genera. The Middle Eastern Doryctinae currently includes 96 species and subspecies in 29 genera and seven tribes (Doryctini Foerster, 1863, Ecphylini Hellén, 1957, Hecabolini Foerster, 1863, Heterospilini Fischer, 1981, Holcobraconini Cameron, 1905;

Rhaconotini Fahringer, 1928, Spathiini Marshall, 1872). In the present checklist, one species *Syngaster lepidus* Brullé, 1846 is excluded from the fauna of Iran because of misidentification.

List of species of the subfamily Doryctinae recorded in the Middle East

Subfamily Doryctinae Foerster, 1863

Tribe Doryctini Foerster, 1863

Genus *Dendrosoter* Wesmael, 1838

Dendrosoter (Caenopachys) hartigii (Ratzeburg, 1848)

Catalogs with Iranian records: This species is new record for the fauna of Iran.

Distribution in Iran: Mazandaran province, Chalus (Faraj-Abad), 2♀, August 2012.

Distribution in the Middle East: Iran (new record), Israel—Palestine (Halperin, 1986; Mendel, 1986; Mendel & Halperin, 1981; Papp, 2012), Turkey (Mancini et al., 2003).

Extralimital distribution: Algeria, Armenia, Austria, Belarus, Belgium, Bosnia-Herzegovina, Bulgaria, Croatia, Czech Republic, Finland, France, Germany, Hungary, Italy, Lithuania, Morocco, Norway, Poland, Portugal, Russia, Serbia, Slovakia, South Africa (introduced), Spain, Sweden, Tunisia, Ukraine, United Kingdom, former Yugoslavia.

Host records: Summarized by Yu et al. (2016) as being a parasitoid of mostly the following curculionids: *Carphoborus minimus* (Fabricius), *Ceutorhynchus quadridens* Panzer, *Ernoporus caucasicus* Lindemann, *Ips acuminatus* (Gyllenhal), *Lixus bidens* Capiomont, *Onthotomicus proximus* Eichhoff, *Onthotomicus suturalis* (Gyllenhal), *Pityogenes* spp., *Polygraphus poligraphus* (L.), *Pityophthorus lichtensteini* Eichhoff, *Scolytus mali* (Bechstein), *Tomicus destruens* (Wollaston), and *Tomicus minor* (Hartig); and the tenebrionid *Corticeus linearis* (Fabricius).

Dendrosoter (Dendrosoter) middendorffii (Ratzeburg, 1848)

Catalogs with Iranian records: Farahani et al. (2016), Yu et al. (2016), Gadallah & Ghahari (2017), Samin, Coronado-Blanco, Kavallieratos et al. (2018), Samin, Coronado-Blanco, Fischer et al. (2018).

Distribution in Iran: Ardabil (Basiri et al., 2012, 2013), Iran (no specific locality cited) (Belokobylskij & Tobias, 1986; Belokobylskij et al., 2019).

Distribution in the Middle East: Iran (see references above), Israel—Palestine (Halperin, 1986; Mendel, 1986; Mendel & Halperin, 1981), Turkey (Schimitschek, 1941, 1944).

Extralimital distribution: Austria, Belarus, Bulgaria, Czech Republic, Finland, France, Georgia, Germany, Hungary, India, Italy, Japan, Latvia, Lithuania, Moldova, Poland,

Russia, Slovakia, Spain, Sweden, Switzerland, Ukraine, United Kingdom.

Host records: Summarized by Yu et al. (2016) as being a parasitoid of the following curculionids: *Cryphalus piceae* (Ratzeburg), *Dendroctonus micans* (Kugelann), *Hylesinus fraxini* (Panzer), *Hylurgops palliatus* Gyllenhal, *Hylurgus micklitzi* Wachtl, *Ips* spp., *Magdalis memnonia* (Gyllenhal in Faldermann), *Orthotomicus erosus* (Wollaston), *Orthotomicus laricis* (Fabricius), *Pissodes castaneus* (DeGeer), *Pityogenes* spp., *Pityokteines curvidens* (Germar), *Pityokteines vorontzowi* (Jakobson), *Polygraphus grandiclava* (Thomson), *Polygraphus poligraphus* (L.), *Scolytus* spp., *Trypodendron lineatum* (Olivier), *Tomicus destruens* (Wollaston), *Tomicus minor* (Hartig), and *Tomicus pineperda* (L.). In Iran, this species has been reared from the curculionid *Scolytus rugulosus* Müller on *Malus domestica* and *Prunus domestics* (Rosaceae) (Basiri et al., 2012, 2013).

Dendrosoter (Dendrosoter) protuberans (Nees von Esenbeck, 1834)

Catalogs with Iranian records: Farahani et al. (2016), Yu et al. (2016), Gadallah and Ghahari (2017), Samin, Coronado-Blanco, Kavallieratos et al. (2018), Samin, Coronado-Blanco, Fischer et al. (2018).

Distribution in Iran: Alborz (Farahani et al., 2014), Kordestan (Samin et al., 2016).

Distribution in the Middle East: Egypt (Abd El-Latif et al., 2009), Iran (Farahani et al., 2014; Samin et al., 2016), Israel—Palestine (Halperin, 1986; Mendel, 1986; Mendel & Halperin, 1981), Turkey (Schimitschek, 1941).

Extralimital distribution: Widely distributed in the Nearctic (introduced) and Palaearctic regions.

Host records: Summarized by Yu et al. (2016) as being a parasitoid of several coleopteran species of the families Buprestidae, Cerambycidae, Chrysomelidae, Curculionidae, and the hymenopterous family Xiphydriidae.

Genus *Dendrosotinus* Telenga, 1941

Dendrosotinus (Dendrosotinus) ferrugineus (Marshall, 1888)

Catalogs with Iranian records: No catalog.

Distribution in Iran: Razavi Khorasan (Ghahari & Gadallah, 2019).

Distribution in the Middle East: Iran (Ghahari & Gadallah, 2019), Israel-Palestine (Halperin, 1986), Saudi Arabia (Edmardash et al., 2020), Turkey (Beyarslan, 2015), United Arab Emirates (Belokobylskij & van Achterberg, 2021).

Extralimital distribution: Armenia, Azerbaijan, Bosnia-Herzegovina, France, Greece, Italy, Russia, Spain, former Yugoslavia.

Host records: Summarized by Yu et al. (2016) as being a parasitoid of the bostrichids *Scobicia chevrieri* (Villa et

Villa), and *Sinoxylon sexdendatum* (Olivier); the curculionids *Chaetoptelius vestitus* (Mulsant et Rey), and *Phloeotribus scarabaeoides* (Bernard).

Dendrosotinus (Gildoria) maculipennis Belokobylskij, 2021
Distribution in the Middle East: Yemen (Belokobylskij & van Achterberg, 2021).
Extralimital distribution: None.
Host records: Unknown.

Dendrosotinus (Gildoria) similis Bouček, 1955
Distribution in the Middle East: Israel—Palestine (Papp, 2012).
Extralimital distribution: Germany, Poland, Switzerland, United Kingdom.
Host records: Summarized by Yu et al. (2016) as being a parasitoid of the following curculionids: *Cryphalus abietis* Ratzeburg, *Cryphalus piceae* Ratzeburg, *Pityophthorus polonicus* Karpinski, and *Pityophthorus pubescens* (Marsham).

Dendrosotinus (Gildoria) subelongatus Belokobylskij, 2021
Distribution in the Middle East: United Arab Emirates (Belokobylskij & van Achterberg, 2021).
Extralimital distribution: None.
Host records: Unknown.
Remarks: This species was described as new species by Belokobylskij and van Achterberg (2021) from two female specimens (holotype ad paratype) collected from Fujairah in the United Arab Emirates; however, they incorrectly mentioned "Yemen" in its distribution.

Dendrosotinus (Gildoria) titubatus Papp, 1985
Catalogs with Iranian records: Farahani et al. (2016 as *Gildoria titubata* (Papp, 1985)), Yu et al. (2016), Gadallah and Ghahari (2017 as *Gildoria titubata*), Samin, Coronado-Blanco, Kavallieratos et al. (2018), Samin, Coronado-Blanco, Fischer et al. (2018).
Distribution in Iran: West Azarbaijan (Ghahari & Fischer, 2011 as *Gildoria titubata* under the subfamily Cheloninae), Kuhgiloyeh and Boyerahmad (Samin et al., 2015 as *Gildoria titubata*).
Distribution in the Middle East: Iran (Ghahari & Fischer, 2011; Samin et al., 2015), Israel—Palestine (Halperin, 1986; Papp, 1989, 2012).
Extralimital distribution: France, Greece, Italy, Spain.
Host records: Summarized by Yu et al. (2016) as being a parasitoid of the cerambycids *Nathrius brevipennis* (Mulsant), and *Stenopterus rufus syriacus* Pic.

Genus *Doryctes* Haliday, 1836
Doryctes (Doryctes) inopinatus Belokobylskij, 1984
Catalogs with Iranian records: Farahani et al. (2016), Yu et al. (2016), Gadallah and Ghahari (2017), Samin, Coronado-Blanco, Kavallieratos et al. (2018), Samin, Coronado-Blanco, Fischer et al. (2018).
Distribution in Iran: Hormozgan (Ameri et al., 2014).
Distribution in the Middle East: Iran.
Extralimital distribution: Tajikistan.
Host records: Unknown.
Comments: Record of *Doryctes inopinatus* from Iran is doubtful, and the specimen must be re-examined (S.A. Belokobylskij).

Doryctes (Doryctes) leucogaster (Nees von Esenbeck, 1834)
Catalogs with Iranian records: Fallahzadeh and Saghaei (2010), Farahani et al. (2016), Yu et al. (2016), Gadallah and Ghahari (2017), Samin, Coronado-Blanco, Kavallieratos et al. (2018), Samin, Coronado-Blanco, Fischer et al. (2018).
Distribution in Iran: Alborz, Qazvin (Farahani et al., 2014), Hormozgan (Ameri et al., 2014), Ilam (Ghahari et al., 2011b), West Azarbaijan (Samin et al., 2014), Iran (no specific locality cited) (Belokobylskij & Tobias, 1986; Belokobylskij et al., 2019; Shenefelt & Marsh, 1976; Telenga, 1941; Tobias, 1976).
Distribution in the Middle East: Iran (see references above), Israel—Palestine (Halperin, 1986; Papp, 2012), Turkey (Beyarslan, 2015).
Extralimital distribution: Widespread in many Western Palaearctic countries.
Host records: Summarized by Yu et al. (2016) as being a parasitoid of the bostrichid *Bostrichus capucinus* (L.); the buprestids *Agrilus biguttatus* Fabricius, *Anthaxia aurulenta* (Fabricius), *Anthaxia manca* (L.), *Chrysobothris affinis* (Fabricius), *Chrysobothris solieri* Gory and Laporte, and *Lampra mirifica* (Mulsant); the following cerambycids: *Acanthocinus aedilis* (L.), *Chlorophorus pilosus* (Forster), *Exocentrus lusitanus* (L.), *Hesperophanes cinereus* Fairmaire et Germain, *Hylotrupes bajulus* (L.), *Penichroa fasciata* (Stephens), *Phymatodes testaceus* (L.), *Plagionotus arcuatus* (L.), *Rhagium bifasciatum* Fabricius, *Rhagium inquisitor* (L.), *Rhagium mordax* (DeGeer), and *Tetropium castaneum* (L.); the chrysomelid *Agelastica alni* (L.); the clerid *Opilo domesticus* (Sturm); and the ptinid *Anobium punctatum* DeGeer.

Doryctes (Doryctes) molorchi Fischer, 1971
Distribution in the Middle East: Turkey (Beyarslan, 2015).
Extralimital distribution: Austria, Georgia, Russia.

Host records: Recorded by Fischer (1971) as being a parasitoid of the cerambycid *Molorchus umbellatarum* (Schreber).

Doryctes (Doryctes) obliteratus (Nees von Esenbeck, 1834)

Catalogs with Iranian records: Farahani et al. (2016 as *D. striatellus* (Nees, 1834)), Yu et al. (2016), Gadallah and Ghahari (2017 as *D. striatellus* (Nees, 1834)), Samin, Coronado-Blanco, Kavallieratos et al. (2018), Samin, Coronado-Blanco, Fischer et al. (2018) as *D. striatellus* (Nees, 1834)).

Distribution in Iran: East Azarbaijan (Rastegar et al., 2012, as *Doryctes striatellus* f. *petrovskii* Kokujev), West Azarbaijan (Rastegar et al., 2012, as *Doryctes striatellus* f. *petrovskii*; Samin et al., 2014).

Distribution in the Middle East: Iran.

Extralimital distribution: Austria, Belgium, China, Finland, France, Germany, Hungary, Italy, Japan, Poland, Russia, Sweden, Switzerland, Ukraine, United Kingdom.

Host records: Summarized by Yu et al. (2016) as being a parasitoid of the following cerambycids: *Callidium violaceum* (L.), *Monochamus galloprovincialis* (Olivier), *Obrium brunneum* (Fabricius), *Pogonocherus hispidus* (L.), *Pyrrhidium sanguineum* L.; the curculionids *Hylurgops palliatus* (Gyllenhal), *Hylurgus ligniperda* (Fabricius), *Magdalis rufa* Germar, *Magdalis violacea* (L.), and *Pissodes castaneus* (DeGeer); and the ptinids *Dorcatoma dresdensis* Herbst, *Ernobius abietis* (Fabricius), and *Ernobius mollis* (L.). It was also recorded by Hedwig (1958) as being a parasitoid of the xiphydriid wood wasp *Xiphydria camelus* (L.).

Doryctes (Doryctes) striatellus (Nees von Esenbeck, 1834)

Distribution in the Middle East: Turkey (Beyarslan, 2015).

Extralimital distribution: Austria, Belgium, China, Czech Republic, Finland, France, Germany, Hungary, Italy, Japan, Poland, Portugal, Russia, Serbia, Slovakia, Sweden, Switzerland, Ukraine, United Kingdom.

Host records: Summarized by Yu et al. (2016) as being a parasitoid of the following cerambycids: *Callidium violaceum* (L.), *Monochamus galloprovincialis* (Olivier), *Obrium brunneum* (Fabricius), *Pogoncherus hispidus* (L.), and *Pyrrhidium sanguineum* L.; the following curculionids: *Hylurgops palliatus* (Gyllenhal), *Hylurgops liguiperda* (Fabricius), *Magdalis rufa* Germar, *Magdalis violacea* (L.), and *Pissodes notatus* (Fabricius); the following ptinids: *Dorcatoma dresdensis* Herbst, *Ernobius abietis* (Fabricius), and *Ernobius mollis* (L.); and the xiphydeiid *Xiphydria camelus* (L.).

Doryctes (Doryctes) undulatus (Ratzeburg, 1852)

Catalogs with Iranian records: Yu et al. (2016), Gadallah and Ghahari (2017), Samin, Coronado-Blanco, Kavallieratos et al. (2018), Samin, Coronado-Blanco, Fischer et al. (2018).

Distribution in Iran: Khuzestan (Samin et al., 2016), Iran (no specific locality cited) (Mamedov et al., 2015).

Distribution in the Middle East: Iran (Mamedov et al., 2015; Samin et al., 2016), Turkey (Beyarslan, 2015).

Extralimital distribution: China, Japan, Kazakhstan, Korea, Mongolia, Russia.

Host records: Summarized by Yu et al. (2016) as being a parasitoid of the buprestids *Agrilus* spp., and *Anthaxia tuerki* Ganglbauer; the cerambycids *Axinopalpis gracilis* (Krynicki), *Grammoptera ruficornis* (Fabricius), *Molorchus kiesenwetter* Mulsant and Rey, *Molorchus umbellatarum* (Schreber), *Pogonocherus* sp., and *Tetrops praeustus* (L.); and the cuculionids *Magdalis armigera* (Geoffroy), *Magdalis ruficornis* (L.), and *Pityogenes bidentatus* (Herbst).

Doryctes (Neodoryctes) arrujumi Belokobylskij, 2021

Distribution in the Middle East: Saudi Arabia, Yemen (Belokobylskij & van Achterberg, 2021).

Extralimital distribution: None.

Host records: Unknown.

Genus Doryctophasmus Enderlein, 1912

Doryctophasmus ferrugineus (Granger, 1949)

Distribution in the Middle East: United Arab Emirates, Yemen (Belokobylskij, 2015; Belokobylskij & van Achterberg, 2021).

Extralimital distribution: Djibouti, Madagascar, Senegal.

Host records: Unknown.

Genus Euscelinus Westwood, 1882

Euscelinus sarawacus Westwood, 1882

Distribution in the Middle East: Israel–Palestine (Halperin, 1986; Papp, 1989).

Extralimital distribution: Australia, India, Malaysia, Myanmar, Pakistan, Philippines, Thailand, United States of America (Hawaii).

Host records: Summarized by Yu et al. (2016) as being a parasitoid of the following bostrichids: *Amphicerus bimaculatus* (Olivier), *Dinoderus minutus* (Fabricius), *Heterobostrychus aequalis* (Waterhouse), *Sinoxylon ceratoniae* (L.), *Sinoxylon conigerum* Gerstäcker, and *Sinoxylon sexdentatum* (Olivier).

Genus Hemispathius Belokobylskij and Quicke, 2000

Hemispathius pilosus (Granger, 1949)

Distribution in the Middle East: Yemen (Belokobylskij & van Achterberg, 2021).

Extralimital distribution: Madagascar.

Host records: Unknown.

Genus *Mimodoryctes* Belokobylskij, 2001

Mimodoryctes arabicus Edmardash, Gadallah and Soliman, 2020

Distribution in the Middle East: Saudi Arabia (Edmardash et al., 2020), Yemen (Belokobylskij & van Achterberg, 2021).
Extralimital distribution: None.
Host records: Unknown.

Mimodoryctes proprius Belokobylskij, 2001

Distribution in the Middle East: Saudi Arabia (Edmardash et al., 2020), Yemen (Belokobylskij & van Achterberg, 2021).
Extralimital distribution: Algeria.
Host records: Unknown.

Genus *Ontsira* Cameron, 1900

Ontsira antica (Wollaston, 1858)

Catalogs with Iranian records: Gadallah and Ghahari (2017), Samin, Coronado-Blanco, Kavallieratos et al. (2018), Samin, Coronado-Blanco, Fischer et al. (2018).
Distribution in Iran: Isfahan (Sakenin et al., 2008), Mazandaran (Samin et al., 2019).
Distribution in the Middle East: Iran (Sakenin et al., 2008; Samin et al., 2019), Turkey (Beyarslan, 2015).
Extralimital distribution: Azerbaijan, China, Croatia, former Czechoslovakia, Finland, France, Georgia, Germany, Hungary, Italy, Lithuania, Madeira Islands, Moldova, New Zealand (introduced), Romania, Russia, Slovakia, Spain, Sweden, Switzerland, United States of America, Ukraine, United Kingdom, former Yugoslavia.
Host records: Summarized by Yu et al. (2016) as being a parasitoid of the buprestistids *Agrilus viridis* (L.), and *Chrysobothris igniventris* Ritter, the following cerambycids: *Clytus arietis* (L.), *Exocentrus punctipennis* Mulsant and Guillebeu, *Phymatodes testaceus* (L.), *Plagionotus arcuatus* (L.), *Plagionotus floralis* (Pallas), *Pogonocherus hispidulus* (Piller and Mitterpacher), and *Pyrrhidium sanguineum* L., the curculionids *Ips acuminatus* (Gyllenhal), *Magdalis frontalis* (Pallas), *Scolytus pygmaeus* (Fabricius), and *Scolytus nrugulosus* (Müller), the eucnemid *Melasis buprestoides* (L.), and the ptinids *Ernobius mollis* (L.) and *Ptilinus pectinicornis* (L.).

Ontsira ignea (Ratzeburg, 1852)

Catalogs with Iranian records: Farahani et al. (2016), Ghahari and Gadallah (2017), Samin, Coronado-Blanco, Kavallieratos et al. (2018), Samin, Coronado-Blanco, Fischer et al. (2018).
Distribution in Iran: Ilam (Ghahari et al., 2011b), Iran (no specific locality cited) (Belokobylskij et al., 2012).
Distribution in the Middle East: Iran (Belokobylskij et al., 2012; Ghahari et al., 2011b), Israel—Palestine (Papp, 1989).

Extralimital distribution: Bulgaria, China, Croatia, former Czechoslovakia, Finland, France, Georgia, Germany, Greece, Hungary, Italy, Japan, Korea, Russia, Sweden, former Yugoslavia.
Host records: Summarized by Yu et al. (2016) as being a parasitoid of the bostrichid *Bostrichus capucinus* (L.), the buprestids *Buprestis strigosa* Gebbler and *Dicerea berolinensis* (Herbst), the cerambycids *Pogonocherus fasciculatus* (DeGeer), *Pogonocherus hispidulus* (Piller & Mitterpacher), and *Tetropium castaneum* (L.).

Ontsira imperator (Haliday, 1836)

Catalogs with Iranian records: Fallahzadeh and Saghaei (2010), Farahani et al. (2016), Yu et al. (2016), Gadallah and Ghahari (2017), Samin, Coronado-Blanco, Kavallieratos et al. (2018), Samin, Coronado-Blanco, Fischer et al. (2018).
Distribution in Iran: Kordestan (Samin et al., 2016), Qazvin (Ghahari et al., 2011a), Iran (no specific locality cited) (Telenga, 1941 as *Doryctodes iranicus* Telenga, Hedwig, 1957 as *Coeloides niger* Hedwig, Shenefelt & Marsh, 1976, Belokobylskij et al., 2012, 2013, 2019).
Distribution in the Middle East: Iran (see references above), Israel—Palestine (Yu et al., 2016), Turkey (Beyarslan, 2015; Beyarslan et al., 2017).
Extralimital distribution: Nearctic, Oriental, Palaearctic regions.
Host records: Summarized by Yu et al. (2016) as being a parasitoid of the bostrichid *Bostrichus capucinus* (L.), the buprestids *Buprestis strigosa* Gebler and *Dicerea borolinensis* (Herbest), the following cerambycids: *Acanthocinus aedilis* (L.), *Acanthocinus griseus* (Fabricius), *Anisarthron barbipes* (Schrank), *Hylotrupes bajulus* (L.), *Leiopus nebulosus* (L.), *Pogonocherus fasciculatus* (DeGeer), *Pogonocherus hispidus* (L.), *Rhagium bifasciatum* Fabricius, *Rhagium mordax* (DeGeer), *Strictoleptura rubra* (L.), *Stromatium fulvum* (Villers), and *Tetropium castaneum* (L.), and the curculionids *Scolytus scolytus* (Fabricius) and *Tomicus minor* (Hartig).

Ontsira longicaudis (Giraud, 1857)

Catalogs with Iranian records: Farahani et al. (2016), Yu et al. (2016), Gadallah and Ghahari (2017), Samin, Coronado-Blanco, Kavallieratos et al. (2018), Samin, Coronado-Blanco, Fischer et al. (2018).
Distribution in Iran: Mazandaran (Farahani et al., 2014).
Distribution in the Middle East: Iran.
Extralimital distribution: Austria, Croatia, France, Germany, Hungary, Italy, Romania, Russia, former Yugoslavia.
Host records: Summarized by Yu et al. (2016) as being a parasitoid of the cerambycids *Aegosoma scabricorne* (Scopoli), *Cerambyx cerdo* L., *Prinobius myardi* Mulsant, and *Saperda punctata* (L.).

Genus *Rhoptrocentrus* Marshall, 1897

Rhoptrocentrus cleopatra Belokobylskij, 2001
Distribution in the Middle East: Egypt (Belokobylskij, 2001).
Extralimital distribution: None.
Host records: Unknown.

Rhoptrocentrus piceus Marshall, 1897
Catalogs with Iranian records: Farahani et al. (2016), Yu et al. (2016), Gadallah and Ghahari (2017), Samin, Coronado-Blanco, Kavallieratos et al. (2018), Samin, Coronado-Blanco, Fischer et al. (2018).
Distribution in Iran: Qazvin (Ghahari et al., 2011a).
Distribution in the Middle East: Iran (Ghahari et al., 2011a), Israel—Palestine (Halperin, 1986), Turkey (Beyarslan, 2015).
Extralimital distribution: Armenia, Austria, Bulgaria, Croatia, former Czechoslovakia, Germany, Greece, Hungary, Italy, Japan, Mexico, Moldova, Poland, Russia, Serbia, Slovakia, Spain, Sweden, Turkmenistan, United States of America, Ukraine, Vietnam, former Yugoslavia.
Host records: Summarized by Yu et al. (2016) as being a parasitoid of the bostrichids *Heterobostrychus brunneus* (Murray), and *Scobicia chevrier* (Villa and Villa), the buprestids *Anthaxia corynthia* Reiche and Saulcy, and *Buprestis haemorrhoidalis araratica* Marseul, the following cerambycids: *Acanthocinus griceus* (Fabricius), *Chlorophorus annularis* (Fabricius), *Chlorophorus glabromaculatus* (Goeze), *Hylotrupes bajulus* (L.), *Penichroa fasciata* (Stephens), *Psacothea hilaris* (Psacoe) and *Stromatuem fulvum* (Villers), the following curculionids: *Hypothenemus eruditus* Westwood, *Phloeotribus scarabaeoides* (Bernard), *Pissodes castaneus* (DeGeer), *Scolytus scolytus* (Fabricius) and *Tomicus destruens* (Wollastone), and the ptinid *Gastrallus corsicus* Schilsky. It was also recorded by Tobias (1976) and Belokobylskij and Tobias (1986) as being a parasitoid of the tortricid *Eupoecilia ambiguella* (Hübner), and by Belokobylskij and Tobias (1986) as being a parasitoid of the xiphydriid woodwasp *Xiphydria camelus* (L.).

Tribe Ecphylini Hellén, 1957
Genus *Aivalykus* Nixon, 1938

Aivalykus microaciculatus Ranjith and Belokobylskij, 2020
Distribution in the Middle East: United Arab Emirates (Belokobylskij & van Achterberg, 2021; Ranjith et al., 2020).
Extralimital distribution: India (Ranjith et al., 2020).
Host records: Unknown.

Genus *Allorhogas* Gahan, 1912

Allorhogas semitemporalis (Fischer, 1960)
Distribution in the Middle East: Iraq (Shenefelt & Marsh, 1976).
Extralimital distribution: None.
Host records: Unknown.

Remarks: Very likely this species was collected in the Neotropical Region, its handwritten geographical label was incorrectly read.

Genus *Ecphylus* Foerster, 1863
Ecphylus (*Ecphylus*) *silesiacus* (Ratzeburg, 1848)
Catalogs with Iranian records: Fallahzadeh and Saghaei (2010), Farahani et al. (2016), Yu et al. (2016), Gadallah and Ghahari (2017), Samin, Coronado-Blanco, Kavallieratos et al. (2018), Samin, Coronado-Blanco, Fischer et al. (2018).
Distribution in Iran: Alborz (Karadj) (Davatchi & Shojai, 1969; Hedqvist, 1967 as *Ecphylus carinatus*; Hedqvist, 1967; Shojai, 1998), Markazi, Semnan, Tehran (Radjabi, 1976, 1991; Shojai, 1968, 1998; Modarres Awal, 1997, 2012 as *E. carinatus*), Qom (Davatchi & Shojai, 1969), Iran (no specific locality cited) (Aubert, 1966; Behdad, 1991; Modarres Awal, 1997 as *E. carinatu*; Shenefelt & Marsh, 1976s), Northern forests of Iran (Amini et al., 2016).
Distribution in the Middle East: Iran (see references above), Israel—Palestine (Halperin, 1986; Mendel, 1986; Papp, 1989, 2012), Turkey (Beyarslan, 2015).
Extralimital distribution: Armenia, Austria, Belarus, Bulgaria, Croatia, Czech Republic, Finland, France, Georgia, Germany, Hungary, Italy, Kazakhstan, Lithuania, Moldova, Montenegro, Poland, Romania, Russia, Serbia, Slovakia, Spain, Sweden, Switzerland, Ukraine, United Kingdom, former Yugoslavia.
Host records: Summarized by Yu et al. (2016) as being a parasitoid of the bostricihid *Bostrichus binodulus* Ratzeburg, as well as many curculionid scolytine bark beetles. In Iran, this species has been reared from the curculionids *Scolytus rugulosus* (Müller) (Behdad, 1991; Davatchi & Shojai, 1969; Hedqvist, 1967; Modarres Awal, 1997, 2012; Shojai, 1998 as *Ruguloscolytus mediterraneus* (Eggers)), and *Taphrorychus lenkoranus* (Reitter, 1913) (Amini et al., 2016).

Genus *Sycosoter* Picard and Lichtenstein, 1917
Sycosoter caudatus (Ruschka, 1916)
Distribution in the Middle East: Israel—Palestine (Halperin, 1986; Mendel, 1986; Papp, 1989).
Extralimital distribution: Algeria, Austria, China, Croatia, France, Italy, Japan, Korea, Malta, Morocco, Romania, Russia, Spain, Tunisia.
Host records: Summarized by Yu et al. (2016) as being a parasitoid of the bostrichid *Sinoxylon sexdentata* (Olivier), and the following curculionids: *Chaetoptelius vestitus* (Mulsant and Rey), *Cryphalus piceae* (Ratzrburg), *Hypoborus ficus* Erichson, *Liparthrum colchicum* Semonov, *Phloeotribus scarabaeoides* (Bernard), and *Pityokteines vorontzovi* (Jacobson).

Tribe Hecabolini Foerster, 1863

Genus *Hecabalodes* Wilkinson, 1929

Hecabalodes anthaxiae Wilkinson, 1929
Distribution in the Middle East: Saudi Arabia (Edmardash et al., 2020), United Arab Emirates, Yemen (Belokobylskij & van Achterberg, 2021).
Extralimital distribution: Sudan.
Host records: Recorded by Wilkinson (1929) as being a parasitoid of the buprestid *Anthaxia congregate* (Klug).

Hecabalodes maculatus Belokobylskij, 2021
Distribution in the Middle East: United Arab Emirates (Belokobylskij & van Achteberg, 2021).
Extralimital distribution: None.
Host records: Unknown.

Hecabalodes radialis Tobias, 1962
Catalogs with Iranian records: Yu et al. (2016), Gadallah and Ghahari (2017), Samin, Coronado-Blanco, Kavallieratos et al. (2018), Samin, Coronado-Blanco, Fischer et al. (2018).
Distribution in Iran: Hormozgan (Ameri et al., 2014).
Distribution in the Middle East: Iran (Ameri et al., 2014), Turkey (Beyarslan, 2015), United Arab Emirates, Yemen (Belokobylskij & van Achterberg, 2021).
Extralimital distribution: Greece, India, Tajikistan, Turkmenistan.
Host records: Summarized by Yu et al. (2016) as being a parasitoid of the buprestids *Acmaeoderella glasunovi* (Semonov), and *Anthaxia judinae* Stepanov, the cerambycid *Turanium scabrum* (Kraatz), and the curculionids *Magdalis egregia* J. Faust, *Magdalis myochroa* Reichardt, and *Scolytus gretschkini* Sokanovskiy.

Hecabalodes xylophagi Fischer, 1962
Catalogs with Iranian records: Farahani et al. (2016), Gadallah and Ghahari (2017), Samin, Coronado-Blanco, Kavallieratos et al. (2018), Samin, Coronado-Blanco, Fischer et al. (2018).
Distribution in Iran: Alborz (Modarres Awal, 1997, 2012; Shojai, 1998).
Distribution in the Middle East: Iran (Modarres Awal, 1997, 2012; Shojai, 1998), United Arab Emirates, Yemen (Belokobylskij & van Achterberg, 2021).
Extralimital distribution: Algeria, Chad, Mauritania.
Host records: In Iran, this species has been reared from the curculionid *Scolytus rugulosus* (Müller) (Shojai, 1998, Modarres Awal, 1997, 2012 as *Ruguloscolytus mediterraneans* (Eggers)).

Genus *Hecabolus* Curtis, 1834

Hecabolus sulcatus Curtis, 1834
Catalogs with Iranian records: Farahani et al. (2016), Yu et al. (2016), Gadallah and Ghahari (2017), Samin, Coronado-Blanco, Kavallieratos et al. (2018), Samin, Coronado-Blanco, Fischer et al. (2018).
Distribution in Iran: Mazandaran (Farahani et al., 2014).
Distribution in the Middle East: Iran (Farahani et al., 2014), Israel—Palestine (Halperin, 1986).
Extralimital distribution: Azerbaijan, Belgium, Croatia, Czech Republic, Finland, France, Georgia, Germany, Hungary, Italy, Latvia, Moldova, Morocco, Poland, Russia, Serbia, Slovakia, Sweden, Switzerland, Ukraine, United Kingdom, former Yugoslavia.
Host records: Summarized by Yu et al. (2016) as being a parasitoid of bostrichid *Lyctus brunneus* (Stephens), the buprestid *Anthaxia* sp., the chrysomelid *Agelastica alni* (L.), the curculionids *Hylesinus fraxini* (Panzer), *Phloeosinus bicolor* (Brullé), and *Phloeosinus thujae* (Perris), and the ptinids *Anobium punctatum* DeGeer, *Anobium rufipes* Fabricius, *Anobium thomsoni* (Kraatz), *Ochina ptinoides* (Marsham), *Ptilinus fuscus* Geoffroy, *Ptilinus pectinicornis* (L.), and *Ptilinus fur* (L.).

Genus *Hemidoryctes* Belokobylskij, 1992

Hemidoryctes carbonarius postfurcalis Belokobylskij, 2021
Distribution in the Middle East: Yemen (Belokobylskij & van Achterberg, 2021).
Extralimital distribution: None.
Host records: Unknown.

Genus *Leluthia* Cameron, 1887

Leluthia (*Euhecabolodes*) *asiatica* (Tobias, 1980)
Catalogs with Iranian records: Farahani et al. (2016), Yu et al. (2016), Gadallah and Ghahari (2017), Samin, Coronado-Blanco, Kavallieratos et al. (2018), Samin, Coronado-Blanco, Fischer et al. (2018).
Distribution in Iran: Mazandaran (Farahani et al., 2014).
Distribution in the Middle East: Iran (Farahani et al., 2014), Turkey (Beyarslan, 2015).
Extralimital distribution: Kazakhstan, Mongolia.
Host records: Reported by Tobias (1980) and Belokobylskij and Tobias (1986) as being a parasitoid of the curculionid *Scolytus schevyrewi* (Semenov-Tian-Shanskij).

Leluthia (*Euhecabolodes*) *ruguloscolyti* (Fischer, 1962)
Catalogs with Iranian records: Fallahzadeh and Saghaei (2010), Farahani et al. (2016), Yu et al. (2016),

Gadallah and Ghahari (2017), Samin, Coronado-Blanco, Kavallieratos et al. (2018), Samin, Coronado-Blanco, Fischer et al. (2018).

Distribution in Iran: Alborz (Aubert, 1966; Davatchi & Shojai, 1969; Fischer, 1962; Shojai, 1998 as *Hecabalodes ruguloscolyti* Fischer), Isfahan, Kordestan, Markazi, Zanjan (Modarres Awal, 1997, 2012 as *H. ruguloscolyti*), Qom (Davatchi & Shojai, 1969), Tehran (Davatchi & Shojai, 1969; Shojai, 1998; Modarres Awal, 1997, 2012 as *H. ruguloscolyti*), Iran (no specific locality cited) (Behdad, 1991; Belokobylskij & Tobias, 1986).

Distribution in the Middle East: Iran (see references above), Turkey (Beyarslan, 2015).

Extralimital distribution: Tajikistan, Turkmenistan.

Host records: Summarized by Yu et al. (2016) as being a parasitoid of the curculionids *Phloeosinus bicolor* (Brullé), and *Scolytus rugulosus* (Müller). In Iran, this species has been reared from the curculionids *Phloeosinus bicolor* (Brullé), *Scolytus rugulosus* (Müller) (as *Ruguloscolytus mediterraneus*), and *Scolytus multistriatus* (Marsham) (Aubert, 1966; Behdad, 1991; Davatchi & Shojai, 1969; Modarres Awal, 1997, 2012; Radjabi, 1976, 1991; Shojai, 1968, 1998).

Leluthia (*Euhecabolodes*) *transcaucasica* (Tobias, 1976)

Catalogs with Iranian records: Farahani et al. (2016), Yu et al. (2016), Gadallah and Ghahari (2017), Samin, Coronado-Blanco, Kavallieratos et al. (2018), Samin, Coronado-Blanco, Fischer et al. (2018).

Distribution in Iran: Hormozgan (Ameri et al., 2014).

Distribution in the Middle East: Iran (Ameri et al., 2014), Turkey (Beyarslan, 2015).

Extralimital distribution: Former Czechoslovakia, Georgia, Kazakhstan, Mongolia, Russia.

Host records: Summarized by Yu et al. (2016) as being a parasitoid of the curculionids *Phloeosinus bicolor* (Brullé), *Scolytus butovitschi* Eggers, and *Scolytus japonicus* Chapius.

Comments: Record of *Leluthia transcaucasica* from Iran is doubtful, and the specimen must be re-examined (S.A. Belokobylskij).

Leluthia (*Leluthia*) *abnormis* Belokobylskij, 2020

Distribution in the Middle East: Yemen (Belokobylskij, 2020a).

Extralimital distribution: None.

Host records: Unknown.

Leluthia (*Leluthia*) *accepta* (Belokobylskij, 1986)

Distribution in the Middle East: Israel−Palestine (Belokobylskij & Tobias, 1986).

Extralimital distribution: Georgia.

Host records: Unknown.

Leluthia (*Leluthia*) *brevitergum* Belokobylskij, 2020

Distribution in the Middle East: Yemen (Belokobylskij, 2020a).

Extralimital distribution: None.

Host records: Unknown.

Leluthia (*Leluthia*) *paradoxa* (Picard, 1938)

Catalogs with Iranian records: Farahani et al. (2016), Yu et al. (2016), Gadallah and Ghahari (2017), Samin, Coronado-Blanco, Kavallieratos et al. (2018), Samin, Coronado-Blanco, Fischer et al. (2018).

Distribution in Iran: Golestan (Sakenin et al., 2012), Ilam (Ghahari et al., 2011b).

Distribution in the Middle East: Iran (Ghahari et al., 2011b; Sakenin et al., 2012), Turkey (Beyarslan, 2015).

Extralimital distribution: Algeria, France, Hungary, Italy, North Macedonia, Spain, Tunisia, former Yugoslavia.

Host records: Summarized by Yu et al. (2016) as being a parasitoid of the curculionids *Pissodes piceae* (Illiger), *Scolytus multistriatus* (Marsham), and *Trypodendron signatum* (Fabricius).

Genus *Monolexis* Foerster, 1863
Monolexis fuscicornis Foerster, 1863

Catalogs with Iranian records: No catalog.

Distribution in Iran: West Azarbaijan (Ghahari et al., 2020).

Distribution in the Middle East: Iran (Ghahari et al., 2020), Israel−Palestine (Halperin, 1986; Papp, 1989, 2012), Turkey (Özgen et al., 2018).

Extralimital distribution: Argentina, Australia, Austria, Azerbaijan, Brazil, Bulgaria, Canada, Costa Rica, former Czechoslovakia, France, Georgia, Germany, Hungary, Italy, Japan, Malaysia, Spain, Tunisia, United Kingdom, United States of America.

Host records: Summarized by Yu et al. (2016) as being a parasitoid of the following bostrichids: *Amphicerus bimaculatus* (Olivier), *Enneadesmus trispinosus* (Olivier), *Heterobostrychus brunneus* (Murray), *Lyctus brunneus* (Stephens), *Lyctus linearus* (Goeze), *Lyctus parallelocollis* Blackburn, *Lyctus planicollis* LeConte, *Minthea rugicollis* (Walker), *Scobicia chevrieri* (Villa and Villa), *Scobicia pustulata* (Fabricius), *Sinoxylon ceratoniae* (L.), *Sinoxylon sexdentatum* (Olivier), and *Trogoxylon parallelopipedum* Melscheimer, the cermabycids *Mesosa curculionoides* (L.), and *Sternidius alpha* (Say), the cucujid *Laemophloeus capensis* (Waltl), and the curculionids *Chaetoptelius vestitus* (Mulsant and Rey), and *Phloeotribus scarabaeoides* (Bernard). In Iran, this species has been reared from the bostrichid *Lyctus linearis* (Goeze) (Ghahari et al., 2020).

Genus *Parallorhogas* Marsh, 1993

Parallorhogas testaceus (Szépligeti, 1914)

Distribution in the Middle East: United Arab Emirates (Belokobylski & van Achterberg, 2021).
Extralimital distribution: Togo.
Host records: Unknown.

Genus *Polystenus* Foerster, 1863

Polystenus rugosus Foerster, 1863

Catalogs with Iranian records: Yu et al. (2016), Gadallah and Ghahari (2017), Samin, Coronado-Blanco, Kavallieratos et al. (2018), Samin, Coronado-Blanco, Fischer et al. (2018).
Distribution in Iran: Kerman (Samin et al., 2016), Khuzestan (Ameri et al., 2020).
Distribution in the Middle East: Iran (Samin et al., 2016; Ameri et al., 2020), Turkey (Beyarslan, 2015).
Extralimital distribution: Austria, China, Czech Republic, Germany, Hungary, Italy, Japan, Kazakhstan, Korea, Liechtenstein, Poland, Russia, Serbia, Slovakia, Switzerland, Tajikistan, Taiwan, Ukraine, former Yugoslavia.
Host records: Summarized by Yu et al. (2016) as being a parasitoid of the bostrichid *Sinoxylon sexdentatum* (Olivier), and the following buprestids: *Agrilus angustulus* (Illiger), *Agrilus auricollis* Kiesenwetter, *Agrilus sulcicollis* Lacordaire, *Agrilus viridis* (L.), *Anthaxia manca* (L.), and *Coraebus florentinus* (Herbst).

Tribe Heterospilini Fischer, 1981

Genus *Heterospilus* Haliday, 1836

Heterospilus (*Eoheterospilus*) *rubrocinctus* (Ashmead, 1905)

Distribution in the Middle East: United Arab Emirates, Yemen (Belokobylskij & van Achterberg, 2021).
Extralimital distribution: China, Japan, Philippines, Russia, Vietnam.
Host records: Recorded by Belokobylskij and Maetô (2009) as being a parasitoid of the anthribid *Choragus sheppardi* Kirby.

Heterospilus (*Heterospilus*) *austriacus* (Szépligeti, 1906)

Catalogs with Iranian records: No catalog.
Distribution in Iran: Zanjan (Samin, Sakenin Chelav et al., 2020).
Distribution in the Middle East: Iran.
Extralimital distribution: Austria, China, Germany, Kazakhstan, Korea, Latvia, Moldova, Russia.

Host records: Summarized by Yu et al. (2016) as being a parasitoid of the cerambycid *Phymatodes maaki* (Kraatz), and the curculionids *Phloeosinus thujae* (Perris), *Pissodes notatus* Duftschmidt, and *Xylocleptes bispinus* Duftschmidt.

Heterospilus (*Heterospilus*) *cephi* Rohwer, 1925

Catalogs with Iranian records: Yu et al. (2016), Gadallah and Ghahari (2017), Samin, Coronado-Blanco, Kavallieratos et al. (2018), Samin, Coronado-Blanco, Fischer et al. (2018).
Distribution in Iran: Guilan (Samin et al., 2016).
Distribution in the Middle East: Iran (Samin et al., 2016), Israel—Palestine (Papp, 1970, 1989, 2012), Turkey (Beyarslan, 2015, 2019 as H. magnastimata Beyarslan).
Extralimital distribution: Armenia, China, Italy, Japan, Kazakhstan, Korea, Mongolia, Russia, Tunisia, Turkmenistan, United States of America.
Host records: Summarized by Yu et al. (2016) as being a parasitoid of the cephid sawflies *Cephus pygmeus* L., and *Trachelus tabidus* (Fabricius).

Heterospilus (*Heterospilus*) *divisus* (Wollaston, 1858)

Distribution in the Middle East: Israel—Palestine (Papp, 2012).
Extralimital distribution: Madeira Islands, Ukraine.
Host records: Unknown.

Heterospilus (*Heterospilus*) *genalis* Tobias, 1976

Catalogs with Iranian records: Samin, Coronado-Blanco, Kavallieratos et al. (2018).
Distribution in Iran: Chaharmahal and Bakhtiari (Samin, Coronado-Blanco, Kavallieratos et al., 2018).
Distribution in the Middle East: Iran (Samin, Coronado-Blanco, Kavallieratos et al., 2018), Turkey (Beyarslan, 2015).
Extralimital distribution: Azerbaijan, Hungary.
Host records: Unknown.

Heterospilus (*Heterospilus*) *hemipterus* (Thomson, 1892)

Catalogs with Iranian records: Farahani et al. (2016), Yu et al. (2016), Gadallah and Ghahari (2017), Samin, Coronado-Blanco, Kavallieratos et al. (2018), Samin, Coronado-Blanco, Fischer et al. (2018).
Distribution in Iran: Qazvin (Ghahari et al., 2011a).
Distribution in the Middle East: Iran (Ghahari et al., 2011a), Turkey (Beyarslan, 2015).
Extralimital distribution: Austria, Bulgaria, France, Germany, Lithuania, Mongolia, Serbia, Slovenia, Sweden.
Host records: Unknown.

Heterospilus (Heterospilus) leptosoma Fischer, 1960

Catalogs with Iranian records: Yu et al. (2016), Gadallah and Ghahari (2017), Samin, Coronado-Blanco, Kavallieratos et al. (2018), Samin, Coronado-Blanco, Fischer et al. (2018).

Distribution in Iran: Mazandaran (Samin et al., 2016).

Distribution in the Middle East: Iran (Samin et al., 2016), Turkey (Beyarslan, 2015).

Extralimital distribution: Austria, Bulgaria, China, Croatia, former Czechoslovakia, Greece, Hungary, Japan, Kazakhstan, Korea, Moldova, Mongolia, Russia, Serbia.

Host records: Unknown.

Heterospilus (Heterospilus) rubicola Fischer, 1968

Catalogs with Iranian records: No catalog.

Distribution in Iran: Khuzestan (Ameri et al., 2020).

Distribution in the Middle East: Iran (Ameri et al., 2020), Turkey (Beyarslan, 2015).

Extralimital distribution: Bulgaria, Georgia, Germany, Hungary, Kazakhstan, Korea, Moldova, Russia, Serbia, Uzbekistan.

Host records: Unknown.

Heterospilus (Heterospilus) tadzhicus Belokobylskij, 1983

Catalogs with Iranian records: Samin, Coronado-Blanco, Kavallieratos et al. (2018).

Distribution in Iran: Isfahan (Samin, Coronado-Blanco, Kavallieratos et al., 2018).

Distribution in the Middle East: Iran (Samin, Coronado-Blanco, Kavallieratos et al., 2018), Israel—Palestine (Papp, 2012), Turkey (Beyarslan, 2015).

Extralimital distribution: Tajikistan.

Host records: Unknown.

Heterospilus (Heterospilus) tauricus Telenga, 1941

Catalogs with Iranian records: Farahani et al. (2016), Yu et al. (2016), Gadallah and Ghahari (2017), Samin, Coronado-Blanco, Kavallieratos et al. (2018), Samin, Coronado-Blanco, Fischer et al. (2018).

Distribution in Iran: Alborz, Guilan, Qazvin, Tehran (Farahani et al., 2014), Mazandaran (Ghahari, 2019).

Distribution in the Middle East: Iran (Farahani et al., 2014; Ghahari, 2019), Israel—Palestine (Papp, 2012), Turkey (Beyarslan, 2015).

Extralimital distribution: Armenia, Azerbaijan, China, France, Germany, Italy, Japan, Kazakhstan, Korea, Lithuania, Moldova, Russia, Serbia, Ukraine.

Host records: Recorded by Fischer (1960) as being a parasitoid of the mordellid *Mordellistena pentas* Mulsant.

Genus *Neoheterospilus* Belokobylskij, 2006

Neoheterospilus (Neoheterospilus) alkowdi Belokobylskij, 2020

Distribution in the Middle East: Saudi Arabia (Edmardash et al., 2020, as *Neoheterospilus* sp.), Yemen (Belokobylskij, 2020b; Belokobylskij & van Achterberg, 2021).

Extralimital distribution: None.

Host records: Unknown.

Neoheterospilus (Neoheterospilus) yemenus Belokobylskij, 2020

Distribution in the Middle East: Yemen (Belokobylskij, 2020b; Belokobylskij & van Achterberg, 2021).

Extralimital distribution: None.

Host records: Unknown.

Tribe Holcobraconini Cameron, 1905

Genus *Zombrus* Marshall, 1897

Zombrus anisopus Marshall, 1897

Distribution in the Middle East: Egypt, Saudi Arabia (Fahringer, 1930; Fischer, 1980).

Extralimital distribution: Morocco.

Host records: Unknown.

Zombrus flavipennis (Brullé, 1846)

Catalogs with Iranian records: Farahani et al. (2016), Yu et al. (2016), Gadallah and Ghahari (2017), Samin, Coronado-Blanco, Kavallieratos et al. (2018), Samin, Coronado-Blanco, Fischer et al. (2018).

Distribution in Iran: Golestan (Sakenin et al., 2008 as *Syngaster lepida* Brullé, 1846—misidentification of *Syngaster flavipennis* Brullé, 1846), Iran (no specific locality cited) (Belokobylskij & Samartsev, 2011; Fallahzadeh & Saghaei, 2010; Shenefelt & Marsh, 1976).

Distribution in the Middle East: Iran.

Extralimital distribution: Pakistan.

Host records: In Iran, this species has been reared from the cerambycid *Phytoecia croceipes* Reiche and Saulcy (Sakenin et al., 2008).

Tribe Rhaconotini Fahringer, 1928

Genus *Platyspathius* Viereck, 1911

Platyspathius (Platyspathius) brevis Belokobylskij, 2021

Distribution in the Middle East: Yemen (Belokobylskij & van Achterberg, 2021).

Extralimital distribution: None.

Host records: Unknown.

Platyspathius (Platyspathius) longicaudis Belokobylskij, 2021

Distribution in the Middle East: Yemen (Belokobylskij & van Achterberg, 2021).
Extralimital distribution: None.
Host records: Unknown.

Genus *Rhaconotinus* Hedqvist, 1965

Rhaconotinus (Rhaconotinus) albosetosus Belokobylskij, 2021

Distribution in the Middle East: Yemen (Belokobylskij & van Achterberg, 2021).
Extralimital distribution: None.
Host records: Unknown.

Rhaconotinus (Rhaconotinus) menippus (Nixon, 1939)

Distribution in the Middle East: Yemen (Belokobylskij & van Achterberg, 2021).
Extralimital distribution: Benin, China, India, Madagascar, Malaysia, South Africa, Thailand, Uganda.
Host records: Summarized by Yu et al. (2016 as *Rhaconotus menippus* Nixon) as being a parasitoid of the brentid *Cylas puncticollis* Boheman; as well as the following curculionids: *Hypolixus truncatulus* (Fabricius), *Lixus* sp., *Peloropus batatae* Marshall, and *Pempheres affinis* Faust.

Genus *Rhaconotus* Ruthe, 1854

Rhaconotus (Rhaconotus) aciculatus Ruthe, 1854

Catalogs with Iranian records: Farahani et al. (2016), Yu et al. (2016), Gadallah and Ghahari (2017), Samin, Coronado-Blanco, Kavallieratos et al. (2018), Samin, Coronado-Blanco, Fischer et al. (2018).
Distribution in Iran: Alborz, Guilan, Mazandaran, Qazvin, Tehran (Farahani et al., 2014), Hormozgan (Ameri et al., 2014), Markazi, Khuzestan (Ameri et al., 2020), West Azarbaijan (Samin et al., 2014; Shahand & Karimpour, 2017), Yazd (Mohammadi-Khoramabadi, 2016).
Distribution in the Middle East: Iran (see references above), Israel—Palestine (Halperin, 1986; Papp, 1989, 2012), Turkey (Beyarslan, 2015).
Extralimital distribution: Azerbaijan, China, Czech Republic, France, Germany, Hungary, Italy, Kazakhstan, Korea, Kyrgyzstan, Moldova, Mongolia, Russia, Serbia, Slovakia, Spain, Tajikistan, Turkmenistan, Ukraine, United Kingdom, Uzbekistan, former Yugoslavia.
Host records: Summarized by Yu et al. (2016) as being a parasitoid of the buprestids *Agrilus viridis* (L.) and *Anthaxia lgockii* Obenberger, the chrysomelid *Caryedon serratus* (Olivier), and the curculionid *Lixus* sp. In Iran, this species has been reared from the curculionid *Lixus (Dilixellus) fasciculatus* Boheman (Shahand & Karimpour, 2017).

Rhaconotus (Rhaconotus) arabicus Belokobylskij, 2001

Distribution in the Middle East: Saudi Arabia (Belokobylskij, 2001; Edmardash et al., 2020), United Arab Emirates, Yemen (Belokobylskij & van Achterberg, 2021).
Extralimital distribution: None.
Host records: None.

Rhaconotus (Rhaconotus) brevicellularis Belokobylskij, 2021

Distribution in the Middle East: United Arab Emirates (Belokobylskij & van Achterberg, 2021).
Extralimital distribution: None.
Host records: Unknown.

Rhaconotus (Rhaconotus) elegans (Foerster, 1863)

Catalogs with Iranian records: No catalog.
Distribution in Iran: Guilan (Ghahari & Sakenin, 2018).
Distribution in the Middle East: Iran (Ghahari & Sakenin, 2018), Turkey (Beyarslan, 2015).
Extralimital distribution: Georgia, Germany, Hungary, Kazakhstan, Poland, Russia, Tajikistan, Turkmenistan, Ukraine, Uzbekistan.
Host records: Unknown.

Rhaconotus (Rhaconotus) kerzhneri Belokobylskij, 1985

Catalogs with Iranian records: Farahani et al. (2016), Yu et al. (2016), Gadallah and Ghahari (2017), Samin, Coronado-Blanco, Kavallieratos et al. (2018), Samin, Coronado-Blanco, Fischer et al. (2018).
Distribution in Iran: Hormozgan (Ameri et al., 2014), Yazd (Mohammadi-Khoramabadi, 2016).
Distribution in the Middle East: Iran (Ameri et al., 2014; Mohammadi-Khoramabadi, 2016), Israel-Palestine (Papp, 2012 as *Rhaconotus asiaticus* Belokobylskij), Turkey (Beyarslan, 2015 both as *R. asiatica* and *R. kerzhneri*).
Extralimital distribution: Kazakhstan, Tajikistan Turkmenistan, Uzbekistan.
Host records: Recorded by Belokobylskij (1990, as *R. asiaticus*) as being a parasitoid of the curculionid *Lixus kiritshenkoi* Ter-Minasian.

Rhaconotus (Rhaconotus) longulus Belokobylskij, 1994

Distribution in the Middle East: Turkey (Beyarslan, 2015).
Extralimital distribution: Russia.
Host records: Unknown.

Rhaconotus (Rhaconotus) magniareolus Belokobylskij, 2021

Distribution in the Middle East: Yemen (Belokobylskij & van Achterberg, 2021).
Extralimital distribution: None.
Host records: Unknown.

Rhaconotus (Rhaconotus) manolus Nixon, 1941

Distribution in the Middle East: United Arab Emirates, Yemen (Belokobylskij & van Achterberg, 2021).
Extralimital distribution: South Africa.
Host records: Unknown.

Rhaconotus (Rhaconotus) microexcavatus Belokobylskij, 2021

Distribution in the Middle East: United Arab Emirates (Belokobylskij & van Achterberg, 2021).
Extralimital distribution: None.
Host records: Unknown.

Rhaconotus (Rhaconotus) ollivieri (Giraud, 1869)

Distribution in the Middle East: Cyprus, Syria (Belokobylskij, 1990; Fahringer, 1932).
Extralimital distribution: Algeria.
Host records: Recorded by Belokobylskij (1990) as being a parasitoid of the gelechiid *Oecocecis guyonella* Guenée.

Rhaconotus (Rhaconotus) pictipennis (Reinhard, 1885)

Catalogs with Iranian records: Gadallah and Ghahari (2017), Samin, Coronado-Blanco, Kavallieratos et al. (2018), Samin, Coronado-Blanco, Fischer et al. (2018).
Distribution in Iran: Hamadan (Gadallah & Ghahari, 2017).
Distribution in the Middle East: Iran (Gadallah & Ghahari, 2017), Turkey (Beyarslan, 2015).
Extralimital distribution: Azerbaijan, Germany, Greece, Hungary, Kazakhstan, Kyrgyzstan, Korea, Russia, Tajikistan, Turkmenistan, Ukraine.
Host records: Unknown.

Rhaconotus (Rhaconotus) scaber Kokujev, 1900

Catalogs with Iranian records: Farahani et al. (2016), Yu et al. (2016), Gadallah and Ghahari (2017), Samin, Coronado-Blanco, Kavallieratos et al. (2018), Samin, Coronado-Blanco, Fischer et al. (2018).

Distribution in Iran: Alborz, Tehran (Farahani et al., 2014).
Distribution in the Middle East: Iran.
Extralimital distribution: Bulgaria, Hungary, Kazakhstan, Moldova, Mongolia, Russia, Spain, Tajikistan, Ukraine, Uzbekistan.
Host records: Unknown.

Rhaconotus (Rhaconotus) scirpophagae Wilkinson, 1927

Distribution in the Middle East: Yemen (Belokobylskij & van Achterberg, 2021).
Extralimital distribution: Bangladesh, China, Ghana, India, Indonesia, Ivory Coast, Kenya, Mauritius, Nigeria, Pakistan, Senegal, Sierra Leone, Tanzania, Trinidad and Tobago, Vietnam.
Host records: Recorded by Belokobylskij and van Achterberg (2021) as being a parasitoid of the following crambids: *Chilo auricilus* Dudgeon, *C. partellus* (Swinhoe), *Scirpophaga excerptalis* (Walker), and *S. nivella* (Fabricius); the noctuid *Busseola fusca* (Fuller), and the pyralid *Maliarpha separatella* Ragonot.

Rhaconotus (Rhaconotus) sudanensis Wilkinson, 1927

Distribution in the Middle East: Saudi Arabia (Edmardash et al., 2020 as *R. carinatus* Polaszek), United Arab Emirates, Yemen (Belokobylskij & van Achterberg, 2021).
Extralimital distribution: Senegal, Sudan.
Host records: Recorded by Belokobylskij and van Achterberg (2021) as being a parasitoid of the buprestid *Sphenoptera gossypii* Kerremans, as well as the crambid *Coniesta ignefusalis* (Hampson).

Rhaconotus (Rhaconotus) testaceus (Szépligeti, 1908)

Catalogs with Iranian records: Farahani et al. (2016), Yu et al. (2016), Gadallah and Ghahari (2017), Samin, Coronado-Blanco, Kavallieratos et al. (2018), Samin, Coronado-Blanco, Fischer et al. (2018).
Distribution in Iran: Hormozgan (Ameri et al., 2014 as *Rh. flavistigma* Telenga), Mazandaran (Amooghli-Tabari & Ghahari, 2021).
Distribution in the Middle East: Cyprus (Beyarslan et al., 2017), Iran (Ameri et al., 2014; Amooghli-Tabari & Ghahari, 2021), Israel—Palestine (Papp, 1989).
Extralimital distribution: China, India, Indonesia, Japan, Korea, Tajikistan, Vietnam.
Host records: Recorded by Yasumatsu (1967) as being a parasitoid of the crambid *Chilo suppressalis* (Walker), and the pyraustid *Scirpophaga incertulas* (Walker). In Iran, this species has been reared from *C. suppressalis* (Amooghli-Tabari & Ghahari, 2021).

Rhaconotus (Rhaconotus) vanharteni Belokobylskij, 2021

Ditribution in the Middle East: Yemen (Belokobylskij & van Achterberg, 2021).
Extralimital distribution: None.
Host records: Unknown.

Rhaconotus (Rhaconotus) zarudnyi Belokobylskij, 1990

Catalogs with Iranian records: Farahani et al. (2016), Yu et al. (2016), Gadallah and Ghahari (2017), Samin, Coronado-Blanco, Kavallieratos et al. (2018), Samin, Coronado-Blanco, Fischer et al. (2018).
Distribution in Iran: Kerman (Belokobylskij, 1990), Iran (no specific locality cited) (Belokobylskij & Chen, 2004; Fallahzadeh & Saghaei, 2010).
Distribution in the Middle East: Iran.
Extralimital distribution: China, Vietnam.
Host records: Unknown.

Tribe Spathiini Marshall, 1872

Genus *Parana* Nixon, 1943

Parana arabica Belokobylskij, 2021

Distribution in the Middle East: United Arab Emirates, Yemen (Belokobylskij & van Achterberg, 2021).
Extralimital distribution: None.
Host records: Unknown.

Genus *Spathiomorpha* Tobias, 1976

Spathiomorpha varinervis Tobias, 1976

Catalogs with Iranian records: No catalog.
Distribution in Iran: Ardabil (Samin, Sakenin Chelav et al., 2020).
Distribution in the Middle East: Iran (Samin, Sakenin Chelav et al., 2020), Turkey (Beyarslan, 2015).
Extralimital distribution: Azerbaijan, Georgia, North Macedonia, Russia, Serbia.
Host records: Unknown.

Genus *Spathius* Nees von Esenbeck, 1819

Spathius alkadanus Belokobylskij, 2021

Distribution in the Middle East: Yemen (Belokobylskij & van Achterberg, 2021).
Extralimital distribution: None.
Host records: Unknown.

Sapthius austroarabicus Belokobylskij, 2021

Distribution in the Middle East: Yemen (Belokobylskij & van Achterberg, 2021).
Extralimital distribution: None.
Host records: Unknown.

Spathius brevicaudis Ratzeburg, 1844

Catalogs with Iranian records: Farahani et al. (2016), Yu et al. (2016), Gadallah and Ghahari (2017), Samin, Coronado-Blanco, Kavallieratos et al. (2018), Samin, Coronado-Blanco, Fischer et al. (2018).
Distribution in Iran: Guilan, Mazandaran (Farahani et al., 2014).
Distribution in the Middle East: Iran (Farahani et al., 2014), Turkey (Beyarslan et al., 2017).
Extralimital distribution: Austria, Azerbaijan, Bulgaria, China, Czech Republic, Denmark, France, Georgia, Germany, Hungary, Italy, Japan, Kazakhstan, Korea, Moldova, Mongolia, Montenegro, Poland, Romania, Russia, Serbia, Slovakia, Sweden, Switzerland, Taiwan, former Yugoslavia.
Host records: Summarized by Yu et al. (2016) as being a parasitoid of several coleopteran species of the families Bostrichidae, Buprestidae, Cerambycidae, Curculionidae and the hymenopteran family Xiphydriidae. It was also recently recorded by Cao et al. (2019) as being a parasitoid of the buprestid *Agrilus mali* Matsumara.

Spathius exarator (Linnaeus, 1758)

Catalogs with Iranian records: Farahani et al. (2016), Yu et al. (2016), Gadallah and Ghahari (2017), Samin, Coronado-Blanco, Kavallieratos et al. (2018), Samin, Coronado-Blanco, Fischer et al. (2018).
Distribution in Iran: East Azarbaijan (Samin, Beyarslan et al., 2020), Guilan (Farahani et al., 2014), Hormozgan (Ameri et al., 2014).
Distribution in the Middle East: Iran (see references above), Turkey (Beyarslan, 2015).
Extralimital distribution: Mostly Palaearctic in distribution, but also recorded for the Nearctic, Oceanic, and Oriental regions.
Host records: Summarized by Yu et al. (2016) as being a parasitoid of several species of the orders Coleoptera (Anobiidae, Buprestidae, Cerambycidae, Chrysomelidae, Curculionidae, Xiphydriidae), Diptera (Tephritidae), Hymenoptera (Xyelidae), and Lepidoptera (Tortricidae). In Iran, this species has been reared from the curculionid *Ips typographus* (L.) (Samin, Beyarslan et al., 2020).
Comments: Record of *Spathius exarator* from Iran is doubtful, and the specimen must be re-examined (S.A. Belokobylskij).

Spathius lahji Belokobylskij, 2021

Distribution in the Middle East: Yemen (Belokobylskij & van Achterberg, 2021).
Extralimital distribution: None.
Host records: Unknown.

Spathius maderi Fahringer, 1930

Catalogs with Iranian records: Farahani et al. (2016), Yu et al. (2016), Gadallah and Ghahari (2017), Samin, Coronado-Blanco, Kavallieratos et al. (2018), Samin, Coronado-Blanco, Fischer et al. (2018).

Distribution in Iran: Ilam (Ghahari et al., 2011b), Mazandaran (Sakenin et al., 2012).

Distribution in the Middle East: Iran.

Extralimital distribution: Albania, Russia, Serbia, Spain.

Host records: Unknown.

Spathius nixoni Belokobylskij and Maetô, 2009

Distribution in the Middle East: United Arab Emirates, Yemen (Belokobylskij & van Achterberg, 2021).

Extralimital distribution: China ? Japan.

Host records: Unknown.

Spathius pedestris Wesmael, 1838

Distribution in the Middle East: Israel–Palestine (Halperin, 1986).

Extralimital distribution: Albania, Belgium, former Czechoslovakia, Denmark, France, Georgia, Germany, Hungary, Italy, Madeira Islands, Netherlands, New Zealand, Poland, Romania, Russia, Slovenia, Spain, Switzerland, Ukraine, United Kingdom, former Yugoslavia.

Host records: Summarized by Yu et al. (2016) as being a parasitoid of the following anobiids: *Anobium fulvicorne* (Sturm), *Anobium punctatum* (De Geer), *Gastrallus immarginatus* Mull, *Gastrallus laevigatus* (Olivier), and *Stegobium paniceum* (L.), as well as the curculionid *Pentarthrum huttoni* Wollaston.

Spathius polonicus Niezabitowski, 1910

Catalogs with Iranian records: Fallahzadeh and Saghaei (2010), Farahani et al. (2016), Yu et al. (2016), Gadallah and Ghahari (2017), Samin, Coronado-Blanco, Kavallieratos et al. (2018), Samin, Coronado-Blanco, Fischer et al. (2018).

Distribution in Iran: Alborz (Fischer, 1970; Radjabi, 1989 as *Spathius radjabii* (Fischer, 1970); Farahani et al., 2014), East Azarbaijan (Ghahari, Fischer et al., 2009), Khuzestan (Ameri et al., 2020), Kordestan, Markazi, Semnan, Zanjan (Modarres Awal, 1997, 2012 as *S. radjabii* Fischer; Radjabi, 1976, 1989), Mazandaran (Ghahari, 2017—around rice fields), Qazvin (Radjabi, 1989 as *S. radjabii* Fischer, Farahani et al., 2014), Tehran (Fischer, 1970; Modarres Awal, 1997, 2012 as *S. radjabii* Fischer), Iran (no specific locality cited) (Belokobylskij, 1989; Modarres Awal, 1997; Shenefelt & Marsh, 1976).

Distribution in the Middle East: Iran (see references above), Turkey (Beyarslan, 2015).

Extralimital distribution: Armenia, Azerbaijan, Belarus, Germany, Italy, Netherlands, Poland, Russia, Serbia, Slovakia, Spain, Tajikistan, Turkmenistan, Ukraine, Uzbekistan.

Host records: Summarized by Yu et al. (2016) as being a parasitoid of the following buprestids: *Agrilus constantini* Obenberger, *Agrilus lineola* Kiesenwetter, *Agrilus planipennis* Fairmaire, *Agrilus viridis* (L.), *Anthaxia* sp., *Coraebus florentinus* (Herbst), *Lampra mirifica* (Mulsant), *Melanophila decastigma* Fabricius, *Sphenoptera kaznakovi* Jakovlev and *Trachypteris picta* (Pallas), and the curculionid *Scolytus* sp. In Iran, this species has been reared from the buprestid *Agrilus viridis* (L.) on *Populus nigra* (Salicaceae) (Ghahari, Fischer et al., 2009), *Sphenoptera davatchii* Descarpentries, 1960, *Sphenoptera kambyses* Obenberger, and *Chrysobothris affinis* (Fabricius) (Modarres Awal, 1997, 2012; Radjabi, 1989).

Spathius rubidus (Rossi, 1794)

Catalogs with Iranian records: Farahani et al. (2016), Yu et al. (2016), Gadallah and Ghahari (2017), Samin, Coronado-Blanco, Kavallieratos et al. (2018), Samin, Coronado-Blanco, Fischer et al. (2018).

Distribution in Iran: Guilan (Farahani et al., 2014), Hormozgan (Ameri et al., 2014), Mazandaran (Farahani et al., 2014; Ghahari, Gadallah et al., 2009).

Distribution in the Middle East: Cyprus (Nixon, 1943), Iran (see references above), Turkey (Beyarslan, 2015).

Extralimital distribution: Mostly distributed in the Palaearctic region, but also in the Nearctic and Oriental regions.

Host records: Summarized by Yu et al. (2016) as being a parasitoid of several coleopteran species of the families Anobiidae, Bostrichidae, Buprestidae, Cerambycidae, Chrysomelidae, Colydiidae, Curculionidae, as well as the hymenopteran family Xiphyriidae.

Spathius subafricanus Belokobylskij, 2021

Distribution in the Middle East: United Arab Emirates (Belokobylskij & van Achterberg, 2021).

Extralimital distribution: None.

Host records: Unknown.

Remarks: Two female specimens (holotype and a paratype) from which the species is described were collected from Fujairah (United Arab Emirates); however, the authors incorrectly cited Yemen in the distribution of the species.

Spathius umbratus (Fabricius, 1798)

Catalogs with Iranian records: Farahani et al. (2016 as *Spathius curvicaudis* Ratzeburg, 1844), Gadallah and Ghahari (2017 as *Spathius erythrocephalus* Wesmael, 1838, and *Spathius radzayanus* Ratzeburg, 1848).

Distribution in Iran: Golestan (Samin et al., 2016), Guilan, Mazandaran (Farahani et al., 2014 as *Spathius curvicaudis*),

West Azarbaijan (Samin et al., 2014 as *Spathius curvicaudis*).

Distribution in the Middle East: Iran (see references above), Israel—Palestine (Halperin, 1986), Turkey (Beyarslan, 2015).

Extralimital distribution: Mostly Palaearctic and sometimes the Oriental region.

Host records: Summarized by Yu et al. (2016) as being a parasitoid of several coleopteran species of the families Anobiidae, Buprestidae, Cerambycidae, Curculionidae, Ptinidae, the hymenopteran family Xiphydriidae, and the lepidopteran families Sesiidae and Tineidae.

Spathius vulnificus Wilkinson, 1931

Distribution in the Middle East: Israel—Palestine (Halperin, 1986).

Extralimital distribution: India.

Host records: Summarized by Yu et al. (2016) as being a parasitoid of the following bostrichids: *Dinoderus brevis* Horn, *Dinoderus minutus* (Fabricius), *Dinoderus ocellaris* Stephens, and *Heterobostrychus aequalis* (Waterhouse).

Species excluded from the fauna of Iran

Tribe Siragrini Belokobylskij, 1994

Genus *Syngaster* Brullé, 1846

Syngaster lepida Brullé, 1846

Catalogs with Iranian records: Gadallah and Ghahari (2017), Samin, Coronado-Blanco, Kavallieratos et al. (2018), Samin, Coronado-Blanco, Fischer et al. (2018).

Records from Iran: Golestan (Sakenin et al., 2008).

General distribution: Australia, United States of America (introduced).

Host records: Summarized by Yu et al. (2016) as being a parasitoid of the cerambycids *Phoracantha recurva* Newman, and *Phoracantha semipunctata* (Fabricius).

Comments: Reexamination of specimens proved Sakenin et al. (2008) identification of *Syngaster lepida* Brullé, 1846 was a misidentification of *Syngaster flavipennis* Brullé, 1846 (=*Zombrus flavipennis* (Brullé, 1846)), and we therefore exclude *Syngaster lepida* from the fauna of Iran and record the data under *Zombrus flavipennis*.

Conclusion

Ninety-six species and subspecies of the subfamily Doryctinae in 29 genera and seven tribes (Doryctini Foerster, 1863, Ecphylini Hellén, 1957, Hecabolini Foerster, 1863, Heterospilini Fischer, 1981, Holcobraconini Cameron, 1905, Rhaconotini Fahringer, 1928, and Spathiini Marshall, 1872) have been reported from the Middle Eastern countries. The genus *Rhaconotus* is the most diverse, with 17 recorded species. The Iranian fauna of the braconids of the subfamily Doryctinae comprises 47 species (2.3% of the world species) in 16 genera and six tribes (Doryctini, Ecphylini, Hecabolini, Heterospilini, Holcobraconini, and Spathiini), which makes Iran the most species rich country among the Middle Eastern countries. *Heterospilus* (with eight species), *Rhaconotus* (with seven species), and *Spathius* (with six species) are more diverse than the other genera, followed by *Doryctes*, *Leluthia*, *Ontsira* (each with four species), *Dendrosoter* (three species), *Dendrosotinus*, *Hecabalodes* (each with two species), *Ecphylus*, *Hecabolus*, *Monolexis*, *Polystenus*, *Rhoptrocentrus*, *Spathiomorpha*, and *Zombrus* (each with one species). Iranian Doryctinae have been recorded from 24 provinces, which Mazandaran (with 14 species) has the highest number of species, followed by Hormozgan (10 species), Alborz (nine species), Guilan and Qazvin with eight and seven species, respectively (Fig. 12.1). Host species have been detected for only eight Doryctinae species, which these hosts belong to five families of Coleoptera, Bostrichidae (two species), Buprestidae (five species), Cerambycidae (two species), Cucujidae (one species) and Curculionidae (seven species), and Crambidae (one species) of Lepidoptera. Comparison of the Doryctinae fauna of Iran with the Middle East countries and adjacent to Iran indicates the fauna of Russia with 110 species in 28 genera (Belokobylskij & Lelej, 2019) is more diverse than Iran, followed by Turkey (39 species), Kazakhstan (31 species), Yemen (30 species), Israel—Palestine (26 species), Azerbaijan (22 species), United Arab Emirates (18 species), Turkmenistan (12 species), Armenia (11 species), Saudi Arabia (nine species), Cyprus, Egypt and Pakistan (each with three species), Afghanistan, Iraq, Syria (each with one species). No species have recorded from Bahrain, Jordan, Kuwait, Lebanon, Oman and Qatar (Yu et al., 2016). Russia is indicated as having the most diverse fauna, which probably should be expected because it is the largest country, but the greater number of reported species undoubtedly also reflects a history of more extensive taxonomic research on Doryctinae in that country. The much lower number of species recorded from most other countries compared to Russia probably not only reflects a smaller land area but also less comprehensive surveys in these countries. Among the 23 countries of the Middle East and adjacent to Iran, Turkey shares 34 species with Iran (87% of total species), followed by Russia (30 species), Kazakhstan (19 species), Israel—Palestine (18 species), Azerbaijan (15 species), Armenia and Turkmenistan (both with nine species), Cyprus and United Arab Emirates (two species), and Egypt, Pakistan and Yemen (one species). There is no doryctine species shared between Iran and the 11 other countries.

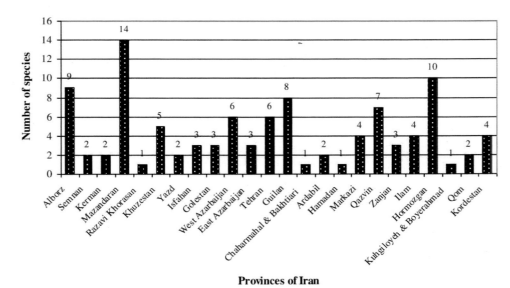

Provinces of Iran

FIGURE 12.1   Number of reported species of Iranian Doryctinae by province.

References

Abd El-Latif, A. N., Soliman, R. H. A., & Abd El-Gayed, A. A. (2009). Ecological and biological studies on some parasitoid species associated with *Scolytus amygdali* Guer. (Coleoptera: Scolytidae) in Fayoum Governorate, Egypt. *Egyptian Journal of Biological Pest Control, 19*(1), 1–4.

Ameri, A., Ebrahimi, E., & Talebi, A. A. (2020). Additions to the fauna of Braconidae (Hym., Ichneumonoidea) of Iran based on specimens housed in Hayk Mirazayansn Insect Museum with six new records for Iran. *Journal of Insect Biodiversity and Systematics, 6*(4), 353–364.

Ameri, A., Talebi, A. A., Rakhshani, E., Beyarslan, A., & Kamali, K. (2014). Taxonomic study of the subfamily Doryctinae (Hymenoptera: Braconidae) in Hormozgan province, southern Iran. *Zoology and Ecology, 24*(1), 40–54.

Amini, S., Nozari, J., Rahati, R., & Etemad, V. (2016). Investigation on parasitoids of bark beetles with new host Record (*Taphrorychus lenkoranus* Reitter, 1913 (Curculionidae: Scolytinae)) from northern forests of Iran. *Acta Phytopathologica et Entomologica Hungarica, 51*(2), 235–246.

Amooghli-Tabari, M., & Ghahari, H. (2021). A study on the transitional population of overwintering larvae of *Chilo suppressalis* Walker (Lepidoptera: Crambidae) with percent rice infestation in main crop, and their natural enemies in Mazandaran, Iran. *Experimental Animal Biology, 9*(4), 39–54 (in Persian, English summary).

Aubert, J. F. (1966). Liste d'identification No. 6 (présentée par le service d'identification des Entomophages). *Entomophaga, 11*(1), 115–134.

Basiri, N., Lotfalizadeh, H., & Kazemi, M. H. (2012). Introduction to parasitoids of bark beetles (Coleoptera: Scolytidae) in Meshkin-Shahr, Ardebil province, Iran. *Journal of Field Crop Entomology, 2*(1), 53–64 (in Persian, English summary).

Basiri, N., Lotfalizadeh, H., & Kazemi, M. H. (2013). *Dendrosoter middendorffii* (Ratzeburg, 1848) (Hymenoptera: Braconidae) a parasitoid of the fruit bark beetles in Iran. *Biharean Biologist, 7*(2), 104–105.

Beyarslan, A. (2015). A faunal study of the subfamily Doryctinae in Turkey (Hymenoptera: Braconidae). *Turkish Journal of Zoology, 39*(1), 126–143.

Behdad, E. (1991). *Pests of fruit crops in Iran*. Isfahan Neshat Publication, 822 pp. (in Persian).

Belokobylskij, S. A. (1989). Palearctic species of braconids of the genus *Spathius* Nees: of the group of species *S. labdacus, S. urios* and *S. leucippus* (Hymenoptera, Braconidae, Doryctinae). *Trudy Zoologicheskogo Instituta Leningrad, 188*, 39–57 (in Russian).

Belokobylskij, S. A. (1990). Review of braconid wasps of the genus *Rhaconotus* Ruthe (Hymenoptera, Braconidae) of the Palearctic region. *Entomologicheskoye Obozreniye, 69*(1), 144–163 (*Entomological Review, 70*(2), 47–67).

Belokobylskij, S. A. (2001). New taxa of the braconid subfamilies Doryctinae and Exothecinae (Hymenoptera, Braconidae) from West Palaearctic. *Entomologicheskoe Obozrenie, 80*(2), 451–471.

Belokobylskij, S. A. (2015). Review of species of the Old World genus *Doryctophasmus* Enderlein, 1912 (Hymenoptera: Braconidae: Doryctinae). *Zootaxa, 3985*(4), 541–564.

Belokobylskij, S. A. (2020a). Two new species of the genus *Leluthia* (Hymenoptera: Braconidae: Doryctinae) from Yemen. *Zoosystematica Rossica, 29*(2), 284–295.

Belokobylskij, S. A. (2020b). Two new species of the genus *Neoheterospilus* Belokobylskij, 2006 (Hymenoptera: Braconidae: Doryctinae) from Yemen. *Zootaxa, 4853*(4), 591–600.

Belokobylskij, S. A., & Chen, X.-X. (2004). The species of the genus *Rhaconotus* Ruthe, 1854 (Hymenoptera: Braconidae: Doryctinae) from China with a key to species. *Annales Zoologici, 54*(2), 319–359.

Belokobylskij, S. A., & Lelej, A. S. (2019). Annotated catalogue of the Hymenoptera of Russia. In, *Proceedings of the Zoological Institute of the Russian Academy of Sciences, suppl. 8: Vol. II. Apocrita: Parasitica*, 594 pp.

Belokobylskij, S. A., & Maetô, K. (2009). *Doryctinae (Hymenoptera, Braconidae) of Japan. Fauna mundi 1*. Warszawa: Warszawska Drukarnia Naukowa, 806 pp.

Belokobylskij, S. A., & Tobias, V. I. (1986). Doryctinae, pp. 21–72. In G. S. Medvedev (Ed.), *Opredelitel Nasekomych Evrospeiskoi Tsasti SSSR 3, Pereponchatokrylye 4. Opredelitel' po Faune SSSR* (Vol. 145). Leningrad: Nauka Press, 501 pp.

Belokobylsij, S. A., & van Achterberg, C. (2021). Review of the braconid parasitoid subfamily Doryctinae (Hymenoptera, Braconidae) from the United Arab Emirates and Yemen. *European Journal of Taxonomy, 765,* 1–134.

Belokobylskij, S. A., Kotenko, A. G., & Samartsev, K. G. (2019). Family Braconidae, pp. 200–329. In S. A. Belokobylskij, K. G. Samartsev, & A. S. Il'inskaya (Eds.), *Apocrita: Parasitica. Proceedings of the Zoological Institute Russian Academy of Sciences. Suppl. 8: Vol. II. Annotated catalogue of the Hymenoptera of Russia.* St. Petersburg: Zoological Institute RAS, 594 pp.

Belokobylskij, S. A., & Samartsev, K. G. (2011). First records of the tribe Holcobraconini and the genus *Zombrus* Marshall, 1897 (Hymenoptera: Braconidae: Doryctinae) in Europe. *Zoosystematica Rossica, 20*(2), 310–318.

Belokobylskij, S. A., Tobias, V. I., Kotenko, A. G., & Proshchalikin, M. Y. (2012). Fam. Braconidae – braconidy, pp. 300–389. In A. S. Lelej (Ed.), *Annotated catalogue of insects of the Russian Far East* (Vol. 1). Hymenoptera. Vladivostok: Dal'nauka, 635 pp.

Belokobylskij, S. A., Zaldívar-Riverón, A., & Quicke, D. L. J. (2004). Phylogeny of the genera of the parasitic wasps subfamily Doryctinae (Hymenoptera: Braconidae) based on morphological evidence. *Zoological Journal of the Linnean Society, 142*(3), 369–404.

Belokobylskij, S. A., Tang, P., & Chen, X.-X. (2013). The Chinese species of the genus *Ontsira* Cameron (Hymenoptera, Braconidae, Doryctinae). *ZooKeys, 345,* 73–96.

Beyarslan, A. (2017). Checklist of the Turkish Doryctinae (Hymenoptera, Braconidae). *Linzer biologische Beiträge, 49/1,* 415–446.

Beyarslan, A. (2019). A new species, *Heterospilus magnastigmata* sp. nov. (Hymenoptera: Braconidae: Doryctinae) from Turkey. *Munis Entomology & Zoology, 14*(1), 36–41.

Beyarslan, A., Gözüaçik, C., Güllü, M., & Konuskal, A. (2017). Taxonomical investigation on Braconidae (Hymenoptera: Ichneumonoidea) fauna in northern Cyprus, with twenty six new records for the country. *Journal of Insect Biodiversity and Systematics, 3*(4), 319–334.

Chen, X.-X., & van Achterberg, C. (2019). Systematics, phylogeny, and evolution of braconid wasps: 30 years of progress. *Annual Review of Entomology, 64,* 1–24.

Davatchi, A., & Shojai, M. (1969). *Les hyménoptères entomophages de l'Iran (études faunistiques).* Tehran University, Faculty of Agriculture, 88 pp.

Edmardash, Y. A., Abu El-Ghiet, U. M., Soliman, A. M., Al-Fifi, Z. I. A., & Gadallah, N. S. (2020). First contribution to the doryctine fauna (Hymenoptera, Braconidae, Doryctinae) of Farasan Archipelago, Saudi Arabia, with new records and the description of a new species. *ZooKeys, 977,* 41–74.

Fahringer, J. (1930). *Opuscula braconologica. Band 3. Palaearktischen region. Lieferung 1-2. Opuscula braconologica, 1–160.* Fritz Wagner, Wien.

Fahringer, J. (1932). *Opuscula braconologica. Band 3. Palaearktischen Region. Lieferung 3. Opuscula braconologica, (1931)* (pp. 161–240). Fritz Wagner, Wien.

Fallahzadeh, M., & Saghaei, N. (2010). Checklist of Braconidae (Insecta: Hymenoptera) from Iran. *Munis Entomology & Zoology, 5*(1), 170–186.

Farahani, S., Talebi, A. A., & Rakhshani, E. (2014). Wasps of the subfamily Doryctinae (Hymenoptera: Braconidae) in Iran. *Zoology in the Middle East, 60*(1), 65–81.

Farahani, S., Talebi, A. A., & Rakhshani, E. (2016). Iranian Braconidae (Insecta: Hymenoptera: Ichneumonoidea), diversity, distribution and host association. *Journal of Insect Biodiversity and Systematics, 2*(1), 1–92.

Fischer, M. (1960). Revision der paläarktischen Arten der Gattung *Heterospilus* Haliday (Hymenoptera, Braconidae). *Polskie Pismo Entomologiczne, 30,* 33–64.

Fischer, M. (1962). Neue Braconiden-Parasiten von schadlichen insecten (Hymenoptera). *Zeitschrift für Angewandte Entomologie, 49,* 297–312.

Fischer, M. (1970). Eine neue *Spathius*-Art aus dem Iran (Hym. Braconidae). *Annales de la Société Entomologique de France, 6,* 705–708.

Fischer, M. (1971). Zwei gezogene *Doryctes*-Arten aus Kärnten (Hymenoptera, Braconidae). *Entomophaga, 16*(1), 101–109.

Fischer, M. (1980). Taxonomische Untersuchungen über Doryctinae aus der *Odontobracon*-Verwandtschaft (Hymenoptera, Braconidae). *Annalen des Naturhistorischen Museums in Wien, 83,* 547–572.

Gadallah, N. S., & Ghahari, H. (2017). An annotated catalogue of the Iranian Doryctinae and Exothecinae (Hymenoptera: Braconidae). *Transactions of the American Entomological Society, 143,* 669–691.

Ghahari, H. (2017). Species diversity of Ichneumonoidea (Hymenoptera) from rice fields of Mazandaran province, northern Iran. *Journal of Animal Environment, 9*(3), 371–378 (in Persian, English summary).

Ghahari, H. (2019). Faunistic survey of parasitoid wasps (Hymenoptera) in forest areas of Mazandaran province, northern Iran. *Iranian Journal of Forest, 11*(1), 61–79 (in Persian, English summary).

Ghahari, H., & Sakenin, H. (2018). Species diversity of Chalcidoidea and Ichneumonoidea (Hymenoptera) in some paddy fields and surrounding grasslands of Mazandaran and Guilan provinces, northern Iran. *Applied Plant Protection, 7*(1), 11–19.

Ghahari, H., Fischer, M., Çetin Erdoğan, Ö., Beyarslan, A., & Havaskary, M. (2009). A contribution to the knowledge of the braconid-fauna (Hymenoptera, Ichneumonoidea, Braconidae) of Arasbaran, northwestern Iran. *Entomofauna, 30,* 329–336.

Ghahari, H., Gadallah, N. S., Çetin Erdoğan, Ö., Hedqvist, K. J., Fischer, F., Beyarslan, A., & Ostovan, H. (2009). Faunistic note on the Braconidae (Hymenoptera: Ichneumonoidea) in Iranian cotton fields and surrounding grasslands. *Egyptian Journal of Biological Pest Control, 19*(2), 115–118.

Ghahari, H., & Fischer, M. (2011). A contribution to the Braconidae (Hymenoptera: Ichneumonoidea) from north-western Iran. *Calodema, 134,* 1–6.

Ghahari, H., Fischer, M., & Papp, J. (2011a). A study on the Braconidae (Hymenoptera: Ichneumonoidea) from Qazvin province, Iran. *Entomofauna, 32,* 197–208.

Ghahari, H., Fischer, M., & Papp, J. (2011b). A study on the Braconidae (Hymenoptera: Ichneumonoidea) from Ilam province, Iran. *Calodema, 160,* 1–5.

Ghahari, H., & Gadallah, N. S. (2019). New records of braconid wasps (Hymenoptera: Ichneumonoidea: Braconidae) from Iran. *Entomological News, 129*(3), 384–394.

Ghahari, H., Beyarslan, A., & Kavallieratos, N. G. (2020). New records of Braconidae (Hymenoptera: Ichneumonoidea) from Iran, and in Memoriam Dr. Maximilian Fischer (7 June 1929 – 15 June 2019). *Scientific Bulletin of Uzhhorod National University (Series: Biology), 48,* 48–55.

Halperin, J. (1986). Braconidae (Hymenoptera) associated with forest and ornamental trees and shrubs in Israel. *Phytoparasitica, 14*(2), 119–135.

Hedqvist, K. J. (1967). Notes on *Ecphylus* Först. and description of two new species (Ichneumonoidea, Braconidae, Doryctinae). *Entomologisk Tidskrift, 88*(1–2), 66–71.

Hedwig, K. (1957). Ichneumoniden und Braconiden aus den Iran 1954 (Hymenoptera). *Jahresheft des Vereins für vaterlaendische Naturkunde, 112*(1), 103−117.

Hedwig, K. (1958). Mitteleuropäische Schlupfwespen und ihre Wirte. *Nachrichten des Naturwissenschaftlichen Museums der Stadt Aschaffenburg, 58*, 21−37.

Jasso-Martínez, J. M., Belokobylskij, S. A., & Zaldívar-Riverón, A. (2019). Molecular phylogenetics and evolution of generic diagnostic morphological features in the doryctine wasp tribe Rhaconotini (Hymenoptera: Braconidae). *Zoologischer Anzeiger, 279*, 164−171.

Mamedov, Z. M., Shirinova, L. A., Atayeva, R. S., & Safarova, E. F. (2015). The main parasites of xylophagous pests (Coleoptera) of trees and shrubs in Absheron Peninsula and their usage as ecological control agents. *AMEA-mn Xabarlai (biologiya va tibb elmlari), 70*(2), 62−67.

Mancini, D., Priore, R., Battaglia, D., & van Achterberg, C. (2003). *Caenopachys hartigii* (ratzeburg) (Hymenoptera: Braconidae: Doryctinae) confirmed for Italy, with notes on the status of the genus *Caenopachys* Foerster. *Zoologische Mededelingen, 77*(26), 459−470.

Marsh, P. M. (2002). The Doryctinae of Costa Rica (excluding the genus *Heterospilus*). *Memoirs of the American Entomological Institute, 70*, 1−319.

Marsh, P., Wild, A., & Whitfield, J. (2013). The Doryctinae (Braconidae) of Costa Rica: genera and species of the tribe Heterospilini. *ZooKeys, 347*, 1−474.

Martínez, J. J., Lázaro, R. N. M., Pedraza-Lara, C., & Zaldívar-Riverón, A. (2016). *Sergey* gen. n., a new doryctine genus from temperate forests of Mexico and Cuba (Hymenoptera, Braconidae). *ZooKeys, 589*, 143−164.

Mendel, Z. (1986). Hymenopterous parasitoids of bark beetles (Scolytidae) in Israel: relationships between host and parasitoid size and sex ratio. *Entomophaga, 31*(2), 127−137.

Mendel, Z., & Halperin, J. (1981). Parasites of the bark beetles (Col.: Scolytidae) on pine and cypress in Israel. *Entomophaga, 26*, 375−379.

Modarres Awal, M. (1997). Family Braconidae (Hymenoptera), pp. 265−267. In M. Modarres Awal (Ed.), *List of agricultural pests and their natural enemies in Iran* (2nd ed.). Ferdowsi University Press, 429 pp.

Modarres Awal, M. (2012). Family Braconidae (Hymenoptera), pp. 483−486. In M. Modarres Awal (Ed.), *List of agricultural pests and their natural enemies in Iran* (3rd ed.). Ferdowsi University of Mashhad Press, 759 pp.

Mohammadi-Khoramabadi, A. (2016). A preliminary study on the seed beetle *Caryedon angeri* (Col., Chrysomelidae, Bruchinae) feeding on *Prosopis farcta* (Fabaceae) in Yaz province. In *Proceedings of the 22ⁿᵈ Iranian plant protection congress, congress, 27−30 August 2016* (p. 656).

Nixon, G. E. J. (1943). A revision of the Spathiinae of the Old World (Hymenoptera: Braconidae). *Transactions of the Royal Entomological Society of London, 93*, 173−456.

Özgen, I., Beyarslan, A., Ruizzer, E., & Topdemir, A. (2018). The important record of *Monolexis fuscicornis* Foerster, 1862 (Hymenoptera, Braconidae, Doryctinae) in Turkey, with notes on *Trogoxylon impressum* (Comoti, 1837). *Cercetari Agronomie in Moldova, LI*(175), 125−130.

Papp, J. (1970). A contribution to the braconid fauna of Israel (Hymenoptera). *Israel Journal of Entomology, 5*, 63−76.

Papp, J. (1989). A contribution to the braconid fauna of Israel (Hymenoptera), 2. *Israel Journal of Entomology, 22*, 45−59.

Papp, J. (2012). A contribution to the braconid fauna of Israel (Hymenoptera: Braconidae), 3. *Israel Journal of Entomology, 41−42*, 165−219.

Quicke, D. L. J. (2015). *The Braconid and Ichneumonid Wasps: Biology, systematics, evolution and ecology, xv*. Oxford: Wiley Blackwell, 681 pp.

Quicke, D. L. J., Ficken, L. C., & Fitton, M. G. (1993). New diagnostic ovipositor characters for doryctine wasps (Hymenoptera: Braconidae). *Journal of Natural History, 26*, 1035−1046.

Radjabi, G. H. (1976). *Xylophagous insects of roseaceous trees in Iran*. Tehran, Iran: Plant Pests and Diseases Research Institute, 241pp.

Radjabi, G. H. (1989). *Insects attacking rosaceous fruit trees in Iran* (Vol. 3, 1−256). Tehran, Iran: Plant Pests and Diseases Research Institute.

Radjabi, G. H. (1991). *Insects attacking rosaceous fruit trees in Iran, 1*. Tehran: Coleoptera. Plant Pests and Diseases Research Institute, 221 pp.

Ranjith, A. P., Belokobylskij, S. A., Sureshan, P. M., & Nasser, M. (2020). The genus *Aivalykus* Nixon, 1938 (Hymenoptera: Braconidae: Doryctinae) with description of a new species from India and Arabian Peninsula. *Zootaxa, 4822*(2), 269−276.

Rastegar, J., Sakenin, H., Khodaparast, S., & Havaskary, M. (2012). On a collection of Braconidae (Hymenoptera) from East Azarbaijan and vicinity, Iran. *Calodema, 226*, 1−4.

Sakenin, H., Eslami, B., Samin, N., Imani, S., Shirdel, F., & Havaskary, M. (2008). A contribution to the most important trees and shrubs as the hosts of woodboring beetles in different regions of Iran and identification of many natural enemies. *Plant and Ecosystem, 16*, 27−46.

Sakenin, H., Naderian, H., Samin, N., Rastegar, J., Tabari, M., & Papp, J. (2012). On a collection of Braconidae (Hymenoptera) from northern Iran. *Linzer biologische Beiträge, 44*(2), 1319−1330.

Samin, N., Coronado-Blanco, J. M., Kavallieratos, N. G., Fischer, M., & Sakenin, H. (2018). Recent findings on Braconidae (Hymenoptera: Ichneumonoidea) of Iran with an updated checklist. *Acta Biologica Turcica, 31*(4), 160−173.

Samin, N., Coronado-Blanco, J. M., Fischer, M., van Achterberg, C., Sakenin, H., & Davidian, E. (2018). Updated checklist of Iranian Braconidae (Hymenoptera: Ichneumonoidea) with twenty-three new records. *Natura Somogyiensis, 32*, 21−36.

Samin, N., Ghahari, H., Gadallah, N. S., & Davidian, E. (2014). A study on the Braconidae (Hymenoptera: Ichneumonoidea) from West Azarbaijan province, northwestern Iran. *Linzer biologische Beiträge, 46*(2), 1447−1478.

Samin, N., van Achterberg, K., & Ghahari, H. (2015). A fauistic study of Braconidae (Hymenoptera: Ichneumonoidea) from southern Iran. *Linzer biologische Beiträge, 47*(2), 1801−1809.

Samin, N., van Achterberg, C., & Çetin Erdoğan, Ö. (2016). A faunistic study on some subfamilies of Braconidae (Hymenoptera: Ichneumonoidea) from Iran. *Arquivos Entomoloxicos, 15*, 153−161.

Samin, N., Fischer, M., Sakenin, H., Coronado-Blanco, J. M., & Tabari, M. (2019). A faunistic study on Agathidinae, Alysiinae, Doryctinae, Helconinae, Microgastrinae, and Rogadinae (Hymenoptera: Braconidae), with eight new country records. *Calodema, 734*, 1−7.

Samin, N., Beyarslan, A., Ranjith, A. P., Ahmad, Z., Sakenin Chelav, H., & Hosseini Boldaji, S. A. (2020). A faunistic study on Braconidae

(Hymenoptera: Ichneumonoidea) from Ardebil and East Azarbayjan provinces, northwestern Iran. *Egyptian Journal of Plant Protection Research Institute, 3*(4), 955–963.

Samin, N., Sakenin Chelav, H., Ahmad, Z., Penteado-Dias, A. M., & Samiuddin, A. (2020). A faunistic study on the family Braconidae (Hymenoptera: Ichneumonoidea) from Iran. *Scientific Bulletin of Uzhhorod National University (Series: Biology), 48*, 14–19.

Schimitschek, E. (1941). Beiträge zur Forestentomologie der Türkei. III. Die Massenvermhrung des *Ips sexdendatus* Börner in Gebiete der orientalischen Fichte. II. Teil. *Zeitschrift für Angewandte Entomologie, 27*, 84–113.

Schimitschek, E. (1944). *Forstinsekten der Tuerkei und ihre Umwelt. Prague* (pp. 273–279), 371 pp.

Shahand, S., & Karimpour, Y. (2017). Biology of mugwort weevil, *Lixus fasciculatus* (Col.: Curculionidae) on *Artemisia vulgaris* (Asteraceae) in Urmia region. *Biocontrol in Plant Protection, 5*(1), 45–57.

Shaw, S. R., & Edgerly, J. S. (1985). A new braconid genus (Hymenoptera) parasitizing webspinners (Embiidina) in Trinidad. *Psyche, 92*(4), 505–512.

Shaw, M. R., & Huddleston, T. (1991). Classification and biology of braconid wasps. In, *Vol. 7, Part 11. Handbooks for the identification of British insects*. London: Royal Entomological Society of London, 126 pp.

Shenefelt, R. D., & Marsh, P. M. (1976). Braconidae 9. Doryctinae. *Hymenopterorum Catalogus (nova editio). Pars 13* (pp. 1263–1424).

Shojai, M. (1968). Résultats de l'étude faunistique des Hyménoptères parasites (Terebrants) en Iran et l'importance de leur utilisation dans la lutte biologique. In *Proceedings of the first Iranian plant Protection congress, September 14–19, 1968* (pp. 25–35). Karaj: Karaj Junior College of Agriculture.

Shojai, M. (1998). *Entomology (etholgy, Social Life and natural enemies) (biological control)* (3rd ed., Vol. III). Tehran University Publications, 550 pp. (in Persian).

Telenga, N. A. (1941). *Family Braconidae, subfamily Braconinae (continuation) and Sigalphinae*. Fauna USSR. *Hymenoptera, 5*(3). Moskva-Leningrad: Akademiya nauk SSSR, 466 pp.

Tobias, V. I. (1976). *Braconids of the Caucasus (Hymenoptera, Braconidae). Opred. Faune SSSR* (Vol. 110). Leningrad: Nauka Press, 286 pp. (in Russian).

Tobias, V. I. (1980). New species of braconids (Hymenoptera, Braconidae)-parasites of bark beetles from Mongolia and the USSR. *Nasekomye Mongolii, 7*, 289–295.

van Achterberg, C. (1993). Illustrated key to the subfamilies of the Braconidae (Hymenoptera: Ichneumonidae). *Zoologische Verhandelingen, 283*, 1–189.

Wilkinson, D. S. (1929). New parasitic Hymenoptera and notes on other species. *Bulletin of Entomological Research, 20*(1), 103–114.

Yasumatsu, K. (1967). Distribution and bionomics of natural enemies of rice stem borers. *Mushi, 39*(Supplement), 33–44.

Yu, D. S., van Achterberg, C., & Horstmann, K. (2016). *Taxapad 2016, Ichneumonoidea 2015, Database on flash-drive*. Ottawa, Ontario, Canada.

Zaldívar-Riverón, A., & Belokobylskij, S. A. (2009). The parasitic wasp genus *Hecabolus* (Hymenoptera: Braconidae: Doryctinae), with the description of a new species from Mexico. *Revista Mexicana de Biodiversidad, 80*, 419–429.

Zaldívar-Riverón, A., Belokobylskij, S. A., León-Regagnon, V., Martínez, J. J., Brinceňo, R., & Quicke, D. L. J. (2007). A single origin of gall association in a group of parasitic wasps with disparate morphologies. *Molecular Phylogenetics and Evolution, 44*, 981–992.

Zaldívar-Riverón, A., Belokobylskij, S. A., León-Regagnon, V., Briceno, G. R., & Quicke, D. L. J. (2008). Molecular phylogeny and historical biogeography of the cosmopolitan parasitic wasp subfamily Doryctinae (Hymenoptera: Braconidae). *Invertebrate Systematics, 22*(3), 345–363.

Zaldívar-Riverón, A., Belokobylskij, S. A., Meza-Lázaro, R., Pedraza-Lara, C., García-París, M., & Meseguer, A. S. (2018). Species delimitation, global phylogeny and historical biogeography of the parasitoid wasp genus *Spathius* (Braconidae: Doryctinae) reveal multiple Oligocene-Miocene intercontinental dispersal events. *Zoological Journal of the Linnean Society, 182*(4), 723–734.

Zaldívar-Riverón, A., Mori, M., & Quicke, D. L. J. (2006). Systematics of the cyclostome subfamilies of braconid parasitic wasps (Hymenoptera: Ichneumonoidea), a simultaneous molecular and morphological Bayesian approach. *Molecular Phylogenetics and Evolution, 38*(1), 130–145.

Zaldívar-Riverón, A., Martínez, J. J., Belokobylskij, S. A., Pedraza-Lara, C., Shaw, S. R., Hanson, P. E., & Verla-Hernández, F. (2014). Systematics and evolution of gallformation in the plant-associated genera of the wasp subfamily Doryctinae (Hymenoptera: Braconidae). *Systematic Entomology, 39*(4), 633–659.

Dinocampus coccinellae (Schrank, 1802) (Euphorinae), ♀, lateral habitus. *Photo prepared by S.R. Shaw.*

Wesmaelia petiolata (Wollaston, 1858) (Euphorinae), ♀, lateral habitus. *Photo prepared by S.R. Shaw, 2020.*

Chapter 13

Subfamily Euphorinae Foerster, 1863

Michael J. Sharkey[1], Neveen Samy Gadallah[2], Hassan Ghahari[3], Donald L.J. Quicke[4] and Scott Richard Shaw[5]

[1]UW Insect Museum, Department of Entomology, University of Kentucky, Lexington, KY, United States; [2]Entomology Department, Faculty of Science, Cairo University, Giza, Egypt; [3]Department of Plant Protection, Yadegar-e Imam Khomeini (RAH) Shahre Rey Branch, Islamic Azad University, Tehran, Iran; [4]Integrative Ecology Laboratory, Department of Biology, Faculty of Science, Chulalongkorn University, Pathumwan, Bangkok, Thailand; [5]UW Insect Museum, Department of Ecosystem Science and Management, University of Wyoming, Laramie, WY, United States

Introduction

The cosmopolitan Euphorinae Foerster, 1863 is a large and heterogenous subfamily of Braconidae (Chen & van Achterberg, 1997; Shaw, 1997a). It comprises about 1270 species classified into 59 genera and 15 tribes (Chen & van Achterberg, 2019; Stigenberg et al., 2015), of which more than 456 species are known from the Palaearctic region (Yu et al., 2016). Euphorines are mainly diagnosed by the open first brachial cell apically (Shaw, 1985, 1995; Tobias, 1986), but they are not the only subfamily with that character, and their morphological variation means that they cannot be keyed out in just one or two couplets.

They have a wider host range than any other braconid subfamily, and, with the exception of the aphidiines, are the only ones that have members parasitizing the adult stages of their hosts. These are probably the major factors that have led to their great morphological diversity including features such as ovipositor length and form, remarkable variations of antennal shape, the size and position of eyes, and exceptional variation in the morphology in the first metasomal segment (Koldaş et al., 2013; Shaw, 1985, 1997a). Some species of the tribe Helorimorphini have remarkably elongate-slender tubular first metasomal segments, which are the thinnest and most elongated petioles among the whole family (Shaw, 1997). Most euphorines are fully winged as adults but some species occurring in windy grassland habitats are known to have flightless macropterous females (Aguirre et al., 2014).

Members of the tribe Meteorini are typical koinobiont endoparasitoids of Lepidoptera caterpillars and of some Coleoptera (especially of the families Chrysomelidae and Curculionidae). Members of the remaining tribes are unusual in that they are koinobiont endoparasitoids of adult (or late instar nymphal) insects (Gómez Durán & van Achterberg, 2011; Shaw, 1985, 1988). Shaw (2004) proposed the term "imagobiosis" to describe the behavior of euphorines that attack adult stages of insects, to differentiate them more typical koinobionts that normally attack larval insects, because although the host continues to feed, it cannot metamorphose to another life stage. At least seven orders of insects serve as hosts for euphorines: Orthoptera, Psocoptera, Hemiptera, Neuroptera, Coleoptera, Lepidoptera, and Hymenoptera (Belokobylskij, 1995, 1996; Shaw, 1985; Shenefelt, 1969; Stigenberg & Ronquist, 2011; van Achterberg 1985). With the exception of many Meteorini, most are solitary. Some euphorine species are potential biological control agents against harmful insects (Fox et al., 2004; Shaw, 2004; Shaw et al., 2001; Yilmaz et al., 2010).

The Meteorini has sometimes been treated as a tribe among Euphorinae (Aguirre et al., 2015; Chen & van Achterberg, 2019; Huddleston & Short, 1978; Perrichot et al., 2009; Stigenberg & Ronquist, 2011; Stigenberg et al., 2015; van Achterberg, 1984), or as a separate subfamily (Shaw, 1985, 1997; Wharton et al., 1997). Molecular analysis recently carried out by (Stigenberg et al., 2015), supported the former hypothesis and recommended treating Meteorini as a tribe in the Euphorinae as was done by some previous authors (Belshaw & Quicke, 2002; Li et al., 2003; Sharanowski et al., 2011) and which we follow in this book.

Members of Neoneurini are exclusively parasitoids of adult ants (Gómez Durán & van Achterberg, 2011; Shaw, 1992, 2007), and they have been formerly treated as a separate subfamily, the Neoneurinae, among the microgastroid complex (Quicke & van Achterberg, 1990; van Achterberg, 1984) because of shared morphological similarities. However, molecular data (Sharanowski et al., 2011; Stigenberg et al., 2015) have conclusively demonstrated that they are just a derived tribe among Euphorinae, thus corroborating the earlier hypothesis of Tobias (1966).

Braconidae of the Middle East (Hymenoptera). https://doi.org/10.1016/B978-0-323-96099-1.00019-4

Checklists of Regional Euphorinae. Modarres Awal (1997): four species in one genus (*Meteorus* Haliday, 1835); Fallahzadeh and Saghaei (2010): 12 species in six genera (without precise localities); Modarres Awal (2012): six species in two genera (*Meteorus* and *Microctonus* Wesmael, 1835); Ameri et al. (2014): 38 species in 10 genera; Sedighi and Madijdzadeh (2015): 42 species in 12 genera; Gadallah et al. (2016) and Farahani et al. (2016): with 54 and 53 species, respectively, in 16 genera; Yu et al. (2016): 55 species in 18 genera; Samin, Coronado-Blanco, Kavallieratos et al. (2018): 58 species in 18 genera; Samin, Coronado-Blanco, Fischer et al. (2018): 62 species in 20 genera. The present checklist includes 120 species in 22 genera and 11 tribes in the Middle East (Centistini, Dinocampini, Euphorini, Helorimorphini, Meteorini, Myiocephalini, Neoneurini, Perilitini, Pygostolini, Syntretini, and Townesilitini). Six species, *Centistes edentatus* (Haliday, 1835), *Meteorus abdominator* (Nees, 1811), *Meteorus cespitator* (Thunberg, 1822), *Pygostolus falcatus* (Nees, 1834), *P. sticticus* (Fabricius, 1798), and *Streblocera macroscapa* (Ruthe, 1856), are recorded here for the first time for the Iranian fauna. Here we follow Stigenberg et al. (2015) for the tribal classification, and Yu et al. (2016) for the general distribution of species, in addition to recent available references in some places.

Key to genera of the subfamily Euphorinae from the Middle East (modified from Belokobylskij, 1995; Chen & van Achterberg, 1997)

1. Marginal cell of fore wing lacking accessory cell; maxillary palp 4-6-segmented, labial palp always more than one-segmented 2
— Marginal cell of fore wing with accessory cell, sometimes vein dividing them not developed, then veins distinctly sclerotized and radio-medial cell distinct; maxillary palp 2-3-segmented, labial palp 1-2-segmented 20
2. Mesosternum of female densely setose and flattened; first metasomal tergite broadly sessile, with large dorsope; tarsi with long setae ventrally, comparatively slender; ovipositor strongly compressed; apex of antenna with a spine .... *Pygostolus* Haliday
— Mesosternum of female normally setose and convex; first metasomal tergite petiolate, if broadly sessile, then dorsope absent; tarsi normally setose ventrally, usually less slender; ovipositor variable, usually without distinct spine 3
3. First metasomal tergite broadly sessile, laterope deep; ovipositor sheath length less than 3.0× its maximum width, if more than 3.0×, then first metasomal tergite with a large dorsope; marginal cell of fore wing long; vein M+CU1 of fore wing largely reduced, unsclerotized, weakly pigmented 4
— First metasomal tergite distinctly petiolate, if subpetiolate or rather sessile, then laterope absent and/or marginal cell of fore wing short, or length of ovipositor sheath more than 3.0× its maximum width; vein M+CU1 of fore wing variable 5
4. Notauli absent; mesoscutum smooth and bulged .... *Centistes* Haliday
— Notauli present, deep; mesoscutum sometimes smooth medially .... *Allurus* Foerster
5. Tarsal claws bifid, abruptly bent submedially; vein 1-M of hind wing shorter than vein 1-rm or absent; vein cu-a of hind wing usually reduced; vein 1-SR+M of fore wing absent; vein 1-SR + 3-SR of fore wing slightly or not curved, resulting in long marginal cell; first metasomal tergite closed ventrally, tube-shaped either completely or on basal half; vein 2-1A of hind wing absent .... *Syntretus* Foerster
— Tarsal claws simple and submedially evenly curved, not twisted; vein 1-M of hind wing usually as long as vein 1r-m or longer; vein cu-a of hind wing, veins 1-SR+M and SR1+3-SR of fore wing variable; basal half of first metasomal tergite variable, if closed ventrally, then vein 1-SR+M of fore wing usually present; vein 2-1A of hind wing usually present 6
6. Antennal scape enlarged, longer than first flagellomere, protruding above top level of vertex or just reaching it, if intermediate, then dorsope present .... *Streblocera* Westwood
— Antennal scape normal, slightly or not enlarged and subequal to or shorter than first flagellomere, not reaching top level of vertex, if about reaching top level of vertex, then dorsope absent 7
7. Laterope deep and submedially situated in a very slender first metasomal tergite; female metasoma strongly compressed; female hypopygium with long setae apically; dorsal aspect of head usually slightly concave anteriorly; clypeus rather narrow .... *Myiocephalus* Marshall
— Laterope absent or shallow, if distinct, then situated subbasally; female metasoma more or less depressed; hypopygium with short setae apically; dorsal aspect of head flat or convex anteriorly; clypeus rather wide 8

8. First metasomal tergite wider submedially than apically, 7.2—9.4× as long as its apical width; vein r-m of fore wing absent .... *Wesmaelia* Foerster

— First metasomal tergite as wide as submedially as apically or narrower, if parallel-sided, then vein r-m of fore wing present, or length of first metasomal tergite usually less than 5.0× as long as its apical width 9

9. Vein M+CU1 of fore wing largely unsclerotized; vein r-m of fore wing absent; ovipositor usually strongly curved downwards and shorter than hind basitarsus; length of ovipositor sheath 3.0× its maximum width or less; marginal cell of fore wing small or absent 10

— Vein M+CU1 of fore wing completely sclerotized, or if rarely unsclerotized, then vein r-m of fore wing present; ovipositor straight or only apically curved and longer than hind basitarsus; ovipositor sheath longer than 5.0× its maximum width; marginal cell of fore wing medium-sized to large 11

10. First discal cell of fore wing densely setose than basal cell, completely glabrous and darker than basal cell; vein cu-a of hind wing variable, if present, then first metasomal tergite opened ventrally; first metasomal tergite hardly widened apically; occipital carina usually widely interrupted dorsally; mesosternum usually smooth medio-posteriorly .... *Leiophron* Nees

— First discal and basal cells of fore wing similarly setose, both are subhyaline; vein cu-a of hind wing present; vein 2-CU1 of fore wing sclerotized; first metasomal tergite usually widened apically; occipital carina usually complete or narrowly interrupted dorsally; mesosternum usually distinctly sculptured medio-posteriorly .... *Peristenus* Foerster

11. First metasomal tergite long, cylindrical, and closed ventrally, smooth; vein r-m of fore wing present 12

— First metasomal tergite shorter, its posterior half depressed and usually opened ventrally; vein r-m of fore wing variable 13

12. Ovipositor distinctly protruding beyond apex of metasoma; mesoscutum punctate to finely granulate; anterior margin of marginal cell as long as or slightly shorter than pterostigma; femora thin; notauli distinct; ovipositor sheath setose and slender; vein 1r-m of hind wing medium-sized .... *Chrysopophthorus* Goidanich

— Ovipositor short, concealed, scarcely protruding from metasomal apex; mesoscutum reticulate or reticulate—areolate; anterior margin of marginal cell distinctly shorter than pterostigma; femora thickened; notauli absent; ovipositor sheath glabrous except for a few setae; vein 1r-m of hind wing short .... *Aridelus* Marshall

13. Vein r-m of fore wing usually present, if absent then mandible with fine medio-longitudinal carina; propodeum often with curved transverse carina anteriorly or submedially, and a more or less developed median carina; vein 1-R1 of fore wing usually longer than pterostigma 14

— Vein r-m of fore wing absent; mandible without fine medio-longitudinal carina, but usually with distinct ventral carina; vein 1-R1 of fore wing usually shorter than pterostigma; propodeum often rugose or reticulate, usually without carina 15

14. Marginal cell of hind wing widened apically, sometimes with a faint vein r; propodeum with transverse carina anteriorly; metasomal fourth and fifth tergites densely setose .... *Zele* Curtis

— Marginal cell of hind wing narrowed apically, rarely subparallel-sided, without vein r-m; propodeum often without transverse carina anteriorly; metasomal fourth and fifth tergites largely glabrous, rarely rather setose in male .... *Meteorus* Haliday

15. First metasomal tergite closed at least baso-ventrally, tube-shaped; clypeus narrower and almost flat, 2.0—2.5× as wide as high; vein 1-M of hind wing shorter than vein 1-rm 16

— First metasomal tergite completely opened ventrally; clypeus broader and relatively convex, 1.4—2.2× as wide as high; vein 1-M of hind wing about as long as or longer than vein 1-rm, but sometimes shorter 17

16. Second to fourth antennal flagellomeres cylindrical, normally setose, setae not flattened apically .... *Townesilitus* Haeselbarth and Loan

— Second to fourth antennal flagellomeres of female wide, flattened (male unknown), densely setose, with long setae, flattened apically .... *Marshiella* Shaw

17. Antennal scape not enlarged or slightly enlarged, not longer than distance from antennal base to anterior ocellus 18

— Antennal scape distinctly enlarged, much longer than distance from it to anterior ocellus .... *Ecclitura* Kokujev

18. Head distinctly transverse, 2.0× as wide as long, distinctly narrow behind eyes; scape as long as distance from it to anterior ocellus; temple one-third as long as eye; propodeum steeply sloping, its posterior vertical part much larger than its anterior horizontal part; discal cell of fore wing small, about as long as brachial cell, recurrent vein antefurcal; occiput punctate; scutellum rugose-punctate .... *Dinocampus* Foerster

— Head less transverse, not distinctly narrowed behind eyes; temple long; scape 0.5× as long as its distance from it to anterior ocellus; propodeum uniformly rounded, less steeply sloping, posterior vertical part about the same as its horizontal surface; discal cell about 2.0× as large as brachial cell; recurrent vein weakly antefurcal or interstitial; occiput and scutellum smooth 19

19. Vein 1-SR+M of fore wing absent, exceptionally partly developed, but not completely sclerotized .... *Microctonus* Wesmael

— Vein 1-SR+M of fore wing present, completely sclerotized .... *Perilitus* Nees

20. Antenna somewhat shorter than body, 16-segmented; basal flagellomeres 2.0—3.0× as long as wide .... *Neoneurus* Haliday

— Antenna short, not longer than head and mesosoma combined, 13-segmented in female and 14-segmented in male; basal flagellomeres somewhat longer than wide .... *Elasmosoma* Ruthe

List of species of the subfamily Euphorinae recorded in the Middle East

Subfamily Euphorinae Foerster, 1863

Tribe Centistini Čapek, 1970

Genus *Allurus* Foerster, 1863

Allurus lituratus (Haliday, 1835)

Catalogs with Iranian records: Sedighi and Madjdzadeh (2015), Gadallah et al. (2016), Farahani et al. (2016), Yu et al. (2016), Samin, Coronado-Blanco, Kavallieratos et al. (2018), Samin, Coronado-Blanco, Fischer et al. (2018).

Distribution in Iran: Golestan (Samin et al., 2015), Hormozgan (Ameri et al., 2014), Mazandaran (Ghahari, Fischer, Erdoğan et al., 2010; Ghahari, 2017—around rice fields).

Distribution in the Middle East: Iran (see references above), Turkey (Güclü & Özbek, 2011; Koldaş et al., 2007, 2013).

Extralimital distribution: Armenia, Azerbaijan, Belgium, Bulgaria, Canada, China, Finland, France, Georgia, Germany, Greece, Hungary, Ireland, Kazakhstan, Lithuania, Moldova, Poland, Russia, Sweden, Ukraine, United Kingdom.

Host records: Summarized by Yu et al. (2016) as being a parasitoid of the curculionids *Sitona crinita* Herbst, *Sitona inops* Gyllenhal, *Sitona lineata* L., *Sitona obsoleta* Gmelin, and *Sitona scissifrons* Say.

Allurus muricatus (Haliday, 1833)

Catalogs with Iranian records: Sedighi and Madjdzadeh (2015), Gadallah et al. (2016), Farahani et al. (2016), Yu et al. (2016), Samin, Coronado-Blanco, Kavallieratos et al. (2018), Samin, Coronado-Blanco, Fischer et al. (2018).

Distribution in Iran: Qazvin (Farahani, Talebi, & Rakhshani, 2013).

Distribution in the Middle East: Iran (Farahani, Talebi, & Rakhshani, 2013), Israel—Palestine (Papp, 2012), Turkey (Güclü & Özbek, 2011; Koldaş et al., 2007, 2013; Yilmaz et al., 2010).

Extralimital distribution: Armenia, Azerbaijan, Belgium, Croatia, former Czechoslovakia, Finland, France, Georgia, Germany, Greece, Hungary, Ireland, Italy, Kazakhstan, North Macedonia, Moldova, Poland, Serbia, Spain, Sweden, Switzerland, United Kingdom.

Host records: Summarized by Yu et al. (2016) as being a parasitoid of the curculionids *Sitona hispidula* (Fabricius), *Sitona humeralis* Stephens, *Sitona lineata* L., *Sitona regensteinensis* (Herbst), and *Sitona sulcifrons* (Thunberg). This species was also recorded by Szócs (1979) as being a parasitoid of the nepticulids *Stigmella aceris* (Frey), *Stigmella obliquella* (Heinemann), and *Stigmella torminalis* (Wood). However, these later records need to be corroborated.

Genus *Centistes* Haliday, 1835

Centistes (Ancylocentrus) ater (Nees von Esenbeck, 1834)

Catalogs with Iranian records: Gadallah et al. (2016), Farahani et al. (2016), Yu et al. (2016), Samin, Coronado-Blanco, Kavallieratos et al. (2018), Samin, Coronado-Blanco, Fischer et al. (2018).

Distribution in Iran: Kermanshah (Gadallah et al., 2016).

Distribution in the Middle East: Iran (Gadallah et al., 2016), Turkey (Yilmaz et al., 2010).

Extralimital distribution: Belgium, Bulgaria, Canada, China, Finland, France, Germany, Hungary, Ireland, Italy, Lithuania, Netherlands, Poland, Russia, Sweden, United Kingdom (Yu et al., 2016), United States of America (S.R. Shaw, unpublished data).

Host records: Summarized by Yu et al. (2016) as being a parasitoid of the curculionid *Sitona scissifrons* Say; and the nepticulids *Stigmella basiguttella* Heinemann, *Stigmella poterii* (Stainton). The nepticulid records are doubtful and need to be verified.

Centistes (Ancylocentrus) collaris (Thomson, 1895)

Catalogs with Iranian records: No catalog.

Distribution in Iran: Markazi (Ghahari & Gadallah, 2019).

Distribution in the Middle East: Iran (Ghahari & Gadallah, 2019), Turkey (Yilmaz et al., 2010).
Extralimital distribution: Czech Republic, Korea, Lithuania, Russia, Slovenia, Sweden, former Yugoslavia.
Host records: Unknown.

Centistes (Ancylocentrus) edentatus (Haliday, 1835)
Catalogs with Iranian records: This species is new record for the fauna of Iran.
Distribution in Iran: East Azarbaijan province, Horand, 2♂, April 2013.
Distribution in the Middle East: Iran (new record), Turkey (Beyarslan, 2021).
Extralimital distribution: Azerbaijan, Bulgaria, Croatia, Czech Republic, Denmark, Finland, Germany, Greece, Hungary, Ireland, Italy, Kazakhstan, Lithuania, Madeira Islands, Moldova, Morocco, Norway, Poland, Russia, Sweden, Switzerland, United Kingdom, former Yugoslavia.
Host records: Unknown.

Centistes (Ancylocentrus) nasutus (Wesmael, 1838)
Catalogs with Iranian records: No catalog.
Distribution in Iran: Mazandaran (Gadallah et al., 2021).
Distribution in the Middle East: Iran.
Extralimital distribution: Belgium, Finland, France, Georgia, Germany, Hungary, Italy, Kazakhstan, Montenegro, Netherland, Poland, Russia, Sweden, Switzerland, Ukraine, United Kingdom.
Host records: Summarized by Yu et al. (2016) as being a parasitoid of the carabid *Amara apricaria* (Paykull), and the chrysomelid *Galerucella viburni* (Paykull).

Centistes (Ancylocentrus) subsulcatus (Thomson, 1895)
Catalogs with Iranian records: No catalog.
Distribution in Iran: East Azarbaijan (Gadallah et al., 2021).
Distribution in the Middle East: Iran.
Extralimital distribution: Azerbaijan, Belgium, Finland, Hungary, Russia, Serbia, Sweden, Switzerland, United Kingdom.
Host records: Summarized by Yu et al. (2016) as being a parasitoid of the coccinellid *Propylaea quatuordecimpunctata* (L.), and the nepticulid *Trifurcula cryptella* (Stainton).

Centistes (Centistes) cuspidatus (Haliday, 1833)
Catalogs with Iranian records: Gadallah et al. (2016), Farahani et al. (2016), Yu et al. (2016), Samin, Coronado-Blanco, Kavallieratos et al. (2018), Samin, Coronado-Blanco, Fischer et al. (2018).
Distribution in Iran: Mazandaran (Gadallah et al., 2016).

Distribution in the Middle East: Iran (Gadallah et al., 2016), Turkey (Koldaş et al., 2007, 2013; Yilmaz, 2010).
Extralimital distribution: Austria, Belgium, Bulgaria, former Czechoslovakia, Denmark, Finland, France, Germany, Hungary, Italy, Kazakhstan, Korea, Kyrgyzstan, Latvia, Lithuania, Moldova, Poland, Russia, Serbia, Slovenia, Sweden, Switzerland, Ukraine, United Kingdom.
Host records: Recorded by Fulmek (1968) as being a parasitoid of the cecidomyiid *Asphondylia conglomerata* de Stefani (but this record is doubtful and needs to be verified). It was also summarized by Yu et al. (2016) as being a parasitoid of the curculionid *Leperisinus varius* (Fabricius); and the staphylinids *Tachyporus chrysomelinus* (L.), *Tachyporus hypnorum* (Fabricius), *Tachyporus obtusus* (L.), and *Tachyporus solutus* (Erichson).

Centistes (Centistes) fuscipes (Nees von Esenbeck, 1834)
Catalogs with Iranian records: Sedighi and Madjdzadeh (2015), Gadallah et al. (2016), Farahani et al. (2016), Yu et al. (2016), Samin, Coronado-Blanco, Kavallieratos et al. (2018), Samin, Coronado-Blanco, Fischer et al. (2018).
Distribution in Iran: Ardabil (Ghahari & Fischer 2011a).
Distribution in the Middle East: Iran (Ghahari & Fischer, 2011a), Turkey (Koldaş et al., 2007, 2013; Yilmaz et al., 2010).
Extralimital distribution: Azerbaijan, Belgium, Bulgaria, Finland, France, Georgia, Germany, Greece, Hungary, Italy, Latvia, Lithuania, Moldova, Poland, Russia, Slovakia, Spain, Sweden, Switzerland, United Kingdom.
Host records: Recorded by Jacobs (1904) as being a parasitoid of the elachistid *Elachista trapeziella* Stainton. This record is questionable and needs to be corroborated.

Tribe Dinocampini Shaw, 1985
Genus *Dinocampus* Foerster, 1863
Dinocampus coccinellae (Schrank, 1802)
Catalogs with Iranian records: Sedighi and Madjdzadeh (2015), Gadallah et al. (2016), Farahani et al. (2016), Yu et al. (2016), Samin, Coronado-Blanco, Kavallieratos et al. (2018), Samin, Coronado-Blanco, Fischer et al. (2018).
Distribution in Iran: Guilan, Qazvin, Tehran (Farahani, Talebi, & Rakhshani, 2013), Hamadan (Soleimani & Madadi, 2015; Tavoosi Ajvad et al., 2012, 2014), Hormozgan (Ameri et al., 2014), Isfahan (Bagheri, 1998; Modarres Awal, 2012 as *Perilitus coccinellae*; Samin, 2015), Kerman (Abdolalizadeh et al., 2017; Alimohammadi et al., 2012), Lorestan (Biranvand et al., 2020), Mazandaran (Ghahari, Fischer, Erdoğan et al., 2010), Northern Khorasan (Ghahari, 2015), Yazd (Farahani, Talebi, & Rakhshani, 2013).

Distribution in the Middle East: Cyprus, Egypt, Iran, Syria (Yu et al., 2016), Turkey (Koldaş et al., 2007, 2013; Yilmaz et al., 2010).

Extralimital distribution: Nearly cosmopolitan (Shaw, 1985).

Host records: Summarized by Yu et al. (2016) as being a parasitoid of several coleopteran hosts of the families Chrysomelidae, Coccinellidae, and Curculionidae. The records from Chrysomelidae and Curculionidae are questionable and needs to be verified. In Iran, this species has been reared from the coccinellids *Coccinella septempunctata* L. (Bagheri, 1998; Farahani, Talebi, & Rakhshani, 2013; Modarres Awal, 2012; Soleimani & Madadi, 2015), and *Hippodamia variegata* (Goeze) (Tavoosi Ajvad et al., 2012, 2014).

Genus *Ecclitura* Kokujev, 1902
Ecclitura primoris Kokujev, 1902
Catalogs with Iranian records: Fallahzadeh and Saghaei (2010 as *Ecclitura primoris* Kokujev, 1902), Sedighi and Madjdzadeh (2015), Gadallah et al. (2016), Farahani et al. (2016), Yu et al. (2016), Samin, Coronado-Blanco, Kavallieratos et al. (2018), Samin, Coronado-Blanco, Fischer et al. (2018).

Distribution in Iran: Sistan and Baluchestan (Hedwig, 1957 as *Blacus pallens* (Hedwig), Iran (no specific locality cited) (Haeselbarth, 1983; Tobias, 1986; van Achterberg, 1980).

Distribution in the Middle East: Iran (see references above), Turkey (Belokobylskij et al., 2013; Koldaş et al., 2013).

Extralimital distribution: Albania, Azerbaijan, Italy, Russia, Turkmenistan.

Host records: Unknown.

Tribe Euphorini Foerster, 1863
Genus *Leiophron* Nees von Esenbeck, 1819
Leiophron (Leiophron) cubocephalus Tobias, 1986
Distribution in the Middle East: Israel—Palestine (Papp, 2012).

Extralimital distribution: Moldova.

Host records: Unknown.

Leiophron (Euphoriella) deficiens (Ruthe, 1856)
Catalogs with Iranian records: Sedighi and Madjdzadeh (2015), Gadallah et al. (2016), Farahani et al. (2016), Yu et al. (2016), Samin, Coronado-Blanco, Kavallieratos et al. (2018), Samin, Coronado-Blanco, Fischer et al. (2018).

Distribution in Iran: Alborz, Tehran (Farahani, Talebi, van Achterberg, & Rakhshani, 2013), Guilan (Ghahari, 2015), Kerman (Abdolalizadeh et al., 2017), Lorestan (Ghahari et al., 2012), Mazandaran (Ghahari & Fischer, 2011b; Sakenin et al., 2012, in citrus orchards), Razavi Khorasan

(Ghahari, 2020 as *Leiophron decifiens* [sic]), Northern Khorasan (Samin, 2015).

Distribution in the Middle East: Iran (see references above), Turkey (Efil & Beyarslan, 2009; Güclü & Özbek, 2011).

Extralimital distribution: Finland, Germany, Greece, Hungary, Kazakhstan, Korea, Moldova, Poland, Russia, Sweden, Ukraine.

Host records: Summarized by Yu et al. (2016) as being a parasitoid of the mirid bugs *Campylomma diversicornis* (Reuter), *Creontiades pallidus* (Rambur), and *Polymerus cognatus* (Fieber).

Leiophron (Euphoriana) heterocordyli Richards, 1967
Catalogs with Iranian records: Sedighi and Madjdzadeh (2015), Gadallah et al. (2016), Farahani et al. (2016), Yu et al. (2016), Samin, Coronado-Blanco, Kavallieratos et al. (2018), Samin, Coronado-Blanco, Fischer et al. (2018).

Distribution in Iran: Guilan (Ghahari & Fischer, 2011b; Sakenin et al., 2012).

Distribution in the Middle East: Iran.

Extralimital distribution: Mongolia, Russia, United Kingdom.

Host records: Summarized by Yu et al. (2016) as being a parasitoid of the mirid bugs *Asciodema obsoleta* (Fieber), *Heterocordylus tibialis* (Hahn), and *Orthotylus adenocarpi* (Perris).

Leiophron (Euphorus) basalis (Curtis, 1833)
Catalogs with Iranian records: Yu et al. (2016), Samin, Coronado-Blanco, Kavallieratos et al. (2018), Samin, Coronado-Blanco, Fischer et al. (2018).

Distribution in Iran: Guilan (Farahani, Talebi, van Achterberg, & Rakhshani, 2013 as *Leiophron (Euphorus) similis* Curtis).

Distribution in the Middle East: Iran.

Extralimital distribution: Belgium, Croatia, former Czechoslovakia, Finland, France, Germany, Hungary, Ireland, Italy, Kazakhstan, Netherlands, Norway, Poland, Sweden, Switzerland, United Kingdom, former Yugoslavia.

Host records: Summarized by Yu et al. (2016) as being a parasitoid of the following chrysomelids: *Longitarsus longipennis* Kutschera, *Longitarsus pellucidus* (Foudras), *Phyllotreta nigripes* (Fabricius), *Phyllotreta undulata* Kutschera, and *Psylliodes attenuata* (Koch). It has been also recorded by New (1970) as being a parasitoid of the caeciliusid psocopteran *Valenzuela flavidus* (Stephens).

Leiophron (Euphorus) pallidistigma Curtis, 1833
Catalogs with Iranian records: Sedighi and Madjdzadeh (2015), Gadallah et al. (2016), Farahani et al. (2016), Yu et al. (2016), Samin, Coronado-Blanco, Kavallieratos et al. (2018), Samin, Coronado-Blanco, Fischer et al. (2018).

Distribution in Iran: Guilan, Mazandaran (Farahani, Talebi, van Achterberg, & Rakhshani, 2013).

Distribution in the Middle East: Iran (Farahani, Talebi, van Achterberg, & Rakhshani, 2013), Turkey (Koldaş et al., 2007; Yilmaz et al., 2010).

Extralimital distribution: Azerbaijan, Belgium, Czech Republic, Finland, France, Germany, Hungary, Ireland, Italy, Kazakhstan, Korea, Moldova, Netherlands, Norway, Poland, Russia, Sweden, Switzerland, United Kingdom.

Host records: Recorded by Rudow (1918) as being a parasitoid of the cecidomyiid *Rabdophaga rigidae* (Osten Sacken), however, this record is questionable. It was also recorded by Belokobylskij (1993) as being a parasitoid of the psocopteran caeciliusid species *Velenzuela flavidus* (Stephens) and the peripsocid *Peripscocus phaeopterus* (Stephens).

Leiophron (Leiophron) apicalis Haliday, 1833
Catalogs with Iranian records: No catalog.
Distribution in Iran: Ardabil (Samin et al., 2019).
Distribution in the Middle East: Iran.
Extralimital distribution: Azerbaijan, former Czechoslovakia, France, Germany, Hungary, Ireland, Kazakhstan, Korea, Netherlands, Poland, Russia, Sweden, Ukraine, United Kingdom, Uzbekistan.
Host records: Recorded by Niezabitowski (1910) as being a parasitoid of the red wood ant *Formica rufa* L., although this is almost certainly erroneous. It summarized by Yu et al. (2016) as being a parasitoid of the mirids *Orthotylus adenocarpi* (Perris), *Orthotylus concolor* (Kirschbaum), *Orthotylus virescens* (Douglas and Scott), and *Platycranus bicolor* (Douglas and Scott).

Leiophron (Leiophron) fascipennis (Ruthe, 1856)
Catalogs with Iranian records: Sedighi and Madjdzadeh (2015), Gadallah et al. (2016), Farahani et al. (2016), Yu et al. (2016), Samin, Coronado-Blanco, Kavallieratos et al. (2018), Samin, Coronado-Blanco, Fischer et al. (2018).
Distribution in Iran: Guilan (Farahani, Talebi, van Achterberg, & Rakhshani, 2013).
Distribution in the Middle East: Iran.
Extralimital distribution: former Czechoslovakia, France, Germany, Hungary, Korea, Russia, Sweden, Switzerland, United Kingdom.
Host records: Unknown.

Leiophron (Leiophron) reclinator (Ruthe, 1856)
Catalogs with Iranian records: No catalog.
Distribution in Iran: Ardabil (Gadallah et al., 2018).
Distribution in the Middle East: Iran (Gadallah et al., 2018), Turkey (Koldaş et al., 2013).
Extarlimital distribution: Azerbaijan, Hungary, Germany, Italy, Kazakhstan, Poland, Russia, Sweden.
Host records: Unknown.

Genus *Peristenus* Foerster, 1863
Peristenus accinctus (Haliday, 1835)
Catalogs with Iranian records: No catalog.
Distribution in Iran: Lorestan (Ghahari & Gadallah, 2019).
Distribution in the Middle East: Iran (Ghahari & Gadallah, 2019), Turkey (Güclü & Özbek, 2011; Yilmaz et al., 2010).
Extralimital distribution: Belgium, Bosnia-Herzegovina, France, Germany, Hungary, Korea, Moldova, Poland, United Kingdom, former Yugoslavia.
Host records: Summarized by Yu et al. (2016) as being a parasitoid of the nitidulid *Meligethes aeneus* (Fabricius) (doubtful record), and the mirid *Lygocoris pabulinus* (L.).

Peristenus adelphocoridis Loan, 1979
Distribution in the Middle East: Turkey (Yilmaz et al., 2010).
Extralimital distribution: Denmark, France, Germany, Hungary.
Host records: Recorded by Loan (1979) as being a parasitoid of the mirid bug *Adelphocoris lineolatus* (Goeze), and *Adelphocoris* sp.

Peristenus digoneutis Loan, 1973
Catalogs with Iranian records: No catalog.
Distribution in Iran: East Azarbaijan (Samin, Papp, & Coronado-Blanco, 2018).
Distribution in the Middle East: Iran (Samin, Papp, & Coronado-Blanco, 2018), Turkey (Koldaş et al., 2007; Yilmaz et al., 2010).
Extralimital distribution: Canada (introduced), Italy, Moldova, Poland, the United States of America (introduced).
Host records: Summarized by Yu et al. (2016) as being a parasitoid of the cicadellid bug *Empoasca solana* (DeLong) (questionable), as well as the mirids *Adelphocoris lineolatus* (Goeze), *Closterotomus norwegicus* (Gmelin), *Leptopterna dolabrata* (L.), *Lygus lineolaris* (Palisot de Beauvois), and *Lygus rugulipennis* Poppius. In Iran, this species has been recorded from *Lygus rugulipennis* Poppius (Samin, Papp, & Coronado-Blanco, 2018).

Peristenus facialis (Thomson, 1892)
Catalogs with Iranian records: Samin, Coronado-Blanco, Fischer et al. (2018).
Distribution in Iran: Razavi Khorasan (Samin, Coronado-Blanco, Fischer et al., 2018).
Distribution in the Middle East: Iran (Samin, Coronado-Blanco, Fischer et al., 2018), Turkey (Güclü & Özbek, 2011; Koldaş et al., 2007; Yilmaz et al., 2010).
Extralimital distribution: Finland, Hungary, Korea, Lithuania, Mongolia, Poland, Russia, Serbia, Sweden, United Kingdom.

Host records: Recorded by Richards (1967) and Tobias (1971) as being a parasitoid of the mirids *Orthotylus marginalis* (Reuter), and *Psallus varians* (Herrich-Schäffer).

Peristenus grandiceps (Thomson, 1892)
Catalogs with Iranian records: Gadallah et al. (2016), Farahani et al. (2016), Yu et al. (2016), Samin, Coronado-Blanco, Kavallieratos et al. (2018), Samin, Coronado-Blanco, Fischer et al. (2018).
Distribution in Iran: Hormozgan (Ameri et al., 2014).
Distribution in the Middle East: Iran (Ameri et al., 2014), Turkey (Güclü & Özbek, 2011; Koldaş et al., 2007, 2013; Yilmaz et al., 2010).
Extralimital distribution: Germany, Greece, Hungary, Ireland, Kazakhstan, North Macedonia, Moldova, Netherlands, Norway, Poland, Sweden, Switzerland, United Kingdom, former Yugoslavia.
Host records: Unknown.

Peristenus kazak (Tobias, 1986)
Distribution in the Middle East: Turkey (Beyarslan, 2021).
Extralimital distribution: Russia.
Host records: Unknown.

Peristenus nitidus (Curtis, 1833)
Catalogs with Iranian records: Gadallah et al. (2016), Farahani et al. (2016), Yu et al. (2016), Samin, Coronado-Blanco, Kavallieratos et al. (2018), Samin, Coronado-Blanco, Fischer et al. (2018).
Distribution in Iran: West Azarbaijan (Gadallah et al., 2016).
Distribution in the Middle East: Iran (Gadallah et al., 2016), Turkey (Koldaş et al., 2007, 2013; Güclü & Özbek, 2011; Yilmaz et al., 2010).
Extralimital distribution: Czech Republic, Germany, Hungary, Mongolia, Norway, United Kingdom.
Host records: Recorded by Haye et al. (2006) as being a potential parasitoid of the mirid *Closterotomus norwegicus* (Gmelin).

Peristenus orchesiae (Curtis, 1833)
Catalogs with Iranian records: No catalog.
Distribution in Iran: West Azarbaijan (Sakenin et al., 2020).
Distribution in the Middle East: Iran.
Extralimital distribution: Azerbaijan, former Czechoslovakia, Germany, Hungary, Italy, Kazakhstan, Mongolia, Norway, United Kingdom.
Host records: Summarized by Yu et al. (2016) as being a parasitoid of the mirid *Closterotomus norwegicus* (Gmelin), and the melandryid *Orchesia micans* (Panzer).

Peristenus pallipes (Curtis, 1833)
Catalogs with Iranian records: Sedighi and Madjdzadeh (2015), Gadallah et al. (2016), Farahani et al. (2016), Yu et al. (2016), Samin, Coronado-Blanco, Kavallieratos et al. (2018), Samin, Coronado-Blanco, Fischer et al. (2018).
Distribution in Iran: Golestan (Ghahari, 2015), Guilan, Mazandaran (Farahani, Talebi, & Rakhshani, 2013), Iran (no specific locality cited) (Khanjani, 2006a,b).
Distribution in the Middle East: Iran (see references above), Turkey (Koldaş et al., 2007, 2013; Yilmaz et al., 2010).
Extralimital distribution: Nearctic (introduced to the United States of America), Neotropical, Oriental, Palaearctic regions.
Host records: Summarized by Yu et al. (2016) as being a parasitoid of the chrysomelid *Timarcha tenebricosa* (Fabricius); the tetratomid *Eustrophus dermestoides* (Fabricius) (doubtful records); the following mirids: *Adelphocoris lineolatus* (Goeze), *Adelphocoris rapidus* (Say), *Capsus ater* (L.), *Closterotomus norwegicus* (Gmelin), *Labops hirtus* Knight, *Leptopterna dolabrata* (L.), *Lygus* spp., *Notostira erratica* (L.), *Plagiognathus medicagus* (Provancher), and *Trigonotylus coelestialium* (Kirkaldy). In Iran, this species has been reared from the mirid *Adelphocoris lineolatus* (Goeze) (Khanjani, 2006a,b).

Peristenus picipes (Curtis, 1833)
Catalogs with Iranian records: Sedighi and Madjdzadeh (2015), Gadallah et al. (2016), Farahani et al. (2016), Yu et al. (2016), Samin, Coronado-Blanco, Kavallieratos et al. (2018), Samin, Coronado-Blanco, Fischer et al. (2018).
Distribution in Iran: Fars (Lashkari Bod et al., 2010, 2011, in alfalfa field).
Distribution in the Middle East: Iran (Lashkari Bod et al., 2010, 2011), Turkey (Koldaş et al., 2007, 2013; Yilmaz et al., 2010).
Extralimital distribution: Azerbaijan, China, Czech Republic, Finland, Germany, Hungary, Ireland, Italy, Kazakhstan, Lithuania, Moldova, Norway, Poland, Russia, Serbia, Sweden, Switzerland, Ukraine, United Kingdom, Uzbekistan.
Host records: Recorded by Pansa et al. (2012) as being a parasitoid of the mirid *Adelphocoris lineolatus* (Goeze).

Peristenus relictus (Ruthe, 1856)
Catalogs with Iranian records: Sedighi and Madjdzadeh (2015), Gadallah et al. (2016), Farahani et al. (2016), Yu et al. (2016), Samin, Coronado-Blanco, Kavallieratos et al. (2018), Samin, Coronado-Blanco, Fischer et al. (2018).
Distribution in Iran: Kerman (Abdolalizadeh et al., 2017), Qazvin (Farahani, Talebi, & Rakhshani, 2013).

Distribution in the Middle East: Iran (Abdolalizadeh et al., 2017; Farahani, Talebi, & Rakhshani, 2013), Turkey (Drea et al., 1973; Koldaş et al., 2007; Loan & Bilewicz-Pawinska, 1973; Yilmaz et al., 2010).

Extralimital distribution: France, Germany, Hungary, Italy, North Macedonia, Poland, Russia, Serbia, Switzerland.

Host records: Summarized by Yu et al. (2016) as being a parasitoid of the mirids *Adelphocoris lineolatus* (Goeze), *Closterotomus norwegicus* (Gmelin), *Lygus* spp., *Polymerus unifasciatus* (Fabricius), and *Trigonotylus coelestialium* (Kirkaldy).

Peristenus rubricollis (Thomson, 1892)
Distribution in the Middle East: Turkey (Güclü & Özbek, 2011; Koldaş et al., 2007; Yilmaz et al., 2010).

Extralimital distribution: Bulgaria, Canada, Finland, Germany, Hungary, Kazakhstan, Moldova, Netherlands, Poland, Sweden, Switzerland, the United States of America.

Host records: Summarized by Yu et al. (2016) as being a parasitoid of the cicadellid *Empoasca solana* Dozier; and the mirids *Adelphocoris lineolatus* (Goeze), *Adelphocoris seticornis* (Fabricius), *Lygus lineolaris* (Palisot de Beauvois), and *Lygus rugulipennis* Poppius.

Tribe Helorimorphini Schmiedeknecht, 1907
Genus *Aridelus* Marshall, 1887
Aridelus cameroni (Szépligeti, 1914)
Distribution in the Middle East: Turkey (Papp, 1974; Yilmaz et al., 2010).

Extralimital distribution: Cameron, Tanzania, Uganda.

Host records: Sumarized by Yu et al. (2016) as being a parasitoid of the pentatomids *Antestia faceta* (Germar), *Antestia lineaticollis* Stål, and *Palomena prasine* (L.).

Aridelus egregius (Schmiedeknecht, 1907)
Catalogs with Iranian records: No catalog.

Distribution in Iran: Markazi (Gadallah et al., 2018).

Distribution in the Middle East: Iran (Gadallah et al., 2018), Turkey (Koldaş et al., 2007; Yilmaz et al., 2010).

Extralimital distribution: Azerbaijan, Bulgaria, China, Czech Republic, France, Germany, Hungary, Korea, Moldova, Russia, Slovakia, Ukraine.

Host records: Summarized by Yu et al. (2016) as being a parasitoid of the pentatomids *Aelia acuminata* L., *Aelia cognata* Fieber, *Dolycoris baccarum* (L.), *Eurydema ornatum* (L.), *Peribalus strictus* (Fabricius), and *Palomena prasina* (L.); the plataspidids *Coptosoma mucronatum* (L.), and *Coptosoma scutellatum* (Geoffroy); and the scutellerids *Eurygaster austriaca* (Schrank), and *Eurygaster maura* (L.). In Iran, this species has been reared from the pentatomid *Dolycoris baccarum* (L.) (Gadallah et al., 2018).

Genus *Chrysopophthorus* Goidanich, 1948
Chrysopophthorus hungaricus (Zilahi-Kiss, 1927)
Catalogs with Iranian records: Sedighi and Madjdzadeh (2015), Gadallah et al. (2016), Farahani et al. (2016), Yu et al. (2016), Samin, Coronado-Blanco, Kavallieratos et al. (2018), Samin, Coronado-Blanco, Fischer et al. (2018).

Distribution in Iran: East Azarbaijan (Ghahari, Fischer, Erdoğan, Beyarslan, & Havaskary, 2009).

Distribution in the Middle East: Iran (Ghahari, Fischer, Erdoğan, Beyarslan, & Havaskary, 2009), Turkey (Yilmaz et al., 2010).

Extralimital distribution: Azerbaijan, Bulgaria, Czech Republic, France, Germany, Greece, Hungary, Italy, Korea, Madeira Islands, Malta, Moldova, Netherlands, Poland, Romania, Russia, Slovakia, Spain, Switzerland, United Kingdom.

Host records: Summarized by Yu et al. (2016) as being a parasitoid of the chrysopids *Chrysopa flavifrons* Braeur, *Chrysoperla carnea* (Stephens), *Mallada bononensis* Okamoto, *Pseudomallada clathratus* (Schneider), *Pseudomallada genei* (Rambur), *Pseudomallada ibericus* (Navás), *Pseudomallada pictati* (McLachlan), and *Pseudomallada parsinus* (Curtis).

Genus *Wesmaelia* Foerster, 1863
Wesmaelia petiolata (Wollaston, 1858)
Catalogs with Iranian records: Fallahzadeh and Saghaei (2010), Sedighi and Madjdzadeh (2015), Gadallah et al. (2016), Farahani et al. (2016), Yu et al. (2016), Samin, Coronado-Blanco, Kavallieratos et al. (2018), Samin, Coronado-Blanco, Fischer et al. (2018).

Distribution in Iran: Hormozgan (Ameri et al., 2014), Kerman (Abdolalizadeh et al., 2017), Iran (no specific locality cited) (Belokobylskij, 1992 as *Wesmaelia pendula* Foerster).

Distribution in the Middle East: Iran (see references above), Israel—Palestine (Papp, 1989), Turkey (Koldaş et al., 2007, 2013; Yilmaz et al., 2010).

Extralimital distribution: Nearctic, Neotropical, Oceanic, Oriental, Palaearctic.

Host records: Recorded by Stoner (1973) and Marsh (1979) as being a parasitoid of the nabids *Nabis alternatus* Parshley, *Nabis americoferus* Carayon, and *Nabis capsiformis* Germar.

Tribe Meteorini Cresson, 1887

Genus *Meteorus* Haliday, 1835

Meteorus abdominator (Nees von Esenbeck, 1811)

Catalogs with Iranian records: This species is a new record for the fauna of Iran.

Distribution in Iran: Zanjan province, Abhar (Soltanieh), 1♀, 1♂, September 2007.

Distribution in the Middle East: Iran (new record), Turkey (Koldaş et al., 2013; Yilmaz et al., 2010).

Extralimital distribution: Armenia, Azerbaijan, Belgium, Bosnia-Herzegovina, Bulgaria, Croatia, former Czechoslovakia, Denmark, Finland, France, Germany, Hungary, Ireland, Italy, Kazakhstan, Korea, Latvia, Lithuania, Moldova, Montenegro, Netherlands, Norway, Poland, Romania, Russia, Serbia, Sweden, Switzerland, Ukraine, United Kingdom.

Host records: Summarized by Yu et al. (2016) as being a parasitoid of the geometrids *Colostygia pectinataria* (Knoch), *Eupithecia lariciata* (Freyer), *Eupithecia pusillata* (Denis and Schiffermüller), *Operophtera brumata* (L.), and *Xanthorhoe fluctuata* (L.); and the noctuid *Cucullia argentea* (Hufnagel).

Meteorus abscissus Thomson, 1895

Distribution in the Middle East: Turkey (Beyarslan, 2021).

Extralimital distribution: Albania, Austria, Bulgaria, China, Croatia, France, Germany, Hungary, Ireland, Italy, Korea, Norway, Romania, Russia, Slovakia, Slovenia, Spain, Switzerland, Slovenia, United Kingdom, former Yugoslavia.

Host records: Summarized by Yu et al. (2016) as being a parasitoid of the curculionid *Ips acuminatus* (Gyllenhal); the crambid *Endonia truncicolella* (Stainton); the erebids *Paidia murina* (Hübner), and *Setina irrorella* (L.); the geometrids *Agriopis leucophaearia* (Denis and Schiffermüller), *Alsophila aescularia* (Denis and Schiffermüller), *Epirrita dilutata* (Denis and Schiffermüller), and *Operophtera brumata* (L.); the noctuids *Anaplectoides prasine* (Denis and Schiffermüller), *Lycophatia porphyrea* (Denis and Schiffermüller), *Orthosia stabilis* (Fabricius), and *Xestia agathina* (Esper); and the oecophorid *Alabonia geoffrella* (Linnaeus).

Meteorus affinis (Wesmael, 1835)

Catalogs with Iranian records: Gadallah et al. (2016), Farahani et al. (2016), Yu et al. (2016), Samin, Coronado-Blanco, Kavallieratos et al. (2018), Samin, Coronado-Blanco, Fischer et al. (2018).

Distribution in Iran: Hamadan (Gadallah et al., 2016).

Distribution in the Middle East: Iran (Gadallah et al., 2016), Israel−Palestine (Papp, 2012), Turkey (Yilmaz et al., 2010).

Extralimital distribution: Armenia, Belgium, Bulgaria, China, Croatia, former Czechoslovakia, Finland, France, Germany, Hungary, Ireland, Italy, Korea, Madeira Islands, Netherlands, Norway, Romania, Russia, Spain, Sweden, Switzerland, United Kingdom, former Yugoslavia.

Host records: Recorded by Stigenberg and Shaw (2013) as being a parasitoid of the gelechiid *Bryotropha senectella* (Zeller); the geometrid *Xanthorhoe fluctuata* (L.); the oecophorids *Aplota palpella* (Haworth), *Crassa unitella* (Hübner), *Denisia albimaculea* (Haworth)? *Endrosis sarcitrella* (L.), *Esperia sulphurella* (Fabricius), and *Metalampra italica* Baldizzone; the psychids *Bruandia comitella* (Bruand), *Dahlica inconspicuella* (Stainton), *Dahlica lichenella* (L.), *Narycia duplicella* (Goeze), *Psyche casta* (Pallas), and *Solenobia* sp.; the pyralids *Dipleurina lacustrata* (Panzer), *Eudonia angustea* (Curtis), *Eudonia mercuriella* (L.), *Eudonia murana* (Curtis), *Eudonia truncicolella* (Stainton), *Scoparis ambigualis* (Treitschke), and *Scoparis basistrigalis* Knaggs; and the tineids *Infurcitinea argentimaculella* (Stainton), and *Monopis laevigella* (Denis and Schiffermüller).

Meteorus alborossicus Lobodenko, 2000

Catalogs with Iranian records: Sedighi and Madjdzadeh (2015), Gadallah et al. (2016), Farahani et al. (2016), Yu et al. (2016), Samin, Coronado-Blanco, Kavallieratos et al. (2018), Samin, Coronado-Blanco, Fischer et al. (2018).

Distribution in Iran: Mazandaran (Farahani & Talebi, 2012; Farahani et al., 2012).

Distribution in the Middle East: Iran.

Extralimital distribution: Belarus, Norway, Sweden, United Kingdom.

Host records: Unknown.

Meteorus arabica Ghramh, 2012

Distribution in the Middle East: Saudi Arabia (Ghramh, 2012).

Extralimital distribution: None.

Host records: Unknown.

Meteorus breviantennatus Tobias, 1986

Catalogs with Iranian records: Sedighi and Madjdzadeh (2015), Gadallah et al. (2016), Farahani et al. (2016), Yu et al. (2016), Samin, Coronado-Blanco, Kavallieratos et al. (2018), Samin, Coronado-Blanco, Fischer et al. (2018).

Distribution in Iran: Guilan, Mazandaran (Farahani & Talebi, 2012).

Distribution in the Middle East: Iran.

Extralimital distribution: Georgia, Russia.

Host records: Recorded by Martikainen and Koponen (2001) as being a parasitoid of the curculionids *Ips acuminatus* Gyllenhal, and *Tomicus minor* (Hartig).

Meteorus breviterebratus Ameri, Talebi and Beyarslan, 2014

Catalogs with Iranian records: Gadallah et al. (2016), Farahani et al. (2016), Yu et al. (2016), Samin, Coronado-Blanco, Kavallieratos et al. (2018), Samin, Coronado-Blanco, Fischer et al. (2018).

Distribution in Iran: Hormozgan (Ameri et al., 2014).

Distribution in the Middle East: Iran.

Extralimital distribution: None.

Host records: Unknown.

Meteorus cespitator (Thunberg, 1822)

Catalogs with Iranian records: This species is new record for the fauna of Iran.

Distribution in Iran: Fars province, Kazeroon (Khorramzar), 2♀, June 2004.

Distribution in the Middle East: Iran (new record), Turkey (Yilmaz et al., 2010).

Extralimital distribution: Algeria, Azerbaijan, Belgium, Bulgaria, China, Denmark, Faeroe Islands, Finland, France, Georgia, Germany, Hungary, Iceland, Ireland, Italy, Japan, Latvia, Lithuania, Luxembourg, Netherlands, New Zealand, Norway, Poland, Romania, Russia, Spain, Sweden, Switzerland, United Kingdom.

Host records: Recorded by Stigenberg and Shaw (2013) as being a parasitoid of the tineids *Monopis laevigella* (Denis and Schiffermüller), *Tinea trinotella* Thunberg, and *Tineola bisselliella* (Hummel). In addition, several specimens were reared from bird nests, owl pellets, old blankets etc. resulting in the above as well as other tineids along with oecophorids (Stigenberg & Shaw, 2013).

Meteorus cinctellus (Spinola, 1808)

Catalogs with Iranian records: Sedighi and Madjdzadeh (2015), Gadallah et al. (2016), Farahani et al. (2016), Yu et al. (2016), Samin, Coronado-Blanco, Kavallieratos et al. (2018), Samin, Coronado-Blanco, Fischer et al. (2018).

Distribution in Iran: Guilan, Mazandaran (Farahani & Talebi, 2012), Qazvin (Ghahari, 2015).

Distribution in the Middle East: Iran.

Extralimital distribution: Albania, Austria, Belgium, Bulgaria, China, Czech Republic, Denmark, Finland, France, Germany, Greece, Hungary, Ireland, Italy, Japan, Korea, Latvia, Madeira Islands, Netherlands, New Zealand, Norway, Poland, Portugal, Romania, Russia, Serbia, Slovenia, Sweden, Switzerland, Ukraine, United Kingdom.

Host records: Summarized by Yu et al. (2016) as being a parasitoid of the crambids *Crambus uliginosellus* Zeller, *Nomophila noctuella* (Denis and Schiffermüller), *Pyrausta purpuralis* (L.), and *Scoparia ambigualis* (Treitscke); the erebid *Orgyia antiqua* (L.); the geometrid *Thera juniperata* (L.); the noctuid *Apamea scolopacina* (Esper); and the tortricids *Acleris hastiana* (L.), *Acleris rhombana* (Denis and Schiffermüller), *Apoctena flavescens* (Buttler), *Archips xylosteana* (L.), *Catamacta gavisana* (Walker), *Ctenopseustis obliquana* (Walker), *Epalxiphora axenana* Meyrick, *Epiphyas postvittana* (Walker), *Gypsonoma oppressana* (Treitscke), *Gypsonoma sociana* (Haworth), *Hedya pruniana* (Hübner), *Notocelia uddmanniana* (L.), *Pandemis heparana* (Denis and Schiffermüller), and *Tortrix viridana* L. It was also recorded by Stigenberg and Shaw (2013) as being a parasitoid of the pyralids *Crambus uliginosellus* Zeller, *Eudonia angustea* (Curtis), *Eudonia truncicolella* (Stainton), *Nomophila noctuella* (Denis and Schiffermüller), *Pyrausta purpuralis* (L.), and *Scoparia ambigualis* (Treitschke).

Meteorus cis (Bouché, 1834)

Catalogs with Iranian records: No catalog.

Distribution in Iran: East Azarbaijan (Drogvalenko & Ghahari, 2021).

Distribution in the Middle East: Iran.

Extralimital distribution: Austria, China, France, Germany, Ireland, Italy, Japan, Korea, Netherlands, Russia, Sweden, United Kingdom.

Host record: *Meteorus cis* has been recorded by Stigenberg et al. (2013) as being a parasitoid of the ciid *Cis boleti* (Scopoli). In Iran, this species has been reared from *Cis boleti* (Scopoli) and *Cis comptus* Gyllenhal (Coleoptera: Ciidae) (Drogvalenko & Ghahari, 2021).

Meteorus colon (Haliday, 1835)

Catalogs with Iranian records: Sedighi and Madjdzadeh (2015), Gadallah et al. (2016), Farahani et al. (2016), Yu et al. (2016), Samin, Coronado-Blanco, Kavallieratos et al. (2018), Samin, Coronado-Blanco, Fischer et al. (2018).

Distribution in Iran: Kermanshah (Ghahari, Erdoğan et al., 2010 as *Meteorus luridus* Ruthe), Guilan (Farahani & Talebi, 2012), Mazandaran (Kian et al., 2020).

Distribution in the Middle East: Iran (see references above), Israel—Palestine (Papp, 2012), Turkey (Okyar et al., 2012; Yilmaz et al., 2010).

Extralimital distribution: Armenia, Austria, Azerbaijan, Belgium, Bulgaria, China, Croatia, Czech Republic, Denmark, Finland, France, Germany, Ireland, Italy, Japan, Lithuania, Moldova, Montenegro, Netherlands, Norway, Poland, Russia, Slovakia, Sweden, Switzerland, United Kingdom.

Host records: Summarized by Yu et al. (2016) as being a parasitoid of the erebids *Leucoma salicis* (L.), *Lithosia quadra* (L.), and *Lymantria monacha* (L.); the geometrids *Anticollix sparsata* (Treitschke), *Eupithecia venosata* (Fabricius), *Hydriomena furcata* (Thunberg), and *Xanthorhoe biriviata* (Barkhausen); the noctuids *Allophyes oxyacanthae* (L.), *Antitype chi* (L.), *Cosmia diffinis* (L.), *Cosmia trapezina* (L.), *Cucullia argentea* (Hufnagel), *Diarsia brunnea* (Denis and Schiffermüller), *Euxoa nigrofusca* L., *Noctua fimbriata* Schreber, *Orthosia gothica* (L.), *Orthosia stabilis* (Fabricius), *Polia nebulosa* (Hufnagel), and *Spodoptera exigua* (Hübner); the nolids *Bena prasinana* L., *Nola cucullatella* (L.), and *Pseudoips prasinana* (L.); the notodontid *Phalera bucephala* (L.); the nymphalid *Limenitis camilla* (L.); the pyralids *Acrobasis consociella* (Hübner), and *Acrobasis pyrivorella* (Matsumura); the saturnid *Saturnia pavonia* (L.); and the tortricids *Archips rosana* (L.), *Pammene giganteana* (Peyerimhoff), and *Sparganothis pilleriana* (Denis and Schiffermüller). It was also recorded by Stigenberg and Shaw (2013) as being a parasitoid of the geometrids *Anticollix sparsata* (Treitschke), *Hydriomena furcata* (Thunberg), and *Xanthorohoe biriviata* (Borkhausen); the noctuids *Diarsia brunnea* (Denis and Schiffermüller), *Orthosia cerasi* (Fabricius), *Orthosia gathica* (L.) *Antitype chi* (L.), *Pseudoips prasinana* (L.), and *Xestia agathina* (Ford); and the nymphalid *Limenitis camilla* (L.).

Meteorus consimilis (Nees von Esenbeck, 1834)
Catalogs with Iranian records: Sedighi and Madjdzadeh (2015), Gadallah et al. (2016), Farahani et al. (2016), Yu et al. (2016), Samin, Coronado-Blanco, Kavallieratos et al. (2018), Samin, Coronado-Blanco, Fischer et al. (2018).
Distribution in Iran: Guilan, Mazandaran (Farahani & Talebi, 2012; Sedighi & Madjdzadeh, 2015).
Distribution in the Middle East: Iran (Farahani & Talebi, 2012; Sedighi & Madjdzadeh, 2015), Turkey (Yilmaz et al., 2010).
Extralimital distribution: Belgium, Croatia, Czech Republic, Denmark, Finland, France, Germany, Hungary, Ireland, Italy, Latvia, Netherlands, Norway, Poland, Romania, Slovenia, Sweden, Switzerland, Turkmenistan, Ukraine, United Kingdom, former Yugoslavia.
Host records: Recorded by Čapek and Hofmann (1997) as being a parasitoid of the curculionid *Scolytus multistriatus* (Marsham).

Meteorus eadyi Huddleston, 1980
Distribution in the Middle East: Turkey (Yilmaz et al., 2010).
Extralimital distribution: Armenia, Belarus, Bulgaria, former Czechoslovakia, France, Germany, Hungary, Korea, Lithuania, Mongolia, Romania, Russia, Sweden, Switzerland, United Kingdom, former Yugoslavia.
Host records: Unknown.

Meteorus filator (Haliday, 1835)
Catalogs with Iranian records: No catalog.
Distribution in Iran: Chaharmahal and Bakhtiari (Gadallah et al., 2018).

Distribution in the Middle East: Iran (Gadallah et al., 2018), Turkey (Yilmaz et al., 2010).
Extralimital distribution: Austria, Azerbaijan, Belgium, Bulgaria, Denmark, Finland, France, Georgia, Germany, Hungary, Ireland, Italy, Korea, Lithuania, Mongolia, Netherlands, Norway, Poland, Russia, Slovakia, Sweden, Switzerland, United Kingdom.
Host records: Unknown.

Meteorus graciliventris Muesebeck, 1956
Distribution in the Middle East: Turkey (Koldaş et al., 2013).
Extralimital distribution: China, Japan, Korea, Russia.
Host records: Recorded by Muesebeck (1950) as being a parasitoid of the gelechiid *Pectinophora gossypiella* (Saunders).

Meteorus hirsutipes Huddleston, 1980
Distribution in the Middle East: Turkey (Beyarslan, 2021).
Extralimital distribution: China, Croatia, Finland, Germany, Hungary, Ireland, Japan, Norway, Russia, Sweden, Switzerland, United Kingdom.
Host records: Unknown.

Meteorus ictericus (Nees von Esenbeck, 1811)
Catalogs with Iranian records: Sedighi and Madjdzadeh (2015), Gadallah et al. (2016), Farahani et al. (2016), Yu et al. (2016), Samin, Coronado-Blanco, Kavallieratos et al. (2018), Samin, Coronado-Blanco, Fischer et al. (2018).
Distribution in Iran: Mazandaran (Farahani & Talebi, 2012).
Distribution in the Middle East: Iran (Farahani & Talebi, 2012), Israel—Palestine (Papp, 1970), Turkey (Aydoğdu, 2014; Beyarslan et al., 2004; Koldaş et al., 2013).
Extralimital distribution: Australasian, Nearctic, Oceanic, Oriental, Palaearctic.
Host records: Summarized by Yu et al. (2016) as being a parasitoid of the sawflies of Diprionidae (very unlikely, but possible, needs verification), as well as a large number of lepidopteran moths and butterflies of the families Blastodacnidae, Choreutidae, Erebidae, Gelechiidae, Geometridae, Gracillariidae, Lycaenidae, Momphidae, Noctuidae, Nymphalidae, Pyralidae, Thaumetopoeidae, Tortricidae, Yponomeutidae, Zygaenidae. It was also recorded by Stigenberg and Shaw (2013) as being a parasitoid of the gracillariids *Caloptilia syringella* (Fabricius), and *Povolnya leucapenella* (Stephens); the tortricids *Acleris ferrugana* (Denis and Schiffermüller), *Acleris hastiana* (L.), *Acleris rufana* (Denis and Schiffermüller), *Acleris umbrana* (Hübner)? *Acleris variegana* (Denis and Schiffermüller)? *Ancylis upupana* (Treitscke), *Archips podana* (Geoffroy), *Archips rosana* (L.), *Cacoecimorpha pronubana* (Hübner), *Croesia forsskaleana* (L.), and

Ephiphyas postvittana (Walker), in addition to large number of rearings from unidentified tortricids feeding on a wide range of trees and bushes.

Meteorus jaculator (Haliday, 1835)

Catalogs with Iranian records: Samin, Coronado-Blanco, Fischer et al. (2018).

Distribution in Iran: Isfahan (Sakenin et al., 2018).

Distribution in the Middle East: Iran (Sakenin et al., 2018), Turkey (Shenefelt, 1969; Yilmaz et al., 2010).

Extralimital distribution: Austria, Bulgaria, Croatia, Denmark, Finland, France, Germany, Hungary, Ireland, Italy, Netherlands, Norway, Poland, Russia, Slovakia, Sweden, Switzerland, United Kingdom, former Yugoslavia.

Host records: Summarized by Yu et al. (2016) as being a parasitoid of the gelechiid *Caryocolum amaurella* (Hering); the psychids *Siederia alpicolella* (Rebel), and *Siederia pineti* (Zeller); and the tineids *Nemapogon granella* (L.), and *Trichophaga tapetzella* (L.). It was also recorded by Stigenberg and Shaw (2013) as being a parasitoid of the tineid *Nemopogon cloacella* (Haworth) (not certain of being its host).

Meteorus lionotus Thomson, 1895

Catalogs with Iranian records: Samin, Coronado-Blanco, Fischer et al. (2018).

Distribution in Iran: Fars, Isfahan (Samin, Coronado-Blanco, Fischer et al., 2018).

Distribution in the Middle East: Iran (Samin, Coronado-Blanco, Fischer et al., 2018), Turkey (Özbek & Coruh, 2012; Yilmaz et al., 2010).

Extralimital distribution: Austria, China, Finland Germany, Greece, Norway, Poland, Slovakia, Sweden, Switzerland, United Kingdom.

Host records: Summarized by Yu et al. (2016) as being a parasitoid of the diprionid *Neodiprion sertifer* (Geoffroy) (doubtful record, needs verification); the geometrids *Eupithecia indigata* (Hübner), *Eupithecia tantillaria* Boisduval, *Operophtera brumata* (L.), *Thera britannica* Turner, *Thera junipera* (L.), *Thera obelicata* (Hübner), and *Thera variata* (Denis and Schiffermüller); and the lasiocampid *Malacosoma neustria* (L.).

Meteorus longicaudis (Ratzeburg, 1848)

Catalogs with Iranian records: Samin, Coronado-Blanco, Kavallieratos et al. (2018).

Distribution in Iran: East Azarbaijan (Samin, Coronado-Blanco, Kavallieratos et al., 2018).

Distribution in the Middle East: Iran (Samin, Coronado-Blanco, Kavallieratos et al., 2018), Turkey (Yilmaz et al., 2010).

Extralimital distribution: Croatia, Finland, Germany, Netherlands, Poland, Sweden, former Yugoslavia.

Host records: Recorded by Huddleston (1980) and Tobias (1986) as being a parasitoid of the melandryid *Orchesia micans* (Panzer), and the tenebrionid *Eledonoprius armatus* (Panzer).

Meteorus micropterus (Haliday, 1833)

Distribution in the Middle East: Turkey (Koldaş et al., 2013).

Extralimital distribution: Czech Republic, Denmark, Finland, Germany, Hungary, Ireland, Japan, Slovakia, Sweden, United Kingdom.

Host records: Summarized by Yu et al. (2016) as being a parasitoid of the hepialids *Hepialus humuli* (L.), and *Pharmacis fusconebulosa* (DeGeer).

Meteorus obfuscatus (Nees von Esenbeck, 1811)

Catalogs with Iranian records: No catalog.

Distribution in Iran: Ardabil (Gadallah et al., 2021).

Distribution in the Middle East: Iran.

Extralimital distribution: Belgium, Finland, France, Germany, Hungary, Italy, Japan, Montenegro, Norway, Poland, Romania, Russia, Slovakia, Sweden, Switzerland, United Kingdom.

Host records: Summarized by Yu et al. (2016) as being a parasitoid of the cerambycids *Pogonocherus hispidus* (L.), and *Tetrops praeustus* (L.); the curculionid *Ips acuminatus* (Gyllenhal); the erotylid *Triplax russica* (L.); the melandryids *Orchesia micans* (Panzer), and *Orchesia minor* Walker; and the tenebrionid *Mycetochara axillaris* (Paykull).

Meteorus obsoletus (Wesmael, 1835)

Catalogs with Iranian records: Fallahzadeh and Saghaei (2010), Sedighi and Madjdzadeh (2015), Gadallah et al. (2016), Farahani et al. (2016), Yu et al. (2016), Samin, Coronado-Blanco, Kavallieratos et al. (2018), Samin, Coronado-Blanco, Fischer et al. (2018).

Distribution in Iran: East Azarbaijan (Nickdel et al., 2004, 2008).

Distribution in the Middle East: Iran (Nickdel et al., 2004, 2008), Turkey (Beyarslan et al., 2004; Hudleston, 1980; Koldaş et al., 2013; Yilmaz et al., 2010).

Extralimital distribution: Austria, Belgium, Bulgaria, Czech Republic, France, Germany, Ireland, Italy, Japan, Korea, Moldova, Montenegro, Netherlands, Russia, Sweden, Switzerland, Ukraine, United Kingdom.

Host records: Recorded by Balevski (1989) as being a parasitoid of the tephritid *Carpomya schineri* (Loew); however, this seems doubtful. Summarized by Yu et al. (2016) as being a parasitoid of the noctuid *Agrotis ipsilon* (Hufnagel); and the tortricids *Choristoneura jezoensis* Yasuda and Suzuki, *Epinotia cruciana* (L.), *Gypsonoma dealbana* (Frölich), *Hedya nubiferana* Haworth, *Rhopobota ustomaculana* (Curtis), *Tortrix viridana* L.,

Zeiraphera griseana (Hübner), and *Zeiraphera rufimitrana* (Herrich-Schäffer). In Iran, this species has been reared from the erebid *Euproctis chrysorrhoea* (L.) (Nickdel et al., 2004, 2008).

Meteorus oculatus Ruthe, 1862

Catalogs with Iranian records: No catalog.
Distribution in Iran: Lorestan (Naderian et al., 2020).
Distribution in the Middle East: Iran (Naderian et al., 2020), Turkey (Yilmaz et al., 2010).
Extralimital distribution: Austria, Bulgaria, former Czechoslovakia, Finland, Germany, Hungary, Kyrgyzstan, Lithuania, Norway, Poland, Russia, Sweden, Switzerland, United Kingdom.
Host records: Recorded by Hauser (1994) as being a parasitoid of larvae of the psychid *Taleporia tubulosa* (Retzius). It was also recorded by Achtelig (1974) as being a parasitoid of the raphidiid snakefly *Xanthostigma xanthostigma* (Schummel), but this record needs further verification.

Meteorus pendulus (Müller, 1776)

Catalogs with Iranian records: Fallahzadeh and Saghaei (2010), Sedighi and Madjdzadeh (2015), Gadallah et al. (2016), Gadallah et al. (2016), Farahani et al. (2016), Yu et al. (2016), Samin, Coronado-Blanco, Kavallieratos et al. (2018), Samin, Coronado-Blanco, Fischer et al. (2018).
Distribution in Iran: Golestan (Ghahari, 2015), Guilan, Qazvin, West Azarbaijan (Farahani & Talebi, 2012), Isfahan (Ghahari et al., 2011 as *Meteorus gyrator* (Thunberg)), Mazandaran (Abbasipour, 2001, 2004; Farahani & Talebi, 2012; Modarres Awal, 2012 as *M. gyrator*; Kian et al., 2020), Iran (no specific locality cited) (Abbasipour, 2001; Abbasipour & Taghavi, 2002, 2004 as *M. gyrator*; Abbasipour et al., 2004 as *M. gyrator*; Khanjani, 2006a).
Distribution in the Middle East: Cyprus, Egypt, Iran, Israel—Palestine, Turkey (Yu et al., 2016).
Extralimital distribution: Nearctic, Oceanic, Oriental, Palaearctic.
Host records: Summarized by Yu et al. (2016) as being a parasitoid of a large number of insects hosts of the orders Coleoptera (doubtful) (Curculionidae), Diptera (Cecidomyiidae) (doubtful), and Lepidoptera (Arctiidae, Blastodacnidae, Erebidae, Gelechiidae, Geometridae, Gracillaridae, Lasiocampidae, Lycaenidae, Noctuidae, Notodontidae, Pyralidae, Tortricidae). It was also recorded by Stigenberg and Shaw (2013) as being a parasitoid of the noctuids *Apamea unaminis* (Hübner), *Agrocola lota* (Clerck), *Brachylomia viminalis* (Fabricius), *Ceramica pisi* (L.), *Cerapteryx graminis* (L.), *Conistra vaccinii* (L.), *Cosmia trapezina* (L.), *Dryobotodes eremita* (Fabricius), *Eremobia ochroleuca* (Denis and Schiffermüller), *Euplexia lucipara* (L.), *Euplexia transversa* (Hufnagel), *Lacanobia*

oleracea (L.), *Mythimna conigera* (Denis and Schiffermüller), *Mythimna farrago* (Fabricius), *Mythimna impura* (Huebner), *Mythimna* sp., *Noctua fimbriata* (Schreber), *Noctua orbona* (Hufnagel), *Orthosia gracilis* (Denis and Schiffermüller), *Phlogophora meticulosa* (L.), *Thalpophila matura* (Hufnagel), and *Xestia xanthographa* (Denis and Schiffermüller). In Iran, this species has been reared from the noctuids *Mythimna unipuncta* (Haworth) (Abbasipour, 2001, 2004; Khanjani, 2006a as *Pseudaletia unipunctata* Haworth; Modarres Awal, 2012), and *Spodoptera exigua* (Hübner) in rice fields (Farahani & Talebi, 2012).

Meteorus politutele Shenefelt, 1969

Catalogs with Iranian records: No catalog.
Distribution in Iran: Golestan (Samin, Papp, & Coronado-Blanco, 2018).
Distribution in the Middle East: Iran (Samin, Papp, & Coronado-Blanco, 2018), Turkey (Beyarslan et al., 2004; Yilmaz et al., 2010).
Extralimital distribution: Uzbekistan.
Host records: Unknown.

Meteorus pulchricornis (Wesmael, 1835)

Catalogs with Iranian records: Fallahzadeh and Saghaei (2010), Sedighi and Madjdzadeh (2015), Gadallah et al. (2016), Farahani et al. (2016), Yu et al. (2016), Samin, Coronado-Blanco, Kavallieratos et al. (2018), Samin, Coronado-Blanco, Fischer et al. (2018).
Distribution in Iran: East Azarbaijan, Northern Khorasan (Herard et al., 1979), Guilan (Ghahari, 2018 - around rice fields), Mazandaran (Farahani & Talebi, 2012; Ghahari, Fischer, Erdoğan, Tabari et al., 2009; Herard et al., 1979), Iran (no specific locality cited) (Modarres Awal, 1997, 2012).
Distribution in the Middle East: Cyprus, Iran, Israel—Palestine, Turkey (Yu et al., 2016).
Extralimital distribution: Australasian, Oceanic, Oriental, Palaearctic, United States of America (introduced).
Host records: Summarized by Yu et al. (2016) as being a parasitoid of a large number of lepidopteran species of the families Arctiidae, Erebidae, Gelechiidae, Geometridae, Hesperiidae, Lasiocampidae, Lycaenidae, Lyonetiidae, Noctuidae, Nolidae, Nymphalidae, Papilionidae, Plutellidae, Psychidae, Pterophoridae, Pyralidae, Simaethidae, Tineidae, and Tortricidae. It was also recorded by Stigenberg and Shaw (2013) as being a parasitoid of the choreutid *Anthophila fabriciana* (L.); the gelechiid *Hypatima rhomboidella* (L.); the geometrids *Agriopis aurantiaria* (Hübner), *Cyclophora* sp., *Eupithecia* sp., *Operophtera brumata* (L.), and *Operophtera fragata* (Scharfenberg); the lasiocampid *Lasiocampa trifolii* (Denis and Schiffermüller); the noctuids *Anata myrtilli* (L.), *Eupsilia transversa* (Hufnagel), *Lycophotia porphyrea* (Denis and Schiffermüller), and

Orthosia cruda (Denis and Schiffermüller); the nymphalid *Charaxes jasius* (L.); the pterophorid *Amblyptilia acanthadactyla* (Hübner); the pyralid *Uresiphita gilvata* (Fabricius); and the zygaenid *Zygaena viciae* (Denis and Schiffermüller). In Iran, this species has been reared from the erebid *Lymantria dispar* (L.) (Herard et al., 1979; Modarres Awal, 1997, 2012).

Meteorus radialis Tobias, 1986
Distribution in the Middle East: Turkey (Beyarslan, 2021).
Extralimital distribution: Russia.
Host records: Unknown.

Meteorus rubens (Nees von Esenbeck, 1811)
Catalogs with Iranian records: Fallahzadeh and Saghaei (2010), Sedighi and Madjdzadeh (2015), Gadallah et al. (2016), Farahani et al. (2016), Yu et al. (2016), Samin, Coronado-Blanco, Kavallieratos et al. (2018), Samin, Coronado-Blanco, Fischer et al. (2018).
Distribution in Iran: Alborz (Davatchi & Shojai, 1969 as *Meteorus mesopotamicus* Fischer; Farahani & Talebi, 2012; Shojai, 1968, 1998), Mazandaran, Qazvin, Tehran (Farahani & Talebi, 2012), Hormozgan (Ameri et al., 2014), Kerman (Abdolalizadeh et al., 2017), Khuzestan (Shojai et al., 1995 - in corn field; Modarres Awal, 1997, 2012; Razavi Khorasan (Darsouei & Karimi, 2015), Tehran (Modarres Awal, 1997, 2012 as *M. mesopotamicus* and as *M. rubens*), West Azarbaijan (Ghahari, 2019), Iran (no specific locality cited) (Aubert, 1966 as *Meteorus mesopotamicus*; Khanjani, 2006a,b; Stigenberg & Ronquist, 2011).
Distribution in the Middle East: Cyprus, Egypt, Iran, Iraq, Israel—Palestine, Turkey (Yu et al., 2016).
Extralimital distribution: Nearctic, Neotropical, Oriental, Palaearctic.
Host records: Summarized by Yu et al. (2016) as a parasitoid of a large number of insects hosts of the orders Diptera (Chloropidae) (doubtful), and Lepidoptera (Coleophoridae, Erebidae, Gelechiidae, Geometridae, Lasiocampidae, Noctuidae, Nymphalidae, Pieridae, Pyralidae, Thaumatopoeidae, Tortricidae, Yponomeutidae). Recorded by Stigenberg and Shaw (2013) as being a parasitoid of the noctuid *Noctua pronuba* (L.). In Iran, this species has been reared from the noctuids *Agrotis segetum* (Denis and Schiffermüller) (Davatchi & Shojai, 1969; Khanjani, 2006a; Modarres Awal, 1997, 2012; Shojai 1968, 1998), *Spodoptera exigua* (Hübner) (Darsouei & Karimi, 2015; Davatchi & Shojai, 1969; Farahani & Talebi, 2012; Ghahari, 2019; Khanjani, 2006a,b; Modarres Awal, 2012; Shojai 1968, 1998), and *Sesamia nonagrioides* (Lefèbvre) (Khanjani, 2006a; Modarres Awal, 1997, 2012); and the gelechiid moth *Phthorimaea operculella* (Zeller) (Khanjani, 2006b).

Meteorus ruficeps (Nees von Esenbeck, 1834)
Catalogs with Iranian records: No catalog.
Distribution in Iran: Kordestan (Naderian et al., 2020).
Distribution in the Middle East: Iran (Naderian et al., 2020), Israel—Palestine (Papp, 2012).
Extralimital distribution: Armenia, Austria, Belgium, Bulgaria, China, France, Germany, Hungary, Ireland, Isle of Man, Japan, Korea, Latvia, Lithuania, North Macedonia, Moldova, Netherlands, Norway, Poland, Russia, Serbia, Slovakia, Sweden, Switzerland, United Kingdom.
Host records: Summarized by Yu et al. (2016) as being a parasitoid of the erebid *Hyphoraia aulica* (L.); the gelechiids *Dichelia histrionana* (Frölich), *Recurvaria nanella* (Denis and Schiffermüller), and *Teleiodes saltuum* (Zeller); the geometrids *Abraxas grossulariata* (L.), and *Operophtera brumata* (L.); the noctuid *Cosmia trapezina* (L.); the tineid *Nemopogon cloacella* (Haworth); and the following tortricids: *Archips oporana* (L.), *Archips rosana* (L.), *Archips xylosteana* (L.), *Argyroploce uddmanniana* (L.), *Choristoneura fumiferana* (Clemens), *Choristoneura murinana* (Hübner), *Epinotia aciculana* Falkovitsh, *Epinotia pusillana* (Peyerimhoff), *Epinotia sordidana* (Hübner), *Exapate duratella* Heyden, *Gravitarmata margarotana* (von Heinemann), *Notocelia cynosbatella* (L.), *Notocelia roborana* (Denis and Schiffermüller), *Pandemis cerasana* (Hübner), *Pandemis corylana* (Fabricius), *Ptycholomoides aerifernaus* (Herrich-Schäffer), *Rhyacionia buoliana* (Denis and Schiffermüller), *Tortrix viridana* L., and *Zeiraphera griseana* (Hübner); and the yponomeutid *Yponomeuta plumbella* (Denis and Schiffermüller).

Meteorus rufus (DeGeer, 1778)
Catalogs with Iranian records: Gadallah et al. (2016), Farahani et al. (2016), Yu et al. (2016), Samin, Coronado-Blanco, Kavallieratos et al. (2018), Samin, Coronado-Blanco, Fischer et al. (2018).
Distribution in Iran: Chaharmahal and Bakhtiari (Gadallah et al., 2016).
Distribution in the Middle East: Cyprus (Ingram, 1981), Iran (Gadallah et al., 2016), Israel—Palestine (Schwartz et al., 1980), Turkey (Aydoğdu, 2014; Beyarslan et al., 2004; Koldaş et al., 2013; Okyar et al., 2012; Steiner, 1936; Stigenberg & Shaw, 2013; Yilmaz et al., 2010).
Extralimital distribution: Austria, Azores, Belgium, Croatia, Czech Republic, France, Germany, Hungary, India, Ireland, Italy, Poland, Romania, Russia, Switzerland, United Kingdom, former Yugoslavia.
Host records: Summarized by Yu et al. (2016) as being a parasitoid of the crambid *Scirpophaga incertulas* (Walker); the erebids *Leucoma salicis* (L.), *Lymantria dispar* (L.), and *Lymantria monacha* (L.); the geometrids *Abraxas pantaria* (L.), and *Thera cembrae* (Kitt); the noctuids *Agrocola lota* (Clerck), *Cosmia diffinis* (L.), *Cucullia*

argentea (Hufnagel), *Leucania loreyi* (Duponchel), *Orthosia cruda* (Denis and Schiffermüller), *Spodoptera exigua* (Hübner), and *Spodoptera littoralis* Boisduval; the pierid *Colias croceus* (Geoffroy); the tortricid *Archips rosana* (L.); and the zygaenids *Zygaena* spp. It was also recorded by Stigenberg and Shaw (2013) as being a parasitoid of the zygaenids *Zygaena carniolica* (Scopoli), *Zygaena lonicerae* (Scheven), *Zygaena trifolii* (Esper), and *Zygaena* sp.

Meteorus salicorniae Schmiedeknecht, 1897
Catalogs with Iranian records: No catalog.
Distribution in Iran: Northern Khorasan (Samin, Papp, & Coronado-Blanco, 2018).
Distribution in the Middle East: Iran (Samin, Papp, & Coronado-Blanco, 2018), Turkey (Fahringer, 1922; Yilmaz et al., 2010).
Extralimital distribution: Croatia, Japan, Korean, Russia.
Host records: Recorded by Patetta and Manino (1990) as being a parasitoid of the pyralids *Achroia grisella* (Fabricius), *Aphamia sociella* (L.), and *Galleria mellonella* (L.).

Meteorus tabidus (Wesmael, 1835)
Catalogs with Iranian records: No catalog.
Distribution in Iran: Lorestan (Ghahari & Gadallah, 2018c).
Distribution in the Middle East: Iran (Ghahari & Gadallah, 2018c), Turkey (Yilmaz et al., 2010).
Extralimital distribution: Austria, Belgium, Bosnia-Herzegovina, Bulgaria, Croatia, France, Georgia, Germany, Greece, Hungary, Ireland, Italy, Korea, Lithuania, Mongolia, Netherlands, Norway, Poland, Romania, Serbia, Slovakia, Sweden, Switzerland, United Kingdom.
Host records: Summarized by Yu et al. (2016) as being a parasitoid of the cerambycids *Leiopus nebulosus* (L.), *Saperda populnea* (L.), and *Saperda scalaris* (L.) (beetle records are doubtful); the coleophorid *Coleophora ledi* Stainton; the geometrid *Eupithecia absinthiata* (Clerck); the psychids *Epichnopterix sieboldin* (Reutti), *Proutia betulina* (Zeller); and the tortricid *Cydia strobidella* (L.).

Meteorus versicolor (Wesmael, 1835)
Catalogs with Iranian records: Fallahzadeh and Saghaei (2010), Sedighi and Madjdzadeh (2015), Gadallah et al. (2016), Farahani et al. (2016), Yu et al. (2016), Samin, Coronado-Blanco, Kavallieratos et al. (2018), Samin, Coronado-Blanco, Fischer et al. (2018).
Distribution in Iran: Alborz (Shojai, 1998 as *Meteorus decoloratus* Ruthe, 1862), East Azarbaijan (Abdinbekova, 1975, 1995; Nickdel et al., 2004, 2008), Fars, Lorestan, Razavi Khorasan (Ghahari, 2020), Golestan (Ghahari, Fischer, Erdoğan, Tabari et al., 2009), Guilan (Farahani & Talebi, 2012), Mazandaran (Ghahari, 2015; Kian et al., 2020), Tehran (Modarres Awal, 1997, 2012 as *Meteorus decoloratus*).

Distribution in the Middle East: Cyprus (Beyarslan et al., 2017), Iran (see references above), Israel—Palestine (Papp, 1989), Turkey (Aydoğdu, 2014; Beyarslan et al., 2004; Steiner, 1936, 2013; Stigenberg & Shaw, 2013; Yilmaz et al., 2010).
Extralimital distribution: Armenia, Austria, Azerbaijan, Belgium, Bulgaria, China, Croatia, Czech Republic, Finland, France, Germany, Greece, Hungary, Ireland, Italy, Japan, Korea, Latvia, Lithuania, North Macedonia, Madeira Islands, Moldova, Mongolia, Netherlands, Norway, Poland, Portugal, Romania, Russia, Serbia, Slovakia, Spain, Sweden, Switzerland, Tajikistan, Ukraine, United Kingdom, Uzbekistan.
Host records: Summarized by Yu et al. (2016) as being a parasitoid of a large number of lepidopteran moths and butterflies of the families Arctiidae, Argyresthiidae, Erebidae, Gelechiidae, Geometridae, Lasiocampidae, Lycaenidae, Noctuidae, Nolidae, Notodontidae, Nymphalidae, Pieridae, Pyralidae, Thaumetopoeidae, and Tortricidae. It was recorded by Stigenberg and Shaw (2013) as being a parasitoid of the erebids *Calliteara pudibunda* (L.), *Euproctis chrysorrhoea* (L.), *Orgyia antiquoides* (Hübner), and *Orgyia dubia* (King); the gelechiid *Dichomeris ustalella* (Fabricius); the following geometrids: *Agriopis aurantiaria* (Hübner), *Agriopis marginaria* (Fabricius), *Apocheima pilosaria* (Denis and Schiffermüller), *Ematurga atomaria* (L.), *Epirrita* sp., *Eulithis testata* (L.), *Hydriomena ruberata* (Freyer), *Pachycnemia hippocastanaria* (Hübner), and *Thera juniperata* (L.); the lasiocampid *Macrothylacia rubi* (L.); the lycaenid *Callophrys rubi* (L.); the noctuids *Agrochola haematidea* (Duponchel), *Anarta myrtilli* (L.), *Lycophotia porphyrea* (Denis and Schiffermüller), *Orthosia miniosa* (Denis and Schiffermüller), *Nycteola revayana* (Scopoli); and the thaumetopoeid *Thaumetopoea pityocampa* (L.). In Iran, this species has been reared from the erebid *Euproctis chrysorrhoea* (L.) (Modarres Awal, 1997, 2012; Nickdel et al., 2004, 2008; Shojai, 1998).

Meteorus vexator (Haliday, 1835)
Catalogs with Iranian records: Sedighi and Madjdzadeh (2015), Gadallah et al. (2016), Farahani et al. (2016), Yu et al. (2016), Samin, Coronado-Blanco, Kavallieratos et al. (2018), Samin, Coronado-Blanco, Fischer et al. (2018).
Distribution in Iran: Guilan (Farahani & Talebi, 2012).
Distribution in the Middle East: Iran (Farahani & Talebi, 2012), Turkey (Yilmaz et al., 2010).
Extralimital distribution: Armenia, Austria, Azerbaijan, Bulgaria, Croatia, former Czechoslovakia, Denmark, Finland, France, Georgia, Germany, Hungary, Ireland, Latvia, Netherlands, Russia, Slovenia, Sweden, Switzerland, United Kingdom, former Yugoslavia.
Host records: Recorded by Huddleston (1980) and Čapek and Hofmann (1997) as being a parasitoid of the biphyllid beetle *Biphyllus lunatus* Fabricius.

***Meteorus* sp.**

Distribution in Iran: West Azarbaijan (Alizadeh & Javan Moghaddam, 2004).

Host records: In Iran, reared from the noctuid moth *Agrotis segetum* Denis and Schiffermüller (Alizadeh & Javan Moghaddam, 2004).

Genus *Zele* Curtis, 1832

Zele albiditarsus Curtis, 1832

Catalogs with Iranian records: Sedighi and Madjdzadeh (2015), Gadallah et al. (2016), Farahani et al. (2016), Yu et al. (2016), Samin, Coronado-Blanco, Kavallieratos et al. (2018), Samin, Coronado-Blanco, Fischer et al. (2018).

Distribution in Iran: Alborz (Farahani & Talebi, 2012).

Distribution in the Middle East: Iran (Farahani & Talebi, 2012), Israel−Palestine (Papp, 1970), Turkey (Stigenberg & Mark, 2013; Yilmaz et al., 2010).

Extralimital distribution: Nearctic, Neotropical, Oriental, Palaearctic.

Host records: Summarized by Yu et al. (2016) as being a parasitoid of the diprionid *Diprion pini* (L.) (doubtful); the crambid moths *Cnaphalocrocis medinalis* (Guenée); *Loxostege sticticalis* L., and *Ostrinia nubilalis* (Hübner); the erebid *Hypena proboscidalis* (L.); the following geometrids: *Abraxas grossulariata* (L.), *Bupalus piniarius* (L.), *Epirrita dilutata* (Denis and Schiffermüller), *Erannis defoliaria* (Clerck), *Hydriamena furcata* (Thunberg), *Macaria continuaria* (Eversmann), *Macaria notata* (L.), *Operophtera brumata* (L.), and *Rheumaptera hastata* (L.); the following noctuids: *Anarta mertilli* (L.), *Cosmia trapezina* (L.), *Dichonia aeruginea* (Hübner), *Dryobotodes eremita* (Fabricius), *Eupsilia transversa* (Hufnagel), *Lycophotia perphyrea* (Denis and Schiffermüller), *Mamestra brassicae* (L.), *Mniotype adusta* (Esper), *Naranza aenescens* Moore, *Orthosia* spp., *Panolis flammea* (Denis and Schiffermüller), *Polia nebulosa* (Hufnagel), *Tiliacea citrago* (L.), and *Xestia triangulum* (Hufnagel); the nymphalid butterfly *Eurodryas aurinia* (Rottemburg); the pyralid *Phycita roborella* (Denis and Schiffermüller); and the tortricids *Acleris variana* (Fernald), *Archips rosana* (L.), *Pammene regiana* (Zeller), and *Tortrix viridana* L.

Zele chlorophthalmus (Spinola, 1808)

Catalogs with Iranian records: Sedighi and Madjdzadeh (2015), Gadallah et al. (2016), Farahani et al. (2016), Yu et al. (2016), Samin, Coronado-Blanco, Kavallieratos et al. (2018), Samin, Coronado-Blanco, Fischer et al. (2018).

Distribution in Iran: Guilan (Ghahari, Fischer, Erdoğan, Tabari et al., 2009), Mazandaran (Farahani & Talebi, 2012).

Distribution in the Middle East: Cyprus, Egypt, Iran, Israel−Palestine (Yu et al., 2016).

Extralimital distribution: Afrotropical, Oceanic, Oriental, Palaearctic.

Host records: Summarized by Yu et al. (2016) as being a parasitoid of the cephid sawfly *Cephus pygmeus* L. (doubtful); the following crambids: *Anania coronata* (Hufnagel), *Anania hortulata* (L.), *Anania terrealis* (Treitschke), *Evergestis forficalis* (L.), *Ostrinia nubilalis* (Hübner), *Patania ruralis* (Scopoli), *Pyrausta aurata* (Scopoli), and *Udea prunalis* (Denis and Schiffermüller); the depressariid moth *Agonopterix conterminella* (Zeller); the erebids *Lymantria monacha* (L.), and *Rhyparia purpurata* (L.); the geometrids *Angerona prunaria* (L.), *Crocallis elinguaria* (L.), *Macaria signaria* (Hübner), *Odontopera bidentata* (Clerck), and *Operophtera brumata* (L.); the lasiocampid *Malacosoma neustria* (L.); the noctuids *Acronicta leporina* (L.), *Condica capensis* (Guenée), *Heliothis viriplaca* (Hufnagel), *Loxostege sticticalis* L., *Mythimna l-album* (L.), *Spodoptera exigua* (Hübner), *Spodoptera littoralis* (Boisduval), and *Spodoptera litura* (Fabricius); the pyralids *Acrobasis consociella* (Hübner), *Acrobasis tumidana* (Denis and Schiffermüller), *Dioryctria abietella* (Denis and Schiffermüller), *Phycita roborella* (Denis and Schiffermüller), *Rhodophaea formosa* (Haworth), *Sciota hostilis* (Stephens), and *Trachycera suavella* (Zincken); the tortricids *Acleris rhombana* (Denis and Schiffermüller), *Archips rosana* (L.), *Cydia pomonella* (L.), *Pandemis heparana* (Denis and Schiffermüller), and *Tortrix viridana* L.; and the zygaenids *Zygaena lonicerae* (Scheven), and *Zygaena minos* (Denis and SChiffermüller). It was also recorded by Stigenberg and Shaw (2013) as being a parasitoid of the pyralids *Algedonia terrealis* (Treitschke), *Eurrhypora hortulata* (L.), and *Trachycera advenella* (Zincken).

Zele deceptor (Wesmael, 1835)

Catalogs with Iranian records: Samin, Coronado-Blanco, Kavallieratos et al. (2018).

Distribution in Iran: Isfahan (Samin, Coronado-Blanco, Kavallieratos et al., 2018).

Distribution in the Middle East: Iran (Samin, Coronado-Blanco, Kavallieratos et al., 2018), Turkey (Koldaş et al., 2013).

Extralimital distribution: Nearctic, Neotropical, Oriental, Palaearctic.

Host records: Summarized by Yu et al. (2016) as being a parasitoid of the crambids *Loxostege sticticalis* (L.), and *Ostrinia nubilalis* (Hübner); the depressariid *Ethmia dodecea* (Haworth); the following geometrids: *Agriopis leucophaeria* (Denis and Schiffermüller), *Alcis repandata* (L.), *Anticlea badiata* (Denis and Schiffermüller), *Campaea margaritaria* (L.), *Catarhoe cuculata* (Hufnagel), *Chesias legatella* (Denis and Schiffermüller), *Clarada limitaria* (Walker), *Colotois pennaria* (L.), *Crocallis elinguaria* (L.), *Ematurga atomnaria* (L.), *Eutephria fla-vicinctata* (Hübner), *Enypia griseata* (Grossbeck), *Epirrhoe galiata* (Denis and Schiffermüller), *Epirrhoe dilutata* (Denis and Schiffermüller), *Eupithecia* spp., *Hydriomena furcata* (Thunberg), *Macaria* spp., *Odontopera bidentata* (Clerck), *Rheumaptera hastata* (L.), and *Xanthorhoe fluctuata* (L.); the noctuids *Anarta myrtilli* (L.), *Hoplodrina octogenarian* (Goeze), and *Lacanobia oleracea* (L.); and the pyralid *Nephopterix angustella* (Hübner). It was also recorded by Stigenberg and Shaw (2013) as being a parasitoid of the ethmiid *Ethmia dodocea* (Haworth); the following geometrids: *Agriopis leucophearaea* (Denis and Schiffermüller), *Alcis repandata* (L.), *Anticlea badiata* (Denis and Schiffermüller), *Campaea margaritata* (L.), *Catarhoe cuculata* (Hufnagel), *Ematurga atomaria* (L.), *Entephria flavicinctata* (Hübner), *Epirrhoe galiata* (Denis and Schiffermüller), *Epirrita autumnata* (Borkhausen), *Epirrita diluata* (Denis and Schiffermüller), *Epirrita* sp., *Eupithecia nanata* (Hübner), *Eupithecia pygmeata* (Hübner), *Eupithecia simpliciata* (Haworth), *Eupithecia tantillaria* Boisduval, *Eupithecia* sp., *Macaria liturata* (Clerck), and *Odontoptera bedentata* (Clerck); the noctuid *Anarta myrtilli* (L.); as well as many rearings from unidentified Geometridae.

Zele nigricornis (Walker, 1871)
Distribution in the Middle East: Egypt (Brues, 1926; Kamal Bey, 1951; Szépligeti, 1904), Saudi Arabia (Brues, 1926; Szépligeti, 1904).
Extralimital distribution: Eritrea, Ethiopia, Kenya, Tanzania.
Host records: Summarized by Yu et al. (2016) as being a parasitoid of the noctuids *Agrotis ipsilon* (Hufnagel), *Spodoptera exigua* (Hübner), and *Spodoptera litura* (Fabricius).

Tribe Myiocephalini Chen and van Achterberg, 1997
Genus *Myiocephalus* Marshall, 1898
Myiocephalus boops (Wesmael, 1835)
Catalogs with Iranian records: Gadallah et al. (2016), Farahani et al. (2016), Yu et al. (2016), Samin, Coronado-Blanco, Kavallieratos et al. (2018), Samin, Coronado-Blanco, Fischer et al. (2018)).
Distribution in Iran: Razavi Khorasan (Samin et al., 2011).
Distribution in the Middle East: Iran.
Extralimital distribution: Belgium, Bulgaria, Canada, China, Czech Republic, Finland, France, Georgia, Germany, Hungary, Ireland, Korea, Lithuania, Mongolia, Norway, Poland, Russia, Switzerland, United Kingdom, United States of America.

Host records: Recorded by Donisthorpe (1927) (as *Spilomma falcovibrans* (Morley)) as being associated as a "myrmeco-phile" with the formicid ant *Formica fusca* L. However, this record has not since being verified and *Myiocephalus* has not yet definitively been reared from any ant.

Tribe Neoneurini Bengtsson, 1918
Genus *Elasmosoma* Ruthe, 1858
Elasmosoma berolinense Ruthe, 1858
Catalogs with Iranian records: Sedighi and Madjdzadeh (2015), Gadallah et al. (2016), Farahani et al. (2016), Yu et al. (2016), Samin, Coronado-Blanco, Kavallieratos et al. (2018), Samin, Coronado-Blanco, Fischer et al. (2018).
Distribution in Iran: Ardabil (Ghahari, Gadallah et al., 2009), Khuzestan (Samin et al., 2019 under Neoneurinae).
Distribution in the Middle East: Iran (Ghahari, Gadallah et al., 2009; Samin et al., 2019), Turkey (Yilmaz et al., 2010).
Extralimital distribution: Albania, Austria, Bulgaria, Croatia, Denmark, Finland, France, Germany, Greece, Hungary, Italy, Japan, Kazakhstan, North Macedonia, Moldova, Mongolia, Netherland, Poland, Russia, Slovakia, Slovenia, Sweden, Tajikistan, United Kingdom, former Yugoslavia.
Host records: Summarized by Yu et al. (2016) as being a parasitoid of the formicids *Camponotus vagus* (Scopoli), *Formica fusca* L., *Formica pratensis* Retzius, *Formica rufa* L., *Formica sanguinea* Latreille, and *Lasius niger* (L.).

Elasmosoma calcaratum Tobias, 1986
Distribution in the Middle East: Turkey (Koldaş et al., 2013).
Extralimital distribution: Moldova.
Host records: Unknown.

Elasmosoma geylanae Beyarslan, 2016
Distribution in the Middle East: Turkey (Beyarslan, 2016).
Extralimital distribution: None.
Host records: Unknown.

Elasmosoma luxembergense Wasmann, 1909
Catalogs with Iranian records: Gadallah et al. (2016), Farahani et al. (2016), Yu et al. (2016), Samin, Coronado-Blanco, Kavallieratos et al. (2018), Samin, Coronado-Blanco, Fischer et al. (2018).
Distribution in Iran: Alborz (Farahani et al., 2014).
Distribution in the Middle East: Iran.
Extralimital distribution: Albania, Hungary, Luxembourg, Netherlands, Russia, Serbia, Spain.

Host records: Recorded by Gómez Durán and van Achterberg (2011) as being a parasitoid of the formicid *Formica rufibarbis* Fabricius.

Genus *Kollasmosoma* van Achterberg and Argaman, 1993

Kollasmosoma platamonensis (Huddleston, 1976)
Distribution in the Middle East: Egypt (Huddleston, 1976), Israel—Palestine (van Achterberg & Argaman, 1993).
Extralimital distribution: Greece, Spain.
Host records: Recorded by Huddleston (1976) as being a parasitoid of the formicid *Cataglyphis bicolor* (Fabricius).

Genus *Neoneurus* Haliday, 1838

Neoneurus auctus (Thomson, 1895)
Catalogs with Iranian records: No catalog.
Distribution in Iran: Hamadan (Ghahari, 2016).
Distribution in the Middle East: Iran (Ghahari, 2016), Turkey (Koldas et al., 2013).
Extralimital distribution: Austria, Bulgaria, Czech Republic, Finland, France, Germany, Hungary, Kazakhstan, Lithuania, Mongolia, Netherlands, Norway, Poland, Russia, Sweden, Turkmenistan, Ukraine, United Kingdom.
Host records: Summarized by Yu et al. (2016) as being a parasitoid of the formicids *Formica pratensis* Retzius, and *Formica rufa* L.

Neoneurus clypeatus (Foerster, 1863)
Catalogs with Iranian records: Yu et al. (2016), Samin, Coronado-Blanco, Kavallieratos et al. (2018), Samin, Coronado-Blanco, Fischer et al. (2018).
Distribution in Iran: Kuhgiloyeh and Boyerahmad (Samin et al., 2016).
Distribution in the Middle East: Iran.
Extralimital distribution: Austria, Czech Republic, Finland, Germany, Hungary, Italy, Kazakhstan, Korea, Lithuania, Moldova, Mongolia, Netherlands, Norway, Poland, Russia, Serbia, Sweden, Ukraine.
Host records: Recorded by Shaw (1992) as being a parasitoid of the formicid *Formica rufa* L.

Tribe Perilitini Foerster, 1863
Genus *Microctonus* Wesmael, 1835
Microctonus aethiops (Nees von Esenbeck, 1834)
Catalogs with Iranian records: Fallahzadeh and Saghaei (2010 as *Perilitus* (*Microctonus*) *aethiopoides Perilitus* (*Microctonus*) *aethiopoides*), Sedighi and Madjdzadeh (2015 as *Perilitus aethiopoides*), Gadallah et al. (2016), Farahani et al. (2016), Yu et al. (2016), Samin, Coronado-Blanco, Kavallieratos et al. (2018), Samin, Coronado-Blanco, Fischer et al. (2018).

Distribution in Iran: Alborz (Mirabzadeh, 1968; Farahani, Talebi, & Rakhshani, 2013 as *Perilitus aethiops*), Guilan, Mazandaran (Farahani, Talebi, & Rakhshani, 2013 as *Perilitus aethiops*), Hamadan (Arbab & McNiell, 2001 as *Microctonus aethiopoides*; Ghahari, Fischer, Hedqvist et al., 2010; Modarres Awal, 2012 as *Microctonus aethiopoides*), Kerman (Abdolalizadeh et al., 2017), Kermanshah (Ghahari, Fischer, Hedqvist et al., 2010 as *Perilitus* (*Microctonus*) *aethiops*), Kordestan (Samin, 2015), Lorestan (Ghahari et al., 2012 as *Perilitus* (*Microctonus*) *aethiops*), Qazvin (Arbab, 2011; Arbab & McNiell, 2001 both as *Microctonus aethiopoides*; Farahani, Talebi, & Rakhshani, 2013 as *Perilitus aethiops*; Ghahari, Fischer, Hedqvist et al., 2010; Modarres Awal, 2012 as *Microctonus aethiopoides*), Semnan (Ghahari, 2015).
Distribution in the Middle East: Iran (see references above), Israel—Palestine (Papp, 2012), Turkey (Güclü & Özbek, 2011; Koldas et al., 2007, 2013; Yilmaz et al., 2010).
Extralimital distribution: Australasian (introduced), Nearctic (introduced), Neotropical, Oceanic, Palaearctic.
Host records: Summarized by Yu et al. (2016) as being a parasitoid of the following chrysomelids: *Phyllotreta nemorum* L., and *Phyllotreta vittula* (Redtenbacher); the curculionids *Charagmus gressorius* (Fabricius), *Charagmus griseus* (Fabricius), *Hypera* spp., *Ireninus* spp., *Listronotus bonariensis* (Kuschel), *Nicaena cervine* Broun, *Pentomorus cervinus* (Boheman), *Rhinocyllus conicus* Frölich, *Sitona* spp., and *Thylacites incanus* (L.). In Iran, this species has been reared from the curculionid *Hypera postica* (Gyllenhal) (Arbab, 2011; Arbab & McNiell, 2001; Mirabzadeh, 1968; Modarres Awal, 2012).

Microctonus colesi Drea, 1968
Catalogs with Iranian records: Fallahzadeh and Saghaei (2010), Sedighi and Madjdzadeh (2015), Gadallah et al. (2016), Farahani et al. (2016), Yu et al. (2016), Samin, Coronado-Blanco, Kavallieratos et al. (2018), Samin, Coronado-Blanco, Fischer et al. (2018).
Distribution in Iran: Iran (no specific locality cited) (Bartlett et al., 1978).
Distribution in the Middle East: Iran.
Extralimital distribution: Armenia, Belarus, Bulgaria, Canada (introduced), Denmark, Germany, Greece, Ireland, North Macedonia, Norway, Spain, United Kingdom, United States of America (introduced), former Yugoslavia.
Host records: Recorded by Drea (1968) as being a parasitoid of the curculionid *Hypera postica* (Gyllenhal). In Iran, this species has been reared from *Hypera postica* (Bartlett et al., 1978).

Microctonus melanopus Ruthe, 1856

Catalogs with Iranian records: Sedighi and Madjdzadeh (2015 as *Perilitus melanopus* (Ruthe, 1856)), Gadallah et al. (2016), Farahani et al. (2016), Yu et al. (2016), Samin, Coronado-Blanco, Kavallieratos et al. (2018), Samin, Coronado-Blanco, Fischer et al. (2018).

Distribution in Iran: Sistan and Baluchestan (Sedighi et al., 2014 as *Perilitus melanopus*).

Distribution in the Middle East: Iran (Sedighi et al., 2014), Turkey (Güclü & Özbek, 2011).

Extralimital distribution: Austria, Azerbaijan, Bulgaria, Canada (Fox et al., 2004), former Czechoslovakia, Denmark, France, Germany, Hungary, Ireland, Italy, Kazakhstan, Kyrgyzstan, Lithuania, North Macedonia, Moldova, Mongolia, Netherlands, Poland, Russia, Spain, Ukraine, United Kingdom, United States of America, Uzbekistan, former Yugoslavia.

Host records: Summarized by Yu et al. (2016) as being a parasitoid of the following curculionids: *Ceutorhynchus assimilis* (Paykull), *Ceutorhynchus leprieuri* Brisout, *Ceutorhynchus obstrictus* (Marsham), *Ceutorhynchus pallidactylus* (Marsham), *Ceutorhynchus pleurostigma* (Marsham), and *Hypera meles* (Fabricius). Fox et al. (2004) discuss this species as an effective biocontrol agent for the cabbage seedpod weevil, *Ceutorhynchus obstrictus*, in Canada.

Microctonus morimi (Ferrière, 1931)

Catalogs with Iranian records: Gadallah et al. (2016), Farahani et al. (2016), Yu et al. (2016), Samin, Coronado-Blanco, Kavallieratos et al. (2018), Samin, Coronado-Blanco, Fischer et al. (2018).

Distribution in Iran: East Azarbaijan (Sakenin et al., 2008).

Distribution in the Middle East: Iran.

Extralimital distribution: Italy.

Host records: recorded by Loan (1967) as being a parasitoid of the cerambycid *Morimus asper* (Sulzer). In Iran, this species has been reared from the cerambycid *Agapanthia* (*Smaragdula*) *violacea* (Fabricius) (Sakenin et al., 2008).

Microctonus stenocari (Haeselbarth, 2008)

Catalogs with Iranian records: Sedighi and Madjdzadeh (2015), Gadallah et al. (2016), Farahani et al. (2016), Yu et al. (2016), Samin, Coronado-Blanco, Kavallieratos et al. (2018), Samin, Coronado-Blanco, Fischer et al. (2018).

Distribution in Iran: Tehran (Haeselbarth, 2008).

Distribution in the Middle East: Iran.

Extralimital distribution: Belarus, Italy, United Kingdom.

Host records: Recorded by Haeselbarth (2008) as being a parasitoid of the curculionids *Ceutorhynchus maculaalba* Germar, *Stenocarus cardui* (Herbst), and *Stenocarus ruficornis* (Stephens).

Genus Perilitus Nees von Esenbeck, 1819

Perilitus annettae Haeselbarth, 2008

Distribution in the Middle East: Cyprus, Israel–Palestine (Haeselbarth, 2008).

Extralimital distribution: France, Italy, Morocco, Spain.

Host records: Summarized by Yu et al. (2016) as being a parasitoid of the curculionids *Conorhynchus mendicus* (Gyllenhal), and *Stephanocleonus excoriatus* (Gyllenhal).

Perilitus apiophaga (Loan, 1974)

Distribution in the Middle East: Turkey (Koldas et al., 2013).

Extralimital distribution: United Kingdom.

Host records: Summarized by Yu et al. (2016) as being a parasitoid of the apionids *Apion assimile* Kirby, *Apion curtirostre* Germar, *Apion flavipes* (Paykull), *Apion miniatum* Germar, and *Apion violaceum* Kirby.

Perilitus brevicollis Haliday, 1835

Catalogs with Iranian records: Gadallah et al. (2016 as *Microtonus brevicollis* (Haliday)), Farahani et al. (2016 as *M. brevicollis*), Yu et al. (2016), Samin, Coronado-Blanco, Kavallieratos et al. (2018), Samin, Coronado-Blanco, Fischer et al. (2018).

Distribution in Iran: Hamadan (Gadallah et al., 2016 as *M. brevicollis*).

Distribution in the Middle East: Iran (Gadallah et al., 2016), Turkey (Güclü & Özbek, 2011; Yilmaz et al., 2010).

Extralimital distribution: Algeria, Armenia, Austria, Belgium, Czech Republic, Finland, France, Germany, Hungary, Ireland, Italy, Kazakhstan, Lithuania, Moldova, Netherlands, Poland, Spain, Sweden, Switzerland, United Kingdom, Uzbekistan.

Host records: Summarized by Yu et al. (2016) as being a parasitoid of the following chrysomelids: *Altica ampelophaga* Guérin-Méneville, *Phratora vitellinae* (L.), *Phratora vulgatissima* (L.), *Plagiodera versicolora* (Laicharting), *Psylliodes chrysocephala* (L.), and *Psylliodes napi* (Fabricius).

Perilitus cerealium Haliday, 1835

Catalogs with Iranian records: No catalog.

Distribution in Iran: Kordestan (Gadallah et al., 2018).

Distribution in the Middle East: Iran (Gadallah et al., 2018), Turkey (Çikman & Beyarslan, 2009; Yilmaz et al., 2010).

Extralimital distribution: Belgium, Bulgaria, Czech Republic, Finland, France, Germany, Hungary, Ireland, Italy, Kazakhstan, Mongolia, Netherlands, Poland, Romania, Russia, Sweden, Switzerland, Tajikistan, United Kingdom.

Host records: Summarized by Yu et al. (2016) as being a parasitoid of chrysomelids *Chaetocnema hortensis* (Geoffroy), *Phyllotreta vittula* (Redtenbacher), and *Psylliodes attenuate* (Koch); and the curculionid *Hypera variabilis* (Herbst). In Iran, this species has been reared from *Chaetocnema* (*Chaetocnema*) *hortensis* (Geoffroy) (Gadallah et al., 2018).

Perilitus dubius (Wesmael, 1838)

Catalogs with Iranian records: No catalog.
Distribution in Iran: Kordestan (Samin et al., 2019).
Distribution in the Middle East: Iran (Samin et al., 2019), Turkey (Beyarslan et al., 2021).
Extralimital distribution: Belgium, former Czechoslovakia, Finland, France, Germany, Greece, Hungary, Kazakhstan, Moldova, Netherlands, Poland, Russia, Sweden, Switzerland, United Kingdom.
Host records: Recorded by Waloff (1961) and Richards and Waloff (1961) as being a parasitoid of the chrysomelid *Gonioctena olivacea* (Förster).

Perilitus eugenii Haeselbarth, 1999

Distribution in the Middle East: Israel—Palestine (Papp, 2012).
Extralimital distribution: Hungary, Mongolia.
Host records: Unknown.

Perilitus falciger (Ruthe, 1856)

Catalogs with Iranian records: Gadallah et al. (2016 as *Microtonus falciger* Ruthe, 1856), Farahani et al. (2016 as *M. falciger*), Yu et al. (2016), Samin, Coronado-Blanco, Kavallieratos et al. (2018), Samin, Coronado-Blanco, Fischer et al. (2018).
Distribution in Iran: Northern Khorasan (Gadallah et al., 2016 as *M. falciger*).
Distribution in the Middle East: Iran (Gadallah et al., 2016), Israel—Palestine (Papp, 2012), Turkey (Koldas et al., 2013; Yilmaz et al., 2010).
Extralimital distribution: Austria, Belarus, Belgium, Denmark, France, Germany, Hungary, Kazakhstan, Moldova, Mongolia, Poland, Russia, Slovakia, Sweden, Switzerland, Ukraine, United Kingdom, Uzbekistan.
Host records: Summarized by Yu et al. (2016) as being a parasitoid of the chrysomelids *Chrysolina banksi* (Fabricius), *Timarcha goettingensis* (L.), *Timarcha coriaria* (Leicharting), *Timarcha laevigata* (L.), and *Timarcha tenebriosa* Fabricius.

Perilitus foveolatus Reinhard, 1862

Catalogs with Iranian records: Sedighi and Madjdzadeh (2015), Gadallah et al. (2016), Farahani et al. (2016), Yu et al. (2016), Samin, Coronado-Blanco, Kavallieratos et al. (2018), Samin, Coronado-Blanco, Fischer et al. (2018).
Distribution in Iran: Guilan, Qazvin (Farahani, Talebi, & Rakhshani, 2013).
Distribution in the Middle East: Iran (Farahani, Talebi, & Rakhshani, 2013), Turkey (Yilmaz et al., 2010).

Extralimital distribution: Belgium, France, Germany, Hungary, Italy, Kazakhstan, Montenegro, Poland, Russia, Serbia, Switzerland, United Kingdom.
Host records: Summarized by Yu et al. (2016) as being a parasitoid of the chrysomelids *Timarcha goettingensis* (L.), *Timarcha laevigata* (L.), *Timarcha maritima* Perris, and *Timarcha tenebriosa* Fabricius.

Perilitus kokujevi Tobias, 1986

Catalogs with Iranian records: Samin, Coronado-Blanco, Fischer et al. (2018).
Distribution in Iran: East Azarbaijan (Sakenin et al., 2018).
Distribution in the Middle East: Iran (Sakenin et al., 2018), Turkey (Güclü & Özbek, 2011; Koldas et al., 2013).
Extralimital distribution: China, Norway, Russia.
Host records: Unknown.

Perilitus longiradialis Tobias, 1986

Distribution in the Middle East: Turkey (Beyarslan, 2021).
Extralimital distribution: Moldova.
Host records: Unknown.

Perilitus marci Haeselbarth, 1999

Distribution in the Middle East: Turkey (Güclü & Özbek, 2011).
Extralimital distribution: United Kingdom.
Host records: Recorded by Haeselbarth (1999) as being a parasitoid of the tenebrionid *Phylan gibbus* (Fabricius).

Perilitus moldavicus (Tobias, 1986)

Distribution in the Middle East: Turkey (Koldas et al., 2013).
Extralimital distribution: Moldova, Serbia.
Host records: Unknown.

Perilitus parcicornis (Ruthe, 1856)

Catalogs with Iranian records: Samin, Coronado-Blanco, Kavallieratos et al. (2018).
Distribution in Iran: Chaharmahal and Bakhtiari (Samin, Coronado-Blanco, Kavallieratos et al., 2018).
Distribution in the Middle East: Iran (Samin, Coronado-Blanco, Kavallieratos et al., 2018), Turkey (Koldas et al., 2013; Yilmaz et al., 2010).
Extralimital distribution: Belarus, Bulgaria, France, Germany, Greece, Hungary, Italy, Moldova, Mongolia, Portugal, Spain, Switzerland, United Kingdom.
Host records: Unknown.

Perilitus regius Haseselbarth, 1999

Distribution in the Middle East: Turkey (Güclü & Özbek, 2011).
Extralimital distribution: Austria, Bulgaria, Croatia, France, Germany, Hungary, Italy, Moldova, Romania, Slovakia.
Host records: Unknown.

Perilitus retusus (Ruthe, 1856)

Distribution in the Middle East: Turkey (Koldas et al., 2013).

Extralimital distribution: Austria, Bulgaria, Finland, Germany, Greece, Hungary, Ireland, Italy, Japan, North Macedonia, Moldova, Netherlands, Serbia, Spain, Sweden, Switzerland, United Kingdom.

Host records: Summarized by Yu et al. (2016) as being a parasitoid of the carabid *Harpalus rufipes* (DeGeer); and the cerambycid *Herophila tristis* (L.).

Perilitus riphaeus (Tobias, 1986)

Distribution in the Middle East: Turkey (Beyarslan, 2021).
Extralimital distribution: Russia.
Host records: Unknown.

Perilitus rutilus (Nees von Esenbeck, 1811)

Catalogs with Iranian records: Sedighi and Madjdzadeh (2015), Gadallah et al. (2016), Farahani et al. (2016), Yu et al. (2016), Samin, Coronado-Blanco, Kavallieratos et al. (2018), Samin, Coronado-Blanco, Fischer et al. (2018).

Distribution in Iran: Guilan, Mazandaran, Qazvin (Farahani, Talebi, & Rakhshani, 2013).

Distribution in the Middle East: Cyprus (Haeselbarth, 1999), Iran (Farahani, Talebi, & Rakhshani, 2013), Turkey (Koldas et al., 2013; Yilmaz et al., 2010).

Extralimital distribution: Belgium, Bulgaria, Canada (introduced), Czech Republic, Finland, France, Georgia, Germany, Hungary, Ireland, Italy, Kazakhstan, Korea, Lithuania, Moldova, Mongolia, Netherlands, Poland, Russia, Serbia, Switzerland, Tunisia, Ukraine, United Kingdom, United States of America (introduced).

Host records: Recorded by Gyoerfi (1959) as being a parasitoid of the coccinellid *Tytthaspis sedecimpunctata* (L.). Summarized by Yu et al. (2016) as being a parasitoid of the following curculionids: *Hylobius abietis* (L.), *Hypera postica* Gyllenhal, *Hypera variabilis* (Herbst), *Pityokteines curvidens* (Germar), *Sitona crinite* (Herbst), *Sitona cylindricollis* Fåhraeus, *Sitona hispidula* (Fabricius), *Sitona humeralis* Stephens, *Sitona lineata* (L.), *Sitona longulus* Gyllenhal, *Sitona puncticollis* Stephens, and *Sitona sulcifrons* (Thunberg).

Perilitus stelleri (Loan, 1972)

Catalogs with Iranian records: Sedighi and Madjdzadeh (2015 as *Microtonus stelleri* Loan), Gadallah et al. (2016 as *M. stelleri*), Farahani et al. (2016 as *M. stelleri*), Yu et al. (2016), Samin, Coronado-Blanco, Kavallieratos et al. (2018), Samin, Coronado-Blanco, Fischer et al. (2018).

Distribution in Iran: Golestan (Samin et al., 2015), Guilan (Farahani, Talebi, & Rakhshani, 2013), Isfahan (Ghahari, Fischer, Hedqvist et al., 2010), Mazandaran (Ghahari, 2015).

Distribution in the Middle East: Iran (see references above), Israel—Palestine (Papp, 2012), Turkey (Koldas et al., 2007; Yilmaz et al., 2010).

Extralimital distribution: Bulgaria, Denmark, France, Germany, Hungary, Lithuania, Mongolia, Russia, Sweden, Switzerland, United States of America (introduced).

Host records: Summarized by Yu et al. (2016) as being a parasitoid of the curculionids *Hypera nigrirostris* (Fabricius), *Hypera postica* Gyllenhal, and *Hypera variabilis* (Herbst).

Tribe Pygostolini Belokobylskij, 2000

Genus *Pygostolus* Haliday, 1833

Pygostolus falcatus (Nees von Esenbeck, 1834)

Catalogs with Iranian records: This species is new record for the fauna of Iran.

Distribution in Iran: Guilan province, Lahijan, 2♂, July 2012, ex *Sitona humeralis* Stephens (Coleoptera: Curculionidae).

Distribution in the Middle East: Iran (new record).

Extralimital distribution: Nearctic, Palaearctic [Adjacent countries to Iran: Armenia, Azerbaijan, Kazakhstan, Russia].

Host records: Summarized by Yu et al. (2016) as being a parasitoid of the chrysomelid *Cryptocephalus punctatus* (L.); and the following curculionids: *Hypera variabilis* (Herbst), *Otiorhynchus ovatus* (L.), *Phyllobius* spp., *Polydrusus pilosus* Greller, *Sitona* spp., *Strophosoma capitatum* (DeGeer), and *Thylacites incanus* (L.). In Iran, this species has been reared from *Sitona humeralis* Stephens (Coleoptera: Curculionidae) (present work).

Pygostolus multiarticulatus (Ratzeburg, 1852)

Catalogs with Iranian records: No catalog.

Distribution in Iran: Northern Khorasan, Semnan (Samin, Papp, & Coronado-Blanco, 2018).

Distribution in the Middle East: Iran (Samin, Papp, & Coronado-Blanco, 2018), Turkey (Beyarslan, 2021).

Extralimital distribution: Belarus, Belgium, Bulgaria, Croatia, former Czechoslovakia, Finland, France, Germany, Hungary, Ireland, Italy, Japan, Moldova, Netherlands, Norway, Russia, Switzerland.

Host records: Summarized by Yu et al. (2016) as being a parasitoid of the curculionids *Otiorhynchus laevigatus* (Fabricius), *Otiorhynchus niger* (Fabricius) and *Thylacites incanus* (L.).

Pygostolus sticticus (Fabricius, 1798)

Catalogs with Iranian records: This species is new record for the fauna of Iran.

Distribution in Iran: Kuhgiloyeh and Boyerahmad province, Landeh (Sarasiab), 2♀, April 2013.

Distribution in the Middle East: Iran (new record), Turkey (Kolarov, 1989).

Extralimital distribution: Argentina, Armenia, Austria, Belgium, Bulgaria, Croatia, former Czechoslovakia, Denmark, Finland, France, Georgia, Germany, Greece, Hungary, Ireland, Italy, Kazakhstan, Latvia, Lithuania, Netherlands, Norway, Poland, Romania, Russia, Slovenia, Spain, Sweden, Switzerland, Tunisia, Ukraine, United Kingdom.

Host records: Summarized by Yu et al. (2016) as being a parasitoid of the curculionids *Barynotus moerens* (Fabricius), *Barynotus obscurus* (Fabricius), *Otiorhynchus* spp., and *Thylacites incanus* (L.).

Tribe Syntretini Shaw, 1985

Genus *Syntretus* Foerster, 1863

Syntretus (Syntretus) daghestanicus Tobias, 1976

Distribution in the Middle East: Turkey (van Achterberg & Haeselbarth, 2003; Yilmaz et al., 2010).

Extralimital distribution: Hungary, Italy, Moldova, Russia.

Host records: Unknown.

Syntretus (Syntretus) elegans (Ruthe, 1856)

Catalogs with Iranian records: Samin, Coronado-Blanco, Fischer et al. (2018).

Distribution in Iran: West Azarbaijan (Sakenin et al., 2018).

Distribution in the Middle East: Iran (Sakenin et al., 2018), Turkey (van Achterberg & Haeselbarth, 2003; Yilmaz et al., 2010).

Extralimital distribution: Austria, Azerbaijan, Bulgaria, former Czechoslovakia, Finland, France, Georgia, Germany, Hungary, Ireland, Italy, Kazakhstan, Lithuania, Moldova, Montenegro, Netherlands, Poland, Romania, Russia, Serbia, Sweden, Switzerland, Ukraine.

Host records: Unknown.

Syntretus (Syntretus) idalius (Haliday, 1833)

Catalogs with Iranian records: Sedighi and Madjdzadeh (2015), Gadallah et al. (2016), Farahani et al. (2016), Yu et al. (2016), Samin, Coronado-Blanco, Kavallieratos et al. (2018), Samin, Coronado-Blanco, Fischer et al. (2018).

Distribution in Iran: Lorestan (Ghahari et al., 2012).

Distribution in the Middle East: Cyprus (van Achterberg & Haeselbarth, 2003), Iran (Ghahari et al., 2012).

Extralimital distribution: Austria, Belgium, Bulgaria, former Czechoslovakia, Denmark, Finland, France, Georgia, Germany, Hungary, Ireland, Italy, Kazakhstan, Lithuania, Madeira Islands, Moldova, Mongolia, Netherlands, Norway, Poland, Romania, Sweden, Switzerland, Ukraine, United Kingdom.

Host records: Unknown.

Syntretus (Syntretus) ocularis van **Achterberg and Haeselbarth, 2003**

Catalogs with Iranian records: Sedighi and Madjdzadeh (2015), Gadallah et al. (2016), Farahani et al. (2016), Yu et al. (2016), Samin, Coronado-Blanco, Kavallieratos et al. (2018), Samin, Coronado-Blanco, Fischer et al. (2018).

Distribution in Iran: Guilan, Mazandaran (Farahani et al., 2012).

Distribution in the Middle East: Iran (Farahani et al., 2012), Turkey (van Achterberg & Haeselbarth, 2003; Yilmaz et al., 2010).

Extralimital distribution: Austria, Bulgaria, France, Germany, Hungary, Italy, Netherlands, Ukraine, United Kingdom.

Host records: Unknown.

Syntretus (Syntretus) xanthocephalus (Marshall, 1887)

Catalogs with Iranian records: Sedighi and Madjdzadeh (2015), Gadallah et al. (2016), Farahani et al. (2016), Yu et al. (2016), Samin, Coronado-Blanco, Kavallieratos et al. (2018), Samin, Coronado-Blanco, Fischer et al. (2018).

Distribution in Iran: Guilan (Farahani et al., 2012).

Distribution in the Middle East: Iran.

Extralimital distribution: Austria, Bulgaria, China, Denmark, Germany, Hungary, Ireland, Italy, Korea, Lithuania, Netherlands, Romania, Russia, Sweden, Switzerland, United Kingdom.

Host records: Recorded by Cole (1959) as being a parasitoid of adults of the ichneumonid wasp *Dirophanes invisor* (Thunberg).

Tribe Townesilitini Shaw, 1985

Genus *Marshiella* Shaw, 1985

Marshiella plumicornis (Ruthe, 1856)

Catalogs with Iranian records: Samin, Coronado-Blanco, Fischer et al. (2018).

Distribution in Iran: Chaharmahal and Bakhtiari (Samin, Coronado-Blanco, Fischer et al., 2018).

Distribution in the Middle East: Iran (Samin, Coronado-Blanco, Fischer et al., 2018), Turkey (Koldas et al., 2013; Yilmaz et al., 2010).

Extralimital distribution: Germany, Greece, Hungary, Kazakhstan, Moldova, United States of America.

Host records: Recorded by Görmitz (1937) as being a parasitoid of the anthicid *Notoxus monoceros* (L.). Specimens of *Marshiella* can be attracted to traps baited with cantharadin, as are its anthicid beetle hosts (Shaw, 1985; Shaw & Marsh, 2000; Young, 1984a,b).

Genus *Streblocera* Westwood, 1833

Streblocera (Streblocera) antennata Jakimavicius, 1973

Distribution in the Middle East: Turkey (Koldas et al., 2013).

Extralimital distribution: Germany, Lithuania.

Host records: Unknown.

Streblocera (Streblocera) fulviceps Westwood, 1833

Catalogs with Iranian records: Samin, Coronado-Blanco, Fischer et al. (2018).

Distribution in Iran: Khuzestan (Samin, Coronado-Blanco, Fischer et al., 2018).

Distribution in the Middle East: Iran (Samin, Coronado-Blanco, Fischer et al., 2018), Turkey (Yilmaz et al., 2010).

Extralimital distribution: China, Germany, Russia, United Kingdom.

Host records: Recorded by He (1984) as being a parasitoid of the chrysomelid *Chaetocnema cilindrica* Baly.

Streblocera (Eutanycerus) macroscapus (Ruthe, 1856)

Catalogs with Iranian records: This species is new record for the fauna of Iran.

Distribution in Iran: Hamadan province, Nahavand (Hadi-Abad), 2♀, April 2012.

Distribution in the Middle East: Iran (new record), Turkey (Koldas et al., 2013; Yilmaz et al., 2010).

Extralimital distribution: Bulgaria, Czech Republic, Finland, germany, Hungary, Kazkhstan, Korea, Lithuania, Mongolia, Netherlands, Poland, Russia, Sweden, Switzerland, United Kingdom.

Host records: Unknown.

Genus *Townesilitus* Haeselbarth and Loan, 1983

Townesilitus aemulus (Ruthe, 1856)

Catalogs with Iranian records: No catalog.

Distribution in Iran: Mazandaran (Ghahari & Gadallah, 2022).

Distribution in the Middle East: Iran (Ghahari & Gadallah, 2022), Turkey (Güclü & Özbek, 2011).

Extralimital distribution: Austria, former Czechoslovakia, Germany, Hungary, Ireland, Italy, Japan, North Macedonia, Netherlands, Russia, Slovenia, Spain, Switzerland, United Kingdom, former Yugoslavia.

Host records: Recorded by Watanabe (1955) as being a parasitoid of the chrysomelid *Psylliodes punctifrons* Baly. In Iran, this species has been reared from the chrysomelid *Psylliodes cuprea* (Koch) (Ghahari & Gadallah, 2022).

Townesilitus bicolor (Wesmael, 1835)

Catalogs with Iranian records: Sedighi and Madjdzadeh (2015), Gadallah et al. (2016), Farahani et al. (2016), Yu et al. (2016), Samin, Coronado-Blanco, Kavallieratos et al. (2018), Samin, Coronado-Blanco, Fischer et al. (2018).

Distribution in Iran: Golestan (Samin et al., 2015 as *Perilitus (Townesilitus) bicolor* (Wesmael); Ghahari, 2015), Guilan (Farahani, Talebi, & Rakhshani, 2013), Isfahan (Ghahari et al., 2011), Mazandaran (Farahani, Talebi, & Rakhshani, 2013, 2014), Semnan (Samin et al., 2011).

Distribution in the Middle East: Iran (see references above), Turkey (Koldas et al., 2013; Yilmaz et al., 2010).

Extralimital distribution: Albania, Armenia, Austria, Azerbaijan, Belgium, Bulgaria, Croatia, Czech Republic, Denmark, France, Germany, Greece, Hungary, Ireland, Italy, Kazakhstan, Latvia, Lithuania, Moldova, Netherlands, Norway, Poland, Romania, Russia, Serbia, Spain, Sweden, Switzerland, United Kingdom, Uzbekistan.

Host records: Summarized by Yu et al. (2016) as being a parasitoid of the following chrysomelids: *Aphthona euphorbiae* (Schrank), *Aphthona violacea* (Koch), *Chaetocnema aridula* Gyllenhal, *Chaetocnema hortensis* Geoffroy, *Longitarsus ballotae* (Marsham), and *Phyllotreta* spp.

Townesilitus deceptor (Wesmael, 1835)

Distribution in the Middle East: Turkey (Güclü & Özbek, 2011).

Extralimital distribution: Belgium, China, Czech Republic, Hungary, Ireland, Italy, Japan, Korea, Moldova, Netherlands, Poland, Russia, Sweden, Switzerland, Ukraine, United Kingdom.

Host records: Summarized by Yu et al. (2016) as being a parasitoid of the chrysomelids *Altica deserticola* (Weice), *Altica quercetorum* Foudras, and *Melasoma aenea* (L.).

Conclusion

One hundred and twenty valid species of the subfamily Euphorinae in 22 genera and 11 tribes (Centistini, Dinocampini, Euphorini, Helorimorphini, Meteorini, Myiocephalini, Neoneurini, Perilitini, Pygostolini, Syntretini, and Townesilitini) have been reported from eight of the Middle East countries: Cyprus, Egypt, Iran, Iraq, Israel—Palestine, Saudi Arabia, Syria, and Turkey, of which Iran with 92 species in 21 genera (7.2% of the world species) (Fig. 13.1) is the richest. It was found that the most species rich genera in the Euphorinae of the Middle East are

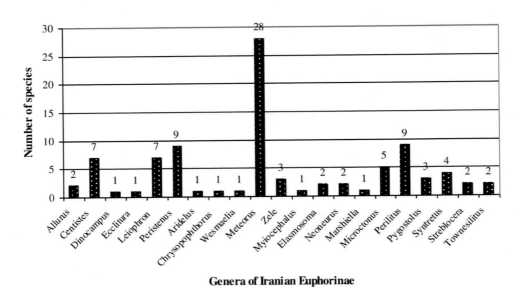

FIGURE 13.1 Number of Euphorinae species within genera known from Iran.

Meteorus (with 35 species), followed by *Perilitus* (18 species), *Peristenus* (12 species), *Leiophron* (eight species), *Centistes* (seven species), *Microctonus* and *Syntretus* (each with five species), *Elasomosoma*, and *Zele* (each with four species), *Pygostolus*, *Streblocera*, and *Townesilitus* (each with three species), *Allurus*, *Aridelus*, and *Neoneurus* (each with two species), and *Chrysopophthorus*, *Dinocampus*, *Ecclitura*, *Kollasmosoma*, *Marshiella*, *Myiocephalus*, and *Wesmaelia* (each with one species). Three euphorine species are known to be endemic to the Middle East fauna, *Elasmosoma geylanae* Beyarslan (2012, 2016) (Turkey), *Meteorus arabica* Ghramh (2012) (Saudi Arabia), and *Meteorus brevitrebratus* Ameri, Talebi and Beyarslan (2016) (Iran).

Distributions of the Iranian Euphorinae have been recorded from 27 provinces which among them, Guilan and Mazandaran (both with 24 species) have the highest number of species (Fig. 13.2). Host species have been detected in Iran for 15 Euphorinae species, and these hosts belong to three orders, Coleoptera (Cerambycidae: one species; Chrysomelidae: two species; Ciidae: two species; Coccinellidae: two species; Curculionidae: two species), Hemiptera (Miridae: two species; Pentatomidae: one species), and Lepidoptera (Erebidae: two species; Gelechiidae: one species; Noctuidae: four species). Comparison of the Euphorinae fauna of Iran with Middle East and adjacent countries indicates the fauna of Russia with 281 recorded species in 29 genera (Belokobylskij & Lelej,

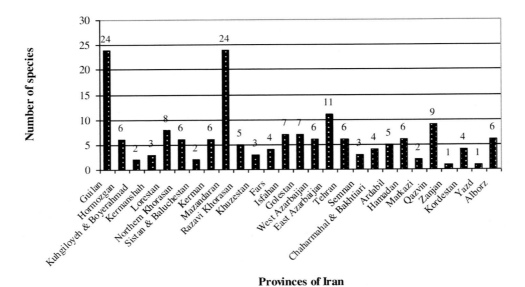

Provinces of Iran

FIGURE 13.2 Number of reported species of Iranian Euphorinae by province.

2019) is more diverse than Iran; followed by Turkey (with 91 species), Kazakhstan (53 species), Azerbaijan (34 species), Armenia (22 species), Israel—Palestine (20 species), Cyprus (10 species), Egypt (six species), Turkmenistan (five species), Saudi Arabia (two species), Afghanistan, Iraq, and Syria (each with one species). No species have been recorded from Bahrain, Jordan, Kuwait, Lebanon, Oman, Pakistan, Qatar, United Arab Emirates, Yemen (Yu et al., 2016). Among the Middle East and adjacent countries to Iran, Russia shares the largest number of species with Iran (with 68 species), followed by Turkey (56 species), Kazakhstan (37 species), Azerbaijan (31 species), Armenia (19 species), Israel—Palestine (16 species), Cyprus (nine species), Turkmenistan (five species), Egypt (four species), Afghanistan, Iraq, and Syria (each with one species).

References

Abbasipour, H. (2001). Report of endoparasitoid wasp, *Meteorus gyrator* (Thunberg) (Hym.: Braconidae) on rice armyworm, *Mythimna unipuncta* (Howarth) from Iran. *Journal of Entomological Society of Iran, 20*(2), 101—102.

Abbasipour, H., & Taghavi, A. H. (2002). Parasitoids of rice armyworm, *Mythimna unipuncta* (Haworth) (Lepidoptera: Noctuidae) in the rice fields of western Mazandaran. In *Proceedings of the 15th Iranian Plant Protection Congress, 7—11 September 2002* (p. 24). Razi University of Kermanshah.

Abbasipour, H., & Taghavi, A. H. (2004). Introduction of the rice armyworm, *Mythimna unipuncta* (Haworth) (Lepidoptera: Noctuidae) parasitoids in western Mazandaran rice fields and preliminary study on their efficiency to control the pest. *Journal of Agricultural Sciences, 1*(1), 19—28 (in Persian, English summary).

Abbasipour, H., Amini, A., & Taghavi, A. (2004). Evaluation of parasitoids efficiency to control the cereal armyworm, *Mythimna unipuncta* (Haworth) (Lep.: Noctuidae) in the rice fields of Iran. In *Proceedings of the 16th Iranian Plant Protection Congress, 28 August — 1 September 2004* (p. 86). University of Tabriz.

Abdinbekova, A. A. (1975). *Brakonidi (Hymenoptera: Braconidae) Azerbaijana*. Azerbaijan, Baku: Akademii Nauk Aserbaid SSR, Institute of Zoology, ELM, 323 pp. (in Russian).

Abdinbekova, A. A. (1995). *Azerbaijinin zarganadli jujuleri (Hymenoptera: Braconidae)*. Baku, 470 pp.

Abdolalizadeh, F., Madjdzadeh, S. M., Farahani, S., & Askari Hesni, M. (2017). A survey of braconid wasps (Hymenoptera: Braconidae: Euphorinae, Homolobinae, Macrocentrinae, Rogadinae) in Kerman province, southeastern Iran. *Journal of Insect Biodiversity and Systematics, 3*(1), 33—40.

Achtelig, M. (1974). Beschreibung des Männchens von *Meteorus pachypus* (Braconidae, Hymenoptera); ein unbekannter Parasit von *Raphidia xanthostigma* Schummel (Raphidioptera). *Nachrichtenblatt der Bayerischen Entomologen, 23*(1), 1—5.

Aguirre, H., Shaw, S. R., Berry, J. A., & de Sassi, C. (2014). Description and natural history of the first micropterous *Meteorus* species: *M. orocrambivorus* sp. n. (Hymenoptera, Braconidae, Euphorinae),

endemic to New Zealand. *Journal of Hymenoptera Research, 38*, 45—57.

Aguirre, H., Almeida, L. F. de, Shaw, S. R., & Sarmiento, C. E. (2015). An illustrated key to neotropical species of the genus *Meteorus* Haliday (Hymenoptera: Braconidae: Euphorinae). *ZooKeys, 489*, 33—94.

Alimohammadi, N., Samih, M. A., & Izadi, H. (2012). First report of *Dinocampus coccinella* (Hym.: Braconidae) as a parasitoid of *Hippodamia variegata* (Col.: Coccinellidae). In *Proceedings of the 17th National and 5th International Iranian Biology Conference, Kerman, Iran* (p. 47).

Alizadeh, S., & Javan Moghaddam, H. (2004). Introduction of some natural enemies of common cutworm (*Agrotis segetum* Schiff). In Miyandoab. *Proceedings of the 16th Iranian Plant Protection Congress, 28 August — 1 September 2004* (p. 45). University of Tabriz.

Ameri, A., Talebi, A. A., Rakhshani, E., Beyarslan, A., & Kamali, K. (2014). A survey of Euphorinae (Hymenoptera: Braconidae) of southern Iran, with description of a new species. *Zootaxa, 3900*(3), 415—428.

Arbab, A. (2011). Introducing *Microctonus aethiopoides* Loan (Hym.: Braconidae) parasitoid of alfalfa root and leaves weevils and some of its behavioural and biological characters. In *Proceedings of the biological control Development Congress in Iran, 27-28 July 2011* (p. 479). Tehran: Iranian Research Institute of Plant Protection.

Arbab, A., & McNiell, M. (2001). New report of *Microctonous aethiopoides* (Hym.: Braconidae) in Iran. *Journal of Entomological Society of Iran, 21*(1), 111—112.

Aubert, J. F. (1966). Liste d'identification No.6 (presentée par le service d'identification des Entomophages). *Entomophaga, 11*(1), 115—134.

Aydoğdu, M. (2014). Parasitoid abundance of *Archips rosana* (Linnaeus, 1758) (lepidoptera: Tortricidae) in organic cherry orchards. *North Western Journal of Zoology, 10*(1), 42—47.

Bagheri, M. R. (1998). The first report of *Perilitus coccinellae* (Hym.: Braconidae) a parasitoid of *Coccinella septempunctata* in Isfahan. In *Proceedings of the 13th Iranian Plant Protection Congress, 23—27 August 1998* (p. 200). Karaj Junior College of Agriculture.

Balevski, N. A. (1989). Species composition and hosts of family Braconidae (Hymenoptera) in Bulgaria. *Acta Zoologica Bulgarica, 38*, 24—45.

Bartlett, B. R., Clausen, C. P., DeBach, P., Goeden, R. D., Legner, E. F., McMurtry, J. A., & Oatman, E. R. (1978). *Introduced parasites and predators of arthropod pests and weeds: a world review* (p. 545). Agricultural Research Service. United States Department of Agriculture Handbook, No. 480.

Belokobylskij, S. A. (1992). Revision of the genus *Centistes* Haliday (Hymenoptera: Braconidae: Euphorinae) of the USSR far East and neighbouring territories. *Zoologische Mededelingen, 66*, 199—237.

Belokobylskij, S. A. (1993). *The braconids of the genus Leiophron (Leiophron) Nees (Hymenoptera: Braconidae: Euphorinae) of the fauna of the Russian far East* (Vol. 251, pp. 61—100). Leningrad: Trudy Zoologicheskogo Instituta (in Russian).

Belokobylskij, S. A. (1995). A new genus and ten species of the subfamily Euphorinae (Hymenoptera: Braconidae) from the Russian Far East. *Zoosystematica Rossica, 3*(2), 293—312.

Belokobylskij, S. A. (1996). New and rare species of the subfamily Euphorinae (Hymenoptera, Braconidae) from the Russian far East. *Zoologische Mededlingen, 70*(20), 275—296.

Belokobylskij, S. A., & Lelej, A. S. (2019). Annotated catalogue of the Hymenoptera of Russia. Volume II. Apocrita: Parasitica. In *Proceedings of the Zoological Institute of the Russian Academy of Sciences, Supplement No. 8* (p. 594).

Belokobylskij, S. A., Loni, A., Lucchi, A., & Bernardo, U. (2013). First records of the genera *Histeromerus* Wesmael (Hymenoptera: Braconidae: Histerominae) and *Ecclitura* Kokujev (Hymenoptera: Braconidae: Euphorinae) in Italy. *ZooKeys, 310*, 29–40.

Belshaw, R., & Quicke, D. L. J. (2002). Robustness of ancestral state estimates: Evolution of life history strategy in ichneumonoid parasitoids. *Systematic Biology, 50*(3), 450–477.

Beyarslan, A. (2016). A new species of *Elasmosoma* Ruthe, 1858 from Turkey (Hymenoptera, Braconidae, Euphorinae). *Journal of the Entomological Research Society, 18*(1), 113–118.

Beyarslan, A. (2021). New species of Euphorinae (Hymenoptera, Braconidae) for the fauna of Turkey. *Acta Biologica Turcica, 34*(1), 38–45.

Beyarslan, A., Aydoğdu, M., & Inanç, F. (2004). A survey of *Meteorus* Haliday, 1835 of Turkey (Hymenoptera, Braconidae, Euphorinae). *Entomofauna, 25*(1), 1–20.

Beyarslan, A., Gozuacik, C., & Ozgen, I. (2013). A contribution on the subfamilies Helconinae, Homolobinae, Macrocentrinae, Meteorinae, and Orgilinae (Hymenoptera: Braconidae) of southeastern Anatolia with new records from other parts of Turkey. *Turkish Journal of Zoology, 37*(4), 501–505.

Beyarslan, A., Gözüaçiki, C., Güllü, M., & Konuksal, A. (2017). Taxonomical investigation on Braconidae (Hymenoptera: Ichneumonoidea) fauna in northern Cyprus, with twenty-six new records for the country. *Journal of Insect Biodiversity and Systematic, 3*(4), 319–334.

Biranvand, A., Nedvĕd, O., Karimi, S., Vahedi, H., Hesami, S., Lotfalizadeh, H., Ajamhasani, M., & Ceryngier, P. (2020). Parasitoids of the ladybird beetles (Coleoptera: Coccinellidae) in Iran: An update. *Annales de la Société Entomologique de France, 56*(2), 106–114.

Brues, C. I. (1926). Studies on the Ethiopian Braconidae with a catalogue of the African species. *Proceedings of the American Academy of Arts and Sciences, 6*, 206–436.

Čapek, M., & Hofmann, C. (1997). The Braconidae (Hymenoptera) in the collections of the Musée Cantonal de Zoologie. *Lausanne Litterae Zoologica (Laussane), 2*, 25–162.

Chen, X. X., & van Achterberg, C. (1997). Revision of the subfamily Euphorinae (excluding the tribe Meteorini Cresson) (Hymenoptera: Braconidae) from China. *Zoologische Verhandelingen, 313*, 1–217.

Chen, X. X., & van Achterberg, C. (2019). Systematics, phylogeny, and evolution of braconid wasps: 30 years of progress. *Annual Review of Entomology, 64*, 1–24.

Çikman, E., & Beyarslan, A. (2009). Four new parasitoid records of the subfamilies Euphorinae and Opiinae (Hymenoptera: Braconidae) from the Adiyaman province of Turkey. *Turkish Journal of Zoology, 33*(3), 367–370.

Cole, L. R. (1959). On a new species of *Syntretus* Foerster (Hymenoptera, Braconidae) parasitic on an adult ichneumonid, with a description of the larva and notes on its life history and that of its hosts, *Phaeogenes invisor* (Thunberg). *Entomologist's Monthly Magazine, 95*, 18–21.

Darsouei, R., & Karimi, J. (2015). Natural enemies of sugar beet armyworm, *Spodoptera exigua* (Lep.: Noctuidae). In *Proceedings of the 1st Iranian International Congress of Entomology, 29.31 August 2015* (p. 98). Tehran: Iranian Research Institute of Plant Protection.

Davatchi, A., & Shojai, M. (1969). *Les hymenopteres entomophages de l'Iran (Etudes Faunestiques)* (Vol. 107, pp. 1–88). Tehran University, Faculty of Agriculture.

Donisthorpe, H. (1927). *The guests of British ants. London* (p. 244) (Ichneumonoidea in pp. 85–91).

Drea, J. J. (1968). A new species of *Microctonus* (Hymenoptera: Braconidae) parasitizing the alfalfa weevil. *Entomological News, 79*, 97–102.

Drea, J. J., Dureseau, L., & Rivet, E. (1973). Biology of *Peristenus stygicas* from Turkey, a potential natural enemy of *Lygus* bugs in North America. *Environmenta Entomology, 2*(2), 278–280.

Drogvalenko, A. N., & Ghahari, H. (2021). An annotated checklist of Ciidae (Coleoptera: Tenebrionoidea) of Iran. *Zootaxa, 4981*(2), 317–330.

Efil, L., & Beyarslan, A. (2009). Factors affecting distribution of two mirid bugs, *Creontiades pallidus* (Rambur) (Hemiptera: Miridae) and notes on on the parasitoid *Leiophron deficiens* Ruthe (Hymenoptera: Braconidae). *Entomologica Fennica, 20*(1), 9–17.

Fahringer, J. (1922). Hymenopterologische Ergebnisse einer wissenschaftlichen Studienreise nach der Türkei und Kleinasien (mit Ausschluβ des Amanusgebirges). *Archiv für naturgeschichte, A88*(9), 149–222.

Fallahzadeh, M., & Saghaei, N. (2010). Checklist of Braconidae (Insecta: Hymenoptera) from Iran. *Munis Entomology & Zoology, 5*(1), 170–186.

Farahani, S., & Talebi, A. A. (2012). A review of the tribe Meteorini (Cresson, 1887) (Hymenoptera: Braconidae, Euphorinae) in northern Iran, with eight new records. *Iranian Journal of Animal Biosystematics, 8*(2), 135–157.

Farahani, S., Talebi, A. A., van Achterberg, C., & Rakhskani, E. (2012). *Syntrenus*, a genus of Euphorinae (Hymenoptera: Braconidae) new for Iran, with first record of two species. *Journal of Crop Protection, 1*(3), 173–179.

Farahani, S., Talebi, A. A., van Achterberg, C., & Rakhshani, E. (2013). Three new records of the genera *Leiophron* and *Euphorus* (Hym.: Braconidae: Euphorinae) from Iran. *Journal of Entomological Society of Iran, 33*(2), 73–79.

Farahani, S., Talebi, A. A., & Rakhshani, E. (2013). A contribution to the knowledge of Euphorinae (Hymenoptera: Braconidae), with six new records from Iran. *Journal of Entomological and Acarological Research, 45*, 43–51.

Farahani, S., Talebi, A. A., & Rakhskani, E. (2014). First report of *Elasmosoma luxembergense* Wasmann, 1909 (Hymenoptera: Braconidae: Euphorinae) from Iran. In *Proceedings of the 21st Iranian Plant Protection Congress, 23–26 August 2014* (p. 496). Urmia University.

Farahani, S., Talebi, A. A., & Rakhshani, E. (2016). Iranian Braconidae (Insecta: Hymenoptera: Ichneumonoidea): Diversity, distribution and host association. *Journal of Insect Biodiversity and Systematics, 2*(1), 1–92.

Fox, A. S., Shaw, S. R., Dosdall, L. M., & Lee, B. (2004). *Microctonus melanopus* (Ruthe) (Hymenoptera: Braconidae), a parasitoid of adult cabbage seedpod weevil (Coleoptera: Curculionidae): Distribution in Southern Alberta and female diagnosis. *Journal of Entomological Science, 39*(3), 350–361.

Fulmek, L. (1968). Parasitinsekten der Insektengallen Europas. *Beiträge zur Entomologie, 18*(7/8), 719—952.

Gadallah, N. S., Ghahari, H., & van Achterberg, C. (2016). An annotated catalogue of the Iranian Euphorinae, Gnamptodontinae, Helconinae, Hormiinae and Rhysipolinae (Hymenoptera: Braconidae). *Zootaxa, 4072*(1), 1—38.

Gadallah, N. S., Ghahari, H., Papp, J., & Beyarslan, A. (2018). New records of Braconidae (Hymenoptera) from Iran. *Wuyi Science Journal, 34*, 43—48.

Gadallah, N. S., Ghahari, H., & Quicke, D. L. J. (2021). Further addition to the braconid fauna of Iran (Hymenoptera: Braconidae). *Egyptian Journal of Biological Pest Control, 31*, 32. https://doi.org/10.1186/s41938-021-00376-8

Ghahari, H. (2015). A faunistic study on the subfamily Euphorinae (Hymenoptera: Ichneumonoidea, Braconidae) from Iran. *Arquivos Entomoloxicos, 14*, 149—156.

Ghahari, H. (2016). Five new records of Iranian Braconidae (Hymenoptera: Ichnemonoidea) for Iran and annotated catalogue of the subfamily Homolobinae. *Wuyi Science Journal, 32*, 35—43.

Ghahari, H. (2017). Species diversity of Ichneumonoidea (Hymenoptera) from rice fields of Mazandaran province, northern Iran. *Journal of Animal Environment, 9*(3), 371—378 (in Persian, English summary).

Ghahari, H. (2018). Species diversity of the parasitoids in rice fields of northern Iran, especially parasitoids of rice stem borer. *Journal of Animal Environment, 9*(4), 289—298 (in Persian, English summary).

Ghahari, H. (2019). Study on the natural enemies of agricultural pests in some sugar beet fields, Iran. *Journal of Sugar Beet, 35*(1), 91—102 (in Persian, English summary).

Ghahari, H. (2020). A study on the fauna of predator and parasitoid arthropod of saffron fields (*Crocus sativus* L.). *Journal of Saffron Research, 7*(2), 203—215 (in Persian, English summary).

Ghahari, H., & Gadallah, N. S. (2022). Additional records to the braconid fauna (Hymenoptera: Ichneumonoidea) of Iran, with new host reports. *Entomological News* (in press).

Ghahari, H., Fischer, M., Erdoğan, Ö.Ç., Tabari, M., Ostovan, H., & Beyarslan, A. (2009). A contribution to Braconidae (Hymenoptera) from rice fields and surrounding grasslands of northern Iran. *Munis Entomology & Zoology, 4*(2), 432—435.

Ghahari, H., Fischer, M., Erdoğan, Ö.Ç., Beyarslan, A., & Havaskary, M. (2009). A contribution to the knowledge of the braconid-fauna (Hymenoptera, Ichneumonoidea, Braconidae) of Arasbaran, northwestern Iran. *Entomofauna, 30*, 329—336.

Ghahari, H., Gadallah, N. S., Erdoğan, Ö.Ç., Hedqvist, K. J., Fischer, F., Beyarslan, A., & Ostovan, H. (2009). Faunistic note on the Braconidae (Hymenoptera: Ichneumonoidea) in Iranian cotton fields and surrounding grasslands. *Egyptian Journal of Biological Pest Control, 19*(2), 115—118.

Ghahari, H., Fischer, M., Erdoğan, Ö.Ç., Beyarslan, A., & Ostovan, H. (2010). A contribution to the braconid wasps (Hymenoptera: Braconidae) from the forests of northern Iran. *Linzer biologische Beiträge, 42*(1), 621—634.

Ghahari, H., Fischer, M., Hedqvist, K. J., Erdoğan, Ö.Ç., van Achterberg, C., & Beyarslan, A. (2010). Some new records of Braconidae (Hymenoptera) for Iran. *Linzer biologische Beiträge, 42*(2), 1395—1404.

Ghahari, H., Erdoğan, Ö.Ç., Šedivý, J., & Ostovan, H. (2010). Survey of the Ichneumonoidea and Chalcidoidea (Hymenoptera) parasitoids of Saturniidae (Lepidoptera) in Iran. *Efflatounia, 10*, 1—6.

Ghahari, H., & Fischer, M. (2011a). A contribution to the Braconidae (Hymenoptera: Ichneumonoidea) from north-western Iran. *Calodema, 134*, 1—6.

Ghahari, H., & Fischer, M. (2011b). A study on the Braconidae (Hymenoptera: Ichneumonoidea) from some regions of northern Iran. *Entomofauna, 32*(8), 181—196.

Ghahari, H., Fischer, M., & Papp, J. (2011). A study on the braconid wasps (Hymenoptera: Braconidae) from Isfahan province, Iran. *Entomofauna, 32*(16), 261—272.

Ghahari, H., Fischer, M., Papp, J., & Tobias, V. (2012). A contribution to the knowledge of braconids (Hymenoptera: Braconidae) from Lorestan province Iran. *Entomofauna, 7*, 65—72.

Ghahari, H., & Gadallah, N. S. (2019). New records of braconid wasps (Hymenoptera: Ichneumonoidea: Braconidae) from Iran. *Entomological News, 129*(3), 384—392.

Ghramh, H. A. (2012). A new species *Meteorus arabica* sp. nov. (Hymenoptera: Braconidae) from Saudi Arabia. *Egyptian Academic Journal of Biological Science, 5*(2), 117—120.

Gómez Durán, J.-M., & van Achterberg, C. (2011). Oviposition behaviour of four ant parasitoids (Hymenoptera, Braconidae, Euphorinae, Neoneurini and Ichneumonidae, Hybrizontinae), with the description of three new European species. *ZooKeys, 125*, 59—106.

Görmitz, K. (1937). Cantharadin als Gift- und Anlockungsmittel für Insekten. *Arbeiten über Physiologische und Angewandte Entomologie, 4*, 116—157.

Güçlü, C., & Özbek, H. (2011). A contribution to the knowledge of Euphorinae (Hymenoptera: Braconidae) from Turkey. *Journal of the Entomological Research Society, 13*(2), 61—70.

Gyoerfi, J. (1959). Beiträge zur Kenntnis der Wirte verschiedener Braconiden-Arten (Hymenoptera, Braconidae). *Acta Zoologica Hungarica, 5*, 49—65.

Haeselbarth, E. (1983). Determination list of entomophagous insects. 9. *Bulletin Section regionale Ouest Palaearctique, Organisation Internationale de Lutte Biologique, 6*(1), 22—23.

Haeselbarth, E. (1999). Zur Braconiden-Gattung *Perilitus* Nees, 1818. 2. Beitrag Die Arten mit ausgebildetem ersten Cubitus-Abschnitte (Insecta, Hymenoptera, Braconidae). *Mitteilungen Munchener Entomologischen Gesellschaft, 89*, 11—46.

Haeselbarth, E. (2008). Zur Braconiden-Gattung *Perilitus* Nees, 1818. 3. Beitrag die Arten ohne ausgebildeten ersten Cubitus- Abscnitt (Hymenoptera, Braconidae). *Linzer biologische Beiträge, 40*(2), 1013—1152.

Hauser, E. (1994). Ökologie der Parasitoide von *Taleporia tubulosa* (Hymenoptera: Ichneumonoidea/Lepidoptera: Psychidae). *Entomologia Generalis, 18*(3—4), 227—233.

Haye, T., van Achterberg, C., Goulet, H., Barratt, B. T. P., & Kuhlmann, U. (2006). Potential for classical biological control of the potato bug *Closterotomus norwegicus* (Hemiptera: Miridae): Description, parasitism and host specificity of *Peristenus closterotomae* sp. n. (Hymenoptera: Braconidae). *Bulletin of Entomological Research, 96*, 421—431.

He, J. H. (1984). Six new species records of the Braconidae (Hymenoptera) to China. *Acta Agriculturae Universitatis Zhejiangensis, 1012*, 199—205 (in Chinese with English summary).

Hedwig, K. (1957). Ichneumoniden und Braconiden aus Iran 1954 (Hymenoptera). *Jharschefte des Vereins für Vaterländische Naturkunde Württemberg, 112*, 103—117.

Herard, F., Mercadier, G., & Abai, M. (1979). Situation de *Lymantria dispar* (Lep.: Lymantriidae) et son complexe parasitaire en Iran, en 1976. *Entomophaga, 24*(4), 371−384.

Huddleston, T. (1976). A revision of *Elasmosoma* Ruthe (Hymenoptera: Braconidae) with two new species from Mongolia. *Annales Historico-Naturalis Musei Nationalis Hungarici, 68*, 215−225.

Huddleston, T. (1980). A revision of the western Palaearctic species of the genus *Meteorus* (Hymenoptera: Braconidae). *Bulletin of the British Museum (Natural History), Entomology series, 41*(1), 1−58.

Huddleston, T., & Short, J. R. T. (1978). A new genus of Euphorinae (Hymenoptera: Braconidae) from Australia with a description of final instar larva of one species. *Journal of the Australian Entomological Society, 17*, 317−321.

Ingram, W. R. (1981). The parasitoids of *Spodoptera littoralis* (Lep.: Noctuidae) and their role in the population control in Cyprus. *Entomophaga, 26*, 23−37.

Jacobs, J. C. (1904). Hyménoptères parasites obtenus de quelques nymphes de microlépidoptères et d'autres nymphes par Mile Baron de Crombrugghe. *Annales de la Société Entomologique de Belgique, 48*, 308.

Kamal Bey, M. (1951). The biological control of the cotton leaf-worm (*Prodenia litura* F.) in Egypt. *Bulletin de la Société Fouad 1er d'Entomologie, 35*, 221−270.

Khanjani, M. (2006a). *Field crop pests in Iran* (3rd ed., p. 719). Bu-Ali Sina University (in Persian).

Khanjani, M. (2006b). *Vegetable pests in Iran* (2nd ed., p. 467). Bu-Ali Sina University (in Persian).

Kian, N., Goldasteh, S., & Farahani, S. (2020). A survey on abundance and species diversity of Braconid wasps in forest of Mazandaran province. *Journal of Entomological Research, 12*(1), 61−69 (in Persian, English summary).

Kolarov, J. A. (1989). Ichneumonidae (Hymenoptera) from Balkan Peninsula and some adjacent regions. II. Lissonotinae, Ctenopelmatinae, Tersilochinae, Cremastinae and Campopleginae. *Turkiye Entomoloji Dergisi, 13*(2), 67−84.

Koldaş, T., Aydoğdu, M., & Beyarslan, A. (2007). Euphorinae (Hymenoptera: Braconidae) fauna from the Thrace Region of Turkey. *Linzer biologische Beiträge, 39*(1), 441−450.

Koldaş, T., Aydoğdu, M., & Beyarslan, A. (2013). New taxonomic and faunistic data on the subfamily Euphorinae Förster, 1862 of Turkey (Hymenoptera: Braconidae). *Journal of the Entomological Research Society, 15*(2), 21−35.

Lashkari Bod, A., Rakhshani, E., Talebi, A. A., & Lozan, A. (2010). Introduction of twelve newly recorded species of Braconidae (Hymenoptera) from Iran. In *Proceedings of the 19th Iranian Plant Protection Congress, 31 July − 3 August 2010* (p. 161). Iranian Research Institute of Plant Protection.

Lashkari Bod, A., Rakhshani, E., Talebi, A. A., Lozan, A., & Žikič, V. (2011). A contribution to the knowledge of Braconidae (Hym., Ichneumonoidea) of Iran. *Bihorean Biologist, 5*(2), 147−150.

Li, F. F., Chen, X. X., Piao, M. H., He, J. H., & Ma, Y. (2003). Phylogenetic relationships of the Euphorinae (Hymenoptera: Braconidae) based on the D2 variable region of 28S ribosomal RNA. *Entomotaxonomia, 25*(3), 217−226.

Loan, C. C. (1967). Studies on the taxonomy and biology of the Euphorinae (Hymenoptera: Braconidae). II. Host relations of six *Microctonus* species. *Annals of the Entomological Society of America, 60*(1), 236−240.

Loan, C. C. (1979). Three new species of *Peristenus* Foerster from Canada and western Europe (Hymenoptera: Braconidae, Euphorinae). *Le Naturaliste, 106*, 387−391.

Loan, C. C., & Bilewicz-Pawinsta, T. (1973). Systematics and biology of four Polish species of *Peristenus* Foerster (Hymenoptera: Braconidae: Euphorinae). *Environmental Entomology, 2*(2), 271−278.

Marsh, P. M. (1979). Braconidae. Aphidiidae. Hybrizontidae, pp. 144−313. In K. V. Krombein, P. D. Hurd, D. R. Smith, & B. D. Burks (Eds.), *Catalog of Hymenoptera in America north of Mexico. Volume 1: Symphyta and Apocrita (Parasitica)* (p. xvi+1198). Washington: Smithsinian Institution Press.

Martikainen, P., & Koponen, M. (2001). *Meteorus corax* Marshall, 1898 (Hymenoptera: Braconidae), a new species to Finland and Russian Karelia, with an overview of northern species of *Meteorus* parasitizing beetles. *Entomologica Fennica, 12*, 169−172.

Mirabzadeh, A. (1968). Investigation of integrated control against alfalfa weevil (*Hypera postica*). In *Proceedings of the 8th Iranian Plant Protection Congress, 30 August − 4 September 1986, Isfahan* (p. 28).

Modarres Awal, M. (1997). Family Braconidae (Hymenoptera), pp. 265−267. In M. Modarres Awal (Ed.), *List of agricultural pests and their natural enemies in Iran* (2d ed., p. 429). Ferdowsi University of Mashhad Press.

Modarres Awal, M. (2012). Family Braconidae (Hymenoptera), pp. 483−486. In M. Modarres Awal (Ed.), *List of agricultural pests and their natural enemies in Iran* (3rd ed., p. 759). Ferdowsi University of Mashhad Press.

Muesebeck, C. F. W. (1950). Some braconid parasites of the pink bollworm *Pectinophora gossypiella* (Saunders). *Bollettino del Laboratoria di Zoologia Generale e Agraria. Portici, 33*, 57−68.

Naderian, H., Penteado-Dias, A. M., Sakenin Chelav, H., & Samin, N. (2020). A faunistic study on Braconidae and Ichneumonidae (Hymenoptera, Ichneumonoidea) of Iran. *Calodema, 844*, 1−9.

New, T. R. (1970). The life histories of two species of *Leiophron* Nees (Hymenoptera, Braconidae) parasitic on Psocoptera in southern England. *Entomologist's Gazette, 21*, 39−48.

Nickdel, M., Sadeghian, B., & Dordaei, M. (2004). Collection and identification of brown-tail moth's natural enemies in Arasbaran forest. In *The Joint Agriculture and Natural Resources Symposium* (pp. 1−5).

Nickdel, M., Sadeghian, B., Dordaei, M., & Askari, H. (2008). Identification, distribution and evaluation of natural enemies associated with *Euproctis chrysorrhoea* (Lep.: Lymantriidae) in Arasbaran forests of Iran. *Iranian Journal of Forest and Range Protection Research, 5*(2), 14−125 (in Persian, English summary).

Niezabitowski, E. L. (1910). Materyaly do fauny Brakonidow Polski. Braconidae, zebrane w Galicyi. *Sprawozdania Akademii Umiejetnosci w Krakowie, 44*, 47−106.

Okyar, Z., Yurtcan, M., Beyarslan, A., & Aktac, N. (2012). The parasitoid complex of white-spotted pinion *Cosmia dffinis* (Linnaeus, 1767) (Lepidoptera: Noctuidae) on *Ulmus minor* Miller (Ulmaceae) in Edirne province (European Turkey). *Journal of the Kansas Entomological Society, 85*(2), 91−96.

Özbek, H., & Coruh, S. (2012). Larval parasitoids and larval diseases of *Malacosoma neustria* L. (Lepidoptera: Lasiocampidae) detected in Erzurum Province, Turkey. *Turkish Journal of Zoology, 36*(4), 447−459.

Pansa, M. G., Guidone, L., & Tavella, L. (2012). Distribution and abundance of nymphal parasitoids of *Lygus rugulipennis* and *Adelphocoris lineolatus* in northwestern Italy. *Bulletin of Insectology, 65*(1), 81—87.

Papp, J. (1970). A contribution to the braconid fauna of Israel (Hymenoptera). *Israel Journal of Entomology, 5,* 63—76.

Papp, J. (1974). *Arideloides niger* gen. and sp. n. from new Guinea (Hymenoptera: Braconidae: Euphorinae). *Proceedings of the Hawaiian Entomological Society, 21*(3), 443—446.

Papp, J. (2012). A contribution to the braconid fauna of Israel (Hymenoptera: Braconidae), 3. *Israel Journal of Entomology, 41—42,* 165—219.

Patetta, A., & Manino, A. (1990). Tarme della cora. *Apicoltore Moderno, 80*(6), 265—274.

Perrichot, V., Nel, A., & Quicke, D. L. J. (2009). New braconid wasps from French Cretaceous amber (Hymenoptera: Braconidae): synonymization with Eoichneumonidae and implications for the phylogeny of Ichneumonoidea. *Zoologica Scripta, 38,* 79—88.

Quicke, D. L. G., & van Achterberg, C. (1990). Phylogeny of the subfamilies of the family Braconidae (Hymenoptera: Ichneumonidae). *Zoologische Verhandlungen, 258,* 1—95.

Richards, O. W. (1967). Some British species of *Leiophron* Nees (Hymenoptera: Braconidae, Euphorinae), with the description of two new species. *Transactions of the Royal Entomological Society of London, 119,* 171—186.

Richards, O. W., & Waloff, N. (1961). A study of a natural population of *Phytodecta olivacea* (Forster) (Coleoptera, Chrysomelidae). *Philosophical Transactions, 244,* 205—257.

Rudow, F. (1918). Braconiden und ihre Wirte. *Entomologische Zeitschrift, 32*(4), 7—8, 11—12, 15—16.

Sakenin, H., Eslami, B., Samin, N., Imani, S., Shirdel, F., & Havaskary, M. (2008). A contribution to the most important trees and shrubs as the hosts of wood-boring beetles in different regions of Iran and identification of many natural enemies. *Plant and Ecosystem, 16,* 27—46.

Sakenin, H., Naderian, H., Samin, N., Rastegar, J., Tabari, M., & Papp, J. (2012). On a collection of Braconidae (Hymenoptera) from northern Iran. *Linzer biologische Beiträge, 44*(2), 1319—1330.

Sakenin, H., Coronado-Blanco, M., Samin, N., & Fischer, M. (2018). New records of Braconidae (Hymenoptera) from Iran. *Far Eastern Entomologist, 362,* 13—16.

Sakenin, H., Samin, N., Beyarslan, A., Coronado-Blanco, J. M., Navaeian, M., Fischer, M., & Hosseini Boldaji, S. A. (2020). A faunistic study on braconid wasps (Hymenoptera: Braconidae) from Iran. *Boletin de la Sociedad Andaluza de Entomologia, 30,* 96—102.

Samin, N. (2015). A faunistic study on the Braconidae of Iran (Hymenoptera: Ichneumonoidea). *Arquivos Entomoloxicos, 13,* 339—345.

Samin, N., Sakemin, H., Imani, S., & Shojai, M. (2011). A study on the Braconidae (Hymenoptera) of Khorasan province and vicinity, Northeastern Iran. *Phagea, 39*(4), 137—143.

Samin, N., Ghahari, H., Gadallah, N. S., & Monaem, R. (2015). A study on the braconid wasps (Hymenoptera: Braconidae), from Golestan province, northern Iran. *Linzer biologische Beiträge, 47*(1), 731—739.

Samin, N., van Achterberg, C., & Erdoğan, Ö.Ç. (2016). A faunistic study on some subfamilies of Braconidae (Hymenoptera: Ichneumonoidea) from Iran. *Arquivos Entomoloxicos, 15,* 153—161.

Samin, N., Coronado-Blanco, J. M., Kavallieratos, N. G., Fischer, M., & Sakenin, H. (2018). Recent findings on Braconidae (Hymenoptera: Ichneumonoidea) of Iran with an updated checklist. *Acta Biologica Turcica, 31*(4), 160—173.

Samin, N., Coronado-Blanco, J. M., Fischer, M., van Achterberg, C., Sakenin, H., & Davidian, E. (2018). Updated checklist of Iranian Braconidae (Hymenoptera: Ichneumonoidea) with twenty-three new records. *Natura Somogyiensis, 32,* 21—36.

Samin, N., Papp, J., & Coronado-Blanco, J. M. (2018c). A faunistic study on braconid wasps (Hymenoptera: Ichneumonoidea: Braconidae) of Iran. *Scientific Bulletin of the Uzhgorod University (Series Biology), 45,* 15—19.

Samin, N., Coronado-Blanco, J. M., Hosseini, A., Fischer, M., & Sakenin Chelav, H. (2019). A faunistic study on the braconid wasps (Hymenoptera: Braconidae) of Iran. *Natura Somogyiensis, 33,* 75—80.

Sedighi, S., Madjdzadeh, S. M., Khaje, N., & Rakhshani, E. (2014). New records of Opiinae and Euphorinae (Hym.: Braconidae) from Iran. In *Proceedings of the 3rd Integrated Pest Management Conference (IPMC), 21—22 January 2014, Kerman* (pp. 399—404) (in Persian, English summary).

Sedighi, S., & Madjdzadeh, M. (2015). Updated checklist of Iranian Euphorinae (Hymenoptera: Ichneumonoidea: Braconidae). *Biharean Biologist, 9*(2), 98—104.

Sharanowski, B. J., Dowling, A. P. G., & Sharkey, M. J. (2011). Molecular phylogenetics of Braconidae (Hymenoptera: Ichneumonoidea), based on multinuclear genes, and amplifications for classification. *Systematic Entomology, 36,* 549—572.

Shaw, M. R. (1985). A phylogenetic study of the subfamilies Meteorinae and Euphorinae (Hymenoptera: Braconidae). *Entomography, 3,* 277—370.

Shaw, S. R. (1988). Euphorine phylogeny: The evolution of diversity in host-utilization by parasitoid wasps (Hymenoptera: Braconidae). *Ecological Entomology, 13*(3), 323—335.

Shaw, S. R. (1992). Seven new North American species of *Neoneurus* (Hymenoptera: Braconidae). *Proceedings of the Entomological Society of Washington, 94*(1), 26—47.

Shaw, S. R. (1995). A new species of *Centistes* from Brazil (Hymenoptera: Braconidae: Euphorinae) parasitizing adults of *Diabroctica* (Coleoptera: Chrysomelidae), with a key to new world species. *Proceedings of the Entomological Society of Washington, 97*(1), 153—160.

Shaw, S. R. (1997a). Subfamily Euphorinae, pp. 234—254. In R. A. Wharton, P. M. Marsh, & M. J. Sharkey (Eds.), *Manual of the new world genera of the family Braconidae (Hymenoptera). Special Publication No. 1* (p. 439). Washington, DC: International Society of Hymenopterists.

Shaw, S. R. (1997b). Subfamily Meteorinae, pp. 326—330. In R. A. Wharton, P. M. Marsh, & M. J. Sharkey (Eds.), *Manual of the new world genera of the family Braconidae (Hymenoptera). Special Publication No. 1* (p. 439). Washington, DC: International Society of Hymenopterists.

Shaw, S. R. (1997c). The Costa Rican species of *Wesmaelia* Foerster with description of a new species (Hymenoptera: Braconidae: Euphorinae). *Pan-Pacific Entomologist, 73*(2), 103—109.

Shaw, S. R. (2004). Essay on the evolution of adult-parasitism in the subfamily Euphorinae (Hymenoptera, Braconidae). *Trudy Russkogo Entomologicheskogo Obshchestva, 75,* 82—95.

Shaw, S. R. (2007). A new species of *Elasmosoma* Ruthe (Hymenoptera: Braconidae: Neoneurinae) from the northwestern United States associated with western thatching ants, *Formica obscuripes* Forel, and *Formica obscuriventris* clivea Creighton (Hymenoptera: Formicidae). *Proceedings of the Entomological Society of Washington, 109*(1), 1—8.

Shaw, S. R., & Marsh, P. M. (2000). Revision of the enigmatic genus *Marshiella* Shaw in the new world with description of three new species (Hymenoptera: Braconidae, Euphorinae). *Journal of Hymenoptera Research, 9*(2), 277–287.

Shaw, S. R., Salerno, G., Colazza, S., & Peri, E. (2001). First record of *Aridelus rufotestaceus* Tobias (Hymenoptera: Braconidae, Euphorinae) parasitizing *Nezara viridula* nymphs (Heteroptera: Pentatomidae) with observations on its immature stages and development. *Journal of Hymenoptera Research, 10*(2), 131–137.

Shenefelt, R. D. (1969). *Braconidae. 1. Hybrizoninae, Euphorinae, Cosmophorinae, Neoneurinae, Macrocentrinae. Hymenopterorum Catlogus (nova edithio). Part 4. Dr. W. Junk N.V.'s-Gravenhage* (p. 176).

Shojai, M. (1968). Resultats de l'étude faunestiques des hyménoptères parasites (Terebrants) en Iran et l'importance de leur utilization dans la lutte biologique. In *Proceedings of the 1st Iranian Plant Protection Congress, 14–19 September 1968* (pp. 25–35). Karaj: Karaj Junior College of Agriculture.

Shojai, M. (1998). *Entomology (ethology, social life and natural enemies) (biological control)* (Vol. III). Tehran University Publications, 550 pp.

Shojai, M., Abbaspour, H., Nasrollahi, A., & Labbafi, Y. (1995). Technology and biocenotic aspects of integrated biocontrol of corn stem borer: *Sesamia cretica* Led. (Lep.: Noctuidae). *Journal of Agricultural Sciences (Islamic Azad University), 1*(2), 5–32 (in Persian, English summary).

Soleimani, S., & Madadi, H. (2015). Seasonal dynamics of: the pea aphid, *Acyrthosiphon pisum* (Harris), its natural enemies the seven spotted lady beetle *Coccinella septempunctata* Linnaeus and variegated lady beetle *Hippodamia variegata* Goeze, and their parasitoid *Dinocampus coccinellae* (Schrank). *Journal of Plant Protection Research, 55*, 421–428.

Steiner, P. (1936). Beiträge zur Kenntnise der Schädlingsfauna Kleinasiens. III. *Laphygma exigua* Hb., ein Großschädling der Zuckerrübe in Anatolien. *Zeitschrift für Angewandte Entomologie, 23*, 177–222.

Stigenberg, J., & Ronquist, F. (2011). Revision of the western Palearctic *Meteorini* (Hymenoptera, Braconidae), with a molecular characterization of hidden Fennoscandian species diversity. *Zootaxa, 3084*, 1–95.

Stigenberg, J., & Shaw, M. R. (2013). Western Palaearctic Meteorinae (Hymenoptera: Braconidae) in the National Museums of Scotland, with rearing, phenological and distributional data, including six species new to Britain, and a discussion of potential route to speciation. *Entomologist's Gazette, 64*, 251–268.

Stigenberg, J., Boring, C. A., & Ronquist, F. (2015). Phylogeny of the parasitic wasp subfamily Euphorinae (Braconidae) and evolution of its host preferences. *Systematic Entomology, 40*, 570–591.

Stoner, A. (1973). Incidence of *Wesmaelia pendula* (Hymenoptera: Braconidae), a parasite of male *Nabis* species in Arizona. *Annals of the Entomological Society of America, 66*(2), 471–473.

Schwacrtz, A., Gerling, D., & Rossler, Y. (1980). Preliminary notes on the parasites of *Spodoptera exigua* (Huebner) (Lepidoptera: Noctuidae) in Israel. *Phytoparasitica, 8*(2), 93–97.

Szépligeti, G. (1904). Hymenoptera, family Braconidae, pp. 78-79. In P. Wytsman (Ed.), *Genera Insectorum* (Vol. 22, pp. 1–253). V. Verteneuil and L. Desmet-Verteneuil Bruxelles.

Szócs, J. (1979). Angaben zu den Parasiten der minierende Motten (Hymenoptera: Braconidae). *Folia Entomologica Hungarica, 32*(2), 199–206.

Tavoosi Ajvad, F., Madadi, H., Kazazi, M., & Sobhani, M. (2012). Seasonal changes of *Hippodamia variegata* populations and its parasitism by *Dinocampus coccinellae* in alfalfa fields of Hamedan. *Biological Control of Pests and Plant Diseases, 1*, 11–18 (in Persian, English abstract).

Tavoosi Ajvad, F., Madadi, H., Sobhani, M., & van Achterberg, C. (2014). First report of *Dinocampus coccinellae* (Hym.: Braconidae) from Iran. *Journal of Entomological Society of Iran, 33*(4), 77.

Tobias, V. I. (1966). Generic groupings and evolution of parasitic Hymenoptera of the subfamily Euphorinae (Hymenoptera: Braconidae). II. Entomologicheskoe Obozrenie. *Entomological Review, 45*, 612–633, 348–358].

Tobias, V. (1971). Review of the Braconidae (Hymenoptera) of the USSR. *Trudy Vsesoyuznogo Entomologicheskogo Obshchestva, 54*, 156–268.

Tobias, V. (1986). Otryad Hymenoptera-Pereponchatokrylye, Semeistov Braconidae. In G. S. Medvedev (Ed.), *Hymenoptera, part 4: Vol. 3. Opredelitelj nasekomykh Evropejskoj tchasty SSSR* (pp. 7–500). Leningrad: Nauka.

van Achterberg, C. (1984). Notes on some species of Braconidae (Hymenoptera) described by Hedwig from Iran and Afghanistan. *Entomologische Berichten, Amsterdam, 40*, 25–31.

van Achterberg, C. (1984). Essay on the phylogeny of Braconidae (Hymenoptera: Ichneumonoidea). *Entomologisk Tidskrift, 105*, 41–58.

van Achterberg, C. (1985). The genera and subgenera of *Centistini*, with description of two new taxa from the Nearctic region (Hymenoptera: Braconidae: Euphorinae). *Zoologische Mededelingen, 59*, 348–362.

van Achterberg, C., & Argaman, Q. (1993). *Kollasmosoma* gen. nov. and a key to the genera of the subfamily Neoneurinae (Hymenoptera: Braconidae). *Zoologische Mededelingen, 67*(5), 63–74.

van Achterberg, C., & Haeselbarth, E. (2003). Revision of the genus *Syntretus* Foerster (Hymenoptera: Braconidae: Euphorinae) from Europe. *Zoologische Mededelingen, 77*(2), 9–78.

Waloff, N. (1961). Observations on the biology of *Perilitus dubius* (Wesmael) (Hymenoptera: Braconidae), a parasite of the chrysomelid beetle *Phytodecta olivacea* (Forster). *Proceedings of the Royal Entomological Society of London (A), 36*(7–9), 96–102.

Watanabe, C. (1955). On Japanese species of the genus *Microctonus* Wesmael with description of a new species (Hymenoptera: Braconidae). *Mushi, 29*, 51–55.

Wharton, R. A., Marsh, P. M., & Sharkey, M. J. (1997). *Manual of the new world genera of the family Braconidae (Hymenoptera)* (p. 439). Washington, DC: International Society of Hymenopterists.

Yilmaz, T., Aydogdu, M., & Beyarslan, A. (2010). The distribution of the Euphorinae wasps (Hymenoptera: Braconidae) in Turkey, with phytogeographical notes. *Turkish Journal of Zoology, 34*(2), 181–194.

Young, D. K. (1984a). Field records and observations of insects associated with cantharidin. *Great Lakes Entomologist, 17*, 195–199.

Young, D. K. (1984b). Cantharidin and insects: An historical review. *Great Lakes Entomologist, 17*, 187–194.

Yu, D. S., van Achterberg, C., & Horstmann, K. (2016). *Taxapad 2016, Ichneumonoidea 2015, Database on flash-drive.* Nepean, Ontario, Canada.

Colastes braconius Haliday, 1833 (Exothecinae), ♀, dorsal habitus. *Photo prepared by S.R. Shaw.*

Shawiana catenator (Haliday 1836) (Exothecinae), ♀, lateral habitus. *Photo prepared by S.R. Shaw.*

Chapter 14

Subfamily Exothecinae Foerster, 1863

Neveen Samy Gadallah[1], Hassan Ghahari[2], Donald L.J. Quicke[3] and Scott Richard Shaw[4]

[1]Entomology Department, Faculty of Science, Cairo University, Giza, Egypt; [2]Department of Plant Protection, Yadegar-e Imam Khomeini (RAH) Shahre Rey Branch, Islamic Azad University, Tehran, Iran; [3]Integrative Ecology Laboratory, Department of Biology, Faculty of Science, Chulalongkorn University, Pathumwan, Bangkok, Thailand; [4]UW Insect Museum, Department of Ecosystem Science and Management, University of Wyoming, Laramie, WY, United States

Introduction

The subfamily Exothecinae Foerster, 1863 is a rather small cosmopolitan subfamily, currently comprising more than 92 species in eight genera (Yu et al., 2016). Their species are widespread but often overlooked due to their small body sizes and comparatively little diversity.

Members of the subfamily Exothecinae are idiobiont ectoparasitoids of the larvae of leaf mining or gall-forming holometabolous larvae mainly of the order Lepidoptera, and rarely attacking some mining Diptera, Hymenoptera, and Coleoptera (Chen & van Achterberg, 2019; van Achterberg, 1993). In rare cases, the host was found to be a stem borer, as in the case of the exothecine *Shawiana foveolator* (Thomson) that was reported as a host of the sawfly *Blasticotoma filiceti* Klug (van Achterberg & Shaw, 2008). Some species are categorized as koinobionts (Shaw, 1983), as they permit the host to feed for some time after having been attacked. On the other hand, other species are recorded as idiobionts, as they permanently paralyze the host (van Achterberg & Shaw, 2008). Van Achterberg and Shaw (2008) reported larvae of two exothecine species living in bracket fungi in the United Kingdom, the Netherlands, and Finland. Based on such habitats, a new subgenus *Fungivenator* van Achterberg and Shaw, of the genus *Colastes*, was designated (van Achterberg & Shaw, 2008). Three other *Colastes* species, that are believed to occur in the same habitat occupied by *Dercatoma* species (Coleoptera: Ptinidae), were also included under the new subgenus *Fungivenator* (van Achterberg & Shaw, 2008). The genera *Xenarcha* and *Shawiana* were formerly treated by Belokobylskij (1998) and Belokobylskij et al. (2003) as subgenera of the genus *Colastes*; however, van Achterberg and Shaw (2008) disagreed and reinstated them as valid genera.

Exothecines were formerly included as a tribe in Rogadinae (Shaw & Huddleston, 1991) but later were elevated to a separate subfamily (Chen & van Achterberg, 2019).

Checklists of Regional Exothecinae. Yu et al. (2016) listed two species in two genera, Gadallah and Ghahari (2017) cataloged three species in one genus, and Samin, Coronado-Blanco, Kavallieratos, et al. (2018), Samin, Coronado-Blanco, Fischer, et al. (2018) represented three species in three genera found in the Middle East. The present checklist includes eight species in three genera. Here we follow Chen and van Acherberg (2019) for classification, Yu et al. (2016) for global distribution of species, and in addition to some recent references whenever available.

Key to genera of the subfamily Exothecinae of the Middle East (modified from van Achterberg, 1983)

1. Pronope usually large and deep; vein r of fore wing variably placed in pterostigma; male fore wing with sometimes enlarged pterostigma 2

— Pronope absent; vein r of fore wing received on pterostigma near basal third, rarely submedially; male fore wing without enlarged pterostigma .... *Colastes* Haliday

Braconidae of the Middle East (Hymenoptera). https://doi.org/10.1016/B978-0-323-96099-1.00023-6

2. Notauli incomplete; vein r of fore wing usually received between base and basal third of pterostigma; if received near basal 0.4 of pterostigma, then vein m-cu of fore wing distinctly converging to vein 1-M posteriorly ....*Shawiana* van Achterberg
— Notauli nearly complete, usually partly crenulate, and with a short medio-posterior carina, its posterior part distinctly impressed; vein r of fore wing received on basal 0.4—0.6 of pterostigma, if from 0.4, then vein m-cu of fore wing only slightly converging to vein 1-M posteriorly .... *Xenarcha* Foerster

List of species of the subfamily Exothecinae recorded in the Middle East

Subfamily Exothecinae Foerster, 1863

Genus *Colastes* Haliday, 1833

Colastes (Colastes) braconius Haliday, 1833

Catalogs with Iranian records: Yu et al. (2016), Gadallah and Ghahari (2017), Samin, Coronado-Blanco, Kavallieratos, et al. (2018), Samin, Coronado-Blanco, Fischer, et al. (2018).
Distribution in Iran: Kuhgiloyeh and Boyerahmad (Samin et al., 2016).
Distribution in the Middle East: Iran (Samin et al., 2016), Turkey (Beyarslan, 2015, 2017).
Extralimital distribution: Japan, Kazakhstan, Korea, Russia, Tunisia, Ukraine.
Host records: Summarized by Yu et al. (2016) as being a parasitoid of several insect pests belonging to the orders Coleoptera (Curculionidae), Diptera (Agromyzidae), Hymenoptera (Tenthredinidae), and Lepidoptera (Coleophoridae, Cosmopterigidae, Elachistidae, Eriocraniidae, Gracillariidae, Heliozelidae, Lycaenidae, Lyonetiidae, Momphidae, Nepticulidae, Pyralidae, Tischeriidae, Tortricidae, Ypsolophidae).

Colastes (Colastes) pubicornis (Thomson, 1892)

Catalogs with Iranian records: No catalog.
Distribution in Iran: Kordestan (Gadallah et al., 2021).
Distribution in the Middle East: Iran.
Extralimital distribution: Bulgaria, China, Finland, Germany, Hungary, Japan, Korea, Lithuania, Russia, Sweden, Ukraine.
Host records: Recorded by Godfray and McGavin (1985) as being a parasitoid of the anthomyiid *Chirosia histricina* (Rondani).

Colastes (Colastes) vividus Papp, 1975

Distribution in the Middle East: Turkey (Beyarslan, 2015, 2017).
Extralimital distribution: Austria, Germany, Hungary, Moldova, Poland, Tunisia.
Host records: Unknown.

Colastes (Fungivenator) aciculatus Tobias, 1986

Distribution in the Middle East: Turkey (Beyarslan, 2015, 2017).
Extralimital distribution: Czech Republic, Russia.
Host records: Unknown.

Genus *Shawiana* van Achterberg, 1983

Shawiana catenator (Haliday, 1836)

Catalogs with Iranian records: No catalog.
Distribution in Iran: Chaharmahal and Bakhtiari (Ghahari & Gadallah, 2019).
Distribution in the Middle East: Iran (Ghahari & Gadallah, 2019), Turkey (Beyarslan, 2017).
Extralimital distribution: Austria, Belarus, Bulgaria, former Czechoslovakia, Georgia, Germany, Hungary, Ireland, Italy, Japan, Latvia, Lithuania, Mongolia, Netherlands, Poland, Russia, Serbia, Switzerland, Ukraine, United Kingdom.
Host records: Summarized by Yu et al. (2016) as being a parasitoid of a number of leaf miners including the following tenthredinids: *Fenella nigrita* Westwood, *Fenusa dohrnii* (Tischbein), *Fenusa pumila* Leach, *Fenusa pusilla* (Lepeletier), *Heterarthus aceris* (Kaltenbach), *Heterarthus vagans* (Fallén), *Messa hortulana* (Klug), *Messa nana* (Klug), *Metallus albipes* (Cameron), *Metallus pumilus* (Klug), *Parna tenella* (Klug), *Profenusa pygmaea* (Klug), and *Scolioneura betuleti* (Klug); the eriocraniid *Eriocrania semipurpurella* (Stephens); and the gracillariid *Phyllonorycter quercifoliella* (Zeller).

Shawiana laevis (Thomson, 1892)

Catalogs with Iranian records: Yu et al. (2016), Gadallah and Ghahari (2017 as *Colastes (Shawiana) laevis* (Thomson, 1892)), Samin, Coronado-Blanco, Kavallieratos, et al. (2018), Samin, Coronado-Blanco, Fischer, et al. (2018).
Distribution in Iran: Razavi Khorasan (Samin et al., 2016).
Distribution in the Middle East: Iran (Samin et al., 2016), Turkey (Beyarslan, 2015).
Extralimital distribution: Belarus, Czech Republic, Finland, Germany, Korea, Lithuania, Mongolia, Netherlands, Russia, Sweden, Ukraine, United Kingdom.
Host records: Summarized by Yu et al. (2016) as being a parasitoid of the following tenthredinids: *Euura mucronate* (Hartig), *Fenusa dohrnii* (Tischbein), *Heterarthus aceris*

(Kaltenbach), *Heterarthus microcephalus* (Klug), *Heterarthus vagans* (Fallén), *Pontania glaucae* Kopelke, *Pontania lapponicola* Kopelke, *Pontania nigricantis* Kopelke, *Pontania vesicator* (Bremi), *Scolioneura betuleti* (Klug), and *Scolioneura vicina* Konow.

Genus *Xenarcha* Foerster, 1863

Xenarcha laticarpus (Thomson, 1892)

Catalogs with Iranian records: No catalog.
Distribution in Iran: East Azarbaijan (Gadallah et al., 2021).
Distribution in the Middle East: Iran.
Extralimital distribution: Austria, Azerbaijan, Finland, Georgia, Germany, Lithuania, Russia, Sweden.
Host records: Unknown.

Xenarcha lustrator (Haliday, 1836)

Catalogs with Iranian records: Gadallah and Ghahari (2017), Samin, Coronado-Blanco, Kavallieratos, et al. (2018), Samin, Coronado-Blanco, Fischer, et al. (2018).
Distribution in Iran: Northern Khorasan (Gadallah & Ghahari, 2017 as *Colastes* (*Xenarcha*) *lustator*).
Distribution in the Middle East: Iran (Gadallah & Ghahari, 2017), Turkey (Beyarslan, 2015, 2017).
Extralimital distribution: Russia.
Host records: Summarized by Yu et al. (2016) as being a parasitoid of the coleophorid *Coleophora* sp.; and the following tenthredinids: *Fenella nigrita* Westwood, *Fenusa pumila* Leach, *Fenusa pusilla* (Lepeletier), *Fenusa ulmi* Sundeval, *Metallus albipes* (Cameron), and *Metallus pumilus* (Klug).

Conclusion

Eight species of the subfamily Exothecinae in three genera, *Colastes*, *Shawiana*, and *Xenarcha*, are reported from only two of the Middle East countries, Iran and Turkey. Exothecinae of Iran, known so far, comprises six species (4.3% of the world species) in three genera, *Colastes* (two species), *Shawiana* (two species), and *Xenarcha* (two species). Iranian Exothecinae species have been recorded from six provinces: Chaharmahal and Bakhtiari, East Azarbaijan, Kordestan, Kuhgiloyeh and Boyerahmad, Northern Khorasan and Razavi Khorasan. No host species have been recorded for these parasitoids in Iran. Among the countries of the Middle East and those adjacent to Iran the subfamily Exothecinae has been recorded from five of them: Russia (78 species in 16 genera (Belokobylskij & Lelej, 2019)), Turkey (six species), Kazakhstan (five species), Armenia, and Azerbaijan (both with three species). Additionally, Russia shares six species with Iran, and Azerbaijan, Kazakhstan, and Turkey each share one species.

References

Belokobylskij, S. A. (1998). Exothecinae, pp. 111–162. *Operdelitel nasekomych dalnego bostoka Rosii, 4*(3), 1–707. Vladivostok.

Belokobylskij, S. A., & Lelej, A. S. (2019). Annotated catalogue of the Hymenoptera of Russia. Volume II. Apocrita: Parasitica. In *Proceedings of the Zoological Institute of the Russian Academy of Sciences*. Supplement No. 8, 594 pp.

Belokobylskij, S. A., Taeger, A., van Achterberg, C., Haeselbarth, E., & Riedel, M. (2003). Checklist of the Braconidae of Germany. *Beitrag Entomologie, 53*, 341–435.

Beyarslan, A. (2015). A faunal study of the subfamily Doryctinae in Turkey (Hymenoptera: Braconidae). *Turkish Journal of Zoology, 39*(1), 126–143.

Beyarslan, A. (2017). Checklist of Turkish Doryctinae (Hymenoptera: Braconidae). *Linzer biologische Beiträge, 49*(1), 415–440.

Chen, X. X., & van Achterberg, C. (2019). Systematics, phylogeny, and evolution of braconid wasps: 30 years of progress. *Annual Review of Entomology, 64*, 1–24.

Gadallah, N. S., & Ghahari, H. (2017). An annotated catalogue of the Iranian Doryctinae and Exothecinae (Hymenoptera: Braconidae). *Transactions of the American Entomological Society, 143*, 669–691.

Gadallah, N. S., Ghahari, H., & Quicke, D. L. J. (2021). Further addition to the braconid fauna of Iran (Hymenoptera: Braconidae). *Egyptian Journal of Biological Pest Control, 31*, 32. https://doi.org/10.1186/s41938-021-00376-8

Ghahari, H., & Gadallah, N. S. (2019). New records of braconid wasps (Hymenoptera: Ichneumonoidea: Braconidae) from Iran. *Entomological News, 129*(3), 384–392.

Godfray, H. C. J., & McGavin, G. C. (1985). *Colastes pubicornis* (Thomson) (Hym., Braconidae, Exothecini) new to Britain, with a first host record. *Entomologist's Monthly Magazine, 121*, 109–110.

Samin, N., van Achterberg, C., & Çetin Erdoğan, Ö. (2016). A faunistic study on some subfamilies of Braconidae (Hymenoptera: Ichneumonoidea) from Iran. *Arquivos Entomoloxicos, 15*, 153–161.

Samin, N., Coronado-Blanco, J. M., Kavallieratos, N. G., Fischer, M., & Sakenin, H. (2018). Recent findings on Braconidae (Hymenoptera: Ichneumonoidea) of Iran with an updated checklist. *Acta Biologica Turcica, 31*(4), 160–173.

Samin, N., Coronado-Blanco, J. M., Fischer, M., van Achterberg, C., Sakenin, H., & Davidian, E. (2018). Updated checklist of Iranian Braconidae (Hymenoptera: Ichneumonoidea) with twenty-three new records. *Natura Somogyiensis, 32*, 21–36.

Shaw, M. R. (1983). *On evolution of endoparasitism: the biology of some genera of Rogadinae (Braconidae)*. Contr (Vol. 20, pp. 307–328). American Enterprise Institute.

Shaw, M. R., & Huddleston, T. (1991). Classification and biology of braconid wasps (Hymenoptera: Braconidae). *Handbook of Identification of British Insects, 7*(11), 1–126.

van Achterberg, C. (1983). Revisionary notes on the Palaearctic genera and species of the tribe Exothecini Foerster (Hymenoptera: Braconidae). *Zoologische Midedelingen, 57*, 339–355.

van Achterberg, C. (1993). Illustrated key to the subfamilies of the Braconidae (Hymenoptera: Ichneumonidae). *Zoologische Verhandelingen, 283,* 1—189.

van Achterberg, C., & Shaw, M. R. (2008). A new subgenus of the genus *Colastes* Haliday (Hymenoptera: Braconidae: Exothecinae) from

species reared from bracket fungi, with description of two new species from Europe. *Journal of Natural History, 42*(27), 1849—1860.

Yu, D. S., van Achterberg, C., & Horstmann, K. (2016). *Taxapad 2016, Ichneumonoidea 2015, Database on flash-drive.* Ottawa, Ontario, Canada.

Gnamptodon sp. (Gnamptodontinae), ♀, lateral habitus. *Photo prepared by S.R. Shaw.*

Pseudognaptodon shawi Williams, 2004 (Gnamptodontinae), ♀, lateral habitus - Nearctic. *Photo prepared by S.R. Shaw.*

Chapter 15

Subfamily Gnamptodontinae Fischer, 1970

Neveen Samy Gadallah[1], Hassan Ghahari[2], Scott Richard Shaw[3] and Donald L.J. Quicke[4]

[1]Entomology Department, Faculty of Science, Cairo University, Giza, Egypt; [2]Department of Plant Protection, Yadegar-e Imam Khomeini (RAH) Shahre Rey Branch, Islamic Azad University, Tehran, Iran; [3]UW Insect Museum, Department of Ecosystem Science and Management, University of Wyoming, Laramie, WY, United States; [4]Integrative Ecology Laboratory, Department of Biology, Faculty of Science, Chulalongkorn University, Pathumwan, Bangkok, Thailand

Introduction

Gnamptodontinae is a small cosmopolitan subfamily of Braconidae with 90 species classified into four genera and three tribes (Exodontiellini Wharton, 1978; Gnamptodontini Fischer, 1970; Telengaiini Tobias, 1962) (Chen & van Achterberg, 2019; Yu et al., 2016). It was formerly placed in Opiinae (Fischer, 1987). Based on a molecular study carried out by Zaldívar-Riverón et al. (2006), the genus *Telengaia* (formerly placed in a separate subfamily, Telengaiinae by Tobias, 1962) is treated as a tribe in Gnamptodontinae. The same was done with the Nearctic exodont genus *Exodontiella* Wharton, 1977 (Wharton et al., 2006), which was formerly included in Opiinae (Wharton, 1978).

Members of this subfamily are small sized (less than 3 mm in length), being characterized by the presence of a transverse groove, delimiting a more or less rectangular, medio-basal area of second metasomal tergite; the non-sculptured propodeum; and the absence of both epicnemial and occipital carinae (Sharkey, 1993; Wharton et al., 1997; Williams, 2004). Gnaptodontines are presumably mostly idiobiont ectoparasitoids of leaf-mining larvae of Nepticulidae (Lepidoptera) (Chen & van Achterberg, 2019; Sharkey, 1993; van Achterberg, 1983), especially of the genus *Stigmella* Schrank, from which they have been reared (Sharkey, 1993; van Achterberg, 1983). Although long suspected of being koinobiont parasitoids by Shaw and Huddleston (1991), their detailed biology is still unknown (Cirelli et al., 2002; Williams, 2004). The genus *Gnamptodon* Haliday, 1833 is the largest in the subfamily Gnamptodontinae, with about 54 described species (Yu et al., 2016). These species are distributed almost worldwide but they are particularly diverse in the Holarctic region (Belokobylskij, 1987; Chen et al., 2002; van Achterberg, 1988; Yu et al., 2016).

Checklists of Regional Gnamptodontinae. Farahani et al. (2016), Gadallah et al. (2016), Yu et al. (2016) and Samin, Coronado-Blanco, Kavallieratos, et al. (2018), Samin, Coronado-Blanco, Fischer, et al. (2018): all with four species in one genus *Gnamptodon* Haliday, 1833. The present Middle Eastern checklist includes four species in one genus. Here we follow Chen and van Achterberg (2019) for the higher classification, and Yu et al. (2016) for the global distribution of species.

List of species of the subfamily Gnamptodontinae recorded in the Middle East

Subfamily Gnamptodontinae Fischer, 1970

Tribe Gnamptodontini Fischer, 1970

Genus *Gnamptodon* Haliday, 1833

Gnamptodon breviradialis (Fischer, 1959)

Catalogs with Iranian records: Gadallah et al. (2016), Farahani et al. (2016), Yu et al. (2016), Samin, Coronado-Blanco, Kavallieratos, et al. (2018), Samin, Coronado-Blanco, Fischer, et al. (2018).

Distribution in Iran: Guilan, Mazandaran (Farahani et al., 2014).

Distribution in the Middle East: Iran.

Extralimital distribution: Former Czechoslovakia, France, Greece, Hungary, Italy, Moldova, Russia.

Host records: Summarized by Yu et al. (2016) as being a parasitoid of the nepticulids *Acalyptris loranthella*

(Klimesch), *Ectoedemia mahalebella* (Klimesch), *Ectoedemia picturata* Puplesis, *Stigmella amygdali* (Klimesch), *Stigmella eberhardi* (Johansson), *Stigmella nostrata* Puplesis, and *Stigmella ulmivora* (Fologne).

Gnamptodon decoris (Foerster, 1863)

Catalogs with Iranian records: Gadallah et al. (2016), Farahani et al. (2016), Yu et al. (2016), Samin, Coronado-Blanco, Kavallieratos, et al. (2018), Samin, Coronado-Blanco, Fischer, et al. (2018).

Distribution in Iran: Guilan, Mazandaran (Farahani et al., 2014).

Distribution in the Middle East: Iran.

Extralimital distribution: Austria, Bulgaria, Czech Republic, Finland, Germany, Greece, Hungary, Italy, Kazakhstan, Korea, Kyrgyzstan, Moldova, Mongolia, Netherlands, Poland, Russia, Switzerland, Ukraine, United Kingdom.

Host records: Summarized by Yu et al. (2016) as being a parasitoid of the nepticiculids *Bohemannia pulverosella* (Stainton), *Ectoedema* spp., *Parafomoria helianthenella* (Herrich-Schäffer), *Stigmella* spp., and *Trifurcula cryptella* (Stainton).

Gnamptodon georginae (van Achterberg, 1983)

Catalogs with Iranian records: Gadallah et al. (2016), Farahani et al. (2016), Yu et al. (2016), Samin, Coronado-Blanco, Kavallieratos, et al. (2018), Samin, Coronado-Blanco, Fischer, et al. (2018).

Distribution in Iran: Alborz, Guilan, Qazvin, Tehran (Farahani et al., 2014), Golestan (Ghahari et al., 2010), Khuzestan (Samin et al., 2015), Mazandaran (Kian et al., 2020).

Distribution in the Middle East: Iran (Yu et al., 2016), Turkey (Beyarslan, 2021).

Extralimital distribution: Algeria, Bulgaria, China, Germany, Hungary, Italy, Moldova, Mongolia, Poland, Russia, Switzerland, Ukraine.

Host records: Summarized by Yu et al. (2016) as being a parasitoid of the gracillariid *Parornix anguliferella* (Zeller); and the nepticulids *Stigmella hybnerella* (Hübner), *Stigmella incognitella* (Herrich-Schäffer), *Stigmella lemniscella* (Zeller), *Stigmella malella* (Stainton), *Stigmella lediella* (Schleich), *Stigmella ruficapitella* (Haworth), and *Stigmella tityrella* (Stainton).

Gnamptodon pumilio (Nees ab Esenbeck, 1834)

Catalogs with Iranian records: Gadallah et al. (2016), Farahani et al. (2016), Yu et al. (2016), Samin, Coronado-Blanco, Kavallieratos, et al. (2018), Samin, Coronado-Blanco, Fischer, et al. (2018).

Distribution in Iran: Guilan, Mazandaran (Farahani et al., 2014).

Distribution in the Middle East: Iran.

Extralimital distribution: Austria, Belgium, Bulgaria, Czech Republic, Denmark, Finland, France, Germany, Greece, Hungary, Ireland, Italy, Korea, Lithuania, Moldova, Netherlands, Norway, Poland, Russia, Slovakia, Sweden, Switzerland, Ukraine, United Kingdom.

Host records: Summarized by Yu et al. (2016) as being a parasitoid of the nepticulids *Bohemannia piorta* Puplesis, *Ectoedema* spp., and *Stigmella* spp.

Conclusion

Four valid species of the subfamily Gnamptodontinae within one genus, *Gnamptodon* have been reported from only two countries of the Middle East (Iran and Turkey). Of these two, Iran was found to be the most diverse country, comprising all the four species (4.4% of the world species), while Turkey is represented by only a single, recently recorded species, *Gnamptodon georginae* (Beyarslan, 2021; van Achterberg, 1983). However, others may have been overlooked so far due to their minute size and cryptic habits. The recorded species are known from seven provinces: Guilan and Mazandaran (both with four species), Alborz, Golestan, Khuzestan, Qazvin, and Tehran (each with one species). No host species has been so far recorded for these parasitoids in Iran. Among the 23 countries of the Middle East and adjacent to Iran, the subfamily Gnamptodontinae has been recorded from only three of them, Russia (nine species), Kazakhstan (two species), and Turkey (one species). No species have been recorded from the remaining Middle Eastern countries (Yu et al., 2016). All the Iranian species are shared with the fauna of Russia and one species each with Kazakhstan and Turkey.

References

Belokobylskij, S. A. (1987). Subfamily Gnaptodontinae (Hymenoptera, Braconidae) of the USSR far East. In *Taxonomy of the insects of the USSR Siberia and far East* (pp. 78–84). Vladivostok: Collection of scientific papers (in Russia).

Beyarslan, A. (2021). Gnamptodontinae Fischer, 1970 a new record subfamily of Braconidae (Hymenoptera) in Turkey. *Acta Biologica Turcica, 34*(2), 55–58.

Chen, X. X., & van Achterberg, C. (2019). Systematics, phylogeny, and evolution of braconid wasps: 30 years of progress. *Annual Review of Entomology, 64*, 1–24.

Chen, X., Whitfield, J. B., & He, J. (2002). The discovery of the genus *Gnamptodon* Haliday (Hymenoptera: Braconidae) in China, with description of one new species. *Pan-Pacific Entomologist, 78*(3), 184–187.

Cirelli, K. R. N., Braga, S. M. P., & Penteado-Dias, A. M. (2002). New species of *Pseudognamptodon* Fischer (Hymenoptera: Braconidae: Gnamptodontinae) from Brazil. *Zoologische Mededelingen, 76*(9), 89–95.

Farahani, S., Talebi, A. A., & Rakhshani, E. (2014). A study of the rare genus *Gnamptodon* Haliday (Hymenoptera: Braconidae: Gnamptodontinae) in Iran, with three new records. *Journal of the Entomological Research Society, 16*(3), 87−93.

Farahani, S., Talebi, A. A., & Rakhshani, E. (2016). Iranian Braconidae (Insecta: Hymenoptera: Ichneumonoidea): diversity, distribution and host association. *Journal of Insect Biodiversity and Systematics, 2*(1), 1−92.

Fischer, M. (1987). Hymenoptera Braconidae (Opiinae III)-äthiopische, orientalische, australische und ozeanische region. *Tierreich, 104*, 1−734.

Gadallah, N. S., Ghahari, H., & van Achterberg, C. (2016). An annotated catalogue of the Iranian Euphorinae, Gnamptodontinae, Helconinae, Hormiinae and Rhysipolinae (Hymenoptera: Braconidae). *Zootaxa, 4072*(1), 1−38.

Ghahari, H., Fischer, M., Hedqvist, K. J., Erdoğan, Ö.Ç., van Achterberg, C., & Beyarslan, A. (2010). Some new records of Braconidae (Hymenoptera) for Iran. *Linzer biologische Beiträge, 42*(2), 1395−1404.

Kian, N., Goldasteh, S., & Farahani, S. (2020). A survey on abundance and species diversity of Braconid wasps in forest of Mazandaran province. *Journal of Entomological Research, 12*(1), 61−69 (in Persian, English summary).

Samin, N., van Achterberg, C., & Ghahari, H. (2015). A faunistic study of Braconidae (Hymenoptera; Ichneumonoidea) from southern Iran. *Linzer biologische Beiträge, 47*(2), 1801−1809.

Samin, N., Coronado-Blanco, J. M., Kavallieratos, N. G., Fischer, M., & Sakenin, H. (2018). Recent findings on Braconidae (Hymenoptera: Ichneumonoidea) of Iran with an updated checklist. *Acta Biologica Turcica, 31*(4), 160−173.

Samin, N., Coronado-Blanco, J. M., Fischer, M., van Achterberg, C., Sakenin, H., & Davidian, E. (2018). Updated checklist of Iranian Braconidae (Hymenoptera: Ichneumonoidea) with twenty-three new records. *Natura Somogyiensis, 32*, 21−36.

Sharkey, M. J. (1993). Family Braconidae, pp. 362−395. In H. Goulet, & J. V. Huber (Eds.), *Hymenoptera of the world: an identification guide to families*. Ottawa: Agriculture Canada Publications, 688 pp.

Shaw, M. R., & Huddleston, T. (1991). Classification and biology of braconid wasps (Hymenoptera: Braconidae). *Handbook of Identification of the British Insects, 7*(11), 1−126.

Tobias, V. I. (1962). A new subfamily of braconids (Hymenoptera: Braconidae) from Central Asia. *Trudy Zoologicheskogo Instituta. Akedemie Nauk SSSR, 30*, 268−270.

van Achterberg, C. (1983). Revisionary notes on the subfamily Gnaptodontinae, with description of eleven new species (Hymenoptera: Braconidae). *Tijdschrift voor Entomologie, 126*, 25−57.

van Achterberg, C. (1988). A new species of the genus *Gnamptodon* from Italy (Hymenoptera: Braconidae). *Entomologische Berichten, Amsterdam, 48*(10), 159−161.

Wharton, R. A. (1978). Exodontiellini, a new tribe of Opiinae with exodont mandibles. *Pan-Pacific Entomology, 531*, 297−303.

Wharton, R. A., Marsh, P. M., & Sharkey, M. J. (1997). *Manual of the New World genera of the family Braconidae* (Vol. 1, pp. 1−439). Special Publications of International Society of Hymenopterists.

Wharton, R. A., Yoder, M. J., Gillespie, J. J., Patton, J. C., & Honeycutt, R. L. (2006). Relationships of *Exodontiella*, a non-alysiine, exdont member of the family Braconidae (Insecta, Hymenoptera). *Zoologica Scripta, 35*, 323−340.

Williams, D. L. J. (2004). Revision of the genus *Pseudognaptodon* Fischer (Hymenoptera: Braconidae: Gnamptodontinae). *Journal of Hymenoptera Research, 13*(1), 149−206.

Yu, D. S., van Achterberg, C., & Horstmann, K. (2016). *Taxapad 2016, Ichneumonoidea 2015, database on flash-drive*. Nepean, Ontario, Canada.

Zaldívar-Riverón, A., Mori, M., & Quicke, D. L. J. (2006). Systematics of the cyclostome subfamilies of braconid parasitic wasps (Hymenoptera: Ichneumonoidea): a simultaneous molecular and morphological Bayesian approach. *Molecular Phylogeny and Evolution, 38*(1), 130−145.

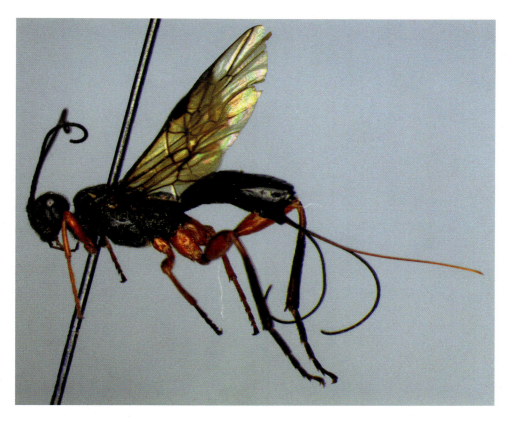

Helcon tardator Nees, 1812 (Helconinae), ♀, lateral habitus. *Photo prepared by S.R. Shaw.*

Wroughtonia dentator Fabricius, 1804 (Helconinae), ♀, lateral habitus. *Photo prepared by S.R. Shaw.*

Chapter 16

Subfamily Helconinae Foerster, 1863

Hassan Ghahari[1], Scott Richard Shaw[2], Neveen Samy Gadallah[3] and Michael J. Sharkey[4]

[1]Department of Plant Protection, Yadegar-e Imam Khomeini (RAH) Shahre Rey Branch, Islamic Azad University, Tehran, Iran; [2]UW Insect Museum, Department of Ecosystem Science and Management, University of Wyoming, Laramie, WY, United States; [3]Entomology Department, Faculty of Science, Cairo University, Giza, Egypt; [4]UW Insect Museum, Department of Entomology, University of Kentucky, Lexington, KY, United States

Introduction

Helconinae is a cosmopolitan subfamily, comprising 119 species into 18 genera and four tribes, Chelonohelconinae Tobias, 1987 (fossil), Helconini Foerster, 1863, Ussurohelconini van Achterberg, 1994, and Vervootihelconini van Achterberg, 1998 (Chen & van Achterberg, 2019). Helconini is by far the most diverse tribe, comprising most of the species and genera of the subfamily (see Yu et al., 2016). On the other hand, the highly aberrant tribe Vervootihelconinae comprises only a single Chilean species (van Achterberg, 1998). Ussurohelconini was formerly placed within Cenocoeliinae, due to the high metasomal insertion on the propodeum as well as the presence of complete postpectal carina (van Achterberg, 1994). Sharanowski et al. (2011) reclassified Blacinae, which was formerly considered as a tribe of Helconinae (Sharkey, 1993; van Achterberg, 1975). Historically, the limits of the subfamilies Blacinae and Helconinae have never been well defined because of the variably included genera within the two groups (Martin, 1956; van Achterberg, 1988). Until recently, Helconinae included the tribes Blacini and Brachistini. Using molecular data, Sharanowski et al. (2011) reconsidered their placement and elevated both to subfamily status.

Members of this subfamily are very easily recognized by the shape of the second submarginal cell of the fore wing.

Additionally, they are characterized by the following combination of characters: scutellar disc not margined by carinae; dorsope absent to shallowly impressed; fore wing with RS+M vein present; and origin of the m-cu vein distinctly separated from base by 2 RS (Arias-Penna, 2007). Helconines are solitary koinobiont endoparasitoids of coleopteran larvae (Chen & van Achterberg, 2019; Sharkey, 1997), especially attacking the families Cerambycidae, Buprestidae, and perhaps other wood—boring beetles (Shaw & Huddleston, 1991; Yu et al., 2016). Despite the high diversity of this subfamily in different parts of the world, few comprehensive taxonomic studies have been done (Chen & van Achterberg, 2019).

Checklists of Regional Helconinae. Fallahzadeh and Saghaei (2010) represented three species in two genera (without precise localities); Farahani et al. (2016) and Gadallah et al. (2016) cataloged nine species in five genera (included the tribe Diospilini Foerster with three species in the genus *Diospilus* Haliday, 1833 and one species in *Taphaeus* Wesmael, 1835 under Helconinae); Yu et al. (2016) and Samin, Coronado-Blanco, Kavallieratos et al. (2018), Samin, Coronado-Blanco, Fischer et al. (2018) listed five species in three genera. The subfamily Helconinae in Middle East is currently represented by six species belonging to two genera, *Helcon* Nees and *Wroughtonia* Cameron.

Key to genera of the subfamily Helconinae in the Middle East (modified from Watanabe, 1972; Yan et al., 2017)

1. Hind femur simple, unarmed, at most slightly widened ventrally; occipital carina distinctly curved ventrally; female antenna entirely dark brown to black; metasomal sternite 1 longer than wide and sculptured at base .... *Helcon* Nees
— Hind femur serrate ventrally, armed with a tooth-like protuberance; occipital carina straight to inconspicuously curved ventrally; female antenna not entirely dark brown to black; metasomal sternite 1 about as long as wide or transverse and smooth at base .... *Wroughtonia* Cameron

Braconidae of the Middle East (Hymenoptera). https://doi.org/10.1016/B978-0-323-96099-1.00012-1

List of species of the subfamily Helconinae recorded in the Middle East

Subfamily Helconinae Foerster, 1863

Tribe Helconini Foerster, 1863

Genus *Helcon* Nees von Esenbeck, 1812

Helcon angustator Nees von Esenbeck, 1812

Catalogs with Iranian records: No catalog.
Distribution in Iran: Northern Khorasan (Samin et al., 2019).
Distribution in the Middle East countries: Iran.
Extralimital distribution: Austria, Belgium, Bulgaria, former Czechoslavakia, Finland, France, Germany, Hungary, Italy, Japan, Lithuania, Moldova, Mongolia, Netherlands, Norway, Poland, Romania, Russia, Serbia, Sweden, Switzerland.
Host records: Summarized by Yu et al. (2016) as being a parasitoid of the buprestids *Agrilus sulcicollis* Lacordaire, and *Chrysobothris igniventris* Reitter; and the cerambycids *Callidium violaceum* (L.), *Molorchus minor* (L.), *Oberea pupillata* (Gyllenhal), *Phymatodes pusillus* (Fabricius), *Phymatodes testaceus* (L.), *Pyrrhidium sanguineum* L., *Ropalopus clavipes* (Fabricius), *Ropalopus macropus* (Germar), *Saperda perforata* (Pallas), *Tetropium fuscum* (Fabricius), and *Tetropium gabrieli* Weise. It was also recorded by Tiensuu (1954) (probably erroneously) as being a parasitoid of the tortricid *Cydia pomonella* (L.).

Helcon claviventris Wesmael, 1835

Catalogs with Iranian records: Fallahzadeh and Saghaei (2010), Gadallah et al. (2016), Farahani et al. (2016), Yu et al. (2016), Samin, Coronado-Blanco, Kavallieratos et al. (2018), Samin, Coronado-Blanco, Fischer et al. (2018).
Distribution in Iran: Tehran (Hedwig, 1957).
Distribution in the Middle East: Iran.
Extralimital distribution: Austria, Belgium, China, Croatia, France, Germany, Hungary, Poland, Sweden, Switzerland, United Kingdom.
Host records: Recorded by Tobias (1986) as being a parasitoid of the melandryid *Melandrya caraboides* (L.).

Helcon heinrichi Hedqvist, 1967

Catalogs with Iranian records: Fallahzadeh and Saghaei (2010), Gadallah et al. (2016), Farahani et al. (2016), Yu et al. (2016), Samin, Coronado-Blanco, Kavallieratos et al. (2018), Samin, Coronado-Blanco, Fischer et al. (2018).
Distribution in Iran: Tehran (Elburs—Holotype) (Hedqvist, 1967).
Distribution in the Middle East: Iran.
Extralimital distribution: None.
Host records: Unknown.

Helcon tardator Nees von Esenbeck, 1812

Catalogs with Iranian records: No catalog.
Distribution in Iran: East Azarbaijan (Samin et al., 2019).
Distribution in the Middle East: Iran.
Extralimital distribution: Algeria, Austria, Azerbaijan, Belgium, Finland, France, Germany, Hungary, Italy, Korea, Moldova, Netherlands, Norway, Poland, Romania, Serbia, Slovakia, Sweden, Switzerland, Ukraine, United Kingdom.
Host records: Summarized by Yu et al. (2016) as being a parasitoid of the buprestid *Anthaxia manca* (L.); and the cerambycids *Callidium aeneum* (DeGeer), *Callidium violaceum* (L.), *Clytus arietis* (L.), *Clytus lama* Mulsant, *Leioderus kollari* Kovács, *Leiopus nebulosus* (L.), *Monochamus* sp., *Oplosia cinerea* (Mulsant), *Phymatodes testaceus* (L.), *Plagionotus arcuatus* (L.), *Pyrrhidium sanguineum* (L.), *Tetropium castaneum* (L.), and *Xylotrechus rusticus* (L.).

Genus *Wroughtonia* Cameron, 1899

Wroughtonia dentator (Fabricius, 1804)

Catalogs with Iranian records: Gadallah et al. (2016) as *Helconidea dendator*, Farahani et al. (2016) as *H. dentator*, Yu et al. (2016) as *H. dentator*, Samin, Coronado-Blanco, Kavallieratos et al. (2018), Samin, Coronado-Blanco, Fischer et al. (2018) as *H. dentator*.
Distribution in Iran: Iran (no specific locality cited) (Sakenin et al., 2008 as *H. dentator*).
Distribution in the Middle East: Iran.
Extralimital distribution: Austria, Belgium, China, Czech, Denmark, Finland, France, Georgia, Germany, Hungary, Italy, Japan, Latvia, Lithuania, Montenegro, Norway, Poland, Romania, Russia, Slovakia, Sweden, Switzerland, United Kingdom.
Host records: Summarized by Yu et al. (2016) as being a parasitoid of the cerambycids *Acanthocinus aedilis* (L.), *Acanthocinus griseus* (Fabricius), *Asemum striatum* (L.), *Callidium aeneum* (DeGeer), *Callidium violaceum* (L.), *Cerambyx scopolii* Fuessly, *Monochamus sartor* (Fabricius), *Monochamus sartor ussurovi* (Fischer), *Monochamus sutor* (L.), *Plagionotus arcuatus* (L.), *Tetropium castaneum* (L.), *Tetropium fuscum* (Fabricius), and *Tetropium gabrieli* Weise. In Iran, this species has been reared from an unknown cerambycid species (Sakenin et al., 2008).

Wroughtonia ruspator (Linnaeus, 1758)

Catalogs with Iranian records: No catalog.
Distribution in Iran: Lorestan (Samin et al., 2019 as *Helconidea ruspator*).
Distribution in the Middle East: Iran.
Extralimital distribution: Austria, Belgium, former Czechoslovakia, Finland, France, Georgia, Germany, Hungary,

Italy, Japan, Korea, Latvia, Lithuania, Netherlands, Norway, Poland, Romania, Russia, Serbia, Slovakia, Sweden, Switzerland, Ukraine, United Kingdom (Yu et al., 2016 as *Helconidea ruspator*).

Host records: Summarized by Yu et al. (2016) as being a parasitoid of the cerambycids *Acanthocinus aedilis* (L.), *Anastrangalia dubia* (Scopoli), *Callidium violaceum* (L.), *Leptura ochraceofasciata* Motschulsky, *Monochamus sutor* (L.), *Strangalia quadrifasciata* L., and *Xylotrechus capricornis* Gleber.

Conclusion

Six valid species of the subfamily Helconinae in two genera, *Helcon* Nees von Esenbeck (four species) and *Wroughtonia* Cameron (two species) in the tribe Helconini have been reported from one of the Middle East countries, Iran (5% of the world species). *Helcon heinrichi* Hedqvist (1967) is so far only known from Iran (endemic to Iran). These species have been recorded from four provinces: Tehran (two species), East Azarbaijan, Lorestan, and Northern Khorasan (each with one species). Also, the exact locality of *Wroughtonia dentator* is unknown in Iran. Only one unknown species of Cerambycidae (Coleoptera) has been recorded as the host of an Iranian helconine species. Among the Middle East and adjacent countries to Iran, the subfamily Helconinae has been recorded only from Russia (with 28 species), Azerbaijan and Kazakhstan (both with one species), of which Russia and Azerbaijan share three and one species with Iran, respectively. No helconine species have been recorded from the other Middle Eastern countries.

References

Arias-Penna, D. C. (2007). New geographical records of the genus *Urosigalphus* Ashmead, 1889 (Hymenoptera: Braconidae, Helconinae) for Colombia. *Boletin del Museo de Entomologia de la Universidad del Valee, 8*(1), 1−9.

Chen, X. X., & van Achterberg, C. (2019). Systematics, phylogeny, and evolution of braconid wasps: 30 years of progress. *Annual Review of Entomology, 64*, 1−24.

Fallahzadeh, M., & Saghaei, N. (2010). Checklist of Braconidae (Insecta: Hymenoptera) from Iran. *Munis Entomology & Zoology, 5*(1), 170−186.

Farahani, S., Talebi, A. A., & Rakhshani, E. (2016). Iranian Braconidae (Insecta: Hymenoptera: Ichneumonoidea): Diversity, distribution and host association. *Journal of Insect Biodiversity and Systematics, 2*(1), 1−92.

Gadallah, N. S., Ghahari, H., & van Achterberg, C. (2016). An annotated catalogue of the Iranian Euphorinae, Gnamptodontinae, Helconinae, Hormiinae and Rhysipolinae (Hymenoptera: Braconidae). *Zootaxa, 4072*(1), 1−38.

Hedqvist, K.-J. (1967). Notes on Helconini (Ichneumonoidea, Braconidae, Helconinae) Part I. *Entomologisk Tidskrift, 88*, 133−143.

Hedwig, K. (1957). Ichneumoniden und Braconiden aus Iran 1954 (Hymenoptera). *Jahrschefte des Vereins für Vaterländische Naturkunde Württemberg, 112*, 103−117.

Martin, J. C. (1956). A taxonomic revision of the triaspidine braconid wasps of the Nearctic America (Hymenoptera). *Canadian Department of Agriculture*, Ottawa.

Sakenin Chelav, H., Eslami, B., Samin, N., Imani, S., Shirdel, F., & Havaskary, M. (2008). A contribution to the most important trees and shrubs as the hosts of wood-boring beetles in different regions of Iran and identification of many natural enemies. *Plant and Ecosystem, 16*, 27−46 (in Persian, English summary).

Samin, N., Coronado-Blanco, J. M., Kavallieratos, N. G., Fischer, M., & Sakenin, H. (2018). Recent findings on Braconidae (Hymenoptera: Ichneumonoidea) of Iran with an updated checklist. *Acta Biologica Turcica, 31*(4), 160−173.

Samin, N., Coronado-Blanco, J. M., Fischer, M., van Achterberg, C., Sakenin, H., & Davidian, E. (2018). Updated checklist of Iranian Braconidae (Hymenoptera: Ichneumonoidea) with twenty-three new records. *Natura Somogyiensis, 32*, 21−36.

Samin, N., Fischer, M., Sakenin, H., Coronado-Blanco, J. M., & Tabari, M. (2019). A faunistic study on Agathidinae, Alysiinae, Doryctinae, Helconinae, Microgastrinae, and Rogadinae (Hymenoptera: Braconidae), with eight new country records. *Calodema, 734*, 1−7.

Sharanowski, B. J., Dowling, A. P. G., & Sharkey, M. J. (2011). Molecular phylogenetics of Braconidae (Hymenoptera: Ichneumonoidea), based on multiple nuclear genes and implications for classification. *Systematic Entomology, 36*, 549−572.

Sharkey, M. J. (1993). Family Braconidae, pp. 362−395. In H. Goulet, & J. T. Huber (Eds.), *Hymenoptera of the world: an identification guide to families*. Ottawa, Ontario: Agriculture Canada, 680 pp.

Sharkey, M. J. (1997). Subfamily Helconinae, pp. 260−273. In R. A. Wharton, P. M. Marsh, & M. Sharkey (Eds.), *Manual of the New World Genera of the Family Braconidae (Hymenoptera)* (Vol. 1, pp. 1−439). Special Publication of the International Society of Hymenopterists.

Shaw, M. R., & Huddleston, T. (1991). Classification and Biology of braconid wasps (Hymenoptera: Braconidae). *Handbooks for the Identification of British Insects, 7*, 1−126.

van Achterberg, C. (1975). A revision of the tribus Blacini (Hymenoptera: Braconidae: Helconinae). *Tijdschrift voor Entomologie, 118*, 159−322.

van Achterberg, C. (1988). revision of the subfamily Blacinae Foerster (Hymenoptera, Braconidae). *Zoologische Verhandlingen Leiden, 249*, 1−324.

van Achterberg, C. (1994). Generic revision of the subfamily Cenocoeliinae Szépligeti (Hymenoptera: Braconidae). *Zoologische Verhandelingen, 292*, 1−52.

van Achterberg, C. (1998). *Vervootihelcon*, a new genus of the subfamily Helconinae Förster (Hymenoptera: Braconidae) from Chile. *Zoologische Verhandelingen, 322*, 401−405.

Watanabe, C. (1972). A revision of the Helconini of Japan and review of Helconinae genera of the world (Hymeoptera: Braconidae). *Insecta Matsumura, 35*, 1−18.

Yan, C. J., Achterberg, C. van, He, J. J., & Chen, X. X. (2017). Review of the tribe Helconini Foerster s.s. from China, with description of 18 new species. *Zootaxa, 4291*(3), 401−457.

Yu, D. S., van Achterberg, C., & Horstmann, K. (2016). *Taxapad 2016, Ichneumonoidea 2015, Database on flash-drive*. Nepean, Ontario, Canada.

Homolobus infumator (Lyle, 1914), (Homolobinae), ♀, lateral habitus. *Photo prepared by S.R. Shaw.*

Homolobus truncator (Say, 1829) (Homolobinae), ♀, lateral habitus. *Photo prepared by S.R. Shaw.*

Chapter 17

Subfamily Homolobinae van Achterberg, 1979

Neveen Samy Gadallah[1], Hassan Ghahari[2], Donald L.J. Quicke[3], Michael J. Sharkey[4] and Scott Richard Shaw[5]

[1]Entomology Department, Faculty of Science, Cairo University, Giza, Egypt; [2]Department of Plant Protection, Yadegar-e Imam Khomeini (RAH) Shahre Rey Branch, Islamic Azad University, Tehran, Iran; [3]Integrative Ecology Laboratory, Department of Biology, Faculty of Science, Chulalongkorn University, Pathumwan, Bangkok, Thailand; [4]UW Insect Museum, Department of Entomology, University of Kentucky, Lexington, KY, United States; [5]UW Insect Museum, Department of Ecosystem Science and Management, University of Wyoming, Laramie, WY, United States

Introduction

Homolobinae van Achterberg, 1979 is a relatively small, cosmopolitan subfamily in the helconoid group of braconid genera (van Achterberg, 1979, 1993; Yu et al., 2016), occurring from sea level to up to 2163 m elevation (in Andes mountains of eastern Equador) (van Achterberg & Shaw, 2009). The subfamily Homolobinae comprises about 67 described species classified into three genera, *Exasticolus* van Achterberg, 1979, *Homolobus* Foerster, 1863, and *Westwoodiella* Szépligeti, 1904, which are arranged into two tribes (Homolobini van Achterberg, 1979, and Westwoodiellini van Achterberg, 1992) (Chen & van Achterberg, 2019). About 25 species are present in the Palaearctic region, with all of them belonging to the genus *Homolobus* (Yu et al., 2016).

Homolobines are relatively large-sized braconids, 5.0–9.0 mm in body length, light-colored (often yellowish brown) insects (Tobias, 1986; Shaw & Huddleston, 1991). They are diagnosed by the following combination of characters: metasomal T1 sessile (broadly attached to propodeum and not petiolate), distinctly narrowed behind spiracles that are located in front of middle of the tergite; T1 lacks longitudinal ridges and has a deep, large, subbasal laterope; the occipital carina is well developed, connected with the hypostomal carina above the mandibular base; the hind tibial spurs are long (0.5× or slightly longer than hind basitarsus); the prepectal ridge is complete; antenna long and thin, last flagellomere ending with a spine; forewing longer than body, with two radio-medial veins, anal cross veins absent; and the ovipositor is straight or nearly so, short, not more than 0.5× as long as metasomal length, with a small subapical notch (Shaw & Huddleston, 1991; Tobias, 1986; van Achterberg, 1979, 1993). Members of the subfamily Homolobinae are solitary koinobiont nocturnal endoparasitoids of exposed lepidopterous caterpillars mainly of the families Noctuidae and Geometridae, many of which are agricultural pests (Quicke & van Achterberg, 1990; Sharanowski et al., 2014; Tobias, 1986; van Achterberg, 1979, 1993). Homolobines are nocturnally active and some species may be collected at lights, while others are better sampled with Malaise traps and yellow bowl traps (van Achterberg & Shaw, 2009).

Checklists of Regional Homolobinae. Fallahzadeh and Saghaei (2010) represented one species, *Homolobus* (*Apatia*) *ophioninus* (Vachal, 1907) (without precise locality cited). Farahani et al. (2016) listed three species in one genus, *Homolobus* Foerster, 1863. Ghahari (2016), Yu et al. (2016), and Samin, Coronado-Blanco, Kavallieratos et al. (2018), Samin, Coronado-Blanco, Fischer et al. (2018) all listed four species in one genus. The present checklist includes seven species from the Middle East belonging to the genus *Homolobus*. Here we follow Chen and van Achterberg (2019) for higher classification, and Yu et al. (2016) for the global distribution of species.

List of species of the subfamily Homolobinae recorded in the Middle East

Subfamily Homolobinae van Achterberg, 1979

Tribe Homolobini van Achterberg, 1979

Genus *Homolobus* Foerster, 1863

Homolobus (Phylacter) annulicornis (Nees von Esenbeck, 1834)

Catalogs with Iranian records: Ghahari (2016), Yu et al. (2016), Samin, Coronado-Blanco, Kavallieratos et al. (2018), Samin, Coronado-Blanco, Fischer et al. (2018).

Braconidae of the Middle East (Hymenoptera). https://doi.org/10.1016/B978-0-323-96099-1.00010-8

Distribution in Iran: Chaharmahal and Bakhtiari (Samin et al., 2016).

Distribution in the Middle East: Cyprus (Beyarslan et al., 2017), Iran (Samin et al., 2016), Turkey (Beyarslan & Aydoğdu, 2013, 2014).

Extralimital distribution: Austria, Azerbaijan, Belarus, Belgium, China, Croatia, Czech Republic, Denamark, Finland, France, Germany, Hungary, Ireland, Italy, Japan, Korea, Latvia, Lithuania, Moldova, Netherlands, Poland, Romania, Russia, Serbia, Sweden, Switzerland, United Kingdom.

Host records: Summarized by Yu et al. (2016) as being a parasitoid of the following noctuids: *Autographa gamma* (L.), *Cosmia diffinis* (L.), *Cosmia trapezina* (L.), *Dichonia convergens* (Denis and Schiffermüller), *Dryobotodes ermita* (Fabricius), *Fissipunctia ypsillon* (Denis and Schiffermüller), *Lithophane lamada* (Fabricius), *Mamestra brassicae* (L.), *Orthosia populeti* (Fabricius), *Panolis flammea* (Denis and Schiffermüller), and *Xestia triangulum* (Hufnagel).

Homolobus (Apatia) arabicus Ghramh, 2012

Distribution in the Middle East: Saudi Arabia (Ghramh, 2012).

Extraalimital distribution: None.

Host records: Unknown.

Homolobus (Chartolobus) infumator (Lyle, 1914)

Catalogs with Iranian records: Farahani et al. (2016), Ghahari (2016), Yu et al. (2016), Samin, Coronado-Blanco, Kavallieratos et al. (2018), Samin, Coronado-Blanco, Fischer et al. (2018).

Distribution in Iran: Guilan (Farahani et al., 2012), Mazandaran (Farahani et al., 2012; Kian et al., 2020).

Distribution in the Middle East: Egypt (Abu El-Ghiet et al., 2014), Iran (see references above), Turkey (Beyarslan & Aydoğdu, 2013, 2014).

Extralimital distribution: Nearctic, Neotropical, Oriental, Palaearctic [Adjacent countries to Iran: Armenia, Azerbaijan, Kazakhstan, Russia].

Host records: Summarized by Yu et al. (2016) as being a parasitoid of the depressariid *Agonopterix alstromeriana* (Clerck); the following geometrids: *Agriopis aurantiaria* (Hübner), *Alcis repandata* (L.), *Bupalus piniarius* (L.), *Cleora cinctaria* (Denis and Schiffermüller), *Compsoptera jourdanaria* (Serres), *Ematurga atonaria* (L.), *Lambdina fiscellaria* Guenée, *Lycia hirtaria* (Clerck), *L. zonaria* (Denis and Schiffermüller), *Nepytia canosaria* Walker, *Pungeleria capreolaria* (Denis and Schiffermüller), and *Sabulodes caberata* Guenée; and the lasiocampid *Dendrolimus superans* (Butler).

Homolobus (Phylacter) meridionalis van Achterberg, 1979

Catalogs with Iranian records: No catalog.

Distribution in Iran: Hamadan (Rajabi Mazhar et al., 2018).

Distribution in the Middle East: Cyprus (van Achterberg, 1979), Iran (Rajabi Mazhar et al., 2018).

Extralimital distribution: Croatia, France, Greece, Italy, Morocco, Spain (Yu et al., 2016), Malta (Mifsud et al., 2019).

Host records: Recorded by Shaw (2015) as being a parasitoid of the noctuid *Dryobota labecula* (Esper).

Homolobus (Apatia) ophioninus (Vachal, 1907)

Catalogs with Iranian records: Fallahzadeh and Saghaei (2010), Farahani et al. (2016), Ghahari (2016), Yu et al. (2016), Samin, Coronado-Blanco, Kavallieratos et al. (2018), Samin, Coronado-Blanco, Fischer et al. (2018).

Distribution in Iran: Fars (van Achterberg, 1979), Kerman (Asadizadeh et al., 2014), Sistan and Baluchestan (Danesh, 2015).

Distribution in the Middle East: Iran.

Extralimital distribution: Australia, Democratic Republic of the Congo, Ethiopia, Kenya, Madagascar, New Caledonia, Norfolk Island, Rwanda, Tanzania, Zambia, Zimbabwe.

Host records: Summarized by Yu et al. (2016) as being a parasitoid of the following noctuids: *Agrotis segetum* (Denis and Schiffermüller), *Mythimna convecta* (Walker), and *Spodoptera exempta* (Walker).

Homolobus (Apatia) truncatoides van Achterberg, 1979

Distribution in the Middle East: Cyprus, Egypt, Iraq, Saudi Arabia (van Achterberg, 1979), Israel—Palestine (Papp, 2012), Oman, United Arab Emirates, Yemen (van Achterberg, 2014).

Extralimital distribution: Cape Verde, China, Ethiopia, Italy, Japan, Madagascar, Malaysia, Nigeria, South Africa, Spain, Sri Lanka, Vietnam.

Host records: Recorded by Caballero et al. (1990) as being a parasitoid of the noctuid *Spodoptera littoralis* Boisduval.

Homolobus (Apatia) truncator (Say, 1829)

Catalogs with Iranian records: Farahani et al. (2016), Ghahari (2016), Yu et al. (2016), Samin, Coronado-Blanco, Kavallieratos et al. (2018), Samin, Coronado-Blanco, Fischer et al. (2018).

Distribution in Iran: Alborz, Guilan, Qazvin, Tehran (Farahani et al., 2012), East Azarbaijan (Ghahari, Fischer, Erdoğan, Tabari et al., 2009; Ghahari, Fischer, Erdoğan,

Beyarslan et al., 2009), Hamadan (Rajabi Mazhar et al., 2018), Kerman (Abdolalizadeh et al., 2017), Mazandaran (Farahani et al., 2012; Kian et al., 2020).

Distribution in the Middle East: Cyprus (van Achterberg, 1979), Egypt (Kamal Bey, 1951), Iran (see references above), Israel—Palestine (Halperin, 1986; Papp, 2012), Turkey (Beyarslan, 2016; Beyarslan & Aydoğdu, 2013, 2014; Beyarslan et al., 2013).

Extralimital distribution: Distributed worldwide (except Australia) [Adjacent countries to Iran: Afghanistan, Kazakhstan, Russia, Turkmenistan].

Host records: Summarized by Yu et al. (2016) as being a parasitoid of several lepidopteran insect pests of the families Cosmopterigidae, Gelechiidae, Geometridae, Lasiocampidae, Noctuidae, Pyralidae.

Homolobus sp.

Distribution in Iran: Khuzestan (Modarres Awal, 1997; Siahpoush et al., 1993).

Host records in Iran: *Mythimna unipuncta* (Haworth, 1809) (Lepidoptera: Noctuidae) (Siahpoush et al., 1993; Modarres Awal, 1997 as *Mythimna loreyi*).

Conclusion

Seven valid species of the subfamily Homolobinae in the genus *Homolobus* Foerster, 1863 (Tribe Homolobini van Achterberg, 1979) have been reported from ten of the Middle East countries. Iranian Homolobinae with five species (7.4% of the world species) is the richest. Species have been recorded from 12 provinces, Guilan, Hamadan, Kerman, Mazandaran (each with two species), Alborz, Chaharmahal and Bakhtiari, East Azarbaijan, Fars, and Sistan and Baluchestan (each with one species), and Khuzestan (with one unknown species). One species of Noctuidae (Lepidoptera) has been recorded as the host of the Iranian Homolobinae. Comparison of the Homolobinae fauna of Iran with the Middle East and adjacent countries indicates the fauna of Russia with 10 species is more diverse than Iran, followed by Cyprus (four species), Egypt and Turkey (each with three species), Azerbaijan, Israel—Palestine, Kazakhstan and Saudi Arabia (each with two species), Afghanistan, Armenia, Iraq, Oman, Turkmenistan, United Arab Emirates, and Yemen (each with one species). On the other hand, no homolobine species are reported from Bahrain, Jordan, Kuwait, Lebanon, Qatar, Pakistan, and Syria. Additionally, among the 23 countries of the Middle East and adjacent to Iran, Cyprus, and Russia share three species with Iran, followed by Azerbaijan, Egypt, Kazakhstan (each with two species), Afghanistan, Armenia, Israel—Palestine, and Turkmenistan (each with one species).

References

Abdolalizadeh, F., Madjdzadeh, S. M., Farahani, S., & Askari Hesni, M. (2017). A survey of braconid wasps (Hymenoptera: Braconidae: Euphorinae, Homolobinae, Macrocentrinae, Rogadinae) in Kerman province, southeastern Iran. *Journal of Insect Biodiversity and Systematics, 3*(1), 33—40.

Abu El-Ghiet, U. M., Edmardash, Y. A., & Gadallah, N. S. (2014). Braconidae diversity (Hymenoptera: Ichneumonoidea) in alfalfa fields, *Medicago sativa* L., of some Western Desert Oases, Egypt. *Journal of Crop Protection, 3*(4), 543—556.

Asadizadeh, A., Mahriyan, K., Talebi, A. A., & Esfandiarpour, I. (2014). Faunistic survey of parasitoid wasps family of Braconidae from Anar Region, Kerman province. In *Proceedings of the 3rd Integrated Pest Management Conference (IPMC), 21—22 January 2014 Kerman* (p. 629).

Beyarslan, A. (2016). Taxonomical investigations of the fauna of Helconinae, Homolobinae, and Ichneutinae (Hymenoptera, Braconidae) in provinces Ardahah, Erzurum, Igdir, and Kars of Turkish north-eastern Anatolia region. *Entomofauna, 37*(25), 413—420.

Beyarslan, A., & Aydoğdu, M. (2013). Additions to the rare species of Braconidae fauna (Hymenoptera: Braconidae) from Turkey. *Munis Entomology & Zoology, 8*(1), 369—374.

Beyarslan, A., & Aydoğdu, M. (2014). Additions to the rare species of Braconidae fauna (Hymenoptera: Braconidae) from Turkey. *Munis Entomology & Zoology, 9*(1), 103—108.

Beyarslan, A., Gozuacik, C., & Ozgen, I. (2013). A contribution on the subfamilies Helconinae, Homolobinae, Macrocentrinae, Meteorinae, and Orgilinae (Hymenoptera: Braconidae) of southeastern Anatolia with new records from other parts of Turkey. *Turkish Journal of Zoology, 37*(4), 501—505.

Beyarslan, A., Gözüaçik, C., Güllü, M., & Konuksal, A. (2017). Taxonomical investigations on Braconidae (Hymenoptera: Ichneumonoidea) fauna in northern Cyprus, with twenty-six new records for the country. *Journal of Insect Biodiversity and Systematics, 3*(4), 319—334.

Caballero, P., Vargas-Osuna, E., Aldebis, H. K., & Santiago-Alvarez, C. (1990). Parasitoids associated to natural populations of *Spodoptera littoralis* Boisduval and *S. exigua* Hb. (Lepidoptera: Noctuidae). *Boletín de Sanidad Vegetal. Plagas, 16*(1), 91—96.

Chen, X. X., & van Achterberg, C. (2019). Systematics, phylogeny, and evolution of braconid wasps: 30 years of progress. *Annual Review of Entomology, 64*, 1—24.

Danesh, M. (2015). *Fauna of Braconidae (Hym., Ichneumonoidea) in the South of Sistan & Baluchestan province* (M. Sc. thesis). University of Zabol, Faculty of Agriculture, Department of Plant Protection.

Fallahzadeh, M., & Saghaei, N. (2010). Checklist of Braconidae (Insecta: Hymenoptera) from Iran. *Munis Entomology & Zoology, 5*(1), 170—186.

Farahani, S., Talebi, A. A., Rakhshani, E., & van Achterberg, C. (2012). First record of *Homolobus infumator* (Lyle, 1914) (Insecta: Hymenoptera: Braconidae: Homolobinae) from Iran. *Check List, 8*(6), 1350—1352.

Farahani, S., Talebi, A. A., & Rakhshani, E. (2016). Iranian Braconidae (Insecta: Hymenoptera: Ichneumonoidea): diversity, distribution and host association. *Journal of Insect Biodiversity and Systematics, 2*(1), 1—92.

Ghahari, H. (2016). Five new records of Iranian Braconidae (Hymenoptera: Ichnemonoidea) for Iran and annotated catalogue of the subfamily Homolobinae. *Wuyi Science Journal, 32,* 35—43.

Ghahari, H., Fischer, M., Erdoğan, Ö.Ç., Tabari, M., Ostovan, H., & Beyarslan, A. (2009). A contribution to Braconidae (Hymenoptera) from rice fields and surrounding grasslands of Northern Iran. *Munis Entomology & Zoology, 4*(2), 432—435.

Ghahari, H., Fischer, M., Erdoğan, Ö.Ç., Beyarslan, A., & Havaskary, M. (2009). A contribution to the Knowledge of the braconid-fauna (Hymenoptera, Ichneumonoidea, Braconidae) of Arasbaran, Northwestern Iran. *Entomofauna, 30,* 329—336.

Ghramh, H. A. (2012). Records of the genus *Homolobus* Foerster (Hymenoptera: Braconidae), with description of a new species from Saudi Arabia. *Trends in Biosciences, 5*(3), 166—167.

Halperin, J. (1986). Braconidae (Hymenoptera) associated with forest and ornamental trees and shrubs in Israel. *Phytoparasitica, 14*(2), 119—135.

Kamal Bey, M. (1951). The biological control of the cotton leaf-worm (*Prodenia litura* F.) in Egypt. *Bulletin de la Société Fouad 1er d'Entomologie, 35,* 221—270.

Kian, N., Goldasteh, S., & Farahani, S. (2020). A survey on abundance and species diversity of braconid wasps in forest of Mazandaran province. *Journal of Entomological Research, 12*(1), 61—69 (in Persian, English summary).

Mifsud, D., Farrugia, L., & Shaw, M. R. (2019). Braconid and ichneumonid (Hymenoptera) parasitoid wasps of Lepidoptera from the Maltese Islands. *Zootaxa, 4567*(1), 47—60.

Modarres Awal, M. (1997). Family Braconidae (Hymenoptera), pp. 265—267. In M. Modarres Awal (Ed.), *List of agricultural pests and their natural enemies in Iran* (Second ed.). Ferdowsi University of Mashhad Press, 429 pp.

Papp, J. (2012). A contribution to the braconid fauna of Israel (Hymenoptera: Braconidae), 3. *Israel Journal of Entomology, 41—42,* 165—219.

Quicke, D. L. G., & van Achterberg, C. (1990). Phylogeny of the subfamilies of the family Braconidae (Hymenoptera: Ichneumonoidea). *Zoologische Verhandlungen, 258,* 1—95.

Rajabi Mazhar, A. R., Goldasteh, S., Farahani, S., & Azizkhani, E. (2018). *Homolobus (Phylacter) meridionalis* van Achterberg, 1979 (Braconidae, Homolobinae) in Iran: the first record from Asia. *Far Eastern Entomologist, 349,* 28—32.

Samin, N., van Achterberg, C., & Erdoğan, Ö.Ç. (2016). A faunistic study on some subfamilies of Braconidae (Hymenoptera: Ichneumonoidea) from Iran. *Arquivos Entomoloxicos, 15,* 153—161.

Samin, N., Coronado-Blanco, J. M., Kavallieratos, N. G., Fischer, M., & Sakenin, H. (2018). Recent findings on Braconidae (Hymenoptera: Ichneumonoidea) of Iran with an updated checklist. *Acta Biologica Turcica, 31*(4), 160—173.

Samin, N., Coronado-Blanco, J. M., Fischer, M., van Achterberg, C., Sakenin, H., & Davidian, E. (2018). Updated checklist of Iranian Braconidae (Hymenoptera: Ichneumonoidea) with twenty-three new records. *Natura Somogyiensis, 32,* 21—36.

Sharanowski, B. J., Zhaw, Y. M., & Wanigasekara, R. W. M. U. M. (2014). Annotated checklist of Braconidae (Hymenoptera) in the Canadian Prairies Ecozone. In D. J. Giberson, & H. A. Cércamo (Eds.), Arthropods of Canadian Grasslands *(Volume 4): Biodiversity and Systematics Part 2* (pp. 399—425). Biological Survey of Canada.

Shaw, M. R., & Hudleston, T. (1991). Classification and biology of Braconid wasps (Hymenoptera: Braconidae). *Handbooks for the Identification of British Insects, 7*(11), 1—126.

Shaw, M. R. (2015). A rearing of *Homolobus (Phylacter) meridionalis* van Achterberg (Hymenoptera: Braconidae, Homolobinae) in the south of France. *Entomologist's Gazette, 66,* 245—247.

Siahpoush, A., Rabee, R., Nazemi, B., Azimi, A., & Mozaffari, M. (1993). Introduction of five hymenopterous wasps parasitizing the larvae of *Mythimna loreyi* (Dup.) (Lep.: Noctuidae) at Khuzestan corn fields. In *Proceedings of the 10th Iranian Plant Protection Congress, 1—5 September 1991* (p. 96). Shahid Bahonar University of Kerman.

Tobias, V. I. (1986). Homolobinae, Orgilinae (Mimagathidinae, Microtypinae). In G. S. Medvedev (Ed.), `Opredelitel Nasekomych Evrospeiskoi Tsasti SSSR 3, Peredpontdatokrylye 4. Opr. Faune SSSR' 145: 1-501 (pp. 263—274).

van Achterberg, C. (1979). A revision of the subfamily *Zelinae auct.* (Hymenoptera, Braconidae). *Tijdschrift voor Entomologie, 122,* 241—479.

van Achterberg, C. (1993). Illustrated key to subfamilies of the Braconidae (Hymenoptera: Ichneumonoidea). *Zoologische Verhandelingen, 283,* 1—189.

van Achterberg, C. (2014). Order Hymenoptera, family Braconidae: the subfamily Homolobinae from the United Arab Emirates, with a review of the fauna of the Arabian Peninsula. *Arthropod Fauna of the UAE, 5,* 426—433.

van Achterberg, C., & Shaw, S. R. (2009). A new species of the genus *Homolobus* from Ecuador (Hymenoptera: Braconidae: Homolobinae). *Zoologische Mededelingen, 83*(24), 805—810.

Yu, D. S., van Achterberg, C., & Horstmann, K. (2016). *Taxapad 2016, Ichneumonoidea 2015, Database on flash-drive.* Nepean, Ontario, Canada.

Hormius sp. (Hormiinae), ♀, lateral habitus. *Photo prepared by S.R. Shaw.*

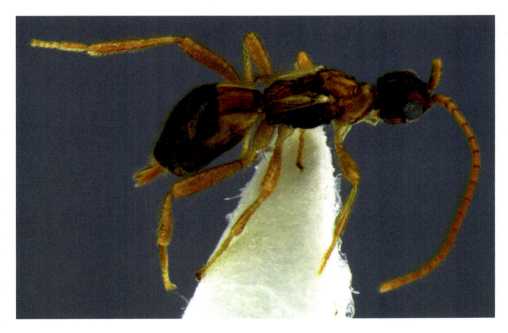

Hormius dispar (Brues, 1907) (Hormiinae), ♀ (brachypterous), lateral habitus—Nearctic. *Photo prepared by S.R. Shaw.*

Chapter 18

Subfamily Hormiinae Foerster, 1863

Hassan Ghahari[1], Neveen Samy Gadallah[2], Donald L.J. Quicke[3] and Scott Richard Shaw[4]

[1]Department of Plant Protection, Yadegar-e Imam Khomeini (RAH) Shahre Rey Branch, Islamic Azad University, Tehran, Iran; [2]Entomology Department, Faculty of Science, Cairo University, Giza, Egypt; [3]Integrative Ecology Laboratory, Department of Biology, Faculty of Science, Chulalongkorn University, Pathumwan, Bangkok, Thailand; [4]UW Insect Museum, Department of Ecosystem Science and Management, University of Wyoming, Laramie, WY, United States

Introduction

Hormiinae is a relatively small cosmopolitan subfamily which is closely related to the Rogadinae (van Achterberg, 1991; Zaldívar-Revirón et al., 2006, 2008); and comprises approximately 25 genera classified into two tribes (Hormiini Foerster and Lysitermini Tobias) (Yu et al., 2016). The limits of the subfamily have varied widely because of the lack of clear diagnostic morphological characters (Wharton, 1993; Whitfield & Wharton, 1997).

In a recent phylogenetic study carried out by Jasso-Martínez et al. (2021), using a generated ultraconserved element, it was recovered that Hormiinae and Lysiterminae are not reciprocally monophyletic; thus, they proposed to unite their members under Hormiinae. The Lysiterminini is tentatively retained as a tribe (Lysitermini) based on morphological characters, although the exact relationships of its constituent elements will require further investigation. It comprises the subtribes Cedriina, Chremylina, Lysitermina, Pentatermina, and Tetratermina. At present only members of the tribe Hormiini have been recorded from Iran although it is highly likely that some of the others also occur there.

Little is known about the biology of most Hormiini, however, those with known biology are idiobiont ectoparasitoids (often gregarious), and mostly attacking lepidopteran families such as Coleophoridae, Depressariidae, Gelechiidae, Gracillariidae, Oecophoridae, Pyralidae, Scythrididae, and Tortricidae (Shaw & Huddleston, 1991; Belokobylskij, 1993; Whitfield & Wharton, 1997). The majority are parasitoids of concealed hosts such as leaf-miners, leaf-rollers, leaf-tiers, bagworms, stem-borers, and seed-borers (Whitfield & Wharton, 1997).

At least one species of *Cedria* (Lysitermini: Cedriina) shows parental care (Beeson & Chaterjee, 1935; Yu et al., 2016). The Chremylina attack case-bearing Tineidae (Lepidoptera) larvae (van Achtertberg, 1995) and the Lysitermina have been reared from Psychidae cases and Depressariidae (Lepidoptera) (van Achterberg, 1995; Yu et al., 2016; Gupta & Quicke, 2018). Species of the genus *Katytermus* (within the Tetratermina), attack leaf-rolling gryllacridids (Orthoptera: Gryllacrididae) (van Achterberg & Steiner, 1996).

It is exceedingly difficult to diagnose the subfamily Hormiinae as a whole even though a close relationship between Hormiini and Lysitermini has been recognized for a long time (Wharton, 1993).

Hormiini are diagnosed by having the metasomal tergites largely membranous medially; the antenna slender in females, with the antennal pedicellus large in relation to the scape, but slightly shorter than it; and having the prepectal carina well-developed, usually complete, but if weakly developed then the fore wing is without the r-m vein (van Achterberg, 1995). Due to the largely soft and membranous metasomal tergites, specimens of Hormiini are best preserved in alcohol vials, or they may be critical-point dried or chemically dried to preserve the shape and characters of the abdomen. In contrast, members of Lysitermini possess strongly sclerotized metasomal tergites and these often form a three or four-segmented carapace (Wharton, 1993; van Achterberg, 1995).

Some species have recorded under the genus name *Hormisca* Telenga, 1941. However, this genus was synonymized with *Hormius* by Tobias (1974), because its recognition was based on a short fore wing marginal cell, but there is a continuum of variation as recognized by Tobias. Belokobylskij (1993) suggested, nevertheless, retaining *Hormisca* as a subgeneric name.

Checklists of Regional Hormiinae. Modarres Awal (1997, 2012) listed one species, *Hormius sculpturatus* Tobias, 1967. Fallahzadeh and Saghaei (2010) represented two species in two genera (without precise localities). Ameri et al. (2015) listed seven species in three genera (considered the genus *Pseudohormius* Tobias and Alexeev, 1973 (under the subfamily Hormiinae). Yu et al. (2016) represented six species in two genera. Gadallah et al. (2016) and Farahani et al. (2016) considered two tribes Hormiini and Avgaini (now in Pambolinae *s.l.*) under the subfamily Hormiinae, and accordingly cataloged eight and 10 species, respectively, in four genera. Samin, Coronado-Blanco, Kavallieratos et al. (2018), Samin, Coronado-Blanco, Fischer et al. (2018) listed six species in two genera. The present checklist comprises 11 species in three genera, *Chremylus* Haliday, *Hormius* Nees, and *Pseudohormius* Tobias and Alexeev reported in the Middle East.

Key to genera of the subfamily Hormiinae of the Middle East

1. Antenna 11-14-segmented; propodeum with a pair of minute tubercles .... *Chremylus* Haliday
— Antenna more than 15-segmented; propodeum without tubercles 2
2. Mesoscutum almost smooth .... *Hormius* Nees
— Mesoscutum with a wide crenulate depression between notauli .... *Pseudohormius* Tobias and Alexeev

List of species of the subfamily Hormiinae recorded in the Middle East

Subfamily Hormiinae Foerster, 1863

Tribe Hormiini Foerster, 1863

Genus *Hormius* Nees von Esenbeck, 1819

Hormius (*Hormisca*) *extima* (Tobias, 1964)

Catalogs with Iranian records: No catalog.
Distribution in Iran: Chaharmahal and Bakhtiari (Samin et al., 2019 as *Hormisca extima* (Tobias)), Northern Khorasan (Samin et al., 2020).
Distribution in the Middle East: Iran.
Extralimital distribution: Azerbaijan, Mongolia, Russia, Spain, Tajikistan, Turkmemistan.
Host records: Unknown.

Hormius (*Hormisca*) *pseudomitis* (Hedwig, 1957)

Catalogs with Iranian records: Fallahzadeh and Saghaei (2010 as *Leiophron* (*Leiophron*) *pseudomitis* (Hedwig) under Euphorinae), Sedighi and Madjdzadeh (2015 as *Leiophron pseudomitis* under Euphorinae), Gadallah et al. (2016 as *Hormisca pseudomitis* (Hedwig)), Farahani et al. (2016 as *Hormisca pseudomitis*), Yu et al. (2016 as *Hormisca pseudomitis*), Samin, Coronado-Blanco, Kavallieratos et al. (2018), Samin, Coronado-Blanco, Fischer et al. (2018) both as *Hormisca pseudomitis*.
Distribution in Iran: Sistan and Baluchestan (Hedwig, 1957 as *Euphorus pseudomitis* under Euphorinae).
Distribution in the Middle East: Iran.
Extralimital distribution: Afghanistan.
Host records: Unknown.

Hormius (*Hormisca*) *tatianae* (Telenga, 1941)

Catalogs with Iranian records: Fallahzadeh and Saghaei (2010 as *Hormisca tatianae* Telenga), Gadallah et al. (2016 as *Hormisca tatianae*), Ameri et al. (2015 as *Hormisca tatianae*), Farahani et al. (2016 as *Hormisca tatianae*), Yu et al. (2016 as *Hormisca tatianae*), Samin, Coronado-Blanco, Kavallieratos et al. (2018), Samin, Coronado-Blanco, Fischer et al. (2018) both as *Hormisca tatianae*.
Distribution in Iran: Guilan (Ghahari, Fischer, Çetin Erdoğan, Tabari, et al., 2009) Iran (no specific locality cited) (Telenga, 1941; Shenefelt, 1975; Belokobylskij & Tobias, 1986; Papp, 1989).
Distribution in the Middle East: Iran (see references above), Israel—Palestine (Papp, 1989, 2012).
Extralimital distribution: Afghanistan, Greece, Kazakhstan, Mongolia, Morocco, Tajikistan, Tunisia, Turkmenistan, Uzbekistan.
Host records: Recorded by Shenefelt (1975) as being a parasitoid of the pyralid *Ancylosis fulvobasella* Ragonot.

Hormius (*Hormius*) *moniliatus* (Nees von Esenbeck, 1811)

Catalogs with Iranian records: Gadallah et al. (2016), Ameri et al. (2015), Farahani et al. (2016), Yu et al. (2016), Samin, Coronado-Blanco, Kavallieratos et al. (2018), Samin, Coronado-Blanco, Fischer et al. (2018).
Distribution in Iran: East Azarbaijan (Ghahari, Fischer, Çetin Erdoğan, Beyarslan, et al., 2009), Hormozgan (Ameri et al., 2015), Isfahan (Ghahari et al., 2011a), Mazandaran (Ghahari, Fischer, Çetin Erdoğan, Tabari, et al., 2009; Ghahari, 2018; Kian et al., 2020; Amooghli-Tabari & Ghahari, 2021), West Azarbaijan (Samin et al., 2014), Northern Iran (Pasandideh Saqalaksari et al., 2020).
Distribution in the Middle East: Egypt (Edmardash et al., 2020), Iran (see references above), Israel—Palestine (Bodenheimer, 1930; Papp, 1970, 2012), Turkey (Beyarslan & Aydoğdu, 2013, 2014; Beyarslan, 2015).
Extralimital distribution: Nearctic, Oceanic, Oriental, Palaearctic [Adjacent countries to Iran: Afghanistan, Armenia, Azerbaijan, Kazakhstan, Russia, Turkmenistan].

Host records: Summarized by Yu et al. (2016) as being a parasitoid of the coleophorid *Coleophora trifariella* Zeller; the crambids *Achyra nudalis* (Hübner), *Hellula undalis* (Fabricius), *Paratalanta hyalinalis* (Hübner), *Pyrausta aurata* (Scopoli), *Pyrausta purpuralis* (L.), and *Pyrausta sanguinalis* (L.); the depressariids *Agonopterix adspersella* (Kollar), *Agonopterix assimilella* (Treitschke), and *Depressaria pulcherrimella* Krulikovsky; the gelechiids *Dichomeris marginella* (Fabricius), and *Pexicopia malvella* (Hübner); the scythridid *Scythris inspersella* (Hübner); and the tortricids *Archips crataegana* (Hübner), and *Pandemis corylana* (Fabricius).

Hormius (*Hormius*) *propodealis* (Belokobylskij, 1989)
Distribution in the Middle East: Egypt (Edmardash et al., 2020), Turkey (Belokobylskij, 2001).
Extralimital distribution: Australia.
Host records: Unknown.

Hormius (*Hormius*) *radialis* Telenga, 1941
Catalogs with Iranian records: Gadallah et al. (2016), Ameri et al. (2015), Farahani et al. (2016), Yu et al. (2016), Samin, Coronado-Blanco, Kavallieratos et al. (2018), Samin, Coronado-Blanco, Fischer et al. (2018).
Distribution in Iran: Hormozgan (Ameri et al., 2015), Isfahan (Ghahari et al., 2011a).
Distribution in the Middle East: Iran.
Extralimital distribution: Afghanistan, Azerbaijan, Greece, Kazakhstan, North Macedonia, Spain, Turkmenistan, former Yugoslavia.
Host records: Unknown.

Hormius (*Hormius*) *sculpturatus* Tobias, 1967
Catalogs with Iranian records: Fallahzadeh and Saghaei (2010), Gadallah et al. (2016), Ameri et al. (2015), Farahani et al. (2016), Yu et al. (2016), Samin, Coronado-Blanco, Kavallieratos et al. (2018), Samin, Coronado-Blanco, Fischer et al. (2018).
Distribution in Iran: Fars (Al-e-Mansour & Moustafavi, 1993; Modarres Awal, 1997, 2012), Hormozgan (Ameri et al., 2015).
Distribution in the Middle East: Egypt (Edmardash et al., 2020), Iran (see references above).
Extralimital distribution: Turkmenistan.
Host records: Unknown.
Plant associations in Iran: Forest and pasture plants (Modarres Awal, 1997, 2012).

Hormius (*Hormius*) *similis* Szépligeti, 1896
Catalogs with Iranian records: Gadallah et al. (2016), Ameri et al. (2015), Farahani et al. (2016), Yu et al. (2016), Samin, Coronado-Blanco, Kavallieratos et al. (2018), Samin, Coronado-Blanco, Fischer et al. (2018).

Distribution in Iran: Qazvin (Ghahari et al., 2011b), West Azarbaijan (Samin et al., 2014).
Distribution in the Middle East: Egypt (Edmardash et al., 2020), Iran (Ghahari et al., 2011b; Samin et al., 2014), Turkey (Beyarslan, 2015).
Extralimital distribution: Azerbaijan, Bulgaria, China, Croatia, Czech Republic, Germany, Greece, Hungary, Japan, Kazakhstan, Korea, North Macedonia, Mongolia, Montenegro, Russia.
Host records: Unknown.

Genus *Pseudohormius* Tobias and Alexeev, 1973
Pseudohormius flavobasalis (Hedwig, 1957)
Catalogs with Iranian records: Fallahzadeh and Saghaei (2010 as *Perilitus* (*Perilitus*) *flavobasalis* Hedwig), Gadallah et al. (2016), Ameri et al. (2015), Farahani et al. (2016), Samin, Coronado-Blanco, Kavallieratos et al. (2018), Samin, Coronado-Blanco, Fischer et al. (2018).
Distribution in Iran: Hormozgan (Ameri et al., 2015), Sistan and Baluchestan (Hedwig, 1957 as *Perilitus flavobasalis* Hedwig).
Distribution in the Middle East: Iran.
Extralimital distribution: Turkmenistan.
Host records: Unknown.

Pseudohormius turkmenus Tobias and Alexeev, 1973
Catalogs with Iranian records: Ameri et al. (2015), Farahani et al. (2016).
Distribution in Iran: Hormozgan (Ameri et al., 2015).
Distribution in the Middle East: Iran (Ameri et al., 2015), Turkey (Beyarslan, 2015).
Extralimital distribution: Russia, Turkmenistan.
Host records: Recorded by Durdyev (1990) as being a parasitoid of the bucculatricid *Bucculatrix bechsteinella* (Bechstein and Scharfenberg).

Pseudohormius sp.
Distribution in Iran: Mazandaran (Kian et al., 2020).

Tribe Lysitermini Tobias, 1968
Genus *Chremylus* Haliday, 1833
Chremylus elaphus Haliday, 1833
Catalogs with Iranian records: No catalog.
Distribution in Iran: West Azarbaijan (Gadallah et al., 2021).
Distribution in the Middle East: Iran.
Extralimital distribution: Argentina, Austria, Azores, Belgium, Czech Republic, Finland, France, Georgia, Germany, Hungary, Ireland, Italy, Japan, Latvia, Lithuania, Netherlands, New Zealand, Poland, Russia, Spain, Sweden, Switzerland, Ukraine, United Kingdom, United States of America.

Host records: Summarized by Yu et al. (2016) as being a parasitoid of the chrysomelids *Bruchidius seminarius* (L.), *Bruchus atomarius* (L.), *Bruchus lentis* Frölich, and *Bruchus rufimanus* Boheman; the curculionids *Pityogenes bidentatus* (Herbst), *Sitophilus granarius* (L.), and *Sitophilus oryzae* (L.); the ptinids *Ernobius abietis* (Fabricius), *Ernobius angusticollis* (Ratzeburg), *Ernobius longicornis* (Sturm), *Ernobius mollis* (L.), and *Stegobium paniceum* (L.); the gelechiid *Pseudotelphusa secalella* (Scopoli); the pyralids *Ephestia kuehniella* Zeller, and *Pyralis farinalis* (L.); the tineids *Nemapogon cloacella* (Haworth), *Nemapogon granella* (L.), *Niditinea fuscella* (L.), and *Tineola bisselliella* (Hummel); and the tortricids *Archips xylosteana* (L.), and *Rhyacionia buoliana* (Denis and Schiffermüller). In Iran, this species has been reared from the tortricid *Archips xylosteana* (L.) (Gadallah et al., 2021).

Conclusion

Eleven valid species of the subfamily Hormiinae in three genera and two tribes (Hormiini and Lysiterminae) have been reported from four of the Middle East countries (Egypt, Iran, Israel—Palestine, and Turkey). It was found that Iran is the richest, with 10 species, followed by Egypt and Turkey (each with four species), then Israel—Palestine (with two species). The subfamily Hormiinae of Iran have been recorded from 11 provinces, Hormozgan (with five species), West Azarbaijan (with three species), Isfahan, Sistan andand Baluchestan (with two species), Chaharmahal and Bakhtiari, East Azarbaijan, Fars, Guilan, Mazandaran, Northern Khorasan, Qazvin (each with one species). Only one host record, *Archips xylosteana* (L.) (Lepidoptera: Tortricidae) has been recorded for a single species, *Chremylus elaphis* Haliday in Iran. Among the countries of the Middle East and those adjacent to Iran, Turkmenistan shares seven species with Iran, followed by Russia (with five species), Afghanistan, Azerbaijan, Kazakhstan (with four species), Egypt and Turkey (both with three species), Israel—Palestine (two species), and Armenia (with one species).

References

Al-e-Mansour, H., & Moustafavi, M. S. (1993). The first record of Braconidae bees on forest and range vegetation in the Fars province. In *Proceedings of the 11th Iranian plant protection congress, 28 August − 2 September 1993* (p. 236). University of Guilan.

Ameri, A., Talebi, A. A., Rakhshani, E., & Beyarslan, A. (2015). A review of the subfamily *Hormiinae* (Hymenoptera: Braconidae) from Iran. *Journal of Insect Biodiversity and Systematics, 1*(2), 111−123.

Amooghli-Tabari, M., & Ghahari, H. (2021). A study on the transitional population of overwintering larvae of *Chilo suppressalis* Walker (Lepidoptera: Crambidae) with percent rice infestation in main crop, and their natural enemies in Mazandaran, Iran. *Experimental Animal Biology, 9*(4), 39−54 [in Persian, English summary].

Beeson, C. F. C., & Chattarjee, S. N. (1935). On the biology of the Braconidae [Hymenoptera]. *Indian forest records - Entomology series, 1*, 105−138.

Belokobylskij, S. A. (1993). On the classification and phylogeny of the braconid wasps of subfamilies Doryctinae and Exothecinae (Hymenoptera: Braconidae). Part I. On the classification, 2. *Entomologicheskoe Obozrenie, 72*, 143−164 [in Russian].

Belokobylskij, S. A. (2001). New taxa of the braconid subfamilies Doryctinae and Exothecinae (Hymenoptera, Braconidae) from west palaearctic. *Entomologicheskoe Obozrenie, 80*(2), 451−471.

Belokobylskij, S. A., & Tobias, V. I. (1986). Doryctinae. In G. S. Medvedev (Ed.), *Opredelitel Nasekomych Evrospeiskoi Tsasti SSSR 3, Peredpontdatokrylye 4. Opr. Faune SSSR* (vol. 145, pp. 21−72). Leningrad: Nauka Press, 501 pp.

Beyarslan, A. (2015). A faunal study of the subfamily Doryctinae in Turkey (Hymenoptera: Braconidae). *Turkish Journal of Zoology, 39*(1), 126−143.

Beyarslan, A., & Aydoğdu, M. (2013). Additions to the rare species of Braconidae fauna (Hymenoptera: Braconidae) from Turkey. *Munis Entomology & Zoology, 8*(1), 369−374.

Beyarslan, A., & Aydoğdu, M. (2014). Additions to the rare species of Braconidae fauna (Hymenoptera: Braconidae) from Turkey. *Munis Entomology & Zoology, 9*(1), 103−108.

Bodenheimer, F. S. (1930). Die schädlingsfauna palästinas. *Monographien zur Angewandten Entomologie, 10*, 438.

Durdyev, S. K. (1990). First raising of *Pseudohormius turkmenus* Tobias et Alexeev (Hymenoptera, Braconidae) from *Bucculatrix crataegi* (Lepidoptera, Bucculatricidae) in orchards of the south Turkmenistan. *Izvestiya Akademii Nauk Turkmenskoi Ssr Seriya Biologicheskikh Nauk, 1990*(2), 75−77.

Edmardash, Y. A., Abu El-Ghiet, U. M., & Gadallah, N. S. (2020). First record of Hormiini Förster, 1863, and Macrocentrinae Förster, 1863 (Hymenoptera: Braconidae) for the fauna of Egypt, with the description of a new species. *Zootaxa, 4722*(6), 555−570.

Fallahzadeh, M., & Saghaei, N. (2010). Checklist of Braconidae (Insecta: Hymenoptera) from Iran. *Munis Entomology and Zoology, 5*(1), 170−186.

Farahani, S., Talebi, A. A., & Rakhshani, E. (2016). Iranian Braconidae (Insecta: Hymenoptera: Ichneumonoidea): diversity, distribution and host association. *Journal of Insect Biodiversity and Systematics, 2*(1), 1−92.

Gadallah, N. S., Ghahari, H., & van Achterberg, C. (2016). An annotated catalogue of the Iranian Euphorinae, Gnamptodontinae, Helconinae, Hormiinae and Rhysipolinae (Hymenoptera: Braconidae). *Zootaxa, 4072*(1), 1−38.

Gadallah, N. S., Ghahari, H., & Quicke, D. L. J. (2021). Further addition to the braconid fauna of Iran (Hymenoptera: Braconidae). *Egyptian Journal of Biological Pest Control, 31*, 32. https://doi.org/10.1186/s41938-021-00376-8

Ghahari, H. (2018). Species diversity of the parasitoids in rice fields of northern Iran, especially parasitoids of rice stem borer. *Journal of Animal Environment, 9*(4), 289−298 [in Persian, English summary].

Ghahari, H., Fischer, M., Çetin Erdoğan, Ö., Tabari, M., Ostovan, H., & Beyarslan, A. (2009). A contribution to Braconidae (Hymenoptera) from rice fields and surrounding grasslands of northern Iran. *Munis Entomology & Zoology, 4*(2), 432−435.

Ghahari, H., Fischer, M., Çetin Erdoğan, Ö., Beyarslan, A., & Havaskary, M. (2009). A contribution to the knowledge of the braconid-fauna (Hymenoptera, Ichneumonoidea, Braconidae) of Arasbaran, northwestern Iran. *Entomofauna, 30*(20), 329−336.

Ghahari, H., Fischer, M., & Papp, J. (2011a). A study on the braconid wasps (Hymenoptera: Braconidae) from Isfahan province, Iran. *Entomofauna, 32*(16), 261–272.

Ghahari, H., Fischer, M., & Papp, J. (2011b). A study on the Braconidae (Hymenoptera: Ichneumonoidea) from Qazvin province, Iran. *Entomofauna, 32*(9), 197–208.

Gupta, A., & Quicke, D. L. J. (2018). A new species of *Acanthormius* (Braconidae: Lysiterminae) reared as a gregarious parasitoid of psychid caterpillar (Lepidoptera: Psychidae) from India. *Zootaxa, 4388*(3), 425–430.

Hedwig, K. (1957). Ichneumoniden und Braconiden aus Iran 1954 (Hymenoptera). *Jharschefte des Vereins für Vaterländische Naturkunde Württemberg, 112*, 103–117.

Jasso-Martínez, J. M., Quicke, D. L. J., Belokobylskij, S. A., Meza-Lázaron, R. N., & Zaldívar-Riverón, A. (2021). Phylogenomics of the lepidopteran endoparasitoid wasp subfamily Rogadinae (Hymenoptera: Braconidae) and related subfamilies. *Systematic Entomology, 46*, 83–95.

Kian, N., Goldasteh, S., & Farahani, S. (2020). A survey on abundance and species diversity of Braconid wasps in forest of Mazandaran province. *Journal of Entomological Research, 12*(1), 61–69 [in Persian, English summary].

Modarres Awal, M. (1997). Family Braconidae (Hymenoptera). In M. Modarres Awal (Ed.), *List of agricultural pests and their natural enemies in Iran* (Second edition, pp. 265–267). Ferdowsi University of Mashhad Press, 429 pp.

Modarres Awal, M. (2012). Family Braconidae (Hymenoptera). In M. Modarres Awal (Ed.), *List of agricultural pests and their natural enemies in Iran* (Third edition, pp. 483–486). Ferdowsi University of Mashhad Press, 759 pp.

Papp, J. (1970). A contribution to the braconid fauna of Israel (Hymenoptera). *Israel Journal of Entomology, 5*, 63–76.

Papp, J. (1989). A contribution to the braconid fauna of Israel (Hymenoptera), 2. *Israel Journal of Entomology, 22*, 45–59.

Papp, J. (2012). A contribution to the braconid fauna of Israel (Hymenoptera: Braconidae), 3. *Israel Journal of Entomology, 41–42*, 165–219.

Pasandideh Saqalasari, M., Talebi, A. A., & van de Kamp, T. (2020). MicroCT 3D reconstruction of three described braconid species (Hymenoptera: Braconidae). *Journal of Insect Biodiversity and Systematics, 6*(4), 331–342.

Samin, N., Ghahari, H., Gadallah, N. S., & Davidian, E. (2014). A study on the Braconidae (Hymenoptera: Ichneumonoidea) from West Azarbaijan province, northern Iran. *Linzer Biologische Beiträge, 46*(2), 1447–1478.

Samin, N., Coronado-Blanco, J. M., Kavallieratos, N. G., Fischer, M., & Sakenin, H. (2018). Recent findings on Braconidae (Hymenoptera: Ichneumonoidea) of Iran with an updated checklist. *Acta Biologica Turcica, 31*(4), 160–173.

Samin, N., Coronado-Blanco, J. M., Fischer, M., van Achterberg, C., Sakenin, H., & Davidian, E. (2018). Updated checklist of Iranian Braconidae (Hymenoptera: Ichneumonoidea) with twenty-three new records. *Natura Somogyiensis, 32*, 21–36.

Samin, N., Fischer, M., Sakenin, H., Coronado-Blanco, J. M., & Tabari, M. (2019). A faunistic study on Agathidinae, Alysiinae, Doryctinae, Helconinae, Microgastrinae and Rogadinae (Hymenoptera: Braconidae), with eight new country records. *Calodema, 734*, 1–7.

Samin, N., Sakenin Chelav, H., Ahmad, Z., Penteado-Dias, A. M., & Samiuddin, A. (2020). A faunistic study on the family Braconidae (Hymenoptera: Ichneumonoidea) from Iran. *Scientific Bulletin of Uzhhorod National University (Series: Biology), 48*, 14–19.

Sedighi, S., & Madjdzadeh, M. (2015). Updated checklist of Iranian Euphorinae (Hymenoptera: Ichneumonoidea: Braconidae). *Biharean Biologist, 9*(2), 98–104.

Shaw, M. R., & Huddleston, T. (1991). Classification and biology of braconid wasps (Hymenoptera: Braconidae). *Handbooks for the identification of British insects* (vol. 7, pp. 1–126). London: Royal Entomological Society of London.

Shenefelt, R. D. (1975). Braconidae 8. Exothecinae, Rogadinae. *Hymenopterorum Catalogus (Nova Editio), 12*, 1115–1262.

Telenga, N. A. (1941). Family braconidae, subfamily Braconinae (continuation) and Sigalphinae. Fauna USSR. *Hymenoptera, 5*(3), 466. Moskva-Leningrad: Akademiya nauk SSSR.

Tobias, V. I. (1974). Contribution to the fauna of Braconidae (Hymenoptera) of Mongolia. *Insects of Mongolia, 2*, 261–274 [in Russian].

van Achterberg, C. (1991). Revision of the genera of the Afrotropical and W. Palaearctic Rogadinae Foerster (Hymenoptera: Braconidae). *Zoologische Verhandelingen, 273*, 1–102.

van Achterberg, C. (1995). Generic revision of the subfamily Betylobraconinae (Hymenoptera: Braconidae) and other groups with modified fore tarsus. *Zoologische Verhandlingen, 298*, 1–242.

van Achterberg, C., & Steiner, H. (1996). A new genus of Tetratermini (Hymenoptera: Braconidae: Lysiterminae) parasitic on grasshoppers (Gryllacrididae). *Zoologische Mededelingen, 70*(17), 249–259.

Wharton, R. A. (1993). Review of the Hormiini (Hymenoptera: Braconidae) with a description of new taxa. *Journal of Natural History, 27*, 107–171.

Whitfield, J. B., & Wharton, R. A. (1997). Subfamily Hormiinae. In R. A. Wharton, P. M. Marsh, & M. J. Sharkey (Eds.), *Manual of the New world genera of the family Braconidae (Hymenoptera)* (vol. 1, pp. 284–301). Special Publications International Society of Hymenopterists (Special publication), 1–439.

Yu, D. S., van Achterberg, C., & Horstmann, K. (2016). *Taxapad 2016, Ichneumonoidea 2015, database on flash-drive*. Nepean, Ontario, Canada.

Zaldívar-Riverón, A., Mori, M., & Quicke, D. L. J. (2006). Systematics of the cyclostome subfamilies of braconid parasitic wasps (Hymenoptera: Ichneumonoidea); a simultaneous molecular and morphological Bayesian approach. *Molecular Phylogenetics and Evolution, 38*, 130–145.

Zaldívar-Riverón, A., Shaw, M. R., Sáez, A. G., Mori, M., Belokobylskij, S. A., Shaw, S. R., & Quicke, D. L. J. (2008). Evolution of the parasitic wasp subfamily Rogadinae (Braconidae): phylogeny and evolution of lepidopteran host ranges and mummy characteristics. *BMC Evolutionary Biology, 8*, 329.

Ichneutes bicolor Cresson, 1872 (Ichneutinae), ♀, lateral habitus—Nearctic. *Photo prepared by S.R. Shaw.*

Ichneutes bicolor Cresson, 1872 (Ichneutinae), ♂, lateral habitus—Nearctic. *Photo prepared by S.R. Shaw.*

Chapter 19

Subfamily Ichneutinae Foerster, 1863

Neveen Samy Gadallah[1], Hassan Ghahari[2], Scott Richard Shaw[3], Michael J. Sharkey[4] and Donald L.J. Quicke[5]

[1]Entomology Department, Faculty of Science, Cairo University, Giza, Egypt; [2]Department of Plant Protection, Yadegar-e Imam Khomeini (RAH) Shahre Rey Branch, Islamic Azad University, Tehran, Iran; [3]UW Insect Museum, Department of Ecosystem Science and Management, University of Wyoming, Laramie, WY, United States; [4]UW Insect Museum, Department of Entomology, University of Kentucky, Lexington, KY, United States; [5]Integrative Ecology Laboratory, Department of Biology, Faculty of Science, Chulalongkorn University, Pathumwan, Bangkok, Thailand

Introduction

Ichneutinae are a relatively small cosmopolitan subfamily of the family Braconidae, with about 81 currently valid species in 10 genera (Fischer et al., 2015; Yu et al., 2016), and two tribes (Ichneutini Foerster, 1863, and Muesebeckiini Mason, 1969) (Chen & van Achterberg, 2019). Sharkey and Wharton (1994) included Proteropinae as a tribe within Ichneutinae; however, molecular studies (Li et al., 2016; Sharanowski et al., 2011) exclude it from Ichneutinae. The position of the tribe Muesebeckiini is still uncertain as it was not included in the molecular study by Sharanowski et al. (2011); however, the morphological and molecular phylogenetic study by Dowton et al. (2002) found that it did not nest within Ichneutinae. The subfamily was established by Foerster (1863) for the genera of *Ichneutes* Nees, 1816 and *Proterops* Wesmael, 1835 (Sharkey & Wharton, 1994). Ichneutinae has received some attention because of its confused taxonomic history (He et al., 1997; Ranjith et al., 2020; Sharkey & Wharton, 1994). Members of the subfamily Ichneutinae are medium-sized and rather stout braconids, which have the forewing with vein 1-M vein abruptly curved at anterior end (Shaw & Huddleston, 1991; van Achterberg, 1993). Ichneutines are unique since they are one of a few braconid subfamilies that include species known as koinobiont ovo-larval endoparasitoids of sawfly larvae especially of the families Tenthredinidae and Argidae (He et al., 1997; Sharanowski & Sharkey, 2007; Sharkey & Wharton, 1994; Shaw & Huddleston, 1991; Tobias, 1986).

Members of Musesebeckiini parasitize leaf-mining lepidopteran hosts (He et al., 1997; Sharkey & Wharton, 1994). Muesebeckiines have not yet been detected in Iran (possibly because of their minute size and cryptic habits); however, they were recently discovered in India (Ranjith et al., 2020). A key to the Old World species of *Paraligoneurus*, and a checklist of species, was provided by Ranjith et al. (2020).

Ichneutinae *s.l.*, including Proteropinae, has been asserted to be paraphyletic by several studies and has been determined to be a "grade" basal to remaining subfamilies of the microgastroid complex (Belshaw & Quicke, 2002; Belshaw et al., 1998, 2000; Dowton et al., 2002; Murphy et al., 2008; Pitz et al., 2007; Quicke & van Achterberg, 1990; Shi et al., 2005), a proposition that is supported by Sharanowski et al. (2011) based upon the lack of polydnaviruses.

Checklists of Regional Ichneutinae. Farahani et al. (2016) represented one species in one genus (*Ichneutes* Nees, 1816). Yu et al. (2016), Samin, Coronado-Blanco, Kavallieratos et al. (2018) and Samin, Coronado-Blanco, Fischer et al. (2018) listed two species in two genera. Gadallah et al. (2019) cataloged three species in three genera (including the genus *Proterops*, which is now treated under the subfamily Proteropinae). The present Middle Eastern checklist includes four species in two genera (*Ichneutes* Nees von Esenbeck, 1816 and *Pseudichneutes* Belokobylskij, 1996). Here we follow Yu et al. (2016) for the global distribution of species.

Key to genera of the subfamily Ichneutinae in the Middle East (modified from Belokobylskij, 1996; He et al., 1997)

1. Anterior tentorial pits small, not deep; notauli fused posteriorly; vein r of fore wing arising behind the middle of pterostigma; marginal cell of fore wing long and acute apically; vein cu-a of hind wing relatively long, distinctly inclivous or subvertical; precoxal sulcus sculptured .... *Ichneutes* Nees

− Anterior tentorial pits very large and deep; notauli not fused posteriorly; vein r of fore wing arising from middle of pterostigma; marginal cell of fore wing narrower and subtruncate apically; vein cu-a of hind wing relatively short, reclivous or almost vertical; precoxal sulcus smooth .... *Pseudichneutes* Belokobylskij

List of species of the subfamily Ichneutinae recorded in the Middle East

Subfamily Ichneutinae Foerster, 1863

Genus *Ichneutes* Nees von Esenbeck, 1816

Ichneutes brevis Wesmael, 1835

Catalogs with Iranian records: No catalog.
Distribution in Iran: Ardabil (Samin et al., 2020).
Distribution in the Middle East: Iran.
Extralimital distribution: Austria, Belgium, former Czechoslovakia, Finland, France, Germany, Hungary, Ireland, Italy, Kazakhstan, Korea, Mongolia, Poland, Russia, Switzerland, United Kingdom.
Host records: Summarized by Yu et al. (2016) as being a parasitoid of the following tenthredinids: *Endophytus anemones* (Hering), *Euura lanatae* Malaise, *Euura mucronata* (Hartig), *Euura pedunculi* (Hartig), *Euura venusta* (Brischke), *Euura viminalis* (L.), *Fenusa pusilla* (Lepeletier), *Nematus ribesii* (Scopoli), *Nematus salicis* (L.), and *Phyllocolpa* sp. In Iran, this species has been reared from the tenthredinid *Nematus fagi* Zaddach (Samin et al., 2020).

Ichneutes reunitor Nees von Esenbeck, 1816

Catalogs with Iranian records: Yu et al. (2016), Samin, Coronado-Blanco, Kavallieratos et al. (2018), Samin, Coronado-Blanco, Fischer et al. (2018), Gadallah et al. (2019).
Distribution in Iran: Chaharmahal and Bakhtiari (Samin et al., 2016).
Distribution in the Middle East: Iran (Samin et al., 2016), Turkey (Beyarslan, 2016; Beyarslan & Aydoğdu, 2013, 2014).
Extralimital distribution: Azerbaijan, Belgium, former Czechoslovakia, Finland, France, Georgia, Germany, Hungary, Ireland, Italy, Japan, Kazakhstan, Lithuania, Mongolia, Netherlands, Norway, Poland, Romania, Russia, Serbia, Sweden, Switzerland, Ukraine, United Kingdom, United States of America.
Host records: Summarized by Yu et al. (2016) as being a parasitoid of the diprionid *Neodiprion sertifer* (Geoffroy); and the following tenthredinids: *Aneugmenus padi* (L.), *Cladius pallipes* Serville, *Croesus septentrionalis* (L.), *Euura viminalis* (L.), *Hemichroa crocea* Geoffroy in Fourcroy, *Namatus* spp., *Phyllocolpa* sp., *Pontania proxima* Lepeletier, *Pristophora* spp., and *Trichiocampus viminalis* (Fallén).

Genus *Pseudichneutes* Belokobylskij, 1996

Pseudichneutes atanassovae van Achterberg, 1997

Catalogs with Iranian records: Farahani et al. (2016), Gadallah et al. (2019).
Distribution in Iran: Alborz (Farahani et al., 2012).
Distribution in the Middle East: Iran.
Extralimital distribution: Bulgaria, Montenegro.
Host records: Unknown.

Pseudichneutes levis (Wesmael, 1835)

Catalogs with Iranian records: No catalog.
Distribution in Iran: Northern Khorasan (Gadallah et al., 2021).
Distribution in the Middle East: Iran (Gadallah et al., 2021), Turkey (Beyarslan & Aydoğdu, 2013, 2014).
Extralimital distribution: Belgium, Finland, France, Germany, Hungary, Italy, Kazakhstan, Netherlands, Poland, Russia, Sweden, Turkey, Ukraine, United Kingdom.
Host records: Summarized by Yu et al. (2016) as being a parasitoid of the following tenthredinids: *Fenusa dohrnii* (Tischbein), *Fenusa ulmi* Sundevall, *Metallus pumilus* (Klug), *Pontania proxima* (Serville), *Pontania viminalis* (L.), and *Scolioneura vicina* Konow.

Conclusion

Four valid species of the subfamily Ichneutinae in two genera, *Ichneutes* Nees and *Pseudichneutes* Belokobylskij, have been reported from only two of the Middle Eastern countries, Iran and Turkey, with four and two reported species, respectively. Ichneutinae of Iran represents 4.9% of the world species. These species have been recorded from four provinces (Alborz, Ardabil, Chaharmahal and Bakhtiari, and Northern Khorasan). One host record has been reported for an Iranian ichneutine species, *Ichneutes brevis*. Among the 23 countries of the Middle East and adjacent to Iran, Kazakhstan and Russia share three species with Iran, followed by Turkey (with two species), and Azerbaijan (with one species).

References

Belokobylskij, S. A. (1996). A new Palaearctic genus of the subfamily ichneutinae (Hymenoptera: Braconidae). *Zoosystematica Rossica, 4*(2), 307−310.

Belshaw, R., & Quicke, D. L. J. (2002). Robustness of ancestral state estimates: Evolution of life history strategy in ichneumonoid parasitoids. *Systematic Biology, 51*, 450−477.

Belshaw, R., Fitton, M., Herniou, E., Gimeno, C., & Quicke, D. L. J. (1998). A phylogenetic reconstruction of the Ichneumonoidea (Hymenoptera) based on the D2 variable region of 28S ribosomal RNA. *Systematic Entomology, 23*, 109−123.

Belshaw, R., Dowton, M., Quicke, D. L. J., & Austin, A. (2000). Estimating ancestral geographical distributions: A Gondwanan origin for aphid parasitoids? *Proceedings of the Royal Society of London B, 267*, 491−496.

Beyarslan, A. (2016). Taxonomical investigations of the fauna of Helconinae, Homolobinae, and Ichneutinae (Hymenoptera, Braconidae) in provinces Ardahah, Erzurum, Igdir, and Kars of Turkish north-eastern Anatolia region. *Entomofauna, 37*(25), 413−420.

Beyarslan, A., & Aydoğdu, M. (2013). Additions to the rare species of *Braconidae fauna* (Hymenoptera: Braconidae) from Turkey. *Munis Entomology and Zoology, 8*(1), 369−374.

Beyarslan, A., & Aydoğdu, M. (2014). Additions to the rare species of Braconidae fauna (Hymenoptera: Braconidae) from Turkey. *Munis Entomology & Zoology, 9*(1), 103−108.

Chen, X. X., & van Achterberg, C. (2019). Systematics, phylogeny, and evolution of braconid wasps: 30 years of progress. *Annual Review of Entomology, 64*, 1−24.

Dowton, M., Belshaw, R., Austin, A. D., & Quicke, D. L. J. (2002). Simultaneous molecular and morphological analysis of braconid relationships (Insecta: Hymenoptera: Braconidae) indicates independent mt-tRNA gene inversions within a single wasp family. *Journal of Molecular Evolution, 54*, 210−226.

Farahani, S., Talebi, A. A., Rakhshani, E., & van Achterberg, C. (2012). New record of *Pseudichneutes atanassovae* (Hymenoptera: Braconidae: Ichneutinae) from Iran. In *Proceedings of the 20th Iranian plant protection congress, 26−29 August 2012* (p. 118). University of Shiraz.

Farahani, S., Talebi, A. A., & Rakhshani, E. (2016). Iranian Braconidae (Insecta: Hymenoptera: Ichneumonoidea): diversity, distribution and host association. *Journal of Insect Biodiversity and Systematics, 2*(1), 1−92.

Fischer, J., Tucker, E., & Sharkey, M. (2015). *Colemanus keeleyrum* (Braconidae: Ichneutinae s.l.): a new genus and species of *Eocene* wasp from the Green River Formation of western North America. *Journal of Hymenoptera Research, 44*, 57−67.

Foerster, A. (1863). Synopsis der Familien und Gattungen der Braconiden. *Verhandlungen des Naturhistorischen Vereins der Preussischen Rheinland und Westfalens, 19*(1862), 225−288.

Gadallah, N. S., Ghahari, H., & Kavallieratos, N. G. (2019). An annotated catalogue of the Iranian Charmontinae, Ichneutinae, Macrocentrinae and Orgilinae (Hymenoptera: Braconidae). *Journal of the Entomological Research Society, 21*(3), 333−354.

Gadallah, N. S., Ghahari, H., & Quicke, D. L. J. (2021). Further addition to the braconid fauna of Iran (Hymenoptera: Braconidae). *Egyptian Journal of Biological Pest Control, 31*, 32. https://doi.org/10.1186/s41938-021-00376-8

He, J., Chen, X., & van Achterberg, C. (1997). Five new species of the subfamily Ichneutinae (Hymenoptera: Braconidae) from China and Europe. *Zoologische Mededelingen, 71*(2), 9−23.

Li, Q., Wei, S. J., Tang, P., Wu, Q., Shi, M., Sharkey, M. J., & Chen, X. X. (2016). Multiple lines of evidence from mitochondrial genomes resolve phylogeny and evolution of parasitic wasp in Braconidae. *Genome Biology and Evolution, 8*(9), 2651−2662.

Murphy, N., Banks, J., Whitfield, J. B., & Austin, A. (2008). Phylogeny of the parasitic microgastroid subfamilies (Hymenoptera: Braconidae) based on sequence data from seven genes, with an improved time estimate of the origin of the lineage. *Molecular Phylogenetics and Evolution, 47*, 378−395.

Pitz, K., Dowling, A. P. G., Sharanowski, B. J., Boring, C. A. B., Seltmann, K. C., & Sharkey, M. J. (2007). Phylogenetic relationships among the Braconidae (Hymenoptera: Ichneumonoidea) as proposed by Shi et al.: A reassessment. *Molecular Phylogenetics and Evolution, 43*, 338−343.

Quicke, D. L. J., & van Achterberg, C. (1990). Phylogeny of the subfamilies of the family Braconidae (Hymenoptera: Ichneumonoidea). *Zoologische Verhandelingen, 258*, 1−95.

Ranjith, A. P., van Achterberg, C., Sankararaman, H., & Nassar, M. (2020). Discovery of the ichneutine genus *Paroligoneurus* Muesebeck (Hymenoptera: Braconidae) from the Indian subcontinent with the description of a new species from Northwest India. *Zootaxa, 4786*(3), 396−408.

Samin, N., van Achterberg, C., & Erdoğan, Ö.Ç. (2016). A faunistic study on some subfamilies of Braconidae (Hymenoptera: Ichneumonoidea) from Iran. *Arquivos Entomolóxicos, 15*, 153−161.

Samin, N., Coronado-Blanco, J. M., Kavallieratos, N. G., Fischer, M., & Sakenin, H. (2018). Recent findings on Braconidae (Hymenoptera: Ichneumonoidea) of Iran with an updated checklist. *Acta Biologica Turcica, 31*(4), 160−173.

Samin, N., Coronado-Blanco, J. M., Fischer, M., van Achterberg, C., Sakenin, H., & Davidian, E. (2018). Updated checklist of Iranian Braconidae (Hymenoptera: Ichneumonoidea) with twenty-three new records. *Natura Somogyiensis, 32*, 21−36.

Samin, N., Beyarslan, A., Ranjith, A. P., Ahmad, Z., Sakenin Chelav, H., & Hosseini Boldaji, S. A. (2020). A faunistic study on Braconidae (Hymenoptera: Ichneumonoidea) from Ardebil and East Azarbayjan provinces, Northwestern Iran. *Egyptian Journal of Plant Protection Research Institute, 3*(4), 955−963.

Sharanowski, B. J., & Sharkey, M. J. (2007). Description of three new species of *Helconichia* Sharkey and Wharton (Hymenoptera: Braconidae: Ichneutinae) with a revised key to all species. *Zootaxa, 1502*, 45−57.

Sharanowski, B. J., Dowling, A. P. G., & Sharkey, M. J. (2011). Molecular phylogenetics of Braconidae (Hymenoptera: Ichneumonoidea), based on multiple molecular genes, and implications for classification. *Systematic Entomology, 36*, 549−572.

Sharkey, M. J., & Wharton, R. A. (1994). A revision of the genera of the world Ichneutinae (Hymenoptera: Braconidae). *Journal of Natural History, 28*, 873—912.

Shaw, M. R., & Huddleston, T. (1991). Classification and biology of *Braconid wasps* (Hymenoptera: Braconidae). *Handbooks for the Identification of British Insects, 7*(11), 1—126.

Shi, M., Chen, X. X., & van Achterberg, C. (2005). Phylogenetic relationships among the Braconidae (Hymenoptera: Ichneumonoidea) inferred from partial 16S rDNA D2, 18S rDNA gene sequences and morphological characters. *Molecular Phylogenetics and Evolution, 37*, 104—116.

Tobias, V. I. (1986). Ichneutinae, pp. 291—293. In G. S. Medvedev (Ed.), *Opredelitel Nasekomych Evrospeiskoi Tsasti SSSR 3, Peredpontdatokrylye 4* (Vol. 145, pp. 1—501). Opr. Faune SSSR.

van Achterberg, C. (1993). Illustrated key to the subfamilies of the Braconidae (Hymenoptera: Ichneumonoidea). *Zoologische Verhandelingen, 283*, 1—189.

Yu, D. S., van Achterberg, C., & Horstmann, K. (2016). *Taxapad 2016, Ichneumonoidea 2015, Database on flash-drive.* Ottawa, Ontario, Canada.

Macrocentrus cingulum Brischke, 1882 (Macrocentrinae), ♀, lateral habitus. *Photo prepared by S.R. Shaw.*

Macrocentrus resinellae Linnaeus, 1758 (Macrocentrinae), ♀, lateral habitus. *Photo prepared by S.R. Shaw.*

Chapter 20

Subfamily Macrocentrinae Foerster, 1863

Hassan Ghahari[1], Scott Richard Shaw[2], Neveen Samy Gadallah[3], Donald L.J. Quicke[4] and James B. Whitfield[5]

[1]Department of Plant Protection, Yadegar-e Imam Khomeini (RAH) Shahre Rey Branch, Islamic Azad University, Tehran, Iran; [2]UW Insect Museum, Department of Ecosystem Science and Management, University of Wyoming, Laramie, WY, United States; [3]Entomology Department, Faculty of Science, Cairo University, Giza, Egypt; [4]Integrative Ecology Laboratory, Department of Biology, Faculty of Science, Chulalongkorn University, Pathumwan, Bangkok, Thailand; [5]Department of Entomology, University of Illinois at Urbana-Champaign, Urbana, IL, United States

Introduction

Macrocentrinae Foerster, 1863 is a subfamily of slender-bodied delicate Braconidae, with a worldwide distribution (Yu et al., 2016). Currently, it comprises 237 species classified into eight genera (Yu et al., 2016). Among them, the genus *Macrocentrus* Curtis, 1833 is the largest and the most diverse genus in the Old World, with 191 described species (80.5% of the total number of species) (Akhtar et al., 2014; Yu et al., 2016). The subfamily Macrocentrinae was placed among the helconoid complex together with 13 other subfamilies by Wharton (1993). This hypothesis was corroborated by a phylogenetic study based on molecular data carried out by Sharanowski et al. (2011). Macrocentrines include medium- to large-sized gracile wasps, often pale and nocturnally active, and usually with quite long antennae and long ovipositors. They are most definitively identified by the presence of subapical comb of short pegs or spines on the anterior sides of the trochantelli. Additionally, the following characters are useful for identifying Macrocentrinae: the head is conspicuously wide and transverse; the occipital carina is lacking; the median lobe of the mesoscutum protrudes above lateral lobes; the metasoma is connected to propodeum somewhat above the hind coxae; and the ovipositor is longitudinally ridged (Chen & van Achterberg, 2019; Shaw & Huddleston, 1991; van Achterberg, 1993a). Species of Macrocentrinae are koinobiont solitary or gregarious endoparasitoids of both macro- and microlepidopteran larvae (Sharanowski et al., 2011, 2014; Shi et al., 2005). Numerous species have been reported from multiple hosts (Yu et al., 2016). Many macrocentrines have been suggested as biological control agents for different pests (Krugnera et al., 2005; Onstad et al., 1991; Orr & Pleasnats, 1996). The European species of the genus *Macrocentrus* were revised by van Achterberg &

Haeselbarth (1983), while Macrocentrinae of the Palaearctic region have been keyed by van Achterberg (1993b).

Checklists of Regional Macrocentrinae. Modarres Awal (1997, 2012) represented one species (*Macrocentrus collaris* (Spinola, 1808)). Fallahzadeh & Saghaei (2010) listed two species in one genus (without precise localities). Farahani et al. (2016) listed seven species in one genus. Yu et al. (2016), Samin, Coronado-Blanco, Fischer et al. (2018) and Samin, Coronado-Blanco, Kavallieratos et al. (2018) all represented 10 species in one genus. Gadallah et al. (2019) cataloged 13 species in one genus. The present Middle Eastern checklist includes 17 species—all belonging to the genus *Macrocentrus* Curtis, 1833, of which *M. nitidus* (Wesmael, 1835) is here recorded for the first time for the Iranian fauna. We follow Yu et al. (2016) regarding the global distribution of species.

List of species of the subfamily Macrocentrinae recorded in the Middle East

Subfamily Macrocentrinae Foerster, 1863

Genus *Macrocentrus* Curtis, 1833

Macrocentrus bicolor Curtis, 1833

Catalogs with Iranian records: Farahani et al. (2016), Yu et al. (2016), Samin, Coronado-Blanco, Fischer et al. (2018), Samin, Coronado-Blanco, Kavallieratos et al. (2018), Gadallah et al. (2019).

Distribution in Iran: Fars (Ghahari et al., 2010; Samin, 2015), Guilan (Farahani et al., 2012), Mazandaran (Kian et al., 2020), northern Iran (Saqalaksari et al., 2020).

Distribution in the Middle East: Iran (see references above), Turkey (Beyarsaln et al., 2013).

Braconidae of the Middle East (Hymenoptera). https://doi.org/10.1016/B978-0-323-96099-1.00028-5

Extralimital distribution: Albania, Andorra, Austria, Azerbaijan, Belarus, Bulgaria, China, Czech Republic, France, Georgia, Germany, Greece, Hungary, Ireland, Italy, Japan, Korea, Lithuania, Moldova, Netherlands, Norway, Poland, Romania, Russia, Serbia, South Korea, Spain, Sweden, Switzerland, Ukraine, United Kingdom.

Host records: Summarized by Yu et al. (2016) as being a parasitoid of the depressariid *Agonopterix ferulae* (Zeller); the gelechiid *Anacampsis populella* (Clerck); the gracillariid *Phyllonorycter scopariella* (Zeller); the lyonetiid *Leucoptera lustratella* (Herrich-Schäffer); the lypusid *Diurnea lipsiella* (Denis and Schiffermüller); the pyralid *Acrobasis consociella* (Hübner); the tineids *Morophaga choragella* Denis and Schiffermüller, *Nemapogon cloacella* (Haworth), *Nemaxera betulinella* (Paykull), and *Triaxomera parasitella* (Hübner); and the tortricids *Archips rosana* (L.), *Archips xylosteana* (L.), and *Tortricodes alternella* (Denis and Schiffermüller).

Macrocentrus blandus Eady and Clark, 1964

Catalogs with Iranian records: Farahani et al. (2016), Yu et al. (2016), Samin, Coronado-Blanco, Fischer et al. (2018), Samin, Coronado-Blanco, Kavallieratos et al. (2018), Gadallah et al. (2019).

Distribution in Iran: Alborz, Guilan, Mazandaran (Farahani et al., 2012).

Distribution in the Middle East: Iran (Farahani et al., 2012), Turkey (Beyarslan & Aydoğdu, 2012).

Extralimital distribution: Andorra, Austria, Belarus, Bulgaria, former Czechoslovakia, Finland, France, Germany, Hungary, Ireland, Japan, Kazakhstan, Korea, Lithuania, Moldova, Mongolia, Netherlands, Norway, Russia, South Korea, Serbia, Sweden, Switzerland, United Kingdom.

Host records: Summarized by Yu et al. (2016) as being a parasitoid of the following noctuids: *Agrotis segetum* (Denis and Schiffermüller), *Dasypolia templi* (Thunberg), *Hydraecia micacea* (Esper), *Hydraecia petasitis* Double-day, and *Mesapamea secalis* (L.); and the tortricid *Zeiraphera griseana* (Hübner).

Macrocentrus cingulum Brischke, 1882

Catalogs with Iranian records: Farahani et al. (2016), Yu et al. (2016), Samin, Coronado-Blanco, Fischer et al. (2018), Samin, Coronado-Blanco, Kavallieratos et al. (2018), Gadallah et al. (2019).

Distribution in Iran: Guilan (Farahani et al., 2012), Golestan (Samin et al., 2019 as *Macrocentrus grandii* Goidanish, 1937), Mazandaran (Farahani et al., 2012; Kian et al., 2020).

Distribution in the Middle East: Iran (see references above), Turkey (Beyarslan & Aydoğdu, 2012).

Extralimital distribution: Azerbaijan, Belarus, Bulgaria, Canada, China, Czech Republic, France, Georgia, Germany, Hungary, India (introduced), Italy, Japan, Lithuania, Moldova, Netherlands, Norway, Poland, Russia, Slovakia, South Africa (introduced), South Korea, Switzerland, Ukraine, United Kingdom.

Host records in Iran: Summarized by Yu et al. (2016) as being a parasitoid of the following crambids: *Anania hortulata* (L.), *Drataea crambidoides* (Grote), *Ostrinia furnacalis* (Guenée), *Ostrinia nubilalis* (Hübner), *Pantania ruralis* (Scopoli), and *Sitochroa verticalis* (L.); the erebid *Orgyia antiqua* (L.); the noctuids *Anadevidia peponis* (Fabricius), and *Sesamia inferens* (Walker); the notodontid *Clostera anachoreta* (Denis and Schiffermüller); and the nymphalid *Vanessa atalanta* (L.). In Iran, this species has been reared from *Ostrinia nubilalis* (Hübner) on *Zea mays* (Samin et al., 2019).

Macrocentrus collaris (Spinola, 1808)

Catalogs with Iranian records: Fallahzadeh and Saghaei (2010), Farahani et al. (2016), Yu et al. (2016), Samin, Coronado-Blanco, Fischer et al. (2018), Samin, Coronado-Blanco, Kavallieratos et al. (2018), Gadallah et al. (2019).

Distribution in Iran: Alborz (Davatchi & Shojai, 1969; Shojai, 1998; Farahani et al., 2012), Ardabil, Zanjan (Ameri et al., 2020), Fars (Al-e-Mansour & Moustafavi, 1993; Modarres Awal, 1997, 2012), Guilan, Qazvin (Farahani et al., 2012), Kerman (Abdolalizadeh et al., 2017; Ameri et al., 2020; Asadizade et al., 2014), Mazandaran (Farahani et al., 2012; Ghahari et al., 2009), Tehran (Farahani et al., 2012; Modarres Awal, 1997; 2012), Iran (no specific locality cited) (Aubert, 1966; Beyarslan & Aydoğdu, 2012).

Distribution in the Middle East: Cyprus (Ingram, 1981), Iran (see references above), Israel—Palestine (Papp, 1970, 2012), Turkey (Beyarslan & Aydoğdu, 2012; Beyarslan et al., 2006; 2013), Egypt (Edmardash et al., 2020), Yemen (van Achterberg, 1993a).

Extralimital distribution: Afghanistan, Albania, Andorra, Argentina, Austria, Azerbaijan, Azores, Belarus, Belgium, Bulgaria, China, Croatia, Czech Republic, Ethiopia, Finland, France, North Macedonia, Germany, Greece, Hungary, India, Italy, Kazakhstan, Latvia, Libya, Lithuania, Moldova, Mongolia, Montenegro, Morocco, Netherlands (introduced), New Zealand (introduced), Norway, Poland, Portugal, Romania, Russia, Serbia, Slovakia, Slovenia, South Korea, Spain, Sweden, Switzerland, Tajikistan, Tunisia, Turkmenistan, Ukraine, United Kingdom, Uzbekistan.

Host records: Summarized by Yu et al. (2016) as being a parasitoid of the elaterid *Agriotes lineatus* L.; the ptinid

Anobium punctatum DeGeer; the erebid *Lymantria monacha* (L.); the geometrid *Lycia hirtaria* (Clerck); the following noctuids: *Agrotis clavis* (Hufnagel), *Agrotis exclamationis* (L.), *Agrotis ipsilon* (Hufnagel), *Agrotis segetum* (Denis and Schiffermüller), *Apamea sordens* (Hufnagel), *Chalciope mygdon* (Cramer), *Diloba caeruleocephala* (L.), *Euxoa cursoria* Hufnagel, *Helicoverpa armigera* (Hübner), *Heliothis viriplaca* (Hufnagel), *Mamestra brassicae* (L.), *Noctua pronuba* (L.), *Polymixis xanthomista* (Hübner), *Spodoptera littoralis* (Boisduval), and *Spodoptera litura* (Fabricius); the nymphalid *Polygonia c-album* (L.); the tortricids *Eupoecilia ambiguella* (Hübner), *Notocelia roborana* (Denis and Schiffermüller), and *Tortrix viridana* L.; and the yponomeutid *Yponomeuta malinella* (Zeller). In Iran, this species has been reared from the noctuid *Agrotis segetum* (Denis and Schiffermüller) (Davatchi & Shojai, 1969; Modarres Awal, 1997, 2012; Shojai, 1998).

Macrocentrus equalis Lyle, 1914

Catalogs with Iranian records: Farahani et al. (2016), Yu et al. (2016), Samin, Coronado-Blanco, Fischer et al. (2018), Samin, Coronado-Blanco, Kavallieratos et al. (2018), Gadallah et al. (2019).
Distribution in Iran: Mazandaran (Farahani et al., 2012).
Distribution in the Middle East: Iran (Farahani et al., 2012), Turkey (Beyarslan & Aydoğdu, 2012).
Extralimital distribution: Belarus, Bulgaria, Finland, Germany, Hungary, Japan, Korea, Lithuania, Mongolia, Netherlands, Russia, United Kingdom.
Host records: Summarized by Yu et al. (2016) as being a parasitoid of the noctuids *Agrotis segetum* (Denis and Schiffermüller), *Xestia ditrapezium* (Denis and Schiffermüller), and *Xestia triangulum* (Hufnagel); the tortricids *Adoxophyes orana* (Fischer), *Orthotaenia undulana* Denis and Schiffermüller, and *Pandemis heparana* (Denis and Schiffermüller).

Macrocentrus flavus Snellen van Vollenhoven, 1878

Catalogs with Iranian records: Fallahzadeh and Saghaei (2010), Farahani et al. (2016), Yu et al. (2016), Samin, Coronado-Blanco, Fischer et al. (2018), Samin, Coronado-Blanco, Kavallieratos et al. (2018), Gadallah et al. (2019).
Distribution in Iran: Kerman (Abdolalizadeh et al., 2017), Iran (no specific locality cited) (van Achterberg, 1993a; Beyarslan & Aydoğdu, 2012; Farahani et al., 2012).
Distribution in the Middle East: Iran (see references above), Turkey (Beyarslan & Aydoğdu, 2012).
Extralimital distribution: Armenia, Austria, Azerbaijan, Belarus, Bulgaria, Czech Republic, France, Germany, Greece, Hungary, Italy, Kazakhstan, Moldova, Netherlands, Poland, Russia, Slovakia, Tajikistan, Ukraine (Yu et al., 2016), Spain, United Kingdom (Shaw, 2020).

Host records: Summarized by Yu et al. (2016) as being a parasitoid of the gelechiid *Teleiodes paripunctella* (Thunberg); the pyralids *Acrobasis consociella* (Hübner), *Acrobasis glaucella* Staudinger, *Acrobasis sodalella* Zeller, and *Acrobasis tumidana* (Denis and Schiffermüller); and the tortricids *Apotomis lutosana* (Kennel), and *Exapate congelatella* (Thunberg). This species was also recorded by Shaw (2020) as being probably a parasitoid of the pyralids *Acrobasis suavella* (Zincken), and *Bazaria ruscinenella* Ragonot.

Macrocentrus infirmus (Nees von Esenbeck, 1834)

Catalogs with Iranian records: Yu et al. (2016), Samin, Coronado-Blanco, Fischer et al. (2018), Samin, Coronado-Blanco, Kavallieratos et al. (2018), Gadallah et al. (2019).
Distribution in Iran: East Azarbaijan (Samin et al., 2020), Kuhgiloyeh and Boyerahmad (Samin et al., 2016).
Distribution in the Middle East: Iran (Samin et al., 2016, 2020), Turkey (Beyarslan & Aydoğdu, 2012).
Extralimital distribution: Austria, Belarus, Belgium, Bulgaria, China, Croatia, Czech Republic, Denmark, Faeroe Islands, Finland, France, Germany, Hungary, Ireland, Italy, Kazakhstan, Korea, Lithuania, Moldova, Mongolia, Netherlands, Norway, Poland, Romania, Russia, Sweden, Switzerland, United Kingdom, former Yugoslavia.
Host records: Summarized by Yu et al. (2016) as being a parasitoid of the cossid *Zeuzera pyrina* (L.); the noctuids *Apamea monoglypha* (Hufnagel), *Hydraecia micacea* (Esper), and *Hydraecia petasitis* Doubleday; and the tortricids *Blastesthia mughiana* (Zeller), *Clavigesta sylvestrana* (Curtis), *Cydia pactolana* (Zeller), *Gypsonoma aceriana* (Duponchel), *Retima resinella* (L.). In Iran, this species has been reared from *Zeuzera pyrina* (L.) (Samin et al., 2020).

Macrocentrus kurnakovi Tobias, 1976

Catalogs with Iranian records: Gadallah et al. (2019).
Distribution in Iran: Guilan (Ghahari, 2016).
Distribution in the Middle East: Iran (Ghahari, 2016), Turkey (Beyarslan & Aydoğdu, 2012).
Extralimital distribution: Azerbaijan, former Czechoslovakia, Georgia, Germany, Hungary, Italy, Japan, Korea, Netherlands, Poland, Russia.
Host records: Summarized by Yu et al. (2016) as being a parasitoid of the tineids *Archinemapogon yildizae* Koçak, *Morophaga choragella* Denis and Schiffermüller, and *Morophagoides ussuriensis* (Caradga).

Macrocentrus linearis (Nees von Esenbeck, 1811)

Catalogs with Iranian records: No catalog.
Distribution in Iran: Ardabil (Samin et al., 2019).

Distribution in the Middle East: Iran (Samin et al., 2019), Turkey (Aydoğdu, 2014).

Extralimital distribution: Afrotropical, Nearctic, Oriental, Palaearctic [Adjacent countries to Iran: Kazakhstan].

Host records: Summarized by Yu et al. (2016) as being a parasitoid of several lepidopteran species belonging to the families Coleophoridae, Depressariidae, Drepanidae, Gelechiidae, Geometridae, Gracillaridae, Lasiocampidae, Lymantriidae, Noctuidae, Oecophoridae, Plutellidae, Psychidae, Pyralidae, Thyatiridae, Tortricidae, Yponomeutidae, Ypsolophidae, Zygaenidae.

Macrocentrus marginator (Nees von Esenbeck, 1811)

Catalogs with Iranian records: Farahani et al. (2016), Yu et al. (2016), Samin, Coronado-Blanco, Fischer et al. (2018), Samin, Coronado-Blanco, Kavallieratos et al. (2018), Gadallah et al. (2019).

Distribution in Iran: Guilan (Farahani et al., 2012).

Distribution in the Middle East: Iran (Farahani et al., 2012), Turkey (Beyarslan & Aydoğdu, 2012).

Extralimital distribution: Austria, Azerbaijan, Belarus, Belgium, Bulgaria, Canada, China, Croatia, Czech Republic, Denmark, Finland, France, Georgia, Germany, Hungary, Italy, Japan, Kazakhstan, Latvia, Lithuania, Moldova, Mongolia, Netherlands, Norway, Poland, Romania, Russia, Serbia, Slovakia, Slovenia, South Korea, Switzerland, Sweden, Ukraine, United Kingdom, United States of America.

Host records: Summarized by Yu et al. (2016) as being a parasitoid of several insect pests of the orders Coleoptera (Curculionidae), Hymenoptera (Cynipidae) and Lepidoptera (Depressariidae, Geometridae, Lycaenidae, Notodontidae, Sesiidae, Tortricidae, Ypsolophidae). The coleopteran and hymenopteran hosts are doubtful.

Macrocentrus nidulator (Nees von Esenbeck, 1834)

Catalogs with Iranian records: Gadallah et al. (2019).

Distribution in Iran: Mazandaran (Gadallah et al., 2019).

Distribution in the Middle East: Iran (Gadallah et al., 2019), Turkey (Beyarslan et al., 2018).

Extralimital distribution: Armenia, Austria, Azerbaijan, Denmark, Finland, France, Germany, Hungary, Ireland, Italy, Japan, Lithuania, Moldova, Mongolia, Montenegro, Netherlands, Norway, Poland, Russia, Slovakia, Spain, Switzerland, Ukraine, United Kingdom.

Host records: Summarized by Yu et al. (2016) as being a parasitoid of the gelechiid *Metzneria metzneriella* (Stainton); the oecophorid *Batia lambdella* (Donovan); the tortricid *Eucosma hoheenwartiana* (Denis and Schiffermüller), and the yponomeutid *Yponomeuta malinella* (Zeller).

Macrocentrus nitidus (Wesmael, 1835)

Catalogs with Iranian records: This species is new record for the fauna of Iran.

Distribution in Iran: Hamadan province, Nahavand, 1♀, 2♂, September 2011, ex *Gypsonoma aceriana* (Duponchel) (Lepidoptera: Tortricidae) in apple orchard.

Distribution in the Middle East: Iran (new record).

Extralimital distribution: Austria, Belgium, Bulgaria, Croatia, former Czechoslovakia, Finland, France, Germany, Hungary, Italy, Japan, Latvia, Mongolia, Netherlands, Norway, Poland, Romania, Russia, Serbia, Sweden, Switzerland, United Kingdom, former Yugoslavia.

Host records: Summarized by Yu et al. (2016) as being a parasitoid of the epermeniid *Epermenia chaerophylella* (Goeze); the tortricids *Acleris hastiana* (L.), *Epinotia abbreviana* (Fabricius), *Gravitarmata dealbana* (Frölich), *Gypsonoma sociana* (Haworth), and *Zeiraphera griseana* (Hübner).

Macrocentrus oriens van Achterberg and Belokobylskij, 1987

Catalogs with Iranian records: Gadallah et al. (2019).

Distribution in Iran: Fars (Hasanshahi et al., 2016).

Distribution in the Middle East: Iran.

Extralimital distribution: Russia.

Host records: Unknown.

Comments: Hasanshahi et al. (2016) has erroneously recorded *M. oriens* in association with pistachio gall aphids, *Forda hirsuta* and *Slavum* sp. (Hemiptera: Aphididae) on *Pistacia atlantica* (Gadallah et al., 2019).

Macrocentrus pallipes (Nees von Esenbeck, 1811)

Catalogs with Iranian records: No catalog.

Distribution in Iran: West Azarbaijan (Ghahari & Gadallah, 2022).

Distribution in the Middle East: Iran (Ghahari & Gadallah, 2022), Turkey (Beyarslan & Aydoğdu, 2012).

Extralimital distribution: Austria, Bulgaria, China, Croatia, Czech Republic, Finland, France, Germany, Hungary, Japan, Korea, Lithuania, Moldova, Netherlands, Poland, Romania, Russia, Slovakia, Switzerland, United Kingdom.

Host records: Summarized by Yu et al. (2016) as being a parasitoid of the crambid *Anania funebris* (Ström); the depressariids *Agonopterix hypericella* (Hübner), and *Agonopterix liturosa* (Haworth); the erebid *Arctia caja* (L.); the gelechiids *Athrips mouffetella* (L.), and *Recurvaria leucatella* (Clerck); the nymphalid *Vanessa atalanta* (L.); and the following tortricids: *Acleris variegana* (Denis and Schiffermüller), *Archips oporana* (L.), *Archips rosana* (L.), *Archips xylosteana* (L.), *Celypha lacunana* (Denis and Schiffermüller), *Celypha siderana* (Treitschke), *Clepsis rogana* (Guenée), *Clepsis spectrana* (Treitschke), *Cydia pactolana* (Zeller), *Cydia zebeana* (Ratzeburg), *Hedya nubiferana* (Haworth), *Hedya pruniana* (Hübner), *Neosphaleroptera nubilana* (Hübner), *Pandemis cerasana* (Hübner), *Pandemis heparana* (Denis and Schiffermüller), *Rhyacionia resinella* (L.), *Spilonota ocellana* (Denis and Schiffermüller), and *Triedris paleana* (Hübner). In Iran, this species has been reared from the tortricid *Archips rosana* (L.) (Ghahari & Gadallah, 2022).

Macrocentrus resinellae (Linnaeus, 1758)

Catalogs with Iranian records: Yu et al. (2016), Samin, Coronado-Blanco, Fischer et al. (2018), Samin, Coronado-Blanco, Kavallieratos et al. (2018), Gadallah et al. (2019).

Distribution in Iran: Chaharmahal and Bakhtiari (Samin et al., 2016).

Distribution in the Middle East: Iran.

Extralimital distribution: Andorra, Austria, Azerbaijan, Belarus, Bulgaria, Belgium, China, Czech Republic, Finland, France, Georgia, Germany, Greece, Hungary, Italy, Japan, Kazakhstan, Latvia, Lithuania, Moldova, Netherlands, Poland, Romania, Russia, Slovakia, Spain, Sweden, Switzerland, United Kingdom.

Host records: Summarized by Yu et al. (2016) as being a parasitoid of several species in the orders: Coleoptera (Melandryidae), Hymenoptera (Pamphiliidae), and Lepidoptera (Gelechiidae, Geometridae, Lasiocampidae, Noctuidae, Pyralidae, Sesiidae, Tortricidae).

Macrocentrus sylvestrellae van Achterberg, 2001

Distribution in the Middle East: Turkey (Beyarslan & Aydoğdu, 2012).

Extralimital distribution: France, Italy.

Host records: Recorded by van Achterberg (2001) as being a prasitoid of the pyralid *Dioryctria sylvestrella* Ratzeburg.

Macrocentrus tessulatanae Hedwig, 1959

Distribution in the Middle East: Turkey (Hedwig, 1959; van Achterberg, 1993a).

Extralimital distribution: None.

Host records: Recorded by van Achterberg (1993) as being a parasitoid of the tortricid *Pseudococcyx tessulatana* (Staudinger).

Macrocentrus thoracicus (Nees von Esenbeck, 1811)

Catalogs with Iranian records: Yu et al. (2016), Samin, Coronado-Blanco, Fischer et al. (2018), Samin, Coronado-Blanco, Kavallieratos et al. (2018), Gadallah et al. (2019).

Distribution in Iran: Chaharmahal and Bakhtiari, East Azarbaijan (Samin et al., 2016).

Distribution in the Middle East: Iran (Samin et al., 2016), Turkey (Beyarslan & Aydoğdu, 2012).

Extralimital distribution: Albania, Armenia, Austria, Azerbaijan, Belarus, Belgium, Bulgaria, China, Croatia, Finland, France, Georgia, Germany, Greece, Hungary, Italy, Japan, Kazakhstan, Lithuania, Moldova, Netherlands, Poland, Russia, Serbia, Slovakia, Spain, Sweden, Switzerland, United States of America (introduced), Ukraine, United Kingdom.

Host records: Summarized by Yu et al. (2016) as being a parasitoid of many insect pest species belonging to the orders Hymenoptera (Cynipidae), and Lepidoptera (Depressariidae, Gelechiidae, Oecophoridae, Pyralidae, Tineidae, Tortricidae, Yponomeutidae). The cynipid host records are doubtful.

Macrocentrus turkestanicus (Telenga, 1950)

Catalogs with Iranian records: No catalog.

Distribution in Iran: Fars (Ameri et al., 2020).

Distribution in the Middle East: Iran.

Extralimital distribution: India, Tajikistan, Turkmenistan, Uzbekistan.

Host records: Summarized by Yu et al. (2016) as being a parasitoid of the noctuids *Sesamia cretica* Lederer, and *Sesamia inferens* (Walker).

Conclusion

Nineteen valid species of the subfamily Macrocentrinae in the genus *Macrocentrus* Curtis, 1833 have been reported from six of the Middle East countries (Cyprus, Egypt, Iran, Israel−Palestine, Turkey, and Yemen). It was found that Iran is the richest country, with 17 species (7.1% of the world species). Iranian Macrocentrinae species have been recorded from 16 provinces, Guilan and Mazandaran (both with six species), Fars (with four species), Alborz, Ardabil, Chaharmahal and Bakhtiari, East Azarbaijan Kerman (each with two species), Golestan, Hamadan, Kuhgiloyeh and Boyerahmad, Qazvin, Tehran, West Azarbiajan, and Zanjan (each with one species). Four host species of the families Cossidae, Crambidae, Noctuidae, and Tortricidae (Lepidoptera) have been recorded for Macrocentrinae of Iran. Comparison of the Macrocentrinae fauna of Iran with the Middle East and adjacent countries to Iran indicates the fauna of Russia with 34 species in three genera (Belokobylskij & Lelej, 2019) is more diverse than Iran, followed by Turkey (15 species), Azerbaijan (11 species), Kazakhstan (eight species), Armenia (four species), Turkmenistan (three species), Pakistan (two species), Afghanistan, Cyprus, Egypt, Israel−Palestine, and Yemen (each with one species). No macrocentrine species have been recorded from Bahrain, Iraq, Jordan, Kuwait, Lebanon, Oman, Qatar, Saudi Arabia, Syria, and United Arab Emirates. Among the countries of the Middle East and those adjacent to Iran, Russia shares 15 species with Iran, followed by Turkey (11 species), Azerbaijan (nine species),

Kazakhstan (eight species), Armenia (three species), Turkmenistan (two species), and Afghanistan, Cyprus, Egypt, Israel—Palestine, and Yemen (each with one species).

References

Abdolalizadeh, F., Madjdzadeh, S. M., Farahani, S., & Askari Hesni, M. (2017). A survey of braconid wasps (Hymenoptera: Braconidae: Euphorinae, Homolobinae, Macrocentrinae, Rogadinae) in Kerman province, southeastern Iran. *Journal of Insect Biodiversity and Systematics, 3*(1), 33—40.

Akhtar, M. S., Singh, L. R., & Ramcmurthy, V. V. (2014). New species of the genus *Macrocentrus* Curtis, 1833 (Hymenoptera: Braconidae) from India. *The Pan-Pacific Entomologist, 90*(1), 11—15.

Al-e-Mansour, H., & Moustafavi, M. S. (1993). The first record of Braconidae bees on forest and range vegetation in the Fars province. In *Proceedings of the 11th Iranian plant protection congress, 28 August — 2 September 1993* (p. 236). University of Guilan.

Ameri, A., Ebrahimi, E., & Talebi, A. A. (2020). Additions to the fauna of Braconidae (Hym., Ichneumonoidea) of Iran based on specimens housed in Hayk Mirazayansn Insect Museum with six new records for Iran. *Journal of Insect Biodiversity and Systematics, 6*(4), 353—364.

Asadizade, A., Mahriyan, K., Talebi, A. A., & Esfandiarpour, I. (2014). Faunistic survey of parasitoid wasps family of Braconidae from Anar region, Kerman province. In *Proceedings of the 3rd integrated pest management conference (IPMC), 21—22 January 2014, Kerman* (p. 629).

Aubert, J. F. (1966). Liste d'identification No. 6 (présentée par le service d'identification des Entomophages). *Entomophaga, 11*(1), 115—134.

Aydoğdu, M. (2014). Parasitoid abundance of *Archips rosana* (Linnaeus, 1758) (Lepidoptera: Tortricidae) in organic cherry orchards. *North-Western Journal of Zoology, 10*(1), 42—47.

Belokobylskij, S. A., & Lelej, A. S. (2019). Annotated catalogue of the Hymenoptera of Russia. Volume II. Apocrita: Parasitica. In *Proceedings of the Zoological Institute of the Russian Academy of Sciences*. Supplement No. 8, 594 pp.

Beyarslan, A., & Aydoğdu, M. (2012). A preliminary study of the *Macrocentrus* Curtis, 1833 (Hymenoptera: Braconidae: Macrocentrinae) fauna of Turkey, with zoogeographical remarks. *Journal of Entomological Research Society, 14*(1), 83—90.

Beyarslan, A., Yurtcan, M., Çetin Erdoğan, Ö., & Aydoğdu, M. (2006). A study on the Braconidae and Ichneumonidae from Ganos Mountains (Therace Region, Turkey) (Hymenoptera: Braconidae, Ichneumonidae). *Linzer biologische Beiträge, 38/1*, 409—422.

Beyarslan, A., Gözüaçik, C., & Özgen, I. (2013). A contribution on the subfamilies Helconinae, Homolobinae, Macrocentrinae, Meteorinae, and Orgilinae (Hymenoptera: Braconidae) of southeastern Anatolia with new records from other parts of Turkey. *Turkish Journal of Zoology, 37*(4), 501—505.

Beyarslan, A., Arisoy, M., & Çolak, D. (2018). Records of some Braconidae species (Hymenoptera: Ichneumonoidea) from Gürün (Sivas) and Ilgaz Montain (Kastamonu). *Acta Biologica Turcica, 31*(4), 178—192.

Chen, X. X., & van Achterberg, C. (2019). Systematics, phylogeny, and evolution of braconid wasps: 30 years of progress. *Annual Review of Entomology, 64*, 1—24.

Davatchi, A., & Shojai, M. (1969). *Les hymenopteres entomophages de l'Iran (etudes faunestiques)*. Tehran University, Faculty of Agriculture, 88 pp.

Edmardash, Y. A., Abu El-Ghiet, U. M., & Gadallah, N. S. (2020). First record of Hormiini Förster, 1863 (Hymenoptera: Braconidae) for the fauna of Egypt, with the description of a new species. *Zootaxa, 4722*(6), 555—570.

Fallahzadeh, M., & Saghaei, N. (2010). Checklist of Braconidae (Insecta: Hymenoptera) from Iran. *Munis Entomology & Zoology, 5*(1), 170—186.

Farahani, S., Talebi, A. A., & Rakhshani, E. (2012). First records of *Macrocentrus* Curtis 1833 (Hymenoptera: Braconidae: Macrocentrinae) from Northern Iran. *Zoology and Ecology, 22*(1), 41—50.

Farahani, S., Talebi, A. A., & Rakhshani, E. (2016). Iranian Braconidae (Insecta: Hymenoptera: Ichneumonoidea): diversity, distribution and host association. *Journal of Insect Biodiversity and Systematics, 2*(1), 1—92.

Gadallah, N. S., Ghahari, H., & Kavallieratos, N. G. (2019). An annotated catalogue of the Iranian Charmontinae, Ichneutinae, Macrocentrinae and Orgilinae (Hymenoptera: Braconidae). *Journal of the Entomological Research Society, 21*(3), 333—354.

Ghahari, H. (2016). Five new records of Iranian Braconidae (Hymenoptera: Ichnemonoidea) for Iran and annotated catalogue of the subfamily Homolobinae. *Wuyi Science Journal, 32*, 35—43.

Ghahari, H., & Gadallah, N. S. (2022). Additional records to the braconid fauna (Hymenoptera: Ichneumonoidea) of Iran, with new host reports. *Entomological News* (in press).

Ghahari, H., Fischer, M., Çetin Erdoğan, Ö., Tabari, M., Ostovan, H., & Beyarslan, A. (2009). A contribution to Braconidae (Hymenoptera) from rice fields and surrounding grasslands of Northern Iran. *Munis Entomology & Zoology, 4*(2), 432—435.

Ghahari, H., Fischer, M., Hedqvist, K. J., Çetin Erdoğan, Ö., van Achterberg, C., & Beyarslan, A. (2010). Some new records of Braconidae (Hymenoptera) for Iran. *Linzer biologische Beiträge, 42*(2), 1395—1404.

Hasanshahi, G., Gharaei, A. M., Mohammadi-Khoramadi, A., Abbasipour, H., & Papp, J. (2016). First record of parasitoid wasp *Macrocentrus oriens* (Hym.: Braconidae, Macrocentrinae) from Iran. *Journal of Entomological Society of Iran, 36*(1), 77—78.

Hedwig, K. (1959). Über einige neue Schlupfwespen, zum teil aus Süditalien und der Türkei. *Nachrichten des Naturwissenschaftlichen Museums der Stadt Aschaffenburg, 62*, 95—102.

Ingram, W. R. (1981). The parasitoids of *Spodoptera littoralis* (Lep.: Noctuidae) and their role in the population control in Cyprus. *Entomophaga, 26*, 23—37.

Kian, N., Goldasteh, S., & Farahani, S. (2020). A survey on abundance and species diversity of Braconid wasps in forest of Mazandaran province. *Journal of Entomological Research, 12*(1), 61—69 [in Persian, English summary].

Krugnera, R., Daanea, K. M., Lawsonc, A. B., & Yokotoa, G. Y. (2005). Biology of *Macrocentrus iridescens* (Hymenoptera: Braconidae): a parasitoid of the obliquebanded leafroller (Lepidoptera: Tortricidae). *Environmental Entomology, 34*(2), 336—343.

Modarres Awal, M. (1997). Family Braconidae (Hymenoptera). In M. Modarres Awal (Ed.), *List of agricultural pests and their natural enemies in Iran* (2nd ed., pp. 265—267). Ferdowsi University of Mashhad Press, 429 pp.

Modarres Awal, M. (2012). Family Braconidae (Hymenoptera). In M. Modarres Awal (Ed.), *List of agricultural pests and their natural enemies in Iran* (3rd ed., pp. 483−486). Ferdowsi University of Mashhad Press, 759 pp.

Onstad, D. W., Siegel, J. P., & Maddox, J. V. (1991). Distribution of parasitism by *Macrocentrus grandii* (Hymenoptera: Braconidae) in maize infested by *Ostrinia nubilalis* (Lepidoptera: Pyralidae). *Environmental Entomology, 20*, 156−159.

Orr, D. B., & Pleasants, J. M. (1996). The potential of native prairie plant species to enhance the effectiveness of the *Ostrinia nubilalis* parasitoid *Macrocentrus grandii*. *Journal of the Kansas Entomological Society, 69*(2), 133−143.

Papp, J. (1970). A contribution to the braconid fauna of Israel (Hymenoptera). *Israel Journal of Entomology, 5*, 63−76.

Papp, J. (2012). A contribution to the braconid fauna of Israel (Hymenoptera: Braconidae), 3. *Israel Journal of Entomology, 41−42*, 165−219.

Samin, N. (2015). A faunistic study on the Braconidae of Iran (Hymenoptera: Ichneumonoidea). *Arquivos Entomoloxicos, 13*, 339−345.

Samin, N., van Achterberg, C., & Çetin Erdoğan, Ö. (2016). A faunistic study on some subfamilies of Braconidae (Hymenoptera: Ichneumonoidea) from Iran. *Arquivos Entomoloxicos, 15*, 153−161.

Samin, N., Coronado-Blanco, J. M., Kavallieratos, N. G., Fischer, M., & Sakenin, H. (2018). Recent findings on Braconidae (Hymenoptera: Ichneumonoidea) of Iran with an updated checklist. *Acta Biologica Turcica, 31*(4), 160−173.

Samin, N., Coronado-Blanco, J. M., Fischer, M., van Achterberg, C., Sakenin, H., & Davidian, E. (2018). Updated checklist of Iranian Braconidae (Hymenoptera: Ichneumonoidea) with twenty-three new records. *Natura Somogyiensis, 32*, 21−36.

Samin, N., Coronado-Blanco, J. M., Hosseini, A., Fischer, M., & Sakenin Chelav, H. (2019). A faunistic study on the braconid wasps (Hymenoptera: Braconidae) of Iran. *Natura Somogyiensis, 33*, 75−80.

Samin, N., Beyarslan, A., Ranjith, A. P., Ahmad, Z., Sakenin Chelav, H., & Hosseini Boldaji, S. A. (2020). A faunistic study on Braconidae (Hymenoptera: Ichneumonoidea) from Ardebil and East Azarbayjan provinces, Northwestern Iran. *Egyptian Journal of Plant Protection Research Institute, 3*(4), 955−963.

Saqalaksari, M. P., Talebi, A. A., & van de Kamp, T. (2020). MicroCT 3D reconstruction of three described braconid species (Hymenoptera: Braconidae). *Journal of Insect Biodiversity and Systematics, 6*(4), 331−342.

Sharanowski, B. J., Dowling, A. P. G., & Sharkey, M. J. (2011). Molecular phylogenetics of Braconidae (Hymenoptera: Ichneumonoidea), based on multiple molecular genes, and amplification for classification. *Systematic Entomology, 36*, 549−572.

Sharanowski, B. J., Zhang, Y. M., & Wanigasekara, R. W. U. M. (2014). Annotated checklist of Braconidae (Hymenoptera) in the Canadian Prairies Ecozone. In D. J. Giberson, & H. A. Cárcamo (Eds.), *Arthropods of Canadian grasslands. Volume 4. Biodiversity and systematics. Part 1* (pp. 399−425). Biological Survey of Canada, 479 pp.

Shaw, M. R. (2020). Rearing record of two species of Hymenoptera: Braconidae (Macrocentrinae and Orgilinae) new to Britain. *Entomologist's Gazette, 71*, 205−209.

Shaw, M. R., & Huddleston, T. (1991). Classification and biology of braconid wasps (Hymenoptera: Braconidae). *Handbooks for the Identification of British Insects, 7*(11), 1−126.

Shi, M., Chen, X. X., & van Achterberg, C. (2005). Phylogenetic relationships among the Braconidae (Hymenoptera: Ichneumonoidea) inferred from partial 16S rDNA, 28S rDNA D2, 18S rDNA gene sequences and morphological characters. *Molecular Phylogenetics and Evolution, 37*(1), 104−116.

Shojai, M. (1998). *Entomology (ethology, social life and natural enemies) (biological control)* (3rd ed., Vol. III). Tehran University Publications, 550 pp. [in Persian].

van Achterberg, C. (1993a). Revision of the subfamily Macrocentrinae Foerster (Hymenoptera: Braconidae) from the Palaearctic region. *Zoologische Verhandelingen, 286*, 110 pp.

van Achterberg, C. (1993b). Illustrated key to the subfamilies of the Braconidae (Hymenoptera: Ichneumonoidea). *Zoologische Verhandelingen, 283*, 1−189.

van Achterberg, C., & Haeselbarth, E. (1983). Revisionary notes on the European species of *Macrocentrus* Curtis sensu stricto (Hymenoptera, Braconidae). *Entomofauna, 4*, 37−59.

Yu, D. S., van Achterberg, C., & Horstmann, K. (2016). *Taxapad 2016, Ichneumonoidea 2015, Database on flash-drive*. Ottawa, Ontario, Canada.

Sathon neomexicanus (Muesbeck, 1921) (Microgastrinae), ♀, lateral habitus—Nearctic. *Photo prepared by S.R. Shaw.*

Cotesia ofella (Nixon, 1974) (Microgastrinae), ♀, lateral habitus. *Photo prepared by S.R. Shaw.*

Chapter 21

Subfamily Microgastrinae Foerster, 1863

James B. Whitfield[1], Neveen Samy Gadallah[2], Hassan Ghahari[3] and Scott Richard Shaw[4]

[1]Department of Entomology, University of Illinois at Urbana-Champaign, Urbana, IL, United States; [2]Entomology Department, Faculty of Science, Cairo University, Giza, Egypt; [3]Department of Plant Protection, Yadegar-e Imam Khomeini (RAH) Shahre Rey Branch, Islamic Azad University, Tehran, Iran; [4]UW Insect Museum, Department of Ecosystem Science and Management, University of Wyoming, Laramie, WY, United States

Introduction

The cosmopolitan subfamily Microgastrinae is perhaps the most hyperdiverse of all the braconid wasps, as well as being the one of greatest economic importance (Fernandez-Triana et al., 2020). At the time of this writing, the subfamily ranks second to the Braconinae in terms of diversity, comprising 2999 extant valid species classified into 81 genera worldwide (Abdoli & Pourhaji, 2019; Fernández-Triana & Boudreault, 2018; Fernández-Triana & Ward, 2017; Fernandez-Triana et al., 2020; Liu et al., 2019; Whitfield et al., 2018; Yu et al., 2016; Zargar et al., 2019a-c). The actual world species richness of the Microgastrinae has been estimated to be between 20,000 and roughly 50,000 species, with a broad range of uncertainty because these estimates are based on data primarily from temperate regions with the exception of Costa Rica (Fernandez-Triana et al., 2020; Rodriguez et al., 2013). Although microgastrines are easily distinguished as a group, their enormous diversity and extremely conservative morphology makes their taxonomic study very difficult, especially at the generic level (Fernandez-Triana, 2020; Shaw & Huddleston, 1991; Smith et al., 2008). The subfamily has previously been divided into six tribes (Apantelini Viereck, 1918; Cotesiini Mason, 1981; Forniciini Mason, 1981; Microgastrini Foerster, 1863; Microplitini Mason, 1981; Semionini Tobias, 1987) (Chen & van Achterberg, 2019). However, these are now generally abandoned and Fernandez-Triana et al. (2020) classified the microgastrine genera into three informal groups: the *Microplitis* group, the *Cotesia* group, and the *Apanteles* group and then the remainder which are currently treated as a group of unplaced genera.

Rich data concerning genetics, ecology, physiology, and behavior have been accumulated for quite a number of microgastrine species (Whitfield et al., 2018). Molecular phylogenetic studies confirmed placement of the subfamily within a larger grouping comprising the Cardiochilinae, Cheloninae (including Adeliini), Khoikoiinae, Mendesellinae, and Miracinae (Banks & Whitfield, 2006; Belshaw et al., 1998, 2000; Dowton & Austin, 1998; Dowton et al., 1998; Murphy et al., 2008; Sharanowski et al., 2011; Whitfield, 1997). This set of subfamilies is collectively referred to as the "microgastroid" lineage. Some molecular analyses further suggest that the Ichneutinae s.l. (i.e., Ichneutinae + Proteropinae) may also belong with the microgastroids, but a robust phylogeny of the group is still lacking. The same applies to relationships within the Microgastrinae, and the limits of some genera are still not well defined.

Despite their small body size, microgastrines are very familiar to both agriculturalists and ecologists as being one of the most critically important groups for understanding insect parasitism of Lepidoptera in many areas of agricultural, ecological, and basic science (Whitfield et al., 2018). Microgastrines are virtually exclusively parasitoids of larval Lepidoptera, and all are koinobiont endoparasitoids (Smith et al., 2008). They are important natural enemies of many pest caterpillars that feed on a wide range of agricultural and forestry "crop" plants in all parts of the world (Fernandez-Triana et al., 2020; Whitfield et al., 2018) and have been the focus of many ecological and agricultural efforts to control and manage caterpillar pests (Fernández-Triana et al., 2011, 2016; Hrcek et al., 2013; Janzen et al., 2009; Le Corff et al., 2000; Smith et al., 2008; Whitfield, 1997; Whitfield et al., 1999, 2009). Indeed, more than 100 microgastrine species have been used successfully in biocontrol programs (Austin & Dangerfield, 1992; Fernandez-Flores et al., 2013; Whitfield, 1997). The enormous diversity of this group as well as its economic importance has caused it to become one of the most studied parasitic wasp groups for establishing DNA barcodes (Fernandez-Floes et al., 2013; Rodriguez et al., 2013; Smith et al., 2013; Stahlhut et al., 2013).

Braconidae of the Middle East (Hymenoptera). https://doi.org/10.1016/B978-0-323-96099-1.00006-6

The Microgastrinae has been recognized as a distinct subfamily for more than 150 years, since the time of Foerster (1863) (Fernández-Triana & Ward, 2017). More than 757 species are currently known from the Palaearctic region, the least diverse zoogeographic realm (Fernandez-Triana et al., 2020). Of these, 529 are known from the western Palaearctic and 496 from the eastern Palaearctic (Fernández-Triana et al., 2014; Yu et al., 2016). A large proportion of the microgastrine species were formerly placed under the genus *Apanteles* by Nixon (1965), who divided the genus into 44 species groups. Subsequent revision of the subfamily's classification by Mason (1981) led to recognition that *Apanteles* was not a monophyletic entity and he divided its members into a number of genera (among his 51 genera and five tribes). However, the genus *Apanteles* Foerster, 1863, even in the narrowed sense, is still one of the most diverse and widespread genera of the subfamily. Currently, *Apanteles* comprises at least 633 described species from all biogeographical regions, representing 21% of the total number of microgastrine species in the world (Fernandez-Triana et al., 2020). The reason why so many species were placed by earlier workers in *Apanteles* was that the definition of the genus was based almost entirely on the basis of the absence of a small wing vein, and Mason showed that the vein in question (1r-m) must have been lost several times independently.

The last published keys for World microgastrine genera were provided by Nixon (1965), who recognized 19 genera, and Mason (1981) who treated 51 genera, although several other regional keys exist. The latest research on the group (Fernandez-Triana et al., 2020) recognizes 81 genera worldwide but as yet these have not been fully treated in an identification key.

Members of the subfamily Microgastrinae are easily recognized from other braconid subfamilies by the following combination of characters: the antenna with 18 segments (except for the endemic New Zealand genus, *Kiwigaster* Fernandez-Triana, Whitfield and Ward, in which the females have 17 (Fernandez-Triana et al., 2011)); the fore wing without well-defined venation on apical third, vein 2-SR of fore wing connected to vein r; first metasomal segment with well-defined mediotergite separated from laterotergite containing the spiracles; the metasoma short compared with other braconid subfamilies (hence the subfamily name); and the scutellar sulcus being more or less developed (Fernández-Triana & Ward, 2017; van Achterberg, 1993). It should be noted that the location of the first tergite spiracles in the membanous lateral tergite is an important key character but often extremely hard to see, so most specialists do not rely on it, they recognize the subfamily (after a lot of practice) by its general gestalt and the rather limited range of fore wing venation.

Checklists of Regional Microgastrinae. A series of increasingly more comprehensive checklists of Iranian Microgastrinae have been produced: Farahbakhsh (1961): one species (*Apanteles glomeratus* (Linnaeus, 1758)); Modarres Awal (1997): 16 species in three genera (*Apanteles* Foerster, *Diolcogaster* Ashmead, *Microplitis* Foerster); Fallahzadeh & Saghaei (2010): 30 species in six genera; Modarres Awal (2012): 21 species in three genera (*Apanteles*, *Cotesia* Cameron, 1891, *Diolcogaster*); Gadallah et al. (2015): 99 species in eight genera; Farahani et al. (2016) and Yu et al. (2016): 102 and 103 species, respectively, in eight genera; Samin, Coronado-Blanco, Kavallieratos et al. (2018) and Samin, Coronado-Blanco, Fischer et al. (2018): 104 and 111 species, respectively, in eight genera. The present checklist includes 292 species in 23 genera in the Middle East. Here we follow Fernandez-Triana et al. (2020) in the classification of the Microgastrinae genera and groups. Global distribution of species is based mainly on Yu et al. (2016) and Fernandez-Triana et al. (2020).

List of species of the subfamily Microgastrinae recorded in the Middle East

Subfamily Microgastrinae Foerster, 1863

Apanteles group (= Apantelini + Microgastrini *sensu* Mason, 1981)

Genus *Apanteles* Foerster, 1863

Apanteles angaleti Muesebeck, 1956

Distribution in the Middle East: Iraq (Al Maliky & Al Izizi, 1990).

Extralimital distribution: China, India, Indonesia, Kenya, Mexico (introduced), Pakistan, United States of America (introduced), Vietnam.

Host records: Summarized by Yu et al. (2016) as being a parasitoid of the cosmopterigid *Anatrachyntis simplex* (Walsingham); the crambid *Haritalodes derogata* (Fabricius); the gelechiid *Pectinophora gossypiella* (Saunders); the nolids *Earias insulana* (Boisduval), and *Earias vittella* Fabricius; the pyralid *Apomyelois ceratomiae* (Zeller); the saturniids *Antheraea paphia* (L.), *Antheraea polyphemus* (Cramer), and *Antheraea roylei* Moore; and the tortricid *Archips occidentalis* (Walsingham).

Apanteles aragatzi Tobias, 1976

Catalogs with Iranian records: Samin, Coronado-Blanco, Fischer et al. (2018).

Distribution in Iran: Kordestan (Samin, Coronado-Blanco, Fischer et al., 2018).

Distribution in the Middle East: Iran (Samin, Coronado-Blanco, Fischer et al., 2018), Turkey (Beyarslan et al., 2002a, b; Inanç & Beyarslan, 2001a, b; Papp, 1984).

Extralimital distribution: Armenia, Russia, Sweden.

Host records: Unknown.

Apanteles articas Nixon, 1965
Distribution in the Middle East: Israel–Palestine (Fernandez-Triana et al., 2020; Papp, 2012), Turkey (Beyarslan et al., 2006; Papp, 1980).
Extralimital distribution: Senegal.
Host records: Recorded by Nixon (1965) as being a parasioid of the tortricid *Eccopsis wahlbergiana* Zeller.

Apanteles bajariae Papp, 1975
Distribution in the Middle East: Turkey (Beyarslan, 1988; Beyarslan et al., 2006; Papp, 1984).
Extralimital distribution: Bulgaria, Greece, Hungary, Montenegro, Spain.
Host records: Unknown.

Apanteles biroicus Papp, 1973
Catalogs with Iranian records: Gadallah et al. (2015), Farahani et al. (2016), Yu et al. (2016), Samin, Coronado-Blanco, Kavallieratos et al. (2018) and Samin, Coronado-Blanco, Fischer et al. (2018).
Distribution in Iran: Sistan and Baluchestan (Khajeh et al., 2014 as *Illidops biroicus* Papp, 1973).
Distribution in the Middle East: Iran.
Extralimital distribution: Hungary, Romania, Tunisia.
Host records: Unknown.

Apanteles brunnistigma Abdinbekova, 1969
Catalogs with Iranian records: Gadallah et al. (2015), Farahani et al. (2016), Yu et al. (2016), Samin, Coronado-Blanco, Kavallieratos et al. (2018) and Samin, Coronado-Blanco, Fischer et al. (2018).
Distribution in Iran: East Azarbaijan (Gadallah et al., 2015).
Distribution in the Middle East: Iran (Gadallah et al., 2015), Turkey (Beyarslan et al., 2002a; Inanç & Beyarslan, 2001a-c).
Extralimital distribution: Azerbaijan, Canada, Czech Republic, Finland, France, Germany, Hungary, Italy, Korea, Lithuania, Russia, Sweden, Switzerland, United Kingdom, Ukraine.
Host records: Summarized by Yu et al. (2016) as being a parasitoid of the crambid *Pyrausta aurata* (Scopoli); the depressariid *Agonopterix umbellana* (Fabricius); the epermiid *Epermenia chaerophylella* (Goeze); and the tortricids *Archips rosana* (L.), *Aphelia viburnana* (Denis and Schiffermüller), *Eucosma catoptrana* (Rebel), *Gynnidomorpha vectisana* (Humphreys and Westwood), and *Rhopobota naevana* (Hübner).

Apanteles carpatus (Say, 1836)
Catalogs with Iranian records: Gadallah et al. (2015), Farahani et al. (2016), Yu et al. (2016), Samin, Coronado-Blanco, Kavallieratos et al. (2018) and Samin, Coronado-Blanco, Fischer et al. (2018).

Distribution in Iran: Isfahan (Ghahari et al., 2011c), Mazandaran (Ghahari, Fischer, Çetin Erdoğan et al., 2010), Sistan and Baluchestan (Samin, 2015), Tehran (Modarres Awal, 1997; 2012; Shojai, 1998 as *Apanteles sarcitorius* Telenga, 1955).
Distribution in the Middle East: Iran (see references above), Israel–Palestine (Papp, 2012), Turkey (Inanç, 1997; Inanç & Beyarslan, 2001a).
Extralimital distribution: Nearly cosmopolitan species.
Host records: Summarized by Yu et al. (2016) as being a parasitoid of several lepidopteran pests of the families Erebidae, Gelechiidae, Gracillariidae, Lasiocampidae, Lecithoceridae, Notodontidae, Pyralidae, Tineidae, Tortricidae, Zygaenidae. In Iran, this species was erroneously recorded from the coccid *Lecanium* sp. by Modarres Awal (1997, 2012) and Shojai (1998).

Apanteles evanidus Papp, 1975
Distribution in the Middle East: Turkey (Beyarslan et al., 2006; Inanç, 1997; Inanç & Beyarslan, 2001a).
Extralimital distribution: Finland, Hungary, Moldova, Russia, Sweden, Ukraine.
Host records: Recorded by Papp (1981) as being a parasitoid of the plutellid *Scythropia cratagella* (L.).

Apanteles firmus Telenga, 1949
Catalogs with Iranian records: No catalog.
Distribution in Iran: Semnan (Samin, Sakenin Chelav, Ahmad, et al., 2020).
Distribution in the Middle East: Iran.
Extralimital distribution: Armenia, Azerbaijan, France, Hungary, Kazakhstan, Korea, Mongolia, Romania, Russia, Tajikistan, Ukraine, former Yugoslavia.
Host records: Recorded by Tobias (1971) as being a parasitoid of the tortricid *Acleris quercinana* (Zeller).

Apanteles galleriae Wilkinson, 1932
Catalogs with Iranian records: Fallahzadeh and Saghaei (2010), Gadallah et al. (2015), Farahani et al. (2016), Yu et al. (2016), Samin, Coronado-Blanco, Kavallieratos et al. (2018) and Samin, Coronado-Blanco, Fischer et al. (2018).
Distribution in Iran: Golestan, Razavi Khorasan (Ghahari et al., 2011a), Guilan (Goldansaz et al., 1996; Modarres Awal, 1997, 2012).
Distribution in the Middle East: Iran (see references above), Turkey (Beyarslan et al., 2002a; Inanç & Beyarslan, 2001a).
Extralimital distribution: Argentina, Armenia, Brazil, Bulgaria, Canada, Cape Verde Islands, China, France, Greece, Hawaiian Islands, Hungary, India, Italy, Japan, Kenya, Malta, Mauritius, New Zealand, Pakistan, Romania, Réunion, Russia, Spain, United Kingdom, United States of America.

Host records: Summarized by Yu et al. (2016) as being a parasitoid of the pyralids *Achroia grisella* (Fabricius), *Achroia innotata* (Walker), *Galleria mellonella* (L.), and *Vitula edmandsii* (Packard). In Iran, this species has been reared from the pyralid *Achroia grisella* (Fabricius) (Modarres Awal, 1997, 2012).

Apanteles hemara Nixon, 1965
Catalogs with Iranian records: Gadallah et al. (2015), Farahani et al. (2016), Yu et al. (2016), Samin, Coronado-Blanco, Kavallieratos et al. (2018) and Samin, Coronado-Blanco, Fischer et al. (2018).
Distribution in Iran: Qazvin (Ghahari et al., 2011b).
Distribution in the Middle East: Cyprus (Papp, 1980), Egypt, Oman, Saudi Arabia, United Arab Emirates, Yemen (Fernandez-Triana et al., 2020), Iran (Ghahari et al., 2011b), Israel—Palestine (Papp, 2012), Turkey (Beyarslan et al., 2002a; Inanç, 1997; Inanç & Beyarslan, 2001a).
Extralimital distribution: Australia, Bulgaria, Cape Verde Islands, China, France, Greece, India, Italy, Kenya, Madagascar, Madeira Islands, Mauritius, Pakistan, Portugal, Republic of Congo, Russia, Senegal, South Africa, Spain, Vietnam, former Yugoslavia.
Host records: Summarized by Yu et al. (2016) as being a parasitoid of the choreutid *Tebenna micalis* (Mann); and the crambids *Herpetogramma stultale* (Walker), *Hydriris ornatalis* (Duponchel), *Spoladera recurvalis* (Fabricius), and *Udea ferrugalis* (Hübner).

Apanteles kubensis Abdinbekova, 1969
Catalogs with Iranian records: No catalog.
Distribution in Iran: Guilan (Samin, Sakenin Chelav, Ahmad, et al., 2020).
Distribution in the Middle East: Iran (Samin, Sakenin Chelav, Ahmad, et al., 2020), Turkey (Inanç, 1997; Inanç & Beyarslan, 2001a).
Extralimital distribution: Azerbaijan, Hungary, Korea, Moldova, Mongolia, Russia.
Host records: Recorded by Ku et al. (2001) as being a parasitoid of the tortricid *Adoxophyes orana* (Fischer).

Apanteles lacteus (Nees von Esenbeck, 1834)
Catalogs with Iranian records: Fallahzadeh and Saghaei (2010), Gadallah et al. (2015), Farahani et al. (2016), Yu et al. (2016), Samin, Coronado-Blanco, Kavallieratos et al. (2018) and Samin, Coronado-Blanco, Fischer et al. (2018).
Distribution in Iran: Iran (no specific locality cited) (Telenga, 1955).
Distribution in the Middle East: Iran (Telenga, 1955), Israel—Palestine (Aubert, 1966; Bodenheimer, 1930; Gothilf, 1969; Halperin, 1986; Kugler, 1966), Turkey (Inanç & Beyarslan, 1990; Inanç, 1997).

Extralimital distribution: Armenia, Azerbaijan, Finland, Germany, Greece, Kazakhstan, Moldova, Poland, Romania, Russia, Slovakia, Sweden, Tajikistan, Tunisia, Ukraine, United Kingdom, Uzbekistan.
Host records: Summarized by Yu et al. (2016) as being a parasitoid of the crambid *Hellula undalis* (Fabricius); the gelechiid *Ptocheuusa inopella* (Zeller); the gracillariid *Phyllonorycter junoniella* (Zeller), the lyonetiid *Lyonetia prunifoliella* (Hübner); the pyralids *Acrobasis sodalella* Zeller, *Dioryctria abietella* (Denis and Schiffermüller), *Ectomyelois ceratoniae* (Zeller), *Etiella zinckenella* (Treitschke), *Homoeosoma nebulella* Denis and Schiffermüller, *Homoeosoma nimbella* (Duponchel), and *Phycitodes maritima* (Tengström); the tineid *Triaxomera parasitella* (Hübner); and the tortricids *Cacoecimorpha pronubana* (Hübner), and *Cydia pactolana* (Zeller).

Apanteles lenea Nixon, 1976
Catalogs with Iranian records: No catalog.
Distribution in Iran: Qazvin (Sakenin et al., 2020).
Distribution in the Middle East: Iran (Sakenin et al., 2020), Turkey (Inanç, 1997).
Extralimital distribution: Austria, Bulgaria, Czech Republic, France, Germany, Hungary, Italy, Korea, Romania, Russia, Serbia, Slovakia, Spain, Sweden, Switzerland, United Kingdom.
Host records: Summarized by Yu et al. (2016) as being a parasitoid of the cosmopterigid *Pancalia schwarzella* (Fabricius); the pyralid *Onocera semirubella* (Scopoli); and the tortricids *Argyroploce arbutella* (L.), and *Sparganothis pilleriana* (Denis and Schiffermüller).

Apanteles metacarpalis (Thomson, 1895)
Catalogs with Iranian records: Gadallah et al. (2015), Farahani et al. (2016), Yu et al. (2016), Samin, Coronado-Blanco, Kavallieratos et al. (2018) and Samin, Coronado-Blanco, Fischer et al. (2018).
Distribution in Iran: Guilan (Ghahari, Fischer, & Tobias et al., 2012), Ilam (Ghahari et al., 2011d).
Distribution in the Middle East: Iran.
Extralimital distribution: Azerbaijan, China, Czech Republic, Finland, France, Germany, Greece, Hungary, Ireland, Italy, Korea, Malta, Moldova, Mongolia, Romania, Russia, Serbia, Spain, Sweden, Tajikistan, Tunisia, United Kingdom, Ukraine, Uzbekistan.
Host records: Summarized by Yu et al. (2016) as being a parasitoid of the following gelechiids: *Scrobipalpa atriplicella* (Fischer), *Scrobipalpa instabilella* (Douglas), *Scrobipalpa nitentella* (Fuchs), *Scrobipalpa obsoletella* (Fischer), *Scrobipalpa samadensis* (Pfaffehzeller), and *Scrobipalpa suaedella* (Richardson); the gracillariid

Caloptilia semifascia (Haworth); and the lycaenid *Drupadia theda* (Felder and Felder).

Apanteles obscurus (Nees von Esenbeck, 1834)

Catalogs with Iranian records: Fallahzadeh and Saghaei (2010), Gadallah et al. (2015), Farahani et al. (2016), Yu et al. (2016), Samin, Coronado-Blanco, Kavallieratos et al. (2018) and Samin, Coronado-Blanco, Fischer et al. (2018).

Distribution in Iran: East Azarbaijan (Ghahari & van Achterberg, 2016), Fars (Lashkari Bod et al., 2011), Sistan and Baluchestan (Khajeh et al., 2014), West Azarbaijan (Samin et al., 2014), Iran (no specific locality cited) (Čapek & Hofmann, 1997; Güçlü & Özbek, 2011).

Distribution in the Middle East: Iran (see references above), Israel−Palestine (Papp, 2012), Turkey (Beyarslan, 1988; Beyarslan et al., 2002a; Beyarslan et al., 2006; Fahringer, 1922; Inanç, 1997; Inanç & Beyarslan, 1990; 2001a; Nixon, 1976).

Extralimital distribution: Albania, Armenia, Azerbaijan, Belgium, Bulgaria, Croatia, Denmark, Finland, France, Georgia, Germany, Greece, Hungary, Ireland, Italy, Kazakhstan, Lithuania, North Macedonia, Moldova, Mongolia, Montenegroa, Netherlands, Poland, Romania, Russia, Serbia, Slovakia, Slovenia, Spain, Sweden, Switzerland, Tunisia, United Kingdom.

Host records: Summarized by Yu et al. (2016) as being a parasitoid of the choreutid *Tebenna bjerkandrella* (Thunberg); the crambids *Algedonia terrealis* (Treitschke), *Anania crocealis* (Hübner), *Pyrausta cingulata* (L.), and *Udea ferrugalis* (Hübner); the epermeniid *Epermenia chaerophylella* (Goeze); the geometrid *Chlorissa viridata* (L.); the gracillariid *Gracillaria syringella* (Fabricius); the nymphalid *Eurodryas aurinia* (Rottemberg)); the tortricid *Clepsis pallidana* (Fabricius); and the zygaenid *Zygaena filipendulae* (L.).

Apanteles pilosus Telenga, 1955

Catalogs with Iranian records: Gadallah et al. (2015), Farahani et al. (2016), Yu et al. (2016), Samin, Coronado-Blanco, Kavallieratos et al. (2018) and Samin, Coronado-Blanco, Fischer et al. (2018).

Distribution in Iran: Sistan and Baluchestan (Khajeh et al., 2014 as *Illidops pilosus* Telenga, 1955).

Distribution in the Middle East: Iran.

Extralimital distribution: Kazakhstan, Turkmenistan, Uzbekistan.

Host records: Unknown.

Apanteles sodalis (Haliday, 1834)

Catalogs with Iranian records: Gadallah et al. (2015), Farahani et al. (2016), Yu et al. (2016), Samin, Coronado-Blanco, Kavallieratos et al. (2018) and Samin, Coronado-Blanco, Fischer et al. (2018).

Distribution in Iran: Kermanshah (Ghahari & Fischer, 2012 as *Apanteles ater* (Ratzeburg, 1852)).

Distribution in the Middle East: Iran (Ghahari & Fischer, 2012), Turkey (Aydoğdu, 2014; Beyarslan et al., 2002a; Beyarslan et al., 2006; Inanç, 1997; Inanç & Beyarslan, 2001a).

Extralimital distribution: Armenia, Azerbaijan, Bulgaria, Canada (introduced), Cape Verde Islands, China, Czech Republic, France, Germany, Greece, Hungary, Ireland, Italy, Japan, Kazakhstan, Korea, Latvia, Lithuania, Moldova, Netherlands, Poland, Romania, Russia, Serbia, Slovakia, Sweden, Switzerland, Ukraine, United Kingdom.

Host records: Summarized by Yu et al. (2016) as being a parasitoid of lepidopteran hosts of the families Crambidae, Erebidae, Geometridae, Plutellidae, Psychidae, Pyralidae, and Yponomeutidae.

Apanteles syleptae Ferrière, 1925

Distribution in the Middle East: Egypt (El-Sherif & Kaschef, 1977; Risbec, 1960).

Extralimital distribution: Chad, Democratic Republic of Congo, Kenya, Nigeria, Senegal, Sudan, Tanzania, Togo.

Host records: Summarized by Yu et al. (2016) as being a parasitoid of the crambids *Chilo zacconius* Bleszyński, *Haritalodes derogata* (Fabricius), and *Palpita vitrealis* (Rossi); and the noctuids *Earias insulana* (Boisduval), *Spodoptera exempta* (Walker), and *Spodoptera exigua* (Hübner).

Apanteles xanthostigma (Haliday, 1834)

Catalogs with Iranian records: Farahani et al. (2016).

Distribution in Iran: East Azarbaijan, Razavi Khorasan, West Azarbaijan (Shojai et al., 2002), Iran (no specific locality cited) (Esmaili, 1983; Modarres Awal, 1997, 2012, both as *Apanteles xanthostigma anarisae*).

Distribution in the Middle East: Iran (see references above), Turkey (Avci, 2009).

Extralimital distribution: Afrotropical, Nearctic, Oceanic and Palaearctic regions; Canada (introduced).

Host records: Summarized by Yu et al. (2016) as being a parasitoid of several lepidopteran species of the families Choreutidae, Gelechiidae, Geometridae, Gracillariidae, Lymantriidae, Nepticulidae, Noctuidae, Oecophoridae, Pyralidae, Tortricidae and Yponomeutidae. In Iran, this species has been reared from the gelechiid *Anarsia lineatella* Zeller (Esmaili, 1983; Modarres Awal, 1997; 2012; Shojai et al., 2002).

Apanteles sp.

Distribution in Iran: Alborz (Shojai, 1998), Kordestan (Kamangar et al., 2017), Razavi Khorasan, West Azarbaijan (Shojai et al., 2000).

Host records in Iran: *Diaspidiotus pronurum* (Laing) (Hemiptera: Diaspididae) (Shojai, 1998), *Cydia pomonella* L. (Lepidoptera: Tortricidae) (Shojai et al., 2000), and the tortricid moth *Tortrix viridana* L. (Kamangar et al., 2017).

Genus *Choeras* Mason, 1981

Choeras afrotropicalis Fernandez-Triana and van Achterberg, 2017

Distribution in the Middle East: Yemen (Fernandez-Triana & van Acherberg, 2017).
Extralimital distribution: None.
Host records: Unknown.

Choeras dorsalis (Spinola, 1808)

Catalogs with Iranian records: Gadallah et al. (2015), Farahani et al. (2016), Yu et al. (2016), Samin, Coronado-Blanco, Kavallieratos et al. (2018) and Samin, Coronado-Blanco, Fischer et al. (2018) [all as *Apanteles (Choeras) dorsalis* (Spinola, 1808)].
Distribution in Iran: Ardabil (Ghahari et al., 2011a), Kermanshah (Ghafouri Mogaddam et al., 2018), Iran (no specific locality cited) (Abdoli et al., 2019a).
Distribution in the Middle East: Cyprus (Nixon, 1965), Egypt, Jordan (Fernandez-Triana et al., 2020), Iran (see references above), Israel—Palestine (Papp, 2012), Turkey (Beyarslan, 1988; Beyarslan et al., 2002a, b; Beyarslan et al., 2006; Inanç & Beyarslan, 1990, 2001a, b).
Extralimital distribution: Armenia, Austria, Azerbaijan, Belgium, Bulgaria, France, Georgia, Germany, Greece, Hungary, Italy, Lithuania, Madeira Islands, Malta, Moldova, Poland, Romania, Russia, Slovakia, Spain, Switzerland, Tunisia, Ukraine, United Kingdom, Uzbekistan.
Host records: Recorded by Fulmek (1968) as being a parasitoid of the choreutid *Tebenna bjerkandrella* (Thunberg); the crambid *Pyrausta aurata* (Scopoli); the depressariid *Ethmia terminella* (Fletcher); the noctuid *Orthosia miniosa* (Denis and Schiffermüller); the oecophorid *Schiffermuelleria schaefferella* (L.); the pyralid *Palpita unionalis* (Hübner); and the tortricids *Avaria hyerana* (Millière), *Cacoecimorpha pronubana* (Hübner), and *Pseudohermenias abietana* (Fabricius).

Choeras formosus Abdoli and Fernandez-Triana, 2019

Catalogs with Iranian records: No catalog.
Distribution in Iran: Alborz, Mazandaran (Abdoli et al., 2019a).
Distribution in the Middle East: Iran.
Extralimital distribution: None.
Host records: Unknown.

Choeras fulviventris Fernandez-Triana and Abdoli, 2019

Catalogs with Iranian records: No catalog.
Distribution in Iran: Mazandaran (Abdoli et al., 2019a).

Distribution in the Middle East: Iran.
Extralimital distribution: None.
Host records: Unknown.

Choeras parasitellae (Bouché, 1834)

Catalogs with Iranian records: No catalog.
Distribution in Iran: Mazandaran (Ghahari & Sakenin, 2018 as *Apanteles (Choeras) parasitellae* (Bouché); Ghahari, 2019a), Tehran (Ghahari & Beyarslan, 2019).
Distribution in the Middle East: Iran (see references above), Israel—Palestine (Halperin, 1986), Turkey (Beyarslan et al., 2002b; Inanç & Beyarslan, 2001b).
Extralimital distribution: Austria, Belgium, Canada, Czech Republic, Finland, France, Georgia, Germany, Hungary, Italy, Korea, Latvia, Moldova, Netherlands, Poland, Romania, Russia, Serbia, Slovakia, Spain, Sweden, Switzerland, Ukraine, United Kingdom, Uzbekistan (Yu et al., 2016 as *Apanteles (Choeras) parasitellae* (Bouché)).
Host records: Summarized by Yu et al. (2016) as being a parasitoid of the gelechiids *Anacampsis timidella* (Wocke), and *Scrobipalpa samadensis* (Pfaffenzeller); the noctuid *Acronicta rumicis* (L.); the prodoxid *Lampronia morose* Zeller; the pterophorids *Adaina microdactyla* (Hübner), and *Hellinsia osteodactyla* (Zeller); the tineids *Archinemapogon yildizae* (Koçak), *Morophaga choragella* Denis and Schiffermüller, *Nemapogon cloacella* (Haworth), *Nemapogon granella* (L.), *Nemapogon orientalis* Petersen, *Nemapogon variatella* (Clemens), *Nemaxera betulinella* (Paykull), *Scardia boletella* (Fabricius), and *Triaxomera parasitella* (Hübner); the tortricids *Gypsonoma sociana* (Haworth), *Pseudohermenias abietana* (Fabricius), *Spilonota prognathana* (Snellen), and *Zeiraphera isertana* (Fabricius); and the yponomeutid *Yponomeuta padella* (L.). In Iran, this species has been reared from *Yponomeuta padella* (L.) on apricot (Ghahari & Sakenin, 2018; Ghahari & Beyarslan, 2019).

Choeras qazviniensis Fernandez-Triana and Talebi, 2019

Catalogs with Iranian records: No catalog.
Distribution in Iran: Qazvin (Abdoli et al., 2019a).
Distribution in the Middle East: Iran.
Extralimital distribution: None.
Host records: Unknown.

Choeras ruficornis (Nees von Esenbeck, 1834)

Catalogs with Iranian records: No catalog.
Distribution in Iran: Chaharmahal and Bakhtiari (Ghahari et al., 2019).
Distribution in the Middle East: Iran.
Extralimital distribution: Belgium, Finland, France, Georgia, Germany, Hungary, Italy, Latvia, Netherlands, Norway, Poland, Romania, Russia, Slovakia, Sweden, Switzerland, United Kindgom (Yu et al., 2016 as *Apanteles ruficornis* (Nees)).

Host records: Summarized by Yu et al. (2016) as being a parasitoid of the crambid *Loxostege sticticalis* (L.); the lasiocampid *Dendrolimus pini* (L.); and the yponomeutid *Yponomeuta padella* (L.).

Choeras semele (Nixon, 1965)

Distribution in the Middle East: Israel—Palestine (Papp, 2012).

Extralimital distribution: Greece, Italy, Malta, Morocco, Spain.

Host records: Summarized by Yu et al. (2016) as being a parasitoid of the choreutid *Choreutis nemorana* (Hübner); the gelechiid *Tuta absoluta* (Meyrick); the plutellid *Plutella xylostella* (L.); and the tortricids *Archips rosana* (L.), and *Cacoecimorpha pronubana* (Hübner).

Choeras taftanensis Ghafouri Moghaddam and van Achterberg, 2018

Catalogs with Iranian records: No catalog.

Distribution in Iran: Guilan, Mazandaran (Abdoli et al., 2019a), Kerman, Kermanshah, Sistan and Baluchestan (Ghafouri Moghaddam et al., 2018).

Distribution in the Middle East: Iran.

Extralimital distribution: None.

Host records: Unknown.

Choeras tedellae (Nixon, 1961)

Catalogs with Iranian records: Gadallah et al. (2015 as *Apanteles* (*Choeras*) *tedellae* Nixon, 1961), Farahani et al. (2016 as *Apanteles* (*Choeras*) *tedellae* Nixon, 1961), Yu et al. (2016 as *Apanteles* (*Choeras*) *tedellae* Nixon, 1961), Samin, Coronado-Blanco, Kavallieratos et al. (2018) and Samin, Coronado-Blanco, Fischer et al. (2018) as *Apanteles* (*Choeras*) *tedellae*).

Distribution in Iran: Golestan (Ghahari & Fischer, 2011b as *Apanteles* (*Choeras*) *tedellae* Nixon, 1961), Iran (no specific locality cited) (Abdoli et al., 2019a).

Distribution in the Middle East: Iran (see references above), Israel—Palestine (Papp, 2012).

Extralimital distribution: Austria, Bulgaria, Croatia, Czech Republic, Denmark, Finland, Germany, Greece, Hungary, Korea, Madeira Islands, Moldova, Netherlands, Poland, Romania, Russia, Slovakia, Sweden, Switzerland, United Kingdom (Yu et al., 2016 as *Apanteles* (*Choeras*) *tedellae* Nixon).

Host records: Summarized by Yu et al. (2016) as being a parasitoid of the tortricids *Cydia laricana* (Busck), *Epinotia pinicola* Kuznetzov, *Epinotia tedella* (Clerck), *Ptycholomoides aeriferanus* (Herrich-Schäffer), and *Zeirafera griseana* (Hübner).

Choeras tiro (Reinhard, 1880)

Catalogs with Iranian records: Gadallah et al. (2015), Farahani et al. (2016), Yu et al. (2016 as), Samin, Coronado-Blanco, Kavallieratos et al. (2018) and Samin, Coronado-Blanco, Fischer et al. (2018) [all as *Apanteles* (*Choeras*) *tiro* (Reinhard, 1880)].

Distribution in Iran: Guilan (Abdoli et al., 2019a), Qazvin (Ghahari et al., 2011b as *Apanteles* (*Choeras*) *tiro* (Reinhard, 1880), Abdoli et al., 2019a), Kermanshah (Ghafouri Moghaddam et al., 2018), Iran (no specific locality cited) (Abdoli et al., 2019a).

Distribution in the Middle East: Iran (see references above), Israel—Palestine (Papp, 2012).

Extralimital distribution: Austria, Bulgaria (introduced), Canada (introduced), France, Germany, Greece, Hungary, Poland, Romania, Russia, Slovakia, Spain, Switzerland (Yu et al., 2016 as *Apanteles* (*Choeras*) *tiro* (Reinhard, 1880); Fernandez-Triana et al., 2020).

Host records: Summarized by Yu et al. (2016) as being a parasitoid of the geometrids *Alsophila aescularia* (Denis and Schiffermüller), *Lycia hirtaria* (Clerck), and *Phigalia pilosaria* (Denis and Schiffermüller); and the tortricids *Choristoneura fumiferana* (Clemens), *Cnephasia asseclana* (Denis and Schiffermüller), and *Cnephasia chrysantheana* (Hübner).

Genus Dolichogenidea Viereck, 1911

Dolichogenidea agilla (Nixon, 1972)

Catalogs with Iranian records: Gadallah et al. (2015), Farahani et al. (2016), Yu et al. (2016), Samin, Coronado-Blanco, Kavallieratos et al. (2018) and Samin, Coronado-Blanco, Fischer et al. (2018).

Distribution in Iran: East Azarbaijan (Ghahari et al., 2011a).

Distribution in the Middle East: Iran.

Extralimital distribution: Finland, Greece, Hungary, Mongolia, Russia, United Kingdom.

Host records: Recorded by Shaw (2012) as being a parasitoid of the tortricid *Dichrorampha plumbagana* (Treitschke).

Dolichogenidea albipennis (Nees von Esenbeck, 1834)

Catalogs with Iranian records: This species is new record for the fauna of Iran.

Distribution in Iran: Kuhgiloyeh and Boyerahmad province, Pataveh, 3♀, September 2015, ex *Etiella zinckenella* (Treitschke) (Lepidoptera: Pyralidae).

Distribution in the Middle East: Iran (new record), Turkey (Papp, 1978).

Extralimital distribution: Afghanistan, Albania, Azerbaijan, Belarus, Denmark, France, Georgia, Germany, Hungary, Italy, Kazakhstan, Kyrgyzstan, Lithuania, Moldova, Mongolia, Netherlands, Poland, Romania, Russia, Sweden, Turkmenistan, Ukraine, United Kingdom (Yu et al., 2016 as *Apanteles* (*Dolichogenidea*) *albipennis* (Nees); Fernandez-Triana et al., 2020).

Host records: Recorded by Yu et al. (2016) as being a parasitoid of the agonoxid *Blastodacna atra* (Haworth); the argysthiid *Argyresthia brockeella* (Hübner); the coleophorids *Coleophora laricella* (Hübner), *Coleophora prunifoliae* Doets, and *Coleophora serinipenella* Christoph; the cosmopterigid *Etebalea albiapicella* (Duponchel); the cossid *Zeuzera pyrina* (L.); the douglasiid *Tinagma ocnerostomellum* (Stainton); the erebid *Rhyparia purpurata* (L.); the gelechiids *Anacampsis timidella* (Wocke), *Aristotelia brizella* (Treitschke), and *Caryocolum tricolorella* (Haworth); the geometrid *Operophtera brumata* (L.); the gracillariid *Caloptilia semifascia* (Haworth), and *Parornix anglicella* (Stainton); the lasiocampid *Malacosoma neustria* (L.); the lyonetiid *Leucoptera malifoliella* (O. Costa); the plutellid *Plutella xylostella* (L.); the pterophorid *Adeina microdactyla* (Hübner); the pyralid *Etiella zinckenella* (Treitschke); the sesiid *Synanthedon tipuliformis* (Clerck); the following tortricids: *Aethes francillana* (Fabricius), *Archips xylosteana* (L.), *Choristoneura murinana* (Hübner), *Cnephasia sedana* (Constant), *Epiblema costipunctata* (Haworth), *Eucosma aemulana* (Schläger), *Falseucaria ruficiliana* (Haworth), *Pammene amygdalana* (Duponchel), *Sparganothis pilleriana* (Denis and Schiffermüller), and *Tortrix viridana* L. In Iran, this species has been reared from the pyralid *Etiella zinckenella* (Treitschke) (the current study).

Dolichogenidea anarsiae (Faure and Alabouvette, 1924)

Catalogs with Iranian records: No catalog.

Distribution in Iran: Golestan (Samin, Fischer, Sakenin et al., 2019 as *Apanteles anarsia* Faure and Alabouvette, 1924).

Distribution in the Middle East: Iran (Samin, Fischer, Sakenin et al., 2019), Turkey (Inanç & Beyarslan, 2001a).

Extralimital distribution: Azerbaijan, Bulgaria, France, Georgia, Hungary, Italy, Moldova, Romania, Russia, Switzerland (Yu et al., 2016 as *Apanteles anarsia* Faure and Alabouvette, 1924; Fernandez-Triana, 2020).

Host records: Summarized by Yu et al. (2016) as being a parasitoid of the gelechiids *Anarsia lineatella* Zeller, *Helcystogramma triannulella* (Herrich-Schäffer); and the tortricids *Cydia funebrana* (Treitschke), and *Grapholita molesta* (Busck).

Dolichogenidea appellator (Telenga, 1949)

Catalogs with Iranian records: Gadallah et al. (2015), Farahani et al. (2016 as *Apanteles (Dolichogenidea) apellerator* Telenga, 1949), Yu et al. (2016), Samin, Coronado-Blanco, Kavallieratos et al. (2018) and Samin, Coronado-Blanco, Fischer et al. (2018).

Distribution in Iran: Ardabil (Samin, Papp, Coronado-Blanco, 2018 as *Apanteles (Dolichogenidea) appellator* Telenga, 1949), Isfahan (Kazemzadeh et al., 2014), West

Azarbaiajan (Samin, Coronado-Blanco, Hosseini et al., 2019 as *Apanteles litae* Nixon, 1972).

Distribution in the Middle East: Cyprus (Ingram, 1981; Nixon, 1972; Papp, 1978), Egypt (Abbas, 1989; Papp, 1978), Iran (see references above), Israel–Palestine (Halperin, 1986), Jordan (Papp, 1978), Turkey (Beyarslan, 1988; Beyarslan et al., 2002a, b; Beyarslan et al., 2006; Inanç, 1997; Inanç & Beyarslan, 1997, 2001a, b; Papp, 1978).

Extralimital distribution: Afghanistan, Armenia, Azerbaijan, Belarus, Bulgaria, Cape Verde Islands, China, Croatia, England, Germany, Ghana, Hungary, Italy, Kazakhstan, Malta, Moldova, Mongolia, Romania, Russia, Northwestern Selvagens Islands, Spain, Switzerland, Tajikistan, Tunisia, Turkmenistan, Ukraine, United Kingdom, Uzbekistan, former Yugoslavia.

Host records: Summarized by Yu et al. (2016) as being a parasitoid of the cosmopterigid *Ascalenia vanelloides* Gerasimov; the crambid *Pyrausta purpuralis* (L.); the gelechiids *Gnorimoschema blapsigona* (Meyrick), *Phthorimaea operculella* (Zeller), *Scrobiplapa nitentella* (Fuchs), *Scrobipalpa salinella* (Zeller), and *Tuta absoluta* (Meyrick); the noctuid *Spodoptera littorlais* Boisduval; the plutellid *Plutella xylostella* (L.); the pyralid *Etiella zinckenella* (Treitschke); and the tortricids *Cryptophlebia leucotreta* (Meyrick), *Eccopsis wahlbergiana* Zeller, and *Spilonota ocellana* (Denis and Schiffermüller). In Iran, this species has been reared from *Plutella xylostella* on cabbage (Kazemzadeh et al., 2014; Samin, Coronado-Blanco, Hosseini et al., 2019).

Dolichogenidea atreus (Nixon, 1973)

Catalogs with Iranian records: Samin, Coronado-Blanco, Fischer et al. (2018).

Distribution in Iran: Khuzestan (Samin, Coronado-Blanco, Fischer et al., 2018 as *Apanteles atreus* Nixon, 1973).

Distribution in the Middle East: Iran (Samin, Coronado-Blanco, Fischer et al., 2018), Turkey (Inanç & Beyarslan, 2001a; Papp, 1984).

Extralimital distribution: Bulgaria, Czech Republic, Denmark, Germany, Greece, Hungary, Italy, Russia, United Kingdom (Yu et al., 2016 as *Apanteles atreus*; Fernandez-Triana et al., 2020).

Host records: Recorded by Nixon (1973) and Tobias (1986) as being a parasitoid of the momphids *Mompha locupletella* (Denis and Schiffermüller), *Mompha propinquella* (Stainton), and *Mompha sturnipenella* (Treitscheke).

Dolichogenidea bilecikensis Inanç and Çetin Erdoğan, 2004

Distribution in the Middle East: Turkey (Inanç et al., 2004).

Extralimital distribution: None.

Host records: Unknown.

Dolichogenidea breviventris (Ratzeburg, 1848)
Catalogs with Iranian records: This species is new record for the fauna of Iran.

Distribution in Iran: Mazandaran province, Tonekabon (Jangal-e 3000), 1♀, August 2014.

Distribution in the Middle East: Egypt (Papp, 1978), Iran (new record), Turkey (Inanç & Beyarslan, 2001a).

Extralimital distribution: Canada, China, Czech Republic, Finland, Germany, Hungary, Ireland, Italy, Korea, Moldova, Netherlands, Poland, Romania, Russia, Serbia, Slovakia, Sweden, Switzerland, United Kingdom.

Host records: Summarized by Yu et al. (2016) as being a parasitoid of the following coleophorids: *Coleophora anatipennella* (Hübner), *Coleophora betulella* Heinemann and Wocke, *Coleophora flavipennella* (Duponchel), *Coleophora gryphipennella* (Hübner), *Coleophora lutipennella* (Zeller), *Coleophora prunifoliae* Doets, and *Coleophora serratella* (L.); and the noctuid *Orthosia miniosa* (Denis and Schiffermüller).

Dolichogenidea britannica (Wilkinson, 1941)
Catalogs with Iranian records: Samin, Coronado-Blanco, Kavallieratos et al. (2018) and Samin, Coronado-Blanco, Fischer et al. (2018).

Distribution in Iran: Ardabil (Ghahari & van Achterberg, 2016).

Distribution in the Middle East: Iran (Ghahari & van Achterberg, 2016), Israel—Palestine (Papp, 2012).

Extralimital distribution: Armenia, Canada, Greece, Hungary, Malta, Russia, Slovakia, Tajikistan, Ukraine, United Kingdom (Yu et al., 2016 as *Apanteles (Dolichogenidea) britannicus* Wilkinson, 1941).

Host records: Summarized by Yu et al. (2016) as being a parasitoid of the cosmopterigid *Limnaecia phragmitella* Stainton; the gelechiid *Isophrictis striatella* (Denis and Schiffermüller), and *Ptocheuusa paupella* (Zeller)); and the tortricid *Enarmonia formosana* (Scopoli).

Dolichogenidea candidata (Haliday, 1834)
Catalogs with Iranian records: Gadallah et al. (2015), Farahani et al. (2016), Yu et al. (2016), Samin, Coronado-Blanco, Kavallieratos et al. (2018) and Samin, Coronado-Blanco, Fischer et al. (2018).

Distribution in Iran: East Azarbaijan (Ghahari & van Achterberg, 2016), Ilam (Ghahari et al., 2011d), Lorestan (Ghahari, Fischer, Papp et al., 2012), Semnan (Samin, Fischer, & Ghahari, 2015), Zanjan (Ghahari et al., 2019 as *Apanteles candidatus*).

Distribution in the Middle East: Iran.

Extralimital distribution: Azerbaijan, Bulgaria, Germany, Greece, Hungary, Ireland, North Macedonia, Mongolia, Romania, Russia, Serbia, Sweden, Tajikistan, Turkmenistan, United Kingdom, Uzbekistan (Yu et al., 2016 as *Apanteles*

(Dolichogenidea) candidatus (Haliday, 1834); Fernandez-Triana et al., 2020).

Host records: Summarized by Yu et al. (2016) as being a parasitoid of the gelechiids *Coleotechnites milleri* (Busck), and *Coleotechnites starki* (Freeman); and the tortricids *Epinotia meritana* Heinrich, and *Eucosma bobana* (Kearfott).

Dolichogenidea cerialis (Nixon, 1976)
Distribution in the Middle East: Israel—Palestine (Halperin, 1986; Nixon, 1976; Wysoki & Izhar, 1981).

Extralimital distribution: Bulgaria, Hungary, Italy, Kazakhstan, Russia, Spain.

Host records: Recorded by Nixon (1976) as being a parasitoid of the geometrid *Ascotis selenaria* (Denis and Schiffermüller).

Dolichogenidea cheles (Nixon, 1972)
Distribution in the Middle East: Turkey (Beyarslan et al., 2002b; Inanç & Beyarslan, 2001b).

Extralimital distribution: Finland, Hungary, Poland, Russia, Sweden.

Host records: Summarized by Yu et al. (2016) as being a parasitoid of the gracillariid *Caloptilia rufipennella* Hübner; and the tortricid *Croesia holmiana* (L.).

Dolichogenidea coleophorae (Wilkinson, 1938)
Catalogs with Iranian records: This species is new record for the fauna of Iran.

Distribution in Iran: Ardabil province, Namin, 4♀, July 2011, ex *Coleophora serratella* (Linnaeus) (Lepidoptera: Coleophoridae).

Distribution in the Middle East: Iran (new record), Turkey (Inanç & Beyarslan, 2001a).

Extralimital Distribution: Azerbaijan, Canada, Finland, Hungary, Poland, Romania, Russia, Slovakia, Switzerland, Tajikistan, Tunisia, United Kingdom, Uzbekistan (Yu et al., 2016 as *Apanteles (Dolichogenidea) coleophorae* Wilkinson; Fernandez-Triana et al., 2020).

Host records: In Iran, this species has been reared from the coleophorid *Coleophora serratella* (L.) (new record).

Dolichogenia coniferoides (Papp, 1972)
Distribution in the Middle East: Turkey (Inanç & Beyarslan, 1990).

Extralimital distribution: Hungary, Sweden.

Host records: Unknown.

Dolichogenidae corvina (Reinhard, 1880)
Catalogs with Iranian records: Gadallah et al. (2015), Farahani et al. (2016), Yu et al. (2016), Samin, Coronado-Blanco, Kavallieratos et al. (2018) and Samin, Coronado-Blanco, Fischer et al. (2018).

Distribution in Iran: Ilam (Ghahari et al., 2011d), Kerman, Markazi (Ghahari & Beyarslan, 2017), West Azarbaijan (Žikić et al., 2014).

Distribution in the Middle East: Iran.

Extralimital distribution: Azerbaijan, Bulgaria, Canada, China, Czech Republic, Finland, Georgia, Germany, Greece, Hungary, Ireland, Japan, Kazakhstan, Lithuania, Moldova, Mongolia, Netherlands, Poland, Romania, Russia, Sweden, Tajikistan, Turkmenistan, Ukrania, United Kingdom, Uzbekistan.

Host records: Summarized by Yu et al. (2016) as being a parasitoid of the bucculatricid *Bucculatrix bechsteiniella* (Bechstein and Scharfender); the coleophorids *Coleophora coracipennella* (Hübner), *Coleophora nigricella* (Stephens), and *Coleophora serratella* (L.); the gracillariids *Phyllonorycter leucographella* (Zeller), and *Phyllonorycter oxyacanthae* (Frey); the lyonetiid *Lyonetia clerkella* (L.); the tortricid *Hedya nubiferana* Haworth; and the yponomeutid *Paraswammerdamia lutarea* (Howarth). In Iran, this species has been reared from the coleophorid *Coleophora serratella* (L.), and the lyonetiid *Lyonetia clerkella* (L.) (Ghahari & Beyarslan, 2017).

Dolichogenidea cytherea (Nixon, 1972)

Catalogs with Iranian records: Gadallah et al. (2015), Farahani et al. (2016), Yu et al. (2016), Samin, Coronado-Blanco, Kavallieratos et al. (2018) and Samin, Coronado-Blanco, Fischer et al. (2018).

Distribution in Iran: Golestan (Ghahari & Fischer, 2011b).

Distribution in the Middle East: Iran.

Extralimital distribution: Greece, Hungary, Mongolia, Poland, Russia, Serbia, Slovakia, Switzerland, Ukraine, United Kingdom (Yu et al., 2016 as *Apanteles (Dolichogenidea) cytherea* Nixon, 1972; Fernandez-Traina et al., 2020).

Host records: Reported by Čapek et al. (1982) as a parasitoid of the ypsolophid *Ypsolopha alpella* Denis and Schiffermüller.

Dolichogenidea decora (Haliday, 1834)

Catalogs with Iranian records: Gadallah et al. (2015), Farahani et al. (2016), Yu et al. (2016), Samin, Coronado-Blanco, Kavallieratos et al. (2018) and Samin, Coronado-Blanco, Fischer et al. (2018) [all as *Apanteles decorus* (Haliday, 1834)].

Distribution in Iran: Qazvin (Ghahari et al., 2011b), Khuzestan (Samin, Fischer, Sakenin et al., 2019 as *Apanteles decora* (Haliday)).

Distribution in the Middle East: Iran.

Extralimital distribution: Bulgaria, China, Czech Republic, Estonia, Finland, Georgia, Germany, Greece, Hungary, Ireland, Kazakhstan, Lithuania, Poland, Romania, Russia, Slovakia, Spain, Sweden, Turkmenistan (Yu et al., 2016 as

Apanteles (Dolichogenidea) decora; Fernandez-Triana et al., 2020).

Host records: Summarized by Yu et al. (2016) as being a parasitoid of the argyresthiid *Argyresthia geodartella* (L.); the erebid *Lymantria dispar* (L.); the tineid *Nemapogon cloacella* (Haworth); and the tortricids *Aethes dilucidana* (Stephens), *Eudemis profundana* (Denis and Schiffermüller), *Gypsonoma minutana* (Hübner), *Rhyacionia duplana* (Hübner), and *Spilonota lariciana* (Heinemann).

Dolichogenidea dilecta (Haliday, 1834)

Catalogs with Iranian records: No catalog.

Distribution in Iran: Mazandaran (Ghahari et al., 2019 as *Apanteles dilectus* (Haliday)).

Distribution in the Middle East: Iran (Ghahari et al., 2019), Turkey (Inanç & Beyraslan, 2001c).

Extralimital distribution: Armenia, Austria, Bulgaria, China, Czech Republic, Finland, France, Germany, Hungary, Ireland, Italy, Japan, Korea, Moldova, Netherlands, Poland, Romania, Russia, Slovakia, Sweden, Switzerland, Ukraine, United Kingdom (Yu et al., 2016 as *Apanteles dilectus*; Fernandez-Triana et al., 2020).

Host records: Summarized by Yu et al. (2016) as being a parasitoid of the coleophorids *Coleophora* spp.; the erebid *Leucoma salicis* (L.); the gracillariids *Acrocercops brongniardella* Fabricius, *Caloptilia elongella* (L.), and *Garcillaria syringella* (Fabricius); the noctuid *Melancha persicariae* (L.); the plutellids *Prays fraxinella* Bjerkander, and *Prays oleae* Bernard; the tineid *Tineola bisselliella* (Himmel); the tortricids *Archips xylosteana* (L.), *Choristoneura murinana* (Hübner), *Dichelia histrionana* (Frölich), *Grapholita inopinata* (Heinrich), *Hedya pruniana* (Hübner), *Tortrix viridana* L., *Zeiraphera griseana* (Hübner), and *Zeiraphera rufimitrana* (Herrich-Schäffer), and the yponomeutids *Yponomeuta* spp.

Dolichogenidea drusilla (Nixon, 1972)

Catalogs with Iranian records: No catalog.

Distribution in Iran: Zanjan (Samin, Beyarslan, Coronado-Blanco et al., 2020 as *Apanteles (Dolichogenidea) drusilla* Nixon).

Distribution in the Middle East: Iran (Samin, Beyarslan, Coronado-Blanco et al., 2020), Turkey (Inanç & Beyarslan, 2001a).

Extralimital distribution: Bulgaria, Germany, Hungary, Italy, Mongolia, Russia, Slovakia, Sweden, Ukraine, United Kingdom (Fernandez-Triana et al., 2020; Yu et al., 2016).

Host records: Unknown.

Dolichogenidea emarginata (Nees von Esenbeck, 1834)

Catalogs with Iranian records: Gadallah et al. (2015), Farahani et al. (2016), Yu et al. (2016), Samin, Coronado-Blanco,

Kavallieratos et al. (2018) and Samin, Coronado-Blanco, Fischer et al. (2018).

Distribution in Iran: Ardabil (Ghahari, Fischer, Çetin Erdoğan et al., 2009).

Distribution in the Middle East: Iran (Ghahari, Fischer, Çetin Erdoğan et al., 2009), Israel–Palestine (Papp, 1970), Turkey (Inanç et al., 2004).

Extralimital distribution: Armenia, Austria, Azerbaijan, Belgium, Bulgaria, France, Germany, Hungary, Ireland, Italy, Kazakhstan, Lithuania, Moldova, Poland, Romania, Russia, Slovakia, Sweden, Switzerland, United Kingdom (Yu et al., 2016 as *Apanteles (Dolichogenidea) emarginatus* (Nees, 1834); Fernandez-Triana et al., 2020).

Host records: Summarized by Yu et al. (2016) as being a parasitoid of the depressarids *Agonopterix* spp., and *Depressaria* spp.; the gelechiids *Anacampsis timidella* Zeller, *Anarsia lineatella* Zeller, and *Sophronia grandii* Hertig; the gracillariids *Caloptilia* spp.; the nymphalid *Vanessa cardui* (L.); the plutellid *Plutella xylostella* (L.); the pyralid *Gallaria mellonella* (L.); the tortricids *Acleris forsskaleana* (L.), *Aethes dilucidana* (Stephens), *Aleimma loeflingiana* (L.), *Ancylis laetana* (Fabricius), *Choristoneura diversana* (Hübner), *Cnephasia* spp., *Cydia pomonella* (L.), *Dichelia histrionana* (Frölich), *Epinotia caprana* (Fabricius), *Ptycholomoides aeriferana* (Herrich-Schäffer), *Tortrix viridana* L., and *Zeiraphera isertana* (Fabricius); and the yponomeutids *Yponomeuta* spp.

Dolichogenidea ensiformis (Ratzeburg, 1844)

Catalogs with Iranian records: No catalog.

Distribution in Iran: East Azarbaijan (Samin, Beyarslan, Ranjith et al., 2020).

Distribution in the Middle East: Iran.

Extralimital distribution: Germany, Hungary, Italy, Latvia, Mongolia, Poland, Romania, Russia, Slovakia, Spain, Tunisia.

Host records: Summarized by Yu et al. (2016) as being a parasitoid of the acrolepiid *Acrolepia autumnitella* Curtis; the erebid *Euproctis similis* (Füssli), and the plutellid *Plutella xylostella* (L.). In Iran, this species has been reared from *Plutella xylostella* (L.) (Samin, Beyarslan, Ranjith et al., 2020).

Dolichogenidea erevanica (Tobias, 1976)

Catalogs with Iranian records: No catalog.

Distribution in Iran: East Azarbaijan (Samin, Beyarslan, Coronado-Blanco et al., 2020 as *Apanteles erevanicus* Tobias).

Distribution in the Middle East: Iran.

Extralimital distribution: Armenia, Bulgaria, Serbia.

Host records: recorded by Balveski (1999) as being a parasitoid of the tortricid *Gypsonoma aceriana* (Duponchel).

Dolichogenidae evonymellae (Bouché, 1834)

Catalogs with Iranian records: This species is new record for the fauna of Iran.

Distribution in Iran: Guilan province, Talesh (Gisoum Forest Reserve), 2♀, September 2013.

Distribution in the Middle East: Iran (new record), Lebanon (Nixon, 1972; Papp, 1978).

Extralimital distribution: Armenia, Azerbaijan, Belarus, Bulgaria, Czech Republic, Germany, Hungary, Italy, Netherlands, Portugal, Romania, Russia, Serbia, United Kingdom (Fernandez-Triana et al., 2020; Yu et al., 2016).

Host records: Summarized by Yu et al. (2016) as being a parasitoid of the sesiids *Paranthrene tabaniformis* (Rottemberg), and *Synanthedon* spp.; the tortricids *Cydia funebrana* (Treitschke), *Spilonota ocellana* (Denis and Schiffermüller), and *Tortrix viridana* L.; and the yponomeutids *Yponomeuta* spp.

Dolichogenidea fernandeztrianai Abdoli and Talebi, 2019

Catalogs with Iranian records: No catalog.

Distribution in Iran: Mazandran (Abdoli et al., 2019c).

Distribution in the Middle East: Iran.

Extralimital distribution: None.

Host records: Unknown.

Dolichogenidea flavostriata (Papp, 1977)

Catalogs with Iranian records: Farahani et al. (2016).

Distribution in Iran: Sistan and Baluchestan (Khajeh et al., 2014).

Distribution in the Middle East: Iran.

Extralimital distribution: Greece, Hungary (Yu et al., 2016 as *Apanteles (Dolichogenidea) flavostriatus* Papp, 1977; Fernandez-Triana et al., 2020).

Host records: Unknown.

Dolicogenidea gagates (Nees von Esenbeck, 1834)

Catalogs with Iranian records: No catalog.

Distribution in Iran: Guilan (Samin, Beyarslan, Ranjith et al., 2020).

Distribution in the Middle East: Iran.

Extralimital distribution: Austria, Belgium, Bulgaria, Estonia, Finland, France, Georgia, Germany, Hungary, Italy, Latvia, Lithuania, Poland, Romania, Russia, Spain, Sweden, Switzerland, United Kingdom.

Host records: Summarized by Yu et al. (2016) as being a parasitoid of the adelid *Nemophora minimella* (Denis and Schiffermüller); the geometrid *Abraxas grosuulariata* (L.); the pterophorids *Cnaemidophorus rhododactyla* (Denis and Schiffermüller), and *Stenoptilia bipunctidactyla* (Scopoli); and the tortricids *Pandemis heparana* (Denis and Schiffermüller), and *Spilonota ocellana* (Denis and Schiffermüller). In Iran, this species has been reared from the tortricid *Pandemis corylana* (Fabricius) (Samin, Beyarslan, Ranjith et al., 2020).

Dolichogenidea gracilariae (Wilkinson, 1940)
Catalogs with Iranian records: No catalog.
Distribution in Iran: West Azarbaijan (Ghahari et al., 2019 as *Apanteles gracilariae* Wilkinson, 1940)).
Distribution in the Middle East: Iran (Ghahari et al., 2019), Turkey (Çetin Erdoğan & Beyarslan, 2005; Inanç & Beyarslan, 2001c).
Extralimital distribution: Armenia, Austria, Azerbaijan, Bulgaria, Czech Republic, Germany, Hungary, Kazakhstan, Moldova, Poland, Romania, Russia, Serbia, Slovakia, Spain, Sweden, Switzerland, United Kingdom (Yu et al., 2016 as *Apanteles gracilariae*; Fernandez-Triana et al., 2020).
Host records: Summarized by Yu et al. (2016) as being a parasitoid of the gracillariids *Caloptilia cuculipennella* (Hübner), and *Gracillaria syringella* (Fabricius); the lyontiid *Lyonetia rajella* (L.); and the yponomeutid *Euhyponomeutoides albithoracellus* Gaj.

Dolichogenidia halidayi (Marshall, 1872)
Catalogs with Iranian records: Gadallah et al. (2015), Farahani et al. (2016), Yu et al. (2016), Samin, Coronado-Blanco, Kavallieratos et al. (2018) and Samin, Coronado-Blanco, Fischer et al. (2018).
Distribution in Iran: Isfahan (Ghahari et al., 2011c).
Distribution in the Middle East: Iran.
Extralimital distribution: Armenia, Croatia, Finland, Germany, Greece, Hungary, Ireland, North Macedonia, Madeira Islands, Romania, Russia, Sweden, Ukraine, United Kingdom, former Yugoslavia (Yu et al., 2016 as *Apanteles (Dolichogenidea) halidayi* Marshall, 1872; Fernandez-Triana et al., 2020).
Host records: Summarized by Yu et al. (2016) as being a parasitoid of the coleophorids *Coleophora glaucicolella* Wood, and *Goniodoma limoniella* (Stainton); the gelechiids *Ptocheuusa inopella* (Hübner), and *Teleiodes luculella* (Hübner); the glyphipterigid *Glyphipterix simpliciella* (Stephens); and the gracillariid *Parectopa ononidis* (Zeller).

Dolichogenidea immissa (Papp, 1977)
Distribution in the Middle East: Turkey (Inanç & Beyarslan, 2001a).
Extralimital distribution: Germany, Hungary, Slovakia.
Host records: Recorded by M. Shaw (2012) as being a parasitoid of the gracillariid *Caloptilia rufipennella* Hübner.

Dolichogenidea imperator (Wilkinson, 1939)
Catalogs with Iranian records: No catalog.
Distribution in Iran: West Azarbaijan (Gadallah et al., 2021).

Distribution in the Middle East: Iran.
Extralimital distribution: Armenia, Austria, Azerbaijan, Czech Republic, Germany, Hungary, Italy, Kazakhstan, Lithuania, Moldova, Netherland, Romania, Russia, Switzerland, Turkmenistan, Tajikistan, United Kingdom, Uzbekistan.
Host records: Summarized by Yu et al. (2016) as being a parasitoid of the acrolepiid *Acrolepia antumnitella* Curtis; the depressariids *Agonopterix assimilella* (Treitscke), *Agonopterix heracliana* (L.), *Agonopterix nervosa* (Haworth), *Depressaria daucella* (Denis and Schiffermüller), and *Depressaria radiella* (Goeze); the epermeniids *Epermenia aequidentellus* (E. Hofmann), and *Epermenia chaerophylella* (Goeze); the geometrid *Chesias legatella* (Denis and Schiffermüller); the gracillariid *Euspilapteryx auroguttellus* Stephens; and the plutellids *Plutella porrectella* (L.), and *Plutella xylostella* (L.). In Iran, this species has been reared from the plutellid *Plutella xylostella* (L.) on *Brassica oleraceae* (Gadallah et al., 2021).

Dolichogenidea impura (Nees von Esenbeck, 1834)
Catalogs with Iranian records: Samin, Coronado-Blanco, Fischer et al. (2018).
Distribution in Iran: Ardabil (Ghahari, 2016), East Azarbaijan (Samin, Coronado-Blanco, Fischer et al., 2018).
Distribution in the Middle East: Iran.
Extralimital distribution: Austria, Azerbaijan, Belgium, Bulgaria, Czech Republic, France, Germany, Greece, Hungary, Ireland, Italy, Latvia, Lithuania, Mongolia, Poland, Russia, Sweden, Switzerland (Yu et al., 2016 as *Apanteles (Dolichogenidea) impurus* (Nees); Fernandez-Triana et al., 2020).
Host records in Iran: Summarized by Yu et al. (2016) as being a parasitoid of the choreutid *Choreutis pariana* (Clerck); the coleophorid *Coleophora paripennella* Zeller; the elachistid *Elachista lastrella* Chrétien; the gelechiids *Metzneria aestivella* (Zeller), and *Scrobipalpa atriplicella* (Fischer); the gracillariid *Gracillaria syringella* (Fabricius); the lycaenid *Lysandra coridon* (Poda); the momphid *Mompha langiella* (Hübner); the pterophorids *Cnaemidophorus rhododactylus* (Denis and Schiffermüller), and *Porrittia galactodactyla* (Denis and Schiffermüller); the tortricids *Archips oporana* (L.), *Choristoneura murinana* (Hübner), *Cydia pactolana* (Zeller), *Lozotaenia forsterana* (Fabricius), *Spilonota ocellana* (Denis and Schiffermüller), and *Zeiraphera grisean* (Hübner); and the yponomeutid *Euhyponomeutoides ribesiella* (de Joannis). In Iran, this species has been reared from the gelechiid *Scrobipalpa* sp. (Samin, Coronado-Blanco, Fischer et al., 2018).

Dolichogenidea infima (Haliday, 1834)
Catalogs with Iranian records: No catalog.
Distribution in Iran: Ardabil (Samin, Fischer, Sakenin et al., 2019 as *Apanteles infimus* (Haliday, 1834)).

Distribution in the Middle East: Iran (Samin, Fischer, Sakenin et al., 2019), Turkey (Inanç & Beyarslan, 2001a). Extralimital distribution: Azerbaijan, Czech Republic, Finland, France, Georgia, Germany, Hungary, Ireland, Italy, Kazakhstan, Lithuania, North Macedonia, Mongolia, Netherlands, Poland, Romania, Russia, Sweden, Switzerland, United Kingdom, Uzbekistan, former Yugoslavia (Yu et al., 2016 as *Apanteles (Dolichogenidea) infimus*; Fernandez-Triana et al., 2020).
Host records: Summarized by Yu et al. (2016) as being a parasitoid of the acrolepiid *Acrolepia autumnitella* Curtis; the coleophorids *Coleophora albitarsella* Zeller; *Coleophora alticolella* Zeller, *Coleophora caespititiella* Zeller, *Coleophora glaucicolella* (Wood), and *Coleophora serratella* (L.); the epermennid *Epermenia chaerophylella* (Goeze); and the plutellid *Plutella porrectella* (L.).

Dolichogenidea iranica (Telenga, 1955)
Catalogs with Iranian records: Fallahzadeh and Saghaei (2010), Gadallah et al. (2015), Farahani et al. (2016), Yu et al. (2016 as *Apanteles (Dolichogenidea) iranicus* Telenga, 1955), Samin, Coronado-Blanco, Kavallieratos et al. (2018), Samin, Coronado-Blanco, Fischer et al. (2018) as *Apanteles iranicus*).
Distribution in Iran: Kerman, Sistan and Baluchestan (Shenefelt, 1972 as *Dolichogenidea iranica*), Iran (no specific locality cited) (Telenga, 1955; Tobias, 1986).
Distribution in the Middle East: Iran.
Extralimital distribution: Kazakhstan, Mongolia.
Host records: Unknown.

Dolichogenidea lacteicolor (Viereck, 1911)
Catalogs with Iranian records: Fallahzadeh and Saghaei (2010), Gadallah et al. (2015), Farahani et al. (2016), Yu et al. (2016 as *Apanteles (Dolichogenidea) lacticolor* Viereck, 1911), Samin, Coronado-Blanco, Kavallieratos et al. (2018), Samin, Coronado-Blanco, Fischer et al. (2018) as *Apanteles lacticolor*).
Distribution in Iran: Alborz (Hérard et al., 1979), Iran (no specific locality cited) (Modarres Awal, 1997, 2012).
Distribution in the Middle East: Iran (see references above), Israel−Palestine (Beyarslan et al., 2002b; Inanç & Beyarslan, 2001a, b).
Extralimital distribution: Nearctic, Oriental, Palaearctic; Canada (introduced).
Host records: recorded by Yu et al. (2016) as being a parasitoid of the depressariid *Carcina quercana* (Fabricius); the erebids *Euproctis* spp., *Hyphantria cunea* (Drury), *Lymantria dispar* (L.), *Orgyia* spp., and *Spilosoma luteum* (Hufnagel); the geometrid *Theria rupicapraria* (Denis and Schiffermüller); the lasiocampids *Malacosoma americanum* (Fabricius), and *Malacosoma neustria* (L.); the limacodod *Heterogenea asella* (Denis and Schiffermüller); the noctuids *Acronicta* spp., *Bena prasinana* (L.), and *Colocasia coryli* (L.); the nolids *Nola cucullatella* (L.), and *Nycteola asiatica* (Krulikovsky); the notodontids *Cerura vinula* (L.), and *Dicranura ulmi* (Denis and Schiffermüller)); the psychid *Sterrhopterix fusca* (Haworth); the pyralid *Palpita unionalis* (Hübner); the tortricid *Cydia funebrana* (Treitschke); and the zygaenid *Theresimima ampellophaga* (Bayle-Barelle). In Iran, this species has been reared from the erebid *Lymantria dispar* (L.) (Hérard et al., 1979; Modarres Awal, 1997, 2012 as *Porthtria dispar*).

Dolichogenidea laevigata (Ratzeburg, 1848)
Catalogs with Iranian records: No catalog.
Distribution in Iran: Lorestan (Ghahari et al., 2019 as *Apanteles laevigatus* (Ratzeburg).)
Distribution in the Middle East: Iran (Ghahari et al., 2019), Israel−Palestine (Halperin, 1986; Kugler, 1966), Turkey (Schimitschek, 1944; Inanç, 1997).
Extralimital distribution: Armenia, Azerbaijan, Bulgaria, China, Finland, France, Georgia, Germany, Hungary, Italy, Kazakhstan, Korea, Latvia, Lithuania, Moldova, Netherlands, Poland, Romania, Russia, Serbia, Slovakia, Spain, Sweden, Switzerland, Ukraine, United Kingdom, Uzbekistan (Yu et al., 2016 as *Apanteles laevigatus*; Fernandez-Triana et al., 2020).
Host records: Summarized by Yu et al. (2016) as being a parasitoid of the coleophorid *Coleophora nigricella* (Stephens); the cossid *Zeuzera pyrina* (L.); the crambid *Palpita vitrealis* (Rossi); the erebid *Euproctis chrysorrhoea* (L.); the gelechiids *Anacampsis* spp., and *Gelechia* spp.; the prodoxid *Lampronia fuscatella* (Tengström); the pterophorid *Adaina microdactyla* (Hübner); the pyralid *Gallaria mellonella* (L.); the rhynchitid *Byctiscus betulae* (L.); the sesiids *Paranthrene tabaniformis* (Rottemburg), and *Sesia apiformis* (Clerck); the tineid *Triaxomera parasitella* (Hübner); and the tortricids *Acleris forsskaleana* (L.), *Ancylis laetana* (Fabricius), *Archips* spp., *Enarmonia formosana* (Scopoli), *Gypsonoma* spp., *Hedya salicella* (L.), *Pandemis heparana* (Denis and Schiffermüller), *Spilonota ocellana* (Denis and Schiffermüller), and *Tortrix viridana* L.

Dolichogenidea laspeyresiella (Papp, 1972)
Catalogs with Iranian records: Gadallah et al. (2015), Farahani et al. (2016), Yu et al. (2016), Samin, Coronado-Blanco, Kavallieratos et al. (2018) and Samin, Coronado-Blanco, Fischer et al. (2018).
Distribution in Iran: Qom (Goldansaz et al., 2011; Norouzi et al., 2009 as *Apanteles laspeyresiella* Papp, 1972).
Distribution in the Middle East: Iran (Goldansaz et al., 2011; Norouzi et al., 2009), Turkey (Inanç & Beyarslan, 2001a, c).

Extralimital distribution: Austria, Azerbaijan, Belarus, Bulgaria, China, Hungary, Romania, Russia.

Host records: Summarized by Yu et al. (2016) as being a parasitoid of the gracillariid *Phyllonorycter* sp.; the pyralid *Ectomyelois ceratoniae* (Zeller); and the tortricid *Cydia funebrana* (Treitschke). In Iran, this species has been reared from the pyralid *Ectomyelois ceratoniae* (Zeller) (Goldansaz et al., 2011; Norouzi et al., 2009).

Dolichogenidea lineipes (Wesmael, 1837)

Catalogs with Iranian records: No catalog.

Distribution in Iran: Guilan (Ghahari, Çetin Erdoğan, Šedivý et al., 2010 as *Dolichogenidea aethiopicus* - misidentification).

Distribution in the Middle East: Iran (Ghahari, Çetin Erdoğan, Šedivý et al., 2010), Israel−Palestine (Halperin, 1986).

Extralimital distribution: Austria, Belgium, Czech Republic, Finland, France, Germany, Italy, Latvia, Poland, Russia, Slovakia, Switzerland, United Kingdom.

Host records: Summarized by Yu et al. (2016) as being a parasitoid of the argyresthiids *Argyresthia conjugella* Zeller, and *Argyresthia goedartella* (L.); the gelechiid *Anacampsis timidella* (Wocke); the gracillriids *Caloptilia semifascia* (Haworth), and *Acleris bergmanniana* (L.); the psychid *Luffia lapidella* Goeze; and the following tortricids: *Archips oporana* (L.), *Choristoneura murinana* (Hübner), *Clavigesta sylvestrana* (Curtis), *Cydia laricana* (Busck), *Cydia nigricana* (Fabricius), *Epinotia abbreviana* (Fabricius), *Epinotia pusillana* (Peyerimhoff), *Epinotia pygmaeana* (Hübner), *Epinotia sordidana* (Hübner), *Epinotia tedella* (Clerck), *Exapate duratella* Heyden, *Rhyacionia buoliana* (Denis and Schiffermüller), *Rhyacionia pinivorana* (Zeller), *Tortricodes alternella* (Denis and Schiffermüller), *Tortrix viridana* L., *Zeiraphera griseana* (Hübner), *Zeiraphera griseana* (Hübner), *Zeiraphera isertana* (Fabricius), and *Zeiraphera rufimitrana* (Herrich-Schäffer). In Iran, this species has been reared from a saturniid larva (Ghahari, Çetin Erdoğan, Šedivý et al., 2010 as *Dolichogenidea aethiopicus*).

Dolichogenidea longicauda (Wesmael, 1837)

Catalogs with Iranian records: No catalog.

Distribution in Iran: Lorestan (Gadallah et al., 2018).

Distribution in the Middle East: Iran (Gadallah et al., 2018), Turkey (Beyarslan et al., 2002b; Inanç, 1997; Inanç & Beyarslan, 1990, 1997, 2001a, b).

Extralimital distribution: Afghanistan, Armenia, Austria, Azerbaijan, Belarus, Belgium, Bulgaria, Canada, Czech Republic, Estonia, Finland, France, Georgia, Germany, Hungary, Italy, Korea, Latvia, Lithuania, Moldova, Mongolia, Netherlands, Poland, Romania, Russia, Serbia, Slovakia, Slovenia, Spain, Switzerland, Turkmenistan, Ukraine, United Kingdom (Yu et al., 2016 as *Apanteles* (*Dolichogenidea*) *longicauda* (Wesmael, 1837); Fernandez-Triana et al., 2020).

Host records: Summarized by Yu et al. as being a parasitoid of several lepidopteran pest species of the families Blastodacnidae, Bucculatrigidae, Choreutidae, Depressariidae, Gelechiidae, Lasiocampidae, Lymantridae, Momphidae, Oecophoridae, Plutellidae, Psychidae, Simaethidae, Tortricidae, Yponomeutidae, and Ypsolophidae.

Dolichogenidea longipalpis (Reinhard, 1880)

Catalogs with Iranian records: Gadallah et al. (2015), Farahani et al. (2016), Yu et al. (2016), Samin, Coronado-Blanco, Kavallieratos et al. (2018) and Samin, Coronado-Blanco, Fischer et al. (2018).

Distribution in Iran: Lorestan (Modarres Awal, 2012; Pirhadi et al., 2008).

Distribution in the Middle East: Iran (Modarres Awal, 2012; Pirhadi et al., 2008), Turkey (Inanç & Beyarslan, 1990; 2001a; Beyarslan et al., 2002a).

Extralimital distribution: Armenia, China, Finland, Germany, Hungary, Poland, Romania, Russia, Slovakia, Tajikistan, Ukraine, United Kingdom (Yu et al., 2016 as *Apanteles* (*Dolichogenidea*) *longipalpis* Reinhard, 1880; Fernandez-Triana et al., 2020).

Host records: Summarized by Yu et al. (2016) as a parasitoid of the plutellid *Plutella xylostella* (L.); the psychid *Epichnopterix plumella* (Denis and Schiffermüller); and the tortricid *Thiodia citrana* (Hübner). In Iran, this species has been reared from the pterolonchid *Syringopais temperatella* (Lederer) (Modarres Awal, 2012; Pirhadi et al., 2008).

Dolichogenidea murinanae (Čapek and Zwölfer, 1957)

Distribution in the Middle East: Turkey (Fernandez-Triana et al., 2020).

Extralimital distribution: Austria, Czech Republic, Finland, France, Germany, Italy, Lithuania, Mongolia, Morocco, Poland, Romania, Russia, Slovakia, Switzerland, United Kingdom.

Host records: Summarized by Yu et al. (2016) as being a parasitoid of the following tortricids: *Archips oporana* (L.), *Choristoneura murinana* (Hübner), *Epinotia nigricana* (Herrich-Schäffer), *Epinotia pusillana* (Peyerimhoff), *Rhyacionia buoliana* (Denis and Schiffermüller), *Zeiraphera griseana* (Hübner), and *Zeiraphera rufimitrana* (Herrich-Schäffer).

Dolichogenidea mycale (Nixon, 1972)

Distribution in the Middle East: Turkey (Inanç & Beyarslan, 2001a).

Extralimital distribution: Bulgaria, China, Czech Republic, Finland, Hungary, Poland, Slovakia, Sweden, Tunisia.

Host records: Unknown.

Dolichogenidea myron (Nixon, 1973)

Distribution in the Middle East: Turkey (Beyarslan et al., 2002a; Inanç & Beyarslan, 2001a).

Extralimital distribution: Austria, Finland, Germany, Greece, Switzerland, United Kingdom.

Host records: Recorded by M. Shaw (2012) as being a parasitoid of the argyresthiid *Argyresthia* sp.

Dolichogenidea nixosiris Papp, 1976

Catalogs with Iranian records: No catalog.

Distribution in Iran: Mazandaran (Naderian et al., 2020).

Distribution in the Middle East: Iran.

Extralimital distribution: China, Finland, Hungary, Mongolia, Russia, Turkmenistan (Fernandez-Triana et al., 2020).

Host records: Unknown.

Dolichogenidea olivierellae (Wilkinson, 1936)

Distribution in the Middle East: Israel−Palestine (Halperin, 1986; Lupo & Gerling, 1984; Papp, 2012).

Extralimital distribution: Algeria, Morocco.

Host records: Recorded by Wilkinson (1936) as being a parasitoid of the gelechiid *Amblypalpis olivierella* Ragonot.

Dolichogenidea phaloniae (Wilkinson, 1940)

Catalogs with Iranian records: No catalog.

Distribution in Iran: Mazandaran (Samin, Beyarslan, Coronado-Blanco et al., 2020 as *Apanteles* (*Dolichogenidea*) *phaloniae* Wilkinson).

Distribution in the Middle East: Iran (Samin, Beyarslan, Coronado-Blanco et al., 2020), Israel−Palestine (Halperin, 1986; Papp, 1970).

Extralimital distribution: Azerbaijan, Finland, Georgia, Germany, Hungary, Ireland, Italy, Lithuania, Madeira Islands, Moldova, Poland, Romania, Russia, Slovakia, United Kingdom (Fernandez-Triana et al., 2020; Yu et al., 2016).

Host records: Summarized by Yu et al. (2016) as being a parasitoid of the cossid *Zeuzera pyrina* (L.); the gelechiid *Isophrictis striatella* (Denis and Schiffermüller); and the tortricid *Aethes smeathmanniana* (Fabricius). In Iran, this species has been reared from *Zeuzera pyrina* (L.) (Samin, Beyarslan, Coronado-Blanco et al., 2020).

Dolichogenidea praetor (Marshall, 1885)

Catalogs with Iranian records: No catalog.

Distribution in Iran: Ardabil (Samin, Beyarslan, Ranjith et al., 2020).

Distribution in the Middle East: Iran.

Extralimital distribution: Armenia, Finland, France, Hungary, Mongolia, Romania, Russia, Slovakia, Sweden, Switzerland, United Kingdom.

Host records: Recorded by Wilkinson (1936) as a parasitoid of the tortricid *Eucosma aemulana* (Schläger).

Dolichogenidea princeps (Wilkinson, 1941)

Catalogs with Iranian records: No catalog.

Distribution in Iran: Guilan (Naderian et al., 2020), Kordestan (Samin, Sakenin Chelav, Ahmad, et al., 2020).

Distribution in the Middle East: Iran (Naderian et al., 2020; Samin, Sakenin Chelav, Ahmad, et al., 2020), Turkey (Beyarslan et al., 2006; Inanç & Beyarlan, 2001c).

Extralimital distribution: Azerbaijan, Hungary, Italy, Korea, Malta, Mongolia, Romania, Russia, Serbia, Slovakia, Spain, Tunisia, Ukraine, United Kingdom.

Host records: Recorded by Nixon (1972) as being a parasitoid of the coleophorids *Coleophora lutipennella* (Zeller), and *Coleophora obscenella* Herrich-Schäffer.

Dolichogenidea punctiger (Wesmael, 1837)

Catalogs with Iranian records: No catalog.

Distribution in Iran: Guilan (Samin, Beyarslan, Coronado-Blanco et al., 2020 as *Apanteles* (*Dolichogenidea*) *punctiger*).

Distribution in the Middle East: Iran.

Extralimital distribution: Belgium, Croatia, Czech Republic, Denmark, France, Germany, Hungary, Ireland, Italy, Netherlands, Poland, Russia, Slovakia, Sweden, Switzerland, Ukraine, United Kingdom.

Host records: Summarized by Yu et al. (2016) as being a parasitoid of the gracillariid *Caloptilia semifascia* (Haworth); the lasiocampid *Dendrolimus pini* (L.); and the tortricid *Eupoecilia* sp.

Dolichogenidea purdus (Papp, 1977)

Distribution in the Middle East: Turkey (Inanç & Beyarslan, 2001a).

Extralimital distribution: China, Hungary.

Host records: Recorded by Papp (1977) as being a parasitoid of the coleophorid *Coleophora glaucicolella* Wood.

Dolichogenidea seriphia (Nixon, 1972)

Catalogs with Iranian records: Gadallah et al. (2015), Farahani et al. (2016), Yu et al. (2016), Samin, Coronado-Blanco, Kavallieratos et al. (2018) and Samin, Coronado-Blanco, Fischer et al. (2018).

Distribution in Iran: Golestan (Ghahari & Fischer, 2011b).

Distribution in the Middle East: Iran (Ghahari & Fischer, 2011b), Turkey (Beyarslan et al., 2002b; Inanç & Beyarslan, 1997; 2001b).

Extralimital distribution: Greece, Hungary, Italy, Montenegro, Poland, Russia, Slovakia, Spain, Tunisia (Yu et al., 2016 as *Apanteles* (*Dolichogenidea*) *seriphia* Nixon, 1972; Fernandez-Triana et al., 2020).

Host records: Summarized by Yu et al. (2016) as being a parasitoid of the bedelliid *Bedellia ehikella* Szöcs; the bucculatricid *Bucculatrix cantabricella* Chrétien; the gelechiid *Metzneria ehikeella* Gezmány; the gracillariid

Parectopa onomidis (Zeller); and the nepticulid *Stigmella eberhardi* (Johansson).

Dolichogenidea sicaria (Marshall, 1885)
Catalogs with Iranian records: Gadallah et al. (2015), Farahani et al. (2016), Yu et al. (2016), Samin, Coronado-Blanco, Kavallieratos et al. (2018) and Samin, Coronado-Blanco, Fischer et al. (2018).
Distribution in Iran: Ilam (Ghahari et al., 2011d).
Distribution in the Middle East: Iran (Ghahari et al., 2011d), Turkey (Beyarslan, 1988; Beyarslan et al., 2002b; Beyarslan et al., 2006; Inanç, 1997; Inanç & Beyarslan, 1990, 1997, 2001a, b).
Extralimital distribution: Austria, Azerbaijan, Belarus, China, Czech Republic, Estonia, Finland, France, Georgia, Germany, Greece, Greenland, Hungary, Kazakhstan, Kyrgyzsatn, North Macedonia, Moldova, Mongolia, Montenegro, Morocco, Netherlands, New Zealand, Poland, Romania, Russia, Serbia, Slovakia, Spain, Switzerland, Tunisia, Ukraine, United Kindom (Yu et al., 2016 as *Apanteles (Dolichogenidea) sicarius* Marshall, 1885; Fernandez-Triana et al., 2020).
Host records: Summarized by Yu et al. (2016) as being a parasitoid of the acrolepiid *Acrolepia autumnitella* Curtis; the choreutid *Choreutis nemorana* (Hübner); the crambid *Diasemia reticularis* (L.); the depressariids *Agonopterix arenella* (Denis and Schiffermüller), and *Depressaria discipunctella* Herrich-Schäffer; the elachistid *Elachistia megerlella* (Hübner); the gelechiids *Epiphthora melanombra* Meyrick, *Isophrictis anthemidella* (Wocke), *Isophrictis striatella* (Denis and Schiffermüller), *Pexicopia malvella* (Hübner), and *Scrobipalpa instabilella* (Douglas); the hesperiids *Carcharodus alceae* (Esper), and *Pyrgus cinarae* (Rambur); the momphid *Mompha sturnipenella* (Treitscke); the nymphalid *Vanessa cardui* (L.); the plutellids *Plutella porrectella* (L.), and *Plutella xylostella* (L.); and the tortricids *Adoxophyes orana* (Fischer), *Aethes francillana* (Fabricius), *Cochylis posterana* Zeller, *Cydia delineana* (Walker), *Cydia nigricana* (Fabricius), *Eucosma aemulana* (Schläger), *Euxanthoides straminea* (Denis and Schiffermüller), *Hedya nubiferana* Haworth, *Lobesia botrana* (Denis and Schiffermüller), *Lobesia littoralis* (Westwood and Humphreys), *Sparganothis pilleriana* (Denis and Schiffermüller), and *Spilonota ocellana* (Denis and Schiffermüller).

Dolichogenidea soikai (Nixon, 1972)
Catalogs with Iranian records: No catalog.
Distribution in Iran: West Azarbaijan (Naderian et al., 2020).
Distribution in the Middle East: Iran (Naderian et al., 2020), Turkey (Inanç, 1997; Inanç & Beyarslan, 2001a).
Extralimital distribution: Bulgaria, Greece, Hungary, Italy, Russia, Switzerland, Tunisia, United Kingdom.
Host records: Unknown.

Dolichogenidea sophiae (Papp, 1972)
Catalogs with Iranian records: No catalog.
Distribution in Iran: Mazandaran (Naderian et al., 2020).
Distribution in the Middle East: Iran (Naderian et al., 2020), Turkey (Inanç & Beyarslan, 1990).
Extralimital distribution: Armenia, China, Georgia, Hungary, Moldova, Russia, Slovakia, Ukraine (Fernandez-Triana et al., 2020; Yu et al., 2016).
Host records: Unknown.

Dolichogenidea subemarginata (Abdinbekova, 1969)
Catalogs with Iranian records: No catalog.
Distribution in Iran: Mazandaran (Samin, Papp, Coronado-Blanco, 2018 as *Apanteles (Dolichogenidea) subemarginatus* Abdinbekova, 1969).
Distribution in the Middle East: Iran (Samin, Papp, Coronado-Blanco, 2018), Turkey (Beyarslan et al., 2002b; Inanç & Beyarslan, 2001b).
Extralimital distribution: Armenia, Azerbaijan, Hungary (Yu et al., 2016 as *Apanteles (Dolichogenidea) subemarginatus* Abdinbekova, 1969; Fernandez-Triana et al., 2020).
Host records: Unknown.

Dolichogenidea trachalus (Nixon, 1965)
Distribution in the Middle East: Syria (Lababidi & Hammoudi, 2008).
Extralimital distribution: Hungary, Ireland, United Kingdom.
Host records: Summarized by Yu et al. (2016) as being a parasitoid of the crambid *Palpita vitrealis* (Rossi); the Oecophorids *Endrosis sarcitrella* (L.) and *Hofmannophila pseudospretella* (Stainton); and the pyralids *Ephestia kuehniella* Zeller, and *Plodia interpunctella* (Hübner).

Dolichogenidea turkmenus (Telenga, 1955)
Catalogs with Iranian records: No catalog.
Distribution in Iran: Fars (Jahan et al., 2016).
Distribution in the Middle East: Iran (Jahan et al., 2016), Jordan (Papp, 1978), Turkey (Inanç & Beyarslan, 1997).
Extralimital distribution: Armenia, China, Kazakhstan, Turkmenistan, Uzbekistan (Yu et al., 2016 as *Apanteles (Dolichogenidea) turkmenus* Fernandez-Triana et al., 2020; Telenga, 1955).
Host records: In Iran, this species was erroneously reported in association with the aphidid species *Forda hirsuta* Mordvilko, and *Slavum mordvilcovi* Kreutz on *Pistacia atlantica* (Anacardiaceae) (Jahan et al., 2016).

Dolichogenidea ultor (Reinhard, 1880)
Catalogs with Iranian records: No catalog.
Distribution in Iran: East Azarbaijan (Samin, Beyarslan, Ranjith et al., 2020).
Distribution in the Middle East: Iran.

Extralimital distribution: Azerbaijan, Czech Republic, Georgia, Germany, Hungary, Italy, Poland, Romania, Russia, Serbia, Slovakia, Slovenia, Switzerland, Ukraine, United Kingdom.

Host records: Summarized by Yu et al. (2016) as being a parasitoid of the erebids *Euproctis chrysorrhoea* (L.), and *Orgyia antiqua* (L.); the lasiocampids *Eriogaster lanestris* (L.), and *Malacosoma neustria* (L.); the noctuid *Acronicta psi* (L.); the notodontid *Notodonta ziczac* (L.); the tortricid *Rhyacionia resinella* (L.); and the zygaenids *Illiberis sinensis* (Walker), and *Theresimima ampellophaga* (Bayle-Barelle). In Iran, this species has been reared from the lasiocampid *Malacosoma neustria* (Samin, Beyarslan, Ranjith et al., 2020).

Genus *Hygroplitis* Thomson, 1895

Hygroplitis rugulosus (Nees von Esenbeck, 1834)

Distribution in the Middle East: Turkey (Inanç & Beyarslan, 1990).

Extralimital distribution: Czech Republic, Germany, Hungary, Ireland, Italy, Netherlands, Poland, Russia, Sweden, Switzerland, Ukraine, United Kingdom.

Host records: Summarized by Yu et al. (2016) as being a parasitoid of the crambid *Elophila nymphaeata* (L.); and the noctuids *Acronicta rumicis* (L.), and *Acronicta tridens* (Denis and Schiffermüller).

Hygroplitis russatus (Haliday, 1834)

Catalogs with Iranian records: No catalog.

Distribution in Iran: Guilan (Ghahari & Sakenin, 2018), Mazandaran (Amooghli-Tabari & Ghahari, 2021).

Distribution in the Middle East: Iran (Amooghli-Tabari & Ghahari, 2021; Ghahari & Sakenin, 2018), Turkey (Beyarslan et al., 2002a; Inanç & Beyarslan, 2001c).

Extralimital distribution: Belgium, China, Finland, France, Germany, Hungary, Ireland, Japan, Korea, Moldova, Netherlands, Poland, Russia, Sweden, Ukraine, United Kingdom, Vietnam.

Host records: Summarized by Yu et al. (2016) as being a parasitoid of the crambids *Calomotropha paludella* (Hübner), *Chilo suppresalis* (Walker), and *Scirpophaga incertulas* (Walker); the noctuid *Sesamia inferens* (Walker); and the tortricid *Adoxophyes orana* (Fischer). In Iran, this species has been reared from the crambid *Chilo suppressalis* (Walker) (Amooghli-Tabari & Ghahari, 2021; Ghahari & Sakenin, 2018).

Genus *Iconella* Mason, 1981

Iconella aeolus (Nixon, 1965)

Distribution in the Middle East: Turkey (Inanç & Beyarslan, 2001c).

Extralimital distribution: Armenia, Germany, Russia, Ukraine, United Kingdom.

Host records: Recorded by M. Shaw (2012) as being a parasitoid of the pyralid *Ortholepis betulae* (Goeze).

Iconella albinervis (Tobias, 1964)

Catalogs with Iranian records: No catalog.

Distribution in Iran: East Azarbaijan (Samin, Papp, Coronado-Blanco, 2018 as *Apanteles* (*Iconella*) *albinervis* Tobias, 1964).

Distribution in the Middle East: Iran (Samin, Papp, Coronado-Blanco, 2018), Turkey (Beyarslan et al., 2002a, b; Inanç, 1997; Inanç & Beyarslan, 2001a, b).

Extralimital distribution: Azerbaijan, Hungary, Kazakhstan, Moldova, Russia, Ukraine (Yu et al., 2016 as *Apanteles* (*Iconella*) *albinervis* Tobias, 1964; Fernandez-Triana et al., 2020).

Host records: Unknown.

Iconella brachyradiata Abdoli and Talebi, 2021

Catalogs with Iranian records: No catalog.

Distribution in Iran: Alborz, Qazvin (Holotype: Zereshk Road) (Abdoli et al., 2021b).

Distribution in the Middle East: Iran.

Extralimital distribution: None.

Host records: Unknown.

Iconella isus (Nixon, 1965)

Catalogs with Iranian records: Gadallah et al. (2015), Farahani et al. (2016), Yu et al. (2016), Samin, Coronado-Blanco, Kavallieratos et al. (2018) and Samin, Coronado-Blanco, Fischer et al. (2018).

Distribution in Iran: Sistan and Baluchestan (Khajeh et al., 2014).

Distribution in the Middle East: Iran (Khajeh et al., 2014), Israel—Palestine (Papp, 2012).

Extralimital distribution: Armenia, Hungary, Kazakhstan, Russia, Serbia, Spain, Uzbekistan (Yu et al., 2016 as *Apanteles* (*Iconella*) *isus*; Fernandez-Triana et al., 2020; Nixon, 1965).

Host records: Recorded by Nixon (1965) and Tobias (1976, 1986) as being a parasitoid of the pyralid *Etiella zinckenella* (Treitschke).

Iconella lacteoides (Nixon, 1965)

Catalogs with Iranian records: No catalog.

Distribution in Iran: Alborz (Abdoli et al., 2021b), Isfahan (Gadallah et al., 2021).

Distribution in the Middle East: Iran (Abdoli et al., 2021b; Gadallah et al., 2021), Turkey (Beyarslan et al., 2002a; Inanç, 1997; Inanç & Beyarslan, 2001a; Papp, 1982).

Extralimital distribution: Armenia, Azerbaijan, Germany, Greece, Hungary, Italy, Kazakhstan, Mongolia, Poland, Russia, Slovakia, Sweden, Turkmenistan, Ukraine, Uzbekistan.

Host records: Summarized by Yu et al. (2016) as being a parasitoid of the pyralids *Acrobasis sodalella* Zeller, and *Homoeosoma nebulella* Denis and Schiffermüller. In Iran, this species has been reared from the pyralid *Homoeosoma nebulella* on *Helianthus* sp. (Gadallah et al., 2021).

Iconella merula (Reinhard, 1880)

Distribution in the Middle East: Israel—Palestine (Papp, 2012), Turkey (Beyarslan et al., 2002a).

Extralimital distribution: Austria, Belgium, Bulgaria, Finland, Germany, Hungary, Poland, Romania, Russia, Slovakia, Ukraine.

Host records: Unknown.

Iconella meruloides (Nixon, 1965)

Catalogs with Iranian records: No catalog.

Distribution in Iran: Alborz (Abdoli et al., 2021b), Khuzestan (Zargar et al., 2019a).

Distribution in the Middle East: Iran (Abdoli et al., 2021b; Zargar et al., 2019a), Israel—Palestine (Papp, 2012), Jordan (Papp, 1982), Turkey (Nixon, 1965; Papp, 1982).

Extralimital distribution: Hungary, Malta, Romania (Yu et al., 2016 as *Apanteles* (*Iconella*) *meruloides*; Fernandez-Triana et al., 2020; Nixon, 1965).

Host records: Recorded by Nixon (1965) and Tobias (1976, 1986) as being a parasitoid of the tortricid *Lobesia botrana* (Denis and Schiffermüller).

Iconella mongashtensis Zargar and Gupta, 2019

Catalogs with Iranian records: No catalog.

Distribution in Iran: Khuzestan (Zargar et al., 2019a).

Distribution in the Middle East: Iran.

Extralimital distribution: None.

Host records: Unknown.

Iconella myeloenta (Wilkinson, 1937)

Catalogs with Iranian records: Gadallah et al. (2015), Farahani et al. (2016), Yu et al. (2016), Samin, Coronado-Blanco, Kavallieratos et al. (2018) and Samin, Coronado-Blanco, Fischer et al. (2018).

Distribution in Iran: Isfahan (Nobakht et al., 2015; Sobhani et al., 2012), Kerman (Mehrnejad, 2010), Khuzestan (Zargar et al., 2019a), Markazi, Tehran (Kishani Farahani et al., 2008a, 2010; Modarres Awal, 2012 as *Apanteles myeloenta*), Qazvin (Ghahari et al., 2011b), Qom (Goldansaz et al., 2011; Kishani Farahani et al., 2008a, 2010, 2012, 2013; Modarres Awal, 2012 as *Apanteles myeloenta*).

Distribution in the Middle East: Cyprus (Nixon, 1965, 1976; Wilkinson, 1937), Iran (see references above), Israel—Palestine (Papp, 2012), Turkey (Beyarslan et al., 2002a; Beyarslan et al., 2006; Haeselbarth, 1983; Inanç, 1997).

Extralimital distribution: Greece, Moldova, Russia, Spain, Turkmenistan (Yu et al., 2016 as *Apanteles* (*Iconella*) *myeloenta*; Fernandez-Triana et al., 2020; Wilkinson, 1937).

Host records: Summarized by Yu et al. (2016) as being a parasitoid of the pyralid *Ectomyelois ceratoniae* (Zeller). In Iran, this species has been reared from the pyralids *Arimania komaroffi* (Ragonot) (Mehrnejad, 2010), and *Ectomyelois ceratoniae* (Zeller) (Goldansaz et al., 2011; Kishani Farahani et al., 2008a, b, c, 2010, 2012, 2013; Modarres Awal, 2012; Nobakht et al., 2015; Sobhani et al., 2012).

Iconella nagyi (Papp, 1975)

Catalogs with Iranian records: Gadallah et al. (2015), Farahani et al. (2016), Yu et al. (2016), Samin, Coronado-Blanco, Kavallieratos et al. (2018) and Samin, Coronado-Blanco, Fischer et al. (2018).

Distribution in Iran: Sistan and Baluchestan (Khajeh et al., 2014 as *Iconella nagyi* (Papp, 1975)).

Distribution in the Middle East: Iran.

Extralimital distribution: Romania.

Host records: Unknown.

Iconella similus Zargar and Gupta, 2019

Catalogs with Iranian records: No catalog.

Distribution in Iran: Khuzestan (Zargar et al., 2019a).

Distribution in the Middle East: Iran.

Extralimital distribution: None.

Host records: Unknown.

Iconella subcamilla (Tobias, 1976)

Catalogs with Iranian records: Gadallah et al. (2015), Farahani et al. (2016), Yu et al. (2016 as *Apanteles* (*Iconella*) *subcamilla*; Tobias, 1976), Samin, Coronado-Blanco, Kavallieratos et al. (2018) and Samin, Coronado-Blanco, Fischer et al. (2018) as *Apanteles subcamilla*).

Distribution in Iran: Khuzestan (Zargar et al., 2019a), Sistan and Baluchestan (Khajeh et al., 2014).

Distribution in the Middle East: Iran (Khajae et al., 2014; Zargar et al., 2019a), Israel—Palestine (Papp, 2012).

Extralimital distribution: Azerbaijan, Cape Verde Islands (Yu et al., 2016 as *Apanteles* (*Iconella*) *subcamilla*; Fernandez-Triana et al., 2020; Tobias, 1976).

Host records: Unknown.

Iconella vindicius (Nixon, 1965)

Distribution in the Middle East: Turkey (Inanç, 1997; Inanç & Beyarslan, 2001a).

Extralimital distribution: Bulgaria, Georgia, Hungary, Italy, Korea, Russia, Ukraine.

Host records: Unknown.

Genus *Illidops* Mason, 1981

Illidops albostigmalis van Achterberg and Fernandez-Triana, 2017

Distribution in the Middle East: United Arab Emirates, Yemen (Fernandez-Triana & van Achterberg, 2017).

Extralimital distribution: None.
Host records: Unknown.

Illidops butalidis (Marshall, 1889)
Distribution in the Middle East: Turkey (Inanç & Beyarslan, 2001a).
Extralimital distribution: Bulgaria, Croatia, Germany, Hungary, Mongolia, Romania, Russia, Serbia, Slovakia, Spain, Sweden, Tunisia, Ukraine, United Kingdom.
Host records: Summarized by Yu et al. (2016) as being a parasitoid of the scythridids *Scythris fuscoaenea* Haworth, and *Scythris picaepennis* (Haworth); and the yponomeutid *Roeslerstammia erxlebella* (Fabricius).

Illidops mutabilis (Telenga, 1955)
Catalogs with Iranian records: No catalog.
Distribution in Iran: Kordestan (Naderian et al., 2020).
Distribution in the Middle East: Iran (Naderian et al., 2020), Turkey (Inanç & Beyarslan, 2001a).
Extralimital distribution: Bulgaria, Georgia, Hungary, Kazakhstan, Mongolia, Romania, Russia, Serbia, Slovakia, Spain, Tunisia, Ukraine.
Host records: Recorded by Tobias (1986) as being a parasitoid of the pyralid *Etielle zinckenella* (Treitschke).

Illidops naso (Marshall, 1885)
Catalogs with Iranian records: Gadallah et al. (2015), Farahani et al. (2016), Yu et al. (2016), Samin, Coronado-Blanco, Kavallieratos et al. (2018) and Samin, Coronado-Blanco, Fischer et al. (2018).
Distribution in Iran: Isfahan (Ghahari et al., 2011c), West Azarbaijan (Rastegar et al., 2012).
Distribution in the Middle East: Iran (Ghahari et al., 2011c; Rastegar et al., 2012), Turkey (Beyarslan et al., 2002a; Inanç & Beyarslan, 2001a; Nixon, 1976; Papp, 1981).
Extralimital distribution: Afghanistan, Armenia, Azerbaijan, Bulgaria, Croatia, Finland, Georgia, Greece, Hungary, Kazakhstan, Korea, Kyrgyzstan, North Macedonia, Moldova, Mongolia, Romania, Russia, Serbia, Slovakia, Switzerland, Turkmenistan, United Kingdom, Uzbekistan.
Host records: Unknown.

Illidops scutellaris (Muesebeck, 1921)
Catalogs with Iranian records: Gadallah et al. (2015), Farahani et al. (2016), Yu et al. (2016 as *Apanteles* (*Illidops*) *scutellaris* Muesebeck, 1921), Samin, Coronado-Blanco, Kavallieratos et al. (2018) and Samin, Coronado-Blanco, Fischer et al. (2018) as *Apanteles scutellaris*).
Distribution in Iran: Ilam (Ghahari et al., 2011d), Khuzestan (Samin, van Achterberg, & Ghahari, 2015), West Azarbaijan (Ghahari & Fischer, 2011a).
Distribution in the Middle East: Cyprus (Papp, 1981), Iran (see references above).

Extralimital distribution: Bulgaria, Greece, Hawaiian Islands, Hungary, Mexico, United States of America (Yu et al., 2016 as *Apanteles* (*Illidops*) *scutellaris*; Fernandez-Triana et al., 2020; Muesebeck, 1921).
Host records: Summarized by Yu et al. (2016) as being a parasitoid of the gelechiids *Kieferia lycopersicella* (Wasingham), and *Phthorimaea operculella* (Zeller).

Illidops suevus (Reinhard, 1880)
Catalogs with Iranian records: Gadallah et al. (2015), Farahani et al. (2016), Yu et al. (2016), Samin, Coronado-Blanco, Kavallieratos et al. (2018) and Samin, Coronado-Blanco, Fischer et al. (2018).
Distribution in Iran: Golestan (Ghahari & Fischer, 2011b as *Apanteles suevus* Reinhard, 1880), Mazandaran (Ghahari, 2017 as *Apanateles suevus* - around rice fields).
Distribution in the Middle East: Iran.
Extralimital distribution: Armenia, Austria, Britain, Bulgaria, Croatia, Czech Republic, France, Germany, Greece, Hungary, Kazakhstan, Korea, North Macedonia, Malta, Moldova, Mongolia, Montenegro, Poland, Romania, Russia, Serbia, Slovakia, Switzerland.
Host records: Recorded by Nixon (1976) and Tobias (1986) as being a parasitoid of the psychid *Epichnopterix* sp.

Illidops urgo (Nixon, 1965)
Catalogs with Iranian records: No catalog.
Distribution in Iran: East Azarbaijan (Samin, Coronado-Blanco, Hosseini et al., 2019 as *Apanteles urgo*; Nixon, 1965).
Distribution in the Middle East: Iran (Samin, Coronado-Blanco, Hosseini et al., 2019), Turkey (Beyarslan et al., 2002a; Inanç, 1997; Inanç & Beyarslan, 1990; 2001a).
Extralimital distribution: Azerbaijan, Croatia, Greece, Hungary, Mongolia, Russia, Slovakia.
Host records: Unknown.

Genus *Microgaster* Latreille, 1804
Microgaster alebion Nixon, 1968
Distribution in the Middle East: Turkey (Beyarslan et al., 2002a; Inanç & Beyarslan, 2001a).
Extralimital distribtution: Czech Republic, Finland, Germany, Hungary, Italy, Poland, Romania, Russia, Serbia, Switzerland, United Kingdom.
Host records: Summarized by Yu et al. (2016) as being a parasitoid of the crambid *Patania ruralis* (Scopoli); the nymphalid *Vanessa cardui* (L.); and the pterophorid *Platyptilia gonodactyla* (Denis and Schiffermüller).

Microgaster asramenes Nixon, 1968
Distribution in the Middle East: Turkey (Nixon, 1968; Tobias, 1986).

Extralimital distribution: China, Georgia, Hungary, Italy, Korea, Poland, Romania, Russia.

Host records: Recorded by Tobias (1986) as being a parasitoid of the tortricid *Eudemis porphyrana* (Hübner).

Microgaster australis Thomson, 1895

Catalogs with Iranian records: Fallahzadeh and Saghaei (2010), Gadallah et al. (2015), Farahani et al. (2016), Yu et al. (2016), Samin, Coronado-Blanco, Kavallieratos et al. (2018) and Samin, Coronado-Blanco, Fischer et al. (2018).

Distribution in Iran: Fars (Lashkari Bod et al., 2011), Iran (no specific locality cited) (Tobias, 1986).

Distribution in the Middle East: Iran (Lashkari Bod et al., 2011; Tobias, 1986), Turkey (Fernandez-Triana et al., 2020).

Extralimital distribution: Georgia, Germany, Greece, Hungary, Italy, Kazakhstan, Latvia, Moldova, Mongolia, Montenegro, Poland, Russia, Slovenia, Spain, Turkmenistan.

Host records: Summarized by Yu et al. (2016) as being a parasitoid of the hesperiids *Carcharodus alceae* (Esper), *Pyrgus armoricanus* (Oberthür), *Pyrgus onopardi* (Kirby), and *Pyrgus serratulae* (Rambur).

Microgaster famula Nixon, 1968

Distribution in the Middle East: Turkey (Inanç & Beyraslan, 2001c).

Extralimital distribution: Austria, Croatia, Hungary, Moldova, Romania, Russia, Serbia, Slovakia, Switzerland.

Host records: Recorded by Tobias (1976) as being a parasitoid of the noctuid *Polychrysia moneta* (Fabricius).

Microgaster filizinancae Koçak and Kemal, 2013

Distribution in the Middle East: Turkey (Koçak and Kemal, 2013).

Extralimital distribution: None.

Host records: Unknown.

Microgaster fischeri Papp, 1960

Distribution in the Middle East: Turkey (Inanç & Beyarslan, 2001a, c).

Extralimital distribution: Austria, Hungary, Moldova, Mongolia, Russia.

Host records: Unknown.

Microgaster fulvicrus Thomson, 1895

Distribution in the Middle East: Turkey (Inanç & Beyarslan, 2001c).

Extralimital distribution: Finland, Germany, Hungary, Ireland, Japan, Korea, Moldova, Montenegro, Romania, Russia, Serbia, Slovakia, Sweden, United Kingdom, Uzbekistan.

Host records: Recorded by Nixon (1968) as being a parasitoid of the depressariid *Agonopterix ocellana* (Fabricius).

Microgaster hospes Marsahll, 1885

Catalogs with Iranian records: No catalog.

Distribution in Iran: East Azarbaijan (Samin, Beyarslan, Ranjith et al., 2020).

Distribution in the Middle East: Iran.

Extralimital distribution: Austria, Azerbaijan, Hungary, Kyrgyzstan, Moldova, Mongolia, Romania, Russia, Ukraine.

Host records: Summarized by Yu et al. (2016) as being a parasitoid of the argyresthiid *Argyresthia calliphanes* Meyrick; the erebid *Lymantria dispar* (L.); the gelechiid *Gelechia hippophaella* (Schrank); the noctuids *Colocasia coryli* (L.), and *Polychrysia moneta* (Fabricius); the pyralids *Acrobasis sodalella* Zeller, and *Conobothra profundana* (Denis and Schiffermüller); and the tortricids *Acleris asperana* (Hübner), *Acleris comariana* (Lienig and Zeller), *Acleris hastiana* (L.), *Acleris variegana* (Denis and Schiffermüller), *Aleimma loeflingiana* (L.), *Ancylis comptana* (Frölich), *Ancylis upupana* (Treitschke), *Archips rosana* (L.), *Archips xylosteana* (L.), *Eudemis profundana* (Denis and Schiffermüller), *Hedya nubiferana* Haworth, *Pandemis heparana* (Denis and Schiffermüller), *Ptycholoma lecheana* (L.), and *Tortrix viridana* L. In Iran, this species has been reared from *Archips rosana* (L.) (Samin, Beyarslan, Ranjith et al., 2020).

Microgaster luctuosa Haliday, 1834

Catalogs with Iranian records: Farahani et al. (2016), Yu et al. (2016), Samin, Coronado-Blanco, Kavallieratos et al. (2018) and Samin, Coronado-Blanco, Fischer et al. (2018).

Distribution in Iran: East Azarbaijan (Samin, Beyarslan, Ranjith et al., 2020), Golestan (Samin, Coronado-Blanco, Hosseini et al., 2019 as *M. curvicrus*), West Azarbaijan (Samin et al., 2014).

Distribution in the Middle East: Iran (see references above), Israel—Palestine (Papp, 2012), Turkey (Fernandez-Triana et al., 2020).

Extralimital distribution: Austria, Azerbaijan, Bulgaria, Croatia, Finland, Germany, Greece, Hungary, Ireland, Moldova, Mongolia, Poland, Romania, Russia, Serbia, Sweden, Switzerland, Tunisia, Turkmenistan, United Kingdom, Uzbekistan.

Host records: Summarized by Yu et al. (2016) as being a parasitoid of the depressariid *Agonopterix pallorella* (Zeller); and the tortricids *Cnephasia asseclana* (Denis and Schiffermüller), and *Olethreutes arbutella* (L.).

Microgaster meridiana Haliday, 1834

Distribution in the Middle East: Turkey (Fernandez-Triana et al., 2020).

Extralimital distribution: Bulgaria, Czech Republic, Finland, Germany, Hungary, Ireland, Italy, Kazakhstan, Latvia, Lithuania, Moldova, Poland, Romania, Russia, Slovakia, Spain, Sweden, Switzerland, Ukraine, United Kingdom.

Host records: Summarized by Yu et al. (2016) as being a parasitoid of the depressariid *Agonopterix petesitis* (Standfuss); the gelechiid *Dichomeres derasella* (Denis and Schiffermüller); and the tortricids *Aphelia viburnana* (L.), *Archips rosana* (L.), and *Cnephasia chrysantheana* (Duponchel).

Microgaster messoria Haliday, 1834

Catalogs with Iranian records: This species is new record for the fauna of Iran.

Distribution in Iran: Golestan province, Gonbad, 2♀, June 2014, ex. *Ostrinia nubilalis* (Hübner) (Lepidoptera: Crambidae).

Distribution in the Middle East: Iran (new record), Turkey (Inanç, 1997).

Extralimital distribution: Armenia, Austria, Azerbaijan, Bulgaria, Canada, China, Croatia, Czech Republic, Finland, France, Georgia, Germany, Hungary, Ireland, Italy Japan, Kazakhstan, Latvia, North Macedonia, Malta, Moldova, Montenegro, Netherlands, Poland, Romania, Russia, Serbia, Spain, Sweden, Switzerland, Turkmenistan, Ukraine, United Kingdom, Uzbekistan.

Host records: Summarized by Yu et al. (2016) as being a parasitoid of the choreutid *Tebenna micalis* (Mann); the crambids *Anania crocealis* (Hübner), and *Ostrinia nubilalis* (Hübner); the depressariid *Agonopterix atomella* (Denis and Schiffermüller); the erebid *Lymantria dispar* (L.); the gelechiid *Acampsis populella* (Clerck); the geometrids *Eupithecia denotata* (Hübner), *Eupithecia linariata* (Denis and Schiffermüller), and *Perizoma flavofsciata* (Thunberg); the gracillariid *Aspilapterix atomella* (Denis and Schiffermüller); the nymphalids *Aglais urticae* (L.), and *Vanessa indica* (Herbst); the sphingid *Smerinthus ocellatus* (L.); and the tortricids *Acleris aspersana* (Hübner), *Archips rosana* (L.), *Clepsis unicolorana* (Duponchel), *Lathronympha strigana* (Fabricius), and *Pelatea klugiana* (Freyer). In Iran, this species has been reared from the crambid *Ostrinia nubilalis* (this study).

Microgaster opheltes Nixon, 1968

Distribution in the Middle East: Turkey (Inanç & Beyarslan, 2001a).

Extralimital distribution: Ireland, Italy, North Macedonia, Romania, former Yugoslavia.

Host records: Unknown.

Microgaster parvistriga Thomson, 1895

Catalogs with Iranian records: Gadallah et al. (2015), Farahani et al. (2016), Yu et al. (2016), Samin, Coronado-Blanco, Kavallieratos et al. (2018) and Samin, Coronado-Blanco, Fischer et al. (2018).

Distribution in Iran: Guilan (Ghahari, Fischer, & Tobias et al., 2012), Semnan (Samin, Fischer, & Ghahari, 2015), West Azarbaijan (Samin et al., 2014).

Distribution in the Middle East: Iran.

Extralimital distribution: Armenia, Bulgaria, Finland, Germany, Greece, Hungary, Korea, Mongolia, Poland, Romania, Russia, Slovakia, Sweden, Switzerland, United Kingdom.

Host records: Summarized by Yu et al. (2016) as being a parasitoid of the argyresthiid *Argyresthia goedartella* (L.); the cosmopterigids *Cosmopterix scribaiella* Zeller, and *Cosmopterix zieglerella* (Hübner); the gelechiid *Anacampsis timidella* (Wocke); and the tortricids *Acleris kochiella* (Goeze), *Epinotia tetraquetrana* (Haworth), and *Tortrix viridana* L.

Microgaster rufipes Nees von Esenbeck, 1834

Catalogs with Iranian records: Fallahzadeh and Saghaei (2010), Gadallah et al. (2015), Farahani et al. (2016), Yu et al. (2016), Samin, Coronado-Blanco, Kavallieratos et al. (2018) and Samin, Coronado-Blanco, Fischer et al. (2018).

Distribution in Iran: Ardabil (Samin, Beyarslan, Ranjith et al., 2020), Guilan (Ghahari, Fischer, Çetin Erdoğan et al., 2010), Iran (no specific locality cited) (Telenga, 1955).

Distribution in the Middle East: Iran (see references above), Turkey (Fernandez-Triana et al., 2020).

Extralimital distribution: Widely distributed in the Palaearctic region.

Host records: Summarized by Yu et al. (2016) as being a parasitoid of the geometrid *Eupithesia immundata* (Lienig), *Epithesia linariata* (Denis and Schiffermüller); the gracillariid *Caloptilia alchimiella* (Scopoli); the pterophorids *Platyptilia gonodactyla* (Denis and Schiffermüller), and *Platylia nemoralis* Zeller; the pyralid *Acrobasis consociella* (Hübner); and the tortricid *Archips rosana* (L.). In Iran, this species has been reared from the yponomeutid *Yponomeuta padella* (L.) (Samin, Beyarslan, Ranjith et al., 2020).

Microgaster stictica Ruthe, 1858

Distribution in the Middle East: Turkey (Beyarslan et al., 2002a; Inanç & Beyarslan, 2001c).

Extralimital distribution: Bulgaria, Croatia, Czech Republic, Finland, Germany, Hungary, Ireland, Italy, Korea, Mongolia, Netherlands, Poland, Romania, Russia, Slovakia, Spain, Sweden, Switzerland, United Kingdom, former Yugoslavia.

Host records: Summarized by Yu et al. (2016) as being a parasitoid of the crambid *Patania ruralis* (Scopoli); the gelechiid *Anacampsis populella* (Clerck); the nymphalids *Vanessa atalanta* (L.), and *Vanessa cardui* (L.); and the tortricids *Archips xylosteana* (L.), *Argyroploa lediana* (L.), and *Sparganothis pilleriana* (Denis and Schiffermüller).

Microgaster subcompleta Nees von Esenbeck, 1834

Catalogs with Iranian records: This species is new record for the fauna of Iran.

Distribution in Iran: Markazi province, Khomein, 3♀, August 2015, ex *Vanessa cardui* (Linnaeus) (Lepidoptera: Nymphalidae).

Distribution in the Middle East: Iran (new record), Turkey (Beyarslan et al., 2002a; Inanç, 1997; Inanç & Beyarslan, 2001a).

Extralimital distribution: Armenia, Austria, Azerbaijan, Belarus, Belgium, Bulgaria, China, Croatia, Czech Republic, France, Georgia, Germany, Hungary, Ireland, Italy, Japan, Korea, Lithuania, North Macedonia, Moldova, Netherlands, Poland, Romania, Russia, Slovakia, Spain, Switzerland, United States of America, Ukraine, United Kingdom.

Host records: Summarized by Yu et al. (2016) as being a parasitoid of the crambids *Anania hortulata* (L.), *Loxostege sticticalis* (L.), *Patania ruralis* (Scopoli), and *Sitochroa verticalis* (L.); the erebid *Hypena proboscidalis* (L.); the noctuids *Acronicta rumicis* (L.), *Acronicta tridens* (Denis and Schiffermüller), *Cucullia verbasci* (L.), *Euchalcia viriabilis* (Piller and Mitterpacher), *Polychrisia moneta* (Fabricius), and *Shargacucullia scrophulariae* (Denis and Schiffermüller); the nymphalids *Aglais io* (L.), *Aglais urticae* (L.), *Polygonia c-album* (L.), *Vanessa atalanta* (L.), and *Vanessa cardui* (L.); the tortricids *Ancylis unculana* (Haworth), *Hedya nubiferana* Haworth, *Phalonidia manniana* (Fischer), and *Tortrix viridana* L. In Iran, this species has been reared from *Vanessa cardui* (this study).

Microgaster subtilipunctata Papp, 1959

Distribution in the Middle East: Turkey (Tobias, 1986).

Extralimital distribution: Austria, Germany, Hungary, Moldova, Romania, Russia, Switzerland.

Host records: Unknown.

Genus *Napamus* Papp, 1993

Napamus vipio (Reinhard, 1880)

Catalogs with Iranian records: No catalog.

Distribution in Iran: Northern Khorasan (Ghafouri Moghaddam et al., 2021).

Distribution in the Middle East: Iran (Ghafouri Moghaddam et al., 2021), Israel—Palestine (Papp, 2012), Turkey (Inanç & Beyarslan, 1990, 2001a; Nixon, 1976).

Extralimital distribution: Armenia, Austria, Croatia, France, Germany, Hungary, Italy, Romania, Russia, Spain.

Host records: Summarized by Yu et al. (2016) as being a solitary parasitoid of the scythridid *Scythris knochella* (Fabricius), and the tineid *Haplotinea insectella* (Fabricius).

Genus *Parapanteles* Ashmead, 1900

Parapanteles aethiopicus (Wilkinson, 1931)

Distribution in the Middle East: Egypt (Fernandez-Triana et al., 2020).

Extralimital distribution: Cameroon, Democratic Republic of Congo, Ethiopia, Ivory Coast, Kenya, Ruwanda, Senegal, Sierra Leone, Somalia, South Africa, Sudan, Tanzania, Uganda.

Host records: Summarized by Yu et al. (2016) as being a parasitoid of the crambids *Antigastra catalaunalis* (Duponchel), and *Chilo zacconius* Bleszyński; the erebids *Achaea catella* Guenée, and *Utetheisa pulchella* (L.); the lasiocampids *Catalebeda* sp., and *Streblote graberi* (Dewitz); the noctuids *Nyodes prasinodes* Prout, and *Spodoptera exigua* (Hübner); the nymphalid *Hamanumida daedalus* (Fabricius); the saturniids *Holocerina angulata* (Aurivillius), and *Gonibrasia tyrrhea* (Cramer); and the zygaenid *Saliuanca homochroa* (Holland).

Genus *Pholetesor* Mason, 1981

Pholetesor arisba (Nixon, 1973)

Catalogs with Iranian records: No catalog.

Distribution in Iran: Golestan (Sakenin et al., 2020 as *Apanteles* (*Pholetesor*) *arisba* Nixon).

Distribution in the Middle East: Egypt (Papp, 1983), Iran (Sakenin et al., 2020), Israel—Palestine (Papp, 2012), Turkey (Fernandez-Triana et al., 2020).

Extralimital distribution: Austria, Bulgaria, China, Czech Republic, Denmark, Germany, Greece, Hungary, Italy, Netherlands, New Zealand, Norway, Russia, Serbia, Spain, Ukraine, United Kingdom.

Host records: Summarized by Yu et al. (2016) as being a parasitoid of the bucculatricid *Bucculatrix thoracella* (Thunberg); the coleophorid *Goniodoma limoniella* (Stainton); the elachistid *Stephensia brunnichella* (L.); and the gracillarids *Parornix finitimella* (Zeller), *Parornix tarquillella* (Zeller), *Phyllonorycter blancardella* (Fabricius), *Phyllonorycter comparella* (Duponchel), *Phyllonorycter corylifoliella* (Hübner), *Phyllonorycter malella* (Gerasimov), *Phyllonorycter oxyacanthae* (Frey), and *Phyllonorycter pyrifoliella* (Gerasimov). In Iran, this species has been reared from *Phyllonorycter salicicolella* (Sircom) (Sakenin et al., 2020).

Pholetesor bicolor (Nees von Esenbeck, 1834)

Catalogs with Iranian records: Gadallah et al. (2015, as *Pholetesor pedias* (Nixon, 1973)), Farahani et al. (2016, as *Apanteles bicolor* (Nees, 1834), and *Pholetesor bicolor* (Nees, 1834)), Samin, Coronado-Blanco, Kavallieratos et al. (2018) and Samin, Coronado-Blanco, Fischer et al. (2018) as *P. schillei* Niezabotowski, 1910).

Ditribution in Iran: Fars (Amiri et al., 2008; Modarres Awal, 2012), Ilam (Ghahari et al., 2011d), Markazi (Radjabi, 1986 as *Apanteles bicolor*, Modarres Awal, 1997, 2012, as *A. bicolor*), Gadallah et al. (2015, as *Pholetesor pedias*).

Distribution in the Middle East: Iran (see references above), Israel—Palestine (Papp, 2012).

Extralimital distribution: Belgium, Bulgaria, Canada (introduced), China, Croatia, Finland, France, Georgia, Germany, Greece, Hungary, Ireland, Italy, Japan, Kyrgyzstan, Lithuania, Moldova, Mongolia, New Zealand (introduced), Poland, Romania, Russia, Serbia, Slovakia, Spain, Switzerland, Tunisia, Turkmenistan, Ukraine, United Kingdom (Yu et al., 2016 as *Apanteles* (*Pholetesor*) *schillei* Niezabetowski, 1910; Fernandez-Triana et al., 2020).

Host records: Summarized by Yu et al. (2016) as being a parasitoid of several lepidopteran insect pests of the families Lymantridae, Lyonetiidae, Noctuidae, Psychidae, Pyralidae, Tineidae, Tischeriidae, Tortricidae). In Iran, this species has been reared from the gracillariid moth *Phyllonorycter corylifoliella* (Hübner) (Amiri et al., 2008; Modarres Awal, 1997, 2012).

Pholetesor circumscriptus (Nees von Esenbeck, 1834)

Catalogs with Iranian records: Gadallah et al. (2015), Farahani et al. (2016), Yu et al. (2016 as *Apanteles* (*Pholetesor*) *circumscriptus* (Nees, 1834)), Samin, Coronado-Blanco, Kavallieratos et al. (2018) and Samin, Coronado-Blanco, Fischer et al. (2018).

Distribution in Iran: Alborz (Shojai, 1998 as *Apanteles lautellus* Marshall, 1885), Fars, Kuhgiloyeh and Boyer-ahmad (Samin, van Achterberg, & Ghahari, 2015 as *Apanteles circumscriptus* (Nees, 1834)), East Azarbaijan (Ghahari & van Achterberg, 2016), Isfahan (Ghahari, Fischer, Hedqvist et al., 2010 as *Apanteles circumscriptus*), Markazi (Rajabi, 1986 as *Apanteles circumscriptus*), Tehran (Modarres Awal, 1997, 2012 as *Apanteles lautellus*), West Azarbaijan (Samin et al., 2014), Iran (no specific locality cited) (Behdad, 1991 as *Apanteles circumscriptus*; Gadallah et al., 2015).

Distribution in the Middle East: Iran (see references above), Israel–Palestine (Halperin, 1986).

Extralimital distribution: Australasian, Nearctic, Oriental, Palaearctic (Yu et al., 2016 as *Apanteles* (*Pholetesor*) *circumscriptus* (Nees, 1834); Fernandez-Triana et al., 2020).

Host records: Summarized by Yu et al. (2016) as being a parasitoid of several lepidopteran species of the families Argyresthiidae, Bucculatrigidae, Choreutidae, Coleophoridae, Elachistidae, Gracillariidae, Lyonetiidae, Noctuidae, Tortricidae. In Iran, this species has been reared from the gracillariid larvae of *Phyllonorycter blancardella* (Fabricius) (Behdad, 1991; Ghahari, Fischer, Hedqvist et al., 2010), *Phyllonorycter corylifoliella* (Hübner) (Radjabi, 1986), and *Phyllonorycter platani* Staudinger (Modarres Awal, 1997; 2012; Shojai, 1998).

Pholetesor elpis (Nixon, 1973)

Catalogs with Iranian records: No catalog.

Distribution in Iran: East Azarbaijan (Samin, Coronado-Blanco, Hosseini et al., 2019 as *Apanteles elpis* Nixon, 1937).

Distribution in the Middle East: Iran.

Extralimital distribution: Austria, Azerbaijan, Bulgaria, Croatia, Finland, Germany, Greece, Hungary, Korea, Mongolia, Netherlands, Poland, Russia, Serbia, Slovakia, Ukraine, United Kingdom (Yu et al., 2016 as *Apanteles* (*Pholetesor*) *elpis*; Fernandez-Triana et al., 2020).

Host records: Summarized by Yu et al. (2016) as being a parasitoid of the coleophorid *Coleophora serratella* (L.); the elachistids *Elachista cingillella* (Herrich-Schäffer), and *Elachista subnigrella* Douglas; the gracillariids *Caloptilia rufipemella* Hübner, *Euspilapteryx auroguttellus* (Stephens), *Phyllonorycter blancardella* (Fabricius), and *Phyllonorycter comparella* (Duponchel).

Pholetesor ingenuoides (Papp, 1971)

Catalogs with Iranian records: Gadallah et al. (2015), Farahani et al. (2016), Yu et al. (2016), Samin, Coronado-Blanco, Kavallieratos et al. (2018) and Samin, Coronado-Blanco, Fischer et al. (2018).

Distribution in Iran: Kordestan (Gadallah et al., 2015).

Distribution in the Middle East: Iran (Gadallah et al., 2015), Turkey (Beyarslan et al., 2006; Çetin Erdoğan & Beyarslan, 2005; Inanç, 1997; Inanç & Beyarslan, 1990, 2001a; Papp, 1984).

Extralimital distribution: Armenia, Bulgaria, Croatia, France, Germany, Greece, Hungary, Korea, Mongolia, Montenegro.

Host records: Summarized by Yu et al. (2016) as being a parasitoid of the gracillariids *Euspilapteryx auroguttellus* (Stephens), *Phyllonorycter agilella* (Zeller), and *Phyllonorycter spinicolella* (Zeller).

Pholetesor pseudocircumscriptus Abdoli, 2019

Catalogs with Iranian records: No catalog.

Distribution in Iran: East Azarbaijan (Abdoli & Pourhaji, 2019).

Distribution in the Middle East: Iran.

Extralimital distribution: None.

Host records: In Iran, this species has been reared from the gracillariid *Phyllonorycter corylifoliella* (Hübner) (Abdoli & Pourhaji, 2019).

Pholetesor rufulus (Tobias, 1964)

Distribution in the Middle East: Turkey (Papp, 1983).

Extralimital distribution: Azerbaijan, Hungary, Kazakhstan, Uzbekistan.

Host records: Unknown.

Pholetesor viminetorum (Wesmael, 1837)

Catalogs with Iranian records: Fallahzadeh and Saghaei (2010, as *Apanteles viminetorum*), Gadallah et al. (2015), Farahani et al. (2016), Yu et al. (2016 as *Apanteles*

(*Pholetesor*) *viminetotum* (Wesmael, 1837)), Samin et al. (2018a, b as *Apanteles viminetotum*).

Distribution in Iran: Golestan (Gadallah et al., 2015; Telenga, 1955).

Distribution in the Middle East: Iran.

Extralimital distribution: Nearctic, Oriental, and Palaearctic regions (Yu et al., 2016 as *Apanteles* (*Pholetesor*) *viminetorum* (Wesmael, 1837); Fernandez-Triana et al., 2020).

Host records: Summarized by Yu et al. (2016) as being a parasitoid of several lepidopteran species of the families Coleophoridae, Gelechiidae, Gracillariidae, Hesperiidae, Lymantridae, Lyonetiidae, Noctuidae, Plutellidae, Pyralidae, Tortricidae, Ypsolophidae, and Zygaenidae.

Cotesia group (= most of Cotesiini *sensu* Mason, 1981)

Genus *Cotesia* Cameron, 1891

Cotesia abjecta (Marshall, 1885)

Catalogs with Iranian records: Gadallah et al. (2015), Farahani et al. (2016), Yu et al. (2016), Samin, Coronado-Blanco, Kavallieratos et al. (2018) and Samin, Coronado-Blanco, Fischer et al. (2018).

Distribution in Iran: Lorestan (Ghahari, Fischer, Papp et al., 2012).

Distribution in the Middle East: Iran (Ghahari, Fischer, Papp et al., 2012), Israel–Palestine (Papp, 2012).

Extralimital distribution: Croatia, Finland, France, Germany, Hungary, Ireland, Italy, Mongolia, Poland, Romania, Russia, Slovakia, Switzerland, United Kingdom, former Yugoslavia.

Host records: Summarized by Yu et al. (2016) as being a parasitoid of the drepanid *Drepana falcataria* (L.); the geometrids *Ennomos autumnaria* (Werneburg), and *Pseudoterpna pruinata* (Hufnagel); the notodontids *Cerura vinula* (L.), *Drymonia ruficornis* (Hufnagel), *Notodonta dictaeoides* (Esper), *Notodonta dromedarius* (L.), *Notodonta ziczac* (L.), *Phoesia gnoma* (Fabricius), *Phoesia tremula* (Clerck), *Ptilodon capucina* (L.); and the sphingid *Smerinthus ocellatus* (L.).

Cotesia acuminata (Reinhard, 1880)

Catalogs with Iranian records: No catalog.

Distribution in Iran: West Azarbaijan (Gadallah et al., 2021).

Distribution in the Middle East: Iran (Gadallah et al., 2021), Israel–Palestine (Papp, 1987, 2012).

Extralimital distribution: Armenia, Austria, China, Czech Republic, Finland, France, Georgia, Germany, Hungary, Romania, Russia, Slovakia, Spain, Sweden, Tajikistan, Ukraine, Uzbekistan (Fernandez-Triana et al., 2020), Iran (Gadallah et al., 2021).

Host records: Summarized by Yu et al. (2016) as being a parasitoid of the geometrid *Epidesmia chilonaria*

Herrich-Schäffer; the following nymphalids: *Aporia crataegi* (L.), *Euphydryas maturna* (L.), *Melitaea didyma* (Esper), *Melitaea leucippe* Schneider, *Melitaea phoebe* (Denis and Schiffermüller), *Melitaea scotosia* Butler, *Melitaea telona* Fruhstorfer, and *Mellicta athalia* (Rottemburg); and the pierid *Aporia crataegi* (L.). In Iran, this species has been reared from the nymphalid *Melitaea didyma* (Esper) on *Helianthus* sp. (Gadallah et al., 2021).

Cotesia affinis (Nees von Esenbeck, 1834)

Catalogs with Iranian records: No catalog.

Distribution in Iran: East Azarbaijan (Samin, Beyarslan, Ranjith et al., 2020).

Distribution in the Middle East: Iran.

Extralimital distribution: Armenia, Austria, Cape Verde Islands, China, France, Germany, Hungary, Italy, Japan, Kazakhstan, Korea, Latvia, Poland, Romania, Russia, Serbia, Slovakia, Spain, Sweden, Switzerland, Ukraine, United Kingdom.

Host records: Summarized by Yu et al. (2016) as being a parasitoid of the crambid *Loxostege sticticalis* (L.); the incurvariid *Phylloporia bistrigella* (Howarth); the noctuids *Cucullia artimisiae* (Hufnagel), and *Lacanobia oleracea* (L.); the notodontids *Cerura felina* Butler, *Cerura menciana* Moore, *Cerura przewalskyii* (Alpheraky), and *Cerura vinula* (L.); and the sphingids *Hyles euphorbiae* (L.), *Laethoe populi* (L.), and *Smerinthus planus* Walker.

Cotesia ancilla (Nixon, 1974)

Catalogs with Iranian records: Gadallah et al. (2015), Farahani et al. (2016), Yu et al. (2016), Samin, Coronado-Blanco, Kavallieratos et al. (2018) and Samin, Coronado-Blanco, Fischer et al. (2018).

Distribution in Iran: Isfahan (Ghahari et al., 2011c), Kermanshah (Ghahari & Fischer, 2012).

Distribution in the Middle East: Iran (Ghahari et al., 2011c; Ghahari & Fischer, 2012), Israel–Palestine (Papp, 1987), Turkey (Beyarslan et al., 2002a; Beyarslan et al., 2006; Inanç, 1997; Inanç & Beyarslan, 2001a; Papp, 1987).

Extralimital distribution: Armenia, Austria, Bulgaria, Croatia, Germany, Greece, Hungary, Italy, Japan, North Macedonia, Mongolia, Netherlands, Russia, Slovakia, Spain, Switzerland, former Yugoslavia.

Host records: Recorded by Shaw et al. (2009) as being a parasitoid of the following pierids: *Colias alfacariensis* Ribbe, *Colias chrysotheme* Esper, *Colias croceus* (Geoffroy), *Colias hyale* (L.), *Colias myrmidone* Esper, *Colias palaeno* (L.), and *Euchloe charlonia* (Donzel).

Cotesia astrarches (Marshall, 1889)

Distribution in the Middle East: Cyprus (Ingram, 1981).

Extralimital distribution: Afghanistan, Azerbaijan, Croatia, Finland, France, Georgia, Germany, Greece, Hungary, Kazakhstan, North Macedonia, Moldova, Norway, Poland, Russia, Slovakia, Slovenia, Spain, United Kingdom, former Yugoslavia.

Host records: Summarized by Yu et al. (2016) as being a parasitoid of the crambid *Evergestis forficalis* (L.), the following lycaenids: *Aricia agestis* (Denis and Schiffermüller), *Aricia artaxerxes* Fabricius, *Cupido minimus* (Fuessley), *Lysandra coridon* (Poda), *Polyommatus admetus* (Esper), *Polyommatus thersites* Cantener, and *Tomares ballus* Fabricius; the noctuids *Helicoverpa armigera* (Hübner), and *Helicoverpa zea* (Boddie); the pierid *Colias croceus* (Geoffroy); and the pyralid *Etiella zinckenella* (Treitschke).

Cotesia bignellii (Marshall, 1885)

Distribution in the Middle East: United Arab Emirates (Porter, 1979).

Extralimital distribution: Finland, France, Germany, Greece, Hungary, Ireland, Italy, Romania, Russia, Spain, Sweden, United Kingdom, former Yugoslavia.

Host records: Summarized by Yu et al. (2016) as being a parasitoid of the nymphalids *Euphydryas maturua* (L.), and *Eurodryas aurinia* (Rottemburg).

Cotesia brevicornis (Wesmael, 1837)

Catalogs with Iranian records: This species is new record for the fauna of Iran.

Distribution in Iran: Chaharmahal and Bakhtiari province, Borojen, 2♀, September 2015.

Distribution in the Middle East: Iran (new record), Turkey (Inanç & Beyarslan, 2001a, c).

Extralimital distribution: Azerbaijan, Belgium, Canada, Croatia, Finland, Germany, Hungary, Iceland, Ireland, Korea, Lithuania, Poland, Romania, Russia, Slovakia, Sweden, Switzerland, Ukraine, United Kingdom, former Yugoslavia.

Host records: Summarized by Yu et al. (2016) as being a parasitoid of the erebids *Euproctis chrysorrhoea* (L.), *Euproctis similis* (Füssly), and *Lymantria dispar* (L.); the geometrids *Epirrita autumnata* (Borkhausen), *Hydriomena furcate* (Thunberg), and *Rheumaptera hastata* (L.); the noctuids *Brachylomia viminalis* (Fabricius), *Enargia decolor* (Walker), *Ipimorpha retusa* (L.), *Ipimorpha subtusa* (Denis and Schiffermüller), *Orthosia miniosa* (Denis and Schiffermüller), and *Xanthi* sp.; and the tortricids *Ptycholoma lecheana* (L.), and *Syndemis musculata* (Hübner).

Cotesia callimone (Nixon, 1974)

Catalogs with Iranian records: Gadallah et al. (2015), Farahani et al. (2016), Yu et al. (2016), Samin, Coronado-Blanco, Kavallieratos et al. (2018) and Samin, Coronado-Blanco, Fischer et al. (2018).

Distribution in Iran: Lorestan (Ghahari, Fischer, Papp et al., 2012), Mazandaran (Sakenin et al., 2012), Semnan (Naderian et al., 2012).

Distribution in the Middle East: Iran (see references above), Turkey (Inanç, 1997; Özbek & Calmasur, 2010).

Extralimital distribution: Bulgaria, Finland, Hungary, Ireland, Mongolia, Russia, Serbia, Slovakia, Switzerland, Ukraine, United Kingdom.

Host records: Summarized by Yu et al. (2016) as being a parasitoid of the erebid *Callimorpha dominula* (L.); and the geometrid *Abraxas pantaria* (L.).

Cotesia calodetta (Nixon, 1974)

Distribution in the Middle East: Turkey (Inanç & Beyarslan, 2001a).

Extralimital distribution: Russia, Sweden.

Host records: Summarized by Yu et al. (2016) as being a parasitoid of the lasiocampids *Eriogaster arbusculae* Freyer, and *Eriogaster lanestris* (L.).

Cotesia chilonis (Munakata, 1912)

Catalogs with Iranian records: Fallahzadeh and Saghaei (2010), Gadallah et al. (2015), Farahani et al. (2016), Yu et al. (2016), Samin, Coronado-Blanco, Kavallieratos et al. (2018) and Samin, Coronado-Blanco, Fischer et al. (2018).

Distribution in Iran: Guilan (Rassipour, 1983; Modarres Awal, 1997; 2012, both as *Apanteles chilonis* Munakata, 1912), Mazandaran (Rassipour, 1983; Modarres Awal, 1997; 2012, both as *A. chilonis*; Amooghli-Tabari & Ghahari, 2021).

Distribution in the Middle East: Iran.

Extralimital distribution: Benin (introduced), China, France (introduced), India, Indonesia, Japan, Korea, Myanmar, Pakistan (introduced), South Africa (introduced).

Host records: Summarized by Yu et al. (2016) as being a parasitoid of the crambids *Chilo infuscatellus* Snellen, *Chilo luteellus* (Motschulsky), *Chilo partellus* (C. Swinha), and *Chilo suppressalis* (Walker); and the noctuid *Condica capensis* (Guenée). In Iran, this species has been recorded in association with the crambid *Chilo suppressalis* (Walker) (Amooghli-Tabari & Ghahari, 2021; Modarres Awal, 1997, 2012; Rassipour, 1983).

Cotesia cuprea (Lyle, 1925)

Catalogs with Iranian records: Gadallah et al. (2015), Farahani et al. (2016), Yu et al. (2016), Samin, Coronado-Blanco, Kavallieratos et al. (2018) and Samin, Coronado-Blanco, Fischer et al. (2018).

Distribution in Iran: Lorestan (Ghahari, Fischer, Papp et al., 2012), Semnan (Samin, Fischer, & Ghahari, 2015), West Azarbaijan (Samin et al., 2014).

Distribution in the Middle East: Iran (see references above), Turkey (Beyarslan, 1988; Beyarslan et al., 2002a, b; Inanç, 1997; Inanç & Beyarslan, 2001a, b).

Extralimital distribution: Azerbaijan, Bulgaria, Finland, France, Germany, Greece, Hungary, Lithuania, Mongolia, Netherlands, Poland, Romania, Slovakia, Spain, Switzerland, United Kingdom.

Host records: Summarized by Yu et al. (2016) as being a parasitoid of the lycaenids *Agriades optilete* (Knoch), *Lampidis borticus* (L.), *Lycaena dispar* (Haworth), *Lycaena helle* (Denis and Schiffermüller), *Lycaena phaeas* (L.), *Lycaena thersamon* Esper, *Plebejus argus* (L.), *Polymmatus icarus* (Rottemburg), and *Zizeeria Knysna* (Trimen).

Cotesia cynthiae (Nixon, 1974)

Catalogs with Iranian records: No catalog.

Distribution in Iran: Khuzestan (Zargar et al., 2019b).

Distribution in the Middle East: Iran (Zargar et al., 2019b), Turkey (Inanç & Beyarsaln, 1990).

Extralimital distribution: Austria, Bulgaria, France, Hungary, Switzerland.

Host records: recorded by Shaw et al. (2009) as being a parasitoid of the nymphalid *Euphydryas cynthia* (Denis and Schiffermüller).

Cotesia elongata Zargar and Gupta, 2019

Catalogs with Iranian records: No catalog.

Distribution in Iran: Khuzestan (Zargar et al., 2019b).

Distribution in the Middle East: Iran.

Extralimital distribution: None.

Host records: Unknown.

Cotesia euryale (Nixon, 1974)

Catalogs with Iranian records: Gadallah et al. (2015), Farahani et al. (2016), Yu et al. (2016), Samin, Coronado-Blanco, Kavallieratos et al. (2018) and Samin, Coronado-Blanco, Fischer et al. (2018).

Distribution in Iran: Ilam (Ghahari et al., 2011d).

Distribution in the Middle East: Iran.

Extralimital distribution: Bulgaria, Czech Republic, France, Greece, Hungary, North Macedonia, Mongolia, Netherlands, Switzerland, former Yugoslavia.

Host records: Summarized by Yu et al. (2016) as being a parasitoid of the erebid *Ctocala fraxini* (L.); and the geometrids *Aplocera efformata* (Guenée), *Aplocera plagiata* (L.), and *Biston betularia* L.

Cotesia evagata (Papp, 1973)

Distribution in the Middle East: Jordan (Papp, 1987).

Extralimital distribution: Turkmenistan.

Host records: Unknown.

Cotesia ferruginea (Marshall, 1885)

Distribution in the Middle East: Turkey (Papp, 1986).

Extralimital distribution: Belgium, Germany, Hungary, Italy, Korea, Lithuania, Netherlands, Romania, Russia, Slovakia, Switzerland, Ukraine, United Kingdom.

Host records: Summarized by Yu et al. (2016) as being a parasitoid of the cossid *Phragmataecia castaneae* (Hübner); the crambid *Chilo phragmitella* (Hübner); and the noctuid *Archanara geminipuncta* (Howarth).

Cotesia flavipes Cameron, 1891

Catalogs with Iranian records: Gadallah et al. (2015), Farahani et al. (2016), Yu et al. (2016), Samin, Coronado-Blanco, Kavallieratos et al. (2018) and Samin, Coronado-Blanco, Fischer et al. (2018).

Distribution in Iran: East Azarbaijan (Ghahari & van Achterberg 2016), Mazandaran (Amooghli-Tabari & Ghahari, 2021; Ghahari, Tabari, Haji-Amiri et al., 2009).

Distribution in the Middle East: Iran.

Extralimital distribution: Afrotropical, Australasian, Oceanic, Nearctic, Neotropical, Oriental and Palaearctic (Eastern) regions.

Host records: Summarized by Yu et al. (2016) as being a parasitoid of several lepidopteran pest species of the families Arctiidae, Brachodidae, Lymantriidae, Noctuidae, Pyralidae, and Tortricidae. In Iran, this species has been reared from the crambid moth, *Chilo suppressalis* Walker (Amooghli-Tabari & Ghahari, 2021).

Cotesia gastropachae (Bouché, 1834)

Catalogs with Iranian records: No catalog.

Distribution in Iran: Hamadan (Naderian et al., 2020).

Distribution in the Middle East: Iran (Naderian et al., 2020), Israel—Palestine (Papp, 1970), Turkey (Telenga, 1955).

Extralimital distribution: Azerbaijan, Bulgaria, China, Czech Republic, Finland, France, Germany, Hungary, Japan, Kazakhstan, Lithuania, Moldova, Mongolia, Montenegro, Poland, Romania, Russia, Slovakia, United Kingdom, Uzbekistan.

Host records: Summarized by Yu et al. (2016) as being a parasitoid of the crambid *Loxostege sticticalis* (L.); the erebid *Lymantria dispar* (L.); the lasiocampids *Eriogaster lanestris* (L.), *Gastropocha quercifolia* (L.), and *Macrothylacia rubi* (L.); the noctuid *Schinia scutosa* (Denis and Schiffermüller); the nymphalis *Vanessa atalanta* (L.), and *Vanessa cardui* (L); and the pierid *Aporia crataegi* (L.).

Cotesia geryonis (Marshall, 1885)

Catalogs with Iranian records: Gadallah et al. (2015), Farahani et al. (2016), Yu et al. (2016), Samin, Coronado-Blanco, Kavallieratos et al. (2018) and Samin, Coronado-Blanco, Fischer et al. (2018).

Distribution in Iran: Kermanshah (Ghahari & Fischer, 2012).

Distribution in the Middle East: Iran (Ghahari & Fischer, 2012), Turkey (Inanç, 1997; Inanç & Beyarslan, 2001a).

Extralimital distribution: Bulgaria, Germany, Hungary, Italy, Korea, Mongolia, Poland, Romania, Russia, Slovakia, Spain, Switzerland, Ukraine, United Kingdom.

Host records: Summarized by Yu et al. (2016) as being a parasitoid of the following zygaenids: *Adscita albanica* (Naufock), *Adscita geryon* (Hübner), *Adscita schmidti* (Naufock), *Adscita statices* (L.), *Jordanita globulariae* (Hübner), and *Rhagadesygaena laeta* (Hübner).

Cotesia glabrata (Telenga, 1955)
Catalogs with Iranian records: No catalog.
Distribution in Iran: Khuzestan (Zargar et al., 2019b).
Distribution in the Middle East: Iran (Zargar et al., 2019b), Israel—Palestine (Halperin, 1986; Papp, 1987), Turkey (Inanç & Beyarslan, 2001a).
Extralimital distribution: Bulgaria, Georgia, Germany, Hungary, Kazakhstan, Russia, Turkmenistan, Ukraine.
Host records: Summarized by Yu et al. (2016) as being a parasitoid of the geometrid *Biston betularia* L.; the hesperiids *Carcharodus alceae* (Esper), *Carcharodus tripolinus* (Verity), and *Pyrgus cirsii* (Rambur); and the pierid *Aporia crataegi* (L.).

Cotesia glomerata (Linnaeus, 1758)
Catalogs with Iranian records: Fallahzadeh and Saghaei (2010), Gadallah et al. (2015), Farahani et al. (2016), Yu et al. (2016), Samin, Coronado-Blanco, Kavallieratos et al. (2018) and Samin, Coronado-Blanco, Fischer et al. (2018).
Distribution in Iran: Alborz (Farahbakhsh, 1961 as *Apanteles glomeratus* (Linnaeus, 1758)), East Azarbaijan (Ghahari & van Achterberg, 2016), Fars (Ghahari & Beyarslan, 2017; Lashkari Bod et al., 2011), Golestan (Ghahari & Jussila, 2015), Guilan (Ghahari, Çetin Erdoğan, Šedivý et al., 2010), Khuzestan (Shojai et al., 1995 as *A. glomeratus*, Ghahari & Beyarslan, 2017), Mazandaran (Shojai, 1968, 1998; Davatchi & Shojai, 1969; Radjabi, 1986; Modarres Awal, 1997, 2012, all as *A. glomeratus*), Tehran (Farahbakhsh, 1961; Shojai, 1968, 1998; Davatchi & Shojai, 1969; Radjabi, 1986; Modarres Awal, 1997, 2012 as *A. glomeratus*, Hasanshahi et al., 2014b), West Azarbaijan (Alizadeh & Javan Moghaddam, 2004; Khanjani, 2006a, b; Razmi et al., 2011; Modarres Awal, 2012 as *A. glomeratus*, Mirfakhraie & Dey, 2013; Ghahari, 2019b).
Distribution in the Middle East: Cyprus, Egypt, Iran, Israel—Palestine, Jordan, Syria, Turkey (Fernandez-Triana et al., 2020).
Extralimital distribution: Nearly cosmopolitan species (except Afrotropical region).
Host records: Summarized by Yu et al. (2016) as being a parasitoid of many lepidopteran pest species of the families Arctiidae, Bombycidae, Geometridae, Lasiocampidae, Lycaenidae, Lymantriidae, Noctuidae, Nymphalidae, Pieridae, Pyralidae, Saturniidae, Sesiidae, Sphingidae, Thyatiridae, Tortricidae, Yponomeutidae, and Zygaenidae. In Iran, this species has been reared from the noctuid

Acronicta sp., and the pierid *Aporia crataegi* (L.) (Modarres Awal, 1997, 2012; Radjabi, 1986), and the noctuids *Agrotis segetum* (Denis and Schiffermüller) (Alizadeh & Javan Moghaddam, 2004; Khanjani, 2006a, b; Modarres Awal, 1997, 2012 as *Scotia segetum*; Ghahari & Beyarslan, 2017; Ghahari, 2019b), *Sesamia cretica* Lederer (Shojai et al., 1995), and *Sesamia nonagrioides botanephaga* Tams and Bowden (Khanjani, 2006a, b); the saturniid *Acronicta aceris* (L.) (Modarres Awal, 1997, 2012 as *Apatele aceris*); the pierids *Pieris brassicae* (L.) (Farahbakhsh, 1961; Modarres Awal, 1997; 2012; Razmi et al., 2011; Mirfakhraie & Dey, 2013; Ghahari & Beyarslan, 2017), and *Pieris rapae* (L.) (Farahbakhsh, 1961; Hasanshahi et al., 2014b; Modarres Awal, 1997; 2012).
Comments: *Lysibia nana* (Gravenhorst, 1829) (Hymenoptera: Ichneumonidae) was recorded by Ghahari and Jussila (2015) as the hyperparasitoid of *Cotesia glomerata* via *Aporia* sp. (Lepidoptera: Pieridae).

Cotesia gonopterygis (Marshall, 1897)
Catalogs with Iranian records: Samin, Coronado-Blanco, Fischer et al. (2018).
Distribution in Iran: Kordestan (Samin, Coronado-Blanco, Fischer et al., 2018).
Distribution in the Middle East: Iran (Samin, Coronado-Blanco, Fischer et al., 2018), Turkey (Çetin Erdoğan & Beyarslan, 2005; Inanç, 1997).
Extralimital distribution: Germany, Hungary, Japan, Romania, Russia, Slovakia, Switzerland, United Kingdom.
Host records: Summarized by Yu et al. (2016) as being a parasitoid of the noctuid *Acronicta rumicis* (L.); the nymphalids *Eurodryas aurinia* (Rottemburg), and *Limenitis camilla* (L.); and the pierid *Gonepteryx cleopatra* (L.).

Cotesia hyphantriae (Riley, 1887)
Catalogs with Iranian records: Gadallah et al. (2015), Farahani et al. (2016), Yu et al. (2016), Samin, Coronado-Blanco, Kavallieratos et al. (2018) and Samin, Coronado-Blanco, Fischer et al. (2018).
Distribution in Iran: Qazvin (Ghahari et al., 2011b).
Distribution in the Middle East: Iran (Ghahari et al., 2011b), Turkey (Inanç & Beyarslan, 2001a).
Extralimital distribution: Nearctic, Neotropical, Oriental, Palaearctic.
Host records: Summarized by Yu et al. (2016) as being a parasitoid of the erebids *Estigmene acaea* (Drury), *Hyphantria cunea* (Drury), *Loptocampa argentata* (Packard), *Orgyia leucostigma* (JE Smith), and *Spilarctia subcarnea* (Walker); the galacticid *Homadaula anisocentra* Meyrick; the geometrid *Operothera brumata* (L.); the hesperiid *Thymelicus lineola* (Ochsenheimer); and the noctuids *Cosmia trapezina* (L.), *Lacanobia suasa* (Denis

and Schiffermüller), *Leucoma salicis* (L.), *Morrisonia confuse* Hübner, *Orthosia* spp., and *Panolis flammea* (Denis and Schiffermüller).

Cotesia icipe Fernandez-Triana and Fiaboe, 2017
Distribution in the Middle East: Saudi Arabia, Yemen (Fiaboe et al., 2017).
Extralimital distribution: Kenya, Madagascar, South Africa.
Host records: Unknown.

Cotesia inducta (Papp, 1973)
Distribution in the Middle East: Israel–Palestine (Papp, 2012), Turkey (Papp, 1986).
Extralimital distribution: Bulgaria, Hungary, Ireland, Korea, Moldova, Russia, Slovakia, Ukraine, United Kingdom, Uzbekistan.
Host records: Summarized by Yu et al. (2016) as being a parasitoid of the lycaenids *Callophrys avis* Chapman, *Celastrina argiolus* (L.), *Glaucopsyche melanops* (Boisduval), *leptotes pirithous* (L.), *Satyrium w-album* (Knoch), *Tomares ballus* Fabricius, and *Zizeeria knysna* (Trimen).

Cotesia jucunda (Marshall, 1885)
Catalogs with Iranian records: Gadallah et al. (2015), Farahani et al. (2016), Yu et al. (2016), Samin, Coronado-Blanco, Kavallieratos et al. (2018) and Samin, Coronado-Blanco, Fischer et al. (2018).
Distribution in Iran: Guilan (Ghahari, Fischer, & Tobias et al., 2012), Markazi (Ghahari et al., 2011a), Khuzestan (Zargar et al., 2019b).
Distribution in the Middle East: Iran (see references above), Turkey (Beyarslan, 1988; Inanç & Beyarslan, 2001a, c).
Extralimital distribution: Armenia, Austria, Azerbaijan, Bulgaria, Croatia, Denmark, Estonia, Finland, France, Germany, Greece, Hungary, Ireland, Moldova, Mongolia, Poland, Romania, Russia, Serbia, Slovakia, Spain, Sweden, Switzerland, United Kingdom.
Host records: Summarized by Yu et al. (2016) as being a parasitoid of the geometrids *Abraxas grossulariata* (L.), *Alsophila aceraria* (Denis and Schiffermüller), *Earophila badiata* (Denis and Schiffermüller), *Arichanna melanaria* (L.), *Campaea margaritata* (L.), *Cyclophora* spp., *Epirrita autumnata* (Borkhausen), *Erannis defoliaria* (Clerck), *Eupithesia dodoneata* Guenée, *Eupithesia pimpinellata* (Hübner), *Operophtera brumata* (L.), *Operophtera fagata* (Scharfenberg), and *Phigalia pilosaria* (Denis and Schiffermüller); the nymphalid *Vanessa atalanta* (L.); the pierids *Pieris brassicae* (L.), and *Pieris rapae* (L.); and the pterophorid *Stenoptilia veronica* Karvonen.

Cotesia judaica (Papp, 1970)
Distribution in the Middle East: Israel–Palestine (Halperin, 1986; Papp, 1970, 1987).

Extralimital distribution: Hungary, Italy, Kazakhstan, Ukraine.
Host records: Summarized by Yu et al. (2016) as being a parasitoid of the noctuids *Orgyia dubia* (Tauscher), and *Orgyia dubia judaeea* Satudinger.

Cotesia kazak (Telenga, 1949)
Catalogs with Iranian records: Fallahzadeh and Saghaei (2010), Gadallah et al. (2015), Farahani et al. (2016), Yu et al. (2016), Samin, Coronado-Blanco, Kavallieratos et al. (2018) and Samin, Coronado-Blanco, Fischer et al. (2018).
Distribution in Iran: Golestan (Ghadiri Rad & Ebrahimi, 2010; Afshari et al., 2014), Tehran (Shojai, 1968; 1998; Davatchi & Shojai, 1969; Modarres Awal, 1997; 2012 as *Apanteles kazak* Telenga, 1949; Ghahari & Gadallah, 2022).
Distribution in the Middle East: Iran (see references above), Israel–Palestine (Papp, 1986, 1987, 2012), Turkey (Beyarslan et al., 2002a, b; Inanç, 1997; Inanç & Beyarslan, 1990, 2001a, b; Papp, 1986, 1987).
Extralimital distribution: Armenia, Australia, Azerbaijan, Bulgaria, China, Croatia, Greece, India, Kazakhstan, Mongolia, Morocco, New Zealand, Portugal, Russia, Spain, Tajikistan, Tunisia, Turkmenistan, Uzbekistan.
Host records: Summarized by Yu et al. (2016) as being a parasitoid of the geometrid *Apocheima cinerarius* (Erschoff); the noctuids *Helicoverpa armigera* (Hübner), *Helicoverpa zea* (Boddie), *Heliothis peltigera* (Denis and Schiffermüller), and *Heliothis viriplaca* (Hufnagel). In Iran, this species has been reared from the noctuids *Heliothis viriplaca* (Hufnagel) (Shojai, 1968), *Heliothis* sp. (Davatchi & Shojai, 1969; Modarres Awal, 1997; 2012; Shojai, 1998), and *Helicoverpa armigra* Hübner in tomato fields (Afshari et al., 2014; Ghadiri Rad & Ebrahimi, 2010).
Comments: *Cotesia kazak* is a hyperparasitoid of *Hyposoter didymator* (Thunberg) (Hymenoptera: Ichneumonidae) (Ghahari & Gadallah, 2022).

Cotesia khuzestanensis Zargar and Gupta, 2019
Catalogs with Iranian records: No catalog.
Distribution in Iran: Khusestan (Zargar et al., 2019b).
Distribution in the Middle East: Iran.
Extralimital distribution: None.
Host records: Unknown.

Cotesia kurdjumovi (Telenga, 1955)
Distribution in the Middle East: Israel–Palestine (Papp, 2012), Turkey (Beyarslan, 1988; Inanç, 1997; Inanç & Beyarslan, 2001a; Papp, 1986, 1987).
Extralimital distribution: Bulgaria, Germany, Hungary, Lithuania, Moldova, Mongolia, Russia, Spain, Turkmenistan, Ukraine, United Kingdom.
Host records: Summarized by Yu et al. (2016) as being a parasitoid of the crambid *Pyrausta aurata* (Scopoli), and the pyralid *Moitrelia obductella* (Zeller).

Cotesia lineola (Curtis, 1830)

Catalogs with Iranian records: No catalog.
Distribution in Iran: Hamadan (Sakenin et al., 2020).
Distribution in the Middle East: Iran (Sakenin et al., 2020), Turkey (Beyarslan et al., 2002a; Beyarslan et al., 2006; Inanç, 1997; Inanç & Beyarslan, 2001a; Papp, 1987).
Extralimital distribution: Armenia, Bulgaria, Czech Republic, Finland, France Germany, Hungary, Latvia, Moldova, Romania, Russia, Spain, United Kingdom.
Host records: Summarized by Yu et al. (2016) as being a parasitoid of the crambids *Evergestis extimalis* (Scopoli), *Evergestis forficalis* (L.), *Evergestis pallidata* (Hufnagel), and *Ostrinia nubilalis* (Hübner); the lasiocampids *Eriogaster lanestris* (L.), and *Malacosoma neustria* (L.); the notodontid *Thaumetopoea processionea* (L.); the pierid *Aporia crataegi* (L.); the pterophorid *Hellinsia osteodactyla* (Zeller); and the tortricid *Selania leplastriana* (Curtis). In Iran, this species has been reared from crambid *Hellula undalis* (Fabricius) (Sakenin et al., 2020).

Cotesia lycophron (Nixon, 1974)

Distribution in the Middle East: Israel—Palestine (Papp, 2012).
Extralimital distribution: France, Hungary, Netherlands.
Host records: Summarized by Yu et al. (2016) as being a parasitoid of the nymphalids *Melitaea didyma* (Esper), *Melitaea trivia* Denis and Schiffermüller, and *Mellicta athalia* (Rottemburg).

Cotesia melanoscelus (Ratzeburg, 1844)

Catalogs with Iranian records: Fallahzadeh and Saghaei (2010), Gadallah et al. (2015), Farahani et al. (2016), Yu et al. (2016), Samin, Coronado-Blanco, Kavallieratos et al. (2018) and Samin, Coronado-Blanco, Fischer et al. (2018).
Distribution in Iran: East Azarbaijan (Ghahari, Çetin Erdoğan, Šedivý et al., 2010), Golestan (Samin, Beyarslan, Coronado-Blanco et al., 2020), Guilan (Hérard et al., 1979 as *Apanteles melanoscelus* (Ratzeburg, 1844); Radjabi, 1986), Mazandaran (Hérard et al., 1979 as *Apanteles melanoscelus*; Radjabi, 1986; Ghahari, 2019a), Iran (no specific locality cited) (Modarres Awal 1997, 2012 as *A. melanoscela*).
Distribution in the Middle East: Iran (see references above), Turkey (Inanç, 1997; Inanç & Beyarslan, 2001a).
Extralimital distribution: Nearctic, Oriental, Palaearctic; Canada, United States of America (introduced to many parts of them).
Host records: Summarized by Yu et al. (2016) as being a parasitoid of the crambid *Ostrinia nubilalis* (Hübner); the erebids *Euproctis chrysorrhoea* (L.), *Leucoma salicis* (L.), *Orgyia antiqua* (L.), and *Orgyia leucostigma* (JE Smith); the gelechiid *Anacampsis populella* (Clerck); the geometrids *Abraxas grossulariata* (L.), *Cyclophora linearia*

(Hübner), *Eulithis testata* (L.), *Macaria aemulataria* Walker, and *Phigalia titea* (Cramer); the lasiocampid *Malacosoma neustria* (L.); the lymantriids *Ivela auripes* (Butler), and *Lymantria* spp.; the noctuids *Noctua fimbriata* Schreber, *Orthosia miniosa* (Denis and Schiffermüller), and *Orthosia stabilis* (Fabricius); the nolids *Nycteola asiatica* (Krulikovsky), and *Nycteola frigidana* (Walker); the saturniid *Hemileuca maia* (Drury); and the tortricids *Lathronympha strigana* (Fabricius), and *Tortrix viridana* (L.). In Iran, this species has been reared from the erebids *Euproctis chrysorrhoea* (L.) (Samin, Beyarslan, Coronado-Blanco et al., 2020), and *Lymantria dispar* (L.) (Hérard et al., 1979; Modarres Awal, 1997, 2012).

Cotesia melitaearum (Wilkinson, 1937)

Catalogs with Iranian records: No catalog.
Distribution in Iran: Mazandaran (Gadallah et al., 2021).
Distribution in the Middle East: Iran (Gadallah et al., 2021), Turkey (Papp, 1986).
Extralimital distribution: Armenia, Azerbaijan, China, Estonia, Finland, France, Germany, Hungary, Ireland, Italy, Kazakhstan, Korea, Moldova, Poland, Romania, Russia, Slovakia, Spain, Sweden, United Kingdom, Uzbekistan.
Host records: Summarized by Yu et al. (2016) as being a parasitoid of the erebid *Lymantria dispar* (L.); and the following nymphalids: *Euphydryas davidi* (Oberthur), *Euphydryas desfontainii* (Godart), *Euphydryas maturna* (L.), *Melitaea cinxia* (L.), *Melitaea diamina* (Lang), *Melitaea didyma* (Esper), *Melitaea leucippe* Schneider, *Melitaea parthenoides* Keferstein, *Melitaea trivia* (Denis and Schiffermüller), *Mellicta athalia* (Rottemburg), and *Nymphalis antiope* (L.). In Iran, this species has been reared from *Lymantria dispar* (L.) on *Ulmus* sp. (Gadallah et al., 2021).

Cotesia neustriae (Tobias, 1986)

Distribution in the Middle East: Turkey (Çetin Erdoğan & Beyarslan, 2005).
Extralimital distribution: Kazakhstan, Moldova, Russia, Ukraine.
Host records: Recorded by Tobias (1986) as being a parasitoid of the lasiocampid *Malacosoma neustria* (L.).

Cotesia nothus (Marshall, 1885)

Catalogs with Iranian records: Gadallah et al. (2015), Farahani et al. (2016), Yu et al. (2016), Samin, Coronado-Blanco, Kavallieratos et al. (2018) and Samin, Coronado-Blanco, Fischer et al. (2018).
Distribution in Iran: Golestan (Ghahari & Fischer, 2011b), Lorestan (Ghahari, Fischer, Papp et al., 2012 *Cotesia nothus* (Marshall, 1885)).
Distribution in the Middle East: Iran Ghahari and Fischer, 2011b; Ghahari, Fischer, Papp et al., 2012), Turkey (Fernandez-Triana et al., 2020).

Extralimital distribution: Germany, Greece, Hungary, Italy, Korea, Mongolia, Romania, Russia, Slovakia, United Kingdom.

Host records: Recorded by Yu et al. (2016) as being a parasitoid of the coleophorids *Coleophora milvipennis* Zeller, *Coleophora nigricella* (Stephens); the erebid *Spilosoma lubricepedum* (L.); the geometrid *Abraxas grossulariata* (L.), *Earophila badiata* (Denis and Schiffermüller), *Catarhoe rubidata* (Denis and Schiffermüller), *Epirrhoe galiata* (Denis and Schiffermüller), *Eulithis pyraliata* (Denis and Schiffermüller), *Horisme vitalbata* (Denis and Schiffermüller), *Hydriomena furcate* (Thunberg), and *Phigalia pilosaria* (Denis and Schiffermüller); the gracillariid *Phyllonorycter lantanella* (Schrank); the noctuids *Cosmia pyralina* (Denis and Schiffermüller), *Ipimorpha retusa* (L.), *Lithophane semibrunnea* (Haworth), and *Mormo maura* (L.); the nymphalids *Maniola jurtina* (L.), and *Melanargia galathea* (L.); and the zygaenid *Jordanita subsolana* (Staudinger).

Cotesia numen (Nixon, 1974)

Distribution in the Middle East: Turkey (Inanç & Beyarslan, 2001a).

Extralimital distribution: Czech Republic, Denmark, France, Germany, Hungary, Mongolia, Slovakia, United Kingdom.

Host records: Summarized by Yu et al. (2016) as being a parasitoid of the geometrids *Epithecia centaureata* (Denis and Schiffermüller), *Epithecia intricata* (Zetterstedt), and *Epithecia nanata* (Hübner); and the noctuid *Hadena confusa* Hufnagel.

Cotesia ofella (Nixon, 1974)

Catalogs with Iranian records: Fallahzadeh and Saghaei (2010), Gadallah et al. (2015), Farahani et al. (2016) (all as *Cotesia ofella* (Nixon, 1974)), Yu et al. (2016), Samin, Coronado-Blanco, Kavallieratos et al. (2018) and Samin, Coronado-Blanco, Fischer et al. (2018).

Distribution in Iran: Mazandaran (Ghahari, Fischer, Çetin Erdoğan et al., 2010), West Azarbaijan (Modarres Awal, 2012; Karimpour et al., 2001; Asem et al., 2016).

Distribution in the Middle East: Iran (see references above), Israel−Palestine (Papp, 2012), Turkey (Beyarslan, 1988; Beyarslan et al., 2002a; Inanç, 1997; Inanç & Beyarslan, 1990; 2001a).

Extralimital distribution: Belgium, Finland, Germany, Italy, Netehrlands, Poland, Serbia, Slovakia, Spain, Switzerland, Ukraine, United Kingdom.

Host records: Summarized by Yu et al. (2016) as being a parasitoid of the erebid *Spilosoma lubricipeda* (L.); the lasiocampid *Euthrix potatoria* (L.); and the noctuids *Acronicta aceris* (L.), *Acronicta rumicis* (L.), and *Simyra*

dentinosa Freyer. In Iran, this species has been reared from the noctuid *Simyra dentinosa* Freyer (Karimpour et al., 2001; Modarres Awal, 2012).

Cotesia ordinaria (Ratzeburg, 1844)

Catalogs with Iranian records: Gadallah et al. (2015), Farahani et al. (2016), Yu et al. (2016), Samin, Coronado-Blanco, Kavallieratos et al. (2018) and Samin, Coronado-Blanco, Fischer et al. (2018).

Distribution in Iran: Golestan (Sakenin et al., 2012 as *Cotesia ordinarius* (Ratzeburg, 1844)), Lorestan (Ghahari, Fischer, Papp et al., 2012 as *C. ordinarius*).

Distribution in the Middle East: Iran (Ghahari, Fischer, Papp et al., 2012; Sakenin et al., 2012), Israel−Palestine (Papp, 1970, 2012), Turkey (Inanç & Beyarslan, 2001c).

Extralimital distribution: China, Czech Republic, Germany, Hungary, Italy, Japan (introduced), Korea, Mongolia, Poland, Romania, Russia, Ukraine.

Host records: Summarized by Yu et al. (2016) as being a parasitoid of the erebid *Amata mestralii* (Bugnion); the lasiocampids *Dendrolimus pini* (L.), *Dendrolimus punctatus* Walker, *Dendrolimus spectabilis* (Butler), *Dendrolimus superans* (Butler), *Dendrolimus tabulaeformis* Tsai and Liu, and *Macrothylacia rubi* (L.); and the pierid *Pieris brassicae* (L.).

Cotesia orestes (Nixon, 1974)

Distribution in the Middle East: Turkey (Beyarslan et al., 2002a; Inanç & Beyarslan, 1990, 2001a, b).

Extralimital distribution: Finland, Germany, Hungary, Korea, Netherlands, Russia, United Kingdom.

Host records: Recorded by Quicke and M. Shaw (2004) as being a parasitoid of the lasiocampid *Euthrix potatoria* (L.).

Cotesia pappi Inanç, 2002

Distribution in the Middle East: Turkey (Inanç, 2002).

Extralimital distribution: None.

Host records: Unknown.

Cotesia pieridis (Bouché, 1834)

Catalogs with Iranian records: No catalog.

Distribution in Iran: Lorestan (Gadallah et al., 2021).

Distribution in the Middle East: Iran (Gadallah et al., 2021), Israel−Palestine (Papp, 2012), Turkey (Bayram et al., 1998).

Extralimital distribution: Armenia, Azerbaijan, China, Georgia, Germany, Hungary, Kazakhstan, Lithuania, Moldova, Mongolia, Romania, Russia, Slovakia, Tajikistan, Uzbekistan.

Host records: Summarized by Yu et al. (2016) as being a parasitoid of the erebids *Euproctis chrysorrhoea* (L.), and *Orgyia splendida* (Rambur); the lasiocampids *Malacosoma castrense* (L.), *Malcosoma neustria* (L.), and *Malcosoma parallellum* Staudinger; the noctuid *Helicoverpa armigera* (Hübner); the pierids *Aporia crataegi* (L.), *Colias* sp., *Gonepteryx rhamni* (L.), and *Pieris brassicae* (L.); and the tortricid *Pandemis chondrillana* (Herrich-Schäffer). In Iran, this species has been reared from lasiocampid *Malacosoma neustria* (L.) on *Poplar* sp. (Gadallah et al., 2021).

Cotesia pilicornis (Thomson, 1895)
Distribution in the Middle East: Turkey (Papp, 1986).
Extralimital distribution: Bulgaria, Croatia, Finland, Germany, Hungary, Italy, Moldova, Romania, Russia, Slovakia, Sweden, Switzerland, United Kingdom.
Host records: Summarized by Yu et al. (2016 as being a parasitoid of the geometrids *Eupithecia pulchellata* Stephens, and *Perizoma alchemillata* (L.); the pterophorids *Amblyptilia acanthadactyla* (Hübner), *Amblyptilia punctidactyla* (Haworth), and *Platyptilia cosmodactyla* Hübner; and the tortricid *Epinotia nigricana* (Herrich-Schaffer).

Cotesia praepotens (Haliday, 1834)
Catalogs with Iranian records: Gadallah et al. (2015), Farahani et al. (2016), Yu et al. (2016), Samin, Coronado-Blanco, Kavallieratos et al. (2018) and Samin, Coronado-Blanco, Fischer et al. (2018).
Distribution in Iran: Qazvin (Ghahari et al., 2011b), Khuzestan (Zargar et al., 2019b).
Distribution in the Middle East: Iran (Ghahari et al., 2011b; Zargar et al., 2019b), Turkey (Beyarslan, 1988; Beyarslan et al., 2002a, b; Inanç & Beyarslan, 1990, 2001a, b; Inanç, 1997; Papp, 1986).
Extralimital distribution: Afghanistan, Armenia, Azerbaijan, Bulgaria, Croatia, Czech Republic, Finland, Germany, Greece, Hungary, Ireland, Italy, Kazakhstan, Lithuania, North Macedonia, Moldova, Mongolia, Poland, Romania, Russia, Slovakia, Spain, Sweden, Switzerland, Tajikistan, Turkmenistan, United Kingdom, Uzbekistan, former Yugoslavia.
Host records: Summarized by Yu et al. (2016) as being a parasitoid of the elachistid *Elachista gleichenella* (Fabricius); the erebids *Euproctis chrysorrhoea* (L.), and *Lymantria dispar* (L.); the geometrids *Agriopis* spp., *Alsophila aescularia* (Denis and Schiffermüller), *Erannis defoliaria* (Clerck), *Eupithecia* spp., *Operophtera* spp., *Phigalia pilosoria* (Denis and Schiffermüller), and *Istrugia arenacearia* (Denis and Schiffermüller); the gracillariid *Phyllonorycter spinicolella* (Zeller); the lasiocampid *Poecilocampa populi* (L.); the lycaenid *Iolana iolas* Ochsenheimer; the noctuids *Abrostola triplasia* (L.), *Eupsitia transversa* (Hufnagel), and *Orthosia* spp.; the notodontid *Drymonia* spp.; the pterophorids *Cnaemidophorus*

rhododactylus (Denis and Schiffermüller), and *Stenoptilia veronicae* Karvonen; the tortricids *Archips crataegana* (Hübner), and *Tortrix viridana* (L.); and the yponomeutid *Yponomeuta cagnagella* (Hübner). No yet recorded in Iran.

Cotesia risilis (Nixon, 1974)
Catalogs with Iranian records: Gadallah et al. (2015), Farahani et al. (2016), Yu et al. (2016), Samin, Coronado-Blanco, Kavallieratos et al. (2018) and Samin, Coronado-Blanco, Fischer et al. (2018).
Distribution in Iran: Ilam (Ghahari et al., 2011d), West Azarbaijan (Ghahari & Fischer, 2011a; Samin, 2015), Khuzestan (Zargar et al., 2019b).
Distribution in the Middle East: Iran (see references above), Turkey (Inanç & Çetin Erdoğan, 2004).
Extralimital distribution: Greece, Hungary, Italy, Mongolia, Montenegro, Netherlands, Romania, Slovakia, United Kingdom.
Host records: Recorded by Shaw et al. (2009) as being a parasitoid of the pierids *Gonepteryx cleopatra* (L.), and *Gonepteryx rhamni* (L.).

Cotesia rubecula (Marshall, 1885)
Catalogs with Iranian records: Fallahzadeh and Saghaei (2010), Gadallah et al. (2015), Farahani et al. (2016), Yu et al. (2016), Samin, Coronado-Blanco, Kavallieratos et al. (2018) and Samin, Coronado-Blanco, Fischer et al. (2018).
Distribution in Iran: Hamadan (Khanjani, 2006b as *Apanteles rebecula* Marshall, 1885), Khuzestan (Siahpoush et al., 1993).
Distribution in the Middle East: Iran.
Extralimital distribution: Europe; Nearctic, Ocenic, Oriental and Palaearctic regions (Yu et al., 2016). Introduced to Australia, Canada, New Zealand, United States of America (some states).
Host records: Summarized by Yu et al. (2016) as being a parasitoid of the crambid *Hellula rogatalis* (Hulst); the lasiocampid *Macrothylacia rubi* (L.); the noctuids *Autographa gamma* (L.), and *Mamestra brassicae* (L.); the pierids *Pieris brassicae* (L.), *Pieris rapae* (L.), and *Pontia protodice* (Boisduval and Leconte); the plutellid *Plutella xylostella* (L.); the sphingids *Hemaris fuciformis* (L.), *Hyles euphorbiae* (L.), and *Laothoe populi* (L.); and the tortrix *Tortrix viridana* (L.). In Iran, this species has been reared from the noctuid *Leucania loreyi* (Duponchel) (Siahpoush et al., 1993 as *Mythimna loreyi*), and the pierid *Pieris rapae* (L.) (Khanjani, 2006b).

Cotesia rubripes (Haliday, 1834)
Catalogs with Iranian records: No catalog.
Distribution in Iran: Kermanshah (Sakenin et al., 2020).

Distribution in the Middle East: Iran (Sakenin et al., 2020), Israel—Palestine (Halperin, 1986; Papp, 1970), Turkey (Fernandez-Triana et al., 2020).

Extralimital distribution: Belarus, Bulgaria, Croatia, Czech Republic, France, Germany, Hungary, Italy, Japan, Kazakhstan, Korea, Lithuania, Mongolia, Morocco, Poland, Romania, Russia, Serbia, Switzerland, Ukraine, United Kingdom, Uzbekistan (Fernandez-Triana et al., 2020; Yu et al., 2016).

Host records: Summarized by Yu et al. (2016) as being a parasitoid of the erebids *Euproctis chrysorrhoea* (L.), *Lymantria dispar* (L.), *Ocnogyna boetica* (Rambur), and *Orgyia dubia judaeea* Stgr.; the geometrids *Abraxas grossulariata* (L.), *Cabera pusaria* (L.), *Geometra papilionaria* (L.), and *Jodis lactearia* (L.); the lasiocampids *Dendrolimus pini* (L.), *Dendrolimus sibiricus* Tschetverickov, and *Malacosoma neustria* (L.); the notodontid *Notodonta ziczac* (L.); the nymphalid *Aglais urticae* (L.); the pierids *Aporia crataegi* (L.), *Pieris brassicae* (L.), and *Pieris rapae* (L.); the plutellid *Plutella xylostella* (L.); the pterophorid *Cnaemidophorus rhododactyla* (Denis and Schiffermüller); and the sphingid *Hemaris fuciformis* (L.).

Cotesia ruficrus (Haliday, 1834)

Catalogs with Iranian records: Fallahzadeh and Saghaei (2010), Gadallah et al. (2015), Farahani et al. (2016), Yu et al. (2016), Samin, Coronado-Blanco, Kavallieratos et al. (2018) and Samin, Coronado-Blanco, Fischer et al. (2018).

Distribution in Iran: East Azarbaijan (Ghahari & van Achterberg, 2016), Fars (Lashkari Bod et al., 2011), Golestan (Daniali, 1993; Shojai et al., 1997 as *Apanteles ruficrus* Haliday, 1834), Khuzestan (Shojai et al., 1995 as *Apanteles ruficrus*; Siahpoush et al., 1993; Modarres Awal, 1997, 2012, as *Apanteles ruficrus*), Mazandaran (Shojai et al., 1997 as *Apanteles ruficrus*; Abbasipour & Taghavi, 2002, 2004a, b; Abbasipour et al., 2004; Modarres Awal, 2012; Amooghli-Tabari & Ghahari, 2021—around rice fields), Sistan and Baluchestan (Khajeh et al., 2014), Iran (no specific locality cited) (Khanjani, 2006a).

Distribution in the Middle East: Cyprus (Ingram, 1981), Egypt (El-Heneidy and Hassanein, 1987; Ismail & Swailem, 1976), Iran (see references above), Iraq (Fernandez-Triana et al., 2020), Israel—Palestine (Halperin, 1986; Papp, 1970, 2012), Turkey (Beyarslan et al., 2002a, b; Inanç & Beyarslan, 1990; 2001a, b; Inanç, 1997; Sertkaya & Beyarslan, 2005; Sertkaya et al., 2004), Yemen (Fernandez-Triana et al., 2020).

Extralimital distribution: Afrotropical, Australasian, Neotropical, Oceanic, Oriental, Palaearctic (Yu et al., 2016). Introduced into Australia (Queenslands), Cook Islands, New Zealand, United States of America.

Host records: Summarized by Yu et al. (2016) as being a parasitoid of many lepidopteran species of the families Arctiidae, Brachodidae, Dilobidae, Erebidae, Gelechiidae, Geometridae, Hesperiidae, Lasiocampidae, Lycaenidae, Noctuidae, Nymphalidae, Pieridae, Plutellidae, Pyralidae, and Tortricidae. In Iran, this species has been reared from the noctuids *Leucania loreyi* (Duponchel) (Siahpoush et al., 1993; Modarres Awal, 1997, 2012 as *Mythimna loreyi*), *Sesamia cretica* Lederer (Shojai et al., 1995), and *Mythimna unipuncta* (Haworth) (Abbasipour & Taghavi, 2004a, b; Abbasipour et al., 2004; Khanjani, 2006a as *Pseudaletia unipunctata* Haworth; Modarres Awal, 2012).

Cotesia salebrosa (Marshall, 1885)

Catalogs with Iranian records: Gadallah et al. (2015), Farahani et al. (2016), Yu et al. (2016), Samin, Coronado-Blanco, Kavallieratos et al. (2018) and Samin, Coronado-Blanco, Fischer et al. (2018).

Distribution in Iran: Kermanshah (Ghahari & Fischer, 2012).

Distribution in the Middle East: Iran (Ghahari & Fischer, 2012), Turkey (Inanç & Beyarslan, 2001c).

Extralimital distribution: Bulgaria, Finland, France, Germany, Hungary, Italy, Korea, Lithuania, Mongolia, Norway, Poland, Russia, Sweden, Switzerland, Ukraine, United Kingdom.

Host records: Summarized by Yu et al. (2016) as being a parasitoid of the erebid *Leucoma salicis* (L.); the geometrids *Earophila badiata* (Denis and Schiffermüller), *Epirrita autumnata* Borkhausen, *Epirrita dilutata* (Denis and Schiffermüller), *Epirrita defoliaria* (Clerck), *Eupithecia intricata* (Zetterstedt), *Eupithesia valerianata* (Hübner), and *Operphtera brumata* (L.); and the noctuid *Anarta myrtilli* (L.).

Cotesia saltator (Thunberg, 1922)

Catalogs with Iranian records: Fallahzadeh and Saghaei (2010), Gadallah et al. (2015), Farahani et al. (2016), Yu et al. (2016), Samin, Coronado-Blanco, Kavallieratos et al. (2018) and Samin, Coronado-Blanco, Fischer et al. (2018).

Distribution in Iran: Sistan and Baluchestan (Khajeh et al., 2014), Iran (no specific locality cited) (Telenga, 1955; Papp, 1987).

Distribution in the Middle East: Iran (see references above), Israel—Palestine (Papp, 1987; Halperin, 1986), Lebanon (Papp, 1987), Turkey (Beyarslan, 1988).

Extralimital distribution: Armenia, Bulgaria, France, Germany, Hungary, Mongolia, Poland, Russia, Slovakia, Sweden, Ukraine, former Yugoslavia.

Host records: Summarized by Yu et al. (2016) as being a parasitoid of the acrolepiid *Acrolepia autumnitella* Curtis; the erebid *Lymantria dispar* (L.); the noctuid *Heliothis peltigera* (Denis and Schiffermüller); the pierids *Anthocharis cardamines* (L.), *Anthocharis euphonoides*

Staudinger, and *Euchloe crameri* (Butler); the sphingids *Daphnis nerii* (L.), and *Hyles euphorbiae* (L.); and the yponomeutid *Yponomeuta padella* (L.).

Cotesia saltatoria (Balevski, 1980)

Distribution in the Middle East: Turkey (Beyarslan et al., 2002a; Inanç, 1997; Inanç & Beyarslan, 1990, 2001a; Papp, 1987).

Extralimital distribution: Bulgaria, Croatia, France, Germany, Hungary, Macedonia, Mongolia, Serbia, Slovakia, Spain, United Kingdom.

Host records: Summarized by Yu et al. (2016) as being a parasitoid of the lycaenids *Aricia agestis* (Denis and Schiffermüler), *Aricia atraxerxes* (Fabricius), *Aricia cramera* Eschscholtz, *Callophrys rubi* (L.), *Lysandra coridon* (Poda), *Polyommatus icarus* (Rottemburg), and *Zizeeria Knysna* (Trimen).

Cotesia scabricula (Reinhard, 1880)

Catalogs with Iranian records: Gadallah et al. (2015), Farahani et al. (2016), Yu et al. (2016), Samin, Coronado-Blanco, Kavallieratos et al. (2018) and Samin, Coronado-Blanco, Fischer et al. (2018).

Distribution in Iran: Guilan (Ghahari, Fischer, & Tobias et al., 2012).

Distribution in the Middle East: Iran.

Extralimital distribution: Armenia, Austria, China, Germany, Hungary, Italy, Korea, North Macedonia, Moldova, Mongolia, Romania, Russia, Serbia, Slovakia, Switzerland.

Host records: Summarized by Yu et al. (2016) as being a parasitoid of the crambid *Ostrinia nubilalis* (Hübner); the erebids *Anomis flava* Fabricius, and *Leucoma salicis* (L.); the geometrid *Erannis defoliaria* (Clerck); the lasiocampid *Malacosoma neustria* (L.); the noctuid *Orthosia* sp.; the nolids *Earias clorana* (L.), *Earias cupreoviridis* (Walker), and *Earias vittella* (Fabricius); and the tortricid *Apotomis capreana* (Hübner).

Cotesia sesamiae (Cameron, 1906)

Catalogs with Iranian records: No catalog.

Distribution in Iran: Razavi Khorasan (Samin, Fischer, Sakenin et al., 2019).

Distribution in the Middle East: Iran.

Extralimital distribution: Barbados (introduced), Benin (introduced), Burkina Faso, Cameroon, Central African Republic, Democratic Republic of Congo, Eritrea, Ethiopia, Ghana, India, Ivory Coast, Kenya, Lesotho, Madagascar (introduced), Malawi, Mauritius (introduced), Mozambique, Nigeria, Réunion (introduced), Senegal, South Africa, Sudan, Tanzania, Uganda, Zambia, Zimbabwe.

Host records: Summarized by Yu et al. (2016) as being a parasitoid of the crambids *Chilo* spp., and *Coniesta ignefusalis* (Hampson); the noctuids *Busseola* spp., *Poeonoma serrata* (Hampson), *Sciomesa mesophaea* (Hampson),

Sciomesa piscator Fletcher, *Sesamia* spp., and *Spodoptera littoralis* (Boisduval); and the pyralids *Eldana saccharina* Walker, and *Maliarpha separatella* Ragonot.

Cotesia sessilis (Geoffroy, 1785)

Catalogs with Iranian records: Fallahzadeh and Saghaei (2010, as *Apanteles juniperatus*), Gadallah et al. (2015), Farahani et al. (2016), Yu et al. (2016), Samin, Coronado-Blanco, Kavallieratos et al. (2018) and Samin, Coronado-Blanco, Fischer et al. (2018).

Distribution in Iran: Alborz (Shojai, 1968, 1998; Davatchi & Shojai, 1969 as *Apanteles juniperatus* (Bouché, 1834), Davatchi & Shojai, 1969), Ardabil (Ghahari et al., 2011a as *Cotesia tetrica* (Reinhard, 1880)), Mazandaran (Ghahari, Çetin Erdoğan, Šedivý et al., 2010 as *Cotesia juniperatae* (Bouché, 1834)), Tehran (Modarres Awal, 1997, 2012 as *A. juniperatae*).

Distribution in the Middle East: Iran (see references above), Turkey (Inanç & Beyarslan, 2001a, c; Inanç & Çetin Erdoğan, 2004).

Extralimital distribution: Armenia, Austria, Azerbaijan, Belarus, Belgium, Croatia, Czech Republic, Estonia, Finland, France, Germany, Greece, Hungary, Ireland, Italy, Kazakhstan, Latvia, Lithuania, Moldova, Norway, Poland, Romania, Russia, Serbia, Sweden, Switzerland, Tajikistan, United Kingdom, Uzbekistan.

Host records: Summarized by Yu et al. (2016) as being a parasitoid several species of lepidopteran pests of the families Arctiidae, Dilobidae, Erebidae, Geometridae, Lasiocampidae, Lyonetiidae, Noctuidae, Nymphalidae, Pieridae, Tortricidae, and Zygaenidae. In Iran, this species has been reared from the lycaenid *Lampides boeticus* (L.) (Davatchi & Shojai, 1969; Modarres Awal, 1997, 2012; Shojai, 1968, 1998).

Comments: Farahani et al. (2016) considered *Saturnia pavonia* (Linnaeus, 1758) (Lepidoptera: Saturniidae) as the host of *Cotesia melanoscela* based on Ghahari et al. (2010c). However, this saturniid species has not been recorded from Iran by Ghahari et al. (2010c), and the host is quoted from *C. glomerata* based on Mason (1981) and Papp (1990).

Remarks: This species is considered as a *Numen dubium* (Fernandez-Triana et al., 2020) as the type and depository unknown, and the country of the type locality is unknown.

Cotesia setebis (Nixon, 1974)

Catalogs with Iranian records: Gadallah et al. (2015), Farahani et al. (2016), Yu et al. (2016), Samin, Coronado-Blanco, Kavallieratos et al. (2018) and Samin, Coronado-Blanco, Fischer et al. (2018).

Distribution in Iran: Golestan (Ghahari et al., 2011a).

Distribution in the Middle East: Iran (Ghahari et al., 2011a), Turkey (Beyarslan et al., 2002b; Inanç & Beyarslan, 2001a, b).

Extralimital distribution: Bulgaria, Czech Republic, Greece, Hungary, Mongolia, Russia, Slovakia, Sweden, Switzerland.
Host records: Recorded by Vidal (1997) as being a parasitoid of the geometrid *Eupithecia veratraria* (Herrich-Schäffer).

Cotesia specularis (Szépligeti, 1896)

Catalogs with Iranian records: Fallahzadeh and Saghaei (2010), Gadallah et al. (2015), Farahani et al. (2016), Yu et al. (2016), Samin, Coronado-Blanco, Kavallieratos et al. (2018) and Samin, Coronado-Blanco, Fischer et al. (2018).
Distribution in Iran: Qazvin (Ghahari et al., 2011b), Iran (no specific locality cited) (Papp, 1986 as *Apanteles specularis* Szépligeti, 1896; Papp, 2007).
Distribution in the Middle East: Iran (see references above), Israel—Palestine (Papp, 2012), Jordan, Turkey (Papp, 1986).
Extralimital distribution: Bulgaria, Germany, Greece, Hungary, Kyrgyzstan, Moldova, Romania, Russia, Spain, Tajikistan, Uzbekistan.
Host records: Summarized by Yu et al. (2016) as being a parasitoid of the geometrid *Chloroclystis v-ata* (Haworth); the lycaenids *Glaucopsyche alexis* Poda, *Iolana iolas* Ochesenheimer, and *Lampides boeticus* (L.); the noctuid *Mythimna unipuncta* (Haworth); and the pierid *Aporia crataegi* (L.).

Cotesia spuria (Wesmael, 1837)

Catalogs with Iranian records: Gadallah et al. (2015), Farahani et al. (2016), Yu et al. (2016), Samin, Coronado-Blanco, Kavallieratos et al. (2018) and Samin, Coronado-Blanco, Fischer et al. (2018).
Distribution in Iran: Qazvin (Ghahari et al., 2011b).
Distribution in the Middle East: Iran (Ghahari et al., 2011b), Israel—Palestine (Papp, 2012), Turkey (Fernandez-Triana et al., 2020).
Extralimital distribution: Afghanistan, Armenia, Austria, Azerbaijan, Belgium, Bulgaria, China, Croatia, Finland, France, Germany, Greece, Hungary, Ireland, Italy, Japan, Kazakhstan, Korea, Latvia, Lithuania, Moldova, Poland, Romania, Russia, Serbia, Slovakia, Slovenia, Sweden, Switzerland, Tajikistan, Ukraine, United Kingdom, Uzbekistan.
Host records: Summarized by Yu et al. (2016) as being a parasitoid of several lepidopteran species of the families Arctiidae, Dilobidae, Erebidae, Gelechiidae, Geometridae, Hepialidae, Hesperiidae, Lasiocampidae, Lycaenidae, Noctuidae, Notodontidae, Nymphalidae, Pieridae, Pterophoridae, Pyralidae, Thyatiridae, Tortricidae, and Zygaenidae.

Cotesia telengai (Tobias, 1972)

Catalogs with Iranian records: Fallahzadeh and Saghaei (2010), Gadallah et al. (2015), Farahani et al. (2016), Yu et al. (2016), Samin, Coronado-Blanco, Kavallieratos et al. (2018) and Samin, Coronado-Blanco, Fischer et al. (2018).

Distribution in Iran: Ilam (Ghahari et al., 2011d), Iran (no specific locality cited) (Papp, 1986).
Distribution in the Middle East: Iran (Papp, 1986; Ghahari et al., 2011d), Israel—Palestine (Papp, 2012), Turkey (Beyarslan, 1988; Beyarslan et al., 2002a; Beyarslan et al., 2006; Inanç, 1997; Inanç & Beyarslan, 1990; 2001a).
Extralimital distribution: Afghanistan, Albania, Algeria, Armenia, Azerbaijan, Bosnia-Herzegovina, Bulgaria, Croatia, Georgia, Germany, Greece, Hungary, Italy, Kazakhstan, Moldova, Morocco, Netherlands, Poland, Russia, Slovakia, Spain, Switzerland, Tajikistan, Tunisia, Turkmenistan, United Kingdom, Uzbekistan.
Host records: Summarized by Yu et al. (2016) as being a parasitoid of the crambid *Pediasia aridella* (Thunberg); the erebid *Catacola elocata* (Esper); the noctuids *Agrotis segetum* (Denis and Schiffermüller), *Athetis hospes* (Freyer), *Autographa gamma* (L.), *Euxoa temera* Hübner, *Helicoverpa armigera* (Hübner), *Heliothis viriplaca* (Hufnagel), *Noctua pronuba* (L.), *Cornutiplusia circumflexa* (L.), and *Xestia c-nigrum* (L.); and the nymphalid *Erebia* sp.

Cotesia tenebrosa (Wesmael, 1837)

Catalogs with Iranian records: Gadallah et al. (2015), Farahani et al. (2016), Yu et al. (2016), Samin, Coronado-Blanco, Kavallieratos et al. (2018) and Samin, Coronado-Blanco, Fischer et al. (2018).
Distribution in Iran: Lorestan (Ghahari, Fischer, Papp et al., 2012).
Distribution in the Middle East: Iran (Ghahari, Fischer, Papp et al., 2012), Israel—Palestine (Kugler, 1966; Halperin, 1986; Papp, 2012), Turkey (Beyarslan, 1988; Inanç, 1997; Inanç & Beyarslan, 1990; 2001a).
Extralimital distribution: Andorra, Belgium, Croatia, Finland, France, Georgia, Germany, Greece, Hungary, Kazakhstan, Korea, North Macedonia, Moldova, Mongolia, Poland, Russia, Serbia, Spain, Sweden, Switzerland, Tajikistan, Ukraine, United Kingdom, Uzbekistan.
Host records: Summarized by Yu et al. (2016) as being a parasitoid of the aceolepiid *Acrolepia autumnitella* Curtis; the erebids *Euproctis chrysorrhoea* (L.), and *Lymantria dispar* (L.); the geometrids *Biston betularia* L., and *Phigalia pilosaria* (Denis and Schiffermüller); the lycaenids *Aricia morronensis* (Ribbe), *Cupido alcetas* (Hoffmannsegg), *Lampides boeticus* (L.), *Lysandra coridon* (Poda), *Plebejus argus* (L.), and *Polyomatus* spp.; the noctuids *Spodoptera littoralis* (Boisduval), and *Spodoptera litura* (Fabricius); the pterophorid *Stenoptilia pterodactyla* (L.); and the yponomeutid *Yponomeuta padella* (L.).

Cotesia tibialis (Curtis, 1830)

Catalogs with Iranian records: Fallahzadeh and Saghaei (2010), Gadallah et al. (2015), Farahani et al. (2016),

Yu et al. (2016), Samin, Coronado-Blanco, Kavallieratos et al. (2018) and Samin, Coronado-Blanco, Fischer et al. (2018).

Distribution in Iran: Alborz (Shojai, 1968, 1998; Davatchi & Shojai, 1969 as *Apanteles congestus* (Nees, 1834), Davatchi & Shojai, 1969), Golestan, Mazandaran (Ghahari et al., 2011a), Tehran (Modarres Awal, 1997, 2012, as *A. congestus*), Iran (no specific locality cited) (Aubert, 1966; Khanjani, 2006a as *Cotesia tibialis* (Curtis, 1830)).

Distribution in the Middle East: Iran (see references above), Israel−Palestine (Halperin, 1986; Papp, 2012), Turkey (Beyarslan et al., 2002a, b; Inanç, 1997; Inanç & Beyarslan, 2001a, b).

Extralimital distribution: Europe, Palaearctic.

Host records: Summarized by Yu et al. (2016) as being a parasitoid of several lepidopteran pests of the families Arctiidae, Bombycidae, Cossidae, Erebidae, Gelechiidae, Geometridae, Lasiocampidae, Lycaenidae, Noctuidae, Notodontidae, Nymphalidae, Pieridae, Plutellidae, Pyralidae, Tortricidae, Yponomeutidae, and Zygaenidae. In Iran, this species has been reared from the noctuid *Agrotis segetum* (Denis and Schiffermüller) (Shojai, 1968, 1998; Davatchi & Shojai, 1969; Modarres Awal, 1997, 2012 as *Scotia segetum*); and the crambid *Ostrinia nubilalis* (Hübner) (Khanjani, 2006a).

Cotesia vanessae (Reinhard, 1880)

Catalogs with Iranian records: Fallahzadeh and Saghaei (2010), Gadallah et al. (2015), Farahani et al. (2016), Yu et al. (2016), Samin, Coronado-Blanco, Kavallieratos et al. (2018) and Samin, Coronado-Blanco, Fischer et al. (2018).

Distribution in Iran: East Azarbijan (Nikdel et al., 2012), Isfahan, Northern Khorasan (Ghahari & Gadallah, 2022), Mazandaran (Ghahari, 2019a), Sistan and Baluchestan (Khajeh et al., 2014), West Azarbaijan (Asem et al., 2016; Karimpour et al., 2001; Modarres Awal, 2012; Samin et al., 2014).

Distribution in the Middle East: Iran (see references above), Israel−Palestine (Papp, 2012), Turkey (Beyarslan, 1988; Çetin Erdoğan & Beyarslan, 2005; Doganlar, 1982; Inanç, 1997; Inanç & Beyarslan, 1990, 2001a; Özbek & Coruh, 2012).

Extralimital distribution: Afghanistan, Armenia, Austria, Azerbaijan, Bulgaria, Canada, China, Czech Republic, Ethiopia, Finland, France, Georgia, Germany, Greece, Hungary, Ireland, Italy, Japan, Kazakhstan, Korea, Latvia, Moldova, Montenegro, Morocco, Netherlands, Poland, Romania, Russia, Serbia, Spain, Tunisia, Ukraine, United Kingdom, Uzbekistan.

Host records: Summarized by Yu et al. (2016) as being a parasitoid of the crambid *Loxostege sticticalis* (L.); the lasiocampid *Lasiocampa trifolii* (Denis and Schiffermüller); the following noctuids: *Acontia lucida* (Hufnagel), *Actebia praecox* (L.), *Apamea sordens* (Hufnagel), *Autographa gamma* (L.), *Calophasia opaline* (Esper), *Chrysodeixis chalcites* (Esper), *Cornotiplusia circumflexa* (L.), *Discestra trifolii* (Hufnagel), *Helicoverpa armigera* (Hübner), *Heliothis viriplaca* (Hufnagel), *Lacanobia oleracea* (L.), *Macdunnoughia confusa* (Stephens), *Mamestra brassciae* (L.), *Parexarnis fugax* (Treitschke), *Simyra dentinosa* Freyer, *Spodoptera exigua* (Hübner), and *Trichoplusia ni* (Hübner); the nolid *Nycteola revayana* (Scopoli); the notodontids *Cerura vinula* (L.), and *Eligmodonta ziczac* (L.); the nymphalids *Aglais* spp., *Argynnis aglaja* (L.), *Limenitis camilla* (L.), *Nymphalis polychloros* (L.), *Vanessa atalanta* (L.), and *Vanessa cardui* (L.); the petrophorid *Cnaemidophorus rhododactylus* (Denis and Schiffermüller); and the thaumetopoeid *Thaumetopoea processionea* (L.). In Iran, this species has been reared from the noctuids *Simyra dentinosa* Freyer (Karimpour et al., 2001; Modarres Awal, 2012; Nikdel et al., 2012), *Apamea sordens* Hufnagel (Ghahari, 2019a), *Helicoverpa armigera* (Hübner), and *Spodoptera exigua* (Hübner) (Ghahari, 2019a; Ghahari & Gadallah, 2022).

Plant associates in Iran: *Euphorbia* sp. (Euphorbiaceae) (Nikdel et al., 2012).

Cotesia vestalis (Haliday, 1834)

Catalogs with Iranian records: Fallahzadeh and Saghaei (2010), Gadallah et al. (2015), Farahani et al. (2016), Yu et al. (2016), Samin et al. (2018a, b as *Cotesia plutellae* (Kurdjumov, 1912)), Samin, Coronado-Blanco, Kavallieratos et al. (2018) and Samin, Coronado-Blanco, Fischer et al. (2018).

Distribution in Iran: Alborz (Golizadeh et al., 2008, Ghahari, Fischer, & Jussila, 2012, both as *Cotesia plutellae* (Kurdjumov, 1912)), East Azarbaijan (Rastegar et al., 2012 as *C. plutellae*, Ghahari, Fischer, & Jussila, 2012), Guilan (Ghahari, Fischer, & Tobias et al., 2012), Isfahan (Afiunizadeh Isfahani & Karimzadeh Isfahani 2010 as *C. plutellae*, Afiunizadeh et al., 2011, Ghahari et al., 2011c as *C. plutellae*, Rezaei et al., 2014, Rabiei et al., 2017), Khuzestan (Zargar et al., 2019b), Tehran (Modarres Awal, 2012 as *C. plutellae*, Hasanshahi et al., 2012a, b, 2014a, 2015 as *C. plutellae*, Rostami et al., 2014 as *C. plutellae*), West Azarbaijan (Karimpour et al., 2005 as *C. plutellae*), Iran (no specific locality cited) (Telenga, 1955).

Distribution in the Middle East: Iran (see references above), Israel−Palestine (Papp, 1970, 2012), Turkey (Fernandez-Triana et al., 2020).

Extralimital distribution: Nearly cosmopolitan species.

Host records: Summarized by Yu et al. (2016) as being a parasitoid of the erebids *Arctia tigrina* (Villers), *Epicallia villica* (L.), *Hyphoraia aulica* (L.), *Ocnogyna clathrate* (Lederer), and *Spiris striata* (L.); the noctuids *Conistra staundengeri* (Graslin), and *Noctua fimbriata* Schreber; the nymphalids *Aglais urticae* (L.), *Bolaria eunomia* Esper, *Eurodryas aurinia* (Rottemburg), *Hipparchia semele* (L.),

Maniola jurtina (L.), *Nymphalis polychloros* (L.), and *Vanessa cardui* (L.); and the plutellid *Plutella xylostella* (L.). In Iran, this species has been reared from the noctuid *Simyra dentinosa* Freyer (Karimpour et al., 2005); the plutellid diamondback *Plutella xylostella* (L.) (Afiunizadeh Isfahani & Karimzadeh Isfahani, 2010; Ghahari, Fischer, & Jussila, 2012; Golizadeh et al., 2008; Hasanshahi et al., 2012a, b; Karimzadeh et al., 2016; Modarres Awal, 2012; Rabiei et al., 2017; Rezaei et al., 2014; Rostami et al., 2014); and the pierid *Pieris rapae* (L.) (Hasanshahi et al., 2015).

Plant associates in Iran: *Brassica oleracea* (Brassicaceae) (Rostami et al., 2014; Hasanshahi et al., 2014a).

Cotesia villana (Reinhard, 1880)

Catalogs with Iranian records: Gadallah et al. (2015), Farahani et al. (2016), Yu et al. (2016), Samin, Coronado-Blanco, Kavallieratos et al. (2018) and Samin, Coronado-Blanco, Fischer et al. (2018).

Distribution in Iran: Qazvin (Ghahari et al., 2011b as *Cotesia villanus* (Reinhard, 1880)).

Distribution in the Middle East: Iran (Ghahari et al., 2011b), Turkey (Inanç & Beyarslan, 2001a).

Extralimital distribution: Croatia, Finland, France, Germany, Greece, Hungary, Mongolia, Poland, Romania, Russia, Slovakia, Switzerland, United Kingdom.

Host records: Summarized by Yu et al. (2016) as being a parasitoid of the erebids *Arctia tigrine* (Villers), *Coscinia cibraria* (L.), *Epicallia villica* (L.), *Hyphoraia aulica* (L.), and *Spiris striata* (L.); the noctuids *Agrotis segetum* (Denis and Schiffermüller), *Conistra staudingeri* (Grasli), and *Noctua fimbriata* Schreber; and the nymphalid *Melitaea didyma* (Esper).

Cotesia zagrosensis Zargar and Gupta, 2019

Catalogs with Iranian records: No catalog.

Distribution in Iran: Khusestan (Zargar et al., 2019b).

Distribution in the Middle East: Iran.

Extralimital distribution: None.

Host records: Unknown.

Cotesia zygaenarum (Marshall, 1885)

Catalogs with Iranian records: Gadallah et al. (2015), Farahani et al. (2016), Yu et al. (2016), Samin, Coronado-Blanco, Kavallieratos et al. (2018) and Samin, Coronado-Blanco, Fischer et al. (2018).

Distribution in Iran: Golestan (Samin, Ghahari, Gadallah et al., 2015a), Ilam (Ghahari et al., 2011d), Mazandaran (Ghahari, Fischer, Çetin Erdoğan et al., 2010).

Distribution in the Middle East: Iran (see references above), Israel−Palestine (Papp, 2012), Turkey (Beyarslan, 1988; Beyarslan et al., 2002a; Çetin Erdoğan & Beyarslan, 2005; Inanç & Çetin Erdoğan, 2004).

Extralimital distribution: Albania, Armenia, Austria, Azerbaijan, China, Czech Republic, Finland, France, Germany, Greece, Hungary, Italy, Japan, Kazakhstan, Korea, North Macedonia, Moldova, Mongolia, Poland, Romania, Russia, Serbia, Slovakia, Switzerland, Tunisia, United Kingdom.

Host records: Summarized by Yu et al. (2016) as being a parasitoid of the geometrid *Phigalia pilosaria* (Denis and Schiffermüller); the lycaenid *Polyommatus icarus* (Rottemburg); the noctuid *Diloba caeruleocephala* (L.); the nymphalid *Eurodryas aurinia* (Rottemburg); and the pierid *Colias hyale* (L.).

Genus *Deuterixys* Mason, 1981

Deuterixys carbonaria (Wesmael, 1837)

Catalogs with Iranian records: No catalog.

Distribution in Iran: Guilan (Samin, Beyarslan, Coronado-Blanco et al., 2020).

Distribution in the Middle East: Iran.

Extralimital distribution: Austria, Belgium, Czech Republic, Finland, France, Germany, Hungary, Italy, Japan, Korea, Lithuania, Mongolia, Netherlands, Poland, Romania, Russia, Slovenia, United Kingdom, former Yugoslavia (Fernandez-Triana et al., 2020; Yu et al., 2016).

Host records: Summarized by Yu et al. (2016) as being a parasitoid of the bucculatricids *Bucculatrix cristatella* (Zeller), and *Bucculatrix nigricomella* (Zeller); the erebid *Leucoma salicis* (L.); the geometrid *Operophtera brumata* (L.); and the lasiocampids *Euthrix potatoria* (L.), and *Macrothylacia rubi* (L.).

Deuterixys rimulosa (Niezabitowski, 1910)

Catalogs with Iranian records: Gadallah et al. (2015), Farahani et al. (2016), Yu et al. (2016), Samin, Coronado-Blanco, Kavallieratos et al. (2018) and Samin, Coronado-Blanco, Fischer et al. (2018).

Distribution in Iran: Isfahan (Ghahari et al., 2011c).

Distribution in the Middle East: Iran.

Extralimital distribution: Azerbaijan, Croatia, Germany, Greece, Hungary, Kazakhstan, Mongolia, Poland, Russia, Slovakia, Spain, Turkmenistan, United Kingdom, Uzbekistan.

Host records: Summarized by Yu et al. (2016) as being a parasitoid of the bucculatricids *Bucculatrix artemisiella* Herrich-Schäffer, *Bucculatrix benacicolella* Hartig, *Bucculatrix cristatella* (Zeller), *Bucculatrix gnaphaliella* (Treitschke), and *Bucculatrix nigrocomella* (Zetter); and the nepticulid *Stigmella anomalella* (Goeze).

Deuterixys tenuiconvergens Zargar and Gupta, 2020

Catalogs with Iranian records: No catalog.

Distribution in Iran: Khuzestan (Zargar et al., 2020).

Distribution in the Middle East: Iran.

Extralimital distribution: None.

Host records: Unknown.

Genus *Diolcogaster* Ashmead, 1900

Diolcogaster abdominalis (Nees von Esenbeck, 1834)

Catalogs with Iranian records: No catalog.

Distribution in Iran: Golestan (Naderian et al., 2020).

Distribution in the Middle East: Iran (Naderian et al., 2020), Israel—Palestine (Papp, 2012).

Extralimital distribution: Azerbaijan, Belgium, France, Georgia, Germany, Hungary, Ireland, Italy, Kazakhstan, Korea, North Macedonia, Moldova, Mongolia, Montenegro, Poland, Romania, Russia, Serbia, Slovakia, Spain, Switzerland, United Kingdom.

Host records: Recorded by Shaw et al. (2009) as being a parasitoid of the nymphalids *Coenonympha oedippus* (Fabricius), and *Coenonympha tullia* (Müller). No yet recorded in Iran.

Diolcogaster alvearia (Fabricius, 1798)

Catalogs with Iranian records: Gadallah et al. (2015), Farahani et al. (2016), Yu et al. (2016), Samin, Coronado-Blanco, Kavallieratos et al. (2018) and Samin, Coronado-Blanco, Fischer et al. (2018).

Distribution in Iran: Qazvin (Ghahari et al., 2011b).

Distribution in the Middle East: Iran (Ghahari et al., 2011b), Israel—Palestine (Papp, 2012), Turkey (Inanç, 1997; Inanç & Beyarslan, 2001a).

Extralimital distribution: Austria, Bulgaria, China, Croatia, France, Germany, Greece, Hungary, Italy, Moldova, Netherlands, Romania, Russia, Slovakia, Spain, Switzerland, United Kingdom, former Yugoslavia.

Host records: Summarized by Yu et al. (2016) as being a parasitoid of the geometrids *Alcis repandata* (L.), *Hypomecis* sp., *Menophra abruptaria* (Thunberg), *Opisthograptis luteolata* (L.), *Ourapteryx sambucaria* (L.), and *Peribatodes rhomboidaria* (Denis and Schiffermüller).

Diolcogaster claritibia (Papp, 1959)

Catalogs with Iranian records: Gadallah et al. (2015), Farahani et al. (2016), Yu et al. (2016), Samin, Coronado-Blanco, Kavallieratos et al. (2018) and Samin, Coronado-Blanco, Fischer et al. (2018).

Distribution in Iran: Golestan (Ghahari & Fischer, 2011b), Kerman, Northern Khorasan, Sistan and Baluchestan (Ghafouri Moghaddam et al., 2019).

Distribution in the Middle East: Iran (Ghahari & Fischer, 2011b; Ghafouri Moghaddam et al., 2019), Jordan (Fernandez-Triana et al., 2014), Turkey (Inanç & Beyarslan, 1990; 1997; 2001a; Inanç, 1997).

Extralimital distribution: Afghanistan, Armenia, Austria, Azerbaijan, Belarus, Canada, Finland, Georgia, Greece, Hungary, Kazakhstan, Lithuania, North Macedonia, Moldova, Russia, Spain, Tunisia, Turkmenistan, Ukraine, formerYugoslavia.

Host records: Recorded by Tobias (1971) and Papp (1981) as being a parasitoid of the plutellid *Plutella xylostella* (L.). Plant association in Iran: *Cucumis sativus* (Cucurbitaceae), *Medicago sativa* (Fabaceae) (Ghafouri Moghaddam et al., 2019).

Diolcogaster mayae (Shestakov, 1932)

Catalogs with Iranian records: Fallahzadeh and Saghaei (2010), Gadallah et al. (2015), Farahani et al. (2016), Yu et al. (2016), Samin, Coronado-Blanco, Kavallieratos et al. (2018) and Samin, Coronado-Blanco, Fischer et al. (2018).

Distribution in Iran: Fars (Al-e Mansour & Mostafavi, 1993; Modarres Awal, 1997, 2012), Sistan and Baluchestan (Hedwig, 1957 as *Microgaster iranensis* (Hedwig, 1957)); Ghafouri Moghaddam et al., 2019), Southern Khorasan (Ghafouri Moghaddam et al., 2019), West Azarbaijan (Telenga, 1955), Iran (no specific locality cited) (Güçlü & Özbek, 2011; Tobias, 1976, 1986).

Distribution in the Middle East: Iran (see references above), Israel—Palestine, Turkey (Fernandez-Triana et al., 2020), Yemen (Fernandez-Triana et al., 2020; Ghafouri Moghadam et al., 2019).

Extralimital distribution: Afghanistan, Algeria, Armenia, Azerbaijan, Kazakhstan, Mongolia, Romania, Russia, Tajikistan, Turkmenistan, Uzbekistan.

Host records: Unknown.

Plant association in Iran: Forest and pasture plants (Modarres Awal, 1997, 2012), *Phoenix* sp. (Arecaceae), *Pistacia atlantica* (Anacardiaceae) (Ghafouri Moghaddam et al., 2019).

Diolcogaster minuta (Reinhard, 1880)

Catalogs with Iranian records: No catalog.

Distribution in Iran: Mazandaran (Naderian et al., 2020).

Distribution in the Middle East: Iran (Naderian et al., 2020), Turkey (Fernandez-Triana et al., 2020).

Extralimital distribution: Armenia, Czech Republic, Finland, Germany, Lithuania, Poland, Romania, Russia, Switzerland, Turkmenistan, Ukraine, United Kingdom.

Host records: Summarized by Yu et al. (2016) as being a parasitoid of the geometrid *Alcis jubata* (Thunberg); and the noctuid *Brachylomia viminalis* (Fabricius).

Diolcogaster spreta (Marshall, 1885)

Catalogs with Iranian records: Gadallah et al. (2015), Farahani et al. (2016), Yu et al. (2016), Samin, Coronado-Blanco, Kavallieratos et al. (2018) and Samin, Coronado-Blanco, Fischer et al. (2018).

Distribution in Iran: East Azarbaijan (Ghahari & Fischer, 2011a), Southern Khorasan (Ghahari & Beyarslan, 2017), Zanjan (Samin, 2015).

Distribution in the Middle East: Iran (see references above), Turkey (Inanç & Çetin Erdoğan, 2004).

Extralimital distribution: Bulgaria, China, Denmark, Germany, Greece, Hungary, Moldova, Romania, Russia, Slovakia, United Kingdom.

Host records: Summarized by Yu et al. (2016) as being a parasitoid of the pyralids *Acrobasis advenella* Zincken, *Acrobasis consociella* (Hübner), *Acrobasis repandana* (Fabricius), and *Pempelia palumbella* (Denis and Schiffermüller).

Genus *Distatrix* Mason, 1981

Distatrix flava (Fernandez-Triana & van Achterberg, 2017)

Distribution in the Middle East: Yemen (Fernandez-Triana & van Achterberg, 2017).

Extralimital distribution: None.

Host records: Unknown.

Distatrix yemenitica van Achterberg and Fernandez-Triana, 2017

Distribution in the Middle East: Yemen (Fernandez-Triana & van Achterberg, 2017).

Extralimital distribution: None.

Host records: Unknown.

Genus *Glyptapanteles* Ashmead, 1904

Glyptapanteles acasta (Nixon, 1973)

Catalogs with Iranian records: Samin, Coronado-Blanco, Fischer et al. (2018).

Distribution in Iran: Northern Khorasan (Samin, Coronado-Blanco, Fischer et al., 2018).

Distribution in the Middle East: Iran (Samin, Coronado-Blanco, Fischer et al., 2018), Turkey (Beyarslan et al., 2002a; Inanç & Beyarslan, 2001b).

Extralimital distribution: Bulgaria, Finland, Germany, Greece, Hungary, Poland, Russia, Slovakia, Switzerland, United Kingdom (Fernandez-Triana et al., 2020; Yu et al., 2016).

Host records: Summarized by Yu et al. (2016) as being a parasitoid of the noctuids *Allophyes oxyacanthae* (L.), *Amphipyra perflua* (Fabricius), *Amphipyra pyramidea* (L.), and *Diarsia mendica* (Fabricius).

Glyptapanteles aliphera (Nixon, 1973)

Catalogs with Iranian records: No catalog.

Distribution in Iran: Guilan (Samin, Coronado-Blanco, Hosseini et al., 2019 as *Protapanteles aliphera* (Nixon, 1973)).

Distribution in the Middle East: Iran (Samin, Coronado-Blanco, Hosseini et al., 2019), Israel−Palestine (Papp, 1983).

Extralimital distribution: Armenia, Azerbaijan, Finland, France, Georgia, Germany, Greece, Hungary, Netherlands, Poland, Romania, Russia, Slovakia, Sweden, Switzerland, United Kingdom.

Host records: Unknown.

Glyptapanteles antinoe (Nixon, 1973)

Distribution in the Middle East: Turkey (Inanç & Beyarslan, 2001a).

Extralimital distribution: Austria, Germany, Hungary.

Host records: Unknown.

Glyptapanteles callidus (Haliday, 1834)

Catalogs with Iranian records: No catalog.

Distribution in Iran: Isfahan (Naderian et al., 2020).

Distribution in the Middle East: Iran (Naderian et al., 2020), Israel−Palestine (Čapek & Hofmann, 1997), Turkey (Inanç & Beyarslan, 2001a).

Extralimital distribution: Armenia, Austria, Belgium, Bulgaria, Czech Republic, Finland, France, Georgia, Germany, Hungary, Ireland, Lithuania, Netherlands, Poland, Romania, Russia, Slovakia, Sweden, Switzerland, Ukraine, United Kingdom.

Host records: Summarized by Yu et al. (2016) as being a parasitoid of the erebid *Parasemia plantaginis* (L.); the geometrid *Abraxas glossulariata* (L.); and the noctuid *Noctua orbona* (Hufnagel).

Glyptapanteles compressiventris (Muesebeck, 1921)

Distribution in the Middle East: Turkey (Inanç, 1997).

Extralimital distribution: Armenia, Azerbaijan, Canada, Croatia, Czech Republic, Finland, Germany, Hungary, Italy, Kazakhstan, Lithuania, North Macedonia, Moldova, Netherlands, Romania, Russia, Serbia, Slovakia, Spain, Switzerland, United Kingdom.

Host records: Summarized by Yu et al. (2016) as being a parasitoid of the erebids *Amata phegea* (L.), *Diaphora mendica* (Clerck), *Epicallia villica* (L.), *Ocnogyna baetica* (rambur), *Ocnogyna loeewii* (Zeller), *Ocnogyna pierreti* Rambur, *Phragmatobia fuliginosa* (L.), and *Spilosoma lubricipedum* (L.); the noctuid *Amphipyra perflua* (Fabricius), and the tortricid *Archips xylosteana* (L.).

Glyptapanteles fulvipes (Haliday, 1834)

Catalogs with Iranian records: No catalog.

Distribution in Iran: Markazi (Sakenin et al., 2020).

Distribution in the Middle East: Iran (Sakenin et al., 2020), Turkey (Beyarslan, 1988; Beyarslan et al., 2002a; Inanç & Beyarslan, 2001a).

Extralimital distribution: Armenia, Austria, Azerbaijan, Belarus, Belgium, Bulgaria, Canada, Croatia, Czech Republic, Faeroe Islands, Finland, France, Georgia,

Germany, Hungary, Iceland, Ireland, Italy, Japan, Kazakhstan, Korea, Lithuania, North Macedonia, Moldova, Mongolia, Netherlands, Poland, Romania, Russia, Serbia, Slovakia, Slovenia, Spain, Sweden, Switzerland, Ukraine, United Kingdom.

Host records: Summarized by Yu et al. (2016) as being a parasitoid of several lepidopteran hosts of the families: Arctiidae, Choreutidae, Dilobidae, Erebidae, Gelechiidae, Geometridae, Lasiocampidae, Noctuidae, Notodontidae, Nymphalidae, Pieridae, and Pyralidae.

Glyptapanteles inclusus (Ratzeburg, 1844)

Catalogs with Iranian records: No catalog.
Distribution in Iran: Ardabil (Samin, Beyarslan, Ranjith et al., 2020).
Distribution in the Middle East: Iran.
Extralimital distribution: Austria, Azerbaijan, Bulgaria, Cape Verde, China, Denmark, France, Germany, Ireland, Italy, Japan, Kazakhstan, Korea, Mongolia, Poland, Romania, Russia, Slovakia, Switzerland, Ukraine, United Kingdom.
Host records: Summarized by Yu et al. (2016) as being a parasitoid of the erebids *Euproctis chrysorrhoea* (L.), *Euproctis similis* (Füssli), *Lithosia quadra* (L.), *Lymantria dispar* (L.), and *Lymantria monacha* (L.); the lasiocampid *Dendrolimus pini* (L.); the noctuid *Acronicta intermedia* Warren; and the zygaenid *Theresimima ampellophaga* (Bayle-Barelle). In Iran, this species has been reared from the erebid *Euproctis chrysorrhoea* (Samin, Beyarslan, Ranjith et al., 2020).

Glyptapanteles indiensis (Marsh, 1979)

Catalogs with Iranian records: No catalog.
Distribution in Iran: East Azarbaijan (Nikdel, 2015).
Distribution in the Middle East: Iran.
Extralimital distribution: India, United States of America.
Host records: Summarized by Yu et al. (2016) as being a parasitoid of the erebids *Lymantria dispar* (L.), and *Lymantria obfuscata* Walker. In Iran, this species has been reared from *Lymantria dispar* (L.) (Nikdel, 2015).

Glyptapanteles liparidis (Bouché, 1834)

Catalogs with Iranian records: Fallahzadeh and Saghaei (2010), Gadallah et al. (2015), Farahani et al. (2016), Yu et al. (2016), Samin, Coronado-Blanco, Kavallieratos et al. (2018) and Samin, Coronado-Blanco, Fischer et al. (2018) (all as *Protapanteles liparidis* (Bouché, 1843)).
Distribution in Iran: Guilan (Hérard et al., 1979 as *Apanteles liparidis* (Bouché, 1834), Radjabi, 1986), West Azarbaijan (Žikić et al., 2014 as *Protapanteles liparidis*),

Iran (no specific locality cited) (Modarres Awal, 1997, 2012 as *Protapanteles liparidis*).
Distribution in the Middle East: Iran.
Extralimital distribution: Oriental and Palaearctic regions (Yu et al., 2016 as *Protapanteles liparidis*; Fernandez-Triana et al., 2020).
Host records: Summarized by Yu et al. (2016) as being a parasitoid of the crambid *Ostrinia nubilalis* (Hübner); the following erebids: *Calliteara abietis* (Denis and Schiffermüller), *Calliteara pudibunda* (L.), *Dasychira pseudabietis* Butler, *Epicallia villica* (L.), *Euproctis chrysorrhoea* (L.), *Euproctis taiwana* (Shiraki), *Ivela auripes* (Butler), *Lymantria dispar* (L.), *Lymantria obfuscata* Walker, and *Orgyia* spp.; the lasiocampids *Dendrolimus* spp., *Eriogaster lanestris* (L.), and *Malacosoma neustria* (L.); the noctuid *Acronicta rumicis* (L.); the notodontid *Clostera anastomosis* (L.); and the tortricid *Rhyacionia buoliana* (Denis and Schiffermüller). In Iran, this species has been reared from the erebid *Lymantria dispar* (L.) (Modarres Awal, 1997, 2012), and the gracillariid *Phyllonorycter populifoliella* (Treitschke) (Žikić et al., 2014).

Glyptapanteles mygdonia (Nixon, 1973)

Catalogs with Iranian records: Gadallah et al. (2015), Farahani et al. (2016), Yu et al. (2016), Samin, Coronado-Blanco, Kavallieratos et al. (2018) and Samin, Coronado-Blanco, Fischer et al. (2018) (all as *Protapanteles mygdonia* (Nixon, 1973)).
Distribution in Iran: East Azarbaijan (Ghahari, Gadallah, Çetin Erdoğan et al., 2009), Golestan (Samin, Ghahari, Gadallah et al., 2015a as *Protapanteles mygdonia*).
Distribution in the Middle East: Iran (Ghahari, Gadallah, Çetin Erdoğan et al., 2009; Samin, Ghahari, Gadallah et al., 2015a), Turkey (Inanç & Çetin Erdoğan, 2004).
Extralimital distribution: Bulgaria, Finland, France, Germany, Hungary, Ireland, Italy, Korea, Madeira Islands, Russia, Slovakia, Spain, Switzerland, United Kingdom (Yu et al., 2016 as *Protapanteles mygdonia*; Fernandez-Triana et al., 2020).
Host records: Summarized by Yu et al. (2016) as being a parasitoid of the choreutid *Choreutis pariana* (Clerck); the geometrids *Operophtera brumata* (L.), *Phigalia pilosaria* (Denis and Schiffermüller), and *Pungeleria capreolaria* (Denis and Schiffermüller).

Glyptapanteles popovi (Telenga, 1955)

Distribution in the Middle East: Turkey (Inanç & Beyarslan, 1990).
Extralimital distribution: Turkmenistan.
Host records: Unknown.

Glytapanteles porthetriae (Muesebeck, 1928)

Catalogs with Iranian records: Gadallah et al. (2015), Farahani et al. (2016), Yu et al. (2016), Samin, Coronado-Blanco, Kavallieratos et al. (2018) and Samin, Coronado-Blanco, Fischer et al. (2018).

Distribution in Iran: East Azarbaijan (Nikdel, 2015 as *Glyptapanteles porthetriae* (Muesebeck, 1928)), Markazi (Ghahari et al., 2011a).

Distribution in the Middle East: Iran (Ghahari et al., 2011a; Nikdel, 2015), Israel—Palestine (Papp, 2012), Turkey (Beyarslan, 1988; Beyarslan et al., 2002a, b; Inanç & Beyarslan, 2001b).

Extralimital distribution: Armenia, Austria, Azerbaijan, Bulgaria, China, Croatia, Finland, France, Georgia, Germany, Greece, Hungary, India, Italy, Korea, Moldova, Morocco, Poland, Portugal, Romania, Russia, Serbia, Slovakia, Spain, Switzerland, Ukraine, United Kingdom.

Host records: Summarized by Yu et al. (2016) as being a parasitoid of the erebid *Lymantria dispar* (L.); the geometrids *Alcis repandata* (L.), and *Phigalia titea* (Cramer); and the noctuids *Calophasia lunula* (Hufnagel), *Craniphora ligustri* (Denis and Schiffermüller), and *Cucullia lanceolata* (Villers). In Iran, this species has been reared from *Lymantria dispar* (L.) (Nikdel, 2015).

Glyptapanteles ripus (Papp, 1983)

Catalogs with Iranian records: No catalog.

Distribution in Iran: Mazandaran (Sakenin et al., 2020).

Distribution in the Middle East: Iran.

Extralimital distribution: Germany, Hungary, Korea, Lithuania, North Macedonia, Poland, Russia, Slovakia, Spain, former Yugoslavia.

Host records: Recorded by Papp (1983, 1996) as being a parasitoid of the noctuid *Cucullia xeranthemi* Boisduval, and the notodontid *Clostera anastomosis* (L.).

Glyptapanteles rubens (Reinhard, 1880)

Distribution in the Middle East: Israel—Palestine (Papp, 2012).

Extralimital distribution: Germany, Russia, Ukraine.

Host records: recorded by Telenga (1955) as being a parasitoid of the noctuid *Acronicta rumicis* (L.).

Glyptapanteles thompsoni (Lyle, 1927)

Catalogs with Iranian records: Fallahzadeh and Saghaei (2010), Gadallah et al. (2015), Farahani et al. (2016), Yu et al. (2016), Samin, Coronado-Blanco, Kavallieratos et al. (2018) and Samin, Coronado-Blanco, Fischer et al. (2018) (all as *Protapanteles thompsoni* (Lyle, 1927)).

Distribution in Iran: Iran (no specific locality cited) (Khanjani, 2006a as *Protapanteles thompsoni*).

Distribution in the Middle East: Iran.

Extralimital distribution: Belgium, Cameroon, China, France, Hungary, Japan, Korea, Moldova, Romania, Russia (Yu et al., 2016 as *Protapanteles thompsoni*; Fernandez-Triana et al., 2020).

Host records: Summarized by Yu et al. (2016) as being a parasitoid of the crambid *Ostrinia nubilalis* (Hübner); the erebid *Autoba admota* (Felder and Rogenhofer); and the gelechiids *Anacampsis timidella* (Wocke), and *Helcystogramma triannulella* (Herrich-Schäffer). In Iran, this species has been reared from *Ostrinia nubilalis* (Hübner) (Khanjani, 2006a).

Glyptapanteles vitripennis (Curtis, 1830)

Catalogs with Iranian records: Farahani et al. (2016), Yu et al. (2016), Samin, Coronado-Blanco, Kavallieratos et al. (2018) and Samin, Coronado-Blanco, Fischer et al. (2018).

Distribution in Iran: West Azarbaijan (Samin et al., 2014).

Distribution in the Middle East: Iran (Samin et al., 2014), Turkey (Doganlar, 1982).

Extralimital distribution: Azerbaijan, Belgium, Bulgaria, Czech Republic, Finland, France, Georgia, Germany, Greece, Hungary, India, Ireland, Italy, Kazakhstan, Kyrgyzstan, Latvia, Pakistan, Poland, Romania, Russia, Serbia, Slovakia, Spain, Sweden, Switzerland, Tajikistan, Ukraine, United Kingdom, Uzbekistan.

Host records: Recorded by Yu et al. (2016) as being a parasitoid of several lepidopteran species of the families Crambidae, Erebidae, Gelechiidae, Geometridae, Lasiocampidae, Lycaenidae, Noctuidea, Nymphalidae, Pieridae, Plutellidae, Pterophoridae, Tortricidae, and Yponomeutidae.

Genus *Keylimepie* Fernandez-Triana, 2016

Keylimepie hadhramautensis van Achterberg and Fernandez-Triana, 2017

Distribution in the Middle East: Yemen (Fernandez-Triana & van Achterberg, 2017).

Extralimital distribution: None.

Host records: Unknown.

Keylimepie sanaaensis van Achterberg and Fernandez-Triana, 2017

Distribution in the Middle East: Yemen (Fernandez-Triana & van Achterberg, 2017).

Extralimital distribution: None.

Host records: Unknown.

Genus *Protapanteles* Ashmead, 1898

Protapanteles albigena Abdoli, Fernandez-Triana and Talebi, 2021

Catalogs with Iranian records: No catalog.

Distribution in Iran: Guilan, Mazandaran (Abdoli et al., 2021).

Distribution in the Middle East: Iran.
Extralimital distribution: None.
Host records: Unknown.

Protapanteles anchisiades (Nixon, 1973)

Distribution in the Middle East: Turkey (Inanç & Beyarslan, 2001a).
Extralimital distribution: Bulgaria, Czech Republic, Finland, Germany, Hungary, Italy, Korae, Mongolia, Netherlands, Norway, Poland, Russia, Slovakia, Sweden, Switzerland, Ukraine, United Kingdom.
Host records: Summarized by Yu et al. (2016) as being a parasitoid of the geometrids *Alcis repandata* (L.), *Epirrita autumnata* (Borkhausen), *Epirrita dilutata* (Denis and Schiffermüller), *Eupithecia lariciata* (Freyer), *Odontopera bidentata* (Clerck), and *Operophtera brumata* (L.); and the pierid *Leptidea sinapis* (L.).

Protapanteles (Protapanteles) immunis (Haliday, 1834)

Catalogs with Iranian records: Gadallah et al. (2015), Farahani et al. (2016), Yu et al. (2016), Samin, Coronado-Blanco, Kavallieratos et al. (2018) and Samin, Coronado-Blanco, Fischer et al. (2018).
Distribution in Iran: West Azarbaijan (Ghahari, Çetin Erdoğan, Šedivý et al., 2010).
Distribution in the Middle East: Iran.
Extralimital distribution: Armenia, Austria, Bulgaria, Croatia, Estonia, Finland, Germany, Greenland, Hungary, Ireland, Italy, Kazakhstan, Korea, Lithuania, Moldova, Netherlands, Norway, Poland, Romania, Russia, Serbia, Slovakia, Sweden, Switzerland, Tunisia, Ukraine, United Kingdom.
Host records: Summarized by Yu et al. (2016) as being a parasitoid of the choreutid *Choreutis diana* (Hübner); the erebids *Hypena proboscidalis* (L.), and *Orgyia antiqua* (L.); the following geometrids: *Agriopis aurantiaria* (Hübner), *Agriopis leucophaearia* (Denis and Schiffermüller), *Asthena albulata* (Hufnagel), *Bupalus piniarius* (L.), *Cabera pusaria* (L.), *Campaea margaritaria* (L.), *Colotois pennaria* (L.), *Cyclophora albipunctata* (L.), *Dysstroma truncata* (Hufnagel), *Electrophaes corylata* (Thunberg), *Ematurga atomaria* (L.), *Ennomos quercinaria* (Hufnagel), *Epirrita autumnata* (Borkhausen), *Epirrita dilutata* (Denis and Schiffermüller), *Erannis defoliaria* (Clerck), *Eupithecia tantillaria* Boisduval, *Eupithecia vulgata* (Haworth), *Isturga arenacearia* (Denis and Schiffermüller), *Isturga limbaria* (Fabricius), *Lycia isabellae* (Harrison), *Operophtera brumata* (L.), and *Pseudoterpna pruinata* (Hufnagel); the lycaenid *Neozephyrus quercus* (L.); the noctuids *Autographa gamma* (L.), *Cosmia trapezina* (L.), and *Dichonia convergens* (Denis and Schiffermüller); the tortricids *Tortrix viridana*

(L.), and *Zeiraphera griseana* (Hübner); and the ypsolophid *Ypsolopha falcella* (Denis and Schiffermüller).
Comments: Farahani et al. (2016) considered *Saturnia pavonia* (L.) (Lepidoptera: Saturniidae) and *Orgyia antiqua* (L.) (Lepidoptera: Erebidae) as the hosts of *Protapanteles* (*Protapanteles*) *immunis* based on Ghahari et al. (2010c) by mistake. They were mentioned by Ghahari et al. (2010c) according to Thompson (1944), Rougeot (1971), Papp (1990) and Mason (1981).

Protapanteles incertus (Ruthe, 1859)

Catalogs with Iranian records: No catalog.
Distribution in Iran: Ardabil (Sakenin et al., 2020).
Distribution in the Middle East: Iran.
Extralimital distribution: Austria, Azerbaijan, Georgia, Germany, Hungary, Iceland, Italy, Mongolia, Poland, Romania, Russia, Slovakia, Sweden, Switzerland, Ukraine, United Kingdom, former Yugoslavia.
Host records: Summarized by Yu et al. (2016) as being a parasitoid of the gelechiid *Exoteleia dodecella* (L.); the following geometrids: *Agriopis marginaria* (Fabricius), *Alsophila aescularia* (Denis and Schiffermüller), *Anticlea derivata* (Denis and Schiffermüller), *Arichanna melanaria* (L.), *Biston betularia* L., *Bupalus piniarius* (L.), *Cabera exanthemata* (Scopoli), *Cabera pusaria* (L.), *Campaea margaritaria* (L.), *Electrophaes corylata* (Thunberg), *Ennomos fuscantaria* (Haworth), *Epirrita dilutata* (Denis and Schiffermüller), *Jodis lactearia* (L.), *Macaria liturata* (Clerck), and *Selenia dentaria* (Fabricius); the lycaenid *Thecla betulae* (L.); the noctuid *Cucullia argentea* (Hufnagel); the nolid *Bena prasinana* (L.); the nymphalid *Pararge aegeria* (L.); and the tortricid *Zeiraphera griseana* (Hübner).

Genus Sathon Mason, 1981

Sathon eugeni (Papp, 1972)

Distribution in the Middle East: Turkey (Beyarslan et al., 2006; Inanç & Beyarslan, 2001a).
Extralimital distribution: Austria, Bulgaria, Finland, Germany, Hungary, Italy, Latvia, Lithuania, Netherlands, Russia, Slovakia, Sweden, Switzerland, United Kingdom.
Host records: Recorded by Papp (1972) as being a parasitoid of the choreutid *Anthophila fabriciana* (L.).

Sathon falcatus (Nees von Esenbeck, 1834)

Catalogs with Iranian records: No catalog.
Distribution in Iran: Northern Khorasan (Ghahari et al., 2019).
Distribution in the Middle East: Iran (Ghahari et al., 2019), Egypt, Turkey (Fernandez-Triana et al., 2020).
Extralimital distribution: Oriental, Palaearctic regions.
Host records: Summarized by Yu et al. (2016) as being a parasitoid of the hepialid *Hepialis humuli* (L.); the noctuids

Actinotia polyodon (Clerck), *Apamea laeritia* (Hufnagel), and *Apamea monoglypha* (Hufnagel); the pterophorids *Adaina microdactyla* (Hübner), and *Emmelina monodactyla* (L.); the sesiid *Synanthedon tipuliformis* (Clerck); and the tortricids *Rhyacionia buoliana* (Denis and Schiffermüller), and *Zeiraphera griseana* (Hübner).

Sathon lateralis (Haliday, 1834)

Catalogs with Iranian records: This species is new record for the fauna of Iran.

Distribution in Iran: Alborz province, Taleghan 2♀, 1♂, August 2015.

Distribution in the Middle East: Iran (new record), Turkey (Beyarslan et al., 2002a; Inanç & Beyarslan, 2001c).

Extralimital distribution: Armenia, Azerbaijan, Belgium, Finland, France, Georgia, Germany, Ireland, Italy, Kazakhstan, Lithuania, Madeira Islands, Moldova, Netherlands, Romania, Russia, Serbia, Slovakia, Spain, Sweden, Switzerland, Ukraine, United Kingdom.

Host records: Summarized by Yu et al. (2016) as being a parasitoid of the choreutid *Anthophila fabriciana* (L.); the geometrid *Eupithecia assimilata* Doubleday; the gracillariids *Callisto denticulella* (Thunberg), and *Gracillaria syringella* (Fabricius); and the yponomeutid *Yponomeuta padella* (L.).

Genus Venanides Mason, 1981

Venanides caspius Abdoli, Fernández-Triana and Talebi, 2019

Catalogs with Iranian records: No catalog.

Distribution in Iran: Guilan (Abdoli et al., 2019d).

Distribution in the Middle East: Iran.

Extralimital distribution: None.

Host records: Unknown.

Venanides longifrons Fernandez-Triana and van Achterberg, 2017

Distribution in the Middle East: Yemen (Fernandez-Triana & van Achterberg, 2017).

Extralimital distribution: None.

Host records: Unknown.

Venanides supracompressus Fernandez-Triana and van Achterberg, 2017

Distribution in the Middle East: Yemen (Fernandez-Triana & van Achterberg, 2017).

Extralimital distribution: None.

Host records: Unknown.

Venanides tenuitergitus Fernandez-Triana and van Achterberg, 2017

Distribution in the Middle East: Yemen (Fernandez-Triana & van Achterberg, 2017).

Extralimital distribution: None.

Host records: Unknown.

Venanides vanharteni Fernandez-Triana and van Achterberg, 2017

Distribution in the Middle East: Yemen (Fernandez-Triana & van Achterberg, 2017).

Extralimital distribution: None.

Host records: Unknown.

Genus Wilkinsonellus Mason, 1981

Wilkinsonellus arabicus van Achterberg and Fernandez-Triana, 2017

Distribution in the Middle East: Yemen (Fernandez-Triana & van Achterberg, 2017).

Extralimital distribution: None.

Host records: Unknown.

Microplitis group (=Microplitini *sensu* Mason, 1981)

Genus Microplitis Foerster, 1863

Microplitis aduncus (Ruthe, 1860)

Catalogs with Iranian records: Fallahzadeh and Saghaei (2010), Gadallah et al. (2015), Farahani et al. (2016), Yu et al. (2016), Samin, Coronado-Blanco, Kavallieratos et al. (2018) and Samin, Coronado-Blanco, Fischer et al. (2018).

Distribution in Iran: Iran (no specific locality cited) (Papp, 1984 as *Microgaster adunca* Ruthe, 1860).

Distribution in the Middle East: Iran.

Extralimital distribution: Bulgaria, Finland, Georgia, Germany, Hungary, Korea, Mongolia, Netherlands, Poland, Russia, Selvagens Islands, Serbia, Sweden, Switzerland, Tunisia, Turkmenistan, United Kingdom.

Host records: Summarized by Yu et al. (2016) as being a parasitoid of the saturniid *Saturnia pavonia* (L.); and the tortricids *Eupoecilia ambiguella* (Hübner), and *Lobesia botrana* (L.).

Microplitis albipennis Abdinbekova, 1969

Catalogs with Iranian records: Samin, Coronado-Blanco, Fischer et al. (2018).

Distribution in Iran: Khuzestan (Samin, Coronado-Blanco, Fischer et al., 2018).

Distribution in the Middle East: Iran (Samin, Coronado-Blanco, Fischer et al., 2018), Turkey (Fernandez-Triana et al., 2020).

Extralimital distribution: Azerbaijan, Hungary, Mongolia, Poland, Russia.

Host records: Unknown.

Microplitis alborziensis Abdoli and Talebi, 2021

Catalogs with Iranian records: No catalog.

Distribution in Iran: Alborz (Abdoli et al., 2021a).

Distribution in the Middle East: Iran.
Extralimital distribution: None.
Host records: Unknown.

Microplitis beyarslani Inanç, 2002
Distribution in the Middle East: Turkey (Inanç, 2002).
Extralimital distribution: None.
Host records: Unknown.

Microplitis cebes Nixon, 1970
Catalogs with Iranian records: No catalog.
Distribution in Iran: Alborz (Abdoli et al., 2021a), Tehran (Abdoli et al., 2016; 2021a).
Distribution in the Middle East: Iran (Abdoli et al., 2016; 2021a), Turkey (Beyarslan, 1988; Beyarslan et al., 2002a, b; Inanç & Beyarslan, 1990, 1997, 2001a, b; Inanç, 1997).
Extralimital distribution: Austria, Croatia, Germany, Greece, Hungary, Mongolia, Poland, Serbia, Slovakia, Spain, Switzerland.
Host records: Unknown.

Microplitis decens Tobias, 1964
Distribution in the Middle East: Turkey (Beyarslan, 1988; Beyarslan et al., 2002a; Inanç, 1997; Inanç & Beyarslan, 1990; 1997; 2001a).
Extralimital distribution: Finland, Germany, Hungary, Italy, Kazakhstan, Korea, Mongolia, Montenegro, Netherlands, Russia, Serbia, Spain, Switzerland, United Kingdom.
Host records: Unknown.

Microplitis decipiens Prell, 1925
Catalogs with Iranian records: Gadallah et al. (2015), Farahani et al. (2016), Yu et al. (2016), Samin, Coronado-Blanco, Kavallieratos et al. (2018) and Samin, Coronado-Blanco, Fischer et al. (2018).
Distribution in Iran: East Azerbaijan (Gadallah et al., 2015).
Distribution in the Middle East: Iran (Gadallah et al., 2015), Turkey (Beyarslan et al., 2002a; Inanç, 1997; Inanç & Beyarslan, 1990; 2001a).
Extralimital distribution: Azerbaijan, Germany, Hungary, Kazakhstan, Lithuania, Moldova, Poland, Russia.
Host records: Recorded by Telenga (1955) as being a parasitoid of the noctuid *Panolis flammea* (Denis and Schiffermüller).

Microplitis deprimator (Fabricius, 1798)
Catalogs with Iranian records: Fallahzadeh and Saghaei (2010), Gadallah et al. (2015), Farahani et al. (2016), Yu et al. (2016), Samin, Coronado-Blanco, Kavallieratos et al. (2018) and Samin, Coronado-Blanco, Fischer et al. (2018).

Distribution in Iran: Iran (no specific locality cited) (Nixon, 1968 as *Microgaster deprimator*).
Distribution in the Middle East: Cyprus, Iran (Nixon, 1968), Turkey (Beyarslan and Nixon, 2002a; Inanç & Beyarslan, 2001a; Nixon, 1968).
Extralimital distribution: Armenia, Austria, Azerbaijan, Belgium, Bulgaria, Czech Republic, Finland, France, Georgia, Germany, Hungary, Ireland, Italy, Kazakhstan, Korea, Latvia, Moldova, Mongolia, Netherlands, Norway, Poland, Romania, Russia, Serbia, Spain, Switzerland, Ukraine, United Kingdom.
Host records: Summarized by Yu et al. (2016) as being a parasitoid of the erebid *Euproctis similis* (Füssli); the hesperiids *Carcharodus floccider* (Zeller), and *Ochlodes venatus* (Bremer and Gray); the noctuids *Acronicta* spp., *Autographa gamma* (L.), *Euchalcia variabilis* (Piller and Mitterpacher), *Cucullia verbasci* (L.), *Eupsilia transversa* (Hufnagel), *Shargacucullia scrophulariae* (Denis and Schiffermüller), and *Xylena exsoleta* (L.); the pyralids *Aphomia sociella* (L.), and *Galaria melonella* (L.); the sphingid *Smerinthus ocellatus* (L.); and the tortricids *Adoxophyes orana* (Fischer), and *Epinotia tetraquetrana* (Haworth).

Microplitis docilis Nixon, 1970
Catalogs with Iranian records: No catalog.
Distribution in Iran: Qazvin (Abdoli et al., 2019b; 2021a).
Distribution in the Middle East: Iran (Abdoli et al., 2019b; 2021a), Turkey (Beyarslan et al., 2002a, b; Inanç & Beyarslan, 2001a, b).
Extralimital distribution: Bulgaria, Croatia, Finland, Germany, Hungary, Russia, Serbia, Sweden.
Host records: Unknown.

Microplitis eremitus Reinhard, 1880
Catalogs with Iranian records: No catalog.
Distribution in Iran: Guilan, Mazandaran, Qazvin (Abdoli et al., 2021a), Lorestan (Naderian et al., 2020).
Distribution in the Middle East: Iran (Abdoli et al., 2021a; Naderian et al., 2020), Turkey (Kohl, 1905).
Extralimital distribution: Armenia, Austria, Azerbaijan, Croatia, Finland, France, Germany, Hungary, Kazakhstan, Korea, Lithuania, Mongolia, Netherlands, Poland, Russia, Serbia, Spain, Sweden, Switzerland, Ukraine, Uzbekistan.
Host records: Summarized by Yu et al. (2016) as being a parasitoid of the gelechiid *Athrips mouffetella* (L.); the noctuids *Acronicta psi* (L.), *Acronicta tridens* (Denis and Schiffermüller), *Autographa gamma* (L.), *Calliergis ramosa* (Esper), *Chloantha hyperici* (Denis and Schiffermüller), *Euxoa lidia* (Stoll), *Helicoverpa armigera* (Hübner), and *Mythimna unipuncta* (Haworth); the notodontid *Notodonta dromadarius* (L.); and the sphingid *Laothoe populi* (L.).

Microplitis erythrogaster Abdinbekova, 1967

Catalogs with Iranian records: No catalog.
Distribution in Iran: West Azarbaijan (Sakenin et al., 2020).
Distribution in the Middle East: Iran.
Extralimital distribution: Azerbaijan, Denmark, Germany, Hungary, Russia, Tajikistan, Turkmenistan.
Host records: Unknown.

Microplitis faifaicus Ghramh and Ahmad, 2020

Distribution in the Middle East: Saudi Arabia (Ghramh et al., 2020).
Extralimital distribution: None.
Host records: Unknown.

Microplitis flavipalpis (Brullé, 1832)

Catalogs with Iranian records: This species is new record for the fauna of Iran.
Distribution in Iran: Chaharmahal and Bakhtiari province, Lordegan, 1♂, June 2015.
Distribution in the Middle East: Iran (new record), Israel—Palestine (Papp, 2012), Turkey (Inanç & Beyarslan, 2001a).
Extralimital distribution: Algeria, Armenia, Bulgaria, Finland, France, Germany, Greece, Hungary, Kazakhstan, Korea, Lithuania, Moldova, Mongolia, Poland, Russia, Serbia, Slovakia, Spain, Switzerland, Tunisia, United Kingdom.
Host records: Summarized by Yu et al. (2016) as being a parasitoid of the noctuids *Amphipyra berbera* Rungs, *Anarta myrtilli* (L.), *Calophasia lunula* (Hufnagel), and *Lithophane lamda* (Fabricius).

Microplitis fordi Nixon, 1970

Catalogs with Iranian records: No catalog.
Distribution in Iran: East Azarbaijan (Samin, Coronado-Blanco, Hosseini et al., 2019).
Distribution in the Middle East: Iran (Samin, Coronado-Blanco, Hosseini et al., 2019), Israel—Palestine (Papp, 1984), Jordan (Papp, 1980), Turkey (Beyarslan et al., 2002a; Inanç & Beyarslan, 1997; 2001a).
Extralimital distribution: Austria, Bulgaria, Croatia, Czech Republic, Germany, Greece, Hungary, Italy, North Macedonia, Mongolia, Russia, Switzerland, Tunisia, United Kingdom, former Yugoslavia (Fernandez-Triana et al., 2020; Yu et al., 2016).
Host records: Summarized by Yu et al. (2016) as being a parasitoid of the erebid *Ocnogyna parasita* (Hübner); and the geometrids *Chesias legatella* (Denis and Schiffermüller), *Chesias rufata* (Fabricius), and *Thera juniperata* (L.).

Microplitis fulvicornis (Wesmael, 1837)

Catalogs with Iranian records: Gadallah et al. (2015), Farahani et al. (2016), Yu et al. (2016), Samin, Coronado-Blanco, Kavallieratos et al. (2018) and Samin, Coronado-Blanco, Fischer et al. (2018).
Distribution in Iran: Razavi Khorasan (Karimi-Malati et al., 2014).
Distribution in the Middle East: Iran (Karimi-Malati et al., 2014), Turkey (Inanç & Beyarslan, 2001a).
Extralimital distribution: Belgium, Croatia, Czech Republic, Finland, Germany, Hungary, Ireland, Netherlands, Poland, Romania, Russia, Serbia, Slovakia, Switzerland, United Kingdom.
Host records: In Iran, this species has been recorded by Karimi et al. (2014) as being a parasitoid of the noctuid *Spodoptera exigua* (Hübner).

Microplitis heterocerus (Ruthe, 1860)

Distribution in the Middle East: Israel—Palestine (Papp, 1970), Turkey (Inanç & Beyarslan, 2001a).
Extralimital distribution: Croatia, Germany, Hungary, Italy, Korea, Poland, Romania, Russia, Slovakia, Spain, Ukraine, former Yugoslvia.
Host records: Summarized by Yu et al. (2016) as being a parasitoid of the arctiid *Ocnogyna loewi* (Zeller); the erebid *Eublemma pannonica* Freyer; and the noctuids *Acronicta psi* (L.), *Dicycla oo* (L.), and *Dryobotodes roboris* (Boisduval).

Microplitis idia Nixon, 1970

Distribution in the Middle East: Israel—Palestine (Papp, 1984, 2012), Turkey (Inanç & Beyarslan, 2001a).
Extralimital distribution: Germany, Hungary, Russia, Sweden.
Host records: Recorded by Papp (2012) as being a parasitoid of the pierid *Euchloe belemia* (Esper).

Microplitis kaszabi Papp, 1980

Catalogs with Iranian records: No catalog.
Distribution in Iran: Alborz, Qazvin, Tehran (Abdoli et al., 2018; 2021a).
Distribution in the Middle East: Iran.
Extralimital distribution: Korea, Mongolia, Russia.
Host records: Unknown.

Microplitis khamisicus Ghramh and Ahmad, 2020

Distribution in the Middle East: Saudi Arabia (Ghramh et al., 2020).
Extralimital distribution: None.
Host records: Unknown.

Microplitis lugubris (Ruthe, 1860)

Catalogs with Iranian records: No catalog.
Distribution in Iran: Ardabil (Gadallah et al., 2018).
Distribution in the Middle East: Iran (Gadallah et al., 2018), Turkey (Doganlar, 1982).

Extralimital distribution: Armenia, Canada, Finland, Germany, Greenland, Hungary, Ireland, Lithuania, Mongolia, Poland, Russia, Serbia, Sweden, Switzerland, United Kingdom.

Host records: Summarized by Yu et al. (2016) as being a parasitoid of the erebids *Euclidia glyphica* (L.), and *Ocnogyna loewii* (Zeller).

Microplitis mandibularis (Thomson, 1895)

Catalogs with Iranian records: Samin, Coronado-Blanco, Fischer et al. (2018).

Distribution in Iran: Alborz (Abdoli et al., 2021a), Fars (Samin, Coronado-Blanco, Fischer et al., 2018).

Distribution in the Middle East: Iran (Abdoli et al., 2021a; Samin, Coronado-Blanco, Fischer et al., 2018), Turkey (Beyarslan, 1988; Beyarslan et al., 2002a; Beyarslan et al., 2006; Inanç, 1997; Inanç & Beyarslan, 1990; 2001a).

Extralimital distribution: Armenia, Azerbaijan, Croatia, Finland, Georgia, Germany, Greenland, Hungary, Macedonia, Mongolia, Netherlands, Russia, Serbia, Slovakia, Spain, Sweden, Switzerland, Tunisia, United Kingdom.

Host records: Summarized by Yu et al. (2016) as being a parasitoid of the following noctuids: *Conistra vaccinii* (L.), *Jodia croceago* (Denis and Schiffermüller), *Lasionycta proxima* (Hübner), *Lithophane ornitopus* (Hufnagel), *Noctua fimbriata* Schreber, and *Xanthia ocellaris* (Borkhausen).

Microplitis marshallii Kokujev, 1898

Catalogs with Iranian records: Gadallah et al. (2015), Farahani et al. (2016), Yu et al. (2016), Samin, Coronado-Blanco, Kavallieratos et al. (2018) and Samin, Coronado-Blanco, Fischer et al. (2018).

Distribution in Iran: Ardabil (Gadallah et al., 2015).

Distribution in the Middle East: Iran (Gadallah et al., 2015), Turkey (Beyarslan et al., 2002a, b; Beyarslan et al., 2006; Inanç & Beyarslan, 2001a, b; Papp, 1984).

Extralimital distribution: Armenia, Azerbaijan, China, Finland, Georgia, Hungary, Moldova, Romania, Russia, Spain.

Host records: Unknown.

Microplitis mediator (Haliday, 1834)

Catalogs with Iranian records: No catalog.

Distribution in Iran: East Azerbaijan (Samin, Fischer, Sakenin et al., 2019), Iran (no specific locality cited) (Belokobylskij et al., 2019).

Distribution in the Middle East: Iran (Belokobylskij et al., 2019; Samin, Fischer, Sakenin et al., 2019), Turkey (Beyarslan, 1988; Beyarslan et al., 2002a; Inanç & Beyarslan, 1997; 2001a).

Extralimital distribution: Palaearctic, Nearctic, Neotropical, Oriental regions.

Host records: Summarized by Yu et al. (2016) as being a parasitoid of the geometrids *Epirrhoe galiata* (Denis and Schiffermüller), and *Eupithecia icterata* (de Villers); the following noctuids: *Agrotis* spp., *Apamea unanimis* (Hübner), *Arcte coerula* (Guenée), *Autographa gamma* (L.), *Conistra vaccinii* (L.), *Cucullia verbasci* (L.), *Helicoverpa* spp., *Heliothis peltigera* (Denis and Schiffermüller), *Jodia croceago* (Denis and Schiffermüller), *Lacanobia oleracea* (L.), *Lithophane furcifera* (Hufnagel), *Mamestra* spp., *Mythimna straminea* (Treitschke), *Naenia typica* (L.), *Noctua* spp., *Orthosia* spp., *Polymixis flavicincta* (Denis and Schiffermüller), *Rileyiana fovea* (Treitschke), *Sarcopolia illoba* (Butler), *Schinia scutosa* (Denis and Schiffermüller), and *Xestia* spp.; the pierid *Aporia crataegi* (L.); and the plutellid *Plutella xylostella* (L.).

Microplitis mongolicus Papp, 1967

Distribution in the Middle East: Israel—Palestine, Jordan (Papp, 1980).

Extralimital distribution: Hungary, Mongolia, Russia.

Host records: Unknown.

Microplitis naenia Nixon, 1970

Distribution in the Middle East: Turkey (Inanç & Beyarslan, 2001a).

Extralimital distribution: Hungary, Russia, Slovakia, United Kingdom.

Host records: Summarized by Yu et al. (2016) as being a parasitoid of the erebid *Conistra vaccinii* (L.); and the noctuids *Cosmia trapezina* (L.), *Eupsilia transversa* (Hufnagel), *Orthosia cruda* (Denis and Schiffermüller), *Orthosia stabilis* (Fabricius), and *Rileyiana fovea* (Treitschke).

Microplitis ochraceus Szépligeti, 1896

Catalogs with Iranian records: Fallahzadeh and Saghaei (2010), Gadallah et al. (2015), Farahani et al. (2016), Yu et al. (2016), Samin, Beyarslan, Coronado-Blanco et al. (2020), Samin, Sakenin Chelav, Ahmad, et al., 2020.

Distribution in Iran: Alborz (Farahani et al., 2014), Ardabil (Telenga, 1955), Golestan (Ghahari et al., 2011a), Tehran (Abdoli et al., 2021a), Iran (no specific locality cited) (Papp, 1984; Tobias, 1976, 1986).

Distribution in the Middle East: Iran.

Extralimital distribution: Azerbaijan, Greece, Hungary, Kazakhstan, Moldova, Mongolia, Romania, Russia, Ukraine, Uzbekistan.

Host records: Recorded by Györfi (1959) as being a parasitoid of the noctuid *Orthosia populeti* (Fabricius).

Microplitis pallidipennis Tobias, 1964

Catalogs with Iranian records: No catalog.
Distribution in Iran: Alborz (Abdoli et al., 2021a).
Distribution in the Middle East: Iran.
Extralimital distribution: Kazakhstan, Mongolia, Russia.
Host records: Unknown.

Microplitis plutellae Muesebeck, 1922

Distribution in the Middle East: Egypt (Hassanein, 1958).
Extralimital distribution: Canada, China, Russia.
Host records: Recorded by Muesebeck (1922) as being a parasitoid of the plutellid *Plutella xylostella* (L.).

Microplitis pseudomurinus Abdinbekova, 1969

Distribution in the Middle East: Turkey (Inanç & Beyarslan, 2001c).
Extralimital distribution: Azerbaijan, Bulgaria, Georgia, Greece, Hungary, Kazakhstan Moldova, Russia.
Host records: Recorded by Tobias (1986) as being a parasitoid of the noctuid *Pyrrhia umbra* (Hufnagel).

Microplitis ratzeburgii (Ruthe, 1858)

Catalogs with Iranian records: No catalog.
Distribution in Iran: Zanjan (Samin, Beyarslan, Coronado-Blanco et al., 2020).
Distribution in the Middle East: Iran (Samin, Beyarslan, Coronado-Blanco et al., 2020), Israel–Palestine (Halperin, 1986).
Extralimital distribution: Armenia, Bulgaria, Denmark, Finland, France, Germany, Italy, Japan, Poland, Russia, Serbia, Ukraine (Fernandez-Triana et al., 2020; Yu et al., 2016).
Host records: Summarized by Yu et al. (2016) as being a parasitoid of the erebid *Euproctis similis* (Füssli); the noctuid *Acronicta auricoma* (Denis and Schiffermüller); and the notodontids *Cerura menciana* Moore, *Cerura vinula* (L.), and *Furcula palaestinensis* Bartel-Gaede. In Iran, this species has been reared from the erebid *Euproctis similis* (Samin, Beyarslan, Coronado-Blanco et al., 2020).

Microplitis rufiventris Kokujev, 1914

Catalogs with Iranian records: Gadallah et al. (2015), Farahani et al. (2016), Yu et al. (2016), Samin, Coronado-Blanco, Kavallieratos et al. (2018) and Samin, Coronado-Blanco, Fischer et al. (2018).
Distribution in Iran: Alborz (Farahani et al., 2014; Abdoli et al., 2021a), Qazvin (Farahani et al., 2016), Tehran (Farahani et al., 2014).
Distribution in the Middle East: Cyprus (Ingram, 1981), Egypt (Hegazi et al., 1973; 1977; Tawfik et al., 1977; Abou-Zaid et al., 1978; Kolaib et al., 1980; Hassanein & El-Heneidy, 1984, 1991; Papp, 1984; Moawad et al., 1987; Mahmoud et al., 2009), Iran (see references above), Israel–Palestine (Gerling, 1969, 1971; Halperin, 1986;

Papp, 1970; 1984; 2012; Schwartz et al., 1980), Jordan (Papp, 1984), Turkey (Papp, 1984; Sertkaya et al., 2004).
Extralimital distribution: Afghanistan, China, Romania, Turkmenistan, Uzbekistan.
Host records: Summarized by Yu et al. (2016) as being a parasitoid of the noctuids *Helicoverpa armigera* (Hübner), *Spodoptera cilium* Guenée, *Spodoptera exigua* (Hübner), and *Spodoptera littoralis* (Boisduval). In Iran, this species has been reared from *Spodoptera exigua* (Hübner) (Farahani et al., 2016).

Microplitis scrophulariae Szépligeti, 1898

Catalogs with Iranian records: Fallahzadeh and Saghaei (2010), Gadallah et al. (2015), Farahani et al. (2016), Yu et al. (2016), Samin, Coronado-Blanco, Kavallieratos et al. (2018) and Samin, Coronado-Blanco, Fischer et al. (2018).
Distribution in Iran: Isfahan (Ghahari et al., 2011c), Iran (no specific locality cited) (Tobias, 1976, 1986).
Distribution in the Middle East: Iran (see references above), Turkey (Beyarslan et al., 2002a; Inanç & Beyarslan, 2001a).
Extralimital distribution: Armenia, Azerbaijan, Bulgaria, Croatia, Czech Republic, France, Georgia, Greece, Hungary, Kazakhstan, Korea, Mongolia, Romania, Russia, Serbia, Slovakia, Sweden, United Kingdom.
Host records: Summarized by Yu et al. (2016) as being a parasitoid of the noctuids *Autographa gamma* (L.), *Cucullia absinthii* (L.), *Cucullia achilleae* Guenée, *Cucullia chamomillae* (Denis and Schiffermüller), *Cucullia blattariae* (Esper), *Cucullia lanceolata* (Villers), *Cucullia verbasci* (L.), *Shargacucullia scrophulariae* (Denis and Schiffermüller), and *Xylena exsoleta* (L.).

Microplitis sofron Nixon, 1970

Catalogs with Iranian records: No catalog.
Distribution in Iran: Qazvin (Samin, Coronado-Blanco, Hosseini et al., 2019).
Distribution in the Middle East: Iran (Samin, Coronado-Blanco, Hosseini et al., 2019), Turkey (Beyarslan, 1988; Inanç & Beyarslan, 2001a, c; Papp, 1984).
Extralimital distribution: Armenia, Azerbaijan, Bulgaria, Denmark, Finland, France, Germany, Greece, Greenland, Hungary, Ireland, Italy, Kazakhstan, Netherlands, Norway, Russia, Serbia, Spain, Sweden, Switzerland, Turkmenistan, United Kingdom.
Host records: Recorded by Nixon (1970) as being a parasitoid of the noctuid *Tholera cespitis* (Denis and Schiffermüller).

Microplitis sordipes (Ziegler, 1834)

Distribution in the Middle East: Turkey (Beyarslan, 1988).
Extralimital distribution: Albania, Armenia, Austria, Azerbaijan, Belgium, Czech Republic, Finland, France, Georgia, Germany, Hungary, Italy, Kazakhstan, Lithuania,

Moldova, Mongolia, Poland, Romania, Russia, Slovakia, Sweden, Switzerland, Turkmenistan, Ukraine, United Kingdom, Uzbekistan, former Yugoslavia.
Host records: Summarized by Yu et al. (2016) as being a parasitoid of the tortricid *Archips rosana* (L.).

Microplitis spectabilis (Haliday, 1834)
Catalogs with Iranian records: Gadallah et al. (2015), Farahani et al. (2016), Yu et al. (2016), Samin, Coronado-Blanco, Kavallieratos et al. (2018) and Samin, Coronado-Blanco, Fischer et al. (2018).
Distribution in Iran: Alborz, Guilan, Mazandaran, Qazvin (Abdoli et al., 2021a), East Azarbaijan (Rastegar et al., 2012), Isfahan (Ghahari et al., 2011c).
Distribution in the Middle East: Iran (see references above), Israel–Palestine (Papp, 2012), Turkey (Beyarslan, 1988; Beyarslan et al., 2002a; Inanç & Beyarslan, 1990).
Extralimital distribution: Armenia, Austria, Azerbaijan, Belgium, Bulgaria, Croatia, Finland, France, Germany, Greece, Hungary, Ireland, Italy, Kazakhstan, Latvia, Lithuania, Madeira Islands, Malta, Moldova, Mongolia, Morocco, Pakistan, Poland, Romania, Russia, Slovakia, Sweden, Switzerland, Tajikistan, Tunisia, Turkmenistan, Ukraine, United Kingdom, Uzbekistan.
Host records: Summarized by Yu et al. (2016) as being a parasitoid of the geometrids *Eulithis testata* (L.), and *Eupithecia succenturiata* (L.); the following noctuids: *Agrotis* spp., *Apamea unanimis* (Hübner), *Charanyca trigrammica* (Hufnagel), *Cosmia affinis* (L.), *Eupsilia transversa* (Hufnagel), *Hadena bicruris* Hufnagel, *Mamestra brassicae* (L.), *Polychrisia moneta* (Fabricius), *Spodoptera exigua* (Hübner), *Tiliacea citrago* (L.), and *Xylena exsoleta* (L.); the nymphalid *Nymphalis polychloras* (L.); and the yponomeutid *Yponomeuta evonymella* (L.).

Microplitis spinolae (Nees von Esenbeck, 1834)
Catalogs with Iranian records: Gadallah et al. (2015), Farahani et al. (2016), Yu et al. (2016), Samin, Coronado-Blanco, Kavallieratos et al. (2018) and Samin, Coronado-Blanco, Fischer et al. (2018).
Distribution in Iran: Alborz, Guilan, Mazandaran, Tehran (Abdoli et al., 2021a), Iran (no specific locality cited) (Nixon, 1970; Shenefelt, 1972; Inanç & Beyarslan, 2001d).
Distribution in the Middle East: Iran (see references above), Turkey (Beyarslan et al., 2002a, b; Inanç & Beyarslan, 1990; 2001a, b).
Extralimital distribution: Armenia, Austria, Azerbaijan, Belgium, Bosnia-Herzegovina, Bulgaria, Croatia, Czech Republic, Finland, France, Georgia, Germany, Greece, Hungary, Ireland, Italy, Japan, Kazakhstan, Korea, Kyrgyzstan, Lithuania, North Macedonia, Moldova, Netherlands, Poland, Romania, Russia, Serbia, Slovakia, Sweden, Switzerland, Tajikistan, Ukraine, United Kingdom, Uzbekistan.

Host records: Summarized by Yu et al. (2016) as being a parasitoid of the geometrids *Ennomos quercinaria* (Hufnagel), and *Isturgia arenacearia* (Denis and Schiffermüller); and the following noctuids: *Abrostola triplasia* (L.), *Acronicta alni* (L.), *Allophyes oxyacanthae* (L.), *Autographa gamma* (L.), *Calophasia lunula* (Hufnagel), *Heliothis peltigera* (Denis and Schiffermüller), *Heliothis viriplaca* (Hufnagel), *Lacanobia oleracea* (L.), *Orthosia gothica* (L.), *Cornutiplusia circumflexa* (L.), and *Tholera cespitis* (Denis and Schiffermüller).

Microplitis stigmaticus (Ratzeburg, 1844)
Catalogs with Iranian records: No catalog.
Distribution in Iran: Ardabil (Samin, Beyarslan, Ranjith et al., 2020).
Distribution in the Middle East: Iran.
Extralimital distribution: Armenia, Azerbaijan, Finland, Germany, Italy Kazakhstan, Latvia, Poland, Romania, Russia, Serbia, Turkmenistan, Ukraine, Uzbekistan.
Host records: Yu et al. (2016) as being a parasitoid of the erebid *Arctia caja* (L.); the noctuids *Autographa gamma* (L.), and *Cosmia trapezina* (L.); and the yponomutids *Yponomeuta cagnagella* (Hübner), and *Yponomeuta padella* (L.). In Iran, this species has been reared from the *Yponomeuta padella* (Samin, Beyarslan, Ranjith et al., 2020).

Microplitis strenuus Reinhard, 1880
Catlogues with Iranian records: No catalog.
Distribution in Iran: Chaharmahal and Bakhtiari (Samin, Sakenin Chelav, Ahmad, et al., 2020).
Distribution in the Middle East: Iran (Samin, Sakenin Chelav, Ahmad, et al., 2020), Turkey (Inanç & Beyarslan, 2001a; Nixon, 1970).
Extralimital distribution: Afghanistan, Armenia, Azerbaijan, China, Croatia, Czech Republic, Germany, Hungary, Kazakhstan, Moldova, Mongolia, Netherlands, Poland, Russia, Serbia, Sweden, Switzerland, Ukraine, United Kingdom, Uzbekistan.
Host records: Summarized by Yu et al. (2016) as being a parasitoid of the following noctuids: *Acronicta psi* (L.), *A. tridens* (Denis and Schiffermüller), *Agrotis segetum* (Denis and Schiffermüller), *Allophyes oxyacanthae* (L.), and *Diloba caeruleocephala* (L.).

Microplitis tadzhicus Telenga, 1949
Catalogs with Iranian records: No catalog.
Distribution in Iran: Kerman, Northern Khorasan (Gadallah et al., 2021).
Distribution in the Middle East: Iran.
Extralimital distribution: Afghanistan, Azerbaijan, China, France, Hungary, Kazakhstan, Korea, Russia, Tajikistan, Turkmenistan, Uzbekistan.
Host records: Unknown.

Microplitis tihamicus Ghramh and Ahmad, 2020
Distribution in the Middle East: Saudi Arabia (Ghramh et al., 2020).
Extralimital distribution: None.
Host records: Unknown.

Microplitis tristis (Nees von Esenbeck, 1834)
Distribution in the Middle East: Turkey (Inanç & Beyarslan, 1990; 2001a).
Extralimital distribution: Azerbaijan, Belgium, Croatia, France, Germany, Hungary, Kazakhstan, Kyrgyzstan, Lithuania, Moldova, Mongolia, Netherlands, Poland, Romania, Russia, Slovakia, Switzerland, Ukraine, United Kingdom.
Host records: Summarized by Yu et al. (2016) as being a parasitoid of the erebids *Calophasia opalina* (Esper), *Cucullia argentea* (Hufnagel), *Cucullia artemisiae* (Hufnagel), and *Cucullia verbasci* L.; the following noctuids: *Cerapteryx graminis* (L.), *Euchalcis consona* Fabricius, *Euchalcia modestoides* Poope, *Euchalcia variabilis* (Piller and Mitterpacher), *Hadena confusa* Hufnagel, *Plusia modesta* Walker, *Polychysia moneta* (Fabricius), and *Sideridis rivularis* Fabricius; and the nymphalid *Aglais urticae* (L.).

Microplitis tuberculatus (Bouché, 1834)
Catalogs with Iranian records: No catalog.
Distribution in Iran: Ardabil (Samin, Beyarslan, Ranjith et al., 2020), Hamadan (Samin, Sakenin Chelav, Ahmad, et al., 2020).
Distribution in the Middle East: Iran (Samin, Sakenin Chelav, Ahmad, et al., 2020, Samin, Beyarslan, Ranjith et al., 2020), Israel−Palestine (Papp, 2012).
Extralimital distribution: Azerbaijan, Finland, Georgia, Germany, Hungary, Ireland, Italy, Moldova, Mongolia, Poland, Romania, Russia, Slovakia, Sweden, Switzerland, Ukraine, United Kingdom.
Host records: Summarized by Yu et al. (2016) as being a parasitoid of the erebid *Euproctis chrysorrhoea* (L.); the noctuids *Acronicta auricoma* (Denis and Schiffermüller), *Acronicta euphorbiae* (Denis and Schiffermüller), *Acronicta menynthidis* (Esper), *Acronicta psi* (L.), *Acronicta rumicis* (L.), *Eupisilia transversa* (Hufnagel), and *Orthosia miniosa* (Denis and Schiffermüller); the notodontids *Cerura bifida* (Brahm), and *Cerura vinula* (L.); and the tortricid *Archips rosana* (L.). In Iran, this species has been reared from the erebid *Euproctis chrysorrhoea* (Samin, Beyarslan, Ranjith et al., 2020).

Microplitis tuberculifer (Wesmael, 1837)
Catalogs with Iranian records: Gadallah et al. (2015), Farahani et al. (2016), Yu et al. (2016), Samin, Coronado-Blanco, Kavallieratos et al. (2018) and Samin, Coronado-Blanco, Fischer et al. (2018).

Distribution in Iran: Alborz, Guilan, Mazandaran, Tehran (Abdoli et al., 2021a), Ilam (Ghahari et al., 2011d), Razavi Khorasan (Sakenin et al., 2021—around cotton fields).
Distribution in the Middle East: Iran (see references above), Israel−Palestine (Papp, 2012), Turkey (Beyarslan, 1988; Beyarslan et al., 2002a; Inanç & Beyarslan, 1990, 1997, 2001a, b; Inanç, 1997).
Extralimital distribution: Armenia, Austria, Azerbaijan, Belarus, Belgium, Bulgaria, China, Croatia, Czech Republic, Estonia, Finland, France, Georgia, Germany, Greece, Hungary, India, Ireland, Italy, Japan, Kazakhstan, Korea, Kyrgyzstan, Latvia, Lithuania, Moldova, Mongolia, Morocco, Netherlands, Poland, Romania, Russia, Serbia, Slovakia, Spain, Sweden, Switzerland, Ukraine, United Kingdom, Uzbekistan, Vietnam.
Host records: Summarized by Yu et al. (2016) as being a parasitoid of the coleophorid *Coleophora follicularis* (Vallot); the erebid *Lesmone irregularis* (Hübner); the geometrids *Eulithis testata* (L.), *Eupithecia* spp., and *Ligdia adustata* (Denis and Schiffermüller); the lasiocampid *Dendrolimus pini* (L.); the following noctuids: *Coranarta cordigera* (Thunberg), *Apamea sordens* (Hufnagel), *Autographa gamma* (L.), *Autographa nigrisigna* (Walker), *Diachrysia chrysitis* (L.), *Griposia aprilina* (L.), *Hadena irregularis* Hufnagel, *Lacanobia oleracea* (L.), *Lithomoia solidaginis* (Hübner), *Mamestra brassicae* (L.), *Noctua fimbriata* Schreber, *Orthosia* spp., *Phlogophora meticulosa* (L.), *Polymixis flavicincta* (Denis and Schiffermüller), *Spodoptera exigua* (Hübner), *Spodoptera litura* (Fabricius), *Stilbia anomala* (Haworth), and *Xestia xanthographa* (Denis and Schiffermüller); the sphingid *Laothoe populi* (L.); the tortricid *Eupoecilia ambiguella* (Hübner); and the ypsolophid *Ypsilopha nemorella* (L.).

Microplitis varipes (Ruthe, 1860)
Catalogs with Iranian records: No catalog.
Distribution in Iran: Alborz, Guilan, Qazvin, Tehran (Abdoli et al., 2021a), East Azarbaijan (Samin, Beyarslan, Ranjith et al., 2020), Hamadan (Samin, Sakenin Chelav, Ahmad, et al., 2020).
Distribution in the Middle East: Iran (Abdoli et al., 2021a; Samin, Sakenin Chelav, Ahmad, et al., 2020, c), Turkey (Beyarslan, 1988; Beyarslan et al., 2002a; Beyarslan et al., 2006; Inanç & Beyarslan, 1990; 2001a; Inanç, 1997).
Extralimital distribution: Austria, Azerbaijan, China, Finland, Georgia, Germany, Hungary, Italy, Kazakhstan, Malta, Moldova, Mongolia, Montenegro, Netherlands, Poland, Russia, Serbia, Slovakia, Switzerland, Ukraine.
Host records: Summarized by Yu et al. (2016) as being a parasitoid of the crambid *Loxostege sticticalis* (L.); and the noctuids *Cucullia scopariae* (Dorfmeister), *Noctua serena* (Hufnagel), and *Omphalophana antirrhinii* (Hübner). In Iran, this species has been reared from *Noctua pronuba* (Samin, Beyarslan, Ranjith et al., 2020).

Microplitis viduus (Ruthe, 1860)

Catalogs with Iranian records: Gadallah et al. (2015), Farahani et al. (2016), Yu et al. (2016), Samin, Coronado-Blanco, Kavallieratos et al. (2018) and Samin, Coronado-Blanco, Fischer et al. (2018).

Distribution in Iran: Qazvin (Ghahari et al., 2011b as *Microplitis vidua* (Ruthe, 1860)).

Distribution in the Middle East: Cyprus (Ingram, 1981; Nixon, 1970), Iran (Ghahari et al., 2011b), Israel—Palestine (Nixon, 1970), Turkey (İnanç & Beyarslan, 2001a, c).

Extralimital distribution: Armenia, Azerbaijan, Croatia, Czech Republic, Finland, Germany, Greece, Hungary, Italy, Kazakhstan, North Macedonia, Moldova, Mongolia, Netherlands, Poland, Romania, Russia, Serbia, Switzerland, Ukraine, United Kingdom, Uzbekistan.

Host records: Summarized by Yu et al. (2016) as being a parasitoid of erebids *Callistege mi* (Clerck), *Euplagia quadripunctaria* (Poda), and *Gramma ornata* (Packard); the geometrids *Chesias rufata* (Fabricius), *Macaria liturata* (Clerck), and *Thera variata* (Denis and Schiffermüller); the noctuids *Acronicta psi* (L.), *Acronicta tridens* (Denis and Schiffermüller), *Apamea sordens* (Hufnagel), *Autographa gamma* (L.), *Cucullia gnaphalii* (Hübner), *Cucullia verbasci* (L.), *Hecatera bicolorata* (Hufnagel), *Helicoverpa armigera* (Hübner), and *Orthosia incerta* (Hufnagel); the pyralid *Dioryctria abietella* (Denis and Schiffermüller); and the yponomeutid *Yponomeuta evonymella* (L.).

Microplitis xanthopus (Ruthe, 1860)

Catalogs with Iranian records: No catalog.

Distribution in Iran: Iran (No specific locality cited) (Belokobylskij et al., 2019).

Distribution in the Middle East: Iran.

Extralimital distribution: Belarus, Bulgaria, Croatia, Czech Republic, Finland, Georgia, Germany, Hungary, Ireland, Italy, Kazakhstan, Moldova, Poland, Romania, Russia, Serbia, Sweden, Switzerland, Ukraine, United Kingdom.

Host records: Summarized by Yu et al. (2016) as being a parasitoid of the crambid *Ostrinia nubilalis* (Hübner); the erebids *Euproctis chrysorrhoea* (L.), and *Euproctis similis* (Füssli); the gelechiid *Aristotelia brizella* (Treitschke); the noctuids *Acronicta aceris* (L.), *Autographa gamma* (L.), *Heliothis peltigera* (Denis and Schiffermüller), and *Orthosia populeti* (Fabricius); and the notodontid *Notodonta ziczac* (L.).

Microplitis sp.

Distribution in Iran: Khuzestan (Modarres Awal, 1997).

Host records: In Iran, reared from the noctuid *Leucania loreyi* (Duponchel) (Modarres Awal, 1997 as *Mythimna loreyi*).

Unplaced genera

Genus *Beyarslania* Koçak and Kemal, 2009

Beyarslania insolens (Wilkinson, 1930)

Distribution in the Middle East: Yemen (Fernandez-Triana et al., 2020).

Extralimital distribution: Rwanda, South Africa.

Host records: Unknown.

Genus *Miropotes* Nixon, 1965

Miropotes inexpectatus van Achterberg and Fernandez-Triana, 2017

Distribution in the Middle East: Yemen (Fernandez-Triana & van Achterberg, 2017).

Extralimital distribution: None.

Host records: Unknown.

Species excluded from the fauna of Iran

Dolichogenidea aethiopicus (Wilkinson, 1931)

Catalogs with Iranian records: Gadallah et al. (2015), Farahani et al. (2016), Yu et al. (2016), Samin, Coronado-Blanco, Kavallieratos et al. (2018) and Samin, Coronado-Blanco, Fischer et al. (2018) [all as *Apanteles aethiopicus* Wilkinson, 1931].

Distribution in Iran: Guilan (Ghahari, Çetin Erdoğan, Šedivý et al., 2010).

General distribution: Cameroon, Democratic Republic of Congo, Ethiopia, Iran, Ivory Coast, Kenya, Rwanda, Senegal, Sierra Leone, Somalia, South Africa, Sudan, Tanzania, Uganda (Yu et al., 2016).

Comments: Reexamination of specimens proved Ghahari et al. (2010c) identification of *Dolichogenidea aethiopicus* was a misidentification of *Dolichogenidea lineipes* and we therefore exclude *D. aethiopicus* from the fauna of Iran and record those data under *D. lineipes*.

Conclusion

Two hundred and ninety-two valid species of the subfamily Microgastrinae in 23 genera have been reported from most of the Middle East countries, with the exception of Bahrain, Kuwait, and Qatar. Among them, Iran is the richest, with 208 species (7% of the world species). The most species rich genera in the Microgastrinae of the Middle East are *Cotesia* with 66 species, followed by *Dolichogenidea* (56 species), *Microplitis* (46 species), *Apanteles* (20 species), *Glyptapanteles* (16 species), *Microgaster* (17 species), *Iconella* (13 species), *Choeras* (11 species), *Pholetesor* (eight species), *illidops* (seven species), *Diolcogaster* (six species), *Venanides* (five species), *Protapanteles* (four species), *Deuterixys*, and *Sathon* (each with three species), *Distatrix*, *Hygroplitis*, and *Keylimepie* (each with two

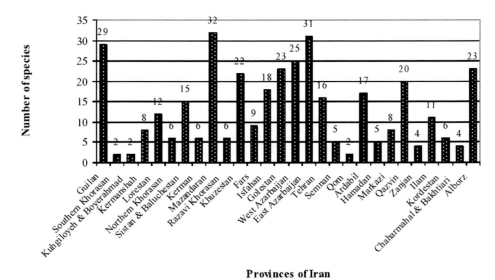

FIGURE 21.1   Number of reported species of Iranian Microgastrinae by province.

species), and finally, *Beyarslania, Miropotes, Napamus, Parapanteles*, and *Wilkinsonellus* (each with one species). Thirty-two microgastrine species are known to be endemic to the Middle East fauna, from which 15 species are so far only known from Iran (being endemic to Iran): *Choeras formosus* Abdoli and Fernandez-Triana, 2019, *Choeras fulviventris* Fernandez-Triana and Abdoli, 2019, *Choeras qazviniensis* Fernandez-Triana and Talebi, 2019, *Choeras taftanensis* Ghafouri Moghaddam and van Achtereberg, 2018, *Cotesia elongata* Zargar and Gupta, 2019, *Cotesia khuzestanensis* Zargar and Gupta, 2019, *Cotesia zagrosensis* Zargar and Gupta, 2019, *Deuterixys tenuiconvergens* Zargar and Gupta, 2019, *Dolichogenidea fernandeztrianai* Abdoli and Talebi, 2019, *Iconella brachyradiata* Abdoli and Talebi, 2021, *Iconella mongashtensis* Zargar and Gupta, 2019, *Iconella similus* Zargar and Gupta, 2019, *Microplitis alborziensis* Abdoli and Talebi, 2021, *Pholetesor pseudocircumscriptus* Abdoli, 2019, *Venanides caspius* Abdoli, Fernandez-Triana and Talebi, 2019, and *Protapanteles albigena* Abdoli, Fernandez-Triana and Talebi, 2021. Distributions of the Iranian Microgastrinae have been recorded from 28 provinces; among them, Mazandaran and East Azarbaijan with 32 and 31 species, respectively, have the highest number of species (Fig. 21.1). Host species have been reported for 53 Iranian Microgastrinae species, of which three of the host records in Aphididae (two species) and Coccidae (one species) (both Hemiptera), have erroneously been reported. The valid hosts belong to 15 families of Lepidoptera (Coleophoridae: one species; Crambidae: two species; Erebidae: three species; Gelechiidae: two species; Gracillariidae: four species; Lasiocampidae: two species,

Lycaenidae: one species; Lyonetiidae: one species; Noctuidae: 13 species; Nymphalidae: one species; Pieridae: three species; Plutellidae: three species; Pterolonchidae: one species; Pyralidae: five species; Saturniidae: two species; Tortricidae: three species; Yponomeutidae: two species). Comparison of the Microgastrinae fauna of Iran with Middle East and adjacent countries indicates the fauna of Russia with 367 species in 20 genera (Belokobylskij & Lelej, 2019) is more diverse than Iran, followed by Turkey (with 173 species), Azerbaijan, and Kazakhstan (both with 112 species), Armenia (93 species), Israel–Palestine (with 72 species), Turkmenistan (54 species), Pakistan (19 species), Afghanistan (18 species), Yemen (17 species), Egypt (12 species), Cyprus and Jordan (both with 11 species), Saudi Arabia (five species), United Arab Emirates (three species), Iraq, Lebanon, and Syria (each with two species), then Oman (with one species) (Yu et al., 2016). Among the Middle East and adjacent countries to Iran, Russia shares the most number of species with Iran (165 species), followed by Turkey (118 species), Azerbaijan (89 species), Armenia (77 species), Kazakhstan (72 species), Israel–Palestine (59 species), Turkmenistan (34 species), Afghanistan (15 species), Cyprus and Egypt (both with 10 species), Jordan (eight species), Pakistan (seven species), Yemen (three species), Lebanon (both with two species), Oman, Saudi Arabia, Syria, and United Arab Emirates (each with one species).

References

Abbas, M. S. T. (1989). Studies on *Apanteles litae* var. *operculella*, a parasite of the diamond-back moth, *Plutella xylostella* in Egypt. *Mitteilungen aus dem Zoologischen Museum in Berlin, 65*(1), 157–160.

Abbasipour, H., & Taghavi, A. (2002). Parasitoids of rice armyworm, *Mythimna unipuncta* (Lepidoptera: Noctuidae) in the rice fields of western Mazandaran. In *Proceedings of the 15th Iranian Plant Protection Congress, 7—11 September 2002* (p. 24). Razi University of Kermanshah.

Abbasipour, H., & Taghavi, A. (2004a). Introduction of the rice armyworm, *Mythimna unipuncta* (Haworth) (Lepidoptera: Noctuidae) parasitoids in western Mazandaran rice fields and preliminary study on their efficiency to control the pest. *Journal of Agricultural Science, 1*(1), 19—28 [in Persian, English summary].

Abbasipour, H., & Taghavi, A. (2004b). Morphological and biological studies of parasitoid wasp, *Cotesia rufricus* (Haliday) (Hymenoptea: Braconidae), a gregarious endoparasitoid of the rice armyworm larvae, *Mythimna unipuncta*. *Pazhohesh va sazandegi, 62*, 18—24 [in Persian, English summary].

Abbasipour, H., Amini Dehaghi, M., & Taghavi, A. (2004). Evaluation of parasitoids efficiency to control the cereal armyworm, *Mythimna unipuncta* (Haworth) (Lep.: Noctuidae) in the rice fields of Iran. In *Proceedings of the 16th Iranian Plant Protection Congress, 28 August — 1 September 2004* (p. 86). University of Tabriz.

Abdoli, P., Talebi, A. A., & Farahani, S. (2016). First record of *Microplitis cebes* (Braconidae: Microgastrinae) from Iran. In *Proceedings of the 22nd Iranian Plant Protection Congress, 27—30 August 2016, College of Agriculture and Natural Resources* (p. 466). Karaj: University of Tehran.

Abdoli, P., Talebi, A. A., & Farahani, S. (2018). First record of *Microplitis kaszabi* (Braconidae: Microgastrinae) from the west Palearctic and Iran. In *Proceedings of the 23rd Iranian Plant Protection Congress, 27—30 August 2016, College of Agriculture and Natural Resources* (p. 934). Karaj: University of Tehran.

Abdoli, P., & Pourhaji, A. (2019). Description of a new species of the genus *Pholetesor* Mason, 1981 and host association from Iran (Braconidae Microgastrinae). *Redia, 102*, 55—90.

Abdoli, P., Talebi, A. A., Farahani, S., & Fernandez-Triana, J. (2019a). Three new species of the genus *Choeras* Mason, 1981 (Hymenoptera: Braconidae, Microgastrinae) from Iran. *Zootaxa, 4545*(1), 77—92.

Abdoli, P., Talebi, A. A., & Farahani, S. (2019b). First record of *Microplitis docilis* (Braconidae: Microgastrinae) from Iran. In *Proceedings of the 3rd Iranian International Congress of Entomology, 17—19 August 2019* (pp. 246—251). Tabriz: University of Tabriz.

Abdoli, P., Talebi, A. A., & Farahani, S. (2019c). *Dolichogenidea fernadeztrianai* sp. nov. (Hymenoptera: Braconidae, Microgastrinae) from Iran. *Journal of Agricultural Science and Technology, 21*(3), 647—658.

Abdoli, P., Talebi, A. A., Farahani, S., & Fernández-Triana, J. (2019d). *Venanides caspius* sp. nov. from Iran, the first species of *Venanides* (Hymenoptera: Braconidae) described from the Palaearctic region. *Acta Entomologica Musei Nationalis Pragae, 59*(2), 543—548.

Abdoli, P., Talebi, A. A., Fernandez-Triana, J., & Farahani, S. (2021a). Taxonomic study of the genus *Microplitis* Förster, 1862 (Hymenoptera, Braconidae, Microgastrinae) from Iran. *European Journal of Taxonomy, 744*, 83—118.

Abdoli, P., Talebi, A. A., & Farahani, S. (2021b). Additional review of the genus *Iconella* (Hymenoptera: Braconidae, Microgastrinae) from Iran with the description of a new species. *North-Western Journal of Zoology, 2021*, e211202 (in press).

Abdoli, P., Talebi, A. A., Fernandez-Triana, J., & Farahani, S. (2021). Description of a new species of the genus *Protapanteles* Ashmead, 1898 (Hymenoptera: Braconidae: Microgastrinae) from Iran. *Annales Zoologici (Warszawa), 71*(2), 289—295.

Abou Zaid, N. A., El-Dakroury, M. S. I., El-Heneidy, A. H., & Abbas, M. S. T. (1978). Biology of *Microplitis rufiventris* Kok. Parasitizing *Heliothis armigera* Hb. in Egypt. *Agricultural Research Review (Cairo), 56*, 31—36.

Afiunizadeh Isfahani, M., & Karimzadeh Isfahani, J. (2010). Larval and pupal parasitoids of *Plutella xylostella* (Lep.: Plutellidae) in Isfahan province, Iran. *Plant Protection Journal, 2*(2), 79—97 [in Persian, English summary].

Afiunizadeh, M., Karimzadeh, J., & Shojai, M. (2011). Naturally-occurring parasitism of diamondback moth in central Iran. In *Proceedings of the 6th International Workshop on Management of the Diamondback Moth and other Crucifer Insect Pests, 23—27 March, Bangalore, India* (pp. 93—96).

Afshari, A., Yazdanian, M., Shabanipour, M., & Ghadiri-Rad, S. (2014). Natural parasitism of tomato fruitworm (*Helicoverpa armigra* Hübner) in tomato fields of Golestan province, northern Iran. In *Proceedings of the 22nd Iranian Plant Protection Congress, 27—30 August 2016, College of Agriculture and Natural Resources* (p. 503). Karaj: University of Tehran.

Al-e Mansour, H., & Mostafavi, M. S. (1993). The first record of Braconidae bees on forest and range vegetations in Fars province. In *Proceedings of the 11th Iranian Plant Protection Congress, 28 August — 2 September 1993* (p. 236). University of Guilan.

Alizadeh, S., & Javan Moghaddam, H. (2004). Introduction of some natural enemies of common cutworm (*Agrotis segetum* Schiff. In *Miyandoab. Proceedings of the 16th Iranian Plant Protection Congress, 28 August — 1 September 2004* (p. 45). University of Tabriz.

Al-Maliky, S. K., & Al Izizi, M. A. J. (1990). *Apanteles angaleti* (Muesebeck), Hymenoptera Braconidae. A new record species in Iraq. *Bulletin of the Iraq Natural History Museum, 3*(3), 193—194.

Amiri, A., Talebi, A. A., & Kamali, K. (2008). Parasitoid complex of *Phyllonorycter corylifoliella* (Hübnr) (Lepidoptera: Gracillariidae) in Fars province of Iran. In *Proceedings of the 18th Iranian Plant Protection Congress, 24—27 August 2008* (p. 76). University of Bu-Ali Sina Hamedan.

Amooghli-Tabari, M., & Ghahari, H. (2021). A study on the transitional population of overwintering larvae of *Chilo suppressalis* Walker (Lepidoptera: Crambidae) with percent rice infestation in main crop, and their natural enemies in Mazandaran, Iran. *Experimental Animal Biology, 9*(4), 39—54 [in Persian, English summary].

Asem, A. R., Eimanifar, A., & Wink, M. (2016). Update of biodiversity of the hypersaline Urmia lake National Park (NW Iran). *Diversity, 8*(6), 9.

Aubert, J. F. (1966). Liste d'identification No. 7 (Présentée par le Service d'Identification des Entomophages). *Entomophaga, 1*, 135—151.

Austin, A. D., & Dangerfield, P. C. (1992). Synopsis of Australasian Microgastrinae (Hymenoptera: Braconidae), with a key to genera and description of new taxa. *Invertebrate Taxonomy, 6*(1), 1—76.

Avci, M. (2009). Parasitoid complex and new host plant of the gypsy moth, *Lymantria dispar* L. in the Lakes district, Turkey. *Journal of the Animal and Vetrinary Advances, 8*(7), 1402—1405.

Aydoğdu, M. (2014). Parasitoid abundance of *Archips rosana* (Linnaeus, 1758) (Lepidoptera: Tortricidae) in organic cherry orchards. *North-Western Journal of Zoology, 10*(1), 42−47.

Banks, J. C., & Whitfield, J. B. (2006). Dissecting the ancient rapid radiation of microgastrine wasp genera using additional nuclear genes. *Molecular Phylogenetics and Evolution, 41*, 690−703.

Balveski, N. A. (1999). *Catalogue of the braconid parasitoids (Hymenoptera: Braconidae) isolated from various phytophagous insect hosts in Bulgaria*. Sofia & Moscow: Pensoft. vi + 126 pp.

Bayram, S., Ulogenturk, S., & Toros, S. (1998). Researches on the insects causing galls on dog rose (*Rosa* spp.) and their parasitoids in Ankara province. *Turkiye Entomoloji Dergisi, 22*(4), 259−268.

Behdad, E. (1991). *Pests of fruit crops in Iran* (p. 822). Isfahan: Isfahan Neshat Publication [in Persian].

Belokobylskij, S. A., & Lelej, A. S. (2019). Annotated catalogue of the Hymenoptera of Russia. Volume II. Apocrita: Parasitica. In *Proceedings of the Zoological Institute of the Russian Academy of Sciences* (p. 594). Supplement No. 8.

Belokobylskij, S. A., Samartsev, K. G., & Il'inskaya, A. S. (2019). Annotated catalogue of the Hymenoptera of Russia. Volume II. Apocrita: Parasitica. *Proceedings of the Zoological Institute RAS, 323*(Suppl. 8), 1−594.

Belshaw, R., Fitton, M., Herniou, E., Gimeno, C., & Quicke, D. L. J. (1998). A phylogenetic reconstruction of the Ichneumonoidea (Hymenoptera) based on the D2 variable region of 28S ribosomal RNA. *Systematic Entomology, 23*, 109−123.

Belshaw, R., Dowton, M., & Quicke, D. L. J. (2000). Estimating ancestral geographical distributions: A gondwanan origin for aphid parasitoids? *Proceedings of the Royal Society of London B, 267*, 491−496.

Beyarslan, A. (1988). Zwei neue Arten der familie Braconidae (Hymenoptera) aus der Türkei. *Zeitschrift der Arbeitsgemeinschaft Österreichischer Entomologen, 39*(3/4), 71−76.

Beyarslan, A., Inanç, F., Çetin Erdoğan, Ö., & Aydoğdu, M. (2002a). Braconidae species of Turkish Aegean region (Hymenoptera). In Melika, G., Thuroezy, C. (saba [Eds.]. *Parasitic Wasps: evolution, systematics, biodiversity and biological control*. Internal symposium: "Parasitic Hymenoptera: Taxonomy and Biological Control" (14−17 May 2001, Koszek, Hungary) Agroinform Kiado & Nyomda Kft, Budapest 2002: i−xx, 1−480. Chapter pagination: 285−290.

Beyarslan, A., Inanç, F., Çetin Erdoğan, Ö., & Aydoğdu, M. (2002b). Braconiden von den tuerkischen Inseln Imbros und Tenedos (Hymenoptera, Braconidae: Agathidinae, Braconinae, Cheloninae, Microgastrinae). *Entomofauna, 23*(15), 173−188.

Beyarslan, A., Yurtcan, M., Çetin Erdoğan, Ö., & Aydoğdu, M. (2006). A study on the Braconidae and Ichneumonidae from Ganos Mountain (Therace Region, Turkey) (Hymenoptera: Braconidae, Ichneumonidae). *Linzer biologische Beiträge, 38/1*, 409−422.

Bodenheimer, F. S. (1930). Die Schädlingsfauna Palästinas. *Monographien zur Angewandten Entomologie, 10*, 438.

Čapek, M., Hladil, J., & Sedivy, J. (1982). Verzeichnis der aus verschiedenen Insekten erzogenen parasitschen Hymenopteren-Teil VI. *Entomological Problems, 17*, 325−371.

Čapek, M., & Hofmann, C. (1997). The Braconidae (Hymenoptera) in the collections of the Musée Cantonal de Zoologie, Lausanne. *Litterae Zoologicae (Lausanne), 2*, 25−162.

Çetin Erdoğan, Ö., & Beyarslan, A. (2005). Microgastrinae species collected from some provinces of Erzurum and Kars in eastern Anatolia of Turkey. *Linzer biologische Beiträge, 37/1*, 393−397.

Chen, X. X., & van Achterberg, C. (2019). Systematics, phylogeny, and evolution of braconid wasps: 30 years of progress. *Annual Review of Entomology, 64*, 1−24.

Daniali, M. (1993). Biology and methods of mass production of *Bracon hebetor* ectoparasite of cotton boll worm, *Heliothis armigera*. In *Gorgan and Gonbad. Proceedings of the 11th Plant Protection Congress of Iran, 28 August − 2 September 1993* (p. 109). University of Guilan.

Davatchi, A., & Shojai, M. (1969). *Les hymenoptères entomophages de l'Iran (etudes faunestiques). Université de Teheran, Faculté d'Agronomia, Publication No* (Vol. 107, pp. 1−88).

Doganlar, M. (1982). Hymenopterous parasites of some lepidopterous pests in eastern Anatolia. *Turkiye Bitkikoruma Dergisi, 6*, 197−205.

Dowton, M., Austin, A. D., & Antolin, M. F. (1998). Evolutionary relationships among the Braconidae (Hymenoptera: Ichneumonoidea) inferred from partial 16S rDNA gene sequences. *Insect Molecular Biology, 7*, 129−150.

El-Heneidy, A. H., & Hassanein, F. A. (1987). Survey of the parasitoids of the grease cutworm *Agrotis ipsilon* Rott. (Lepidoptera: Noctuidae), in Egypt. *Anzeiger für Schädlingskunde Pflanzenschutz Umweltschulz, 60*(8), 155−157.

El-Sherif, L., & Kaschef, A. H. (1977). Morphological and biological studies on *Apanteles syleptae* F. (Hymenoptera: Braconidae) recovered from the jasmineum moth *Palpita unionalis* Hb. *Zeitschrift für Angewandte Entomologie, 84*(4), 419−424.

Esmaili, M. (1983). *The important pests of fruit trees* (p. 578). Tehran: Sepehr Publication.

Fahringer, J. (1922). Hymenopterologische Ergebnisse einer wissenschaftlichen Studienreise nach der Türkei und Kleinasien (mit Ausschluß des Amanusgebirges). *Archiv für naturgeschichte, A88*(9), 149−222.

Fallahzadeh, M., & Saghaei, N. (2010). Checklist of Braconidae (Insecta: Hymenoptera) from Iran. *Munis Entomology & Zoology, 5*(1), 170−186.

Farahani, S., Talebi, A. A., van Achterberg, C., & Rakhshani, E. (2014). First record of *Microplitis rufiventris* Kokujev, 1914 (Braconidae: Microgastrinae) from Iran. *Check List, 10*(2), 441−444.

Farahani, S., Talebi, A. A., & Rakhshani, E. (2016). Iranian Braconidae (Insecta: Hymenoptera: Ichneumonoidea): Diversity, distribution and host association. *Journal of Insect Biodiversity* and Systematics, 2(1), 1−92.

Farahbakhsh, G. (1961). Family Braconidae (Hymenoptera). In G. Farahbakhsh (Ed.), *The Ministry of Agriculture, Department Plant Protection, No: Vol. 1. A checklist of economically important insects and other enemies of plants and agricultural products in Iran* (p. 124), 153 pp.

Fernandez-Triana, J., Shaw, M. R., Boudrault, C., Beaudin, M., & Broad, G. R. (2020). Annotated and illustrated world checklist of Microgastrinae parasitoid wasps (Hymenoptera, Braconidae). *ZooKeys, 920*, 1−1089.

Fernandez-Triana, J., & Boudreault, C. (2018). Seventeen new genera of microgastrine parasitoid wasps (Hymenoptera, Braconidae) from tropical areas of the world. *Journal of Hymenoptera Research, 64*, 25−140.

Fernández-Triana, J., Smith, M. A., Boudreault, C., Goulet, H., Herbert, P. D. N., Smith, A. C., & Roughley, R. (2011). A poorly-known high-latitude parasitoid wasp community: Unexpected diversity and dramatic changes through time. *PLoS One, 6*, e23719.

Fernández-Flores, S., Fernández-Triana, J. L., Martínez, J. J., & Zaldivár-Riverón, A. (2013). DNA barcoding species inventory of Microgastrinae wasps (Hymenoptera: Braconidae) from a Mexican tropical dry forest. *Molecular Ecology Resources, 13*(6), 1146−1150.

Fernández-Triana, J. L., Shaw, M. R., Cardinal, S., & Mason, P. G. (2014). Contributions to the study of the Holarctic fauna of Microgastrinae (Hymenoptera: Braconidae). I. Introduction and first results of transatlantic comparisons. *Journal of Hymenoptera Research, 37*, 61−76.

Fernandez-Triana, J., Ward, D. F., & Whitfield, J. B. (2011). *Kiwigaster* gen. nov. (Hymenoptera: Braconidae) from New Zealand: The first Microgastrinae with sexual dimorphism in number of antennal segments. *Zootaxa, 2932*, 24–32.

Fernández-Triana, J., Boudreault, C., Buffam, J., & Mclean, R. (2016). A biodiversity hotspot for Microgastrinae (Hymenoptera, Braconidae) in North America: Annotated species checklist for Ottawa, Canada. *ZooKeys, 633*, 1–93.

Fernández-Triana, J. L., & Ward, D. (2017). *Microgastrinae wasps of the world.* http://microgastrinae.myspecies.info/.

Fernández-Triana, J. L., & van Achterberg, C. (2017). Order Hymenoptera, family Braconidae, subfamily Microgastrinae from the Arabian Peninsula. In A. van Harten (Ed.), *Arthropod fauna of the UAE* (Vol. 6, pp. 275–321). Abu Dhabi: Department of the President's Affairs, 775 pp.

Fiaboe, K. K. M., Fernandez-Triana, J., Nyamu, F. W., & Agbodzavu, K. M. (2017). *Cotesia icipe* sp. n., a new microgastrine wasp (Hymenoptera: Braconidae) of importance in the biological control of lepidopteran pests in Africa. *Journal of Hymenoptera Research, 61*, 49–64.

Fulmek, L. (1968). Parasitinsekten der Insektengallen Europas. *Beiträge zur Entomologie, 18*(7/8), 719–952.

Gadallah, N. S., Ghahari, H., & Peris-Felipo, F. J. (2015). Catalogue of the Iranian Microgastrinae (Hymenoptera: Braconidae). *Zootaxa, 4043*(1), 1–69.

Gadallah, N. S., Ghahari, H., Papp, J., & Beyarslan, A. (2018). New records of Braconidae (Hymenoptera) from Iran. *Wuyi Science Journal, 34*, 43–48.

Gadallah, N. S., Ghahari, H., & Quicke, D. L. J. (2021). Further addition to the braconid fauna of Iran (Hymenoptera: Braconidae). *Egyptian Journal of Biological Pest Control, 31*, 32. https://doi.org/10.1186/s41938-021-00376-8

Gerling, D. (1969). The parasites of *Spodoptera littoralis* Bois. (Lep., Noctuidae) eggs and larvae in Israel. *Israel Journal of Entomology, 4*, 73–81.

Gerling, D. (1971). Occuraence, abundance and efficiency of some local parasitoids attacking *Spodoptera littoralis* (Lepidoptera: Noctuidae) in selected cotton fields in Israel. *Annals of the Entomological Society of America, 64*(2), 492–499.

Ghadiri Rad, S., & Ebrahimi, E. (2010). Analysis regression of the effects of *Cotesia kazak* Telenga (Hym.: Braconidae) on population densities of *Helicoverpa armigera* Hubner (Lep.: Noctuidae) and crop injury indices. In *Proceedings of the 19th Plant Protection Congress of Iran, 31 July – 3 August 2010* (p. 67). Iranian Research Institute of Plant Protection.

Ghafouri Moghaddam, M., Rakhshani, E., van Achterberg, C., & Mokhtari, A. (2018). A study of the Iranian species of *Choeras* Mason (Hymenoptera: Braconidae: Microgastrinae), with the description of a new species. *Zootaxa, 4446*(4), 455–476.

Ghafouri Moghaddam, M., Rakhshani, E., van Achterberg, C., & Mokhtari, A. (2019). A taxonomic review of the genus *Diolcogaster* Ashmead (Hymenoptera, Braconidae, Microgastrinae) in Iran, distribution and morphological variability. *Zootaxa, 4590*(1), 95–124.

Ghafouri Moghaddam, M., Rakhshani, E., van Achterberg, C., & Mokhtari, A. (2021). Revision of the genus *Napamus* Papp (Hymenoptera, Braconidae, Microgastrinae). *International Journal of Tropical Insect Science.* https://doi.org/10.1007/s42690-021-00433-7

Ghahari, H. (2016). Five new records of Iranian Braconidae (Hymenoptera: Ichneumonoidea) for Iran and annotated catalogue of the subfamily Homolobinae. *Wuyi Science Journal, 32*, 35–43.

Ghahari, H. (2017). Species diversity of Ichneumonoidea (Hymenoptera) from rice fields of Mazandaran province, northern Iran. *Journal of Animal Environment, 9*(3), 371–378 [in Persian, English summary].

Ghahari, H. (2019a). Faunistic survey of parasitoid wasps (Hymenoptera) in forest areas of Mazandaran province, northern Iran. *Iranian Journal of Forest, 11*(1), 61–79 [in Persian, English summary].

Ghahari, H. (2019b). Study on the natural enemies of agricultural pests in some sugar beet fields, Iran. *Journal of Sugar Beet, 35*(1), 91–102 [in Persian, English summary].

Ghahari, H., & Jussila, R. (2015). Faunistic notes on the Ichneumonid wasps (Hymenoptera: Ichneumonidae) in alfalfa fields in some regions of Iran. *Entomofauna, 36*(12), 185–192.

Ghahari, H., & Sakenin, H. (2018). Species diversity of Chalcidoidea and Ichneumonoidea (Hymenoptera) in some paddy fields and surrounding grasslands of Mazandaran and Guilan provinces, northern Iran. *Applied Plant Protection, 7*(1), 11–19 [in Persain, English summary].

Ghahari, H., & Gadallah, N. S. (2022). Additional records to the braconid fauna (Hymenoptera: Ichneumonoidea) of Iran, with new host reports. *Entomological News* (in press).

Ghahari, H., Fischer, M., Çetin Erdoğan, Ö., Beyarslan, A., Hedqvist, K. J., & Ostovan, H. (2009). Faunistic note on the Braconidae (Hymenoptera: Ichneumonoidea) in Iranian alfalfa fields and surrounding grasslands. *Entomofauna, 30*(24), 437–444.

Ghahari, H., Gadallah, N. S., Çetin Erdoğan, Ö., Hedqvist, K. J., Fischer, F., Beyarslan, A., & Ostovan, H. (2009). Faunistic note on the Braconidae (Hymenoptera: Ichneumonoidea) in Iranian cotton fields and surrounding grasslands. *Egyptian Journal of Biological Pest Control, 19*(2), 115–118.

Ghahari, H., Tabari, M., Haji-Amiri, M., Sakenin, H., & Ostovan, H. (2009). Population fluctuation of rice stem borer, *Chilo suppressalis* Walker (Lepidoptera: Pyralidae) in paddy fields of northern Amol in Mazandaran Province. *Journal of Plant Protection, 23*(1), 41–49 [in Persian, English summary].

Ghahari, H., Fischer, M., Çetin Erdoğan, Ö., Beyarslan, A., & Ostovan, H. (2010). A contribution to the braconid wasps (Hymenoptera: Braconidae) from the forests of northern Iran. *Linzer biologische Beiträge, 42*(1), 621–634.

Ghahari, H., Fischer, M., Hedqvist, K. J., Çetin Erdoğan, Ö., van Achterberg, C., & Beyarslan, A. (2010). Some new records of Braconidae (Hymenoptera) for Iran. *Linzer biologische Beiträge, 42*(2), 1395–1404.

Ghahari, H., Çetin Erdoğan, Ö., Šedivý, J., & Ostovan, H. (2010). Survey of the Ichneumonoidea and Chalcidoidea (Hymenoptera) parasitoids of Saturniidae (Lepidoptera) in Iran. *Efflatounia, 10*, 1–6.

Ghahari, H., & Fischer, M. (2011a). A contribution to the Braconidae (Hymenoptera: Ichneumonoidea) from north-western Iran. *Calodema, 134*, 1–6.

Ghahari, H., & Fischer, M. (2011b). A contribution to the Braconidae (Hymenoptera) from Golestan National Park, northern Iran. *Zeitschrift Arbeitsgemeinschaft Österreichischer Entomologen, 63*, 77–80.

Ghahari, H., Fischer, M., Sakenin, H., & Imani, S. (2011a). A contribution to the Agathidinae, Alysiinae, Aphidiinae, Braconinae, Microgastrinae and Opiinae (Hymenoptera: Braconidae) from cotton fields and surrounding grasslands of Iran. *Linzer biologische Beiträge, 43*(2), 1269–1276.

Ghahari, H., Fischer, M., & Papp, J. (2011b). A study on the Braconidae (Hymenoptera: Ichneumonoidea) from Qazvin province, Iran. *Entomofauna, 32*(9), 197–208.

Ghahari, H., Fischer, M., & Papp, J. (2011c). A study on the braconid wasps (Hymenoptera: Braconidae) from Isfahan province, Iran. *Entomofauna, 32*(16), 261−272.

Ghahari, H., Fischer, M., & Papp, J. (2011d). A study on the Braconidae (Hymenoptera: Ichneumonoidea) from Ilam province, Iran. *Calodema, 160*, 1−5.

Ghahari, H., & Fischer, M. (2012). A faunistic survey on the braconid wasps (Hymenoptera: Braconidae) from Kermanshah province, Iran. *Entomofauna, 33*(20), 305−312.

Ghahari, H., Fischer, M., & Jussila, R. (2012). Braconid and ichneumonid wasps (Hymenoptera, Ichneumonoidea) as the parasitoids of *Plutella xylostella* (L.) (Lepidoptera: Plutellidae) in Iran. *Entomofauna, 33*(18), 281−288.

Ghahari, H., Fischer, M., Papp, J., & Tobias, V. (2012). A contribution to the knowledge of braconids (Hymenoptera: Braconidae) from Lorestan province Iran. *Entomofauna, 33*(7), 65−72.

Ghahari, H., Fischer, M., & Tobias, V. (2012). A study on the Braconidae (Hymenoptera: Ichneumonoidea) from Guilan province, Iran. *Entomofauna, 33*(22), 317−324.

Ghahari, H., & van Achterberg, C. (2016). A contribution to the study of subfamilies Microgastrinae and Opiinae (Hymenoptera: Braconidae) from the Arasbaran Biosphere Reserve and vicinity, Northwestern Iran. *Natura Somogyiensis, 28*, 23−32.

Ghahari, H., & Beyarslan, A. (2017). A faunistic study on Braconidae (Hymenoptera: Ichneumonoidea) from Iran. *Natura Somogyiensis, 30*, 39−46.

Ghahari, H., & Beyarslan, A. (2019). A faunistic study on Braconidae (Hymenoptera: Ichneumonoidea) from Iran, and in Memoriam Dr. Jenő Papp (20 May 1933-11 December 2017). *Acta Biologica Turcica, 32*(4), 248−254.

Ghahari, H., Fischer, M., Beyarslan, A., Navaeian, M., & Hosseini Boldaji, S. A. (2019). New records of Brachistinae, Braconinae, Cheloninae, and Microgastrinae (Hymenoptera: Braconidae) from Iran. *Wuyi Science Journal, 35*(2), 135−141.

Ghramh, H. A., Ahmad, Z., Khan, K. A., & Khan, F. (2020). Three new species of the genus *Microplitis* Förster, 1862 (Hymenoptera: Braconidae: Microgastrinae) from Saudi Arabia. *Pakistan Journal of Zoology, 52*(2), 2185−2192.

Goldansaz, S. H., Esmaili, M., & Ebadi, R. (1996). Lesser wax moth, *Achroia grisella* F., and its parasitic wasps. In *Proceedings of the XX International Congress of Entomology, Firenze, Italy* (p. 663).

Goldansaz, S. H., Kishani, H., Talaei, L., Poorjavad, N., & Sobhani, M. (2011). Biological control of carob moth, past, present and future. In *Proceedings of the Biological Control Development Congress in Iran, 27-28 July 2011* (pp. 55−63). Iranian Research Institute of Plant Protection [in Persian, English summary].

Golizadeh, A., Kamali, K., Fathipour, Y., Abbasipour, H., & Lozan, A. (2008). Report of the parasitoid wasp, *Cotesia plutellae* (Hym.: Braconidae): From Iran. *Journal of Entomological Society of Iran, 27*(2), 19−20.

Gothilf, S. (1969). Natural enemies of the carob moth *Ectomyelois ceratoniae* (Zeller). *Entomophaga, 14*(2), 195−202.

Güçlü, C., & Özbek, H. (2011). A contribution to the knowledge of Microgastrinae (Hymenoptera: Braconidae) from Turkey. *Journal of Biology and Life Sciences, 2*(2), 1−5.

Györfi, J. (1959). Beiträge zur Kenntnis der Wirte verschiedener Braconiden-Arten (Hymenoptera, Braconidae). *Acta Zoologica Hungarica, 5*, 49−65.

Haeselbarth, E. (1983). Determination list of entomophagous insects. Nr. 9. *Bulletin. Section Regionale Ouest Palaearctique, Organisation Internationale de Lutte Biologiqu, 6*(1), 1−49.

Halperin, J. (1986). Braconidae (Hymenoptera) associated with forest and ornamental trees and shrubs in Israel. *Phytoparasitica, 14*(2), 119−135.

Hasanshahi, G., Askarianzadeh, A., Abbasipour, H., & Karimi, J. (2012a). Natural parasitism of diamondback moth, *Plutella xylostella* L. (Lep.: Plutellidae) on different cultivars of cauliflower. In *Proceedings of the 20th Iranian Plant Protection Congress, 26−29 August 2012* (p. 11). University of Shiraz.

Hasanshahi, G., Askarianzadeh, A., Abbasipour, H., & Karimi, J. (2012b). Identification of parasitoids of diamondback moth, *Plutella xylostella* L. (Lep.: Plutellidae) and their parasitism rate in cauliflower fields of south of Tehran. In *Proceedings of the 20th Iranian Plant Protection Congress, 26−29 August 2012* (p. 116). University of Shiraz.

Hasanshahi, G., Abbasipour, H., Jussila, R., Jahan, F., & Dosti, Z. (2014a). First record of the genus and species, *Syrphophilus bizonarius* from Iran. *Biocontrol in Plant Protection, 1*(2), 111−113.

Hasanshahi, G., Abbasipour, H., Jussila, R., Jahan, F., & Dosti, Z. (2014b). Host report of *Hyposoter clausus* (Brischke, 1880) (Ichneumonidae: Campopleginae), a larval parasitoid of the cabbage white butterfly, *Pieris rapae* from cauliflower fields in Tehran. *Biocontrol in Plant Protection, 2*(1), 95−97.

Hasanshahi, G., Abbasipour, H., Moghbeli Gharaei, A., Jussila, R., & Mohammadi-Khoramabadi, A. (2015). First report of *Hyposoter ebeninus* a larval parasitoid of small cabbage butterfly, *Pieris rapae* from Tehran province. *Applied Entomology and Phytopathology, 82*(2), 185−186.

Hassanein, M. H. (1958). Biological studies on the diamond-back moth, *Plutella maculipennis* Curtis (Lepidoptera: Plutellidae). *Bulletin de la Société Entomologique d'Egypte, 42*, 325−337.

Hassanein, F. A., & El-Heneidy, A. H. (1984). On the parasitism on the cotton leafworm *Spodoptera littoralis* (Boisd.) on cabbage in Egypt. *Bulletin of the Entomological Society of Egypt Economic Series, 14*, 257−262.

Hassanein, F. A., & El-Heneidy, A. H. (1991). Comparative study of the parasitism by *Microplitis rufiventris* Kok. (Hymenoptera, Braconidae) and *Periboea (Periboea) orbata* Wied. (Diptera: Tachinidae) on main lepidopterous pests in vegetable crop fields in Egypt. *Bulletin of the Entomological Society of Egypt Economic Series, 17*, 127−135.

Hedwig, K. (1957). Ichneumoniden und Braconiden aus den Iran 1954 (Hymenoptera). *Jahresheft des Vereins für Vaterlaendische Naturkunde, 112*(1), 103−117.

Hegazi, E. M., Hammad, S. M., Altahtawy, M., & El-Sawaf, S. K. (1973). Parasites of the larval stage of cotton leaf-worm *Spodoptera littoralis* (Boisd.) (Noctuidae, Lepidoptera) in Alexandria region. *Zeitschrift für Angewandte Entomologie, 74*, 332−336.

Hegazi, E. M., Hammad, S. M., & Minshawy, A. M. (1977). Field and laboratory observations on the parasitoids of *Spodoptera littoralis* (Boisd.) (Lep., Noctuidae) in Alexandria. *Zeitschrift für Angewandte Entomologie, 84*, 316−321.

Hérard, F., Mercadier, G., & Abai, M. (1979). Situation de *Lymantria dispar* (Lep.: Lymantriidae) et son complexe parasitaire en Iran, en 1976. *Entomophaga, 24*(4), 371−384.

Hrcek, J., Miller, S. E., Whitfield, J. B., Shima, H., & Novotny, V. (2013). Parasitism rate, parasitoid community composition and host specificity on exposed and semi-concealed caterpillars from a tropical rainforest. *Oecologia, 173*, 521−532.

Inanç, F. (1997). The Microgastrinae (Hymenoptera: Braconidae) fauna of the Thrace Region of Turkey. *Turkish Journal of Zoology, 21*(2), 135−165.

Inanç, F. (2002). *Cotesia pappi* sp. n. (Hymenoptera, Braconidae, Microgastrinae) from Turkey. *Acta Zoologica Academiae Scientiarum Hungaricae, 48*(2), 157−160.

Inanç, F., & Beyarslan, A. (1990). Instranca dağlannin *Apanteles* Förster (Hym., Braconidae, Microgastrinae) türteri. Doğa. *Turkish Journal of Zoology, 14*, 281−300.

Inanç, F., & Beyarslan, A. (1997). Microgastrinae (Hymenoptera, Braconidae) species collected from some provinces of Gaziantep and Sanliurfa. *Turkiye Entomoloji Dergisi, 21*(3), 213−223.

Inanç, F., & Beyarslan, A. (2001a). Die Microgastrinae-Fauna der ostmarmara region der Turkei (Hymenotera: Braconidae). *Entomofauna, 22*(11), 221−244.

Inanç, F., & Beyarslan, A. (2001b). A study on Microgastrinae (Hymenoptera: Braconidae) species in Gokceada and Bozcaada. *Turkish Journal of Zoology, 25*(3), 287−296.

Inanç, F., & Beyarslan, A. (2001c). New records of Microgastrinae (Hymenoptera: Braconidae) from Turkish Thrace in Turkey. *Turkiye Entomoloji Dergisi, 25*(3), 205−216.

Inanç, F., & Beyarslan, A. (2001d). Untersuchungen ueber Microgastrinae fauna der Ost Marmara region der Türkei (Hymenoptera: Braconidae). *Entomofauna, 22*(11), 221−244.

Inanç, F., & Çetin Erdoğan, Ö. (2004). Contribution to the Microgastrinae (Hymenoptera: Braconidae) fauna of Turkey, with description of a new species of *Dolichogenidea*. *Biologia, 59*(5), 547−551.

Ingram, W. R. (1981). The parasitoids of *Spodoptera littoralis* (Lep.: Noctuidae) and their role in the population control in Cyprus. *Entomophaga, 26*, 23−37.

Ismail, I. I., & Swailem, S. M. (1976). On the biology of the bollworm *Heliothis armigera* (Hubner) (Lepidoptera: Noctuidae). *Bulletin de la Société Entomologique d'Egypte, 59*(1975), 207−216.

Jahan, F., Moghbeli Gharaei, A., Hasanshahi, G., Mohammadi-Khoramabadi, A., Abbasipour, H., & Papp, J. (2016). First report of the parasitoid wasp *Dolichogenidea turkmena* (Hym.: Braconidae, Microgastrinae) from Iran. *Applied Entomology and Phytopathology, 83*(2), 276.

Janzen, D. H., Hallwachs, W., Blandin, P., Burns, J. M., Cadiou, J. M., Chacon, I., Dapkey, T., Deans, A. R., Epstein, M. E., Espinoza, B., Franclemont, J. G., Haber, W. A., Hajibabaei, M., Hall, J. P., Hebert, P. D., Gauld, I. D., Harvey, D. J., Hausmann, A., Kitching, I. J., Lafontaine, D., & Wilson, J. J. (2009). Integration of DNA barcoding into an ongoing inventory of tropical complex biodiversity. *Molecular Ecology Resources, 9*(Suppl. 1), 1−26.

Kamangar, S., Lotfalizadeh, H., Mohammadi-Khoramabadi, A., & Seyedi-Sahebari, F. (2017). Parasitoids of *Tortrix viridana* L. in Kurdistan province. *Iranian Journal of Forest and Range Protection Research, 15*(2), 176−186 [in Persian, English summary].

Karimi-Malati, A., Fathipour, Y., Talebi, A. A., & Lozan, A. (2014). The first report of *Microplitis fulvicornis* (Hym.: Braconidae: Microgastrinae) as a parasitoid of *Spodoptera exigua* (Lep.: Noctuidae) from Iran. *Journal of Entomological Society of Iran, 33*(4), 71−72.

Karimpour, Y., Fathipour, Y., Talebi, A. A., & Moharramipour, S. (2001). Report of two endoparasitoid wasps, *Cotesia ofella* (Nixon) and *Cotesia vanessae* (Hym.: Braconidae) on larvae of *Simara dentinosa* Freyer (Lep.: Noctuidae) from Iran. *Journal of Entomological Society of Iran, 21*(2), 106.

Karimpour, Y., Fathipour, Y., Talebi, A. A., Moharramipour, S., Horstman, K., & Papp, J. (2005). New records of two parasitoid wasps of *Simyra dentinosa* Freyer (Lep., Noctuidae) larvae from Iran. *Applid Entomology and Phytopathology, 73*(1), 133.

Karimzadeh, J., Rabiei, A., Shakarami, J., & Jafari, S. H. (2016). Promotion of mass-rearing procedure of *Cotesia vestalis* (Haliday) and its quality control measured by fetility. In *Proceedings of the 22nd Iranian Plant Protection Congress, 27−30 August 2016, College of Agriculture and Natural Resources* (p. 659). Karaj: University of Tehran.

Kazemzadeh, Z., Shaw, M. R., & Karimzadeh, J. (2014). A new record for Iran of *Dolichogenidea appellator* (Hym.: Braconidae: Microgastrinae), a larval endoparasitoid of diamondback moth, *Plutella xylostella* (Lep.: Plutellidae). *Journal of Entomological Society of Iran, 33*(4), 81−82.

Khajeh, N., Rakhshani, E., Arjmandi, A. A., & Barahoei, H. (2014). A faunistic study on Microgastrinae in Sistan region. In *Proceedings of the 21st Iranian Plant Protection Congress, 23−26 August 2014* (p. 575). Urmia University.

Khanjani, M. (2006a). *Field crop pests in Iran* (3rd ed., p. 719). Bu-Ali Sina University [in Persian].

Khanjani, M. (2006b). *Vegetable pests in Iran* (2nd ed., p. 467). Bu-Ali Sina University [in Persian].

Kishani Farahani, H., Goldansaz, H., & Sabahi, G. (2008a). Study on the larval parasitoids of carob moth, *Ectomyelois ceratoniae* Zeller (Lep.: Pyralidae) in Saveh, Qom, and Varamin. In *Proceedings of the 18th Iranian Plant Protection Congress, 24−27 August 2008* (p. 72). University of Bu-Ali Sina Hamedan.

Kishani Farahani, H., Goldansaz, S. H., Sabahi, G., Ziaddini, M., & Haghani, S. (2008b). Effect of host range on immature developmental time, sex ratio, and parasitism percentage of *Apanteles myloenta* Wilkinson (Hym.: Braconidae). In *Proceedings of the 18th Iranian Plant Protection Congress, 24−27 August 2008* (p. 397). University of Bu-Ali Sina Hamedan.

Kishani Farahani, H., Goldansaz, H., Sabahi, G., Ziaaddini, M., & Haghani, S. (2008c). Biology of *Apanteles myeloenta* a parasitoid of carob moth, *Ectomyelois ceratoniae* Zeller (Lep.: Pyralidae). In *Proceedings of the 18th Iranian Plant Protection Congress, 24−27 August 2008* (p. 474). University of Bu-Ali Sina Hamedan.

Kishani Farahani, H., Goldansaz, H., Sabahi, G., & Shakeri, M. (2010). Study on the larval parasitoids of carob moth *Ectomyelois ceratoniae* Zeller (Lepidoptera: Pyralidae) in Varamin, Qom and Saveh. *Iranian Journal of Plant Protection Science, 41*(2), 337−344 [in Persian, English summary].

Kishani Farahani, H., Bell, H., & Goldansaz, S. H. (2012). Biology of *Apanteles myeloenta* (Hymenoptera: Braconidae), a larval parasitoid of carob moth *Ectomyelais ceratoniae* (Lepidoptera: Pyralidae). *Journal of Asia-Pacific Entomology, 15*, 607−610.

Kishani Farahani, H., & Goldansaz, S. H. (2013). Is host age an important factor in the bionomics of *Apanteles myeloenta* (Hymenoptera: Braconidae)? *European Journal of Entomology, 110*(2), 277−283.

Koçak, A. O., & Kemal, M. (2013). Nomenclatural notes on the Asiatic Ichneumonoidea (Hymenoptera). *Centre for Entomological Studies Miscellaneous Papers, 160*, 7−8.

Kohl, F. F. (1905). Hymenoptera. In A. Penther, & E. Zederbauer (Eds.), *Annalen des Naturhistorischen Museum in Wien: Vol. 20. Ergebnisse einer naturwissenschatlischen Reise zur Erdschias-Dagh (Kleinasien)* (pp. 220–246).

Kolaib, M. O., Hegazi, E. M., & Abd El Fattah, M. I. (1980). Parasitoids of the Egyptian cotton leafworm in Menoufia Governorate, Egypt. *Zeitschrift für Angewandte Entomologie, 89*(2), 193–198.

Ku, D. S., Belokobylskij, S. A., & Cha, J. Y. (2001). Hymenoptera (Braconidae). Economic Insects of Korea 16. *Insecta Koreana, Supplement 23* (p. 283).

Kugler, J. (1966). A list of parasites of Lepidoptera from Israel. *Israel Journal of Entomology, 1*, 75–88.

Lababidi, M. S., & Hammoud, D. H. (2008). Biological and ecological studies on the parasitoid *Dolichogenidea trachalus* (Nixon) (Hymenoptera: Braconidae) collected from the olive moth (jasmine moth) *Palpita unionalis* Hubner (Lepidoptera: Pyralidae) in Syria. *Arab Journal of Plant Protection, 26*(1), 1–6.

Lashkari Bod, A., Rakhshani, E., Talebi, A. A., Lozan, A., & Zikic, V. (2011). A contribution to the knowledge of Braconidae (Hym., Ichneumonoidea) of Iran. *Biharean Biologist, 5*(2), 147–150.

Le Corff, J., Marquis, R. J., & Whitfield, J. B. (2000). Temporal and spatial variation in a parasitoid community associated with the herbivores that feed on Missouri *Quercus*. *Environmental Entomology, 29*, 181–194.

Liu, Z., He, J.-H., Chen, X.-X., Gupta, A., & Ghafouri Moghaddam, M. (2019). The *ultor*-group of the genus *Dolichogenidea* Viereck (Hymenoptera, Braconidae, Microgastrinae) from China with the descriptions of thirty-nine new species. *Zootaxa, 4710*(1), 1–134.

Lupo, A., & Gerling, D. (1984). Bionomics of the tamarix spindle-gall moth *Amblypalpis olivierella* (Lepidoptera: Gelechiidae) and its natural enemies. *Bolletino Del laboratorio Di Entomologia Agraria Filippo Silvestri, 41*, 71–90.

Mahmoud, S. M., El-Heneidy, A. H., Gadallah, N. S., & Ahmed, R. S. (2009). Survey and abundance of common ichneumonoid parasitoid species in Suez Canal Region, Egypt. *Egyptian Journal of Biological Pest Control, 19*(2), 185–190.

Mason, W. R. M. (1981). The polyphyletic nature of *Apanteles* Förster (Hymenoptera: Braconidae): A phylogeny and reclassification of Microgastrinae. *Memoirs of the Entomological Society of Canada, 115*, 1–147.

Mehrnejad, M. R. (2010). The parasitoids of the pistachio fruit hull borer moth, *Arimania komaroffi*. *Journal of Applied Entomology and Phytopathology, 78*(1), 129–130 [in Persian, English summary].

Mirfakhraie, S. H., & Dey, D. (2013). New records of Hymenoptera associated with *Pieris brassicae* L. (Lepidoptera: Pieridae) from Iran. In *Indian Society of Vegetable Science, National Symposium on Abiotic and Biotic Stress Management in Vegetable Crops, 12–14 April 2013*. India: IIVR.

Moawad, G. M., Topper, C. P., El-Husseini, M. M., & Kamel, A. M. (1987). Seasonal abundance of parasites and predators of *Spodoptera littoralis* Boisd. in clover fields in Egypt. *Agricultural Research Review (Cairo), 63*(1), 45–55.

Modarres Awal, M. (1997). Family Braconidae (Hymenoptera). In M. Modarres Awal (Ed.), *List of agricultural pests and their natural enemies in Iran* (2nd ed., pp. 265–267). Ferdowsi University of Mashhad Press, 429 pp.

Modarres Awal, M. (2012). Family Braconidae (Hymenoptera). In M. Modarres Awal (Ed.), *List of agricultural pests and their natural enemies in Iran* (3rd ed., pp. 483–486). Ferdowsi University of Mashhad Press, 759 pp.

Muesebeck, C. F. W. (1922). A revision of the North American ichneumon-flies, belonging to the subfamilies Neoneurinae and Microgastrinae. In *Proceedings of the United States national Museum* (Vol. 61, pp. 1–76), 2436.

Murphy, N., Banks, J., Whitfield, J. B., & Austin, A. (2008). Phylogeny of the parasitic microgastroid subfamilies (Hymenoptera: Braconidae) based on sequence data from seven genes, with an improved estimate of the origin of the lineage. *Molecular Phylogenetics and Evolution, 47*, 378–395.

Naderian, H., Ghahari, H., & Asgari, S. (2012). Species diversity of natural enemies in corn fields and surrounding grasslands of Semnan province, Iran. *Calodema, 217*, 1–8.

Naderian, H., Penteado-Dias, A. M., Sakenin Chelav, H., & Samin, N. (2020). A faunistic study on Braconidae and Ichneumonidae (Hymenoptera, Ichneumonoidea) of Iran. *Calodema, 844*, 1–9.

Nikdel, M. (2015). Complementary identification of biological control agents of gypsy moth, *Lymantria dispar* (Lep.: Lymantriidae) in Arasbaran forests. *Agricultural Pest Management, 2*(1), 30–38 [in Persian, English summary].

Nikdel, M., Dordaei, A. A., & Pezeshki, M. H. (2012). Introduction of weedy spurge leaf defoliators and their parasitoids in East Azarbijan province rangelands. *Iranian Journal of Forest and Range Protection Research, 9*(1), 46–52 [in Persian, English summary].

Nixon, G. E. J. (1965). A reclassification of the tribe Microgasterini (Hymenoptera: Braconidae). *Bulletin of the British Museum (Natural History), Entomology series*, (Supplement 2), 1–284.

Nixon, G. E. J. (1968). A revision of the genus *Microgaster* Latreille (Hymenoptera: Braconidae). *Bulletin of the British Museum, 22*, 33–72.

Nixon, G. E. J. (1970). A revision of the N. W. European species of *Microplitis* Förster (Hymenoptera: Braconidae). *Bulletin of the British Museum (Natural History), Entomology series, 25*(1), 1–30.

Nixon, G. E. J. (1972). A revision of the north-western European species of *laevigatus*-group of *Apanteles* Foerster (Hymenoptera, Braconidae). *Bulletin of Entomological Research, 61*, 701–743.

Nixon, G. E. J. (1973). A revision of the north-western European species of the *vitripennis, pallipes, octonarius, triangulator, fraternus, formosus, parasitellas, metacarpalis* and *circumscriptus*-groups of *Apanteles* Förster (Hymenoptera: Braconidae). *Bulletin of Entomological Research, 63*, 169–228.

Nixon, G. E. J. (1976). A revision of the north-western European of *merula, laetus, vipio, ultor, ater, butalidis, popularis, carbonarius* and *validus*-groups of *Apanteles* Förster (Hym.: Braconidae). *Bulletin of Entomological Research, 65*, 687–732.

Nobakht, Z., Karimzadeh, J., Shakaram, J., & Jafari, S. H. (2015). Identification of parasitoids of *Apomyelois ceratoniae* (Zeller) (Lepidoptera, Pyralidae) on pomegranate in Isfahan province. *Journal of Entomology and Zoology Studies, 3*(1), 287–289.

Norouzi, A., Talebi, A. A., Fathipour, Y., & Lozan, A. (2009). *Apanteles laspeyresiellus* (Hymenoptera: Braconidae), a new record for Iran insect fauna. *Journal of Entomological Society of Iran, 28*(2), 79–80.

Özbek, H., & Calmasur, O. (2010). Spotted ash looper, *Abraxas pantaria* (L.) (lepidoptera: Geometridae), a new ash pest in Turkey. *Turkish Journal of Zoology, 34*(3), 351–358.

Özbek, H., & Coruh, S. (2012). Larval parasitoids and larval diseases of *Malacosoma neustria* L. (Lepidoptera: Lasiocampidae) detected in Erzurum Province, Turkey. *Turkish Journal of Zoology, 36*(4), 447–459.

Papp, J. (1972). New *Apanteles* Först. species from Hungary (Hymenoptera, Braconidae: Microgastrinae). I. *Annales Historico-Naturales Musei Nationalis Hungarici, 64*, 335–345.

Papp, J. (1970). A contribution to the braconid fauna of Israel (Hymenoptera). *Israel Journal of Entomology, 5*, 63–76.

Papp, J. (1977). New *Apanteles* Först. Species from Hungary (Hymenoptera, Braconidae: Microgastrinae), V. *Annales Historico-Naturales Musei Nationalis Hungarici, 69*, 201–217.

Papp, J. (1978). A survey of the European species of *Apanteles* Först. (Hymenoptera: Braconidae: Microgastrinae) II. The *laevigatus*-group, 1. *Annales Historico-Naturales Musei Nationalis Hungarici, 70*, 265–301.

Papp, J. (1980). Braconidae (Hymenoptera) from Mongolia VIII. *Acta Zoologica Hungarica, 26*, 401–413.

Papp, J. (1981). Contributions to the braconid fauna of Hungary, III. Opiinae and Microgastrinae (Hymenoptera: Braconidae). *Folia Entomologica Hungarica, 42*(34), 127–141 (2).

Papp, J. (1982). A survey of the European species of *Apanteles* Först. (Hymenoptera, Braconidae: Microgastrinae), VI. The *laspeyresiella*-, *merula*-, *falcatus*- and *validus*-group. *Annales Historico-Naturales Musei Nationalis Hungarici, 74*, 255–267.

Papp, J. (1983). A survey of the European species of *Apanteles* Först. (Hymenoptera, Braconidae: Microgastrinae), VII. The *carbonarius*-, *circumscriptus*-, *fraternus*-, *pallipes*-, *parasitellae*-, *vitripennis*-, *liparides*-, *octanarius*-, and *thompsoni*-group. *Annales Historico Naturalis Musi Nationalis Hungarici, 75*, 247–283.

Papp, J. (1984). A survey of the European species of *Apanteles* Först. (Hymenoptera, Braconidae: Microgastrinae), III. The *metacarpalis*-, *formosus*-, *popularis*- and *suevus*-group. *Annales Historico-Naturales Musei Nationalis Hungarici, 76*, 265–295.

Papp, J. (1986). A survey of the European species of *Apanteles* Först. (Hymenoptera, Braconidae: Microgastrinae). IX. The *glomeratus*-group, 1. *Annales Historico-Naturales Musei Nationalis Hungarici, 78*, 225–247.

Papp, J. (1987). A survey of the European species of *Apanteles* Foerster (Hymenoptera: Braconidae: Microgastrinae). X. The *glomeratus* group 2. *Annales Historico-Naturales Musei Nationalis Hungarici, 79*, 207–258.

Papp, J. (1990). A survey of the European species of *Apanteles* Förster (Hymenoptera, Braconidae: Microgastrinae) XII. Supplement to the key of the *glomeratus* group. Parasitoid/host list 2. *Annales Historico-Naturales Musei Nationalis Hungarici, 81*, 159–203.

Papp, J. (1996). The braconid wasps (Hymenoptera, Braconidae) of the Bükk National Park (NE Hungary). In S. Mahunka (Ed.), *The fauna of the Bükk National Park* (Vol. 2). Budapest: Hungarian Natural History Museum, 1–655. Chapter pagination: 453–476.

Papp, J. (2007). Braconidae (Hymenoptera) from Greece, 6. *Notes Fauniques de Gembloux, 60*(3), 99–127.

Papp, J. (2012). A contribution to the braconid fauna of Israel (Hymenoptera: Braconidae), 3. *Israel Journal of Entomology, 41–42*, 165–219.

Pirhadi, A., Rajabi, G., Ebrahimi, E., Ostovan, H., Shekarian Moghaddam, B., Mohiseni, A. A., Mozaffarian, F., & Ghavami, S. (2008). Natural enemies of cereal leaf miner, *Syringopais temperatella* Led. (Lep.: Elachistidae) in Lorestan province. In *Proceedings of the 18th Iranian Plant Protection Congress, 24–27 August 2008* (p. 71). University of Bu-Ali Sina Hamedan.

Porter, K. (1979). A third generation of *Apanteles bignellii* Marsh (Hym., Braconidae). *Entomologist's Monthly Magazine, 114*, 214.

Quicke, D. L. J., & Shaw, M. R. (2004). Cocoon silk chemistry in parasitic wasps (Hymenoptera, Ichneumonoidea) and their hosts. *Biological Journal of the Linnean Society, 81*(2), 161–170.

Rabiei, A., Shakarami, J., Karimzadeh Isfahani, J., & Jafari, S. (2017). Study on some biological characteristics of parasitoid wasp *Cotesia vestalis* in different mass-rearing conditions. *Biocontrol in Plant Protection, 4*(2), 99–108.

Radjabi, G.h. (1986). *Insects attacking rosaceous fruit trees in Iran* (Vol. II, p. 209). Tehran, Iran: Lepidoptera. Plant pests and diseases research institute [in Persian].

Rassipour, A. (1983). Etude biologique d'*Apanteles chilonis* Mun. (Hym.: Braconidae) en vue de la lute biologique contre la pyrale du riz, *Chilo suppressalis* Walk. (Lep.: Pyralidae). *Bulletin Plant Protection Organization, 29*, 1–24.

Rastegar, J., Sakenin, H., Khodaparast, S., & Havaskary, M. (2012). On a collection of Braconidae (Hymenoptera) from East Azarbaijan and vicinity, Iran. *Calodema, 226*, 1–4.

Razmi, M., Karimpour, Y., Safaralizadeh, M. H., & Safavi, S. A. (2011). Parasitoid complex of cabbage large white butterfly *Pieris brassicae* (L.) (Lepidoptera: Pieridae) in Urmia with new records from Iran. *Journal of Plant Protection Research, 51*(3), 248–251.

Rezaei, M., Karimzadeh, J., Shakarami, J., & Jafari, S. (2014). Side effects of insecticides on the adult longevity of *Cotesia vestalis*, a larval parasitoid of the diamondback moth, *Plutella xylostella*. *Journal of Entomology and Zoology Studies, 2*(4), 49–51.

Risbec, J. (1960). Les parasites des insects d'importance en Afrique tropicale et à Madagascar. *Agronomie Tropicale, 15*, 624–656.

Rodriguez, J., Oltra-Moscrdó, T., Peris-Felipo, F. J., & Jiménez-Peydró, R. (2013). Microgastrinae (Hymenoptera: Braconidae) in the forest state of Artikutza (Navarra: Spain): Diversity and community structure. *Insects, 4*(3), 493–505.

Rostami, E., Hasanshahi, G.h., Abbasipour, H., Askarianzadeh, A., & Karimi, J. (2014). Spatial distribution of the diamondback moth, *Plutella xylostella* and its parasitoid, *Cotesia plutellae* on the cauliflower in south of Tehran region. In *Proceedings of the 3rd Integrated Pest Management (IPMC), 21–22 January 2014, Kerman, Iran* (p. 521).

Sakenin, H., Naderian, H., Samin, N., Rastegar, J., Tabari, M., & Papp, J. (2012). On a collection of Braconidae (Hymenoptera) from northern Iran. *Linzer biologische Beiträge, 44*(2), 1319–1330.

Sakenin, H., Samin, N., Beyarslan, A., Coronado-Blanco, J. M., Navaeian, M., Fischer, M., & Hosseini Boldaji, S. A. (2020). A faunistic study on braconid wasps (Hymenoptera: Braconidae) from Iran. *Boletin de la Sociedad Andaluza de Entomologia, 30*, 96–102.

Sakenin, H., Ghahari, H., & Navaeian, M. (2021). A study on the predator and parasitoid insects in some cotton fields of Iran. *Journal of Animal Environment, 13*(1), 397–406 [in Persian, English summary].

Samin, N. (2015). A faunistic study on the Braconidae of Iran (Hymenoptera: Ichneumonoidea). *Arquivos Entomoloxicos, 13*, 339–345.

Samin, N., Ghahari, H., Gadallah, N. S., & Davidian, E. (2014). A study on the Braconidae (Hymenoptera: Ichneumonoidea) from West Azarbaijan province, northern Iran. *Linzer biologische Beiträge, 46*(2), 1447–1478.

Samin, N., Ghahari, H., Gadallah, N. S., & Monaem, R. (2015). A study on the braconid wasps (Hymenoptera: Ichneumonoidea: Braconidae) from Golestan province, northern Iran. *Linzer biologische Beiträge, 47*(1), 731–739.

Samin, N., van Achterberg, C., & Ghahari, H. (2015). A faunistic study of Braconidae (Hymenoptera: Ichneumonoidea) from southern Iran. *Linzer biologische Beiträge, 47*(2), 1801–1809.

Samin, N., Fischer, M., & Ghahari, H. (2015). A contribution to the study on the fauna of Braconidae (Hymenoptera: Ichneumonoidea) from the province of Semnan, Iran. *Arquivos Entomoloxicos, 13*, 429–433.

Samin, N., Coronado-Blanco, J. M., Kavallieratos, N. G., Fischer, M., & Sakenin, H. (2018). Recent findings on Braconidae (Hymenoptera: Ichneumonoidea) of Iran with an updated checklist. *Acta Biologica Turcica, 31*(4), 160–173.

Samin, N., Coronado-Blanco, J. M., Fischer, M., van Achterberg, C., Sakenin, H., & Davidian, E. (2018). Updated checklist of Iranian Braconidae (Hymenoptera: Ichneumonoidea) with twenty-three new records. *Natura Somogyiensis, 32*, 21–36.

Samin, N., Papp, J., & Coronado-Blanco, J. M. (2018). A faunistic study on braconid wasps (Hymenoptera: Ichneumonoidea: Braconidae) of Iran. *Scientific Bulletin of the Uzhgorod University (Series: Biology), 45*, 15–19.

Samin, N., Coronado-Blanco, J. M., Hosseini, A., Fischer, M., & Sakenin Chelav, H. (2019). A faunistic study on the braconid wasps (Hymenoptera: Braconidae) of Iran. *Natura Somogyiensis, 33*, 75–80.

Samin, N., Fischer, M., Sakenin, H., Coronado-Blanco, J. M., & Tabari, M. (2019). A faunistic study on Agathidinae, Alysiinae, Doryctinae, Helconinae, Microgastrinae, and Rogadinae (Hymenoptera: Braconidae), with eight new country records. *Calodema, 734*, 1–7.

Samin, N., Beyarslan, A., Coronado-Blanco, J. M., Navaeian, M., & Kavallieratos, N. (2020). A contribution to the braconid wasps (Hymenoptera: Braconidae) from Iran. *Natura Somogyiensis, 35*, 25–28.

Samin, N., Sakenin Chelav, H., Ahmad, Z., Penteado-Dias, A. M., & Samiuddin, A. (2020). A faunistic study on the family Braconidae (Hymenoptera: Ichneumonoidea) from Iran. *Scientific Bulletin of Uzhhorod National University (Series: Biology), 48*, 14–19.

Samin, N., Beyarslan, A., Ranjith, A. P., Ahmad, Z., Sakenin Chelav, H., & Hosseini Boldaji, S. A. (2020). A faunistic study on Braconidae (Hymenoptera: Ichneumonoidea) from Ardebil and East Azarbayjan provinces, Northwestern Iran. *Egyptian Journal of Plant Protection Research Institute, 3*(4), 955–963.

Schimitschek, E. (1944). *Forstinesketen der Tuerkei und ihre Umwelt, Prague* (Vol. 371, pp. 273–279).

Schwartz, A., Gerling, D., & Rossler, Y. (1980). Preliminary notes on the parasites of *Spodoptera exigua* (Huebner) (Lepidoptera: Noctuidae) in Israel. *Phytoparasitica, 8*(2), 93–97.

Sertkaya, E., & Bayram, A. (2005). Parasitoid community of the lorrei leafworm *Mythimna (Acantholeucania) loreyi* novel host-parasitoid associations and their efficiency in the eastern Mediterranean Region of Turkey. *Phytoparasitica, 33*(5), 441–449.

Sertkaya, E., Bayram, A., & Kornosor, S. (2004). Egg and larval parasitoids of the beet armyworm *Spodoptera exigua* on maize in Turkey. *Phytoparasitica, 32*(3), 305–312.

Shaw, M. R. (2012). Notes on some European Microgastrinae (Hymenoptera: Braconidae) with the National Museums of Scotland, with twenty species new to Britain, new host data, taxonomic changes and remarks, and descriptions of two new species of *Microgaster* Latreille. *Entomologist's Gazette, 63*, 173–201.

Shaw, M. R., & Huddleston, T. (1991). Classification and biology of braconid wasps (Hymenoptera: Braconidae). *Handbooks for the Identification of British Insects, 7*(11), 1–126.

Shaw, M. R., Stefanescu, C., & van Saskya, N. (2009). Parasitoids of European butterflies. In J. Settele, T. Shreeve, M. Konvicka, & H. van Dyck (Eds.), *Ecology of butterflies in Europe* (pp. 130–156). Cambridge University Press, 526 pp.

Sharanowski, B. J., Dowling, A. P. G., & Sharkey, M. J. (2011). Molecular phylogenetics of Braconidae (Hymenoptera: Ichneumonoidea), based on multiple nuclear genes, and implication for classification. *Systematic Entomology, 36*, 549–572.

Shenefelt, R. D. (1972). Braconidae 4. Microgastrinae *Apanteles* Foerster. *Hymenopterorum Catalogus [S'Gravenhage]* (Vol. 4, pp. 429–668).

Shojai, M. (1968). Resultats de létude faunestiques des Hyménopteres parasites (Terebrants) en Iran et l'importance de leur utilization des la lute biologique. *The First National Congress of Plant Medicine of Iran* (pp. 25–35).

Shojai, M. (1998). *Entomology (ethology, social life and natural enemies) (biological control)* (Third Edition, Vol. III, p. 550). Tehran University Publications [in Persian].

Shojai, M., Abbaspour, H., Nasrollahi, A., & Labbafi, Y. (1995). Technology and biocenotic aspects of integrated biocontrol of corn stem borer: *Sesamia cretica* Led. (Lep.: Noctuidae). *Journal of Agricultural Science, 1*(2), 5–32 [in Persian, English summary].

Shojai, M., Nasrollahi, A., Labafi, Y., Azma, M., Amiri, B., Baiat, H., Daniali, M., & Maghsodi, A. (1997). The biocenotic aspects of the Iranian subspecies corn stem borer, *Ostrinia nubilalis persica* (Lepidoptera: Pyraustididae) and its role in the increasing efficiency of *Trichogramma* wasps in IPM program in corn field of northern Iran. *Journal of Agricultural Sciences, 3*, 5–48 [in Persian, English summary].

Shojai, M., Esmaili, M., Ostovan, H., Khodaman, A., Daniali, M., Hosseini, M., Assadi, Y., Sadighfar, M., Korosh-Najad, A., Nasrollahi, A., Labbafi, Y., Azma, M., Ghavam, F., & Honarbakhsh, S. (2000). Integrated pest management of codling moth and other important pests of Pomoidea fruit trees. *Journal of Agricultural Sciences, 6*(2), 15–45 [in Persian, English summary].

Shojai, M., Ostovan, H., Hosseini, M., Sadighfar, M., Khodaman, A., Labbafi, Y., Nasrolahi, A., Ghavam, F., & Honarbakhsh, S. (2002). Biocenotic potentials of apple orchards IPM in organic crop production programme. *Journal of Agricultural Sciences, 8*(1), 1–27 [in Persian, English summary].

Siahpoush, A., Azimi, A., Rabee, R., & Mozaffari, M. (1993). Introduction of three species of *Mythimna* (Lep.: Noctuidae) at Khuzestan corn fields. In *Proceedings of the 11th Iranian Plant Protection Congress of Iran, 28 August – 2 September 1993* (p. 94). University of Guilan.

Smith, M. A., Rodriguez, J. J., Whitfield, J. B., Deans, A., Janzen, D. H., Hallwachs, W., & Herbert, P. D. N. (2008). Extreme diversity of tropical parasitoid wasps exposed by integration of natural history, DNA barcoding, morphology, and collections. *Proceedings of the National Academy of Sciences of the United States of America, 105*(34), 12359–12364.

Smith, M. A., Fernández-Triana, J. L., Eveleigh, E., Gómez, J., Guclu, C., Hallwachs, W., Hebert, P. D. N., et al. (2013). DNA barcoding and the taxonomy of Microgastrinae wasps (Hymenoptera: Braconidae): impacts after 8 years and nearly 20 000 sequences. *Molecular Ecology Resources, 13*, 168–176.

Sobhani, M., Goldansaz, S. H., & Hatami, B. (2012). Study of larval parasitoids of carob moth *Ectomyelois ceratoniae* (Lep.: Pyralidae) in Kashan region. In *Proceedings of the 20th Iranian Plant Protection Congress, 26–29 August 2012* (p. 83). University of Shiraz.

Stahlhut, J. K., Fernández-Triana, J., Adamowicz, S. J., Buck, M., Goulet, H., Herbert, P. D. N., Huber, J. T., Merilo, M. T., Sheffield, C. S., Woodcock, T., & Smith, M. A. (2013). DNA barcoding reveals diversity of Hymenoptera and dominance of parasitoids in a sub-Arctic environment. *BMC Ecology, 13*(2), 1–13.

Tawfik, M. F. S., Hafez, M., & Ibrahim, A. A. (1977). On the bionomics of *Microplitis rufiventris* Kok. (Hym., Braconidae). *Bulletin de la Société Entomologique d'Egypte, 6*(1978), 123–135.

Telenga, N. A. (1955). Fam. Braconidae, subfamilies Microgastrinae, Agathidinae. *Fauna Rossili (Hymenoptera), 5*(4), 311.

Tobias, V. I. (1971). Review of the Braconidae (Hymenoptera) of the U.S.S.R. *Trudy Vsesoyuznogo Entomologicheskogo Obshchestva, 54*, 156–268.

Tobias, V. I. (1976). *Braconids of the Caucasus (Hymenoptera, Braconidae). Opred. Faune SSSR* (p. 286). Leningrad: Nauka Press.

Tobias, V. I. (1986). Acaeliinae, Cardiochelinae, Microgastrinae, Miracinae. Supplement, pp. 336–501. In G. S. Medvedev (Ed.), *Opredelitel nasekomyeh Evropeiskoi Tsasti SSSR 3, Peredpontdatokrylye 4. Opr. Faune SSSR* (Vol. 145, pp. 1–501). Leningrad: Nauka [in Russian].

van Achterberg, C. (1993). Illustrated key to the subfamilies of the Braconidae (Hymenoptera: Ichneumonoidea). *Zoologische Verhandelingen, 283*, 1–189.

Vidal, S. (1997). Determination list of entomophagous insects, Nr. 13. *IOBC-WPRS Bulletin, 20*(2), 1–8.

Whitfield, J. B. (1997). Subfamily Microgastrinae. In R. A. Wharton, P. M. Marsh, & M. J. Sharkey (Eds.), *Manual of the New World Genera of the Family Braconidae (Hymenoptera)* (pp. 333–366). Washington: The International Society of Hymenopterists, 439 pp.

Whitfield, J. B., Marquis, R. J., & Le Corff, J. (1999). Host associations of braconid parasitoids (Hymenoptera: Braconidae) reared from Lepidoptera feeding on oaks (*Quercus* spp.) in the Missouri Ozarks. *Entomological News, 110*, 225–230.

Whitfield, J. B., Rodriguez, J. J., & Masonick, P. K. (2009). Reared microgastrine wasps (Hymenoptera: Braconidae) from Yanayacu Biological Station and environs (Napo province, Ecuador): Diversity and host specialization. *Journal of Insect Science, 9*, 31.

Whitfield, J. B., Austin, A. D., & Fernández-Triana, J. (2018). Systematics, biology and evolution of microgastrine parasitoid wasps. *Annual Review of Entomology, 63*, 389–406.

Wilkinson, D. S. (1936). A list of Lepidoptera from which parasites are particularly desired. *Entomologist, 69*, 81–84.

Wilkinson, D. S. (1937). A new species of *Apanteles* (Hym. Brac.) bred from *Myelois ceratoniae* attacking carobs in Cyprus. *Bulletin of Entomological Research, 28*, 463–466.

Wysoki, M., & Izhar, V. (1981). Biological data on *Apanteles cerialis* Nixon (Hymenoptera: Braconidae), a parasite of *Boarmia (Ascotis) selenaria* Schiff. (Lepidoptera: Geometridae). *Phytoparasitica, 9*, 19–25.

Yu, D. S., van Achterberg, C., & Horstmann, K. (2016). *Taxapad 2016, Ichneumonoidea 2015, Database on flash-drive*. Nepean, Ontario, Canada.

Zargar, M., Gupta, A., Talebi, A. A., & Farahani, S. (2019a). A review of the Iranian species of genus *Iconella* Mason (Hymenoptera: Braconidae: Microgastrinae) with description of two new species. *Zootaxa, 4586*(3), 491–504.

Zargar, M., Gupta, A., Talebi, A. A., & Farahani, S. (2019b). Three new species and two new records of the genus *Cotesia* Cameron (Hymenoptera: Braconidae) from Iran. *European Journal of Taxonomy, 571*, 1–25.

Zargar, M., Gupta, A., Talebi, A. A., & Farahani, S. (2020). Description of a new species of the genus *Deuterixys* (Hymenoptera: Braconidae: Microgastrinae) from Iran. *Biologia, 75*, 267–272.

Žikić, V., Lotfalizadeh, H., Sadeghi, S., Petrović, A., Janković, M., & Tomanović, Z. (2014). New record and new associations two leaf miner parasitoids (Hymenoptera: Braconidae: Microgastrinae) from Iran. *Archives of Biological Science Belgrade, 66*(4), 1591–1594.

Microtypus desertorum Shestakov, 1932 (Microtypinae), ♀, lateral habitus. *Photo prepared by S.R. Shaw.*

Microtypus wesmaelii Ratzeburg, 1848 (Microtypinae), ♂, lateral habitus. *Photo prepared by S.R. Shaw.*

Chapter 22

Subfamily Microtypinae Szépligeti, 1908

Scott Richard Shaw[1], Hassan Ghahari[2], Neveen Samy Gadallah[3] and Donald L.J. Quicke[4]

[1]*UW Insect Museum, Department of Ecosystem Science and Management, University of Wyoming, Laramie, WY, United States;* [2]*Department of Plant Protection, Yadegar-e Imam Khomeini (RAH) Shahre Rey Branch, Islamic Azad University, Tehran, Iran;* [3]*Entomology Department, Faculty of Science, Cairo University, Giza, Egypt;* [4]*Integrative Ecology Laboratory, Department of Biology, Faculty of Science, Chulalongkorn University, Pathumwan, Bangkok, Thailand*

Introduction

Microtypinae Szépligeti, 1908 is a small subfamily of Braconidae. Even so, microtypines are widely distributed in almost all parts of the world except the Australian region (Yu et al., 2016). The subfamily Microtypinae is revised by van Achterberg (1992), where he provided a key to genera, and proposed the genus *Neomicrotypus* van Achterberg. The subfamily comprises 23 known species classified into three genera (*Microtypus* Ratzeburg, *Neomicrotypus* van Achterberg, and *Plesiotypus* van Achterberg) (Čapek & van Achterberg, 1992; Edmardash et al., 2017; Sabahatullah et al., 2015; van Achterberg, 1992, 2010; Yu et al., 2016). Most of the species belong to the genus *Microtypus*, with 17 species (of which seven species are extant and the other 10 are known only from fossils) (Edmardash et al., 2017; Yu et al., 2016).

Microtypines were formerly included in Orgilinae or Homolobinae (van Achterberg, 1984, 1987). Based on larval and adult morphology and biology, a sister relationship, Orgilinae (Homolobinae + Microtypinae), is proposed by some authors (e.g., Čapek, 1970; van Achterberg, 1984, 1992). This relationship is also strongly supported by the phylogenetic analysis of Sharanowski et al. (2011) who included Microtypinae among the macrocentroid subcomplex (helconid complex). They are thought to utilize lepidopteran hosts.

Members of the Microtypinae are mainly diagnosed by the small triangular or trapezoidal second submarginal cell of the fore wing; the first metasomal tergite having weak carinae or none; having the scutellum depressed medioposteriorly; the presence of a slit-like or triangular pronope; and the absence of hind tibial pegs apically (van Achterberg, 1987, 1992; Wharton et al., 1997).

Although known only from a few *Microtypus* species (*Microtypus trigonus* (Nees) and *Microtypus wesmaelii* Ratzeburg), microtypines are considered to be koinobiont endoparasitoids of concealed larvae of Microlepidoptera of the families Pyralidae, Gelechiidae, Tortricidae, and Yponomeutidae (van Achterberg, 1993; Yu et al., 2016).

Checklists of Regional Microtypinae. Fallahzadeh and Saghai (2010) represented one species (without precise locality cited). Farahani et al. (2016) and Yu et al. (2016) listed two and three species, respectively, in one genus. Beyarslan et al. (2017), Samin, Coronado-Blanco, Kavallieratos et al. (2018), and Samin, Coronado-Blanco, Fischer et al. (2018) both cataloged four species in one genus. The present checklist includes six species in one genus (*Microtypus* Ratzeburg, 1848).

List of species of the subfamily Microtypinae recorded in the Middle East

Subfamily Microtypinae Szépligeti, 1908

Genus *Microtypus* Ratzeburg, 1848

Microtypus aegypticus Edmardash et al., 2017

Distribution in the Middle East: Egypt (Edmardash et al., 2017).
Extralimital distribution: None.
Host records: Unknown.

Microtypus algiricus Szépligeti, 1908

Catalogs with Iranian records: Yu et al. (2016), Beyarslan et al. (2017), Samin, Coronado-Blanco, Kavallieratos et al. (2018), Samin, Coronado-Blanco, Fischer et al. (2018).

Distribution in Iran: Khuzestan (Samin et al., 2016).
Distribution in the Middle East: Egypt (Edmardash et al., 2017), Iran (Samin et al., 2016), Jordan (Čapek & van Achterberg, 1992).
Extralimital distribution: Algeria, China.
Host records: Unknown.

Microtypus desertorum Shestakov, 1932

Catalogs with Iranian records: Fallahzadeh and Saghaei (2010), Farahani et al. (2016), Yu et al. (2016), Beyarslan et al. (2017), Samin, Coronado-Blanco, Kavallieratos et al. (2018), Samin, Coronado-Blanco, Fischer et al. (2018).
Distribution in Iran: Alborz (Farahani et al., 2014), Sistan and Baluchestan (Hedwig, 1957), Iran (no specific locality cited) (Shenefelt, 1970 under Mimagathidinae; Tobias, 1976, 1986 under Orgilinae).
Distribution in the Middle East: Egypt (Edmardash et al., 2017), Iran (see references above), Israel−Palestine (Papp, 2012).
Extralimital distribution: Algeria, China, Kazakhstan, Mongolia, Turkmenistan, Uzbekistan.
Host records: Unknown.

Microtypus trigonus (Nees von Esenbeck, 1843)

Catalogs with Iranian records: Beyarslan et al. (2017), Samin, Coronado-Blanco, Kavallieratos et al. (2018), Samin, Coronado-Blanco, Fischer et al. (2018).
Distribution in Iran: Mazandaran (Beyarslan et al., 2017).
Distribution in the Middle East: Iran (Beyarslan et al., 2017), Turkey (Beyarslan & Çetin Erdoğan, 2011).
Extralimital distribution: Canada, Mongolia, Russia, Tajikistan.
Host records: Summarized by Yu et al. (2016) as being a parasitoid of the gelechiids *Chionodes tragicella* (Heyden), and *Teleiodes saltuum* (Zeller); the pyralid *Acrobasis consociella* (Hübner); and the tortricids *Cydia illutana* (Herrich-Schäffer), and *Zeiraphera griseana* (Hübner).

Microtypus vanharteni van Achterberg, 2010

Distribution in the Middle East: Egypt (Edmardash et al., 2017), United Arab Emirates (van Achterberg, 2010).
Extralimital distribution: None.
Host records: Unknown.

Microtypus wesmaelii Ratzeburg, 1848

Catalogs with Iranian records: Farahani et al. (2016), Yu et al. (2016), Beyarslan et al. (2017), Samin, Coronado-Blanco, Kavallieratos et al. (2018), Samin, Coronado-Blanco, Fischer et al. (2018).
Distribution in Iran: Fars (Al-e Mansour & Mostafavi, 1993), Guilan (Ghahari & Fischer, 2011; Sakenin et al., 2012; Farahani et al., 2014), Kuhgiloyeh and Boyerahmad (Samin et al., 2015), Mazandaran (Kian et al., 2020).
Distribution in the Middle East: Iran (see references above), Turkey (Beyarslan & Çetin Erdoğan, 2011).
Extralimital distribution: Bulgaria, Canada, China, former Czechoslovakia, Germany, Hungary, Netherlands, Russia, Slovakia, United Kingdom, United States of America (Yu et al., 2016), Tajikistan (Čapek & van Achterberg, 1992).
Host records: Summarized by Yu et al. (2016) as being a parasitoid of the crambid *Loxostege sticticalis* (L.); the gelechiid *Coleotechnites atrupictellus* (Dietz); the pyralids *Acrobasis* spp., *Dioryctria auranticella* (Grote), and *Pococera asperatella* (Clemens); the tortricid *Grapholita molesta* (Busck); and the yponomeutid *Zelleria haimbachi* Busck.

Conclusion

Six valid species of the subfamily Microtypinae in the genus *Microtypus* Ratzeburg, 1848 have been reported from six of the Middle Eastern countries (Egypt, Iran, Israel−Palestine, Jordan, Turkey, and United Arab Emirates), of which Iran and Egypt (each with four species) are the richest. Microtypinae of Iran with four reported species represents 17.4% of the world species, which are recorded from seven provinces, Alborz, Fars, Guilan, Khuzestan, Kuhgiloyeh and Boyerahmad, Mazandaran, and Sistan and Baluchestan. No host species have been recorded for these parasitoid wasps in Iran. Comparison of the Microtypinae fauna of Iran with the Middle East and adjacent countries to Iran indicates the fauna of Iran and Egypt is the most diverse (both with four species), followed by Russia and Turkey (each with two species), Armenia, Israel−Palestine, Jordan, Turkmenistan, and United Arab Emirates (each with one species). No species have been recorded from Afghanistan, Azerbaijan, Bahrain, Cyprus, Iraq, Kazakhstan, Kuwait, Lebanon, Oman, Pakistan, Qatar, Saudi Arabia, Syria, and Yemen (Yu et al., 2016). Additionally, among the 23 countries of the Middle East and adjacent to Iran, Egypt, and Russia share two species with Iran, and Israel−Palestine, Jordan, Kazakhstan, Turkey, and Turkmenistan share one species.

References

Al-e Mansour, H., & Mostafavi, M. S. (1993). The first record of Braconidae bees on forest and range vegetations in Fars province. In *Proceedings of the 11th Iranian plant protection congress, 28 August − 2 September 1993* (p. 236). Rasht: University of Guilan.

Beyarslan, A., & Çetin Erdoğan, Ö. (2011). A study of Orgilinae and Microtypinae (Hymenoptera: Braconidae) from Turkey, with the description of a new species. *Biologia (Bratislava), 66*(1), 121−129.

Beyarslan, A., Gadallah, N. S., & Ghahari, H. (2017). An annotated catalogue of the Iranian Microtypinae and Rogadinae (Hymenoptera: Braconidae). *Zootaxa, 4291*(1), 99–116.

Čapek, M. (1970). A new classification of the Braconidae (Hymenoptera) based on the cephalic structures of the final instar larva and biological evidence. *The Canadian Entomologist, 102*, 846–875.

Čapek, M., & van Achterberg, C. (1992). A revision of the genus *Microtypus* Ratzeburg (Hymenoptera: Braconidae). *Zoologische Mededelingen, 66*(21), 323–338.

Edmardash, Y. A., Gadallah, N. S., & van Achterberg, K. (2017). Discovery of the subfamily Microtypinae (Hymenoptera: Braconidae) in Egypt, with the description of a new species. *Zoology in the Middle East, 63*(3), 239–249.

Fallahzadeh, M., & Saghaei, N. (2010). Checklist of Braconidae (Insecta: Hymenoptera) from Iran. *Munis Entomology & Zoology, 5*(1), 170–186.

Farahani, S., Talebi, A. A., van Achterberg, C., & Rakhshani, E. (2014). A taxonomic study of Orgilinae and Microtypinae from Iran (Hymenoptera, Braconidae). *Spixiana, 37*(1), 93–102.

Farahani, S., Talebi, A. A., & Rakhshani, E. (2016). Iranian Braconidae (Insecta: Hymenoptera: Ichneumonoidea): diversity, distribution and host association. *Journal of Insect Biodiversity and Systematics, 2*(1), 1–92.

Ghahari, H., & Fischer, M. (2011). A study on the Braconidae (Hymenoptera: Ichneumonoidea) from some regions of northern Iran. *Entomofauna, 32*, 181–196.

Hedwig, K. (1957). Ichneumoniden und Braconiden aus den Iran 1954 (Hymenoptera). *Jahresheft des Vereins für Vaterlaendische Naturkunde, 112*(1), 103–117.

Kian, N., Goldasteh, S., & Farahani, S. (2020). A survey on abundance and species diversity of braconid wasps in forest of Mazandaran province. *Journal of Entomological Research, 12*(1), 61–69 (in Persian, English summary).

Papp, J. (2012). A contribution to the Braconid fauna of Israel (Hymenoptera: Braconidae), 3. *Israel Journal of Entomology, 41–42*, 165–219.

Sabahatullah, M., Inayatullah, M., & Tahira, Q. A. (2015). First record of Microtypinae (Hymenoptera: Braconidae) from Pakistan with the description of a new species. *Journal of Entomology and Zoology Studies, 3*(2), 127–230.

Sakenin, H., Naderian, H., Samin, N., Rastegar, J., Tabari, M., & Papp, J. (2012). On a collection of Braconidae (Hymenoptera) from northern Iran. *Linzer biologische Beiträge, 44*(2), 1319–1330.

Samin, N., van Achterberg, C., & Ghahari, H. (2015). A faunistic study of Braconidae (Hymenoptera: Ichneumonoidea) from southern Iran. *Linzer biologische Beiträge, 47*(2), 1801–1809.

Samin, N., van Achterberg, C., & Çetin Erdoğan, Ö. (2016). A faunistic study on some subfamilies of Braconidae (Hymenoptera: Ichneumonoidea) from Iran. *Arquivos Entomolóxicos, 15*, 153–161.

Samin, N., Coronado-Blanco, J. M., Kavallieratos, N. G., Fischer, M., & Sakenin, H. (2018). Recent findings on Braconidae (Hymenoptera: Ichneumonoidea) of Iran with an updated checklist. *Acta Biologica Turcica, 31*(4), 160–173.

Samin, N., Coronado-Blanco, J. M., Fischer, M., van Achterberg, C., Sakenin, H., & Davidian, E. (2018). Updated checklist of Iranian Braconidae (Hymenoptera: Ichneumonoidea) with twenty-three new records. *Natura Somogyiensis, 32*, 21–36.

Sharanowski, B. J., Dowling, A. P. G., & Sharkey, M. J. (2011). Molecular phylogenetics of Braconidae (Hymenoptera: Ichneumonoidea), based on multiple nuclear genes, and implications for classification. *Systematic Entomology, 36*, 549–572.

Shenefelt, R. D. (1970). *Braconidae 2. Helconinae, Calyptinae, Mimagathidinae, Triapinae. Hymenopterorum Catalogus* (Nova ed., pp. 177–306) Pars 5.

Tobias, V. I. (1976). Braconids of the Caucasus (Hymenoptera, Braconidae). Opred. In *Faune SSSR* (Vol. 110, pp. 1–286). Leningrad: Nauka Press.

Tobias, V. I. (1986). Subfamily Rogadinae, pp. 72–85. In G. S. Medvedev (Ed.), *Keys to the insects of the European part of the USSR. Vol. III. Hymenoptera Part IV* (pp. 1–883). New Delhi: Amerind Publishing Co.

van Achterberg, C. (1984). Essay on the phylogeny of Braconidae (Hymenoptera: Ichneumonoidea). *Entomologisk Tidskrift, 105*, 41–58.

van Achterberg, C. (1987). Revisionary notes of the subfamily Orgilinae (Hymenoptera: Braconidae). *Zoologische Verhandelingen, 242*, 1–111.

van Achterberg, C. (1992). Revision of the genera of the subfamily Microtypinae (Hymenoptera: Braconidae). *Zoologische Mededelingen, 66*(26), 369–380.

van Achterberg, C. (1993). Illustrated key to the subfamilies of the Braconidae. *Zoologische Verhandelingen, 283*, 1–189.

van Achterberg, C. (2010). Order Hymenoptera, family Braconidae, genus *Microtypus* Ratzeburg (Hymenoptera: Braconidae: Microtypinae). *Arthropod Fauna of the UAE, 3*, 381–387.

Wharton, R. W., Marsh, P. M., & Sharkey, M. J. (1997). *The New World Genera of the Family Braconidae (Hymenoptera)* (pp. 1–439). Special Publication of the International Society of Hymenopterists.

Yu, D. S., van Achterberg, C., & Horstmann, K. (2016). *Taxapad 2016, Ichneumonoidea 2015, Database on flash-drive, Nepean, Ontario, Canada.*

Mirax sp. (Miracinae), ♀, lateral habitus. *Photo prepared by S.R. Shaw.*

Mirax sp. (Miracinae), ♀, dorsal habitus. *Photo prepared by S.R. Shaw.*

Chapter 23

Subfamily Miracinae Viereck, 1918

Neveen Samy Gadallah[1], Hassan Ghahari[2], Scott Richard Shaw[3] and Donald L.J. Quicke[4]

[1]Entomology Department, Faculty of Science, Cairo University, Giza, Egypt; [2]Department of Plant Protection, Yadegar-e Imam Khomeini (RAH) Shahre Rey Branch, Islamic Azad University, Tehran, Iran; [3]UW Insect Museum, Department of Ecosystem Science and Management, University of Wyoming, Laramie, WY, United States; [4]Integrative Ecology Laboratory, Department of Biology, Faculty of Science, Chulalongkorn University, Pathumwan, Bangkok, Thailand

Introduction

Miracinae Viereck, 1918 is a small cosmopolitan microgastroid subfamily (Chen & van Achterberg, 2019), currently with 48 described species classified into three genera, *Centistidea* Rohwer, 1914, *Miracoides* Brues, 1933 (fossil genus), and *Mirax* Haliday, 1833 (Cauich-Kumul et al., 2014; Farahani et al., 2014; Papp, 2013; Yu et al., 2016), of which 17 species are known from the Palaearctic region (Farahani et al., 2014; Yu et al., 2016). It has been formerly included in the subfamily Microgastrinae by several authors (e.g., Čapek, 1970; Huddleston, 1978; Nixon, 1965) until being raised to a subfamily level by van Achterberg (1984). The genus *Centistidea* has been synonymized with *Mirax* by some authors (Muesebeck, 1922; Papp, 2013), but is treated as a separate genus by many others (Chen et al., 1997; Farahani et al., 2014; Penteado-Dias, 1999; van Achterberg & Mehrnejad, 2002; Yu et al., 2016). Miracinae is placed in the microgastroid complex (Murphy et al., 2008; Sharanowski et al., 2011), and this is supported in their molecular analyses. It was recovered as a sister group of Cardiochilinae + Microgastrinae (Dowton & Austin, 1998; Shi et al., 2005), a relationship that was corroborated by Sharanowski et al. (2011). In other studies, it was predominantly a sister group of Cardiochilinae (Banks & Whitfield, 2006; Belshaw et al., 1998; Murphy et al., 2008).

Miracines are small, compact insects recognized by their 14-segmented antenna; the absence of an occipital carina; the vein 2-SR of the fore wing being connected with the pterostigma or nearly so; the apical abscissa of the RS vein of the fore wing being either present as depressed line or completely absent; the hind wing with a strongly oblique cu-a vein; the absence of a prepectal carina; the dorsum of metasomal T1 being strongly narrowed toward apex and medially; the absence of a scutellar sulcus; and the metasoma with an inverted Y-shaped structure formed by the sclerotized part of the first metasomal tergites surrounded by lateral membranous areas (Farahani et al., 2014; Shaw & Huddleston, 1991; van Achterberg, 1984, 1993). They are koinobiont endoparasitoids attacking small plant-mining lepidopterous caterpillars of different families, mostly Nepticulidae, but also Heliozelidae and Lyonetiidae (Farahani et al., 2014; van Achterberg, 1993; Yu et al., 2016).

Checklists of Regional Miracinae. Fallahzadeh and Saghaei (2010) represented one species (without precise localities). Farahani et al. (2016), Gadallah and Ghahari (2016), Yu et al. (2016), Samin, Coronado-Blanco, Kavallieratos et al. (2018), and Samin, Coronado-Blanco, Fischer et al. (2018) all listed three species in two genera. The present Middle Eastern checklist includes six species in two genera, of which two species, *Centistidea pistaciella* and *Mirax caspiana*, are only known from the Iranian fauna. On the other hand, one species, *Centistidea tihamica*, is only known from Saudi Arabia, and one, *Mirax striata*, is only known from the Turkish fauna.

Key to genera of the subfamily Miracinae in the Middle East

1. Propodeum with a distinct median carina; notauli present, at least partly .... *Centistidea* Rohwer
— Propodeum without median carina; notauli absent, or shortly present anteriorly .... *Mirax* Haliday

List of species of the subfamily Miracinae recorded in the Middle East

Subfamily Miracinae Viereck, 1918

Genus *Centistidea* Rohwer, 1914

Centistidea (Paracentistidea) pistaciella van Achterberg and Mehrnejad, 2002

Catalogs with Iranian records: Fallahzadeh and Saghaei (2010), Yu et al. (2016), Gadallah and Ghahari (2016), Farahani et al. (2016), Samin, Coronado-Blanco, Kavallieratos et al. (2018), Samin, Coronado-Blanco, Fischer et al. (2018).

Distribution in Iran: Kerman (Mehrnejad & Basirat, 2009; van Achterberg & Mehrnejad, 2002).

Distribution in the Middle East: Iran.

Extralimital distribution: None.

Host records: In Iran, it was reared from the pistachio twig borer tineid *Kermania pistaciella* Amsel in pistachio orchards (Mehrnejad and Basirat, 2009; van Achterberg and Mehrnejad, 2002).

Centistidea sculpturator (Belokobylskij, 1989)

Distribution in the Middle East: Turkey (Beyarslan, 2009).

Extralimital distribution: Russia.

Host records: Unknown.

Centistidea tihamica Ghramh and Ahmad, 2019

Distribution in the Middle East: Saudi Arabia (Gharmh et al., 2019).

Extralimital distribution: None.

Host records: Unknown.

Genus *Mirax* Haliday, 1833

Mirax caspiana Farahani, Talebi, van Achterberg and Rakhshani, 2014

Catalogs with Iranian records: Yu et al. (2016), Gadallah and Ghahari (2016), Farahani et al. (2016), Samin, Coronado-Blanco, Kavallieratos et al. (2018), Samin, Coronado-Blanco, Fischer et al. (2018).

Distribution in Iran: Guilan (Farahani et al., 2014), Mazandaran (Farahani et al., 2014; Kian et al., 2020).

Distribution in the Middle East: Iran.

Extralimital distribution: None.

Host records: Unknown.

Mirax rufilabris Haliday, 1833

Catalogs with Iranian records: Yu et al. (2016), Gadallah and Ghahari (2016), Farahani et al. (2016), Samin, Coronado-Blanco, Kavallieratos et al. (2018), Samin, Coronado-Blanco, Fischer et al. (2018).

Distribution in Iran: Chaharmahal and Bakhtiari (Samin et al., 2016), Isfahan (Ghahari et al., 2011 as *Mirax dryochares* Marshall, 1898).

Distribution in the Middle East: Iran (see references above), Israel—Palestine (Papp, 2012), Turkey (Beyarslan, 2009; Inanç et al., 2012).

Extralimital distribution: Armenia, Azerbaijan, Bulgaria, Croatia, Czech Republic, Finland, France, Germany, Greece, Ireland, Italy, Lithuania, Malta, Moldova, Poland, Portugal, Romania, Russia, Slovakia, Spain, Sweden, Ukraine, United Kingdom.

Host records: Summarized by Yu et al. (2016) as being a parasitoid of several lepidopteran insect pests of the families Coleophoridae, Nepticulidae, and Tineidae.

Mirax striata Beyarslan, 2009

Distribution in the Middle East: Turkey (Beyarslan, 2009).

Extralimital distribution: None.

Host records: Unknown.

Conclusion

Six species of the subfamily Miracinae (12% of the world species) in two genera, *Cetistidea* Rohwer and *Mirax* Haliday (each with three recorded species), have been reported from the Middle East countries. In total, three species of Miracinae in two genera, *Centistidea* (one species) and *Mirax* (two species), have been recorded from Iran (6% of the world described species). Two of them, *Centistidea (Paracentistidea) pistaciella* and *Mirax caspiana*, are only known from Iran (endemic to Iran). The species of Iranian Miracinae have been recorded from five provinces, Chaharmahal and Bakhtiari, Guilan, Isfahan, Kerman, and Mazandaran. Host species of Iranian Miracinae has been detected for only one species which belongs to the family Tineidae (Lepidoptera) (van Achterberg & Mehrnejad, 2002). Comparison of the Miracinae fauna of Iran with the Middle East and adjacent countries to Iran indicates the fauna of Russia with eight species in three genera (Belokobylskij & Lelej, 2019) is more diverse than Iran, followed by Turkey (three species), Armenia, Azerbaijan, Israel—Palestine, and Saudi Arabia (each with one species). Among the Middle East and adjacent countries to Iran, both of Russia and Turkey share two species with Iran, and then Armenia, Azerbaijan, and Israel—Palestine (each with one species). No species have been recorded from the other Middle East countries or adjacent ones to Iran.

References

Banks, J. C., & Whitfield, J. B. (2006). Dissecting the ancient rapid radiation of microgastrine wasps genera using additional nuclear genes. *Molecular Phylogenetics and Evolution, 41*, 690—703.

Belokobylskij, S. A., & Lelej, A. S. (2019). Annotated catalogue of the Hymenoptera of Russia. Volume II. Apocrita: Parasitica. *Proceedings of the Zoological Institute of the Russian Academy of Sciences*, (Suppl. 8), 594 pp.

Belshaw, R., Fitton, M., Herniou, E., Gimeno, C., & Quicke, D. L. J. (1998). A phylogenetic reconstruction of the Ichneumonoidea (Hymenoptera) based on the D2 variable region of 28S ribosomal RNA. *Systematic Entomology, 23*(2), 109−123.

Beyarslan, A. (2009). A survey of Turkish Miracinae, with the description of a new species, *Mirax striacus* (Hymenoptera: Braconidae). *Entomological News, 120*(3), 291−296.

Čapek, M. (1970). A new classification of the Braconidae (Hymenoptera) based on the cephalic structures of the final instar larva and biological evidence. *The Canadian Entomologist, 102*, 846−875.

Cauich-Kumul, R., López-Martínez, V., & de García-Ramírez, M. J. (2014). Two new species of braconid wasps (Hymenoptera: Braconidae: Miracinae: *Mirax* and Rogadinae: *Choreborogas*) from Mexico. *Florida Entomologist, 97*(3), 902−910.

Chen, X.-X., He, J.-H., & Ma, Y. (1997). Two new species of the subfamily Miracinae (Hym.: Braconidae) from China. *Wuyi Science Journal, 13*, 63−69.

Dowton, M., & Austin, A. D. (1998). Phylogenetic relationships among the microgastroid wasps (Hymenoptera: Braconidae): combined analysis of 16S and 28S rDNA genes. *Molecular Phylogenetics and Evolution, 10*(3), 354−366.

Fallahzadeh, M., & Saghaei, N. (2010). Checklist of Braconidae (Insecta: Hymenoptera) from Iran. *Munis Entomology & Zoology, 5*(1), 170−186.

Farahani, S., Talebi, A. A., van Achterberg, C., & Rakhshani, E. (2014). A new species of the genus *Mirax* Haliday, 1833 (Hymenoptera: Braconidae: Miracinae) from Iran. *Annales Zoologici, 64*(4), 677−682.

Farahani, S., Talebi, A. A., & Rakhshani, E. (2016). Iranian Braconidae (Insecta: Hymenoptera: Ichneumonoidea), diversity, distribution and host association. *Journal of Insect Biodiversity and Systematics, 2*(1), 1−92.

Gadallah, N. S., & Ghahari, H. (2016). An updated checklist of the Iranian Miracinae, Pambolinae and Sigalphinae (Hymenoptera: Braconidae). *Orsis, 30*, 51−61.

Ghahari, H., Fischer, M., & Papp, J. (2011). A study of the braconid wasps (Hymenoptera: Braconidae) from Isfahan province, Iran. *Entomofauna, 32*, 261−270.

Ghramh, H. A., Ahmad, Z., & Pandey, K. (2019). Three new species of the genus *Centistidea* Rohwer, 1914 (Hymenoptera, Braconidae, Miracinae) from India and Saudi Arabia. *ZooKeys, 889*, 37−47.

Huddleston, T. (1978). Braconidae [and] Aphidiidae. In G. S. Kloet, & W. D. A. Hinks (Eds.), *Handbooks for the identification of British Insects: Vol. 11(4). A checklist of British insects (Second edition). Part 4: Hymenoptera* (pp. 46−62).

Inanç, O., Halil, B., & Beyarslan, A. (2012). *Chelonus flavipalpis* Szépligeti, 1896 and *Mirax rufilabris* Haliday, 1833 (Hymenoptera: Braconidae): Two larva-pupa parasitoids of Pistachio twig borer *Kirmania pistaciella* Amsel, 1964 (Lepidoptera: Oinophilidae) with the parasitization ratios from Turkey. *Munis Entomology & Zoology, 7*(1), 238−242.

Kian, N., Goldasteh, S., & Farahani, S. (2020). A survey on abundance and species diversity of Braconid wasps in forest of Mazandaran province. *Journal of Entomological Research, 12*(1), 61−69 (in Persian, English summary).

Mehrnejad, M. R., & Basirat, M. (2009). Parasitoid complex of the pistachio twig borer moth, *Kermania pistaciella*, in Iran. *Biocontrol Science and Technology, 19*(5), 499−510.

Muesebeck, C. F. W. (1922). A revision of the North American Ichneumonflies belonging to the subfamilies Neoneurinae and Microgasterinae. *Proceedings of the United States National Museum, 61*(2436), 1−76.

Murphy, N., Banks, J., Whitfield, J. B., & Austin, A. (2008). Phylogeny of the parasitic microgastroid subfamilies (Hymenoptera: Braconidae) based on sequence data from seven genes, with an improved time estimate of the origin of the lineage. *Molecular Phylogenetics and Evolution, 47*, 378−395.

Nixon, G. E. J. (1965). A reclassification of the tribe Microgasterini (Hymenoptera: Braconidae). *Bulletin of the British Museum (Natural History), Entomology series (Supplement), 2*, 1−284.

Papp, J. (2013). Eleven new *Mirax* Haliday, 1833 species, from Colombiand Honduras and key to the sixteen Neotropical *Mirax* species (Hymenoptera: Braconidae: Miracinae). *Acta Zoologica Academiae Scientiarum Hungaricae, 59*(2), 97−129.

Penteado-Dias, A. M. (1999). New species of parasitoids on *Perileucoptera coffeella* (Guérin-Meneville) (Lepidoptera: Lyonetiidae) from Brazil. *Zoologische Mededelingen, 72*(10), 189−197.

Samin, N., van Achterberg, C., & Çetin Erdoğan, O. (2016). A faunistic study on some subfamilies of Braconidae (Hymenoptera: Ichneumonoidea) from Iran. *Arquivos Entomolóxicos, 15*, 153−161.

Samin, N., Coronado-Blanco, J. M., Kavallieratos, N. G., Fischer, M., & Sakenin, H. (2018). Recent findings on Braconidae (Hymenoptera: Ichneumonoidea) of Iran with an updated checklist. *Acta Biologica Turcica, 31*(4), 160−173.

Samin, N., Coronado-Blanco, J. M., Fischer, M., van Achterberg, C., Sakenin, H., & Davidian, E. (2018). Updated checklist of Iranian Braconidae (Hymenoptera: Ichneumonoidea) with twenty-three new records. *Natura Somogyiensis, 32*, 21−36.

Sharanowski, B. J., Dowling, A. P. G., & Sharkey, M. J. (2011). Molecular phylogenetics of Braconidae (Hymenoptera: Ichneumonoidea), based on multiple nuclear genes, and implications for classification. *Systematic Entomology, 36*, 549−572.

Shaw, M. R., & Huddleston, T. (1991). Classification and biology of braconid wasps (Hymenoptera: Braconidae). *Handbooks for the Identification of British Insects, 7*, 1−126.

Shi, M., Chen, X. X., & van Achterberg, C. (2005). Phylogenetic relationships among the Braconidae (Hymenoptera: Ichneumonoidea) inferred from partial 16S rDNA, 28S rDNA D2, 18S rDNA gene sequences and morphological characters. *Molecular Phylogenetics and Evolution, 37*(1), 104−116.

van Achterberg, C. (1984). Essay on the phylogeny of Braconidae (Hymenoptera: Ichneumonoidea). *Entomologisk Tidskrift, 105*, 41−58.

van Achterberg, C. (1993). Illustrated key to the subfamilies of the Braconidae. *Zoologische Verhandelingen, 283*, 1−189.

van Achterberg, C., & Mehrnejad, M. R. (2002). The braconid parasitoids (Hymenoptera: Braconidae) of *Kermania pistaciella* Amsel (Lepidoptera: Tineidae: Hieroxestinae) in Iran. *Zoologische Mededelingen, 76*, 27−39.

Yu, D. S., van Achterberg, C., & Horstmann, K. (2016). *Taxapad 2016, Ichneumonoidea 2015, Database on flash-drive, Nepean, Ontario, Canada.*

Diachasmimorpha longicaudata (Ashmead, 1905) (Opiinae), ♀, lateral habitus. *Photo prepared by S.R. Shaw.*

Phaedrotoma cingulatus (Wesmael, 1835) (Opiinae), ♀, lateral habitus. *Photo prepared by S.R. Shaw.*

Chapter 24

Subfamily Opiinae Blanchard, 1845

Neveen Samy Gadallah[1], Francisco Javier Peris Felipo[2], Hassan Ghahari[3], Donald L.J. Quicke[4], Scott Richard Shaw[5] and Maximilian Fischer[6]

[1]*Entomology Department, Faculty of Science, Cairo University, Giza, Egypt;* [2]*Basel, Switzerland;* [3]*Department of Plant Protection, Yadegar-e Imam Khomeini (RAH) Shahre Rey Branch, Islamic Azad University, Tehran, Iran;* [4]*Integrative Ecology Laboratory, Department of Biology, Faculty of Science, Chulalongkorn University, Pathumwan, Bangkok, Thailand;* [5]*UW Insect Museum, Department of Ecosystem Science and Management, University of Wyoming, Laramie, WY, United States;* [6]*Naturhistorisches Museum, Zoologische Abteilung, Wien, Austria*

Introduction

Within the Braconidae, the Opiinae is one of the largest subfamilies with about 2063 described species classified into 39 genera and two tribes (Biosterini Fischer, 1970; Opiini Blanchard, 1845) (Chen & van Achterberg, 2019; Yu et al., 2016). By far the largest genus is *Opius*, with over 1280 species (Yu et al., 2016). The first world Opiinae revisions were carried out by Fischer (1972, 1977, 1986, 1987), who, in 1987, established the subgeneric classification of *Opius*, comprising 38 subgenera. The validity of this subgeneric classification has not been widely accepted. Some of them have now been synonymized and others raised to genus level (Li et al., 2013; Wharton, 1997). All species of the Opiinae are solitary koinobiont endoparasitoids of fly (Diptera) larvae, especially of mining or fruit infesting species (Li et al., 2013; Quicke & van Achterberg, 1990; Yu et al., 2016). Opiines therefore play important roles in controlling dipteran pests of the families Agromyzidae and Tephritidae (Li et al., 2013; Schuster & Wharton, 1993; Wharton, 1984, 1997). Most species for which biological observations are available are larval parasitoids, but a few oviposit into the egg of the hosts and are egg-larval parasitoids (Wang et al., 2004).

Most Opiinae are small, stout, and weakly sculptured wasps. Many species are uniformly brownish to blackish in color, just a few having brighter markings such as with orange or yellow patterns. Opiines are diagnosed by the absence of the occipital carina medio-dorsally; and also by having the clypeus usually broadly emarginate, with a relatively narrow space or opening that could be seen between it and mandibles when closed (leading to a confusion between them and other cyclostomes). Additionally, the 3-SR vein of the fore wing is usually distinctly longer than the 2-SR, and vein M+CU1 is usually largely unsclerotized (not tubular, only pigmented). In cases where M+CU1 is fully sclerotized, then the laterope is distinct on first metasomal tergite; and the second metasomal tergite is mostly smooth with at most a few striations (Shaw & Huddleston, 1991; van Achterberg, 1993). A rare exception is the genus *Ademon* Haliday, which has the second tergum distinctly granulate.

All phylogenetic studies to date dealing with the opiinae (e.g., Belshaw et al., 1998; Dowton et al., 1998; Sharanowski et al., 2011; Shi et al., 2005; Wharton et al., 2006; Zaldívar-Revirón et al., 2006), the Opiinae are recovered most closely related to the Alysiinae, another morphologically similar group that also are entirely endoparasitoids of Diptera. These studies provide support for what has been termed the alysioid subcomplex, comprising the subfamilies Opiinae, Alysiinae and Exothecinae, and Gnamptodontinae. This complex, in turn, appears to be a sister group of the Braconinae. The great majority of species of all these subfamilies lack a prepectal carina. Opiines differ from Exothecinae by having the labrum more or less flat (rather than being strongly concave), and most opiines have distinctly broader and more triangular fore wings, often with a narrow and elongated pterostigma (Shaw & Huddleston, 1991). Opiines differ from alysiines in that the latter have exodont mandibles. The Gnamptodontinae have a distinctive raised, usually polished, anterior zone of the second metasomal tergite.

Checklists of Regional Opiinae. Modarres Awal (1997, 2012) listed one species (*Opius* (*Misophthora*) *monilicornis* Fischer, 1962), and two species, respectively, in two genera. Fallahzadeh and Saghaei (2010) represented seven species in four genera (without precise localities). Khajeh et al. (2014) listed 68 species in eight genera. Farahani et al. (2016) and

Yu et al. (2016) listed 95 and 99 species, respectively, in 13 genera. Gadallah et al. (2016) cataloged 101 species in 11 genera. Samin, Coronado-Blanco, Kavallieratos et al. (2018), Samin, Coronado-Blanco, Fischer et al. (2018) listed 101 and 104 species, respectively, in 13 genera. The present checklist includes 238 species in 23 genera and two tribes (Biosterini and Opiini) in the Middle East. In this chapter, we follow Yu et al. (2016), and Chen and van Achterberg (2019) in classification, and Yu et al. (2016) for general distribution of species, in addition to most recent references in some cases.

List of species of the subfamily Opiinae recorded in the Middle East

Subfamily Opiinae Blanchard, 1845

Tribe Biosterini Fischer, 1970

Genus *Biosteres* Foerster, 1863

Biosteres (*Biosteres*) *adanaensis* Fischer & Beyarslan, 2005
Distribution in the Middle East: Turkey (Beyarslan & Fischer, 2011b; Fischer & Beyarslan, 2005).
Extralimital distribution: None.
Host records: Unknown.

Biosteres (*Biosteres*) *analis* (Wesmael, 1835)
Distribution in the Middle East: Turkey (Beyarslan, 2015).
Extralimital distribution: Austria, Belgium, Bulgaria, Finland, France, Germany, Hungary, Ireland, Italy, Russia, Spain, Sweden, United Kingdom.
Host records: Summarized by Yu et al. (2016) as being a parasitoid of the anthomyiid *Chirosia histricina* (Rondani); the cecidomyiid *Wachtliella rosarum* (Hardy); and the scatophagid *Parallelomma vittatum* (Meigen).

Biosteres (*Biosteres*) *carbonarius* (Nees von Esenbeck, 1834)
Distribution in the Middle East: Turkey (Beyarslan, 2015; Beyarslan & Fischer, 2011a, b).
Extralimital distribution; Albania, Austria, Belgium, Bulgaria, Canada, Denmark, Estonia, Faeroe islands, Finland, France, Germany, Hungary, Iceland, Ireland, Italy, Japan, Korea, Lithuania, Netherlands, Norway, Poland, Russia, Slovakia, Sweden, Switzerland, United Kingdom, United States of America, former Yugoslavia.
Host records: Summarized by Yu et al. (2016) as being a parasitoid of the following anthomyiids: *Delia antiqua* (Meigen), *Delia radicum* (L.), *Pegomya bicolor* (Wiedemann), *Pegomya hyoscyami* (Panzer), *Pegomya nigrisquama* Stein, *Pegomya solennis* (Meigen), and *Pegomya steini* Hendel.

Biosteres (*Biosteres*) *kayapinarensis* Fischer & Beyarslan, 2005
Distribution in the Middle East: Turkey (Beyarslan & Fischer, 2011a, b; Fischer & Beyarslan, 2005, 2013).

Extralimital distribution: None.
Host records: Unknown.

Biosteres (*Biosteres*) *lentulus* Papp, 1979
Distribution in the Middle East: Turkey (Beyarslan, 2015).
Extralimital distribution: Romania.
Host records: Unknown.

Biosteres (*Biosteres*) *longicauda* (Thomson, 1895)
Catalogs with Iranian records: Gadallah et al. (2016), Farahani et al. (2016), Yu et al. (2016).
Distribution in Iran: Northern Khorasan (Samin, 2015), Razavi Khorasan (Samin et al., 2011).
Distribution in the Middle East: Iran (Samin, 2015; Samin et al., 2011), Turkey (Beyarslan, 2015; Beyarslan & Fischer, 2011a, b; Fischer & Beyarslan, 2013).
Extralimital distribution: Austria, China, former Czechoslovakia, Denmark, Finland, France, Germany, Italy, Lithuania, Moldova, Russia, Slovenia, Sweden, Switzerland, former Yugoslavia.
Host records: Unknown.

Biosteres (*Biosteres*) *remigii* Fischer, 1971
Catalogs with Iranian records: Gadallah et al. (2016), Farahani et al. (2016), Yu et al. (2016), Samin, Coronado-Blanco, Kavallieratos et al. (2018), Samin, Coronado-Blanco, Fischer et al. (2018).
Distribution in Iran: Guilan (Gadallah et al., 2016).
Distribution in the Middle East: Iran (Gadallah et al., 2016), Turkey (Beyarslan & Fischer, 2011a, b).
Extralimital distribution: Armenia, Hungary, Korea, Mongolia, Russia.
Host records: Unknown.

Biosteres (*Biosteres*) *spinaciae* (Thomson, 1895)
Catalogs with Iranian records: Gadallah et al. (2016), Farahani et al. (2016), Yu et al. (2016), Samin, Coronado-Blanco, Kavallieratos et al. (2018), Samin, Coronado-Blanco, Fischer et al. (2018).
Distribution in Iran: Kerman (Safahani et al., 2018 on *Medicago sativa*-Fabaceae), Kermanshah (Ghahari & Fischer, 2012), Khuzestan (Ameri et al., 2020).
Distribution in the Middle East: Iran (see references above), Turkey (Beyarslan & Fischer, 2011a, b; Fischer & Beyarslan, 2005, 2013).
Extralimital Distribution: Austria, Bulgaria, Canada, former Czechoslovakia, Denmark, Finland, Germany, Hungary, Italy, Lithuania, Mongolia, Netherlands, Norway, Poland, Russia, Spain, United Kingdom, United States of America.
Host records: Summarized by Yu et al. (2016) as being a parasitoid of the anthomyiids *Delia platura* (Meigen), *Pegomya atriplicis* Goureau, *Pegomya betae* (Curtis), *Pegomya hyoscyami* (Panzer), and *Pegomya solennis* (Meigen).

Biosteres (*Biosteres*) *spinaciaeformis* Fischer, 1971
Catalogs with Iranian records: No catalog.
Distribution in Iran: Khuzestan (Ameri et al., 2020).

Distribution in the Middle East: Iran.
Extralimital distribution: Mongolia.
Host records: Unknown.

Biosteres (Chilotrichia) advectus Papp, 1979

Catalogs with Iranian records: No catalog.
Distribution in Iran: Mazandaran (Dolati et al., 2021).
Distribution in the Middle East: Iran.
Extralimital distribution: Hungary, North Macedonia, Serbia.
Host records: Unknown.

Biosteres (Chilotrichia) arenarius (Stelfox, 1959)

Catalogs with Iranian records: Khajeh et al. (2014), Gadallah et al. (2016), Farahani et al. (2016), Yu et al. (2016), Samin, Coronado-Blanco, Kavallieratos et al. (2018), Samin, Coronado-Blanco, Fischer et al. (2018).
Distribution in Iran: Lorestan (Ghahari, Fischer, Papp et al., 2012).
Distribution in the Middle East: Cyprus (Beyarslan et al., 2017), Iran (Ghahari, Fischer, Papp et al., 2012), Turkey (Beyarslan, 2015; Beyarslan & Fischer, 2011a, b; Çikman & Beyarslan, 2009).
Extralimital distribution: France, Germany, Ireland, Moldova, Russia.
Host records: Unknown.

Biosteres (Chilotrichia) blandus (Haliday, 1837)

Catalogs with Iranian records: Gadallah et al. (2016), Farahani et al. (2016), Yu et al. (2016), Samin, Coronado-Blanco, Kavallieratos et al. (2018), Samin, Coronado-Blanco, Fischer et al. (2018).
Distribution in Iran: Kermanshah (Ghahari & Fischer, 2012), Mazandaran (Dolati et al., 2021).
Distribution in the Middle East: Iran (Ghahari & Fischer, 2012; Dolati et al., 2021), Turkey (Beyarslan, 2015; Beyarslan & Fischer, 2011a, b; Fischer & Beyarslan, 2005).
Extralimital distribution: Austria, Bulgaria, Czech Republic, Denmark, Finland, France, Germany, Hungary, Ireland, Italy, Lithuania, Mongolia, Russia, Serbia, Sweden, Switzerland, Ukraine, United Kingdom, Uzbekistan.
Host records: Recorded by Fischer (1969) as being a parasitoid of the agromyzid *Agromyza* sp.

Biosteres (Chilotrichia) brevipalpis (Thomson, 1895)

Catalogs with Iranian records: Khajeh et al. (2014), Gadallah et al. (2016), Farahani et al. (2016), Yu et al. (2016), Samin, Coronado-Blanco, Kavallieratos et al. (2018), Samin, Coronado-Blanco, Fischer et al. (2018).
Distribution in Iran: Guilan (Ghahari, Fischer, & Tobias, 2012).
Distribution in the Middle East: Iran (Ghahari, Fischer, & Tobias, 2012), Turkey (Beyarslan, 2015; Beyarslan & Fischer, 2011a, b; Fischer & Beyarslan, 2005).

Extralimital distribution: Austria, Bulgaria, Czech Republic, Denmark, Finland, Germany, Hungary, Italy, Lithuania, Romania, Russia, Spain, Sweden, Switzerland, United Kingdom.
Host records: Recorded by Fischer and Koponen (1999) as being a parasitoid of the agromyzid *Agromyza hendeli* Griffiths; and the anthomyiid *Delia quadripila* (Stein).

Biosteres (Chilotrichia) cumatus Zaykov & Fischer, 1983

Distribution in the Middle East: Turkey (Beyarslan & Fischer, 2011b; Fischer & Beyarslan, 2013).
Extralimital distribution: Bulgaria.
Host records: Unknown.

Biosteres (Chilotrichia) haemorrhoeus (Haliday, 1837)

Catalogs with Iranian records: Khajeh et al. (2014), Gadallah et al. (2016), Farahani et al. (2016), Yu et al. (2016), Samin, Coronado-Blanco, Kavallieratos et al. (2018), Samin, Coronado-Blanco, Fischer et al. (2018).
Distribution in Iran: East Azarbaijan (Ghahari & van Achterberg, 2016), Guilan (Ghahari, Fischer, & Tobias, 2012), West Azarbaijan (Rastegar et al., 2012).
Distribution in the Middle East: Iran (see references above), Turkey (Beyarslan, 2015; Beyarslan & Fischer, 2011a, b; Fischer & Beyarslan, 2005).
Extralimital distribution: Austria, Belgium, Bulgaria, Canada, Croatia, former Czechoslovakia, Denmark, Finland, France, Germany, Hungary, Ireland, Italy, Malta, Netherlands, Poland, Romania, Russia, Sweden, Switzerland, United Kingdom, United States of America, former Yugoslavia.
Host records: Recorded by Fischer (1969) as being a parasitoid of the anthomyiids *Pegomya bicolor* (Wiedemann), *Pegomya holosteae* (Hering), and *Pegomya solennis* (Meigen).

Biosteres (Chilotrichia) punctiscuta (Thomson, 1895)

Catalogs with Iranian records: Gadallah et al. (2016), Farahani et al. (2016), Yu et al. (2016), Samin, Coronado-Blanco, Kavallieratos et al. (2018), Samin, Coronado-Blanco, Fischer et al. (2018).
Distribution in Iran: Kordestan (Gadallah et al., 2016).
Distribution in the Middle East: Iran (Gadallah et al., 2016), Turkey (Beyarslan & Fischer, 2011a; Fischer & Beyarslan, 2013).
Extralimital distribution: Austria, former Czechoslovakia, Denmark, Finland, France, Germany, Hungary, Korea, Moldova, Russia.
Host records: Unknown.

Biosteres (Chilotrichia) rusticus (Haliday, 1837)

Catalogs with Iranian records: Khajeh et al. (2014), Gadallah et al. (2016), Farahani et al. (2016), Yu et al.

(2016), Samin, Coronado-Blanco, Kavallieratos et al. (2018), Samin, Coronado-Blanco, Fischer et al. (2018).

Distribution in Iran: Guilan (Dolati et al., 2018), Lorestan (Ghahari, Fischer, Papp et al., 2012).

Distribution in the Middle East: Iran (Dolati et al., 2018; Ghahari, Fischer, Papp et al., 2012), Turkey (Beyarslan, 2015; Beyarslan & Fischer, 2011a, b; Fischer & Beyarslan, 2005).

Extralimital distribution: Austria, Bulgaria, former Czechoslovakia, Denamark, Finland, France, Germany, Hungary, Ireland, Italy, Lithuania, Poland, Russia, Switzerland, United Kingdom, Uzbekistan.

Host records: Recorded by Fischer (1962, 1969) as being a parasitoid of the anthomyiids *Delia echinata* Seguy, *Pegomya* sp., and *Phorbia* sp.

Biosteres (*Chilotrichia*) scabriculus (Wesmael, 1835)
Catalogs with Iranian records: Samin, Coronado-Blanco, Kavallieratos et al. (2018), Samin, Coronado-Blanco, Fischer et al. (2018).

Distribution in Iran: Fars (Ghahari & Beyarslan, 2017).

Distribution in the Middle East: Iran (Ghahari & Beyarslan, 2017), Turkey (Beyarslan & Fischer, 2011a, b).

Extralimital distribution: Belgium, former Czechoslovakia, Finland, France, Germany, Hungary, Ireland, Italy, Norway, Poland, United Kingdom.

Host records: Unknown.

Biosteres (*Chilotrichia*) sylvaticus (Haliday, 1837)
Distribution in the Middle East: Turkey (Beyarslan, 2015).

Extralimital distribution: Armenia, Austria, Belgium, Denmark, Finland, France, Germany, Hungary, Ireland, Latvia, Lithuania, Norway, Poland, Russia, Sweden, Switzerland, United Kingdom, Uzbekistan.

Host records: Summarized by Čapek and Hofmann (1997) as being a parasitoid of the anthomyiid *Pegomya hyoscyami* (Panzer).

Biosteres (*Chilotrichia*) ultor (Foerster, 1863)
Catalogs with Iranian records: Yu et al. (2016), Samin, Coronado-Blanco, Kavallieratos et al. (2018), Samin, Coronado-Blanco, Fischer et al. (2018).

Distribution in Iran: Kermanshah (Ghahari & Fischer, 2012).

Distribution in the Middle East: Iran (Ghahari & Fischer, 2012), Turkey (Beyarslan & Fischer, 2011a, b; Fischer & Beyarslan, 2005).

Extralimital distribution: Austria, Belgium, former Czechoslovakia, Denmark, Finland, Germany, Hungary, Ireland, United Kingdom.

Host records: Unknown.

Comments: Gadallah et al. (2016) and Farahani et al. (2016) listed erroneously *Biosteres* (*Chilotrichia*)

wesmaelii (Haliday, 1837) with reference to Ghahari and Fischer (2012).

Biosteres (*Chilotrichia*) wesmaelii (Haliday, 1837)
Catalogs with Iranian records: Khajeh et al. (2014 as *Biosteres carbonarius* (Nees, 1834)), Gadallah et al. (2016 as *Biosteres carbonarius*), Farahani et al. (2016 as *Biosteres carbonarius*), Yu et al. (2016), Samin, Coronado-Blanco, Kavallieratos et al. (2018), Samin, Coronado-Blanco, Fischer et al. (2018).

Distribution in Iran: Kerman (Ranjbar et al., 2016), Lorestan (Ghahari, Fischer, Papp et al., 2012).

Distribution in the Middle East: Iran (Ghahari, Fischer, Papp et al., 2012; Ranjbar et al., 2016), Turkey (Beyarslan, 2015; Beyarslan & Fischer, 2011a, b; Fischer & Beyarslan, 2005, 2013).

Extralimital distribution: Austria, Belgium, Bulgaria, China, former Czechoslovakia, Denmark, Estonia, Finland, France, Germany, Hungary, Ireland, Italy, Lithuania, Netherlands, Poland, Russia, Sweden, Switzerland, United Kingdom, Uzbekistan.

Host records: Summarized by Yu et al. (2016) as being a parasitoid of the anthomyiids *Delia quadripila* (Stein), *Fucellia* sp., *Pegomya atriplicis* Goureau, *Pegomya betae* (Curtis), *Pegomya setaria* (Meigen) and *Pegomya solennis* (Meigen).

Plant associations in Iran: *Descurainia sophia* (Brassicaceae) (Ranjbar et al., 2016).

Genus *Diachasma* Foerster, 1863
Diachasma graeffei (Fischer, 1959)
Distribution in the Middle East: Turkey (Beyarslan, 2015; Beyarslan & Fischer, 2011b; Fischer & Beyarslan, 2005).

Extralimital distribution: Hungary, Italy.

Host records: Unknown.

Tribe Opiini Blanchard, 1845
Genus *Apodesmia* Foerster, 1863
Apodesmia angelus (Fischer, 1971)
Distribution in the Middle East: Turkey (Beyarslan & Fischer, 2011a, b).

Extralimital distribution: Mongolia.

Host records: Unknown.

Apodesmia arenacea (Jakimavicius, 1986)
Distribution in the Middle East: Cyprus (Beyarslan et al., 2017 as *Opius arenaceus*), Turkey (Beyarslan, 2015).

Extralimital distribution: Ukraine.

Host records: Unknown.

Apodesmia austriacus (Fischer, 1958)
Distribution in the Middle East: Turkey (Beyarslan, 2015).

Extralimital distribution: Austria, Finland, Hungary, Italy, Poland, Russia, Slovakia.
Host records: Unknown.

Apodesmia damnosa (Papp, 1980)

Catalogs with Iranian records: Gadallah et al. (2016 as *Opius (Allotypus) damnosus*), Farahani et al. (2016 as *Opius (Allotypus) damnosus*), Yu et al. (2016), Samin, Coronado-Blanco, Kavallieratos et al. (2018), Samin, Coronado-Blanco, Fischer et al. (2018).
Distribution in Iran: Hormozgan (Ameri et al., 2014 as *Opius (Allotypus) damnosus*).
Distribution in the Middle East: Iran.
Extralimital distribution: India, Korea, Russia.
Host records: Unknown.

Apodesmia geniculata (Thomson, 1895)

Distribution in the Middle East: Turkey (Beyarslan, 2015).
Extralimital distribution: Austria, former Czechoslovakia, Denmark, Finland, France, Germany, Hungary, Italy, Japan, Lithuania, Norway, Poland, Romania, Russia, Sweden, Switzerland, United Kingdom.
Host records: Summarized by Yu et al. (2016) as being a parasitoid of the tephritids *Stemonocera cornuta* (Scopoli), and *Trypeta immaculata* (Macquart).

Apodesmia irregularis (Wesmael, 1835)

Catalogs with Iranian records: Khajeh et al. (2014 as *Opius (Allotypus) irregularis* Wesmael, 1835), Gadallah et al. (2016 as *Opius (Allotypus) irregularis*), Farahani et al. (2016 as *Opius (Allotypus) irregularis*), Yu et al. (2016), Samin, Coronado-Blanco, Kavallieratos et al. (2018), Samin, Coronado-Blanco, Fischer et al. (2018).
Distribution in Iran: East Azarbaijan (Ghahari & van Achterberg, 2016 as *Opius (Allotypus) irregularis*), Lorestan (Ghahari, Fischer, Papp et al., 2012 as *O. (Allotypus) irregularis*).
Distribution in the Middle East: Iran (Ghahari, Fischer, Papp et al., 2012; Ghahari & van Achterberg, 2016), Turkey (Beyarslan, 2015; Beyarslan & Fischer, 2011a, b; Fischer & Beyarslan, 2005, 2013).
Extralimital distribution: Austria, Azerbaijan, Belgium, Canada, former Czechoslovakia, Faeroe Islands, Finland, France, Germany, Hungary, Ireland, Italy, Japan, Kazakhstan, Korea, Lithuania, Montenegro, Netherlands, Poland, Russia, Serbia, Spain, Sweden, Switzerland, United Kingdom, United States of America.
Host records: Summarized by Yu et al. (2016) as being a parasitoid of the agromyzid *Chromatomyia primulae* (Robineau-Desvoidy); the anthomyiid *Pegomya solennis* (Meigen); the cecidomyiid *Asphondylia verbasci* (Vallot); the ephydrid *Hydrellia griseola* (Fallén); and the tephritid *Tephritis leontodontis* (DeGeer).

Apodesmia ispartaensis (Fischer & Beyarslan, 2005)

Distribution in the Middle East: Turkey (Fischer & Beyarslan, 2005 as *Opius ispartaensis*; Beyarslan & Fischer, 2011b).
Extralimital distribution: None.
Host records: Unknown.

Apodesmia karesuandensis (Fischer, 1964)

Catalogs with Iranian records: Gadallah et al. (2016 as *Opius (Apodesmia) karesuandensis*), Farahani et al. (2016 as *Opius (Apodesmia) karesuandensis*), Yu et al. (2016), Samin, Coronado-Blanco, Kavallieratos et al. (2018), Samin, Coronado-Blanco, Fischer et al. (2018).
Distribution in Iran: Kermanshah (Ghahari & Fischer, 2012 as *Opius (Apodesmia) karesuandensis*).
Distribution in the Middle East: Iran (Ghahari & Fischer, 2012), Turkey (Beyarslan & Fischer, 2011a, b).
Extralimital distribution: Austria, Canada, Finland, Germany, Italy, Sweden, United States of America.
Host records: Unknown.

Apodesmia novosimilis (Fischer, 1989)

Catalogs with Iranian records: Khajeh et al. (2014 as *Opius (Agnopius) novosimilis*), Gadallah et al. (2016 as *Opius (Agnopius) novosimilis*), Farahani et al. (2016 as *Opius (Agnopius) novosimilis*), Yu et al. (2016), Samin, Coronado-Blanco, Kavallieratos et al. (2018), Samin, Coronado-Blanco, Fischer et al. (2018).
Distribution in Iran: Hormozgan (Ameri et al., 2014 as *Opius (Agnopius) novosimilis*), Kerman (Safahani et al., 2018 on *Medicago sativa*—Fabaceae).
Distribution in the Middle East: Iran (Ameri et al., 2014; Safahani et al., 2018), Turkey (Beyarslan, 2015).
Extralimital distribution: India, Slovakia.
Host records: Unknown.

Apodesmia nowakowskii (Fischer, 1959)

Catalogs with Iranian records: Gadallah et al. (2016 as *Opius (Agnopius) nowakowskii*), Farahani et al. (2016 as *Opius (Agnopius) nowakowskii*), Yu et al. (2016), Samin, Coronado-Blanco, Kavallieratos et al. (2018), Samin, Coronado-Blanco, Fischer et al. (2018).
Distribution in Iran: Hormozgan (Ameri et al., 2014 as *Opius (Agnopius) nowakowskii*).
Distribution in the Middle East: Iran (Ameri et al., 2014), Turkey (Beyarslan & Fischer, 2011a, b).
Extralimital distribution: Hungary, Poland.
Host records: Recorded by Fischer (1959) as being a parasitoid of the agromyzid *Phytomyza thysselini* Hendel.

Apodesmia posticatae (Fischer, 1957)

Catalogs with Iranian records: No catalog.
Distribution in Iran: Guilan (Dolati et al., 2021).

Distribution in the Middle East: Iran (Dolati et al., 2021), Israel−Palestine (Papp, 2012 as *Opius* (*Utetes*) *posticatae*). Extralimital distribution: Austria, Bulgaria, former Czechoslovakia, Finland, Germany, Hungary, Italy, Korea, Lithuania, Netherlands, Poland, Russia, Spain, Sweden, United Kingdom (Yu et al., 2016 as *Opius* (*Utetes*) *posticatae* Fischer).

Host records: Summarized by Yu et al. (2016) as being a parasitoid of the following agromyzids: *Agromyza nigescens* Hendel, *Agromyza nigripes* Meigen, *Cerodontha* spp., *Chromatomyia luzulae* (Hering), *Chromatomyia ramosa* (Hendel), *Liriomyza pusio* Meigen, *Nemoromyza posticata* (Meigen), *Ophiomyza maura* (Meigen), and *Phytomyza* spp.

Apodesmia pseudarenacea (Fischer & Beyarslan, 2005)

Distribution in the Middle East: Turkey (Fischer & Beyarslan, 2005 as *Opius pseudarenacea*; Beyarslan & Fischer, 2011b).

Extralimital distribution: None.

Host records: Unknown.

Apodesmia rugata (Fischer, 1992)

Distribution in the Middle East: Turkey (Beyarslan & Fischer, 2011a, b; Fischer, 1992; Fischer & Beyarslan, 2005).

Extralimital distribution: None.

Host records: Unknown.

Apodesmia saeva (Haliday, 1837)

Catalogs with Iranian records: No catalog.

Distribution in Iran: Mazandaran (Dolati et al., 2019 as *Opius* (*Allotypus*) *saevus* Haliday, 1837).

Distribution in the Middle East: Iran (Dolati et al., 2019), Turkey (Beyarslan, 2015).

Extralimital distribution: Armenia, Austria, Bulgaria, Denmark, Finland, Germany, Hungary, Italy, Korea, Russia, United Kingdom.

Host records: Recorded by Fischer (1967) as being a parasitoid of the agromyzid *Nemorimyza posticata* Meigen.

Apodesmia selimbassai (Fischer, 1992)

Distribution in the Middle East: Turkey (Fischer, 1992 as *Opius selimbassai*; Beyarslan, 2015; Beyarslan & Fischer, 2011a, b; Fischer & Beyarslan, 2005).

Extralimital distribution: Spain.

Host records: Unknown.

Apodesmia sharynensis (Fischer, 2001)

Catalogs with Iranian records: Fallahzadeh and Saghaei (2010 as *Opius* (*Apodesmia*) *sharynensis*), Khajeh et al. (2014 as *Opius* (*Apodesmia*) *sharynensis*), Gadallah et al.

(2016 as *Opius* (*Apodesmia*) *sharynensis*), Farahani et al. (2016 as *Opius* (*Apodesmia*) *sharynensis*), Yu et al. (2016), Samin, Coronado-Blanco, Kavallieratos et al. (2018), Samin, Coronado-Blanco, Fischer et al. (2018).

Distribution in Iran: Alborz (Fischer, 2001 as *Opius* (*Apodesmia*) *sharynensis*).

Distribution in the Middle East: Iran.

Extralimital distribution: Kazakhstan.

Host records: Unknown.

Apodesmia similis (Szépligeti, 1898)

Catalogs with Iranian records: Gadallah et al. (2016 as *Opius* (*Agnopius*) *similis*), Farahani et al. (2016 as *Opius* (*Agnopius*) *similis*), Yu et al. (2016), Samin, Coronado-Blanco, Kavallieratos et al. (2018), Samin, Coronado-Blanco, Fischer et al. (2018).

Distribution in Iran: Alborz (Dolati et al., 2019 as *Opius* (*Agnopius*) *basirufus* Fischer), Guilan (Dolati et al., 2019 as *Opius* (*Agnopius*) *basirufus* Fischer; Dolati et al., 2021), Golestan (Sakenin et al., 2012 as *Opius* (*Agnopius*) *similis*), Kerman (Safahani et al., 2018 on *Mentha pulegium* (Laminaceae)), Lorestan (Ghahari, Fischer, Papp et al., 2012 as *O.* (*Agnopius*) *similis*), Mazandaran (Dolati et al., 2019 as *O. basirufus*, and as *O. periclymenii* Fischer; Dolati et al., 2021).

Distribution in the Middle East: Iran (see references above), Israel−Palestine (Fischer, 1997), Turkey (Beyarslan, 2015; Beyarslan & Fischer, 2011a, b; Fischer, 1958, 1972, 1997; Fischer & Beyarslan, 2005).

Extralimital distribution: Austria, Bulgaria, Croatia, Czech Republic, Estonia, Finland, France, Greece, Hungary, Ireland, Italy, Kazakhstan, Lithuania, Montenegro, Netherlands, Poland, Romania, Russia, Serbia, Slovakia, Spain, Sweden, Switzerland, Ukraine, United Kingdom.

Host records: Summarized by Yu et al. (2016) as being a parasitoid of the agromyzids of the following genera: *Agromyza* Fallén, *Amauromyza* Hendel, *Aulagromyza* Enderlein, *Calycomyza* Hendel, *Cerodontha* Rondani, *Chromatomyia* Hardy, *Galiomyza* Spencer, *Liriomyza* Mik, *Napomyza* Westwood, *Ophiomyia* Brascsnikov, and *Phytomyza* Fallén.

Apodesmia similis parvipunctum (Fischer, 1958)

Distribution in the Middle East: Turkey (Fischer, 1958, 1972).

Extralimital distribution: Greece.

Host records: Unknown.

Apodesmia similoides (Fischer, 1962)

Distribution in the Middle East: Turkey (Beyarslan, 2015).

Extralimital distribution: Austria, Bulgaria, former Czechoslovakia, Finland, France, Germany, Hungary, Italy, Lithuania, Netherlands, Poland, Russia.

Host records: Unknown.

Apodesmia striatula (Fischer, 1957)
Catalogs with Iranian records: No catalog.
Distribution in Iran: Guilan (Dolati et al., 2021).
Distribution in the Middle East: Iran.
Extralimital distribution: Austria, Denmark, Hungary, Russia, Switzerland.
Host records: Unknown.

Apodesmia tirolensis (Fischer, 1958)
Catalogs with Iranian records: Khajeh et al. (2014 as *Opius* (*Agnopius*) *tirolensis*), Gadallah et al. (2016 as *Opius* (*Agnopius*) *tirolensis*), Farahani et al. (2016 as *Opius* (*Agnopius*) *tirolensis*), Yu et al. (2016), Samin, Coronado-Blanco, Kavallieratos et al. (2018), Samin, Coronado-Blanco, Fischer et al. (2018).
Distribution in Iran: Guilan (Ghahari, Fischer, & Tobias, 2012; Dolati et al., 2019 as *Opius* (*Agnopius*) *tirolensis* Fischer, 1958), Hormozgan (Ameri et al., 2014).
Distribution in the Middle East: Iran (see references above), Turkey (Beyarslan & Fischer, 2011a, b; Fischer & Beyarslan, 2005).
Extralimital distribution: Austria, Czech Republic, Denmark, Finland, Germany, Hungary, Italy, Poland, Romania, Russia, Spain.
Host records: Recorded by Fischer (1969) as being a parasitoid of the agromyzid *Phytomyza flavicornis* Fallén.

Apodesmia tuberculata (Fischer, 1959)
Catalogs with Iranian records: Khajeh et al. (2014 as *Opius* (*Allotypus*) *tuberculatus*), Gadallah et al. (2016 as *Opius* (*Allotypus*) *tuberculatus*), Farahani et al. (2016 as *Opius* (*Allotypus*) *tuberculatus*), Yu et al. (2016), Samin, Coronado-Blanco, Kavallieratos et al. (2018), Samin, Coronado-Blanco, Fischer et al. (2018).
Distribution in Iran: Razavi Khorasan (Ghahari, Fischer, Sakenin et al., 2011 as *Opius* (*Allotypus*) *tuberculatus*).
Distribution in the Middle East: Iran (Ghahari, Fischer, Sakenin et al., 2011), Turkey (Fischer, 1959, 1972; Beyarslan & Fischer, 2011a, b; Fischer & Beyarlan, 2013).
Extralimital distribution: Denmark, France, Poland, Russia.
Host records: Unknown.

Apodesmia uttoisimilis (Fischer, 1999)
Catalogs with Iranian records: Khajeh et al. (2014 as *Opius* (*Cryptognathopius*) *uttoisimilis*), Gadallah et al. (2016 as *Opius* (*Cryptognathopius*) *uttoisimilis*), Farahani et al. (2016 as *Opius* (*Cryptognathopius*) *uttoisimilis*), Yu et al. (2016), Samin, Coronado-Blanco, Kavallieratos et al. (2018), Samin, Coronado-Blanco, Fischer et al. (2018).
Distribution in Iran: Isfahan (Ghahari, Fischer, & Papp, 2011 as *Opius* (*Cryptognathopius*) *uttoisimilis*),

Mazandaran (Ghahari, 2018—around rice fields), Qazvin (Dolati et al., 2021).
Distribution in the Middle East: Iran (see references above), Turkey (Beyarslan, 2015; Beyarslan & Fischer, 2011a; Yildirim et al., 2010).
Extralimital distribution: Russia.
Host records: Unknown.

Genus *Atormus* van Achterberg, 1997
Atormus victus (Haliday, 1837)
Catalogs with Iranian records: Gadallah et al. (2016), Farahani et al. (2016), Yu et al. (2016), Samin, Coronado-Blanco, Kavallieratos et al. (2018), Samin, Coronado-Blanco, Fischer et al. (2018).
Distribution in Iran: Kermanshah (Ghahari & Fischer, 2012).
Distribution in the Middle East: Iran (Ghahari & Fischer, 2012), Turkey (Beyarslan & Fischer, 2011a, b; Fischer & Beyarslan, 2005).
Extralimital distribution: Austria, Belgium, Bulgaria, Czech Republic, Denmark, Estonia, Finland, Germany, Hungary, Georgia, Ireland, Italy, Mongolia, Montenegro, Netherlands, Poland, Romania, Russia, Serbia, Slovakia, Slovenia, Sweden, Switzerland, United Kingdom.
Host records: Summarized by Yu et al. (2016) as being a parasitoid of the following agromyzids: *Agromyza* spp., *Amauromyza labiatarum* (Hendel), *Aulagromyza hendeliana* (Hering), *Cerodontha pygmaea* (Meigen), *Chromatomyia periclymeni* (de Meijere), *Liriomyza amoena* (Meigen), *Liriomyza eupatorii* (Kaltenbach), and *Phytomyza* spp.

Genus *Biophthora* Foerster, 1863
Biophthora bajula (Haliday, 1837)
Catalogs with Iranian records: No catalog.
Distribution in Iran: Kerman (Ranjbar et al., 2016).
Distribution in the Middle East: Iran (Ranjbar et al., 2016), Israel—Palestine (Fischer, 1997), Turkey (Beyarslan & Fischer, 2011a; Fischer, 1972; Fischer & Beyarslan, 2005).
Extralimital distribution: Austria, Czech Republic, Germany, Hungary, Ireland, Montenegro, Poland, Serbia, United Kingdom.
Host records: Unknown.
Plant associations in Iran: *Medicago sativa* (Fabaceae) (Ranjbar et al., 2016).

Biophthora rossica (Szépligeti, 1901)
Catalogs with Iranian records: No catalog.
Distribution in Iran: Sistan and Baluchestan (Sedighi et al., 2014).
Distribution in the Middle East: Iran (Sedighi et al., 2014), Turkey (Beyarslan & Fischer, 2011a; Fischer & Beyarslan, 2005).

Extralimital distribution: Hungary, Poland, Russia, Spain.
Host records: Unknown.
Plant associations in Iran: *Hordeum vulgare* (Poaceae), and *Medicago sativa* (Fabaceae) (Ranjbar et al., 2016).

Genus *Bitomus* Szépligeti, 1910

Bitomus (*Bitomus*) *castus* (Zaykov, 1983)
Distribution in the Middle East: Turkey (Beyarslan & Fischer, 2011a, b; Fischer & Beyarslan, 2005).
Extralimital distribution: Bulgaria, Hungary, Russia.
Host records: Unknown.

Bitomus (*Bitomus*) *multipilis* Fischer, 1990
Catalogs with Iranian records: No catalog.
Distribution in Iran: Alborz (Dolati et al., 2021).
Distribution in the Middle East: Iran.
Extralimital distribution: Hungary (Fischer, 1990).
Host records: Unknown.

Bitomus (*Bitomus*) *pamboloides* (Tobias, 1986)
Distribution in the Middle East: Turkey (Beyarslan & Fischer, 2011b; Fischer & Beyarslan, 2005).
Extralimital distribution: Italy, Moldova, Russia, Turkmenistan.
Host records: Unknown.

Bitomus (*Bitomus*) *valdepusillus* Fischer & Beyarslan, 2005
Distribution in the Middle East: Turkey (Beyarslan & Fischer, 2011a, b; Fischer & Beyarslan, 2005).
Extralimital distribution: None.
Host records: Unknown.

Genus *Desmiostoma* Foerster, 1863

Desmiostoma parvulum (Wesmael, 1835)
Catalogs with Iranian records: No catalog.
Distribution in Iran: Qazvin (Dolati et al., 2021).
Distribution in the Middle East: Iran (Dolati et al., 2021), Saudi Arabia (El-Hag & El-Meleigi, 1991).
Extralimital distribution: Austria, Belgium, Bermuda, Canada, Czech Republic, Denmark, Finland, France, Germany, Hungary, Italy, Spain, Switzerland, United States of America, United Kingdom.
Host records: Summarized by Yu et al. (2016) as being a parasitoid of the following agromyzids: *Agromyza nigrella* (Rondani), *Liriomyza pusilla* (Meigen), *Liriomyza sativae* Blanchard, and *Phytomyza* sp.

Genus *Diachasmimorpha* Viereck, 1913

Diachasmimorpha paeoniae (Tobias, 1980)
Distribution in the Middle East: Turkey (Beyarslan, 2015; Beyarslan & Fischer, 2011a, b).
Extralimital distribution: Russia.
Host records: Unknown.

Diachasmimorpha tryoni (Cameron, 1911)
Distribution in the Middle East: Egypt (introduced, Fischer, 1959).
Extralimital distribution: Algeria (introduced), Australia, Brazil (introduced), (introduced), Fiji, Mexico, Puerto Rico, Spain, United States of America.
Host records: Summarized by Yu et al. (2016) as being a parasitoid of several tephritid species of the following genera: *Anastrepha* Schiner, *Bactrocera* Macquart, *Ceratitis* Macleay, *Eutreta* Loew, *Procecichares* Hendel, and *Rhagoletis* Loew.

Genus *Eurytenes* Foerster, 1863

Eurytenes (*Eurytenes*) *abnormis* (Wesmael, 1835)
Catalogs with Iranian records: Gadallah et al. (2016), Farahani et al. (2016), Yu et al. (2016), Samin, Coronado-Blanco, Kavallieratos et al. (2018), Samin, Coronado-Blanco, Fischer et al. (2018).
Distribution in Iran: Mazandaran (Dolati et al., 2021), West Azarbaijan (Gadallah et al., 2016).
Distribution in the Middle East: Iran (Dolati et al., 2021; Gadallah et al., 2016), Turkey (Fischer & Beyarslan, 2005).
Extralimital distribution: Austria, Belgium, Bulgaria, Canada, Croatia, Finland, Germany, Hungary, Ireland, Italy, Korea, Lithuania, Poland, Russia, Ukraine, United Kingdom, United States of America, former Yugoslavia.
Host records: Summarized by Yu et al. (2016) as being a parasitoid of the following agromyzids: *Agromyza* spp., *Amauromyza* spp., *Cerodontha* spp., *Liriomyza* spp., *Phytoliriomyza variegata* Meigen, and *Phytomyza* spp.; and the anthomyiid *Pegomya bicolor* (Wiedemann).

Genus *Fopius* Wharton, 1987

Fopius carpomyiae (Silvestri, 1916)
Catalogs with Iranian records: Fallahzadeh and Saghaei (2010), Khajeh et al. (2014), Gadallah et al. (2016), Farahani et al. (2016), Yu et al. (2016), Samin, Coronado-Blanco, Kavallieratos et al. (2018), Samin, Coronado-Blanco, Fischer et al. (2018).
Distribution in Iran: Bushehr (Farrar & Chou, 2000; Farrar et al., 2002, 2004, 2009, 2012; Golestaneh et al., 2018a, b; Modarres Awal, 2012), Khuzestan (Golestaneh et al., 2018a, b).
Distribution in the Middle East: Iran.
Extralimital distribution: India, Sri Lanka.
Host records: Summarized by Yu et al. (2016) as being a parasitoid of the tephritids *Bactrocera dorsalis* (Hendel), *Bactrocera zonata* (Saunders), and *Carpomya vesuviana* A. Costa. In Iran, this species has been reared from the agromyzid *Carpomya vesuviana* Costa (Farrar et al., 2004, 2009, 2012; Golestaneh et al., 2018a, b; Modarres Awal, 2012).

Genus *Hoplocrotaphus* Telenga, 1950

Hoplocrotaphus hamooniae Peris-Felipo, Belokobylskij & Rakhshani, 2018

Catalogs with Iranian records: No catalog.

Distribution in Iran: Sistan and Baluchestan (Zabol: Hamoon wetland—Holotype) (Peris-Felipo et al., 2018).

Distribution in the Middle East: Iran.

Extralimital distribution: Uzbekistan.

Host records: Unknown.

Host plants in Iran: *Tamarix stricta* (Tamaricaceae) (Peris-Felipo et al., 2018).

Genus *Indiopius* Fischer, 1966

Indiopius cretensis Fischer, 1983

Catalogs with Iranian records: Gadallah et al. (2016), Farahani et al. (2016), Yu et al. (2016), Samin, Coronado-Blanco, Kavallieratos et al. (2018), Samin, Coronado-Blanco, Fischer et al. (2018).

Distribution in Iran: Kerman (Safahani et al., 2018), Sistan and Baluchestan (Peris-Felipo et al., 2014).

Distribution in the Middle East: Iran (Peris-Felipo et al., 2014; Safahani et al., 2018), Israel—Palestine (Papp, 2012), Turkey (Beyarslan & Fischer, 2011b; Fischer & Beyarslan, 2005; Papp, 1990).

Extralimital distribution: Cape Verde Islands, Greece.

Host records: Unknown.

Host plants in Iran: *Medicago sativa* (Fabaceae) (Safahani et al., 2018).

Indiopius saigonensis Fischer, 1966

Distribution in the Middle East: Turkey (Beyarslan & Fischer, 2011b; Fischer & Beyarslan, 2005).

Extralimital distribution: Vietnam.

Host records: Unknown.

Indiopius yilmazae Fischer & Beyarslan, 2011

Distribution in the Middle East: Turkey (Beyarslan & Fischer, 2011b; Fischer & Beyarslan, 2011).

Extralimital distribution: None.

Host records: Unknown.

Genus *Opiognathus* Fischer, 1972

Opiognathus propepactum (Fischer, 1984)

Distribution in the Middle East: Turkey (Beyarslan, 2015; Beyarslan & Fischer, 2011a, b).

Extralimital distribution: Denmark.

Host records: Unknown.

Opiognathus silifkeensis (Fischer & Beyarslan, 2005)

Distribution in the Middle East: Turkey (Beyarslan & Fischer, 2011b; Fischer & Beyarslan, 2005).

Extralimital distribution: None.

Host records: Unknown.

Opiognathus propodealis (Fischer, 1958)

Catalogs with Iranian records: Gadallah et al. (2016), Farahani et al. (2016), Yu et al. (2016), Samin, Coronado-Blanco, Kavallieratos et al. (2018), Samin, Coronado-Blanco, Fischer et al. (2018).

Distribution in Iran: Guilan, Mazandaran (Dolati et al., 2019), Kermanshah (Ghahari & Fischer, 2012 as *Opius* (*Opiognathus*) *propodealis*).

Distribution in the Middle East: Iran (Ghahari & Fischer, 2012; Dolati et al., 2019), Turkey (Beyarslan & Fischer, 2011a, b).

Extralimital distribution: Austria, former Czechoslovakia, Estonia, Finland, France, Germany, Hungary, India, Italy, Korea, Lithuania, Netherlands, Poland, Russia, Serbia, Spain, Switzerland, United Kingdom.

Host records: Summarized by Yu et al. (2016) as being a parasitoid of the following agromyzids: *Agromyza* spp., *Amauromyza frontella* (Rondani), *Chromatomyia lonicerae* (Robineau-Desvoidy), *Liriomyza* spp., and *Phytomyza* spp.

Genus *Opiostomus* Fischer, 1972

Opiostomus campanariae (Fischer, 1959)

Catalogs with Iranian records: Khajeh et al. (2014), Gadallah et al. (2016), Farahani et al. (2016), Yu et al. (2016), Samin, Coronado-Blanco, Kavallieratos et al. (2018), Samin, Coronado-Blanco, Fischer et al. (2018).

Distribution in Iran: Semnan (Ghahari, Fischer, Sakenin et al., 2011).

Distribution in the Middle East: Iran (Ghahari, Fischer, Sakenin et al., 2011), Turkey (Beyarslan & Fischer, 2011a, b).

Extralimital distribution: Armenia, Austria, Finland, Germany, Hungary, Kazakhstan, Lithuania, Mongolia, Poland, Russia, Slovakia, Spain, Ukraine, Uzbekistan.

Host records: Summarized by Yu et al. (2016) as being a parasitoid of the agromyzids *Aulagromyza luteoscutellata* (de Meijere), *Phytoliriomyza melampyga* (Loew), and *Phytomyza campanariae* Nowakowski.

Opiostomus dividus (Tobias, 1998)

Distribution in the Middle East: Turkey (Beyarslan, 2015).

Extralimital distribution: Russia.

Host records: Unknown.

Opiostomus griffithsi (Fischer, 1962)

Distribution in the Middle East: Israel—Palestine (Fischer, 1997), Turkey (Beyarslan, 2015; Beyarslan & Fischer, 2011a, b; Fischer & Beyarslan, 2005).

Extralimital distribution: Belgium, Bulgaria, Croatia, Hungary, United Kingdom.

Host records: Recorded by Fischer (1962, 1967, 1972) as being a parasitoid of the agromyzid *Amauromyza verbasci* Bouché.

Opiostomus (Jucundopius) impatientis (Fischer, 1957)

Catalogs with Iranian records: No catalog.

Distribution in Iran: Markazi (Samin, Sakenin Chelav et al., 2020).

Distribution in the Middle East: Iran.

Extralimital distribution: Armenia, Austria, Czech Republic, Finland, Germany, Hungary, Italy, Kazakhstan, Kyrgyzstan, Mongolia, Poland, Spain, Uzbekistan.

Summarized by Yu et al. (2016) as being a parasitoid of the following agromyzids: *Agromyza johannae* de Meijere, *Aulagromyza luteoscutellata* (de Meijere), *Napomyza xylostei* (Kaltenbach), and *Phytoliriomyza melampyga* (Loew).

Opiostomus (Opiostomus) riphaeus (Tobias, 1986)

Catalogs with Iranian records: Gadallah et al. (2016 as *Opius riphaeus* Tobias, 1986), Farahani et al. (2016 as *O. riphaeus*), Yu et al. (2016), Samin, Coronado-Blanco, Kavallieratos et al. (2018), Samin, Coronado-Blanco, Fischer et al. (2018).

Distribution in Iran: Hormozgan (Ameri et al., 2014 as *O. riphaeus*).

Distribution in the Middle East: Iran.

Extralimital distribution: India, Russia.

Host records: Unknown.

Opiostomus snoflaki (Fischer, 1959)

Catalogs with Iranian records: Gadallah et al. (2016 as *Opius snoflaki* Fischer, 1959), Farahani et al. (2016 as *Opius snoflaki* Fischer, 1959), Yu et al. (2016), Samin, Coronado-Blanco, Kavallieratos et al. (2018), Samin, Coronado-Blanco, Fischer et al. (2018).

Distribution in Iran: Kermanshah (Ghahari & Fischer, 2012), Mazandaran (Dolati et al., 2021).

Distribution in the Middle East: Iran (Dolati et al., 2021; Ghahari & Fischer, 2012), Turkey (Beyarslan & Fischer, 2011a, b).

Extralimital distribution: Bulgaria, Czech Republic, India, Serbia.

Host records: Unknown.

Genus *Opius* Wesmael, 1835

Opius abditiformis Fischer, 1984

Distribution in the Middle East: Turkey (Beyarslan & Fischer, 2011b; Fischer & Beyarslan, 2005).

Extralimital distribution: Former Czechoslovakia, Finland, Hungary, Slovakia.

Host records: Unknown.

Opius abditus Fischer, 1960

Catalogs with Iranian records: Fallahzadeh and Saghaei (2010), Khajeh et al. (2014), Gadallah et al. (2016), Farahani et al. (2016), Yu et al. (2016), Samin, Coronado-Blanco, Kavallieratos et al. (2018), Samin, Coronado-Blanco, Fischer et al. (2018).

Distribution in Iran: Golestan (Sakenin et al., 2012), Lorestan (Ghahari, Fischer, Papp et al., 2012), Semnan (Samin, Fischer et al., 2015), Iran (no specific locality cited) (Fischer, 1960; 1972; Beyarslan & Fischer, 2013).

Distribution in the Middle East: Iran (see references above), Turkey (Beyarslan, 2019; Beyarslan & Fischer, 2011a, b).

Extralimital distribution: None.

Host records: Unknown.

Opius adanacola Fischer & Beyarslan, 2005

Distribution in the Middle East: Turkey (Beyarslan & Fischer, 2011b; Fischer & Beyarslan, 2005).

Extralimital distribution: None.

Host records: Unknown.

Opius adentatus Fischer, 1981

Distribution in the Middle East: Turkey (Beyarslan, 2015).

Extralimital distribution: Austria, Finland, Russia, Uzbekistan.

Host records: Unknown.

Opius agromyzicola Fischer, 1967

Distribution in the Middle East: Turkey (Beyarslan & Fischer, 2011a, b; Fischer & Beyarslan, 2005).

Extralimital distribution: Austria, China, Germany, Hungary, Sweden, United Kingdom.

Host records: Recorded by Fischer (1967) and Fischer and Koponen (1999) as being a parasitoid of the agromyzid *Agromyza potentillae* (Kaltenbach).

Opius areatus Tobias, 1986

Distribution in the Middle East: Turkey (Beyarslan, 2015).

Extralimital distribution: Russia.

Host records: Unknown.

Opius arundinis Fischer, 1964

Catalogs with Iranian records: Khajeh et al. (2014), Gadallah et al. (2016), Farahani et al. (2016), Yu et al. (2016), Samin, Coronado-Blanco, Kavallieratos et al. (2018), Samin, Coronado-Blanco, Fischer et al. (2018).

Distribution in Iran: Kermanshah (Ghahari & Fischer, 2012), Sistan and Baluchestan (Khajeh et al., 2014).

Distribution in the Middle East: Iran (Ghahari & Fischer, 2012; Khajeh et al., 2014), Jordan (Papp, 1982), Turkey (Beyarslan & Fischer, 2011a, b).

Extralimital distribution: Hungary, Italy.

Host records: Unknown.

Plant associations in Iran: *Medicago sativa* (Fabaceae) (Khajeh et al., 2014).

Opius attributus Fischer, 1962

Catalogs with Iranian records: No catalog.
Distribution in Iran: Mazandaran (Dolati et al., 2021).
Distribution in the Middle East: Iran.
Extralimital distribution: Austria, Czech Republic, Finland, Hungary, Lithuania, Slovakia, Ukraine.
Host records: Unknown.

Opius basalis Fischer, 1958

Catalogs with Iranian records: Khajeh et al. (2014), Gadallah et al. (2016), Farahani et al. (2016), Yu et al. (2016), Samin, Coronado-Blanco, Kavallieratos et al. (2018), Samin, Coronado-Blanco, Fischer et al. (2018).
Distribution in Iran: Alborz (Dolati et al., 2019), Mazandaran (Dolati et al., 2019; Ghahari & Fischer, 2011; Sakenin et al., 2012).
Distribution in the Middle East: Egypt (Aamer et al., 2014) Iran (see references above), Turkey (Beyarslan & Fischer, 2011b; Çikman et al., 2006; Fischer & Beyarslan, 2005; 2013).
Extralimital distribution: Former Czechoslovakia, Hungary, Italy.
Host records: Recorded by Çikman et al. (2006) as being a parasitoid of the agromyzids *Agromyza albitarsis* Meigen, and *Liriomyza trifolii* (Burgess). In Iran, this species has been reared from *Agromyza* sp. (Ghahari & Fischer, 2011; Sakenin et al., 2012).

Opius bouceki Fischer, 1958

Catalogs with Iranian records: Khajeh et al. (2014), Gadallah et al. (2016), Farahani et al. (2016), Yu et al. (2016), Samin, Coronado-Blanco, Kavallieratos et al. (2018), Samin, Coronado-Blanco, Fischer et al. (2018).
Distribution in Iran: Kerman (Safahani et al., 2018), Sistan and Baluchestan (Khajeh et al., 2014).
Distribution in the Middle East: Iran (Khajeh et al., 2014; Safahani et al., 2018), Turkey (Beyarslan, 2015; Beyarslan & Fischer, 2011a, b; Fischer & Beyarslan, 2013).
Extralimital distribution: Former Czechoslovakia, Hungary.
Host records: Unknown.
Plant associations in Iran: *Medicago sativa* (Fabaceae) (Khajeh et al., 2014; Safahani et al., 2018).

Opius bulgaricus Fischer, 1959

Distribution in the Middle East: Turkey (Beyarslan, 2015; Çikman & Beyarslan, 2009).
Extralimital distribution: Austria, Bulgaria, Hungary, Italy, Spain, Switzerland.
Host records: Recorded by Priore and Tremblay (1995) as being a parasitoid of the agromyzid *Chromatomyia horticola* (Goureau).

Opius caricivorae Fischer, 1964

Catalogs with Iranian records: Khajeh et al. (2014), Gadallah et al. (2016), Farahani et al. (2016), Yu et al.

(2016), Samin, Coronado-Blanco, Kavallieratos et al. (2018), Samin, Coronado-Blanco, Fischer et al. (2018).
Distribution in Iran: Sistan and Baluchestan (Khajeh et al., 2014).
Distribution in the Middle East: Iran (Khajeh et al., 2014), Turkey (Beyarslan & Fischer, 2011a, b; Fischer & Beyarslan, 2005).
Extralimital distribution: Afghanistan, Austria, China, Czech Republic, Estonia, Finland, Germany, Greece, Hungary, Italy, Korea, Poland, Russia, Spain, Ukraine.
Host records: Summarized by Yu et al. (2016) as being a parasitoid of the following agromyzids: *Agromyza nigriscens* Hendel, *Cerodontha* spp., *Chromatomyia horticola* Goureau, *Liriomyza chinensis* (Kato), *Liriomyza sativae* (Blanchard), *Napomyza salviae* (Hering), *Phytomyza* spp.; and the drosophilid *Scaptomyza flava* (Fallén).
Plant associations in Iran: *Medicago sativa* (Fabaceae) (Khajeh et al., 2014).

Opius ciceris Fischer, 1974

Distribution in the Middle East: Turkey (Beyarslan & Fischer, 2011a, b; Fischer & Beyarslan, 2005).
Extralimital distribution: Austria, Germany, Spain.
Host records: Recorded by Fischer (1974) as being a parasitoid of the agromyzid *Liriomyza cicerinae* (Rondani).

Opius cingutolicus Fischer, 1992

Distribution in the Middle East: Turkey (Beyarslan & Fischer, 2011b; Fischer, 1992; Fischer & Beyarsaln, 2005, 2013).
Extralimital distribution: None.
Host records: Unknown.

Opius circinus Papp, 1979

Catalogs with Iranian records: No catalog.
Distribution in Iran: Kerman (Ranjbar et al., 2016).
Distribution in the Middle East: Iran.
Extralimital distribution: Finland, Hungary, Korea.
Host records: Unknown.
Plant associations in Iran: *Medicago sativa* (Fabaceae) (Ranjbar et al., 2016).

Opius circulator (Nees von Esenbeck, 1834)

Catalogs with Iranian records: Khajeh et al. (2014), Gadallah et al. (2016), Farahani et al. (2016), Yu et al. (2016), Samin, Coronado-Blanco, Kavallieratos et al. (2018), Samin, Coronado-Blanco, Fischer et al. (2018).
Distribution in Iran: Guilan (Ghahari, Fischer, & Tobias, 2012), Mazandaran (Dolati et al., 2021).
Distribution in the Middle East: Iran (Dolati et al., 2021; Ghahari, Fischer, & Tobias, 2012), Turkey (Beyarslan & Fischer, 2011a, b; Fischer & Beyarslan, 2013).

Extralimital distribution: Austria, Belgium, Bulgaria, Czech Republic, Finland, France, Germany, Hungary, Italy, Korea, Lithuania, Poland, Russia, Switzerland.
Host records: Unknown.

Opius coloraticeps Fischer, 1996
Distribution in the Middle East: Turkey (Beyarslan, 2015).
Extralimital distribution: Greece.
Host records: Unknown.

Opius connivens Thomson, 1895
Catalogs with Iranian records: Khajeh et al. (2014), Gadallah et al. (2016), Farahani et al. (2016), Yu et al. (2016), Samin, Coronado-Blanco, Kavallieratos et al. (2018), Samin, Coronado-Blanco, Fischer et al. (2018).
Distribution in Iran: Guilan (Ghahari, Fischer, & Tobias, 2012).
Distribution in the Middle East: Iran (Ghahari, Fischer, & Tobias, 2012), Turkey (Beyarslan & Fischer, 2011a, b).
Extralimital distribution: former Czechoslovakia, Denmark, Estonia, Finland, Hungary, Kazakhstan, Netherlands, Sweden.
Host records: Recorded by Fischer and Koponen (1999) as being a parasitoid of the agromyzid *Aulagromyza similis* (Brischke).

Opius corfuensis Fischer, 1996
Distribution in the Middle East: Turkey (Beyarslan, 2015; Beyarslan & Fischer, 2011b; Fischer & Beyarslan, 2005, 2013).
Extralimital distribution: Greece.
Host records: Unknown.

Opius crassipes Wesmael, 1835
Catalogs with Iranian records: Khajeh et al. (2014), Gadallah et al. (2016), Farahani et al. (2016), Yu et al. (2016), Samin, Coronado-Blanco, Kavallieratos et al. (2018), Samin, Coronado-Blanco, Fischer et al. (2018).
Distribution in Iran: Golestan (Samin, Ghahari et al., 2015, as *Opius paraplasticus* Fischer, 1972—misidentification), Guilan (Ghahari, Fischer, & Tobias, 2012), Isfahan (Ghahari, Fischer, & Papp, 2011), Semnan (Ghahari, Fischer, Hedqvist et al., 2010 as *Opius paraplasticus*—misidentification).
Distribution in the Middle East: Iran (see references above), Turkey (Beyarslan, 2015; Beyarslan & Fischer, 2011a, b; Fischer & Beyarslan, 2005, 2013).
Extralimital distribution: Armenia, Austria, Belgium, Czech Republic, Denmark, Finland, France, Germany, Greece, Hungary, Ireland, Italy, Lithuania, North Macedonia, Poland, Romania, Serbia, Spain, Sweden, Switzerland, United Kingdom.
Host records: Unknown.

Opius curticornis Fischer, 1960
Catalogs with Iranian records: Khajeh et al. (2014), Gadallah et al. (2016), Gadallah et al. (2016), Farahani et al. (2016), Yu et al. (2016), Samin, Coronado-Blanco, Kavallieratos et al. (2018), Samin, Coronado-Blanco, Fischer et al. (2018).
Distribution in Iran: East Azarbaijan (Ghahari et al., 2009), Lorestan (Ghahari, Fischer, Papp et al., 2012).
Distribution in the Middle East: Iran (Ghahari et al., 2009; Ghahari, Fischer, Papp et al., 2012), Turkey (Beyarslan & Fischer, 2011a, b; Fischer, 1960, 1972; Fischer & Betarslan, 2013).
Extralimital distribution: Hungary.
Host records: Unknown.

Opius delipunctis Fischer & Beyarslan, 2005
Distribution in the Middle East: Turkey (Beyarslan & Fischer, 2011b; Fischer & Beyarslan, 2005).
Extralimital distribution: None.
Host records: Unknown.

Opius erzurumensis Fischer, 2004
Distribution in the Middle East: Turkey (Beyarslan & Fischer, 2011b; Fischer, 2004; Fischer & Beyarslan, 2013).
Extralimital distribution: None.
Host records: Unknown.

Opius exiloides Fischer, 1989
Distribution in the Middle East: Turkey (Beyarslan, 2015).
Extralimital distribution: Hungary.
Host records: Unknown.

Opius ficedus Papp, 1979
Catalogs with Iranian records: No catalog.
Distribution in Iran: Kerman (Madjdzadeh et al., 2021).
Distribution in the Middle East: Iran.
Extralimital distribution: Former Czechoslovakia, Poland.
Host records: Unknown.

Opius flammeus Fischer, 1959
Distribution in the Middle East: Turkey (Beyarslan & Fischer, 2011b; Fischer & Beyarslan, 2005).
Extralimital distribution: Austria, Hungary, Poland, United Kingdom.
Host records: Recorded by Fischer (1967, 1969) as being a parasitoid of the agromyzids *Galiomyza morio* (Brischke), and *Liriomyza amoena* (Meigen).

Opius flavipes Szépligeti, 1898
Catalogs with Iranian records: Gadallah et al. (2016), Farahani et al. (2016), Yu et al. (2016), Samin, Coronado-Blanco, Kavallieratos et al. (2018), Samin, Coronado-Blanco, Fischer et al. (2018).

Distribution in Iran: Guilan (Dolati et al., 2021), Hormozgan (Ameri et al., 2014), Kerman (Safahani et al., 2018).
Distribution in the Middle East: Iran.
Extralimital distribution: Hungary, Switzerland.
Host records: Unknown.
Plant associations in Iran: *Medicago sativa* L. (Fabaceae) (Safahani et al., 2018).

Opius fuscipennis Wesmael, 1835
Catalogs with Iranian records: Samin, Coronado-Blanco, Fischer et al. (2018).
Distribution in Iran: Fars (Samin, Coronado-Blanco, Fischer et al., 2018), Mazandaran (Ghahari, 2019).
Distribution in the Middle East: Iran (Ghahari, 2019; Samin, Coronado-Blanco, Fischer et al., 2018), Turkey (Beyarslan, 2015; Beyarslan & Fischer, 2011a, b; Fischer & Beyarslan, 2005, 2013).
Extralimital distribution: Austria, Belgium, former Czechoslovakia, Finland, France, Germany, Greece, Hungary, Italy, Moldova, Poland, Russia, Spain, Switzerland, United Kingdom.
Host records: Summarized by Yu et al. (2016) as being a parasitoid of the agromyzid *Agromyza nigriciliata* Hendel; and the tephritid *Ensina sonchi* (L.).

Opius gigapiceus Fischer, 1989
Distribution in the Middle East: Turkey (Beyarslan & Fischer, 2011b; Fischer & Beyarslan, 2005).
Extralimital distribution: Austria, former Czechoslovakia, Finland, Germany, Hungary.
Host records: Unknown.

Opius gracilis Fischer, 1957
Catalogs with Iranian records: Samin et al. (2018b).
Distribution in Iran: Alborz, Guilan, Tehran (Dolati et al., 2019), Golestan (Gadallah et al., 2018), Kerman (Safahani et al., 2018), Khuzestan (Samin, Coronado-Blanco, Fischer et al., 2018).
Distribution in the Middle East: Iran (see references above), Israel−Palestine (Papp, 2012), Turkey (Beyarslan, 2015; Beyarslan & Fischer, 2011a, b; Fischer & Beyarslan, 2005, 2013).
Extralimital distribution: Austria, Bosnia-Herzegovina, Bulgaria, Croatia, Denmark, Estonia, Finland, France, Germany, Hungary, Italy, Kazakhstan, Korea, Lithuania, Mongolia, Montenegro, Poland, Romania, Russia, Serbia, Slovakia, Spain, Sweden, Switzerland, Tunisia, United Kingdom, Uzbekistan.
Host records: Summarized by Yu et al. (2016) as being a parasitoid of the following agromyzids: *Agromyza* spp., *Amauromyza gyrans* (Fallén), *Chromatomyia lonicerae* (Robineau-Desvoidy), *Chromatomyia syngenesiae* Hardy, *Liriomyza* spp., *Napomyza xylostei* Kaltenbach, and *Phytomyza* spp. In Iran, this species has been reared from

Amuromyza gyrans (Fallén) (Samin, Coronado-Blanco, Fischer et al., 2018).
Plant association in Iran: *Medicago sativa* (Fabaceae), *Rubus* sp. (Rosaceae) (Safahani et al., 2018).

Opius imitabilis Telenga, 1950
Distribution in the Middle East: Iraq, Jordan (Papp, 1982).
Extralimital distribution: Spain, Uzbekistan.
Host records: Unknown.

Opius inancae Fischer & Beyarslan, 2005
Distribution in the Middle East: Turkey (Beyarslan & Fischer, 2011b; Fischer & Beyarslan, 2005).
Extralimital distribution: Montenegro, Serbia.
Host records: Unknown.

Opius instabilis Wesmael, 1835
Catalogs with Iranian records: No catalog.
Distribution in Iran: Chaharmahal and Bakhtiari, Lorestan (Samin, Papp et al., 2018).
Distribution in the Middle East: Iran (Samin, Papp et al., 2018), Turkey (Beyarslan & Fischer, 2011b; Fischer & Beyarslan, 2005).
Extralimital distribution: Algeria, Austria, Belgium, former Czechoslovakia, Denmark, Finland, France, Germany, Hungary, Ireland, Italy, Kyrgyzstan, Lithuania, Poland, Serbia, Spain, Sweden, Switzerland, United Kingdom.
Host records: Recorded by Fischer (1962, 1967) as being a parasitoid of the agromyzids *Agromyza potentillae* (Kaltenbach), and *Chromatomyia ramosa* (Hendel).

Opius izmirensis Fischer & Beyarslan, 2005
Distribution in the Middle East: Turkey (Beyarslan & Fischer, 2011b; Fischer & Beyarslan, 2005).
Extralimital distribution: None.
Host records: Unknown.

Opius kilisanus Fischer & Beyarslan, 2005
Distribution in the Middle East: Turkey (Beyarslan & Fischer, 2011b; Fischer & Beyarslan, 2005).
Extralimital distribution: None.
Host records: Unknown.

Opius kirklareliensis Fischer & Beyarslan, 2005
Distribution in the Middle East: Turkey (Beyarslan & Fischer, 2011b; Fischer & Beyarslan, 2005).
Extralimital distribution: None.
Host records: Unknown.

Opius larissa Fischer, 1968
Catalogs with Iranian records: No catalog.
Distribution in Iran: Alborz, Guilan (Dolati et al., 2019), Razavi Khorasan (Ghahari & Beyarslan, 2019).

Distribution in the Middle East: Iran (Dolati et al., 2019; Ghahari & Beyarslan, 2019), Turkey (Beyarslan, 2015; Beyarslan & Fischer, 2011a, b; Fischer & Beyarslan, 2005). Extralimital distribution: Afghanistan, Greece, Hungary, Korea, Mongolia.

Host records: Unknown.

Opius latidens Fischer, 1990

Catalogs with Iranian records: Gadallah et al. (2016), Farahani et al. (2016), Yu et al. (2016), Samin, Coronado-Blanco, Kavallieratos et al. (2018), Samin, Coronado-Blanco, Fischer et al. (2018).

Distribution in Iran: Hormozgan (Ameri et al., 2014).

Distribution in the Middle East: Iran.

Extralimital distribution: Hungary.

Host records: Unknown.

Opius latipediformis Fischer, 2004

Catalogs with Iranian records: Gadallah et al. (2016), Farahani et al. (2016), Yu et al. (2016), Samin, Coronado-Blanco, Kavallieratos et al. (2018), Samin, Coronado-Blanco, Fischer et al. (2018).

Distribution in Iran: Hormozgan (Ameri et al., 2014).

Distribution in the Middle East: Iran.

Extralimital distribution: Denmark.

Host records: Unknown.

Opius latistigma Fischer, 1960

Catalogs with Iranian records: Khajeh et al. (2014), Gadallah et al. (2016), Farahani et al. (2016), Yu et al. (2016), Samin, Coronado-Blanco, Kavallieratos et al. (2018), Samin, Coronado-Blanco, Fischer et al. (2018).

Distribution in Iran: Razavi Khorasan (Ghahari, Fischer, Sakenin et al., 2011).

Distribution in the Middle East: Iran (Ghahari, Fischer, Sakenin et al., 2011), Turkey (Beyarslan, 2015; Beyarslan & Fischer, 2011a, b; Fischer & Beyarslan, 2013).

Extralimital distribution: Austria, Hungary, North Macedonia, Spain, former Yugoslavia.

Host records: Unknown.

Opius levis Wesmal, 1835

Catalogs with Iranian records: Khajeh et al. (2014 as both Opius filicornis Thomson, 1895 and Opius (Opiothorax) levis), Gadallah et al. (2016 as Opius filicornis), Farahani et al. (2016 as Opius filicornis), Samin, Coronado-Blanco, Kavallieratos et al. (2018), Samin, Coronado-Blanco, Fischer et al. (2018).

Distribution in Iran: East Azarbaijan (Rastegar et al., 2012 as Opius filicornis), Fars (Lashkari Bod et al., 2010, 2011), Guilan (Ghahari, Fischer, & Tobias, 2012 as Opius filicornis), Isfahan (Ghahari, Fischer, & Papp, 2011), Kermanshah (Ghahari & Fischer, 2012), Mazandaran (Dolati et al., 2021), Semnan (Samin, Fischer et al., 2015), Sistan and Baluchestan (Khajeh et al., 2014 as both Opius (Opiothorax) levis and Opius filicornis).

Distribution in the Middle East: Iran (see references above), Israel—Palestine (Papp, 1989), Turkey (Beyarslan, 2015; Beyarslan & Fischer, 2011a, b).

Extralimital distribution: Algeria, Austria, Belgium, Bulgaria, Croatia, Czech Republic, Denmark, Estonia, Ethiopia, Finland, France, Germany, Greece, Hungary, Ireland, Italy, Korea, Liechtenstein, Lithuania, North Macedonia, Montenegro, Netherlands, Norway, Poland, Romania, Russia, Serbia, Spain, Sweden, Switzerland, United Kingdom, Uzbekistan.

Host records: Summarized by Yu et al. (2016) as being a parasitoid of the following agromyzids: Agromyza nana Meigen, Liriomyza congesta (Becker), Liriomyza flaveola (Fallén), Liriomyza strigata (Meigen), Phytomyza crassiseta Zetterstedt, Phytomyza glechomae Kaltenbach, Phytomyza scotina Hendel; the drosophilid Scaptomyza graminum Fallén; and the ephydrid Hydrellia griseola (Fallén).

Plant associations in Iran: Lactuca sp. (Asteraceae), Malva sylvestris (Malvaceae) (Khajeh et al., 2014), Medicago sativa (Fabaceae) (Lashkari Bod et al., 2010, 2011; Khajeh et al., 2014).

Opius longicornis Thomson, 1895

Catalogs with Iranian records: Khajeh et al. (2014), Gadallah et al. (2016), Farahani et al. (2016), Yu et al. (2016), Samin, Coronado-Blanco, Kavallieratos et al. (2018), Samin, Coronado-Blanco, Fischer et al. (2018), Samin, Coronado-Blanco, Kavallieratos et al. (2018), Samin, Coronado-Blanco, Fischer et al. (2018).

Distribution in Iran: Guilan (Ghahari, Fischer, & Tobias, 2012), Kerman (Ranjbar et al., 2016; Safahani et al., 2018).

Distribution in the Middle East: Iran (see references above), Turkey (Beyarslan & Fischer, 2011a, b).

Extralimital distribution: Austria, Croatia, former Czechoslovakia, Finland, France, Germany, Greece, Hungary, Italy, Lithuania, North Macedonia, Montenegro, Romania, Russia, Serbia, Slovenia, Spain, Sweden, United Kingdom.

Host records: Recorded by Fischer (1967) as being a parasitoid of the agromyzid Phytomyza isais Hering.

Opius lonicerae Fischer, 1958

Catalogs with Iranian records: Gadallah et al. (2016), Farahani et al. (2016), Samin, Coronado-Blanco, Kavallieratos et al. (2018), Samin, Coronado-Blanco, Fischer et al. (2018).

Distribution in Iran: Kermanshah (Ghahari & Fischer, 2012).

Distribution in the Middle East: Iran (Ghahari & Fischer, 2012), Turkey (Beyarslan, 2015; Beyarslan & Fischer, 2011a, b; Beyarslan et al., 2006; Çikman & Uygun, 2003; Fischer & Beyarslan, 2005).

Extralimital distribution: Algeria, Austria, former Czechoslovakia, France, Germany, Hungary, Italy, Kazakhstan, Lithuania, Moldova, Poland, Spain, Switzerland, Uzbekistan.

Host records: Summarized by Yu et al. (2016) as being a parasitoid of the agromyzids *Amauromyza morionella* (Zetterstedt), *Chromatomyia lonicerae* (Robineau-Desvoidy), *Liriomyza trifolii* (Burgess), and *Napomyza xylostei* Kaltenbach.

Opius lara Fischer, 1968
Distribution in the Middle East: Turkey (Beyarslan & Fischer, 2011a, b; Fischer & Beyarslan, 2005, 2013).
Extralimital distribution: Mongolia.
Host records: Unknown.

Opius lugens Haliday, 1837
Catalogs with Iranian records: Gadallah et al. (2016), Farahani et al. (2016), Yu et al. (2016), Samin, Coronado-Blanco, Kavallieratos et al. (2018), Samin, Coronado-Blanco, Fischer et al. (2018).
Distribution in Iran: Guilan (Dolati et al., 2021), Kerman (Safahani et al., 2018), Kermanshah (Ghahari & Fischer, 2012).
Distribution in the Middle East: Iran (see references above), Israel—Palestine (Papp, 1989), Turkey (Beyarslan & Fischer, 2011a, b; Beyarslan & Inanç, 1992; Fischer & Beyarslan, 2005, 2013).
Extralimital distribution: Algeria, Austria, Bulgaria, Croatia, Czech Republic, Denmark, Estonia, Ethiopia, Finland, France, Germany, Greece, Hungary, Ireland, Italy, Lithuania, Mongolia, Montenegro, Poland, Russia, Serbia, Spain, Sweden, Switzerland, United Kingdom.
Host records: Summarized by Yu et al. (2016) as being a parasitoid of the following agromyzids: *Chromatomyia synenesiae* Hardy, *Liriomyza congesta* (Becker), *Liriomyza pusilla* (Meigen), *Liriomyza trifolii* (Burgess), *Ophiomyia* sp., and *Phytomyza thymi* Hering; the cecidomyiid *Asphondylia verbasci* (Vallot); and the tephritid *Ensina sonchi* (L.).
Plant associations in Iran: *Medicago sativa* (Fabaceae), *Rubus* sp. (Rosaceae) (Safahani et al., 2018).

Opius macedonicus Papp, 1973
Distribution in the Middle East: Turkey (Beyarslan & Fischer, 2011b; Fischer & Beyarslan, 2013).
Extralimital distribution: Hungary, North Macedonia, former Yugoslavia.
Host records: Unknown.

Opius magnicauda Fischer, 1958
Catalogs with Iranian records: Khajeh et al. (2014), Gadallah et al. (2016), Farahani et al. (2016), Yu et al.

(2016), Samin, Coronado-Blanco, Kavallieratos et al. (2018), Samin, Coronado-Blanco, Fischer et al. (2018).
Distribution in Iran: Golestan (Ghahari, Fischer, Sakenin et al., 2011).
Distribution in the Middle East: Iran (Ghahari, Fischer, Sakenin et al., 2011), Turkey (Beyarslan & Fischer, 2011a, b; Fischer & Beyarslan, 2005).
Extralimital distribution: Austria, former Czechoslovakia, Denmark, Estonia, Finland, Germany, Hungary, Italy, Kazakhstan, Poland, Russia, Sweden, Switzerland.
Host records: Unknown.

Opius mendus Papp, 1982
Catalogs with Iranian records: No catalog.
Distribution in Iran: Semnan (Samin, Papp et al., 2018).
Distribution in the Middle East: Iran (Samin, Papp et al., 2018), Turkey (Beyarslan, 2015; Beyarslan & Fischer, 2011a, b).
Extralimital distribution: Afghanistan, Spain.
Host records: Unknown.

Opius metanivens Fischer, 1992
Distribution in the Middle East: Turkey (Beyarslan & Fischer, 2011b; Fischer, 1992; Fischer & Beyarslan, 2005).
Extralimital distribution: None.
Host records: Unknown.

Opius minusculae Fischer, 1967
Catalogs with Iranian records: Gadallah et al. (2016), Farahani et al. (2016), Yu et al. (2016), Samin, Coronado-Blanco, Kavallieratos et al. (2018), Samin, Coronado-Blanco, Fischer et al. (2018).
Distribution in Iran: Hormozgan (Ameri et al., 2014), Guilan (Dolati et al., 2021).
Distribution in the Middle East: Iran (Ameri et al., 2014; Dolati et al., 2021), Turkey (Beyarslan, 2015; Beyarslan & Fischer, 2011a, b; Fischer & Beyarslan, 2005, 2013).
Extralimital distribution: Finland, Hungary, Spain, United Kingdom.
Host records: Recorded by Fischer (1967) as being a parasitoid of the agromyzid *Phytomyza minuscula* Goureau.

Opius mirabilis Fischer, 1958
Catalogs with Iranian records: Gadallah et al. (2016), Farahani et al. (2016), Yu et al. (2016), Samin, Coronado-Blanco, Kavallieratos et al. (2018), Samin, Coronado-Blanco, Fischer et al. (2018).
Distribution in Iran: Kermanshah (Ghahari & Fischer, 2012).
Distribution in the Middle East: Iran (Ghahari & Fischer, 2012), Turkey (Beyarslan & Fischer, 2011a, b; Beyarslan & Inanç, 1992; Fischer & Beyarslan, 2005, 2013).

Extralimital distribution: Armenia, Austria, China, Croatia, Finland, France, Germany, Hungary, Italy, Kazakhstan, Poland, Romania, Spain, Tunisia, former Yugoslavia.
Host records: Summarized by Yu et al. (2016) as being a parasitoid of the agromyzids *Chromatomyia horticola* Goureau, *Liriomyza balcanica* (Strobl), and *Liriomyza sativae* (Blanchard).

Opius mischa Fischer, 1968
Catalogs with Iranian records: No catalog.
Distribution in Iran: Guilan (Dolati et al., 2021).
Distribution in the Middle East: Iran (Dolati et al., 2021), Turkey (Beyarslan, 2015).
Extralimital distribution: Hungary, Mongolia.
Host records: Unknown.

Opius monilicornis Fischer, 1962
Catalogs with Iranian records: Khajeh et al. (2014), Gadallah et al. (2016), Farahani et al. (2016), Yu et al. (2016), Samin, Coronado-Blanco, Kavallieratos et al. (2018), Samin, Coronado-Blanco, Fischer et al. (2018).
Distribution in Iran: Lorestan (Ghahari, Fischer, Papp et al., 2012), Mazandaran (Ghahari, Fischer, Çetin Erdoğan et al., 2010; Sakenin et al., 2012), West Azarbaijan (Adldoost, 1995; Modarres Awal, 1997, 2012; Khanjani, 2006a, b).
Distribution in the Middle East: Iran (see references above), Jordan (Papp, 1982), Syria (El Bouhssini et al., 2008), Turkey (Beyarslan & Fischer, 2011a, b; Çikman, 2006).
Extralimital distribution: Algeria, Moldova, Spain.
Host records: Recorded by Çikman et al. (2006) as being a parasitoid of the agromyzid *Liriomyza cicerinae* (Rondani). In Iran, this species has been reared from *Liriomyza cicerina* (Rondani) (Khanjani, 2006a, b; Modarres Awal, 1997, 2012).

Opius moravicus Fischer, 1960
Distribution in the Middle East: Turkey (Beyarslan, 2015).
Extralimital distribution: Former Czechoslovakia, Hungary.
Host records: Unknown.

Opius nanosoma Fischer, 1989
Distribution in the Middle East: Turkey (Beyarslan, 2015).
Extralimital distribution: Slovakia.
Host records: Unknown.

Opius neopendulus Fischer, 1989
Distribution in the Middle East: Turkey (Beyarslan & Fischer, 2011a, b).
Extralimital distribution: Austria, Germany, Hungary, Italy.
Host records: Unknown.

Opius nigricolor Fischer, 1960
Distribution in the Middle East: Turkey (Beyarslan, 2015).
Extralimital distribution: Austria, France, Germany, Hungary, Italy, Poland, Russia, Slovakia, Spain, Ukraine.

Host records: Recorded by Fischer (1969, 1995) as being a parasitoid of the agromyzid *Aulagromyza* sp.

Opius nigricoloratus Fischer, 1958
Catalogs with Iranian records: Khajeh et al. (2014), Gadallah et al. (2016), Farahani et al. (2016), Yu et al. (2016), Samin, Coronado-Blanco, Kavallieratos et al. (2018), Samin, Coronado-Blanco, Fischer et al. (2018).
Distribution in Iran: Guilan (Ghahari, Fischer, & Tobias, 2012), Semnan (Ghahari, Fischer, Hedqvist et al., 2010; Samin, 2015).
Distribution in the Middle East: Iran (see references above), Turkey (Beyarslan & Fischer, 2011a, b; Fischer & Beyarslan, 2005, 2013).
Extralimital distribution: Austria, Denmark, Finland, France, Germany, Greece, Hungary, Italy, Mongolia, Serbia, Spain, Switzerland.
Host records: Recorded by Drea et al. (1982) as being a parasitoid of the agromyzids *Agromyza frontella* (Rondani), and *Agromyza nana* Meigen.

Opius occulisus Telenga, 1950
Catalogs with Iranian records: Khajeh et al. (2014), Gadallah et al. (2016), Farahani et al. (2016), Yu et al. (2016), Samin, Coronado-Blanco, Kavallieratos et al. (2018), Samin, Coronado-Blanco, Fischer et al. (2018).
Distribution in Iran: Tehran (Ghahari, Fischer, Sakenin et al., 2011).
Distribution in the Middle East: Iran (Ghahari, Fischer, Sakenin et al., 2011), Turkey (Beyarslan & Fischer, 2011a, b).
Extralimital distribution: Ukraine.
Host records: Unknown.

Opius ocuvergens Papp, 2012
Distribution in the Middle East: Israel—Palestine (Papp, 2012).
Extralimital distribution: None.
Host records: Unknown.

Opius opacus Fischer, 1968
Catalogs with Iranian records: Khajeh et al. (2014), Gadallah et al. (2016), Farahani et al. (2016), Yu et al. (2016), Samin, Coronado-Blanco, Kavallieratos et al. (2018), Samin, Coronado-Blanco, Fischer et al. (2018).
Distribution in Iran: Kerman (Safahani et al., 2018), Sistan and Baluchestan (Khajeh et al., 2014).
Distribution in the Middle East: Iran (Khajeh et al., 2014; Safahani et al., 2018), Turkey (Beyarslan & Fischer, 2011a, b; Fischer & Beyarslan, 2005).
Extralimital distribution: Finland, Greece, Hungary, Moldova, Mongolia, Spain.
Host records: Unknown.

Plant associations in Iran: *Cardaria draba* (Brassicaceae) (Khajeh et al., 2014), *Medicago sativa* (Fabaceae) (Safahani et al., 2018).

Opius orbiculator (Nees von Esenbeck, 1811)

Distribution in the Middle East: Turkey (Beyarslan, 2015; Beyarslan & Fischer, 2011b; Fischer & Beyarslan, 2013).

Extralimital distribution: Austria, former Czechoslovakia, Estonia, Finland, Germany, Hungary, Ireland, Italy, Lithuania, Mongolia, Montenegro, Netherlands, Poland, Russia, Serbia, Spain, Sweden.

Host records: Summarized by Yu et al. (2016) as being a parasitoid of the following agromyzids *Phytomyza affinis* Fallén, *Phytomyza ranunculi* (Schrank), and *Phytomyza solidaginis* Hendel.

Opius osogovoensis Fischer, 1964

Distribution in the Middle East: Turkey (Beyarslan & Fischer, 2011a, b; Çikman & Uygun, 2003; Fischer & Beyarslan, 2005, 2013).

Extralimital distribution: Austria, Bulgaria, Hungary, Moldova, Poland, Russia.

Host records: Recorded by Çikman and Uygun (2003) as being a parasitoid of the agromyzid *Chromatomyia horticola* (Goureau).

Opius pallipes Wesmael, 1835

Catalogs with Iranian records: Khajeh et al. (2014), Gadallah et al. (2016), Farahani et al. (2016), Yu et al. (2016), Samin, Coronado-Blanco, Kavallieratos et al. (2018), Samin, Coronado-Blanco, Fischer et al. (2018).

Distribution in Iran: Guilan (Ghahari, Fischer, & Tobias, 2012), Kerman (Ranjbar et al., 2016; Safahani et al., 2018), Sistan and Baluchestan (Khajeh et al., 2014 as *Opius exilis* Haliday, 1837).

Distribution in the Middle East: Iran (see references above), Israel—Palestine (Papp, 2012), Turkey (Beyarslan, 2015; Beyarslan & Fischer, 2011a, b).

Extralimital distribution: Austria, Belgium, Bulgaria, Canada, China, Croatia, Czech Republic, Denmark, Estonia, Finland, France, Georgia, Germany, Greece, Hungary, Ireland, Italy, Kazakhstan, Korea, Lithuania, Mongolia, Montenegro, Netherlands, Norway, Poland, Romania, Russia, Serbia, Slovenia, Spain, Sweden, Switzerland, Ukraine, United Kingdom, United States of America, Uzbekistan.

Host records: Summarized by Yu et al. (2016) as being a parasitoid of several dipteran families including, Agromyzidae, Anthomyiidae, Drosophilidae, Scatophagidae, and Tephritidae.

Plant associations in Iran: *Brassica napus* (Brassicaceae), *Lactuca oleracea* (Asteraceae), *Malva sylvestris* (Malvaceae) (Khajeh et al., 2014), *Medicago sativa* (Fabaceae) (Khajeh et al., 2014; Ranjbar et al., 2016; Safahani et al., 2018), *Rubus* sp. (Rosaceae) (Safahani et al., 2018).

Opius paranivens Fischer, 1990

Catalogs with Iranian records: No catalog.
Distribution in Iran: Kerman (Ranjbar et al., 2016).
Distribution in the Middle East: Iran.
Extralimital distribution: Finland, Hungary, Spain.
Host records: Unknown.
Plant associations in Iran: *Medicago sativa* (Fabaceae) (Ranjbar et al., 2016).

Opius paraqvisti Fischer, 2004

Distribution in the Middle East: Turkey (Fischer, 2004; Beyarslan & Fischer, 2011a, b).
Extralimital distribution: None.
Host records: Unknown.

Opius pendulus Haliday, 1837

Catalogs with Iranian records: Khajeh et al. (2014), Gadallah et al. (2016), Farahani et al. (2016), Yu et al. (2016), Samin, Coronado-Blanco, Kavallieratos et al. (2018), Samin, Coronado-Blanco, Fischer et al. (2018).

Distribution in Iran: Hormozgan (Ameri et al., 2014), Lorestan (Ghahari, Fischer, Papp et al., 2012), Semnan (Naderian et al., 2012), Sistan and Baluchestan (Khajeh et al., 2014).

Distribution in the Middle East: Iran (see references above), Turkey (Beyarslan & Fischer, 2011a, b; Fischer & Beyarslan, 2005).

Extralimital distribution: Austria, Belgium, Bulgaria, Canada, Croatia, Denmark, Estonia, Finland, France, Germany, Greece, Hungary, Ireland, Italy, Kazakhstan, Kyrgyzstan, Lithuania, Moldova, Montenegro, Poland, Romania, Russia, Serbia, Slovakia, Spain, Sweden, Switzerland, United Kingdom, United States of America.

Host records: Summarized by Yu et al. (2016) as being a parasitoid of the agromyzids *Liriomyza congesta* (Becker), *Phytomyza cineracea* Hendel, *Phytomya evanescens* Hendel, and *Phytomya nigritula* Zetterstedt.

Plant associations in Iran: *Cyperus rotundus* (Cyperaceae), *Hordeum vulgare* (Poaceae), *Matricaria recutita* (Asteraceae), *Triticum aestivum* (Poaceae) (Khajeh et al., 2014).

Opius peterseni Fischer, 1964

Distribution in the Middle East: Turkey (Beyarslan & Fischer, 2011a, b; Fischer & Beyarslan, 2005).
Extralimital distribution: Austria, Denmark, Finland, Hungary, Montenegro, Serbia.
Host records: Unknown.

Opius phytobiae Fischer, 1959

Catalogs with Iranian records: No catalog.
Distribution in Iran: Guilan (Dolati et al., 2021).

Distribution in the Middle East: Iran.
Extralimital distribution: Georgia, Greece, Hungary, Italy, Korea, Poland, Russia, Spain, United Kingdom.
Host records: Recorded by Fischer (1959) as being a parasitoid of the agromyzid *Phytoliriomyza hilarella* (Zetterstedt).

Opius piloralis Fischer, 1989
Distribution in the Middle East: Israel–Palestine (Fischer, 1997).
Extralimital distribution: Slovakia.
Host records: Unknown.

Opius podomelas Fischer, 1972
Distribution in the Middle East: Turkey (Beyarslan, 2019).
Extralimital distribution: South Africa.
Host records: Unknown.

Opius ponticus Fischer, 1958
Catalogs with Iranian records: Gadallah et al. (2016), Yu et al. (2016), Samin, Coronado-Blanco, Kavallieratos et al. (2018), Samin, Coronado-Blanco, Fischer et al. (2018).
Distribution in Iran: Golestan, Mazandaran (Gadallah et al., 2016).
Distribution in the Middle East: Iran (Gadallah et al., 2016), Turkey (Beyarslan, 2015; Beyarslan & Fischer, 2011b; Fischer & Beyarslan, 2005).
Extralimital distribution: Austria, Kazakhstan, Kyrgyzstan, Turkmenistan, Uzbekistan.
Host records: Unknown.

Opius pulicariae Fischer, 1969
Catalogs with Iranian records: No catalog.
Distribution in Iran: Guilan (Dolati et al., 2019).
Distribution in the Middle East: Iran (Dolati et al., 2019), Turkey (Beyarslan, 2019).
Extralimital distribution: Armenia, Austria, former Czechoslovakia, Finland, Germany, Hungary, Italy, Poland, Russia, Spain, Switzerland.
Host records: Summarized by Yu et al. (2016) as being a parasitoid of the agromyzids *Chromatomyia syngenesiae* Hardy, *Ophiomyia pulicaria* (Meigen), *Phytomyza ranunculi* (Schrank), and *Phytomya virgaureae* Hering.

Opius pygmaeator (Nees von Esenbeck, 1811)
Catalogs with Iranian records: Gadallah et al. (2016), Yu et al. (2016), Samin, Coronado-Blanco, Kavallieratos et al. (2018), Samin, Coronado-Blanco, Fischer et al. (2018).
Distribution in Iran: East Azarbaijan (Gadallah et al., 2016).
Distribution in the Middle East: Iran (Gadallah et al., 2016), Turkey (Beyarslan, 2015; Beyarslan & Fischer, 2011b; Civelek et al., 2002).

Extralimital distribution: Austria, Belgium, Czech Republic, Denmark, Estonia, Finland, France, Georgia, Germany, Greece, Hungary, Ireland, Italy, Lithuania, Montenegro, Norway, Poland, Russia, Serbia, Slovenia, Sweden, Switzerland, United Kingdom.
Host records: Summarized by Yu et al. (2016) as being a parasitoid of the agromyzids *Liriomyza congesta* (Becker), *Liriomyza huidobrensis* (Blanchard), *Phytomyza cineracea* Hendel, and *Phytomyza evanescens* Hendel, and the tephritid *Chaetostomella cylindrica* (Robineau-Desvoidy).

Opius pygmaeus Fischer, 1962
Catalogs with Iranian records: Khajeh et al. (2014), Gadallah et al. (2016), Farahani et al. (2016), Yu et al. (2016), Samin, Coronado-Blanco, Kavallieratos et al. (2018), Samin, Coronado-Blanco, Fischer et al. (2018).
Distribution in Iran: Alborz, Guilan, Tehran (Dolati et al., 2021), Fars (Lashkari Bod et al., 2010, 2011), Kerman (Ranjbar et al., 2016; Safahani et al., 2018), Lorestan (Ghahari, Fischer, Papp et al., 2012).
Distribution in the Middle East: Iran (see references above), Turkey (Beyarslan, 2015; Beyarslan & Fischer, 2011a, b; Fischer & Beyarslan, 2005, 2013).
Extralimital distribution: Algeria, Austria, former Czechoslovakia, Estonia, Finland, France, Germany, Hungary, Italy, Kazakhstan, Lithuania, Russia, Spain, United Kingdom.
Host records: Recorded by Fischer (1962, 1967) as being a parasitoid of the agromyzids *Liriomyza cicerinae* (Rondani) and *Phytomyza adjuncta* Hering.
Plant associations in Iran: *Descurainia sophia* (Brassicaceae), *Glycirhiza glabra* (Fabaceae), *Medicago sativa* (Fabaceae), *Rubus* sp. (Rosaceae) (Lashkari Bod et al., 2010; 2011; Ranjbar et al., 2016; Safahani et al., 2018).

Opius quasilatipes Fischer & Beyarslan, 2005
Distribution in the Middle East: Turkey (Beyarslan & Fischer, 2011b; Fischer & Beyarslan, 2005).
Extralimital distribution: None.
Host records: Unknown.

Opius quasipulvis Fischer, 1989
Distribution in the Middle East: Turkey (Beyarslan & Fischer, 2011a, b; Çikman et al., 2006).
Extralimital distribution: Hungary.
Host records: Unknown.

Opius radialis Fischer, 1957
Distribution in the Middle East: Turkey (Beyarslan, 2015).
Extralimital distribution: Austria, former Czechoslovakia, Estonia, Finland, Germany, Hungary, Spain.
Host records: Recorded by Fischer (1967) as being a parasitoid of the agromyzid *Phytomyza spondylii* Robineau-Disvoidy.

Opius repentinus Papp, 1980
Catalogs with Iranian records: No catalog.
Distribution in Iran: Mazandaran (Dolati et al., 2019).
Distribution in the Middle East: Iran.
Extralimital distribution: Croatia, Hungary, Korea.
Host records: Unknown.

Opius robustus Telenga, 1950
Catalogs with Iranian records: Khajeh et al. (2014), Gadallah et al. (2016), Gadallah et al. (2016), Farahani et al. (2016), Yu et al. (2016), Samin, Coronado-Blanco, Kavallieratos et al. (2018), Samin, Coronado-Blanco, Fischer et al. (2018).
Distribution in Iran: Alborz (Dolati et al., 2021), Fars (Lashkari Bod et al., 2011).
Distribution in the Middle East: Iran.
Extralimital distribution: Kazakhstan, Ukraine, Uzbekistan.
Host records: Unknown.
Plant associations in Iran: *Medicago sativa* (Fabaceae) (Lashkari Bod et al., 2010, 2011).

Opius rudiformis Fischer, 1958
Distribution in the Middle East: Turkey (Beyarslan, 2015; Beyarslan & Fischer, 2011b; Fischer & Beyarslan, 2005).
Extralimital distribution: Austria, Canada, former Czechoslovakia, Denmark, Finland, France, Hungary, Italy, Netherlands, Poland, Russia, Spain, United Kingdom.
Host records: Unknown.

Opius rufimixtus Fischer, 1958
Catalogs with Iranian records: No catalog.
Distribution in Iran: Mazandaran (Dolati et al., 2021).
Distribution in the Middle East: Iran.
Extralimital distribution: Austria, Greece, Hungary, Italy, Russia.
Host records: Unknown.

Opius russalka Fischer, 1968
Distribution in the Middle East: Syria (Papp, 1982).
Extralimital distribution: Mongolia.
Host records: recorded by Papp (1982) as being a parasitoid of the agromyzid *Liriomyza cicerinae* (Rondani).

Opius seductus Fischer, 1959
Catalogs with Iranian records: Khajeh et al. (2014), Gadallah et al. (2016), Farahani et al. (2016), Yu et al. (2016), Samin, Coronado-Blanco, Kavallieratos et al. (2018), Samin, Coronado-Blanco, Fischer et al. (2018).
Distribution in Iran: Isfahan (Ghahari, Fischer, & Papp, 2011), Kerman (Safahani et al., 2018).
Distribution in the Middle East: Iran (Ghahari, Fischer, & Papp, 2011; Safahani et al., 2018), Turkey (Beyarslan & Fischer, 2011b; Fischer & Beyarslan, 2005).
Extralimital distribution: Greece, Turkmenistan, Uzbekistan.

Host records: Unknown.
Plant associations in Iran: *Mentha pulegium* (Lamiaceae) (Safahani et al., 2018).

Opius sigmodus Papp, 1981
Distribution in the Middle East: Cyprus (Beyarslan et al., 2017), Turkey (Beyarslan, 2015; Beyarslan & Fischer, 2011b; Fischer & Beyarslan, 2013).
Extralimital distribution: Finland, Germany, Hungary.
Host records: Unknown.

Opius singularis Wesmael, 1835
Catalogs with Iranian records: Khajeh et al. (2014), Gadallah et al. (2016), Farahani et al. (2016), Yu et al. (2016), Samin, Coronado-Blanco, Kavallieratos et al. (2018), Samin, Coronado-Blanco, Fischer et al. (2018).
Distribution in Iran: East Azarbaijan (Ghahari & Gadallah, 2022), Guilan (Ghahari, Fischer, & Tobias, 2012; Dolati et al., 2019), Mazandaran (Dolati et al., 2019), Semnan (Samin, Fischer et al., 2015).
Distribution in the Middle East: Iran (see references above), Turkey (Beyarslan & Fischer, 2011a, b).
Extralimital distribution: Armenia, Austria, Belgium, Bulgaria, Croatia, Czech Republic, Denmark, Finland, France, Georgia, Germany, Greece, Hungary, Ireland, Italy, Korea, Lithuania, Mongolia, Montenegro, Netherlands, Poland, Romania, Russia, Serbia, Spain, Sweden, Switzerland, Tunisia, United Kingdom.
Host records: Summarized by Yu et al. (2016) as being a parasitoid of the following agromyzids: *Agromyza* spp., *Amauromyza labiatarum* (Hendel), *Aulagromyza hendeliana* (Hering), *Cerodontha pygmaea* (Meigen), *Chromatomyza periclymeni* (de Meijere), *Liriomyza amoena* (Meigen), *Liriomyza eupatorii* (Kaltenbach), and *Phytomyza* spp. In Iran, this species has been reared from the agromyzid *Cerodontha pygmaea* (Meigen) (Ghahari & Gadallah, 2022).

Opius soenderupianus Fischer, 1967
Distribution in the Middle East: Turkey (Beyarslan & Fischer, 2011a, b).
Extralimital distribution: Former Czechoslovakia, Finland, Germany, Russia.
Host records: Recorded by Fischer (1967) and Fischer and Koponen (1999) as being a parasitoid of the agromyzids *Cerodontha caricicola* (Hering) and *Phytomyza soenderupi* Hering.

Opius solymosae Fischer, 1989
Distribution in the Middle East: Turkey (Beyarslan, 2019).
Extralimital distribution: Hungary.
Host records: Unknown.

Opius sonja Fischer, 1968
Distribution in the Middle East: Turkey (Beyarslan & Fischer, 2011a, b).
Extralimital distribution: Mongolia.
Host records: Unknown.

Opius tabificus Papp, 1979
Catalogs with Iranian records: Khajeh et al. (2014), Gadallah et al. (2016), Farahani et al. (2016), Yu et al. (2016), Samin, Coronado-Blanco, Kavallieratos et al. (2018), Samin, Coronado-Blanco, Fischer et al. (2018).
Distribution in Iran: Hormozgan (Ameri et al., 2014), Kerman (Safahani et al., 2018 on *Medicago sativa*—Fabaceae), Sistan and Baluchestan (Khajeh et al., 2014).
Distribution in the Middle East: Iran (see references above), Turkey (Beyarslan, 2015; Beyarslan & Fischer, 2011a, b; Fischer & Beyarslan, 2013).
Extralimital distribution: Italy, Spain, Tunisia.
Host records: Unknown.

Opius tekirdagensis Fischer & Beyarslan, 2005
Distribution in the Middle East: Turkey (Beyarslan & Fischer, 2011b; Fischer & Beyarslan, 2005).
Extralimital distribution: None.
Host records: Unknown.

Opius tenellae Fischer, 1969
Distribution in the Middle East: Turkey (Fischer & Beyarslan, 2005, Beyarslan & Fischer, 2011a, b).
Extralimital distribution: Austria, Croatia, Germany, Ireland, Mongolia.
Host records: Recorded by Fischer (1969, 1972) as being a parasitoid of the agromyzid *Phytomyza tenella* Meigen.

Opius tersus (Foerster, 1863)
Catalogs with Iranian records: Khajeh et al. (2014), Gadallah et al. (2016), Farahani et al. (2016), Yu et al. (2016), Samin, Coronado-Blanco, Kavallieratos et al. (2018), Samin, Coronado-Blanco, Fischer et al. (2018).
Distribution in Iran: Guilan (Ghahari, Fischer, & Tobias, 2012), Kerman (Ranjbar et al., 2016).
Distribution in the Middle East: Iran (Ghahari, Fischer, & Tobias, 2012; Ranjbar et al., 2016), Israel—Palestine (Papp, 2012), Turkey (Fischer & Beyarslan, 2005, 2013, Beyarslan & Fischer, 2011a, b).
Extralimital distribution: Algeria, Austria, Croatia, former Czechoslovakia, Estonia, Finland, France, Germany, Italy, Korea, Montenegro, Poland, Russia, Serbia, Spain, Turkmenistan, Uzbekistan.
Host records: Summarized by Yu et al. (2016) as being a parasitoid of the following agromyzids: *Amauromyza*

flavifrons (Meigen), *Chromatomyia horticola* Goureau, *Liriomyza* spp., and *Phytomyza scotina* Hendel.
Plant associations in Iran: *Medicago sativa* (Fabaceae) (Ranjbar et al., 2016; Safahani et al., 2018), *Mentha longifolia* (Lamiaceae) (Ranjbar et al., 2016), *Anethum graveolens* (Apiaceae) (Safahani et al., 2018).

Opius truncatulus Fischer, 1963
Distribution in the Middle East: Turkey (Atay et al., 2019).
Extralimital distribution: Austria, Hungary.
Host records: Unknown.

Opius tuberculifer Fischer, 1958
Distribution in the Middle East: Turkey (Beyarslan, 2015).
Extralimital distribution: Austria, former Czechoslovakia, Finland, Germany, Hungary, Russia, Switzerland.
Host records: Unknown.

Opius tunensis Fischer, 1962
Distribution in the Middle East: Turkey (Beyarslan & Fischer, 2011b; Fischer & Beyarslan, 2005).
Extralimital distribution: Tunisia.
Host records: Unknown.

Opius turcicus Fischer, 1960
Catalogs with Iranian records: Khajeh et al. (2014), Gadallah et al. (2016), Farahani et al. (2016), Yu et al. (2016), Samin, Coronado-Blanco, Kavallieratos et al. (2018), Samin, Coronado-Blanco, Fischer et al. (2018).
Distribution in Iran: Markazi (Ghahari, Fischer, Sakenin et al., 2011).
Distribution in the Middle East: Iran (Ghahari, Fischer, Sakenin et al., 2011), Iraq (Fischer & Koponen, 1999), Israel—Palestine (Fischer & Koponen, 1999; Papp, 1989), Jordan (Fischer & Koponen, 1999; Papp, 1982), Turkey (Beyarslan, 2015; Beyarslan & Fischer, 2011a, b; Fischer & Beyarslan, 2005; Fischer & Koponen, 1999).
Extralimital distribution: Algeria, Bulgaria, Croatia, former Czechoslovakia, Estonia, Finland, France, Germany, Hungary, India, Mongolia, Netherlands, Spain, Ukraine.
Host records: Summarized by Yu et al. (2016) as being a parasitoid of the agromyzids *Agromyza lathyri* Hendel, *Agromyza rondensis* Strobl, and *Chromatomyia horticola* Goureau.

Opius viennensis Fischer, 1959
Distribution in the Middle East: Turkey (Beyarslan & Fischer, 2011b; Fischer & Beyarslan, 2005, 2013).
Extralimital distribution: Austria, former Czechoslovakia, Finland, Hungary, Italy, Spain.
Host records: Unknown.

Genus *Phaedrotoma* Foerster, 1863

Phaedrotoma bajariae (Fischer, 1989)

Catalogs with Iranian records: Gadallah et al. (2016), Farahani et al. (2016), Yu et al. (2016), Samin, Coronado-Blanco, Kavallieratos et al. (2018), Samin, Coronado-Blanco, Fischer et al. (2018) (all as *Opius (Ilicopius) bajariae* Fischer).

Distribution in Iran: Hormozgan (Ameri et al., 2014 as *O. bajariae*).

Distribution in the Middle East: Cyprus (Beyarslan et al., 2017 as *Opius bajarae*), Iran (Ameri et al., 2014), Turkey (Beyarslan, 2015; Beyarslan & Fischer, 2011a, b; Fischer & Beyarslan, 2013).

Extralimital distribution: Hungary, India.

Host records: Unknown.

Phaedrotoma benignus (Papp, 1981)

Distribution in the Middle East: Turkey (Beyarslan, 2015; Beyarslan & Fischer, 2011a, b; Fischer & Beyarslan, 2013).

Extralimital distribution: Korea, Russia.

Host records: Unknown.

Phaedrotoma aethiops (Haliday, 1837)

Catalogs with Iranian records: Khajeh et al. (2014), Gadallah et al. (2016), Farahani et al. (2016), Yu et al. (2016), Samin, Coronado-Blanco, Kavallieratos et al. (2018), Samin, Coronado-Blanco, Fischer et al. (2018) (all as *Opius (Phaedrotoma) aethiops* Haliday).

Distribution in Iran: Ardabil (Rastegar et al., 2012 as *O. aethiops*), Guilan (Ghahari, Fischer, & Tobias, 2012 as *O. Aethiops*, Dolati et al., 2021).

Distribution in the Middle East: Iran (see references above), Israel–Palestine (Fischer, 1997; Papp, 2012), Turkey (Beyarslan & Fischer, 2011a, b; Fischer & Beyarslan, 2013).

Extralimital distribution: Austria, Bulgaria, Czech Republic, Denmark, Estonia, Finland, France, Germany, Hungary, Ireland, Italy, Lithuania, Montenegro, Poland, Russia, Serbia, Spain, Sweden, United Kingdom, Uzbekistan.

Host records: Recorded by Fischer (1964) as being a parasitoid of the agromyzids *Cerodontha denticornis* (Panzer) and *Chromatomyia milii* (Kaltenbach).

Phaedrotoma ambiguus (Wesmael, 1835)

Catalogs with Iranian records: Khajeh et al. (2014), Gadallah et al. (2016), Farahani et al. (2016), Yu et al. (2016), Samin, Coronado-Blanco, Kavallieratos et al. (2018), Samin, Coronado-Blanco, Fischer et al. (2018) (all as *Opius (Nosopoea) ambiguus* Wesmael).

Distribution in Iran: Isfahan (Ghahari, Fischer, & Papp, 2011 as *O. ambiguus*), Mazandaran (Dolati et al., 2019 as *O. ambiguus*).

Distribution in the Middle East: Iran (Dolati et al., 2019; Ghahari, Fischer, & Papp, 2011), Turkey (Beyarslan &

Fischer, 2011b; Çikman & Uygun, 2003; Fischer & Beyarslan, 2005, 2013).

Extralimital distribution: Austria, Belgium, Bulgaria, China, Czech Republic, Denmark, Estonia, Finland, France, Germany, Greece, Hungary, Ireland, Italy, Korea, Mongolia, Montenegro, Netherlands, Norway, Poland, Russia, Serbia, Sweden, Switzerland, Ukraine, United Kingdom.

Host records: Summarized by Yu et al. (2016) as being a parasitoid of the following agromyzids: *Agromyza* spp., *Aulagromyza buhri* (de Meijere), *Cerodontha* spp., *Chromatomyia horticola* Goureau, *Cerodontha milii* (Kaltenbach), *Liriomyza pusio* (Meigen), *Nemorimyza posticata* Meigen, and *Phytomyza* spp.

Phaedrotoma biroi (Fischer, 1960)

Catalogs with Iranian records: Khajeh et al. (2014), Gadallah et al. (2016), Farahani et al. (2016), Yu et al. (2016), Samin, Coronado-Blanco, Kavallieratos et al. (2018), Samin, Coronado-Blanco, Fischer et al. (2018) (all as *Opius (Phaedrotoma) biroi* Fischer).

Distribution in Iran: Golestan (Ghahari, Fischer, Sakenin et al., 2011 as *Opius (Phaedrotoma) biroi*), Kerman (Ranjbar et al., 2016 as *O. biroi*), Sistan and Baluchestan (Khajeh et al., 2014 as *O. biroi*).

Distribution in the Middle East: Iran (see references above), Israel–Palestine (Papp, 2012), Turkey (Beyarslan, 2015; Beyarslan & Fischer, 2011a, b; Fischer & Beyarslan, 2005, 2013).

Extralimital distribution: China, Hungary, Italy, Spain.

Host records: Recorded by Ku et al. (2001) as being a parasitoid of the agromyzids *Chromatomyia horticola* Goureau, *Liriomyza bryoniae* (Kaltenbach), and *Liriomyza trifolii* (Burgess).

Plant associations in Iran: *Cardaria draba* (Brassicaceae) (Khajeh et al., 2014), *Descurainia sophia* (Brassicaceae), *Medicago sativa* (Fabaceae) (Ranjbar et al., 2016).

Phaedrotoma biroicus (Fischer & Beyarslan, 2005)

Catalogs with Iranian records: Khajeh et al. (2014), Gadallah et al. (2016 as *O. biroicus*), Farahani et al. (2016 as *O. biroicus*), Yu et al. (2016 as *O. biroicus*), Samin, Coronado-Blanco, Kavallieratos et al. (2018), Samin, Coronado-Blanco, Fischer et al. (2018 as *Opius biroicus*).

Distribution in Iran: Sistan and Baluchestan (Khajeh et al., 2014).

Distribution in the Middle East: Iran (Khajeh et al., 2014), Turkey (Beyarslan & Fischer, 2011b; Fischer & Beyarslan, 2005).

Extralimital distribution: None.

Host records: Unknown.

Plant associations in Iran: *Medicago sativa* (Fabaceae) (Khajeh et al., 2014).

Phaedrotoma caesus (Haliday, 1837)

Catalogs with Iranian records: No catalog.

Distribution in Iran: Fars (Ghahari & Beyarslan, 2019 as *Opius (Tolbia) caesus* Haliday).

Distribution in the Middle East: Iran (Ghahari & Beyarslan, 2019), Israel−Palestine (Papp, 2012), Syria (Papp, 1982), Turkey (Beyarslan & Fischer, 2011b; Fischer & Beyarslan, 2005).

Extralimital distribution: Austria, Bulgaria, former Czechoslovakia, Denmark, Finland, France, Germany, Hungary, Iceland, Ireland, Italy, Lithuania, Nepal, Poland, Romania, Russia, Serbia, Spain, Sweden, Switzerland, Tajikistan, United States of America (Yu et al., 2016 as *Opius caesus*).

Host records: Summarized by Yu et al. (2016) as being a parasitoid of the agromyzid *Chromatomyia primulae* (Robineau-Desvoidy), the drosophilid *Scatomyza graminum* Fallén, and the ephydrids *Hydrellia cochleariae* Haliday, *Hydrellia griseola* (Fallén), *Hydrellia nigripes* Zetterstedt, and *Hydrellia tarsata* Haliday.

Phaedrotoma caucasi (Tobias, 1986)

Catalogs with Iranian records: Khajeh et al. (2014), Gadallah et al. (2016), Farahani et al. (2016), Yu et al. (2016), Samin, Coronado-Blanco, Kavallieratos et al. (2018), Samin, Coronado-Blanco, Fischer et al. (2018) (all as *Opius (Gastrosema) caucasi* Tobias).

Distribution in Iran: East Azarbaijan (Ghahari & van Achterberg, 2016; Rastegar et al., 2012, both as *Opius caucasi*), Guilan (Ghahari, Fischer, & Tobias, 2012 as *O. caucasi*), Hormozgan (Ameri et al., 2014 as *O. caucasi*).

Distribution in the Middle East: Cyprus (Beyarslan et al., 2017 as *O. caucasi*), Iran (see references above), Turkey (Beyarslan, 2015; Beyarslan & Fischer, 2011a, b; Fischer & Beyarslan, 2005, 2013).

Extralimital distribution: Japan, Russia (Yu et al., 2016 as *O. caucasi*).

Host records: Recorded by Ku et al. (2001) as being a parasitoid of the agromyzid *Chromatomyia horticola* Goureau.

Phaedrotoma cingulatus (Wesmael, 1835)

Catalogs with Iranian records: Gadallah et al. (2016), Farahani et al. (2016), Yu et al. (2016), Samin, Coronado-Blanco, Kavallieratos et al. (2018), Samin, Coronado-Blanco, Fischer et al. (2018) (all as *Opius (Nosopoea) cingulatus* Wesmael).

Distribution in Iran: Kermanshah (Ghahari & Fischer, 2012 as *O. cingulatus*), Khuzestan (Ghahari & Beyarslan, 2017 as *O. cingulatus*), Mazandaran (Ghahari, 2019 as *O. cingulatus*).

Distribution in the Middle East: Iran (see references above), Israel−Palestine (Papp, 2012), Turkey (Beyarslan, 2015; Beyarslan & Fischer, 2011a, b; Fischer & Beyarslan, 2005).

Extralimital distribution: Armenia, Austria, Belgium, Bulgaria, former Czechoslovakia, Denmark, Estonia, Finland, France, Germany, Greece, Hungary, Ireland, Italy, Lithuania, Netherlands, Norway, Poland, Russia, Serbia, Slovenia, Spain, Sweden, Switzerland, United Kingdom.

Host records: Summarized by Yu et al. (2016) as being a parasitoid of the agromyzids *Agromyza* spp., *Amauromyza labiatarum* (Hendel), *Cerodontha incisa* (Meigen), *Galiomyza moria* (Brischke), *Ophiomyza maura* Meigen, and *Phytomyza evanescens* Hendel, the tephritids *Euleia heracleid* (L.), and *Trypeta immaculata* (Macquart). In Iran, this species has been reared from *Phytomyza plantaginis* Goureau (Ghahari & Beyarslan, 2017).

Phaedrotoma depeculator (Foerster, 1863)

Catalogs with Iranian records: Gadallah et al. (2016), Farahani et al. (2016), Yu et al. (2016), Samin, Coronado-Blanco, Kavallieratos et al. (2018), Samin, Coronado-Blanco, Fischer et al. (2018) (all as *Opius (Phaedrotoma) depeculator* (Foerster)).

Distribution in Iran: Ardabil (Ghahari & van Achterberg, 2016 as *O. depeculator*), Guilan, Mazandaran (Dolati et al., 2018 as *O. depeculator*), Kermanshah (Ghahari & Fischer, 2012 as *O. depeculator*).

Distribution in the Middle East: Iran (see references above), Turkey (Beyarslan, 2015; Beyarslan & Fischer, 2011a, b; Beyarslan & Inanç, 1992; Fischer & Beyarslan, 2005, 2013).

Extralimital distribution: Armenia, Austria, Czech Republic, Denmark, Estonia, Finland, France, Georgia, Germany, Hungary, Ireland, Italy, Kyrgyzstan, Montenegro, Netherlands, Poland, Romania, Russia, Serbia, Spain, Switzerland, United Kingdom, Uzbekistan.

Host records: Summarized by Yu et al. (2016) as being a parasitoid of the following agromyzids: *Cerodontha denticornis* (Panzer), *Chromatomyia ramosa* (Hendel), *Ophiomyia* sp., *Phytomyza griffithsi* Spencer, and *Phytomyza plantaginis* Goureau.

Phaedrotoma diversiformis (Fischer, 1960)

Catalogs with Iranian records: Khajeh et al. (2014), Gadallah et al. (2016), Farahani et al. (2016), Yu et al. (2016), Samin, Coronado-Blanco, Kavallieratos et al. (2018), Samin, Coronado-Blanco, Fischer et al. (2018) (all as *Opius (Phaedrotoma) diversiformis* Fischer).

Distribution in Iran: Golestan (Sakenin et al., 2012 as *O. diversiformis*), Lorestan (Ghahari, Fischer, Papp et al., 2012 as *O. diversiformis*), Sistan and Baluchestan (Sedighi et al., 2014 as *O. diversiformis*).

Distribution in the Middle East: Iran (see references above), Israel−Palestine (Papp, 1989), Turkey (Beyarslan & Fischer, 2011a, b; Fischer & Beyarslan, 2005).

Extralimital distribution: Armenia, Austria, Georgia, Greece, Hungary, Italy, Korea, North Macedonia, Montenegro, Serbia, Spain.
Host records: Unknown.
Plant associations in Iran: *Medicago sativa* (Fabaceae) (Sedighi et al., 2014).

Phaedrotoma diversus (Szépligeti, 1898)

Catalogs with Iranian records: Khajeh et al. (2014), Gadallah et al. (2016), Farahani et al. (2016), Yu et al. (2016), Samin, Coronado-Blanco, Kavallieratos et al. (2018), Samin, Coronado-Blanco, Fischer et al. (2018) (all as *Opius* (*Phaedrotoma*) *diversus* Szépligeti).
Distribution in Iran: Ardabil (Rastegar et al., 2012 as *O. diversus*), East Azarbaijan (Ghahari & van Achterberg, 2016 as *O. diversus*), Guilan (Ghahari, Fischer, & Tobias, 2012 as *O. diversus*), Kerman (Ranjbar et al., 2016; Safahani et al., 2018), Semnan (Samin, Fischer et al., 2015 as *O. diversus*), Sistan and Baluchestan (Khajeh et al., 2014 as *O. diversus*).
Distribution in the Middle East: Iran (see references above), Israel—Palestine (Fischer, 1997; Papp, 1970), Turkey (Beyarslan, 2015; Beyarslan & Fischer, 2011a, b; Fischer & Beyarslan, 2005, 2013).
Extralimital distribution: Austria, Bulgaria, Cape Verde Islands, former Czechoslovakia, Denmark, Estonia, Finland, France, Germany, Greece, Hungary, Italy, Montenegro, Poland, Romania, Russia, Serbia, Sweden, United Kingdom, Uzbekistan.
Host records: Summarized by Yu et al. (2016) as being a parasitoid of the following agromyzids *Calycomyza solidaginis* (Kaltenbach), *Chromatomyia scabiosae* (Hendel), *Chromatomyia succisae* (Hering), *Chromatomyia syngenesiae* Hardy, *Liriomyza centaureae* Hering, *Phytomyza gentianae* Hendel, and *Phytomyza plantaginis* Goureau.
Plant associations in Iran: *Medicago sativa* (Fabaceae) (Khajeh et al., 2014; Ranjbar et al., 2016), and *Rubus* sp. (Rosaceae) (Safahani et al., 2018).

Phaedrotoma exigua (Wesmael, 1835)

Catalogs with Iranian records: Fallahzadeh and Saghaei (2010), Khajeh et al. (2014), Gadallah et al. (2016), Farahani et al. (2016), Yu et al. (2016), Samin, Coronado-Blanco, Kavallieratos et al. (2018), Samin, Coronado-Blanco, Fischer et al. (2018) (all as *Opius* (*Phaedrotoma*) *exigua* Wesmael).
Distribution in Iran: Alborz, Qazvin (Dolati et al., 2018), Ardabil, East Azarbaijan (Ghahari & van Achterberg, 2016 as *O. exigua*), Kerman (Ranjbar et al., 2016; Safahani et al., 2018 both as *O. Exigua*, Madjdzadeh et al., 2021), Lorestan (Ghahari, Fischer, Papp et al., 2012 as *Opius* (*Phaedrotoma*) *exiguus*), Mazandaran (Ghahari 2017—around rice fields, Dolati et al., 2018 both as *O. exigua*), Sistan and

Baluchestan (Khajeh et al., 2014 as *O. exigua*), Tehran (Dolati et al., 2018; Fischer, 1990 both as *O. exigua*).
Distribution in the Middle East: Egypt (Yu et al., 2016), Iran (see references above), Israel—Palestine (Papp, 1989, 2012), Turkey (Beyarslan, 2015; Beyarslan & Fischer, 2011a, b; Beyarslan & Inanç, 1992; Çikman, 2006; Çikman & Uygun, 2003; Çikman et al., 2006).
Extralimital distribution: Afghanistan, Austria, Belgium, Bulgaria, Czech Republic, Denmark, Estonia, Ethiopia, Finland, France, Germany, Greece, Hungary, India, Ireland, Italy, Kazakhstan, Korea, Liechtenstein, Lithuania, Mongolia, Montenegro, Poland, Portugal (Madeira Islands), Russia, Serbia, South Africa, Spain, Sweden, Switzerland, Turkmenistan, United Kingdom, Uzbekistan.
Host records: Summarized by Yu et al. (2016) as being a parasitoid of the agromyzids *Agromyza nana* Meigen, *Agromyza rondensis* Strobl, *Calycomyza solidaginis* (Kaltenbach), *Chromatomyia* spp., *Liriomyza* spp., and *Phytomyza* spp., and the drosophilid *Scaptomyza graminum* Fallén.
Plant associations in Iran: *Brassica napus* (Brassicaceae), *Melilotus officinalis* (Fabaceae), *Trigonella* sp. (Fabaceae) (Khajeh et al., 2014), *Cicer arietinum* (Fabaceae) (Ranjbar et al., 2016), *Medicago sativa* (Fabaceae) (Ranjbar et al., 2016; Safahani et al., 2018).

Phaedrotoma gafsaensis (Fischer, 1964)

Catalogs with Iranian records: No catalog.
Distribution in Iran: Kerman (Safahani et al., 2018 as *Opius* (*Phaedrotoma*) *gafsaensis* Fischer).
Distribution in the Middle East: Iran (Safahani et al., 2018), Turkey (Beyarslan & Fischer, 2011b; Fischer & Beyarslan, 2005; Papp, 1982).
Extralimital distribution: Spain, Tunisia.
Host records: Unknown.
Plant associations in Iran: *Medicago sativa* (Fabaceae) (Safahani et al., 2018).

Phaedrotoma laetatorius (Fischer, 1958)

Catalogs with Iranian records: No catalog.
Distribution in Iran: Chaharmahal and Bakhtiari (Samin, Papp et al., 2018 as *Opius* (*Merotrachys*) *laetatorius* Fischer).
Distribution in the Middle East: Iran (Samin, Papp et al., 2018), Turkey (Beyarslan & Fischer, 2011b; Fischer & Beyarslan, 2005).
Extralimital distribution: Former Czechoslovakia, France, Germany, Hungary, Moldova, Russia, Spain.
Host records: Unknown.

Phaedrotoma leclytus (Fischer, 1996)

Distribution in the Middle East: Turkey (Beyarslan & Fischer, 2011a; Fischer & Beyarslan, 2005).

Extralimital distribution: Greece.
Host records: Unknown.

Phaedrotoma mirabundus (Papp, 1982)

Catalogs with Iranian records: Khajeh et al. (2014), Gadallah et al. (2016), Farahani et al. (2016), Yu et al. (2016), Samin, Coronado-Blanco, Kavallieratos et al. (2018), Samin, Coronado-Blanco, Fischer et al. (2018) (all as *Opius (Phaedrotoma) mirabundus* Papp).
Distribution in Iran: Sistan and Baluchestan (Khajeh et al., 2014 as *O. mirabunda*).
Distribution in the Middle East: Iran (Khajeh et al., 2014), Turkey (Beyarslan, 2015; Bayarslan & Fischer, 2011a, b; Fischer & Beyarslan, 2013).
Extralimital distribution: Greece.
Host records: Unknown.
Plant associations in Iran: *Medicago sativa* (Fabaceae), *Taraxacum* sp. (Asteraceae) (Khajeh et al., 2014).

Phaedrotoma nitidulator (Nees von Esenbeck, 1834)

Catalogs with Iranian records: Gadallah et al. (2016), Yu et al. (2016 as *Opius (Phaedrotoma) nitidulator* (Nees)), Samin, Coronado-Blanco, Kavallieratos et al. (2018), Samin, Coronado-Blanco, Fischer et al. (2018 as *O. nitidulator*).
Distribution in Iran: Ardabil (Gadallah et al., 2016).
Distribution in the Middle East: Egypt (Abd El Wahab et al., 1998), Iran (Gadallah et al., 2016), Turkey (Beyarslan, 2015; Beyarslan & Fischer, 2011b; Fischer & Beyarslan, 2005).
Extralimital distribution: Austria, Belgium, Bulgaria, Canada, former Czechoslovakia, Denmark, Finland, France, Germany, Greece, Hungary, Ireland, Italy, Korea, Libya, Lithuania, Poland, Russia, Selvagens Islands, Spain, Sweden, Tunisia, United Kingdom, United States of America, Uzbekistan.
Host records: Summarized by Yu et al. (2016) as being a parasitoid of the following anthomyiids: *Pegomya atriplicis* Goureau, *Pegomya betae* (Curtis), *Pegomya conformis* (Fallén), *Pegomya hyoscyami* (Panzer), *Pegomya mixta* Villeneuve, and *Pegomya solennis* Meigen, the calliphorids *Calliphora vomitoria* (L.), and *Lucilia caesar* (L.), and the muscid *Musca domestica* L.

Phaedrotoma ochrogaster (Wesmael, 1835)

Catalogs with Iranian records: No catalog.
Distribution in Iran: Alborz, Guilan, Mazandaran (Dolati et al., 2019 as *Opius (Phaedrotoma) ochrogaster* Wesmael).
Distribution in the Middle East: Iran (Dolati et al., 2019), Turkey (Beyarslan, 2015; Beyarslan & Fischer, 2011a, b; Fischer & Beyarslan, 2005, 2013).

Extralimital distribution: Austria, Belgium, Bulgaria, Croatia, former Czechoslovakia, France, Germany, Hungary, Ireland, Italy, Kazakhstan, Lithuania, Mongolia, Montenegro, Netherlands, Poland, Romania, Serbia, Spain, Switzerland, United Kingdom.
Host records: Summarized by Yu et al. (2016) as being a parasitoid of the following agromyzids: *Agromyza phragmitidis* Hendel, *Amauromyza flavifrons* (Meigen), *Amauromyza gyrans* (Fallén), *Amauromyza labiatarum* (Hendel), *Amauromyza verbasci* (Bouché), and *Phytomyza campanulae* Hendel.

Phaedrotoma pseudonitidus (Fahringer, 1943)

Catalogs with Iranian records: No catalog.
Distribution in Iran: Tehran (Dolati et al., 2021).
Distribution in the Middle East: Iran.
Extralimital distribution: Bulgaria, Kazakhstan.
Host records: Recorded by Fischer (1972) as being a parasitoid of the anthomyiid *Pegomya hyoscyami* (Panzer).

Phaedrotoma pulchriceps (Szépligeti, 1898)

Phaedrotoma pulchriventris (Fischer, 1958) (Synonymy by van Achterberg, 2014).
Catalogs with Iranian records: Khajeh et al. (2014), Gadallah et al. (2016), Farahani et al. (2016), Yu et al. (2016), Samin, Coronado-Blanco, Kavallieratos et al. (2018), Samin, Coronado-Blanco, Fischer et al. (2018) (all as *Opius (Phaedrotoma) pulchriceps* Szépligeti).
Distribution in Iran: Alborz, Qazvin, Tehran (Dolati et al., 2018 as *Phaedrotoma pulchriventris* (Fischer, 1958)), East Azarbaijan (Rastegar et al., 2012 as *Opius pulchriventris*), Guilan (Ghahari, Fischer, & Tobias, 2012 as *Opius (Phaedrotoma) pulchriventris*), Kerman (Madjdzadeh et al., 2021).
Distribution in the Middle East: Iran (see references above), Turkey (Beyarslan, 2015; Beyarslan & Fischer, 2011a, b; Fischer & Beyarslan, 2005).
Extralimital distribution: Austria, Canada, China, former Czechoslovakia, Finland, France, Germany, Hungary, India, Ireland, Italy, Kazakhstan, Korea, Lithuania, Moldova, Netherlands, Poland, Russia, Spain, Switzerland, Ukraine, United Kingdom.
Host records: Summarized by Yu et al. (2016) as being a parasitoid of the following agromyzids: *Amauromyza verbasci* (Bouché), *Chromatomyia horticola* Goureau, *Liriomyza bryoniae* Kaltenbach, *Liriomyza congesta* (Becker), *Liriomyza strigata* (Meigen), *Phytoliriomyza hilarella* (Zetterstedt), *Phytomyza agromyzina* Meigen, and *Phytomyza ilicis* Curtis.

Phaedrotoma pumilio (Wesmael, 1835)

Catalogs with Iranian records: Khajeh et al. (2014 as *Opius pumilio* Wesmael), Gadallah et al. (2016 as *O. pumilio*),

Farahani et al. (2016 as *O. pumilio*), Yu et al. (2016 as *O. pumilio*), Samin, Coronado-Blanco, Kavallieratos et al. (2018), Samin, Coronado-Blanco, Fischer et al. (2018 as *O. pumilio*).

Distribution in Iran: Fars (Lashkari Bod et al., 2010, 2011 as *O. pumilio*).

Distribution in the Middle East: Iran (Lashkari Bod et al., 2010, 2011), Turkey (Fischer & Beyarslan, 2005 as *Opius pumilio*; Beyarslan & Fischer, 2011b).

Extralimital distribution: Belgium, Bulgaria, France, Germany, Hungary, Korea, Poland, Russia, Serbia, Spain.

Host records: Recorded by Fischer (1958) as being a parasitoid of the agromyzid *Amauromyza verbasci* (Bouché), and the anthomyiid *Pegomya bicolor* (Wiedemann).

Plant associations in Iran: *Medicago sativa* (Fabaceae) (Lashkari Bod et al., 2010, 2011).

Phaedrotoma quasiqvisti (Fischer, 1991)

Distribution in the Middle East: Turkey (Beyarslan & Fischer, 2011b; Fischer & Beyarslan, 2005, 2013).

Extralimital distribution: Austria, Montenegro, Serbia, Slovakia.

Host records: Unknown.

Phaedrotoma rex (Fischer, 1958)

Catalogs with Iranian records: Khajeh et al. (2014 as *Opius (Agnopius) rex*), Gadallah et al. (2016 as *Opius (Agnopius) rex*), Farahani et al. (2016 as *Opius (Agnopius) rex*), Yu et al. (2016), Samin, Coronado-Blanco, Kavallieratos et al. (2018), Samin, Coronado-Blanco, Fischer et al. (2018).

Distribution in Iran: Ardabil (Ghahari & Fischer, 2011 as *Opius (Agnopius) rex*), Guilan (Dolati et al., 2021), Kerman (Safahani et al., 2018), Mazandaran (Ghahari, 2018; Ghahari, Fischer, Sakenin et al., 2011—around rice fields, as *Opius (Agnopius) rex*).

Distribution in the Middle East: Iran (see references above), Turkey (Beyarslan & Fischer, 2011a, b; Yildirim et al., 2010).

Extralimital distribution: Austria, Czech Republic, Denmark, Finland, France, Germany, Greece, Hungary, Italy, Korea, Mongolia, Netherlands, Poland, Romania, Russia, Sweden, Switzerland, United Kingdom, former Yugoslavia.

Host records: Summarized by Yu et al. (2016) as being a parasitoid of the following agromyzids: *Agromyza* spp., *Aulagromyza heringi* (Hendel), *Chromatomyia milii* (Kaltenbach), *Liriomyza* spp., and *Phytomyza* spp.

Phaedrotoma rhodopicola (Zaykov & Fischer, 1986)

Distribution in the Middle East: Turkey (Beyarslan & Fischer, 2011a, b).

Extralimital distribution: Bulgaria, Montenegro, Serbia.

Host records: Unknown.

Phaedrotoma rudis (Wesmael, 1835)

Catalogs with Iranian records: Khajeh et al. (2014), Gadallah et al. (2016), Farahani et al. (2016), Yu et al. (2016), Samin, Coronado-Blanco, Kavallieratos et al. (2018), Samin, Coronado-Blanco, Fischer et al. (2018) (all as *Opius (Phaedrotoma) rudis* Wesmael).

Distribution in Iran: Guilan (Dolati et al., 2021), Kerman (Madjdzadeh et al., 2021), Kermanshah (Ghahari & Fischer, 2012 as *O. rudis*), Sistan and Baluchestan (Khajeh et al., 2014 as *O. rudis*).

Distribution in the Middle East: Iran (see references above), Turkey (Beyarslan, 2015; Beyarslan & Fischer, 2011a, b; Fischer & Beyarslan, 2005, 2013).

Extralimital distribution: Algeria, Armenia, Austria, Belgium, Bulgaria, Canada, Former Czechoslovakia, Denmark, Finland, France, Germany, Hungary, Ireland, Korea, Lithuania, Madeira Islands, Montenegro, Netherlands, Poland, Russia, Serbia, Spain, Sweden, United Kingdom, United States of America, Uzbekistan.

Host records: Summarized by Yu et al. (2016) as being a parasitoid of the agromyzids *Agromyza megalopsis* Hering, and *Agromyza nigripes* Meigen.

Plant associations in Iran: *Daucus carota* (Apiaceae), *Medicago sativa* (Fabaceae), *Tamarix* sp. (Tamaricaceae), *Triticum aestivum* (Poaceae) (Khajeh et al., 2014).

Phaedrotoma scaptomyzae (Fischer, 1967)

Catalogs with Iranian records: No catalog.

Distribution in Iran: Kerman (Safahani et al., 2018 as *Opius scaptomyzae* (Fischer)).

Distribution in the Middle East: Iran (Safahani et al., 2018), Turkey (Beyarslan & Fischer, 2011a, b; Beyarslan & Inanç, 1992; Fischer & Beyarslan, 2013).

Extralimital distribution: Croatia, North Macedonia, Russia, former Yugoslavia.

Host records: Recorded by Fischer (1967) as being a parasitoid of the drosophilid *Scaptomyza graminum* Fallén.

Plant associations in Iran: *Medicago sativa* (Fabaceae) (Safahani et al., 2018).

Phaedrotoma schmidti (Fischer, 1960)

Catalogs with Iranian records: No catalog.

Distribution in Iran: Guilan (Dolati et al., 2019 as *Opius (Merotrachys) schmidti* Fischer).

Distribution in the Middle East: Iran (Dolati et al., 2019), Turkey (Beyarslan & Fischer, 2011a, b).

Extralimital distribution: Hungary.

Host records: Unknown.

Phaedrotoma seiunctus (Fischer, 1959)

Distribution in the Middle East: Turkey (Beyarslan & Fischer, 2011a, b; Fischer & Beyarslan, 2005, 2013).

Extralimital distribution: Mongolia, Turkmenistan, Uzbekistan.
Host records: Unknown.

Phaedrotoma speciosus (Fischer, 1959)
Catalogs with Iranian records: Gadallah et al. (2016), Yu et al. (2016), Samin, Coronado-Blanco, Kavallieratos et al. (2018), Samin, Coronado-Blanco, Fischer et al. (2018) (all as *Opius* (*Nosopoea*) *speciosus* Fischer).
Distribution in Iran: East Azarbaijan (Gadallah et al., 2016 as *O. speciosus*), Fars (Ghahari & Beyarslan, 2017 as *O. speciosus*).
Distribution in the Middle East: Cyprus (Beyarslan et al., 2017), Iran (Gadallah et al., 2016; Ghahari & Beyarslan, 2017), Turkey (Beyarslan, 2015; Beyarslan & Fischer, 2011a, b; Fischer & Beyarslan, 2005).
Extralimital distribution: Armenia, Austria, former Czechoslovakia, Finland, France, Hungary, Italy, Spain.
Host records: Unknown.

Phaedrotoma staryi (Fischer, 1958)
Catalogs with Iranian records: Gadallah et al. (2016), Yu et al. (2016), Samin, Coronado-Blanco, Kavallieratos et al. (2018), Samin, Coronado-Blanco, Fischer et al. (2018) (all as *Opius* (*Phaedrotoma*) *staryi* Fischer).
Distribution in Iran: Guilan, Mazandaran (Dolati et al., 2021), Hormozgan (Ameri et al., 2014 as *Opius* (*Allophlebus*) *staryi* Fischer, 1958).
Distribution in the Middle East: Iran (Ameri et al., 2014; Dolati et al., 2021), Turkey (Beyarslan, 2015; Beyarslan & Fischer, 2011a, b).
Extralimital distribution: Austria, Bosnia-Herzegovina, Czech Republic, Estonia, Finland, France, Germany, Hungary, Italy, Korea, Montenegro, Netherlands, Romania, Russia, Serbia, Slovakia, Spain, Sweden, Switzerland, United Kingdom, Uzbekistan.
Host records: Recorded by Fischer (1962, 1967) as being a parasitoid of the agromyzids *Aulagromyza tremulae* (Hering), *Phytomyza aquilegiae* Hardy, and *Phytomyza minusculae* Goureau.

Phaedrotoma teheranensis (Fischer, 1990)
Catalogs with Iranian records: Fallahzadeh and Saghaei (2010), Khajeh et al. (2014), Gadallah et al. (2016), Farahani et al. (2016), Yu et al. (2016), Samin, Coronado-Blanco, Kavallieratos et al. (2018), Samin, Coronado-Blanco, Fischer et al. (2018) (all as *Opius* (*Nosopoea*) *teheranesis* Fischer).
Distribution in Iran: Tehran (Fischer, 1990 as *O. teheranensis*).
Distribution in the Middle East: Iran.
Extralimital distribution: None.
Host records: Unknown.

Phaedrotoma umlalaziensis (Fischer, 1996)
Distribution in the Middle East: Turkey (Çikman & Beyarslan, 2009).
Extralimital distribution: South Africa.
Host records: Unknown.

Phaedrotoma variegatus (Szépligeti, 1896)
Catalogs with Iranian records: No catalog.
Distribution in Iran: Markazi (Gadallah et al., 2018 as *Opius variegatus*), Mazandaran (Ghahari & Sakenin, 2018), Tehran (Dolati et al., 2018; Ghahari & Beyarslan, 2019 both as *O. variegatus*).
Distribution in the Middle East: Iran (see references above), Turkey (Beyarslan & Fischer, 2011a, b; Fischer & Beyarslan, 2013).
Extralimital distribution: Austria, Croatia, former Czechoslovakia, Denmark, Estonia, Finland, France, Germany, Hungary, Ireland, Italy, Kazakhstan, Mongolia, Netherlands, Poland, Russia, Spain, Sweden, Switzerland, United Kingdom.
Host records: Summarized by Yu et al. (2016) as being a parasitoid of the agromyzids *Agromyza* spp., *Amauromyza flavifrons* (Meigen), *Amauromyza verbasci* (Bouché), *Cerodontha* spp., and *Chromatomyia horticola* Goureau, and the anthomyiid *Pegomya hyoscyami* (Panzer). In Iran, this species has been reared from the agromyzid *Chromatomyia horticola* (Goureau) on Polygonaceae (Ghahari & Beyarslan, 2019).

Phaedrotoma zomborii (Papp, 1982)
Distribution in the Middle East: Turkey (Beyarslan, 2015).
Extralimital distribution: Korea, Russia.
Host records: Unknown.

Genus *Pokomandya* Fischer, 1959
Pokomandya curticornis Fischer, 1959
Catalogs with Iranian records: Khajeh et al. (2014), Farahani et al. (2016), Yu et al. (2016), Samin, Coronado-Blanco, Kavallieratos et al. (2018), Samin, Coronado-Blanco, Fischer et al. (2018).
Distribution in Iran: Alborz (Dolati et al., 2021), East Azarbaijan (Ghahari et al., 2009).
Distribution in the Middle East: Iran (Dolati et al., 2021; Ghahari et al., 2009), Israel–Palestine (Papp, 1989), Turkey (Fischer, 1959, 1972; Papp, 1982).
Extralimital distribution: Greece, Hungary, Kazakhstan, Slovakia.
Host records: Unknown.

Genus *Psyttalia* Walker, 1860
Psyttalia (*Psyttalia*) *carinata* (Thomson, 1895)
Catalogs with Iranian records: No catalog.

Distribution in Iran: Ardabil (Samin, Beyarslan et al., 2020), Zanjan (Samin, Sakenin Chelav et al., 2020).

Distribution in the Middle East: Iran.

Extralimital distribution: Armenia, Austria, Bulgaria, former Czechoslovakia, Finland, France, Germany, Hungary, Italy, Kazakhstan, Kyrgyzstan, Lithuania, Malta, Moldova, Norway, Poland, Sweden, Switzerland, Uzbekistan, former Yugoslavia.

Host records: Summarized by Yu et al. (2016) as being a parasitoid of the following tephritids: *Carpomya schineri* (Loew), *Myoleja lucida* (Fallén), *Rhagoletis alternata* (Fallén), *Rhagoletis batava* Hering, *Rhagoletis cerasi* (L.), *Rhagoletis flavicincta* Enderlein, *Rhagoletis magniterebra* (Rohdendorf), and *Rhagoletis meigenii* (Loew). In Iran, this species has been reared from *Rhagoletis cerasi* (L.) (Samin, Beyarslan et al., 2020).

Psyttalia (Psyttalia) concolor (Szépligeti, 1910)

Catalogs with Iranian records: Khajeh et al. (2014), Gadallah et al. (2016), Farahani et al. (2016), Yu et al. (2016), Samin, Coronado-Blanco, Kavallieratos et al. (2018), Samin, Coronado-Blanco, Fischer et al. (2018).

Distribution in Iran: Lorestan (Ghahari, Fischer, Papp et al., 2012), Mazandaran (Dolati et al., 2021), Semnan (Naderian et al., 2012).

Distribution in the Middle East: Egypt (Abdel-Galil et al., 2014; Mahmoud et al., 2009), Iran (see references above), Israel−Palestine (Bodenheimar, 1930; Halperin, 1986; Papp, 1970, 1989, 2012), Turkey (Fischer & Beyarslan, 2005, 2013, Beyarslan & Fischer, 2011a, b).

Extralimital distribution: Algeria, Australia (introduced), Bermuda, Bulgaria, Cook Islands (introduced), Fiji (introduced), Finland, France (introduced), Greece (introduced), Guam (introduced), Iran, Italy (introduced), Kenya, Libya, Madagascar (introduced), Malta, Morocco, New Caledonia (introduced), South Africa, Spain (introduced), Tanzania, Tunisia, United States of America (introduced).

Host records: Summarized by Yu et al. (2016) as being a parasitoid of the following tephritids: *Anastrepha suspensa* (Loew), *Bactrocera oleae* (Rossi), *Capparimyia savastani* (Martelli), *Carpomya incompleta* (Becker), *Ceratitis capitata* (Wiedemann), *Dacus ciliatus* Loew, *Euphranta connexa* Fabricius, and *Rhagoletis cerasi* L.

Psyttalia nilotica (Schmideknecht, 1900)

Distribution in the Middle East: Egypt (Brues, 1926; Fischer, 1958, 1972, 1987; Schmiedecknecht, 1900; Szépligeti, 1904), Israel−Palestine (Papp, 1989).

Extralimital distribution: None.

Host records: Unknown.

Psyttalia subsulcata (Granger, 1949)

Distribution in the Middle East: Jordan (Papp, 1982).

Extralimital distribution: Madagascar, Réunion.

Host records: Recorded by Fischer (2008) as being a parasitoid of the tephritid *Spathulina acroleuca* (Schiner).

Genus *Psyttoma* van Achterberg & Li, 2012

Psyttoma wachsmanni (Szépligeti, 1898)

Catalogs with Iranian records: Khajeh et al. (2014 as *Opius wachsmanni* Szépligeti, 1898), Gadallah et al. (2016 as *O. wachsmanni*), Farahani et al. (2016 as *O. wachsmanni*), Yu et al. (2016), Samin, Coronado-Blanco, Kavallieratos et al. (2018), Samin, Coronado-Blanco, Fischer et al. (2018).

Distribution in Iran: Alborz, Guilan, Qazvin (Dolati et al., 2019 as *Opius (Mesophthora) wachsmanii* Szépligeti, 1898), Lorestan (Ghahari, Fischer, Papp et al., 2012).

Distribution in the Middle East: Iran (Dolati et al., 2019; Ghahari, Fischer, Papp et al., 2012), Turkey (Beyarslan & Fischer, 2011a; Beyarslan & Inanç, 1992; Fischer & Beyarslan, 2005).

Extralimital distribution: Azerbaijan, Hungary, Kazakhstan, Korea, Moldova, Uzbekistan.

Host records: Unknown.

Genus *Rhogadopsis* Brèthes, 1913

Rhogadopsis parvungula (Thomson, 1895)

Distribution in the Middle East: Turkey (Beyarslan & Fischer, 2011a, b; Fischer & Beyarslan, 2005, 2013).

Extralimital distribution: Austria, Bulgaria, former Czechoslovakia, Denmark, Estonia, Finland, France, Germany, Hungary, Hungary, Italy, Korea, Montenegro, Norway, Poland, Russia, Serbia, Spain, Sweden, United Kingdom.

Host records: Summarized by Yu et al. (2016) as being a parasitoid of the following agromyzids: *Agromyza albipennis* Meigen, *Agromyza ambigua* Fallén, *Agromyza graminicola* Hendel, *Agromyza lucida* Hendel, *Agromyza nigrella* (Rondani), *Agromyza reptans* Fallén, and *Cerodontha denticornis* (Panzer).

Rhogadopsis reconditor (Wesmael, 1835)

Catalogs with Iranian records: Gadallah et al. (2016), Farahani et al. (2016), Yu et al. (2016), Samin, Coronado-Blanco, Kavallieratos et al. (2018), Samin, Coronado-Blanco, Fischer et al. (2018).

Distribution in Iran: Kermanshah (Ghahari & Fischer, 2012 as *Opius (Gastrosema) docilis* Haliday, 1837), Mazandaran (Dolati et al., 2019 as *O. (Gastrosema) docilis*).

Distribution in the Middle East: Iran (Dolati et al., 2019; Ghahari & Fischer, 2012), Turkey (Beyarslan & Fischer, 2011a, b; Beyarslan & Fischer, 2015).

Extralimital distribution: Austria, Belgium, Denmark, Finland, France, Georgia, Germany, Greece, Hungary, Ireland, Kyrgyzstan, Montenegro, Poland, Russia, Serbia, Slovakia, Spain, Sweden, Switzerland, United Kingdom, Uzbekistan.

Host records: Summarized by Yu et al. (2016) as being a parasitoid of the following agromyzids: *Agromyza albipennis* Meigen, *Agromyza ambigua* Fallén, *Agromyza phragmitidis* Hendel, *Agromyza reptans* Fallén, *Amauromyza labiatarum* (Hendel), *Amauromyza verbasci* (Bouché), *Phytomyza affinis* Fallén, and *Phytomyza xylostei* Kaltenbach.

Genus *Sternaulopius* Fischer, 1965

Sternaulopius edirneanus Fischer & Beyarslan, 2005

Distribution in the Middle East: Turkey (Fischer & Beyarslan, 2005).
Extralimital distribution: None.
Host records: Unknown.

Genus *Utetes* Foerster, 1863

Utetes aemuloides (Fischer, 1958)

Distribution in the Middle East: Turkey (Beyarslan & Fischer, 2011b; Fischer & Beyarslan, 2005).
Extralimital distribution: United Kingdom.
Host records: Unknown.

Utetes aemulus (Haliday, 1837)

Distribution in the Middle East: Turkey (Beyarslan, 2015; Beyarslan & Fischer, 2011b; Fischer & Beyarslan, 2005, 2013).
Extralimital distribution: Austria, former Czechoslovakia, Finland, France, Germany, Hungary, Ireland, Italy, Lithuania, Netherlands, Poland, Spain, United Kingdom.
Host records: Summarized by Yu et al. (2016) as being a parasitoid of the agromyzids *Liriomyza congesta* (Becker), and *Phytomyza stolonigena* Hering.

Utetes coracinus (Thomson, 1895)

Catalogs with Iranian records: Khajeh et al. (2014), Gadallah et al. (2016), Farahani et al. (2016), Yu et al. (2016), Samin, Coronado-Blanco, Kavallieratos et al. (2018), Samin, Coronado-Blanco, Fischer et al. (2018).
Distribution in Iran: Guilan (Ghahari, Fischer, & Tobias, 2012 as *Opius* (*Utetes*) *coracinus*).
Distribution in the Middle East: Iran (Ghahari, Fischer, & Tobias, 2012), Turkey (Beyarslan, 2015; Beyarslan & Fischer, 2011a, b; Fischer & Beyarslan, 2005).
Extralimital distribution: Austria, Bulgaria, former Czechoslovakia, Denmark, Finland, France, Germany, Hungary, Italy, Montenegro, Poland, Russia, former Yugoslavia.
Host records: Unknown.

Utetes curtipectus (Fischer, 1958)

Catalogs with Iranian records: No catalog.
Distribution in Iran: Mazandaran (Dolati et al., 2021).
Distribution in the Middle East: Iran (Dolati et al., 2021), Turkey (Beyarslan, 2019).
Extralimital distribution: Austria, former Czechoslovakia, Finland, Lithuania, Spain.
Host records: Recorded by Docavo et al. (1987) as being a parasitoid of the agromyzid *Aulagromyza populi* (Kaltenbach).

Utetes fasciatus (Thomson, 1895)

Distribution in the Middle East: Turkey (Beyarslan, 2015).
Extralimital distribution: Former Czechoslovakia, Denmark, Finland, Germany, Italy, Lithuania, Russia, Sweden, Switzerland, United Kingdom.
Host record: Unknown.

Utetes ferrugator (Goureau, 1862)

Catalogs with Iranian records: Khajeh et al. (2014 under *Utetes magnus*), Gadallah et al. (2016 as *Opius* (*Utetes*) *ferrugator*), Farahani et al. (2016 as *O. ferrugator*), Yu et al. (2016 as *O. ferrugator*), Samin, Coronado-Blanco, Kavallieratos et al. (2018), Samin, Coronado-Blanco, Fischer et al. (2018 as *O. ferrugator*).
Distribution in Iran: Ardabil (Ghahari & Fischer, 2011 as *Opius* (*Utetes*) *magnus* Fischer, 1958), Guilan (Ghahari, Fischer, & Tobias, 2012 as *O.* (*Utetes*) *magnus*), West Azarbaijan (Rastegar et al., 2012 as *O.* (*Utetes*) *magnus*).
Distribution in the Middle East: Iran (see references above), Turkey (Beyarslan & Fischer, 2011a, b; Yildirim et al., 2010).
Extralimital distribution: Austria, Finland, France, Georgia, Germany, Hungary, Italy, Kazakhstan, Poland, Russia, Switzerland, Ukraine.
Host records: Summarized by Yu et al. (2016) as being a parasitoid of the following tephritids: *Anomoia purmunda* (Harris), *Carpomya schineri* (Loew), *Rhagoletis alternata* Fallén, *Rhagoletis berberis* Jermy, *Rhagoletis cerasi* L., *Rhagoletis meigenii* (Loew), and *Trypeta* sp. In Iran, this species has been reared from *Rhagoletis* sp. (Ghahari & Fischer, 2011).

Utetes fulvicollis (Thomson, 1895)

Catalogs with Iranian records: No catalog.
Distribution in Iran: Khuzestan (Ameri et al., 2020).
Distribution in the Middle East: Iran.
Extralimital distribution: Belgium, Canada, Denmark, Finland, Germany, Lithuania, Netherlands, Poland, Russia, Sweden, United States of America, United Kingdom, Uzbekistan (Yu et al., 2016 as *O.* (*Utetes*) *fulvicollis* Thomson).

Host records: Summarized by Yu et al. (2016) as being a parasitoid of the anthomyiids *Pegomya betae* (Curtis) and *Pegomya hyoscyami* (Panzer).

Utetes propecoracium (Fischer, 1999)
Distribution in the Middle East: Turkey (Beyarslan & Fischer, 2011a, b).
Extralimital distribution: Ukraine.
Host records: Unknown.

Utetes pygmisoma (Fischer, 1968)
Distribution in the Middle East: Turkey (Beyarslan & Fischer, 2011b; Fischer & Beyarslan, 2013).
Extralimital distribution: Democratic Republic of Congo.
Host records: Unknown.

Utetes rhodopensis (Zaykov, 1983)
Distribution in the Middle East: Turkey (Beyarslan, 2015).
Extralimital distribution: Bulgaria.
Host records: Unknown.

Utetes rotundiventris (Thomson, 1895)
Catalogs with Iranian records: Samin, Coronado-Blanco, Fischer et al. (2018).
Distribution in Iran: Guilan, Mazandaran (Dolati et al., 2018), Markazi (Samin, Coronado-Blanco, Fischer et al., 2018 as *Opius rotundiventris*).
Distribution in the Middle East: Iran (Dolati et al., 2018; Samin, Coronado-Blanco, Fischer et al., 2018), Turkey (Beyarslan, 2015; Beyarslan & Fischer, 2011a, b; Fischer & Beyarslan, 2005, 2013).
Extralimital distribution: Austria, Bulgaria, former Czechoslovakia, Denmark, Finland, France, Germany, Hungary, Italy, Japan, Korea, Lithuania, Netherlands, Poland, Russia, Sweden, Switzerland, United Kingdom, former Yugoslavia (Yu et al., 2016 as *O. (Utetes) rotundiventris*).
Host records: Recorded by Fischer (1969) as being a parasitoid of the agromyzids *Agromyza albitarsis* Meigen, *Agromyza rufipes* Meigen, *Phytomyza diversicornis* Hendel, and *Phytomyza sedicola* Hering.

Utetes ruficeps (Wesmael, 1835)
Distribution in the Middle East: Syria (Fischer, 1997), Turkey (Beyarslan & Fischer, 2011b; Fischer & Beyarslan, 2013).
Extralimital distribution: Austria, Belgium, former Czechoslovakia, Denmark, Finland, France, Hungary, Italy, Korea, Lithuania, Poland, Russia, Spain, Sweden, Switzerland, United Kingdom, former Yugoslavia.
Host records: Summarized by Yu et al. (2016) as being a parasitoid of the following anthomyiids: *Pegomya betae* (Curtis), *Pegomya bicolor* (Wiedemann), *Pegomya conformis* (Fallén), *Pegomya hyoscyami* (Panzer), *Pegomaya solennis* (Meigen), and *Pegomya terebrans* (Rondani), and the agromyzids *Agromyza abiens* Zetterstedt, and *Agromyza rufipes* Macquart.

Utetes testaceus (Wesmael, 1838)
Catalogs with Iranian records: Khajeh et al. (2014), Gadallah et al. (2016), Farahani et al. (2016), Yu et al. (2016), Samin, Coronado-Blanco, Kavallieratos et al. (2018), Samin, Coronado-Blanco, Fischer et al. (2018) (all as *Opius (Utetes) testaceus* Wesmael).
Distribution in Iran: Alborz (Dolati et al., 2021), Guilan (Ghahari, Fischer, & Tobias, 2012 as *Opius (Utetes) truncates*).
Distribution in the Middle East: Iran (Dolati et al., 2021; Ghahari, Fischer, & Tobias, 2012), Turkey (Beyarslan & Fischer, 2011a, b).
Extralimital distribution: Austria, Belgium, Croatia, former Czechoslovakia, England, France, Germany, Italy, Netherlands, former Yugoslavia.
Host records: Recorded by Fischer (1972) and Tobias (1977) as being a parasitoid of the anthomyiids *Pegomya hyoscyami* (Panzer), and *Pegomya solennis* Meigen, and the tephritids *Euphranta connexa* Fabricius, *Goniglossum wiedemanni* (Meigen), and *Rhagoletis cerasi* L.

Utetes truncatus (Wesmael, 1835)
Catalogs with Iranian records: Khajeh et al. (2014), Gadallah et al. (2016), Farahani et al. (2016), Yu et al. (2016), Samin, Coronado-Blanco, Kavallieratos et al. (2018), Samin, Coronado-Blanco, Fischer et al. (2018) (all as *Opius (Utetes) truncatus* Wesmael).
Distribution in Iran: East Azarbaijan (Ghahari & Gadallah, 2022), Lorestan (Ghahari, Fischer, Papp et al., 2012 as *O. truncatus*), Mazandaran (Ghahari, 2019 as *O. truncatus*).
Distribution in the Middle East: Iran (see references above), Turkey (Beyarslan & Fischer, 2011a, b).
Extralimital distribution: Armenia, Austria, Belgium, Bulgaria, former Czechoslovakia, Denmark, Finland, France, Germany, Hungary, Lithuania, Mongolia, Poland, Russia, Sweden, Switzerland.
Host records: Recorded by Fischer (1972) and Tobias (1977) as being a parasitoid of the tephritids *Trypeta artemisiae* (Fabricius), and *Trypeta zoe* Meigen. In Iran, this species has been reared from the tephritid *Bactrocera* sp. (Ghahari & Gadallah, 2022).

Utetes ussuriensis (Tobias, 1977)
Distribution in the Middle East: Turkey (Beyarslan, 2015).
Extralimital distribution: Russia.

Host records: Recorded by Tobias (1977) as being a parasitoid of the tephritid *Rhagoletis reducta* Hering.

Genus *Xynobius* Foerster, 1863

Xynobius (*Xynobius*) *aciculatus* (Thomson, 1895)
Catalogs with Iranian records: Gadallah et al. (2016), Farahani et al. (2016), Yu et al. (2016), Samin, Coronado-Blanco, Kavallieratos et al. (2018), Samin, Coronado-Blanco, Fischer et al. (2018).
Distribution in Iran: Kermanshah (Ghahari & Fischer, 2012 as *Opius* (*Xynobius*) *aciculatus* Thomson, 1895), Mazandaran (Dolati et al., 2021).
Distribution in the Middle East: Iran (Dolati et al., 2021; Ghahari & Fischer, 2012), Turkey (Beyarslan, 2015; Beyarslan & Fischer, 2011a).
Extralimital distribution: Austria, Denmark, Russia, Spain, Sweden, Switzerland.
Host records: Unknown.

Xynobius (*Xynobius*) *caelatus* (Haliday, 1837)
Catalogs with Iranian records: No catalog.
Distribution in Iran: Mazandaran, Qazvin (Dolati et al., 2018).
Distribution in the Middle East: Iran (Dolati et al., 2018), Israel−Palestine, Syria (Fischer, 1997), Turkey (Beyarslan & Fischer, 2011b; Fischer & Beyarslan, 2005, 2013).
Extralimital distribution: Austria, Denmark, Germany, Hungary, Ireland, Lithuania, Norway, Poland, Spain, Switzerland, United Kingdom.
Host records: Recorded by Fischer (1962) as being a parasitoid of the anthomyiid *Pegomya seitenstettensis* (Strobl).

Xynobius (*Xynobius*) *christenseni* (Papp, 1982)
Distribution in the Middle East: Turkey (Beyarslan & Fischer, 2011b; Fischer & Beyarslan, 2005).
Extralimital distribution: Georgia, Greece, Ukraine.
Host records: Unknown.

Xynobius (*Xynobius*) *curtifemur* (Fischer, 1961)
Catalogs with Iranian records: No catalog.
Distribution in Iran: Guilan (Dolati et al., 2018).
Distribution in the Middle East: Iran (Dolati et al., 2018), Turkey (Beyarslan, 2015; Beyarslan & Fischer, 2011a, b; Fischer & Beyarslan, 2005).
Extralimital distribution: Bulgaria, Croatia, Czech Republic, Germany, Hungary, Moldova, Romania, Slovakia, Ukraine.
Host records: Recorded by Fischer (1964, 1967) as being a parasitoid of the agromyzid *Agromyza nana* Meigen.

Xynobius (*Xynobius*) *decoratus* (Stelfox, 1949)
Catalogs with Iranian records: Khajeh et al. (2014), Gadallah et al. (2016), Farahani et al. (2016), Yu et al. (2016), Samin, Coronado-Blanco, Kavallieratos et al. (2018), Samin, Coronado-Blanco, Fischer et al. (2018).
Distribution in Iran: East Azarbaijan (Rastegar et al., 2012 as *Opius* (*Xynobius*) *decorates* Stelfox, 1949), Guilan (Ghahari, Fischer, & Tobias, 2012 as *O.* (*Xynobius*) *decorates*), Semnan (Samin, Fischer et al., 2015).
Distribution in the Middle East: Iran (see references above), Turkey (Beyarslan & Fischer, 2011a, b).
Extralimital distribution: Armenia, former Czechoslovakia, England, Finland, Hungary, Ireland, Lithuania, Russia.
Host records: Unknown.

Xynobius (*Xynobius*) *discoidalis* (Fischer, 1957)
Distribution in the Middle East: Turkey (Beyarslan, 2015).
Extralimital distribution: Sweden.
Host records: Unknown.

Xynobius (*Xynobius*) *macrocerus* (Thomson, 1895)
Catalogs with Iranian records: Khajeh et al. (2014), Gadallah et al. (2016), Farahani et al. (2016), Yu et al. (2016), Samin, Coronado-Blanco, Kavallieratos et al. (2018), Samin, Coronado-Blanco, Fischer et al. (2018).
Distribution in Iran: East Azarbaijan (Ghahari & van Achterberg, 2016), Golestan (Ghahari & Gadallah, 2022), Guilan (Sakenin et al., 2012 as *Opius macrocerus* Thomson, 1895), Lorestan (Ghahari, Fischer, Papp et al., 2012 as *Opius macrocerus*), Semnan (Samin, Fischer et al., 2015).
Distribution in the Middle East: Iran (see references above), Turkey (Beyarslan & Fischer, 2011a).
Extralimital distribution: Austria, Bulgaria, former Czechoslovakia, Finland, France, Germany, Hungary, Ireland, Italy, Japan, Montenegro, Netherlands, Norway, Poland, Romania, Russia, Serbia, Switzerland, United Kingdom, United States of America.
Host records: Summarized by Yu et al. (2016) as being a parasitoid of the agromyzid *Agromyza prespana* Spencer, and the tephritids *Euleia heraclei* (L.), and *Trypeta immaculata* (Macquart). In Iran, this species has been reared from the agromyzid *Agromyza ambigua* Fallén (Ghahari & Gadallah, 2022).

Xynopius (*Xynopius*) *maculipes* (Wesmael, 1835)
Catalogs with Iranian records: Fallahzadeh and Saghaei (2010 as *Opius maculipes* Wesmael, 1835), Khajeh et al. (2014 as *O. maculipes*), Gadallah et al. (2016 as *O. maculipes*), Farahani et al. (2016 as *O. maculipes*), Yu et al. (2016), Samin, Coronado-Blanco, Kavallieratos et al. (2018), Samin, Coronado-Blanco, Fischer et al. (2018).

Distribution in Iran: East Azarbaijan (Rastegar et al., 2012), Guilan (Ghahari, Fischer, & Tobias, 2012 as *O. maculipes*, Ghahari & Sakenin, 2018 as *Opius (Nosopoea) maculipes* (Wesmael), Dolati et al., 2019 as *O. (Nosopoea) maculipes*), Mazandaran (Dolati et al., 2019, Sakenin et al., 2021—around cotton fields, both as *O. (Nosopoea) maculipes*), Tehran (Fischer, 1990 as *O. maculipes*, Khajeh et al., 2014 as *O. maculipes*), Iran (no specific locality cited) (Beyarslan & Fischer, 2013 as *O. maculipes*).

Distribution in the Middle East: Iran (see references above), Israel—Palestine (Fischer, 1997; Papp, 2012), Turkey (Beyarslan & Fischer, 2011a, b; Fischer & Beyarslan, 2005, 2013).

Extralimital distribution: Austria, Belgium, Bulgaria, former Czechoslovakia, Denmark, Finland, France, Germany, Greece, Hungary, India, Ireland, Italy, Moldova, Netherlands, Norway, Poland, Spain, Sweden, Switzerland, Poland, Tunisia, Turkmenistan, United Kingdom, United States of America (introduced), Uzbekistan.

Host records: Summarized by Yu et al. (2016) as being a parasitoid of several agromyzids of the genus *Agromyza* Fallén.

Xynobius (Xynobius) notabilis (Fischer, 1958)

Distribution in the Middle East: Turkey (Beyarslan, 2015; Beyarslan & Fischer, 2011b; Fischer & Beyarslan, 2013).
Extralimital distribution: Finland, Hungary, Lithuania, Romania.
Host records: Unknown.

Xynobius (Xynobius) thomsoni (Fischer, 1971)

Distribution in the Middle East: Turkey (Beyarslan & Fischer, 2011a, b).
Extralimital distribution: Austria, former Czechoslovakia, France, Hungary, Italy, Switzerland, United Kingdom.
Host records: Unknown.

Xynobius (Xynobiotenes) scutellatus (Fischer, 1962)

Catalogs with Iranian records: Fallahzadeh & Saghaei (2010), Khajeh et al. (2014), Gadallah et al. (2016), Farahani et al. (2016), Yu et al. (2016), Samin, Coronado-Blanco, Kavallieratos et al. (2018), Samin, Coronado-Blanco, Fischer et al. (2018).
Distribution in Iran: Fars (Lashkari Bod et al., 2011), Tehran (Fischer, 1990; Khajeh et al., 2014).
Distribution in the Middle East: Iran (see references above), Israel—Palestine (Fischer, 1997; Papp, 2012).
Extralimital distribution: Spain, Tunisia.
Host records: Unknown.
Plant associations in Iran: *Medicago sativa* (Fabaceae) (Lashkari Bod et al., 2011).

Xynobius (Xynobiotenes) stigmaticus (Fischer, 1958)

Distribution in the Middle East: Turkey (Beyarslan & Fischer, 2011b; Fischer, 1958, 1972).

Extralimital distribution: Armenia.
Host records: Unknown.

Species excluded from the Middle Eastern countries.

Two species, *Opius paraplasticus* Fischer, 1972 and *Phaedrotoma penetrator* (Fischer, 1966) were doubtfully recorded for the Iranian fauna as they are Afrotropical and Australasian species, respectively (Fischer, 1972, 1987).

Opius paraplasticus Fischer, 1972

Catalogs with Iranian records: Khajeh et al. (2014), Gadallah et al. (2016), Farahani et al. (2016), Yu et al. (2016), Samin, Coronado-Blanco, Kavallieratos et al. (2018), Samin, Coronado-Blanco, Fischer et al. (2018).
Records from Iran: Golestan (Samin, Ghahari et al., 2015), Semnan (Ghahari, Fischer, Hedqvist et al., 2010).
General distribution: South Africa.
Host records: Unknown.
Comments: Reexamination of specimens proved Ghahari, Fischer, Hedqvist et al. (2010), and Samin, Ghahari et al. (2015) identification of *Opius paraplasticus* was a misidentification of *Opius crassipes* Wesmael, 1835 and we therefore exclude *O. paraplasticus* from the fauna of Iran and record the data under *O. crassipes*.

Phaedrotoma penetrator (Fischer, 1966)

Catalogs with Iranian records: Gadallah et al. (2016), Farahani et al. (2016), Yu et al. (2016), Samin, Coronado-Blanco, Kavallieratos et al. (2018), Samin, Coronado-Blanco, Fischer et al. (2018) (all as *Opius (Merotrachys) penetrator* Fischer).
Records from Iran: Hormozgan (Ameri et al., 2014 as *Opius penetrator*).
General distribution: Australia.
Host records: Unknown.
Comments: *Phaedrotoma penetrator* is an Australasian species, and its presence in Iran is strongly doubtful, we therefore exclude this species from the fauna of Iran.

Conclusion

A total of 238 species of the subfamily Opiinae (11.5% of the world species) in 23 genera and two tribes (Biosterini Fischer; Opiini Blanchard) have been reported from the Middle East countries. The genus *Opius* is the most diverse, with 98 recorded species. The Turkish fauna is the most diverse, with 204 species in 23 genera, followed by Iran, with 137 species in 20 genera. The genus *Opius* is more diverse than the other genera, represented by 49 species, followed by *Phaedrotoma* (28 species), *Biosteres* (14 species), *Apodesmia* (13 species), *Utetes*, *Xynobius* (both with seven species), *Opiostomus* (four species), *Biophthora* and *Psyttalia* (both with two species), *Atormus*, *Bitomus*, *Desmiostoma*, *Eurytenes*, *Fopius*, *Hoplocrotaphus*, *Indiopius*, *Opiognathus*, *Pokomandya*, *Psyttoma*, and *Rhogadopsis*

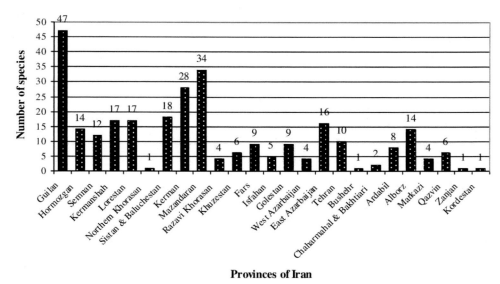

FIGURE 24.1 Number of reported species of Iranian Opiinae by province.

(each with one species). Within all species, two species, *Opius* (*Nosopoea*) *teheranensis* Fischer, 1990, and *Hoplocrotaphus hamooniae* Peris-Felipo, Belokobylskij, & Rakhshani, 2018, are so far only known from Iran (endemic or subendemic to Iran). Opiinae have been recorded from 25 Iranian provinces which among them, Guilan with 47 species has the highest number of species, followed by Mazandaran and Kerman with 34 and 28 species, respectively (Fig. 24.1). From the biological point of view, host species have been detected only for nine opiine species belonging to three families of Diptera, Agromyzidae (seven species), Anthomyiidae (one species), and Tephritidae (four species). Comparison of the Opiinae fauna of Iran with the Middle East countries and adjacent ones to Iran indicates the faunas of Russia with 330 species in 18 genera (Belokobylskij & Lelej, 2019), and Turkey with 204 species (Yu et al., 2016) are more diverse than Iran, followed by Kazakhstan (32 species), Israel−Palestine (26 species), Armenia (23 species), Turkmenistan (10 species), Afghanistan, Cyprus and Egypt (each with six species), Jordan and Syria (both with five species), Azerbaijan (three species), Iraq (two species), Pakistan, Saudi Arabia (both with one species), Bahrain, Kuwait, Lebanon, Oman, Qatar, United Arab Emirates, and Yemen (no species) (Yu et al., 2016). Russia is indicated as having the most diverse fauna, which probably should be expected because it is the largest country, but the greater number of reported species undoubtedly, also reflects a history of more extensive taxonomic research on Opiinae in that country. The much lower number of species recorded from most other countries compared to Russia, Turkey, and Iran probably not only

reflects a smaller land area but also less comprehensive surveys in those countries. Among the countries of the Middle East and adjacent to Iran, Turkey shares 104 species with Iran, followed by Russia (68 species), Kazakhstan (24 species), Israel−Palestine (22 species), Armenia (16 species), Turkmenistan (five species), Afghanistan, Cyprus, Egypt (each with four species), Jordan, Syria (both with three species), Azerbaijan (two species), and Iraq and Saudi Arabia (one species). There is no opiine species shared between Iran and nine other Middle Eastern countries.

References

Abdel-Galil, F. A., Moustafa, M. A., Temerak, S. A. H., & Dalia, Y. A. (2014). Studies on the parasitoid *Opius concolor* Szépligeti (Hymenoptera: Braconidae) associated with ziziphus fruit fly, *Carpomyia incompleta* Becker (Diptera: Tephritidae). *Archives of Phytopathology and Plant Protection, 47*(6), 665−674.

Abd El Wahab, A. A., El Adl, F. E., Gabre, A. M., & Abd El Ghaffar, F. M. (1998). Studies on the beet-fly *Pegomyia mixta* Vill (Diptera: Agromyzidae) and its parasite *Opius nitidulator* (Nees) (Hymenoptera Braconidae). *Journal of the Egyptian German Society of Zoology, 27*(E), 43−56.

Adldoost, H. (1995). Study of population dynamics of rainfed chickpea leafminer in West Azarbaijan. In *Proceedings of the 12th Iranian Plant Protection Congress, 2−7 September 1995* (p. 140). Karaj: Karaj Junior College of Agriculture.

Aamer, N. A., & Hegazi, E. M. (2014). Parasitoids of the leaf miners *Liriomyza* spp. (Diptera: Agromyzidae) attacking faba bean in Alexandria, Egypt. *Egyptian Journal of Biological Pest Control, 24*(2), 301−305.

Ameri, A., Ebrahimi, E., & Talebi, A. A. (2020). Additions to the fauna of Braconidae (Hym., Ichneumonoidea) of Iran based on specimens housed in Hayk Mirzayans Insect Museum with six new records for Iran. *Journal of Insect Biodiversity and Systematics, 6*(4), 353−364.

Ameri, A., Talebi, A. A., Rakhshani, E., Beyarslan, A., & Kamali, K. (2014). Study of the genus *Opius* Wesmael (Hymenoptera: Braconidae: Opiinae) in southern Iran, with eleven new records. *Zootaxa, 3884*(1), 1−26.

Atay, C., Çetin Erdoğan, Ö., & Beyarslan, A. (2019). The Braconidae (Hymenoptera-Apocrita) of Gala Lake National Park and the surrounding area. *Turkish Journal of Zoology, 43*, 131−141.

Belokobylskij, S. A., & Lelej, A. S. (2019). Annotated catalogue of the Hymenoptera of Russia. Volume II. Apocrita: Parasitica. In *Proceedings of the Zoological Institute of the Russian Academy of Sciences*. Supplement No. 8, 594 pp.

Belshaw, R., Fitton, M., Herniou, E., Gimeno, C., & Quicke, D. L. J. (1998). A phylogenetic reconstruction of the Ichneumonoidea (Hymenoptera) based on the D2 variable region of 28S ribosomal RNA. *Systematic Entomology, 23*(2), 109−123.

Beyarslan, A. (2015). A faunal study of the subfamily Doryctinae in Turkey (Hymenoptera: Braconidae). *Turkish Journal of Zoology, 39*(1), 126−143.

Beyarslan, A. (2019). Taxonomic studies on the Opiinae (Hymenoptera, Braconidae) fauna of the the Turkish central part of Eastern Anatolia region (Bingöl, Bitlis, Muş and Van). *Acta Biologica Turcica, 33*(1), 1−7.

Beyarslan, A., & Fischer, M. (2011a). Contribution to the Opiinae fauna of Turkey (Hymenoptera: Braconidae). *Turkish Journal of Zoology, 35*(3), 293−305.

Beyarslan, A., & Fischer, M. (2011b). Checklist of Turkish Opiinae (Hymenoptera: Braconidae). *Zootaxa, 3721*(5), 401−454.

Beyarslan, A., & Fischer, M. (2013). Checklist of Turkish Opiinae (Hymenoptera: Braconidae). *Zootaxa, 3721*(5), 401−454.

Beyarslan, A., & Inanç, F. (1992). Turkish Opiinae (Hymenoptera: Braconidae). *Ulusal Biydoji Kongresi, XI*, 61−68.

Beyarslan, A., Gözüaçik, C., Güllü, M., & Konuksal, A. (2017). Taxonomical investigation on Braconidae (Hymenoptera: Ichneumonoidea) fauna in the Northern Cyprus, with twenty-six new records for the country. *Journal of Insect Biodiversity and Systematics, 3*(4), 319−334.

Beyarslan, A., Yurtcan, M., Çetin Erdoğan, Ö., & Aydoğdu, M. (2006). A study on the Braconidae and Ichneumonidae from ganos Mountain (Thrace Region, Turkey) (Hymenoptera, Braconidae, Ichneumonidae). *Linzer biologische Beiträge, 38/1*, 409−422.

Bodenheimer, F. S. (1930). Die Schädlingsfauna Palästinas. *Monographien zur Angewandten Entomologie, 10*, 438.

Brues, C. I. (1926). Studies on the Ethiopian Braconidae with a catalogue of the African species. *Proceedings of the American Academy of Arts and Sciences, 6*, 206−436.

Čapek, M., & Hofmann, C. (1997). The Braconidae (Hymenoptera) in the collections of the Musée Cantonal de Zoologie, Lausanne. *Litterae Zoologicae (Lausanne), 2*, 25−162.

Chen, X. X., & van Achterberg, C. (2019). Systematics, phylogeny, and evolution of braconid wasps: 30 years of progress. *Annual Review of Entomology, 64*, 1−24.

Çikman, E. (2006). Parasitoids of the leaf miners (Diptera: Agromyzidae) from Adiyaman province. *Türkiye Entomoloji Dergisi, 30*(2), 99−111.

Çikman, E., & Beyarslan, A. (2009). Four new parasitoid records of the subfamilies Euphorinae and Opiinae (Hymenoptera: Braconidae) from the Adiyaman province of Turkey. *Turkish Journal of Zoology, 33*(3), 367−370.

Çikman, E., & Uygun, N. (2003). The determination of leaf miners (Diptera: Agromyzidae) and their parasitoids in cultivated and non-cultivated areas in Sanliurfa province, southern Turkey. *Türkiye Entomoloji Dergisi, 27*(4), 305−318.

Çikman, E., Beyarslan, A., & Civelek, H. S. (2006). Parasitoids of leaf-miners (Diptera: Agromyzidae) from southeast Turkey with 3 new records. *Turkish Journal of Zoology, 30*, 167−173.

Civelek, H. S., Yoldas, Z., & Weintraub, P. (2002). The parasitoid complex of *Liriomyza huidobrensis* in cucumber greenhouses in Izmir Province, western Turkey. *Phytoparasitica, 30*(3), 285−287.

Docavo, I., Jiménez, R., Tormos, J., & Verdu, M. J. (1987). [Braconidae and Chalcidoidea (Hymenoptera, Apocrita, Terebrantia) parasites of Agromyzidae (Diptera, Cyclorrhapha) in Valencia (Spain)]. *Investigation Agraria Production Y. Protection Vegetales, 2*(2), 195−209.

Dolati, S., Talebi, A. A., Farahani, S., & Khayrandish, M. (2018). New finding in Opiinae (Hymenoptera: Braconidae) from north of Iran. *Journal of Insect Biodiversity and Systematics, 4*(3), 163−182.

Dolati, S., Talebi, N. N., Farahani, S., & Khyrandish, M. (2019). New data of the genus *Opius* Wesmael (Hymenoptera: Braconidae, Opiinae) from northern Iran. *Journal of Agricultural Science and Technology, 21*(Supplement), 1871−1887.

Dolati, S., Talebi, A. A., Peris-Felipo, F. J., Farahani, S., & Khayrandish, M. (2021). New data on the subfamily Opiinae (Hymenoptera: Braconidae) from Iran. *Zootaxa, 4903*(3), 331−352.

Dowton, M., & Austin, A. D. (1998). Phylogenetic relationships among the microgastroid wasps (Hymenoptera: Braconidae): Combined analysis of 16S and 28S rDNA genes. *Molecular Phylogenetics and Evolution, 10*(3), 354−366.

Drea, J. J., Jeandel, D., & Gruber, F. (1982). Parasites of agromyzid leafminers (Diptera: Agromyzidae) on alfalfa in Europe. *Annals of the Entomological Society of America, 75*(3), 297−310.

El Bouhssini, M., Mardini, K., Malhotra, R. S., m Joubi, A., & Kagka, N. (2008). Effect of planting date, varieties and insecticides on chickpea leaf miner *Liriomyza cicerinae* R. infestation and the parasitoid *Opius monilicornis* F. *Crop Protection, 27*(6), 915−919.

El-Hag, E. T. A., & El-Meleigi, M. A. (1991). Bionomics of the wheat leafminer, *Agromyza* sp. (Diptera: Agromyzidae) in central Saudi Arabia. *Crop Protection, 10*(1), 70−73.

Fallahzadeh, M., & Saghaei, N. (2010). Checklist of Braconidae (Insecta: Hymenoptera) from Iran. *Munis Entomology & Zoology, 5*(1), 170−186.

Farahani, S., Talebi, A. A., & Rakhshani, E. (2016). Iranian Braconidae (Insecta: Hymenoptera: Ichneumonoidea): Diversity, distribution and host association. *Journal of Insect Biodiversity and Systematics, 2*(1), 1−92.

Farrar, N., & Chou, L. Y. (2000). Introduction of *Fopius carpomyiae* (Silvestri, 1916) (Braconidae: Opiinae), as a parasitoid of ber fruit fly larvae *Carpomya vesuviana* Costa (Tephritidae), in Iran. *Applied Entomology and Phytopathology, 67*(1−2), 27−28.

Farrar, N., Askary, H., Asadi, Gh., & Minaei, K. (2002). *Fopius carpomyiae* (Silvestri, 1916) (Hym.: Braconidae) as a biological control agent of the ber fruit fly by the wasp in Boushehr province, Iran. In *Proceedings of the 15th Iranian Plant Protection Congress, 7−11 September 2002* (p. 115). Razi University of Kermanshah.

Farrar, N., Askari, H., Golestaneh, R., Karampour, F., & Haghani, M. (2012). Study on parasitism of *Carpomyia vesuviana* Costa (Diptera: Techritidae) by *Fopius carpomyie* (Silvestri) (Hymenoptera:

Braconidae) in Bushehr province. *Journal of Iranian Plant Pests Research, 1*(1), 1—9.

Farrar, N., Golestaneh, R., Askari, H., & Assareh, M. H. (2009). Studies on parasitism of *Fopius carpomyie* (Silvestri) (Hymenoptera: Braconidae), an egg-pupal parasitoid of ber (Konar) fruit fly, *Carpomyia vesuviana* Costa (Diptera: Techritidae), in Bushehr-Iran. *Acta Horticulturae, 84*, 431—438.

Farrar, N., Golestaneh, S. R., & Sadeghi, S. E. (2004). Relation between diapause of *Carpomya vesuviana* Costa (Diptera: Tephritidae) and *Thiacidas postica* Walker (Lepidoptera: Noctuidae) and their survival and evolution. In *Proceedings of the 16th Iranian Plant Protection Congress, 28 August—1 September 2004* (p. 457). University of Tabriz.

Fischer, M. (1958). Die europäischen Arten der Gattung *Opius* Wesm. Teil 1a (Hymenoptera, Braconidae). *Annali del Museo Civico di Storia Naturale di Genova, 70*, 33—70.

Fischer, M. (1959). Neue *Opius*-Arten aus Polen (Hymenoptera, Braconidae). *Annales Zoologici Warszawa, 18*, 81—87.

Fischer, M. (1960). Die europäischen Arten der Gattung *Opius* Wesmael, Teil IVa. *Annales Zoologici Warszawa, 19*, 33—112.

Fischer, M. (1962). Beitrag zur Kenntnis der Wirte von Opius-Arten (Hymenoptera, Braconidae). *Entomophaga, 7*(2), 79—90.

Fischer, M. (1964). Gezüchtete Opiinae aus dem Zoologischen Museum der Humboldt-Universität zu Berlin (Hym., Braconidae). *Zeitschrift für Angewandte Entomologie, 55*, 55—70.

Fischer, M. (1967). Über gezüchtete Opiinae aus Europa (Hymenoptera: Braconidae). *Zeitschrift für Angewandte Entomologie, 60*(3), 318—350.

Fischer, M. (1969). Die von Dr. H. Buhr gezüchteten Opiinae (Hymenoptera, Braconidae). *Zeitschrift für Angewandte Zoologie, 56*, 56—88.

Fischer, M. (1972). Hymenoptera Braconidae (Opiinae I). (Paläarktische Region). *Das Tierreich, 91*(1973), 1—620.

Fischer, M. (1974). Eine neue *Opius*-Arten Spanien (Hymenoptera, Braconidae, Opiinae). *Zeitschrift der Arbeitsgemeinschaft Österreichischer Entomologen, 24*, 113—115.

Fischer, M. (1977). Hymenoptera Braconidae (Opiinae II). *Amerika) Das Tierreich, 96*, 1—1001.

Fischer, M. (1986). Neue Bestimmungsschlussel für paläarktische Opiinae, neue Subgenera, Redeskriptionen und eine neue Art (Hymenoptera, Braconidae). *Annalen des Naturhistorischen Museums in Wien, 88/89*, 607—662.

Fischer, M. (1987). Hymenoptera Opiinae III — Aethiopische, orientalische, australische und ozeanische Region. *Das Tierreich, 104*, 1—734.

Fischer, M. (1990). Paläarktische Opiinae (Hymenoptera, Braconidae): Neue Arten und neue Funde aus dem Ungarischen Naturwissenschaftlichen Museum in Budapest. *Annales Historico-Naturales Musei Nationalis Hungarici, 81*(1989), 205—238.

Fischer, M. (1992). Neue *Opius*-Arten aus der Turkei (Hymenoptera, Braconidae, Opiinae). *Zeitschrift der Arbeitsgemeinschaft Österreichischer Entomologen, 44*(3—4), 79—86.

Fischer, M. (1995). Korrekturen und Erganzungun zur Taxonomie altweltlicher Opiinae und Neufassung eines Bestimmungsschlussels für die paläarktischen Arten des Subgenus *Opiothorax* Fischer, 1972 des Genus *Opius* Wesmael, 1835 (Hymenoptera: Braconidae). *Entomofauna, 16*(9), 217—242.

Fischer, M. (1997). Die paläarktischen Opiinae (Madenwespen) der Zoologischen Stäatssammlang München (Hymenoptera: Braconidae). *Entomofauna, 18*(14), 137—194.

Fischer, M. (2001). Mitteilungen über neue und schon bekannte *Opius* Wesmael-Arten der Alten Welt (Hymenoptera, Braconidae, Opiinae). *Linzer biologische Beiträge, 33*(1), 5—33.

Fischer, M. (2004). Einige neue Brackwespen (Insecta: Hymenoptera: Braconidae) und weitere Formen der Kiefer und Madenwespen (Alysiinae, Opiinae). *Annalen des Naturhistorischen Museums in Wien. Series B Botanik und Zoologie, 10B*(2003), 277—318.

Fischer, M. (2008). Eine neue *Bitomus*-Arten aus Bulgarien (Hymenoptera: Braconidae: Opiinae). *Zeitschrift der Arbeitsgemeinschaft Österreichischer Entomologen, 60*, 55—58.

Fischer, M., & Beyarslan, A. (2005). A survey of Opiinae (Hymenoptera: Braconidae) of Turkey. *Fragmenta Faunistica (Warsaw), 48*(1), 27—62.

Fischer, M., & Beyarslan, A. (2011). *Indiopius yilmazae* sp. n. (Hymenoptera: Braconidae: Opiinae), a new species from Turkey. *Zeitschrift der Arbeitsgemeinschaft Österreichischer Entomologen, 63*(2), 123—125.

Fischer, M., & Beyarslan, A. (2013). Additional contributions to the Opiinae fauna of Turkey (Hymenoptera: Braconidae). *Turkish Journal of Zoology, 37*(5), 525—583.

Fischer, M., & Koponen, M. (1999). A survey of Opiinae (Hymenoptera, Braconidae) of Finland, part 2. *Entomologica Fennica, 10*, 129—160.

Gadallah, N. S., Ghahari, H., Papp, J., & Beyarslan, A. (2018). New records of Braconidae (Hymenoptera) from Iran. *Wuyi Science Journal, 34*, 43—48.

Gadallah, N. S., Ghahari, H., Peris-Felipo, F. J., & Fischer, M. (2016). Updated checklist of Iranian Opiinae (Hymenoptera: Braconidae). *Zootaxa, 4066*(1), 1—40.

Ghahari, H. (2017). Species diversity of Ichneumonoidea (Hymenoptera) from rice fields of Mazandaran province, northern Iran. *Journal of Animal Environment, 9*(3), 371—378 (in Persian, English summary).

Ghahari, H. (2018). Species diversity of the parasitoids in rice fields of northern Iran, especially parasitoids of rice stem borer. *Journal of Animal Environment, 9*(4), 289—298 (in Persian, English summary).

Ghahari, H. (2019). Faunistic survey of parasitoid wasps (Hymenoptera) in forest areas of Mazandaran province, northern Iran. *Iranian Journal of Forest, 11*(1), 61—79 (in Persian, English summary).

Ghahari, H., & Beyarslan, A. (2017). A faunistic study on Braconidae (Hymenoptera: Ichneumonoidea) from Iran. *Natura Somogyiensis, 30*, 39—46.

Ghahari, H., & Beyarslan, A. (2019). A faunistic study on Braconidae (Hymenoptera: Ichneumonoidea) from Iran, and in Memoriam Dr. Jenő Papp (20 May 1933-11 December 2017). *Acta Biologica Turcica, 32*(4), 248—254.

Ghahari, H., & Fischer, M. (2011). A study on the Braconidae (Hymenoptera: Ichneumonoidea) from some regions of northern Iran. *Entomofauna, 32*(8), 181—196.

Ghahari, H., & Fischer, M. (2012). A faunistic survey on the braconid wasps (Hymenoptera: Braconidae) from Kermanshah province, Iran. *Entomofauna, 33*(20), 305—312.

Ghahari, H., & Gadallah, N. S. (2022). Additional records to the braconid fauna (Hymenoptera: Ichneumonoidea) of Iran, with new host reports. *Entomological News* (in press).

Ghahari, H., & Sakenin, H. (2018). Species diversity of Chalcidoidea and Ichneumonoidea (Hymenoptera) in some paddy fields and surrounding grasslands of Mazandaran and Guilan provinces, northern Iran. *Applied Plant Protection, 7*(1), 11–19 (in Persian, English summary).

Ghahari, H., & van Achterberg, C. (2016). A contribution to the study of subfamilies Microgastrinae and Opiinae (Hymenoptera: Braconidae) from the Arasbaran Biosphere Reserve and vicinity, northwestern Iran. *Natura Somogyiensis, 28*, 23–32.

Ghahari, H., Fischer, M., Çetin Erdoğan, Ö., Beyarslan, A., & Havaskary, M. (2009). A contribution to the knowledge of the braconid-fauna (Hymenoptera, Ichneumonoidea, Braconidae) of Arasbaran, northwestern Iran. *Entomofauna, 30*(20), 329–336.

Ghahari, H., Fischer, M., Çetin Erdoğan, Ö., Beyarslan, A., & Ostovan, H. (2010). A contribution to the braconid wasps (Hymenoptera: Braconidae) from the forests of northern Iran. *Linzer biologische Beiträge, 42*(1), 621–634.

Ghahari, H., Fischer, M., Hedqvist, K. J., Çetin Erdoğan, Ö., van Achterberg, C., & Beyarslan, A. (2010). Some new records of Braconidae (Hymenoptera) for Iran. *Linzer biologische Beiträge, 42*(2), 1395–1404.

Ghahari, H., Fischer, M., & Papp, J. (2011). A study on the braconid wasps (Hymenoptera: Braconidae) from Isfahan province, Iran. *Entomofauna, 32*(16), 261–272.

Ghahari, H., Fischer, M., Papp, J., & Tobias, V. (2012). A contribution to the knowledge of braconids (Hymenoptera: Braconidae) from Lorestan province Iran. *Entomofauna, 33*(7), 65–72.

Ghahari, H., Fischer, M., Sakenin, H., & Imani, S. (2011). A contribution to the Agathidinae, Alysinae, Aphidiinae, Braconinae, Microgastrinae and Opiinae (Hymenoptera: Braconidae) from cotton fields and surrounding grasslands of Iran. *Linzer biologische Beiträge, 43*(2), 1269–1276.

Ghahari, H., Fischer, M., & Tobias, V. (2012). A study on the Braconidae (Hymenoptera: Ichneumonoidea) from Guilan province, Iran. *Entomofauna, 33*(22), 317–324.

Golestaneh, S. R., Kocheili, F., Rasekh, A., Esfandiari, M., & Farashiani, M. E. (2018a). A survey on foraging behavior of *Fopius carpomyiae* (Hymenoptera: Braconidae), a parasitoid of ber fruit fly *Carpomyia vesuviana* (Diptera: Tephritidae). *Plant Protection (Scientific Journal of Agriculture), 41*(3), 73–85 (in Persian, English summary).

Golestaneh, S. R., Kocheili, F., Rasekh, A., Esfandiari, M., & Farashiani, M. E. (2018b). Study on foraging behavior of *Fopius carpomyiae*, a parasitoid of ber fruit fly *Carpomyia vesuviana*. In *Proceedings of the 23th Iranian Plant Protection Congress, 27–30 August 2018* (p. 1039). Gorgan University of Agricultural Sciences and Natural Resources.

Halperin, J. (1986). Braconidae (Hymenoptera) associated with forest and ornamental trees and shrubs in Israel. *Phytoparasitica, 14*(2), 119–135.

Khajeh, N., Rakhshani, E., Peris-Felipo, F. J., & Žikić, V. (2014). Contributions to the Opiinae (Hymenoptera: Braconidae) of Eastern Iran with updated checklist of Iranian species. *Zootaxa, 3784*(2), 131–147.

Khanjani, M. (2006a). *Field crop pests in Iran* (3rd ed). Bu-Ali Sina University, 719 pp. (in Persian).

Khanjani, M. (2006b). *Vegetable pests in Iran* (2nd ed). Bu-Ali Sina University, 467 pp. (in Persian).

Ku, D. S., Belokobylskij, S. A., & Cha, J. Y. (2001). Hymenoptera (Braconidae). Economic insects of Korea 16. *Insecta Koreana (Supplement), 23*, 283.

Lashkari Bod, A., Rakhshani, E., Talebi, A. A., & Lozan, A. (2010). Introduction of twelve newly recorded species of Braconidae (Hymenoptera) from Iran. In *Proceedings of the 19th Iranian Plant Protection Congress, 31 July–3 August 2010* (p. 161). Tehran: Iranian Research Institute of Plant Protection.

Lashkari Bod, A., Rakhshani, E., Talebi, A. A., Lozan, A., & Žikić, V. (2011). A contribution to the knowledge of Braconidae (Hym., Ichneumonoidea) of Iran. *Biharean Biologist, 5*(2), 147–150.

Li, X. Y., van Achterberg, C., & Tan, J. C. (2013). Revision of the subfamily Opiinae (Hymenoptera, Braconidae) from Hunan (China), including thirty-six new species and two new genera. *ZooKeys, 268*, 1–168.

Madjdzadeh, S. M., Parrezaali, M., Dolati, S., & Ghassemi-Khademi, T. (2021). New data on the braconid wasps (Hymenoptera: Braconidae: Cheloninae, Opiinae, Rogadinae) of South-Eastern Iran. *Faunistic Entomology, 74*, 1–7.

Mahmoud, S. M., El-Heneidy, A. H., Gadallah, N. S., & Ahmed, R. S. (2009). Survey and abundance of common ichneumonoid parasitoid species in Suez Canal Region, Egypt. *Egyptian Journal of Biological Pest Control, 19*(2), 185–190.

Modarres Awal, M. (1997). Family Braconidae (Hymenoptera), pp. 265–267. In M. Modarres Awal (Ed.), *List of agricultural pests and their natural enemies in Iran* (2nd ed). Ferdowsi University of Mashhad Press, 429 pp.

Modarres Awal, M. (2012). Family Braconidae (Hymenoptera), pp. 483–486. In M. Modarres Awal (Ed.), *List of agricultural pests and their natural enemies in Iran* (3rd ed). Ferdowsi University of Mashhad Press, 759 pp.

Naderian, H., Ghahari, H., & Asgari, S. (2012). Species diversity of natural enemies in corn fields and surrounding grasslands of Semnan province, Iran. *Calodema, 217*, 1–8.

Papp, J. (1970). A contribution to the braconid fauna of Israel (Hymenoptera). *Israel Journal of Entomology, 5*, 63–76.

Papp, J. (1982). Taxonomical and faunistical novelities of the Opiinae in the Arctogaea (Hymenoptera, Braconidae). *Annales Historico-Naturales Musei Nationalis Hungarici, 74*, 241–253.

Papp, J. (1989). A contribution to the braconid fauna of Israel (Hymenoptera), 2. *Israel Journal of Entomology, 22*, 45–59.

Papp, J. (1990). Braconidae (Hymenoptera) from Greece, 3. *Annales Musei Goulandris, 8*, 269–290.

Papp, J. (2012). A contribution to the braconid fauna of Israel (Hymenoptera: Braconidae), 3. *Israel Journal of Entomology, 41–42*, 165–219.

Peris-Felipo, F. J., Belokobylskij, S. A., Derafshan, H. A., & Rakhshani, E. (2018). Revision of the genus *Hoplocrotaphus* Telenga, 1950 (Hymenoptera, Braconidae, Opiinae). *Journal of Hymenoptera Research, 62*, 55–72.

Peris-Felipo, F. J., Rahmani, Z., Belokobylskij, S. A., & Rakhshani, E. (2014). Genus *Indiopius* Fischer, 1966 (Hymenoptera, Braconidae, Opiinae) in Iran with a key to the world species. *ZooKeys, 368*, 37–44.

Priore, R., & Tremblay, E. (1995). Parassitoidi (Hymenoptera: Braconidae) di alcuni ditteri fillomina tori (Diptera: Agromyzidae). *Bolletino de Laboratorio di Entomologia Agraria Filippo Silvetsri, 50*, 109–120.

Quicke, D. L. J., & van Achterberg, C. (1990). Phylogeny of the subfamilies of the family Braconidae (Hymenoptera: Ichneumonoidea). *Zoologische Vehandelingen, 258*, 1–180.

Ranjbar, M., Madjdzadeh, S. M., Peris-Felipo, F. J., Askari, M., & Rakhshani, E. (2016). A contribution to the fauna of Opiinae (Hym.: Braconidae) of Kerman province South-Eastern Iran. *Journal of the Entomological Research Society, 18*(1), 19–26.

Rastegar, J., Sakenin, H., Khodaparast, S., & Havaskary, M. (2012). On a collection of Braconidae (Hymenoptera) from East Azarbaijan and vicinity, Iran. *Calodema, 226,* 1–4.

Safahani, S., Madjdzadeh, S. M., & Peris-Felipo, F. J. (2018). Contribution to the fauna and phenological knowledge of high mountains Opiinae (Hymenoptera, Braconidae) in Kerman province (Iran). *Journal of Insect Biodiversity and Systematics, 4*(2), 73–83.

Sakenin, H., Naderian, H., Samin, N., Rastegar, J., Tabari, M., & Papp, J. (2012). On a collection of Braconidae (Hymenoptera) from northern Iran. *Linzer biologische Beiträge, 44*(2), 1319–1330.

Sakenin, H., Ghahari, H., & Navaeian, M. (2021). A study on the predator and parasitoid insects in some cotton fields of Iran. *Journal of Animal Environment, 13*(1), 397–406 (in Persian, English summary).

Samin, N. (2015). A faunistic study on Braconidae of Iran (Hymenoptera: Ichneumonoidea). *Arquivos Entomolóxicos, 13,* 339–345.

Samin, N., Beyarslan, A., Ranjith, A. P., Ahmad, Z., Sakenin Chelav, H., & Hosseini Boldaji, S. A. (2020). A faunistic study on Braconidae (Hymenoptera: Ichneumonoidea) from Ardebil and East Azarbayjan provinces, northwestern Iran. *Egyptian Journal of Plant Protection Research Institute, 3*(4), 955–963.

Samin, N., Coronado-Blanco, J. M., Fischer, M., van Achterberg, C., Sakenin, H., & Davidian, E. (2018). Updated checklist of Iranian Braconidae (Hymenoptera: Ichneumonoidea) with twenty-three new records. *Natura Somogyiensis, 32,* 21–36.

Samin, N., Coronado-Blanco, J. M., Kavallieratos, N. G., Fischer, M., & Sakenin, H. (2018). Recent findings on Braconidae (Hymenoptera: Ichneumonoidea) of Iran with an updated checklist. *Acta Biologica Turcica, 31*(4), 160–173.

Samin, N., Fischer, M., & Ghahari, H. (2015). A contribution to the study on the fauna of Braconidae (Hymenoptera, Ichneumonoidea) from the province of Semnan, Iran. *Arquivos Entomoloxicos, 13,* 429–433.

Samin, N., Ghahari, H., Gadallah, N. S., & Monaem, R. (2015). A study on the braconid wasps (Hymenoptera: Ichneumonoidea: Braconidae) from Golestan province, northern Iran. *Linzer biologische Beiträge, 47*(1), 731–739.

Samin, N., Papp, J., & Coronado-Blanco, J. M. (2018). A faunistic study on braconid wasps (Hymenoptera: Ichneumonoidea: Braconidae) of Iran. *Scientific Bulletin of the Uzhgorod University (Series: Biology), 45,* 15–19.

Samin, N., Sakenin, H., Imani, S., & Shojai, M. (2011). A study on the Braconidae (Hymenoptera) of Khorasan province and vicinity, Northeastern Iran. *Phegea, 39*(4), 137–143.

Samin, N., Sakenin Chelav, H., Ahmad, Z., Penteado-Dias, A. M., & Samiuddin, A. (2020). A faunistic study on the family Braconidae (Hymenoptera: Ichneumonoidea) from Iran. *Scientific Bulletin of Uzhhorod National University (Series: Biology), 48,* 14–19.

Schmiedeknecht, O. (1900). Neue Hymenopteren aus Nord-Africa. *Természetrajzi Füzetek, 23,* 220–247.

Schuster, D. J., & Wharton, R. A. (1993). Hymenopterous parasitoids of leafmining *Liriomyza* spp. (Diptera: Agromyzidae) on tomato in Florida. *Environmental Entomology, 22,* 1188–1191.

Sedighi, S., Madjdzadeh, M., Khajeh, N., & Rakhshani, E. (2014). New records of Opiinae and Euphorinae (Hym.: Braconidae) from Iran. In *Proceedings of the 3rd Integrated Pest Management Conference (IPMC), 21–22 January 2014, Kerman* (pp. 399–404) (in Persian, English summary).

Sharanowski, B. J., Dowling, A. P. G., & Sharkey, M. J. (2011). Molecular phylogenetics of Braconidae (Hymenoptera: Ichneumonoidea), based on multiple nuclear genes, and implications for classification. *Systematic Entomology, 36,* 549–572.

Shaw, M. R., & Huddleston, T. (1991). *Classification and biology of braconid wasps (Hymenoptera: Braconidae), 7.11.* London: Royal Entomological Society, 126 pp.

Shi, M., Chen, X. X., & van Achterberg, C. (2005). Phylogenetic relationships among the Braconidae (Hymenoptera: Ichneumonoidea) inferred from partial 16S rDNA, 28S rDNA D2, 18S rDNA gene sequences and morphological characters. *Molecular Phylogenetics and Evolution, 37*(1), 104–116.

Szépligeti, G. (1904). Hymenoptera. Fam. Braconidae. *Genera Insectorum, 22,* 1–253.

Tobias, V. I. (1977). The genus *Opius* Wesm. (Hymenoptera, Braconidae) as parasites of fruit flies (Diptera, Tephritidae). *Entomologicheskoye Obozreniye, 56,* 420–430.

van Achterberg, C. (1993). Illustrated key to the subfamilies of the Braconidae (Hymenoptera: Ichneumonoidea). *Zoologische Verhandelingen, 283,* 1–189.

van Achterberg, C. (2014). Notes on the checklist of Braconidae (Hymenoptera) from Switzerland. *Mitteilungen der Schweizerischen Entomologischen Gesellschaft, 87,* 191–213.

Wang, X. G., Bokonan-Ganta, A. H., Ramadan, M. M., & Messing, R. H. (2004). Egg-larval opiine parasitoids (Hym., Braconidae) of tephritid fruit fly pests do not attack the flowerheaded-feeder *Trupanea dubautiae* (Dipt., Tephritidae). *Journal of Applied Entomology, 128*(9/10), 716–722.

Wharton, R. A. (1984). The status of certain Braconidae (Hymenoptera) cultured for biological control programs, and description of a new species of *Macrocentrus*. *Proceedings of the Entomological Society of Washington, 86,* 902–912.

Wharton, R. A. (1997). Generic relationships of opiine Braconidae (Hymenoptera) parasitic on fruit-infesting Tephritidae (Diptera). *Contributions of the American Entomological Institute, 30,* 1–53.

Wharton, R. A., Yoder, M. J., Gillespie, J. J., Patton, J. C., & Haoneyatt, R. L. (2006). Relatioships of *Exodontiella*, a non-alysiine, exdont member of the family Braconidae (Insecta, Hymenoptera). *Zoologica Scripta, 35,* 323–340.

Yildirim, E. M., Civelek, H. S., Çikman, E., Dursun, O., & Atay, E. (2010). Contributions to the Turkish Braconidae (Hymenoptera) fauna with seven new records. *Turkiye Entomoloji Dergisi, 34*(1), 29–35.

Yu, D. S., van Achterberg, C., & Horstmann, K. (2016). *Taxapad 2016, Ichneumonoidea 2015, database on flash-drive.* Ottawa, Ontario, Canada.

Zaldívar-Riverón, A., Mori, M., & Quicke, D. L. J. (2006). Systematics of the cyclostome subfamilies of braconid parasitic wasps (Hymenoptera: Ichneumonoidea): A simultaneous molecular and morphological Bayesian approach. *Molecular Phylogeny and Evolution, 38*(1), 130–145.

Orgilus macrurus Muesebeck, 1970 (Orgilinae), ♀, lateral habitus—Nearctic. *Photo prepared by S.R. Shaw.*

Orgilus grapholithae Muesebeck, 1970 (Orgilinae), ♀, lateral habitus—Nearctic. *Photo prepared by S.R. Shaw.*

Chapter 25

Subfamily Orgilinae Ashmead, 1900

Hassan Ghahari[1], Scott Richard Shaw[2], Neveen Samy Gadallah[3], Donald L.J. Quicke[4] and James B. Whitfield[5]

[1]Department of Plant Protection, Yadegar-e Imam Khomeini (RAH) Shahre Rey Branch, Islamic Azad University, Tehran, Iran; [2]UW Insect Museum, Department of Ecosystem Science and Management, University of Wyoming, Laramie, WY, United States; [3]Entomology Department, Faculty of Science, Cairo University, Giza, Egypt; [4]Integrative Ecology Laboratory, Department of Biology, Faculty of Science, Chulalongkorn University, Pathumwan, Bangkok, Thailand; [5]Department of Entomology, University of Illinois at Urbana-Champaign, Urbana, IL, United States

Introduction

Orgilinae Ashmead, 1900 is a small subfamily of Braconidae that is distributed in almost all parts of the world (Yu et al., 2016). Orgilinae comprises about 362 described species belonging to 13 genera and three tribes, which are Antestrigini van Achterberg, 1987 (Neotropical), Mimagathidini Enderlein, 1905 (Afrotropical, Indo-Australian, Neotropical, NE Palaearctic), and Orgilini Ashmead, 1900 (cosmopolitan) (Chen & van Achterberg, 2019; van Achterberg et al., 2017; Yu et al., 2016). Most of the species belong to the genus *Orgilus* Haliday, 1833, which includes about 254 described species (71% of the total number of species) (Yu et al., 2016). A sister relationship, Orgilinae (Homolobinae + Microtypinae), has been suggested by some authors (e.g., van Achterberg, 1984, 1992), based on larval and adult morphology and biology. This relationship has also been corroborated by Sharanowski et al. (2011) through a phylogenetic study. More broadly, Orgilinae has been included within the helconoid complex (macrocentroid subcomplex) (Sharanowski et al., 2011).

Members of the subfamily Orgilinae are diagnosed by having the following combination of characters: slender, medium-sized bodies (4.0–7.0 mm), usually with a somewhat long ovipositor; the occipital carina being reduced dorsally and meeting the hypostomal carina above base of mandible; prepectal carina usually developed but sometimes partly or largely reduced; discoidal cell of forewing sessile, forewing 2-1A vein somewhat developed; head narrow, with face and clypeus strongly protuberant; and hind tibia usually with pegs near base of spurs (Shaw & Huddleston, 1991; Toibas, 1986; van Achterberg, 1987, 1993).

As far known, all of them are koinobiont endoparasitoids in concealed-feeding lepidopteran larvae (Chen & van Achterberg, 2019; Shaw & Huddleston, 1991; van Achterberg et al., 2017) mainly of the families Coleophoridae, Gelechiidae, Gracillariidae, Oecophoridae, Pyralidae, Crambidae, and Tortricidae (Braet & Quicke, 2004; Kula et al., 2009; Muesebeck, 1938, 1970; Sharanowski et al., 2014; van Achterberg, 1987). While most orgilines attack hosts that are concealed in leaf-rolls or silk webbing, some of the smallest Orgilinae species are parasitoids of leaf-miners, especially Gracillariidae and Solanaceae-mining Gelechiidae (Muesebeck, 1970; Whitfield & Wagner, 1991). Some orgiline species are considered to be potential biocontrol agents of pest Lepidoptera species (van Achterberg, 1987). *Stantonia pallida* (Ashmead) *sensu* Braet and Quicke (2004) has been successfully used as a biocontrol agent for suppression of *Neomusotima conspurcatalis* Warren (Lepidoptera: Crambidae), as well as *Diaphania hyalinata* (Linnaeus) and *Diaphania nitidalis* (Stoll) (Lepidoptera: Pyralidae) in Florida (Kula et al., 2009). The genera of the subfamily Orgilinae were revised and keyed by van Achterberg (1987), with a subsequent addition by van Achterberg and Quicke (1992) and van Achterberg (1994). The Palaearctic species of the genera *Kerorgilus* and *Orgilus* have been studied by van Achterberg (1985) and Taeger (1989), respectively. The Nearctic species of *Orgilus* have been revised by Muesebeck (1970).

Checklists of Regional Orgilinae. Modarres Awal (1997, 2012) recorded one species (*Orgilus* (*Orgilus*) *obscurator* (Nees, 1812)). Fallahzadeh and Saghaei (2010) listed five species in one genus (without precise localities). Farahani et al. (2016) recorded 12 species in one genus. Yu et al. (2016), Samin, Coronado-Blanco, Kavallieratos et al. (2018), and Samin, Coronado-Blanco, Fischer et al. (2018) all listed 15 species in two genera. Gadallah et al. (2019) cataloged 16 species in two genera (*Orgilus*, and *Kerorgilus* van Achterberg, 1985). The present checklist includes 34 species in two genera in the Middle Eastern countries. Additionally, *Orgilus jennieae* Marsh, 1979 which has been erroneously reported for the Iranian fauna (Khanjani, 2006) is here excluded from the Iranian fauna.

Key to genera of the subfamily Orgilinae in the Middle East (modified from van Achterberg, 1994)

Clypeus with a pair of upwardly bent tubercles; tarsal claws very slender; hind tarsus long and slender ....*Kerorgilus* van Achterberg
Clypeus without tubercles; tarsal claws simple, less slender; hind tarsus shorter, usually less slender .... *Orgilus* Haliday

List of species of the subfamily Orgilinae in the Middle East

Subfamily Orgilinae Ashmead, 1900

Tribe Orgilini Ashmead, 1900

Genus *Kerorgilus* van Achterberg, 1985

Kerorgilus longicaudis van Achterberg, 1985
Distribution in the Middle East: Turkey (van Achterberg, 1985).
Extralimital distribution: None.
Host records: Unknown.

Kerorgilus zonator (Szépligeti, 1896)
Catalogs with Iranian records: Yu et al. (2016), Samin, Coronado-Blanco, Kavallieratos et al. (2018), Samin, Coronado-Blanco, Fischer et al. (2018), Gadallah et al. (2019).
Distribution in Iran: West Azarbaijan (Samin et al., 2016).
Distribution in the Middle East: Iran (Samin et al., 2016), Turkey (Beyarslan & Çetin Erdoğan, 2011).
Extralimital distribution: Azerbaijan, China, Germany, Greece, Hungary, Korea, Mongolia, Romania.
Host records: Unknown.

Genus *Orgilus* Haliday, 1833

Orgilus (*Orgilus*) *abbreviator* (Ratzeburg, 1852)
Catalogs with Iranian records: Fallahzadeh and Saghaei (2010), Farahani et al. (2016), Yu et al. (2016), Samin, Coronado-Blanco, Kavallieratos et al. (2018), Samin, Coronado-Blanco, Fischer et al. (2018), Gadallah et al. (2019).
Distribution in Iran: Southern Khorasan (Ghahari, 2020), Tehran (Damavand) (Taeger, 1989 as *Orgilus nanellae* Tobias, 1986), Iran (no specific locality cited) (Farahani et al., 2014; Güçlü & Özbek, 2015).
Distribution in the Middle East: Iran (see references above), Turkey (Beyarslan & Çetin Erdoğan, 2011; Güçlü & Özbek, 2015).
Extralimital distribution: Armenia, Bulgaria, Germany, Greece, Hungary.

Host records: Recorded by Taeger (1989) as being a parasitoid of the gelechiids *Recurvaria leucatella* (Clerck), and *Recurvaria nanella* (Denis and Schiffermüller).

Orgilus (*Origlus*) *asper* Taeger, 1989
Distribution in the Middle East: Jordan (Taeger, 1989).
Extralimital distribution: Hungary.
Host records: Unknown.

Orgilus (*Orgilus*) *caliginosus* Taeger, 1989
Distribution in the Middle East: Egypt (Taeger, 1989), Israel—Palestine (Papp, 2012; Taeger, 1989).
Extralimital distribution: None.
Host records: Unknown.

Orgilus (*Orgilus*) *claripennis* Ivanov, 1899
Distribution in the Middle East: Turkey (Beyarslan & Çetin Erdoğan, 2011; Güçlü & Özbek, 2015; Taeger, 1989).
Extralimital distribution: Moldova, Ukraine.
Host records: Recorded by Tobias (1976, 1986) as being a parasitoid of the depressariid *Depressaria depresella* Hübner.

Orgilus (*Orgilus*) *dilleri* Beyarslan, 1996
Distribution in the Middle East: Turkey (Beyarslan, 1996).
Extralimtal distribution: None.
Host records: Unknown.

Orgilus (*Orgilus*) *dovnari* Tobias, 1986
Catalogs with Iranian records: No catalog.
Distribution in Iran: Kordestan (Samin, Papp, & Coronado-Blanco, 2018).
Distribution in the Middle East: Iran (Samin, Papp, & Coronado-Blanco, 2018), Turkey (Beyarslan, 1996).
Extralimital distribution: Former Czechoslovakia, Germany, Hungary, Mongolia, Netherlands, Russia, Slovakia, Sweden, Ukraine.
Host records: Unknown.

Orgilus (Orgilus) festivus Papp, 1975
Distribution in the Middle East: Turkey (Beyarslan & Çetin Erdoğan, 2011; Güçlü & Özbek, 2015).
Extralimital distribution: Croatia, Greece, former Yugoslavia.
Host records: Unknown.

Orgilus (Orgilus) grunini Tobias, 1986
Distribution in the Middle East: Turkey (Beyarslan, 1996; Beyarslan & Çetin Erdoğan, 2011).
Extralimital distribution: Austria, former Czechoslovakia, France, Germany, Hungary, Kazakhstan, Netherlands, Poland, Romania.
Host records: Summarized by Yu et al. (2016) as being a parasitoid of the following coleophorids: *Coleophora coracipennella* (Hübner), *Coleophora hemerobiella* (Scopoli), *Coleophora ibipennella* Zeller, and *Coleophora serratella* (L.).

Orgilus (Orgilus) hungaricus Szépligeti, 1896
Catalogs with Iranian records: Farahani et al. (2016), Yu et al. (2016), Samin, Coronado-Blanco, Kavallieratos et al. (2018), Samin, Coronado-Blanco, Fischer et al. (2018), Gadallah et al. (2019).
Distribution in Iran: East Azarbaijan (Ghahari et al., 2009), Iran (no specific locality cited) (Farahani et al., 2014).
Distribution in the Middle East: Iran (Farahani et al., 2014; Ghahari et al., 2009), Turkey (Beyarslan & Çetin Erdoğan, 2011; Güçlü & Özbek, 2015).
Extralimital distribution: Hungary, Kazakhstan, Romania, Serbia, Slovakia.
Host records: Unknown.

Orgilus (Orgilus) interjectus Taeger, 1989
Distribution in the Middle East: Turkey (Güçlü & Özbek, 2015).
Extralimital distribution: Austria, Germany, Hungary, Italy, Netherlands, Romania, Switzerland, United Kingdom.
Host records: Summarized by Yu et al. (2016) as being a parasitoid of the depressariid *Agonopterix hypericella* (Hübner), and the tortricid *Cochylis atricapitana* (Stephens).

Orgilus (Orgilus) ischnus Marshall, 1898
Catalogs with Iranian records: Farahani et al. (2016), Yu et al. (2016), Samin, Coronado-Blanco, Kavallieratos et al. (2018), Samin, Coronado-Blanco, Fischer et al. (2018), Gadallah et al. (2019).
Distribution in Iran: Alborz, Tehran (Farahani et al., 2014), Ardabil (Samin, Papp, & Coronado-Blanco, 2018).
Distribution in the Middle East: Iran.
Extralimital distribution: Austria, China, Czech Republic, Germany, Hungary, Mongolia, Netherlands, Norway, Poland, Russia, Switzerland, United Kingdom, former Yugoslavia.
Host records: Recorded by Taeger (1989) and Papp (1994) as being a parasitoid of the coleophorids *Coleophora frischella* (L.), *Coleophora paripennella* Zeller, and *Coleophora peisoniella* Kasy; and the tortricid *Spilonota ocellana* (Denis and Schiffermüller).

Orgilus (Orgilus) leptocephalus (Hartig, 1838)
Catalogs with Iranian records: Gadallah et al. (2019).
Distribution in Iran: Guilan (Gadallah et al., 2019).
Distribution in the Middle East: Iran.
Extralimital distribution: Austria, Belgium, Canada (unspecified), Czech Republic, Finland, France, Germany, Hungary, Ireland, Italy, Luxembourg, Mongolia, Netherlands, Poland, Russia, Sweden, Switzerland, United Kingdom, United States of America, former Yugoslavia.
Host records: Recorded by Papp (1994) as being a parasitoid of the tortricid *Rhyacionia buoliana* (Denis and Schiffermüller).

Orgilus (Orgilus) mediterraneus Taeger, 1989
Distribution in the Middle East: Cyprus, Israel–Palestine (Taeger, 1989).
Extralimital distribution: None.
Host records: Unknown.

Orgilus (Orgilus) meyeri Telenga, 1933
Catalogs with Iranian records: Fallahzadeh and Saghaei (2010), Farahani et al. (2016), Yu et al. (2016), Samin, Coronado-Blanco, Kavallieratos et al. (2018), Samin, Coronado-Blanco, Fischer et al. (2018), Gadallah et al. (2019).
Distribution in Iran: Alborz, Guilan (Farahani et al., 2014), Mazandaran (Farahani et al., 2014; Kian et al., 2020), Tehran (Taeger, 1989), Iran (no specific locality cited) (Güçlü & Özbek, 2015).
Distribution in the Middle East: Iran (see references above), Turkey (Güçlü & Özbek, 2015).
Extralimital distribution: Azerbaijan, Mongolia, Uzbekistan.
Host records: Unknown.

Orgilus (Orgilus) nitidior Taeger, 1989
Catalogs with Iranian records: Farahani et al. (2016), Yu et al. (2016), Samin, Coronado-Blanco, Kavallieratos et al. (2018), Samin, Coronado-Blanco, Fischer et al. (2018), Gadallah et al. (2019).
Distribution in Iran: Alborz, Guilan, Qazvin, Tehran (Farahani et al., 2014).
Distribution in the Middle East: Iran.
Extralimital distribution: Azerbaijan.
Host records: Unknown.

Orgilus (Orgilus) nitidus Marshall, 1898
Catalogs with Iranian records: No catalog.
Distribution in Iran: East Azarbaijan (Samin, Papp, & Coronado-Blanco, 2018).
Distribution in the Middle East: Iran.
Extralimital distribution: Azerbaijan, Hungary, Mongolia, Russia, Spain.
Host records: Unknown.

Orgilus (Orgilus) obscurator (Nees von Esenbeck, 1812)

Catalogs with Iranian records: Fallahzadeh and Saghaei (2010), Farahani et al. (2016), Yu et al. (2016), Samin, Coronado-Blanco, Kavallieratos et al. (2018), Samin, Coronado-Blanco, Fischer et al. (2018), Gadallah et al. (2019).

Distribution in Iran: Tehran (Modarres Awal, 1997, 2012), Iran (no specific locality cited) (Farahani et al., 2014; Sabzevari, 1968).

Distribution in the Middle East: Iran (see references above), Turkey (Beyarslan & Çetin Erdoğan, 2011; Beyarslan et al., 2013; Güçlü & Özbek, 2015).

Extralimital distribution: Armenia, Azerbaijan, Canada (introduced), Chile (introduced), China, Croatia, Czech Republic, Finland, France, Germany, Hungary, Ireland, Italy, Kazakhstan, Latvia, Lithuania, North Macedonia, Moldava, Mongolia, Netherlands, Norway, Poland, Russia, Slovakia, Slovenia, Sweden, Switzerland, Ukraine, United Kingdom, United States of America (introduced), former Yugoslavia.

Host records: Summarized by Yu et al. (2016) as being a parasitoid of the coleophorids *Coleophora* spp.; the crambid *Loxostege sticticalis* (L.); the depressariids *Agonopterix conterminella* (Zeller), and *Agonopterix kaekeritziana* (L.); the gelechiids *Aproaerena anthyllidella* (Hübner), *Dichomeris juniperella* (L.), *Exoteleia dodecella* (L.), *Recurvia nanella* (Denis and Schiffermüller), *Scrobipalpa acuminatella* (Sircom), and *Scrobipalpa ocellatella* (Boyd); the lasiocampid *Dendrolimus pini* (L.); the momphids *Mompha epilobiella* (Denis and Schiffermüller), and *Mompha miscella* (Denis and Schiffermüller); the psychid *Phalacropterix graslinella* (Boisduval); the scythridid *Scythris picaepennis* (Haworth); the tortricids *Epinotia cruciana* (L.), *Gypsonoma aceriana* (Duponchel), *Lathronympha strigana* (Fabricius), *Rhyacionia* spp., *Stictea mygindiana* (Denis and Schiffermüller), and *Tortrix viridana* L.; and the yponomeutid *Yponomeuta evonymella* (L.). In Iran, this species has been reared from the gelechiid *Recurvaria nanella* (Denis and Schiffermüller) (Modarres Awal, 1997, 2012; Sabzevari, 1968).

Orgilus (Orgilus) pimpinellae Niezabitowski, 1910

Catalogs with Iranian records: Farahani et al. (2016), Yu et al. (2016), Samin, Coronado-Blanco, Kavallieratos et al. (2018), Samin, Coronado-Blanco, Fischer et al. (2018), Gadallah et al. (2019).

Distribution in Iran: Golestan (Samin et al., 2015), Guilan, Qazvin (Farahani et al., 2014), Mazandaran (Ghahari et al., 2010; Ghahari & Sakenin, 2018).

Distribution in the Middle East: Iran (see references above), Turkey (Beyarslan & Çetin Erdoğan, 2011; Beyarslan et al., 2013; Güçlü & Özbek, 2015).

Extralimital distribution: Afghanistan, Austria, Bulgaria, Czech Republic, Germany, Greece, Hungary, Ireland, Italy, Kazakhstan, Korea, Lithuania, Moldova, Mongolia, Norway, Poland, Romania, Russia, Serbia, Switzerland, Ukraine, United Kingdom, Uzbekistan.

Host records: Summarized by Yu et al. (2016) as being a parasitoid of the acrolepiid *Digitivalva arnicella* (Denis and Schiffermüller); the coleophorid *Coleophora spiraecella* Rebel; the depressariid *Agonopterix bipunctosa* (Curtis); the gelechiids *Anacampsis populella* (Clerck), *Anacampsis temerella* (Lienig and Zeller), *Argolamprotes micella* (Denis and Schiffermüller), *Athrips pruinosella* (Lienig and Zeller), *Caryocolum tricolorella* (Haworth), *Dichomeris juniperella* (L.), *Phthorimaea operculella* (Zeller), *Recurvria nanella* (Denis and Schiffermüller), and *Scrobipalpa ocellatella* (Boyd); the momphid *Mompha miscella* (Denis and Schiffermüller); and the pyralid *Oncocera obductella* (Zeller).

Orgilus (Orgilus) ponticus Tobias, 1986

Catalogs with Iranian records: Farahani et al. (2016), Yu et al. (2016), Samin, Coronado-Blanco, Kavallieratos et al. (2018), Samin, Coronado-Blanco, Fischer et al. (2018), Gadallah et al. (2019).

Distribution in Iran: Golestan (Samin et al., 2015), West Azarbaijan (Ghahari & Fischer, 2011), Iran (no specific locality cited) (Farahani et al., 2014 as *O. puncticus*).

Distribution in the Middle East: Iran (Farahani et al., 2014), Turkey (Güçlü & Özbek, 2015; Taeger, 1989).

Extralimital distribution: Albania, Greece, Hungary, Italy, Russia, Slovenia.

Host records: Unknown.

Orgilus (Orgilus) priesneri Fischer, 1958

Catalogs with Iranian records: Farahani et al. (2016), Yu et al. (2016), Samin, Coronado-Blanco, Kavallieratos et al. (2018), Samin, Coronado-Blanco, Fischer et al. (2018), Gadallah et al. (2019).

Distribution in Iran: Fars (Lashkari Bod et al., 2010, 2011 as *Orgilus kazakhstanicus* Tobias, 1986), Iran (no specific locality cited) (Farahani et al., 2014).

Distribution in the Middle East: Egypt (Fischer, 1958; Taeger, 1989), Iran (see references above), Israel−Palestine (Papp, 2012; Taeger, 1989), Jordan, Saudi Arabia (Taeger, 1989).

Extralimital distribution: Kazakhstan.

Host records: Unknown.

Orgilus (Orgilus) punctiventris Tobias, 1976

Catalogs with Iranian records: Farahani et al. (2016), Yu et al. (2016), Samin, Coronado-Blanco, Kavallieratos et al. (2018), Samin, Coronado-Blanco, Fischer et al. (2018), Gadallah et al. (2019).

Distribution in Iran: Guilan (Farahani et al., 2014).

Distribution in the Middle East: Iran (Farahani et al., 2014), Turkey (Beyarslan & Çetin Erdoğan, 2011).

Extralimital distribution: Armenia, Azerbaijan.

Host records: Unknown.

Orgilus (Orgilus) punctulator (Nees von Esenbeck, 1812)

Catalogs with Iranian records: Yu et al. (2016), Samin, Coronado-Blanco, Kavallieratos et al. (2018), Samin, Coronado-Blanco, Fischer et al. (2018), Gadallah et al. (2019).

Distribution in Iran: Kordestan (Samin et al., 2016).

Distribution in the Middle East: Iran (Samin et al., 2016), Turkey (Beyarslan & Çetin Erdoğan, 2011; Güçlü & Özbek, 2015).

Extralimital distribution: Armenia, Azerbaijan, Bulgaria, Croatia, former Czechoslovakia, France, Germany, Hungary, Italy, Kazakhstan, Lithuania, Moldova, Mongolia, Netherlands, Poland, Russia, Sweden, Switzerland, United Kingdom, former Yugoslavia.

Host records: Summarized by Yu et al. (2016) as being a parasitoid of the following coleophorids: *Coleophora auricella* (Fabricius), *Coleophora follicularis* (Vallot), *Coleophora galbulipennella* Zeller, *Coleophora nigricella* (Stephens), *Coleophora saponariella* Heeger, and *Coleophora serratella* (L.), the pyralid *Apterona helicoidella* (Vallot), the tortricid *Ancyclis apicella* (Denis and Schiffermüller), and the yponomeutids *Yponomeuta padella* (L.), and *Yponomeuta malinellus* (Zeller).

Orgilus (Orgilus) radialiformis Beyarslan, 2011

Distribution in the Middle East: Turkey (Beyarslan & Çetin Erdoğan, 2011).

Extralimital distribution: None.

Host records: Unknown.

Orgilus (Orgilus) rudolphae Tobias, 1976

Catalogs with Iranian records: No catalog.

Distribution in Iran: Chaharmahal and Bakhtiari (Samin, Papp, & Coronado-Blanco, 2018).

Distribution in the Middle East: Iran (Samin, Papp, & Coronado-Blanco, 2018), Turkey (Beyarslan, 1996; Beyarslan & Çetin Erdoğan, 2011; Güçlü & Özbek, 2015).

Extralimital distribution: Former Czechoslovakia, Greece, Hungary, Kazakhstan, Russia, former Yugoslavia.

Host records: Recorded by Taeger (1989) as being a parasitoid of the gelechiids *Phthorimaea operculella* (Zeller), and *Scrobipalpa ocellatella* (Boyd).

Orgilus (Orgilus) rufigaster Tobias, 1964

Distribution in the Middle East: Turkey (Beyarslan & Çetin Erdoğan, 2011).

Extralimital distribution: China, Kazakhstan, Moldova.

Host records: Unknown.

Orgilus (Orgilus) rugosus (Nees von Esenbeck, 1834)

Catalogs with Iranian records: This species is new record for the fauna of Iran.

Distribution in Iran: Golestan province, Minudasht, 2♂, April 2014, ex *Coleophora serratella* (Linnaeus) (Lepidoptera: Coleophoridae).

Distribution in the Middle East: Iran (new record), Turkey (Beyarslan, 1996; Beyarslan & Çetin Erdoğan, 2011).

Extralimital distribution: Andorra, Austria, Belarus, Bulgaria, Croatia, former Czechoslovakia, Finland, France, Germany, Hungary, Italy, Kazakhstan, Lithuania, North Macedonia, Moldova, Poland, Romania, Russia, Slovakia, Slovenia, Ukraine, United Kingdom, former Yugoslavia.

Host records: Summarized by Yu et al. (2016) as being a parasitoid of the choreutid *Choreutes pariana* (Clerck), the coleophorids *Coleophora* spp., the tortricid *Rhyacionia buoliana* (Denis and Schiffermüller), and the zygaenid *Zygaena carniolica* (Scopoli). In Iran, this species has been reared from *Coleophora serratella* (L.) (present work).

Orgilus (Orgilus) similis Szépligeti, 1896

Catalogs with Iranian records: Gadallah et al. (2019).

Distribution in Iran: Kordestan (Ghahari, 2016), West Azarbaijan (Samin, Papp, & Coronado-Blanco, 2018).

Distribution in the Middle East: Iran (Ghahari, 2016; Samin, Papp, & Coronado-Blanco, 2018), Turkey (Beyarslan & Çetin Erdoğan, 2011).

Extralimital distribution: Bulgaria, Croatia, Hungary, Italy, Moldova, Mongolia, Russia.

Host records: Recorded by Györfi (1959) as being a parasitoid the psychid *Bijugis bombycella* (Denis and Schiffermüller).

Orgilus (Orgilus) simillimus Taeger, 1989

Distribution in the Middle East: Turkey (Beyarslan, 1996; Beyarslan & Çetin Erdoğan, 2011; Güçlü & Özbek, 2015).

Extralimital distribution: Germany, Greece, Italy, Switzerland.

Host records: Unknown.

Orgilus (Orgilus) sticticus Taeger, 1989

Distribution in the Middle East: Turkey (Beyarslan, 1996).

Extralimital distribution: Austria, Bulgaria, Germany, Hungary, Romania, Slovakia, Slovenia.

Host records: Summarized by Yu et al. (2016) as being a parasitoid of the coleophorids *Coleophora colutella* (Fabricius), and *Coleophora discordella* Zeller.

Orgilus (Orgilus) temporalis Tobias, 1976

Catalogs with Iranian records: Farahani et al. (2016), Yu et al. (2016), Samin, Coronado-Blanco, Kavallieratos et al. (2018), Samin, Coronado-Blanco, Fischer et al. (2018), Gadallah et al. (2019).

Distribution in Iran: Mazandaran (Farahani et al., 2014).

Distribution in the Middle East: Iran (Farahani et al., 2014), Turkey (Beyarslan, 1996; Beyarslan & Çetin Erdoğan, 2011; Güçlü & Özbek, 2015).

Extralimital distribution: Azerbaijan, Czech Republic, Finland, Germany, Hungary, Mongolia, Romania, Russia, Switzerland.
Host records: Unknown.

Orgilus (Orgilus) tobiasi Taeger, 1989

Catalogs with Iranian records: Fallahzadeh and Saghaei (2010), Farahani et al. (2016), Yu et al. (2016), Samin, Coronado-Blanco, Kavallieratos et al. (2018), Samin, Coronado-Blanco, Fischer et al. (2018), Gadallah et al. (2019).
Distribution in Iran: Tehran (Taeger et al., 2005), Iran (no specific locality cited) (Farahani et al., 2014; Güçlü & Özbek, 2015; Taeger, 1989).
Distribution in the Middle East: Iran (see references above), Turkey (Güçlü & Özbek, 2015).
Extralimital distribution: Albania, Armenia, Czech Republic, Germany, Greece, Hungary, Ireland, Italy, Romania, Serbia, Spain, Switzerland, United Kingdom.
Host records: Unknown.

Orgilus (Orgilus) turkmenus Telenga, 1933

Distribution in the Middle East: Egypt (Taeger, 1989).
Extralimital distribution: Afghanistan, Turkmenistan.
Host records: Unknown.
Species excluded from the fauna of Iran.

Species formerly erroneously reported in the Middle East which are now excluded

Orgilus (Orgilus) jennieae Marsh, 1979

Catalogs with Iranian records: Fallahzadeh and Saghaei (2010), Yu et al. (2016), Samin, Coronado-Blanco, Kavallieratos et al. (2018), Samin, Coronado-Blanco, Fischer et al. (2018).
Distribution in Iran: Iran (no specific locality cited) (Khanjani, 2006).
General distribution: Costa Rica (Marsh, 1979; Yu et al., 2016).
Host records in Iran: *Phthorimaea operculella* (Zeller) (Lepidoptera: Gelechiidae) (Khanjani, 2006).
Comments: *Orgilus (Orgilus) jennieae* was excluded from the fauna of Iran by Farahani et al. (2016) and Gadallah et al. (2019). This species was previously reported from the Neotropical region (Marsh, 1979).

Conclusion

Thirty-four valid species of the subfamily Orgilinae in two genera, *Kerorgilus* van Achterberg, 1985 and *Orgilus* Haliday, 1833, have been reported from seven of the Middle East countries (Cyprus, Egypt, Iran, Israel–Palestine, Jordan, Saudi Arabia, and Turkey), which among them, Turkey with 25 reported species is the richest. Iranian

Orgilinae comprises 20 recorded species in two genera, *Kerorgilus* (one species) and *Orgilus* (19 species). These species have been recorded from 13 provinces: Tehran (with six species), Guilan (with five species), Alborz, Golestan, Kordestan, Mazandaran, West Azarbaijan (each with three species), East Azarbaijan, Qazvin (both with two species), Ardabil, Chaharmahal and Bakhtiari, Fars, and Southern Khorasan (each with one species). Host species have been determined for only two Iranian Orgilinae species, and these hosts belong to two families of Lepidoptera (Coleophoridae and Gelechiidae). Comparison of the Orgilinae fauna of Iran with the Middle East and adjacent countries to Iran indicates the fauna of Russia with 28 species is more diverse than Iran, followed by Turkey (25 species), Kazakhstan (11 species), Azerbaijan (eight species), Armenia (five species), Egypt, Israel–Palestine (each with three species), Afghanistan and Jordan (each with two species), Cyprus, Pakistan, Saudi Arabia, and Turkmenistan (each with one species). No orgiline species are reported from Bahrain, Kuwait, Iraq, Lebanon, Oman, Qatar, Syria, United Arab Emirates, or Yemen. Additionally, among the adjacent countries to Iran, Russia and Turkey share 12 species with Iran, followed by Azerbaijan and Kazakhstan (both with eight species), Armenia (five species), Egypt and Israel–Palestine (both with three species), Jordan (two species), and Afghanistan, Cyprus, and Saudi Arabia (each with one species).

References

Beyarslan, A. (1996). Die *Orgilus*-Arten der Turkei (Hymenoptera: Braconidae: Orgilinae). *Entomofauna, 22*, 353–360.

Beyarslan, A., & Çetin Erdoğan, Ö. (2011). A study of Orgilinae and Microtypinae (Hymenoptera: Braconidae) from Turkey, with the description of a new species. *Biologia, 66*(1), 121–129.

Braet, Y., & Quicke, D. L. J. (2004). A phylogenetic analysis of the Mimagathidini with revisionary notes on the genus *Stantonia* Ashmead, 1904 (Hymenoptera: Braconidae: Orgilinae). *Journal of Natural History, 38*, 1489–1589.

Beyarslan, A., Gözüaçik, C., & Özgen, I. (2013). A contribution on the subfamilies Helconinae, Homolobinae, Macrocentrinae, Meteorinae, and Orgilinae (Hymenoptera: Braconidae) of southeastern Anatolia with new records from other parts of Turkey. *Turkish Journal of Zoology, 37*(4), 501–505.

Chen, X. X., & van Achterberg, C. (2019). Systematics, phylogeny, and evolution of braconid wasps: 30 years of progress. *Annual Review of Entomology, 64*, 1–24.

Fallahzadeh, M., & Saghaei, N. (2010). Checklist of Braconidae (Insecta: Hymenoptera) from Iran. *Munis Entomology & Zoology, 5*(1), 170–186.

Farahani, S., Talebi, A. A., van Achterberg, C., & Rakhshani, E. (2014). A taxonomic study of Orgilinae and Microtypinae from Iran (Hymenoptera, Braconidae). *Spixiana, 37*(1), 93–102.

Farahani, S., Talebi, A. A., & Rakhshani, E. (2016). Iranian Braconidae (Insecta: Hymenoptera: Ichneumonoidea): Diversity, distribution and host association. *Journal of Insect Biodiversity and Systematics, 2*(1), 1–92.

Fischer, M. (1958). Drei neue Braconiden (Hymenoptera). *Entomologisches Nachrichtenblatt Oesterreichischer und Schweizer Entomologen, 10*, 33–37.

Gadallah, N. S., Ghahari, H., & Kavallieratos, N. G. (2019). An annotated catalogue of the Iranian Charmontinae, Ichneutinae, Macrocentrinae and Orgilinae (Hymenoptera: Braconidae). *Journal of the Entomological Research Society, 21*(3), 333–354.

Ghahari, H. (2016). Five new records of Iranian Braconidae (Hymenoptera: Ichnemonoidea) for Iran and annotated catalogue of the subfamily Homolobinae. *Wuyi Science Journal, 32*, 35–43.

Ghahari, H. (2020). A study on the fauna of predator and parasitoid arthropod of saffron fields (*Crocus sativus* L.). *Journal of Saffron Research, 7*(2), 203–215 (in Persian, English summary).

Ghahari, H., & Fischer, M. (2011). A contribution to the Braconidae (Hymenoptera: Ichneumonoidea) from north-western Iran. *Calodema, 134*, 1–6.

Ghahari, H., & Sakenin, H. (2018). Species diversity of Chalcidoidea and Ichneumonoidea (Hymenoptera) in some paddy fields and surrounding grasslands in Mazandaran and Guilan provinces, northern Iran. *Applied Plant Protection, 7*(1), 11–19 (in Persian, English summary).

Ghahari, H., Fischer, M., Çetin Erdoğan, Ö., Beyarslan, A., & Havaskary, M. (2009). A contribution to the knowledge of the braconid-fauna (Hymenoptera, Ichneumonoidea, Braconidae) of Arasbaran, Northwestern Iran. *Entomofauna, 30*, 329–336.

Ghahari, H., Fischer, M., Çetin Erdoğan, Ö., Beyarslan, A., & Ostovan, H. (2010). A contribution to the braconid wasps (Hymenoptera: Braconidae) from the forests of northern Iran. *Linzer biologische Beiträge, 42*(1), 621–634.

Güçlü, C., & Özbek, H. (2015). A study of Orgilinae (Hymenoptera: Braconidae) from Turkey with new records. *Journal of the Entomological Research Society, 17*(1), 61–69.

Györfi, J. (1959). Beiträge zur kenntnis der Wirte verschiendener Braconiden-Arten (Hymenoptera, Braconidae). *Acta Zoologica Hungarica, 5*, 49–65.

Khanjani, M. (2006). *Vegetable pests in Iran* (2nd ed.). Bu-Ali Sina University, 467 pp. (in Persian).

Kian, N., Goldasteh, S., & Farahani, S. (2020). A survey on abundance and species diversity of Braconid wasps in forest of Mazandaran province. *Journal of Entomological Research, 12*(1), 61–69 (in Persian, English summary).

Kula, R. R., Boughton, A. J., & Pemberton, R. W. (2009). *Stantonia pallida* (Ashmead) (Hymenoptera: Braconidae) reared from *Neomusotima conspurcatalis* Warren (Lepidoptera: Crambidae), a classical biological control agent of *Lygodium microphyllum* (Cav.) R. Br. (Polypodiales: Lygodiaceae). *Proceedings of the Entomological Society of Washington, 112*(1), 61–68.

Lashkari, B. A., Rakhshani, E., Talebi, A. A., & Lozan, A. (2010). Introduction of twelve newly recorded species of Braconidae (Hymenoptera) from Iran. In *Proceedings of the 19th Iranian Plant Protection Congress, 31 July – 3 August 2010* (p. 161). Iranian Research Institute of Plant Protection.

Lashkari Bod, A., Rakhshani, E., Talebi, A. A., Lozan, A., & Žikić, V. (2011). A contribution to the knowledge of Braconidae (Hym., Ichneumonoidea) of Iran. *Biharean Biologist, 5*(2), 147–150.

Marsh, P. M. (1979). Description of new Braconidae (Hymenoptera) parasitic on the potato tuberworm and related Lepidoptera from Central and South America. *Journal of the Washington Academy of Sciences, 69*(1), 12–17.

Modarres Awal, M. (1997). Family Braconidae (Hymenoptera). In M. Modarres Awal (Ed.), *List of agricultural pests and their natural enemies in Iran* (2nd ed., pp. 265–267). Ferdowsi University of Mashhad Press, 429 pp.

Modarres Awal, M. (2012). Family Braconidae (Hymenoptera). In M. Modarres Awal (Ed.), *List of agricultural pests and their natural enemies in Iran* (3rd ed., pp. 483–486). Ferdowsi University of Mashhad Press, 759 pp.

Muesebeck, C. F. (1938). Two reared north American species of the genus *Stantonia* Ashmead (Hymenoptera: Braconidae). *Proceedings of the Entomological Society of Washington, 40*, 89–91.

Muesebeck, C. F. W. (1970). The Nearctic species of *Orgilus* Haliday (Hymenoptera: Braconidae). *Smithsonian Contributions to Zoology, 30*, 1–104.

Papp, J. (1994). Contribution to the braconid fauna of Hungary, 10. Homolobinae, Macrocentrinae, Orgilinae, and Microtypinae (Hymenoptera: Braconidae). *Folia Entomologica Hungarica, 55*, 287–304.

Papp, J. (2012). A contribution to the braconid fauna of Israel (Hymenoptera: Braconidae), 3. *Israel Journal of Entomology, 41–42*, 165–219.

Sabzevari, A. (1968). Lepidopterous pest on apricot. In *Proceedings of the 1st Iranian Plant Protection Congress, 14–19 September 1968* (pp. 63–80). Karaj: Karaj Junior College of Agriculture.

Samin, N., Ghahari, H., Gadallah, N. S., & Monaem, R. (2015). A study on the braconid wasps (Hymenoptera: Ichneumonoidea: Braconidae) from Golestan province, northern Iran. *Linzer biologische Beiträge, 47*(1), 731–739.

Samin, N., van Achterberg, C., & Çetin Erdoğan, Ö. (2016). A faunistic study on some subfamilies of Braconidae (Hymenoptera: Ichneumonoidea) from Iran. *Arquivos Entomolóxicos, 15*, 153–161.

Samin, N., Coronado-Blanco, J. M., Kavallieratos, N. G., Fischer, M., & Sakenin, H. (2018). Recent findings on Braconidae (Hymenoptera: Ichneumonoidea) of Iran with an updated checklist. *Acta Biologica Turcica, 31*(4), 160–173.

Samin, N., Coronado-Blanco, J. M., Fischer, M., van Achterberg, C., Sakenin, H., & Davidian, E. (2018). Updated checklist of Iranian Braconidae (Hymenoptera: Ichneumonoidea) with twenty-three new records. *Natura Somogyiensis, 32*, 21–36.

Samin, N., Papp, J., & Coronado-Blanco, J. M. (2018). A faunistic study on braconid wasps (Hymenoptera: Ichneumonoidea: Braconidae) of Iran. *Scientific Bulletin of the Uzhgorod University (Series: Biology), 45*, 15–19.

Sharanowski, B. J., Dowling, A. P. G., & Sharkey, M. J. (2011). Molecular phylogenetics of Braconidae (Hymenoptera: Ichneumonoidea), based on multiple nuclear genes, and implications for classification. *Systematic Entomology, 36*, 549–572.

Sharanowski, B. J., Zhang, Y. M., & Wanigasekara, R. W. U. M. (2014). Annotated checklist of Braconidae (Hymenoptera) in the Canadian Prairies Ecozone. In D. J. Giberson, & H. A. Cárcamo (Eds.), *Arthropods of Canadian grasslands. Volume 4.* Biodiversity and systematics. *Part 1* (pp. 399–425). Biological Survey of Canada, 479.

Shaw, M. R., & Huddleston, T. (1991). Classification and biology of braconid wasps (Hymenoptera: Braconidae). *Handbooks for the Identification of British Insects, 7*(11), 1–126.

Taeger, A. (1989). *Die Orgilus-Arten der Paläarktis (Hymenoptera, Braconidae).* Arbeit aus dem Institut für Pflanzenschutzforschung Kleinmachnow, Bereich Eberswalde. Akademie der Landwirtschaftswissenschaften der DDR, 260 pp.

Taeger, A., Gaedike, H., & Blank, S. M. (2005). Katalog der primären Hymenopteren-Typen des DEI. *Beiträge zur Entomologie, 55*(1), 151−250.

van Achterberg, C. (1985). I. *Kerorgilus* gen. nov. a new genus of Orgilinae (Hym., Braconidae) from the Palaearctic region. *Zoologische Mededelingen, 59*(15), 163−167.

van Achterberg, C. (1987). Revisionary notes on the subfamily Orgilinae (Hymenoptera: Braconidae). *Zoologische Verhandlingen, 242,* 1−111.

van Achterberg, C. (1992). Revisionary notes on the subfamily Homolobinae (Hymenoptera: Braconidae). *Zoologische Mededelingen, 66*(25), 359−368.

van Achterberg, C. (1993). Illustrated key to the subfamilies of the Braconidae. *Zoologische Verhandelingen, 283,* 1−189.

van Achterberg, C. (1994). Two new genera of the tribe Orgilini Ashmead (Hymenoptera: Braconidae: Orgilinae). *Zoologische Mededelingen, 68*(16), 173−190.

van Achterberg, C., & Quicke, D. L. J. (1992). Phylogeny of the subfamilies of the family Braconidae: a reassessment assessed. *Cladistics, 8,* 237−264.

van Achterberg, C., Long, K. D., & Chen, X.-X. (2017). Review of *Stantonia* Ashmead (Hymenoptera, Braconidae, Orgilinae) from Vietnam, China, Japan, and Russia, with description of six new species. *ZooKeys, 713,* 61−119.

Whitfield, J. B., & Wagner, D. L. (1991). Annotated key to the genera of Braconidae (Hymenoptera) attacking leafmining Lepidoptera in the Holarctic region. *Journal of Natural History, 25,* 733−754.

Yu, D. S., van Achterberg, C., & Horstmann, K. (2016). *Taxapad 2016, Ichneumonoidea 2015, Database on flash-drive.* Nepean, Ontario, Canada.

Pambolus sp. (Pambolinae), ♀, lateral habitus. *Photo prepared by S.R. Shaw.*

Pambolus sp. (Pambolinae), ♀, lateral habitus. *Photo prepared by S.R. Shaw.*

Chapter 26

Subfamily Pambolinae Marshall, 1885

Scott Richard Shaw[1], Hassan Ghahari[2], Neveen Samy Gadallah[3] and Donald L.J. Quicke[4]

[1]UW Insect Museum, Department of Ecosystem Science and Management, University of Wyoming, Laramie, WY, United States; [2]Department of Plant Protection, Yadegar-e Imam Khomeini (RAH) Shahre Rey Branch, Islamic Azad University, Tehran, Iran; [3]Entomology Department, Faculty of Science, Cairo University, Giza, Egypt; [4]Integrative Ecology Laboratory, Department of Biology, Faculty of Science, Chulalongkorn University, Pathumwan, Bangkok, Thailand

Introduction

Pambolinae is a relatively small cosmopolitan subfamily of the family Braconidae, with currently 70 valid species in nine genera (Ahmad et al., 2019; Aguilera-Uribe et al., 2018; Yu et al., 2016). They form a part of cyclostome group of braconids (Aguilera-Uribe et al., 2018). They are often included in the Hormiinae (Wharton, 1993; Whitfield & Wharton, 1997); however, the shape of venom apparatus resulted in treating it as a separate subfamily (Quicke et al., 1992). In a systematic study based on molecular and morphological data, carried out by Zaldívar-Riverón et al. (2006), it occupied a basal position near to Rogadinae (s.l.). Pambolines are diagnosed by the following combination of characters: labrum sculptured; propodeum with two lateral spines, if absent, then first metasomal tergite greatly widened posteriorly; hypoclypeal depression present; labrum usually nearly flattened; metasomal first and second tergites with large flap-like epipleura (van Achterberg, 1993, 1995; Wharton, 1993).

Little is known about the biology of Pambolinae; some species of Pambolus have been reared from beetle-infested wood and Chremylus elaphus is a gregarious idiobiont ectoparasitoid of the case bearing larval Lepidoptera (Quicke, 2015; Shaw & Huddleston, 1991). The hosts in most cases are unknown (Braet & van Achterberg, 2003; Martínez et al., 2012; van Achterberg & Braet, 2004). Among Coleoptera, some records indicate that chrysomelid larvae are the main hosts (Shaw & Huddleston, 1991; Zaldívar-Riverón & Quicke, 2002).

The genus Pambolus Haliday is encountered from nearly all biogeographical regions, and is particularly diverse in the Neotropics (Whitfield & Wharton 1997; Yu et al., 2016). It currently comprises 42 recognized species (Martínez et al., 2012; Yu et al., 2016). It has been included in the subfamily Hormiinae by some authors (e.g., Whitfield & Wharton, 1997) or in the small subfamily Pambolinae (Aguilera-Uribe et al., 2018; Braet and van Achterberg, 2003; Martínez et al., 2012; van Achterberg, 1995; van Achterberg & Braet, 2004; Yu et al., 2016), which is presently more preferred based on previous phylogenetic studies using DNA analysis (Belshaw et al., 2000; Zaldívar-Riverón et al., 2006). The studies revealed that the Pambolinae are a more derived group than the group of subfamilies Hormiinae, Lysiterminae, and Betylobraconinae (Belshaw et al., 2000).

We include Avga here, but its true position remains uncertain. It was recently excluded from Rhyssalinae and Hormiinae (Quicke et al., 2020).

Checklists of Regional Pambolinae. Fallahzadeh and Saghaei (2010), Yu et al. (2016), Gadallah and Ghahari (2016), Farahani et al. (2016), and Samin, Coronado-Blanco, Kavallieratos et al. (2018) all represented one species, Pambolus (Phaenodus) pallipes (Foerster, 1863), and Samin, Coronado-Blanco, Fischer et al. (2018) listed two species in two genera. The present Middle Eastern checklist includes five species in three genera, Avga Nixon, Pambolus Haliday, and Phaenodus Foerster.

List of species of the subfamily Pambolinae recorded in the Middle East

Subfamily Pambolinae Marshall, 1885

Genus *Avga* Nixon, 1940

Avga blaciformis (Hedwig, 1961)

Catalogs with Iranian records: Farahani et al. (2016 as *Pseudobiosteres blaciformis* Hedwig).
Distribution in Iran: Kerman, Southern Khorasan (Samin et al., 2011 as *Psudobiosteres blaciformis*).
Distribution in the Middle East: Iran.

Extralimital distribution: Afghanistan.
Host records: Unknown.

Avga imperfectus (Hedwig, 1961)

Catalogs with Iranian records: Gadallah et al. (2016 as *Pseudobiosteres imperfectus* Hedwig), Farahani et al. (2016 as *Pseudobiosteres imperfectus* Hedwig), Samin, Coronado-Blanco, Kavallieratos et al. (2018) and Samin, Coronado-Blanco, Fischer et al. (2018 as *Pseudobiosteres imperfectus*).
Distribution in Iran: Fars (Samin et al., 2015 as *Pseudobiosteres imperfectus*), Razavi Khorasan (Samin et al., 2011 as *Pseudobiosteres imperfectus*).
Distribution in the Middle East: Iran.
Extralimital distribution: Afghanistan.
Host records: Unknown.

Avga sinaitica Edmardash and Gadallah, 2020

Distribution in the Middle East: Egypt (Edmardash et al., 2020).
Extralimital distribution: None.
Host records: Unknown.

Genus Pambolus Haliday, 1836
Pambolus biglumis (Haliday, 1836)

Catalogs with Iranian records: Samin, Coronado-Blanco, Fischer et al. (2018).
Distribution in Iran: Chaharmahal and Bakhtiari, Mazandaran (Sakenin et al., 2018).
Distribution in the Middle East: Iran.
Extralimital distribution: Belgium, France, Germany, Hungary, Kazakhstan, Moldova, Mongolia, Netherlands, Romania, Russia, Spain, United Kingdom.
Host records: Unknown.

Genus Phaenodus Foerster, 1863
Phaenodus pallipes Foerster, 1863

Catalogs with Iranian records: Fallahzadeh and Saghaei (2010), Farahani et al. (2016), Gadallah and Ghahari (2016) [all as *Pambolus* (*Phaenodus*) *pallipes* (Foerster, 1863)], Yu et al. (2016), Samin, Coronado-Blanco, Kavallieratos et al. (2018), Samin, Coronado-Blanco, Fischer et al. (2018).
Distribution in Iran: Iran (no specific locality cited) (Belokobylskij, 1998 as *Pambolus* (*Phaenodus*) *pallipes*).
Distribution in the Middle East: Iran (Belokobylskij, 1998), Turkey (Belokobylskij, 1986; Beyarslan, 2015).
Extralimital distribution: Bulgaria, Czech Republic, Finland, Germany, Hungary, Italy, Japan, Kazakhstan, Korea, Lithuania, Moldova, Russia, Slovenia, Spain, Sweden, Tajikistan, Ukraine, United Kingdom, former Yugoslavia.

Host records: Unknown.

Phaenodus rugulosus Hellén, 1927

Catalogs with Iranian records: No catalog.
Distribution in Iran: West Azarbaijan (Samin et al., 2020).
Distribution in the Middle East: Iran.
Extralimital distribution: Former Czechoslovakia, Finland, Germany, Kazakhstan, Lithuania, Moldova, Mongolia, Russia, Sweden.
Host records: Unknown.

Conclusion

Six valid species of the subfamily Pambolinae in three genera (*Avga* Nixon, 1940; *Pambolus* Haliday, 1936; and *Phaenodus* Foerster, 1863) have been reported from the Middle East countries. Of these, Iran was found to be the most diverse country, comprising five species in three genera. The five species have been recorded from seven provinces, Chaharmahal and Bakhtiari, Fars, Kerman, Mazandaran, Razavi Khorasan, Southern Khorasan, and West Azarbaijan. The exact locality of *Phaenodus pallipes* Foerster, 1863 is not specified. No host species have been recorded for these wasps in Iran. Among the Middle East and adjacent countries to Iran, five countries share species with Iran: Afghanistan, Kazakhstan and Russia (each with two species), Egypt and Turkey (both with one species).

References

Ahmad, Z., Ghramh, H. A., & Ansari, A. (2019). Two new species of braconid wasps (Hymenoptera, Braconidae) from India. *ZooKeys, 889*, 23−35.

Aguilera-Uribe, M., Martínez, J. J., & Zaldívar-Riverón, A. (2018). Three new species of *Pambolus* (Braconidae: Pambolinae) from Mexico, with comments on the variation of *P. oblongispina* Papp. *Zootaxa, 4377*(1), 125−137.

Belokobylskij, S. A. (1986). A review of the Palearctic species of the genera *Pambolus* Hal. and *Dimerus* Ruthe (Hymenoptera: Braconidae). *Trudy Zoologicheskogo Instituta, 159*, 18−37.

Belokobylskij, S. A. (1998). 1. Rhyssalinae, 2. Doryctinae, 3. Histeromerinae, 4. Exothecinae, 7. Gnamptodontinae, 9. Alysiinae (Alysiini), 10. Helconinae, 11. Cenocoeliinae, 12. Brachistinae, 14. Meteorideinae, 16. Xiphozelinae, 17. Homolobinae, 18. Charmontinae, 19. Orgilinae, 20. Ecnomiinae, 21. Sigalphinae, 23. Ichneutinae, 25. Cardiochilinae, 27. Dirrhopinae, 28. Miracinae, 29. Adeliinae. In P. A. Lehr (Ed.), *Neuropteroidea, Mecoptera, Hymenoptera. Part 3: Vol. 4. Key to the insects of Russian far East*. Vladivostok, St. Petersburg: Nauka, 706 pp. (pp. 41−162, 163−298, 411−520, 531−558).

Belshaw, R., Dowton, M., & Quicke, D. L. J. (2000). Estimating ancestral geographic distribution: A Gondwanan origin for a group of principally north temperate aphid parasitoids. *Proceedings of the Royal Society, B267*(1442), 491−496.

Beyarslan, A. (2015). A faunal study of the subfamily Doryctinae in Turkey (Hymenoptera: Braconidae). *Turkish Journal of Zoology, 39*(1), 126–143.

Braet, Y., & van Achterberg, C. (2003). New species of *Pambolus* Haliday and *Phaeonocarpa* Foerster (Hymenoptera: Braconidae: Pambolinae, Alysiinae) from French Guyana, Suriname and Panama. *Zoologische Mededelingen, 77*(7), 153–179.

Edmardash, Y. A., Abu El-Ghiet, U. M., & Gadallah, N. S. (2020). First record of Hormiini Foerster, 1863, and Macrocentrinae Foerster, 1863 (Hymenoptera: Braconidae) for the fauna of Egypt, with the description of a new species. *Zootaxa, 4722*(6), 555–570.

Fallahzadeh, M., & Saghaei, N. (2010). Checklist of Braconidae (Insecta: Hymenoptera) from Iran. *Munis Entomology & Zoology, 5*(1), 170–186.

Farahani, S., Talebi, A. A., & Rakhshani, E. (2016). Iranian Braconidae (Insecta: Hymenoptera: Ichneumonoidea): Diversity, distribution and host association. *Journal of Insect Biodiversity and Systematics, 2*(1), 1–92.

Gadallah, N. S., & Ghahari, H. (2016). An updated checklist of the Iranian Miracinae, Pambolinae and Sigalphinae (Hymenoptera: Braconidae). *Orsis, 30*, 51–61.

Martínez, J. J., Ceccarelli, F. S., & Zaldívar-Riverón, A. (2012). Two new species of *Pambolus* (Hymenoptera: Braconidae) from Jamaica. *Journal of Hymenoptera Research, 24*, 85–93.

Quicke, D. L. J. (2015). *Braconid and Ichneumonoid parasitoid wasps: biology, systematics, Evolution and Ecology.* Hobken, NJ: Wiley Blackwell, 681 pp.

Quicke, D. L. J., Tunstead, J., Falcó, J. V., & Marsh, P. M. (1992). Venom gland and reservoir morphology in the Doryctinae and related braconid wasps (Insects, Hymenoptera, Braconidae). *Zoologica Scripta, 21*, 403–416.

Quicke, D. L. J., Belokobylskij, S. A., Braet, Y., van Achterberg, C., Hebert, P. D. N., Prosser, S. W. J., Austin, A. D., Fagan-Jeffries, E. P., Ward, D. F., Shaw, M. R., & Butcher, B. A. (2020). Phylogenetic reassignment of basal cyclostome braconid parasitoid wasps (Hymenoptera) with description of a new, enigmatic Afrotropical tribe with a highly anomalous 28S D2 secondary structure. *Zoological Journal of the Linnean Society, XX*, 1–18.

Sakenin, H., Coronado-Blanco, M., Samin, N., & Fischer, M. (2018). New records of Braconidae (Hymenoptera) from Iran. *Far Eastern Entomologist, 362*, 13–16.

Samin, N., Sakenin, H., Imani, S., & Shojai, M. (2011). A study on the Braconidae (Hymenoptera) of Khorasan province and vicinity, Northeastern Iran. *Phegea, 39*(4), 137–143.

Samin, N., van Achterberg, C., & Ghahari, H. (2015). A faunistic study of Braconidae (Hymenoptera: Ichneumonoidea) from southern Iran. *Linzer biologische Beiträge, 47*(2), 1801–1809.

Samin, N., Coronado-Blanco, J. M., Kavallieratos, N. G., Fischer, M., & Sakenin, H. (2018). Recent findings on Braconidae (Hymenoptera: Ichneumonoidea) of Iran with an updated checklist. *Acta Biologica Turcica, 31*(4), 160–173.

Samin, N., Coronado-Blanco, J. M., Fischer, M., van Achterberg, C., Sakenin, H., & Davidian, E. (2018). Updated checklist of Iranian Braconidae (Hymenoptera: Ichneumonoidea) with twenty-three new records. *Natura Somogyiensis, 32*, 21–36.

Samin, N., Beyarslan, A., Coronado-Blanco, J. M., & Navaeian, M. (2020). A contribution to the braconid wasps (Hymenoptera: Braconidae) from Iran. *Natura Somogyiensis, 35*, 25–28.

Shaw, M. R., & Huddleston, T. (1991). Classification and biology of braconid wasps (Hymenoptera: Braconidae). *Handbooks for the Identification of British Insects, 7*, 1–126.

van Achterberg, C. (1993). Illustrated key to the subfamilies of the Braconidae. *Zoologische Verhandelingen, 283*, 1–189.

van Achterberg, C. (1995). Generic revision of the subfamily Betylobraconinae (Hymenoptera; Braconidae) and other groups with modified fore tarsus. *Zoologische Verhandelingen, 298*, 1–242.

van Achterberg, C., & Braet, Y. (2004). Two new species of *Pambolus* Haliday (Hymenoptera: Braconidae: Pambolinae) from Argentina. *Zoologische Mededelingen, 78*(22), 337–344.

Wharton, R. A. (1993). Review of Hormiini (Hymenoptera; Braconidae) with a description of new taxa. *Journal of Natural History, 27*, 107–171.

Whitfield, J. B., & Wharton, R. A. (1997). Hormiinae, pp. 285–301. In R. A. Wharton, P. M. Marsh, & M. J. Sharkey (Eds.), *Manual of the New World Genera of the Family Braconidae (Hymenoptera).* Washington, DC: Special Publication of the International Society of Hymenopterists, 439 pp.

Yu, D. S., van Achterberg, C., & Horstmann, K. (2016). *Taxapad 2016, Ichneumonoidea 2015, Database on flash-drive, Nepean, Ontario, Canada.*

Zaldívar-Riverón, A., & Quicke, L. J. (2002). First host record for the parasitic wasp genus *Notiopambolus* Achterberg and Quicke (Hymenoptera: Braconidae: Pambolinae). *Journal of Hymenoptera Research, 11*(2), 370–471.

Zaldívar-Riverón, A., Mori, M., & Quicke, D. L. J. (2006). Systematics of the cyclostome subfamilies of braconid parasitic wasps (Hymenoptera: Ichneumonoidea): A simultaneous molecular and morphological Bayesian approach. *Molecular Phylogenetics and Evolution, 38*(1), 130–145.

Proterops nigripennis Wesmael, 1835 (Proteropinae), ♀, lateral habitus. *Photo prepared by S.R. Shaw.*

Proterops sp. (Proteropinae), ♀, dorsal habitus. *Photo prepared by S.R. Shaw.*

Chapter 27

Subfamily Proteropinae van Achterberg, 1976

Neveen Samy Gadallah[1], Hassan Ghahari[2], Scott Richard Shaw[3] and Michael J. Sharkey[4]

[1]*Entomology Department, Faculty of Science, Cairo University, Giza, Egypt;* [2]*Department of Plant Protection, Yadegar-e Imam Khomeini (RAH) Shahre Rey Branch, Islamic Azad University, Tehran, Iran;* [3]*Department of Ecosystem Science and Management, University of Wyoming, Laramie, WY, United States;* [4]*UW Insect Museum, Department of Entomology, University of Kentucky, Lexington, KY, United States*

Introduction

The Proteropinae is a small subfamily in the family Braconidae (Chen & van Achterberg, 2019). It is rarely collected subfamily, occurring in the New World, Palaearctic, and Oriental regions (van Achterberg, 1976). It currently comprises eight species classified into five genera; however, only *Proterops* Wesmael is found outside the New World (Yu et al., 2016). *Proterops* was traditionally included in the subfamily Ichneutinae and van Achterberg (1976) proposed the tribe Proteropini for *Proterops* and allied genera (Sharkey & Wharton, 1994; van Achterberg, 1976; Wharton & van Achterberg, 2000). However, several studies suggest that proteropines belong to a different clade than the Ichneutinae based on both morphological data (Quicke & van Achterberg, 1990) and molecular analyses (Belshaw & Quicke, 2002; Li et al., 2016; Sharanowski et al., 2011).

Members of the subfamily Proteropinae are characterized by the following combination of characters: occipital carina absent; anterior tentorial pits very large; the presence of a minute, smooth, elliptical tubercle anteriorly on the subalar depression; vein 1-M of fore wing gently curved; radial cell of the fore wing short and comparatively high; disc of T1 not differentiated from the related epipleuron; and first metasomal tergite flattened basally and laterally, without dorsal carinae (van Achterberg, 1976, 1993). As far as is known, proteropines are koinobiont endoparasitoids of argid sawfly larvae (Tenthredinoidea) (Chen & van Achterberg, 2019; van Achterberg, 1976).

Checklists of Regional Proteropinae. In the Middle East, the subfamily Proteropinae comprises only a single known species, *Proterops nigripennis* Wesmael, 1835. In all previous regional studies, proteropines were treated under the subfamily Ichneutinae (Gadallah et al., 2019; Samin et al., 2016; Samin, Coronado-Blanco, Kavallieratos et al., 2018; Samin, Coronado-Blanco, Fischer et al., 2018).

List of species of the subfamily Proteropinae recorded in the Middle East

Subfamily Proteropinae van Achterberg, 1976

Genus *Proterops* Wesmael, 1835

Proterops nigripennis Wesmael, 1835

Catalogs with Iranian records: Yu et al. (2016, under Ichneutinae), Samin, Coronado-Blanco, Kavallieratos et al. (2018), Samin, Coronado-Blanco, Fischer et al. (2018, under Ichneutinae) Gadallah et al. (2019, under Ichneutinae).

Distribution in Iran: Khuzestan (Samin et al., 2016 under Ichneutinae), Semnan (Samin et al., 2020 under the subfamily Ichneutinae), Iran (no specific locality cited) (Belokobylskij & Lelej, 2019).

Distribution in the Middle East: Iran (see references above), Turkey (Beyarslan & Aydoğdu, 2013, 2014).

Extralimital distribution: Austria, Azerbaijan, Belgium, China, former Czechoslovakia, Denmark, Finland, France, Georgia, Germany, Hungary, Ireland, Italy, Kazakhstan, Korea, Mongolia, Netherlands, Norway, Poland, Russia, Sweden, Switzerland, United Kingdom.

Host records: Summarized by Yu et al. (2016) as being a parasitoid of the argids *Arge berberidis* Schrank, *Arge enodis* (L.), and *Arge rustica* (L.); and the tenthredinid *Athalia roasae* (L.). In Iran, this species has been reared from *Arge ochrupus* (Gmelin in Linnaeus) (Samin et al., 2020).

Braconidae of the Middle East (Hymenoptera). https://doi.org/10.1016/B978-0-323-96099-1.00025-X

Conclusion

A single valid species of the subfamily Proteropinae, *Proterops nigripennis* Wesmael, 1835 has been reported from the Middle East countries (Iran and Turkey). In Iran, this species has been recorded from Khuzestan and Semnan provinces. No host species has been so far recorded for this parasitoid in Iran, of which only *Arge* spp. (Hymenoptera: Argidae) are known as hosts of it worldwide (Yu et al., 2016). Among the Middle East and adjacent countries to Iran, this species (*P. nigripennis*) is shared with Azerbaijan, Kazakhstan, Russia, and Turkey.

References

Belokobylskij, S. A., & Lelej, A. S. (2019). Annotated catalogue of the Hymenoptera of Russia. In *Apocrita: Parasitica. Proceedings of the Zoological Institute of the Russian Academy of Sciences* (Vol. II). Suppl. 8, 594 pp.

Belshaw, R., & Quicke, D. L. J. (2002). Robustness of ancestral state estimates: Evolution of life history strategy in ichneumonoid parasitoids. *Systematic Biology, 51*, 450−477.

Beyarslan, A., & Aydoğdu, M. (2013). Additions to the rare species of Braconidae fauna (Hymenoptera: Braconidae) from Turkey. *Munis Entomology & Zoology, 8*(1), 369−374.

Beyarslan, A., & Aydoğdu, M. (2014). Additions to the rare species of Braconidae fauna (Hymenoptera: Braconidae) from Turkey. *Munis Entomology & Zoology, 9*(1), 103−108.

Chen, X. X., & van Achterberg, C. (2019). Systematics, phylogeny, and evolution of braconid wasps: 30 years of progress. *Annual Review of Entomology, 64*, 1−24.

Gadallah, N. S., Ghahari, H., & Kavallieratos, N. G. (2019). An annotated catalogue of the Iranian Charmontinae, Ichneutinae, Macrocentrinae and Orgilinae (Hymenoptera: Braconidae). *Journal of the Entomological Research Society, 21*(3), 333−354.

Li, Q., Wei, S. J., Tang, P., Wu, Q., Shi, M., Sharkey, M. J., & Chen, X. X. (2016). Multiple lines of evidence from mitochondrial genomes resolve phylogeny and evolution of parasitic wasp in Braconidae. *Genome Biology and Evolution, 8*(9), 2651−2662.

Quicke, D. L. J., & van Achterberg, C. (1990). Phylogeny of the subfamilies of the family Braconidae (Hymenoptera: Ichneumonoidea). *Zoologische Verhandlingen, Leiden, 285*, 1−95.

Samin, N., van Achterberg, C., & Çetin Erdoğan, Ö. (2016). A faunistic study on some subfamilies of Braconidae (Hymenoptera: Ichneumonoidea) from Iran. *Arquivos Entomoloxicos, 15*, 153−161.

Samin, N., Coronado-Blanco, J. M., Kavallieratos, N. G., Fischer, M., & Sakenin, H. (2018). Recent findings on Braconidae (Hymenoptera: Ichneumonoidea) of Iran with an updated checklist. *Acta Biologica Turcica, 31*(4), 160−173.

Samin, N., Coronado-Blanco, J. M., Fischer, M., van Achterberg, C., Sakenin, H., & Davidian, E. (2018). Updated checklist of Iranian Braconidae (Hymenoptera: Ichneumonoidea) with twenty-three new records. *Natura Somogyiensis, 32*, 21−36.

Samin, N., Beyarslan, A., Ranjith, A. P., Ahmad, Z., Sakenin Chelav, H., & Hosseini Boldaji, S. A. (2020). A faunistic study on Braconidae (Hymenoptera: Ichneumonoidea) from Ardebil and East Azarbayjan provinces, Northwestern Iran. *Egyptian Journal of Plant Protection Research Institute, 3*(4), 955−963.

Sharanowski, B. J., Dowling, A. P. G., & Sharkey, M. J. (2011). Molecular phylogenetics of Braconidae (Hymenoptera: Ichneumonoidea), based on multiple molecular genes, and implications for classification. *Systematic Entomology, 36*, 549−572.

Sharkey, M. J., & Wharton, R. A. (1994). A revision of the genera of the world Ichneutinae (Hymenoptera: Braconidae). *Journal of Natural History, 28*, 873−912.

van Achterberg, C. (1976). A preliminary key to the subfamilies of the subfamilies of the Braconidae (Hymenoptera). *Tijdschrift voor Entomologie, 119*, 33−78.

van Achterberg, C. (1993). Illustrated key to the subfamilies of the Braconidae (Hymenoptera: Ichneumonoidea). *Zoologische Verhandelingen, 283*, 1−198.

Wharton, R. A., & van Achterberg, C. (2000). Family group names in the Braconidae (Hymenoptera: Ichneumonoidea). *Journal of Hymenoptera Research, 9*(2), 254−270.

Yu, D. S., van Achterberg, C., & Horstmann, K. (2016). *Taxapad 2016, Ichneumonoidea 2015, Database on flash-drive*. Ottawa, Ontario, Canada.

Rhysipolis meditator Haliday, 1836 (Rhysipolinae), ♂, lateral habitus. *Photo prepared by S.R. Shaw.*

Rhysipolis platygaster Spencer, 1999 (Rhysipolinae), ♀, lateral habitus—Nearctic. *Photo prepared by S.R. Shaw.*

Chapter 28

Subfamily Rhysipolinae Belokobylskij, 1984

Neveen Samy Gadallah[1], Hassan Ghahari[2], Scott Richard Shaw[3] and Donald L.J. Quicke[4]

[1]Entomology Department, Faculty of Science, Cairo University, Giza, Egypt; [2]Department of Plant Protection, Yadegar-e Imam Khomeini (RAH) Shahre Rey Branch, Islamic Azad University, Tehran, Iran; [3]UW Insect Museum, Department of Ecosystem Science and Management, University of Wyoming, Laramie, WY, United States; [4]Integrative Ecology Laboratory, Department of Biology, Faculty of Science, Chulalongkorn University, Pathumwan, Bangkok, Thailand

Introduction

Rhysipolinae Belokobylskij, 1984, is a small, nearly cosmopolitan (worldwide except in the Afrotropical region) subfamily, including 51 species classified into eight genera (Yu et al., 2016). As far as is known, they are mostly solitary koinobiont ectoparasitoids of leaf-mining and concealed-feeding microlepidoptera larvae (Chen & van Achterberg, 2019; Scatolini et al., 2002; Spencer & Whitfield, 1999; van Achterberg, 1995; Whitfield & Wagner, 1991). Townsend and Shaw (2009) described the biology of the (presumed) rhysipoline *Andesipolis yanayacu* Townsend and Shaw as being a gregarious koinobiont endoparasitoid of shelter-building Pyralidae feeding on Urticaceae. However, Shimbori et al. (2017) asserted that *Andesipolis* should instead be classified as a part of Mesostoinae. The biology of only two Palaearctic rhysipoline species has been described in detail (Shaw, 1983; Shaw & Sims, 2015).

Rhysipolines have historically been variously placed either within Exothecinae, Hormiinae, or Rogadinae (e.g., Shaw, 1995; Whitfield & Wharton, 1997). Rhysipolines were treated as a separate subfamily by Quicke (1994) and van Achterberg (1995) based on phylogenetic studies. Rhysipolinae was recovered as a basal lineage of the alysioid subcomplex + Braconinae based on molecular and morphological data (Zaldívar-Riverón et al., 2006).

Rhysipolines are diagnosed by the following combination of characters: the antennal pedicellus is distinctly shorter than the scapus; the occipital carina runs straight to base of mandible; the propodeum has a long median carina, with a small or incomplete areola; the hind wing lacks an m-cu vein; and the spiracle of T2 is situated laterally, and not sclerotized (van Achterberg, 1993, 1995).

Checklists of Regional Rhysipolinae. Gadallah et al. (2016) and Farahani et al. (2016) both listed three species in two genera. Yu et al. (2016) represented four species in two genera. Samin, Coronado-Blanco, Kavallierato, et al. (2018) and Samin, Coronado-Blanco, Fischer, et al. (2018) listed four species in two genera. The present checklist includes six species in three genera in the Middle East.

Key to the genera of the subfamily Rhysipolinae in the Middle East

1. Occiptial and hypostomal carinae are parallel to each other (not joined), reaching lower margin of head 2
— Occipital and hypostomal carinae joined above base of mandible .... *Pachystigmus* Hellén
2. Head with conspicuously projecting antennal tubercles; antennal scape with horn-like process .... *Cerophanes* Tobias
— Head with slightly projected antennal tubercles; antennal scape without horn-like process .... *Rhysipolis* Foerster

List of species of the subfamily Rhysipolinae in the Middle East

Subfamily Rhysipolinae Belokobylskij, 1984

Genus *Cerophanes* Tobias, 1971

Cerophanes kerzhneri Tobias, 1971

Catalogs with Iranian records: Gadallah et al. (2016), Farahani et al. (2016), Yu et al. (2016), Samin, Coronado-Blanco, Kavallierato, et al. (2018), Samin, Coronado-Blanco, Fischer, et al. (2018).

Distribution in Iran: East Azarbaijan (Rastegar et al., 2012), Isfahan (Ghahari et al., 2011).

Distribution in the Middle East: Iran.

Extralimital distribution: Armenia, Bulgaria, Kazakhstan, Moldova, Serbia (Yu et al., 2016), Russia (European part) (Samartsev, 2013).

Host records: Unknown.

Genus *Pachystigmus* Hellén, 1927

Pachystigmus nitidulus Hellén, 1927

Distribution in the Middle East: Turkey (Beyarslan, 2015).

Extralimital distribution: Finland, Spain.

Host records: Unknown.

Genus *Rhysipolis* Foerster, 1863

Rhysipolis decorator (Haliday, 1836)

Catalogs with Iranian records: Gadallah et al. (2016), Farahani et al. (2016), Yu et al. (2016), Samin, Coronado-Blanco, Kavallierato, et al. (2018), Samin, Coronado-Blanco, Fischer, et al. (2018).

Distribution in Iran: Isfahan (Ghahari et al., 2011), Khuzestan (Samin et al., 2016).

Distribution in the Middle East: Iran (Ghahari et al., 2011; Samin et al., 2016), Turkey (Beyarslan, 2015).

Extralimital distribution: Nearctic, Palaearctic [Adjacent countries to Iran: Azerbaijan, Kazakhstan, Russia].

Host records: Summarized by Yu et al. (2016) as being a parasitoid of the choreutid *Choreutis pariana* (Clerck); the crambid *Loxostege sticticalis* (L.); the gelechiids *Scrobipalpa atriplicella* (Fischer), and *Syncopacma coronillella* (Treitschke); the gracillariids *Acrocercops brongniardella* Fabricius, *Caloptilia* spp., *Parornix alta* (Braun), *Parornix geminatella* (Packard), and *Phyllonorycter* spp.; the lyonetiid *Leucoptera malifoliella* (Costa); and the momphid *Mompha raschkiella* (Zeller). In their North American revision of *Rhysipolis*, the hosts confirmed by Spencer and Whitfield (1999) were all species of Gracillariidae.

Rhysipolis enukidzei Tobias, 1976

Catalogs with Iranian records: No catalog.

Distribution in Iran: West Azarbaijan (Samin, Sakenin Chelav et al., 2020).

Distribution in the Middle East: Iran (Samin, Sakenin Chelav et al., 2020), Turkey (Beyarslan, 2015).

Extralimtal distribution: Azerbaijan, Georgia, Hungary, Japan, Kazakhstan, Korea, Poland, Russia, Ukraine.

Host records: Unknown.

Rhysipolis hariolator (Haliday, 1836)

Catalogs with Iranian records: Yu et al. (2016), Samin, Coronado-Blanco, Kavallierato, et al. (2018), Samin, Coronado-Blanco, Fischer, et al. (2018).

Distribution in Iran: Chaharmahal and Bakhtiari (Samin et al., 2016), Mazandaran (Samin, Beyarslan, et al., 2020).

Distribution in the Middle East: Iran (Samin et al., 2016; Samin, Beyarslan, et al., 2020), Turkey (Beyarslan, 2015).

Extralimital distribution: Armenia, Azerbaijan, Belgium, Bulgaria, Czech Republic, Finland, France, Germany, Hungary, Italy, Japan, Kazakhstan, Korea, Lithuania, Moldova, Mongolia, Netherlands, Poland, Russia, Slovakia, Sweden, Switzerland, Ukraine, United Kingdom.

Host records: Summarized by Yu et al. (2016) as being a parasitoid of the cosmopterigid *Cosmopterix zieglerella* (Hübner); and the gracillariids *Caloptilia alchimiella* (Scopoli), *Caloptilia elongella* (L.), *Parornix* spp., and *Phyllonorycter* spp. In Iran, this species has been reared from the gracillariid *Phyllonorycter salicicolella* (Sircom) (Samin, Beyarslan, et al., 2020).

Rhysipolis meditator (Haliday, 1836)

Catalogs with Iranian records: Gadallah et al. (2016), Farahani et al. (2016), Yu et al. (2016), Samin, Coronado-Blanco, Kavallierato, et al. (2018), Samin, Coronado-Blanco, Fischer, et al. (2018).

Distribution in Iran: Isfahan (Ghahari et al., 2011 as *Rhysipolis similis*), Kuhgiloyeh and Boyerahmad (Samin et al., 2016), West Azarbaijan (Rastegar et al., 2012 as *Rhysipolis similis*).

Distribution in the Middle East: Iran (see references above), Turkey (Beyarslan, 2015).

Extralimital distribution: Austria, Belgium, China, Czech Republic, Finland, France, Germany, Hungary, Ireland, Italy, Korea, Lithuania, Moldova, Mongolia, Poland, Russia, Serbia, Slovakia, Spain, Sweden, Switzerland, Tajikistan, Turkmenistan, Ukraine, United Kingdom, Uzbekistan, Vietnam.

Host records: Summarized by Yu et al. (2016) as being a parasitoid of the crambid *Loxostege sticticalis* (L.); the elachistid *Elachista* sp.; the gracillariids *Acrocercops brongniardella* Fabricius, *Caloptilia* spp., *Gracillaria syringella* (Fabricius), and *Parornix* sp.; the lyonetiid *Leucoptera laburnella* (Stainton); the momphids *Mompha reschkiella* (Zeller) and *Mompha terminella* (Humphreys and Westwood); the tineid *Agnathosia mendicella* (Denis and Schiffermüller); the tischeriid *Coptotriche gaunacella* (Duponchel); and the tortricids *Ancylis mitterbacheriana* (Denis and Schiffermüller), and *Cydia cosmophorana* (Treitschke).

Conclusion

Six valid species of the subfamily Rhysipolinae (11.7% from the total number of the world species) in three genera (*Cerophanes* Tobias, 1971; *Pachystigmus* Hellén, 1927, and *Rhysipolis* Foerster, 1863) have been reported from only two of the Middle East countries (Iran and Turkey). Each of the two countries is represented by five rhysipoline species. The Iranian Rhysipolinae comprises five species in two genera *Cerophanes* (one species) and *Rhysipolis* (four species) (7.8% of the world species). These species are recorded from seven provinces, Isfahan (three species), West Azarbaijan (two species), Chaharmahal and Bakhtiari, East Azarbaijan, Khuzestan, Kuhgiloyeh and Boyerahmad, Mazandaran (each with one species). Only one host species of Gelechiidae has been recorded for an Iranian rhysipoline species. Comparison of the Rhysipolinae fauna of Iran with the Middle East and adjacent countries to Iran indicates the fauna of Russia (with 11 species) is more diverse than Iran, followed by Kazakhstan (seven species), Turkey (five species), Azerbaijan (three species), Armenia (two species), and Turkmenistan (one species) (Yu et al., 2016). No rhysipoline species have been recorded from the other Middle Eastern countries. Additionally, among the Middle East countries and adjacent ones to Iran, Russia shares five species with Iran, followed by Kazakhstan (four species), Azerbaijan, Turkey (both with three species), Armenia (two species), and Turkmenistan (one species).

References

Beyarslan, A. (2015). A faunal study of the subfamily Doryctinae in Turkey (Hymenoptera: Braconidae). *Turkish Journal of Zoology, 39*(1), 126−143.

Chen, X. X., & van Achterberg, C. (2019). Systematics, phylogeny, and evolution of the braconid wasps: 30 years of progress. *Annual Review of Entomology, 64*, 1−24.

Farahani, S., Talebi, A. A., & Rakhshani, E. (2016). Iranian Braconidae (Insecta: Hymenoptera: Ichneumonoidea): diversity, distribution and host association. *Journal of Insect Biodiversity and Systematics, 2*(1), 1−92.

Gadallah, N. S., Ghahari, H., & van Achterberg, C. (2016). An annotated catalogue of the Iranian Euphorinae, Gnamptodontinae, Helconinae, Hormiinae and Rhysipolinae (Hymenoptera: Braconidae). *Zootaxa, 4072*(1), 1−38.

Ghahari, H., Fischer, M., & Papp, J. (2011). A study on the braconid wasps (Hymenoptera: Braconidae) from Isfahan province, Iran. *Entomofauna, 32*(16), 261−272.

Quicke, D. L. (1994). Phylogenetics and biological transitions in the Braconidae (Hymenoptera: Ichneumonoidea). *Norwegian Journal of Agricultural Sciences*, (Suppl. 16), 155−162.

Rastegar, J., Sakenin, H., Khodaparast, S., & Havaskary, M. (2012). On a collection of Braconidae (Hymenoptera) from East Azarbaijan and vicinity, Iran. *Calodema, 226*, 1−4.

Samartsev, K. G. (2013). On the rare species of cyclostome braconid wasps (Hymenoptera: Braconidae) from the Middle and Lower Volga territories of Russia. *Caucasian Entomological Bulletin, 9*(2), 315−328.

Samin, N., van Achterberg, C., & Erdoğan, Ö.Ç. (2016). A faunistic study on some subfamilies of Braconidae (Hymenoptera: Ichneumonoidea) from Iran. *Arquivos Entomoloxicos, 15*, 153−161.

Samin, N., Coronado-Blanco, J. M., Kavallieratos, N. G., Fischer, M., & Sakenin, H. (2018). Recent findings on Braconidae (Hymenoptera: Ichneumonoidea) of Iran with an updated checklist. *Acta Biologica Turcica, 31*(4), 160−173.

Samin, N., Coronado-Blanco, J. M., Fischer, M., van Achterberg, C., Sakenin, H., & Davidian, E. (2018). Updated checklist of Iranian Braconidae (Hymenoptera: Ichneumonoidea) with twenty-three new records. *Natura Somogyiensis, 32*, 21−36.

Samin, N., Beyarslan, A., Coronado-Blanco, J. M., & Navaeian, M. (2020). A contribution to the braconid wasps (Hymenoptera: Braconidae) from Iran. *Natura Somogyiensis, 35*, 25−28.

Samin, N., Sakenin Chelav, H., Ahmad, Z., Penteado-Dias, A. M., & Samiuddin, A. (2020). A faunistic study on the family Braconidae (Hymenoptera: Ichneumonoidea) from Iran. *Scientific Bulletin of Uzhhorod National University (Series: Biology, 48*, 14−19.

Scatolini, D., Pentiado-Dias, A. M., & van Achterberg, C. (2002). *Pseudorhysipolis* gen. nov. (Hymenoptera: Braconidae: Rhysipolinae), with nine new species from Brazil, Suriname and Panama. *Zoologische Mededelingen, 76*(13), 109−131.

Shaw, M.R. (1983). On evolution of endoparasitism: The biology of some genera of Rogadinae (Braconidae). *Contributions of the American Entomological Institute, 20*, 307−328.

Shaw, M. R., & Sims, I. (2015). Notes on the biology, morphology, nomenclature and classification of *Pseudavga flavicoxa* Tobias, 1964 (Hymenoptera, Braconidae, Rhysipolinae), a genus and species new to Britain parasitizing *Bucculatrix thoracella* (Thunberg) (Lepidoptera, Bucculatricidae). *Journal of Hymenoptera Research, 42*, 21−32.

Shaw, S. R. (1995). Chapter 12.2, Braconidae, pp. 431−463. In P. Hanson, & I. D. Gauld (Eds.), *The Hymenoptera of Costa Rica* (p. 893). Oxford University Press.

Shimbori, E. M., Gessner, C. da S. S., Penteado-Dias, A. M., & Shaw, S. R. (2017). A revision of the genus *Andesipolis* (Hymenoptera: Braconidae: Mesostoinae) and redefinition of the subfamily Mesostoinae. *Zootaxa, 4216*(2), 101−152.

Spencer, L., & Whitfield, J. (1999). Revision of the Nearctic species of *Rhysipolis* Foerster (Hymenoptera: Braconidae). *Transactions of the American Entomological Society, 125*(3), 295−324.

Townsend, A., & Shaw, S. R. (2009). A new species of *Andesipolis* from the eastern Andes of Ecuador with notes on biology and classification (Hymenoptera: Braconidae). *Journal of Insect Science, 9*(36), 1−7.

van Achterberg, C. (1993). Illustrated key to the subfamilies of the Braconidae. *Zoologische Verhandelingen, 283*, 1−189.

van Achterberg, C. (1995). Generic revision of the subfamily Betylo-braconinae (Hymenoptera: Braconidae) and other groups with modified fore tarsus. *Zoologische Verhandlingen, 298*, 1−242.

Whitfield, J. B., & Wagner, D. L. (1991). Annotated key to the genera of Braconidae (Hymenoptera) attacking leaf mining Lepi-doptera in the Holarctic region. *Journal of Natural History, 25*, 733−754.

Whitfield, J. B., & Wharton, R. A. (1997). Subfamily Hormiinae, pp. 284−301. In R. A. Wharton, P. M. Marsh, & M. J. Sharkey (Eds.), *Manual of the New World Genera of the Family Braconidae (Hymenoptera)* (vol. 1, pp. 1−439). International Society of Hyme-noptera, Special Publication.

Yu, D. S., van Achterberg, C., & Horstmann, K. (2016). *Taxapad 2016, Ichneumonoidea 2015, Database on flash-drive*. Nepean, Ontario, Canada.

Zaldívar-Riverón, A., Mori, M., & Quicke, D. L. J. (2006). Systematics of the cyclostome subfamilies of braconid parasitic wasps (Hymenoptera: Ichneumonoidea): a simultaneous molecular and morphological Bayesian approach. *Molecular Phylogenetics and Evolution, 38*(1), 130−145.

Dolopsidea sp. (Rhyssalinae), ♀, lateral habitus. *Photo prepared by S.R. Shaw.*

Oncophanes atriceps (Ashmead, 1889) (Rhyssalinae), ♀, lateral habitus—Nearctic. *Photo prepared by S.R. Shaw.*

Chapter 29

Subfamily Rhyssalinae Foerster, 1863

Hassan Ghahari[1], Scott Richard Shaw[2], Neveen Samy Gadallah[3] and Donald L.J. Quicke[4]

[1]Department of Plant Protection, Yadegar-e Imam Khomeini (RAH) Shahre Rey Branch, Islamic Azad University, Tehran, Iran; [2]UW Insect Museum, Department of Ecosystem Science and Management, University of Wyoming, Laramie, WY, United States; [3]Entomology Department, Faculty of Science, Cairo University, Giza, Egypt; [4]Integrative Ecology Laboratory, Department of Biology, Faculty of Science, Chulalongkorn University, Pathumwan, Bangkok, Thailand

Introduction

The cosmopolitan Rhyssalinae Foerster, 1863 is a relatively small but varied subfamily of cyclostome Braconidae. Rhyssalines comprise about 63 species classified into 14 genera and five tribes (Achaiabraconini Belokobylskij, 2009, Acrisidini Hellén, 1957, Histeromerini Fahringer, 1930, Laibaleini Quicke, Butcher and Belokobylskij, 2020, and Rhyssalini Foerster, 1863) (Chen & van Achterberg, 2019; Quicke et al., 2020; van Achterberg et al., 2017; Yu et al., 2016), of which 35 species are known from the Palaearctic region (Yu et al., 2016). Rhyssalinae, as presently circumscribed, includes several taxa that were formerly included in a variety of other subfamilies (Sharanowski et al., 2011, 2014; van Achterberg et al., 2017).

Rhyssalines are diagnosed by having the following combination of characters: the antenna with more than 14 segments; labrum smooth and slightly concave; the propodeum usually without tubercles (but if tuberculate, then the tubercles are obtuse); metasomal T4 and following tergites usually largely exposed (except notably in *Lysitermoides*); dorsope usually present and metasomal T2 with spiracles present on epipleurite or near lateral fold (van Achterberg, 1993, 1995). So far as is known, rhyssalines are idiobiont ectoparasitoids of holometabolous larvae (Chen & van Achterberg, 2019; van Achterberg, 1995; Yu et al., 2016). However, the only paper treating the biology of a rhyssaline in detail is Shaw (1995) on *Histeromerus*.

The systematic position of tribe Histeromerini was problematic for a long time (van Achterberg, 1992). The single genus *Histeromerus* Wesmael, 1838 was long considered as a member of the subfamily Doryctinae and included in the tribe Histeromerini or Doryctini by Shenefelt and Marsh (1976) as well as Belokobylskij and Tobias (1986). On the other hand, van Achterberg (1976), Tobias (1976) and Quicke (1987) treated it as a member of Braconinae. Histeromines have also been recognized as a separate subfamily by van Achterberg (1984, 1988), Quicke and van Achterberg (1990), Shaw and Huddleston (1991), and Belokobylskij (1998). More recently, phylogenetic studies based on molecular data conducted by Shi et al. (2005), Zaldívar-Riverón et al. (2006), and Sharanowski et al. (2011) included Histeromerini as a tribe in the subfamily Rhyssalinae. Van Acterberg et al. (2017) also placed the genus *Histeromerus* in the tribe Histeromerini among the subfamily Rhyssalinae.

Checklists of Regional Rhyssalinae. Farahani et al. (2016) listed one species (Histeromerinae) + two species in one genus (Rhyssalinae). Yu et al. (2016) and Samin, Coronado-Blanco, Kavallieratos, et al. (2018) both recorded one species (Histeromerinae) + two species in two genera (Rhyssalinae). Samin, Coronado-Blanco, Fischer, et al. (2018) listed one species (Histeromerinae) + three species in three genera (Rhyssalinae), and finally, Gadallah and Ghahari (2019) cataloged four species in three genera. The present checklist includes seven species in five genera and three tribes in the Middle East. Here we follow the classification of van Achterberg et al. (2017), and Chen and van Achterberg (2019) in placing Histeromerini as a tribe of the Rhyssalinae comprising the single genus, *Histeromerus*.

Key to genera of the subfamily Rhyssalinae in the Middle East (modified from Belokobylskij, 2004; Quicke et al., 2020)

1. Postgenal bridge of head wide; propleuron without posterolateral flange; prepectal carina completely absent; fore tibia with cluster of numerous spines; hind basitarsus almost as long as second to fifth tarsomeres combined; metasomal T1 without dorsope .... *Histeromerus* Wesmael
— Postgenal bridge of head narrow or absent; propleuron with posterodorsal flange; prepectal carina present; fore tibia without or with a few spines; hind basitarsus always shorter than second to fifth tarsomeres combined; metasomal T1 usually with a distinct dorsope 2

2. Marginal cell of fore wing open apically; first subdiscal cell of fore wing widely open; vein cu-a of hind wing mainly or entirely absent; male hind tibia conspicuously swollen (claviform); metasomal T2 and T3 enlarged, thus concealing the following tergites; ovipositor sheath short, not protruding beyond apex of metasoma .... *Acrisis* Foerster
— Marginal vein of fore wing closed apically; first subdiscal cell of fore wing closed; vein cu-a of hind wing present; hind tibia of male not swollen (except in many *Rhyssalus*); metasomal T2 and T3 not enlarged or inconspicuously enlarged, almost not concealing the following tergites (except in some genera); ovipositor sheath distinctly protruding beyond apex of metasoma 3

3. Vein r of fore wing arising about from the apical third of pterostigma; vein 2A absent; distal part of ovipositor ventrally with wide serration; fore tibia with more or less distinct spines; male hind tibia usually distinctly swollen (claviform), with granulate sculpture; metasomal T2 entirely smooth; female metasoma usually inconspicuously compressed .... *Rhyssalus* Haliday
— Vein r of fore wing arising from or just before middle of pterostigma; vein 2A present; distal part of ovipositor smooth ventrally; fore tibia without spines; male hind tibia not thickened, without granulate sculpturing 4

4. Female metasomal tergites behind sculptured one weakly sclerotized, rather soft; metasomal T2 striated at least basally; ovipositor sheath not longer than half length of metasoma; propodeum with lateral protuberances; mesoscutum usually almost entirely densely setose .... *Oncophanus* Foerster
— Female metasomal tergites entirely conspicuously sclerotized; metasomal T2 entirely smooth; ovipositor sheath longer than metasoma, usually as long as body; propodeum with distinct lateral protuberances; mesoscutum glabrous, setae present along notauli and marginal .... *Dolopsidea* Hincks

List of species of the subfamily Rhyssalinae in the Middle East

Subfamily Rhyssalinae Foerster, 1863

Tribe Acrisidini Hellén, 1957

Genus *Acrisis* Foerster, 1863

Acrisis brevicornis Hellén, 1957

Catalogs with Iranian records: Farahani et al. (2016), Yu et al. (2016), Samin, Coronado-Blanco, Kavallieratos, et al. (2018), Samin, Coronado-Blanco, Fischer, et al. (2018), Gadallah and Ghahari (2019).

Distribution in Iran: Iran (no specific locality cited) (Belokobylskij, 1998; Papp, 2018).

Distribution in the Middle East: Iran.

Extralimital distribution: Finland, Hungary, Korea, Russia, Spain (Papp, 2018 under Exothecinae).

Host records: Recorded by Belokobylskij (1990) as being a parasitoid of the cecidomyiid *Kaltenbachiola strobi* (Winnertz).

Acrisis suomii Tobias, 1983

Catalogs with Iranian records: Farahani et al. (2016), Gadallah and Ghahari (2019).

Distribution in Iran: Guilan (Farahani et al., 2015).

Distribution in the Middle East: Iran.

Extralimital distribution: Finland.

Host records: Unknown.

Tribe Histeromerini Fahringer, 1930

Genus *Histeromerus* Wesmael, 1838

Histeromerus mystacinus Wesmael, 1838

Catalogs with Iranian records: Farahani et al. (2016), Yu et al. (2016) under Histermerinae, Samin, Coronado-Blanco, Kavallieratos, et al. (2018), Samin, Coronado-Blanco, Fischer, et al. (2018) under Histeromerinae), Gadallah and Ghahari (2019).

Distribution in Iran: Golestan (Samin et al., 2015), Mazandaran (Ghahari & Fischer, 2011; Sakenin et al., 2012), West Azarbaijan (Samin et al., 2016), North of Iran (Belokobylskij et al., 2013).

Distribution in the Middle East: Iran.

Extralimital distribution. Belgium, Bulgaria, Czech Republic, Denmark, France, Georgia, Germany, Hungary, Ireland, Italy, Lithuania, Netherlands, Poland, Russia, Slovakia, Sweden, Ukraine, United Kingdom (Yu et al., 2016), Malta (Papp, 2015), North America (Quicke et al., 2018).

Host records: Summarized by Yu et al. (2016) as being a parasitoid of the buprestid *Dicera alni* (Fischer); and the following cerambycids: *Leptura aurulenta* Fabricius, *Rhagium fasciculatum* Faldermann, *Saphanus piceus* (Laicharting), *Sinodendron cylindricum* (L.), *Stictoleptura scutellata* (Fabricius), and *Xylosteus caucasicola* Plavilstshikoe.

Tribe Rhyssalini Foerster, 1863

Genus *Dolopsidea* Hincks, 1944

Dolopsidea indagator (Haliday, 1836)

Catalogs with Iranian records: Farahani et al. (2016), Yu et al. (2016), Samin, Coronado-Blanco, Kavallieratos, et al. (2018), Samin, Coronado-Blanco, Fischer, et al. (2018), Gadallah and Ghahari (2019).

Distribution in Iran: Kordestan (Samin et al., 2016).

Distribution in the Middle East: Iran (Samin et al., 2016), Turkey (Beyarslan, 2015; Beyraslan & Aydoğdu, 2013, 2014).

Extralimital distribution: Armenia, Austria, Azerbaijan, Belgium, Bulgaria, Czech Republic, Finland, France, Germany, Hungary, Ireland, Italy, Korea, Lithuania, Montenegro, Russia, Sweden, Switzerland, Ukraine, United Kingdom.

Host records: Recorded by Fahringer (1934) as being a parasitoid of the ptinids *Dorcatoma chrysomelina* Sturm and *Dorcatoma dresdensis* Herbst. This species was also recorded by Anonymous (1960) as being a parasitoid of the cerambycid *Pogonocherus hispidus* (L.).

Genus *Oncophanes* Foerster, 1863

Oncophanes (Oncophanes) minutus (Wesmael, 1838)

Catalogs with Iranian records: Samin, Coronado-Blanco, Fischer, et al. (2018).

Distribution in Iran: Guilan, Mazandaran (Samin, Beyarslan, et al., 2020), Lorestan (Samin, Sakenin Chelav, et al., 2020 as *O. lanceolator* (Nees)), West Azarbaijan (Sakenin et al., 2018).

Distribution in the Middle East: Iran (see references above), Turkey (Beyarslan, 2015).

Extralimital distribution: Armenia, Austria, Azerbaijan, Belarus, Belgium, Bulgaria, Czech Republic, Finland, France, Georgia, Germany, Hungary, Ireland, Italy, Japan, Kazakhstan, Korea, Latvia, Lithuania, Moldova, Mongolia, Netherlands, Poland, Romania, Russia, Slovakia, Sweden, Switzerland, Ukraine, United Kingdom.

Host records: Summarized by Yu et al. (2016) as being a parasitoid of several host species of the orders Coleoptera (Bostrichidae, Cerambycidae), Hymenoptera (Tenthredinidae), and Lepidoptera (Choreutidae, Coleophoridae, Depressariidae, Gelechiidae, Geometridae, Gracillariidae, Nepticulidae, Oecophoridae, Pterophoridae, Tortricidae, Yponomeutidae). In Iran, this species has been reared from the erebid *Lymantria dispar* (L.), and the tortricid *Archips rosana* (L.) (Samin, Beyarslan, et al., 2020).

Genus *Rhyssalus* Haliday, 1833

Rhyssalus clavator Haliday, 1833

Distribution in the Middle East: Turkey (Beyarslan, 2015).

Extralimital distribution: Austria, Bulgaria, Croatia, former Czechoslovakia, Finland, France, Germany, Hungary, Ireland, Lithuania, Netherlands, Poland, Russia, Switzerland, United Kingdom, former Yugoslavia.

Host records: Unknown.

Rhyssalus longicaudis (Tobias & Belokobylskij, 1981)

Catalogs with Iranian records: No catalog.

Distribution in Iran: Markazi (Ameri et al., 2020).

Distribution in the Middle East: Iran.

Extralimital distribution: Bosnia-Herzegovina, Finland, Hungary, Mongolia, Russia.

Host records: Unknown.

Conclusion

Seven valid species of the subfamily Rhyssalinae in five genera and three tribes (Acrisidini Foerster, 1863, Histeromerini Fahringer, 1930, and Rhyssalini Foerster, 1863) have been reported from only two countries of the Middle East (Iran and Turkey). Rhyssalinae of Iran comprises six species in five genera, *Acrisis*, *Dolopsidea*, *Histeromerus*, *Oncophanes*, and *Rhyssalus* (9.5% of the world species), and three tribes (Acrisidini, Histeromerini, and Rhyssalini). These species are recorded from only seven provinces (Golestan, Guilan, Kordestan, Lorestan, Markazi, Mazandaran, and West Azarbaijan). Only two host species have been recorded for the Iranian Rhyssalinae: *Lymantria dispar* (L.) (Lepidoptera: Erebidae), and *Archips rosana* (L.) (Lepidoptera: Tortricidae), both as the hosts of *Oncophanes (Oncophanes) minutus*. Comparison of the Rhyssalinae fauna of Iran with the Middle East and adjacent countries of Iran indicates the fauna of Russia with 22 species in seven genera (Belokobylskij & Lelej, 2019) is more diverse than Iran, followed by Turkey (three species), Armenia, Azerbaijan (both with two species), and Kazakhstan (one species) (Yu et al., 2016). No rhyssaline species have been recorded from the other Middle Eastern countries. Additionally, among the adjacent countries, Russia shares five species with Iran, followed by Armenia, Azerbaijan, Turkey (each with two species), and Kazakhstan (one species).

References

Ameri, A., Ebrahimi, E., & Talebi, A. A. (2020). Additions to the fauna of Braconidae (Hym., Ichneumonoidea) of Iran based on specimens housed in Hayk Mirzayans Insect Museum with six new records for Iran. *Journal of Insect Biodiversity and Systematics, 6*(4), 353–364.

Anonymous. (1960). Secatariat du service d'identification des Entomophages. Liste d'identification No.3. *Entomophaga, 5,* 337–373.

Belokobylskij, S. A. (1990). A contribution to the braconid fauna (Hymenoptera) of the Far East (USSR). *Vestnik Zoologii, 6,* 32–39 (in Russian).

Belokobylskij, S. A. (1998). 1. Rhyssalinae, 2. Doryctinae, 3. Histeromerinae, 4. Exothecinae, 7. Gnamptodontinae, 9. Alysiinae (Alysiini), 10. Helconinae, 11. Cenocoeliinae, 12. Brachistinae, 14. Meteorideinae, 16. Xiphozelinae, 17. Homolobinae, 18. Charmontinae, 19. Orgilinae, 20. Ecnomiinae, 21. Sigalphinae, 23. Ichneutinae, 25. Cardiochilinae, 27. Dirrhopinae, 28. Miracinae, 29. Adeliinae. In P. A. Ler (Ed.), *Key to the insects of Russian far East. Vol. 4. Neuropteroidea, Mecoptera, Hymenoptera. Pt 3* (p. 706). Vladivostok: Dal'nauka.

Belokobylskij, S. A. (2004). Taxonomic reclassification of the East Asian species of the genus *Oncophanes* Förster (Hymenoptera: Braconidae, Rhyssalinae). *Proceedings of the Russian Entomological Society. St. Petersburg, 75*(1), 106–117.

Belokobylskij, S. A., & Tobias, V. I. (1986). Doryctinae, pp. 21–72. In G. S. Medvedev (Ed.), *Opredelitel Nasekomych Evrospeiskoi Tsasti SSSR 3, Peredpontdatokrylyae 4* (Vol. 145, pp. 1–501). Opr. Faune SSSR.

Belokobylskij, S. A., & Lelej, A. S. (2019). Annotated catalogue of the Hymenoptera of Russia. Volume II. Apocrita: Parasitica. In *Proceedings of the Zoological Institute of the Russian Academy of Sciences.* Supplement No. 8, 594 pp.

Beyarslan, A. (2015). A faunal study of the subfamily Doryctinae in Turkey (Hymenoptera: Braconidae). *Turkish Journal of Zoology, 39*(1), 126–143.

Beyarslan, A., & Aydoğdu, M. (2013). Additions to the rare species of *Braconidae fauna* (Hymenoptera: Braconidae) from Turkey. *Munis Entomology & Zoology, 8*(1), 369–374.

Beyarslan, A., & Aydoğdu, M. (2014). Additions to the rare species of *Braconidae fauna* (Hymenoptera: Braconidae) from Turkey. *Munis Entomology & Zoology, 9*(1), 103–108.

Chen, X. X., & van Achterberg, C. (2019). Systematics, phylogeny, and evolution of the braconid wasps: 30 years of progress. *Annual Review of Entomology, 64,* 1–24.

Fahringer, J. (1934). *Oposcula braconologica. Band 3. Palaearktischen region. Lieferung 5–8* (pp. 321–594). Wien: Opuscula braconologica, Fritz Wagner.

Farahani, S., Talebi, A. A., & Rakhshani, E. (2015). First record of *Acrisis suomii* Tobias, 1983 (Hymenoptera: Braconidae: Rhyssalinae) from Iran. In *Proceedings of the 1st Iranian International Congress of Entomology, 29–31 August 2015* (p. 72). Tehran: Iranian Research Institute of Plant Protection.

Farahani, S., Talebi, A. A., & Rakhshani, E. (2016). Iranian Braconidae (Insecta: Hymenoptera: Ichneumonoidea): diversity, distribution and host association. *Journal of Insect Biodiversity and Systematics, 2*(1), 1–92.

Gadallah, N. S., & Ghahari, H. (2019). An updated checklist of Iranian Cardiochilinae, Ryssalinae and Blacini (Hymenoptera: Ichneumonoidea: Braconidae). *Oriental Insects, 54*(2), 143–161.

Ghahari, H., & Fischer, M. (2011). A study on the Braconidae (Hymenoptera: Ichneumonoidea) from some regions of northern Iran. *Entomofauna, 32,* 181–196.

Papp, J. (2015). First contribution to the knowledge of the braconid wasps (Hymenoptera, Braconidae) of Malta. *Bulletin of the Entomological Society of Malta, 7,* 93–108.

Papp, J. (2018). Braconidae (Hymenoptera) from Korea, XXIV. Species of thirteen subfamilies. *Acta Zoologica Academiae Scientiarum Hungaricae, 64*(1), 21–50.

Quicke, D. L. J. (1987). The Old World genera of braconine wasps (Hymenoptera: Braconidae). *Journal of Natural History, 21*(1), 43–157.

Quicke, D. L. J., & van Achterberg, C. (1990). Phylogeny of the subfamilies of the family Braconidae (Hymenoptera: Ichneumonoidea). *Zoologische Verhandelingen, 258,* 1–180.

Quicke, D. L. J., Herbert, P. D. N., & Butcher, B. A. (2018). DNA barcoding reveals the Palaearctic species *Histeromerus mystacinus* (Hymenoptera: Braconidae: Rhyssalinae) in eastern North America. *The Canadian Entomologist, 150*(4), 495–498.

Quicke, D. L. J., Belokobylskij, S. A., Braet, Y., van Acheterberg, C., Hebert, P. D. N., Prosser, S. W. J., Austin, A. D., Fagan-Jeffries, E. P., ward, D. F., Shaw, M. R., & Butcher, B. A. (2020). Phylogenetic reassignment of basal cyclostome braconid parasitoid wasps (Hymenoptera) with description of a new enigmatic Afrotropical tribe with a highly anomalous 28S D2 secondary structure. *Zoological Journal of the Linnean Society, XX,* 1–18.

Sakenin, H., Naderian, H., Samin, N., Rastegar, J., Tabari, M., & Papp, J. (2012). On a collection of Braconidae (Hymenoptera) from northern Iran. *Linzer biologische Beiträge, 44*(2), 1319–1330.

Sakenin, H., Coronado-Blanco, M., Samin, N., & Fischer, M. (2018). New records of Braconidae (Hymenoptera) from Iran. *Far Eastern Entomologist, 362,* 13–16.

Samin, N., Ghahari, H., Gadallah, N. S., & Monaem, R. (2015). A study on the braconid wasps (Hymenoptera: Ichneumonoidea: Braconidae) from Golestan province, northern Iran. *Linzer biologische Beiträge, 47*(1), 731–739.

Samin, N., van Achterberg, C., & Çetin Erdoğan, Ö. (2016). A faunistic study on some subfamilies of Braconidae (Hymenoptera: Ichneumonoidea) from Iran. *Arquivos Entomoloxicos, 15,* 153–161.

Samin, N., Coronado-Blanco, J. M., Kavallieratos, N. G., Fischer, M., & Sakenin, H. (2018). Recent findings on Braconidae (Hymenoptera: Ichneumonoidea) of Iran with an updated checklist. *Acta Biologica Turcica, 31*(4), 160–173.

Samin, N., Coronado-Blanco, J. M., Fischer, M., van Achterberg, C., Sakenin, H., & Davidian, E. (2018). Updated checklist of Iranian Braconidae (Hymenoptera: Ichneumonoidea) with twenty-three new records. *Natura Somogyiensis, 32,* 21–36.

Samin, N., Beyarslan, A., Coronado-Blanco, J. M., & Navaeian, M. (2020). A contribution to the braconid wasps (Hymenoptera: Braconidae) from Iran. *Natura Somogyiensis, 35,* 25–28.

Samin, N., Sakenin Chelav, H., Ahmad, Z., Penteado-Dias, A. M., & Samiuddin, A. (2020). A faunistic study on the family Braconidae (Hymenoptera: Ichneumonoidea) from Iran. *Scientific Bulletin of Uzhhorod National University (Series: Biology), 48,* 14–19.

Sharanowski, B. J., Dowling, A. P. G., & Sharkey, M. J. (2011). Molecular phylogenetics of Braconidae (Hymenoptera: Ichneumonoidea), based on multiple nuclear genes, and implications for classification. *Systematic Entomology, 36,* 549–571.

Sharanowski, B. J., Zhang, Y. M., & Wanigasekara, W. M. U. M. (2014). Annotated checklist of Braconidae (Hymenoptera) in the Canadian Prairies ecozone, pp. 399–425. In D. J. Giberson, & H. A. Cárcamo (Eds.), *Arthropods of Canadian grasslands. Volume 4. Biodiversity and systematic. Part 2* (p. 479). Biological Survey of Canada.

Shaw, M. R. (1995). Observations on the adult behaviour and biology of *Histeromerus mystacinus* Wesmael (Hymenoptera: Braconidae). *Entomologist, 114*, 1–13.

Shaw, M. R., & Huddleston, T. (1991). Classification and biology of braconid wasps (Hymenoptera: Braconidae). *Handbook for identification of British Insects, 7*(11), 1–126.

Shenefelt, R. D., & Marsh, P. (1976). *Braconidae, 9. Doryctinae, Hymenopterorum Catalogus* (Nova edition, pp. 1263–1424). Pars 13.

Shi, M., Chen, X. X., & van Achterberg, C. (2005). Phylogenetic relationships among the Braconidae (Hymenoptera: Ichneumonoidea) inferred from partial 16S rDNA, 28S rDNA D2, 18S rDNA gene sequences and morphological characters. *Molecular Phylogenetics and Evolution, 37*(1), 104–116.

Tobias, V. I. (1976). *Braconids of the Caucasus (Hymenoptera: Braconidae). Opred. Faune SSSR* (Vol. 110). Leningrad: Nauka Press, 286 pp.

van Achterberg, C. (1976). A preliminary key to the subfamilies of the Braconidae (Hym.). *Tijdschrift voor Entomologie, 119*, 33–78.

van Achterberg, C. (1984). Essay on the phylogeny of Braconidae (Hymenoptera: Ichneumonoidea). *Entomologisk Tidskrift, 105*, 41–58.

van Achterberg, C. (1988). *Parallelisms in the Braconidae (Hymenoptera) with special reference to the biology. Advances Par* (pp. 85–115). Hymenoptera Research.

van Acterberg, C. (1992). Revision of the genus *Histeromerus* Wesmael (Hymenoptera: Braconidae). *Zoologische Mededelingen, 66*, 189–196.

van Achterberg, C. (1993). Illustrated key to the subfamilies of the Braconidae (Hymenoptera: Ichneumonoidea). *Zoologische Verhandelingen, 238*, 1–189.

van Achterberg, C. (1995). *Glyptoblacus* gen. Nov. (Hymenoptera: Braconidae: Blacinae) from Honduras. *Zoologische Mededelingen, 69*(23), 303–306.

van Achterberg, C., Taeger, A., Blank, S. M., Zwakhlas, K., Viitasaari, M., Yu, D. S., & de Jong, Y. (2017). Fauna Europaea: Hymenoptera - Symphyta and Ichneumonoidea. *Biodiversity Data Journal, 5*, e14650.

Yu, D. S., van Achterberg, C., & Horstmann, K. (2016). *Taxapad 2016, Ichneumonoidea 2015, Database on flash-drive*. Ottawa, Ontario, Canada.

Zaldívar-Riverón, A., Mori, M., & Quicke, D. L. J. (2006). Systematics of the cyclostome subfamilies of braconid parasitic wasps (Hymenoptera: Ichneumonoidea): a simultaneous molecular and morphological Bayesian approach. *Molecular Phylogenetics and Evolution, 38*(1), 130–145.

Aleiodes albitibia (Herrich-Schäffer, 1883) (Rogadinae), ♀, lateral habitus. *Photo prepared by S.R. Shaw.*

Clinocentrus cunctator (Haliday, 1836) (Rogadinae), ♀, lateral habitus. *Photo prepared by S.R. Shaw.*

Chapter 30

Subfamily Rogadinae Foerster, 1863 *s.s.*

Scott Richard Shaw[1], Hassan Ghahari[2], Neveen Samy Gadallah[3] and Donald L.J. Quicke[4]

[1]*UW Insect Museum, Department of Ecosystem Science and Management, University of Wyoming, Laramie, WY, United States;* [2]*Department of Plant Protection, Yadegar-e Imam Khomeini (RAH) Shahre Rey Branch, Islamic Azad University, Tehran, Iran;* [3]*Entomology Department, Faculty of Science, Cairo University, Giza, Egypt;* [4]*Integrative Ecology Laboratory, Department of Biology, Faculty of Science, Chulalongkorn University, Pathumwan, Bangkok, Thailand*

Introduction

Rogadinae Foerster, 1863 is a large cosmopolitan subfamily of Braconidae (van Achterberg, 1993). It is very diverse subfamily, comprising six tribes (Aleiodini, Betylobraconini, Clinocentrini, Rogadini, Stiropiini, and Yeliconini) (Chen & He, 1997; Chen & van Achterberg, 2019; Zaldívar-Riverón et al., 2006). All the tribes are cosmopolitan (Butcher et al., 2012; Shaw et al., 2006; Shimbori & Shaw, 2014; van Achterberg & Shaw, 2016; Yu et al., 2016), except for the Stiropiini which is restricted to the New World (predominantly North and Central America) (van Achterberg, 1993). In some works, the subfamily was also taken to include various other components, notably Hormiinae and Lysiterminae. Characterization of this later Rogadinae *s.l.* is very problematic as being an extended subfamily, because of the lack of synapomorphies (Belokobylskij & Tobias, 1986; van Achterberg, 1995; Wharton, 1993; Whitfield & Wharton, 1997). However, a recent phylogenomic study using UCEs has conclusively shown that the Rogadinae *s.s.* constitute a monophyletic group to the exclusion of Hormiinae (including Lysitermini, Cedriini, and Chymelini) (Jasso-Martínez et al., 2020). Therefore, we treat the later separately in Chapter 18 even though a number of studies based on one or a few gene fragments usually recovered some Hormiinae nested among the Rogadinae s.s (Quicke et al., 2016; Ranjith et al., 2017).

Members of the Rogadinae *s.s.* are all koinobiont endoparasitoids attacking a wide range of lepidopterous larvae, and mummifying the host caterpillar (Shaw, 1983, 2006; Shaw & Huddleston, 1991; Shimbori & Shaw, 2014; Townsend & Shaw, 2009) although the biology of Betylobraconini remains unknown. Such mummified host caterpillars often are very host specific and provide useful information for diagnosing the rogadine species involved, so the host remains should be preserved with the wasp specimen whenever possible (Shaw, 2006; Shimbori &

Shaw, 2014). Monophyly of each the tribes Aleiodini, Clinocentrini, Stiropiini, Rogadini, and Yeliconini has been supported by molecular analysis of 28S ribosomal RNA (Chen et al., 2003; Jasso-Martínez et al., 2020; Quicke & Butcher, 2015; Zaldívar-Riverón et al., 2006).

Worldwide, *Aleiodes* is by far the commonest and most species-rich genus (Shaw et al., 2020). Revisionary studies of the New World *Aleiodes* have advanced in recent years by defining and examining species-groups (Marsh & Shaw, 1998, 1999, 2001, 2003; Shaw et al., 1997, 1998a,b, 2006, 2013, 2020; Shimbori & Shaw, 2014; Townsend & Shaw, 2009). Shaw et al. (1997) divided *Aleiodes* into 15 species-groups, with three additional groups proposed after phylogenetic analyses (Fortier & Shaw, 1999; Shimbori et al., 2016). However, researchers in the Palaearctic Region have tended not to follow these arrangements. In the Old World, *Aleiodes* has often been divided into a few subgenera, the ones relevant to the Iranian fauna being *Aleiodes*, *Chelonorhogas*, and *Neorhogas* (van Achterberg, 1991). However, molecular studies (Quicke et al., in prep; van Achterberg et al., 2020; Zaldívar-Riverón et al., 2008) firmly reject monophyly of the first two and therefore van Achterberg and Shaw (2016) propose instead referring to these as species-groups. Of the species treated here, the following are classified into the *Chelonorhogas* species-group (= *A. apicalis* "sp. Group"): *A. aestuosus*, *A. agilis*, *A. apicalis*, *A. dimidiatus*, *A. eurinus*, *A. gasterator*, *A. pallidicornis*, *A. rufipes*, *A. rugulosis*, *A. schirjajewi*, *A. unipunctator*, and *A. ductor*. The status of *Tetrasphaeropyx* (= *A. pilosus* species group) as a monophyletic group (Fortier, 2009; Fortier & Shaw, 1999) represented in Iran by *A. arcticus*, may be valid but requires molecular testing. The species included in *Neorhogas* are represented by only a single species in molecular studies to date and are represented in Iran only by *A. caucasicus*. Therefore, we refrain from including any species in formal subgenera.

Braconidae of the Middle East (Hymenoptera). https://doi.org/10.1016/B978-0-323-96099-1.00024-8

Checklists of Regional of Rogadinae: Farahbakhsh (1961) and Modarres Awal (1997) recorded an unknown species of *Rogas* as the parasitoid of *Spodoptera exigua* (Hübner, 1808) (Lepidoptera: Noctuidae though this is almost certainly a misidentification of an *Aleiodes* species since *Rogas* more commonly parasitizes Limacodidae); Fallahzadeh and Saghaei (2010): eight species in two genera; Farahani et al. (2015): species in four genera; Yu et al. (2016): 32 species in four genera; Beyarslan, Gadallah, & Ghahari (2017) and Beyarslan, Gözüaçik, Güllü, & Konuksal (2017): 33 species in four genera, and finally Samin, Coronado-Blanco, Kavallieratos et al. (2018) and Samin, Coronado-Blanco, Fischer et al. (2018) listed 37 species in four genera. In the present list, 68 species are reported in six genera and five tribes in the Middle East (Aleiodini Muesebeck, Clinocentrini van Achterberg, Rogadini Foerster, Pentatermini Belokobylskij, and Yeliconini van Achterberg. *Aleiodes dissector* (Nees) is recorded here as a new record for the Iranian fauna.

List of species of the subfamily Rogadinae recorded in the Middle East

Subfamily Rogadinae Foerster, 1863

Tribe Aleiodini Muesebeck, 1928

Genus *Aleiodes* Wesmael, 1838

Aleiodes aestuosus (Reinhard, 1863)

Catalogs with Iranian records: Farahani et al. (2015, 2016), Yu et al. (2016), Beyarslan, Gadallah, and Ghahari (2017), Beyarslan, Gözüaçik, Güllü, and Konuksal (2017), Samin, Coronado-Blanco, Kavallieratos et al. (2018), Samin, Coronado-Blanco, Fischer et al. (2018).

Distribution in Iran: Mazandaran (Farahani et al., 2015).

Distribution in the Middle East: Cyprus (Fahringer, 1932; Marshall, 1890; Shenefelt, 1975; Szépligeti, 1904; Zaldivar-Riverón et al., 2004), Iran (Farahani et al., 2015), Israel−Palestine (Papp, 1989, 2012), Jordan (van Achterberg et al., 2020), Syria (Szépligeti, 1901, 1904, 1906, Fahringer, 1932).

Extralimital distribution: Albania, Azerbaijan, Bulgaria, Georgia, Greece, Russia, Tunisia, Turkmenistan, Uzbekistan (van Achterberg et al., 2020; Yu et al., 2016).

Host records: Recorded by Tobias (1976, 1986) and van Achterberg et al. (2020) as being a parasitoid of the noctuids *Autographa gamma* (L.), and *Heliothis peltigera* (Denis and Schiffermüller).

Aleiodes agilis (Telenga, 1941)

Catalogs with Iranian records: Fallahzadeh and Saghaei (2010), Farahani et al. (2015, 2016), Yu et al. (2016), Beyarslan, Gadallah, and Ghahari (2017), Beyarslan, Gözüaçik, Güllü, and Konuksal (2017), Samin, Coronado-Blanco, Kavallieratos et al. (2018), Samin, Coronado-Blanco, Fischer et al. (2018).

Distribution in Iran: Iran (no specific locality cited) (Shenefelt, 1975; Telenga, 1941 as *Rhogas agilis*; Tobias, 1976, 1986).

Distribution in the Middle East: Iran.

Extralimital distribution: Armenia (van Achterberg et al., 2020; Yu et al., 2016).

Host records: Unknown.

Aleiodes albitibia (Herrich-Schäffer, 1838)

Catalogs with Iranian records: Beyarslan, Gadallah, & Ghahari (2017), Samin, Coronado-Blanco, Kavallieratos et al. (2018), Samin, Coronado-Blanco, Fischer et al. (2018).

Distribution in Iran: Guilan (Beyarslan, Gadallah, & Ghahari, 2017).

Distribution in the Middle East: Iran (Beyarslan, Gadallah, & Ghahari, 2017), Turkey (Beyarslan, 2015).

Extralimital distribution: Belgium, Canada, Finland, France, Germany, Hungary, Ireland, Korea, Lithuania, Netherlands, Poland, Russia, Sweden, Switzerland, United States of America (Shaw et al., 1998b), Ukraine, United Kingdom, Venezuela (Yu et al., 2016).

Host records: Summarized by Yu et al. (2016) as being a parasitoid of the notodontids *Natada gibbosa* (Smith), *Nerice bidentata* Walker, and *Notodonta dromadarius* (L.). It was also reared from the notodontid moths *Clostera pigra* (Hufnagel), *Notodonta ziczac* (L.), *Phalera bucephala* (L.), *Pheosia gnoma* (Fabricius), *P. tremula* (Clerck), *Pterosoma palpina* (Clerck), and *Ptilodon capucina* (L.) (van Achterberg & Shaw, 2016).

Aleiodes alternator (Nees von Esenbeck, 1834)

Catalogs with Iranian records: Farahani et al. (2015, 2016), Yu et al. (2016), Beyarslan, Gadallah, and Ghahari (2017), Beyarslan, Gözüaçik, Güllü, and Konuksal (2017), Samin, Coronado-Blanco, Kavallieratos et al. (2018), Samin, Coronado-Blanco, Fischer et al. (2018).

Distribution in Iran: Guilan (Farahani et al., 2015), Mazandaran (Kian et al., 2020).

Distribution in the Middle East: Iran.

Extralimital distribution: Belgium, China, France, Germany, Ireland, Latvia, Netherlands, Romania, Spain, United Kingdom.

Host records: Summarized by Yu et al. (2016) as being a parasitoid of the erebids *Cymbalophora pucida* (Esper), *Leucoma salicis* (L.), *Orgyia antiqua* (L.), and *Spiris striata* (L.); the noctuid *Noctua fimbriata* Schreber; and the nymphalids *Hippachia alcyone* (Denis and Schiffermüller), and *Hippachia semele* (L.).

Aleiodes apicalis (Brullé, 1832)

Catalogs with Iranian records: Fallahzadeh and Saghaei (2010 as *Aleiodes* (*Neorhogas*) *ductor* (Thunberg)), Farahani et al. (2015, 2016), Yu et al. (2016), Beyarslan, Gadallah, and Ghahari (2017), Beyarslan, Gözüaçik,

Güllü, and Konuksal (2017), Samin, Coronado-Blanco, Kavallieratos et al. (2018), Samin, Coronado-Blanco, Fischer et al. (2018).

Distribution in Iran: Guilan, Mazandaran (Farahani et al., 2015), Isfahan (Ghahari et al., 2011a as *Aleiodes (Neorhogas) ductor* (Thunberg)), Kerman (Abdolalizadeh et al., 2017; Madjdzadeh et al., 2021), Sistan and Baluchestan (Hedwig, 1957 as *Aleiodes ductor*), West Azarbaijan (Samin et al., 2014 as *Aleiodes ductor*), Iran (no specific locality cited) (Tobias, 1976, 1986).

Distribution in the Middle East: Cyprus (van Achterberg et al., 2020; Zaldivar-Riverón et al., 2004), Iran (see references above), Iraq, Israel—Palestine, Oman, Syria (van Achterberg et al., 2020), Turkey (Aydoğdu & Beyarslan, 2006).

Extralimital distribution: Albania, Austria, Bosnia-Herzegovina, Bulgaria, Croatia, Czech Republic, France, Georgia, Germany, Greece, Hungary, Italy, Kazakhstan, North Macedonia, Malta, Moldova, Montenegro, Morocco, Portugal, Romania, Russia, Serbia, Slovakia, Spain, Switzerland, Tunisia, Turkmenistan (van Achterberg et al., 2020; Yu et al., 2016).

Host records: Recorded by van Achterberg et al. (2020) as being a parasitoid of the gelechiid *Anarsia lineatella* Zeller; and the noctuids *Autographa gamma* (L.), and *Sesamia* sp.

Aleiodes arabiensis Butcher and Quicke, 2015

Distribution in the Middle East: Saudi Arabia (Butcher & Quicke, 2015).

Extralimital distribution: None.

Host records: Unknown.

Aleiodes arcticus (Thomson, 1892)

Catalogs with Iranian records: No catalog.

Distribution in Iran: Markazi (Ameri et al., 2020).

Distribution in the Middle East: Iran (Ameri et al., 2020), Turkey (Beyarslan, 2015).

Extralimital distribution: Finland, Germany, Mongolia, Poland, Russia, Sweden, Switzerland, United Kingdom (Yu et al., 2016).

Host records: Summarized by Yu et al. (2016) as a koinobiont endoparasitoid of the geometrid larvae of *Itame wanaria* L., *Macaria brunneata* (Thunberg), and *Macaria fusca* (Thunberg).

Aleiodes arnoldii (Tobias, 1976)

Catalogs with Iranian records: Farahani et al. (2015, 2016), Yu et al. (2016), Beyarslan, Gadallah, and Ghahari (2017), Beyarslan, Gözüaçik, Güllü, and Konuksal (2017), Samin, Coronado-Blanco, Kavallieratos et al. (2018), Samin, Coronado-Blanco, Fischer et al. (2018).

Distribution in Iran: Alborz, Guilan, Tehran (Farahani et al., 2015; Pasandideh Saqalaksari et al., 2020), Mazandaran (Farahani et al., 2015; Kian et al., 2020; Pasandideh Saqalaksari et al., 2020).

Distribution in the Middle East: Iran (see references above), Turkey (van Achterberg et al., 2020).

Extralimital distribution: Azerbaijan (van Achterberg et al., 2020).

Host records: Unknown.

Aleiodes aterrimus (Ratzeburg, 1852)

Catalogs with Iranian records: No catalog.

Distribution in Iran: Guilan (Sakenin et al., 2020).

Distribution in the Middle East: Iran (Sakenin et al., 2020), Turkey (Aydoğdu & Beyarslan, 2005; van Achterberg et al., 2020).

Extralimital distribution: Austria, Czech Republic, France, Germany, Hungary, Italy, Moldova, Norway, Poland, Russia, Serbia, Spain, Ukraine, United Kingdom (Yu et al., 2016).

Host records: Recorded by Shaw (1981) as being a parasitoid of the noctuid *Amphipyra pyramidea* (L.).

Aleiodes bicolor (Spinola, 1808)

Catalogs with Iranian records: Fallahzadeh and Saghaei (2010), Farahani et al. (2015, 2016), Yu et al. (2016), Beyarslan, Gadallah, and Ghahari (2017), Beyarslan, Gözüaçik, Güllü, and Konuksal (2017), Samin, Coronado-Blanco, Kavallieratos et al. (2018), Samin, Coronado-Blanco, Fischer et al. (2018).

Distribution in Iran: Alborz, Guilan, Qazvin, Tehran (Farahani et al., 2015), Fars (Lashkari Bod et al., 2011), Isfahan (Ghahari et al., 2011a), Kerman (Asadizade et al., 2014; Abdolalizadeh et al., 2017; Madjdzadeh et al., 2021), Mazandaran (Ghahari et al., 2009 as *Rogas bicolor* Spinola, Farahani et al., 2015), Sistan and Baluchestan (Hedwig, 1957 as *Rhogas difficilis* Kokujev and *Rhogas incertus* Kokujev), Iran (no specific locality cited) (Telenga, 1941; Shenefelt, 1975 as *Aleiodes basalis* (Costa), as *Aleiodes bicolor* and *Aleiodes difficilis* Kokujev; Papp, 2011—12; Yu et al., 2016).

Distribution in the Middle East: Cyprus (Beyarslan, Gözüaçik, Güllü, and Konuksal, 2017), Israel—Palestine (Papp, 1970), Jordan (Yu et al., 2016), Turkey (Aydoğdu & Beyarslan, 2005, 2006).

Extralimital distribution: Widely distributed in the Palaearctic countries [Adjacent countries to Iran: Afghanistan, Armenia, Azerbaijan, Kazakhstan, Russia, Turkmenistan].

Host records: Summarized by Yu et al. (2016) as being a parasitoid of the crambids *Loxostege sticticalis* (L.), *Pyrausta purpuralis* (L.), and *Pyrausta sanguinalis* (L.); the erebids *Dasychira albodentata* Bremer, and *Leucoma salicis* (L.); the geometrids *Apocheima cinerarius* (Erschoff), *Archiearis parthenias* (L.), *Eupithesia linariata* (Denis and Schiffermüller), and *Operophtera* sp.; the lycaenids *Aricia agestis* (Denis and Schiffermüller), *Aricia Artaxerxes* (Fabricius), *Cupido alcetas* (Hoffmannsegg), *Cupido minimus* (Füssli), *Lysandra coridon* (Poda),

Plebejus idas (L.), and *Polyommatus* spp.; the noctuid *Apamea sordens* (Hufnagel); the nymphalids *Aglais urticae* (L.), and *Maniola jurtina* (L.); the pterophorids *Emmelina monodactyla* (L.), and *Hellinsia tephradactyla* (Hübner); and the zygaenids *Jordanita chloros* (Hübner), *Jordanita graeca* (Jordan), *Rhagades pruni* (Denis and Schiffermüller), and *Zygaena* spp.

Aleiodes caucasicus (Tobias, 1976)
Catalogs with Iranian records: No catalog.
Distribution in Iran: Ardabil (Samin, Beyarslan, Coronado-Blanco et al., 2020).
Distribution in the Middle East: Iran (Samin, Beyarslan, Coronado-Blanco et al., 2020), Turkey (Aydoğdu & Beyarslan, 2005; van Achterberg et al., 2020).
Extralimital distribution: Bulgaria, Russia (van Achterberg et al., 2020).
Host records: Unknown.

Aleiodes circumscriptus (Nees von Esenbeck, 1834)
Catalogs with Iranian records: Fallahzadeh and Saghaei (2010), Farahani et al. (2015, 2016), Yu et al. (2016), Beyarslan, Gadallah, and Ghahari (2017), Samin, Coronado-Blanco, Kavallieratos et al. (2018), Samin, Coronado-Blanco, Fischer et al. (2018).
Distribution in Iran: Fars (Lashkari Bod et al., 2011), Golestan (Ghahari et al., 2009 as *Rogas circumscriptus* Nees, 1834), Guilan (Farahani et al., 2015; Ghahari & Sakenin, 2018 as *Rogas circumscriptus* (Nees)), Kerman (Ghahari, 2020), Mazandaran (Farahani et al., 2015), Ilam (Ghahari et al., 2011c), Kerman (Abdolalizadeh et al., 2017), West Azarbaijan (Rastegar et al., 2012), Iran (no specific locality cited) (Telenga, 1941 as *Rhogas circumscriptus*; Shenefelt, 1975).
Distribution in the Middle East: Iran (see references above), Turkey (Aydoğdu & Beyarslan, 2005, 2006).
Extralimital distribution: Widely distributed in the Palaearctic countries [Adjacent countries to Iran: Armenia, Azerbaijan, Kazakhstan, Russia].
Host records: Summarized by Yu et al. (2016) as being a parasitoid of the crambid *Anania crocealis* (Hübner); the erebids *Atolmis rubricollis* (L.), and *Leucoma salicis* (L.); the gelechiid *Caryocolum amaurella* (Hering); the geometrids *Abraxas grossulariata* (L.), *Bupalus piniarius* (L.), *Cyclophora ruficiliaria* (Herrich-Schäffer), *Epirrhoe galiata* (Denis and Schiffermüller), *Epirrita autumnata* (Borkhausen), *Eupithecia* spp., *Gymnoscelis rufifasciata* (Haworth), *Hylaea fasciaria* (L.), *Lycia isabellae* (Harrison), and *Thera variata* (Denis and Schiffermüller); the noctuids *Apamea crenata* (Hufnagel), *Autographa gamma* L., *Diarsia rubi* (Vieweg), *Helicoverpa armigera* (Hübner), *Hoplodrina octogenaria* (Goeze), *Lithomoia solidaginis* (Hübner), *Mamestra brassicae* (L.), *Noctua fimbriata* (Schreber), *Noctua pronuba* (L.), *Orthosia* spp.,

Penicillaria jocosatrix Guenée, *Pseudaletia unipuncta* (Haworth), *Spodoptera exigua* (Hübner), and *Xestia* spp.; the pterophorid *Pterophorus pentadactyla* (L.); the pyralid *Dioryctria abietella* (Denis and Schiffermüller); and the tortricids *Archips rosana* (L.), *Croesia bergmanniana* (L.), *Croesia holmiana* (L.), *Cydia strobilella* (L.), *Pammene amygdalana* (Duponchel), *Pammene gallicana* (Guenée), *Tortrix viridana* L., and *Zeiraphera griseana* (Hübner). It was also reared from the erebid larva of *Hypena proboscidalis* (L.) (van Achterberg & Shaw, 2016).

Aleiodes compressor (Herrich-Schäffer, 1838)
Catalogs with Iranian records: Farahani et al. (2015, 2016), Yu et al. (2016), Beyarslan, Gadallah, and Ghahari (2017), Beyarslan, Gözüaçik, Güllü, and Konuksal (2017), Samin, Coronado-Blanco, Kavallieratos et al. (2018), Samin, Coronado-Blanco, Fischer et al. (2018).
Distribution in Iran: East Azarbaijan (Samin, Beyarslan, Ranjith et al., 2020), Ilam (Ghahari et al., 2011c as *Petalodes compressor* (Herrich-Schäffer, 1838)).
Distribution in the Middle East: Iran.
Extralimital distribution: Austria, Belgium, Bulgaria, China, Croatia, Czech Republic, Finland, France, Georgia, Germany, Hungary, Kazakhstan, Korea, Lithuania, Montenegro, Netherlands, Norway, Poland, Russia, Serbia, Slovakia, Spain, Sweden, Switzerland, United Kingdom.
Host records: Summarized by Yu et al. (2016) as being a parasitoid of the drepanid *Tethea or* (Denis and Schiffermüller); the erebid *Leucoma salicis* (L.); the geometrids *Apocheima hispidaria* (Denis and Schiffermüller), and *Hydriomena furcata* (Thunberg); the limacodid *Apoda limacodes* (Hufnagel); the nolids *Nycteola asiatica* (Krulikovsky), and *Nycteola revayana* (Scopoli); the notodontid *Clostera anachoreta* (Denis and Schiffermüller), and *Clostera pigra* (Hufnagel); and the tortricids *Acleris hastiana* (L.), *Archips rosana* (L.), and *Gypsonoma minutana* (Hübner). In Iran, this species has been reared from the tortricid *Archips rosana* L. (Samin, Beyarslan, Ranjith et al., 2020).
Comments: This species was treated under the genus *Petalodes* Wesmael. Based on van Achterberg (1991, 2014) and Chen and He (1997), both morphological analysis and DNA data (Fortier & Shaw, 1999; Quicke et al., 2006, respectively), the genus *Petalodes* Wesmal is synonymized with the genus *Aleiodes* (van Achterberg, 2014).

Aleiodes coxalis (Spinola, 1808)
Catalogs with Iranian records: Beyarslan, Gadallah, and Ghahari (2017), Samin, Coronado-Blanco, Kavallieratos et al. (2018), Samin, Coronado-Blanco, Fischer et al. (2018).
Distribution in Iran: Zanjan (Beyarslan, Gadallah, and Ghahari (2017)).
Distribution in the Middle East: Iran.

Extralimital distribution: Widely distributed in the Palaearctic and Oriental realms and introduced to Canada [Adjacent countries to Iran: Kazakhstan, Russia, Turkmenistan].

Host records: Summarized by Yu et al. (2016) as being a parasitoid of the crambids *Cnaphalocrocis medinalis* (Guenée), and *Ostrinia nubilalis* (Hübner); the erebid *Erebia* sp.; the hesperiid *Thymelicus lineola* (Ochsenheimer); the geometrids *Eupithesia pimpinellata* (Hübner), and *Scopula nigropunctata* (Hufnagel); the noctuid *Sideridis rivularis* Fabricius; the nymphalids *Coenonympha pamphilus* L., *Coenonympha tullia* (Müller), *Limenitis populi* (L.), *Maniola jurtina* (L.), *Melanargia Lachesis* Hübner, and *Pyronia tithonus* (L.); and the pterophorid *Emmelina monodactyla* (L.).

Aleiodes crassipes (Thomson, 1892)
Catalogs with Iranian records: Farahani et al. (2015, 2016), Yu et al. (2016), Beyarslan, Gadallah, and Ghahari (2017)), Samin, Coronado-Blanco, Kavallieratos et al. (2018), Samin, Coronado-Blanco, Fischer et al. (2018).
Distribution in Iran: Isfahan (Ghahari et al., 2011a).
Distribution in the Middle East: Iran.
Extralimital distribution: Finland, Germany, Greece, Hungary, Mongolia, Spain, Sweden, Switzerland.
Host records: Unknown.

Aleiodes cruentus (Nees von Esenbeck, 1834)
Catalogs with Iranian records: No catalog.
Distribution in Iran: East Azarbaijan (Samin, Beyarslan, Ranjith et al., 2020).
Distribution in the Middle East: Iran.
Extralimital distribution: Austria, Bulgaria, Croatia, Czech Republic, Finland, France, Germany, Greece, Italy, Moldova, Netherlands, Norway, Romania, Slovakia, Slovenia, Spain, Sweden, Ukraine (Shaw et al., 2020).
Host records: Summarized by Yu et al. (2016) as being a parasitoid of the noctuid *Sideridis rivularis* (Fabricius); and the crambid *Teia antiquiodes* (Hübner). It has also been recorded as being a parasitoid of the noctuid *Hadena confusa* (Hufnagel) (Shaw et al., 2020).

Aleiodes curticornis van Achterberg and Shaw, 2016
Distribution in the Middle East: Turkey (van Achterberg & Shaw, 2016).
Extralimital distribution: Austria, Finland, France, Hungary, Italy, Romania, Slovakia, Slovenia, Spain (van Achterberg & Shaw, 2016).
Host records: Unknown.

Aleiodes dimidiatus (Spinola, 1808)
Catalogs with Iranian records: Fallahzadeh and Saghaei (2010), Farahani et al. (2015, 2016), Yu et al. (2016), Beyarslan, Gadallah, and Ghahari (2017), Samin,

Coronado-Blanco, Kavallieratos et al. (2018), Samin, Coronado-Blanco, Fischer et al. (2018).
Distribution in Iran: Kerman (Asadizade et al., 2014), Iran (no specific locality cited) (Telenga, 1941 as *Rhogas dimidiatus*).
Distribution in the Middle East: Iran (Telenga, 1941; Asadizade et al., 2014), Turkey (Aydoğdu & Beyarslan, 2005, 2006; Beyraslan et al., 2002, 2006).
Extralimital distribution: Widely distributed in the Oriental and Palaearctic countries [Adjacent countries to Iran: Afghanistan, Azerbaijan, Russia, Turkey and Turkmenistan].
Host records: Summarized by Yu et al. (2016) as being a parasitoid of the following erebids: *Arctia caja* (L.), *Diacrisia irene* Butler, *Diacrisia sannio* (L.), *Orgyia antiqua* (L.), *Orgyia splendida* (Rambur), and *Palearctia gratiosa* (Grum-Grshimailo); the lasiocampids *Euthrix potatoria* (L.), *Lasiocampa quercus* (L.), and *Macrothylacia rubi* (L.); the noctuids *Agrotis* spp., *Apamea anceps* (Denis and Schiffermüller), *Caradrina morpheus* (Hufnagel), *Cerapteryx graminis* (L.), *Cosmia subtilis* Staudinger, *Euxoa* spp., *Helicoverpa armigera* (Hübner), *Helicoverpa zea* (Boddie), *Hoplodrina blanda* (Denis and Schiffermüller), *Hoplodrina octogenaria* (Goeze), and *Mythimna separata* Walker; and the thaumetopoeid *Thaumetopoea processionea* (L.).

Aleiodes dissector (Nees von Esenbeck, 1834)
Catalogs with Iranian records: This species is a new record for the fauna of Iran.
Distribution in Iran: Mazandaran province, Ramsar, Dalkhani forest, 1♀, 1♂, July 2013.
Distribution in the Middle East: Iran (new record), Turkey (Beyarslan, 2015; Quicke et al., 2006; van Achterberg et al., 2020).
Extralimital distribution: Belgium, Croatia, Czech Republic, Finland, France, Germany, Hungary, Ireland, Japan, Korea, Latvia, Lithuania, Mongolia, Montenegro, Netherlands, Norway, Poland, Russia, Slovakia, Spain, Sweden, Switzerland, Ukraine, United Kingdom (Yu et al., 2016).
Host records: Summarized by Yu et al. (2016) as being a parasitoid of the noctuids *Acronicta rumicis* (L.), *Acronicta tridens* (Denis and Schiffermüller), and *Orthosia incerta* (Hufnagel).

Aleiodes ductor (Thunberg, 1822)
Catalogs with Iranian records: Yu et al. (2016), Beyarslan, Gadallah, and Ghahari (2017).
Distribution in Iran: Isfahan (Ghahari et al., 2011a), Sistan and Baluchestan (Hedwig, 1957 as *Rhogas ductor* var. *similis* Szépligeti), West Azerbaijan (Samin et al., 2014), Iran (no specific locality cited) (Shenefelt, 1975 as *Rogas ductor*; Tobias, 1986 as *Rogas ductor*).

Distribution in the Middle East: Cyprus, Egypt (Yu et al., 2016), Iran (Beyarslan, Gadallah, & Ghahari, 2017), Israel–Palestine (Papp, 1970), Turkey (Beyarslan, 2015; Beyarslan et al., 2002, 2006).

Extralimital distribution: Widely distributed in the Palaearctic countries [Adjacent countries to Iran: Afghanistan, Armenia, Azerbaijan, Kazakhstan, Russia, Turkmenistan].

Host records: Summarized by Yu et al. (2016) as being a parasitoid of the lasiocampid *Euthrix potatoria* (L.); the noctuids *Autographa gamma* (L.), *Hadula trifolii* (Hufnagel), and *Mamestra brassicae* (L.); the nymphalid *Brenthis ino* Rottemberg; and the sesiid *Synanthedon scoliaeformis* (Borkhausen).

Aleiodes esenbeckii (Hartig, 1838)

Catalogs with Iranian records: Farahani et al. (2016), Yu et al. (2016), Beyarslan, Gadallah, and Ghahari (2017), Beyarslan, Gözüaçik, Güllü, and Konuksal (2017), Samin, Coronado-Blanco, Kavallieratos et al. (2018), Samin, Coronado-Blanco, Fischer et al. (2018).

Distribution in Iran: East Azarbaijan (Samin, 2015), Razavi Khorasan (Samin et al., 2011).

Distribution in the Middle East: Iran (Samin, 2015; Samin et al., 2011), Turkey (Beyarslan, 2015).

Extralimital distribution: Widely distributed in the Palaearctic and Oriental countries [Adjacent countries to Iran: Afghanistan, Russia].

Host records: Summarized by Yu et al. (2016) as being a parasitoid of the endromid *Endromis versicolora* (L.); the lasiocampids *Dendrolimus* spp. and *Cosmotriche lobulina* (Denis and Schiffermüller); and the zygaenid *Zygaena lonicerae* (Scheven).

Aleiodes eurinus (Telenga, 1941)

Catalogs with Iranian records: Farahani et al. (2015, 2016), Yu et al. (2016), Beyarslan, Gadallah, and Ghahari (2017), Beyarslan, Gözüaçik, Güllü, and Konuksal (2017), Samin, Coronado-Blanco, Kavallieratos et al. (2018), Samin, Coronado-Blanco, Fischer et al. (2018).

Distribution in Iran: Guilan (Farahani et al., 2015).

Distribution in the Middle East: Cyprus (Beyarslan, Gözüaçik, Güllü, and Konuksal, 2017), Iran (Farahani et al., 2015), Turkey (van Achterberg et al., 2020).

Extralimital distribution: China, Italy, Mongolia, Russia, Spain (van Achterberg et al., 2020).

Host records: Recorded by Tobias (1986) as a parasitoid of the noctuids *Apamea anceps* (Denis and Schiffermüller) and *Euxoa ochrogaster* (Guenée).

Aleiodes fortipes (Reinhard, 1863)

Distribution in the Middle East: Turkey (van Achterberg et al., 2020).

Extralimital distribution: Austria, British Isles, Bulgaria, Croatia, Czech Republic, Finland, France, Germany, Hungary, Netherlands, Poland, Spain, Sweden, former Yugoslavia (van Achterberg et al., 2020; Yu et al., 2016).

Host records: Recorded by van Achterberg et al. (2020) as being a parasitoid of the geometrid, either *Idaea aversata* (L.) or *Idaea straminata* (Borkhausen).

Aleiodes gasterator (Jurine, 1807)

Catalogs with Iranian records: Farahani et al. (2015, 2016), Yu et al. (2016), Beyarslan, Gadallah, and Ghahari (2017), Beyarslan, Gözüaçik, Güllü, and Konuksal (2017), Samin, Coronado-Blanco, Kavallieratos et al. (2018), Samin, Coronado-Blanco, Fischer et al. (2018).

Distribution in Iran: East Azarbaijan (Rastegar et al., 2012), Isfahan (Ghahari et al., 2011a).

Distribution in the Middle East: Cyprus, Iraq, Jordan, Syria (van Achterberg et al., 2020), Iran (Ghahari et al., 2011a; Rastegar et al., 2012), Turkey (Aydoğdu & Beyarslan, 2005; Papp, 1985; van Achterberg et al., 2020).

Extralimital distribution: Albania, Greece, Italy, Macedonia, Portugal, Spain, Tunisia, Turkmenistan (van Achterberg et al., 2020).

Host records: Summarized by Yu et al. (2016) as being a parasitoid of the erebids *Euproctis chrysorrhoea* (L.), and *Orgyia splendida* (Rambur); and the noctuid *Autographa gamma* (L.). It was also reared from low-feeding noctuids: *Agrotis segetum* (Denis and Schiffermüller), *Agrotis* sp., *Spodoptera littoralis* (Boisduval) (van Achterberg et al., 2020).

Aleiodes gastritor (Thunberg, 1822)

Catalogs with Iranian records: Farahani et al. (2015, 2016), Yu et al. (2016), Beyarslan, Gadallah, and Ghahari (2017), Beyarslan, Gözüaçik, Güllü, and Konuksal (2017), Samin, Coronado-Blanco, Kavallieratos et al. (2018), Samin, Coronado-Blanco, Fischer et al. (2018).

Distribution in Iran: East Azarbaijan (Ghahari et al., 2009 as *Rogas rossicus* Kokujev; Rastegar et al., 2012), Qazvin (Ghahari et al., 2011b).

Distribution in the Middle East: Iran (see references above), Israel–Palestine (Papp, 1970), Turkey (Aydoğdu & Beyarslan, 2005, 2006; Beyarslan, 2015).

Extralimital distribution: Most areas of Nearctic (Shaw, 2006; Townsend & Shaw, 2009), Oriental and Palaearctic regions [Adjacent countries to Iran: Russia] (Yu et al., 2016).

Host records: Summarized by Yu et al. (2016) as being a parasitoid of the crambids *Anania coronata* (Hufnagel) and *Ostrinia nubilalis* (Hübner); the depressariid *Depressaria absinthiella* Herrich-Schäffer; the drepanid *Cilix glaucata* (Scopoli); the erebids *Euproctis chrysorrhoea* (L.), *Euproctis similis* (Füssli), *Hypena scabra* Fabricius, and *Leucoma salicis* (L.); the geometrids *Alsophila pometaria* (Harris), *Apocheima cinerarius* (Erschoff), *Apocheima hispidaria* (Denis and Schiffermüller), *Chiasmia clethrata* (L.), *Chloroclystis v-ata* (Howarth), *Digrammia gnophosaria* (Guenée), *Epirrita automnata* (Borkhausen), *Eupithesia* spp., *Glena cribrataria* (Guenée), *Hylaea fasciaria* (L.), *Hypagyrtis unipunctata* (Haworth), *Isturgia arenacearia* (Denis and Schiffermüller), *Isturgia limbaria* (Fabricius), *Lycia hirtaria* (Clerck), *Lycia pomonaria* (Hübner), *Operophtera brumata* (L.), and *Phthonandria atritilineata* (Butler); the lasiocampid *Malacosoma neustria* (L.); the noctuids *Agrapha agnata* (Staudinger), *Autographa gama* (L.), *Helicoverpa armigera* (Hübner), *Pseudaletia unipuncta* (Haworth), *Spodoptera exigua* (Hübner), and *Trichoplusia ni* (Hübner); the notodontid *Cerura vinula* (L.); the plutellid *Prays oleae* (Bernard); the thaumetopoeid *Thaumetopoea processionea* (L.); and the tortricids *Archips rosana* (L.) and *Lobesia botrana* (Denis and Schiffermüller).

Aleiodes leptofemur van Achterberg and Shaw, 2016

Distribution in the Middle East: Cyprus (van Achterberg & Shaw, 2016).

Extralimital distribution: Andorra, Austria, Belgium, British Isles, Bulgaria, Czech Republic, France, Finland, Germany, Greece, Hungary, Italy, Netherlands, Norway, Slovakia, Spain, Sweden, Switzerland.

Host records: Recorded by van Achterberg and Shaw (2016) as being a parasitoid of the erebids *Cucullia chamomillae* (Denis and Schiffermüller), and *Shargacucullia verbasci* (L.); as well as the following noctuids: *Ammoconia caecimacula* (Denis and Schiffermüller), *Autographa gamma* (L.), ?*Cerastis rubricosa* (Denis and Schiffermüller), ?*Diarsia rubi* (Viweg), *Dicestra trifolii* (Hufnagel), *Euplexia lucipara* (L.), *Lacanobia oleracea* (L.), *Melanchra pisi* (L.), *Mythimna ferrago* (Fabricius), ? *Mythimna impura* (Hübner), ?*Mythimna littoralis* (Curtis), *Noctua comes* Hübner, *Noctua finbriata* (Schreber), *Noctua interjecta* Hübner, *Noctua janthina* (Denis and Schiffermüller), *Nuctua orbana* (Hufnagel), *Noctua pronuba* (L.), *Orthosia gracilis* (Denis and Schiffermüller), *Paradiarsia glareosa* (Esper), *Phlogophora meticulosa* (L.), *Stilbia anomala* (Haworth), *Xestia agathina* (Duponchel), *Xestia baja* (Denis and Schiffermüller), *Xestia castanea* (Esper), and *Xestia xanthographa* (Denis and Schiffermüller).

Aleiodes miniatus (Herrich-Schäffer, 1838)

Catalogs with Iranian records: No catalog.

Distribution in Iran: Ardabil (Samin, Beyarslan, Ranjith et al., 2020).

Distribution in the Middle East: Iran (Samin, Beyarslan, Ranjith et al., 2020), Syria (Fahringer, 1932), Turkey (Beyarslan, 2015).

Extralimital distribution: Austria, Czech Republic, Finland, France, Germany, Hungary, Kazakhstan, Kyrgyzstan, Romania, Russia, Sweden, Ukraine (Shaw et al., 2020).

Host records: Unknown.

Aleiodes modestus (Reinhard, 1863)

Catalogs with Iranian records: No catalog.

Distribution in Iran: East Azarbaijan (Samin, Sakenin Chelav et al., 2020).

Distribution in the Middle East: Iran.

Extralimital distribution: Austria, British Isles, Bulgaria, Czech Republic, Denmark, Finland, France, Germany, Italy, Netherlands, Poland, Romania, Russia, Slovakia, Sweden, Switzerland (van Achterberg & Shaw, 2016).

Host records: Recorded by van Achterberg and Shaw (2016) as being a parasitoid of the geometrids *Eupithecia absinthiata* (Clerck), *Eupithecia gelidata hyperboreata* Staudingar, *Eupithecia goossensiata* Mabille, *Eupithecia innotata* (Hufnagel), *Eupithecia lariciata* (Freyer), *Eupithecia nanata* (Hübner), *Eupithecia satyrata* (Hübner), *Eupithecia subfuscata* (Haworth), *Eupithecia succenturiata* (L.), and *Eupithecia vulgata* (Haworth).

Aleiodes moldavicus Tobias, 1986

Distribution in the Middle East: Turkey (Beyarslan, 2015).

Extralimital distribution: Moldova.

Host records: Unknown.

Aleiodes nigricornis Wesmael, 1838

Catalogs with Iranian records: Farahani et al. (2015, 2016), Yu et al. (2016), Beyarslan, Gadallah, and Ghahari (2017), Beyarslan, Gözüaçik, Güllü, and Konuksal (2017), Samin, Coronado-Blanco, Kavallieratos et al. (2018), Samin, Coronado-Blanco, Fischer et al. (2018).

Distribution in Iran: Ilam (Ghahari et al., 2011c), Mazandaran (Kian et al., 2020).

Distribution in the Middle East: Iran.

Extralimital distribution: Armenia, Azerbaijan, Belgium, Bulgaria, Croatia, Czech Republic, Faeroe Islands, Finland, France, Germany, Greece, Hungary, Ireland, Italy, Korea, Latvia, Lithuania, Netherlands, Poland, Russia, Serbia, Slovakia, Spain, Sweden, Switzerland, United Kingdom.

Host records: Summarized by Yu et al. (2016) as being a parasitoid of the geometrids *Bupalus piniarius* (L.) and *Hylaea fasciaria* (L.); the noctuids *Apamea crenata* (Hufnagel), *Cucullia serraticornis* Lintner, *Lithomoia solidaginis* (Hübner), *Mythimna farrago* (Fabricus), *Orthosia gothica* (L.), and *Orthosia incerta* (Hufnagel); and the pyralid *Dioryctria abietella* (Denis and Schiffermüller). It was also reared from the noctuids *Apamea epomidion* (Haworth), *Apamea*? *remisa* (Hübner), *Polia* sp., and *Xestia xanthographa* (Denis and Schiffermüller) (van Achterberg & Shaw, 2016).

Aleiodes nobilis (Curtis, 1834)

Catalogs with Iranian records: No catalog.

Distribution in Iran: Fars (Sakenin et al., 2020).

Distribution in the Middle East: Iran.

Extralimital distribution: Bulgaria, former Czechoslovakia, Finland, Germany, Hungary, Ireland, Italy, Moldova, Netherlands, Russia, Sweden, United Kingdom (Yu et al., 2016).
Host records: Unknown.

Aleiodes nocturnus (Telenga, 1941)

Catalogs with Iranian records: Fallahzadeh and Saghaei (2010), Farahani et al. (2015), Yu et al. (2016), Beyarslan, Gadallah, and Ghahari (2017), Beyarslan, Gözüaçik, Güllü, and Konuksal (2017), Samin, Coronado-Blanco, Kavallieratos et al. (2018), Samin, Coronado-Blanco, Fischer et al. (2018).
Distribution in Iran: Guilan, Tehran (Farahani et al., 2015), Kerman (Abdolalizadeh et al., 2017; Madjdzadeh et al., 2021), Iran (no specific locality cited) (Papp, 2012; Shenefelt, 1975; Telenga, 1941; Tobias, 1986).
Distribution in the Middle East: Iran (see references above), Israel–Palestine (Papp, 2012), Turkey (Aydoğdu & Beyarslan, 2005, 2006; Beyarslan, 2015).
Extralimital distribution: China, Hungary, Kazakhstan, Mongolia, Slovakia, Tajikistan, Turkmenistan, Uzbekistan.
Host records: Summarized by Yu et al. (2016) as being a parasitoid of the noctuids *Helicoverpa armigera* (Hübner) and *Helicoverpa zea* (Boddie); the notodontids *Cerura menciana* Moore, *Cerura vinula* (L.), and *Harpyia hermelina* (Goeze); and the nymphalid *Polygonia a-album* (L.).

Aleiodes pallescens Hellén, 1927

Catalogs with Iranian records: Farahani et al. (2015, 2016), Yu et al. (2016), Beyarslan, Gadallah, and Ghahari (2017), Beyarslan, Gözüaçik, Güllü, and Konuksal (2017), Samin, Coronado-Blanco, Kavallieratos et al. (2018), Samin, Coronado-Blanco, Fischer et al. (2018).
Distribution in Iran: Ardabil (Rastegar et al., 2012), Isfahan (Ghahari et al., 2011a).
Distribution in the Middle East: Iran.
Extralimital distribution: China, Finland, Mongolia, Montenegro, Russia, Switzerland.
Host records: Summarized by Yu et al. (2016) as being a parasitoid of the notodontids *Cerura menciana* Moore, *Cerura vinula* (L.), and *Furcula bifida* (Brahm).

Aleiodes pallidator (Thunberg, 1822)

Catalogs with Iranian records: Farahani et al. (2015, 2016), Yu et al. (2016), Beyarslan, Gadallah, and Ghahari (2017), Beyarslan, Gözüaçik, Güllü, and Konuksal (2017), Samin, Coronado-Blanco, Kavallieratos et al. (2018), Samin, Coronado-Blanco, Fischer et al. (2018).
Distribution in Iran: Kerman (Ghahari et al., 2010), Khuzestan (Samin et al., 2015), Qazvin (Ghahari et al., 2011b), Iran (no specific locality cited) (Papp, 1989).
Distribution in the Middle East: Iran (see references above), Israel–Palestine (Papp, 1989), Turkey (Aydoğdu & Beyarslan, 2005, 2006; Beyarslan, 2015).

Extralimital distribution: Nearctic, Neotropical, Palaearctic; introduced in the United States of America (Shaw, 2006; Shaw et al., 2013).
Host records: Summarized by Yu et al. (2016) as being a parasitoid of the erebids *Dasichira* sp., *Euproctis chrysorrhoea* (L.), *Leucoma salicis* (L.), *Lymantria dispar* (L.), *Orgyia definita* Packard, and *Orgyia leucostigma* (Smith); the gelechiids *Anarsia lineatella* Zeller and *Caryocolum amaurella* (Hering); the geometrids *Bupalus piniarius* (L.), *Cyclophora* spp., *Dyscia conspersaria* (Denis and Schiffermüller), *Eupethecia pimpinellata* (Hübner), *Hyleae fasciaria* (L.); the noctuid *Agrotis segetum* (Denis and Schiffermüller) and *Spodoptera exigua* (Hübner); the psychid *Megalophanes viciella* (Denis and Schiffermüller); and the tortricid *Apotomis sororculana* (Zetterstedt). It was also reared from the erebid larvae of *Dicallomera fascelina* (L.), *Euproctis similis* (Füssli), and *Orgyia antiqua* (L.) (van Achterberg & Shaw, 2016).

Aleiodes pallidicornis (Herrich-Schäffer, 1838)

Catalogs with Iranian records: Samin, Coronado-Blanco, Kavallieratos et al. (2018), Samin, Coronado-Blanco, Fischer et al. (2018).
Distribution in Iran: Zanjan (Ghahari and Beyarslan, 2017).
Distribution in the Middle East: Iran (Ghahari & Beyarslan, 2017), Turkey (van Achterberg et al., 2020).
Extralimital distribution: Austria, Belarus, British Isles (Scotland), Bulgaria, Croatia, Czech Republic, Hungary, Italy, Montenegro, Netherlands, North Korea, Romania, Russia, Slovakia, Switzerland (van Achterberg et al., 2020).
Host records: Unknown.

Aleiodes periscelis (Reinhard, 1863)

Catalogs with Iranian records: No catalog.
Distribution in Iran: Kordestan (Sakenin et al., 2020).
Distribution in the Middle East: Iran.
Extralimital distribution: Austria, Czech Republic, Germany, Hungary, Korea, Russia, Slovenia, Ukraine, United Kingdom, former Yugoslavia.
Host records: Unknown.

Aleiodes pictus (Herrich-Schäffer, 1838)

Distribution in the Middle East: Turkey (van Achterberg & Shaw, 2016).
Extralimital distribution: Austria, British Isles, Bulgaria, Czech Republic, Finland, France, Germany, Greece, Hungary, Iceland, Italy, Netherlands, Norway, Poland, Portugal, Romania, Russia, Serbia, Slovakia, spain, Sweden.
Host records: Recorded by van Achterberg and Shaw (2016) as being a parasitoid of the following geometrids: *Camptogramma bilineata* (L.), *Epirrhoe alternata* (Müller), *Lithostege griseata* (Denis and Schiffermüller), and *Xanthorhoe fluctuata* (L.); as well as the following noctuids: *Agrotis exclamationis* (L.), *Hoplodrina ambigua*

(Denis and Schiffermüller), *Haplodrina blanda* (Denis and Schiffermüller), *Hoplodrina octogenaria* (Goeze), *Haplodrina superstes* (Ochsenheimer), *Noctua fimbriata* (Schreber), *Phlogophora meticulosa* (L.), *Syngrapha interrogationis* (L.), and *Xestia xanthographa* (Denis and Schiffermüller).

Aleiodes praetor (Reinhard, 1863)

Catalogs with Iranian records: No catalog.
Distribution in Iran: Ardabil (Samin, Beyarslan, Ranjith et al., 2020).
Distribution in the Middle East: Iran (Samin, Beyarslan, Ranjith et al., 2020), Turkey (Aydoğdu & Beyarslan, 2005).
Extralimital distribution: Austria, Belgium, British Isles, Bulgaria, Croatia, Finland, France, Germany, Hungary, Netherlands, Serbia, Spain, Sweden, Switzerland (van Achterberg & Shaw, 2016).
Host records: Summarized by Yu et al. (2016) as being a parasitoid of the erebids *Euproctis chrysorrhoea* (L.), and *Leucoma salicis* (L.); the sphingids *Callambulyx tartarinovii* (Bremer and Gray), *Laohoe populi* (L.), *Mimas tiliae* (L.), *Smerinthis ocellatus* (L.), *Smerinthis planus* Walker, and *Sphinx pinastri* L. In Iran, this species has been reared from the erebid *Euproctis similis* (Füssli) (Samin, Beyarslan, Ranjith et al., 2020).

Aleiodes quadrum (Tobias, 1976)

Catalogs with Iranian records: No catalog.
Distribution in Iran: West Azarbaijan (Samin, Sakenin Chelav et al., 2020).
Distribution in the Middle East: Iran (Samin, Sakenin Chelav et al., 2020), Turkey (van Achterberg et al., 2020).
Extralimital distribution: Bulgaria, Croatia, France, Greece, Hungary, North Macedonia (van Achterberg et al., 2020).
Host records: Unknown.

Aleiodes ruficeps (Telenga, 1941)

Catalogs with Iranian records: No catalog.
Distribution in Iran: Iran (no specific locality cited) (van Achterberg et al., 2020).
Distribution in the Middle East: Iran, Turkey (van Achterberg et al., 2020).
Extralimital distribution: Armenia, Bulgaria, Russia, Ukraine.
Host records: Unknown.

Aleiodes ruficornis (Herrich-Schäffer, 1838)

Catalogs with Iranian records: Beyarslan, Gadallah, and Ghahari (2017), Beyarslan, Gözüaçik, Güllü, and Konuksal (2017), Samin, Coronado-Blanco, Kavallieratos et al. (2018), Samin, Coronado-Blanco, Fischer et al. (2018).
Distribution in Iran: Kerman (Abdolalizadeh et al., 2017).

Distribution in the Middle East: Iran (Abdolalizadeh et al., 2017), Turkey (van Achterberg et al., 2020).
Extralimital distribution: Afghanistan, Andorra, British Isles (England, Wales), Bulgaria, Croatia, Czech Republic, Finland, France, Germany, Kazakhstan, Kyrgyzstan, North Macedonia, Montenegro, Netherlands, Norway, Romania, Russia, Serbia, Slovakia, Sweden, Switzerland, Ukraine (van Achterberg et al., 2020).
Host records: Summarized by Yu et al. (2016) as being a parasitoid of the noctuids *Agrotis ipsilon* (Hufnagel), *Euxoa sibirica* (Boisduval), and *Mythimna separata* Walker. It was also reared from the noctuids *Agrotis clavis* (Hufnagel), *Agrotis segetum* (Denis and Schiffermüller), *Agrotis sp.*, *Euxoa nigricans* (L.), *Euxoa* sp., *Hoplodrina blanda* (Denis and Schiffermüller), *Hoplodrina octogenarian* (Goeze), and *Mythimna impura* (Hübner) (van Achterberg et al., 2020).

Aleiodes rufipes (Thomson, 1892)

Catalogs with Iranian records: Farahani et al. (2015, 2016), Yu et al. (2016), Beyarslan, Gadallah, and Ghahari (2017), Beyarslan, Gözüaçik, Güllü, and Konuksal (2017), Samin, Coronado-Blanco, Kavallieratos et al. (2018), Samin, Coronado-Blanco, Fischer et al. (2018).
Distribution in Iran: Ilam (Ghahari et al., 2011c).
Distribution in the Middle East: Iran (Ghahari et al., 2011c), Turkey (Aydoğdu & Beyarslan, 2005, 2006).
Extralimital distribution: Finland, Norway, Sweden (van Achterberg et al., 2020).
Host records: Unknown.

Aleiodes rugulosus (Nees von Esenbeck, 1811)

Catalogs with Iranian records: No catalog.
Distribution in Iran: West Azarbaijan (Samin et al., 2019).
Distribution in the Middle East: Iran.
Extralimital distribution: Albania, Austria, Belgium, Bulgaria, Canada (Shaw, 2006), Czech Republic, British Isles (England, Wales, Scotland, Ireland), Finland, France, Germany, Hungary, North Macedonia, Netherlands, Moldova, Norway, Poland, Romania, Russia, Slovakia, Spain, Sweden (van Achterberg et al., 2020).
Host records: Summarized by Yu et al. (2016) as being a parasitoid of the noctuids *Acronicta* spp., *Orthosia miniosa* (Denis and Schiffermüller), *Oxicesta geographica* (Fabricius), *Panchrysia deaurata* (Esper), and *Simyra albovenosa* (Goeze). It was also reared from the noctuids *Acronicta auricoma* (Denis and Schiffermüller), *Acronita euphorbiae* (Denis and Schiffermüller), *Acronita cinereal* (Hufnagel), *Acronita menyanthidis* (Esper), *Acronita rumicis* (L.), *Oxicestra geographica* (Fabricius), and *Simyra albovenosa* (Goeze) (van Achterberg et al., 2020). In Iran, this species has been reared from the noctuid *Simyra albovenosa* (Goeze) (Samin et al., 2019).

Aleiodes schewyrewi (Kokujev, 1898)

Catalogs with Iranian records: No catalog.

Distribution in Iran: Iran (no specific locality cited) (van Achterberg et al., 2020).

Distribution in the Middle East: Iran.

Extralimital distribution: Mongolia, Russia.

Host records: Unknown.

Aleiodes schirjajewi (Kokujev, 1898)

Catalogs with Iranian records: No catalog.

Distribution in Iran: Iran (no specific locality cited) (Abdinbekova et al., 2010 as *Rogas schirjaevi* Kokujev).

Distribution in the Middle East: Iran (Abdinbekova et al., 2010), Israel−Palestine (Papp, 2012), Turkey (Aydoğdu & Beyarslan, 2006).

Extralimital distribution: Bulgaria, Hungary, Italy, Kazakhstan, Moldova, Russia, Ukraine (van Achterberg et al., 2020).

Host records: Unknown.

"*Aleiodes seriatus* (Herrich-Schäffer, 1838)" *sensu lato*

Catalogs with Iranian records: Farahani et al. (2015, 2016), Yu et al. (2016), Beyarslan, Gadallah, and Ghahari (2017), Beyarslan, Gözüaçik, Güllü, and Konuksal (2017), Samin, Coronado-Blanco, Kavallieratos et al. (2018), Samin, Coronado-Blanco, Fischer et al. (2018).

Distribution in Iran: Markazi (Ameri et al., 2020), Mazandaran (Farahani et al., 2015; Kian et al., 2020).

Notes: Here we follow van Achterberg and Shaw (2016). Quicke and Shaw (unpublished data) suggest that, according to DNA evidence, an aggregate of two species present in Europe and the Russian Far East are currently placed under *A. seriatus* remains problematic. However, this will be addressed in a future work by van Achterberg, Quicke and Shaw (unpublished data).

Aleiodes sibiricus (Kokujev, 1903)

Distribution in the Middle East: Turkey (van Achterberg et al., 2020).

Extralimital distribution: Albania, Austria, Bulgaria, France, Germany, Greece, Hungary, Italy, North Macedonia, Russia, Sweden (van Achterberg et al., 2020).

Host records: Recorded by Papp (1985) and van Achterberg et al. (2020) as being a parasitoid of the noctuid *Noctua comes* Hübner.

Aleiodes signatus (Nees von Esenbeck, 1811)

Catalogs with Iranian records: Farahani et al. (2015, 2016), Yu et al. (2016), Beyarslan, Gadallah, and Ghahari (2017), Beyarslan, Gözüaçik, Güllü, and Konuksal (2017), Samin, Coronado-Blanco, Kavallieratos et al. (2018), Samin, Coronado-Blanco, Fischer et al. (2018).

Distribution in Iran: Golestan (Samin, Beyarslan, Coronado-Blanco et al., 2020), Qazvin (Ghahari et al., 2011b), West Azarbaijan (Samin et al., 2014).

Distribution in the Middle East: Iran (see references above), Israel−Palestine (Papp, 1989), Turkey (Aydoğdu & Beyarslan, 2005, 2006; Beyarslan, 2015).

Extralimital distribution: Widely distributed in the Palaearctic countries [Adjacent countries to Iran: Armenia, Azerbaijan, Kazakhstan, Russia].

Host records: Summarized by Yu et al. (2016) as being a parasitoid of the erebids *Arctia caja* (L.), *Coscinia cibraria* (L.), *Epicallia villica* (L.), *Euproctis chrysorrhoea* (L.), *Euproctis similis* (Füssli), *Gynaephora selenitica* (Esper), *Ocnogyna boetica* (Rambur), *Orgyia antiqua* (L.), and *Orgyia aurolimbata* Guenée; the lasiocampid *Euthrix potatoria* (L.); the noctuids *Acronicta rumicis* (L.), *Autographa gamma* (L.), *Noctua fimbriata* Schreber, and *Noctua pronuba* (L.); the thaumetopoeids *Thaumetopoea pityocampa* (Denis and Schiffermüller), and *Thaumetopoea processionea* (L.); and the tortricid *Spilonota ocellana* (Denis and Schiffermüller). In Iran, reared from the erebid *Orgyia antiqua* (L.) (Samin, Beyarslan, Coronado-Blanco et al., 2020).

Aleiodes testaceus (Telenga, 1941)

Catalogs with Iranian records: Fallahzadeh and Shojai (2010).

Distribution in Iran: East Azarbaijan (Samin et al., 2019 as *Heterogamus testaceus* (Telenga, 1941)), Golestan (Samin, 2015 as *Aleiodes* (*Heterogamus*) *testaceus* (Telenga, 1941)), Sistan and Baluchestan (Hedwig, 1957).

Distribution in the Middle East: Cyprus (van Achterberg and Shaw, 2016), Iran (see references above).

Extralimital distribution: Austria, Bulgaria, Cyprus, France, Greece, Hungary, Italy, Madeira Islands, Malta, Morocco, Netherlands, Russia, Spain, Switzerland, Turkmenistan, United Kingdom (van Achterberg & Shaw, 2016; Yu et al., 2016).

Host records: Recorded by van Achterberg and Shaw (2016) as being a parasitoid of the geometrids *Chloroclystis v-ata* (Haworth), *Epithecia dodoneata* Guenée, *Eupithesia* sp., and *Gymnoscelis rufifasciata* (Haworth).

Aleiodes turcicus van Achterberg and Shaw, 2020

Distribution in the Middle East: Turkey (van Achterberg & Shaw, 2020).

Extralimital distribution: None.

Host records: Unknown.

Aleiodes ungularis (Thomson, 1892)

Catalogs with Iranian records: No catalog.

Distribution in Iran: East Azarbaijan (Naderian et al., 2020).

Distribution in the Middle East: Iran (Naderian et al., 2020), Turkey (Aydoğdu & Beyarslan, 2006).

Extralimital distribution: Finland, France, Germany, Greece, Hungary, Romania, Slovakia, Sweden, Switzerland, United Kingdom (van Achterberg & Shaw, 2016).

Host records: Recorded by van Achterberg and Shaw (2016) as being a parasitoid of the nolid *Pseudopis prasinana* (L.).

Aleiodes unipunctator (Thunberg, 1822)

Catalogs with Iranian records: Farahani et al. (2015, 2016), Yu et al. (2016), Beyarslan, Gadallah, and Ghahari (2017), Beyarslan, Gözüaçik, Güllü, and Konuksal (2017), Samin, Coronado-Blanco, Kavallieratos et al. (2018), Samin, Coronado-Blanco, Fischer et al. (2018).

Distribution in Iran: East Azarbaijan (Rastegar et al., 2012), Guilan (Sakenin et al., 2012), Qazvin (Ghahari et al., 2011b).

Distribution in the Middle East: Iran.

Extralimital distribution: Austria, British Isles (England, Wales, Scotland, Ireland), Bulgaria, Czech Republic, Denmark, Finland, Germany, Greece, Hungary, Italy, Kazakhstan, Montenegro, Netherlands, Norway, Romania, Sweden, Tajikistan (van Achterberg et al., 2020).

Host records: Summarized by Yu et al. (2016) as being a parasitoid of the lasiocampid *Euthrix potatoria* (L.); the noctuids *Apamea sordens* (Hufnagel), *Apamea unanimis* (Hübner), and *Leucania comma* (L.); and the sesiid *Synanthedon formicaeformis* (Esper). It was also reared from the noctuids *Apamea crenata* (Hufnagel), *Apamea? sordens* (Hufnagel), and *Apamea unanimis* (Hübner) (van Achterberg et al., 2020).

Aleiodes varius (Herrich-Schäffer, 1838)

Catalogs with Iranian records: Yu et al. (2016), Beyarslan, Gadallah, and Ghahari (2017), Beyarslan, Gözüaçik, Güllü, and Konuksal (2017), Samin, Coronado-Blanco, Kavallieratos et al. (2018), Samin, Coronado-Blanco, Fischer et al. (2018).

Distribution in Iran: Chaharmahal and Bakhtiari (Samin et al., 2016).

Distribution in the Middle East: Iran (Samin et al., 2016), Turkey (Beyarslan, 2015).

Extralimital distribution: Austria, Belgium, Czech Republic, Finland, France, Germany, Hungary, Italy, Japan, Lithuania, Netherlands, Russia, Switzerland (Yu et al., 2016; van Achterberg & Shaw, 2016).

Host records: Recorded by van Achterberg and Shaw (2016) as being a parasitoid of the lasiocampid *Euthrix potatoria* (L.).

Aleiodes venustulus (Kokujev, 1905)

Distribution in the Middle East: Turkey (Aydoğdu & Beyarslan, 2005, 2006).

Extralimital distribution: Kazakhstan, Kyrgyzstan, Mongolia.

Host records: Unknown.

Aleiodes zwakhalzi van Achterberg and Shaw, 2020

Distribution in the Middle East: Turkey (van Achterberg & Shaw, 2020).

Extralimital distribution: None.

Host records: Unknown.

Aleiodes sp.

Distribution in Iran: Fars (Modarres Awal, 1997), Kerman (Iranmanesh et al., 2017), Khuzestan (Shojai et al., 1995; Modarres Awal, 1997).

Host records in Iran: *Sesamia cretica* Lederer, 1857 (Lepidoptera: Noctuidae) (Modarres Awal, 1997; Shojai et al., 1995).

Plant associations in Iran: *Medicago sativa* (Fabaceae), *Mentha pulegium* (Lamiaceae), *Rubus* sp. (Rosaceae), *Sophora alopecuroides* (Fabaceae) (Iranmanesh et al., 2017), *Zea mays* (Poaceae) (Shojai et al., 1995).

Genus *Heterogamus* Wesmael, 1838

Heterogamus dispar (Haliday, 1833)

Catalogs with Iranian records: Farahani et al. (2015, 2016), Yu et al. (2016), Beyarslan, Gadallah, and Ghahari (2017), Beyarslan, Gözüaçik, Güllü, and Konuksal (2017), Samin, Coronado-Blanco, Kavallieratos et al. (2018), Samin, Coronado-Blanco, Fischer et al. (2018).

Distribution in Iran: Mazandaran (Farahani et al., 2015).

Distribution in the Middle East: Iran.

Extralimital distribution: Widely distributed in the Oriental and Palaearctic countries.

Host records: Recorded by Čapek and Hofmann (1997) as being a parasitoid of the noctuid *Agrotis segetum* (Denis and Schiffermüller) but this was a result of misidentification of an *Aleiodes* species and there are no host records for the genus.

Tribe Clinocentrini van Achterberg, 1991

Genus *Clinocentrus* Haliday, 1833

Clinocentrus amiri Rakhshani and Farahani, 2020

Catalogs with Iranian records: No catalog.

Distribution in Iran: Sistan and Baluchestan (Derafshan et al., 2020).

Distribution in the Middle East: Iran.

Host records: Unknown.

Clinocentrus caucasicus Tobias, 1976

Catalogs with Iranian records: No catalog.

Distribution in Iran: Ardabil (Samin et al., 2019), Zanjan (Samin, Beyarslan, Coronado-Blanco et al., 2020).

Distribution in the Middle East: Iran.

Extralimital distribution: Azerbaijan, China, Georgia, Hungary, Japan, Korea, Mongolia, Netherlands, Russia, Ukraine.

Host records: Unknown.

Clinocentrus cunctator (Haliday, 1836)

Catalogs with Iranian records: Farahani et al. (2015, 2016), Yu et al. (2016), Beyarslan, Gadallah, and Ghahari (2017), Beyarslan, Gözüaçik, Güllü, and Konuksal (2017), Samin, Coronado-Blanco, Kavallieratos et al. (2018), Samin, Coronado-Blanco, Fischer et al. (2018).
Distribution in Iran: Ardabil (Rastegar et al., 2012), Mazandaran (Sakenin et al., 2012), Qazvin (Ghahari et al., 2011b), West Azarbaijan (Samin et al., 2016).
Distribution in the Middle East: Iran.
Extralimital distribution: Belgium, Bulgaria, Czech Republic, Finland, France, Finland, Georgia, Germany, Hungary, Ireland, Italy, Korea, Montenegro, Netherlands, Poland, Russia, Slovakia, Sweden, United Kingdom.
Host records: Summarized by Yu et al. (2016) as being a parasitoid of the choreutids *Anthophila fabriciana* (L.), *Prochoreutis myllerana* (Fabricius), and *Prochoreutis sehestediana* (Fabricius), and the geometrid *Pseudoterpna pruinata* (Hufnagel).

Clinocentrus excubitor (Haliday, 1836)

Catalogs with Iranian records: Farahani et al. (2015, 2016), Yu et al. (2016), Beyarslan, Gadallah, and Ghahari (2017), Beyarslan, Gözüaçik, Güllü, and Konuksal (2017), Samin, Coronado-Blanco, Kavallieratos et al. (2018), Samin, Coronado-Blanco, Fischer et al. (2018).
Distribution in Iran: Mazandaran (Derafshan et al., 2020; Farahani et al., 2015).
Distribution in the Middle East: Iran.
Extralimital distribution: Albania, Azerbaijan, Belgium, Bulgaria, Finland, France, Georgia, Germany, Hungary, Ireland, Japan, Kazakhstan, Korea, Lithuania, Moldova, Netherlands, Poland, Russia, Slovakia, Sweden, Switzerland, Ukraine, United Kingdom, Vietnam.
Host records: Summarized by Yu et al. (2016) as being a parasitoid of the depressariid *Agonopterix kaekeritziana* (L.), the geometrid *Biston betularia* L., the gracillariid *Parornix scoticella* (Stainton), the noctuid *Xestia ditrapezium* (Denis and Schiffermüller), and the tortricids *Eucosmomorpha albersana* (Hübner), *Gravitarmata margarotana* (von Heinemann), *Pandemis corylana* (Fabricius), and *Pseudosciaphila branderiana* (L.).

Clinocentrus exsertor (Nees von Esenbeck, 1811)

Catalogs with Iranian records: Farahani et al. (2015, 2016), Yu et al. (2016), Beyarslan, Gadallah, and Ghahari (2017), Beyarslan, Gözüaçik, Güllü, and Konuksal (2017), Samin, Coronado-Blanco, Kavallieratos et al. (2018), Samin, Coronado-Blanco, Fischer et al. (2018).
Distribution in Iran: Guilan (Derafshan et al., 2020; Farahani et al., 2015), Isfahan (Ghahari et al., 2011a under subfamily Hormiinae), Mazandaran (Kian et al., 2020).
Distribution in the Middle East: Iran (see references above), Israel–Palestine (Papp, 2012).

Extralimital distribution: Widely distributed in the Nearctic and Palaearctic countries [Adjacent countries to Iran: Armenia, Azerbaijan, Kazakhstan, Russia].
Host records: Summarized by Yu et al. (2016) as being a parasitoid of the momphid *Mompha conturbatella* (Hübner), and the tortricids *Argyrotaenia velutinana* (Walker), *Avaria hyerana* (Millière), *Choristoneura fumiferana* (Clemens), *Cnephasia genitalana* Pierce and Metcalfe, *Grapholita delineana* Walker, *Gypsonoma sociana* (Haworth), and *Rhyacionia buoliana* (Denis and Schiffermüller).

Clinocentrus umbratilis Haliday, 1833

Catalogs with Iranian records: No catalog.
Distribution in Iran: Kordestan (Samin et al., 2019).
Distribution in the Middle East: Iran.
Extralimital distribution: China, former Czechoslovakia, Finland, Germany, Hungary, Ireland, Japan, Korea, Lithuania, Norway, Poland, Russia, Sweden, Switzerland, Ukraine, United Kingdom.
Host records: Summarized by Yu et al. (2016) as being a parasitoid of the crambid *Catoptria falsella* (Denis and Schiffermüller), and the noctuid *Hadena* sp.

Clinocentrus vestigator (Haliday, 1836)

Catalogs with Iranian records: Yu et al. (2016), Samin, Coronado-Blanco, Kavallieratos et al. (2018), Samin, Coronado-Blanco, Fischer et al. (2018).
Distribution in Iran: Kuhgiloyeh and Boyerahmad (Samin et al., 2016).
Distribution in the Middle East: Iran (Samin et al., 2016), Turkey (Beyarslan, 2015).
Extralimital distribution: Denmark, Finland, Germany, Ireland, Italy, Kazakhstan, Lithuania, Norway, Poland, Russia, Switzerland, Ukraine, United Kingdom.
Host records: Summarized by Yu et al. (2016) as being a parasitoid of the epermeniids *Epermenia chaerophylella* (Goeze) and *Epermenia illigerella* (Hübner), and the ypsolophid *Ypsolopha vittella* (L.).

Tribe Pentatermini Belokobylskij, 1990
Genus Pentatermus Hedqvist, 1963
Pentatermus striatus (Szépligeti, 1908)

Catalogs with Iranian records: No catalog.
Distribution in Iran: Ardabil (Gadallah et al., 2021).
Distribution in the Middle East: Iran (Gadallah et al., 2021), Oman (Belokobylskij, 2002).
Extralimital distribution: China, India, Indonesia, Japan, Madagascar, Malaysia, Niger, Nigeria, Somalia, South Africa, Vietnam.
Host records: Recorded by He and Chen (1995) as being a parasitoid of the hesperiid *Panara guttata* (Bremer and Gray).

Tribe Rogadini Foerster, 1863

Genus *Rogas* Nees von Esenbeck, 1819

Rogas luteus Nees von Esenbeck, 1834

Catalogs with Iranian records: No catalog.
Distribution in Iran: Mazandaran (Sakenin et al., 2020).
Distribution in the Middle East: Cyprus (Ingram, 1981), Iran (Sakenin et al., 2020).
Extralimital distribution: Armenia, Austria, Azerbaijan, Belgium, Croatia, Czech Republic, Denmark, Finland, France, Georgia, Germany, Hungary, Ireland, Italy, Japan, Kazakhstan, Latvia, Moldova, Netherlands, Norway, Poland, Romania, Russia, Slovakia, Switzerland, Ukraine, United Kingdom.
Host records: Summarized by Yu et al. (2016) as being a parasitoid of the erebid *Leucoma salicis* (L.), the geometrids *Eupithecia subfuscata* (Haworth), *Hylaea fasciaria* (L.), *Thera fermata* Hübner, and *Thera variata* (Denis and Schiffermüller), the limacodid *Apoda limacodes* (Hufnagel), the noctuid *Simyra albovenosa* (Goeze), the papilionids *Iphiclides podalirius* (L.), and *Papilio machaon* L., and the tortricid *Tortrix testudinana* Hübner.

Tribe Yeliconini van Achterberg, 1991

Genus *Yelicones* Cameron, 1887

Yelicones delicatus (Cresson, 1872)

Distribution in the Middle East: Israel−Palestine (Papp, 1989).
Extralimital distribution: Dominichian Republic, Mexico, Panama, Puerto Rico, United States of America.
Host records: Summarized by Yu et al. (2016) as being a parasitoid of the following pyralids: *caristanius decoloralis* (Walker), *Nephopterix uvinella* Ragonot, and *Psorosina hammondi* (Riley).

Yelicones iranus (Fischer, 1963)

Catalogs with Iranian records: Fallahzadeh and Saghaei (2010), Farahani et al. (2015, 2016), Yu et al. (2016), Beyarslan, Gadallah, and Ghahari (2017), Beyarslan, Gözüaçik, Güllü, and Konuksal (2017), Samin, Coronado-Blanco, Kavallieratos et al. (2018), Samin, Coronado-Blanco, Fischer et al. (2018).
Distribution in Iran: Kerman (Fischer, 1963 as *Pectenopius iranus* Fischer, 1963).
Distribution in the Middle East: Iran (Fischer, 1963), Israel−Palestine (Quicke et al., 2018).
Extralimital distribution: None.
Host records: This species has been reared from the pyralid *Phycita diaphana* (Staudinger) feeding on *Ricinus communis* L. (Euphorbiaceae) in Israel (Quicke et al., 2018).

Yelicones vojnitsi Papp, 1992

Distribution in the Middle East: Egypt, Oman (Quicke & Chishti, 1997).
Extralimital distribution: Kenya, Namibia, Niger, Senegal, Spain, Tanzania.
Host records: Unknown.

Conclusion

In total, 68 valid species of the subfamily Rogadinae in six genera and five tribes (Aleiodini, Clinocentrini, Pentatermini, Rogadini, and Yeliconini) have been reported from most of the Middle East countries, with the exception of Bahrain, Kuwait, Lebanon, Qatar, United Arab Emirates, and Yemen. The most species rich genera in the Rogadinae of the Middle East are *Aleiodes* (with 55 species), followed by *Clinocentrus* (with seven species). Fifty-six species of Rogadinae in six genera have been recorded from Iran so far, followed by Turkey, with 38 species. Among the genera of Iranian Rogadinae, *Aleiodes* with 46 recorded species is more diverse than the other genera, followed by *Clinocentrus* (with seven species), *Heterogamus*, *Rogas*, *Pentatermus*, and *Yelicones* (each with one species). Iranian Rogadinae have been recorded from 20 provinces, which Mazandaran with 14 recorded species, has the highest number of species, followed by Guilan and Kerman with 11 and 10 species, respectively (Fig. 30.1). Only five host

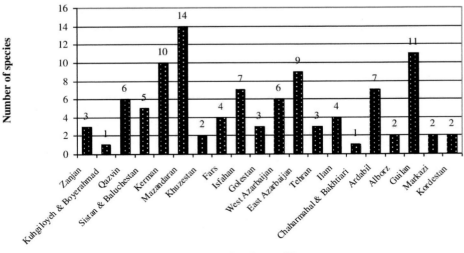

FIGURE 30.1 Number of reported species of Iranian Rogadinae by province.

species have been recorded for Rogadinae in Iran: *Archips rosana* (L.) (Lepidoptera: Tortricidae) (for *Aleiodes compressor*), two species of Erebidae (Lepidoptera), *Orgyia antiqua* (L.) (for *Aleiodes signatus*), *Euproctis similis* (Fuessli) (for *Aleiodes praetor*), and two noctuid species (Lepidoptera), *Sesamia cretica* Lederer (for *Aleiodes* sp.), and *Simyra albovenosa* (Goeze) (for *Aleiodes rugulosus*). Among the 23 countries of the Middle East and adjacent to Iran, Russia shares the highest number of species with Iran (37 species), followed by Turkey (28 species), Kazakhstan (16 species), Azerbaijan (13 species), Israel—Palestine (11 species), Turkmenistan (nine species), Armenia and Cyprus (both with eight species), Afghanistan (five species), Jordan and Syria (each with three species), Iraq and Oman (each with two species), and Egypt (one species).

References

Abdinbekova, A., Huseynova, E., & Kerimova, I. (2010). Braconidae (Hymenoptera) in the collection of the Institute of Zoology, NAS of the Azerbaijan Republic. *Beiträge zur Entomologie, 60*(2), 427—440.

Abdolalizadah, F., Madjdzadeh, S. M., Farahani, S., & Askari Hesni, M. (2017). A survey of braconid wasps (Hymenoptera: Braconidae: Euphorinae, Homolobinae, Macrocentrinae, Rogadinae) in Kerman province, south-eastern Iran. *Journal of Insect Biodiversity and Systematics, 3*(1), 33—40.

Ameri, A., Ebrahimi, E., & Talebi, A. A. (2020). Additions to the fauna of Braconidae (Hym., Ichneumonoidea) of Iran based on specimens housed in Hayk Mirzayans Insect Museum with six new records for Iran. *Journal of Insect Biodiversity and Systematics, 6*(4), 353—364.

Asadizade, A., Mahdiyan, K., Talebi, A. A., & Esfandiarpour, I. (2014). Faunistic surveys of parasitoid wasps family of Braconidae from Anar region, Kerman province. In *Proceedings of the 3rd Integrated Pest Management Conference (IPMC), 21—22 January 2014, Kerman* (p. 629).

Aydoğdu, M., & Beyarslan, A. (2005). The first records of *Aleiodes* Wesmael, 1838 (Hymenoptera: Braconidae: Rogadinae). The fauna of Thrace region of Turkey. *Linzer biologische Beiträge, 37/1*, 185—193.

Aydoğdu, M., & Beyarslan, A. (2006). The first records of *Aleiodes* Wesmael, 1838 species in east Marmara region of Turkey (Hymenoptera, Braconidae, Rogadinae). *Acta Entomologica Slovenica, 14*(1), 81—88.

Belokobylskij, S. A. (2002). On the genus *Pentatermus* Hedqvist (Hymenoptera: Braconidae). *Zoosystematica Rossica, 10*(2), 387—396.

Belokobylskij, S. A., & Tobias, V. I. (1986). Doryctinae. In G. S. Medvedev (Ed.), *Opredelitel Nasekomych Evrospeiskoi Tsasti SSSR 3, Peredpontdatokrylye 4. Opr. Faune SSSR* (Vol. 145, pp. 21—72). Leningrad: Nauka Press, 501 pp.

Beyarslan, A. (2015). Taxonomic survey on the Rogadinae Foerster, 1862 (Hymenoptera, Braconidae) in the northeastern Anatolia region, Turkey. *Turkish Journal of Zoology, 39*(5), 811—819.

Beyarslan, A., Inanç, F., Çetin Erdoğan, Ö., & Aydoğdu, M. (2002). Braconidae species of Turkish Aegean region (Hymenoptera). In G. Melika, & C. Thuroczy (Eds.), *Parasitic Hymenoptera: taxonomy and biological control, 14—17 May, 2001, Koszeg, Hungary, i—xx* (pp. 1—480).

Beyarslan, A., Yurtcan, M., Çetin Erdoğan, Ö., & Aydoğdu, M. (2006). A study on Braconidae and Ichneumonidae from Ganos Mountains (Thrace Region, Turkey) (Hymenoptera, Braconidae, Ichneumonoidea). *Linzer biologische Beiträge, 3811*, 409—422.

Beyarslan, A., Gadallah, N. S., & Ghahari, H. (2017). An annotated catalogue of the Iranian Microtypinae and Rogadinae (Hymenoptera: Braconidae). *Zootaxa, 4291*(1), 99—116.

Beyarslan, A., Gözüaçik, C., Güllü, M., & Konuksal, A. (2017). Taxonomical investigation on Braconidae (Hymenoptera: Ichneumonoidea) fauna in northern Cyprus, with twenty six new records for the country. *Journal of Insect Biodiversity and Systematics, 3*(4), 319—334.

Butcher, B. A., & Quicke, D. L. J. (2015). First record of *Aleiodes* (*Hemigyroneuron*) (Hymenoptera: Braconidae: Rogadinae) from the Arabian Peninsula: description of new species with remarkable wing venation convergence to *Gyroneuron* and *Gyroneurella*. *Zootaxa, 4033*(2), 275—279.

Butcher, B. A., Smith, M. A., Sharkey, M. J., & Quicke, D. L. J. (2012). A turbo-taxonomic study of Thai *Aleiodes* (*Aleiodes*) and *Aleiodes* (*Arcaleiodes*) (Hymenoptera: Braconidae: Rogadinae) based largely on COI barcoded specimens, with rapid descriptions of 179 new species. *Zootaxa, 3457*, 1—232.

Čapek, M., & Hofmann, C. (1997). The Braconidae (Hymenoptera) in the collections of the Musée Cantonal de Zoologie, Lausanne. *Litterae Zoologicae, 2*, 25—162.

Chen, X. X., & He, J. H. (1997). Revision of the subfamily Rogadinae (Hymenoptera: Braconidae) from China. *Zoologische Verhandelingen, 308*, 1—187.

Chen, X. X., & van Achterberg, C. (2019). Systematics, phylogeny, and evolution of braconid wasps: 30 years of progress. *Annual Review of Entomology, 64*, 1—24.

Chen, X. X., Piao, M. H., Whitfield, J. B., & He, J. H. (2003). Phylogenetic relationships within the Rogadinae (Hymenoptera: Braconidae) based on the D2 variable region of 28S ribosomal RNA. *Acta Entomologica Sinica, 46*, 209—217.

Derafshan, H. A., Rakhshani, E., Farahani, S., Moghaddam, M. G., & van Achterberg, C. (2020). The genus *Clinocentrus* Haliday (Hymenoptera, Braconidae, Rogadinae) in Iran, with the description of a new species. *Journal of Natural History, 54*(19—20), 1223—1241.

Fallahzadeh, M., & Saghaei, N. (2010). Checklist of Braconidae (Insecta: Hymenoptera) from Iran. *Munis Entomology & Zoology, 5*(1), 170—186.

Farahani, S., Talebi, A. A., van Achterberg, C., & Rakhshani, E. (2015). A review of the subfamily Rogadinae (Hymenoptera: Braconidae) from Iran. *Zootaxa, 3973*(2), 227—250.

Farahani, S., Talebi, A. A., & Rakhshani, E. (2016). Iranian Braconidae (Insecta: Hymenoptera: Ichneumonoidea): diversity, distribution and host association. *Journal of Insect Biodiversity* and Systematics, 2(1), 1—92.

Farahbakhsh, G. (1961). Family Braconidae (Hymenoptera). In G. Farahbakhsh (Ed.), *A checklist of economically important insects and other enemies of plants and agricultural products in Iran* (p. 124). Tehran, Iran: Department of Plant Protection, Ministry of Agriculture Publication, 133 pp.

Fahringer, J. (1932). *Opuscula braconologica. Band 3. Palaearcktischen Region. Lieferung 3. Opuscula braconologica* (pp. 161—240). Wien: Fritz Wagner.

Fischer, M. (1963). Eine neue *Pectonopius*-Art aus dem Iran (Hymenotera, Braconidae, Opiinae). *Stuttgarter Beiträge zur Naturkunde, 98*, 1—3.

Fortier, J. C. (2009). A revision of the *Tetrasphaeropyx* Ashmead lineage of the genus *Aleiodes* Wesmael (Hymenoptera: Braconidae: Rogadinae). *Zootaxa, 2256*, 1−126.

Fortier, J. C., & Shaw, S. R. (1999). Cladistics of the *Aleiodes* lineage of the subfamily Rogadinae (Hymenoptera: Braconidae). *Journal of Hymenoptera Research, 8*(2), 204−237.

Gadallah, N. S., Ghahari, H., & Quicke, D. L. J. (2021). Further addition to the braconid fauna of Iran (Hymenoptera: Braconidae). *Egyptian Journal of Biological Pest Control, 31*, 32. https://doi.org/10.1186/s41938-021-00376-8

Ghahari, H. (2020). A study on the fauna of predator and parasitoid arthropods in saffron fields (*Crocus sativus* L.). *Journal of Saffron Research, 7*(2), 203−215 (in Persian, English summary).

Ghahari, H., & Beyarslan, A. (2017). A faunistic study on Braconidae (Hymenoptera: Ichneumonoidea) from Iran. *Natura Somogyiensis, 30*, 39−46.

Ghahari, H., & Sakenin, H. (2018). Species diversity of Chalcidoidea and Ichneumonoidea (Hymenoptera) in some paddy fields and surrounding grasslands of Mazandaran and Guilan provinces, northern Iran. *Applied Plant Protection, 7*(1), 11−19 (in Persian, English summary).

Ghahari, H., Fischer, M., Çetin Erdoğan, Ö., Tabari, M., Ostovan, H., & Beyarslan, A. (2009). A contribution to Braconidae (Hymenoptera) from rice fields and surrounding grasslands of Northern Iran. *Munis Entomology & Zoology, 4*(2), 432−435.

Ghahari, H., Fischer, M., Hedqvist, K. J., Çetin Erdoğan, Ö., van Achterberg, C., & Beyarslan, A. (2010). Some new records of Braconidae (Hymenoptera) for Iran. *Linzer biologische Beiträge, 42*(2), 1395−1404.

Ghahari, H., Fischer, M., & Papp, J. (2011a). A study on the braconid wasps (Hymenoptera: Braconidae) from Isfahan province, Iran. *Entomofauna, 32*, 261−272.

Ghahari, H., Fischer, M., & Papp, J. (2011b). A study on the Braconidae (Hymenoptera: Ichneumonoidea) from Qazvin province, Iran. *Entomofauna, 32*, 197−208.

Ghahari, H., Fischer, M., & Papp, J. (2011c). A study on the Braconidae (Hymenoptera: Ichneumonoidea) from Ilam province, Iran. *Calodema, 160*, 1−5.

He, J. H., & Chen, X. X. (1995). [The genus *Pentatermus* Hedqvist (Hymenoptera: Braconidae: Lysiterminae) of China]. *Entomotaxonomia, 17*(2), 225−227 (in Chinese, English summary).

Hedwig, K. (1957). Ichneumoniden und Braconiden aus den Iran 1954 (Hymenoptera). *Jahresheft des Vereins für Vaterlaendische Naturkunde, 112*(1), 103−117.

Ingram, W. R. (1981). The parasitoids of *Spodoptera littoralis* (Lep.: Noctuidae) and their role in the population control in Cyprus. *Entomophaga, 26*, 23−37.

Iranmanesh, M., Madjdzadeh, S. M., & Askari Hesni, M. (2017). Biodiversity of braconid fauna (Hym.: Ichneumonoidea: Braconidae) in Sirch region, Kerman province, Iran. *Journal of Entomological Society of Iran, 37*(1), 1−13 (in Persian, English summary).

Jasso-Martínez, J. M., Quicke, D. L. J., Belokobylskij, S. A., Meza-Lázaro, R. M., & Zaldívar-Riverón, A. (2020). Phylogenomics of the lepidopteran endoparasitoid wasp subfamily Rogadinae (Hymenoptera: Braconidae) and related subfamilies. *Systematic Entomology.* https://doi.org/10.1111/syen.12449

Kian, N., Goldasteh, S., & Farahani, S. (2020). A survey on abundance and species diversity of Braconid wasps in forest of Mazandaran province. *Journal of Entomological Research, 12*(1), 61−69 (in Persian, English summary).

Lashkari Bod, A., Rakhshani, E., Talebi, A. A., Lozan, A., & Žikić, V. (2011). A contribution to the knowledge of Braconidae (Hym., Ichneumonoidea) of Iran. *The Bihar Journal of 'Agricultural Marketing, 5*(2), 147−150.

Madjdzadeh, S. M., Parrezaali, M., Dolati, S., & Ghassemi-Khademi, T. (2021). New data on the braconid wasps (Hymenoptera: Braconidae: Cheloninae, Opiinae, Rogadinae) of south-eastern Iran. *Faunistic Entomology, 74*, 1−7.

Marsh, P. M., & Shaw, S. R. (1998). Revision of North American *Aleiodes* (part 3): the *seriatus* species-group (Hymenoptera: Braconidae: Rogadinae). *Proceedings of the Entomological Society of Washington, 100*(3), 395−408.

Marsh, P. M., & Shaw, S. R. (1999). Revision of North American *Aleiodes* (part 5): the *melanopterus* (Erichson) species-group in North America (Hymenoptera: Braconidae: Rogadinae). *Journal of Hymenoptera Research, 8*, 98−108.

Marsh, P. M., & Shaw, S. R. (2001). Revision of North American *Aleiodes* (part 6): the *gasterator* (Jurine) and *unipunctator* (Thunberg) species groups (Hymenoptera: Braconidae: Rogadinae). *Proceedings of the Entomological Society of Washington, 103*(3), 291−307.

Marsh, P. M., & Shaw, S. R. (2003). Revision of North American *Aleiodes* (part 7): the *compressor* Herrich-Schaeffer, *ufei* (Walley), *gressetti* (Muesebeck) and *procerus* Wesmael species-groups (Hymenoptera: Braconidae: Rogadinae). *Proceedings of the Entomological Society of Washington, 105*(3), 698−707.

Marshall, T. A. (1890). Les Braconides. In E. André (Ed.), *Species des Hyménoptères d'Europe et d'Algerie. Tome* (Vol. 4, p. 609).

Modarres Awal, M. (1997). Family Braconidae (Hymenoptera). In M. Modarres Awal (Ed.), *List of agricultural pests and their natural enemies in Iran* (2nd ed., pp. 265−267). Ferdowsi University of Mashhad Press, 429 pp.

Naderian, H., Penteado-Dias, A. M., Sakenin Chelav, H., & Samin, N. (2020). A faunistic study on Braconidae and Ichneumonidae (Hymenoptera, Ichneumonoidea) of Iran. *Calodema, 844*, 1−9.

Papp, J. (1970). A contribution to the braconid fauna of Israel (Hymenoptera). *Israel Journal of Entomology, 5*, 63−76.

Papp, J. (1985). Braconidae (Hymenoptera) from Greece. 2. *Annales Hostorico-Naturales Musei Nationalis Hungarici, 77*, 217−226.

Papp, J. (1989). A contribution to the braconid fauna of Israel (Hymenoptera), 2. *Israel Journal of Entomology, 22*, 45−59.

Papp, J. (2012). A contribution to the braconid fauna of Israel (Hymenoptera: Braconidae), 3. *Israel Journal of Entomology, 41−42*, 165−219.

Pasandideh Saqalasari, M., Talebi, A. A., & van de Kamp, T. (2020). MicroCT 3D reconstruction of three described braconid species (Hymenoptera: Braconidae). *Journal of Insect Biodiversity and Systematics, 6*(4), 331−342.

Quicke, D. L. J., & Chishti, M. J. K. (1997). A revision of *Yelicones* species (Hymenoptera: Braconidae: Rogadinae) from Africa and the Arabian Peninsula, with the descriptions of four new species. *African Entomology, 5*(1), 77−91.

Quicke, D. L. J., & Butcher, B. A. (2015). Description of a new Betylobraconini-like parasitoid wasps genus and species (Hymenoptera: Braconidae: Rogadinae) from Chile. *Zootaxa, 4021*, 459−466.

Quicke, D. L. J., Mori, M., Zaldivar-Riverón, A., Laurenna, M., & Shaw, M. R. (2006). Suspended mummies in *Aleiodes* species (Hymenoptera: Braconidae: Rogadinae) with descriptions of six new species from western Uganda based largely on DNA sequence data. *Journal of Natural History, 40*(47−48), 2663−2680.

Quicke, D. L. J., Belokobylskij, S. A., Smith, M. A., Rota, J., Hrcek, J., & Butcher, B. A. (2016). A new genus of rhysopoline wasp (Hymenoptera: Braconidae) with modified wing venation from Africa and Papua New Guinea, parasitoid on Choreutidae (Lepidoptera). *Annales Zoologici, 66*, 173−192.

Quicke, D. L. J., Kuslitzky, W., & Butcher, B. A. (2018). First host record for Old World *Yelicones* (Hymenoptera: Braconidae: Rogadinae) adds to evidence that they are strictly parasitoids of Pyralidae (Lepidoptera). *Israel Journal of Entomology, 48*, 33−40.

Ranjith, A. P., Belokobylskij, S. A., Quicke, D. L. J., Kittel, R. N., Butcher, B. A., & Nasser, M. (2017). An enigmatic new genus of Hormiinae (Hymenoptera: Braconidae) from south India. *Zootaxa, 4272*(3), 371−385.

Rastegar, J., Sakenin, H., Khodaparast, S., & Havaskary, M. (2012). On a collection of Braconidae (Hymenoptera) from East Azarbaijan and vicinity, Iran. *Calodema, 226*, 1−4.

Sakenin, H., Naderian, H., Samin, N., Rastegar, J., Tabari, M., & Papp, J. (2012). On a collection of Braconidae (Hymenoptera) from northern Iran. *Linzer biologische Beiträge, 44*(2), 1319−1330.

Sakenin, H., Samin, N., Beyarslan, A., Coronado-Blanco, J. M., Navaeian, M., Fischer, M., & Hosseini Boldaji, S. A. (2020). A faunistic study on braconid wasps (Hymenoptera: Braconidae) from Iran. *Boletin de la Sociedad Andaluza de Entomologia, 30*, 96−102.

Samin, N. (2015). A faunistic study on the Braconidae of Iran (Hymenoptera: Ichneumonoidea). *Arquivos Entomoloxicos, 13*, 339−345.

Samin, N., Sakenin, H., Imani, S., & Shojai, M. (2011). A study on the Braconidae (Hymenoptera) of Khorasan province and vicinity, northeastern Iran. *Phegea, 39*(4), 137−143.

Samin, N., Ghahari, H., Gadallah, N. S., & Davidian, E. (2014). A study on the Braconidae (Hymenoptera: Ichneumonoidea) from West Azarbaijan province, Northwestern Iran. *Linzer biologische Beiträge, 46*(2), 1447−1478.

Samin, N., van Achterberg, C., & Ghahari, H. (2015). A faunistic study of Braconidae (Hymenoptera: Ichneumonoidea) from southern Iran. *Linzer biologische Beiträge, 47*(2), 1801−1809.

Samin, N., van Achterberg, C., & Çetin Erdoğan, Ö. (2016). A faunistic study on some subfamilies of Braconidae (Hymenoptera: Ichneumonoidea) from Iran. *Arquivos Entomolóxicos, 15*, 153−161.

Samin, N., Coronado-Blanco, J. M., Kavallieratos, N. G., Fischer, M., & Sakenin, H. (2018a). Recent findings on Braconidae (Hymenoptera: Ichneumonoidea) of Iran with an updated checklist. *Acta Biologica Turcica, 31*(4), 160−173.

Samin, N., Coronado-Blanco, J. M., Fischer, M., van Achterberg, C., Sakenin, H., & Davidian, E. (2018b). Updated checklist of Iranian Braconidae (Hymenoptera: Ichneumonoidea) with twenty-three new records. *Natura Somogyiensis, 32*, 21−36.

Samin, N., Fischer, M., Sakenin, H., Coronado-Blanco, J. M., & Tabari, M. (2019). A faunistic study on Agathidinae, Alysiinae, Doryctinae, Helconinae, Microgastrinae and Rogadinae (Hymenoptera: Braconidae), with eight new country records. *Calodema, 734*, 1−7.

Samin, N., Beyarslan, A., Coronado-Blanco, J. M., & Navaeian, M. (2020). A contribution to the braconid wasps (Hymenoptera: Braconidae) from Iran. *Natura Somogyiensis, 35*, 25−28.

Samin, N., Sakenin Chelav, H., Ahmad, Z., Penteado-Dias, A. M., & Samiuddin, A. (2020). A faunistic study on the family Braconidae (Hymenoptera: Ichneumonoidea) from Iran. *Scientific Bulletin of Uzhhorod National University (Series: Biology), 48*, 14−19.

Samin, N., Beyarslan, A., Ranjith, A. P., Ahmad, Z., Sakenin Chelav, H., & Hosseini Boldaji, S. A. (2020c). A faunistic study on Braconidae (Hymenoptera: Ichneumonoidea) from Ardebil and east Azarbayjan provinces, Northwestern Iran. *Egyptian Journal of Plant Protection Research Institute, 3*(4), 955−963.

Shaw, M. R. (1981). Possible foodplant differences of *Amphipyra pyramidea* (L.) and *A. berbera svenssoni* Fletcher (Lepidoptera: Noctuidae), and a note on their parasites (Hymenoptera). *Entomologist's Gazette, 32*, 165−167.

Shaw, M. R. (1983). On evolution of endoparasitism: the biology of some genera of Rogadinae (Braconidae). *Contributions of the American Entomological Institute, 20*, 307−328.

Shaw, M. R., & Huddleston, T. (1991). Classification and biology of braconid wasps (Hymenoptera: Braconidae). *Handbooks for the Identification of British Insects, 7*(11), 1−126.

Shaw, S. R. (2006). *Aleiodes wasps of eastern forests: a guide to parasitoids and associated dead caterpillars. Forest Health Technology Enterprise Team (FHTET), FHTET-2006-08*, 126 pp.

Shaw, S. R., Marsh, P. M., & Fortier, J. C. (1997). Revision of North American *Aleiodes* (part 1): the *pulchripes* Wesmael species-group in the new World (Hymenoptera: Braconidae: Rogadinae). *Journal of Hymenoptera Research, 6*, 10−35.

Shaw, S. R., Marsh, P. M., & Fortier, J. C. (1998a). Revision of North American *Aleiodes* (part 2): the *apicalis* Brullé species-group in the new World (Hymenoptera: Braconidae: Rogadinae). *Journal of Hymenoptera Research, 7*(1), 62−73.

Shaw, S. R., Marsh, P. M., & Fortier, J. C. (1998b). Revision of North American *Aleiodes* (part 4): the *albitibia* and *praetor* species-groups in the new World (Hymenoptera: Braconidae: Rogadinae). *Proceedings of the Entomological Society of Washington, 100*(3), 553−565.

Shaw, S. R., Marsh, P. M., & Fortier, J. C. (2006). Revision of North American *Aleiodes* (part 8): the *coxalis* (Spinola) species-group (Hymenoptera: Braconidae: Rogadinae). *Zootaxa, 1314*, 1−30.

Shaw, S. R., Marsh, P. M., & Talluto, M. A. (2013). Revision of North American *Aleiodes* (Part 9): the *pallidator* (Thunberg) species-group with description of two new species (Hymenoptera: Braconidae, Rogadinae). *Zootaxa, 3608*(3), 204−212.

Shaw, S. R., Shimbori, E. M., & Penteado-Dias, A. (2020). A revision of the *Aleiodes bakeri* (Brues) species subgroup of the *A. seriatus* species group with the descriptions of 18 new species from the Neotropical Region (Hymenoptera: Braconidae: Rogadinae). *ZooKeys, 964*, 41−107.

Shenefelt, R. D. (1975). Braconidae 8. Exothecinae, Rogadinae. *Hymenopterorum Catalogus. Nova Editio, Pars, 12*, 1115−1262.

Shimbori, E. M., & Shaw, S. R. (2014). Twenty-four new species of *Aleiodes* Wesmael from the eastern Andes of Ecuador with associated biological information (Hymenoptera, Braconidae, Rogadinae). *ZooKeys, 405*, 1−81.

Shimbori, E. M., Shaw, S. R., Ventura de Almeida, L. F., & Penteado-Dias, A. M. (2016). Eleven new species of *Athacryvac* Braet & van Achterberg from the neotropical region (Hymenoptera, Braconidae, Rogadinae). *Zootaxa, 4138*(1), 83−117.

Shojai, M., Abbas-Pour, H., Nasrollahi, A., & Labbafi, Y. (1995). Technology and biocenotic aspects of integrated biocontrol of corn stem borer: *Sesamia cretica* Led. (Lep., Noctuidae). *Journal of Agricultural Science, 1*(2), 5−32 (in Persian, English summary).

Szépligeti, G. (1901). Tropische Cenocoeliden und Braconiden aus der Sammlung des Ungarischen National-Museums. *Természetrajzi Füsetek, 24,* 354–402.

Szépligeti, G. (1904). Hymenoptera, family Braconidae. In P. Wytsman (Ed.), *Genera Insectorum* (Vol. 22, pp. 78–79), 1–253. V. Verteneuil and L. Desmet-Verteneuil Bruxelles.

Szépligeti, G. (1906). Braconiden aus der Sammlung des ungarischen National-Museums, 1. *Annales Historico-Naturales Musei Nationalis Hungarici, 4,* 547–618.

Telenga, N. A. (1941). *Family Braconidae, subfamily Braconinae (continuation) and Sigalphinae* (pp. 1–466). Hymenoptera: Fauna USSR.

Tobias, V. I. (1976). Braconids of the Caucasus (Hymenoptera, Braconidae). Opred. In *Faune SSSR* (Vol. 110, pp. 1–286). Leningrad: Nauka Press.

Tobias, V. I. (1986). Subfamily Rogadinae. In G. S. Medvedev (Ed.), *Keys to the insects of the European part of the USSR Hymenoptera Part IV* (Vol. III, pp. 72–85). New Delhi: Amerind Publishing Co.. pp. 1–883.

Townsend, A., & Shaw, S. R. (2009). Nine new species of *Aleiodes* reared from caterpillars in the northeastern Andes of Ecuador (Hymenoptera: Braconidae: Rogadinae). *Journal of Insect Science, 9*(33), 1–21.

van Achterberg, C. (1991). Revision of the genera of the Afrotropical and W. Palaearctic Rogadinae Foerster (Hymenoptera: Braconidae). *Zoologische Verhandelingen, 273,* 1–102.

van Achterberg, C. (1993). Illustrated key to the subfamilies of the Braconidae. *Zoologische Verhandelingen, 283,* 1–189.

van Achterberg, C. (1995). Generic revision of the subfamily Betylobraconinae (Hymenoptera: Braconidae) and other groups with modified fore tarsus. *Zoologische Verhandlingen, 298,* 1–242.

van Achterberg, C. (2014). Notes on the checklist of Braconidae (Hymenoptera) from Switzerland. *Mitteilungen der Schweizerischen Entomologischen Gesellschaft, 87,* 191–213.

van Achterberg, C., & Shaw, M. R. (2016). Revision of the western Palaearctic species of *Aleiodes* Wesmael (Hymenoptera: Braconidae: Rogadinae). Part 1: Introduction, key to species groups, outlying distinctive species, and revisionary notes on some further species. *ZooKeys, 639,* 1–164.

van Achterberg, C., Shaw, M. R., & Quicke, D. L. J. (2020). Revision of the western Palaearctic species of *Aleiodes* Wesmael (Hymenoptera, Braconidae, Rogadinae). Part 2: revision of the *A. apicalis* gropup. *ZooKeys, 919,* 1–259.

Wharton, R. A. (1993). Bionomics of the Braconidae. *Annual Review of Entomology, 38,* 121–143.

Whitfield, J. B., & Wharton, R. A. (1997). Subfamily Hormiinae. In R. A. Wharton, P. M. Marsh, & M. J. Sharkey (Eds.), *Manual of the New World Genera of the Family Braconidae (Hymenoptera)* (pp. 284–301). Washington, DC: Special Publication No. 1, International Society of Hymenopterists, 439 pp.

Yu, D. S., van Achterberg, C., & Horstmann, K. (2016). *Taxapad 2016, Ichneumonoidea 2015, Database on flash-drive.* Nepean, Ontario, Canada.

Zaldívar-Riverón, A., Areekul, B., Shaw, M. R., & Quicke, D. L. J. (2004). Comparative morphology of the venom apparatus in the braconid wasp subfamily Rogadinae (Insecta, Hymenoptera, Braconidae) and related taxa. *Zoologica Scripta, 33*(3), 223–237.

Zaldívar-Riverón, A., Mori, M., & Quicke, D. L. J. (2006). Systematics of the cyclostome subfamilies of braconid parasitic wasps (Hymenoptera: Ichneumonoidea): a simultaneous molecular and morphological Bayesian approach. *Molecular Phylogenetics and Evolution, 38*(1), 130–145.

Zaldívar-Riverón, A., Belokobylskij, S. A., Leon Regagnon, V., Briceno, G. R., & Quicke, D. L. J. (2008). Molecular phylogeny and historical biogeography of the cosmopolitan parasitic wasp subfamily Doryctinae (Hymenoptera: Braconidae). *Invertebrate Systematics, 22*(3), 345–363.

Sigalphus irrorator (Fabricius, 1775) (Sigalphinae), ♀, lateral habitus. *Photo prepared by S.R. Shaw.*

Sigalphus bicolor (Cresson, 1880) (Sigalphinae), ♀, lateral habitus—Nearctic. *Photo prepared by S.R. Shaw.*

Chapter 31

Subfamily Sigalphinae Haliday, 1833

Neveen Samy Gadallah[1], Hassan Ghahari[2], Scott Richard Shaw[3] and Donald L.J. Quicke[4]

[1]Entomology Department, Faculty of Science, Cairo University, Giza, Egypt; [2]Department of Plant Protection, Yadegar-e Imam Khomeini (RAH) Shahre Rey Branch, Islamic Azad University, Tehran, Iran; [3]UW Insect Museum, Department of Ecosystem Science and Management, University of Wyoming, Laramie, WY, United States; [4]Integrative Ecology Laboratory, Department of Biology, Faculty of Science, Chulalongkorn University, Pathumwan, Bangkok, Thailand

Introduction

The cosmopolitan Sigalphinae Haliday, 1833 is a small subfamily of the family Braconidae. It comprises less than 50 species classified into eight genera and four tribes (Acampsini van Achterberg, 1992, Minangini de Saeger, 1948, Pselaphanini van Achterberg, 1985, and Sigalphini Haliday, 1833) (Chen & van Achterberg, 2019; van Achterberg, 2014; Yu et al., 2016). Thirteen species in the two genera Acampsis Wesmael, 1835 and Sigalphus Latreille, 1802, are known from the Palaearctic region (Yu et al., 2016).

Because of the form of their metasomal carapace, sigalphines were formerly and traditionally regarded as chelonines. However, they differ from chelonine by having their carapace articulated between first and second tergites (Shaw & Huddleston, 1991). The subfamily is regarded as the sister-group of Agathidinae (Belshaw & Quicke, 2002; Belshaw et al., 1998; Dowton et al., 1998; Quicke et al., 2008; Sharanowski et al., 2011; Shi et al., 2005).

Members of the subfamily Sigalphinae are diagnosed by having the first three metasomal tergites forming a sculptured shield that largely or completely conceals the following tergites (Dudarenko, 1974). Additionally, the ovipositor is short and smooth, being more or less curved, and without teeth, nodus or notch; the ovipositor sheath is usually widened (but slender in *Afrocampsis*); the occipital carina is reduced either completely (as in *Acampsis*) or only medially; the prepectal ridge is present; the notaulices are deep (resulting in the middle lobe of the mesoscutum being as convex as the lateral lobes); the prescutellar depression is absent; the pronotum has a lateral and dorsal pronope; the trochantelli are present and without small pegs; the dorsal carina of first metasomal tergite is strongly developed; and the marginal cell of the fore wing is short (van Achterberg, 1976; 1990, 1993; van Achterberg & Austin, 1992; Tobias, 1986).

Apart from the comprehensively studied case of *Acampsis alternipes* (Nees) (Shaw & Quicke, 2000), little is known in detail about the detailed biology of most Sigalphinae (Braet, 1997; van Achterberg & Austin, 1992). Shaw and Quicke (2000) demonstrated that *Acampsis alternipes* oviposits with precision into larval nerve ganglia. Sigalphines are mostly thought to be koinobiont endoparasitoids of lepidoptera larvae of the families Noctuidae and Geometridae (Quicke & van Achterberg, 1990; van Achterberg, 1984, 1990, 1993; van Achterberg and Austin, 1992; Sharanowski et al., 2014; Sharkey et al. 2019; Tobias, 1986; Yu et al., 2016). It has been suggested (but not proven) that sigalphines might be egg-larval parasitoids because of the carapace-like metasoma (Braet, 1997; van Achterberg and Austin, 1992). However, Cushman (1913) observed oviposition by *Sigalphus* into small caterpillars (not eggs). Shaw and Huddleston (1991) surmised that the idea that sigalphines could be egg-larval parasitoids may just be carried over from earlier times when sigalphines were (incorrectly) classified as Cheloninae.

Checklists of Regional Sigalphinae. Farahani et al. (2016) represented one species. Gadallah and Ghahari (2016), Yu et al. (2016), and Samin, Coronado-Blanco, Kavallieratos, et al. (2018) all listed two species in two genera, and finally, Samin, Coronado-Blanco, Fischer et al. (2018) listed three species in two genera. The present checklist includes two species in two genera and two tribes (Acampsini and Sigalphini) in the Middle East.

Braconidae of the Middle East (Hymenoptera). https://doi.org/10.1016/B978-0-323-96099-1.00013-3

Key to genera of the subfamily Sigalphinae of the Middle East (modified from Tobias et al., 1995)

1. Metasoma broadened toward apex; third metasomal tergite with two denticles ventrally and dense appressed golden hairs; first metasomal tergite strongly elevated in basal third, with two strong carinae and a weak middle one; scutellum flattened .... *Sigalphus* Latreille

− Metasoma oval; third metasomal tergite without denticles ventrally, with sparse hairs; first metasomal tergite slightly elevated in basal third, with weak carinae; scutellum convex .... *Acampsis* Wesmael

List of species of the subfamily Sigalphinae recorded in the Middle East

Subfamily Sigalphinae Haliday, 1833

Tribe Acampsini van Achterberg, 1992

Genus *Acampsis* Wesmael, 1835

Acampsis alternipes (Nees von Esenbeck, 1816)

Catalogs with Iranian records: Yu et al. (2016), Gadallah and Ghahari (2016), Samin, Coronado-Blanco, Kavallieratos, et al. (2018), Samin, Coronado-Blanco, Fischer et al. (2018). Distribution in Iran: West Azarbaijan (Samin et al., 2016). Distribution in the Middle East: Iran.

Extralimital distribution: Austria, Belgium, Bulgaria, France, Germany, Hungary, Italy, North Macedonia, Moldova, Netherlands, Poland, Russia, Slovakia, Switzerland, United Kingdom, former Yugoslavia.

Host records: Summarized by Yu et al. (2016) as being a parasitoid of the geometrids *Alsophila aceraria* (Denis and Schiffermüller), *Alsophila aescularia* (Denis and Schiffermüller), *Erannis defoliaria* (Clerck), and *Operophtera brunata* (L.).

Comments: The biology of *Acampsis alternipes* was studied by Shaw and Quicke (2000).

Tribe Sigalphini Haliday, 1833

Genus *Sigalphus* Latreille, 1802

Sigalphus irrorator (Fabricius, 1775)

Catalogs with Iranian records: Yu et al. (2016), Gadallah and Ghahari (2016), Farahani et al. (2016), Samin, Coronado-Blanco, Kavallieratos, et al. (2018), Samin, Coronado-Blanco, Fischer et al. (2018). Distribution in Iran: Golestan (Ghahari et al., 2010; Samin et al., 2015), Mazandaran (Ghahari, 2017 - around rice fields). Distribution in the Middle East: Iran.

Extralimital distribution: Austria, Belgium, Croatia, Finland, France, Germany, Hungary, Italy, Japan, Korea, Latvia, Netherlands, Poland, Romania, Russia, Slovakia, Spain, Sweden, Switzerland, Ukraine, United Kingdom.

Host records: Summarized by Yu et al. (2016) as being a parasitoid of the noctuids *Acronicta aceris* (L.), *Acronicta psi* (L.), *Acronicta tridens* (Denis and Schiffermüller), *Calophasia lunula* (Hufnagel), and *Ceramica pisi* (L.).

Conclusion

Two valid species of the subfamily Sigalphinae in two genera (*Acampsis* Wesmael, 1835 and *Sigalphus* Latreille, 1802) have been reported from the Middle East (only Iran). Iranian Sigalphinae represents 4.4% of the world species. These species have been recorded from three provinces, Golestan, Mazandaran, and West Azarbaijan. No host species are known for these parasitoids in Iran. Additionally, among the Middle Eastern countries and adjacent to Iran, the subfamily Sigalphinae has been recorded only from Iran and Russia (with five species) and both Iranian species are shared with the fauna of Russia. However, less comprehensive surveys in those countries are a possible reason of the absence of this subfamily.

References

Belshaw, R., & Quicke, D. L. J. (2002). Robustness of ancestral state estimates: evolution of life history strategy in ichneumonoid parasitoids. *Systematic Biology, 51*(3), 450−477.

Belshaw, R., Fitton, M., Herniou, E., Gimeno, C., & Quicke, D. L. J. (1998). A phylogenetic reconstruction of the Ichneumonoidea (Hymenoptera) based on the D2 variable region of 28S ribosomal RNA. *Systematic Entomology, 23*, 109−123.

Braet, Y. (1997). Occurrences of Sigalphinae, (Hymenoptera: Braconidae) in Belgium. *Annales de la Société entomologique de Belgique, 133*, 225−227.

Chen, X. X., & van Achterberg, C. (2019). Systematics, phylogeny, and evolution of braconid wasps: 30 years of progress. *Annual Review of Entomology, 64*, 1−24.

Cushman, R. A. (1913). Biological notes on a few rare or little known parasitic Hymenoptera. *Proceedings of the Entomological Society of Washington, 15*, 153−160.

Dowton, M., Austin, A. D., & Antolin, M. F. (1998). Evolutionary relationships among the Braconidae (Hymenoptera: Ichneumonoidea) inferred from partial 16S rDNA gene sequences. *Insect Molecular Biology, 7*, 129−150.

Dudarenko, G. P. (1974). Formation of the abdominal carapace in braconids (Hymenoptera, Braconidae) and some aspects of the classification of the family. *Entomological Review, 53*, 80−90.

Farahani, S., Talebi, A. A., & Rakhshani, E. (2016). Iranian Braconidae (Insecta: Hymenoptera: Ichneumonoidea): diversity, distribution and host association. *Journal of Insect Biodiversity and Systematics, 2*(1), 1—92.

Gadallah, N. S., & Ghahari, H. (2016). An updated checklist of the Iranian Miracinae, Pambolinae and Sigalphinae (Hymenoptera: Braconidae). *Orsis, 30*, 51—61.

Ghahari, H. (2017). Species diversity of Ichneumonoidea (Hymenoptera) from rice fields of Mazandaran province, northern Iran. *Journal of Animal Environment, 9*(3), 371—378 (in Persian, English summary).

Ghahari, H., Fischer, M., Hedqvist, K. J., Çetin Erdogan, Ö, van Achterberg, C., & Beyarslan, A. (2010). Some new records of Braconidae (Hymenoptera) for Iran. *Linzer biologische Beiträge, 42*(2), 1395—1404.

Quicke, D. L. J., & van Achterberg, C. (1990). Phylogeny of the subfamilies of the family Braconidae (Hymenoptera: Ichneumonoidea). *Zoologische Verhandelingen, 258*, 1—180.

Quicke, D. L. J., Sharkey, M. J., Laurenne, N. M., & Dowling, A. (2008). A preliminary molecular phylogeny of the Sigalphinae (Hymenoptera: Braconidae), including *Pselaphanus* Szépligeti, based on 28S rDNA, with descriptions of new Afrotropical and Madagascan *Minanga and Malasigalphus* and Malasigalphus species. *Journal of Natural History, 42*(43—44), 2703—2719.

Samin, N., Ghahari, H., Gadallah, N. S., & Monaem, R. (2015). A study on the braconid wasps (Hymenoptera: Ichneumonoidea: Braconidae) from Golestan province, northern Iran. *Linzer biologische Beiträge, 47*(1), 731—739.

Samin, N., van Achterberg, C., & Erdoğan, Ö.Ç. (2016). A faunistic study on some subfamilies of Braconidae (Hymenoptera: Ichneumonoidea) from Iran. *Arquivos Entomoloxicos, 15*, 153—161.

Samin, N., Coronado-Blanco, J. M., Kavallieratos, N. G., Fischer, M., & Sakenin, H. (2018). Recent findings on Braconidae (Hymenoptera: Ichneumonoidea) of Iran with an updated checklist. *Acta Biologica Turcica, 31*(4), 160—173.

Samin, N., Coronado-Blanco, J. M., Fischer, M., van Achterberg, C., Sakenin, H., & Davidian, E. (2018). Updated checklist of Iranian Braconidae (Hymenoptera: Ichneumonoidea) with twenty-three new records. *Natura Somogyiensis, 32*, 21—36.

Sharanowski, B. J., Dowling, A. P. G., & Sharkey, M. J. (2011). Molecular phylogenetics of Braconidae (Hymenoptera: Ichneumonoidea), based on multiple nuclear genes, and implication for classification. *Systematic Entomology, 36*, 549—572.

Sharanowski, B. J., Zhang, Y. M., & Wanigasekara, R. W. M. U. M. (2014). Annotated checklist of Braconidae (Hymenoptera) in the Canadian Prairies Ecozone. In D. J. Giberson, & H. A. Carcamo (Eds.), *Arthropods of Canadian Grasslands (volume 4): Biodiversity and systematics Part 2* (pp. 399—425). Ottawa: Biological Survey of Canada.

Sharkey, M. J., penteado-Diaz, A. M., Smith, M. A., Hallwachs, W., & Janzen, D. (2019). Synopsis of New World Sigalphinae (Hymenoptera, Braconidae) with the description of two new species and key to genera. *Journal of Hymenoptera Research, 68*, 1—11.

Shaw, M. R., & Huddleston, T. (1991). Classification and biology of braconid wasps (Hymenoptera: Braconidae). *Handbooks for the Identification of British Insects, 7*, 1—126.

Shaw, M. R., & Quicke, D. L. J. (2000). The biology and early stages of *Acampsis alternipes* (Nees). with comments on the relationships of the Sigalphinae (Hymenoptera: Braconidae). *Journal of Natural History, 34*, 611—628.

Shi, M., Chen, X. X., & van Achterberg, C. (2005). Phylogenetic relationships among the Braconidae (Hymenoptera: Ichneumonoidea) inferred from partial 16S rDNA, 28S rDNA D2, 18S rDNA gene sequences and morphological characters. *Molecular Phylogenetics and Evolution, 37*(1), 104—116.

Tobias, V. I. (1986). Keys to the insects of the European part of USSR. In G. S. Medvedev (Ed.), *[Opredelitel Nasekomykh Evropeiskoi Chasti SSSR. Tom III, Pereponchatokrylye, Chetvertaia Chasf]. Vol. III. Hymenoptera. Part IV*. Braconidae, 883 pp.

Tobias, V. I., Belokobylskij, S. A., & Kotenko, A. G. (1995). *Keys to the insects of the European part of the USSR. III (part 4)*. Institute of Zoology, Academy of Sciences of the USSR, 908 pp.

van Achterberg, C. (1976). A preliminary key to the subfamilies of the Braconidae (Hym.). *Tijdschrift voor Entomologie, 119*(3), 33—78.

van Achterberg, C. (1984). Essay on the phylogeny of Braconidae (Hymenoptera: Ichneumonoidea). *Entomologisk Tidskrift, 105*, 41—58.

van Achterberg, C. (1990). Illustrated key to the subfamilies of the Holarctic Braconidae (Hymenoptera: Ichneumonoidea). *Zoologische Mededelingen, 64*, 1—20.

van Achterberg, C. (1993). Illustrated key to the subfamilies of the Braconidae. *Zoologische Verhandelingen, 283*, 1—189.

van Achterberg, C. (2014). *Sigalphus anjae* spec. nov. (Hymenoptera: Braconidae: Sigalphinae) from southern Vietnam. *Zoologische Mededelingen, 88*(2), 9—17.

van Achterberg, C., & Austin, A. D. (1992). Revision of the genera of the subfamily Sigalphinae (Hymenoptera: Braconidae), including a revision of the Australian species. *Zoologische Verhandelingen, 280*, 1—44.

Yu, D. S., van Achterberg, C., & Horstmann, K. (2016). *Taxapad 2016, Ichneumonoidea 2015, Database on flash-drive*. Nepean, Ontario, Canada.

Hormius sp. (Hormiinae), ♀, lateral habitus. *Photo prepared by S.R. Shaw.*

Doryctes erythromelas (Brullé, 1846) (Doryctinae), ♀, lateral habitus. *Photo prepared by S.R. Shaw.*

Chapter 32

Diversity of Braconidae in the Middle East with an emphasis on Iran

Hassan Ghahari[1], Neveen Samy Gadallah[2], Scott Richard Shaw[3] and Donald L.J. Quicke[4]

[1]Department of Plant Protection, Yadegar-e Imam Khomeini (RAH) Shahre Rey Branch, Islamic Azad University, Tehran, Iran; [2]Entomology Department, Faculty of Science, Cairo University, Giza, Egypt; [3]UW Insect Museum, Department of Ecosystem Science and Management, University of Wyoming, Laramie, WY, United States; [4]Integrative Ecology Laboratory, Department of Biology, Faculty of Science, Chulalongkorn University, Pathumwan, Bangkok, Thailand

Braconid wasps (Hymenoptera: Braconidae) play important ecological roles as parasitoids of agricultural and forest pests by helping to regulate populations of herbivorous insects. They are thus essential for the maintenance of ecological processes and shaping the diversity of other organisms (e.g., by selectively feeding on caterpillars they can shape the distribution and abundance of the host insect's food plants) (Ghahari et al., 2006; Hanson & Gauld, 2006; LaSalle & Gauld, 1993; Quicke, 2015). Faunal lists are a starting point for understanding a country's natural resources and have applications in biodiversity and conservation. Faunal lists are also essential for successful integrated pest management (IPM) by allowing researchers and practitioners to better understand the relationships between insect pests and their associated natural enemies. Here we provide the first comprehensive checklist for braconid wasps for the Middle East countries, with additional data (species diversity and host records upon the constituent provinces) for Iran.

In total, 2037 species in 255 genera and 30 subfamilies of Braconidae have been recorded from the Middle East countries, of which two subfamilies, Braconinae and Microgastrinae, with 347 and 292 recorded species, respectively, are more diverse than the other subfamilies (Table 32.1). Among the Middle Eastern countries, Iran, represented by 1363 recorded species, is more diverse, followed by Turkey (1197 species), and Israel–Palestine (439 species) (Table 32.10). Faunistic knowledge of the family Braconidae in many of the Middle Eastern countries, despite of having arid and semiarid climates, is largely incomplete due to the paucity of regional studies, and to an extent, greater taxonomic complexity compared with well-studied other western Palaearctic countries. Examples are that of Russia, comprising 3272 species in 268 genera

(14.2% of the world species) (Belokobylskij & Lelej, 2019) which has more recorded species than all of the Middle Eastern countries combined.

To compile the faunal lists of Braconidae in the Middle East we consulted all relevant sources published prior to the end of 2021. There are more than 23,062 recognized braconid species worldwide (Yu et al., 2016), and therefore, the Middle Eastern fauna represents 8.9% of the total number of species worldwide. The difference in species per unit area can be attributed largely to the history and relative effort on insect taxonomy in the regions, although climatic and ecosystem factors might also be relevant. Hundreds of new species have been described since Yu et al. (2016). Estimates of 45,000 to 50,000 species of Braconidae will be discovered in the future after faunistic surveys synthetically in all parts of the world.

The fauna of Iranian Braconidae with 1363 species in 203 genera was studied rather well. Among the 30 subfamilies, Braconinae and Microgastrinae with 212 and 209 recorded species, respectively, were represented by the largest number of species, followed by Alysiinae (171 species), Opiinae (137 species), and Cheloninae (113 species) (Table 32.2). Thirty-one species belonging to 12 subfamilies are newly recorded for the Iranian fauna. The checklists include one new country record each for the subfamilies Agathidinae, Aphidiinae, Brachistinae, Doryctinae, Macrocentrinae, Orgilinae, Rogadinae, two for Cheloninae, three for Braconinae, four for Alysiinae, six for Euphorinae, and nine for Microgastrinae. Our checklist excludes eight species previously recorded from the Iranian fauna because of suspected misidentifications. These are *Aphidius nigripes* Ashmead, 1901 (Aphidiinae), *Coeloides brunneri* Viereck, 1911, and *Habrobracon johnson* (Ashmead) (Braconinae), *Peristenus rubricollis* (Thomson, 1892)

TABLE 32.1 Species diversity of Braconidae of the Middle East by subfamily.

Subfamily	Number of genera	Number of species	Subfamily	Number of genera	Number of species
Agathidinae	12	80	Homolobinae	1	7
Alysiinae	32	213	Hormiinae	3	11
Aphidiinae	18	111	Ichneutinae	2	4
Brachistinae	10	100	Macrocentrinae	1	19
Braconinae	35	347	Microgastrinae	23	292
Cardiochilinae	4	21	Microtypinae	1	6
Cenocoelinae	2	2	Miracinae	2	6
Charmontinae	1	2	Opiinae	23	238
Cheloninae	5	219	Orgilinae	2	34
Dirrhopinae	1	1	Pambolinae	3	6
Doryctinae	29	96	Proteropinae	1	1
Euphorinae	22	120	Rhysipolinae	3	6
Exothecinae	3	8	Rhyssalinae	5	7
Gnamptodontinae	1	4	Rogadinae	6	68
Helconinae	2	6	Sigalphinae	2	2

TABLE 32.2 Species diversity of Iranian Braconidae by subfamily.

Subfamily	Number of genera	Number of species	Subfamily	Number of genera	Number of species
Agathidinae	10	47	Homolobinae	1	5
Alysiinae	28	171	Hormiinae	3	10
Aphidiinae	18	92	Ichneutinae	2	4
Brachistinae	9	73	Macrocentrinae	1	17
Braconinae	15	212	Microgastrinae	17	209
Cardiochilinae	3	11	Microtypinae	1	4
Cenocoelinae	2	2	Miracinae	2	3
Charmontinae	1	2	Opiinae	20	137
Cheloninae	5	113	Orgilinae	2	20
Dirrhopinae	1	1	Pambolinae	3	5
Doryctinae	16	47	Proteropinae	1	1
Euphorinae	21	92	Rhysipolinae	2	5
Exothecinae	3	6	Rhyssalinae	5	6
Gnamptodontinae	1	4	Rogadinae	6	56
Helconinae	2	6	Sigalphinae	2	2

(Euphorinae), *Dolichogenidea aethiopicus* (Wilkinson, 1931) (Microgastrinae), *Opius paraplasticus* Fischer, 1972, *Phaedrotoma penetrator* (Fischer, 1966) (Opiinae) and *Orgilus* (*Orgilus*) *jennieae* Marsh, 1979 (Orgilinae). Additionally, this checklist includes 559 more species which are not listed for Iran in Yu et al. (2016).

Endemic species to the Middle East countries. Among the 16 countries of the Middle East, 11 countries (exception Bahrain, Kuwait, Lebanon, Oman, and Qatar) have the species which have so far been known only from there (endemic species). In total, 242 species are endemic to the Middle East countries, of which Turkey with 72 species has the highest number of species, followed by Iran and Yemen with 53 and 45 species, respectively. However, at least some, if not most, of these endemic species might

eventually be found in one or more of the neighboring countries with future collecting and more taxonomic studies.

Cyprus. Three species in the subfamily Cheloninae are endemic for this country: *Chelonus cyprensis* (Tobias, 2001), *Chelonus cypri* (Tobias, 2001), and *Chelonus cyprianus* Fahringer, 1937.

Egypt. Seventeen species in six subfamilies are endemic to this country (10.5% of total species of Egypt). Cardiochilinae with six species is more diverse than other five subfamilies (Table 32.3).

Iran. Fifty-three species are so far only known, or recorded from Iran (3.8% of total species of Iran), of which Microgastrinae with 16 species had the highest number of endemic species (Table 32.4).

TABLE 32.3 List of Braconidae species only known from Egypt (endemic species).

Subfamily	Species	Subfamily	Species
Braconinae	*Iphiaulax* (*Iphiaulax*) *congruus* (Walker, 1871)	Cardiochilinae	*Schoenlandella obscuriceps* (Fischer, 1958)
Braconinae	*Vipio indecisus* (Walker, 1871)	Cardiochilinae	*Schoenlandella pseudoglabra* Edmardash, Gadallah and Sharkey, 2018
Braconinae	*Vipio walkeri* (Dalla Torre, 1898)	Cheloninae	*Phanerotoma elbaiensis* Edmardash, Abdel Dayem and Gadallah, 2011
Cardiochilinae	*Bohayella temporalis* (Fischer, 1958)	Cheloninae	*Phanerotoma ponti* Edmardash, Abdel Dayem and Gadallah, 2011
Cardiochilinae	*Cardiochiles priesneri* Fischer, 1958	Cheloninae	*Phanerotoma sinaitica* Edmardash and Gadallah, 2019
Cardiochilinae	*Cardiochiles weidholzi* Fischer, 1958	Doryctinae	*Rhoptrocentrus cleopatra* Belokobylskij, 2001
Cardiochilinae	*Schoenlandella acrenulata* (Fischer, 1958)	Microtypinae	*Microtypus aegypticus* Edmardash, Gadallah and van Achterberg, 2017
Cardiochilinae	*Schoenlandella glaber* (Fischer, 1958)	Pambolinae	*Avga sinaitica* Edmardash and Gadallah, 2020
Cardiochilinae	*Schoenlandella maculata* (Fischer, 1958)	—	—

TABLE 32.4 List of Braconidae species only known from Iran (endemic species).

Subfamily	Species	Subfamily	Species
Agathidinae	*Aerophilus persicus* (Farahani & Talebi, 2014)	Cheloninae	*Chelonus setaceus* Papp, 1993
Agathidinae	*Cremnops richteri* Hedwig, 1957	Cheloninae	*Chelonus moczari* (Papp, 2014)
Alysiinae	*Aspilota alfalfae* Fischer, Lashkari Bod, Rakhshani and Talebi, 2011	Cheloninae	*Chelonus subpamiricus* Farahani and van Achterberg, 2018

Continued

TABLE 32.4 List of Braconidae species only known from Iran (endemic species).—cont'd

Subfamily	Species	Subfamily	Species
Alysiinae	*Aspilota isfahanensis* Peris-Felipo, 2016	Euphorinae	*Meteorus breviterebratus* Ameri, Talebi and Beyarslan, 2014
Alysiinae	*Aristelix persica* Peris-Felipo, 2015	Helconinae	*Helcon heinrichi* Hedqvist, 1967
Alysiinae	*Chorebus axillaris* Fischer, Lashkari Bod, Rakhshani and Talebi, 2011	Microgastrinae	*Choeras formosus* Abdoli and Fernandez-Triana, 2019
Alysiinae	*Chorebus longiarticulis* Fischer, Lashkari Bod, Rakhshani and Talebi, 2011,	Microgastrinae	*Choeras fulviventris* Fernandez-Triana and Abdoli, 2019
Alysiinae	*Chorebus nigridiremptus* Fischer, Lashkari Bod, Rakhshani and Talebi, 2011	Microgastrinae	*Choeras qazviniensis* Fernandez-Triana and Talebi, 2019
Alysiinae	*Chorebus properesam* Fischer, Lashkari Bod, Rakhshani and Talebi, 2011	Microgastrinae	*Choeras taftanensis* Ghafouri Moghaddam and van Achtereberg, 2018
Alysiinae	*Chorebus zarghanensis* Fischer, Lashkari Bod, Rakhshani and Talebi, 2011	Microgastrinae	*Cotesia elongata* Zargar and Gupta, 2019
Aphidiinae	*Aphidius iranicus* Rakhshani and Starý, 2007	Microgastrinae	*Cotesia khuzestanensis* Zargar and Gupta, 2019
Aphidiinae	*Aphidius stigmaticus* Rakhshani and Tomanović, 2011	Microgastrinae	*Cotesia zagrosensis* Zargar and Gupta, 2019
Aphidiinae	*Tanytrichophorous petiolaris* Mackauer, 1961	Microgastrinae	*Deuterixys tenuiconvergens* Zargar and Gupta, 2019
Aphidiinae	*Trioxys metacarpalis* Rakhshani and Starý, 2012	Microgastrinae	*Dolichogenidea fernandeztrianai* Abdoli and Talebi, 2019
Braconinae	*Bracon persiangulfensis* Ameri, Beyarslan and Talebi, 2013	Microgastrinae	*Iconella brachyradiata* Abdoli and Talebi, 2021
Braconinae	*Habrobracon iranicus* Fischer, 1972	Microgastrinae	*Iconella mongashtensis* Zargar and Gupta, 2019
Braconinae	*Iphiaulax mirabilis* (Hedwig, 1957)	Microgastrinae	*Iconella similus* Zargar and Gupta, 2019
Braconinae	*Megalommum pistacivora* van Achterberg and Mehrnejad, 2011	Microgastrinae	*Microplitis alborziensis* Abdoli and Talebi, 2021
Braconinae	*Pseudovipio nigrirostris* (Kokujev, 1907)	Microgastrinae	*Pholetesor pseudocircumscriptus* Abdoli, 2019
Braconinae	*Pseudovipio schaeuffelei* (Hedwig, 1957)	Microgastrinae	*Protapanteles albigena* Abdoli, Fernandez-Triana and Talebi, 2021
Braconinae	*Vipio xanthurus* (Fahringer, 1926)	Microgastrinae	*Venanides caspius* Abdoli, Fernandez-Triana and Talebi, 2019
Cardiochilinae	*Cardiochiles tibialis* Hedwig, 1957	Miracinae	*Centistidea (Paracentistidea) pistaciella* van Achterberg and Mehrnejad, 2002
Cardiochilinae	*Cardiochiles triplus* Shenefelt, 1973	Miracinae	*Mirax caspiana* Farahani, Talebi, van Achterberg and Rakhshani, 2014
Cardiochilinae	*Pseudcardiochilus abnormipes* Hedwig, 1957	Opiinae	*Opius (Nosopoea) teheranensis* Fischer, 1990
Cardiochilinae	*Schoenlandella angustigena* Kang, 2021	Rogadinae	*Hoplocrotaphus hamooniae* Peris-Felipo, Belokobylskij and Rakhshani, 2018
Cardiochilinae	*Schoenlandella latigena* Kang, 2021	Rogadinae	*Yelicones iranus* (Fischer, 1963)
Cheloninae	*Chelonus iranicus* Tobias, 1972	—	—

Iraq. Three species in three subfamilies are endemic to Iraq: *Trioxys quercicola* Starý, 1969 (Aphidiinae), *Phanerotoma longiradialis* van Achterberg, 1990 (Cheloninae), and *Allorhogas semitemporalis* (Fischer, 1960) (Doryctinae).

Israel–Palestine. Sixteen species in five subfamilies are endemic to this country (3.8% of total species of Israel–Palestine). The subfamily Braconinae with six species is more diverse than other four subfamilies (Table 32.5).

Jordan. Three species in three subfamilies are endemic to Jordan: *Agathis jordanicola* Koçak and Kemal, 2013 (Agathidinae), *Dinotrema* (*Synaldis*) *jordanica* (Fischer, 1993) (Alysiinae), and *Chelonus jordanicus* (Tobias, 2001) (Cheloninae).

Saudi Arabia. Fourteen species in seven subfamilies are endemic to this country (27% of total species of Saudi Arabia) (Table 32.6).

Syria. Four species in two subfamilies are endemic to Syria: *Campyloneurus manni* Fahringer, 1928, *Iphiaulax*

(*Iphiaulax*) *ehrenbergi* Strand, 1912, *Bathyaulax syraensis* (Strand, 1912) (Braconinae), and *Chelonus xanthoscaposus* (Tobias, 2001) (Cheloninae).

Turkey. In total, 71 species in 12 subfamilies are endemic to this country (6% of total species of Turkey). Two subfamilies Braconinae and Opiinae both with 20 endemic species are more diverse than other subfamilies (Table 32.7).

United Arab Emirates. Thirteen species in three subfamilies are endemic to this country (21% of total species): *Agathis luteotegula* van Achterberg, 2011, *Agathis mealnotegula* van Achterberg, 2011, *Coccygidium maculatum* van Achterberg, 2011, *Disophrys angitemporalis* van Achterberg, 2011 (Agathidinae), *Phanerotoma ejuncida* van Achterberg, 2021, *Phanerotoma micrommata* van Achterberg, 2021, *Phanerotoma stenochora* van Achterberg, 2021 (Cheloninae), *Aivalykus microaciculatus* Ranjith and Belokobylskij, 2020, *Dendrosotinus* (*Gildoria*) *subelongatus* Belokobylskij, 2021, *Hecabalodes maculatus* Belokobylskij, 2021, *Rhaconotus* (*Rhaconotus*)

TABLE 32.5 List of Braconidae species only known from Israel–Palestine (endemic species).

Subfamily	Species	Subfamily	Species
Alysiinae	*Dinotrema* (*Synaldis*) *argamani* (Fischer, 1993)	Braconinae	*Bracon* (*Bracon*) *israelicus* Papp, 2015
Alysiinae	*Dinotrema* (*Synaldis*) *israelica* (Fischer, 1993)	Braconinae	*Bracon* (*Glabrobracon*) *propebella* Papp, 2012
Alysiinae	*Dinotrema* (*Dinotrema*) *paucilica* Papp, 2012	Braconinae	*Bracon* (*Lucobracon*) *freidbergi* Papp, 2015
Alysiinae	*Dinotrema* (*Synaldis*) *soederlundi* (Fischer, 2003)	Braconinae	*Rhytimorpha pappi* Quicke and Butcher, 2018
Alysiinae	*Idiasta* (*Idiasta*) *argamani* Papp, 2012	Cheloninae	*Chelonus argamani* Papp, 2012
Aphidiinae	*Lysiphlebus marismotui* Mescheloff and Rosen, 1989	Cheloninae	*Chelonus atrotibia* (Papp, 2012)
Braconinae	*Bracon* (*Bracon*) *furthi* Papp, 2015	Cheloninae	*Chelonus halperini* (Papp, 2012)
Braconinae	*Bracon* (*Bracon*) *heberola* Papp, 2012	Opiinae	*Opius ocuvergens* Papp, 2012

TABLE 32.6 List of Braconidae species only known from Saudi Arabia (endemic species).

Subfamily	Species	Subfamily	Species
Agathidinae	*Camptothlipsis arabica* Ghramh, 2012	Doryctinae	*Rhaconotus* (*Rhaconotus*) *arabicus* Belokobylskij, 2001
Agathidinae	*Coccygidium arabicum* Ghramh, 2011	Euphorinae	*Meteorus arabica* Ghramh, 2012
Agathidinae	*Coccygidium hebabi* Ghramh, 2013	Microgastrinae	*Microplitis faifaicus* Ghramh and Ahmad, 2020
Braconinae	*Bathyaulax fritzeni* Kaartinen and Quicke, 2007	Microgastrinae	*Microplitis khamisicus* Ghramh and Ahmad, 2020
Braconinae	*Bathyaulax juhai* Kaartinen and Quicke, 2007	Microgastrinae	*Microplitis tihamicus* Ghramh and Ahmad, 2020
Braconinae	*Bathyaulax ollilae* Kaartinen and Quicke, 2007	Miracinae	*Centistidea tihamica* Ghramh and Ahmad, 2019
Cheloninae	*Phanerotoma arabica* Ghramh, 2011	Rogadinae	*Aleiodes arabiensis* Butcher and Quicke, 2015

TABLE 32.7 List of Braconidae species only known from Turkey (endemic species).

Subfamily	Species	Subfamily	Species
Agathidinae	*Agathis berkei* Çetin Erdoğan, 2010	Braconinae	*Vipio alpi* Beyarslan, 2002
Agathidinae	*Agathis fischeri* Zettel and Beyarslan, 1992	Braconinae	*Vipio lalapasaensis* (Beyarslan, 1992)
Agathidinae	*Bassus beyarslani* Çetin Erdoğan, 2005	Braconinae	*Vipio spilogaster* (Walker, 1871)
Alysiinae	*Dinotrema (Dinotrema) partimrufa* Fischer, 2009	Cheloninae	*Chelonus beyarslani* Aydoğdu, 2008
Alysiinae	*Dinotrema (Dinotrema) samsunense* Fischer and Sullivan, 2014	Cheloninae	*Chelonus chetini* Bayarslan and Şahan, 2019
Alysiinae	*Eudinostigma subpulvinatum* Fischer, 2009	Cheloninae	*Chelonus turcius* (Tobias, 2008)
Alysiinae	*Idiasta (Idiasta) adanacola* Fischer and Beyarslan, 2012		
Alysiinae	*Idiasta (Idiasta) rugosipleurum* Fischer and Beyarslan, 2012	Euphorinae	*Elasmosoma geylanae* Beyarslan, 2016
Alysiinae	*Orthostigma (Orthostigma) curtiradiale* Fischer, 1995	Microgastrinae	*Microgaster filizinancae* Koçak and Kemal, 2013
Alysiinae	*Orthostigma (Orthostigma) impunctatum* Fischer, 1995	Microgastrinae	*Cotesia pappi* Inanç, 2002
Alysiinae	*Orthostigma (Orthostigma) robusticeps* Fischer, 1995	Miracinae	*Mirax striata* Beyarslan, 2009
Brachistinae	*Blacus (ganychorus) madli* Haeselbarth, 1992	Opiinae	*Biosteres (Biosteres) adanensis* Fischer and Beyarslan, 2005
Brachistinae	*Chelostes robustus* van Achterberg, 1990	Opiinae	*Biosteres (Biosteres) kayapinarensis* Fischer and Beyarslan, 2005
Brachistinae	*Chelostes subrobustus* Yilmaz and Beyarslan, 2009	Opiinae	*Apodesmia ispartaensis* (Fischer and Beyarslan, 2005)
Brachistinae	*Eubazus (Brachistes) aydae* Beyarslan, 2011	Opiinae	*Apodesmia pseudarenacea* (Fischer and Beyarslan, 2005)
Brachistinae	*Schizoprymnus (Schizoprymnus) erzurumus* Belokobylskij, Güclü and Ozbek, 2004	Opiinae	*Apodesmia rugata* (Fischer, 1992)
Brachistinae	*Schizoprymnus (Schizoprymnus) ozlemae* Beyarslan, 1988	Opiinae	*Bitomus (Bitomus) valdepusillus* Fischer and Beyarslan, 2005
Brachistinae	*Diospilus (Diospilus) angorensis* Beyarslan, 2014	Opiinae	*Indiopius yilmazae* Fischer and Beyarslan, 2011
Brachistinae	*Diospilus (Diospilus) belokobylskiji* Beyarslan, 2008	Opiinae	*Opiognathus silifkeensis* (Fischer and Beyarslan, 2005)
Braconinae	*Bracon (Bracon) bachtiae* Beyarslan, 2012	Opiinae	*Opius adanacola* Fischer and Beyarslan, 2005
Braconinae	*Bracon (Bracon) bilecikator* Beyarslan, 1996	Opiinae	*Opius cingutolicus* Fischer, 1992
Braconinae	*Bracon (Bracon) cakili* Beyarslan, 1996	Opiinae	*Opius delipunctis* Fischer and Beyarslan, 2005
Braconinae	*Bracon (Bracon) chagrinicus* Beyarslan, 2002	Opiinae	*Opius erzurumensis* Fischer, 2004
Braconinae	*Bracon (Bracon) selviae* Beyarslan, 2016	Opiinae	*Opius izmerensis* Fischer and Beyarslan, 2005
Braconinae	*Barcon (Glabrobracon) baseflavus* Beyarslan, 2002	Opiinae	*Opius kilisanus* Fischer and Beyarslan, 2005
Braconinae	*Bracon (Glabrobracon) fadiche* Beyarslan, 1996	Opiinae	*Opius kirklareliensis* Fischer and Beyarslan, 2005
Braconinae	*Bracon (Glabrobracon) jenoi* Beyarslan, 2010	Opiinae	*Opius metanivens* Fischer, 1992
Braconinae	*Bracon (Glabrobracon) malatyensis* Beyarslan, 2009	Opiinae	*Opius paraqvisti* Fischer, 2004

Continued

TABLE 32.7 List of Braconidae species only known from Turkey (endemic species).—cont'd

Subfamily	Species	Subfamily	Species
Braconinae	*Bracon (Glabrobracon) surucicus* Beyarslan, 2002	Opiinae	*Opius quasilatipes* Fischer and Beyarslan, 2005
Braconinae	*Bracon (Lucobracon) achterbergi* Beyarslan, 2010	Opiinae	*Opius tekirdagensis* Fischer and Beyarslan, 2005
Braconinae	*Bracon (Lucobracon) attilae* Papp, 2011	Opiinae	*Sternaulopius edirneanus* Fischer and Beyarslan, 2005
Braconinae	*Bracon (Lucobracon) breviradius* Beyarslan, 2011	Orgilinae	*Kerorgilus longicaudis* van Achterberg, 1985
Braconinae	*Bracon filizae* Beyarslan, 2002	Orgilinae	*Orgilus (Orgilus) dilleri* Beyarslan, 1996
Braconinae	*Bracon (Lucobracon) isiklericus* Beyarslan, 2002	Orgilinae	*Orgilus (Orgilus) radialiformis* Bayarslan, 2011
Braconinae	*Bracon (Lucobracon) kuzguni* Beyarslan, 2011	Rogadinae	*Aleiodes turcicus* van Achterberg and Shaw, 2020
Braconinae	*Ceratobracon adaniensis* Beyarslan, 1987	Rogadinae	*Aleiodes zwakhalzi* van Achterberg and Shaw, 2020

brevicellularis Belokobylskij, 2021, *Rhaconotus (Rhaconotus) microexcavatus* Belokobylskij, 2021, and *Spathius subafricanus* Belokobylskij, 2021 (Doryctinae).

Yemen. In total, 44 species in six subfamilies are endemic to this country (39.6% of total species of Yemen). Doryctinae

with 14, Agathidinae and Microgastrinae both with 10 endemic species are more diverse than other subfamilies (Table 32.8).

Species diversity of Iranian Braconidae by province. The distribution of subfamily records across the 31 Iranian

TABLE 32.8 List of Braconidae species only known from Yemen (endemic species).

Subfamily	Species	Subfamily	Species
Agathidinae	*Camptothlipsis breviantennalis* van Achterberg, 2011	Doryctinae	*Leluthia (Leluthia) abnormis* Belokobylskij, 2020
Agathidinae	*Camptothlipsis fuscistigmalis* van Achterberg, 2011	Doryctinae	*Leluthia (Leluthia) brevitergum* Belokobylskij, 2020
Agathidinae	*Camptothlepsis luteostigmalis* van Achterberg, 2011		
Agathidinae	*Coccygidium rugiferum* van Achterberg, 2011	Doryctinae	*Neoheterospilus (Neoheterospilus) yemenus* Belokobylskij, 2020
Agathidinae	*Disophrys punctifera* van Achterberg, 2011	Doryctinae	*Platyspathius (Platyspathius) brevis* Belokobylskij, 2021
Agathidinae	*Lytopylus brevitarsis* van Achterberg, 2011	Doryctinae	*Platyspathius (Platyspathius) longicaudis* Belokobylskij, 2021
Agathidinae	*Therophilus breviscutum* van Achterberg, 2011	Doryctinae	*Rhaconotinus (Rhaconotinus) albosetosus* Belokobylskij, 2021
Agathidinae	*Therophilus longiscutum* van Achterberg, 2011	Doryctinae	*Rhaconotus (Rhaconotus) magniareolus* Belokobylskij, 2021
Agathidinae	*Therophilus nigrator* van Achterberg, 2011	Doryctinae	*Rhaconotus (Rhaconotus) vanharteni* Belokobylskij, 2021
Agathidinae	*Therophilus sulciferus* van Achterberg, 2011	Doryctinae	*Spathius alkadanus* Belokobylskij, 2021
Alysiinae	*Asobara vanharteni* van Achterberg, 2019	Doryctinae	*Sapthius austroarabicus* Belokobylskij, 2021

Continued

TABLE 32.8 List of Braconidae species only known from Yemen (endemic species).—cont'd

Subfamily	Species	Subfamily	Species
Braconinae	*Bathyaulax kossui* Kaartinen and Quicke, 2007	Doryctinae	*Spathius lahji* Belokobylskij, 2021
Braconinae	*Bathyaulax marjae* Kaartinen and Quicke, 2007	Microgastrinae	*Distatrix flava* (Fernandez-Triana and van Achterberg, 2017)
Braconinae	*Cyanopterus (Ipobracon) amorosus* (Kohl, 1906)	Microgastrinae	*Distatrix yemenitica* van Achterberg and Fernandez-Triana, 2017
Cheloninae	*Phanerotoma aspidiota* van Achterberg, 2021		
Cheloninae	*Phanerotoma caudatoides* van Achterberg, 2021	Microgastrinae	*Keylimepie hadhramautensis* van Achterberg and Fernandez-Triana, 2017
Cheloninae	*Phanerotoma latifemorata* van Achterberg, 2021	Microgastrinae	*Keylimepie sanaaensis* van Achterberg and Fernandez-Triana, 2017
Cheloninae	*Phanerotoma longivena* van Achterberg, 2021	Microgastrinae	*Venanides longifrons* Fernandez-Triana and van Achterberg, 2017
Cheloninae	*Phanerotoma signifera* van Achterberg, 2021	Microgastrinae	*Venanides supracompressus* Fernandez-Triana and van Achterberg, 2017
Cheloninae	*Phanerotoma spuriserrata* van Achterberg, 2021	Microgastrinae	*Venanides tenuitergitus* Fernandez-Triana and van Achterberg, 2017
Cheloninae	*Phanerotomella yemenitica* van Achterberg, 2021	Microgastrinae	*Venanides vanharteni* Fernandez-Triana and van Achterberg, 2017
Doryctinae	*Dendrosotinus (Gildoria) maculipennis* Belokobylskij, 2021	Microgastrinae	*Wilkinsonellus arabicus* van Achterberg and Fernandez-Triana, 2017
Doryctinae	*Hemidoryctes carbonarius postfurcalis* Belokobylskij, 2021	Microgastrinae	*Miropotes inexpectatus* van Achterberg and Fernandez-Triana, 2017

provinces (Fig. 32.1) are summarized in Table 32.11. Among the provinces, Guilan with 271 species has the highest reported species diversity, followed by Mazandaran with 263 recorded species (Fig. 32.2). However, these results are probably biased toward these more sampled provinces and the other regions have not yet been sampled as systematically. Based on the known species diversity in countries adjacent to or near Iran, as well as numerous specimens collected in Iran and preserved in different university and private insect collections, we estimate that the total number of species of Braconidae in Iran may reach 2000 or more species eventually. More comprehensive surveys in all areas of the country and additional taxonomic studies are likely to discover more Braconidae species from Iran in the future.

Parasitoid-host relationships of Iranian Braconidae. Iranian braconid species have been reported as parasitoids of five insect orders and 38 families: Coleoptera (Bostrichidae, Buprestidae, Cerambycidae, Chrysomelidae, Ciidae, Coccinellidae, Curculionidae, and Mordellidae), Diptera (Agromyzidae, Anthomyiidae, Chloropidae, and Tephritidae), Hemiptera (Aphididae, Miridae, and Pentatomidae), Hymenoptera (Cynipidae and Tenthredinidae), and Lepidoptera (Batrachedridae, Coleophoridae, Cossidae, Crambidae, Erebidae, Gelechiidae, Gracillariidae, Lasiocampidae, Lycaenidae, Lyonetiidae, Noctuidae, Nolidae, Nymphalidae, Pieridae, Plutellidae, Pterolonchidae, Pyralidae, Tineidae, Tortricidae, Yponomeutidae, and Zygaenidae).

Among the different host families, Aphididae (Hemiptera) with 181 records has the highest number of the recorded braconid parasitoids, followed by Curculionidae (with 18 records), Noctuidae (with 15 records), and Agromyzidae (with 12 records). Additionally, among the 30 subfamilies of Iranian Braconidae, Aphidiinae with 181 host records (all are aphids) has the highest number of known parasitoid—host relationships in Iran; followed by Braconinae (with 58 host records), Microgastrinae (46), Euphorinae (20), Cheloninae (13), Opiinae (12), Doryctinae (12), Agathidinae, Alysiinae (both with seven host records), Macrocentrinae (five), Brachistinae (four), Cardiochilinae, Orgilinae, Rogadinae (three), Rhyssalinae (two), Hormiinae, Ichneutinae, Miracinae, and Rhysipolinae (each with one host record) (Table 32.9). Aphidiines are all solitary endoparasitoids of aphids. They are cosmopolitan and their distribution and abundance follows

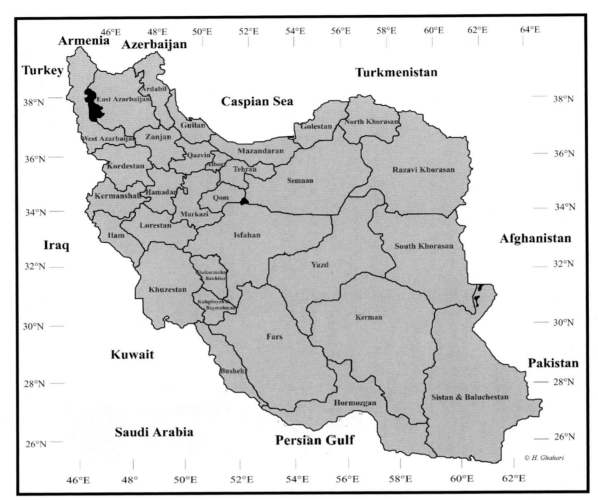

FIGURE 32.1 Map of Iran with provincial boundaries.

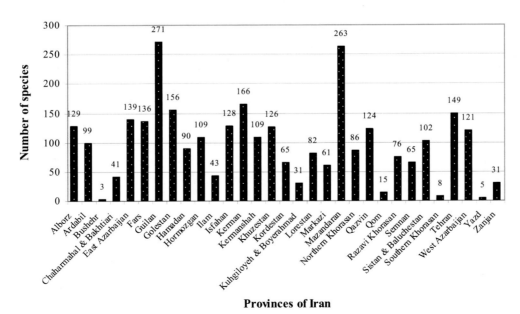

FIGURE 32.2 Number of reported species of Iranian Braconidae by province.

TABLE 32.9 Host—braconid parasitoid relationships in Iran.

Host species	Parasitoid species	Parasitoid subfamily
Order Coleoptera		
Family Bostrichidae		
Lyctus linearis	Monolexis fuscicornis	Doryctinae
Family Buprestidae		
Agrilus viridis	Spathius polonicus	Doryctinae
Chrysobothris affinis	Atanycolus ivanowi; A. initiator	Braconinae
Sphenoptera davatchii	Atanycolus ivanowi; Pseudovipio castrator	Braconinae
	Spathius polonicus	Doryctinae
Sphenoptera kambyses	Spathius polonicus	Doryctinae
Sphenoptera servistana	Atanycolus ivanowi	Braconinae
Sphenoptera tappesi	Atanycolus ivanowi	Braconinae
Trachypteris picta	Atanycolus ivanowi	Braconinae
Family Cerambycidae		
Acanthocerus elegans	Vipio longicauda	Braconinae
Agapanthia violacea	Microctonus morimi	Euphorinae
Calchaenesthes pistacivora	Megalommum pistacivorae	Braconinae
Phytoecia croceipes	Syngaster lepida; Zombrus flavipennis	Doryctinae
Plagionotus arcuatus	Glyptomorpha pectoralis, Pseudovipio castrator	Braconinae
Saperda populnea	Iphiaulax impostor	Braconinae
Family Chrysomelidae		
Bruchus rufimanus	Triaspis thoracica	Brachistinae
Chaetocnema hortensis	Perilitus cerealium	Euphorinae
Gastrophysa viridula viridula	Bracon erraticus	Braconinae
Psylliodes cuprea	Townesilitus aemulus	Euphorinae
Family Coccinellidae		
Coccinella septempunctata	Dinocampus coccinellae	Euphorinae
Hippodamia variegata	Dinocampus coccinellae	Euphorinae
Family Ciidae		
Cis boleti	Meteorus cis	Euphorinae
Cis comptus	Meteorus cis	Euphorinae
Family Curculionidae		
Anthonomus pomorum	Bracon discoideus, B. minutator	Braconinae
Apion sp.	Bracon exhilarator	Braconinae
Archarius crux	Bracon intercessor	Braconinae
Ceutorhynchus assimilis	Diospilus capito, D. morosus	Euphorinae
Hypera postica	Microctonus aethiops, M. colesi	Brachistinae
Ips typographus	Spathius exarator	Doryctinae
Larinus flavescens	Bracon pectoralis	Braconinae
Larinus turbinatus	Bracon illyricus	Braconinae

Continued

TABLE 32.9 Host—braconid parasitoid relationships in Iran.—cont'd

Host species	Parasitoid species	Parasitoid subfamily
Lixus fasciculatus	*Schizoprymnus telengai*	Brachistinae
	Rhaconotus aciculatus	Doryctinae
Lixus incanescens	*Bracon intercessor, B. kozak, Habrobracon hebetor*	Braconinae
Mononychus punctumalbum	*Bracon fulvipes*	Braconinae
Phloeosinus bicolor	*Leluthia ruguloscolyti*	Doryctinae
Sibinia femoralis	*Bracon intercessor*	Braconinae
Scolytus multistriatus	*Leluthia ruguloscolyti*	Doryctinae
	Diospilus oleraceus	Euphorinae
Scolytus rugulosus	*Atanycolus ivanowi*	Braconinae
	Dendrosoter middendorffii, Ecphylus silesiacus, Hecabalodes xylophagi, Leluthia ruguloscolyti	Doryctinae
Sitona humeralis	*Pygostolus falcatus*	Euphorinae
Smicronyx robustus	*Bracon murgabensis*	Braconinae
Taphrorychus lenkoranus	*Ecphylus silesiacus*	Doryctinae
Family Mordellidae		
Mordellistena parvula	*Schizoprymnus pallidipennis*	Brachistinae
Order Diptera		
Family Agromyzidae		
Agromyza ambigua	*Xynobius macrocerus*	Opiinae
Agromyza sp.	*Opius basalis*	Opiinae
Amauromyza gyrans	*Opius gracilis*	Opiinae
Carpomya vesuviana	*Fopius carpomyiae*	Opiinae
Cerodontha pygmaea	*Opius singularis*	Opiinae
Chromatomyia horticola	*Dacnusa laevipectus*	Alysiinae
	Phaedrotoma variegatus	Opiinae
Liriomyza cicerinae	*Opius monilicornis*	Opiinae
Liriomyza congesta	*Chorebus misellus*	Alysiinae
Liriomyza trifolii	*Chorebus axillaris, C. calthae, Dacnusa heringi*	Alysiinae
Phytomyza horticola	*Chorebus aphantus, C. axillaris, C. calthae, C. uliginosus, Dacnusa heringi, D. hospita, D. sibirica*	Alysiinae
Phytomyza plantaginis	*Phaedrotoma cingulatus*	Opiinae
Phytomyza sp.	*Orthostigma pumilum*	Alysiinae
Family Anthomyiidae		
Pegomya hyoscyami	*Phaenocarpa ruficeps*	Alysiinae
Family Chloropidae		
Chlorops pumilionis	*Coelinidea nigra*	Alysiinae
Family Tephritidae		
Acanthiophilus helianthi	*Bracon luteator, Habrobracon brevicornis, H. hebetor*	Braconinae
Bactrocera sp.	*Utetes truncatus*	Opiinae
Carpomya vesuviana	*Fopius carpomyiae*	Opiinae

Continued

TABLE 32.9 Host—braconid parasitoid relationships in Iran.—cont'd

Host species	Parasitoid species	Parasitoid subfamily
Chaetorellia carthami	Habrobracon brevicornis	Braconinae
Rhagolites cerasi	Psyttalia carinata	Opiinae
Rhagolites sp.	Utetes ferrugator	Opiinae
Terellia luteola	Habrobracon brevicornis	Braconinae
Urophora mauritanica	Habrobracon brevicornis	Braconinae
Urophora solstitialis	Bracon luteator	Braconinae
Urophora sp.	Bracon delibator	Braconinae
Order Hemiptera		
Family Aphididae		
Acyrthosiphon bidentis	Lysiphlebus fabarum	Aphidiinae
Acyrthosiphon gossypii	Aphidius ervi, A. matricariae, A. urticae, Lysiphlebus fabarum, Trioxys asiaticus, Praon volucre	Aphidiinae
Acyrthosiphon kondoi	Aphidius ervi, A. smithi	Aphidiinae
Acyrthosiphon lactucae	Lysiphlebus fabarum, Praon volucre, P. yomenae	Aphidiinae
Acyrthosiphon pisum	Aphidius avenae, A. eadyi, A. ervi, A. smithi, A. urticae, Praon barbatum, P. exsoletum, P. volucre, Trioxys asiaticus	Aphidiinae
Amegosiphon platicaudum	Aphidius colemani, A. matricariae, Diaeretiella rapae	Aphidiinae
Amphorophora catharinae	Aphidius popovi, Praon volucre	Aphidiinae
Amphorophora rubi	Aphidius urticae	Aphidiinae
Aphis acetosae	Aphidius colemani, Binodoxys angelicae, Lysiphlebus fabarum	Aphidiinae
Aphis affinis	Aphidius colemani, A. matricariae, Binodoxys acalephae, B. angelicae, Ephedrus persicae, Lysiphlebus confusus, L. fabarum, Praon volucre	Aphidiinae
Aphis alexandrae	Lysiphlebus fabarum	Aphidiinae
Aphis anthemidis	Lysiphlebus fabarum	Aphidiinae
Aphis craccivora	Adialytus veronicaecola, Aphidius colemani, A. matricariae, Binodoxys acalephae, B. angelicae, Diaeretiella rapae, Ephedrus persicae, Lysiphlebus confusus, L. desertorum, L. fabarum, Lysiphlebus testaceipes, Praon necans, P. volucre	Aphidiinae
Aphis crepidis	Aphidius matricariae	Aphidiinae
Aphis davletshinae	Lysiphlebus fabarum	Aphidiinae
Aphis dlabolai	Aphidius matricariae, Praon volucre	Aphidiinae
Aphis epilobii	Lysiphlebus fabarum	Aphidiinae
Aphis euphorbiae	Aphidius matricariae, Binodoxys acalephae, B. angelicae, Lysiphlebus fabarum	Aphidiinae
Aphis eunymi	Lysiphlebus fabarum	Aphidiinae
Aphis euphorbicola	Lysiphlebus fabarum	Aphidiinae
Aphis fabae	Aphidius colemani, A. matricariae, A. smithi, Binodoxys acalephae, B. angelicae, Diaeretiella rapae, Ephedrus persicae, Lysiphlebus cardui, Lysiphlebus confusus, L. fabarum, Praon volucre	Aphidiinae
Aphis fabae cirsiiacanthoides	Aphidius matricariae, Lysiphlebus fabarum	Aphidiinae
Aphis farinosa	Binodoxys angelicae, Lysiphlebus confuses	Aphidiinae

Continued

TABLE 32.9 Host—braconid parasitoid relationships in Iran.—cont'd

Host species	Parasitoid species	Parasitoid subfamily
Aphis gerardianae	*Lysiphlebus fabarum*	Aphidiinae
Aphis gossypii	*Adialytus veronicaecola, Aphidius colemani, A. ervi, A. matricariae, A. platensis, Binodoxys acalephae, B. angelicae, Diaeretiella rapae, Ephedrus persicae, Lysiphlebus confusus, L. fabarum, Praon necans, P. volucre*	Aphidiinae
Aphis hederae	*Binodoxys angelicae*	Aphidiinae
Aphis idaei	*Binodoxys acalephae, B. angelicae, Lysiphlebus confusus, L. fabarum*	Aphidiinae
Aphis intybi	*Aphidius colemani, A. matricariae, A. platensis, Lysiphlebus confusus, L. fabarum*	Aphidiinae
Aphis nasturtii	*Aphidius matricariae, Lysiphlebus fabarum*	Aphidiinae
Aphis nerii	*Aphidius colemani, A. platensis, Binodoxys acalephae, B. angelicae, Ephedrus persicae, Lysiphlebus fabarum, L. testaceipes, Praon necans*	Aphidiinae
Aphis origani	*Lysiphlebus fabarum*	Aphidiinae
Aphis plantaginis	*Aphidius colemani, A. matricareae, Ephedrus persicae, Lysiphlebus fabarum*	Aphidiinae
Aphis polygonata	*Binodoxys angelicae*	Aphidiinae
Aphis pomi	*Aphidius matricariae, Binodoxys acalephae, B. angelicae, Ephedrus persicae, Lysiphlebus fabarum, Praon volucre*	Aphidiinae
Aphis punicae	*Aphidius colemani, A. matricariae, Binodoxys angelicae, Diaeretiella rapae, Ephedrus persicae, Lysiphlebus fabarum, Praon necans*	Aphidiinae
Aphis ruborum	*Lysiphlebus fabarum*	Aphidiinae
Aphis rumicis	*Aphidius colemani, Binodoxys angelicae, Lysiphlebus fabarum*	Aphidiinae
Aphis salviae	*Lipolexis gracilis, Lysiphlebus fabarum*	Aphidiinae
Aphis solanella	*Aphidius colemani, A. matricariae, Binodoxys angelicae, Diaeretiella rapae, Lysiphlebus fabarum, Praon volucre*	Aphidiinae
Aphis spiraecola	*Aphidius matricariae, Binodoxys acalephae, B. angelicae, Lysiphlebus fabarum*	Aphidiinae
Aphis taraxacicola	*Lysiphlebus fabarum*	Aphidiinae
Aphis umbrella	*Aphidius colemani, A. matricariae, Binodoxys acalephae, B. angelicae, Diaeretiella rapae, Lysiphlebus fabarum*	Aphidiinae
Aphis urticata	*Binodoxys acalephae, Lysiphlebus confusus L. fabarum, Praon volucre*	Aphidiinae
Aphis verbasci	*Lysiphlebus confusus, L. volkli*	Aphidiinae
Betulaphis quadrituberculata	*Aphidius aquilus*	Aphidiinae
Brachycaudus amygdalinus	*Aphidius colemani, A. matricariae, Diaeretiella rapae, Ephedrus persicae, Lipolexis gracilis, Praon volucre*	Aphidiinae
Brachycaudus cardui	*Aphidius colemani, A. matricariae, Diaeretiella rapae, Ephedrus persicae, E. plagiator, Lysiphlebus confusus, L. fabarum, Praon abjectum, P. volucre*	Aphidiinae
Brachycaudus divaricatae	*Aphidius matricariae*	Aphidiinae
Brachycaudus helichrysi	*Aphidius colemani, A. matricariae, A. platensis, Binodoxys acalephae, B. angelicae, Ephedrus persicae, E. plagiator, Lysiphlebus confusus, L. fabarum, Praon volucre*	Aphidiinae
Brachycaudus persicae	*Aphidius matricariae, Lysiphlebus fabarum, Praon volucre, Tanytrichophorus petiolaris*	Aphidiinae

Continued

TABLE 32.9 Host—braconid parasitoid relationships in Iran.—cont'd

Host species	Parasitoid species	Parasitoid subfamily
Brachycaudus tragopogonis	Aphidius colemani, A. matricariae, A. platensis, Lysiphlebus fabarum, L. volkli	Aphidiinae
Brachycaudus tragopogonis setosus	Lysiphlebus fabarum	Aphidiinae
Brachyunguis harmalae	Lysiphlebus fabarum	Aphidiinae
Brachyunguis skafi	Lysiphlebus confusus	Aphidiinae
Brachyunguis tamaricis	Ephedrus persicae	Aphidiinae
Brachyunguis zygophylli	Lysiphlebus fabarum	Aphidiinae
Brevicoryne brassicae	Aphidius ervi, Diaeretiella rapae, Ephedrus persicae	Aphidiinae
Capitophorus elaeagni	Aphidius colemani, A. platensis, A. salicis	Aphidiinae
Capitophorus hippophaes	Aphidius matricariae	Aphidiinae
Capitophorus similis	Aphidius matricariae	Aphidiinae
Cavariella aegopodii	Aphidius salicis, Binodoxys brevicornis	Aphidiinae
Cavariella aquatica	Aphidius salicis, Ephedrus helleni	Aphidiinae
Cavariella aspidaphoides	Aphidius salicis, Binodoxys heracleid	Aphidiinae
Cavariella theobaldi	Aphidius salicis	Aphidiinae
Chaitaphis tenuicauda	Trioxys metacarpalis	Aphidiinae
Chaitophorus euphraticus	Adialytus salicaphis	Aphidiinae
Chaitophorus leucomelas	Adialytus salicaphis	Aphidiinae
Chaitophorus pakistanicus	Adialytus salicaphis	Aphidiinae
Chaitophorus populeti	Adialytus salicaphis, Ephedrus chaitophori	Aphidiinae
Chaitophorus populialbae	Adialytus salicaphis	Aphidiinae
Chaitophorus remaudierei	Adialytus salicaphis	Aphidiinae
Chaitophorus salijaponicus	Adialytus salicaphis	Aphidiinae
Chaitophorus salijaponicus niger	Adialytus salicaphis	Aphidiinae
Chaitophorus truncatus	Adialytus salicaphis	Aphidiinae
Chaitophorus vitellinae	Adialytus salicaphis	Aphidiinae
Chromaphis juglandicola	Trioxys pallidus	Aphidiinae
Cinara pinihabitans	Pauesia hazratbalensis	Aphidiinae
Cinara tujafilina	Pauesia hazratbalensis	Aphidiinae
Coloradoa absinthii	Trioxys tanaceticola	Aphidiinae
Coloradoa achilleae	Aphidius arvensis, Lysiphlebus testaceipes	Aphidiinae
Coloradoa heinzei	Trioxys tanaceticola	Aphidiinae
Coloradoa santolinae	Aphidius arvensis	Aphidiinae
Diuraphis noxia	Aphidius colemani, A. ervi, A. matricariae, A. rhopalosiphi, A. uzbekistanicus, Diaeretiella rapae, Ephedrus persicae, E. plagiator, Praon volucre	Aphidiinae
Drepanosiphum platanoidis	Trioxys cirsii	Aphidiinae
Dysaphis crataegi	Ephedrus persicae, E. plagiator	Aphidiinae
Dysaphis devecta	Aphidius matricariae, Diaeretiella rapae, Ephedrus persicae	Aphidiinae

Continued

TABLE 32.9 Host–braconid parasitoid relationships in Iran.—cont'd

Host species	Parasitoid species	Parasitoid subfamily
Dysaphis foeniculus	Ephedrus persicae	Aphidiinae
Dysaphis lappae	Lysiphlebus fabarum	Aphidiinae
Dysaphis plantaginea	Aphidius matricariae, Lysiphlebus fabarum, Ephedrus cerasicola, E. persicae	Aphidiinae
Dysaphis pulverina iranica	Aphidius colemani	Aphidiinae
Dysaphis pyri	Aphidius matricariae, Ephedrus persicus, E. plagiator, Praon volucre	Aphidiinae
Dysaphis radicola	Aphidius colemani, Lysiphlebus fabarum	Aphidiinae
Dysaphis reaumuri	Ephedrus persicae	Aphidiinae
Ephedraphis ephedrae	Lysiphlebus confusus	Aphidiinae
Eriosoma lanuginosum	Areopraon lepelleyi	Aphidiinae
Eucarazzia elegans	Aphidius matricariae	Aphidiinae
Hayhurstia atriplicis	Aphidius colemani, Diaeretiella rapae, Ephedrus helleni, Lysiphlebus fabarum	Aphidiinae
Hoplocallis pictus	Trioxys pallidus	Aphidiinae
Hyadaphis coriandri	Lysiphlebus fabarum, Ephedrus persicae	Aphidiinae
Hyadaphis foeniculi	Aphidius salicis	Aphidiinae
Hayhurstia atriplicis	Diaeretiella rapae, Ephedrus helleni, E. nacheri, E. persicae	Aphidiinae
Hyalopterus amygdali	Aphidius colemani, A. matricariae, A. transcaspicus, Ephedrus persicae, Lysiphlebus fabarum, Praon volucre	Aphidiinae
Hyalopterus pruni	Aphidius colemani, A. matricariae, A. transcaspicus, Diaeretiella rapae, Lysiphlebus fabarum, Praon volucre	Aphidiinae
Hyperomyzus lactucae	Aphidius matricariae, Lysiphlebus confusus, Praon orpheusi, Praon volucre	Aphidiinae
Lipaphis erysimi	Aphidius matricariae, Diaeretiella rapae Lysiphlebus fabarum	Aphidiinae
Lipaphis fritzmuelleri	Lysiphlebus fabarum	Aphidiinae
Lipaphis lepidii	Aphidius matricariae, Binodoxys angelicae, Diaeretiella rapae, Lysiphlebus fabarum	Aphidiinae
Lipaphis pseudobrassicae	Aphidius matricariae, Diaeretiella rapae	Aphidiinae
Macrosiphoniella abrotani	Aphidius absinthii, Ephedrus niger, Praon flavinode	Aphidiinae
Macrosiphoniella absinthii	Praon flavinode	Aphidiinae
Macrosiphoniella artemisiae	Aphidius absinthii, Praon flavinode	Aphidiinae
Macrosiphoniella helichrysi	Aphidius absinthia	Aphidiinae
Macrosiphoniella nr. Macrura	Aphidius absinthia	Aphidiinae
Macrosiphoniella oblonga	Aphidius absinthii, Praon flavinode	Aphidiinae
Macrosiphoniella papilata	Lysiphlebus fabarum	Aphidiinae
Macrosiphoniella pulvera	Aphidius absinthii, Ephedrus niger	Aphidiinae
Macrosiphoniella riedeli	Aphidius absinthia	Aphidiinae
Macrosiphoniella sanborni	Ephedrus niger, Lysiphlebus fabarum, Praon flavinode, P. necans, P. unitum	Aphidiinae
Macrosiphoniella tanacetaria	Aphidius stigmaticus	Aphidiinae
Macrosiphoniella tapuskae	Lysiphlebus confuses	Aphidiinae

Continued

TABLE 32.9 Host—braconid parasitoid relationships in Iran.—cont'd

Host species	Parasitoid species	Parasitoid subfamily
Macrosiphoniella tuberculata	Aphidius absinthii, A. tanacetarius, Trioxys pannonicus	Aphidiinae
Macrosiphum euphorbiae	Aphidius colemani, A. ervi	Aphidiinae
Macrosiphum rosae	Aphidius colemani, A. ervi, A. rosae, Ephedrus plagiator, Lysiphlebus fabarum, Praon rosaecola, P. volucre	Aphidiinae
Mariaella lambersi	Diaeretiella rapae	Aphidiinae
Melanaphis sacchari	Lysiphlebus fabarum	Aphidiinae
Metopolophium dirhodum	Aphidius colemani, A. ervi, A. matricariae, A. popovi, A. rhopalosiphi, A. uzbekistanicus, Diaeretiella rapae, Ephedrus persicae, Lysiphlebus fabarum, Praon volucre	Aphidiinae
Microlophium carnosum	Aphidius ervi, Aphidius urticae	Aphidiinae
Myzus ascalonicus	Aphidius matricariae	Aphidiinae
Myzus beybienkoi	Aphidius matricariae, Diaeretiella rapae, Lysiphlebus fabarum, Praon volucre	Aphidiinae
Myzus cerasi	Aphidius matricariae, Ephedrus cerasicola, E. persicae, E. plagiator	Aphidiinae
Myzus certus	Aphidius matricariae	Aphidiinae
Myzus ornatus	Aphidius matricariae	Aphidiinae
Myzus persicae	Aphidius avenae, A. colemani, A. ervi, A. matricariae, A. platensis, Binodoxys angelicae, Diaeretiella rapae, Lysiphlebus fabarum, Ephedrus cerasicola, E. persicae, Praon necans, P. volucre	Aphidiinae
Nasonovia ribisnigri	Aphidius hieraciorum, A. matricariae, Diaeretiella rapae, Praon pubescens, P. volucre	Aphidiinae
Nearctaphis bakeri	Aphidius smithi, Praon barbatum	Aphidiinae
Neomyzus circumflexus	Praon volucre	Aphidiinae
Ovatus crataegarius	Aphidius matricariae	Aphidiinae
Ovatus insitus	Aphidius matricariae	Aphidiinae
Pemphigus spyrothecae	Monoctonia vesicarii	Aphidiinae
Periphyllus testudinaceus	Aphidius setiger	Aphidiinae
Phorodon humuli	Aphidius colemani, A. matricariae, A. transcaspicus, Lysiphlebus confusus, Lysiphlebus fabarum, Ephedrus cerasicola, E. persicae, Praon volucre	Aphidiinae
Protaphis elongata	Lysiphlebus fabarum	Aphidiinae
Protaphis terricola	Aphidius colemani, Lysiphlebus desertorum, L. fabarum	Aphidiinae
Pterocallis alni	Trioxys pallidus	Aphidiinae
Pterochloroides persicae	Pauesia antennata	Aphidiinae
Pterocomma pilosum	Aphidius cingulatus	Aphidiinae
Pterocomma populeum	Aphidius cingulatus	Aphidiinae
Rhopalosiphum maidis	Aphidius avenae, A. colemani, A. matricariae, A. rhopalosiphi, A. uzbekistanicus, Diaeretiella rapae, Ephedrus persicae, Praon volucre	Aphidiinae
Rhopalosiphum nymphaeae	Aphidius matricariae, A. transcaspicus, Ephedrus cerasicola, Lysiphlebus confusus, L. fabarum, Praon necans	Aphidiinae
Rhopalosiphum padi	Aphidius colemani, A. ervi, A. matricariae, A. rhopalosiphi, A. uzbekistanicus, Diaeretiella rapae, Lysiphlebus fabarum, Ephedrus persicae, E. plagiator, Praon gallicum, P. necans, P. volucre	Aphidiinae
Saltusaphis scirpus	Diaeretiella rapae, Lysiphlebus fabarum	Aphidiinae

Continued

TABLE 32.9 Host—braconid parasitoid relationships in Iran.—cont'd

Host species	Parasitoid species	Parasitoid subfamily
Schizaphis graminum	*Aphidius colemani, A. matricariae, A. rhopalosiphi, A. uzbekistanicus, Diaeretiella rapae, Diaeretus leucopterus, Ephedrus persicae, E. plagiator, Praon volucre*	Aphidiinae
Sipha elegans	*Adialytus ambiguus*	Aphidiinae
Sipha flava	*Adialytus ambiguus*	Aphidiinae
Sipha maydis	*Adialytus ambiguus, Aphidius salicaphis, A. uzbekistanicus, Ephedrus persicae, Praon unitum*	Aphidiinae
Sitobion avenae	*Aphidius avenae, A. colemani, A. ervi, A. matricariae, A. hieraciorum, A. rhopalosiphi, A. smithi, A. uzbekistanicus, Diaeretiella rapae, Ephedrus persicae, E. plagiator, Lysiphlebus fabarum, Praon necans, P. volucre*	Aphidiinae
Sitobion fragariae	*Aphidius matricariae*	Aphidiinae
Thelaxes suberi	*Adialytus thelaxis*	Aphidiinae
Therioaphis khayami	*Trioxys complanatus*	Aphidiinae
Therioaphis riehmi	*Praon exsoletum, Trioxys complanatus*	Aphidiinae
Therioaphis trifolii	*Praon exsoletum, P. volucre, Trioxys complanatus*	Aphidiinae
Therioaphis trifolii maculata	*Praon exsoletum*	Aphidiinae
Tinocallis nevskyi	*Praon flavinode, Trioxys pallidus*	Aphidiinae
Tinocallis saltans	*Betuloxys hortorum*	Aphidiinae
Titanosiphon bellicosum	*Aphidius iranicus, Trioxys tanaceticola*	Aphidiinae
Titanosiphon dracunculi	*Aphidius absinthia*	Aphidiinae
Titanosiphon neoartemisiae	*Trioxys pannonicus*	Aphidiinae
Toxoptera aurantii	*Ephedrus persicae, Lysiphlebus confusus, L. fabarum*	Aphidiinae
Uroleucon acroptilidis	*Aphidius funebris, Ephedrus niger, Praon unitum, P. yomenae*	Aphidiinae
Uroleucon bielawskii	*Aphidius persicus, Ephedrus niger*	Aphidiinae
Uroleucon carthami	*Aphidius persicus, Praon yomenae*	Aphidiinae
Uroleucon chondrillae	*Aphidius funebris, A. persicus, Ephedrus niger, Praon yomenae*	Aphidiinae
Uroleucon cichorii	*Aphidius funebris, Ephedrus niger, Praon volucre, P. yomenae*	Aphidiinae
Uroleucon compositae	*Aphidius funebris, A. persicus, Ephedrus niger, Lysiphlebus fabarum, Praon volucre, P. yomenae*	Aphidiinae
Uroleucon erigeronense	*Aphidius funebris, Ephedrus niger*	Aphidiinae
Uroleucon jaceae	*Aphidius funebris, A. persicus, Ephedrus niger, Lysiphlebus fabarum, Praon dorsale, P. unitum, P. volucre, P. yomenae*	Aphidiinae
Uroleucon ochropus	*Aphidius persicus*	Aphidiinae
Uroleucon sonchi	*Aphidius colemani, A. funebris, A. persicus, Ephedrus niger, Praon unitum, Praon volucre, P. yomenae*	Aphidiinae
Uroleucon tortuosissimae	*Praon yomenae*	Aphidiinae
Wahlgreniella nervata	*Aphidius ervi, A. matricariae, Praon rosaecola*	Aphidiinae
Xerobium cinae	*Lysiphlebus desertorum*	Aphidiinae
Family Miridae		
Adelphocoris lineolatus	*Peristenus pallipes, P. rubricollis*	Euphorinae

Continued

TABLE 32.9 Host–braconid parasitoid relationships in Iran.—cont'd

Host species	Parasitoid species	Parasitoid subfamily
Lygus rugulipennis	Peristenus digoneutis	Euphorinae
Family Pentatomidae		
Dolycoris baccarum	Aridelus egregious	Euphorinae
Order Hymenoptera		
Family Cynipidae		
Biorhiza pallida	Bracon erraticus	Braconinae
Family Tenthredinidae		
Nematus fagi	Ichneutes brevis	Ichneutinae
Pontania vesicator	Bracon picticornis	Braconinae
Order Lepidoptera		
Family Batrachedridae		
Batrachedra amydraula	Habrobracon brevicornis, H. hebetor	Braconinae
	Phanerotoma leucobasis	Cheloninae
Family Coleophoridae		
Coleophora serratella	Therophilus linguarius	Agathidinae
	Dolichogenidea coleophorae, D. corvina	Microgastrinae
	Orgilus rugosus	Orgilinae
Family Cossidae		
Zeuzera pyrina	Macrocentrus infirmus	Macrocentrinae
	Dolichogenidea phaloniae	Microgastrinae
Family Crambidae		
Antigastra catalaunalis	Habrobracon hebetor	Braconinae
Chilo suppressalis	Amyosoma chinense	Braconinae
	Cotesia chilonis, C. flavipes, Hygroplitis russatus	Microgastrinae
	Rhaconotus testaceus	Doryctinae
	Stenobracon deesae	Braconinae
Hellula undalis	Cotesia lineola	Microgastrinae
Ostrinia nubilalis	Pseudovipio inscriptor	Braconinae
	Macrocentrus cingulum	Macrocentrinae
	Cotesia tibialis, Glyptapanteles thompsoni	Microgastrinae
Family Erebidae		
Euproctis chrysorrhoea	Meteorus obsoletus, M. versicolor	Euphorinae
	Cotesia melanoscela, Glyptapanteles inclusus, Microplitis tuberculatus	Microgastrinae
Euproctis similis	Aleiodes praetor	Orgilinae
	Microplitis ratzeburgii	Microgastrinae
Lymantria dispar	Phanerotoma dentata	Cheloninae
	Meteorus pulchricornis	Euphorinae
	Cotesia melanoscela, C. melitaearum, Dolichogenidea lacteicolor, Glyptapanteles indiensis, G. liparidis, G. porthetriae	Microgastrinae
	Oncophanes minutus	Rhyssalinae
Orgyia antiqua	Aleiodes signatus	Rogadinae

Continued

TABLE 32.9 Host—braconid parasitoid relationships in Iran.—cont'd

Host species	Parasitoid species	Parasitoid subfamily
Family Gelechiidae		
Anarsia lineatella	*Bracon variegator*	Braconinae
	Apanteles xanthostigma	Microgastrinae
Phthorimaea operculella	*Habrobracon brevicornis, H. hebetor, H. iranicus, H. radialis*	Braconinae
	Chelonus inanitus	Cheloninae
	Meteorus rubens	Euphorinae
Recurvia nanella	*Orgilus obscurator*	Orgilinae
Scrobipalpa ocellatella	*Bracon intercessor*	Braconinae
	Chelonus contractus, Chelonus subcontractus	Cheloninae
Scrobipalpa sp.	*Dolichogenidea impura*	Microgastrinae
Tuta absoluta	*Agathis fuscipennis*	Agathidinae
	Habrobracon hebetor	Braconinae
Family Gracillariidae		
Caloptilia rufipennella	*Earinus elator*	Agathidinae
Phyllonorycter blancardella	*Pholetesor circumscriptus*	Microgastrinae
Phyllonorycter coryloliella	*Pholetesor bicolor, P. circumscriptus, P. pseudocircumscriptus*	Microgastrinae
Phyllonorycter platani	*Pholetesor circumscriptus*	Microgastrinae
Phyllonorycter populifoliella	*Glyptapanteles liparidis*	Microgastrinae
Phyllonorycter salicicolella	*Rhysipolis hariolator*	Rhysipolinae
	Pholetesor arisba	Microgastrinae
Family Lasiocampidae		
Malacosoma neustria	*Habrobracon nygmiae*	Braconinae
	Cotesia pieridis, Dolichogenidea ulter	Microgastrinae
Family Lycaenidae		
Lampides boeticus	*Cotesia sessilis*	Microgastrinae
Family Lyonetiidae		
Lyonetia clerkella	*Dolichogenidae corvina*	Microgastrinae
Family Noctuidae		
Acronicta aceris	*Cotesia glomerata*	Microgastrinae
Acronicta sp.	*Cotesia glomerata*	Microgastrinae
Agrotis segetum	*Chelonus inanitus*	Cheloninae
	Meteorus rubens	Euphorinae
	Macrocentrus collaris	Macrocentrinae
	Cotesia glomerata, C. tibialis	Microgastrinae
Apamea sordens	*Cotesia vanessae*	Microgastrinae
Helicoverpa armigera	*Habrobracon brevicornis, H. hebetor*	Braconinae
	Cotesia kazak, C. vanessae	Microgastrinae
Heliothis viriplaca	*Habrobracon hebetor, H. iranicus*	Braconinae
	Cotesia kazak	Microgastrinae
Heliothis sp.	*Cotesia kazak*	Microgastrinae
Leucania loreyi	*Habrobracon hebetor*	Braconinae

Continued

TABLE 32.9 Host—braconid parasitoid relationships in Iran.—cont'd

Host species	Parasitoid species	Parasitoid subfamily
	Cotesia rubecula, C. ruficrus	Microgastrinae
Mythimna unipuncta	Meteorus pendulus	Euphorinae
	Cotesia ruficrus	Microgastrinae
Noctua pronuba	Microplitis varipes	Microgastrinae
Sesamia cretica	Habrobracon hebetor	Braconinae
	Cotesia glomerata, C. ruficrus	Microgastrinae
Sesamia nonagrioides	Habrobracon hebetor	Braconinae
	Meteorus rubens	Euphorinae
	Cotesia glomerata	Microgastrinae
Simyra albovenosa	Aleiodes rugulosis	Rogadinae
Simyra dentinosa	Cotesia ofella, C. vanessae, C. vestalis	Microgastrinae
Spodoptera exigua	Habrobracon brevicornis, H. hebetor	Braconinae
	Chelonus corvulus, C. inanitus	Cheloninae
	Meteorus pendulus, M. rubens	Euphorinae
	Cotesia vanessae, Microplitis fulvicornis, M. rufiventris	Microgastrinae
Family Nolidae		
Earias insulana	Bracon lefroyi, Habrobracon brevicornis	Braconinae
Family Nymphalidae		
Melitaea didyma	Cotesia acuminata	Microgastrinae
Family Pieridae		
Aporia crataegi	Cotesia glomerata	Microgastrinae
Pieris brassicae	Cotesia glomerata	Microgastrinae
Pieris rapae	Cotesia glomerata, C. rubecula, C. vestalis	Microgastrinae
Family Plutellidae		
Plutella xylostella	Habrobracon hebetor	Braconinae
	Cotesia vestalis, Diolcogaster claritibia, Dolichogenidea appellator, D. imperator, D. inseformis	Microgastrinae
Family Pterolonchidae		
Syringopais temperatella	Dolichogenidea longipalpis	Microgastrinae
Family Pyralidae		
Achroia grisella	Cardiochiles shestakovi	Cardiochilinae
	Apanteles galleriae	Microgastrinae
Arimania komaroffi	Habrobracon telengai	Braconinae
	Iconella myeloenta	Microgastrinae
Ectomyelois ceratoniae	Habrobracon hebetor, H. iranicus	Braconinae
	Phanerotoma leucobasis	Cheloninae
	Dolichogenidea laspeyresiella, Iconella myeloenta	Microgastrinae
Ephestia kuehniella	Habrobracon hebetor	Braconinae
Epischidia caesariella	Cardiochiles shestakovi	Cardiochilinae
Etiella zinckenella	Cardiochiles fallax	Cardiochilinae

Continued

TABLE 32.9 Host—braconid parasitoid relationships in Iran.—cont'd

Host species	Parasitoid species	Parasitoid subfamily
	Dolichogenidea albipennis	Microgastrinae
Galleria mellonella	*Habrobracon hebetor*	Braconinae
Homoeosoma nebulella	*Habrobracon hebetor*	Braconinae
	Iconella lacteoides	Microgastrinae
Plodia interpunctella	*Habrobracon hebetor*	Braconinae
	Phanerotoma leucobasis	Cheloninae
Family Tineidae		
Kermania pistaciella	*Chelonus kermakiae*	Cheloninae
	Centistidea pistaciella	Miracinae
Family Tortricidae		
Acleris variegana	*Earinus gloriatorius*	Agathidinae
Archips rosana	*Macrocentrus pallipes*	Macrocentrinae
	Microgaster hospes	Microgastrinae
	Aleiodes compressor	Rogadinae
	Oncophanes minutus	Rhyssalinae
Archips xylosteana	*Chremylus elaphus*	Hormiinae
Cydia johanssoni	*Bracon variator*	Braconinae
Cydia pomonella	*Aerophilus rufipes, Cremnops desertor*	Agathidinae
	Habrobracon iranicus	Braconinae
	Ascogaster quadridentate, Phanerotoma dentata, P. tritoma	Cheloninae
Grapholita funebrana	*Phanerotoma planifrons*	Cheloninae
Gypsonoma aceriana	*Macrocentrus nitidus*	Macrocentrinae
Lobesia botrana	*Aerophilus rufipes*	Agathidinae
	Habrobracon hebetor, H. iranicus	Braconinae
	Ascogaster quadridentata	Cheloninae
Pandemis cerasana	*Bracon variegator*	Braconinae
Pandemis corylana	*Dolichogenidea gagates*	Microgastrinae
Spilonota ocellana	*Aerophilus rufipes, Therophilus cingulipes*	Agathidinae
Family Yponomeutidae		
Yponomeuta malinellus	*Habrobracon iranicus*	Braconinae
Yponomeuta padella	*Ascogaster similis*	Cheloninae
	Choeras parasitellae, Microgaster rufipes	Microgastrinae
Family Zygaenidae		
Zygaena loti	*Bracon nigratus*	Braconinae

rather closely that of their aphid hosts, which are more abundant in the temperate and subtropical zones of the northern hemisphere (Starý, 1979, 1989). Hosts of aphidiines are particularly likely to be recorded because many of them are significant in agro-forestry systems. Those not associated with agro-ecosystems are less studied.

In total, there are 380 records of parasitoid—host relationships for Iranian Braconidae (see Table 32.9). These results

TABLE 32.10 Total number of braconid species in the Middle East and nearby countries to Iran, and their land area.

Country	Land area (km²)	Number of recorded species	Country	Land area (km²)	Number of recorded species
Russia	17,098,242	3272	Syria	185,180	52
Kazakhstan	2,724,900	730	Jordan	89,342	41
Saudi Arabia	2,149,690	52	Azerbaijan	86,600	556
Iran	1,648,195	1363	United Arab Emirates	83,600	57
Egypt	1,010,408	162	Armenia	29,749	338
Pakistan	7,96,095	86	Israel—Palestine	28,092	439
Turkey	7,83,562	1196	Kuwait	17,818	0
Afghanistan	6,52,230	107	Qatar	11,586	0
Yemen	555,000	111	Lebanon	10,452	22
Turkmenistan	4,88,100	304	Cyprus	9251	159
Iraq	4,38,317	45	Bahrain	760	0
Oman	3,09,500	9	—	—	—

indicate that the most research on Iranian Braconidae has been done on the basis of collecting of free-flying adults not by rearing them from hosts. Since these parasitoids have efficient role in natural biological control of agricultural and forest pests, determining the hosts of braconids can be important in order to establish successful programs of biological control and IPM (DeBach & Rosen, 1991; Neuenschwander et al., 2003; Peshin & Dhawan, 2009).

The braconine, *Habrobracon hebetor* (Say, 1836), is a highly polyphagous idiobiont ectoparasitoid attacking 19 host species within nine families of three orders Coleoptera (Curculionidae), Diptera (Tephritidae), and Lepidoptera (Batrachedridae, Crambidae, Gelechiidae, Noctuidae, Plutellidae, Pyralidae, Tortricidae). *Habrobracon hebetor* is one of the most important parasitoids of several lepidopteran pests, which is reared in different countries such as Iran and released in various agroecosystems under biological control programs. Since as demonstrated in this catalog, there is a diverse fauna of braconid wasps in Iran, these parasitoids may have many applications in biological control of agricultural and forest pests. Conservation of these efficient parasitoids is necessary for successful control of pests (Hawkins & Sheehan, 1994; Radcliffe et al., 2008). Decreasing pesticide applications is one of the main strategies in order to better conserve braconid wasp species and to increase these parasitoids' survival and efficiency (Croft, 1990; Maredia et al., 2003).

Comparison of Iranian Braconidae to the Middle East and nearby countries. The Middle East region includes 16 countries, Bahrain, Cyprus, Egypt, Iran, Iraq, Israel—Palestine, Jordan, Kuwait, Lebanon, Oman, Qatar,

Saudi Arabia, Syria, Turkey, United Arab Emirates, and Yemen. Iran has land and sea connections with 15 countries, which among the countries adjacent to Iran, eight (Bahrain, Iraq, Kuwait, Oman, Qatar, Saudi Arabia, Turkey, and United Arab Emirates) are located in the Middle East region too (see Fig. 1.6). However, the different states of faunal surveys of braconids in each of these areas are quite variable, from extensive to almost nonexistent, depending on the country in question. Because of the absence of equally up-to-date and comprehensive checklists of Braconidae for all these countries any comparisons must be interpreted with some caution. However, they are offered here as baseline data to stimulate future studies and analyses.

Among the countries of the Middle East and adjacent to Iran, only the fauna of Russia with 3272 recorded species (Belokobylskij & Lelej, 2019) has more recorded species than Iran (Table 32.10), but this is hardly surprising given its vast area (more than 10 × that of Iran) as well as its long history of active insect studies. Russia also has the most species diversity in all the subfamilies except for the Braconinae, a group which is often considered to be more species-rich in tropical regions. Turkey has the most recorded species after Iran, being the only of the other countries with more than 1000 braconid species. Turkmenistan, Egypt, and Kazakhstan have more recorded species of Cardiochilinae than Russia but this is not surprising as this subfamily is typically associated with more arid ecosystems, or at least has a high proportion of species adapted to such conditions. The much lower number of species recorded from some other countries as compared to

TABLE 32.11 Species diversity of each subfamily of Iranian Braconidae by province.

Subf. Prov.	Ag	Al	Ap	Br	Bc	Ca	Ce	Cha	Che	Di	Do	Eu	Ex	Gn	Hl	Hm	Hr	Ic	Ma	Mg	Mt	Mr	Op	Or	Pa	Pr	Rh	Rs	Ro	Si
AL	2	–	29	7	13	1	–	–	15	–	9	6	–	1	–	1	–	–	2	23	1	–	14	3	–	–	–	–	2	–
AR	3	12	16	8	15	–	–	–	2	–	2	5	–	–	–	–	–	–	2	17	–	–	8	1	–	–	–	–	7	–
BU	–	–	–	–	2	–	–	–	–	–	–	–	–	–	–	–	–	–	–	–	–	–	1	–	–	–	–	–	–	–
CH	1	3	–	3	11	–	–	–	1	–	–	4	1	–	–	–	–	–	–	4	–	1	2	1	1	–	1	–	1	–
EA	8	12	8	5	18	–	–	–	7	1	3	11	1	–	–	1	–	–	2	31	1	–	16	2	1	–	1	–	9	–
FA	4	26	40	3	16	1	–	1	10	–	–	4	–	–	–	–	–	–	4	9	1	–	9	1	1	–	–	–	4	–
GL	19	10	31	21	27	1	–	–	22	–	8	24	–	4	–	2	1	–	6	29	1	1	46	5	–	–	–	1	11	1
GO	10	17	32	10	25	1	–	–	6	–	3	7	–	1	–	–	–	–	1	23	–	–	9	3	–	–	–	1	3	–
HA	2	6	25	5	23	–	–	–	14	–	1	6	–	–	–	2	–	–	1	5	–	–	14	–	–	–	–	–	–	–
HO	1	9	15	–	33	2	–	–	14	–	10	6	–	–	–	–	5	–	–	–	–	–	14	–	–	–	–	–	–	–
IL	4	7	2	–	8	–	–	–	3	–	4	–	–	–	–	–	2	–	–	11	–	–	–	–	–	–	–	–	4	–
IS	1	13	31	6	23	1	1	–	5	–	3	7	–	–	–	–	–	–	–	18	1	1	5	–	1	–	3	1	7	–
KE	–	33	35	1	32	1	–	–	6	–	2	6	–	–	–	2	–	–	2	6	1	1	28	–	1	–	–	–	10	–
KH	1	6	26	2	44	1	–	–	3	–	5	3	–	1	–	1	–	–	–	22	1	–	6	3	–	1	1	–	2	–
KO	1	5	21	3	5	–	1	–	6	–	4	4	–	–	–	–	–	–	–	6	–	–	1	–	–	–	–	–	2	–
KS	7	6	34	5	25	–	–	–	5	–	4	4	–	1	–	–	–	–	–	8	–	–	17	3	–	–	–	1	2	–
KU	2	2	7	5	5	–	–	–	2	–	–	2	2	–	–	–	–	–	–	2	1	–	17	–	–	–	–	–	–	–
LO	9	3	5	5	17	–	–	–	7	–	8	8	–	–	1	–	–	–	–	12	–	–	17	–	–	–	–	–	–	–
MA	12	8	27	23	32	–	–	–	18	–	14	24	–	4	–	2	2	–	6	32	1	1	34	3	–	–	1	1	14	1
MR	–	1	35	–	3	–	–	–	1	–	4	2	–	–	–	–	–	–	–	8	–	–	4	–	–	–	–	–	2	–
NK	–	10	33	–	26	–	–	–	1	–	–	6	1	1	1	–	–	1	–	6	–	–	1	–	–	–	–	–	–	–
QA	4	11	15	6	24	–	–	–	9	–	7	9	–	1	–	–	1	–	1	20	–	–	6	2	–	1	–	–	6	–
QO	–	9	9	–	1	–	–	–	1	–	–	–	–	–	–	–	–	–	–	2	–	–	–	–	–	–	–	–	–	–
RK	2	8	23	4	15	–	–	–	4	–	6	5	1	–	–	–	2	–	–	6	–	–	4	–	–	–	–	–	5	–
SB	2	12	19	1	12	3	–	–	9	–	2	2	5	–	–	1	–	–	–	15	1	–	18	–	–	–	–	–	5	–
SE	5	2	3	6	21	–	–	–	5	–	3	3	–	1	–	1	–	–	–	5	–	–	12	–	–	–	–	–	5	–
SK	–	–	–	–	2	–	–	–	1	–	–	–	–	–	–	–	–	–	–	2	–	–	–	1	–	–	–	–	–	–
TE	7	3	52	4	13	3	–	–	13	–	6	6	–	1	2	–	–	–	1	16	–	–	10	6	–	–	2	–	3	–
WA	7	11	10	8	20	–	–	1	5	–	6	6	–	–	–	–	3	–	1	25	–	–	4	3	–	–	–	–	6	1
YZ	–	–	–	–	–	1	–	–	–	–	2	1	–	–	–	–	–	–	–	–	–	–	–	–	–	–	–	–	–	–
ZA	2	2	12	–	1	1	–	–	1	–	2	1	–	–	–	–	–	–	1	4	–	–	1	–	–	–	–	–	3	–

Abbreviations of subfamilies: Ag, Agathidinae; Al, Alysiinae; Ap, Aphidiinae; Br, Brachistinae; Bc, Braconinae; Ca, Cardiochilinae; Ce, Cenocoelinae; Cha, Charmontinae; Che, Cheloninae; Di, Dirrhopinae; Do, Doryctinae; Eu, Euphorinae; Ex, Exothecinae; Gn, Gnamptodontinae; Hl, Helconinae; Hm, Homolobinae; Hr, Horminae; Ic, Ichneutinae; Ma, Macrocentrinae; Mg, Microgastrinae; Mt, Microtypinae; Mr, Miracinae; Op, Opiinae; Or, Orgilinae; Pa, Pambolinae; Pr, Proteropinae; Rh, Rhysalinae; Rs, Rhyssalinae; Ro, Rogadinae; Si, Sigalphinae. Abbreviations of provinces: AL, Alborz; AR, Ardabil; BU, Bushehr; CH, Chaharmahal and Bakhtiari; EA, East Azarbaijan; FA, Fars; GL, Guilan; GO, Golestan; HA, Hamadan; HO, Hormozgan; IL, Ilam; IS, Isfahan; KE, Kerman; KH, Khuzestan; KO, Kordestan; KS, Kermanshah; KU, Kuhgiloyeh and Boyerahmad; LO, Lorestan; MA, Mazandaran; MR, Markazi; NK, Northern Khorasan; QA, Qazvin; QO, Qom; RK, Razavi Khorasan; SB, Sistan and Baluchestan; SE, Semnan; SK, Southern Khorasan; TE, Tehran; WA, West Azarbaijan; YZ, Yazd; ZA, Zanjan.

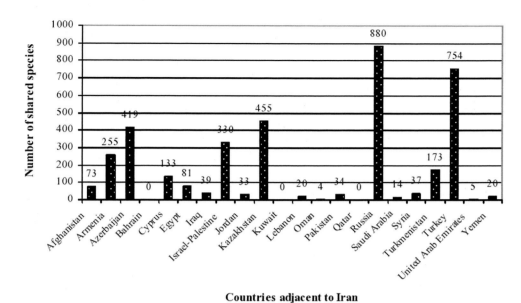

FIGURE 32.3 Number of shared species of Braconidae between Iran and the Middle East and nearby countries to Iran.

Iran probably not only reflects a smaller land area but also few comprehensive surveys in those countries. Some countries, especially Bahrain, Kuwait, and Qatar, have virtually no records for Braconidae species. Also, among the 23 countries of the Middle East and adjacent to Iran, Russia shares the largest number of species with Iran (880 species), followed by Turkey (754 species), Kazakhstan (449 species), and Azerbaijan (419 species) (Fig. 32.3). Comparison of faunal similarity of Iran with the Middle East and nearby countries with their braconid fauna studied rather well shows that 78.5% of the Israel−Palestine fauna of Braconidae is similar to Iran; followed by Armenia (75.4%), Azerbaijan (75.3%), Turkey (63%), Kazakhstan (62.3%), Turkmenistan (56.9%), Russia (26.9%). These results are expected because of significant land borders with Iran for the countries which have the highest percent in faunal similarity; although the highest similarity with the fauna of Israel−Palestine is due to the relatively complete determination of the braconid fauna of this country. Additionally, the fauna of Braconidae of the remaining countries is poorly studied and so comparison of their faunal similarity to Iran is not feasible at present.

Conclusions

Several faunistic studies in Iran showed that a diverse fauna of Braconidae is present in agricultural and natural ecosystems. Taxonomic research on braconid wasps has been accelerated in Iran during the past 10 years. In this book, 31 species were added as new records for the Iranian fauna (about 2.3% of the total number of reported species). The present compilation also includes 559 species that were not

reported from Iran by Yu et al. (2016), as a result of a great deal of recent research on braconids having been conducted after the last update of that catalog. Similar comprehensive cataloging is lacking for some of the Middle East and adjacent countries of Iran and many more records for Iran are likely to be discovered in the future. Iran is a large country that includes a diversity of ecozones and climates. It is near the crossroads of three distinct biogeographical regions, the Palaearctic, Afrotropical, and Oriental regions, which undoubtedly positively impacts its diversity. Most Iranian species likely are Palaearctic in origin, but almost certainly there are faunal elements from all three regions, with Oriental species likely in the southeast. There are some Afrotropical species, although probably not too many because Iran is north of the Arabian Peninsula.

Regarding the importance of braconid wasps for biological control of agricultural and forest pests (Quicke, 2015; Wharton, 1993), and since determining of the fauna of natural enemies is the first step in successful biological control and IPM programs (Bellows et al., 1999; Jervis, 2005), we suggest that researchers and students of the Middle East countries continue the faunistic works toward better understanding the species diversity of Braconidae together with their parasitoid−host relationships. Future studies on various ecosystems will likely result in the discovery of more valuable native parasitoids which are adapted to local environmental conditions. More faunistic surveys are needed in order to find new data about parasitoid−host relationships of Braconidae for establishment of successful biological control programs. We hope that this work will make a sound foundation for future work on the braconids of the whole region and not just for Iran.

Many of the species recorded are likely to be beneficial in some way or another. Some are acting as control agents against pests, either through deliberate human intervention or naturally. The importance of others has probably gone unnoted because they have been preventing or reducing outbreaks of potential pests all along. Although there are not cases for accurate assessment of the parasitism rate of Braconidae in the Middle East countries, conservation of these important natural enemies by decreasing the application of chemical and nonselective pesticides must be a fundamental goal for farmers and gardeners.

References

Bellows, T. S., Fisher, T. W., & Caltagirone, L. E. (1999). *Handbook of biological control: Principles and applications of biological control.* Academic Press, 1046 pp.

Belokobylskij, S. A., & Lelej, A. S. (2019). Annotated catalogue of the Hymenoptera of Russia. In , *Proceedings of the Zoological Institute of the Russian Academy of Sciences, Suppl. No. 8: Vol. II. Apocrita: Parasitica,* 594 pp.

Croft, B. A. (1990). *Arthropod biological control agents and pesticides.* John Wiley and Sons Inc., 723 pp.

DeBach, P., & Rosen, D. (1991). *Biological control by natural enemies.* Cambridge, UK: Cambridge Univ. Press, 440 pp.

Ghahari, H., Yu, D. S., & van Achterberg, C. (2006). *World bibliography of the family Baraconidae (Hymenoptera: Ichneumonoidea) (1964−2003)* (Vol. 8). NNM Technical Bulletin, 293 pp.

Hanson, P., & Gauld, I. (2006). Hymenoptera de la region tropical. *Memoirs of the American Entomological Institute, 77,* 1−994.

Hawkins, B. A., & Sheehan, W. (1994). *Parasitoid community ecology.* Oxford, UK: Oxford University Press, 516 pp.

Jervis, M. A. (2005). *Insect natural enemies. A practical perspective.* Springer, 748 pp.

LaSalle, J., & Gauld, I. D. (1993). *Hymenoptera and biodiversity.* New York: Oxford University Press, 368 pp.

Maredia, K. M., Dakouo, D., & Mota-Sanchez, D. (2003). *Integrated pest management in the global arena.* Trowbridge, UK: Cromwell Press, 512 pp.

Neuenschwander, P., Borgemeister, C., & Langewald, J. (2003). *Biological control in IPM system in Africa.* CABI Publishing, 414 pp.

Peshin, R., & Dhawan, A. K. (2009). *Integrated pest management: Dissemination and impact* (vol. 2). Springer Science+Business Media, 627 pp.

Quicke, D. L. J. (2015). *The braconid and ichneumonid parasitoid wasps: Biology, systematics, evolution and ecology.* Chichester: Wiley Blackwell, 688 pp.

Radcliffe, E. B., Hutchison, W. D., & Cancelado, R. E. (2008). *Integrated pest management: Concepts, tactics, strategies and case studies.* Cambridge University Press, 529 pp.

Starý, P. (1979). *Aphid parasites (Hymenoptera, Aphidiidae) of the central Asian area.* Springer Netherlands, 124 pp.

Yu, D. S., van Achterberg, K., & Horstmann, K. (2016). World Ichneumonoidea 2011. In *Taxonomy, Biology, Morphology and distribution.* Canada: Taxapad.com.

Wharton, R. A. (1993). Bionomics of the Braconidae. *Annual Review of Entomology, 38,* 121−143.

Trachyusa aurora (Haliday, 1838) (Alysiinae), ♂. *Photo courtesy of T. Legrand and C. van Achterberg.*

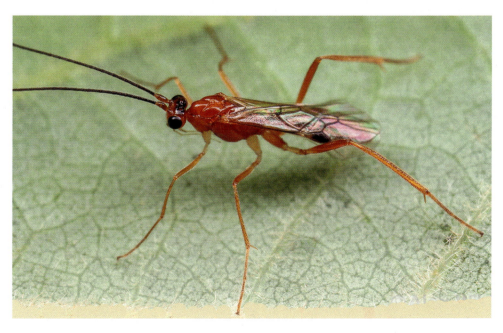

Macrocentrus sp. (Macrocentrinae), ♂. *Photo courtesy of Alex Wild and James B. Whitfield.*

Appendix 1

List of Braconidae (Hymenoptera) of Iran

SUBFAMILY AGATHIDINAE

(1) *Aerophilus persicus* (Farahani & Talebi, 2014)
(2) *A. rufipes* (Nees von Esenbeck, 1812)
(3) *Agathis anglica* Marshall, 1885
(4) *A. assimilis* Kokujev, 1895
(5) *A. breviseta* Nees von Esenbeck, 1812
(6) *A. fulmeki* Fischer, 1957
(7) *A. fuscipennis* (Zetterstedt, 1838)
(8) *A. glaucoptera* Nees von Esenbeck, 1834
(9) *A. griseifrons* Thomson, 1895
(10) *A. levis* Abdinbekova, 1970
(11) *A. lugubris* (Foerster, 1863)
(12) *A. malvacearum* Latreille, 1805
(13) *A. mediator* (Nees von Esenbeck, 1812)
(14) *A. melpomene* Nixon, 1986
(15) *A. montana* Shestakov, 1932
(16) *A. nigra* Nees von Esenbeck, 1812
(17) *A. pedias* Nixon, 1986
(18) *A. pumila* (Ratzeburg, 1844)
(19) *A. rostrata* Tobias, 1963
(20) *A. rufipalpis* Nees von Esenbeck, 1812
(21) *A. semiaciculata* Ivanov, 1899
(22) *A. syngenesiae* Nees von Esenbeck, 1812
(23) *A. tatarica* Telenga, 1933
(24) *A. taurica* Telenga, 1955
(25) *A. tibialis* Nees von Esenbeck, 1812
(26) *A. umbellatarum* Nees von Esenbeck, 1812
(27) *A. varipes* Thomson, 1895
(28) *A. zaisanika* Tobias, 1963
(29) *Bassus calculator* (Fabricius, 1798)
(30) *Camptothlipsis armeniaca* (Telenga, 1955)
(31) *Coccygidium transcaspicum* (Kokujev, 1902)
(32) *Cremnops desertor* (Linnaeus, 1758)
(33) *C. richteri* Hedwig, 1957
(34) *Disophrys caesa* (Klug, 1835)
(35) *D. dissors* Kokujev, 1903
(36) *D. inculcatrix* (Kriechbaumer, 1898)
(37) *Earinus elator* (Fabricius, 1804)

(38) *E. gloriatorius* (Panzer, 1809)
(39) *Euagathis indica* Enderlein, 1920
(40) *Therophilus cingulipes* (Nees von Esenbeck, 1812)
(41) *T. clausthalianus* (Ratzeburg, 1844)
(42) *T. conspicuus* (Wesmael, 1837)
(43) *T. dimidiator* (Nees von Esenbeck, 1834)
(44) *T. linguarius* (Nees von Esenbeck, 1812)
(45) *T. nugax* (Reinhard, 1867)
(46) *T. tegularis* (Thomson, 1895)
(47) *T. tumidulus* (Nees von Esenbeck, 1812)

SUBFAMILY ALYSIINAE

(48) *Adelurola amplidens* (Fischer, 1966)
(49) *A. florimela* (Haliday, 1838)
(50) *Alloea contracta* (Haliday, 1833)
(51) *Alysia alticola* (Ashmead, 1890)
(52) *A. frigida* Haliday, 1838
(53) *A. fuscipennis* (Haliday, 1838)
(54) *A. incongrua* Nees von Esenbeck, 1834
(55) *A. lucicola* Haliday, 1838
(56) *A. luciella* Stelfox, 1941
(57) *A. manducator* (Panzer, 1799)
(58) *A. rufidens* Nees von Esenbeck, 1834
(59) *A. tipulae* (Scopoli, 1763)
(60) *A. truncator* (Nees von Esenbeck, 1812)
(61) *Angelovia elipsocubitalis* Zaykov, 1980
(62) *Aphaereta brevis* Tobias, 1962
(63) *A. difficilis* Nixon, 1939
(64) *A. minuta* (Nees von Esenbeck, 1811)
(65) *A. pallipes* (Say, 1828)
(66) *Aristelix persica* Peris-Felipo, 2015
(67) *Asobara tabida* (Nees von Esenbeck, 1834)
(68) *Aspilota alfalfae* Fischer, Lashkari Bod, Rakhshani & Talebi, 2011
(69) *A. delicata* Fischer, 1973
(70) *A. flagellaris* Fischer, 1973
(71) *A. flagimilis* Fischer, 1996
(72) *A. fuscicornis* (Haliday, 1838)

(73) *A. globipes* (Fischer, 1962)

(74) *A. insolita* (Tobias, 1962)

(75) *A. isfahanensis* Peris-Felipo, 2016

(76) *A. latitemporata* Fischer, 1976

(77) *A. nidicola* Hedqvist, 1972

(78) *A. ruficornis* (Nees von Esenbeck, 1834)

(79) *Carinthilota vichti* van Achterberg, 1988

(80) *Chorebus affinis* (Nees von Esenbeck, 1812)

(81) *C. albipes* (Haliday, 1839)

(82) *C. anasellus* (Stelfox, 1951)

(83) *C. aphantus* (Marshall, 1896)

(84) *C. ares* (Nixon, 1944)

(85) *C. asphodeli* Griffiths, 1968

(86) *C. avestus* (Nixon, 1944)

(87) *C. axillaris* Fischer, Lashkari Bod, Rakhshani & Talebi, 2011

(88) *C. baeticus* Griffiths, 1967

(89) *C. bathyzonus* (Marshall, 1895)

(90) *C. brevicornis* (Thomson, 1895)

(91) *C. caesariatus* Griffiths, 1967

(92) *C. calthae* Griffiths, 1967

(93) *C. compressiiventris* (Telenga, 1935)

(94) *C. cubocephalus* (Telenga, 1935)

(95) *C. dagda* (Nixon, 1943)

(96) *C. diremtus* (Nees von Esenbeck, 1834)

(97) *C. femoratus* (Tobias, 1962)

(98) *C. flavipes* (Goureau, 1851)

(99) *C. fordi* (Nixon, 1954)

(100) *C. fuscipennis* (Nixon, 1937)

(101) *C. gedanensis* (Ratzeburg, 1852)

(102) *C. geminus* (Tobias, 1962)

(103) *C. gnaphalii* Griffiths, 1967

(104) *C. gracilipes* (Thomson, 1895)

(105) *C. groschkei* Griffiths, 1967

(106) *C. heringianus* Griffiths, 1967

(107) *C. iridis* Griffiths, 1968

(108) *C. lar* (Morley, 1924)

(109) *C. larides* (Nixon, 1944)

(110) *C. leptogaster* (Haliday, 1839)

(111) *C. longiarticulis* Fischer, Lashkari Bod, Rakhshani & Talebi, 2011

(112) *C. merellus* (Nixon, 1937)

(113) *C. melanophytobiae* Griffiths, 1968

(114) *C. misellus* (Marshall, 1895)

(115) *C. mucronatus* (Telenga, 1935)

(116) *C. nigridiremptus* Fischer, Lashkari Bod, Rakhshani & Talebi, 2011

(117) *C. nigriscaposus* (Nixon, 1949)

(118) *C. nixoni* Burghele, 1959

(119) *C. ornatus* (Telenga, 1935)

(120) *C. parvungula* (Thomson, 1895)

(121) *C. perkinsi* (Nixon, 1944)

(122) *C. posticus* (Haliday, 1839)

(123) *C. properesam* Fischer, Lashkari Bod, Rakhshani & Talebi, 2011

(124) *C. pseudomisellus* Griffiths, 1968

(125) *C. ruficollis* (Stelfox, 1956)

(126) *C. scabiosae* Griffiths, 1967

(127) *C. senilis* (Nees von Esenbeck, 1812)

(128) *C. solstitialis* (Stelfox, 1951)

(129) *C. spenceri* Griffiths, 1964

(130) *C. stilifer* Griffiths, 1968

(131) *C. tamsi* (Nixon, 1944)

(132) *C. thusa* (Nixon, 1937)

(133) *C. tumidus* (Tobias, 1966)

(134) *C. uliginosus* (Haliday, 1839)

(135) *C. varunus* (Nixon, 1945)

(136) *C. venustus* (Tobias, 1962)

(137) *C. zarghanensis* Fischer, Lashkari Bod, Rakhshani & Talebi, 2011

(138) *Coelinidea elegans* (Curtis, 1829)

(139) *C. gracilis* (Curtis, 1829)

(140) *C. nigra* (Nees von Esenbeck, 1811)

(141) *C. vidua* (Curtis, 1829)

(142) *Coelinius parvulus* (Nees von Esenbeck, 1811)

(143) *Coloneura arestor* (Nixon, 1954)

(144) *C. dice* (Nixon, 1943)

(145) *Cratospila circe* (Haliday, 1838)

(146) *Dacnusa abdita* (Haliday, 1839)

(147) *D. adducta* (Haliday, 1839)

(148) *D. alpestris* Griffiths, 1967

(149) *D. aquilegiae* Marshall, 1896

(150) *D. areolaris* (Nees von Esenbeck, 1811)

(151) *D. aterrima* Thomson, 1895

(152) *D. clematidis* Griffiths, 1967

(153) *D. confinis* Ruthe, 1859

(154) *D. discolor* (Foerster, 1863)

(155) *D. evadne* Nixon, 1937

(156) *D. faeroeensis* (Roman, 1917)

(157) *D. gentianae* Griffiths, 1967

(158) *D. heringi* Griffiths, 1967

(159) *D. hospita* (Foerster, 1863)

(160) *D. laevipectus* Thomson, 1895

(161) *D. metula* (Nixon, 1954)

(162) *D. monticola* (Foerster, 1863)

(163) *D. pubescens* (Curtis, 1826)

(164) *D. sasakawai* Takada, 1977

(165) *D. sibirica* Telenga, 1935

(166) *Dinotrema amoenidens* (Fischer, 1973)

(167) *D. amparoae* Peris-Felipo, 2013

(168) *D. borzhomoii* Tobias, 2004

(169) *D. concinnum* (Haliday, 1838)

(170) *D. concolor* (Nees von Esenbeck, 1812)

(171) *D. contracticorne* (Fischer, 1974)

(172) *D. cratocera* (Thomson, 1895)

(173) *D. cruciform* (Fischer, 1973)

(174) *D. dimidiatum* (Thomson, 1895)

(175) *D. dimorpha* (Fischer, 1976)

(176) *D. distractum* (Nees von Esenbeck, 1834)

(177) *D. intermissum* (Fischer 1974)

(178) *D. maxima* (Fischer, 1962)
(179) *D. megastigma* (Fischer, 1967)
(180) *D. naevium* (Tobias, 1962)
(181) *D. oleraceum* (Tobias, 1962)
(182) *D. perlustrandum* (Fischer, 1973)
(183) *D. significarium* (Fischer, 1973)
(184) *D. sinecarinum* Fischer, 1993
(185) *D. speculum* (Haliday, 1838)
(186) *D. tauricum* (Telenga, 1939)
(187) *D. ultima* (Fischer, 1970)
(188) *D. varipes* (Tobias, 1962)
(189) *Epimicta marginalis* (Haliday, 1839)
(190) *Exotela gilvipes* (Haliday, 1839)
(191) *E. umbellina* (Nixon, 1954)
(192) *Idiasta dichrocera* Königsmann, 1960
(193) *I. picticornis* (Ruthe, 1854)
(194) *I. subannellata* (Thomson, 1895)
(195) *Orthostigma beyarslani* Fischer, 1995
(196) *O. laticeps* (Thomson, 1895)
(197) *O. longicorne* Königsmann, 1969
(198) *O. maculipes* (Haliday, 1838)
(199) *O. mandibulare* (Tobias, 1962)
(200) *O. pumilum* (Nees von Esenbeck, 1834)
(201) *Pentapleura angustula* (Halilday, 1838)
(202) *P. pumilio* (Nees von Esenbeck, 1812)
(203) *Phaenocarpa bicolor* (Foerster, 1863)
(204) *P. brevipalpis* (Thomson, 1895)
(205) *P. canaliculata* Stelfox, 1941
(206) *P. carinthiaca* Fischer, 1975
(207) *P. conspurcator* (Haliday, 1838)
(208) *P. picinervis* (Haliday, 1838)
(209) *P. ruficeps* (Nees von Esenbeck, 1812)
(210) *Polemochartus liparae* (Giraud, 1863)
(211) *Protodacnusa aridula* (Thomson, 1895)
(212) *P. litoralis* Griffiths, 1964
(213) *P. tristis* (Nees von Esenbeck, 1834)
(214) *Pseudopezomachus cursitans* (Ferrière, 1930)
(215) *P. masii* Nixon, 1940
(216) *Tanycarpa bicolor* (Nees von Esenbeck, 1812)
(217) *Trachionus hians* (Nees von Esenbeck, 1816)
(218) *Trachyusa aurora* (Haliday, 1838)

SUBFAMILY APHIDIINAE

(219) *Aclitus obscuripennis* Foerster, 1863
(220) *Adialytus ambiguus* (Haliday, 1834)
(221) *A. salicaphis* (Fitch, 1855)
(222) *A. thelaxis* (Starý, 1961)
(223) *A. veronicaecola* (Starý, 1978)
(224) *Aphidius absinthii* Marshall, 1896
(225) *A. aquilus* Mackauer, 1961
(226) *A. arvensis* (Starý, 1960)
(227) *A. artemisicola* Tizado & Núñez-Perez, 1994
(228) *A. asteris* Haliday, 1834

(229) *A. avenae* Haliday, 1834
(230) *A. cingulatus* Ruthe, 1859
(231) *A. colemani* Viereck, 1912
(232) *A. eadyi* Starý, González & Hall, 1980
(233) *A. ervi* Haliday, 1834
(234) *A. funebris* Mackauer, 1961
(235) *A. hieraciorum* Starý, 1962
(236) *A. hortensis* Marshall, 1896
(237) *A. iranicus* Rakhshani & Starý, 2007
(238) *A. matricariae* Haliday, 1834
(239) *A. persicus* Rakhshani & Starý, 2006
(240) *A. phalangomyzi* Starý, 1963
(241) *A. platensis* Brèthes, 1913
(242) *A. popovi* Starý, 1978
(243) *A. rhopalosiphi* de Stefani-Perez, 1902
(244) *A. rosae* Haliday, 1833
(245) *A. salicis* Haliday, 1834
(246) *A. setiger* (Mackauer, 1961)
(247) *A. smithi* Sharma & Subba Rao, 1959
(248) *A. sonchi* Marshall, 1896
(249) *A. stigmaticus* Rakhshani & Tomanovic, 2011
(250) *A. tanacetarius* Mackauer, 1962
(251) *A. transcaspicus* Telenga, 1958
(252) *A. urticae* Haliday, 1834
(253) *A. uzbekistanicus* Luzhetzki, 1960
(254) *Areopraon lepelleyi* (Waterston, 1926)
(255) *Betuloxys hortorum* (Starý, 1960)
(256) *Binodoxys acalephae* (Marshall, 1896)
(257) *B. angelicae* (Haliday, 1833)
(258) *B. brevicornis* (Haliday, 1833)
(259) *B. centaureae* (Haliday, 1833)
(260) *B. heraclei* (Haliday, 1833)
(261) *Diaeretiella rapae* (McIntosh, 1855)
(262) *Diaeretus leucopterus* (Haliday, 1834)
(263) *Ephedrus cerasicola* Starý, 1962
(264) *E. chaitophori* Gärdenfors, 1986
(265) *E. helleni* Mackauer, 1968
(266) *E. lacertosus* (Haliday, 1833)
(267) *E. laevicollis* (Thomson, 1895)
(268) *E. nacheri* Quilis, 1934
(269) *E. niger* Gautier, Bonnamour & Gaumont, 1929
(270) *E. persicae* Froggatt, 1904
(271) *E. plagiator* (Nees von Esenbeck, 1811)
(272) *Lipolexis gracilis* Foerster, 1863
(273) *Lysiphlebus cardui* (Marshall, 1896)
(274) *L. confusus* Tremblay & Eady, 1978
(275) *L. desertorum* Starý, 1965
(276) *L. fabarum* (Marshall, 1896)
(277) *L. testaceipes* (Cresson, 1880)
(278) *L. volkli* Tomanović & Kavallieratos, 2018
(279) *Monoctonia pistaciaecola* Starý, 1962
(280) *M. vesicarii* Tremblay, 1991
(281) *Monoctonus mali* van Achterberg, 1989
(282) *Pauesia antennata* (Mukerji, 1950)

(283) *P. hazratbalensis* Bhagat, 1981
(284) *Praon abjectum* (Haliday, 1833)
(285) *P. barbatum* Mackauer, 1967
(286) *P. bicolor* Mackauer, 1959
(287) *P. dorsale* (Haliday, 1833)
(288) *P. exsoletum* (Nees, 1811)
(289) *P. flavinode* (Haliday, 1833)
(290) *P. gallicum* (Starý, 1971)
(291) *P. longicorne* Marshall, 1896
(292) *P. necans* Mackauer, 1959
(293) *P. orpheusi* Kavallieratos, Athanassiou & Tomanovic, 2003
(294) *P. pubescens* Starý, 1961
(295) *P. rosaecola* Starý, 1961
(296) *P. unitum* Mescheloff & Rosen, 1989
(297) *P. volucre* (Haliday, 1833)
(298) *P. yomenae* Takada, 1968
(299) *Tanytrichophorus petiolaris* Mackauer, 1961
(300) *Trioxys asiaticus* Telenga, 1953
(301) *T. auctus* (Haliday, 1833)
(302) *T. betulae* Marshall, 1896
(303) *T. cirsii* (Curtis, 1831)
(304) *T. complanatus* Quilis, 1931
(305) *T. curvicaudus* Mackauer, 1967
(306) *T. metacarpalis* Rakhshani & Starý, 2012
(307) *T. pallidus* (Haliday, 1833)
(308) *T. pannonicus* Starý, 1960
(309) *T. pappi* Takada, 1979
(310) *T. tanaceticola* Starý, 1971

SUBFAMILY BRACHISTINAE

(311) *Aspicolpus carinator* (Nees von Esenbeck, 1812)
(312) *A. sibiricus* (Fahringer, 1934)
(313) *Blacus achterbergi* Haeselbarth, 1976
(314) *B. armatulus* Ruthe, 1861
(315) *B. bovistae* Haeselbarth, 1973
(316) *B. conformis* Wesmael, 1835
(317) *B. diversicornis* (Nees von Esenback, 1834)
(318) *B. errans* (Nees von Esenback, 1811)
(319) *B. exilis* (Nees von Esenback, 1811)
(320) *B. filicornis* Haeselbarth, 1973
(321) *B. forticornis* Haeselbarth, 1973
(322) *B. hastatus* Haliday, 1835
(323) *B. humilis* (Nees von Esenbeck, 1811)
(324) *B. instabilis* Ruthe, 1861
(325) *B. interstitialis* Ruthe, 1861
(326) *B. longipennis* (Gravenhorst, 1809)
(327) *B. maculipes* Wesmael, 1835
(328) *B. nixoni* Haeselbarth, 1973
(329) *Blacus paganus* Haliday, 1835
(330) *B. rufescens* Ruthe, 1861
(331) *B. ruficornis* (Nees von Esenbeck, 1811)

(332) *B. stelfoxi* Haeselbarth, 1973
(333) *B. tripudians* Haliday, 1835
(334) *Diospilus capito* (Nees von Esenbeck, 1834)
(335) *D. melanoscelus* (Nees von Esenbeck, 1834)
(336) *D. morosus* Reinhard, 1862
(337) *D. nigricornis* (Wesmael, 1835)
(338) *D. oleraceus* Haliday, 1833
(339) *D. productus* Marshall, 1894
(340) *Eubazus cingulatus* (Szépligeti, 1896)
(341) *E. fasciatus* (Nees von Esenbeck, 1816)
(342) *E. flavipes* (Haliday, 1835)
(343) *E. fuscipes* (Herrich-Schäffer, 1838)
(344) *E. gallicus* (Reinhard, 1867)
(345) *E. lepidus* (Haliday, 1835)
(346) *E. minutus* (Ratzeburg, 1848)
(347) *E. nigricoxis* (Wesmael, 1835)
(348) *E. pallipes* Nees con Esenbeck, 1812
(349) *E. parvulus* (Reinhard, 1867)
(350) *E. ruficoxis* (Wesmael, 1835)
(351) *E. semirugosus* (Nees von Esenbeck, 1816)
(352) *E. tibialis* (Haliday, 1835)
(353) *Foersteria longicauda* van Achterberg, 1990
(354) *Polydegmon foveolatus* (Herrich-Schäffer, 1838)
(355) *P. sinuatus* Foerster, 1863
(356) *Schizoprymnus ambiguus* (Nees von Esenbeck, 1816)
(357) *S. angustatus* (Herrich-Schäffer, 1838)
(358) *S. azerbajdzhanicus* (Abdinbekova, 1967)
(359) *S. bidentulus* (Szépligeti, 1901)
(360) *S. brevicornis* (Herrich-Schäffer, 1838)
(361) *S. crassiceps* (Thomson, 1892)
(362) *S. elongatus* (Szépligeti, 1898)
(363) *S. excisus* (Snoflák, 1953)
(364) *S. hilaris* (Herrich-Schäffer, 1838)
(365) *S. nigripes* (Thomson, 1892)
(366) *S. obscurus* (Nees von Esenbeck, 1816)
(367) *S. pallidipennis* (Herrich-Schäffer, 1838)
(368) *S. parvus* (Thomson, 1892)
(369) *S. pullatus* (Dahlbom, 1833)
(370) *S. tantalus* Papp, 1981
(371) *S. telengai* Tobias, 1976
(372) *S. terebralis* (Snoflák, 1953)
(373) *Taphaeus hiator* (Thunberg, 1822)
(374) *Triapsis armeniaca* Tobias, 1976
(375) *T. caucasica* Abdinbekova, 1969
(376) *T. caudata* (Nees von Esenbeck, 1816)
(377) *T. complanellae* (Hartig, 1847)
(378) *T. floricola* (Wesmael, 1835)
(379) *T. lugubris* Šnoflak, 1953
(380) *T. luteipes* (Thomson, 1874)
(381) *T. obscurella* (Nees, 1816)
(382) *T. pallipes* (Nees, 1816)
(383) *T. thoracica* (Curtis, 1860)

SUBFAMILY BRACONINAE

(384) *Amyosoma chinense* (Szépligeti, 1902)
(385) *Atanycolus denigrator* (Linnaeus, 1768)
(386) *A. fulviceps* (Kriechbaumer, 1898)
(387) *A. genalis* (Thomson, 1892)
(388) *A. initiator* (Fabricius, 1793)
(389) *A. ivanowi* (Kokujev, 1898)
(390) *Baryproctus barypus* (Marshall, 1897)
(391) *B. zarudnianus* Telenga, 1936
(392) *Bracon abbreviator* Nees von Esenbeck, 1834
(393) *B. ahngeri* Telenga, 1936
(394) *B. alutaceus* Szépligeti, 1901
(395) *B. angustiventris* Tobias, 1957
(396) *B. apricus* Schmiedeknecht, 1897
(397) *B. arcuatus* Thomson, 1892
(398) *B. atrator* Nees von Esenbeck, 1834
(399) *B. batis* Papp, 1981
(400) *B. bilgini* Beyarslan, 2002
(401) *B. biroicus* Papp, 1990
(402) *B. brachycerus* Thomson, 1892
(403) *B. brevicalcaratus* Tobias, 1957
(404) *B. brevitemporis* Tobias, 1959
(405) *B. byurakanicus* Tobias, 1976
(406) *B. caudatus* Ratzeburg, 1848
(407) *B. caudiger* Nees von Esenbeck, 1834
(408) *B. chivensis* Telenga, 1936
(409) *B. chrysostigma* Greese, 1928
(410) *B. cingulator* Szépligeti, 1901
(411) *B. ciscaucasicus* Telenga, 1936
(412) *B. cisellatus* Papp, 1989
(413) *B. colpophorus* Wesmael, 1838
(414) *B. concavus* Tobias, 1957
(415) *B. crassiceps* Thomson, 1892
(416) *B. crassungula* Thomson, 1892
(417) *B. curticaudis* (Szépligeti, 1901)
(418) *B. debitor* Papp, 1971
(419) *B. delibator* Haliday, 1833
(420) *B. delusor* Spinola, 1808
(421) *B. densipilosus* Tobias, 1957
(422) *B. dichromus* (Wesmael, 1838)
(423) *B. dilatus* Papp, 1999
(424) *B. discoideus* (Wesmael, 1838)
(425) *B. dolichurus* Marshall, 1897
(426) *B. epitriptus* Marshall, 1885
(427) *B. erraticus* (Wesmael, 1838)
(428) *B. erzurumiensis* Beyarslan, 2002
(429) *B. exhilarator* Nees von Esenbeck, 1834
(430) *B. fallax* Szépligeti, 1901
(431) *B. femoralis* (Brullé, 1832)
(432) *B. filicornis* Thomson, 1892
(433) *B. flagellaris* Thomson, 1892
(434) *B. flamargo* Papp, 2011
(435) *B. fortipes* (Wesmael, 1838)

(436) *B. frater* Tobias, 1957
(437) *B. fulvipes* Nees von Esenback, 1834
(438) *B. fumarius* Szépligeti, 1901
(439) *B. fumatus* Szépligeti, 1901
(440) *B. fumigidus* Szépligeti, 1901
(441) *B. grandiceps* Thomson, 1892
(442) *B. gusaricus* Telenga, 1933
(443) *B. guttiger* (Wesmael, 1838)
(444) *B. helleni* Telenga, 1936
(445) *B. hemiflavus* Szépligeti, 1901
(446) *B. humidus* Tobias, 1976
(447) *B. hungaricus* (Szépligeti, 1896)
(448) *B. hylobii* Ratzeburg, 1848
(449) *B. illyricus* Marshall, 1888
(450) *B. immutator* Nees von Esenbeck, 1834
(451) *B. intercessor* Nees von Esenbeck, 1834
(452) *B. intersessor laetus* (Wesmael, 1838)
(453) *B. iskilipus* Beyarslan & Tobias, 2008
(454) *B. jaroslavensis* Telenga, 1936
(455) *B. kirgisorum* Telenga, 1936
(456) *B. kozak* Telenga, 1936
(457) *B. larvicida* (Wesmael, 1838)
(458) *B. lefroyi* (Dudgeon & Gough, 1914)
(459) *B. leptus* Marshall, 1897
(460) *B. lividus* Telenga, 1936
(461) *B. longicollis* (Wesmael, 1838)
(462) *B. longulus* Thomson, 1892
(463) *B. luteator* Spinola, 1808
(464) *B. mariae* Dalla Torre, 1898
(465) *B. marshalli* Szépligeti, 1901
(466) *B. meyeri* Telenga, 1936
(467) *B. minutator* (Fabricius, 1798)
(468) *B. mirus* Szépligeti, 1901
(469) *B. moczari* Papp, 1969
(470) *B. murgabensis* Tobias, 1957
(471) *B. necator* (Fabricius, 1777)
(472) *B. negativus* Tobias, 1957
(473) *B. nigratus* (Wesmael, 1838)
(474) *B. nigripilosus* Tobias, 1957
(475) *B. nigriventris* (Wesmael, 1838)
(476) *B. nigriventris indubius* Szépligeti, 1901
(477) *B. nocturnus* (Tobias, 1962)
(478) *B. novus* Szépligeti, 1901
(479) *B. obscurator* Nees von Esenbeck, 1811
(480) *B. ochraceus* Szépligeti, 1896
(481) *B. ophtalmicus* Telenga, 1933
(482) *B. osculator* Nees, 1811
(483) *B. otiosus* Marshall, 1885
(484) *B. pallicarpus* Thomson, 1892
(485) *B. parvicornis* Thomson, 1892
(486) *B. parvulus* (Wesmael, 1838)
(487) *B. pectoralis* (Wesmael, 1838)
(488) *B. pelliger rumezensis* Samartsev & Zargar, 2020
(489) *B. peroculatus* (Wesmael, 1838)

(490) *B. persiangulfensis* Ameri, Beyarslan & Talebi, 2013

(491) *B. picticornis* (Wesmael, 1838)

(492) *B. piger* (Wesmael, 1838)

(493) *B. pineti* Thomson, 1892

(494) *B. planinotus* Tobias, 1957

(495) *B. popovi* Telenga, 1936

(496) *B. praecox* (Wesmael, 1838)

(497) *B. punctifer* Thomson, 1892

(498) *B. punctithorax* Tobias, 1959

(499) *B. quadrimaculatus* Telenga, 1936

(500) *B. radiatus* Tobias, 1957

(501) *B. roberti* (Wesmael, 1838)

(502) *B. robustus* Hedwig, 1961

(503) *B. sabulosus* Szépligeti, 1896

(504) *B. santaecrucis* Schmiedeknecht, 1897

(505) *B. scabriusculus* Dalla Torre, 1898

(506) *B. schmidti* Kokujev, 1912

(507) *B. shestakoviellus* Tobias, 1957

(508) *B. speerschneideri* Schmiedecknecht, 1897

(509) *B. spectabilis* (Telenga, 1936)

(510) *B. sphaerocephalus* Szépligeti, 1901

(511) *B. subrugosus* Szépligeti, 1901

(512) *B. suchorukovi* Telenga, 1936

(513) *B. tekkensis* Telenga, 1936

(514) *B. tenuicornis* (Wesmael, 1838)

(515) *B. terebella* (Wesmael, 1838)

(516) *B. thuringiacus* Schmiedeknecht, 1897

(517) *B. titubans* (Wesmael, 1838)

(518) *B. triangularis* (Nees von Esenbeck, 1834)

(519) *B. trucidator* Marshall, 1888

(520) *B. tschitscherini* Kokujev, 1904

(521) *B. urinator* (Fabricius, 1798)

(522) *B. variator* Nees von Esenbeck, 1811

(523) *B. variator bipartitus* (Wesmael, 1838)

(524) *B. variegator* Spinola, 1808

(525) *Ceratobracon stschegolevi* (Telenga, 1933)

(526) *Coeloides abdominalis* (Zetterstedt, 1838)

(527) *C. bostrichorum* Giraud, 1872

(528) *C. filiformis* Ratzeburg, 1852

(529) *C. rossicus* (Kokujev, 1902)

(530) *Cyanopterus flavator* (Fabricius, 1793)

(531) *C. extricator* (Nees von Esenbeck, 1834)

(532) *Doggerella* **spp.**

(533) *Glyptomorpha discolor* Thunberg, 1822

(534) *G. elector* (Kokujev, 1898)

(535) *G. exsculpta* Shestakov, 1926

(536) *G. kasparyani* Tobias, 1976

(537) *G. nachitshevanica* Tobias, 1976

(538) *G. pectoralis* (Brullé, 1832)

(539) *G. sicula* (Marshall, 1888)

(540) *Habrobracon brevicornis* (Wesmael, 1838)

(541) *H. concolorans* (Marshall, 1900)

(542) *H. crassicornis* (Thomson, 1892)

(543) *H. didemie* (Beyarslan, 2002)

(544) *H. excisus* Tobias, 1957

(545) *H. hebetor* (Say, 1836)

(546) *H. iranicus* Fischer, 1972

(547) *H. kopetdagi* Tobias, 1957

(548) *H. lissothorax* Tobias, 1967

(549) *H. nygmiae* Telenga, 1936

(550) *H. ponticus* (Tobias, 1986)

(551) *H. radialis* Telenga, 1936

(552) *H. simonovi* Kokujev, 1914

(553) *H. stabilis* (Wesmael, 1838)

(554) *H. telengai* Muljarskaya, 1955

(555) *H. viktorovi* Tobias, 1961

(556) *Iphiaulax fastidiator* (Fabricius, 1781)

(557) *I. hians* Pérez, 1907

(558) *I. impeditor* (Kokujev, 1898)

(559) *I. impostor* (Scopoli, 1763)

(560) *I. iranicus* Quicke, 1985

(561) *I. jacobsoni* Shestakov, 1927

(562) *I. jakowlewi* (Kokujev, 1898)

(563) *I. mactator* (Klug, 1817)

(564) *I. perezi* (Fahringer, 1926)

(565) *I. potanini* (Kokujev, 1898)

(566) *I. tauricus* Shestakov, 1927

(567) *I. umbraculator* (Nees von Esenbeck, 1834)

(568) *Megalommum pistacivorae* van Achterberg & Mehrnejad, 2011

(569) *Pseudovipio barchanicus* (Telenga, 1936)

(570) *P. castrator* (Fabricius, 1798)

(571) *P. guttiventris* (Thomson, 1892)

(572) *P. inscriptor* (Nees von Esenbeck, 1834)

(573) *P. insubricus* (Fahringer, 1926)

(574) *P. kirmanensis* (Kokujev, 1907)

(575) *P. minutus* (Telenga, 1936)

(576) *P. nigrirostris* (Kokujev, 1907)

(577) *P. schaeuffelei* (Hedwig, 1957)

(578) *P. tataricus* (Kokujev, 1898)

(579) *Rhadinobracon zarudnyi* (Telenga, 1936)

(580) *Stenobracon deesae* (Cameron, 1902)

(581) *S. nicevillei* (Bingham, 1901)

(582) *Vipio appellator* (Nees von Esenbeck, 1834)

(583) *V. humerator* (Costa, 1885)

(584) *V. illusor* (Klug, 1817)

(585) *V. intermedius* Szépligeti, 1896

(586) *V. longicauda* (Boheman, 1853)

(587) *V. mlokossewiczi* Kokujev, 1898

(588) *V. nomioides* Shestakov, 1926

(589) *V. sareptanus* Kawall, 1865

(590) *V. shestakovi* Telenga, 1936

(591) *V. simulator* Kokujev, 1898

(592) *V. striolatus* Telenga, 1936

(593) *V. tentator* (Rossi, 1790)

(594) *V. terrefactor* (Villers, 1789)

(595) *V. xanthurus* (Fahringer, 1926)

SUBFAMILY CARDIOCHILINAE

(596) **Cardiochiles** *fallax* Kokujev, 1895
(597) *C. fumatus* Telenga, 1949
(598) *C. pseudofallax* Telenga, 1955
(599) *C. saltator* (Fabricius, 1781)
(600) *C. shestakovi* Telenga, 1949
(601) *C. tibialis* Hedwig, 1957
(602) *C. triplus* Shenefelt, 1973
(603) **Pseudcardiochilus** *abnormipes* Hedwig, 1957
(604) **Schoenlandella** *angustigena* Kang, 2021
(605) *S. deserta* (Telenga, 1955)
(606) *S. latigena* Kang, 2021

SUBFAMILY CENOCOELIINAE

(607) **Cenocoelius** *analis* (Nees von Esenbeck, 1834)
(608) **Lestricus** *secalis* (Linnaeus, 1758)

SUBFAMILY CHARMONTINAE

(609) **Charmon** *cruentatus* Haliday, 1833
(610) *C. extensor* (Linnaeus, 1758)

SUBFAMILY CHELONINAE

(611) **Adelius** *aridus* (Tobias, 1967)
(612) *A. erythronotus* (Foerster, 1851)
(613) *A. subfasciatus* Haliday, 1833
(614) **Ascogaster** *abdominator* (Dahlbom, 1833)
(615) *A. annularis* (Nees von Esenbeck, 1816)
(616) *A. armata* Wesmael, 1835
(617) *A. bicarinata* (Herrich-Schäffer, 1838)
(618) *A. bidentula* Wesmael, 1835
(619) *A. bimaris* Tobias, 1986
(620) *A. brevicornis* Wesmael, 1835
(621) *A. caucasica* Kokujev, 1895
(622) *A. dispar* Fahringer, 1934
(623) *A. disparilis* Tobias, 1986
(624) *A. excavata* Telenga, 1941
(625) *A. excisa* (Herrich-Schäffer, 1838)
(626) *A. gonocephala* Wesmael, 1835
(627) *A. grahami* Huddleston, 1984
(628) *A. kasparyani* Tobias, 1976
(629) *A. klugii* (Nees, 1816)
(630) *A. quadridentata* Wesmael, 1835
(631) *A. rufipes* (Latreille, 1809)
(632) *A. similis* (Nees von Esenbeck, 1816)
(633) *A. varipes* Wesmael, 1835
(634) **Chelonus** *annulatus* (Nees von Esenbeck, 1816)
(635) *C. annulipes* Wesmael, 1835
(636) *C. areolatus* Cameron, 1906
(637) *C. armeniacus* Tobias, 1976
(638) *C. basalis* Curtis, 1837
(639) *C. bidens* Tobias, 1972
(640) *C. bimaculatus* Szépligeti, 1896
(641) *C. bonellii* (Nees von Esenbeck, 1816)
(642) *C. breviventris* Thomson, 1874
(643) *C. canescens* Wesmael, 1835
(644) *C. caradrinae* Kokujev, 1914
(645) *C. carbonator* Marshall, 1885
(646) *C. cesa* Koçak & Kemal, 2013
(647) *C. cisapicalis* (Tobias, 1989)
(648) *C. contractus* (Nees von Esenbeck, 1816)
(649) *C. corvulus* Marshall, 1885
(650) *C. dauricus* Telenga, 1941
(651) *C. depressus* Thomson, 1874
(652) *C. devius* (Tobias, 1964)
(653) *C. elongatus* Szépligeti, 1898
(654) *C. erosus* Herrich-Schäffer, 1838
(655) *C. erythrogaster* Lucas, 1849
(656) *C. exilis* Marshall, 1885
(657) *C. fenestratus* (Nees von Esenbeck, 1816)
(658) *C. flavipalpis* Szépligeti, 1896
(659) *C. flavoneavulus* Abdinbekova, 1971
(660) *C. inanitus* (Linnaeus, 1767)
(661) *C. incisus* (Tobias, 1986)
(662) *C. iranicus* Tobias, 1972
(663) *C. kermakiae* (Tobias, 2001)
(664) *C. kopetdagicus* (Tobias, 1966)
(665) *C. latrunculus* Marshall, 1885
(666) *C. lissogaster* Tobias, 1972
(667) *C. longiventris* (Tobias, 1964)
(668) *C. luzhetzkji* (Tobias, 1966)
(669) *C. medus* Telenga, 1941
(670) *C. microphtalmus* Wesmael, 1838
(671) *C. microsomus* Tobias, 1964
(672) *C. milkoi* (Tobias, 2003)
(673) *C. mirandus* Tobias, 1964
(674) *C. moczari* (Papp, 2014)
(675) *C. mongolicus* (Telenga, 1941)
(676) *C. mucronatus* Thomson, 1874
(677) *C. nigritibialis* Abdinbekova, 1971
(678) *C. obscuratus* Herrich-Schäffer, 1838
(679) *C. ocellatus* Alexeev, 1971
(680) *C. oculator* (Fabricius, 1775)
(681) *C. olgae* Kokujev, 1895
(682) *C. pannonicus* Szépligeti, 1896
(683) *C. pectinophorae* Cushman, 1931
(684) *C. pellucens* (Nees von Esenbeck, 1816)
(685) *C. productus* Herrich-Schäffer, 1838
(686) *C. pusillus* (Szépligeti, 1908)
(687) *C. retusus* (Nees von Esenbeck, 1816)
(688) *C. risorius* Reinhard, 1867
(689) *C. rostratus* (Tobias, 1966)
(690) *C. scabrator* (Fabricius, 1793)
(691) *C. scabrosus* Szépligeti, 1896
(692) *C. setaceus* Papp, 1993

(693) *C. smirnovi* Telenga, 1953

(694) *C. subarcuatilis* (Tobias, 1986)

(695) *C. subcontractus* Abdinbekova, 1971

(696) *C. submuticus* Wesmael, 1835

(697) *C. subpamiricus* Farahani & van Achterberg, 2018

(698) *C. subpusillus* (Tobias, 1997)

(699) *C. subseticornis* Tobias, 1971

(700) *C. sulcatus* Jurine, 1807

(701) *C. szepligetii* Dalla Torre, 1898

(702) *C. telengai* (Abdinbekova, 1965)

(703) *C. vescus* (Kokujev, 1899)

(704) *C. wesmaeli* Curtis, 1837

(705) *Phanerotoma atra* Šnoflak, 1951

(706) *P. acuminata* Szépligeti, 1908

(707) *P. bilinea* Lyle, 1924

(708) *P. dentata* (Panzer, 1805)

(709) *P. fracta* Kokujev, 1903

(710) *P. intermedia* van Acterberg, 1990

(711) *P. katkowi* Kokujev, 1900

(712) *P. kozlovi* Shestakov, 1930

(713) *P. leucobasis* Kriechbaumer, 1894

(714) *P. masiana* Fahringer, 1934

(715) *P. minuta* Kokujev, 1903

(716) *P. parva* Kokujev, 1903

(717) *P. permixtellae* Fischer, 1968

(718) *P. planifrons* (Nees von Esenbeck, 1816)

(719) *P. rufescens* (Latreille, 1809)

(720) *P. syleptae* Zettel, 1990

(721) *P. tritoma* (Marshall, 1898)

(722) *Phanerotomella bisulcata* (Herrich-Schäffer, 1838)

(723) *P. rufa* (Marshall, 1898)

SUBFAMILY DIRRHOPINAE

(724) *Dirrhope rufa* Foerster, 1851

SUBFAMILY DORYCTINAE

(725) *Dendrosoter hartigii* (Ratzeburg, 1848)

(726) *D. middendorffii* (Ratzeburg, 1848)

(727) *D. protuberans* (Nees von Esenbeck, 1834)

(728) *Dendrosotinus ferrugineus* (Marshall, 1888)

(729) *D. titubatus* Papp, 1985

(730) *Doryctes inopinatus* Belokobylskij, 1984

(731) *D. leucogaster* (Nees von Esenbeck, 1834)

(732) *D. obliteratus* (Nees von Esenbeck, 1834)

(733) *D. undulatus* (Ratzeburg, 1852)

(734) *Ecphylus silesiacus* (Ratzeburg, 1848)

(735) *Hecabalodes radialis* Tobias, 1962

(736) *H. xylophagi* Fischer, 1962

(737) *Hecabolus sulcatus* Curtis, 1834

(738) *Heterospilus austriacus* (Szépligeti, 1906)

(739) *H. cephi* Rohwer, 1925

(740) *H. genalis* Tobias, 1976

(741) *H. hemipterus* (Thomson, 1892)

(742) *H. leptosoma* Fischer, 1960

(743) *H. rubicola* Fischer, 1968

(744) *H. tadzhicus* Belokobylskij, 1983

(745) *H. tauricus* Telenga, 1941

(746) *Leluthia asiatica* (Tobias, 1980)

(747) *L. paradoxa* (Picard, 1938)

(748) *L. ruguloscolyti* (Fischer, 1962)

(749) *L. transcaucasica* (Tobias, 1976)

(750) *Monolexis fuscicornis* Foerster, 1863

(751) *Ontsira antica* (Wollaston, 1858)

(752) *O. ignea* (Ratzeburg, 1852)

(753) *O. imperator* (Haliday, 1836)

(754) *O. longicaudis* (Giraud, 1857)

(755) *Polystenus rugosus* Foerster, 1863

(756) *Rhaconotus aciculatus* Ruthe, 1854

(757) *R. elegans* (Foerster, 1863)

(758) *R. kerzhneri* Belokobylskij, 1985

(759) *R. pictipennis* (Reinhard, 1855)

(760) *R. scaber* Kokujev, 1900

(761) *R. testaceus* (Szépligeti, 1908)

(762) *R. zarudnyi* Belokobylskij, 1990

(763) *Rhoptrocentrus piceus* Marshall, 1897

(764) *Spathiomorpha varinervis* Tobias, 1976

(765) *Spathius brevicaudis* Ratzeburg, 1844

(766) *S. exarator* (Linnaeus, 1758)

(767) *S. maderi* Fahringer, 1930

(768) *S. polonicus* Niezabitowski, 1910

(769) *S. rubidus* (Rossi, 1794)

(770) *S. umbratus* (Fabricius, 1798)

(771) *Zombrus flavipennis* (Brullé, 1846)

SUBFAMILY EUPHORINAE

(772) *Allurus lituratus* (Haliday, 1835)

(773) *A. muricatus* (Haliday, 1833)

(774) *Aridelus egregius* (Schmiedeknecht, 1907)

(775) *Centistes ater* (Nees von Esenbeck, 1834)

(776) *C. collaris* (Thomson, 1895)

(777) *C. cuspidatus* (Haliday, 1833)

(778) *C. edentatus* (Haliday, 1835)

(779) *C. fuscipes* (Nees von Esenbeck, 1834)

(780) *C. nasutus* (Wesmael, 1838)

(781) *C. subsulcatus* (Thomson, 1895)

(782) *Chrysopophthorus hungaricus* (Zilahi-Kiss, 1927)

(783) *Dinocampus coccinellae* (Schrank, 1802)

(784) *Ecclitura primoris* Kokujev, 1902

(785) *Elasmosoma berolinense* Ruthe, 1858

(786) *E. luxembergense* Wasmann, 1909

(787) *Leiophron apicalis* (Haliday, 1833)

(788) *L. basalis* Curtis, 1833

(789) *L. deficiens* (Ruthe, 1856)

(790) *L. fascipennis* (Ruthe, 1856)

(791) *L. heterocordyli* Richards, 1967

(792) *L. pallidistigma* Curtis, 1833
(793) *L. reclinator* (Ruthe, 1858)
(794) *Marshiella plumicornis* (Ruthe, 1856)
(795) *Meteorus abdominator* (Nees von Esenbeck, 1811)
(796) *M. affinis* (Wesmael, 1835)
(797) *M. alborossicus* Lobodenko, 2000
(798) *M. breviantennatus* Tobias, 1986
(799) *M. breviterebratus* Ameri, Talebi & Beyarslan, 2014
(800) *M. cespitator* (Thunberg, 1822)
(801) *M. cinctellus* (Spinola, 1808)
(802) *M. cis* (Bouché, 1834)
(803) *M. colon* (Haliday, 1835)
(804) *M. consimilis* (Nees, 1834)
(805) *M. filator* (Haliday, 1835)
(806) *M. ictericus* (Nees, 1811)
(807) *M. jaculator* (Haliday, 1835)
(808) *M. lionotus* Thomson, 1895
(809) *M. longicaudis* (Ratzeburg, 1848)
(810) *M. obfuscatus* (Nees von Esenbeck, 1811)
(811) *M. obsoletus* (Wesmael, 1835)
(812) *M. oculatus* Ruthe, 1862
(813) *M. pendulus* (Müller, 1776)
(814) *M. politutele* Shenefelt, 1969
(815) *M. pulchricornis* (Wesmael, 1835)
(816) *M. rubens* (Nees, 1811)
(817) *M. ruficeps* (Nees von Esenbeck, 1834)
(818) *M. rufus* (DeGeer, 1778)
(819) *M. salicorniae* Schmiedeknecht, 1897
(820) *M. tabidus* (Wesmael, 1835)
(821) *M. versicolor* (Wesmael, 1835)
(822) *M. vexator* (Haliday, 1835)
(823) *Microctonus aethiops* (Nees von Esenbeck, 1834)
(824) *M. colesi* Drea, 1968
(825) *M. melanopus* Ruthe, 1856
(826) *M. morimi* (Ferrière, 1931)
(827) *M. stenocari* (Haeselbarth, 2008)
(828) *Myiocephalus boops* (Wesmael, 1835)
(829) *Neoneurus auctus* (Thomson, 1895)
(830) *N. clypeatus* (Foerster, 1863)
(831) *Perilitus brevicollis* Haliday, 1835
(832) *P. cerealium* Haliday, 1835
(833) *P. dubius* (Wesmael, 1838)
(834) *P. falciger* (Ruthe, 1856)
(835) *P. foveolatus* Reinhard, 1862
(836) *P. kokujevi* Tobias, 1986
(837) *P. parcicornis* (Ruthe, 1856)
(838) *P. rutilus* (Nees von Esenbeck, 1811)
(839) *P. stelleri* (Loan, 1972)
(840) *Peristenus accinctus* (Haliday, 1835)
(841) *P. digoneutis* Loan, 1973
(842) *P. facialis* (Thomson, 1892)
(843) *P. grandiceps* (Thomson, 1892)

(844) *P. nitidus* (Curtis, 1833)
(845) *P. orchesiae* (Curtis, 1833)
(846) *P. pallipes* (Curtis, 1833)
(847) *P. picipes* (Curtis, 1833)
(848) *P. relictus* (Ruthe, 1856)
(849) *Pygostolus falcatus* (Nees von Esenbeck, 1834)
(850) *P. multiarticulatus* (Ratzeburg, 1852)
(851) *P. sticticus* (Fabricius, 1798)
(852) *Streblocera macroscapa* (Ruthe, 1856)
(853) *S. fulviceps* Westwood, 1833
(854) *Syntretus elegans* (Ruthe, 1856)
(855) *S. idalius* (Haliday, 1833)
(856) *S. ocularis* van Achterberg & Haeselbarth, 2003
(857) *S. xanthocephalus* (Marshall, 1887)
(858) *Townesilitus aemulus* (Ruthe, 1856)
(859) *T. bicolor* (Wesmael, 1835)
(860) *Wesmaelia petiolata* (Wollaston, 1858)
(861) *Zele albiditarsus* Curtis, 1832
(862) *Z. chlorophthalmus* (Spinola, 1808)
(863) *Z. deceptor* (Wesmael, 1835)

SUBFAMILY EXOTHECINAE

(864) *Colastes braconius* Haliday, 1833
(865) *C. pubicornis* (Thomson, 1892)
(866) *Shawiana catenator* (Haliday, 1836)
(867) *S. laevis* (Thomson, 1892)
(868) *Xenarcha laticarpus* (Thomson, 1892)
(869) *X. lustrator* (Haliday, 1836)

SUBFAMILY GNAMPTODONTINAE

(870) *Gnamptodon breviradialis* (Fischer, 1959)
(871) *G. decoris* (Foerster, 1863)
(872) *G. georginae* (van Achterberg, 1983)
(873) *G. pumilio* (Nees, 1834)

SUBFAMILY HELCONINAE

(874) *Helcon angustator* Nees von Esenbeck, 1812
(875) *H. claviventris* Wesmael, 1835
(876) *H. heinrichi* Hedqvist, 1967
(877) *H. tardator* Nees von Esenbeck, 1812
(878) *Wroughtonia dentator* (Fabricius, 1804)
(879) *W. ruspator* (Linnaeus, 1758)

SUBFAMILY HOMOLOBINAE

(880) *Homolobus annulicornis* (Nees von Esenbeck, 1834)
(881) *H. infumator* (Lyle, 1914)
(882) *H. meridionalis* van Achterberg, 1979
(883) *H. ophioninus* (Vachal, 1907)
(884) *H. truncator* (Say, 1829)

SUBFAMILY HORMIINAE

(885) ***Chremylus elaphus*** Haliday, 1833
(886) ***Hormius extima*** Tobias, 1964
(887) *H. moniliatus* (Nees von Esenbeck, 1811)
(888) *H. pseudomitis* (Hedwig, 1957)
(889) *H. radialis* Telenga, 1941
(890) *H. sculpturatus* Tobias, 1967
(891) *H. similis* Szépligeti, 1896
(892) *H. tatianae* (Telenga, 1941)
(893) ***Pseudohormius*** *flavobasalis* (Hedwig, 1957)
(894) *P. turkmenus* Tobias & Alexeev, 1973

SUBFAMILY ICHNEUTINAE

(895) ***Ichneutes brevis*** Wesmael, 1835
(896) *I. reunitor* Nees von Esenbeck, 1816
(897) ***Pseudichneutes atanassovae*** van Achterberg, 1997
(898) *P. levis* (Wesmael, 1835)

SUBFAMILY MACROCENTRINAE

(899) ***Macrocentrus bicolor*** Curtis, 1833
(900) *M. blandus* Eady & Clark, 1964
(901) *M. cingulum* Brischke, 1882
(902) *M. collaris* (Spinola, 1808)
(903) *M. equalis* Lyle, 1914
(904) *M. flavus* Snellen van Vollenhoven, 1878
(905) *M. infirmus* (Nees von Esenbeck, 1834)
(906) *M. kurnakovi* Tobias, 1976
(907) *M. linearis* (Nees von Esenbeck, 1811)
(908) *M. marginator* (Nees von Esenbeck, 1811)
(909) *M. nidulator* (Nees von Esenbeck, 1834)
(910) *M. nitidus* (Wesmael, 1835)
(911) *M. oriens* van Achterberg & Belokobylskij, 1987
(912) *M. pallipes* (Nees von Esenbeck, 1811)
(913) *M. resinellae* (Linnaeus, 1758)
(914) *M. thoracicus* (Nees von Esenbeck, 1811)
(915) *M. turkestanicus* (Telenga, 1950)

SUBFAMILY MICROGASTRINAE

(916) ***Apanteles aragatzi*** Tobias, 1976
(917) *A. biroicus* Papp, 1973
(918) *A. brunnistigma* Abdinbekova, 1969
(919) *A. carpatus* (Say, 1836)
(920) *A. firmus* Telenga, 1949
(921) *A. galleriae* Wilkinson, 1932
(922) *A. hemara* Nixon, 1965
(923) *A. kubensis* Abdinbekova, 1969
(924) *A. lacteus* (Nees von Esenbeck, 1834)
(925) *A. lenea* Nixon, 1976
(926) *A. metacarpalis* (Thomson, 1895)
(927) *A. obscurus* (Nees von Esenbeck, 1834)

(928) *A. pilosus* Telenga, 1955
(929) *A. sodalis* (Haliday, 1834)
(930) *A. xanthostigma* (Haliday, 1834)
(931) ***Choeras dorsalis*** (Spinola, 1808)
(932) *C. formosus* Abdoli & Fernandez-Triana, 2019
(933) *C. fulviventris* Fernandez-Triana & Abdoli, 2019
(934) *C. parasitellae* (Bouché, 1834)
(935) *C. qazviniensis* Fernandez-Triana & Talebi, 2019
(936) *C. ruficornis* (Nees von Esenbeck, 1834)
(937) *C. taftanensis* Ghafouri Moghaddam & van Achterberg, 2018
(938) *C. tedellae* (Nixon, 1961)
(939) *C. tiro* (Reinhard, 1880)
(940) ***Cotesia abjecta*** (Marshall, 1885)
(941) *C. acuminata* (Reinhard, 1880)
(942) *C. affinis* (Nees von Esenbeck, 1834)
(943) *C. ancilla* (Nixon, 1974)
(944) *C. brevicornis* (Wesmael, 1837)
(945) *C. callimone* (Nixon, 1974)
(946) *C. chilonis* (Munakata, 1912)
(947) *C. cuprea* (Lyle, 1925)
(948) *C. cynthiae* (Nixon, 1974)
(949) *C. elongata* Zargar & Gupta, 2019
(950) *C. euryale* (Nixon, 1974)
(951) *C. flavipes* Cameron, 1891
(952) *C. gastropachae* (Bouché, 1834)
(953) *C. geryonis* (Marshall, 1885)
(954) *C. glabrata* (Telenga, 1955)
(955) *C. glomerata* (Linnaeus, 1758)
(956) *C. gonopterygis* (Marshall, 1898)
(957) *C. hyphantriae* (Riley, 1887)
(958) *C. jucunda* (Marshall, 1885)
(959) *C. kazak* (Telenga, 1949)
(960) *C. khuzestanensis* Zargar & Gupta, 2019
(961) *C. lineola* (Curtis, 1830)
(962) *C. melanoscela* (Ratzeburg, 1844)
(963) *C. melitaearum* (Wilkinson, 1937)
(964) *C. nothus* (Marshall, 1885)
(965) *C. ofella* (Nixon, 1974)
(966) *C. ordinaria* (Ratzeburg, 1844)
(967) *C. pieridis* (Bouché, 1834)
(968) *C. praepotens* (Haliday, 1834)
(969) *C. risilis* (Nixon, 1974)
(970) *C. rubecula* (Marshall, 1885)
(971) *C. rubripes* (Haliday, 1834)
(972) *C. ruficrus* (Haliday, 1834)
(973) *C. salebrosa* (Marshall, 1885)
(974) *C. saltator* (Thunberg, 1822)
(975) *C. scabricula* (Reinhard, 1880)
(976) *C. sesamiae* (Cameron, 1906)
(977) *C. sessilis* (Geoffroy, 1785)
(978) *C. setebis* (Nixon, 1974)
(979) *C. specularis* (Szépligeti, 1896)
(980) *C. spuria* (Wesmael, 1837)

(981) *C. telengai* (Tobias, 1972)

(982) *C. tenebrosa* (Wesmael, 1837)

(983) *C. tibialis* (Curtis, 1830)

(984) *C. vanessae* (Reinhard, 1880)

(985) *C. vestalis* (Haliday, 1834)

(986) *C. villana* (Reinhard, 1880)

(987) *C. zagrosensis* Zargar & Gupta, 2019

(988) *C. zygaenarum* (Marshall, 1885)

(989) *Deuterixys carbonaria* (Wesmael, 1837)

(990) *D. rimulosa* (Niezabitowski, 1910)

(991) *D. tenuiconvergens* Zargar & Gupta, 2020

(992) *Diolcogaster abdominalis* (Nees von Esenbeck, 1834)

(993) *D. alvearia* (Fabricius, 1798)

(994) *D. claritibia* (Papp, 1959)

(995) *D. mayae* (Shestakov, 1932)

(996) *D. minuta* (Reinhard, 1880)

(997) *D. spreta* (Marshall, 1885)

(998) *Dolichogenidea agilla* (Nixon, 1972)

(999) *D. albipennis* (Nees von Esenbeck, 1834)

(1000) *D. anarsia* (Faure & Alabouvitte, 1924)

(1001) *D. appellator* (Telenga, 1949)

(1002) *D. atreus* (Nixon, 1973)

(1003) *D. breviventris* (Ratzeburg, 1848)

(1004) *D. britannica* (Wilkinson, 1941)

(1005) *D. candidata* (Haliday, 1834)

(1006) *D. coleophorae* (Wilkinson, 1938)

(1007) *D. corvina* (Reinhard, 1880)

(1008) *D. cytherea* (Nixon, 1972)

(1009) *D. decora* (Haliday, 1834)

(1010) *D. dilecta* (Haliday, 1834)

(1011) *D. drusilla* (Nixon, 1972)

(1012) *D. emarginata* (Nees von Esenbeck, 1834)

(1013) *D. ensiformis* (Ratzeburg, 1844)

(1014) *D. erevanica* (Tobias, 1976)

(1015) *D. evonymellae* (Bouché, 1834)

(1016) *D. fernandeztrianai* Abdoli & Talebi, 2019

(1017) *D. flavostriata* (Papp, 1977)

(1018) *D. gagates* (Nees von Esenbeck, 1834)

(1019) *D. gracilariae* (Wilkinson, 1840)

(1020) *D. halidayi* (Marshall, 1872)

(1021) *D. imperator* (Wilkinson, 1939)

(1022) *D. impura* (Nees von Esenbeck, 1834)

(1023) *D. infima* (Haliday, 1834)

(1024) *D. iranica* (Telenga, 1955)

(1025) *D. lacteicolor* (Viereck, 1911)

(1026) *D. laevigata* (Ratzeburg, 1848)

(1027) *D. laspeyresiella* (Papp, 1972)

(1028) *D. lineipes* (Wesmael, 1837)

(1029) *D. longicauda* (Wesmael, 1837)

(1030) *D. longipalpis* (Reinhard, 1880)

(1031) *D. nixosiris* (Papp, 1976)

(1032) *D. phaloniae* (Wilkinson, 1940)

(1033) *D. praetor* (Marshall, 1885)

(1034) *D. princeps* (Wilkinson, 1941)

(1035) *D. punctiger* (Wesmael, 1837)

(1036) *D. seriphia* (Nixon, 1972)

(1037) *D. sicaria* (Marshall, 1885)

(1038) *D. soikai* (Nixon, 1972)

(1039) *D. sophiae* (Papp, 1972)

(1040) *D. subemarginata* (Abdinbekova, 1969)

(1041) *D. turkmenus* (Telenga, 1955)

(1042) *D. ultor* (Reinhard, 1880)

(1043) *Glyptapanteles acasta* (Nixon, 1973)

(1044) *G. aliphera* (Nixon, 1973)

(1045) *G. callidus* (Haliday, 1834)

(1046) *G. fulvipes* (Haliday, 1834)

(1047) *G. inclusus* (Ratzeburg, 1844)

(1048) *G. indiensis* (Marsh, 1979)

(1049) *G. liparidis* (Bouché, 1834)

(1050) *G. mygdonia* (Nixon, 1973)

(1051) *G. porthetriae* (Muesebeck, 1928)

(1052) *G. ripus* (Papp, 1983)

(1053) *G. thompsoni* (Lyle, 1927)

(1054) *G. vitripennis* (Curtis, 1830)

(1055) *Hygroplitis russatus* (Haliday, 1834)

(1056) *Iconella albinervis* (Tobias, 1964)

(1057) *I. brachyradiata* Abdoli & Talebi, 2021

(1058) *I. isus* (Nixon, 1965)

(1059) *I. lacteoides* (Nixon, 1965)

(1060) *I. meruloides* (Nixon, 1965)

(1061) *I. mongashtensis* Zargar & Gupta, 2019

(1062) *I. myeloenta* (Wilkinson, 1937)

(1063) *I. nagyi* (Papp, 1975)

(1064) *I. similus* Zargar & Gupta, 2019

(1065) *I. subcamilla* (Tobias, 1976)

(1066) *Illidops mutabilis* (Telenga, 1955)

(1067) *I. naso* (Marshall, 1885)

(1068) *I. scutellaris* (Muesebeck, 1921)

(1069) *I. suevus* (Reinhard, 1880)

(1070) *I. urgo* (Nixon, 1965)

(1071) *Microgaster australis* Thomson, 1895

(1072) *M. hospes* Marshall, 1885

(1073) *M. luctuosa* Haliday, 1834

(1074) *M. messoria* Haliday, 1834

(1075) *M. parvistriga* Thomson, 1895

(1076) *M. rufipes* Nees von Essenbeck, 1834

(1077) *M. subcompleta* Nees von Esenbeck, 1834

(1078) *Microplitis aduncus* (Ruthe, 1860)

(1079) *M. albipennis* Abdinbekova, 1969

(1080) *M. alborziensis* Abdoli & Talebi, 2021

(1081) *M. cebes* Nixon, 1970

(1082) *M. decipiens* Prell, 1925

(1083) *M. deprimator* (Fabricius, 1798)

(1084) *M. docilis* Nixon, 1970

(1085) *M. eremitus* Reinhard, 1880

(1086) *M. erythrogaster* Abdinbekova, 1967

(1087) *M. flavipalpis* (Brullé, 1832)

(1088) *M. fordi* Nixon, 1970
(1089) *M. fulvicornis* (Wesmael, 1837)
(1090) *M. kaszabi* Papp, 1980
(1091) *M. lugubris* (Ruthe, 1860)
(1092) *M. mandibularis* (Thomson, 1895)
(1093) *M. marshallii* Kokujev, 1898
(1094) *M. mediator* (Haliday, 1834)
(1095) *M. ochraceus* Szépligeti, 1896
(1096) *M. pallidipennis* Tobias, 1964
(1097) *M. ratzeburgii* (Ruthe, 1858)
(1098) *M. rufiventris* Kokujev, 1914
(1099) *M. scrophulariae* Szépligeti, 1898
(1100) *M. sofron* Nixon, 1970
(1101) *M. spectabilis* (Haliday, 1834)
(1102) *M. spinolae* (Nees von Esenbeck, 1834)
(1103) *M. stigmaticus* (Ratzeburg, 1844)
(1104) *M. strenuus* Reinhard, 1880
(1105) *M. tadzhicus* Telenga, 1949
(1106) *M. tuberculatus* (Bouché, 1834)
(1107) *M. tuberculifer* (Wesmael, 1837)
(1108) *M. varipes* (Ruthe, 1860)
(1109) *M. viduus* (Ruthe, 1860)
(1110) *M. xanthopus* (Ruthe, 1860)
(1111) *Napamus vipio* (Reinhard, 1880)
(1112) *Pholetesor arisba* (Nixon, 1973)
(1113) *P. bicolor* (Nees von Esenbeck, 1834)
(1114) *P. circumscriptus* (Nees von Esenbeck, 1834)
(1115) *P. elpis* (Nixon, 1973)
(1116) *P. ingenuoides* (Papp, 1971)
(1117) *P. pseudocircumscriptus* Abdoli, 2019
(1118) *P. viminetorum* (Wesmael, 1837)
(1119) *Protapanteles albigena* Abdoli, Fernandez-Triana & Talebi, 2021
(1120) *P. anchisiades* (Nixon, 1973)
(1121) *P. immunis* (Haliday, 1834)
(1122) *P. incertus* (Ruthe, 1859)
(1123) *Sathon falcatus* (Nees von Esenbeck, 1834)
(1124) *S. lateralis* (Haliday, 1834)
(1125) *Venanides caspius* Abdoli, Fernandez-Triana & Talebi, 2019

SUBFAMILY MICROTYPINAE

(1126) *Microtypus algiricus* Szépligeti, 1908
(1127) *M. desertorum* Shestakov, 1932
(1128) *M. trigonus* (Nees, 1843)
(1129) *M. wesmaelii* Ratzeburg, 1848

SUBFAMILY MIRACINAE

(1130) *Centistidea pistaciella* van Achterberg & Mehrnejad, 2002
(1131) *Mirax caspiana* Farahani, Talebi, van Achterberg & Rakhshani, 2014
(1132) *M. rufilabris* Haliday, 1833

SUBFAMILY OPIINAE

(1133) **Apodesmia** *damnosa* (Papp, 1980)
(1134) *A. irregularis* (Wesmael, 1835)
(1135) *A. karesuandensis* (Fischer, 1964)
(1136) *A. novosimilis* (Fischer, 1989)
(1137) *A. nowakowskii* (Fischer, 1959)
(1138) *A. posticatae* (Fischer, 1957)
(1139) *A. saeva* (Haliday, 1837)
(1140) *A. sharynensis* (Fischer, 2001)
(1141) *A. similis* (Szépligeti, 1898)
(1142) *A. striatula* (Fischer, 1957)
(1143) *A. tirolensis* (Fischer, 1958)
(1144) *A. tuberculata* (Fischer, 1959)
(1145) *A. uttoisimilis* (Fischer, 1999)
(1146) **Atormus** *victus* (Haliday, 1837)
(1147) **Biophthora** *bajula* (Haliday, 1837)
(1148) *B. rossica* (Szépligeti, 1901)
(1149) **Biosteres** *advectus* Papp, 1979
(1150) *B. arenarius* (Stelfox, 1959)
(1151) *B. blandus* (Haliday, 1837)
(1152) *B. brevipalpis* (Thomson, 1895)
(1153) *B. haemorrhoeus* (Haliday, 1837)
(1154) *B. longicauda* (Thomson, 1895)
(1155) *B. punctiscuta* (Thomson, 1895)
(1156) *B. remigii* Fischer, 1971
(1157) *B. rusticus* (Haliday, 1837)
(1158) *B. scabriculus* (Wesmael, 1835)
(1159) *B. spinaciae* (Thomson, 1895)
(1160) *B. ultor* (Foerster, 1863)
(1161) *B. wesmaelii* (Haliday, 1837)
(1162) **Bitomus** *multipilis* Fischer, 1990
(1163) **Desmiostoma** *parvulum* (Wesmael, 1835)
(1164) **Eurytenes** *abnormis* (Wesmael, 1835)
(1165) **Fopius** *carpomyiae* (Silvestri, 1916)
(1166) **Hoplocrotaphus** *hamooniae* Peris-Felipo, Belokobylskij & Rakhshani, 2018
(1167) **Indiopius** *cretensis* Fischer, 1983
(1168) **Opiognathus** *propodealis* (Fischer, 1958)
(1169) **Opiostomus** *campanariae* (Fischer, 1959)
(1170) *O. impatientis* (Fischer, 1957)
(1171) *O. riphaeus* (Tobias, 1986)
(1172) *O. snoflaki* (Fischer, 1959)
(1173) **Opius** *abditus* Fischer, 1960
(1174) *O. arundinis* Fischer, 1964
(1175) *O. attributus* Fischer, 1962
(1176) *O. bouceki* Fischer, 1958
(1177) *O. caricivorae* Fischer, 1964
(1178) *O. circinus* Papp, 1979
(1179) *O. circulator* (Nees von Esenbeck, 1834)
(1180) *O. connivens* Thomson, 1895
(1181) *O. crassipes* Wesmael, 1835
(1182) *O. curticornis* Fischer, 1960
(1183) *O. flavipes* Szépligeti, 1898

(1184) *O. fuscipennis* Wesmael, 1835
(1185) *O. gracilis* Fischer, 1957
(1186) *O. instabilis* Wesmael, 1835
(1187) *O. larissa* Fischer, 1968
(1188) *O. latidens* Fischer, 1990
(1189) *O. latipediformis* Fischer, 2004
(1190) *O. latistigma* Fischer, 1960
(1191) *O. levis* Wesmael, 1835
(1192) *O. longicornis* Thomson, 1895
(1193) *O. lonicerae* Fischer, 1958
(1194) *O. lugens* Haliday, 1837
(1195) *O. magnicauda* Fischer, 1958
(1196) *O. mendus* Papp, 1982
(1197) *O. minusculae* Fischer, 1967
(1198) *O. mirabilis* Fischer, 1958
(1199) *O. mischa* Fischer, 1968
(1200) *O. monilicornis* Fischer, 1962
(1201) *O. nigricoloratus* Fischer, 1958
(1202) *O. nitidulator* (Nees, 1834)
(1203) *O. occulisus* Telenga, 1950
(1204) *O. opacus* Fischer, 1968
(1205) *O. pallipes* Wesmael, 1835
(1206) *O. paranivens* Fischer, 1990
(1207) *O. pendulus* Haliday, 1837
(1208) *O. phytobiae* Fischer, 1959
(1209) *O. ponticus* Fischer, 1958
(1210) *O. pulicariae* Fischer, 1969
(1211) *O. pygmaeator* (Nees von Esenbeck, 1811)
(1212) *O. pygmaeus* Fischer, 1962
(1213) *O. repentinus* Papp, 1980
(1214) *O. robustus* Telenga, 1950
(1215) *O. rufimixtus* Fischer, 1958
(1216) *O. seductus* Fischer, 1959
(1217) *O. singularis* Wesmael, 1835
(1218) *O. tabificus* Papp, 1979
(1219) *O. teheranensis* Fischer, 1990
(1220) *O. tersus* (Foerster, 1863)
(1221) *O. turcicus* Fischer, 1960
(1222) *Phaedrotoma bajariae* (Fischer, 1989)
(1223) *P. aethiops* (Haliday, 1837)
(1224) *P. ambiguus* (Wesmael, 1835)
(1225) *P. biroi* (Fischer, 1960)
(1226) *P. biroicus* (Fischer & Beyarslan, 2005)
(1227) *P. caesus* (Haliday, 1837)
(1228) *P. caucasi* (Tobias, 1986)
(1229) *P. cingulatus* (Wesmael, 1835)
(1230) *P. depeculator* Foerster, 1863
(1231) *P. diversiformis* (Fischer, 1960)
(1232) *P. diversus* (Szépligeti, 1898)
(1233) *P. exigua* (Wesmael, 1835)
(1234) *P. gafsaensis* (Fischer, 1964)
(1235) *P. laetatorius* (Fischer, 1958)
(1236) *P. mirabunda* (Papp, 1982)
(1237) *P. nitidulator* (Nees von Esenbeck, 1834)

(1238) *P. ochrogaster* (Wesmael, 1835)
(1239) *P. pseudonitidus* (Fahringer, 1943)
(1240) *P. pulchriceps* (Szépligeti, 1898)
(1241) *P. pumilio* (Wesmael, 1835)
(1242) *P. rex* (Fischer, 1958)
(1243) *P. rudis* (Wesmael, 1835)
(1244) *P. scaptomyzae* (Fischer, 1967)
(1245) *P. schmidti* (Fischer, 1960)
(1246) *P. speciosus* (Fischer, 1959)
(1247) *P. staryi* (Fischer, 1958)
(1248) *P. teheranensis* (Fischer, 1990)
(1249) *P. variegatus* (Szépligeti, 1896)
(1250) *Pokomandya curticornis* Fischer, 1959
(1251) *Psyttalia carinata* (Thomson, 1895)
(1252) *P. concolor* (Szépligeti, 1910)
(1253) *Psyttoma wachsmanni* (Szépligeti, 1898)
(1254) *Rhogadopsis reconditor* (Wesmael, 1835)
(1255) *Utetes coracinus* (Thompson, 1895)
(1256) *U. curtipectus* (Fischer, 1958)
(1257) *U. ferrugator* (Goureau, 1862)
(1258) *U. fulvicollis* (Thomson, 1895)
(1259) *U. rotundiventris* (Thomson, 1895)
(1260) *u. testaceus* (Wesmael, 1838)
(1261) *U. truncatus* (Wesmael, 1835)
(1262) *Xynobius aciculatus* (Thomson, 1895)
(1263) *X. caelatus* (Haliday, 1837)
(1264) *X. curtifemur* (Fischer, 1961)
(1265) *X. decoratus* (Stelfox, 1949)
(1266) *X. macrocerus* (Thomson, 1895)
(1267) *X. maculipes* (Wesmael, 1835)
(1268) *X. scutellatus* (Fischer, 1962)

SUBFAMILY ORGILINAE

(1269) *Kerorgilus zonator* (Szépligeti, 1896)
(1270) *Orgilus abbreviator* (Ratzeburg, 1852)
(1271) *O. dovnari* Tobias, 1986
(1272) *O. hungaricus* Szépligeti, 1896
(1273) *O. ischnus* Marshall, 1898
(1274) *O. leptocephalus* (Hartig, 1838)
(1275) *O. meyeri* Telenga, 1933
(1276) *O. nitidior* Taeger, 1989
(1277) *O. nitidus* Marshall, 1898
(1278) *O. obscurator* (Nees von Esenbeck, 1812)
(1279) *O. pimpinellae* Niezabitowski, 1910
(1280) *O. ponticus* Tobias, 1986
(1281) *O. priesneri* Fischer, 1958
(1282) *O. punctiventris* Tobias, 1976
(1283) *O. punctulator* (Nees von Esenbeck, 1812)
(1284) *O. rudolphae* Tobias, 1976
(1285) *O. rugosus* (Nees von Esenbeck, 1834)
(1286) *O. similis* Szépligeti, 1896
(1287) *O. temporalis* Tobias, 1976
(1288) *O. tobiasi* Taeger, 1989

SUBFAMILY PAMBOLINAE

(1289) *Avga blaciformis* (Hedwig, 1961)
(1290) *A. imperfectus* (Hedwig, 1961)
(1291) *Pambolus biglumis* (Haliday, 1836)
(1292) *Phaenodus pallipes* Foerster, 1863
(1293) *P. rugulosus* Hellén, 1927

SUBFAMILY PROTEROPINAE

(1294) *Proterops nigripennis* Wesmael, 1835

SUBFAMILY RHYSIPOLINAE

(1295) *Cerophanes kerzhneri* Tobias, 1971
(1296) *Rhysipolis decorator* (Haliday, 1836)
(1297) *R. enukiazie* Tobias, 1976
(1298) *R. hariolator* (Haliday, 1836)
(1299) *R. meditator* (Haliday, 1836)

SUBFAMILY RHYSSALINAE

(1300) *Acrisis brevicornis* Hellén, 1957
(1301) *A. suomii* Tobias, 1983
(1302) *Dolopsidea indagator* (Haliday, 1836)
(1303) *Histeromerus mystacinus* Wesmael, 1838
(1304) *Oncophanes minutus* (Wesmael, 1838)
(1305) *Rhyssalus longicaudis* (Tobias & Belokobylskij, 1981)

SUBFAMILY ROGADINAE

(1306) *Aleiodes aestuosus* (Reinhard, 1863)
(1307) *A. agilis* (Telenga, 1941)
(1308) *A. albitibia* (Herrich-Schäffer, 1838)
(1309) *A. alternator* (Nees von Esenbeck, 1834)
(1310) *A. apicalis* (Brullé, 1832)
(1311) *A. arcticus* (Thomson, 1892)
(1312) *A. arnoldii* (Tobias, 1976)
(1313) *A. aterrimus* (Ratzeburg, 1852)
(1314) *A. bicolor* (Spinola, 1808)
(1315) *A. caucasicus* (Tobias, 1976)
(1316) *A. circumscriptus* (Nees von Esenbeck, 1834)
(1317) *A. compressor* (Herrich-Schäffer, 1838)
(1318) *A. coxalis* (Spinola, 1808)
(1319) *A. crassipes* (Thomson, 1892)
(1320) *A. cruentus* (Nees von Esenbeck, 1834)

(1321) *A. dimidiatus* (Spinola, 1808)
(1322) *A. dissector* (Nees von Esenbeck, 1834)
(1323) *A. ductor* (Thunberg, 1822)
(1324) *A. esenbeckii* (Hartig, 1838)
(1325) *A. eurinus* (Telenga, 1941)
(1326) *A. gasterator* (Jurine, 1807)
(1327) *A. gastritor* (Thunberg, 1822)
(1328) *A. miniatus* (Herrich-Schäffer, 1838)
(1329) *A. modestus* (Reinhard, 1863)
(1330) *A. nigricornis* Wesmael, 1838
(1331) *A. nobilis* (Curtis, 1834)
(1332) *A. nocturnus* (Telenga, 1941)
(1333) *A. pallescens* Hellén, 1927
(1334) *A. pallidator* (Thunberg, 1822)
(1335) *A. pallidicornis* (Herrich-Schäffer, 1838)
(1336) *A. periscelis* (Reinhard, 1863)
(1337) *A. praetor* (Reinhard, 1863)
(1338) *A. quadrum* (Tobias, 1976)
(1339) *A. ruficeps* (Telenga, 1941)
(1340) *A. ruficornis* (Herrich-Schäffer, 1838)
(1341) *A. rufipes* (Thomson, 1892)
(1342) *A. rugulosis* (Nees von Esenbeck, 1811)
(1343) *A. schirjajewi* (Kokujev, 1898)
(1344) *A. schewyrewi* (Kokujev, 1898)
(1345) *A. seriatus* (Herrich-Schäffer, 1838)
(1346) *A. signatus* (Nees von Esenbeck, 1811)
(1347) *A. testaceus* (Telenga, 1941)
(1348) *A. ungularis* (Thomson, 1892)
(1349) *A. unipunctator* (Thunberg, 1822)
(1350) *A. varius* (Herrich-Schäffer, 1838)
(1351) *Clinocentrus amiri* Rakhshani & Farahani, 2020
(1352) *C. caucasicus* Tobias, 1976
(1353) *C. cunctator* (Haliday, 1836)
(1354) *C. excubitor* (Haliday, 1836)
(1355) *C. exsertor* (Nees von Esenbeck, 1811)
(1356) *C. umbratilis* Haliday, 1833
(1357) *C. vestigator* (Haliday, 1836)
(1358) *Heterogamus dispar* (Haliday, 1833)
(1359) *Pentatermus striatus* (Szépligeti, 1908)
(1360) *Rogas luteus* Nees von Esenbeck, 1834
(1361) *Yelicones iranus* (Fischer, 1963)

SUBFAMILY SIGALPHINAE

(1362) *Acampsis alternipes* (Nees von Esenbeck, 1816)
(1363) *Sigalphus irrorator* (Fabricius, 1775)

Index

Printed in the United States
by Baker & Taylor Publisher Services